NOBEL LAUREATE	COUNTRY*	YEAR OF AWARD	ACHIEVEMENT
Robert Holley	United States	1968	Studies on the genetic codes for amino acids
Har Gobind Khorana	United States (India)		
Marshall W. Niremberg	United States		
Max Delbruck	United States (Germany)	1969	Studies on the mechanism of viral infection of cells
Alfred D. Hershey	United States		
Salvadore E. Luria	United States (Italy)		
Gerald M. Edelman	United States	1972	Elucidation of the nature and structure of antibody molecules
Rodney R. Porter	United Kingdom		
David Baltimore	United States	1975	Studies on the transformation of cells by tumor viruses
Howard Temin	United States		
Renato Dulbecco	United States (Italy)		
Baruch Blumberg	United States	1976	Discovery of Australia antigen
D. Carleton Gajdusek	United States		Description of slow virus disease
Rosalyn Yalow	United States	1977	Development of the radioimmunoassay procedure
Robert C.L. Guillemin	United States (France)		Synthesis of peptide hormones
Andrew V. Schally	United States (Poland)		
Daniel Nathans	United States	1978	Studies on restriction enzymes and their use in genetic engineering
Hamilton Smith	United States		
Werner Arber	Switzerland		
Baruj Benacerral	United States	1980	Discovery of the histocompatibility antigens used in tissue typing
George D. Snell	United States		
Jean Dausser	France		
Barbara McClintock	United States	1983	Studies on transposable genetic elements
Cesar Milstein	United Kingdom (Argentina)	1984	Development of monoclonal antibody technique
Georges J.F. Köhler	Switzerland (Germany)		
Niels K. Jerne	Switzerland (Denmark)		Research in immunology
Susumu Tonegawa	United States (Japan)	1987	Discovery of the genetic principle of antibody diversity
James W. Black	United Kingdom	1988	Elucidation of important principles of drug therapy for disease
Gertrude B. Elion	United States		
George H. Hitchings	United States		
J. Michael Bishop	United States	1989	Studies on the genetic basis of cancer
Harold Varmus	United States		
Joseph E. Murray	United States	1990	Research into transplantation methods
E. Donnall Thomas	United States		
Erwin Neher	Germany	1991	Discovery of ion channels in biological membranes
Bert Sakmann	Germany		
Philip Sharp	United States	1993	Identification of exons and introns in mRNA synthesis
Richard Roberts	United States		

* Indicates country in which work was performed; country of birth in parentheses.

FUNDAMENTALS *of* MICROBIOLOGY

FIFTH EDITION

I. EDWARD ALCAMO, PH.D.

PROFESSOR OF MICROBIOLOGY
STATE UNIVERSITY OF NEW YORK AT FARMINGDALE

An imprint of Addison Wesley Longman, Inc.

Menlo Park, California • Reading, Massachusetts • New York • Harlow, England
Don Mills, Ontario • Sydney • Mexico City • Madrid • Amsterdam

ASSISTANT EDITOR: Leslie A. With
MARKETING MANAGER: Nina Horne
FREELANCE PRODUCTION EDITOR: Catherine E. Lewis
COVER MANAGEMENT: Don Kesner
COVER DESIGNER: Yvo Riezebos
FIFTH EDITION ILLUSTRATORS: Karl Mijajima (technical art), Cyndie C. H. Wooley (cartoons)
COPY EDITOR: Mary Prescott
PROOFREADER: Holly McLean-Aldis
MANUFACTURING SUPERVISOR: Merry Free Osborne

COVER PHOTO: © Telegraph Colour Library/FPG.

TO THE STUDENT

A Study Guide for the textbook is available through your college bookstore under the title, *Study Guide to Accompany Fundamentals of Microbiology*, Fifth Edition, by I. Edward Alcamo. The Study Guide can help you with course material by acting as a review and study aid. If the Study Guide is not in stock, ask the bookstore manager to order a copy for you.

Library of Congress Cataloging-in-Publication Data

Alcamo, I. Edward.
    Fundamentals of microbiology / I. Edward Alcamo.––5th ed.
       p. cm.
    Includes bibliographical references and index.
    ISBN 0-8053-0532-7
    1. Microbiology. 2. Medical microbiology I. Title.
QR41.2.A43 1996
576––dc20   96-43386
                                        CIP

ISBN 0-8053-0532-7
1 2 3 4 5 6 7 8 9 10–RN–00 99 98 97 96

DEDICATION

It gives me great pleasure to dedicate the fifth edition of **Fundamentals of Microbiology** *to my colleagues and friends at the State University of New York at Farmingdale. For the past thirty years, I have been privileged to work alongside you as we sought to provide a rich and rewarding learning environment for our students. Thanks, guys.*

PREFACE

One cold morning in January 1980, I stumbled out of bed and decided that if this book was ever going to get written, I'd better get at it. Three years later, almost to the day, the first edition of *Fundamentals of Microbiology* appeared in our college bookstore. I remember picking it up, leafing through it, and thinking, "Wow!" Then I took a deep breath and walked over to the cafeteria to have a cup of coffee—my steps were a little lighter that winter day.

Early on, my intention had been to write a microbiology book for beginning students of microbiology. At the time I had been teaching for 15 years and I believed they needed something more than the encyclopedias on the market. Their response was enthusiastic beyond my wildest expectations. Students and instructors wrote to let me know that the book was up-to-date, readable, and interesting. The special features of the book included storied illustrations, a wealth of knock-your-socks-off electron micrographs, useful marginal notes, and an occasional tongue-in-cheek look at infectious disease. There was enough repetition of information to reinforce the key concepts and provide a sense of continuity, and the book had sufficient medical orientation to appeal to health science students.

It has been sixteen years since that first winter morning, and now we present the fifth edition of *Fundamentals*. Why do we need new editions? Well, consider the following: When I sat down to write in 1980, AIDS was unknown; hardly anyone understood the significance of T-cells; genetic engineering bore promise but no fruit; and scientists were confidently talking about the end of infectious disease. All that has changed: AIDS has become the epidemic of the century; T-cells are now recognized as key underpinnings of immunity; diabetics routinely take genetically engineered insulin; and the emerging viruses have taught us that infectious disease will be around for many generations to come.

Moreover, writing new editions has given me the opportunity to use the English and grammar skills drummed into my head by countless editors (I now watch my pronouns; I keep my paragraphs short; and I use a comma before *which* but not before *that*). I was able to add some new stories and thought questions to the text, revitalize and update the sections about immunology, and present a number of career essays relating to microbiology. So that's the why and wherefore of this book. Now let me tell you some of the other important details.

Audience

Fundamentals of Microbiology is written for introductory microbiology courses having an emphasis on biology and human disease. It is geared toward students in health and allied health science curricula such as nursing, dental hygiene, medical assistance, sanitary science, and medical laboratory technology. It will also be an asset to students studying food science, agriculture, environmental science, and health administration. In addition, the text should provide a firm foundation for advanced programs in the biological sciences, medicine, dentistry, and other health professions.

Objectives

Fundamentals of Microbiology is an ambitious title for a book because it suggests broad coverage of microorganisms. I have attempted to fulfill that promise by conveying the multiple dimensions of microbiology and by showing that knowledge gained about microorganisms has enriched all segments of biology. For example, there are numerous discussions of the roles microorganisms play in food production, ecology, sewage disposal, and the emerging technologies of our time. Discussions also treat the intimate relationships between structure and function, the variety of life forms within the microbial world, the novel genetic patterns seen in microorganisms, and the basic patterns of human disease and resistance.

One of my most important objectives has always been accessibility. We both know that a book can be long on fact and information, but short on the ability to communicate. I have tried to improve the accessibility to microbiology by telling lots of stories (that was largely how ancient peoples learned); by focusing on the people who worked to make our knowledge base possible; by telling you *why* something is important, not just that it is important; by sharing my fascination with the microbial world; and by trying to elicit from you an occasional "gee whiz." I hope you will find the reading friendly and comfortable.

Organization

Fundamentals of Microbiology is divided into seven major areas of concentration. These areas use basic principles as frameworks to provide a unity and diversity to microbiology. Among the principles explored are the variations in anatomy and physiology of microorganisms, the basis for infectious disease and resistance, and the beneficial effects microorganisms have on the quality of life.

Part 1 deals with foundations for microbiology. It includes chapters on the origins of microbiology and the universal concepts within the science. Part 2 then concentrates on the bacteria, since these are the microorganisms most of us think of when we say "germs." The discussions carry over to Part 3, where the spectrum of bacterial diseases is surveyed. Part 4 then looks briefly at other microorganisms, including viruses, fungi, and protozoa.

In Part 5 of the text, the emphasis is on infectious disease and the body's resistance through the immune system. We study the reasons for disease and the means for surviving it. A logical segue is in Part 6, which presents various mechanisms for controlling microorganisms with sterilization procedures, disinfectants, and antibiotics. Part 7 closes the text with brief discussions of how public health measures interrupt the spread of microorganisms. Some insights are also given to the positive effects microorganisms have on the quality of our lives.

Incidentally, users of former editions will note that we have condensed the book's final four chapters to two in order to help reduce sagging of student backpacks. We have also continued the cover motif used on all the previous editions. Have you noticed it?

Why Pathogens?

Microorganisms perform many useful services for humans when they produce dairy products, manufacture organic materials, and recycle such elements as carbon, nitrogen, sulfur, and phosphorus. The emphasis of this book, however, is on the tiny percentage of microorganisms that cause human disease, the so-called pathogens. Why do we emphasize pathogens? Here are several reasons:

- Pathogens have regularly altered the course of human history
- Pathogens are familiar to audiences of microbiology
- Pathogens add drama to an invisible world of microorganisms
- Pathogens illustrate ecological relationships between humans and microorganisms
- Pathogens point up the diversity of microorganisms

Moreover, the study of pathogens shows that microbiology is not necessarily the same as molecular biology; studying pathogens makes basic science relevant and shows how microbiology interfaces with other disciplines such as sociology, economics, politics, and geography; and the study of classical and emerging pathogens helps us understand contemporary newspaper articles, magazine headlines, and stories on the late news. And in the end, that makes us better citizens.

Indeed, the famous essayist Thomas Mann has written, "All interest in disease . . . is only another expression of interest in life."

Special Features

To help you achieve your learning goals and to reduce the anxiety that comes with approaching a new subject, we have incorporated several features to this book. We hope that these features will enhance the accessibility of the key concepts and generate real enthusiasm for microbiology.

Introductions provide a stimulating thought or historical perspective to set the tone for the chapter and encourage a graceful entry.

Career Essays titled "Microbiology Pathways" discuss career options available to the microbiology student.

Mid-Chapter Summaries entitled "To This Point" allow you to pause and summarize the previous few pages and preview the next few pages.

Boldfaced Terms highlight key terms and phrases in the text so you can focus your energies.

Marginal Pronunciations encourage you to learn the pronunciations of difficult terms as they are encountered in the text.

Marginal Definitions give succinct definitions of notable terms as they enter the discussion or recall them from previous uses.

Marginal Representations contain pictorial presentations of microorganisms discussed nearby to place a vision in your mind.

MicroFocus Boxes explore stimulating topics of microbiological interest.

Notes to the Student at the end of each chapter present an editorial comment within the framework of the chapter.

Chapter Summaries highlight the fundamental principles of the chapter's contents.

Questions for Thought and Discussion encourage the use of textual material to resolve thought-provoking problems.

Reviews contain short answer questions of unconventional types to assist your review of the chapter contents.

Summary Tables pull together the similarities and differences of topics discussed in the chapter.

Vocabulary Boxes allow you to see major terms as they relate to one another and to determine the salient differences of the terms.

Appendixes include the classification of bacteria, temperature and metric conversion charts, a list of incubation periods, and answers to all the questions in the text.

The Glossary lists over a thousand words used in the text together with a concise definition, pronunciation, and chapter location.

Supplements to the Text

To encourage your efforts to learn microbiology, we have prepared a number of supplements that augment and further develop the concepts of this text:

Laboratory Fundamentals of Microbiology is a series of 30 multi-part laboratory exercises that provide basic training in the handling of microorganisms, while exploring microbial properties and uses.

The *Study Guide* to accompany the textbook contains over three thousand (3000) practice exercises, study questions, and puzzles to assist the learning and retention of text information and concepts.

The Microbiology Coloring Book is a set of 125 coloring exercises that allow you to track information and see processes and information in a more visual format and in a planned and meaningful sequence.

The *Instructor's Manual and Test Bank* lists two thousand (2000) questions suitable for testing purposes, discusses issues of contemporary concern, suggests outlines for classroom lectures, and gives insights to improve the instructor's presentations.

A set of 100 *Acetate Overhead Transparencies* highlight in two colors the major line drawings from the textbook.

A set of 35-mm color slides on HIV and other agents of microbial disease is available to instructors to help them augment and update their lectures.

ACKNOWLEDGMENTS

Many talented people have contributed to this book, and I would like to extend a word of appreciation to them. My editors Anne Scanlon-Rohrer and Leslie With lent their considerable expertise to this project during the editorial phase, and Cathy Lewis handled the production phase with a sure-handed touch. The book also benfited from the wisdom of my fellow microbiologists who have reviewed this and past editions. Once again I thank my colleagues from six continents whose photographs have added outstanding visual appeal to this book. And I express my thanks to the building janitor who agreed to leave the lights on while I worked into the wee hours.

I am also indebted to students and instructors like yourself who have written to me since those first days in the 1980s. You have told me how to make this book more useful to your needs—what to add and what to eliminate—and you have encouraged me to take some unusual (and occasionally, frivolous) steps to write a book that you can learn from. Microbiology is an imposing subject, but you have taught me to present it so it can be understood. I thank your predecessors and I hope you will continue to keep me on the straight-and-narrow path.

I could not finish without acknowledging my three favorite people: my children Michael, Elizabeth, and Patricia. As I punch this out on my trusty Mac, Michael is a successful attorney in New York; Elizabeth is completing a doctoral degree (in molecular biology) at Massachusetts Institute of Technology; and Patricia has graduated from Princeton and is enjoying a rewarding career in business, with travels that take her nationally and internationally. Since my wife died of cancer six years ago, they have given a warm glow to my life and have become my best friends. No father could be prouder of his brood.

Farmingdale, New York
Fall 1996

E. A.

TO THE STUDENT

When I was a student, I hardly ever read the "To the Student" sections of my textbooks, and I only discovered later in my years what they were all about. Therefore, I am encouraged that you are reading this, and I welcome the opportunity to let you know what *Fundamentals of Microbiology* is all about.

Let me begin by saying I have yet to meet a student who picked up a microbiology textbook for the sheer joy of it. I am assuming, therefore, that this book has been assigned as part of your microbiology program, and I have worked hard to make it your friend, and relieve some of the stress that comes with learning microbiology. Let me tell you how.

I have kept the chapters to approximately the same length so you can gauge the time necessary for each and plan your study accordingly. I have also avoided lengthy presentations and instead have focused on smaller sections, each with its own heading. These should accommodate short, interrupted study periods while providing flexibility for your instructor.

As you proceed, you will find many historical narratives within the chapters. All too often, significant events become opaque and the names of key investigators fall into obscurity. I have tried to resurrect these events and names to show that real people like ourselves pondered, experimented, and sometimes gave their lives to make the knowledge of microbiology possible. I have also tried to give you a glimpse into the future by explaining current research trends in microbiology.

Careful attention has been given to chapter introductions to let you know what is coming, and I have written mid-chapter summaries to let you reorient yourself before getting too deeply into the material. I hope these pauses will help you to see the forest for the trees and maintain in your mind the objectives of the chapter. In many chapters you will find a Vocabulary of that chapter's major terms to help you organize your thoughts. Several chapters also contain stories of local disease outbreaks expressed as pictorial representations, a medium that may help you remember them better.

At the end of each chapter you will find a Note to the Student. This is where I have an opportunity to do a bit of soapboxing and present an editorial opinion relative to the chapter. A textbook writer has to be noncommittal, and one soon gets tired of "just the facts." I have therefore added a personal thought or two which I offer for your consideration. The end of the chapter also contains a Summary of the major principles of that chapter. It is a good idea to read the summary before delving into the chapter. That way you will have a framework of the chapter contents before filling in the nitty gritty information.

I really enjoyed writing the Questions for Thought and Discussion, and I trust you will see them as challenging applications of microbiology. Perhaps they may even precipitate a classroom discussion or two, in which case they will fulfill my hope that you understand the relevance of microbiology. The Review Questions are a bit unconventional and should provide a refreshing alternative to the traditional type of short-answer questions. More of the traditional questions asking you to choose, list, agree or disagree, summarize, and so forth can be found in the *Study Guide* that accompanies this text. Incidentally, there are more than 3000 questions in the *Study Guide* (plus more puzzles). That translates to over 100 questions per chapter at a cost of less than $1.00 per chapter. I think it is a good investment and hope you will agree.

As you leaf through the pages, you will note that the pronunciations of difficult words are presented alongside the word. Learning the correct pronunciations should increase your familiarity with the terms, make you more confident in using them, and add to your knowledge of microbiology. The pages also contain a running glossary of important terms. These definitions are culled from the neighboring text or from other places in the book where explanations are more complete. They should provide quick reviews and refreshers.

At some point, I would really like to hear from you. Please write and let me know what is good about the book so I can build on it and what is bad so I can change or eliminate it. My goal is to make this text as useful to your study time as your instructor is making the lectures to your class time. Also I would be pleased to hear about any news of microbiology in your community and would be happy to help you locate information not covered in the text. I can be reached at the Department of Microbiology; State University of New York; Farmingdale, New York 11735. My phone number is 516-420-2423 (or you can e-mail me at alcamoie@snyfarva.cc.farmingdale.edu).

In closing, I would offer you a thought that a student named Michael expressed to me several years ago. As I recall, it was a spring afternoon and he was busily peering into his microscope. Quite spontaneously, he looked up and said, "You know, Dr. Alcamo, education is like soup: the more you put into it, the more you get out of it."

Please accept my best wishes for a successful experience in microbiology. And welcome to the wild and wonderful world of microorganisms.

E. Alcamo

CONTENTS IN BRIEF

DETAILED CONTENTS

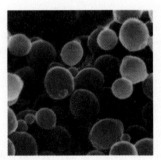

PART 2

THE BACTERIA   84

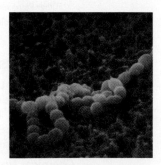

PART 3

BACTERIAL DISEASES OF HUMANS   192

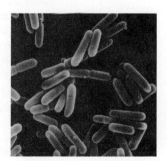

PART 4

OTHER MICROORGANISMS   320

PART 5

DISEASE AND RESISTANCE   518

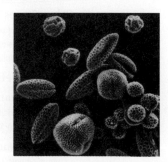

PART 6

CONTROL OF MICROORGANISMS 646

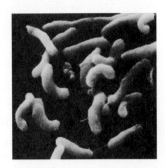

FOUNDATIONS OF MICROBIOLOGY

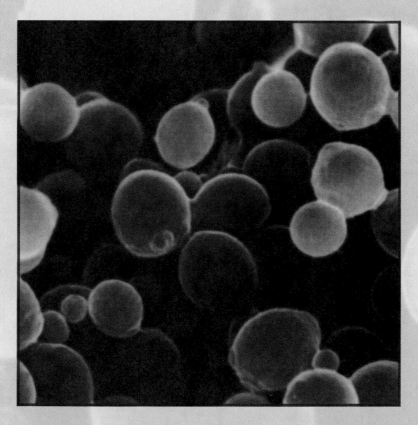

In 1676, a century before the Declaration of Independence, a Dutch merchant named Anton van Leeuwenhoek sent a noteworthy letter to the Royal Society of London. Writing in the vernacular of his home in the Netherlands, van Leeuwenhoek described how he used a primitive microscope to observe vast populations of minute creatures. His report opened a chapter of science that would evolve into the study of microscopic organisms and the discipline of microbiology. At that time, few people, including van Leeuwenhoek, attached any practical significance to the microorganisms, but during the next three centuries, scientists learned how profoundly they influence the quality of our lives. ■ We shall begin our study of the microorganisms by exploring the grass roots developments that led to the establishment of microbiology as a science. These developments are surveyed in Chapter 1, where we focus on the individuals who stood at the forefront of discovery. You will note that Chapter 1 only goes so far as the early 1900s because the remaining chapters of this book describe the discoveries of microbiology since that time. Indeed, our understanding of microorganisms continues to grow even as you read this book. ■ Part I will also contain a chapter on basic chemistry since microorganisms are chemical machines. Moreover, their activities are all related to chemistry. The third and final chapter in Part I will set down some basic concepts that apply to all microorganisms, much as the alphabet applies to word development. In succeeding chapters, we shall formulate words into sentences and sentences into ideas as we survey the types of microorganisms and concentrate on their importance to public health and human welfare.

MICROBIOLOGY PATHWAYS

BEING A SCIENTIST

As you read the next chapter and those beyond, you might wonder why certain individuals have the good fortune to make key discoveries while others never quite reach the pages of textbooks. To be sure, it is sometimes the luck of the draw, but in other cases individuals have a set of characteristics that put them on the trail to success.

Robert S. Root-Bernstein has attempted to identify some of these characteristics in an article published in *The Sciences* (May/June 1988). Root-Bernstein points out that prominent scientists like to goof around, play games, and surround themselves with a type of chaos aimed at revealing the unexpected. Their labs may outwardly appear a disorderly mess, but they know exactly where every tube or bottle belongs. Scientists also identify intimately with the things or creatures they study (it is said that Louis Pasteur actually dreamed about microorganisms), and this identification brings on an intuition, a sense about what an organism will do. Plus, there is the ability to recognize patterns that might eventually reveal hidden truths. (Pasteur had studied art as a teenager, and therefore he had an appreciation of patterns.)

In 1992, I received a letter from a student at Red Deer College in Alberta, Canada asking why I became a microbiologist. Essentially, I answered, it was because I enjoyed my undergraduate microbiology course (thanks, Dr. Marazzella), and when I needed to select a graduate major, microbiology seemed like a good idea. I also think I had some of the characteristics described by Root-Bernstein: I loved to try out different projects; my corner of the world qualified as disaster area; and I was a nut on organization, insisting that all the square pegs fit into the square holes. If all this sounds familiar, then maybe you fit the mold of a scientist. Why not consider pursuing a career in microbiology? Some possibilities are listed in this book, but you should also visit with your instructor. Simply stop by the cafeteria, buy two cups of coffee, and you're on your way . . .

CHAPTER 1

THE DEVELOPMENT OF MICROBIOLOGY

E ach year a group of pilgrims gather in the English countryside outside the village of Eyam to pay homage to the townsfolk who gave their lives three centuries before so that others might live. The pilgrims pause, bow their heads, and remember.

i'am

Bubonic plague erupted in Eyam in the spring of 1666. The rich were the first to flee the town, and soon the commonfolk also contemplated leaving. However, they knew that by doing so they would probably spread the disease to nearby communities. At this point, the village rector made a passionate plea that they stay. After some deep soul-searching, most resolved to remain. A circle of stones was marked off outside the village limits, and people from the adjacent towns nervously brought food and supplies to the barrier, leaving it there for the self-quarantined villagers. In the end, 259 of the town's 350 residents succumbed to the plague.

Bubonic plague:
a highly fatal bacterial disease of the blood, spread by fleas and respiratory secretions.

The memorial service has a poignant moment as the pilgrims somberly recite a rhyme traced to that period:

Ring-a-ring of rosies,
A pocketful of posies
Achoo! Achoo!
We all fall down.

There is no laughter in the group; indeed, some are moved to tears. The ring of rosies refers to the rose-shaped splotches on the chest and armpits of plague victims. Posies were tiny flowers that the people hoped would ward off the evil spirits. "Achoo!" refers to the fits of sneezing that accompanied the disease; and the last line, the saddest of all, suggests the death that befell so many.

Disease has always left people thunderstruck with terror. Before the 1900s, the fear was compounded by ignorance because no one was really sure what caused disease, much less how to deal with it. To be sure, many strange brews and local customs were offered as preventatives or cures, but few were reliable. As we shall see in this chapter, the answers would not start coming until the late 1800s when a link between microorganisms and disease was forged by Pasteur and Koch. There followed an explosion of research and discovery that led to the control of many diseases.

Microorganisms:
microscopic forms of life studied in microbiology.

Beginnings of Microbiology

Ironically, an interest in microscopic objects was developing even as the tragedy and sacrifice of Eyam were unfolding. In 1665, in another part of England, a scientist named **Robert Hooke** published a major work called *Micrographie*. The book contained a miscellany of his thoughts on chemistry as well as a description of the microscope and its uses. Hooke did not invent the microscope (that distinction is generally attributed to Zacharias Janssen, a spectacle-maker from Middleburg, the Netherlands), but he did give it respectability.

Hooke's writings awakened the learned of Europe to the world of very small objects and creatures. Among his illustrations were representations of the eye of a fly, stinger of a bee, structure of a feather, shell of a protozoan, and plantlike form of a mold (MicroFocus: 1.1). He also described a slice of cork and suggested that the cork was composed of compartments, which he called cells. By this account he secured for himself a place in the history of cell biology.

Other scientists were quick to follow Hooke's lead. For example, the Dutch naturalist Jan Swammerdam described tiny bodies, the red blood cells, in samples of blood. Another Dutchman, Regnier de Graaf, discovered cell clusters called follicles in the animal ovary (known today as Graafian follicles). Still another scientist, the Italian physiologist Marcello Malpighi, described the tiny capillaries of an animal's cardiovascular system.

The discoveries by Hooke, Swammerdam, and other scientists showed that the microscope was an important tool for unlocking the secrets of nature. It is not surprising, therefore, that genuine interest was forthcoming when Anton van Leeuwenhoek revealed his descriptions of microorganisms in the 1670s.

Mold:
a type of microorganism composed of threadlike filaments.

mal-pig'ē

1.1

MICROFOCUS

FIRST TO THE MICROORGANISMS

Although Anton van Leeuwenhoek has received considerable credit for his vivid descriptions of microorganisms, he was probably not the first to observe or write about microorganisms. Indeed, Robert Hooke, known for his description of cells, had written about microorganisms 10 years before van Leeuwenhoek.

In 1988, David Bardwell of Kean College reported on his reading on Hooke's *Micrographie* (*ASM News* **54**:183–185). Published in 1665, *Micrographie* contains Hooke's reports of shells of microorganisms among grains of sand. The shells are now believed to be those of foraminiferans, a group of ameboid protozoa that secrete chambered shells about themselves (Chapter 14 has a diagram and explanation of these protozoa). Hooke illustrated the microscopic snail-like shells for his readers and

for posterity and concluded that they were fossilized remains of living microscopic organisms.

Hooke's observations also included a detailed description of molds. His illustrations give structural details of a mold and his writings on the parts of a mold would warm the heart of any microbiology instructor. Hooke's "elongated stalks" are now called sporangiophores and "round knobs" are sporangia. Bardwell concluded that the mold was probably a member of the order Mucorales. Hooke's curiosity also led him to describe a fungus that was parasitizing a rose plant, what mycologists would today call a rust disease of roses. Hooke's illustrations of the mold were clear, concise, and vivid enough to assure us that he was one of history's first microbiologists.

Anton van Leeuwenhoek

Anton van Leeuwenhoek (Figure 1.1a) was a draper and haberdasher and the owner of a dry goods business in Delft, the Netherlands. His father had made the baskets used to pack Delft china off to world markets, and van Leeuwenhoek enjoyed a comfortable living selling silk, wool, cotton, buttons, and other supplies. He was head of the City Council, inspector of weights and measures, and court surveyor. In his spare time, he ground pieces of glass into fine lenses, placing them between two silver or brass plates riveted together, as shown in Figure 1.1b. Later he added an adjustable device for holding tiny specimens. Gradually, he developed a skill that would remain unmatched for many generations.

van-lu'en-hōk"

Van Leeuwenhoek's lenses were no larger than the head of a pin, but they served him well. By most accounts they could magnify an object over 200 times. Initially he used the lenses to inspect the quality of cloth, but as his fascination with microscopic objects developed, he examined hair fibers, skin scales, eye lenses, blood cells, and even samples of his own feces. At one point, van Leeuwenhoek observed tiny sperm cells and speculated that they contain microscopic embryos transferred to the female for nourishment and development.

Van Leeuwenhoek's work did not go unnoticed or unreported. Probably the only scientific group of that period was the Royal Society of London, but correspondence with this group was tenuous because England and the Netherlands were bitter rivals for commercial treasures of the East Indies. Nevertheless, Regnier de Graaf, a fellow of the Royal Society, dispatched a letter to the members in 1673 urging them to contact van Leeuwenhoek. They did so, and van Leeuwenhoek was soon sending along his illustrations and observations. The contact was maintained for the next 50 years.

ren'yā

FIGURE 1.1

Anton van Leeuwenhoek

(a) Van Leeuwenhoek at work in his study. Using a primitive microscope, van Leeuwenhoek was able to achieve magnifications of over 200 times and describe various biological specimens including numerous types of microorganisms. (b) Details of van Leeuwenhoek's lens system. The object is placed on the point of the specimen mount (B). The mount is adjusted by turning the focusing screw (C) and the elevating screw (D). Light is reflected from the specimen through the lens (A), thereby magnifying the specimen.

(a)

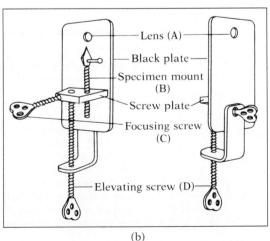

Lens (A)

Black plate

Specimen mount (B)

Screw plate

Focusing screw (C)

Elevating screw (D)

(b)

Animalcules:
Anton von Leeuwenhoek's
term for microorganisms.

In September 1674, van Leeuwenhoek filled a glass with greenish, cloudy water from a marshy lake outside Delft and placed a sample under his lens. The water teemed with tiny microorganisms, which he called **animalcules**. His curiosity aroused, van Leeuwenhoek soon located the animalcules in rainwater, in material from his own teeth and feces, and eventually in most of the specimens he examined. The creatures astonished him at first, then delighted and perplexed him as he pondered their origin and purpose.

In 1676, van Leeuwenhoek sent his eighth letter to the Royal Society. This letter, published the next year, has special significance to microbiology because it contained his first detailed description of the microorganisms:

> May the 26th, I took about $\frac{1}{3}$ of an ounce of whole pepper and having pounded it small, I put it into a Thea-cup with $2\frac{1}{2}$ ounces of Rainwater upon it, stirring it about, the better to mingle the pepper with it, and then suffering the pepper to fall to the bottom. After it had stood an hour or two, I took some of the water, before spoken of, wherein the whole pepper lay, and wherein were so many several sorts of little animals; and mingled it with this water, wherein the pounded pepper had lain an hour or two, and observed that, when there was much of the water of the pounded pepper, with that other, the said animals soon died, but when little they remained alive.
>
> June 2, in the morning, after I had made divers observations since the 26th of May, I could not discover any living thing, but saw some creatures, which tho they had the figures of little animals, yet could I perceive no life in them how attentively I beheld them. . .

Protozoa:
animal-like microorganisms,
often large and motile.

Van Leeuwenhoek's sketches were elegant in detail and clarity. In succeeding letters, he outlined structural details of the familiar protozoa known today as *Paramecium* and *Amoeba*, and he described the threadlike fungi, as well as certain microscopic algae. A particularly noteworthy letter, written in 1683, included drawings of what are now believed to be bacteria in the form of rods, spheres, and spirals. Figure 1.2 illustrates how van Leeuwenhoek perceived them. His work was performed exclusively with single-lens microscopes, of which he constructed about 550. His studies opened the door to a completely new concept of the makeup of all living things.

FIGURE 1.2

van Leeuwenhoek's Descriptions of Bacteria

Examples from a Letter to the Royal Society Dated September 17, 1683. Both (A) and (B) represent rod forms, with (C) and (D) indicating the pathway of motion. A spherical form is shown in (E) and longer types are depicted in (F). The type shown in (G) appears to be a spiral form, and a cluster of spheres is shown in (H). Van Leeuwenhoek found many of these organisms between his teeth.

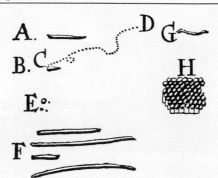

In 1680, van Leeuwenhoek was elected to fellowship in the Royal Society and, with Isaac Newton and Robert Boyle, he became one of the most famous men of his times. Peter the Great of Russia and the Queen of England came to peer into his microscopes. His technique was so fastidious that when some of his slides were discovered 300 years later, scientists found perfectly preserved specimens of cotton seeds, slices of cork, and the optic nerve of a cow. In all, van Leeuwenhoek bombarded the Royal Society with over 200 letters about his findings. He died in 1723 at the age of 90, his advanced age itself a notable achievement.

Van Leeuwenhoek was a very suspicious and secretive person. He invited no one to work with him, nor did he show anyone how to grind lenses or construct a microscope. This is one reason why interest in the microorganisms waned after his death. Another reason, and perhaps of more significance, is that scientists of the day saw the microorganisms merely as curiosities of nature. In the 1700s, disease was still shrouded in magic and mysticism, and few people believed that microorganisms could cause disease in so lofty a creature as the human being. Thus, the substantial development of microbiology would be delayed until the technology for an efficient microscope emerged and until an association was drawn between microorganisms and disease. This was not to happen until the late 1800s.

Royal Society:
an English scientific society of physicists, chemists, and biologists founded in the 1660s and in existence today.

The Transition Period

Biology of the 1700s was a body of knowledge without a focus. Basically, it consisted of observations of plant and animal life and the attempts by scientists to place the organisms in some logical order. New World explorers returned to Europe regularly with specimens to be described and catalogued, and interest in the variety of life forms grew steadily. The dominant figure of the era was **Carolus Linnaeus**, a Swedish botanist who brought all the plant and animal forms together under one great classification scheme (Chapter 3). His book, *Systema Naturae*, was first published in 1735.

lin-ā′us

A few scientists continued to explore the microscopic world. In 1718, for example, Louis Joblot published a treatise on protozoa, and in 1725 Abraham Tremblay described the simple animal known as the hydra. However, scientists generally did not think that tiny living organisms could cause infection. Rather, they believed that an infectious disease spread by an altered chemical quality of the atmosphere, an entity called **miasma**. Miasma arose from decaying or diseased bodies known as miasms. The **miasma theory** figured prominently in medical thinking well into the 1800s.

job-lo′

Hydra:
a multicellular aquatic animal having long fingerlike projections.

mi-az′mah

As the years unfolded, some biologists began to scrutinize the laws of nature and to question the origin of living things as they exist today. The resulting controversy surrounding spontaneous generation could be answered only by experimentation. Explorations of this type reached to the very essence of biology and bridged the gap between van Leeuwenhoek's time and the mid-1800s.

Spontaneous Generation

In the fourth century B.C., Aristotle wrote that flies, worms, and other small animals arose from decaying matter without the need of parent organisms. His observations laid the basis for the doctrine of **spontaneous generation**, a belief that lifeless substances could give rise to living creatures. Indeed, in the early 1600s, the eminent Flemish physician Jan Baptista van Helmont lent credence to the belief when he observed rats "originate" from wheat bran and old rags. Common people embraced the idea, for even they could see slime "breeding" toads and meat "generating" wormlike maggots.

Spontaneous generation:
the doctrine that held that lifeless objects give rise to living organisms.

red'ē

Maggots:
wormlike larvae of certain insects such as houseflies.

nēd'am

spa"lan-zan'e

Among the first to dispute the theory of spontaneous generation was the Florentine scientist **Francesco Redi**. Noting van Leeuwenhoek's complex descriptions of tiny animals, Redi reasoned that flies had reproductive organs. He suggested that flies land on pieces of exposed meat and lay their eggs, which then hatch to maggots. This would explain the "spontaneous" appearance of maggots. In the 1670s, Redi performed a series of tests in which he covered jars of meat with fine lace, thereby preventing the entry of flies. So protected, the meat would not produce maggots and Redi temporarily put to rest the notion of spontaneous generation. His work was one of history's first experiments in biology (Figure 1.3).

Although Redi's work became widely known, the doctrine of spontaneous generation was too firmly entrenched to be abandoned. Science of the 1700s had theological overtones, and some radical philosophers found spontaneous generation to be a useful way of showing that God had no place in creation. Reports of microorganisms were becoming widespread during that period, and in 1748 a British clergyman named **John Needham** put forth the notion that in flasks of mutton gravy microorganisms arise by spontaneous generation. Needham even boiled several flasks of gravy and sealed the flasks with corks, as Redi had sealed his jars. Still, the microorganisms appeared. The Royal Society of London was duly impressed and elected Needham to membership.

However, an Italian cleric and scientist, **Abbé Lazzaro Spallanzani**, criticized Needham's work. In 1767, Spallanzani boiled meat and vegetable broths for long periods of time and then sealed the necks by melting the glass. As control experiments, he left some flasks open to the air, stoppered some loosely with corks, and boiled some briefly, as Needham had done. After two days, he found the control flasks swarming with organisms, but the sealed flasks contained none. Spallanzani's experiment did not settle the issue, though. Needham countered that Spallanzani had destroyed the "vital force" of life with excessive amounts of heat. Other scientists suggested that the air necessary for life had been excluded. When oxygen was discovered in 1774, Spallanzani's opponents pointed to this gas as the vital element eliminated.

FIGURE 1.3

Disputing Spontaneous Generation

In the 1670s Francesco Redi attempted to disprove the belief that maggots (fly larvae) arise from decaying meat. (a) He placed a piece of meat in an open jar and showed that flies lay their eggs, which hatch to form maggots. People of that period believed that the maggots arose spontaneously from the decaying meat. (b) He then placed a second meat sample in a jar covered with cheesecloth. The flies could not reach the meat, and maggots did not appear on the surface. Redi's conclusion was that maggots arise from fly eggs, not spontaneously from the meat. This relatively simple experiment was among the first ever performed in biology.

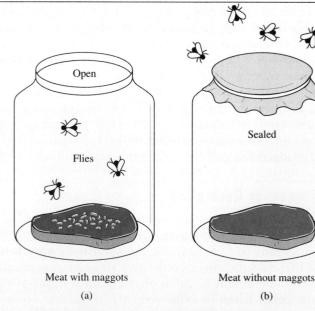

Meat with maggots
(a)

Meat without maggots
(b)

The controversy on spontaneous generation was no closer to being resolved as the 1800s began. Indeed, resolution would not come until Louis Pasteur's work in the 1870s. However, the debate focused attention on the fundamental nature of life and encouraged scientists to experiment in biology.

Disease Transmission

While the doctrine of spontaneous generation was being debated, other scientists were more concerned with how disease was transmitted. As early as 1546, the Italian poet and scientist **Girolamo Fracostoro** wrote: "Contagion is an infection that passes from one thing to another." Fracostoro recognized three forms of passage: contact, lifeless objects, and air (Table 1.1). His ideas were later interwoven with the miasma theory, and "contagion" came to imply a mechanism by which miasma could be passed to a susceptible person.

fra"cos-tor'o

The notion that microorganisms were the substance of contagion received little credibility, however. Thus the German priest Athanasius Kircher was paid little attention when in the mid-1600s he reported "microscopic worms" in the blood of plague victims; nor was Christian Fabricius taken seriously in the 1700s when he suggested that fungi might be the cause of rust and smut diseases in plants. On the contrary, Edward Jenner was accorded many honors in 1798 when he discovered immunization for smallpox (MicroFocus 1.2), despite the fact that he could not explain the cause of the disease.

kir'ker

Smallpox:
a highly fatal viral disease of the skin, accompanied by extensive lesions.

By the mid-1800s enough understanding had accumulated to convince physicians that disease could be transmitted among individuals and that the transmission could be interrupted. The works of two investigators, Semmelweis and Snow, intensified this belief.

Ignaz Semmelweis was a Hungarian physician employed by the Vienna Hospital in Austria. In 1847, he reported that the agent of blood poisoning was transmitted to maternity patients by physicians fresh from performing autopsies in the mortuary. Semmelweis showed that hand washing in chlorine water could stop the spread of disease. However, his call for disinfection practices went largely unheeded because it implied that physicians were at fault.

sem'el-vīs

Blood poisoning:
a colloquial expression for infectious disease of the blood.

TABLE 1.1
Some Early Observations in Microbiology

INVESTIGATOR	TIME FRAME	OBSERVATIONS/CONCLUSIONS
Aristotle	Fourth century B.C.	Living things do not need parents; spontaneous generation apparently occurs
Fracostoro	Mid-1500s	"Contagion" passes among individuals, objects, and air
Kircher	Mid-1600s	"Microscopic worms" are present in blood of plague victims
Redi	Mid-1600s	Fly larvae arise by spontaneous generation
Van Leeuwenhoek	Late 1600s	Microscopic organisms are present in numerous environments
Fabricius	Early 1700s	Fungi cause plant disease
Joblot	Early 1700s	Various forms of protozoa exist
Needham	Mid-1700s	Microorganisms in broth arise by spontaneous generation
Spallanzani	Mid-1700s	Heat destroys microorganisms in broth
Jenner	Late 1700s	Recoverers from cowpox do not contract smallpox
Semmelweis	Mid-1800s	Chlorine disinfection prevents disease spread
Snow	Mid-1800s	Water is involved in disease transmission

FIGURE 1.4

A Streptococcus

An electron microscopic view of a bacterium known as a streptococcus. The bacterium has been caught on the rough surface of a filter, and the cells appear in chains typical of a streptococcus. A species of streptococcus is now known to be one of the causes of the "blood poisoning" studied by Semmelweis. The bar represents a length of one micrometer (1 μm) or 0.001 millimeter.

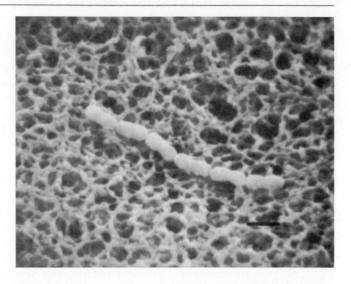

Cholera:
a bacterial disease of the intestine, accompanied by loss of large volumes of fluid.

John Snow, a British physician, traced the source of cholera to the municipal water supply of London during an 1854 outbreak. He reasoned that by avoiding the contaminated water source, people could avoid the disease. Snow's recommendations were adopted, and the spread of disease was halted. In their work, both Semmelweis and Snow also drew attention to the fact that a poison or unseen object in the environment, not a miasma, was responsible for disease (Figure 1.4). Proof, however, was still lacking.

TO THIS POINT
■

We have explored the development of microbiology by focusing on the work of Anton van Leeuwenhoek and illustrating how he called attention to a previously invisible world of microorganisms. In the late 1600s, van Leeuwenhoek communicated with the Royal Society of London and portrayed many commonly recognized microorganisms including protozoa, fungi, and bacteria. After his death, however, there was little interest in microbiology because microscopes were unavailable and because people simply could not conceive that organisms so tiny could bring on something so overpowering as sickness and death.

We then moved on to describe spontaneous generation and show how the controversy surrounding this doctrine stirred people to question the nature of living things and take heed of the existence of microorganisms. Redi, Needham, and Spallanzani achieved distinction during the controversy. The second focus was on disease transmission. We explored the works of Semmelweis and Snow to show how scientists were coming to believe that something tangible in nature was causing disease. This specific agent contrasted with the miasma theory, which pointed to a vague, intangible disease cause.

In the next section we shall see how Louis Pasteur and Robert Koch sought to establish the principle that microorganisms cause infectious disease. We shall discuss their experiments in depth and observe how the proofs for the germ theory of disease evolved. As the foundations of microbiology were strengthened, many other scientists expanded the work of Pasteur and Koch, and worldwide interest in microorganisms emerged. Modern microbiology is a product of that interest.

The Golden Age of Microbiology

The science of microbiology blossomed during a period of about 60 years referred to as the **Golden Age of Microbiology**. The period began in 1857 with the work of Louis Pasteur and continued into the twentieth century until the advent of World War I. During these years, numerous branches of microbiology were established and the foundations were laid for the maturing process that has led to modern microbiology.

Louis Pasteur

In a world ravaged by plague, tuberculosis, typhoid fever, and diphtheria, a large family was often necessary to ensure the next generation. Neither royalty nor commonfolk were immune to disease. There were virtually no cures for disease. Indeed, no one was really sure what caused it.

Such were the times in which **Louis Pasteur** (Figure 1.5) studied at the French school, *École Normale Supérieure*. In 1848, he achieved distinction in organic chemistry for his discovery that tartaric acid, a four-carbon organic compound, forms two different types of crystals. Pasteur successfully separated

---

F I G U R E 1 . 5
Louis Pasteur

Two portraits of Louis Pasteur, one of the founders of the science of microbiology. (a) As a young man while studying chemistry at the *École Normale Supérieure*. (b) As an established scientist in his laboratory at the Pasteur Institute.

(a)

(b)

1.2

MICROFOCUS

AHEAD OF HIS TIME

In the 1700s, smallpox was hardly ever absent from Europe. In England epidemics were so severe that one-third of the children died before reaching the age of three. Many victims were blinded by smallpox, and most people were left pockmarked for life.

But for those who survived the pox, immunity lasted forever. People therefore sought a way of contracting a mild form of the disease. From the Far East came word of inoculation parties. A doctor would make a small wound in the arm and insert a few drops of pus from a smallpox skin lesion. A walnut was then tied over the area. Though mild smallpox often resulted, there was danger of developing a deadly form.

Then Edward Jenner came on the scene. In the 1700s, while apprenticed to a country surgeon, Jenner learned that people who experienced coxpox were apparently immune to small pox. Cowpox was a

disease of the udders of cows. It was common in farmers and milkmaids, and it expressed itself as a mild, smallpoxlike disease.

For years, Jenner pondered whether intentionally giving cowpox to people would protect them against smallpox. He finally decided to put the matter to the test. In 1796, a dairymaid named Sarah Nelmes came to his office, the lesions of cowpox evident on her hand. Jenner took material from her lesions and scratched it into the skin of a boy named James Phipps. Several weeks later he inoculated young Phipps with material from a smallpox lesion. Within days, the boy developed a reaction at the site but failed to show any sign of smallpox.

Jenner continued his experiments for two years, switching to infected cows as a source of cowpox material. In 1798, he published an historic pamphlet on his work

and excited considerable interest. Prominent physicians confirmed his findings, and within a few years, Jenner's method of vaccination spread through the world (*vacca* is Latin for cow). It is estimated that by 1801, 100,000 people in England were vaccinated. In Russia, the first child vaccinated was renamed Vaccinor and was educated by the state.

A hundred years passed before scientists realized that the milder cowpox viruses were setting up a defensive mechanism in the body against the deadlier smallpox viruses. It is thus remarkable that Jenner accomplished what he did. His experimentation methods and interpretation of the vaccination results might well serve as models for accurate and careful laboratory work in modern microbiology.

Yeast cells:
oval or round microorganisms important in fermentation, baking, and other industrial processes.

the crystals while looking through a microscope. In doing so, he developed a skill that aided his later studies of microorganisms. In 1854, at the age of 32, he was appointed Professor of Chemistry at the University of Lille in northern France.

Pasteur believed that the discoveries of science should have practical applications. He therefore grasped the opportunity in 1857 to try and unravel the mystery of why local wines were turning sour, almost immediately stirring controversy. The prevailing theory held that wine fermentation results from the chemical breakdown of grape juice to alcohol. No living thing seemed to be involved. But Pasteur's microscope consistently revealed large numbers of tiny **yeast cells** overlooked by other scientists. Pasteur believed that yeasts played a major role in fermentation. Moreover, he noticed that sour wines contained populations of barely visible sticks and rods, known to physicians then and now as bacteria.

In a classic series of experiments, Pasteur clarified the role of yeasts in fermentation and showed that **bacteria** were responsible for sour wine. First he removed all traces of yeast from a sample of grape juice and set the juice aside to ferment. Nothing happened. Next he added yeasts back to the grape juice, and soon the fermentation was proceeding normally. He then found that if he could remove all traces of bacteria from the grape juice, the wine would not turn sour.

Pasteur's work shook the scientific community. His results demonstrated that yeast cells and bacteria were tiny, living factories where important chemical changes took place. His work also drew attention to microorganisms as the agents of change, because bacteria appeared to make the wine "sick." For years, physicians had interpreted bacteria as an effect of disease; that is, they were thought to arise in the body during illness. Pasteur's work appeared to indicate that they could be a cause of disease, for if they could sour the wine, perhaps they could also make the body ill. In 1857, Pasteur wrote a short paper on souring by bacteria. In the paper, he implied that microorganisms were related to human illness, and in doing so he set down the foundation for the **germ theory of disease**. Essentially this theory holds that microorganisms are responsible for infectious diseases.

Pasteur also recommended a practical solution to the sour wine problem. He suggested that grape juice be heated to destroy all the evidence of life, after which yeasts could be added to begin the fermentation. An alternative was to heat the wine before the bacteria soured it. Acceptance of his technique, known as **pasteurization**, gradually ended the problem. Pasteur's elation was tempered with sadness, however. In 1859, his daughter Jeanne died of typhoid fever.

Germ:
a common expression for a microorganism.

Typhoid fever:
a serious bacterial disease of the human intestine, blood, and other organs, accompanied by high fever.

Pasteur and Disease

Pasteur's interest in microorganisms rose as he learned more about them. He found bacteria in soil, water, air, and the blood of disease victims. Extending his germ theory of disease, he reasoned that if microorganisms were acquired from the environment, their spread could be controlled. If so, perhaps the chain of disease transmission could also be broken. However, many scientists stubbornly stuck to the notion that bacteria arose spontaneously from organic matter and that disease was inevitable so long as the "life force" was present. Pasteur was therefore drawn into the lingering debate on spontaneous generation, and he sought to discredit the doctrine in order to salvage his own germ theory of disease.

Pasteur first showed that where disease was rampant, the air was full of microorganisms, but where disease was uncommon, the air was clean. He opened flasks of nutrient-laden broth to air from the crowded city, then from the countryside, and next, from a high mountain. In each succeeding experiment, fewer flasks became contaminated with microorganisms. Still, his critics were vocal. When he boiled broths and showed that they remained free of life, the critics argued that boiling destroyed the "life force" believed to reside in the air.

Finally, in an elegant series of experiments, Pasteur silenced all but the most extreme supporters of spontaneous generation. In the early 1860s, he prepared broth in a series of **swan-neck flasks**, so named because their S-shaped necks resembled a swan's. Pasteur boiled the flasks of broth, then left them open to the air and any "life force." However, the S-shaped curvature of the neck trapped airborne dust and microorganisms and prevented their entry to the flask. No microorganisms grew in the broth. When the neck was later cut off, however, airborne organisms quickly fell into the broth, and growth appeared within hours. Pasteur's classic experiment is outlined in Figure 1.6.

Pasteur's work brought to an end the long and tenacious debate on spontaneous generation begun two centuries earlier. By now he was a national celebrity. Once more, however, tragedy entered his life; his daughter Camille developed a tumor and died of blood poisoning in 1865. Pasteur realized that he was no closer to solving the riddle of disease.

That same year, 1865, **cholera** engulfed Paris, killing 200 people a day. Pasteur attempted to capture the responsible bacterium by filtering the hospital air and trapping the bacteria in cotton. The cotton was then placed in broth

FIGURE 1.6

Pasteur and the Spontaneous Generation Controversy

(a) When flasks of sterile broth were left open to the air, living organisms appeared. Pasteur suggested that this was because microorganisms entered the flasks. His critics countered that a "life force" entered from the air to give rise to life. (b) When flasks were boiled and sealed, no life appeared. Pasteur believed that the heat killed the microorganisms; his opponents believed the heat had destroyed the "life force." (c) When the flasks were left open and the air was heated, no life appeared. Pasteur proposed that airborne microorganisms were killed by the heat, but his critics suggested that the "life force" had been destroyed. (d) The swan-neck flask solved the controversy. Here the flask was left open to the air, but the neck was curved to trap microorganisms as they fell into the side arm. Pasteur proposed that no life would appear in the flask. His opponents believed the life would arise because the "life force" had free access to the flask. When life failed to appear, Pasteur was proved correct.

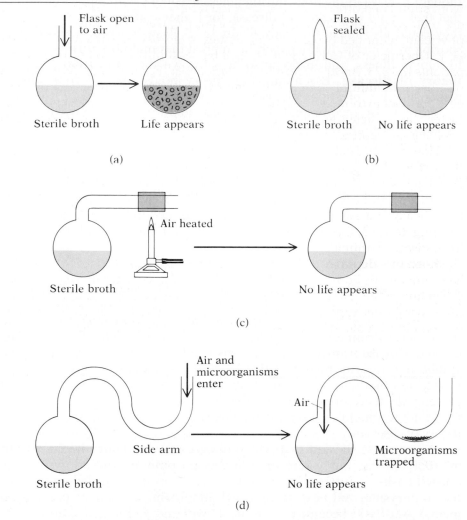

medium where a mixture of bacteria grew. Unfortunately, Pasteur was unable to cultivate one bacterium apart from the others because he was using broth. (In later experiments Koch would use solid culture media instead of broth media.) Although Pasteur demonstrated that bacterial inoculations made animals ill, he could not pinpoint an exact cause. Some of his critics claimed that a poison, or toxin, in the broth was responsible for the disease.

In an effort to help French industry, Pasteur turned his attention to **pébrine**, the disease of silkworms. Late in 1865, he identified a protozoan infesting the sick silkworms and the mulberry leaves fed to them. Next he separated the healthy silkworms from the diseased silkworms and their food, and he managed to quell the spread of disease. The achievement strengthened the germ theory of disease. For Pasteur, however, it was another time of grief. Cecille, another of his daughters, died of typhoid fever in 1866. Again he returned to the study of human disease.

Toxin:
a chemical substance poisonous to the body tissues.

pa-brēn'

Robert Koch

Although Pasteur failed to relate a specific organism to a specific disease, his work stimulated others to investigate the nature of micoorganisms and to ponder their association with disease. For example, Gerhard Hansen, a Norwegian physician, identified bacteria in the tissues of leprosy patients in 1871, and Otto H. F. Obermeier of Germany described bacteria in the blood of relapsing fever patients in 1873. Another German bacteriologist, Ferdinand J. Cohn, discovered that bacteria multiply by dividing into two cells. He also observed that certain ones form an extremely resistant structure called the spore. In England, Joseph Lister was sufficiently impressed with Pasteur's writings on disease to begin applying antiseptics to wound infections.

However, the definitive proof of the germ theory of disease was offered by **Robert Koch** (Figure 1.7a), a country doctor from East Prussia, now part of Germany. Koch's primary interest was **anthrax**, a deadly blood disease in cattle and sheep. Anthrax was a threat to farmers because it ravaged their herds periodically and seemed to reappear time and again in the same district without warning. In 1863, Casimir Davaine, a French physician, had cultivated a bacillus from diseased animals.

Koch was determined to learn more about anthrax. In 1875, in a makeshift laboratory in his home (Figure 1.7b), he injected mice with the blood of diseased sheep and cattle. He then performed meticulous autopsies and noted that the same symptoms appeared regularly. Next, he isolated a few rod-shaped bacilli from a mouse's blood by placing the bacilli in the sterile aqueous humor from

Relapsing fever:
a bacterial disease, accompanied by recurring periods of fever.

Spore:
a highly resistant oval body formed by certain types of bacteria.

Aqueous humor:
the sterile gel-like fluid from behind the lens of the eye.

FIGURE 1.7

Robert Koch

(a) Koch was one of the first to relate a specific organism to a specific disease and thus prove the germ theory of disease. Koch also pioneered pure culture techniques in microbiology. (b) Robert Koch's study. Note the bell jar (A) for cultivating microorganisms and the camera apparatus (B) for photographs. Koch's microscope is evident (C), and the various stains and chemicals used for smear preparations are shown (D).

(a)

(b)

an ox's eye. Koch watched for hours as the bacilli multiplied, formed tangled threads, and finally reverted to the resistant spores described by Cohn. At this point, he took several spores on a sliver of wood and injected them into healthy mice. The symptoms of anthrax appeared within hours. Koch autopsied the animals and found their blood swarming with bacilli. He reisolated the bacilli in sterile aqueous humor. The cycle was now complete.

Koch communicated his findings to Cohn, and in 1876 Cohn invited the 35-year-old physician to present his work at the University of Breslau. Scientists there were astonished. Here was the proof for the germ theory of disease that had eluded Pasteur and for which many of them were waiting. Koch's procedures came to be known as **Koch's postulates**. These techniques, illustrated in Figure 1.8 were quickly adopted as a guide for relating specific organisms to specific diseases. They state that the same microorganism must be identified in all cases of the disease, the microorganism must be isolated and grown in pure culture, and the disease must be reproduced in experimental animals inoculated with these pure cultures. The same microorganism must then be recovered. Originally the postulates had been outlined in 1840 by Jakob Henle, a German researcher. One of his students had been Robert Koch.

Koch's postulates:
a series of procedures by which a specific microorganism can be related to a specific infectious disease.

FIGURE 1.8

Demonstration of Koch's Postulates

Koch's postulates are used to relate a single microorganism to a single disease. (a) Organisms are observed in tissue specimens from a dead animal and (b) isolated and grown in pure culture. (c) A sample of the pure culture is inoculated to a healthy animal, and (d) the disease is reproduced in the animal. (e) Identical organisms are then observed in tissue samples of the dead animal and (f) reisolated in pure culture. The inset shows anthrax rods as Koch viewed them. Many rods are swollen with clear spores.

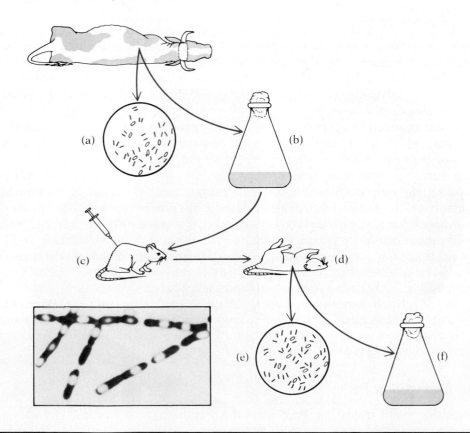

1.3

MICROFOCUS

JAMS, JELLIES, AND MICROORGANISMS

One of the major developments in microbiology was Robert Koch's use of a solid culture medium on which bacteria would grow. He accomplished this by solidifying beef broth with gelatin. When inoculated onto the surface of the nutritious medium, bacteria grew vigorously and produced discrete, visible colonies.

On occasion, however, Koch was dismayed to find that the gelatin-broth medium turned to liquid. It appeared that certain types of bacteria were producing a chemical substance to digest the gelatin. Moreover, gelatin liquefied at the high incubator temperatures commonly used to cultivate bacteria.

Walther Hesse, an associate of Koch, mentioned the problem to his wife and laboratory assistant, Fanny Eilshemius. She had a possible solution. For years, she had

been using a seaweed-derived powder called agar (pronounced ahg'ar) to solidify her jams and jellies. The formula had been passed to her by her mother, who learned it from Dutch friends living in Java. Agar was valuable because it mixed easily with most liquids and, once gelled, it did not liquefy, even at high temperatures.

Hesse was sufficiently impressed to recommend agar to Koch. Soon Koch was using it routinely in his culture media, and in 1884 he first mentioned agar in his paper on the isolation of the tubercle bacillus. It is noteworthy that Fanny Eilshemius may have been among the first Americans to make a significant contribution to microbiology (she was originally from New Jersey).

Another point of interest: the common Petri dish was also invented about this time by Julius Petri, one of Koch's assistants.

Pure Culture Techniques

After the sensation at Breslau subsided, Koch returned to his laboratory and developed numerous staining methods for bacteria. In 1880 he accepted an appointment to the Imperial Health Office, and during this period he happened upon the **cultivation techniques** that sparked the further development of microbiology.

Koch chanced to observe that a slice of potato contained small masses of bacteria, which he termed **colonies**. Colonies contained millions of just one kind of bacteria. Koch concluded that bacteria could grow and multiply on solid surfaces, and he added gelatin to his broth to prepare a solid culture medium. He then inoculated bacteria to the surface and set the medium aside to incubate. Next morning, visible colonies were present on the surface. When colonies of the same bacterium grew together, a **pure culture** formed. Koch could now inoculate laboratory animals with a pure culture of bacteria and be certain that only one species of bacterium was involved. His work also proved that bacteria, not toxins in the broth, were the cause of disease. MicroFocus: 1.3 details a further advance in cultivation techniques.

In 1881, Koch demonstrated his pure culture techniques to the International Medical Congress, meeting at Lister's laboratory in London. Pasteur and several of his co-workers were present as Koch outlined his methods. Several days later, Koch received a personal letter of congratulations from Pasteur.

Colony:
a visible mass of microorganisms, usually of a single type.

Pure culture:
an accumulation of one type of microorganism formed by the growth of colonies of that organism.

The Competition Period

At another point in history, Pasteur and Koch might have become friends and colleagues in their mutual search for the agents of disease. However, the years following the 1870 Franco-Prussian war were accompanied by fierce national pride. Both France and Germany were undergoing unification, and heroes played an important role in the spirit of nationalism. Pasteur became a symbol of French achievement and Koch, his German rival. There sprung up a competition that would last into the next century. (Table 1.2)

Koch's proof of the germ theory was presented in 1876. Within two years, Pasteur had verified the proof and gone a step further. He reported that bacteria were temperature-sensitive because chickens did not acquire anthrax (Figure 1.9) at their normal body temperature of 42° C but did so when the animals were cooled down to 37° C. He also recovered anthrax spores from the soil and pointed out that cattle were probably infected during grazing. This explained the periodic recurrence of the disease. Pasteur suggested that dead animals be burned or buried deeply in soil unfit for grazing.

One of Pasteur's more remarkable discoveries was made in 1880 when a group of inoculated chickens failed to develop chicken cholera. For months he had been working on ways to enfeeble bacteria using heat, different growth media, passages among animals, and virtually anything he thought might weaken them. Finally, he had developed two cultures whose ability to cause disease was reduced. The trick, according to his notebooks, was to suspend the bacteria in a mildly acidic medium and allow the culture to remain undisturbed for a long period of time. When it was inoculated to chickens and later followed by a dose of lethal cholera bacilli, the animals did not become sick. This principle is the basis for the use of many **vaccines** for immunity. Pasteur applied the principle to anthrax in 1881 and found he could protect sheep against the disease.

Pasteur's experiments put France once more in the forefront of science. However, Koch's 1881 announcement of pure culture techniques drew attention back to Germany, and within a year Koch isolated the tubercle bacillus, the cause of **tuberculosis**. In 1884 his associate, Georg Gaffky, cultivated the **typhoid** bacillus, and that same year another co-worker, Friederich Löeffler, isolated the **diphtheria** bacillus. Soon, news was forthcoming from France that Emile Roux and Alexandre Yersin of Pasteur's group had linked diphtheria to a toxin produced in the body. In later years, Koch's co-worker, Emil von Behring, successfully treated diphtheria by injecting antitoxin, a blood product (actually a preparation of antibodies) obtained from animals given injections of the toxin. For his work, von Behring was awarded the first Nobel Prize in Physiology or Medicine (MicroFocus: 1.4).

Vaccine:
a preparation of modified microorganisms, treated toxins, or parts of microorganisms used for immunization purposes.

lef'ler
dif-the'rē-ah
roo

Diphtheria:
a bacterial disease, accompanied by disintegration of the throat tissues in children.

TABLE 1.2

A Comparison of Louis Pasteur and Robert Koch

CHARACTERISTIC	LOUIS PASTEUR	ROBERT KOCH
Country of origin	France	Germany (Prussia)
Preparatory education	Chemistry	Medicine
Initial investigations	Milk souring; beer, wine fermentations	Cause of anthrax
Accomplishments	Proposed germ theory of disease Disproved theory of spontaneous generation Developed immunization techniques Resolved pébrine problem of silkworms Developed rabies vaccine	Proved germ theory of disease Developed cultivation methods for bacteria Isolated bacterium that causes tuberculosis Developed staining methods for bacteria Investigated cholera, malaria, sleeping sickness
Associates	Roux, Yersin, Metchnikoff	Gaffky, Loeffler, von Behring, Kitasato
Nobel Prize	No	Yes

FIGURE 1.9

The Anthrax Bacillus as Seen by Pasteur

A copy of a photomicrograph of the anthrax bacillus taken by Louis Pasteur. Pasteur circled the bacilli in tissue and annotated the photograph "the parasite of Charbonneuse." (Charbonneuse is the French equivalent of anthrax.) He requested that the photograph be brought to the attention of Monsieur Méderlle, dated it 20 March, 1885, and initialed it in the lower right.

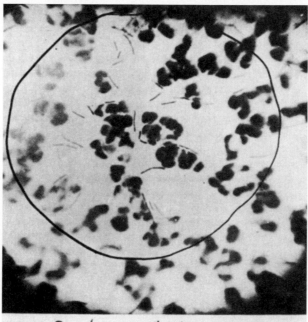

1.4

MICROFOCUS

THE NOBEL PRIZE

The Nobel Prizes are among the world's most venerated awards. Originally, they were a gesture of peace by a man whose discovery had unintentionally added to the destructive forces of warfare.

Alfred Bernhard Nobel was the third son of a Swedish munitions expert. As a young engineer, he developed an interest in nitroglycerine, the oily substance that was 25 times more explosive than gunpowder. In 1863, Nobel obtained a patent for a detonator of mercury fulminate, and within four years he succeeded in solidifying nitroglycerine by mixing it with a type of sandy clay. The mixture was called dynamite, from the Greek *dynamis* for power.

Dynamite had a clear advantage over other explosives because it could be transported easily and handled without fear. It became an overnight success and was adapted to applications in mining, tunnel construction, and bridge and road building. Before long, it was being used in armaments on the battlefield.

Nobel soon amassed a fortune through control of several European companies that produced dynamite. However, toward the end of his life he became a pacifist and began speaking out against the use of dynamite in warfare. In 26 lines of his handwritten will, Nobel directed that the bulk of his estate should be used to award prizes that would promote peace, friendship, and service to humanity.

After his death in 1896, the governments of Sweden and Norway established Nobel Prizes in five categories: chemistry, physics, physiology or medicine, literature, and peace. A sixth category, economics, was added in 1969. Each year Nobel laureates assemble in Oslo or Stockholm on December 10, the anniversary of Nobel's death. Each laureate receives a medallion, a scroll, and all or part of a cash award currently valued at about one million dollars per category.

The first Nobel Prize winners were announced in 1901. Among the recipients were Wilhelm K. Roentgen, the discoverer of X rays; Jean Henri Durant, the founder of the Red Cross; and Emil von Behring, the developer of the diphtheria antitoxin.

1.5

MICROFOCUS

SAVED FROM CERTAIN DEATH

Louis Pasteur was a tireless worker who drove himself and his assistants mercilessly. At heart, however, he was something of a sentimentalist, as the events of July 1885 display.

Pasteur's intuition and scientific technique reached a peak with his development of a vaccine for rabies. To make the vaccine, Pasteur inoculated rabbits with the brain tissue of rabid animals. He then removed the spinal cords and dried them. Next he inoculated experimental animals with 15-day-old cord tissue and followed this up the next day with 14-day-old cord tissue, and so on for two weeks. Animals so treated would not develop rabies.

On July 6, 1885, a nine-year-old boy named Joseph Meister was brought to Pasteur. Two days earlier, the boy had suffered multiple bites from a rabid dog, and physicians assured Pasteur that a horrible death was imminent. The boy's only hope was the vaccine.

Pasteur's vaccine had never been tried on humans, and the scientist, now 63, could not bring himself to make an immediate decision. In his writing, he recalls the incident: "The child's death appeared inevita-

ble. I decided not without acute and harrowing anxiety, as may be imagined, to apply to Joseph Meister the method I had found consistently successful with dogs."

The next day, Pasteur began the treatment. He turned the first sample of vaccine over to two physicians from the Academy of Medicine and watched as they injected the material to the terrified, crying child. Each day his concern lessened as the injections proceeded smoothly. The vaccine appeared to be working.

Joseph Meister survived his ordeal. When he left Paris for his village, Pasteur gave him stamped envelopes so that he could write often. In later years, a grateful Joseph Meister returned to Paris as gatekeeper and caretaker of the Pasteur Institute. He died in 1940 as the German army descended upon the Institute during the occupation of France.

A photograph of the flask used by Pasteur when preparing the rabies vaccine. The flask contains a strip of rabbit spinal cord hanging over a drying agent. When completely dry, the spinal cord was ground up and suspended in fluid to form the vaccine.

---

fag″o-sī-to′sis

Rabies:
a deadly viral disease, accompanied by paralysis and usually acquired by animal bites.

There was also an international flavor in the French and German laboratories. Shibasaburo Kitasato of Japan studied with Koch and successfully cultivated the **tetanus** bacillus, an organism that grows only in the absence of oxygen. One of Pasteur's associates was Elie Metchnikoff, a native of the Ukraine. In 1884, Metchnikoff published an account of **phagocytosis**, a defensive process in which the body's white blood cells engulf and destroy microorganisms.

This period also witnessed two improvements in microscopy, both attributed to the German physicist Ernst Karl Abbé. In 1878, Abbé introduced the **oil immersion lens**, a standard feature of modern microscopes. Eight years later he invented the system of lenses and mirrors known as the **Abbé condenser**. This apparatus concentrates light on objects being viewed and makes increased magnification feasible. Thus, the improved technology for seeing microorganisms dovetailed the interest in microorganisms.

In 1885, Pasteur reached the zenith of his career when he successfully immunized young Joseph Meister against the dreaded disease **rabies** (MicroFocus: 1.5). Although he never saw the agent of rabies, Pasteur was able to cultivate it in the brain of animals and inject the boy with bits of the tissue. The experiment was a triumph for Pasteur because it fulfilled his dream of

FIGURE 1.10

Robert Koch and Paul Ehrlich

Two investigators at the forefront of discovery in microbiology. (a) A photograph of Robert Koch in East Africa in 1906. Koch was 63 years old when he went to Africa to study the cause of sleeping sickness. By this time he had achieved worldwide recognition for his proof of the germ theory of disease. (b) Paul Ehrlich, a chemist who synthesized hundreds of organic compounds while searching for a "magic bullet" that would kill microorganisms in the human body.

(a)

(b)

applying the principles of science to practical problems. Many monetary rewards were also forthcoming, including a generous gift from the Russian government after Pasteur immunized 20 Russians against rabies. The funds were used to help establish the Pasteur Institute in Paris, one of the world's foremost scientific establishments. Pasteur presided over the Institute until his death in 1895.

Koch also reached the height of his influence in the 1880s. His isolation of the tubercle bacillus in 1882 was part of a continuing interest in tuberculosis that continued for the remainder of his years. In 1883, he interrupted his work to lead groups studying **cholera** in Egypt and India. In both countries, Koch isolated a comma-shaped bacillus and confirmed the suspicion first raised by John Snow 30 years earlier that water transmits the disease. In 1885 he was appointed to the University of Berlin, and in 1891 he became Director of the Institute for Infectious Diseases. At various times, Koch studied malaria, plague, and sleeping sickness (Figure 1.10a), but his work with tuberculosis ultimately gained him the 1905 Nobel Prize in Physiology or Medicine. He died of a stroke in 1910 at the age of 66.

Other Pioneers of Microbiology

With the aging of Pasteur and Koch, a new generation of scientists stepped in to expand their work. For example, a Pasteur Institute scientist, Charles Nicolle, proved that **typhus fever** was transmitted by lice. Albert Calmette, also of the Institute, developed a harmless strain of the tubercle bacillus used for immunization. Jules Bordet, another French scientist, isolated the bacillus of **pertussis** (whooping cough) and developed the complement fixation test, a procedure once widely used in the diagnosis of disease.

Typhus fever:
a louseborne bacterial disease of the blood, characterized by a skin rash and high fever.

kal'met
bord-dā'

Meningitis:
disease of the coverings of
the spinal cord and brain.

Syphilis:
a bacterial disease
of multiple body organs,
transmitted by sexual
contact.

Undulant fever:
a bacterial disease,
accompanied by periodic
bouts of fever.

**Rocky Mountain spotted
fever:**
a tickborne bacterial disease
of the blood, characterized
by high fever and skin rash.

Koch's successors included Emil von Behring and Richard Pfeiffer, who isolated one of several organisms that cause **meningitis**. Still another co-worker was Paul Ehrlich, a chemist who explored the mechanisms of immunity and synthesized the "**magic bullet**," an arsenic compound that would seek out and destroy syphilis organisms in the human body (Figure 1.10b).

By the turn of the century, microbiology had moved far beyond the boundaries of France and Germany (MicroFocus: 1.6). Ronald Ross, an English physician working in the Far East, proved that the mosquitoes were the vital link in **malaria** transmission. The discovery earned him the 1902 Nobel Prize. Another Englishman, David Bruce, isolated the cause of undulant fever. While working in Africa, Bruce also showed that tsetse flies transmit **sleeping sickness**. The control of this insect opened the African continent to British colonization. A third British subject, Almroth Wright, described opsonins, the chemical substances that assist phagocytosis in the body. Thirty years later, one of his students, Alexander Fleming, discovered the antibiotic penicillin.

Two Japanese investigators also achieved distinction in these years. In 1897, the Tokyo physician Masaki Ogata reported that rat fleas transmit **bubonic plague**. This discovery solved a centuries-old mystery of how plague spread. A year later, Kiyoshi Shiga isolated the bacterium that causes **bacterial dysentery**, an important intestinal disease. The organism was later named *Shigella*.

The American group of microbiologists was represented by several researchers, among them Daniel E. Salmon and Theobald Smith. These investigators were among the first to use heat-killed bacteria for immunizations. Salmon later studied swine plague and lent his name to *Salmonella*, the cause of **typhoid fever**. Smith showed that **Texas fever**, a disease of cattle, was transmitted by ticks. In addition, the University of Chicago pathologist Howard Taylor Ricketts located the agent of Rocky Mountain spotted fever in the human bloodstream and demonstrated its transmission via ticks. Another American, William Welch, isolated the **gas gangrene bacillus** at his laboratory at Johns Hopkins University. And, Walter Reed led a contingent to Cuba and pinpointed mosquitoes as the insects involved in **yellow fever** transmission. His discovery led to the mosquito eradication programs that made possible the building of the Panama Canal.

Amid the burgeoning interest in medical microbiology, other scientists devoted their research to the environmental importance of microorganisms. The Russian scientist Sergius Winogradsky, for example, discovered that soilborne bacteria could bring about chemical changes in ammonia and in iron and sulfur compounds to obtain life-sustaining energy. One of his more significant observations was that certain bacteria utilize carbon dioxide to synthesize carbohydrates, much as plants do in photosynthesis (Chapter 5).

Another environmental microbiologist was Martinus Beijerinck, a Dutch investigator. Beijerinck isolated bacteria that could trap nitrogen in the soil and make it available to plants for use in constructing amino acids (Chapter 26). He also studied the bacteria prominent in sulfur cycles of the soil and devoted much of his time to the properties of viruses (Chapter 11). Together with Winogradsky, he developed many of the laboratory media essential for the study of environmental microbiology.

End of the Golden Age

The advent of World War I in 1914 signaled a dramatic pause in microbiology research and brought to an end the Golden Age of Microbiology. The

1.6

OF MICROORGANISMS AND CHERRY TREES

In the 1880s, while medical microbiologists were searching out the causes of infectious disease, a Japanese biochemist was busy acquiring enormous experience and wealth as one of the first industrial microbiologists.

The biochemist's name was Jokichi Takamini. In the 1870s, Takamini left his native Japan to study engineering in the West. By the 1890s, he had married an American girl and was an expert on the enzymes of a common mold named *Aspergillus* (as-per-jil′us). Having isolated and purified a certain *Aspergillus* enzyme, he used it to help make whiskey at his fermentation plant in Peoria, Illinois. Local brewers turned against his innovative methods, however, so he patented the enzyme and li-

censed it to a pharmaceutical company. The company mixed it with peppermint and sold it as a digestive aid. Marketed as "taka-diastase," the mold enzyme became the Alka-Seltzer of the 1890s.

Takamini was not only a pioneer of industrial microbiology, he was also one of the first biotechnologists. After moving to New York, he and an assistant set out to isolate the active principle (the hormone) in adrenal glands. For months they separated, precipitated, dissolved, purified, repurified, and hunted for the elusive substance. Then one night, his assistant, too tired to wash the glassware, left it for the morning and went home. The next day, a glass that had contained extract solution was lined with

crystals of epinephrine (adrenalin). The isolation was a success and Takamini's patent rights to epinephrine yielded wealth beyond his imagination.

In later years, Takamini became a patron of the arts and a philanthropist. It so happened that on August 30, 1909 the city of Tokyo announced a gift to the people of the United States: the city would be honored to adorn the Tidal Basin in Washington, DC with dozens of Japanese cherry trees. In the decades following, the cherry trees grew to become a splendid attraction in the nation's capitol and an annual harbinger of spring. What few people know is that the gift was funded anonymously by Jokichi Takamini.

Pasteur Institute was closed as Paris came under siege, and the German laboratories focused on producing antibacterial serums for treating war-related diseases.

The Golden Age witnessed a series of discoveries unparalleled in medicine, most involving identification of the agents of disease (Table 1.3). Although cures for established diseases would not come until the 1940s, scientists developed an awareness that infectious disease was caused by microorganisms and that the chains of transmission could be broken. Their discoveries led to calls for sterile practices in hospitals, pasteurization of milk, purification of water, control of insects, and care in the preparation of food for consumption. Together with improved sanitation and careful personal hygiene, these measures brought about a substantial reduction in the incidence of bacterial disease.

However, the agents of such diseases as measles, mumps, smallpox, yellow fever, and polio continued to elude microbiologists. As the 1920s and 1930s passed, the belief developed that some sort of "virus" was to blame and that discovery of the virus was simply a matter of time. When viruses were finally visualized in the 1940s, they were unlike anything that scientists had imagined. Nevertheless, microbiologists found a way to immunize against their diseases and interrupt their transmission. We shall explore how this took place in Chapter 11.

In our era many viral diseases are coming under control much as many bacterial diseases did at the turn of the century. Smallpox has not occurred in the world since 1977, and polio, rubella, and measles are on a rapid decline. As this trend continues to take place, we shall be watching for the diseases of the future. In this generation, the lexicon of microbiology already has added such terms as Legionnaires' disease, Ebola fever, Kawasaki disease, toxic shock syndrome, and acquired immune deficiency syndrome (AIDS). Illnesses such as these represent the challenges for tomorrow's pioneers of microbiology.

Virus: an infectious particle, composed of nucleic acid and protein, that replicates within living cells.

TABLE 1.3

Some Landmark Accomplishments during the Golden Age of Microbiology, 1857–1914

INVESTIGATOR	COUNTRY	ACCOMPLISHMENT
Gerhard Hansen	Norway	Observed bacteria in leprosy patients
Otto F. H. Obermeier	Germany	Observed bacteria in relapsing fever patients
Ferdinand J. Cohn	Germany	Described life cycle of certain bacteria
Joseph Lister	Great Britain	Developed the principles of aseptic surgery
Georg Gaffky	Germany	Cultivated the typhoid bacillus
Emile Roux and Alexandre Yersin	France	Identified the diphtheria toxin
Emil von Behring	Germany	Developed the diphtheria antitoxin
Shibasaburo Kitasato	Japan	Isolated the tetanus bacillus
Elie Metchnikoff	Ukraine	Described phagocytosis
Ernst Karl Abbé	Germany	Developed the oil immersion lens and Abbé condenser
Charles Nicolle	France	Proved that lice transmit typhus fever
Albert Calmette	France	Developed immunization process for tuberculosis
Jules Bordet	France	Isolated the pertussis bacillus
Richard Pfeiffer	Germany	Identified a cause of meningitis
Paul Ehrlich	Germany	Synthesized a "magic bullet" for syphilis
Ronald Ross	Great Britain	Showed that mosquitoes transmit malaria
David Bruce	Great Britain	Isolated the undulant fever bacillus Proved that tsetse flies transmit sleeping sickness
Almroth Wright	Great Britain	Described opsonins to assist phagocytosis
Masaki Ogata	Japan	Discovered that rat fleas transmit plague
Kiyoshi Shiga	Japan	Isolated a cause of bacterial dysentery
Daniel E. Salmon	United States	Studied swine plague
Theobald Smith	United States	Proved that ticks transmit Texas fever
Howard T. Ricketts	United States	Showed that ticks transmit Rocky Mountain spotted fever
William Welch	United States	Isolated the gas gangrene bacillus
Walter Reed	United States	Proved that mosquitoes transmit yellow fever
Sergius Winogradsky	Russia	Studied biochemistry of soil bacteria
Martinus Beijerinck	Netherlands	Developed discipline of environmental microbiology

NOTE TO THE STUDENT

It would be wrong to believe that Pasteur and Koch were the only ones of their era to think that microorganisms cause infectious disease. Quite the contrary, a substantial number of scientists sensed that different diseases were due to microbial agents, but few were bold enough to hold to this notion without proof. Blaming disease on microorganisms was breaking with centuries-old traditions rooted in beliefs and dogmas. That is why the works of Pasteur and Koch are so special—Pasteur's studies brought microorganisms to the public eye, and Koch provided irrefutable proof of their involvement in disease. There followed a veritable explosion in research that was worldwide in scope. In this chapter I have tried to highlight some of the excitement of the times and variety of studies taking place.

You will note that we have focused on the development of microbiology in Europe. This chapter says little about medical discoveries taking place in the Middle and Far East simply because historians are not really sure what was happening there. However, we know that Eastern peoples were far ahead of Europeans in certain areas. For instance, people in China practiced vaccination against smallpox long before Jenner introduced it in Europe. Perhaps when the history of microbiology in the East is written, it will be necessary to rewrite this chapter substantially.

We have largely ignored twentieth-century microbiology because the remaining chapters in this text cover it in depth. Most of what you will learn is a product of research in the 1900s. Witness, for example, how measles has come under control during our lifetime. If you watch the newspapers, magazines, and other media, you will see the history of microbiology unfolding. Perhaps your grandchildren will be surprised to learn that you once had measles, a disease they could read about in a textbook but would never experience firsthand.

Summary

In the late 1600s, Anton van Leeuwenhoek reported the existence of microorganisms and sparked interest in a previously unknown world of microscopic life. Although strong attention to the microbial world did not continue after his death in 1723, the controversy concerning spontaneous generation drew scientists to explore the origin and nature of living things. In the 1860s, scientists such as Snow and Semmelweis believed that infectious disease could be caused by something transmitted from the environment and that the transmission could be interrupted.

The fermentation experiments of Louis Pasteur, begun in the 1850s, were innovative and imaginative because they focused attention on the chemical changes that microorganisms could induce. Gradually, Pasteur was led to the possibility that human disease could be due to chemical changes brought about by microorganisms in the body, and he proposed a germ theory of disease.

However, Pasteur was unable to cultivate microorganisms in pure culture and prove his germ theory, and it fell to Robert Koch to set down the methods for relating a single microorganism to a single disease. Now the germ theory could be verified. Soon an intense rivalry developed between the associates of Pasteur and Koch as they hunted down the microorganisms of infectious disease. Scientists throughout the world continued and developed the research in microbiology, and they instituted a Golden Age of Microbiology, a period that lasted until the advent of World War I. The science of microbiology has developed from that research.

Questions for Thought and Discussion

1. Abu-Bakr Muhammed al-Razi, better known to history as Rhazes, was a famous Arabic doctor of the 900s. On moving to Bagdad to establish a hospital, he picked a site by hanging pieces of meat throughout the city and choosing the place where the last piece of meat turned rotten. How does this insight show an early appreciation of the relationship between disease and something present in the air?

2. Louis Pasteur received veneration from his contemporaries, and he continues to be held in high esteem to this day as one of the foremost scientists ever to have lived. However, his fellow scientist Robert Koch received no such adulation, nor is his name remembered as fondly as Pasteur's even though the accomplishments of both appear parallel. What do you think has contributed to Pasteur's popularity?

3. Suppose you were a research microbiologist in the year 1900 and you had a consuming interest in the organism that causes measles. What direction would your research be taking and what would be its fate?

4. Why do you think it was important to the changing view of biology in the 1700s for people to believe in spontaneous generation?

5. Many people are fond of pinpointing events that alter the course of history. In your mind, which single event in this chapter had the greatest influence on the development of microbiology? What event would be in second place?

6. Few cures for disease were found between 1857 and 1914, and yet this period has come to be known as the Golden Age of Microbiology. Why is this so?

7. In 1911, the Polish chemist Casimir Funk isolated an active substance from rice husks and proposed that it could prevent beriberi, a disease of the muscular and nervous systems. "Nonsense," he was told. "Spend your time more profitably—look for a microorganism!" Today we know that Funk was correct—beriberi is a nutritional disease due to a vitamin deficiency. But at that time, the scientific community scoffed at his idea. Why do you think that happened?

8. A textbook author writes that the science of microbiology originated from various attempts to solve the mystery of the origin of life. Would you support or reject this concept? Why?

9. The year 1884 was a particularly fruitful one in microbiology. For example, the Danish physician Christian Gram developed the Gram stain technique, which is used to separate bacteria into distinctive groups. In the same year, the German physician Albert Frankel identified an important cause of bacterial pneumonia. How many other events or pieces of research can you locate that occurred or were ongoing that year?

10. The word "microbe" was coined in 1879 by the French investigator Charles E. Sedillot as a term for a microscopic creature. Today the word is used interchangeably with "microorganism." Do you think that Sedillot had something else in mind a hundred years ago? If so, what?

11. Suppose uncooked hamburger meat became smelly after some days in the refrigerator, and imagine that you brought a sample to the laboratory, where microorganisms were revealed. How could you prove that the organisms did not arise by spontaneous generation?

12. Abbé's development of oil immersion microscopy and his invention of the microscope condenser appeared to parallel the interest in microorganisms in the 1880s. Some maintain that the availability of new technologies made the discovery and study of microorganisms possible, but others assert that the intense interest in microorganisms demanded that the technologies be developed. Which argument would you favor? Why?

13. This chapter suggests two factors that retarded the development of microbiology after the death of van Leeuwenhoek. Can you suggest any others?

14. Why do you suppose there was no follow-up work on smallpox after Edward Jenner's development of a smallpox vaccine? And why might you enjoy learning the art of persuasion from Jenner?

15. In downtown Mexico City at the crossroads of Insurgents Avenue and Paseo de la Reforma there is an area called Plaza Louis Pasteur. At the plaza there is an elegant statue of Pasteur given to Mexico City by the French residents of Mexico in 1910. That year marked the centennial of the start of the Mexican War of Independence. What connection can you see among these facts and events?

Review

This chapter has explored the development of microbiology up through the period of World War I. To test your knowledge of the chapter contents, complete the following crossword puzzle. The answers are listed in the appendix.

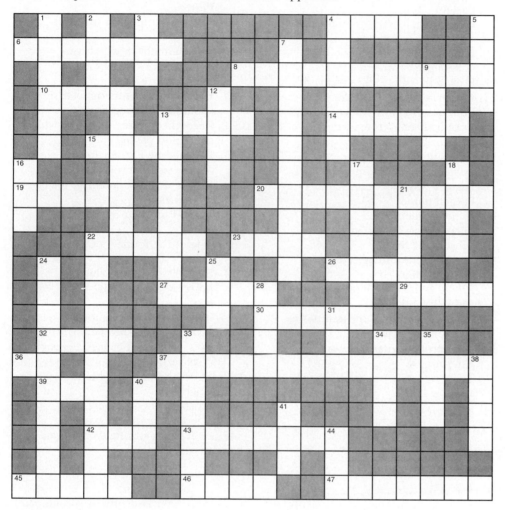

ACROSS

4. Arthropod that transmits plague
6. Microorganisms poorly studied during the Golden Age
8. Van Leeuwenhoek's term for microorganisms
10. Pasteur's flasks had the neck of a _____
13. Related diphtheria to a toxin
14. Mass of bacteria isolated by Koch
15. Attempted to disprove spontaneous generation by covering meat
19. Isolated the gas gangrene bacillus
20. Used chlorine water for hand disinfection in hospital
22. Related a specific arthropod to Texas fever

DOWN

1. Vague airborne principle believed to cause disease
2. Country in which Reed performed his studies
3. Insect whose stinger was described by van Leeuwenhoek
4. Pasteur's country
5. Fungal disease of plants
7. Disease agent isolated by Loeffler
9. Van Leeuwenhoek's microscope had a single _____
11. His broth experiments supported spontaneous generation
12. Type of culture obtained by Koch
13. Related ticks to Rocky Mountain spotted fever

ACROSS

23. Environment where anthrax spores are located
26. Souring attracted attention of Pasteur
27. Isolated the bacterium that causes dysentery
29. Believed by Pasteur to be the cause of spoilage in wine
30. Proved that tsetse flies transmit African sleeping sickness
32. First described by Robert Hooke
36. Tetanus bacilli grow where there is _____ oxygen
37. Dutch investigator who reported existence of microorganisms
39. Pasteur was the _____ of a French tanner
42. In the Golden _____, microbiology experienced substantial growth
43. Searched for a magic bullet to treat syphilis
45. Produced by the bacterium in cases of diphtheria
46. Studied the role of mosquitoes in yellow fever epidemics
47. Causative agent studied by Kitasato

DOWN

16. Van Leeuwenhoek's letters number over _____ hundred
17. Disease of silkworms studied by Pasteur
18. Arthropod that can transmit Texas fever
21. Environment in which van Leeuwenhoek observed microorganisms
22. Broth experiments opposed spontaneous generation
24. Recognized three forms by which contagion passed
25. Believed by Pasteur to be a source of disease microorganisms
28. Developed the condenser for use in microscopy
31. Animal infected by anthrax bacillus
33. One of the founders of the science of microbiology
34. Studied cholera during an outbreak in London
35. Won the Nobel Prize for work on malaria and mosquitoes
38. Proved that a specific microorganism causes a specific disease
40. Source of aqueous humor used by Koch
41. Used by Abbé to increase microscope magnification
44. Article of clothing sold by van Leeuwenhoek

CHAPTER 2

PRINCIPLES OF CHEMISTRY

To many people, the words "microbial life" bring to mind images of microorganisms cavorting in a droplet of pond water; or molds growing in a cup of outdated yogurt; or bacteria multiplying in a festering wound. But microbiologists generally go a step beyond. They commonly ask *why* microorganisms act as they do, *how* they are organized, and *what* they are made of. Their questions reflect a curiosity about aspects of microbial life that are less visible, but perhaps more fundamental.

Untold generations ago, humans first pondered the composition of living things. They reasoned that because living things differed so greatly from nonliving things, there must be a corresponding difference in how they were constructed. Then, during the early 1800s chemists identified a group of basic substances found in the earth and atmosphere, including carbon, hydrogen, and oxygen. Surprisingly, the cells of living things proved to be made of the same materials; the only apparent difference was how the substances were organized in living and nonliving things. Chemists began referring to the materials associated with living things as "organic substances," while all others were termed "inorganic substances." Organic substances could be converted to inorganic substances easily enough, but the reverse appeared impossible. Living things seemed to be unique.

Then in 1821 an important discovery was made by the German chemist Friedrich Wöhler. Wöhler was investigating the properties of cyanides, a group of chemicals generally accepted as inorganic. While heating ammonium cyanate, he found, to his amazement, that crystals of urea were forming. Urea is a major component of urine and is definitely an organic compound. Wöhler's work showed that an inorganic substance could be converted to an organic substance and that it was possible to synthesize the substances of living things. His work encouraged other scientists to tackle the problem of synthesizing the organic substances of living things, and soon scientists realized that knowledge of the chemicals of life was within their grasp.

During the late 1800s and into this century the distinction between living and nonliving things continued to evaporate as chemists formulated numerous

vay′ler

u-rē′ah

Organic compounds:
compounds associated with
living things

organic compounds and began to understand life processes in terms of chemical reactions. Today it is clear that all biology, including microbiology, has a chemical basis. Biology and chemistry are inseparable if scientists are to find answers to the questions of *why*, *how*, and *what*.

In this chapter we shall review the fundamental concepts of chemistry in order to lay a foundation for the chapters ahead. We shall enumerate the elements that make up all known substances and show how these elements combine to form the components of living things. Four major groups of organic substances found in virtually all forms of life will be studied in depth. An understanding of chemistry is vital if we are to comprehend the principles of living systems that researchers are uncovering in our era.

The Elements of the Physical Universe

Element:
any of 92 naturally occurring
basic substances that make
up the physical universe.

So far as scientists know, all matter in the physical universe is composed of a number of basic substances called **elements**. Ninety-two naturally occurring elements have been discovered and characterized. Each is designated by one or two letters that stand for its English or Latin name. For example, H is the symbol for hydrogen, O for oxygen, Cl for chlorine, and Mg for magnesium. Some Latin abbreviations include Na from *natrium* (which translates to sodium), K from *kalium* (which is Latin for potassium), and Fe from *ferrum* (the Latin word for iron). Table 2.1 lists some of the major elements in living things. Note that only six elements—carbon, hydrogen, nitrogen, oxygen, phosphorus, and sulfur—make up 99 percent of the dry weight of a bacterium. (The acronym CHNOPS is helpful in remembering these.)

The Structure of Atoms

Atom:
a unit of an element that
cannot be further subdivided
without losing the quality of
the element

The matter in elements cannot be subdivided without limit. Eventually, the subdivisions lead to units indivisible by ordinary chemical means. These units are called **atoms**. Although sophisticated devices are only beginning to visualize atoms, their existence has been assured by reams of experimental evidence. An atom cannot be broken down further without losing the quality of the element. Thus, chlorine consists of chlorine atoms, carbon of carbon atoms, and so forth.

As early as 1808, the English chemist John Dalton conceived that matter is composed of atoms combined to form more complex substances. For the next century, the atom was considered to be solid and indivisible. However, in 1911, Ernest Rutherford proposed that every atom consists of a positively charged **nucleus** surrounded by a negatively charged system of **electrons**. In the ensuing years scientists found that the nucleus contains most of the atom's mass and is composed of two kinds of particles called **protons** and **neutrons**. These two particles have about the same mass. The protons bear a positive electrical charge, while the neutrons have no charge. Electrons, by contrast, have a negative charge. In any atom, the electrons are equal in number to the protons but each electron has only $\frac{1}{1835}$ of the mass of a proton.

Atomic number:
the number of protons in an
atom.

The number of protons in an atom is the determining factor in the physical and chemical properties of the atom and, therefore, of the element. For example, carbon atoms have six protons, and nitrogen atoms have seven. The number of protons is the **atomic number** of the atom. Thus, carbon with six protons has an atomic number of 6. The **mass number** is determined by the number of protons and neutrons combined. Since carbon atoms have six protons and six neutrons in their nucleus, the mass number of carbon is 12. Figure 2.1 indicates the structures, atomic numbers, and mass numbers of various atoms.

TABLE 2.1
The Major Elements of Living Organisms

ELEMENT	SYMBOL	PERCENTAGE BY WEIGHT IN BODY	ATOMIC NUMBER	ATOMIC WEIGHT
Oxygen	O	65%	8	16
Carbon	C	18	6	12
Hydrogen	H	10	1	1
Nitrogen	N	3	7	14
Calcium	Ca	2	20	40
Phosphorus	P	1	15	31
Potassium	K		19	39
Sulfur	S		16	32
Chlorine	Cl	0.9	17	35
Sodium	Na		11	23
Magnesium	Mg		12	24
Iron	Fe		26	56
Manganese	Mn		25	55
Copper	Cu		29	64
Iodine	I		53	127
Cobalt	Co	0.1	27	59
Zinc	Zn		30	65
Boron	B		5	11

The major elements of a bacterium

Carbon	C	12.14%		
Oxygen	O	73.68		
Nitrogen	N	3.04		
Hydrogen	H	9.94		
Phosphorus	P	0.60		
Sulfur	S	0.32		

In chemistry it is often helpful to express the relative size of a chemical substance. In such instances, scientists use the term **atomic weight**. The atomic weight of an atom is nearly the same as its mass number. Atomic weights are determined by adding together the weights of the protons, neutrons, and electrons in the atom. These are not the real weights of the particles, but only relative weights. Physicists assign the value of 12 to the weight of a carbon atom and determine the weights of other atoms relative to carbon. For example, a hydrogen atom actually weighs 1.67×10^{-24} g, but relative to carbon its weight is 1.008. Therefore the atomic weight of a hydrogen atom is regarded as 1 and, in this case, it is close enough to be the same as the mass number.

Atomic weight: the sum of the relative weights of the protons, neutrons, and electrons of an atom.

Isotopes and Ions

Although the number of protons is the same for all atoms in an element, the number of neutrons may vary. Consequently, the mass number of atoms of an element may vary. For example, most carbon atoms have a mass number of 12,

FIGURE 2.1

The Atomic Structure of Five Important Elements

Note that the number of protons is equal to the number of electrons but not necessarily equal to the number of neutrons. Also note that the electrons are arranged in concentric shells about the nucleus. The atomic number is the proton number, and the mass number is the sum of the numbers of protons and neutrons.

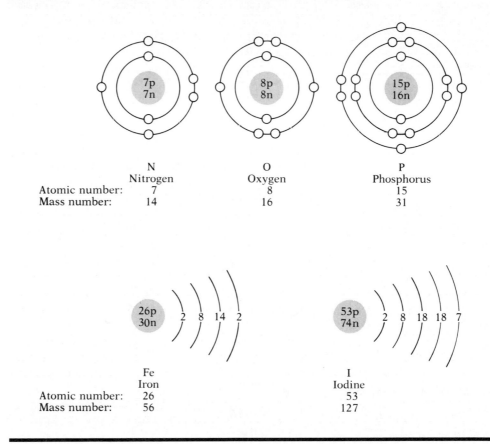

	N Nitrogen	O Oxygen	P Phosphorus
Atomic number:	7	8	15
Mass number:	14	16	31

	Fe Iron	I Iodine
Atomic number:	26	53
Mass number:	56	127

but some atoms of carbon have eight neutrons in the nucleus and, hence, a mass number of 14. Atoms of the same element that have different numbers of neutrons are called **isotopes**. For example carbon-12 and carbon-14 (symbolized as ^{12}C and ^{14}C) are isotopes of carbon. Carbon-14 is especially important because it gives off energy and is radioactive. By attaching it to organic substances, biologists can use carbon-14 as a tracer to follow the fate of the substance.

Atoms are uncharged when they contain equal numbers of electrons and protons. Atoms do not gain or lose protons, but a gain or loss of electrons is possible. When this takes place, the atom is converted to a charged atom, or **ion**. For example, the addition of one electron adds a negative electrical charge to the atom and converts it to a negatively charged ion. By contrast, the loss of one electron would leave the atom with one extra proton and yield a positively charged ion. Ion formation is important in ionic bonding in atoms.

Groups of atoms often join together to form complex ions. Some of the important ions in microbiology are the ammonium ion (NH_4^+), the carbonate ion ($CO_3^=$), the hydrogen ion (H^+), the hydroxyl ion (OH^-), the nitrate ion (NO_3^-), the phosphate ion ($PO_4^=$), and the sulfate ion (SO_4^+). They will occur often in our discussions and an appreciation of their elemental composition is of value.

Electron Placement

In 1913, the Danish physicist Niels Bohr suggested one of the first models of the atom. In his model the atom is perceived as a miniature solar system, with Rutherford's system of electrons revolving around the nucleus in concentric orbits (Figure 2.2). This simplified model has been greatly modified by contemporary physicists. Definite electron pathways have been eliminated and the location of electrons is now regarded as a three-dimensional region of space around the nucleus, as shown in Figure 2.2. This cloudlike pathway is referred to as an **orbital** to distinguish it from Bohr's orbit.

The distance of an electron from the nucleus is a function of the electron's energy. Electrons with higher energy are probably farther from the nucleus than those with lower energy. Therefore while physicists cannot say precisely where an electron is, they can make a statistical estimate of where it is likely to be. The energy level in which an electron is found most often is called the **shell** of the electron. In illustrations of the atom, the shell and the orbital are commonly represented by a ring, with the electron shown as a spot on the ring. Note in Figure 2.2 how this representation is reminiscent of the Rutherford model. The model, though technically wrong, is useful for interpreting chemical phenomena.

The number of concentric electron shells that an atom contains varies with the atom. Each shell can hold a maximum number of electrons. The first shell next to the nucleus can accommodate two electrons, while the second shell can hold eight. Other shells also have maximum numbers, but no atom can have more than eight electrons in its outermost shell. Inner shells are filled first and, if there are not enough electrons to fill all the shells, the outer one is left incomplete.

The shells of atoms of certain elements are filled completely and the elements do not enter into any chemical reactions. An element such as this is said to be an **inert element**. Neon is an inert element because its outer shell contains the maximum eight electrons. By contrast, a carbon atom, with six electrons, has two electrons in its first shell and only four in the second. For this reason carbon is extremely active chemically and forms innumerable combinations with other elements.

Orbital:
the cloudlike pathway describing a three-dimensional region where an electron is located.

Shell:
the energy level of an atom in which a particular electron will most likely be found.

Inert element:
an element whose atoms have an outer shell filled with electrons.

FIGURE 2.2

Two Models of the Atom's Structure

(a) A modern view of the atom with electrons revolving in a cloudlike pathway in a three-dimensional region of space around the nucleus. (b) A simplified view of the atom showing electrons in a ringlike energy level called a shell. The electrons are shown as spots on the ring. Note the similarity to Rutherford's model of the atom.

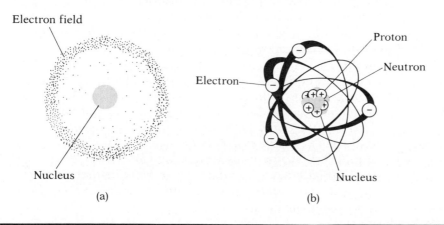

Electron field

Nucleus

(a)

Proton

Neutron

Electron

Nucleus

(b)

Compounds and Molecules

Compound:
a substance composed of two or more elements.

When atoms of two or more elements interact with one another to achieve stability, they form a substance called a **compound**. In a compound, the atoms of different elements combine in specific proportions with a particular pattern of linkages called **chemical bonds**. Each compound, like each element, has a definite formula and set of properties that distinguish it from its components. For example, both hydrogen and oxygen are gases, but the compound they form is water.

Molecule:
the smallest unit of a compound that retains the properties of the compound and cannot be further subdivided without losing the quality of the compound.

The smallest part of a compound that retains the properties of the compound is termed a **molecule**. The compound and molecule are analogous to the element and atom. Molecules may be composed of only one kind of atom, as in oxygen gas (O_2), or they may consist of different kinds of atoms in such things as water (H_2O), ammonia (NH_3), and the sugar glucose ($C_6H_{12}O_6$). The kinds and amounts of atoms in a molecule are symbolized by a designation called the **molecular formula**.

Molecular weight:
the sum of the atomic weights of the atoms in a molecule.

To appreciate the relative size of a molecule it is valuable to know its **molecular weight**. This is determined by adding together the atomic weights of the atoms in the molecule and expressing the sum in units called **daltons**. (A dalton is equal to the mass of one hydrogen atom.) The molecular weight of a water molecule is 18 daltons, while the molecular weight of a glucose molecule is 180 daltons. Some molecular weights reach astonishing proportions. For instance, antibodies produced by the body's immune system may have a molecular weight of 150,000 daltons; and the toxic poison secreted by the bacterium of botulism has a molecular weight of over 900,000 daltons.

Antibodies:
protein molecules produced by the immune system as a defensive measure against certain substances.

The term **mole** is used to express the quantity of a substance whose weight in grams is numerically equivalent to its molecular weight: a mole of water weighs 18 grams, a mole of glucose weighs 180 grams, and a mole of a typical antibody weighs 150,000 grams. The mole concept is important in Chapter 5, where the energy content of a mole of certain substances is discussed.

TO THIS POINT
■

We began studying the principles of chemistry by noting why chemistry is essential to the study of living things. We then outlined the elements of the physical universe and reduced the elements to the level of atoms because atoms join together to form the chemical substances of life. To understand how the joining takes place, we explored atomic structure, with emphasis on the electrons and their placement in the atom. As we shall see in the next section, the atoms may combine in a number of ways to yield the familiar substances of all living things, be they bacteria, trees, or humans.

The section also reviewed some important concepts of chemistry that will recur in many of the later chapters. For example, isotopes are commonly used in diagnostic procedures for disease, and ions are essential to the chemical processes of most microorganisms. We shall also encounter molecules and compounds in the chemistry of life processes, and, in many places, the molecular formula and molecular weight of a compound will be mentioned. It is therefore essential to grasp the meaning of these terms as a prelude to later use.

We shall now turn to the types of bonding that occur in chemical compounds. This is where we construct the substances of living things, including microorganisms. We shall examine chemical reactions from a general standpoint and learn the basis for the chemical processes that go on in all life forms. A short explanation of acids and bases is included because they are very significant in biological chemistry.

Chemical Bonding

Molecules form when atoms interact with one another and link together. Linkage requires that two atoms come close enough for their electron orbitals to overlap. At this point, an energy exchange takes place and each of the participating atoms assumes an electron configuration more stable than its original configuration. The rearrangement can occur in two major ways: one atom can give up some of its electrons to the other, or each atom can share electrons with the other. When two or more atoms are linked together, the force that holds them is called a **chemical bond**. Chemical bonds usually form in order to complete the outer electron shell.

Chemical bond: the force that holds atoms together in molecules.

Ionic Bonding

In **ionic bonding**, one atom gives up its outermost electrons to another. The reaction between sodium and chlorine is illustrative. Sodium has an atomic number of 11, with electrons arranged in groups of 2, 8, and 1, as shown in Figure 2.3. Chlorine, with an atomic number of 17, has its electrons in groups of 2, 8, and 7. To achieve stability, sodium atoms need only lose one electron and chlorine atoms need only gain one. Thus when atoms of the two elements are brought together, the sodium atoms donate one electron to the chlorine atoms.

The transfer of electrons leads to ion formation. By acquiring one electron, chlorine atoms form chloride ions (Cl^-). By contrast, sodium atoms now have an extra proton and therefore are sodium ions (Na^+). Since opposite electrical charges attract each other, the chloride ions and sodium ions come together to form sodium chloride (NaCl) molecules, or table salt (MicroFocus: 2.1). The force that holds the ions together in a molecule is called an **ionic bond**.

Ionic bond: the force that holds ions together in a molecule.

Covalent Bonding

Atoms can also achieve stability by sharing their electrons, a process called **covalent bonding** (Table 2.2). Such bonds are very important in biology because the major elements of life (carbon, oxygen, nitrogen, and hydrogen) almost always enter into covalent bonds with one another.

TABLE 2.2
Three Types of Chemical Bonds in Organic Molecules

TYPE	CHEMICAL BASIS	STRENGTH	EXAMPLE
Ionic	Attraction between oppositely charged ions	Strong	Sodium chloride
Covalent	Sharing of electron pairs between atoms	Strong	Carbon-to-carbon bonds
Hydrogen	Attraction of a hydrogen nucleus (a proton) to negatively charged atoms in neighboring molecules	Weak	Cohesiveness of water

FIGURE 2.3

Formation of the Ionic Bond in Sodium Chloride

(a) Sodium atoms contain one electron in the outer shell, whereas chlorine atoms lack one electron in this shell. (b) The sodium atom loses an electron to the chlorine atom. (c) A sodium ion is thus formed with one positive charge, and a chloride ion is formed with one negative charge. An electrical attraction develops between the ions. (d) The ions are drawn together to form an ionic bond.

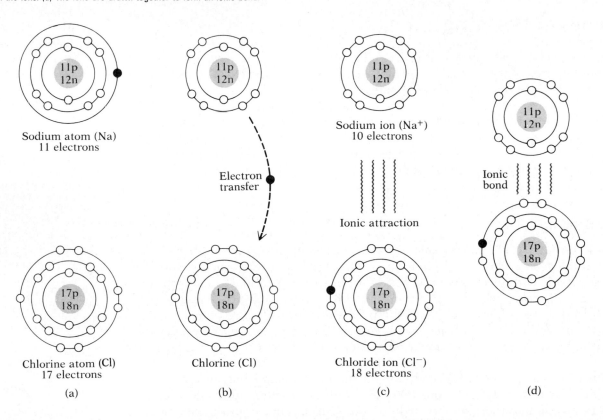

Sodium atom (Na)
11 electrons

Electron transfer

Sodium ion (Na$^+$)
10 electrons

Ionic attraction

Ionic bond

Chlorine atom (Cl)
17 electrons

Chlorine (Cl)

Chloride ion (Cl$^-$)
18 electrons

(a)   (b)   (c)   (d)

Covalent bond:
the force resulting from the sharing of electrons among the atoms in a molecule.

Organic chemistry:
the chemistry of living things.

Hydrocarbons:
a group of organic substances consisting solely of carbon and hydrogen.

The bonding in water molecules is typical of covalent bonding. Oxygen atoms are not strong enough to attract the electrons from hydrogen atoms, nor is a hydrogen atom sufficiently weak to give up its electron. Therefore, the oxygen and hydrogen atoms share their electrons and form a **covalent bond** in water. The electrons circle both the oxygen and hydrogen.

Scientists use one or more lines between chemical symbols to represent covalent bonds. Water is represented as H—O—H. If two pairs of electrons are shared, as in carbon dioxide (CO_2), then two lines are used to indicate the double bond (O=C=O).

Covalent bonding occurs frequently in carbon because this element has four electrons in its outer shell. The carbon atom is not strong enough to acquire four additional electrons, but it is sufficiently strong to retain the four that it has. It therefore enters into a huge variety of covalent bonds with four other atoms, ions, or groups of atoms. So vast is the array of compounds that **organic chemistry**, the chemistry of living things, is basically the chemistry of carbon.

The simplest derivatives of carbon are the **hydrocarbons**, so named because they consist solely of hydrogen and carbon. Methane, or natural gas, is the most fundamental hydrocarbon. It contains one carbon atom and four hydrogen atoms. Other hydrocarbons may consist of chains of carbon atoms and, in some cases, the chains may be closed to form a cyclic molecule.

THE FLAVORING OF HISTORY

About two-thirds of the way through the movie *Gandhi*, a *New York Times* reporter portrayed by Martin Sheen visits Mohandas Gandhi to renew an old friendship. They chat for a spell, and then Gandhi confides his next plan to defy the British. "I'm going to march to the sea and make salt!"

Salt was a commodity whose production in India was controlled by the British. So important was salt to life that by dominating its sale, the British dominated life at its very core. Gandhi's intention was to demonstrate that Indians could and must control their own destiny. His dream was realized after World War II when the British left India.

Throughout history, salt has been an undercurrent to life. Salt may have been discovered when early humans followed cattle to outcrops of salt rock and watched as the animals licked the salt. Probably one of the first human industries was evaporating water from saltwater to obtain the salt. Many early civilizations centered on salt-pans such as the Dead Sea in the Middle East, the Nile River in Egypt, and the Yellow River in China.

Salt's utility as a preservative and seasoning made it a form of money at one time. The early Greeks and Phoenicians, both noted traders, exchanged salt for goods. In addition, the Arabs based their economy in large measure on salt from the Dead Sea. Roman legionnaires were given salt allowances, and later, cash was substituted so they could purchase salt. The word "salary" is the modern descendent of this cash.

Well after the Renaissance, salt continued to have economic value. For example, it was often used in the slave trade in Central Africa, where little salt was available, and we have previously noted Gandhi's intention to "make salt." The Via Saleria (salt road) of Italy remains as a testament to the ancient and current significance of salt. Today the world consumes almost 200 million tons annually. It is not surprising that when we leave for school or work, we often think of "going to the salt mines."

Some carbon compounds consist of carbon atoms covalently bonded to a group of atoms that operate as a unit. Such a group is called a **functional group**. In microbiology, the important functional groups include the hydroxyl group ($-OH$), the amino group ($-NH_2$), the acid, or carboxyl group ($-COOH$), the sulfhydryl group ($-SH$), and the aldehyde group ($-CHO$). When a molecule containing any of these groups reacts with another molecule, the groups generally function as a single entity and may be exchanged for another group.

hi-drox'ill
kar-box'ill
sulf-hi'drill

Hydrogen Bonding

When atoms bond together to form a molecule, they establish a definite geometric relationship determined largely by the electron configuration. This leads to an uneven distribution of electrical charge. In water, the two hydrogen atoms are attached on one side of the oxygen atom. As a result, the protons gather at this side and give the molecule a slightly positive charge. By contrast, the other side of the molecule carries a slightly negative charge owing to the accumulation of electrons. The water molecule therefore has poles and is said to be polar. When attractions develop between the poles of adjacent molecules, a weak polar bond is formed. Such a bond is usually called a **hydrogen bond** because it generally involves hydrogen.

Hydrogen bonds may last only a brief instant, but they are strong enough to prevent water from easily becoming ice or steam. Also, the hydrogen bonds are important in shaping the structures of proteins and nucleic acids, two of the major components of all living things.

Hydrogen bond: the force resulting from the attractions between oppositely charged poles of adjacent molecules.

Chemical Reactions

A **chemical reaction** is a process in which atoms or molecules interact and form new bonds, thereby undergoing a change through electron rearrangement. Different combinations of atoms or molecules result from the reaction.

Chemists use an arrow to separate the original and final substances in a chemical reaction. The atoms or molecules to the left of the arrow are the **reactants**, and those to the right are the **products**. In some cases, the reaction consists of the simple combining of two reactants to form a molecule:

$$\underset{\text{Sodium ion}}{Na^+} \quad + \quad \underset{\text{Chloride ion}}{Cl^-} \quad \longrightarrow \quad \underset{\text{Sodium chloride}}{NaCl}$$

In other cases, reactions involve a switch of reactants:

$$\underset{\substack{\text{Hydrochloric}\\\text{acid}}}{HCl} \quad + \quad \underset{\substack{\text{Sodium}\\\text{hydroxide}}}{NaOH} \quad \longrightarrow \quad \underset{\text{Water}}{H_2O} \quad + \quad \underset{\substack{\text{Table}\\\text{salt}}}{NaCl}$$

Other reactions involve a decomposition in which water functions as a reactant. Such reactions are called **hydrolysis** reactions:

$$\underset{\text{Sucrose}}{C_{12}H_{22}O_{11}} \quad + \quad \underset{\text{Water}}{H_2O} \quad \longrightarrow \quad \underset{\text{Glucose}}{C_6H_{12}O_6} \quad + \quad \underset{\text{Fructose}}{C_6H_{12}O_6}$$

In certain reactions, an oxidation and a reduction take place. In an **oxidation**, one reactant loses electrons. In a **reduction**, a reactant gains electrons. Energy transfer is often involved in such a reaction:

$$Cytochrome\!-\!A\!-\!Fe^{++} + Cytochrome\!-\!B\!-\!Fe^{+++} \xrightarrow{\text{electron transfer}}$$
$$Cytochrome\!-\!A\!-\!Fe^{+++} + Cytochrome\!-\!B\!-\!Fe^{++}$$

In this reaction, cytochrome A has lost an electron to cytochrome B. Note that by losing an electron, cyctochrome A is left with an additional positive charge, while cytochrome B gains the electron that neutralizes one of its positive charges. This type of reaction is essential to bacterial metabolism, as we shall see in Chapter 5.

Acids and Bases

There are many possible definitions for acids and bases. For our purposes, an **acid** is a chemical substance that donates protons or hydrogen ions (H^+) to water or other solution. By contrast, a **base** (or **alkali**) is a substance that accepts hydrogen ions in solution. Since the acceptance of a hydrogen ion increases the concentration of hydroxyl ions (OH^-) in solution, a base may also be considered a substance that increases the amount of hydroxyl ions.

Acids are distinguished by their sour taste. Some common examples are acetic acid in vinegar, citric acid in citrus fruits, and lactic acid in sour milk products. Strong acids are those that donate large amounts of hydrogen ion to a solution. Hydrochloric acid (HCl), sulfuric acid (H_2SO_4), and nitric acid (HNO_3) are examples. Weak acids donate a small amount of hydrogen ion, as typified by carbonic acid (H_2CO_3).

Bases have a bitter taste. Certain ones are strong bases that take up large amounts of hydrogen ion from a solution and leave it with a high concentration of hydroxyl ions. Potassium hydroxide (KOH), a material used to make soap, is among them. Weak bases are illustrated by compounds containing amino ($-NH_2$) groups.

Hydrolysis:
a decomposition chemical reaction in which water functions as an intermediary.

Cytochrome:
a cellular pigment that functions in energy reactions in a living cell.

Acid:
a chemical substance that adds hydrogen ions (protons) to a solution.

Base:
a chemical substance that removes hydrogen ions from a solution and thereby increases the number of hydroxyl ions.

Because of their opposing chemical characteristics, acids and bases frequently react with each other. Such a reaction neutralizes both the acid and base to form water and a salt. For example, when hydrochloric acid (HCl) and sodium hydroxide (NaOH) react, the result is water (H_2O) and ordinary table salt (NaCl).

In 1909, the Danish chemist Søren P. L. Sørensen introduced the symbol pH and the **pH scale** to refer to the strength of an acid or base. The scale extends from 0 to 14 and is based on actual calculations of the number of hydrogen ions present when a substance mixes with water. The strongest acid has a pH of 0; the strongest base has a pH of 14. A substance with a pH of 7, such as water, is said to be neutral. Most bacteria live at neutral pH environments, but many fungi tolerate a more acidic environment. Thus fungi are often found in sour cream, yogurt, and citrus fruit products. Figure 2.4 summarizes the pH values of some common substances.

pH:
a measure of the acidity or alkalinity of a substance.

TO THIS POINT ∎

We have studied atoms as the fundamental units of elements and noted that they achieve stability when their electron shells are complete. We then moved on to discuss how atoms combine with one another in three types of bonding. Ionic bonding, the first type, involves a transfer of electrons between atoms and an attraction of the ions that form. Covalent bonding, by comparison, is based on a sharing of electrons without an actual transfer of electrons. Hydrogen bonding, the third method, involves the linking of groups of atoms in a molecule by weak forces developing from the polarity of the components. Virtually all chemical reactions are based on the interactions of the chemical forces in bonding.

We also reviewed the various types of chemical reactions in which bonds are broken and reconstituted. The reaction may involve a combining of two reactants, a switch among reactants, a hydrolysis, or an oxidation-reduction reaction. The discussion closed with a brief review of acids and bases. Acids are hydrogen ion donors, while bases are hydrogen ion acceptors. By taking up hydrogen ions, bases also increase the level of hydroxyl ions in a solution. We reviewed the pH scale that is used to measure the relative strength or weakness of an acid or base.

In this final section, we shall focus on four important classes of organic compounds: carbohydrates, lipids, proteins, and nucleic acids. We shall examine the structures and functions of these compounds and touch on their importance in microbiology. All living things, including microorganisms, are composed of these substances. The discussions in this chapter's first two sections were largely designed to prepare us for this study.

Major Organic Compounds of Living Things

Living things are composed of numerous organic compounds, most of which fall into four distinct classes having different chemical compositions and properties. In this section we shall briefly explore the four classes.

FIGURE 2.4

A Sample of pH Values

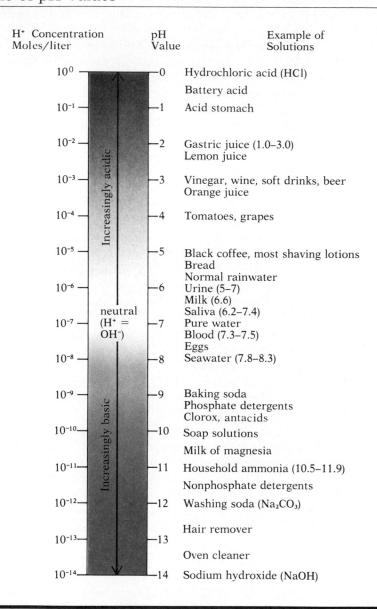

H^+ Concentration Moles/liter	pH Value	Example of Solutions
10^0	0	Hydrochloric acid (HCl)
		Battery acid
10^{-1}	1	Acid stomach
10^{-2}	2	Gastric juice (1.0–3.0)
		Lemon juice
10^{-3}	3	Vinegar, wine, soft drinks, beer
		Orange juice
10^{-4}	4	Tomatoes, grapes
10^{-5}	5	Black coffee, most shaving lotions
		Bread
		Normal rainwater
10^{-6}	6	Urine (5–7)
		Milk (6.6)
		Saliva (6.2–7.4)
10^{-7}	7	Pure water
		Blood (7.3–7.5)
		Eggs
10^{-8}	8	Seawater (7.8–8.3)
10^{-9}	9	Baking soda
		Phosphate detergents
		Clorox, antacids
10^{-10}	10	Soap solutions
		Milk of magnesia
10^{-11}	11	Household ammonia (10.5–11.9)
		Nonphosphate detergents
10^{-12}	12	Washing soda (Na_2CO_3)
		Hair remover
10^{-13}	13	Oven cleaner
10^{-14}	14	Sodium hydroxide (NaOH)

Increasingly acidic — neutral ($H^+ = OH^-$) — Increasingly basic

Carbohydrates

Carbohydrates are organic compounds of carbon, hydrogen, and oxygen. In simple carbohydrates, the ratio of hydrogen to oxygen is 2 to 1, the same as in water. For this reason, the carbohydrates are often considered "hydrated carbon," hence their name. However, the atoms are not present as water molecules bound to carbon, but rather in H—C—OH configurations. This combination occurs frequently in carbohydrate structures.

Carbohydrates function as energy sources in cells. They are also found in several cellular structures such as bacterial capsules. Carbohydrates are synthesized from water and carbon dioxide through the process of photosynthesis (Chapter 5). Certain bacteria and other microorganisms have the necessary

Photosynthesis:
a chemical process in which light energy is used to form the chemical bonds in carbohydrate molecules.

chemical machinery for this process. Often the carbohydrates are termed **saccharides** from the Latin *saccharon* for sugar. The word "sugar" is applied to the simple carbohydrates.

Carbohydrates are generally divided into three classes: monosaccharides, disaccharides, and polysaccharides. **Monosaccharides** are the simplest carbohydrates, and represent the building blocks for disaccharides and polysaccharides. Some monosaccharides, such as glyceraldehyde, have three carbon atoms; others have four, five, six, or more carbon atoms.

Glucose and fructose are among the most widely encountered monosaccharides. Both have a molecular formula typical of carbohydrates: $C_6H_{12}O_6$. However, their structural formulas are different, as shown in Figure 2.5. Such molecules are called **isomers**. Isomers react differently and usually have different properties. For example, fructose is much sweeter than glucose. Glucose serves as the basic supply of energy in the world. Estimates vary, but many scientists suggest that half the world's carbon exists as glucose. Chapter 5 is devoted to the production of glucose in living things and its use for energy.

Disaccharides are double sugars. They are composed of two monosaccharides held together by covalent bonds. Maltose is an example. This disaccharide is constructed from glucose molecules by **dehydration synthesis** (Figure 2.6a). In dehydration synthesis, the products of water are removed during the formation of covalent bonds. Maltose occurs in cereal grains such as barley and is fermented by yeasts to form the alcohol in beer. Another disaccharide, lactose, is composed of the monosaccharides glucose and galactose (Figure 2.6a). Lactose is known as milk sugar because it is the principal carbohydrate in milk. Under controlled industrial conditions, microorganisms digest the lactose to form acid in yogurt, sour cream, and other sour dairy products. Sucrose, or table sugar, is a third disaccharide. It is a starting point in wine fermentations, and it may contribute to tooth decay (MicroFocus: 2.2).

Complex sugars are called **polysaccharides**. These carbohydrates are generally large compounds formed by joining together hundreds or even thousands of glucose units. Covalent bonds resulting from dehydration synthesis link the units. Starch, a common polysaccharide, is used by many bacteria as an energy source. Cellulose, another polysaccharide, is a component of the cell walls of plants and certain molds.

i'so-mer

Isomers:
chemical molecules with the same molecular formulas but different structural formulas.

Dehydration synthesis:
a chemical process in which the products of water are removed during the synthesis of a molecule.

FIGURE 2.5

Structural Formulas for Three Important Monosaccharides

Glucose and fructose are isomers because they have identical molecular formulas but different structural formulas. Glucose is shown in the chain form and its more common ring form.

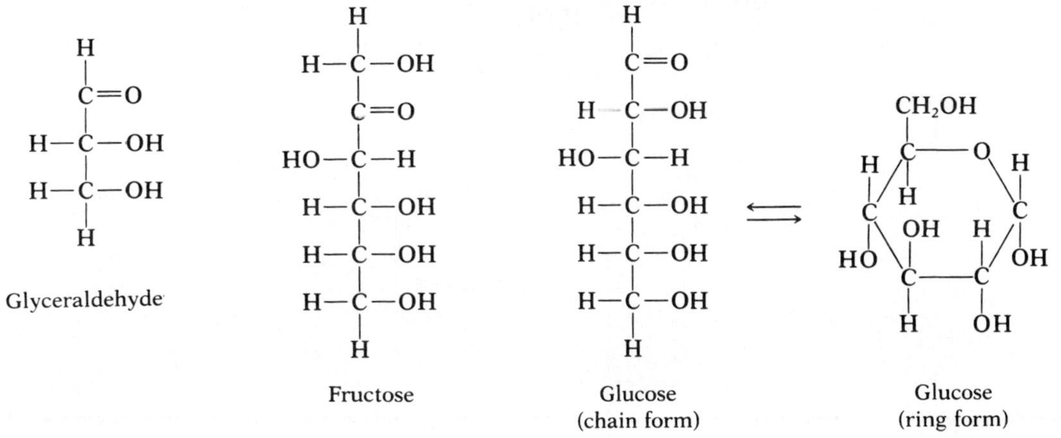

FIGURE 2.6

Structural Formulas for Three Disaccharides and One Polysaccharide

Note that in the synthesis of the disaccharides the products of water are removed from the reactants. This is a dehydration synthesis. The polysaccharide is pictured as a very complex arrangement of glucose molecules covalently bonded to one another.

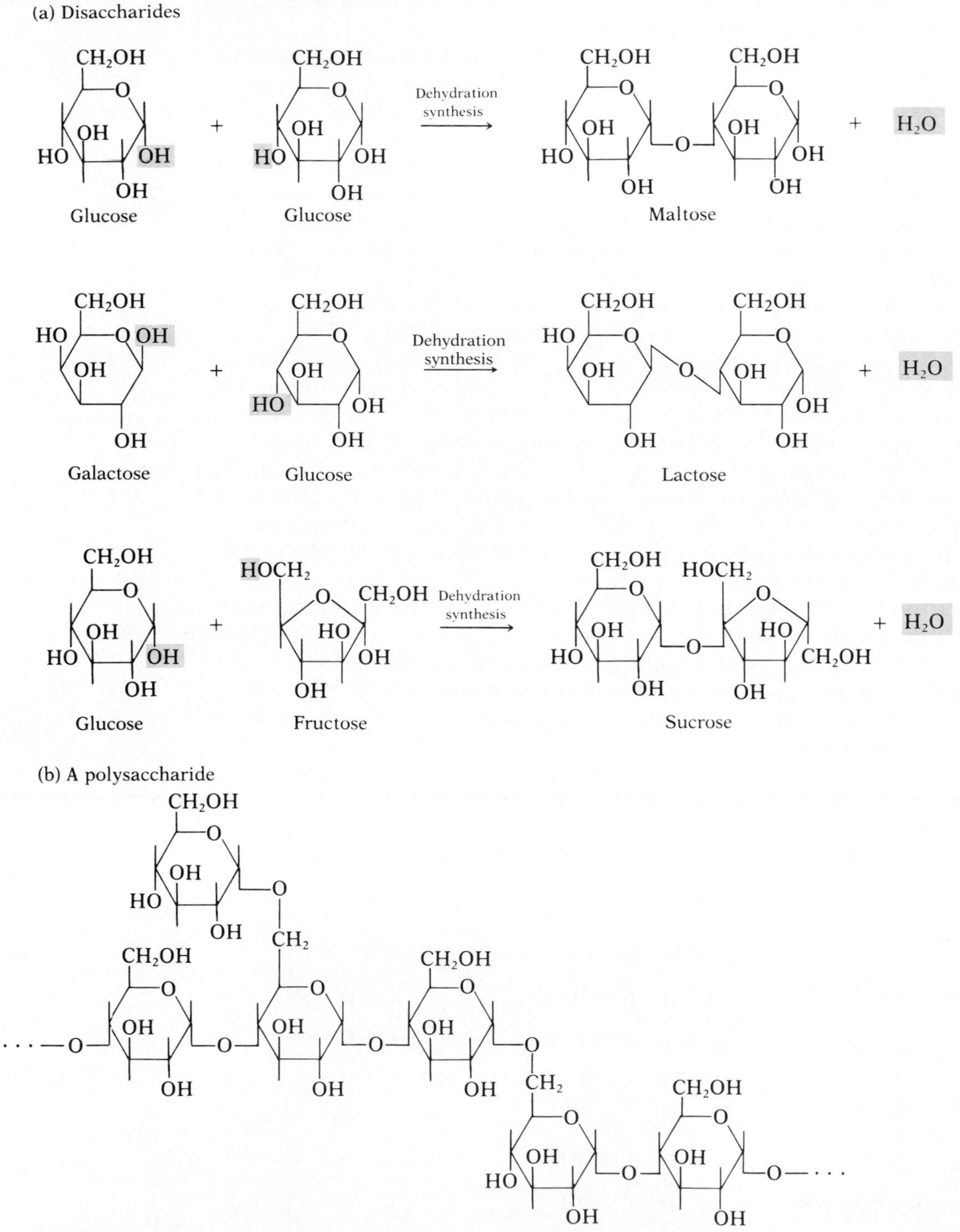

(a) Disaccharides

(b) A polysaccharide

THE REAL CAUSE OF CAVITIES

How many times were we told as children to avoid sweets because they cause cavities? Probably more times than we could count. The advice was essentially good because sweets do contribute to cavities, but only in an indirect way. The real culprits are the bacteria.

What happens in a cavity is this. As bacteria pass through the mouth, certain ones are trapped in the plaque. Plaque is a gummy layer of protein and carbohydrate materials formed from saliva and food particles. It accumulates between the teeth and at the gum line. Ironically, plaque is an oxygen-free environment in a part of the body where air is plentiful. Bacteria that accumulate in the plaque must be able to survive under oxygen-free conditions. Such bacteria are said to be anaerobic.

As the bacteria multiply, they digest the sucrose and other carbohydrates in sweets, producing large amounts of acid. The acid eats away at the tooth enamel, and soon a depression, or cavity, forms. When the soft dental tissues underneath are reached, the bacteria penetrate to this area. Toothache pain arises from exposure of the sensitive nerve endings in the soft tissues, and part of the throb is due to bacterial gases that press against the nerves. When the dentist finally drills through to this area, the foul-smelling gas is released and the pain ebbs.

In retrospect, sweets do not present serious problems so long as the acid-producing bacteria are eliminated. This is where vigorous brushing and flossing enter the picture. Tooth decay is really a disease. It is so widespread that it is considered one of the most prevalent diseases in the world.

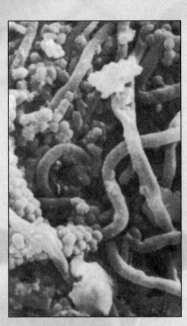

Highly magnified view of bacteria from the surface of the teeth ($\times$ 2500). Many spherical bacteria may be seen clinging to the surface of the rod forms. Organisms such as these exist on the enamel surface and produce the acid that leads to cavities.

Lipids

The **lipids** are a broad group of organic compounds that dissolve in organic solvents such as ether or benzene but generally do not dissolve in water. Like carbohydrates, the lipids are composed of carbon, hydrogen, and oxygen, but the proportion of oxygen is much lower. Lipids also may contain other elements such as sulfur or phosphorus.

The best-known lipids are the **fats**. Fats serve living organisms as important energy sources. They are located in the granules of bacteria and are components of the cellular membrane in most microorganisms. As shown in Figure 2.7, fats consist of a three-carbon glycerol molecule and up to three long-chain fatty acids. Each fatty acid generally has between 16 and 18 carbon atoms in the chain. Bonding to the glycerol molecule occurs by dehydration synthesis.

There are two major types of fatty acids. The **saturated fatty acids** contain the maximum possible number of hydrogen atoms, while the unsaturated fatty acids contain less than the maximum. The **unsaturated fatty acids** appear to be a factor in good health for humans because they lower the levels of cholesterol in the blood and thus reduce the likelihood of clogged blood vessels.

Other types of lipids are the waxes, phospholipids, and steroids. Waxes are composed of long chains of fatty acids, while phospholipids are lipids that contain a phosphate group. Steroids, such as cholesterol, are composed of several rings of carbon atoms with side chains.

Fatty acid:
a long chain of carbon atoms with multiple hydrogen atoms and an acid group.

FIGURE 2.7

The Molecular Structures of Fat Components and Synthesis of a Fat

(a) Fats consist of glycerol and fatty acids. The fatty acids may be saturated or unsaturated, as shown. Unsaturated fats contain less hydrogen. (b) Fat formation by dehydration synthesis. The components of water are removed from the reactants and covalent bonds are formed.

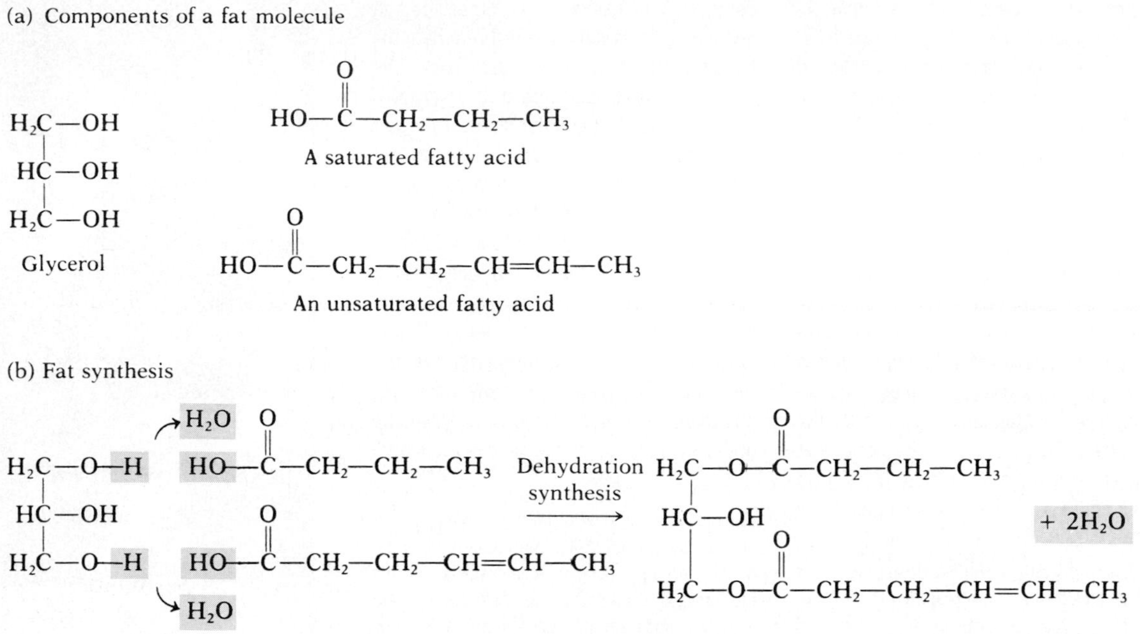

(a) Components of a fat molecule

(b) Fat synthesis

Proteins

Proteins are by far the most abundant organic components of microorganisms and other living things. They function as structural materials as well as **enzymes**, a group of biological compounds that catalyze chemical reactions in living things. Destruction of the proteins in an organism, such as with heat or chemicals, usually spells death for the organism.

Proteins consist essentially of chains of nitrogen-containing compounds called **amino acids**. At the root of each amino acid is a carbon atom. Attached to this atom are an amino group ($-NH_2$), an organic acid group ($-COOH$), and another side group. The side group determines the nature of the amino acid. There are approximately 20 different amino acids in proteins.

In protein formation, amino acids are joined together by covalent bonds as the amino groups are linked to the acid groups by a dehydration synthesis (Figure 2.8). The chain that results is known as a peptide, and the bond is therefore called a **peptide bond**. In some cases the peptide is the total protein, but in other situations, one or more chemical groups are added to constitute the complete protein. How the amino acids are slotted into position in peptide synthesis is a complex process discussed in Chapter 5. It should be noted that the sequence of amino acids is of the utmost importance because a single amino acid improperly positioned may change the character of the protein.

The chain of amino acids in the protein represents the **primary structure** of the protein. Many simple proteins such as the human hormone oxytocin occur in this form. Numerous proteins have a **secondary structure** that forms when the amino acid chain twists itself into a corkscrewlike pattern. Hydrogen

---

Enzymes:
a group of proteins that regulate the rates of many biochemical reactions while themselves remaining unchanged.

Amino acid:
a building block of proteins that consists of an amino group, an acid group, and a specified side group.

Peptide:
a relatively small chain of amino acids.

Oxytocin:
the hormone that stimulates uterine contractions during menstruation and birth.

FIGURE 2.8

Formation of a Small Protein (a Dipeptide) by Dehydration Synthesis

The —OH group from the acid group from alanine combines with the —H ion from the amino group of valine to form water. The open bonds then link together, yielding a peptide bond. The dipeptide in this example is called alanylvaline.

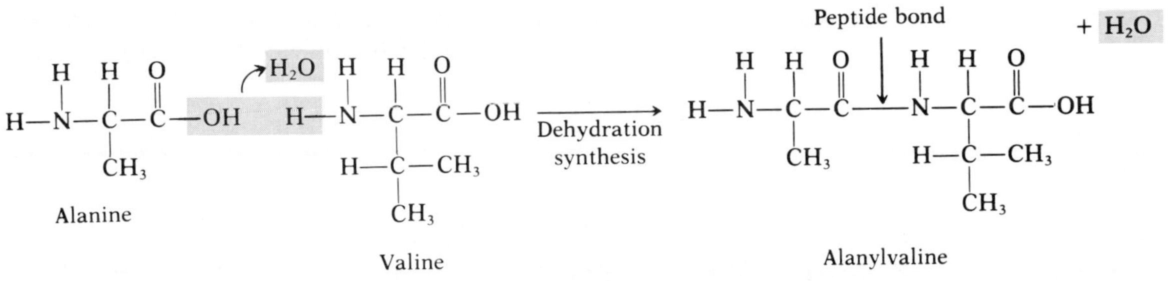

bonds between nearby amino acids and sulfur-to-sulfur (disulfide) linkages help maintain this structure. A secondary structure may also form when amino acid chains line up alongside one another, as in the animal hormone insulin.

In addition, many proteins have a **tertiary structure**. In this case, the protein is folded back on itself much like a telephone cord folded on a table. Hydrogen bonds maintain the protein in its tertiary structure. When subjected to heat or chemicals, the bonds break easily and the protein reverts to its secondary structure. This process is referred to as **denaturation**. Since most enzymes are tertiary proteins, heat or chemicals may be used to denature them and thus interrupt the important chemical reactions they control. Death of the organism

Denaturation:
a chemical process in which the tertiary structure of the protein is destroyed and the protein often solidifies.

TABLE 2.3

Major Organic Compounds of Living Things

ORGANIC COMPOUND	BUILDING BLOCK	SOME MAJOR FUNCTIONS	EXAMPLES
Carbohydrate: Monosaccharide	—	Energy storage; physical structure	Glucose, galactose, fructose
Disaccharide	Monosaccharides	Energy storage; physical structure	Lactose, maltose, sucrose
Polysaccharide	Monosaccharides	Energy storage; physical structure	Starch, cellulose, chitin
Protein	Amino acids	Enzymes; toxins; physical structures	Antibodies; viral surface; flagella; pili
Lipid: Triglycerides	Fatty acids and glycerol	Energy storage; thermal insulation; shock absorption	Fat; oil
Phospholipids	Fatty acids, glycerol, phosphate, and an R group*	Foundation for cell membranes	Plasma (cell) membranes
Steroids	Four-ringed structure†	Membrane stability	Cholesterol
Nucleid acid	Ribonucleotides; Deoxyribonucleotides	Inheritance; instructions for protein synthesis	DNA, RNA

* R group = a variable portion of a molecule.

† Technically, steroids are neither polymers nor macromolecules.

usually follows. Viruses may also be destroyed by denaturing the tertiary proteins found in the outer viral surfaces. The white of a boiled egg is denatured egg protein; cottage cheese is denatured milk protein.

Nucleic Acids

de-ox"e-ri-bo-new-klay'ik

The **nucleic acids** are among the largest molecules found in organisms. Two types function in all living things: deoxyribonucleic acid (DNA) and ribonucleic

---

FIGURE 2.9
The Molecular Structures of Nucleotide Components

The carbohydrates in nucleotides are ribose and deoxyribose. These two small compounds are identical except at C-2. Phosphate ions are formed by the loss of protons from phosphoric acid. The nitrogenous bases include adenine and guanine, which are large purine molecules, and thymine, cytosine, and uracil, which are smaller pyrimidine molecules. Note the similarities in the structures of these bases and the differences in the side groups. A symbol for each molecule is shown to the right. At the bottom of the figure, a DNA nucleotide and an RNA nucleotide have been formed from the component molecules.

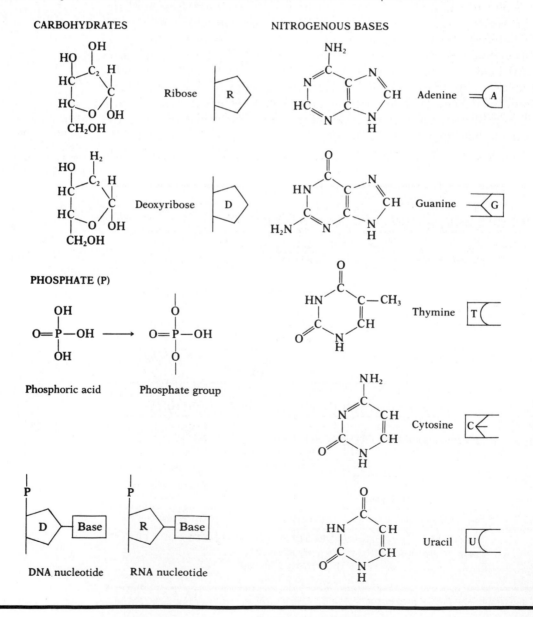

acid (RNA). DNA acts as the genetic material of the chromosome, while RNA functions in the construction of proteins (Table 2.3).

Both DNA and RNA are composed of repeating units called **nucleotides**. Each nucleotide has three components: a carbohydrate molecule, a phosphate group, and a nitrogenous base. The carbohydrate molecule in RNA is ribose, while in DNA it is the closely related compound deoxyribose. The phosphate group is an organic ion formed by the loss of hydrogen ions from phosphoric acid. **Nitrogenous bases** are any of five nitrogen-containing compounds that have excessive numbers of amino groups and therefore act as bases. In DNA, the bases are adenine, guanine, cytosine, and thymine. In RNA, adenine, guanine, and cytosine are also present, but uracil is found instead of thymine. Figure 2.9 displays the structural formulas of these bases. The nucleotides are joined by covalent bonds to form the nucleic acids.

The most familiar location of DNA is in the chromosome of the cell, where it passes on genetic information and directs the synthesis of protein. To form the complete DNA molecule, two single strands of DNA oppose each other in a ladderlike arrangement. Guanine and cytosine line up opposite one another, and thymine and adenine oppose each other. This forms a double strand of DNA. The double strand then twists to form a spiral arrangement called the **double helix**, shown in Figure 2.10.

As with proteins, the nucleic acids cannot be altered without injuring the organism or killing it. Ultraviolet light damages DNA, and thus it can be used to control bacteria on an environmental surface. Chemicals such as formaldehyde alter the nucleic acids of viruses and can be used in the preparation of vaccines. Certain antibiotics interfere with RNA function and thereby kill

Nucleotide:
a building block of nucleic acids consisting of a carbohydrate molecule, a phosphate group, and a nitrogenous base.

ni-troj'en-us

gwan'in

Chromosome:
a molecule consisting of DNA that contains the hereditary information of the cell.

F I G U R E 2.10

DNA Formation from Its Components

Formation of a double-stranded molecule of deoxyribonucleic acid (DNA) (a) Two single-stranded DNA molecules line up next to each other to form the double strand. Adenine molecules always oppose thymine molecules, and guanine molecules always oppose cytosine molecules. (b) The double-stranded DNA molecule then twists as shown to form the double helix, which is characteristic of the DNA in the chromosome of living organisms.

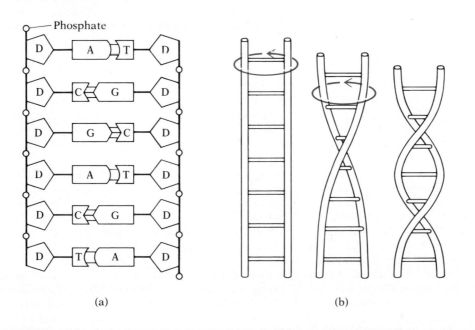

(a) (b)

bacteria. We shall encounter other instances where tampering with nucleic acids is the basis for controlling microorganisms. An entire chapter, Chapter 6, is devoted to the role of nucleic acids in the genetics of bacteria. Together with the proteins, carbohydrates, and lipids, the nucleic acids form the chemical bases for all living things, including microorganisms.

NOTE TO THE STUDENT

You might well ask at this point whether this is going to be a chemistry book or a microbiology book. In a sense, it is going to be both.

In past centuries, biology was the study of nature. Biologists described natural phenomena, classified plants and animals, performed dissections, and marveled at the world of microscopic creatures. Then, in 1828, Friedrich Wöhler synthesized urea and showed that the substances of living things could be produced by scientists. Later, in the 1850s, Pasteur announced his discoveries on fermentation and pointed out that microorganisms were tiny chemical factories in which sugar was converted to alcohol. As a result of these advances, microbiology became irrevocably bound to chemistry.

In future chapters, we shall discuss such topics as the metabolism of microorganisms, the activity of antibodies, the mechanisms of disease, the action of antibiotics on bacteria, the replication of viruses, and the chemical control of microorganisms. None of these processes was understood a hundred years ago, partly because knowledge of their chemical bases was lacking. Such is not the case today.

The chemicals are the nuts and bolts of all living things, including microorganisms. I recommend that you give this chapter a careful reading. In succeeding chapters I shall try to make your time investment worthwhile.

Summary

Ninety-two naturally occurring elements make up the substance of all known living and nonliving things. The simplest unit of these elements is the atom, a particle consisting of a nucleus (with protons and neutrons) and electrons. Electrons are arranged outside the nucleus in shells, with each shell holding a maximum number of electrons. Interactions occur between atoms to fill the shells with electrons. These interactions can consist of a transfer of electrons between two atoms to form an ionic bond, or a sharing of electrons between two atoms to form a covalent bond. The interactions between atoms result in molecules. A mass of molecules is a compound.

Among the important compounds in living things are acids and bases. Acids are compounds that donate hydrogen ions to a solution; bases are compounds that remove hydrogen ions from a solution and, in so doing, increase the number of hydroxyl ions. The pH scale indicates the number of hydrogen ions in a solution and indicates how acidic or basic (alkaline) the solution is.

Other important compounds in living things are the carbohydrates. These compounds contain carbon, hydrogen, and oxygen. They are used primarily as energy sources for life processes, and they include monosaccharides such as glucose and fructose, disaccharides such as lactose and maltose, and polysaccharides such as starch and cellulose. Lipids have the same types of atoms as carbohydrates, but the proportion of oxygen is much lower. Like carbohydrates, lipids serve as energy sources, but they are also used in structural materials. Fats are types of lipids.

The compounds of living things also include proteins. These are chains of amino acids connected by peptide bonds, a type of covalent bond. Proteins are used as enzymes and as structural components of cells. Primary, secondary, and tertiary structures are found in many proteins. The genetic instructions for living things are located in compounds called nucleic acids. A nucleic acid is composed of carbohydrate molecules, phosphate ions, and a series of nitrogenous bases. Two nucleic acids are important in biological systems: ribonucleic acid (RNA) and deoxyribonucleic acid (DNA). With the other organic compounds, nucleic acids form the chemical bases of all living things.

Questions for Thought and Discussion

1. Calcium atoms have two electrons in their outer shells; magnesium atoms also have two. Do you expect that good "chemistry" will occur between these elements when their atoms are brought together? Why or why not?

2. Some individuals unable to drink milk are said to suffer "lactose intolerance." What chemistry do you think their digestive tracts are unable to perform?

3. Certain detergent disinfectants are known to dissolve lipids. How would this activity encourage the destruction of bacterial cells?

4. Why do organic molecules tend to be so large?

5. An atom of carbon is an almost incomprehensible one ten-billionth of a meter in diameter. Put another way, 10 billion carbon atoms could line up along the length of a meter stick. Suppose you were to count each carbon atom along the length of the meter stick at a rate of one atom per second. How long do you suppose it would take to complete your counting?

6. Why is Wöhler's synthesis of urea in 1828 considered to be a landmark achievement in bridging the gap between biology and chemistry?

7. Bacteria do not grow on bars of soap even though the soap is wet and covered with bacteria after one has washed. Can you surmise why this is so?

8. The area occupied by the nucleus of an atom relative to the area of the entire atom is estimated to be similar to the area occupied by a raisin sitting on the 50-yard line of a football stadium relative to the remainder of the stadium. What takes up the remainder of the space in the atom?

9. Oxygen comprises about 65 percent of the weight of a living organism. This means that a 120-pound person contains 78 pounds of oxygen. How can this be so?

10. Suppose you had the choice of destroying one group of chemical substances in bacteria in order to prevent their spread. Which would you choose? Why?

11. Two bottles of acid are taken from the shelf. One is labeled pH 1.5, the other pH 6.78. Which bottle should be handled more carefully? Why?

12. Carbon is so common in living things that the chemistry of carbon is often said to be the chemistry of life. What property of carbon is responsible for its ability to enter a wide variety of chemical combinations?

13. The toxin associated with the foodborne disease botulism is a protein. To avoid botulism, home canners are advised to heat preserved foods to boiling for at least 12 minutes before tasting or consuming them. How does the heat help?

14. Water is generally considered to be the universal solvent of life. In how many places in this chapter does water live up to its reputation?

15. Proteins are made up of chains of amino acids, and yet the proteins themselves are not acidic. Why do you think this is so?

Review

This chapter has focused on the elements of living things, how the elements combine to form molecules, and the major compounds of microorganisms and other life forms. To test your knowledge of the chapter contents, rearrange the scrambled letters to spell out the correct word for the available space. The answers are listed in the appendix.

1. During _____ bonding, an electron or a series of electrons are transferred between atoms.   Ⓞ Ⓒ Ⓘ Ⓘ Ⓝ

2. The chemistry of living things is _____ chemistry.
   Ⓡ Ⓝ Ⓒ Ⓘ Ⓞ Ⓖ Ⓐ

3. An acid is a chemical substance that donates _____ ions to a solution such as water.   Ⓨ Ⓡ Ⓝ Ⓗ Ⓖ Ⓞ Ⓓ Ⓔ

4. The carbohydrate _____ contains the basic supply of the world's
   energy to living things.   Ⓤ Ⓒ Ⓔ Ⓢ Ⓛ Ⓖ Ⓞ

5. The best known lipids are the _____.   Ⓐ Ⓢ Ⓣ Ⓕ

6. In all living organisms, proteins function both as structural materials and as

   _____.   Ⓔ Ⓔ Ⓢ Ⓨ Ⓜ Ⓝ Ⓩ

7. Both DNA and RNA are composed of repeating units called

   _____.   Ⓤ Ⓛ Ⓢ Ⓣ Ⓔ Ⓝ Ⓘ Ⓔ Ⓒ Ⓞ Ⓓ

8. The _____ structure of a protein consists of the amino acid chain.
   Ⓘ Ⓐ Ⓜ Ⓡ Ⓨ Ⓟ Ⓡ

9. In a _____ synthesis, the products of water are removed during the
   formation of covalent bonds.   Ⓨ Ⓓ Ⓐ Ⓘ Ⓡ Ⓔ Ⓞ Ⓓ Ⓗ Ⓣ Ⓝ

10. The pH scale relates the measure of _____ of a chemical substance.
    Ⓐ Ⓓ Ⓨ Ⓣ Ⓘ Ⓒ Ⓘ

11. The smallest part of a compound that retains the property of the compound is a

    _____.   Ⓛ Ⓤ Ⓒ Ⓞ Ⓔ Ⓛ Ⓜ

12. _____ are atoms of the same element that have different numbers
    of neutrons.   Ⓣ Ⓘ Ⓟ Ⓔ Ⓢ Ⓢ Ⓞ Ⓞ

13. A functional group designated —COOH is known as an acid group or a

    _____ group.   Ⓐ Ⓑ Ⓧ Ⓒ Ⓡ Ⓞ Ⓛ Ⓨ

14. Examples of disaccharides include lactose, sucrose, and _____.
    Ⓣ Ⓐ Ⓜ Ⓔ Ⓞ Ⓛ Ⓢ

15. The _____ bond is a weak bond that exists between poles of adjacent
    molecules.   Ⓖ Ⓡ Ⓝ Ⓗ Ⓔ Ⓞ Ⓓ Ⓨ

CHAPTER 3

BASIC CONCEPTS OF MICROBIOLOGY

H istorically, patrons of fast-food restaurants have been remarkably safe from infectious disease, especially in view of the huge volume of food served and the speed of preparation. But occasionally things go wrong, and in the winter of 1993 things went very wrong. That January, close to 500 people in several northwestern states became seriously ill with intestinal disease after eating at their local Jack-In-The-Box restaurants. Three children died during the epidemic and many others required hospitalization, including some who needed kidney dialysis.

The problem came to light when several patrons of Jack-In-The-Box restaurants in Washington State called their physicians to report the onset of bloody diarrhea. Most were experiencing gut-wrenching stomach cramps, and many were sick with fever. Similar reports were soon received from patients in Nevada, Oregon, Idaho, and California; it was obvious that a major epidemic was under way.

Local health departments moved quickly and contacted the federal Centers for Disease Control and Prevention, the national agency for dealing with problems of this magnitude (MicroFocus: 3.1). Almost overnight health officials were on the scene, taking food samples at several restaurants and sending the samples to laboratories for bacterial testing. Within days, the laboratories all reported isolations of a common bacterium called *Escherichia coli*. Normally a benign   esh"er-ik'e-a
inhabitant of the human and animal intestines, this particular strain of *E. coli* was the notorious toxin-producing strain 0157:H7. The toxin was destroying cells of the intestinal lining, which led to the bleeding, and it was causing hemorrhages in the kidneys. All signs pointed to the hamburger meat as the source of the bacteria. Most likely it had been contaminated with feces from the cattle's intestine during slaughter.

Now it was time to prevent any further spread of the epidemic. The management of Jack-In-The-Box restaurants ordered its employees to increase the cooking time for hamburgers by 12.5 percent. Federal guidelines were instituted requiring ground beef to be cooked at 155° F (instead of the previous 140° F). Physicians throughout the country were alerted to watch for additional cases

3.1

MICROFOCUS

THE CENTERS FOR DISEASE CONTROL AND PREVENTION

The Centers for Disease Control and Prevention (CDC) has its headquarters in Atlanta, Georgia, and is one of six major agencies of the U.S. Public Health Service. Originally established as the Communicable Disease Center in 1946, the CDC was the first governmental health organization ever set up to coordinate a national control program against infectious diseases. At first, it was concerned with diseases spread from person to person, from animals, or from the environment to humans. Eventually, though, all communicable diseases came under its aegis. Atlanta was selected as the site for the CDC because it was a convenient central point for the study of malaria, which was then common in the South.

In April 1955, two weeks after release of the Salk vaccine for polio, the CDC received reports of six cases of polio in vaccinated children. Two days later, it established the Polio Surveillance Unit and began collecting data on polio occurrence and summarizing it for health professionals. More than 80 percent of vaccine-associated polio cases were related to a single manufacturer, and its vaccine was withdrawn at once. This incident established the role of the CDC in health emergencies, and soon it became a national resource for the development and dissemination of information on communicable disease. In 1960, the CDC moved to a new headquarters complex adjoining Emory University. The unassuming appearance of the facility belies its importance.

Reorganized with its current name in 1980 (the word Prevention was added in 1992 but the CDC acronym was retained), the CDC is charged with protecting the public health of the United States populace by providing leadership and direction in the prevention and control of infectious disease and other preventable conditions such as cancer. It is concerned with urban rat control, quarantine measures, health education, and the upgrading and licensing of clinical laboratories. The organization also provides international consultation on disease and participates with other nations in the control and eradication of communicable infections. It employs 3500 physicians and scientists, the largest group in the world, and processes 170,000 samples of tissue annually. Its publication *Morbidity and Mortality Weekly Report* is distributed each week to over 100,000 health professionals.

that might indicate spread of the epidemic. Infected children were not permitted to return to day-care centers until two successive tests proved they had no residual bacteria in their intestine, and infected adults were given similar restrictions if they worked in day-care centers or with frail or sick people, such as in hospitals. And consumers were warned to avoid rare hamburgers at fast-food restaurants; instead, they were told, the meat should be gray (the new slogan became "If it's gray, it's OK").

So it was that microorganisms once again influenced the way people think and act. And again, it became painfully clear that although microorganisms are very small, the ramifications of microbiology are very great. Virtually no one has escaped infectious disease, and the search goes on daily for the causes, treatments, and preventatives for microbial disorders. In the quality control laboratory, inspectors are on constant alert to interrupt disease transmission in food and dairy products, and water purification and sewage treatment have become highly sophisticated technologies for preventing the spread of microorganisms through the fluids we drink.

But microbiology also has many positive aspects. Ecologists study microbiology to understand the natural recycling of minerals such as carbon and nitrogen. Evolutionists look to the microorganisms to learn more about the ancestors of contemporary forms, and biochemists use microorganisms as miniature laboratories to discover the chemical systems that underlie all living things. Geneticists observe hereditary material at work in microscopic forms and then apply what they have learned to more complex organisms. Even behaviorists find value in studying how certain microorganisms respond to environmental stimuli.

Our study of microbiology begins with a review of the position microorganisms occupy in the world of living things. Thumbnail sketches of the major groups of microorganisms will be presented, after which we shall explore the methods used to classify microorganisms. Discussions will follow on the origin of their names, their sizes and shapes, and the techniques used to view them. These basic considerations provide a foundation for studying the microorganisms in detail in future chapters.

A Brief Survey of Microorganisms

Despite their microscopic size, microorganisms exhibit numerous features also found in more complex cells. For example, DNA is the hereditary material in microorganisms as well as all other living things, and many of the biochemical patterns of growth are identical. However, a number of significant differences do set microorganisms apart. Before we explore these differences and unique features, we shall examine the total spectrum of living things by distinguishing two broad groups of organisms, the prokaryotes and eukaryotes. Once the characteristics of these groups are defined, we shall see how microorganisms fit into the pattern.

DNA: deoxyribonucleic acid, the organic substance that contains the hereditary information of the cell.

Prokaryotes and Eukaryotes

One of the most important generalizations to emerge in recent decades is that all organisms may be categorized as either **prokaryotes** or **eukaryotes** (terms first proposed in 1937 by Edward Chatton). Both terms are derived from the Greek word *karyon* meaning "nut," a reference to the nucleus of the cell. Prokaryotes lack a nucleus ("pro" implies primitive), whereas eukaryotes possess a well-defined nucleus ("eu" means true). In some texts, the words are spelled procaryote and eucaryote, but we shall retain the original spelling.

pro-kar'e-ōt
u-kar'e-ōt

According to modern biologists, bacteria and cyanobacteria ("blue-green algae") are prokaryotes, while fungi, protozoa, plants, and animals (including humans) are eukaryotes. Knowing this can have practical significance because we can define an organism's characteristics by knowing if its cells are prokaryotic or eukaryotic.

Indeed, scientists have found that the distinctions between prokaryotes and eukaryotes go deeper than the nucleus. The cells of prokaryotes and eukaryotes both have a gel-like cytoplasm (sometimes called the cytosol), but eukaryotic cells have a variety of membrane-bound components lacking in prokaryotes (Figure 3.1). These components are called **organelles**. They include the **endoplasmic reticulum**, a series of membranes extending throughout the cytoplasm; the **Golgi body**, a number of flattened sacs where the cell's proteins and lipids are processed and packaged before release; the **lysosome**, a somewhat circular, droplike sac of digestive enzymes within the cytoplasm; and the **mitochondrion**, the organelle where much chemical energy is released for cell use (which is why mitochondria are called the "powerhouses of the cells").

Organelles: membrane-bound compartments in eukaryotic cells.

Certain microorganisms such as simple algae and green-sulfur bacteria possess chlorophyll, a pigmented substance used in photosynthesis. The chlorophyll is dissolved in the cytoplasm in prokaryotic cells, but in eukaryotic microorganisms (such as the protozoan *Euglena*) it is contained in an organelle called the **chloroplast**. Still another organelle within the eukaryotic cell is the **cytoskeleton**, an interconnected system of fibers, threads, and interwoven molecules that give structure to the cell. The main components of the cytoskeleton are microtubules, microfilaments, and intermediate filaments, all assembled from subunits of protein. Another organelle called the **centriole** is a cylinderlike structure occurring in pairs and functioning in cell division.

A Comparison of Prokaryotic and Eukaryotic Cells

The prokaryotic cell is shown in (a). There are no organelles, nor is there a nucleus. The cell's DNA is a naked molecule that occupies the nuclear region. Ribosomes exist free in the cytoplasm. A cell wall surrounds the prokaryotic cell. The eukaryotic cell is shown in (b). Note the complexity of the cell and the variety of membrane-bound organelles. A well-defined nucleus, nuclear membrane, and nucleolus are evident. The ribosomes are bound to membranes.

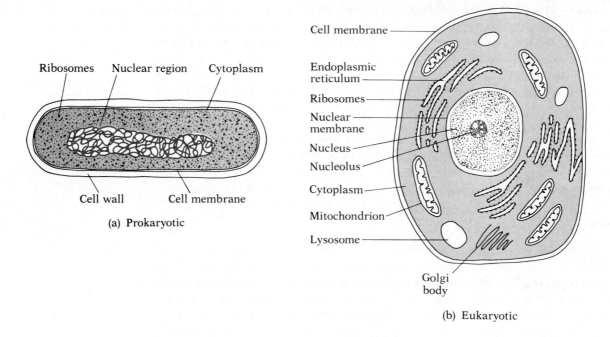

(a) Prokaryotic

(b) Eukaryotic

Ribosomes:

ultramicroscopic bodies of RNA and protein where protein synthesis takes place in cells.

    All prokaryotes and eukaryotes possess **ribosomes**, the RNA-protein bodies that participate in protein synthesis (Chapter 5). Prokaryotic ribosomes (70S ribosomes) are smaller than their counterparts (80S ribosomes) in eukaryotic cells. Moreover, prokaryotic ribosomes exist free in the cytoplasm, while eukaryotic ribosomes are usually bound to membranes of the endoplasmic reticulum.

    Many prokaryotic and eukaryotic cells contain organelles called flagella and cilia. **Flagella** are long and hairlike and extend from the cell to permit cell movement. In prokaryotic cells, such as bacteria, the flagella rotate like the propeller of a motorboat (Chapter 4). In eukaryotic cells, such as certain protozoa and sperm cells, the flagella whip about. **Cilia** are shorter and more numerous than flagella. In moving cells, they wave in synchrony and move the cell forward. *Paramecium* is a well-known ciliated protozoan. Cilia are also found at the surface of several types of cells, such as those lining the human respiratory tract.

    Both prokaryotic and eukaryotic cells have their hereditary characteristics stored in molecules of deoxyribonucleic acid (DNA) organized in **chromosomes**. In prokaryotic cells, the chromosome consists of a single DNA molecule in a closed loop, while in eukaryotic cells the DNA occurs in multiple chromosomes arranged in the nucleus. Eukaryotic chromosomes contain histone proteins; prokaryotic chromosomes do not.

    The **cell membrane** (also known as the **plasma membrane**) lies at the border of all cells. Like other membranes of the cell, it is composed primarily of proteins and lipids, especially phospholipids. The lipid occurs in two layers, referred to as a bilayer, and protein globules appear to float within the lipid.

Therefore, the membrane is constantly in flux and is referred to as a **fluid mosaic structure**. Within the fluid mosaic, proteins carry out most of the functions of the membrane. In order for the cell's cytoplasm to communicate with the external environment, materials move through the cell membrane by a series of processes explored in Chapter 4.

Many prokaryotic and eukaryotic cells contain a structure outside the cell membrane called the **cell wall**. With only a few exceptions, all bacteria have a thick, rigid cell wall. Among the eukaryotes, the algae, fungi, and plants have cell walls. The cell walls are not identical in these organisms, however. In fungi, the cell wall contains a polysaccharide called chitin (Chapter 14), while bacteria possess a different polysaccharide, one called peptidoglycan (Chapter 4). Algae and plants have cellulose as their primary wall components. Cell walls provide support for the cells, give them shape, and help them resist mechanical pressures. A summary of the many differences between prokaryotes and eukaryotes is presented in Table 3.1.

fun'jī
al'je

It is important to remember that all living things are either prokaryotes or eukaryotes. Note, however, that viruses are neither, since they are not living organisms. Viruses are segments of nucleic acid packaged in a protein shell. Nevertheless we consider them "microorganisms" due to their infectious nature. In the following sections we shall summarize other significant properties of the major groups of microorganisms, including the viruses.

Bacteria

The term **bacteria** is a plural form of the Latin *bacterium*, meaning staff or rod. Bacteria (sing., bacterium) are prokaryotes and are among the most abundant organisms on Earth. The vast majority play a positive role in nature: they digest sewage into simple chemicals; they extract nitrogen from the air and make it available to plants for protein production; they break down the remains of all that die and recycle the carbon; and they produce foods for our consumption and products for industrial technology. Many biologists believe that life as we know it would be impossible without the bacteria.

TABLE 3.1

A Comparison of Prokaryotes and Eukaryotes

CHARACTERISTIC	PROKARYOTES	EUKARYOTES
Nucleus	Absent	Present with nuclear membrane
Organelles	Absent	Present in a variety of forms
DNA structure	Single closed loop Naked strand with no protein	Multiple chromosomes Protein associated with DNA
Chlorophyll	When present, dissolved in cytoplasm	When present, contained in chloroplast
Ribosomes	Smaller than eukaryotic ribosomes Free in cytoplasm	Larger than prokaryotic ribosomes Bound to membranes
Cell walls	Generally present Complex chemical composition	Present in some types, absent in others Simple chemical composition
Reproduction	Usually by fission No evidence of mitosis Sexual reproduction unusual	By mitosis Sexual reproduction usual
Examples	Bacteria, rickettsiae, chlamydiae, cyanobacteria	Fungi, protozoa, plants, animals, humans (all other organisms)

Toxin:
a chemical substance
poisonous to body tissues.

Of course, some bacteria are harmful. Certain bacteria multiply within the human body, where they digest the tissues or produce toxins that lead to disease. Other bacteria infect plant crops and animal herds. Disease-causing bacteria are a global threat to all forms of life.

Bacteria have adapted to more different living conditions than any other group of organisms. They inhabit the air, soil, and water, and they exist in enormous numbers on the surfaces of virtually all plants and animals. They can be isolated from Arctic ice, hot springs, the fringes of space, and the tissues of animals (Figure 3.2). Some can withstand the acid in vinegar, the crushing pressures of ocean trenches, and the enzyme activity of digestive juices. Other bacteria survive in oxygen-free environments, boiling water, and extremely dry locations. Bacteria have so completely invaded every part of the earth that the mass of bacterial cells is estimated to outweigh the mass of all plants and animals combined (MicroFocus: 3.2). Chapters 4, 5, and 6 are devoted to the structure, growth, biochemistry, and genetics of bacteria, and Chapters 7 through 10 discuss the diseases they cause.

Rickettsiae, Chlamydiae, and Mycoplasmas

rik-et′se-ē
klah-mid′e-e

Rickettsiae (sing. rickettsia), chlamydiae, and mycoplasmas are also prokaryotes. Up until the early 1970s these microorganisms were considered apart from the bacteria, but contemporary microbiologists now classify them as "small bacteria." Unfortunately, old habits linger and some microbiologists still think of them as distinct groups, which is why we separate them here.

Rickettsiae were first described by Howard Taylor Ricketts in 1909. These tiny bacteria can barely be seen with the light microscope and are transmitted

FIGURE 3.2

Mixed Bacteria

An electron microscopic view of mixed bacteria from the gastro-intestinal tract of a lamb. Rod and spherical forms in various sizes and shapes are visible. Bar = 1 μm.

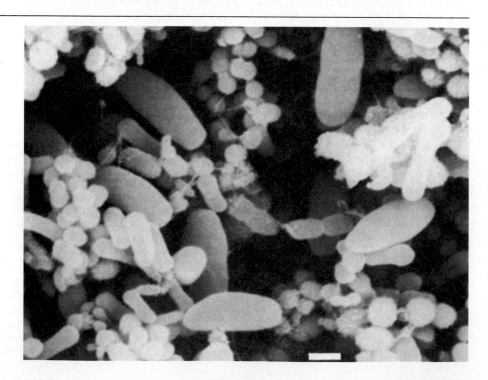

MICROFOCUS

AN UNKNOWN WORLD

Practically everyone knows the term bacteria, but it may surprise you that hardly anyone *knows* the bacteria. Yes, bacteria have been thoroughly studied in medicine, ecology, and molecular genetics, but the vast majority of bacteria remain unknown to science.

Try holding a pinch of rich soil in the palm of your hand. You are now face to face with an estimated billion bacteria representing some 10,000 different species. If you were to attempt to cultivate them using the most sophisticated methods available, perhaps 500 species might grow and multiply. That leaves 9500 species unknown until the right combinations of nutrients and environmental conditions is established in the laboratory.

Through the 1990s, *Bergey's Manual* has been the most authoritative guide to bacterial species. About 4000 species are listed in this book, a far cry from the thousands more believed to exist. Those who believe there are no more worlds to conquer should take note.

among humans primarily by arthropods such as ticks and lice. They are cultivated only in living tissues such as fertilized eggs, and different species cause a number of important diseases, including Rocky Mountain spotted fever and typhus fever. Chapter 9 contains a more thorough description of their properties.

Chlamydiae (sing. chlamydia) are roughly half the size of the rickettsiae, so small that they cannot be seen with the light microscope. The chlamydiae can be cultivated in cultures of living cells, and one species causes the gonorrhealike disease known as chlamydia. This and several other chlamydial diseases are described in Chapters 7 and 10.

Mycoplasmas are another notch smaller than chlamydiae and are possibly the smallest known bacteria (Table 3.2). They can be cultivated outside living tissues in artificial laboratory media and, although they are prokaryotes, they do not have cell walls like other bacteria. One form of pneumonia is due to a mycoplasma (Chapter 7) and a sexually transmitted disease (Chapter 10) is a mycoplasmal illness.

Arthropods:
animals with jointed appendages and a hard outer skeleton; examples are mosquitoes, ticks, and fleas.

Cyanobacteria

In older textbooks, **cyanobacteria** were known as blue-green algae. Today these microorganisms are considered more closely related to bacteria than to algae because of their structural and biochemical properties. However, cyanobacteria are not regarded as true bacteria, because of fundamental differences in how they form carbohydrates in photosynthesis (Chapter 5).

Cyanobacteria are prokaryotes. They possess light-trapping pigments that function in photosynthesis taking place in the cell. Many of the pigments are blue, but some are black, yellow, green, or red. The periodic redness of the Red Sea, for example, is due to a species of cyanobacteria whose members contain large amounts of red pigment.

si"ah-no-bak-tēr'e-ah

al'je
Algae:
plantlike organisms that practice photosynthesis and differ structurally from other land plants.

Photosynthesis:
a chemical process in which light energy is used to form chemical bonds in carbohydrate molecules.

TABLE 3.2

An Overview of the Microorganisms

MICROORGANISMS	CLASSIFICATION	DISTINGUISHING CHARACTERISTICS
Bacteria	Prokaryotic	Extremely abundant; microscopic; many positive roles in nature; some cause disease; all microscopic
Rickettsiae Chlamydiae Mycoplasmas	Prokaryotic	Types of "small bacteria"; many involved in disease
Cyanobacteria	Prokaryotic	Pigmented organisms; formerly "blue-green algae"; similar to true bacteria
Protozoa	Eukaryotic	Animal-like; classified by type of motion; no cell walls; not photosynthetic
Fungi	Eukaryotic	Molds and yeasts; usually filamentous; plantlike but not photosynthetic; unique cell walls
Unicellular algae	Eukaryotic	Plantlike; photosynthetic; most marine forms include diatoms and dinoflagellates
Viruses	(Nonliving)	Fragment of nucleic acid (RNA or DNA) enclosed in protein; envelope in some; noncellular inert particles; replicate only in host living cells; ultramicroscopic

Cyanobacteria may occur as unicellular or filamentous organisms. Many species are able to incorporate atmospheric nitrogen (they "fix" nitrogen) into organic compounds useful to plants, thereby filling an important ecological niche. Cyanobacteria inhabit freshwater as well as marine environments. When ponds or lakes contain a rich supply of nutrients, the organisms may "bloom" and convert the water to a pea-soup green with a foul odor. Swimming pools and aquaria experience this problem if algicide is not used regularly.

Protozoa

Protozoa (sing., protozoan) are single-celled eukaryotes. Many thousands of protozoal species exist, among which about two dozen species cause diseases such as malaria and sleeping sickness. The word "protozoa" is derived from Greek stems that mean "first animals," a term reflecting one view of their evolutionary significance.

Protozoa exhibit a bewildering assortment of shapes, sizes, and structural components (Figure 3.3). A few species have the necessary pigments for photosynthesis and can therefore synthesize their own foods. Most protozoal species, however, must obtain nutrients from preformed organic matter. Protozoa generally exist wherever there is water and a supply of food. The majority of species are microscopic, although a few species may be seen with the unaided eye.

Protozoa are classified into groups according to how they move. Some have whiplike flagella, others possess hairlike cilia, and still others move by means of cytoplasmic extensions called pseudopodia (false feet). We shall study the protozoa in depth in Chapter 15.

Fungi

For many generations **fungi** (sing., fungus) were considered plants. In recent years, however, laboratory researchers have discovered characteristics that set the fungi apart from plants. For example, the cell walls of fungi contain chitin, a carbohydrate not found in plant cell walls.

Malaria:
a mosquito-borne blood disease in which protozoa multiply within and destroy red blood cells.

fun′jī

ki′tin

Chitin:
a carbohydrate substance found in the cell wall of fungi.

FIGURE 3.3
Three Species of Protozoa

Electron micrographs of three species of protozoa illustrating differences in shape and size. (a) The ameba *Naegleria fowleri*, a cause of meningitis in humans. Note the irregular shape of this organism. Bar = 5 μm. (b) A protozoan called a trypanosome. The elongated shape and hairlike appendage (flagellum) are apparent. This organism causes African sleeping sickness. (c) Another flagellated protozoan, *Giardia lamblia* (× 8026). The organism has a flat shape with multiple flagella extending toward the rear. The protozoan on the left shows the sucker device on its lower surface for holding fast to tissue. *Giardia* causes diarrhea in humans.

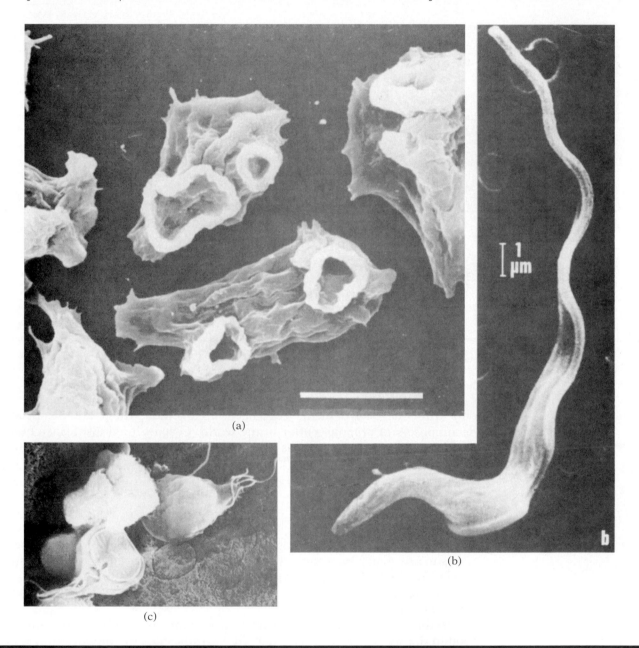

(a)

(b)

(c)

Fungi are eukaryotic organisms subdivided into two groups: the yeasts and the molds. **Yeasts** (Figure 3.4) are unicellular organisms about the size of large bacteria. They play a vital role in the fermentation of wine and beer and the production of bread. **Molds** are long chains of cells often seen as fuzzy masses on bread and other food products, especially acidic products. Commonly the molds assume vivid colors due to the pigments in spores they produce for reproductive purposes.

FIGURE 3.4

Two Microscopic Views of Fungi

(a) A transmission electron micrograph of a yeast cell showing typical eukaryotic features. Note the cell wall (CW), capsule (C), and plasma membrane (PM) on the cell surface. In the cytoplasm, vacuoles (V) and mitochondria (M) may be seen. The nucleus (N) is surrounded by a nuclear membrane (NM). In the upper portion of the photograph, a reproductive structure, the bud (B), is growing out from the parent cell. (b) Two molds, one wrapped around the other, in a parasitic relationship. The long, filamentous form of each mold and the many branches are visible. Bar = 10 μm.

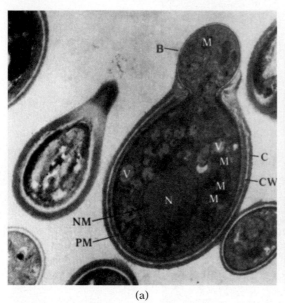

(a)

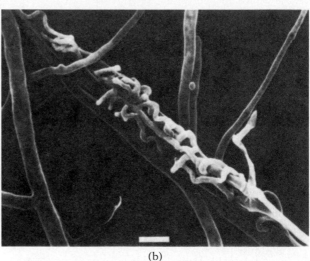

(b)

Fungi lack photosynthetic pigments, so they must obtain preformed nutrients from the environment. Together with bacteria, fungi are the primary decomposers of organic matter in the world. Certain fungi also cause human disease, such as athlete's foot and thrush. Chapter 14 discusses the fungi in detail.

Unicellular Algae

al'je

The word **algae** (sing., alga) has no technical significance in biology. It merely refers to any plantlike organisms that practice photosynthesis and differ structurally from typical land plants such as mosses, ferns, and seed plants. Two types of unicellular algae, the diatoms and the dinoflagellates, are important in microbiology.

di'ah-tomz

Diatoms are eukaryotic marine microorganisms and the main source of food in the world's oceans. Through photosynthesis, they trap the sun's energy and manufacture carbohydrates, which are passed on to other marine organisms as food. The cell walls of diatoms are impregnated with silicon dioxide, a glasslike substance. When diatoms die, their remains accumulate on the sea-

Diatomaceous earth:
the remains of diatoms used in filtering materials.

shore as diatomaceous earth. The latter is gathered and used to produce filters. **Dinoflagellates** are a group of photosynthetic eukaryotes composed of amoebas encased in hard shells (hence the name). Dinoflagellates are important members of the world's food chains. They also cause the periodic red tides occurring in the oceans. Both diatoms and dinoflagellates are discussed in Chapter 25.

Viruses

Viruses are neither prokaryotes nor eukaryotes, as noted previously. Many microbiologists even question whether viruses are living organisms because they are noncellular. They do not grow, nor do they display any nutritional patterns. Viruses have no observable activity except replication, and they accomplish this function only within living cells. Indeed, outside a living cell, a virus is no more alive than a grain of sand.

Viruses consist of an ultramicroscopic fragment of nucleic acid, either DNA or RNA, surrounded by a sheath of protein. Although we shall consider viruses "microorganisms," they appear to be at the threshhold of life. And, despite their simplicity, viruses are responsible for many human diseases including influenza, hepatitis, polio, and chickenpox. We shall pay considerable attention to the nature of viruses in Chapter 11 and discuss the diseases they cause in Chapters 12 and 13.

TO THIS POINT ■

We have begun our study of microbiology by placing microorganisms into the panorama of living things as prokaryotes or eukaryotes. Knowing whether an organism is a prokaryote or a eukaryote is important because it permits us to see microorganisms in relation to plants and animals, while giving us a glimpse of their properties. We then developed brief sketches of different microorganisms to observe the spectrum of forms we shall be studying in the chapters ahead. Viruses were included because they are important agents of disease, even though they are neither prokaryotes nor eukaryotes.

We shall now focus on some general properties that apply to all microorganisms. Included here are such properties as classification, nomenclature, and size. This study will serve as a prelude to studying individual groups of microorganisms in other chapters.

General Properties of Microorganisms

The basic concepts of microbiology include a number of properties that apply to all microorganisms. For example, all microorganisms have scientific names, and all have certain sizes. In addition, there are classification schemes into which most fit. In this section we shall explore the principles on which these properties are based.

Origins of Classification

The science of classification, or **taxonomy**, deals with the systematized arrangements of related microorganisms and other living things into logical categories. Derived from Greek roots meaning "law of arrangements," taxonomy is essential to an understanding of relationships among living things. Taxonomy focuses on unifying concepts among organisms, while providing a basis for communication among biologists.

One of the first taxonomists was the Greek philosopher **Aristotle**. In the fourth century B.C., Aristotle categorized living forms in the world around him and described over 500 species of plants and animals according to their appearance and habits. Without reference books or instruments to help him, Aristotle made some key observations, including the fact that the whale is more like a mammal than a fish.

Taxonomy:
the science that deals with the logical arrangement of living things into categories.

lin-ā′us

plant′ā
an-i-mal′ē-ah

The modern basis of taxonomy was devised by the Swedish botanist Carl von Linné, better known to history as **Carolus Linnaeus**. In his *Systema Naturae*, published in several editions between 1735 and 1759, Linnaeus named thousands of plants and animals and classified them in the kingdoms **Plantae** and **Animalia**. He inspired an unprecedented worldwide program of specimen hunting and sent his students around the globe in search of new life forms. He also devised the binomial scheme of nomenclature, thus resurrecting the use of Latin as a language of the learned (MicroFocus: 3.3). In his otherwise exacting system, Linnaeus reflected the scant knowledge of microorganisms and his general disinterest in them by grouping them separately under the heading of Vermes (as in vermin) in the category Chaos (confusion). In the years after Linnaeus, however, some microorganisms came to be considered plants while others were categorized as animals.

hek′el

Protist:
a name coined by Haeckel to
refer to microorganisms;
now used for a specific
group of microorganisms.

In 1866, the German naturalist **Ernst H. Haeckel** separated the microorganisms from the plant and animal kingdoms. Haeckel coined the term **protist** for microorganisms and placed the fungi, protozoa, and microscopic algae in a third kingdom, **Protista**. As more and more bacteria were discovered, they were included in this group. However, biologists could not fully agree on the placement of bacteria and fungi. For example, Ferdinand Cohn, the prominent bacteriologist of the late 1800s (and Koch's patron), published a three-volume work on bacteria and classified them with the other plants because of the presence of a cell wall (Table 3.3).

3.3

MICROFOCUS

"WHAT WAS WHAT NAME?"

The binomial system of nomenclature appears to be so obvious that it hardly needed to be invented. However, before Carolus Linnaeus devised the scheme in 1735, scientists could not agree on scientific names for the organisms they studied. Different names were invented by different writers to serve as both designation and description. Confusion was rampant.

Linnaeus' great simplifying decision was to use the genus name and a modifying adjective for the label and description of an organism. He had to work quickly lest other naturalists use the same name for different organisms. In a monumental task of linguistic invention, Linnaeus ransacked his Latin for enough terms to make up thousands of labels. Some names he took from an organism's manner of growth, others from the discoverer, others from classical heroes, and still others from vernacular names. Any parent who has had to name a child can appreciate what he was up against.

In a 1753 book on plants, Linnaeus supplied binomial names for over 5900 plant species known at that time. In his tenth edition of *Systema Naturae* (1759), he extended

the scheme to animals. Within decades, his binomial names were adopted by European scientists and were reaching across the world. And just in time, because explorers were returning to Europe from distant corners of the globe with newly discovered life forms (penguins, tobacco, potato, manatees, kangaroos, and others).

How important is the binomial system of nomenclature? It is clear that, to be understood, scientists must avoid ambiguity, and this is what the binomial system accomplishes. In the United States, for example, corn refers to a tall plant that produces yellow kernels on a cob. However, in England, the same plant is called maize. English "corn" is any number of different cereal grains, such as wheat, rye, and barley. Thus, to eliminate confusion, scientists use the term *Zea mays* when they refer to American corn or English maize. The binomial system contains more than a grain of truth for scientists.

TABLE 3.3

A Summary of Microbial "Classifiers"

INVESTIGATOR	TIME FRAME	PROPOSAL
Aristotle	Fourth century B.C.	Described 500 species of plants and animals
Carolus Linnaeus	1750s	Devised plant and animal kingdoms Developed classification system Grouped microorganisms as "Chaos"
Ernst Haeckel	1860s	Separated microorganisms to third kingdom (Protista)
Ferdinand Cohn	1880s	Placed bacteria with plants because of presence of cell wall
David Bergey	1923	Attempted a classification of bacteria in *Bergey's Manual*
Robert Whittaker	1969	Devised five-kingdom classification; microorganisms occupy kingdoms Monera, Protista, and Fungi

Modern Classification

In the twentieth century, advances in the study of cell biology and increasing interest in evolutionary biology raised serious questions about two- or three-kingdom classification schemes. Finally, in 1969, **Robert H. Whittaker** of Cornell University proposed a system that has gained wide acceptance among modern biologists. Further expanded in succeeding years by Lynn Margulis of the University of Massachusetts, the system recognizes five kingdoms for all living things: Monera, Protista, Fungi, Animalia, and Plantae (Figure 3.5).

According to the five-kingdom system, the bacteria and cyanobacteria are classified together in the kingdom **Monera** because they are the only true prokaryotes. The second kingdom, **Protista**, includes eukaryotes such as protozoa and one-celled algae. These organisms are unicellular all their lives; many contain chlorophyll (plantlike), and move (animal-like). The fungi are placed in the third kingdom, **Fungi**, because they are nongreen eukaryotic organisms whose cell walls differ chemically from those of traditional plants. Also the cytoplasm mingles among adjacent cells in the fungus, so that fungi are not considered true multicellular organisms. The final two kingdoms, **Plantae** and **Animalia**, are the traditional plants and animals. Note that the viruses are not to be found in any of these kingdoms because they are neither prokaryotes nor eukaryotes. Viruses appear to be partial or degenerate forms of living systems.

Chlorophyll: the green pigment used by plant and other cells in photosynthesis.

The mechanics of the classification system have remained consistent since first outlined by Linnaeus. The fundamental rank is the **species** (pl., species). For microorganisms, this is a group that has 70 percent biochemical similarity and differs significantly from other species. (For plants and animals, a species is a group of individuals in a population that can breed with one another). Two or more species are grouped together as a **genus** (pl., genera). A collection of genera make up a **family**, and families with similar characteristics make up an **order**. Orders are placed together as a **class**, and classes are assembled into a **phylum** (or **division**, in bacteriology and botany). Two or more phyla are grouped as a **kingdom**. Table 3.4 outlines the classification schemes for three organisms in the five-kingdom system.

Species: the fundamental rank in the classification system.

Genus: a collection of two or more species of organisms.

In bacteriology, a microorganism may belong to a rank below the species level to indicate that a special characteristic exists within a subgroup of the species. Such ranks have no official standing in nomenclature, but they have practical usefulness in helping further to identify an organism. For example,

FIGURE 3.5

The Five-Kingdom System of Classification

Devised by Robert H. Whittaker, this system implies an evolutionary lineage beginning with the Monera and extending to the Protista. Certain of the Protista are proposed to be ancestors of the Plantae, Fungi, and Animalia, respectively. Divergence at each level is based on the three modes of nutrition: photosynthesis, absorption, and ingestion. Unicellular and multicellular organization are also key features in the system.

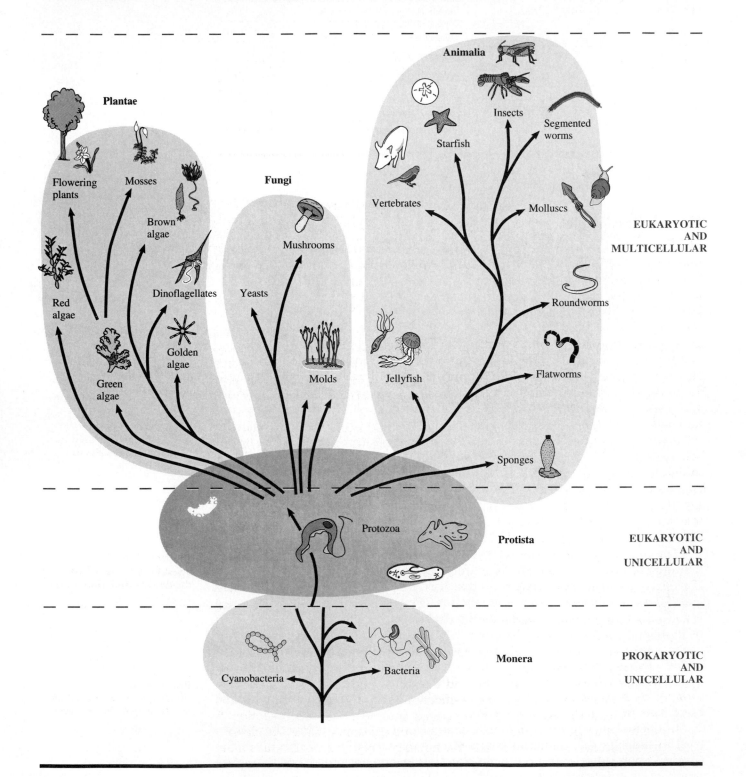

TABLE 3.4
The Taxonomy of Three Common Species of Organisms

	HOMO SAPIENS (HUMAN BEING)	FELIS DOMESTICA (HOUSE CAT)	LEPTOSPIRA INTERROGANS (A BACTERIUM)
Kingdom	Animalia	Animalia	Procaryotae
Phylum (Division)	Chordata	Chordata	Gracilicutes
Class	Mammalia	Mammalia	Scotobacteria
Order	Primata	Carnivora	Spirochaetales
Family	Hominidae	Felidae	Leptospiraceae
Genus	Homo	Felis	Leptospira
Species	H. sapiens	F. domestica	L. interrogans

two biotypes of the cholera bacillus, *Vibrio cholerae*, are known to exist: *Vibrio cholerae* classical biotype and *Vibrio cholerae* El tor biotype. Other designations of ranks include subspecies, serotype, strain, morphotype, and variety. Some bacteriologists recommend changing all these subspecies ranks to "varieties" and then using names such as biovar and serovar.

Cholera:
a bacterial disease of the intestine accompanied by extreme diarrhea.

Bacterial Taxonomy

We shall consider the taxonomy of most groups of microorganisms in their respective chapters. Bacterial taxonomy, however, merits special attention because bacteria occupy an important position in microbiology and have a complex system of taxonomy. We shall therefore discuss it here to show how a classification system develops. Also, we shall use bacterial classification as a prelude to Chapter 4.

One of the first systems of classification for the bacteria was devised in 1923 by David Hendricks Bergey. His book *Bergey's Manual of Determinative Bacteriology* was updated in the decades thereafter and has come to our era as **Bergey's Manual of Systematic Bacteriology**. The first volume of the first edition of this later work was published in 1984, and three succeeding volumes appeared in following years. *Bergey's Manual*, as it is commonly known, is considered the official listing of all recognized bacteria. It is also intended to be a guide to identification.

In 1984, the editors of *Bergey's Manual* noted that there is no official classification of bacteria and that the closest approximation to an official classification is that most widely accepted by the community of microbiologists. They stated that a comprehensive classification may one day be possible but a general scheme cannot currently be perceived due to the volume and complexity of available information.

Bergey's Manual:
the official listing of all recognized bacteria.

Nevertheless, volume I of *Bergey's Manual* contains an interim classification scheme during the years of transition. The editors propose that bacteria be placed in the kingdom **Procaryotae** and that the kingdom be further broken down to four divisions. No further classification is offered beyond the division level, but instead the bacteria are separated into sections. Each section is described by a number of experts in that area, and decisions on further classification are left to the experts. Therefore the taxonomy of certain bacteria is more complete than that of others. Appendix D of this text contains the complete outline of *Bergey's Manual*.

Procaryotae:
the kingdom name used by bacteriologists for bacteria.

Antibodies:
protein molecules produced by the immune system that react specifically with organisms or chemical substances.

The criteria used in the **identification** and classification of bacteria are rigorous and thorough. For example, the organism's shape and size; its oxygen, pH, and temperature requirements; and its laboratory characteristics are considered (Figure 3.6). Staining reactions, sporeforming ability, and type of movement are other important determinants. Pathogenic effects on animals are also noted. The biochemistry of the organism, including its photosynthetic nature and ability to digest certain organic substances, yields additional data. Tests are also performed to see whether it interacts with known antibodies. Even a laser beam can be used to identify an organism. In 1993, for example, researchers discovered that each species of bacteria has an "optical fingerprint" and reflects or absorbs laser light in its own characteristic way.

FIGURE 3.6
A Scratch and a Corneal Infection

This incident happened in Georgia during the winter of 1989. Morphological, staining, and biochemical characteristics were used to identify the pathogenic bacterium.

1. On the morning of January 11, 1989 a woman scratched her left cornea while applying mascara.

2. A day later her eye was painfully swollen and red. The woman was very sensitive to light and decided to seek medical help.

3. That day, the woman visited an ophthalmologist and was treated with gentamicin ointment. However, the symptoms became worse and on January 14 she was admitted to the hospital with a severe corneal abscess.

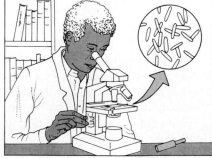

4. Technologists at the hospital laboratory isolated the bacterium *Pseudomonas aeruginosa* from the abscessed cornea. They found the identical organism in a sample of her mascara and concluded that the scratch had probably introduced bacteria into her eye tissues.

In the 1970s, scientists found that the DNA of bacteria could be isolated rather easily and that purified DNA could be centrifuged at a high speed to measure its density, which was found to vary according to the relative amounts of the four nitrogenous bases in DNA. Scientists could then estimate the proportion of guanine and cytosine relative to the total amount of bases in DNA and derive a so-called **GC ratio** for numerous species of bacteria. While this procedure added another criterion for classification, it also caused some consternation because bacteria thought to be related were found to have very different GC ratios. The current trend is to assume a close relationship between two species of bacteria only if their GC ratios differ by less than 10 percent.

Guanine, cytosine: **two of the four nitrogenous bases found in DNA.**

Nomenclature

In addition to giving the descriptions and properties of microorganisms, taxonomies like that in *Bergey's Manual* provide the genus name by which scientists refer to specific microorganisms. All microorganisms have a double name, usually from Latin or Greek stems. The name consists of the genus to which the organism belongs and a species modifier (a modifying adjective) that further describes the genus name. This system of nomenclature is called the **binomial system** from the Latin *bi-nomial*, meaning two names. First suggested by Linnaeus, the binomial system gives biologists throughout the world an international language for life forms and eliminates incalculable amounts of confusion.

Genus: **a group of two or more species.**

When the genus and species modifier of an organism are written, only the first letter of the genus name is capitalized. The remainder of the genus name and species modifier are written in lowercase letters. Both words should be printed in italics, but if this is not possible, they should be underlined. For example, the species of intestinal rod discovered in 1888 by Theodor Escherich is written as *Escherichia coli* or Escherichia coli.

esh'er-ik
esh"er-ik'e-a

Scientists often abbreviate binomial names by writing the first letter of the genus name, or some accepted substitution, together with the full species modifier. The abbreviation should also be italicized or underlined. Thus *Escherichia coli* becomes *E. coli*, and *Bacillus subtilis* is written *B. subtilis*. Where similarities occur, international committees on nomenclature have intervened.

Size Relationships

Another important property of a microorganism is its size. In microbiology, the unit of length most often used is the **micrometer**. This unit is equivalent to a millionth (micro-) of a meter. The micrometer is symbolized as μm, a combination of the Greek letter "μ" (pronounced "mue") and the abbreviation "m" for meter. To appreciate how small a micrometer is, consider this: comparing a micrometer to an inch is like comparing a housefly to New York City's Empire State Building, 1472 feet high.

Micrometer: **a unit of length equal to a millionth of a meter.**

Microorganisms range in size from the relatively large, almost visible protozoa (100 μm) down to the incredibly tiny viruses (0.01 μm), ten thousand times smaller (Figure 3.7). Molds consist of intertwined filaments so long and twisted that they are visible, but the individual mold cells measure only about 40 μm by 10 μm. One group of fungi, the yeasts, are commonly about 8 μm in diameter. Most bacteria are about 1 μm to 5 μm in length, but the largest ones reach approximately 20 μm in length, although in 1993 a notable addition to the microbial world appeared in print (MicroFocus: 3.4). The smaller bacteria known as the rickettsiae may be only about 0.5 μm long. Chlamydiae, among the smallest bacteria, are a scant 0.25 μm in length.

FIGURE 3.7
Size Comparison Among Various Living Things

A comparison of size relationships among common objects and microorganisms. Most microorganisms have a size that falls in the range of 100 to 0.25 μm. Viruses lie below this range.

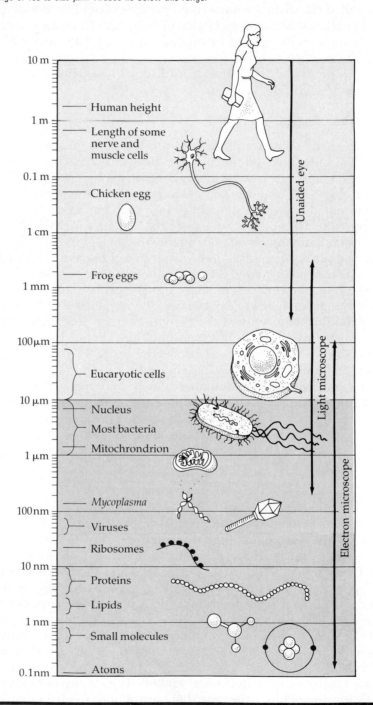

3.4

A BIOLOGICAL OXYMORON

An English professor might define an oxymoron as two mutually exclusive juxtaposed terms. Translated to the vernacular, an oxymoron is simply two words that do not go together. Some typical oxymorons include jumbo shrimp, holy war, negative attraction, and sweet sorrow.

In 1993, researchers at Indiana University reported their discovery of a visible bacterium, and headlines proclaimed it a biological oxymoron. Why? Because a bacterium by definition is an invisible organism. And yet, here was a visible bacterium, an organism so large that a microscope was not needed to see it. The spectacular giant measures over 0.6 mm in length (that's 600 μm compared with 2 μm for *E. coli*) and dwarfs a *Paramecium*. Found in the gut of the surgeonfish near Australian reefs, the bacterium is the largest ever observed.

But is it really a bacterium? Its discoverers maintain it is. They have used biochemical techniques to amplify the genes for ribosomal RNA, and they have determined the sequence of bases in the genes. The base sequence, they report, is consistent with that of bacterial genes. Moreover, cellular analysis has failed to reveal a nucleus or nuclear membrane or subcellular compartmentalization, all of which bacteria characteristically lack. In addition, the DNA

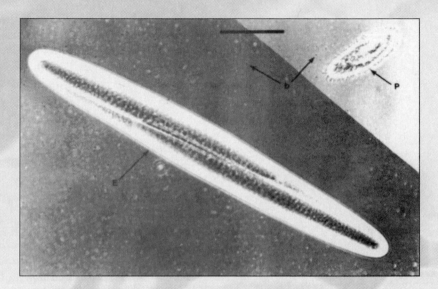

analysis of the organism's chromosome has indicated its relationship to the bacteria of the genera *Epulopiscium* and *Clostridium*.

Will the data stand the scrutiny of time and additional research? If they do not, then the definition of the bacterium is safe. If, however, this giant is indeed a bacterium, then the definition of a bacterium will have to change lest microbiologists be accused of spreading an oxymoron—and no one wants that to happen.

A scanning electron micrograph of an immense bacterium isolated from the gut of a marine surgeonfish in 1993. The organism is labeled E because its name is *Epulopiscium fishelsoni*. Over 0.61 mm (610 μm) long, the bacterium can be compared to *Paramecium* (P), a relatively large protozoan at the upper right. The arrows from b point to typical bacteria added to the mixture for reference. Bar = 0.1 mm.

Because the viruses are so tiny, scientists often express their size in **nanometers** to avoid decimals of micrometers. The nanometer, symbolized as nm, is equivalent to a billionth (nano-) of a meter. Using nanometers the size of a smallpox virus may be written as 250 nm, rather than 0.25 μm. The polio virus, among the smaller viruses, measures 20 nm. Other objects measured in nanometers include the wavelength of radiant energy, such as visible or ultraviolet light, and the size of certain large molecules. To appreciate how incredibly small a nanometer really is, consider what happened a "mere" billion seconds ago—John F. Kennedy was still President of the United States and the idol of a teenaged Bill Clinton. It was 1963. Chances are you were not even born yet.

Nanometer:
a unit of measurement equal to a billionth of a meter.

TO THIS POINT

■

We have discussed the spectrum of microorganisms and focused on some of their properties. One such property, classification, has been studied by Linnaeus and Haeckel, but the currently accepted system is that devised by Whittaker. Microorganisms occupy three of the five kingdoms in Whittaker's system. We noted that the classification systems for different microbial groups are handled in more specific textbooks on the groups (for example, fungi, protozoa, and viruses). For bacteria, the classification system is complex and is outlined in four volumes of Bergey's Manual of Systematic Bacteriology. *We listed several criteria used to classify and identify a bacterium to provide a glimpse of the taxonomist's work.*

The discussion then turned to nomenclature, where the binomial system is the method employed for all living things. A binomial name consists of the genus to which an organism belongs and a modifying species adjective. Both parts of the binomial name are written in italics. We closed the section with a brief review of the size of microorganisms and the units used to measure their length.

In the final section of this chapter, we shall survey the methods used for observing microorganisms. We shall open with a discussion of the common light microscope and then review three specialized types of microscopy. Some remarks on electron microscopy will complete the chapter.

Microscopy

Modern technologists have made available to microbiologists a broad range of instruments for viewing microorganisms. These instruments include the common light microscope, as well as a number of specialized instruments and the highly sophisticated electron microscopes. All operate on the same basic principle: energy is projected toward an object, such as a microorganism. The energy bounces off the object and creates an impression on a sensing device. This device may be a television screen, a photographic film, or the human eye. The image reveals the form, shape, size, and other structural features of the object.

Light Microscopy

The basic microscopic system used in the microbiology laboratory is the **light microscope** (Figure 3.8). This instrument is also called the bright-field microscope because visible light passes directly through its lenses until it reaches the eye. Another common term is the **compound microscope** because it has a two-lens system, with the objective lens nearer the object and the ocular lens nearer the eye.

In light microscopy, visible light is projected through a substage condenser, which focuses the light into a sharp cone. The light then passes through the opening in the stage, into the slide, and bounces off the object. Next, it enters the objective lens to form a magnified image darker than the background. This image is called a **real image** because it can be projected onto a screen. However, the image is not seen by the microscopist. Instead, the image becomes an object for the ocular lens, which magnifies the image a second time to create a **virtual image** in space. Only the observer can see this image. It appears about as distant from the eye as this page is from your eye. Figure 3.9 traces the pathway of light and the formation of the virtual image.

Objective lens:
the lens of a compound microscope nearest the object.

Ocular lens:
the lens of a compound microscope nearest the eye.

FIGURE 3.8

The Light Microscope

The familiar light microscope used in many instructional and clinical laboratories. Note the important features of the microscope that contribute to the visualization of the object.

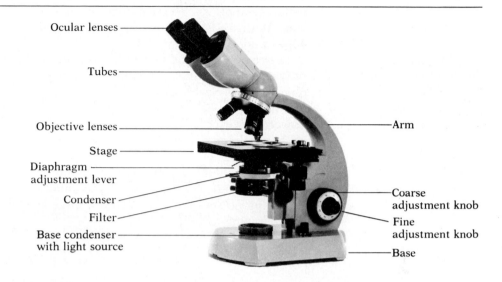

Ocular lenses

Tubes

Objective lenses

Stage

Diaphragm adjustment lever

Condenser

Filter

Base condenser with light source

Arm

Coarse adjustment knob

Fine adjustment knob

Base

FIGURE 3.9

Image Formation in Light Microscopy

Light passes through the objective lens, forming an inverted real image (A). This image serves as an object for the ocular lens, which remagnifies the image and forms the virtual image (B). The lens system of the eye perceives this image and captures it on the retina (C).

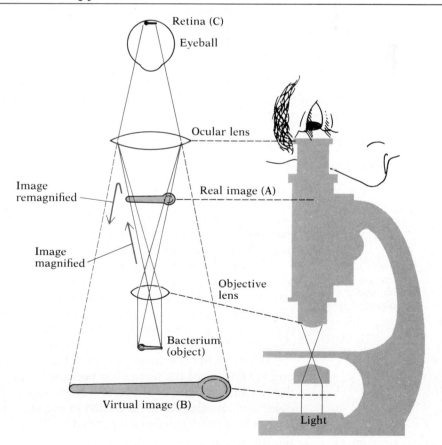

Retina (C)

Eyeball

Ocular lens

Image remagnified

Real image (A)

Image magnified

Objective lens

Bacterium (object)

Virtual image (B)

Light

A light microscope usually has three objective lenses, the low-power, high-power, and oil-immersion lenses. Generally these magnify an object about 10, 40, and 100 times, respectively. The magnification is represented by the multiplication symbol "×." The real image each forms is then remagnified by the ocular lens. With a 10× ocular, the total magnifications achieved would be 100×, 400×, and 1000×, respectively.

For an object to be seen distinctly, the lens system must have good **resolving power**; that is, it must transmit light without variation and allow closely spaced objects to be clearly distinguished. For example, a car seen in the distance at night may appear to have a single headlight because the eyes lack resolving power. However, as the car comes closer, the two headlights can be seen clearly as the resolving power of the eye increases. The headlights now have **resolution**, or clarity.

The resolving power (RP) of a lens system is important in microscopy because it indicates the size of the smallest object that can be seen clearly. Resolving powers vary for each objective lens and are calculated according to the following formula:

Resolving power: the ability of a lens system to transmit light without variation and permit nearby objects to be clearly distinguished.

$$RP = \frac{\lambda}{2 \times NA}$$

In this formula, the Greek letter λ (lambda) represents the wavelength of light and is usually set at 550 nm, the halfway point between the limits of visible light. The symbol NA stands for the **numerical aperture** of the lens. This number is generally printed by the manufacturer on the side of the objective lens. It refers to the size of the cone of light that will enter the objective and the medium in which the lens is suspended, usually air. For a low-power objective with an NA of 0.25, the resolving power may be calculated as follows:

nm: the symbol for nanometer, a billionth of a meter.

$$RP = \frac{550 \text{ nm}}{2 \times 0.25} = \frac{550}{0.50} = 1100 \text{ nm or } 1.1 \ \mu m$$

Since the resolving power for this lens system is 1.1 μm, any object smaller than 1.1 μm could not be seen, but an object larger than 1.1 μm would be visible.

For an oil-immersion lens, the NA is often 1.25. Using this figure in the formula, we can solve for the resolving power as follows:

$$RP = \frac{550 \text{ nm}}{2 \times 1.25} = \frac{550}{2.50} = 220 \text{ nm or } 0.22 \ \mu m$$

In this situation, objects as small as 0.22 μm may be visualized. This would include most bacteria, with the notable exception of chlamydiae and mycoplasmas. Viruses would not be visible.

Another factor of the compound microscope is the **working distance** (Figure 3.10), the amount of clearance between the slide and the bottom of the objective lens. Working distance is related to where the object comes into focus. For the low-power objective, a common working distance is 6.8 millimeters (mm); for the oil-immersion objective it is a scant 0.12 mm, almost 60 times closer.

Working distance: the amount of clearance between the slide and the bottom of the objective lens.

When switching from the low-power lens to the oil-immersion lens, one quickly finds that the image has become fuzzy. The object lacks resolution, and the resolving power of the lens system appears to be poor. This is because the objective lens should be used with immersion oil. The system's resolving power is calculated with the lens suspended in oil rather than air, a factor that increases the numerical aperture to 1.25.

FIGURE 3.10

Aspects of Light Microscopy

(a) The magnifications, numerical apertures, and working distances of the three common objectives of the light microscope are indicated. (b) The use of oil in oil-immersion microscopy. When light rays enter the air (solid arrow), they miss the objective lens. However, they remain on a straight line (dashed arrow) in the oil. This pathway leads them directly into the lens.

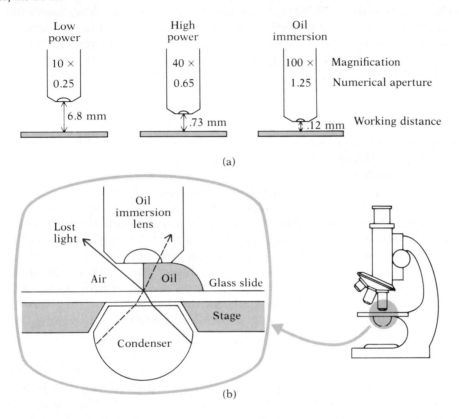

(a)

(b)

Oil is needed for **oil-immersion microscopy** because light bends abruptly as it leaves the glass slide and enters the air (Figure 3.10). Both low-power and high-power objectives are wide enough to capture sufficient light for viewing, but the oil-immersion objective is so narrow that most light bends away and would miss the objective if oil were not used. Immersion oil has an index of refraction of 1.5, which is almost identical to the index of refraction (or refractive index) of glass. The **index of refraction** is a measure of the light-bending ability of a medium. Because it is the same for oil and glass, the light does not bend as it passes from the glass slide into the oil (MicroFocus: 3.5). By comparison, air has an index of refraction of 1.0, which accounts for the abrupt bending as light enters it. The oil thus provides a homogeneous pathway for light from the slide to the objective, and the resolution of the object increases.

Index of refraction: **a measure of the light-bending ability of a medium.**

Staining Techniques

In preparing for light microscopy, microbiologists commonly stain bacteria because the cytoplasm of bacteria lacks color. Several techniques have been developed for this purpose.

Heat fixing:
a procedure in which a slide containing bacteria is subjected to a moment of heat from a flame.

To perform the **simple stain technique**, a small amount of bacteria is placed in a droplet of water on a glass slide and the slide is air-dried. Next, the slide is passed through a flame in a process called **heat fixing**. This bonds the cells to the slide, kills many organisms that may still be alive, and prepares them for staining. Now the slide is flooded with a **basic dye** such as crystal violet or methylene blue. Cytoplasm generally has a negative charge, and since basic dyes have a positive charge, the dye is attracted to the cytoplasm, where staining takes place. Figure 3.11 displays this principle.

The **negative stain technique** works in the opposite manner (Figure 3.11). Bacteria are mixed on a slide with an **acidic dye** such as nigrosin (a black stain) or Congo red (a red dye). The mixture is then smeared across the face of the slide and allowed to air-dry. Because the acidic dye carries a negative charge, it is repelled by the cytoplasm. The stain gathers around the negatively charged cells and the microscopist observes clear or white cells on a colored background. Since this technique avoids chemical reactions and heat fixing, the cells appear less shriveled and less distorted and are closer to their natural condition.

Gram stain technique:
a staining procedure that differentiates bacteria into two separate groups, Gram-positive and Gram-negative.

The **Gram stain technique** allows us to view stained cells while learning something about them. The technique is named for Christian Gram, the Danish physician who first suggested its use in 1884. It is a differential technique because it differentiates bacteria into two groups depending on the results. Certain bacteria are called Gram-positive bacteria; others are Gram-negative.

The first two steps of the technique are straightforward. Air-dried heat-fixed smears are stained with crystal violet (MicroFocus 3.6), then with a special Gram's iodine solution. All bacteria become blue-purple. Next the smear is rinsed with a decolorizer such as 95 percent alcohol or an alcohol–acetone mixture. At this point, certain bacteria lose their color and become transparent. These are the **Gram-negative bacteria**. Other bacteria regain the blue-purple stain. These are the **Gram-positive bacteria**. When safranin, a red dye, is applied to the slide, only the Gram-negative organisms accept the stain. Thus at the technique's conclusion, Gram-positive bacteria are blue-purple while Gram-negative organisms appear orange or red (Figure 3.11). By observing the color of the cells at the conclusion of the process, one may decide the group to which the bacteria belong.

Safranin:
a basic dye with a red color; used in Gram staining.

It is not totally clear why bacteria respond differently to the Gram stain technique. One theory suggests that crystal violet and iodine form a chemical complex in the bacterial cytoplasm. Since Gram-negative bacteria have a high lipid content in their cell walls, some microbiologists maintain that the alcohol dissolves the lipid and allows the crystal violet–iodine complex to leak out of the cytoplasm. Gram-positive bacteria, with less cell-wall lipid, are less susceptible to the alcohol's effects. Another theory points to the heavy concentration of

FIGURE 3.11

Important Staining Reactions in Microbiology

(a) In the simple staining technique, the positive-charged stain is attracted to the negative-charged bacteria and staining takes place. (b) With the negative stain technique, negative-charged dye is repelled by the bacteria and the cells remain clear on a dark background. (c) The Gram stain technique is a differential procedure. All bacteria stain with the crystal violet and iodine, but only the Gram-negative bacteria lose the color when alcohol is applied. Subsequently, these bacteria stain with the safranin dye. The Gram-positive bacteria remain blue-purple.

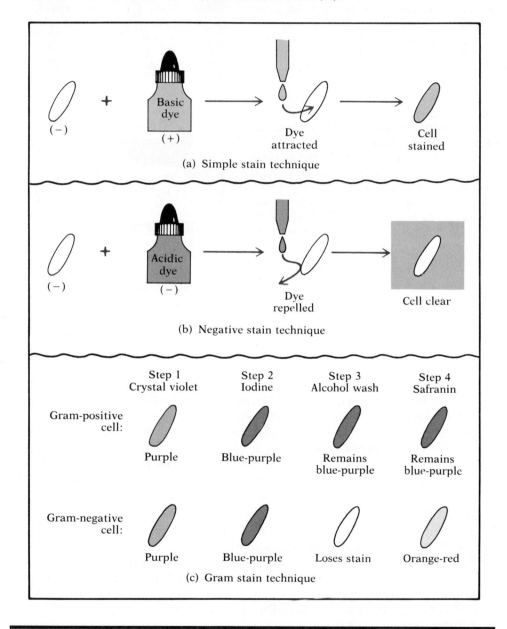

peptidoglycan in the cell wall of Gram-positive bacteria. Peptidoglycan, a complex carbohydrate, is thought to trap the crystal violet–iodine complex in its many cross-linkages. Gram-negative bacteria have considerably less peptidoglycan in their cell walls, hence they would trap less of the complex. It should be noted that the terms "positive" and "negative" are nothing more than convenient expressions and that electrical charges play a minimal role in Gram staining.

pep′ti-do-gli″kan

Peptidoglycan:
a complex carbohydrate present in the cell walls of bacteria.

"A.O. MEANS WHAT?"

In modern bacteriology laboratories the crystal violet solution used for Gram staining is prepared by mixing solid dye particles with ammonium oxalate. This procedure has not changed since 1929, when a graduate student named Thomas Hucker introduced it. How this "Hucker modification" came about is part of the folklore of microbiology.

Hucker was studying bacteriology at Yale University. Early in 1929, his advisor suggested that he contact several hospital and university laboratories to see how they were performing the Gram stain technique. Hucker was to report his findings in a paper presentation at an upcoming scientific meeting in Philadelphia. He dutifully sent out a series of letters and learned that the standard procedures were being used at all laboratories—all, that is, except Dartmouth's.

The reply from Dartmouth College piqued his interest. At the time, the usual procedure was to dissolve crystal violet in aniline oil. But Dartmouth bacteriologists apparently were using ammonium oxalate. Hucker tried ammonium oxalate and found that the stain improved with age and gave clearer results. He prepared his paper for the Philadelphia meeting and sent a draft to Dartmouth's biology department with a note of thanks. Soon thereafter he received a phone call from Dartmouth—they had never heard of ammonium oxalate for Gram staining. Hucker was perplexed.

In the days that followed, Hucker learned that a chemist had intercepted his survey letter and sent the reply. In writing out the method for crystal violet preparation, the chemist had read "A.O." on the bottle of stain in the laboratory and assumed that it meant the dye was dissolved in ammonium oxalate. Aniline oil simply did not occur to him. Moreover, he had not bothered to check with the biology department because it was inventory time and other things were

on his mind. Thus a case of badly interpreted bacteriological shorthand led to the Hucker modification. Hucker became famous; the chemist remained anonymous.

Knowing whether an organism is Gram-positive or Gram-negative is important for several reasons. For instance, microbiologists use results from the Gram stain technique to identify an unknown organism and classify it in *Bergey's Manual*. Gram-positive and Gram-negative bacteria differ in their susceptibility to chemical substances such as antibiotics (Gram-positive bacteria are more susceptible to penicillin, Gram-negatives to tetracycline), and they have different structural components (Gram-negative bacteria have more complex cell walls). They produce different types of toxic poisons as well.

One other differential technique, the **acid-fast technique**, bears mention. This technique is used to identify members of the genus *Mycobacterium*, one species of which causes tuberculosis. These bacteria are normally difficult to stain, but they stain red when treated with carbolfuchsin and heat (or lipid solubilizer). Then they retain their color when washed with a dilute acid-alcohol solution. Other bacteria lose their color easily during the acid-alcohol wash. The *Mycobacterium* species is therefore said to be acid-resistant or "acid-fast." (A blue counterstain is used to give color to nonacid-fast bacteria.) Because they stain red and break sharply when they reproduce, *Mycobacterium* species are euphemistically referred to as "red snappers."

Acid-fast technique:
a staining technique in which stain is forced into bacteria, then retained by the cells even on treatment with a dilute acid-alcohol solution; used to identify *Mycobacterium* species.

Dark-Field Microscopy

Condenser:
a series of lenses mounted under the stage of a microscope.

In **dark-field microscopy**, the background remains dark, and only the object is illuminated (Figure 3.12). A special condenser mounted under the stage of the dark-field microscope scatters the light and causes it to hit the object from

FIGURE 3.12

Variations in Microscopy

The same organism seen under three different microscopes. The cyanobacterium *Gleocapsa* is visualized under (a) the light microscope (× 1568); (b) the phase-contrast microscope (× 1736), and (c) the dark-field microscope (× 1568). Note the different view that each microscope affords.

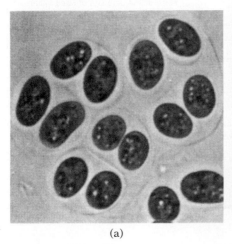 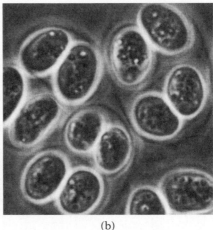 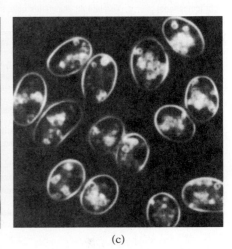

(a)   (b)   (c)

different angles. Some light bounces off the object into the lens to make the object visible, but the surrounding area appears dark because it lacks background light. The effect is similar to seeing the moon at night. In this case, sunlight from behind the earth reflects off the moon to let us see it, but the sky appears dark because the sun is hidden.

Dark-field microscopy helps in the diagnosis of diseases caused by spiral bacteria because these organisms are near the limit of resolution and do not stain well. For example, syphilis is due to *Treponema palladum*, a spiral bacterium with a diameter of about 0.15 μm. This bacterium may be observed in scrapings taken from the lesion of one who has the disease. The unstained organisms are seen moving about with a characteristic rotary motion.

trep"o-nē'-mah
pal'e-dum

Phase-Contrast Microscopy

Special microscope parts are also used in **phase-contrast microscopy**. A series of special condensers and filters split a light beam and throw the light rays slightly out of phase. The separated beams of light then pass through and around microscopic objects, and small differences in the densities of the objects show up as different degrees of brightness and contrast (Figure 3.12). The Dutch scientist Fritz Zernicke received the 1935 Nobel Prize in Physics for his development of the system.

Phase-contrast microscope: one in which light rays are thrown out of phase by special condensers, then used to illuminate details of living cells.

With the phase-contrast microscope microbiologists can see organisms alive and unstained. Internal details not visible with the light microscope may also be observed. The fine structures of yeasts, molds, and protozoa are studied with this instrument (Table 3.5).

Fluorescent Microscopy

During the past generation, **fluorescent microscopy** has emerged to become a major asset to the diagnostic and research laboratory. The technique has been applied to the identification of many microorganisms and is a mainstay of modern microbiology. Microorganisms are coated with a fluorescent dye, such

TABLE 3.5

A Comparison of Various Types of Microscopy

TYPE OF MICROSCOPE	SPECIAL FEATURE	APPEARANCE OF OBJECT	MAGNIFICATION	OBJECTS OBSERVED
Light (compound)	Visible light illuminates object	Stained microorganisms on clear background	1000×	Form, shape, and size of killed micro-organisms (except viruses, spirochetes)
Dark-field	Special condenser scatters light	Unstained microorganisms on dark background	1000×	Live, unstained microorganisms (e.g., spirochetes)
Phase-contrast	Special condenser throws light rays "out of phase"	Unstained microorganisms on dark background	1000×	Internal structures of live, unstained eukaryotic microorganisms
Fluorescent	UV light illuminates fluorescent-coated objects	Fluorescing microorganisms on dark background	1000×	Outline of microorganisms coated with fluorescent-tagged antibodies
Transmission electron microscope (TEM)	Short-wavelength electron beam penetrates sections	Alternating light and dark areas reflecting internal cell structures	20 million ×	Ultrathin slices of microorganisms and internal components of eukaryotes
Scanning electron microscope (SEM)	Short-wavelength electron beam knocks loose electron showers	Microbial surfaces	100,000×	Surfaces and textures of micro-organisms and cell components

Ultraviolet light:
a type of energy whose wavelength is shorter than that of visible light.

as **fluorescein**, then illuminated with ultraviolet light energy. The energy excites electrons in the dye, and they move to higher energy levels. However, the electrons quickly drop back to their original energy levels and give off the excess energy as visible light. The coated microorganisms appear to fluoresce.

An important application of fluorescent microscopy is in the **fluorescent antibody technique** used to identify an unknown organism. In one variation of this procedure, fluorescein is chemically united with antibodies, the protein molecules produced by the body's immune system when stimulated by a specific organism. "Tagged" antibodies result. Next, these antibodies are mixed with a sample of the unknown organism. If the antibodies are specific for that organism they will bind to it and coat the cells with the dye. When subjected to ultraviolet light, the organisms will fluoresce (Figure 3.13). If the organisms fail to fluoresce, antibodies for a different organism are tried.

FIGURE 3.13

Fluorescent Microscopy

A photomicrograph of *Clostridium botulinum*, the cause of botulism. The bacteria were obtained from the feces of an infected child and stained by the fluorescent antibody technique (× 626). In this view, the cells have combined with tagged antibodies and are fluorescing.

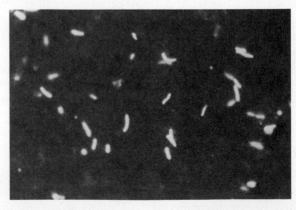

Electron Microscopy

The **electron microscope** grew out of an engineering design made in 1933 by German physicist Ernst Ruska (1986 Nobel Prize winner in Physics). Ruska showed that electrons flow in a sealed tube if a vacuum is maintained to prevent electron wandering. Magnets pinpointed the flow onto an object, where the electrons were absorbed or deflected depending on the density of structures within the object. When projected onto a screen underneath, the electrons formed an image that outlined the structures. By 1941, American engineers at RCA research laboratories had adapted the device to an instrument for microscopy and were publishing photographs with magnifications of 10,000 times.

The key to electron microscopy is the extraordinarily short wavelength of the beam of electrons. Measured at 0.005 nm (as compared to 550 nm for visible light), the short wavelength dramatically increases the resolving power of the system to nanometers and makes possible the visualization of viruses, fine cellular structures, and large molecules such as DNA (Figure 3.14).

Two types of electron microscopes are currently in wide use. The first type, the **transmission electron microscope (TEM)**, is used to photograph detailed structures within cells. Ultrathin sections of the object must be prepared because the electron beam can penetrate matter only a short distance. After embedding the specimen in a suitable mounting medium or freezing it, scientists cut the

Electron microscope: a microscope in which a beam of electrons substitutes for the light energy used in other microscopes.

Transmission electron microscope: one in which an electron beam passes through an ultrathin slice of an object.

FIGURE 3.14

Improving Figure Resolution

How energy of shorter wavelengths increases the resolution of a figure. In diagrams 1 to 7 successively smaller circles outline the figure and the figure becomes clearer. In the same way, different energies of decreasing wavelengths (e.g., visible light–ultraviolet light–electron beam) outline an object more precisely and give an instrument better resolving power and the object better resolution. (Courtesy of John Lennox of Penn State University, Altoona)

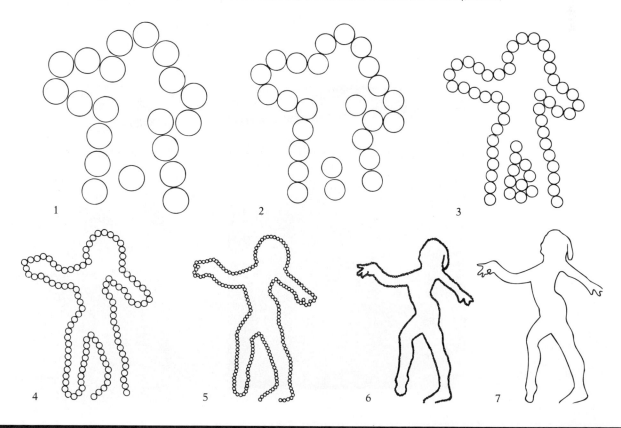

specimen into sections with a diamond knife. In this manner, a single bacterium can be sliced the long way into a hundred or more sections.

Once sectioned, the portions of the object are floated in water and picked up on a wire grid. Then they are stained with a heavy metal such as gold or palladium to make certain parts dense. Next, the microscopist inserts the sections into the vacuum chamber of the instrument and focuses a 10,000-volt electron beam on them using magnetic lenses. An image forms below. A photograph prepared from the image may be enlarged with enough resolution to achieve a total magnification of over 20 million times. Objects as small as 2.0 nanometers can be seen.

The second type is the **scanning electron microscope (SEM)**. This instrument, developed in the late 1960s, permits researchers to see the surfaces of objects in the natural state and without sectioning. The specimens are placed in the vacuum chamber and covered with a thin coat of gold to increase electrical conductivity and decrease blurring. The electron beam then sweeps across the object and knocks loose showers of electrons that are captured by a detector. An image builds line by line much as in a television receiver. Electrons that strike a sloping surface yield fewer electrons, thereby producing a darker contrasting spot and a sense of three dimensions. Magnifications with the scanning electron microscope are limited to about 75,000 to 100,000 times, but the instrument is relatively easy to operate and gives vivid and undistorted views of an organism's surface details. Figure 3.15 displays the same organism viewed with the two different forms of electron microscopy.

Scanning electron microscope:
one in which an electron beam sweeps across the surface of an object.

FIGURE 3.15

Scanning and Transmission Electron Microscopy Compared

The bacterium *Pseudomonas aeruginosa* as seen with two types of electron microscopy. (a) A view of whole cells seen with the scanning electron microscope. Bar = 1.0 μm. (b) A view of sectioned cells seen with the transmission electron microscope. Bar = 0.5 μm. The difference in perspective with the two microscopes is clear. Figure 3.6 in this chapter concerns the eye infections this organism can cause.

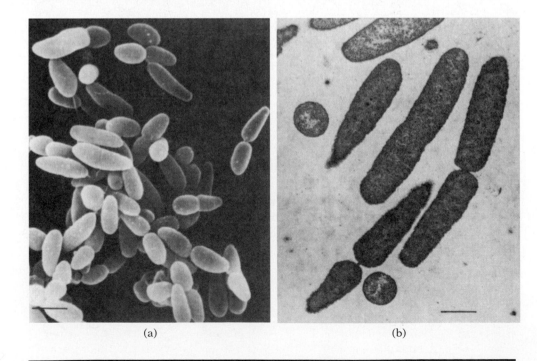

(a)   (b)

The electron microscope has added immeasurably to our understanding of the structure and function of microorganisms by permitting us to penetrate their innermost secrets. In the chapters ahead, we shall study myriad fine structures displayed by the electron microscope, and we will better appreciate microbial physiology as it is defined by microbial structures.

NOTE TO THE STUDENT

As microbiology continues to capture headlines in the public media, it becomes more and more common to see microbial names in print. And all too often the name is misspelled, miswritten, or misinterpreted. For example, I once read in a newspaper that Staphylococcus aureus, *a well-known cause of skin infections, was a recurring problem in a local hospital. In the vernacular of our times, the disease is often called a "staph infection." However, the newspaper reported it as a "staff infection."*

I believe that if one is to write or speak microbiology, one should try to do it correctly. I would recommend, therefore, that you pay particular attention to the microbial names as we encounter them and that you develop the habit of pronouncing them and writing them correctly. You should also insist that magazines and newspapers follow the accepted standards for writing the names of microorganisms. A letter-to-the-editor does much to improve the publication's awareness of its error. You should not be reluctant to send one.

Summary

Some characteristics applied to all microorganisms are surveyed in this chapter. For example, microorganisms fit the pattern wherein all living things are categorized as prokaryotes or eukaryotes, the distinguishing hallmarks being cellular features such as the nucleus, ribosomes, and organelles as well as patterns such as reproductive methods. Bacteria are prokaryotes, while fungi and protozoa are eukaryotes. Brief sketches of microorganisms give insight into microbial anatomy and physiology, and they point to the importance of microorganisms in the natural world and in medical microbiology.

Many systems of classification for living things have been devised. The efforts of Aristotle, Linnaeus, and Haeckel are notable. The currently accepted system is the five-kingdom system proposed by Robert H. Whittaker in 1969. In Whittaker's classification, bacteria are placed in the kingdom Monera, protozoa in Protista, and fungi in a kingdom with the same name, Fungi. Part of an organism's binomial name is the genus name; the remaining part is a modifying adjective that describes the genus name. Another criterion of a microorganism is its size, a characteristic that varies among members of different groups. The micrometer, a millionth of a meter, is used to measure the dimensions of bacteria, protozoa, yeasts, and fungi. The nanometer, a billionth of a meter, is commonly used to express viral measurements.

The instrument most widely used to observe microorganisms is the light microscope. For bacteria, staining generally precedes observation. The simple, negative, Gram, and other staining techniques can be used to impart color to bacteria and determine structural or physiological properties. Microscopes such as the dark-field, phase-contrast, and fluorescent microscopes have specialized uses in microbiology. To increase resolution and achieve extremely high magnification, the electron microscope employs a beam of electrons instead of a beam of light. To see whole objects, the scanning electron microscope is useful; to observe internal details, the transmission electron microscope is most often used. The various microscopes help one to visualize and conceptualize microorganisms, and understanding their basic properties helps one see their place in the scheme of living things.

Questions for Thought and Discussion

1. A student is asked on an examination to write a description of the protozoa. She blanks out. However, she remembers that protozoa are eukaryotes, and she recalls the properties of eukaryotes. How can she use this information to answer the question?

2. A local newspaper once contained an article about "the famous bacteria eecoli." How many things can you find wrong in this phrase?

3. For many years after its description in the mid-1970s, Lyme disease remained a disease without a cause. Finally, in 1985, scientists discovered a spiral bacterium and related the new organism to the disease. What type of laboratory techniques might have been used to make the discovery?

4. A student is performing the Gram stain technique in the laboratory. In reaching for the alcohol bottle in step 3, he inadvertently takes the water bottle and proceeds with the technique. What will be the colors of Gram-positive and Gram-negative bacteria at the conclusion of the technique?

5. In older textbooks the words "micron" and "millimicron" are used where we now use the terms "micrometer" and "nanometer." Why is current terminology preferable?

6. A chapter in a general biology textbook is entitled "The Protists." What might you expect this chapter to cover?

7. It is a midsummer day and a dip in the pool is an inviting prospect until you discover that the water has turned an eerie blue-green. What type of microorganism is probably responsible and how can the problem be solved?

8. A new bacteriology laboratory is opening in your community. What is one of the first books that the laboratory director will want to purchase?

9. While scanning a menu in an Italian restaurant, you note an entire section entitled "Fungi." Among the choices are spaghetti with fungi, stuffed fungi, and fungi parmigiana. What will you receive if you order any of these?

10. A car parked along the street displays a license plate that reads E. COLI. What do you suppose the owner does for a living?

11. The technician in the laboratory suspects that a person's blood contains the plague bacillus. How might this be verified using fluorescein, a preparation of plague antibodies, and the fluorescence microscope?

12. In 1987, in a respected journal of science, an author wrote, "Linnaeus gave each lifeform two Latin names, the first denoting its genus and the second its species." A few lines later, the author wrote, "Man was given his own genus and species *Homo sapiens*." What is conceptually and technically wrong with both statements?

13. A student of general biology observes a microbiology student using immersion oil and asks why the oil is used. "To increase the magnification of the microscope" is the reply. Would you agree? Why?

14. Assume that a small spherical bacterium has a diameter of 1 micrometer. A million of these bacteria would therefore fit in the space occupied by a meter (39.39 inches, or slightly more than 3 feet). Suppose you were to count each bacterium in this space at a rate of one bacterium per second. How long would it take you to count all the bacteria? Does this help you conceptualize how small a bacterium truly is?

15. A 1980s "Far Side" cartoon by Gary Larson was entitled "Single Cell Bar." Assuming you were the artist, how many different cell shapes and sizes would you include in your drawing?

Review

The types of microorganisms; their classification, nomenclature, and size; and the methods for observing microorganisms were the major themes of this chapter. To test your understanding of these themes, match the statement on the left to the term on the right by placing the letter of the term in the available space. The appendix contains the correct answers.

___ 1. System of nomenclature used for microorganisms and other living things.

___ 2. Unit of measurement used for viruses and equal to a billionth of a micrometer.

___ 3. Major group of organisms whose cells have no nucleus or organelles in the cytoplasm.

___ 4. Once known as blue-green algae.

___ 5. Devised the five-kingdom system of classification in which microorganisms are placed.

___ 6. Type of microscope that uses a special condenser to split the light beam.

___ 7. Type of electron microscope for which cell sectioning is not required.

___ 8. Eukaryotes classified into groups according to how they move.

___ 9. Prokaryotic microorganisms that have no cell wall.

___ 10. Neither prokaryotes nor eukaryotes.

___ 11. Kingdom in which the bacteria are classified.

___ 12. Staining technique that differentiates bacteria into two groups.

___ 13. Author of an early system of classification for bacteria.

___ 14. Category into which two or more species of bacteria are grouped.

___ 15. Coined the name Protista for microorganisms.

___ 16. Considered to be unicellular algae.

___ 17. Used to write the binomial name of microorganisms.

___ 18. Unit of measurement for bacteria and equal to a millionth of a meter.

___ 19. Type of unspecialized laboratory microscope having a two-lens system.

___ 20. Staining technique in which the background is colored and the cells are clear.

A. Viruses
B. Italics
C. Gram
D. Mycoplasmas
E. Simple
F. Genus
G. Binomial
H. Boldface
I. Diatoms
J. Micrometer
K. Prokaryotes
L. Cyanobacteria
M. Dark-field
N. Scanning
O. Haeckel
P. Bergey
Q. Negative
R. Protista
S. Eukaryote
T. Nanometer
U. Phase contrast
V. Monera
W. Protozoa
X. Whittaker
Y. Chlamydia
Z. Compound

PART 2

THE BACTERIA

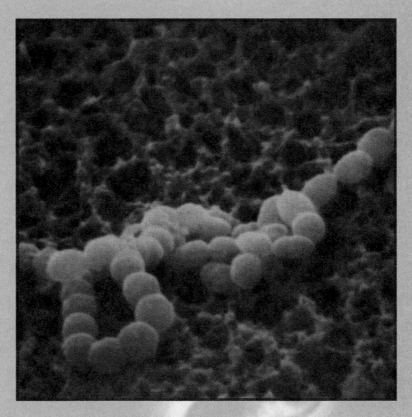

We live at the center of a microbial universe. On all sides, microscopic organisms surround us and make their presence felt for good or ill. The useful species outnumber the harmful ones by thousands and are so valuable we could not live without them. The remaining species are agents of disease and death. ■ Only since the late 1800s have scientists linked microorganisms to events of human importance (not surprisingly, because microorganisms can be seen only with a microscope). Before then, processes now attributed to microorganisms, such as infectious disease, seemed to happen almost spontaneously and without apparent cause. Even thinking about disease and microorganisms could be dangerous because it was heresy to consider disease as anything other than supernatural. ■ In Part II of this text we shall focus on one group of microorganisms, the bacteria. These microorganisms have traditionally occupied an important niche in microbiology because scientists probably know more about bacteria than any other organisms of any kind. Bacteria have been involved in the great plagues of history, and for centuries their effects have captured the imagination of scientists and writers. Bacteria are easily studied in the laboratory, and their chemical activities have been charted and well documented. Also, many helpful ones play key roles in industrial processes. We usually mean bacteria when we talk about "germs," and we need only consider how often we use that term to appreciate the significance of bacteria in our lives. ■ Small as they are, bacteria are endowed with the ability to perform certain acts characteristic of all living things. They take in food, grow, excrete waste products, reproduce, and die. In addition, bacteria are sensitive to external agents and stimuli and they respond in some fashion. In Chapter 4, we shall survey their anatomical frameworks and growth patterns, and in Chapter 5, examine their biochemical activities. Chapter 6, devoted to the genetics of bacteria, includes some of the key findings of modern biotechnology. The discussions in Part II have broad significance not only in medicine, research, and industry, but also in our day-to-day lives.

BIOTECHNOLOGY

During the 1980s, the editors of *Time Magazine* referred to DNA technology as "the most awesome skill acquired by man since the splitting of the atom." Indeed, the work with DNA, begun in the 1950s and continuing today, has opened vistas previously unimagined. Scientists can now remove bits of DNA from organisms, snip and rearrange the genes, and insert them into fresh organisms, where the genes will express themselves. Practical results of these experiments have led to the mass production of hormones, clotting factors, and other pharmaceutical products; diagnostic methods based on DNA fingerprinting; advances in gene therapy; a revolution in agricultural research; barnyard animals producing human hemoglobin; and a colossal attempt to map the entire human genome.

If you would like to be part of what promises to be the next century's great technology, then microbiology is the place to be. You would be well advised to take a course in biochemistry, as well as one in genetics. Courses in physiology and neurobiology are also helpful. Employers will be looking for individuals with good laboratory skills, so be sure to take as many lab courses as you can. And don't be afraid to become a "lab-rat" (the scientific equivalent of basketball's "gym-rat").

You may enter the biotechnology field with an associate's, bachelor's, master's, or doctoral degree. This is because there are so many levels at which individuals are hired. An employer will be looking for work experience, which you can obtain by assisting a senior scientist, doing an internship, working summers (usually for slave wages) in a biotech firm, or doing volunteer work. The campus research lab is another good place to obtain work experience. It might also be a good idea to sharpen your writing skills, since you will be preparing numerous reports. As Chapter 6 explains, the novel and imaginative research that established biotechnology was founded in microbiology and continues to call on microbiology today for its continuing growth.

CHAPTER 4

ANATOMY AND GROWTH OF BACTERIA

I n a 1683 letter to the Royal Society of London, Anton von Leeuwenhoek described microscopic "streaks and threads" among his tiny animals. The streaks and threads remained nameless until 1773, when the Danish scientist Otto Frederick Müller christened them "bacilli." Bacilli is the plural form of the Latin word *bacillus*, meaning little rod.

But not all bacilli were rods. Some were spiral and some were circular, and "bacilli" would not do. Therefore, in the 1850s, the French investigator Casimir Davaine began calling the microscopic creatures "bacteria," even though this derivative of the Greek *bacterion* also means rod. Time has a way of sorting out confusion, and in the next few decades "bacteria" came to refer to all the microorganisms in that group, and the word "bacillus" was reserved for rod forms only.

The terminology problem was resolved just in time because, in the 1850s, bacteria were stimulating substantial interest. Pasteur's work with fermentation showed that bacteria are chemical factories capable of bringing about significant changes in nature, and in the 1870s, Koch's experiments proved their link to disease. In the late 1800s, the rush to locate and isolate the bacterial causes of disease was unlike anything previously experienced in medical science.

As it happened, neither van Leeuwenhoek, nor Davaine, nor Pasteur, nor Koch could see what lay ahead. As the twentieth century unfolded, scientists found that bacteria had structures and growth patterns far beyond what had been imagined in the years before. With the development of the electron microscope in the 1940s and the revelations of biochemistry in that period, bacteria revealed themselves as more than simple sticks and rods. Scientists uncovered a wealth of microscopic and submicroscopic details in bacteria and showed how the very minute bacteria can be as complex as very large, visible organisms. As we shall see in this chapter, a study of the anatomical features of bacteria provides a window to their activities and illustrates how bacteria relate to other living things.

Anatomy of the Bacteria

The anatomy of an organism refers to its size and shape and the structural features that make it distinctive. Anatomy is an inherited trait derived from information stored in the chromosomal DNA. This trait is passed from generation to generation in the genes.

Viewed under the light microscope, most bacteria appear in variations of three different shapes: the rod, the sphere, and the spiral. As suggested by Müller, the rod is known as a **bacillus** (pl., bacilli). In various species of bacteria, a bacillus may be as long as 20 μm or as short as 0.5 μm. Certain rods such as those of typhoid fever are slender; others such as the agents of anthrax are rectangular with squared ends; still others such as diphtheria bacilli are club shaped. Most rods occur singly, but some form long chains called **streptobacilli**. Figure 4.1 illustrates this diversity. It should be noted that the word bacillus is used two ways in microbiology: to denote a rod form, and as a genus name. The organism of anthrax, for instance, is a bacillus having the name *Bacillus anthracis*.

A bacterial sphere is known as a **coccus** (pl., cocci), a term derived from the Greek *kokkos*, for berry. Cocci are approximately 0.5 μm in diameter. They are usually round, but they may also be oval, elongated, or indented on one side (Figure 4.1). Those cocci remaining together in pairs after reproducing are called **diplococci**. The organisms of gonorrhea and one type of bacterial meningitis are examples. Those cocci consisting of chains of diplococci are called **streptococci** (Figure 4.2a). Certain streptococci are involved in strep throat and tooth decay, but many are harmless enough to be used for producing dairy products such as yogurt.

Another variation of cocci is the **sarcina**. The sarcina is a cubelike packet of eight cocci (*sarcina* is Latin for bundle). One species, *Micrococcus luteus*, is a common inhabitant of the skin. Certain cocci divide randomly and form an irregular grapelike cluster of cells called a **staphylococcus** (Figure 4.2b), from *staphyle*, the Greek word for grape. A well-known example, *Staphylococcus aureus*, is a widespread cause of food poisoning (MicroFocus: 4.1), as well as toxic shock syndrome and numerous skin infections. The latter are known in the modern vernacular as "staph" infections.

The third important shape of bacterial organisms is the **spiral**. Certain spiral bacteria called **vibrios** are curved rods that resemble commas. The cholera organism is typical. Other spiral bacteria called **spirilla** (sing., spirillum) have a corkscrew shape with a rigid cell wall and hairlike projections called flagella that assist movement (Figure 4.3). Still others, known as **spirochetes**, have a flexible cell wall but no flagella in the traditional sense. Movement in these organisms occurs by contractions of long filaments (endoflagella) that run the length of the cell. The organism of syphilis typifies a spirochete.

Variations in bacterial anatomy are readily visible when the organisms are magnified a thousand times under the light microscope. When the electron microscope is used, however, a magnification of a million times or more is possible and scientists can observe a world of fine bacterial details not otherwise seen by the casual observer. We shall examine some of these details next.

Flagella

Numerous species of bacterial rods and spirilla and a limited number of species of cocci are capable of independent motion. To achieve motion, they utilize structures called **flagella** (sing., flagellum). Flagella are composed of long, rigid

---

Genes:
segments of DNA that provide the biochemical code for inherited traits.

μm:
the symbol for micrometer, a millionth of a meter.

Streptobacillus:
a chain of bacterial rods.

kok'us
kok'si

Streptococcus:
a chain of bacterial diplococci.

sar-cē'nah

staf'i-lo-kok'us

au're-us

spi'ro-kēt

FIGURE 4.1

Variations in Bacterial Anatomy

(a) Anthrax bacilli. (b) Tetanus bacillus swollen with spores. (c) Diphtheria bacilli displaying a club shape. (d) Streptobacilli. (e) Diplococci of bacterial pneumonia. (f) Diplococci of gonorrhea. (g) Streptococci such as those involved in strep throat. (h) Cubelike packets (sarcinae) of eight cocci. (i) Staphylococci in a grapelike cluster. (j) Vibrios such as the cholera organism. (k) Spirilla. (l) Spirochetes of syphilis.

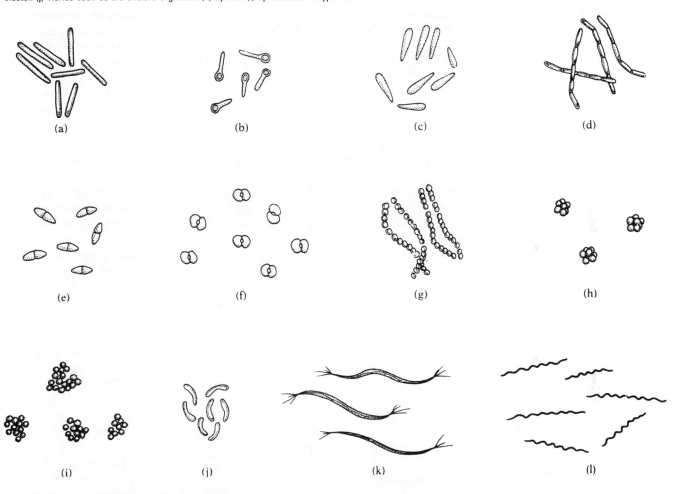

(a)   (b)   (c)   (d)

(e)   (f)   (g)   (h)

(i)   (j)   (k)   (l)

4.1

MICROFOCUS

BIOLOGICAL WARFARE

During World War II a U.S. Navy ship was cruising off the coast of California. One evening the galley served baked ham with no apparent consequences. However, the leftover ham remained overnight at room temperature and soon it was contaminated with staphylococci and their toxins. The next morning's menu consisted of ham and eggs. Within hours, hundreds of sailors developed diarrhea and became severely indisposed.

The story is told that a senior officer radioed the Los Angeles County Department of Health and asked for advice on how to deal with the situation. When little help was forthcoming he became frustrated and exclaimed, "What if the Japanese fleet appears on the horizon?" Equally frustrated, the laboratory director replied, "Shoot over the damned hams!"

FIGURE 4.2

Scanning Electron Micrographs of Three Variations of Cocci

(a) A streptococcus such as is found in the intestine. Note the characteristic string-of-beads appearance of diplococci in the chain. The arrows indicate three of the many positions at which the cocci are undergoing division. (b) *Staphylococcus aureus* showing the typical grapelike cluster of cocci (× 15,000). (c) Clusters of sarcinae in cubelike packets of eight.

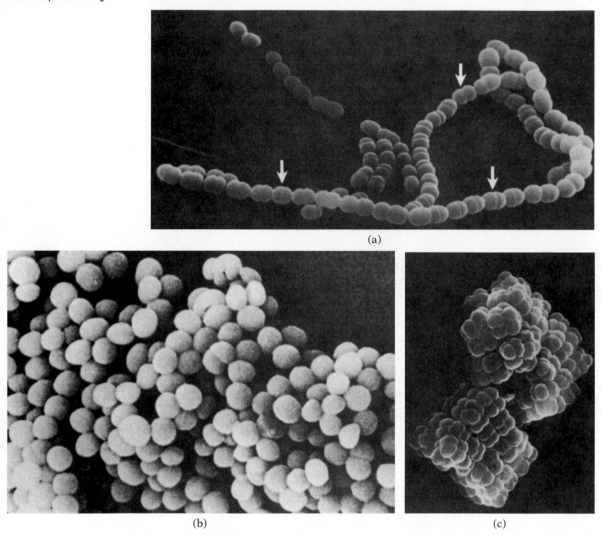

(a)

(b)   (c)

Flagellin:
the protein of which flagella
are composed.

strands of a protein called **flagellin**. Within the strands, the protein exists in ultrathin fibers permanently bent like a coil or helix. This structure permits the flagellum to rotate. (By contrast, the flagella whip about in eukaryotic cells such as protozoa [*flagellum* is Latin for whip]. Here the strands are flexible, and the fibers are elongated and slide past one another.)

Electron microscopy reveals that the flagellum of Gram-negative bacteria is anchored to a hooklike shaft, which penetrates the cell wall and attaches to two ring-shaped bases in the cell membrane illustrated in Figure 4.4. Gram-positive bacteria have only the inner ring. The inner ring rotates while the outer ring, when present, remains in place. This activity creates a propellerlike rotation that drives the bacterium forward much as a motor propels a boat. The boat, however, remains upright while the bacterium rotates in a direction opposite to the flagellar rotation (Figure 4.4).

FIGURE 4.3

Electron Micrographs of a Freshwater Spirillum Showing the Typical Corkscrew Pattern

(a) The spiral shape is seen clearly in the cell, and flagella are visible at the poles of the cell. (b) A second cell is observed, with the background enhanced by negative staining to enable viewing of the cell interior. An electron-dense particle chain (PC) is visible in the cytoplasm. This chain of particles gives magnetic properties to the spirillum. In the 1980 publication where these photographs first appeared, the researchers proposed that the chain be called a "magnetosome." Bar = 1.0 μm.

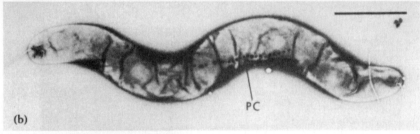

Flagella can vary in number and placement. A **monotrichous** bacterium (a montrichaete) possesses a single flagellum, while a **lophotrichous** organism (a lophotrichaete) has a group of two or more flagella at one pole of the cell. An **amphitrichous** bacterium (an amphitrichaete) has groups of flagella at both ends, and a **peritrichous** organism (a peritrichaete) is covered with flagella. The arrangement of flagella is characteristic of a species and is used in classifying the species in taxonomic schemes.

The flagellum ranges in length from 10 μm to 20 μm and is therefore many times the length of the cell, as Figure 4.4d displays. However, the flagellum is only about 0.2 μm thick and cannot be seen under the light microscope unless coated with dye. In the human body, flagella enable bacteria such as cholera bacilli to move among the tissues and colonize various areas. Some bacteria are known to travel up to 2000 times their own length in an hour.

Pili

Pili (sing., pilus) are bacterial appendages that appear as short flagella (Figure 4.5) but have no function in motility. Instead, certain pili aid the transfer of genetic material among bacteria (Chapter 6), while other pili anchor bacteria to surfaces such as living tissue. By doing so, pili enhance an organism's ability to cause disease (Table 4.1).

Pili are primarily found on Gram-negative bacteria such as *Neisseria gonorrhoeae*, the cause of **gonorrhea**. Because pili are composed of protein, the body's immune system responds to their presence by producing antipili antibodies.

mon″o-trik′us
loff″o-trik′us

am″fi-trik′us

Cholera:
a serious bacterial disease of the intestine, characterized by loss of large volumes of water.

pi′lus

nī-se′re-ah
Gonorrhea:
a sexually transmitted bacterial disease, characterized by colonization of the reproductive and urinary tract tissues.

FIGURE 4.4
Details of the Bacterial Flagellum

(a) In Gram-negative bacteria, the flagellum is attached to the cell wall and membrane by a complex mechanism of structures having two rings. (b) Rotation of the flagellum in one direction causes rotation of the bacterial cell in the opposite direction and the bacterium moves forward. (c) Various types of flagellation occur among bacteria. Monotrichous bacteria possess a single flagellum, amphitrichous bacteria have flagella at both poles of the cell, and lophotrichous organisms have them at one end. Peritrichous bacteria are surrounded by flagella. (d) A transmission electron micrograph of *Pseudomonas marginalis* showing polar flagella (× 38,800). Note that the flagella are many times the length of the bacillus and appear in a characteristic wavy format. This bacterium is lophotrichous.

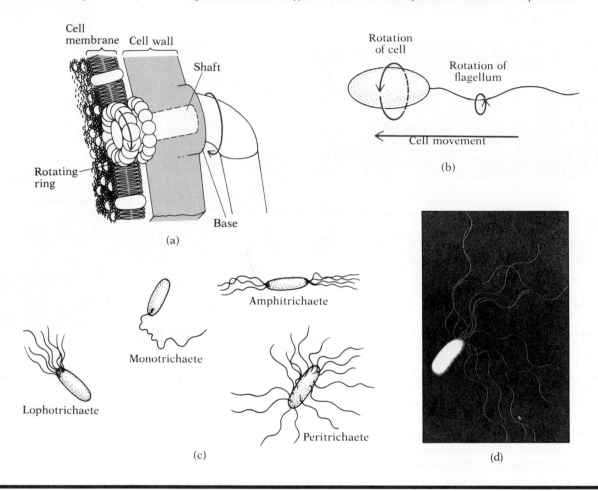

TABLE 4.1
A Comparison of Flagella and Pili

CHARACTERISTIC	FLAGELLA	PILI
Composition	Flagellin, a protein	Pilin, a protein
Size	0.01 to 0.05 μm width; 10 to 20 μm length	0.007 to 0.008 μm width; 0.5 to 2 μm length
Occurrence	Many bacterial rods but few cocci	Many Gram-negative rods and cocci
Structure	Adjacent fibrils with no regular pattern	Wound fibers with a hollow center
Number	Varies according to the organism; one to several hundred	Varies according to the organism; one to several hundred
Function	Motion	Attachment, conjugation
Origin	Hooklike insertion to basal granule inside cell wall	Cell cytoplasm, then through cell membrane, wall, and capsule

FIGURE **4.5**

Two Examples of Pili in Bacteria

(a) A heavily piliated cell of *Neisseria gonorrhoeae*, the agent of gonorrhea. The stain has obscured the diplococcus form of the organism. Bar = 1 μm.
(b) Numerous *Klebsiella pneumoniae* cells displaying pili (× 24,400). Both organisms are pathogens, and the pili enhance pathogenicity by permitting the organism to adhere to the tissue.

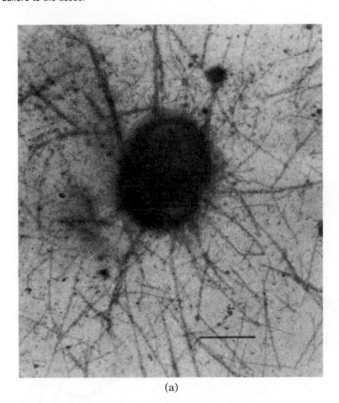

(a)

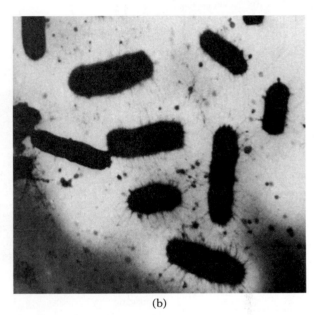

(b)

Therefore, an important avenue of gonorrhea research has been to develop antipili antibodies to neutralize the pili and prevent attachment to the patient's tissues. This would limit the need for antibiotics, especially since drug resistance is increasing among gonorrhea organisms. It should be noted that some microbiologists use the word **fimbriae** (sing., fimbria) to refer to bacterial structures of attachment and reserve the word pili for structures that function in genetic transfers.

fim'bre-ā

Capsule

Many species of bacteria secrete a layer of polysaccharides and small proteins that adheres to the bacterial surface. Commonly known as a **capsule**, this layer is a very sticky, gelatinous structure formed by various species of bacilli and cocci, but not by spiral bacteria.

The capsule serves as a buffer between the cell and its external environment (Figure 4.6). Because of its high water content, the capsule protects the cell against dehydration while preventing nutrients from flowing away. In the body, it also contributes to the establishment of disease because white blood cells that normally engulf and destroy bacteria by phagocytosis cannot perform this function on encapsulated bacteria. For example, a principal cause of bacterial pneumonia, *Streptococcus pneumoniae*, is deadly in its encapsulated form but harmless when the capsule has been experimentally removed.

Phagocytosis:
a defensive measure of the body in which white blood cells engulf and destroy microorganisms.

FIGURE 4.6
A Fictionalized Bacterial Cell

Composite diagram of a fictionalized bacterial cell. The anatomical features of this "ideal" bacterium are seen in perspective to one another to visualize their relationships. Such a bacterium probably does not exist.

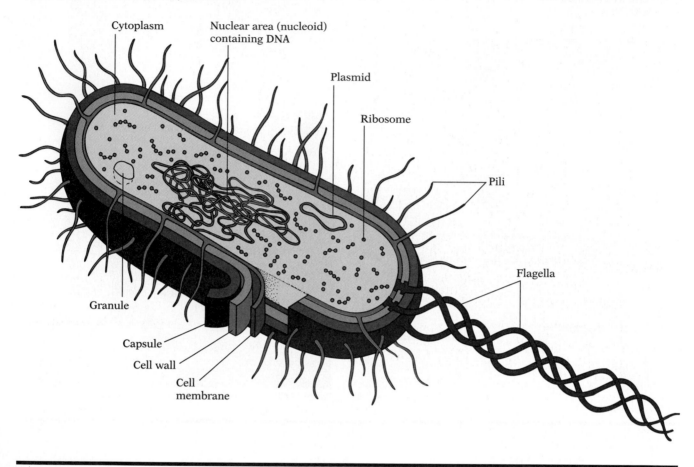

When the capsule has a looser consistency and is less tightly bound to the cell, it is commonly referred to as a slime layer, or **glycocalyx** (Figure 4.7). This structure usually contains a mass of tangled fibers of a polysaccharide called **dextran**. The fibers attach the bacterium to tissue surfaces. A case in point is *Streptococcus mutans*, an important cause of dental caries. This bacterium attaches itself to the surface of the teeth using dextran it synthesizes from sucrose (table sugar). Soon a layer of dental plaque has formed and the streptococci begin breaking down dietary carbohydrates to acids that dissolve the enamel.

In food products, the slime-producing bacteria may cause an unsightly and distasteful experience. For instance, the gluelike slime of *Alcaligenes viscolactis* accumulates in milk, causing it to become thick and stringy. The result is **ropy milk**. Bread may also become ropy if contaminated with capsule-producing *Bacillus subtilis*.

gli″ko-ka′liks

al″kah-lij′ē-nēz

Ropy milk:
thick, stringy milk that contains the gluelike slime of capsule-producing bacteria.

Cell Wall

With the notable exception of mycoplasmas, all bacteria have a **cell wall**. This structure protects the cell and, to a large extent, determines its shape. A century

FIGURE 4.7

The Glycocalyx of *Escherichia coli*

An electron micrograph of *Escherichia coli* from the intestine of an animal. Each bacillus is surrounded by a glycocalyx. Bar = 0.1 μm.

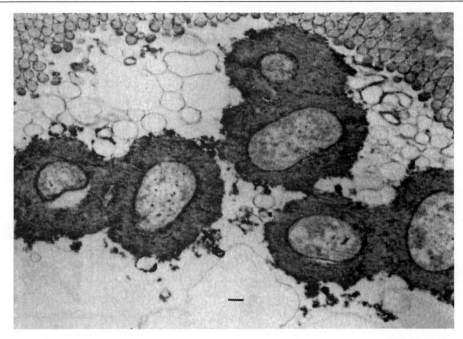

ago, taxonomists classified bacteria as plants because of the presence of a cell wall, but modern biochemists have established that the chemical composition of the bacterial cell wall differs from that in plants.

The important chemical constituent of the bacterial cell wall is **peptidoglycan**. This is a very large molecule composed of alternating units of two amino-containing carbohydrates, *N*-acetylglucosamine and *N*-acetylmuramic acid, joined by cross-bridges of amino acids. Peptidoglycan occurs in multiple layers connected by side chains of four amino acids, as illustrated in Figure 4.8a. Therefore, the many layers comprise one extremely large molecule.

pep′tī-do-gli″kan
acetyl-glucose-amine
acetyl-muramic

The cell walls of Gram-positive and Gram-negative bacteria differ considerably. In Gram-positive bacteria, the peptidoglycan layer is about 25 nm wide and contains an additional polysaccharide called **teichoic acid**. About 60 to 90 percent of the cell wall is peptidoglycan, and the material is so abundant that Gram-positive bacteria are able to retain the crystal violet–iodine complex in **Gram staining** (Chapter 3).

nm:
the symbol for nanometer, a billionth of a meter.

By contrast, Gram-negative bacteria have a peptidoglycan layer only 3 nm wide without any evidence of teichoic acid. The cell wall in these bacteria contains various polysaccharides, proteins, and lipids and so is much more complex than the cell wall of Gram-positive bacteria. Also, the cell wall is surrounded by an outer membrane barely separated from the cell wall by a so-called **periplasmic space** containing a gel-like material called **periplasm**. On the inner side of the cell wall the periplasmic space is wider. Bacterial toxins and enzymes apparently remain in this space and destroy antibacterial substances before they can affect the cell membrane, and other proteins, facilitate passage through the cell membrane. The multiple layers of the Gram-negative cell also afford protection by restricting the passage of chemicals such as antibiotics, salts, and dyes to the cell. The crystal violet–iodine complex in Gram staining is lost partly because of the thinness of the cell wall in Gram-negative bacteria.

Gram stain technique:
a four-part staining procedure that differentiates bacteria into two groups, Gram-positive and Gram-negative.

FIGURE 4.8

A Comparison of the Cell Walls of Gram-positive and Gram-negative Bacteria

(a) The cell wall of a Gram-positive bacterium is composed of peptidoglycan layers combined with teichoic acid molecules. The structure of peptidoglycan is shown as units of NAG and NAM joined laterally by amino acid cross-bridges and vertically by side chains of four amino acids. (b) In the Gram-negative cell wall, the amount of peptidoglycan is much smaller and there is no teichoic acid. Moreover, an outer membrane closely overlies the peptidoglycan layer so that the membrane and layer comprise the cell wall. Note the structure of the cell membrane in this figure. The bacterial membrane conforms to the fluid mosaic model that describes most membranes.

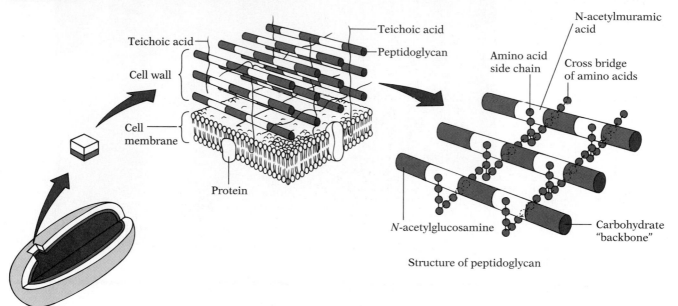

(a) Gram-positive cell wall

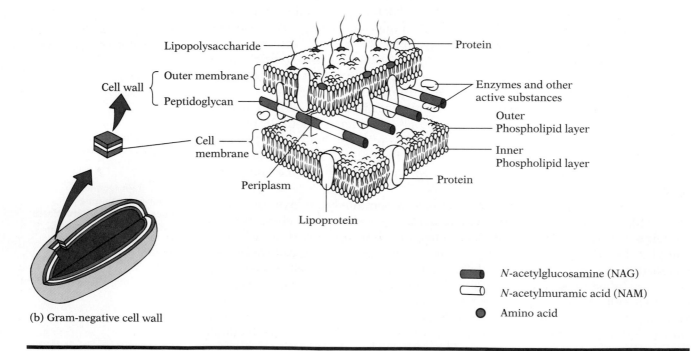

(b) Gram-negative cell wall

FIGURE 4.9
The Effect of Penicillin

A photomicrograph of a *Staphylococcus aureus* cell exploding on exposure to penicillin (× 150,000). The antibiotic has prevented construction of the peptidoglycan layer of the cell wall, and internal pressures have led to weakening and disruption of the cell membrane.

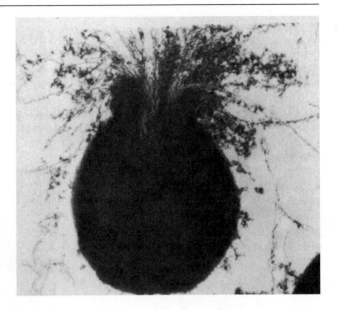

The cell wall holds the cell together. It also prevents the cell from bursting because the internal pressure may be up to 20 times the external pressure due to the high concentration of inorganic salts, carbohydrates, amino acids, and other small molecules within the cell. The antibiotic **penicillin** prevents the construction of the cell wall in new cells, and they quickly burst (Figure 4.9). Where penicillin acts on new cells, lysozyme destroys existing cells. **Lysozyme** is an enzyme in human tears and saliva. It attacks the linkages between carbohydrates in the peptidoglycan layer, thus causing the cell wall to break down and the cell to explode. In both cases, the effect is more dramatic in Gram-positive bacteria because these organisms have more peptidoglycan. Moreover the lipopolysaccharide layer plays a protective role in Gram-negative organisms.

Penicillin:
a mold-derived antibiotic that prevents the construction of the cell wall, especially in Gram-positive bacteria.

li'so-zīm

Cell Membrane

The **cell membrane** (also called the **plasma membrane**) is the boundary layer of the bacterial cell. It exists inside the cell wall and functions in transporting nutrients into the cell and waste materials out of the cell. It also anchors the DNA during replication and is a site for enzymes that function in cell wall synthesis. Moreover, it is the location of enzymes used in energy production by the cell, a factor that makes it the equivalent of the membranes in mitochondria of a eukaryotic cell. Some microbiologists combine the cell membrane, cell wall, and capsule (if present) together as a group and term them the **cell envelope**.

Approximately 60 percent of the cell membrane is composed of protein, and about 40 percent of lipid, mainly phospholipid. The phospholipid molecules are arranged in two parallel layers (a phospholipid bilayer), one at the outside, the other at the inside of the membrane. In contrast, the proteins are arranged as globules floating like icebergs at or near the inner and outer surfaces of the membrane, and some globules extend from one surface of the membrane to the other. This model of the membrane, called the **fluid mosaic model**, accounts

Mitochondria:
organelles of eukaryotic cells in which energy-yielding biochemical reactions occur.

Phospholipids:
molecules of lipids that contain a large number of phosphate groups.

A (NOT SO) FATAL ATTRACTION

To get from place to place, humans usually require the assistance of maps, compasses, and gas station attendants. In the microbial world life is generally more simple, and traveling is no exception.

Consider the bacteria, for example. In the early 1980s, Richard P. Blakmore and his colleagues at the University of New Hampshire observed that certain mud-dwelling bacteria tend to gather at the north end of water droplets. On further study, they found that each bacterium had a chain of magnetic particles acting as a kind of bacterial dipole directing the organism's movements. Bacteria possessing the particles swim toward the north in the northern hemisphere and toward the south in the southern hemisphere. Indeed, when genetic mutations are caused in the bacteria, they head in the wrong direction—and wind up in a hostile environment and die.

This last observation is particularly noteworthy because it appears to give rhyme and reason to the particles. The conventional wisdom is that the so-called magnetotactic bacteria use their traveling skills to locate a favorable environment. In 1992, Dennis A. Bazylinski of Stevens Institute of Technology theorized how this hypothesis might work in nature (Figure B4.2). Certain magnetotactic bacteria are anaerobic; that is, they live in an oxygen-free environment. While swimming toward a pole (north or south), the bacteria are also oriented downward by the Earth's magnetic field. The downward tilt pulls them away from the oxygen-rich water and toward the oxygen-poor mud below. On reaching their type of optimum environment, the bacteria reach a sort of biological nirvana and settle in for a life of anaerobic bliss.

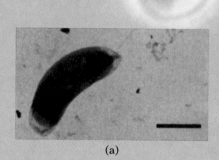

(a)

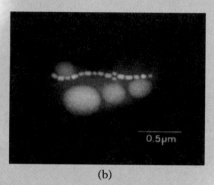

0.5μm

(b)

Two views of magnetotactic bacteria. (a) An electron micrograph of a negatively stained organism showing two chains of magnetite-containing magnetosomes. Bar = 0.5 μm. (b) A dark-field image of an electron micrograph displaying a cell with a chain of magnetite-containing magnetosomes. Other cellular bodies appear as ovals.

for the membrane's appearance under the electron microscope and helps explain how it allows passage of certain substances. For example, lipid-soluble materials dissolve in the phospholipid layer and pass through the membrane, while amino acids and nitrogenous bases, which do not dissolve in lipids, move through the protein passageways.

When antimicrobial substances act on the cell membrane, bacterial death usually follows. Certain detergents, for instance, dissolve the phospholipid layers and cause the cytoplasmic contents to leak out. Ethyl alcohol and some antibiotics such as polymyxin work similarly.

Detergents:
synthetic chemicals used as antiseptics and disinfectants to kill bacteria by altering the structure of the cell membrane.

pol-e-mix'in

Cytoplasm

Inside the cell membrane lies the **cytoplasm**, a gelatinous mass of proteins, carbohydrates, lipids, nucleic acids, salts, and inorganic ions, all dissolved in water. Cytoplasm is the foundation substance of a cell and the center of its growth and biochemistry. It is thick, semitransparent, and elastic.

Several cytoplasmic bodies are of interest. **Ribosomes** are bodies of RNA and protein associated with the synthesis of protein (Chapter 5). Other bodies found in various bacteria include globules of starch, glycogen, or lipid. Often referred to collectively as **inclusion bodies**, these globules store nutrients for later use during periods of starvation. Certain other bodies serve as phosphate

FIGURE 4.10

The Nucleoids of *Escherichia coli*

In this transmission electron micrograph, nucleoids are seen in irregular coralline (coral-shaped) forms. Nucleoids can be observed occupying a large area in a bacterium, which indicates how far the DNA of the bacterium is spread out. Both longitudinal and cross sections of *E. coli* are visible. Bar = 0.5 μm.

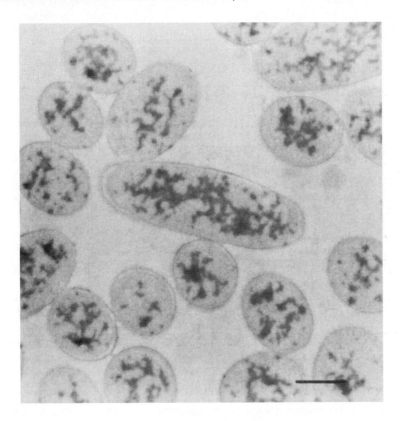

depots. Commonly known as **metachromatic granules**, or volutin, these bodies stain deeply with dyes such as methylene blue. Their presence in diphtheria bacilli assists identification procedures. A recently discovered body, the **magnetosome**, helps certain bacteria orient themselves to the environment (MicroFocus: 4.2). Crystals of an iron-containing compound called magnetite fill the magnetosome and align themselves with the local magnetic field. Scientists believe that the magnetite directs bacteria toward their preferred habitat.

vol'u-tin
dif-the're-ah

The cytoplasm is also the site of the bacterial **chromosome**. This closed loop of DNA contains the hereditary information of the cell. It is suspended in the cytoplasm without a covering or membrane and is not associated with protein. These factors are important in classifying bacteria as prokaryotes. The term **nucleoid** is often applied to the chromosome region (Figure 4.10).

Nucleoid:
the chromosome region of a bacterial cell.

Smaller molecules of DNA exist apart from the chromosome in closed loops called **plasmids**. Although they contain few genes and are not essential for bacterial growth, plasmids are significant because many carry genes for drug resistance. For this reason they are often called **R factors** ("R" for resistance). Plasmids may be transferred between cells during recombination processes (Chapter 6) and are known to multiply during cell reproduction. They are a focus of attention in industrial technologies that utilize genetic engineering. Table 4.2 summarizes the morphological features of bacterial cells.

Recombination:
a process in which the genetic material of a bacterium changes due to the incorporation of new DNA.

TABLE 4.2

A Summary of Anatomical Features of Bacteria

STRUCTURE	CHEMICAL COMPOSITION	FUNCTION	COMMENT
Flagella	Protein	Movement	Present in many rods and spirilla; few cocci; vary in number and placement
Pili	Protein	Attachment to surfaces Genetic transfers	Found in many Gram-negative bacteria Stimulate immune system
Capsule and glycocalyx	Polysaccharides and small proteins	Buffer to environment Contributes to disease Cell protection Attachment to surfaces	Source of ropy milk and bread Found in plaque bacteria
Cell wall	Gram-positives have peptidoglycan with teichoic acid Gram-negatives have peptidoglycan, lipopolysaccharide, phospholipid, lipoprotein	Cell protection Shape determination	Site of activity of penicillin and lysozyme Absent in mycoplasmas Multiple layers in Gram-negatives
Cell membrane	Protein Phospholipid	Cell boundary Transport into/out of cell Site of enzymes	Conforms to fluid mosaic model Susceptible to detergents, alcohols, and some antibiotics
Cytoplasm	Water, proteins, lipids, carbohydrates, nucleic acids, etc.	Foundation substance of cell	Center of biochemistry and growth Semitransparent, thick
Ribosomes	RNA and protein	Protein synthesis	Inhibited by certain antibiotics
Inclusion bodies	Starch, glycogen, or lipid	Nutrient storage	Used as nutrients during starvation periods
Metachromatic granules	Phosphate	Storage	Found in diphtheria bacilli
Magnetosome	Magnetite	Cell orientation	Helps locate preferred habitat
Chromosome	DNA	Site of genetic code Site of inheritance	Exists as single, closed loop Referred to as nucleoid
Plasmids	DNA	Site of some genes	Contains R factors
Spores	Complex chemical composition with dipicolinic acid	Resistance to environment	Produced by *Bacillus* and *Clostridium* species Probably the most resistant living thing known

Spores

klo-strid'e-um

Vegetative cell:
a cell that is growing, maturing, and reproducing.

Certain Gram-positive bacteria are able to produce highly resistant structures called **endospores** or simply, **spores**. Members of the genera *Bacillus* and *Clostridium* are among the best-known sporeformers (Figure 4.11). These bacteria grow, mature, and reproduce for several hours as vegetative cells. Spore formation then begins. The bacterial chromosome replicates, a small amount of cytoplasm gathers with it, and the cell membrane grows in to seal off the developing

FIGURE 4.11

Formation of the Bacterial Spore

The cell metabolizes nutrients and multiplies for many generations in the vegetative cycle. After some time, the cell enters the sporulation cycle. The cellular DNA duplicates (1) and is surrounded by the cell membrane (2). Resistant layers of coat material form around the spore (3) and formation is complete (4). Now the remaining cell disintegrates (5) to liberate the free spore (6). Later, when conditions are suitable, the spore germinates (7) and the bacterium emerges (8) to resume the vegetative cycle.

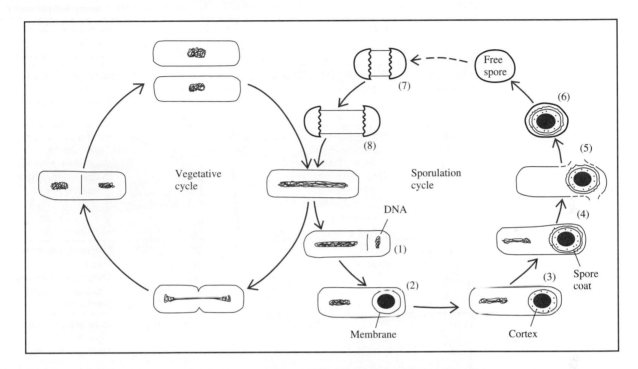

spore. Thick layers of peptidoglycan form and a series of coats are synthesized to protect the contents further. The cell wall of the vegetative cell then disintegrates and the spore is freed.

Endospores may develop at the end of the cell, near the end, or at the center of the cell, depending on the species. They contain little water and exhibit very few chemical reactions. However, they do have a large amount of **dipicolinic acid**, an organic substance that helps stabilize their proteins. When the external environment is favorable, the protective layers break down and the spores germinate to vegetative cells (Figure 4.12). It should be noted that spore formation is not a reproductive process; a vegetative cell forms a single spore, and later the spore germinates to one vegetative cell.

di"pik-o-lin'ik

Bacterial spores are probably the most resistant living things known. For example, most vegetative bacteria die quickly in water over 80° C, but bacterial spores may remain alive in boiling water (100° C) for 2 hours or more. When placed in 70 percent ethyl alcohol, spores have survived for 20 years. Humans can barely withstand 500 rems of radiation, but spores can survive a million rems. Drying has little effect on the spores, and living spores have been recovered from the intestines of Egyptian mummies. In 1983, archaeologists found spores alive in sediment lining Minnesota's Elk Lake. The sediment was 7518 years old.

Ethyl alcohol:
a two-carbon alcohol
compound used as a
disinfectant and antiseptic.

FIGURE 4.12

Three Different Views of Bacterial Spores

(a) A view of *Clostridium* under the light microscope showing terminal spore formation. Note the characteristic drumstick appearance of the cells. (b) The fine structure of a *Bacillus thuringiensis* spore seen under the transmission electron microscope. The visible spore structures include the core membrane (CM), core wall (CW), cortex (C), inner coat layer (ICL), outer coat layer (OCL), and exosporium basal layer (XBL). These layers contribute to spore resistance. (c) A scanning electron microscope view of a germinating spore (× 30,000). Note that the spore coat divides equatorially along the long axis, and as it separates, the vegetative cell emerges.

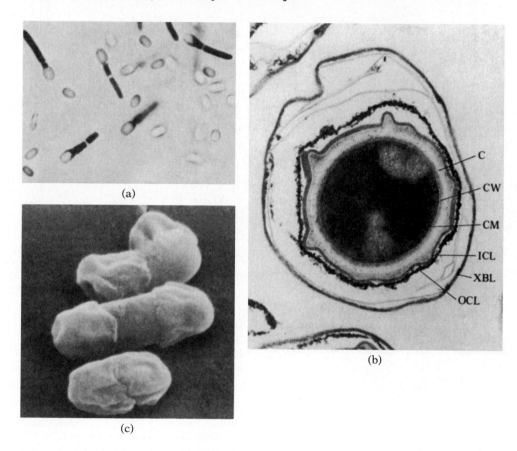

(a)

(b)

(c)

Gas gangrene:
a bacterial disease of the muscles accompanied by large amounts of gas.

Only four serious diseases in humans are known to be caused by sporeformers. The first, anthrax, is due to *Bacillus anthracis* (Chapter 9). This deadly blood disease was studied by Koch and Pasteur. The periodic recurrence of anthrax in the countryside was due to spores remaining alive in the soil where they could be ingested by animals.

The other three diseases are botulism, gas gangrene, and tetanus. These diseases are caused by species of *Clostridium*. Clostridial spores are often found in soil as well as in human and animal intestines. For the spores to germinate to vegetative cells, the environment must be free of oxygen. The dead tissue in a wound provides such an environment for the tetanus and gas gangrene spores (Chapter 9), and a vacuum-sealed can of food is suitable for the spores of botulism (Chapter 8). Some imaginative scientists envision spores as visitors to Earth from a distant galaxy (MicroFocus: 4.3).

4.3

MICROFOCUS

VISITORS?

In 1901, the Swedish chemist Svante Arrhenius (Ar-ren′e-us), taking his cue from the interest in bacteria, proposed that the first form of life on Earth may have been bacterial spores drifting from some distant planet. In the ensuing decades, Arrhenius′ suggestion formed the basis for the panspermia theory. This theory argues that primitive life began with biological substances raining down from space.

In the 1970s, two scientists from the University of Leiden, The Netherlands, took a closer look at Arrhenius′ proposal (and by inference, the panspermia theory). They subjected bacterial spores to laboratory conditions that mimicked conditions found in outer space. The scientists, Peter Weber and J. Mayo Greenberg, found that *Bacillus subtilis* spores could survive in a vacuum chamber made extremely cold by helium. Moreover, the spores would survive when subjected to irradiation by mercury and hydrogen lamps simulating the sun′s ultraviolet rays. Their results intrigued the scientific community.

But the ability to survive a spacelike environment would not matter unless transport to Earth was possible. Arrhenius suggested that biological objects must be lifted into space from a source. Weber and Greenberg answered that concern by theorizing

that a comet or meteorites striking a planet could create the necessary updraft. During space transport, they postulated, a molecular cloud of dust and gas could protect spores against radiation during the in-drift. Arrhenius also wondered how living things could survive for thousands of years. The Dutch scientists projected a 0.1 percent survival of spores for as long as 2500 years and for up to tens of millions of years within a molecular cloud.

Finally, Arrhenius reasoned that the spores would have to enter the Earth environment and land. This was difficult for Weber and Greenberg to explain, because radiation pressures tend to prevent small particles from entering Earth′s environment and solar winds push particles away from Earth.

The prospects for visitations by extraterrestrial spores do not appear bright, according to current thinking. Such a scenario would require several coinciding events, which at present do not appear possible. Still . . .

TO THIS POINT

■

We have explored the anatomical features of bacteria and have noted the three major shapes that a bacterium can take. These are the bacillus, coccus, and spiral shapes. We then analyzed the fine details of a bacterium and discussed how bacterial structures are related to bacterial functions. For example, flagella assist movement in many bacterial rods and spirilla, while pili are used as organs of attachment and in genetic transfers. The bacterial capsule and glycocalyx are sticky, gelatinous structures serving as a protective buffer between the bacterium and its environment. All bacteria, except mycoplasmas, have a rigid cell wall that gives shape to the organism and prevents its disruption by internal pressure. Penicillin prevents cell wall synthesis and lysozyme breaks the cell wall down, both substances contributing to bacterial destruction. The cell membrane conforms to the fluid mosaic model of a membrane and functions primarily in nutrient and waste transport and as an anchor for DNA and a site for key enzymes. The cytoplasm is the center of cell growth and the ground substance in which several structures such as the chromosome are located. The final structure discussed was the bacterial spore, an ultraresistant body formed by species of Bacillus *and* Clostridium.

In the second half of this chapter we shall turn to the growth patterns exhibited by bacteria. We shall study how bacteria reproduce and then explore the dynamics that attend the growth of a bacterial population. Next we shall outline some of the conditions that encourage bacterial growth and discuss certain nutritional patterns. We shall also see how bacteria are cultivated in the laboratory and describe their relationships with other organisms. You might take note of the broad variety of environmental conditions under which bacteria grow. For this reason they can be located in virtually any environment on Earth.

Bacterial Reproduction and Growth

Bacteria reproduce by an asexual process called **binary fission**. In this scquence of events, the chromosome duplicates, the cell elongates, and the plasma membrane pinches inward at the center of the cell. When the nuclear material has been evenly distributed, the cell wall thickens and grows inward to separate the dividing cell. No mitotic structures (e.g., spindle, aster) are present as in eukaryotic cells.

Reproduction by binary fission lends a certain immortality to bacteria because there is never a moment at which the first bacterium has died. Bacteria mature, undergo binary fission, and are young again. In a sense, the original bacterium, though billions of years old, is still among us.

Once the division is complete, bacteria grow and develop the features that make each species unique. The interval of time until the completion of the next division is known as the **generation time**. In some bacteria, the generation time is very short; for others it is quite long. For example, for *Staphylococcus aureus*, the generation time is about 30 minutes; for *Mycobacterium tuberculosis*, the agent of tuberculosis, it is approximately 18 hours; and for the syphilis spirochete, *Treponema pallidum*, it is a long 33 hours. The generation time is a determining factor in the amount of time that passes before disease symptoms appear in an infected individual.

One of the most remarkable generation times is the 20 minutes for *Escherichia coli* growing under optimal conditions. If you were to begin with a single rod at 8:00 A.M. this morning, two would be present by 8:20, four by 8:40, and eight by 9:00 A.M. Sixty-four rods would be present by 10:00 A.M. and 512 by 11:00 A.M. By 6:00 P.M. tonight, the culture would contain just over a billion rods. One enterprising mathematician has calculated that if binary fission were to continue for 36 hours, or until 8:00 P.M. tomorrow night, there would be enough bacteria to cover the face of the earth!

Fortunately, the reproductive potential of a bacterium is never realized because of the limitations of the external environment. Thus we need never worry about being smothered with bacteria. Apparently bacteria are subject to the same controls as all other organisms on Earth, as we shall see next.

Bacterial Growth Curve

A typical **growth curve** for a population of bacterial cells, as shown in Figure 4.13, illustrates some of the dynamics that affect the population over the course of time. The population's history may begin when several bacteria enter the human respiratory tract or are transferred to a tube of growth medium in the laboratory. Four distinct phases of the curve are recognized as follows:

The **lag phase** encompasses the first few hours of the curve. During this time bacteria adapt to their new environment. In the respiratory tract scavenging white blood cells may engulf and destroy some bacteria; in growth medium

Generation time: the time period that passes between binary fissions in bacteria.

esh″er-ik′e-a

Growth medium: material for the cultivation of microorganisms.

FIGURE 4.13

The Growth Curve for a Bacterial Population

During the lag phase (a), the population numbers remain stable as bacteria prepare for division. During the logarithmic phase (b), the numbers double with each generation time. Environmental factors later lead to cell death, and the stationary phase (c) shows a stabilizing population. The decline phase (d) is the period during which cell death becomes substantial.

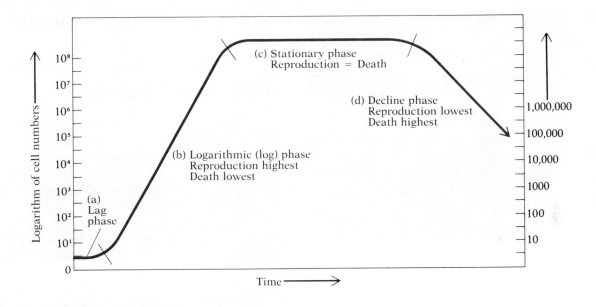

some organisms may die from the shock of transfer. However, the biochemical activity in the remaining bacteria is intense as they store nutrients, synthesize enzymes, and prepare for binary fission. The curve remains at a plateau, balanced by early reproduction in some cells and death in others.

The population then enters an active stage of growth called the **logarithmic phase** (often shortened to "log phase"). The mass of each cell increases rapidly and reproduction follows. As each generation time passes, the number of bacteria doubles and the graph rises in a straight line if logarithms (powers of 10) of the actual numbers are used for the curve.

In humans, disease symptoms usually develop during the log phase because the bacterial population has reached a high enough level to cause tissue damage. Coughing or fever may occur, and fluid may enter the lungs if the air sacs are damaged. If the bacteria produce toxins, tissue destruction may become apparent. In the laboratory, the population growth may be so vigorous that visible colonies appear on solid media, each colony consisting of millions of organisms (Figure 4.14). Broth media may become cloudy with growth. Because the population is at its biochemical optimum, research experiments are generally performed during the log phase.

After some hours or days, the vigor of the population changes and, as the reproductive and death rates equalize, the population enters another plateau, the **stationary phase**. In the respiratory tract, antibodies from the immune system have begun to attack the bacteria and phagocytosis by white blood cells adds to their destruction. Perhaps the person was given an antibiotic to supplement the body's defensive measures. In the culture tube, nutrients have become scarce, waste products have accumulated, and factors such as oxygen or water are in short supply.

Logarithmic phase:
the phase of a bacterial growth curve at which reproduction and growth are at their highest rates.

Antibiotic:
the naturally occurring or synthetic product of a microorganism that inhibits the growth of other microorganisms.

FIGURE 4.14
Two Views of Bacterial Colonies

(a) Colonies of *Bacillus macerans* isolated from sewage and growing in a medium of solidified soybean meal and casein peptone. These colonies are several millimeters in height; they have been euphemistically called "Rockies in a Petri dish." (b) A scanning electron micrograph of the surface of a colony of *Staphylococcus aureus* on a solid medium (× 6000). Note the irregular nature of the surface of the colony, with numerous conical pits.

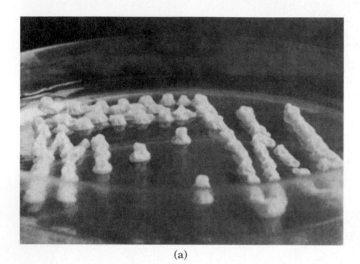

(a)

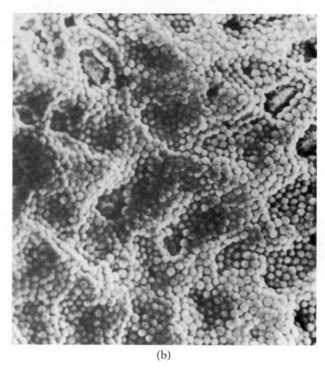

(b)

If these conditions continue, the external environment will exert its limiting powers on the population and the **decline phase** will ensue. Now the number of cells dying exceeds the number of new cells formed. A bacterial capsule may forestall death by acting as a buffer to the environment, and flagella may allow organisms to move to a new location. If the organism is a species of *Bacillus* or *Clostridium*, the vegetative cells will have reverted to spores by this time and the stationary phase may be extended for months or years. For many species of bacteria, though, the history of the population soon comes to an end with the death of the last cell.

Temperature

si'kro-filz

Bacteria inhabit almost every environment on Earth because different species can tolerate the myriad conditions found on the planet. For example, different bacterial species grow at different temperatures (MicroFocus: 4.4). Certain bacteria have their shortest generation times at temperatures in the range of 0° C to 20° C; these bacteria are called **psychrophiles**. Other bacteria, the **mesophiles**, thrive at the middle range of 20° C to 40° C; and **thermophiles** multiply best at high temperatures of 40° C to 90° C or higher.

4 . 4

A VOCABULARY OF BACTERIAL ANATOMY AND GROWTH

flagellum: a long, rigid strand of protein that rotates and imparts motility to a bacterium.

pilus: a short, hairlike protein structure at the surface of a bacterium that functions in attachment and genetic recombinations.

capsule: a gelatinous layer of polysaccharides and small proteins that surrounds a bacterium and protects it from changes in the external environment.

cell wall: the rigid structure of organic materials that encloses a bacterium outside the cell membrane and protects it from bursting.

cell membrane: the flexible structure of protein and phospholipid that encloses the bacterial cytoplasm and determines what enters and leaves the internal environment.

endospore (spore): a highly resistant structure formed by members of the genera *Bacillus* and *Clostridium* that tolerates extreme conditions in the external environment.

nucleoid: the chromosomal region of a bacterium.

ribosome: bodies of protein and RNA that serve at sites of protein synthesis in the cytoplasm of bacterial and eukaryotic cells.

magnetosome: a cytoplasmic body in certain bacterial species that orients the cell toward a favorable environment.

psychrophile: a bacterium that lives best at temperatures ranging from 0° C to 20° C.

mesophile: a bacterium that lives best at temperatures ranging from 20° C to 40° C.

thermophile: a bacterium that lives best at temperatures ranging from 40° C to 90° C or above.

aerobe: a bacterium that utilizes oxygen in its metabolism and prefers an oxygen-rich atmosphere for life.

anaerobe: a bacterium that does not utilize oxygen in its metabolism and prefers an oxygen-free atmosphere for life.

autotrophy: the nutritional pattern in which microorganisms synthesize their own foods from simple carbon compounds.

heterotrophy: the nutritional pattern in which microorganisms obtain their foods from the external environment and utilize complex carbon compounds as food sources.

enriched medium: a cultivation medium that contains special additives to enhance the growth of certain organisms.

selective medium: a cultivation medium that contains special additives to encourage the growth of one species while retarding the growth of other species.

differential medium: a cultivation medium that contains special additives to distinguish the growth of one species from that of another.

Most bacteria appear to be mesophiles. This is especially true of pathogenic bacteria growing in the human body, where the temperature is 37° C. It should be noted that pathogenic bacteria usually grow over a 35° C to 42° C range. Thus, when the body temperature rises to 40° C (104° F), there is a negligible "cooking effect" on bacteria. The common laboratory incubator is set at 37° C to provide the proper environment for mesophiles.

Some mesophiles can grow at temperatures substantially below their normal range. Certain species, for instance, grow in refrigerated foods at 5° C. Here they may produce waste products and cause food spoilage. For example, staphylococci deposit their toxins in cold cuts, salads, and various leftovers. When such foods are consumed without heating, the toxins may cause mild food poisoning. Other examples of mesophiles growing in the cold include *Salmonella* species, which can be infectious (Figure 4.15), and *Proteus vulgaris*, which causes blackening of eggs with a characteristic rotten odor. Since these organisms are not truly psychrophilic, some microbiologists prefer to describe them as **psychrotrophic**. True psychrophiles live in the ocean depths and in Arctic and Antarctic regions where no other form of life is known to exist.

Thermophiles are present in compost heaps and are important contaminants in dairy products because they survive pasteurization temperatures. However, thermophiles pose little threat to human health because they do not grow well at cooler body temperature. In the 1980s scientists isolated thermophilic bacteria from seawater brought up from hot water vents along rifts on the floor of the Pacific Ocean. Using high pressure to keep the water from boiling, they found that the bacteria grew at an astonishing 250° C (482° F).

Pathogenic bacteria: bacteria able to cause disease in plants and animals.

Food poisoning: an intestinal condition due to the consumption of toxins, usually from staphylococci.

si″kro-troph′ik

FIGURE 4.15

A Textbook Case of *Salmonella* Food Infection

This outbreak occurred at a restaurant during October 1991. *Salmonella enteritidis* was isolated from the stools of all 13 patrons of the restaurant who required medical attention.

1. On the morning of October 17, 1991 a restaurant employee prepared Caesar salad by hand-cracking eggs into a large bowl containing olive oil. The eggs were probably contaminated with *Salmonella*.

2. She then added anchovies, garlic, and warm water to the egg-and-oil mixture. The warm water raised the temperature of the mixture slightly.

3. Although the salad dressing was placed in the refrigerator, bacteria from the eggs probably had an opportunity to grow in the temporarily warm environment.

4. Later that day, the Caesar salad dressing was placed at the salad bar in a cooled compartment having a temperature of about 60°F (comfortable room temperature is about 70 to 75°F).

5. The dressing remained in the compartment until the restaurant closed (a period of 8 to 10 hours) while patrons helped themselves to the Caesar salad.

6. Within three days, 15 restaurant patrons fell ill, with diarrhea, fever, abdominal cramps, nausea, and chills. Thirteen sought medical care and eight required intravenous rehydration.

Oxygen

Anaerobic bacteria: bacteria that grow in the absence of oxygen.

The growth of many bacteria also depends on a plentiful supply of oxygen and, in this respect, the **aerobic** bacteria are similar to more complex organisms. **Anaerobic** bacteria, by contrast, must have an oxygen-free environment in order to survive. Some anaerobic bacteria actually die if oxygen is present, while others simply fail to grow and multiply. Certain anaerobic bacteria use sulfur in their chemistry instead of oxygen, and therefore they produce hydrogen sulfide (H_2S) rather than water (H_2O) as a waste product of their metabolism. Others synthesize considerable amounts of methane (CH_4), often called swamp gas. Both of these gases give putrid odors to marshes, swamps, and landfills. Petroleum is a product of anaerobic metabolism.

Certain anaerobic bacteria cause disease in humans. For example, the **Clostridium** species that cause tetanus and gas gangrene multiply in the dead, anaerobic tissue of a wound and produce toxins that lead to tissue damage. Another species of *Clostridium* multiplies in the oxygen-free environment of a vacuum-sealed can of food, where it produces the lethal toxin of botulism. In one bizarre incident, a restaurant owner died of botulism after tasting a piece of fish marinating under a layer of oil. Anaerobic conditions had apparently developed under the oil giving the *Clostridium* an opportunity to grow and produce its toxin.

Anaerobic conditions may be established in the laboratory by a number of methods. One method involves using thioglycollic acid to bind oxygen in thioglycollate medium. Another employs a mixture of pyrogallic acid and sodium hydroxide to absorb oxygen from the environment. Among the most widely used methods is the GasPak system, in which hydrogen reacts with oxygen in the presence of a catalyst to form water, thereby creating an oxygen-free atmosphere.

thi″o-gli′ko-lāt

Some bacteria are neither aerobic nor anaerobic, but **facultative**. Facultative bacteria grow either in the presence or absence of oxygen. This group includes many staphylococci and streptococci, as well as members of the genus *Bacillus* and a variety of intestinal rods, among them *E. coli*. Some microbiologists believe that a majority of bacteria may be facultative organisms. A facultative aerobe prefers aerobic conditions (but grows anaerobically), while a facultative anaerobe prefers oxygen-free conditions (but grows aerobically).

Facultative bacteria: bacteria that grow in the presence or absence of oxygen.

A final group is the **microaerophilic** bacteria. These organisms require a low concentration of oxygen for growth. In the body, certain microaerophiles cause disease of the oral cavity, urinary tract, and gastrointestinal tract. Certain species of bacteria, said to be **capnophilic**, require an atmosphere low in oxygen but rich in carbon dioxide. The CO_2 content can be increased in the laboratory by using a gas-generating apparatus or by burning a candle in a closed jar with the bacteria.

mi″kro-a-ro-fil′ik

cap-no-fil′ik

Acidity/Alkalinity

Because the internal environment of most bacteria has a pH of about 7.0, the majority of species grow best under neutral pH conditions. Although most growth media for laboratory cultivation are set at pH 7.0, bacteria tolerate acidic conditions as low as pH 6.5 and alkaline conditions as high as pH 7.5. Human blood and tissues, with a pH of approximately 7.2 to 7.4, provide a suitable environment for the proliferation of disease-causing bacteria.

pH: a measure of the acidity or alkalinity of a substance or solution.

Certain acid-tolerant bacteria called **acidophiles** are valuable in the food and dairy industries. For example, *Lactobacillus* and *Streptococcus* produce the acid that converts milk to buttermilk, cream to sour cream, and milk curds to cheese. These organisms pose no threat to good health even when consumed in large amounts. The "active cultures" in a cup of yogurt are actually acid-tolerant (or, acidophilic) bacteria.

The vast majority of bacteria, however, do not grow well under acidic conditions. Thus the acidic environment of the stomach helps deter disease in this organ while posing a natural barrier to the organs beyond. In addition, you may have noted that certain acidic foods are hardly ever contaminated with bacteria. Examples include lemons, oranges, and other citrus fruits, as well as vegetables such as cabbage and rhubarb. Traditionally, tomatoes were too acidic to support bacterial growth, but modern technologists have developed the "neutral tomato" and with it a bevy of new problems for consumers, especially those who grow and can their own tomatoes.

Patterns of Nutrition

Bacteria must meet certain nutritional requirements in order to grow. Most bacteria have relatively simple requirements, with **water** an absolute necessity. In addition, bacteria need foods that can serve as energy sources and raw materials for the synthesis of cell components. These foods generally include proteins for structural compounds and enzymes, carbohydrates for energy, and a series of vitamins, minerals, and inorganic salts.

Two different patterns exist for satisfying an organism's nutritional needs. These patterns are called autotrophy and heterotrophy. They are primarily based on the source of carbon used for making cell components.

Organisms that practice **autotrophy** are able to synthesize their own foods from simple carbon sources (Chapter 5). The organisms are said to be autotrophic (literally "self-feeding"). Autotrophs obtain their carbon from inorganic compounds such as carbon dioxide and ions such as carbonate, nitrate, and sulfate. Energy for food synthesis may come from the sun or from chemical reactions taking place in the cytoplasm as Chapter 5 explains.

The second pattern, **heterotrophy**, is employed by heterotrophic organisms (literally "other-feeders"). These organisms obtain preformed organic molecules from the environment and use them for structural components and energy. The heterotrophic bacteria that feed exclusively on dead organic matter such as rotting wood are commonly called **saprobes**. For many years these organisms were known as saprophytes, from Greek stems meaning "rotten" and "plant," but the name has been changed to saprobes to reflect feeding on both plants and animals. Heterotrophs that feed on living organic matter such as human tissues are commonly known as **parasites**. The word **pathogen** (from the Greek *pathos* for "suffering") is used if the parasite causes disease in its host organism.

Bacterial Cultivation

Since the time of Pasteur and Koch, microbiologists have used media such as beef broth for the laboratory cultivation of bacteria (MicroFocus: 4.5). The modern form of this liquid medium, called **nutrient broth**, consists of water, beef extract, and peptone, a protein supplement from plant or animal sources. When agar is added to solidify the medium, the product is called **nutrient agar**. Agar is a polysaccharide derived from marine algae. It adds no nutrients to the medium but only serves to make it solid so that bacteria can be cultivated on the surface. The introduction of agar to bacteriology is described in Chapter 1. Sometimes it is valuable to use a semisolid medium, such as when testing bacterial motility. In this case, a small portion of agar is added to the medium to make it stiff but not as solid as nutrient agar.

Most common bacteria grow well in nutrient broth and nutrient agar, but certain fastidious bacteria may require specially **enriched media**. For example, the streptococci that cause strep throat and scarlet fever grow well when whole blood is added to the nutrient medium. In this instance, the medium is called blood agar. To encourage the growth of *Neisseria* species, blood agar is heated before solidification, a process that disrupts the red blood cells and releases the hemoglobin. The medium is now termed chocolate agar because of its charred brown appearance.

Selective media are those containing ingredients to inhibit the growth of certain bacteria in a mixture while permitting the growth of others. For example, staphylococci are cultivated on mannitol salt agar. This medium contains mannitol, an alcoholic carbohydrate fermented by staphylococci, as well as a high salt concentration that inhibits most other bacteria. Another example is eosin methylene blue (EMB) agar. This selective medium has carbohydrates

---

aw″to-trōph′e
het″er-o-trōph′e

Autotroph:
an organism that synthesizes its foods from simple inorganic carbon compounds.

Heterotroph:
an organism that obtains its foods from preformed organic carbon compounds.

Medium:
a substance used to support the growth of microorganisms (e.g., nutrient agar).

ahg′ar

Fastidious:
having special requirements.

ni-se′re-ah

e′o-sin

ASLEEP FOR 11,000 YEARS

It was 2 feet long, reddish-brown, and tube shaped. It could have been a piece of rusted pipe, but it smelled so bad and had such a convoluted shape that the pipe theory was quickly discarded. Workers had found the object while building a new fourteenth hole at the Burning Tree golf course near Columbus, Ohio. While digging a mere 5 feet into the soil they hit upon the skeletal remains of a mastodon, an elephantlike animal that lived 11,000 years ago. This tube-shaped object was near the animal's rib cage.

Paleontologists theorized that the object was probably part of the mastodon's intestinal tract. It so happened that Gerald Goldstein of Ohio Wesleyan University was visiting the site at the invitation of a friend involved in the excavation. He half-jokingly suggested that something might still be alive in the intestinal contents, and he was given a small bag of the smelly material. Back at his laboratory, he placed a sample of the intestinal contents on ordinary bacteriological medium and surprise, surprise ... the next day the medium teemed with bacteria. The organism was *Enterobacter cloacae*, a well-known resident of a mammal's (including a mastodon's) intestine.

In 1991, Goldstein announced his recovery to a skeptical scientific community. He determined that the *E. cloacae* was not a contaminant from the surrounding soil by searching for the organism in 12 samples from nearby sites. All 12 samples failed to yield *E. cloacae*. An independent "blind" analysis, in which scientists were not told the sources of the samples, confirmed the results.

Assuming Goldstein's discovery remains intact, the bacteria he cultivated would be the oldest known living bacteria recovered from nature (bacterial spores excluded). A 3-foot cap of clay had apparently sealed the site, and the chilly 45° F temperature probably contributed to placing the bacteria in a state of suspended animation. As one writer suggested "Move over Rip van Winkle. There's a new record for slumber time."

---

fermented by *E. coli* and other Gram-negative bacteria, but it also contains eosin and methylene blue, two dyes that inhibit Gram-positive bacteria.

Another type of medium is the **differential medium** (Table 4.3). This medium makes it easy to distinguish colonies of one organism from colonies of other organisms on the same plate. MacConkey agar is typical. It contains the dyes neutral red and crystal violet as well as the carbohydrate lactose. Those bacteria that ferment the lactose take up the dyes and form red colonies; other bacteria show up as colorless colonies. In addition, MacConkey agar contains bile salts that inhibit the growth of Gram-positive bacteria. This medium is thus selective as well as differential.

Differential medium: a medium in which colonies of different bacteria can be distinguished.

TABLE 4.3

A Comparison of Bacterial Media

NAME	COMPONENTS	USES	EXAMPLES
Nutrient broth	Water, beef extract, peptone	General use	—
Nutrient agar	Water, beef extract, peptone, agar	General use	—
Enriched medium	Growth stimulants	Cultivating fastidious bacteria	Blood agar for streptococci; chocolate agar for *Neisseria* species
Selective medium	Growth stimulants Growth inhibitors	Selecting certain bacteria out of mixture	Mannitol salt agar for staphylococci; EMB agar for Gram-negative bacteria
Differential medium	Dyes Growth stimulants Growth inhibitors	Distinguishing different bacteria in a mixture	MacConkey agar for Gram-negative bacteria

Synthetic medium:
a medium in which the
nature and quantity of each
component are known.

Pure culture:
a population containing only
one species.

The media just described represent nonchemically defined or **natural media**. This is because one cannot be certain of the exact components or their quantity. Another type of medium is the chemically defined or **synthetic medium**. Here the nature and amount of each component are known. Such a medium might contain glucose, ammonium phosphate, potassium phosphate, magnesium sulfate, and sodium chloride. The glucose supplies energy to the cell; the ammonium ions are a source of nitrogen for amino acid and nucleic acid formation; the phosphate is used in DNA and RNA synthesis; sulfur from magnesium sulfate is valuable for enzyme formation; and sodium chloride maintains a stable internal environment in the cytoplasm.

Bacteria rarely occur in nature alone. In virtually all cases, they are mixed with species of other bacteria, a so-called **mixed culture** (a swab from the gum area is an example). But to work with bacteria, the laboratory technologist must use a **pure culture**, that is, a population of only one species of bacteria. This is particularly important if an identification of a pathogen is to be made (Micro-Focus: 4.6).

When isolating bacteria from mixed cultures, two standard methods are available. The first method, called the **streak plate isolation method**, uses a single plate of bacteriological medium. An inoculum is taken with a loop or needle, and a series of streaks is made in one area of the plate. The instrument is flamed, touched to the first area, and a second series is made in a second area. Similarly, streaks are made in the third and fourth areas, thereby spreading out the different bacteria so they can form discrete colonies on incubation. The second method is the **pour plate isolation method**. Here, a sample of bacteria is diluted in several tubes of melted agar medium. The agar is then poured into sterile Petri dishes and permitted to harden. On incubation, the bacteria will form discrete colonies where they have been diluted the most. The technologist can then pick samples of the colonies for further testing.

Various types of culture methods are available to the research microbiologist. For example, bacteria can be cultivated in a large **batch method**, then removed from the container for further study; or they may be cultivated by a **continuous method** where nutrients are regularly added to the container and waste products are drawn off. Bacteria remain in the log phase in this method, and they can be used to produce useful industrial products such as vitamins and enzymes. On occasion, it is desirable to have all bacteria at the same stage of population growth (e.g., the stationary phase). A physical treatment such as heat is used in this so-called **synchronous method** of cultivation.

To measure the amount of bacterial growth in a medium, there are numerous methods. For example, the cloudiness, or **turbidity**, of a liquid culture may be used to indicate the number of bacteria it contains. It is also possible to perform a **direct microscopic count** using a known sample of the culture on a specially prepared slide (a Petroff-Hauser counting chamber) containing a counting grid. The **dry weight** of the bacteria gives an indication of the cell mass, and the **oxygen uptake** in metabolism can be measured as an indication of the bacterial number.

It is also possible to perform a **most probable number test** (Chapter 25) or a **standard plate count procedure** (Chapter 25). In the former test, samples of bacteria are added to numerous lactose broth tubes and the presence or absence of gas formed in fermentation gives a rough estimation of the bacterial number. In the latter test, a bacterial culture is diluted, and samples of dilutions are placed in agar plates. The number of colonies appearing after incubation reflects the number of bacteria originally present. This test is desirable because it gives the **viable count** of bacteria (the living bacteria only) as compared to a microscopic count or dry weight test that gives the **total bacterial count** (the living as well as dead bacteria).

4.6

TIME FOR THE ERASER

Richard Pfeiffer was so excited he was jumping for joy. He had found it—the influenza bacillus! There it was under the microscope for all to see—a small Gram-negative rod. Naming it was easy—*Haemophilus influenzae*. His colleagues were equally excited—they simply called it Pfeiffer's bacillus.

That was in 1892. Now it was 26 years later in 1918, and the world was in the midst of an influenza pandemic. Millions were dying, and Pfeiffer's bacillus was not as easy to find as scientists anticipated. Sure, they could locate it in people already sick with the flu, but it did not seem to be causing the influenza. Time and again, inoculations with the bacillus failed to bring on typical flu symptoms in healthy volunteers. The doubts were creeping in. Was it possible? Had Pfeiffer been wrong? Was it time to rewrite the textbooks?

Indeed, it was time. Pfeiffer had isolated a bacillus but it was not the influenza bacillus. By 1919, it had become quite clear that influenza is caused by a virus, not a bacterium. Pfeiffer's cultures had probably been contaminated by a bacterial pathogen from a patient. Once again, a lesson was driven home to students of microbiology—aseptic technique is paramount to success in the lab. Contamination means disaster.

Some bacteria are impossible to cultivate on artificial laboratory media but, instead, require a living tissue medium. Most rickettsiae and chlamydiae are examples of such bacteria. They must be grown in fertilized eggs, tissue cultures, animals, or other environments where living cells are found. The difficulty in cultivation often makes detection and study of these organisms burdensome.

Intermicrobial Relationships

Closely allied to the nutritional needs of bacteria is their relationship with other organisms. The term **symbiosis** (literally, "living together") is applied to the relationship. Symbiosis implies a situation in which two populations of organisms interact in a close and permanent association. The benefits obtained through this interaction may involve food, protection, support, or other life-sustaining factors.

sim"bi-o'sis

If a symbiosis between two populations of organisms benefits both populations, the relationship is further defined as **mutualism**. For example, bacteria live on the roots of pod-bearing plants such as peas and beans where they trap nitrogen from the atmosphere and convert it to ammonium ions. The plant then uses the ammonium ions to synthesize amino acids. The plant, in turn, provides a stable environment for the bacteria and supplies them with essential growth factors. The significance of this relationship is explored more fully in Chapter 25.

Another type of symbiosis is **commensalism**. This relationship occurs when one population receives benefit from the relationship while the other is neither benefited nor harmed (Figure 4.16). An example is found in the populations of bacteria inhabiting the human skin. Such bacteria are called commensals. Another commensal is *Escherichia coli*, the Gram-negative rod that thrives in the human intestine but usually causes no damage. To some investigators this relationship is more symbolic of mutualism, because research suggests that *E. coli* provides certain vitamins for human nutrition and breaks down otherwise indigestible foodstuffs.

Commensalism:
a symbiotic relationship in which one population receives benefit while the other receives neither benefit nor harm.

A third type of symbiosis is **synergism**. In this case, two populations or organisms live together and accomplish what neither population could accomplish alone. In trench mouth, for example, at least two populations of bacteria must be present for infection of the oral cavity to occur. Usually one population consists of rods, the other of spirochetes.

sin'er-jizm

Trench mouth:
a disease of the dental tissues, accompanied by bleeding ulcers, foul odor, and bad taste.

FIGURE 4.16

Commensalism

A scanning electron micrograph of a section from the gastrointestinal tract of a young sheep showing an area inhabited by a mass of curved rods. These organisms are probably commensals because the animal exhibited no signs of disease. Bar = 1.0 μm.

When a symbiosis is beneficial to one organism but harmful to the other, the result is **parasitism**. In this situation, the organism that benefits is the parasite, and the one that suffers injury is the host. The bacteria of human disease are typical parasites. Many microbiologists believe that injury to the host is probably not in the best interest of the parasite, because if the host were to die, the parasite might also die. A relationship such as this is explored more completely in Chapter 17.

Archaeobacteria

We shall close our discussion of bacterial anatomy and growth with insights into a unique type of bacteria called **archaeobacteria** (sometimes spelled without the "o"). Archaeobacteria have existed on Earth longer than any other bacteria or surviving organism of any type. Their fossils have been located in African and Australian rock 3.5 billion years old. The name "archaeobacteria" reflects their status of being the oldest known bacteria. All other bacteria, by contrast, are referred to as "eubacteria," meaning true bacteria (Figure 4.17). In this text we shall use the term bacteria interchangeably with eubacteria, as we have done in this chapter.

Archaeobacteria are distinctive in several respects. Their cell walls do not contain peptidoglycan and their cell membranes have an unusual lipid composition. Moreover, they have a unique type of ribosomal RNA. Indeed, biochemist Carl Woese and his co-workers have determined the sequence of nucleotides in the RNA of archaeobacterial ribosomes and have suggested that the differences are significant enough to separate the archaeobacteria into their own kingdom. However, *Bergey's Manual* classifies the archaeobacteria in Division IV of Procaryotae.

FIGURE 4.17

A Possible Classification of Living Things

A proposed classification of living things beginning with a common universal ancestor and taking into account the division of bacteria into Archaeobacteria and Eubacteria. Note the three major types of archaeobacteria. Also note the theorized place of bacteria in the formation of mitochondria and chloroplasts in eukaryotes.

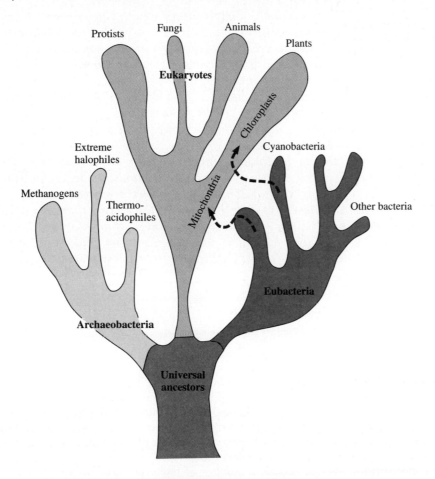

Archaeobacteria fall into three separate categories: methanogens, thermoacidophiles, and extreme halophiles. **Methanogens** are rods that live in strict anaerobic environments and produce large quantities of methane in their metabolism. They may be found in marshes, lake bottoms, and feces (and hence in the intestines of some animals). **Thermoacidophiles** show remarkable resistance to acid and extremely high temperatures. One genus, *Sulfolobus*, grows in hot springs that have temperatures as high as 85° C and acidic soils that have a pH of 1.0. Dense communities of archaeobacteria have also been found in the boiling waters just above the Macdonald Seamount, an active volcano 150 feet below the surface of the Pacific Ocean where the water temperature exceeds 100° C. **Extreme halophiles** thrive in high-salt environments such as in Utah's Great Salt Lake. Additional aspects of archaeobacterial physiology are discussed in Chapter 5.

Archaeobacteria are of particular interest to molecular paleontologists because the bacteria live under conditions that existed on primitive Earth. Certain scientists postulate that an ancient strain of prokaryote may have given rise to archaeobacteria by one route and to eubacteria by another. It is even conceivable that traces of that ancestor are waiting to be discovered in some primordial swamp.

NOTE TO THE STUDENT

It is fairly common to read in biology books about the "higher" and "lower" forms of life. Typically, humans are cast as higher forms, while insects, worms, microorganisms, and similar creatures are considered "lower" forms.

Discerning biologists would probably disagree with this concept. They would point out that each species of organism is a product of evolution and, as such, each is exquisitely adapted to its environment and way of life. Indeed, it is probable that a cockroach is better equipped to survive the rigors of this world than a human being.

Now consider the bacteria. Bacterial species thrive in environments ranging from ice to boiling-hot springs; many species can live with or without oxygen; a large percentage make their own foods from chemicals in the soil; they have no built-in death age; and they double in number every hour or so. We humans, by contrast, must maintain a constant body temperature; we die without oxygen; we eat complex foods; we reach a certain age, then die; and it takes a full 25 years to produce a generation.

It is difficult to believe that microorganisms are "lower" than any other forms of living things. Certainly they are not "lower" than humans. Bacteria were here long before we humans came on the scene, and they will undoubtedly be here long after we "higher" forms have vanished. Sorry for the put down.

Summary

Bacteria occur in variations of three shapes, the rod (bacillus), the sphere (coccus), and the spiral (vibrio, spirochete, or spirillum). Under the microscope, rods and spirals generally appear singly, but cocci are seen in a number of configurations, including the diplococcus, streptococcus, and staphylococcus.

The electron microscope reveals a number of bacterial structures that give insight into bacterial functions. Flagella, for example, occur on many rods and impart motion. Pili are short hairlike appendages that permit attachment to a surface. The capsule is a sticky layer of polysaccharides that buffers a bacterium against the external environment. And the cell wall provides rigidity and structure to the cell, while the cell membrane is a site of transfer into and out of the cell as well as an enzyme site. Various bodies exist in the cell cytoplasm, but there is no bacterial nucleus. A highly resistant structure called the spore is produced by members of the genera *Bacillus* and *Clostridium*.

The reproduction of bacteria takes place by binary fission, a process wherein chromosomal duplication and cytoplasmic separation are major events. Binary fissions occur at intervals called the generation time, which for bacteria may be as short as 20 minutes. Although the potential for incalculable masses of bacteria is great, the dynamics of the growth curve show how a population reaches a certain peak and then levels off and declines.

Bacterial species are able to grow over a vast array of conditions. Certain species, for example, grow at temperatures as low as 0° C, others at 90° C. Bacteria may grow with or without oxygen, depending on the species, and most grow over a range of pH. Different species display different nutritional patterns, such as autotrophy and heterotrophy, and laboratory media are devised to reflect these patterns. In nature, bacterial populations interact with other populations of living things in certain recognizable ways. These interrelationships fit the bacteria into the scheme of living things on Earth.

Questions for Thought and Discussion

1. Suppose a bacterium had the opportunity to form a capsule, a flagellum, or a spore. Which do you think it might choose? Why?

2. For certain bacterial diseases such as strep throat, a brief few days separate entry of the bacterium to the individual and the appearance of symptoms. For other bacterial diseases, such as tuberculosis, the period may be weeks or months. How many factors can you name that may account for this difference?

3. Consumers are advised to avoid stuffing a turkey the night before cooking, even though the turkey is refrigerated. A homemaker questions this advice and points out that the bacteria of human disease grow mainly at warm temperatures not in the refrigerator. What explanation might you offer to counter this argument?

4. Toothbrushes are constantly exposed to bacteria in the mouth. Nevertheless, bacterial colonies do not appear on toothbrushes hanging in a rack. Can you postulate why?

5. In the fall of 1993, public health officials found that the water in a midwestern town was contaminated with sewage bacteria. The officials suggested that homeowners boil their water for a couple of minutes before drinking it. Would this treatment sterilize the water? Why?

6. An organism is described as a peritrichous, anaerobic, heterotrophic, mesophilic streptococcus. How might you translate this complex bacteriological gibberish into a description of the organism?

7. The dynamics of a growth curve for bacteria apply equally well to populations of most other organisms. Begin with a group of Pilgrims arriving at Plymouth, Massachusetts, in 1620 and follow the growth curve of the population through to the modern era while drawing a parallel to a population of bacteria.

8. Is it advantageous for a bacterium to live as a saprobe, or would life as a parasite be preferable? Why?

9. Suppose this chapter on the anatomy and growth of bacteria had been written in 1940, before the electron microscope became available. Which parts of the chapter would probably be missing?

10. Bacterium "X" has been identified as a cause of human disease. Knowing that it is a pathogen, can you guess what structures it might possess? Explain your reasons for each choice.

11. There are thousands of species of bacteria, yet most apparently have variations of one of three shapes: the rod, sphere, or spiral. Do you find this strange? Why or why not?

12. If a bacterium is able to grow in a synthetic (chemically defined) medium, do you suppose it will grow better or worse in a natural medium such as nutrient agar? Why?

13. Although thermophilic bacteria are presumably harmless because they do not grow at body temperatures, they may still pose a hazard to good health. Can you think of a situation in which this might occur?

14. To prevent decay by bacteria and to display the mummified remains of ancient peoples, museum officials place the mummies in glass cases where oxygen has been replaced with nitrogen gas. Why do you think nitrogen is used? Will any bacteria-induced decay occur in this environment?

15. Every state has an official animal, flower, or tree, but only Oregon has a bacterium named in its honor: *Methanohalophilus oregonese*. The species modifier is obvious, but can you decipher the meaning of the genus name?

Review

One of the major topics of this chapter has been a survey of microscopic bacterial structures seen through the light and electron microscopes. To test your recall of these structures, fill in the following crossword puzzle. The answers to the puzzle are in the Appendix.

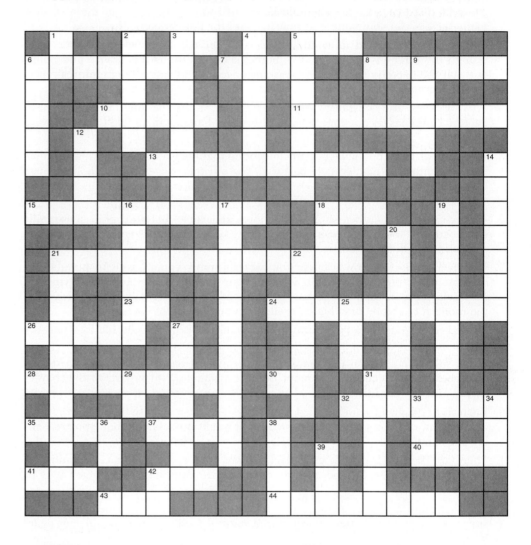

ACROSS

3. Capsule-producing bacterial species (initials)
5. The genetic material of a bacterium
6. A cluster of four or eight cocci
7. The _____ structure of a bacterium is visible with the electron microscope
8. Contains the enzyme lysozyme for bacterial destruction
10. The _____ mosaic model describes the structure of the cell membrane

DOWN

1. Bacterium (initials) that appears as a grapelike cluster
2. A pair of bacterial cocci is a _____-coccus
3. A bacterial rod
4. A curved rod that resembles a comma under the microscope
5. A polysaccharide existing as tangled fibers in the glycocalyx
6. Another name for the glycocalyx is the _____ layer
9. About 40 percent of the cell membrane consists of _____

ACROSS (*continued*)

11. The cell membrane is the structure of _____ into the cell cytoplasm
13. A bacterial capsule having a loose consistency
15. Closed loop of DNA having the bacterial inheritance characteristics
18. Nucleic acid not found in plasmids
21. The important chemical constituent of the bacterial cell wall
23. Unit of measurement (abbr) for width of cell wall
24. Bacterial spiral with a flexible cell wall
26. A short hairlike projection for attachment and genetic transfers
28. Bodies of RNA and protein that function in protein synthesis
30. Bacterium (initials) that occurs as a sarcina
32. Closed loop of DNA apart from the chromosome in the cytoplasm
35. The slime of the capsule of *A. viscolactis* resembles

    _____
37. Teichoic acid is present in the cell walls of _____-positive bacteria
40. Flagella permit a bacterium to _____
41. A monotrichaete has only _____ flagellum
42. A component (abbr) of the principal layer of the bacterial cell wall
43. The cell membrane has _____ important organic constituents
44. The name for the chromosomal region of a bacterium

DOWN (*continued*)

12. The side chains of peptidoglycan consist of _____ amino acids
14. The presence of a capsule contributes to the ability to cause _____
16. A capsule-containing cause of caries is *Streptococcus*

    _____
17. A cytoplasmic body that helps a bacterium orient itself
18. Nucleic acid not found in plasmids
19. Many times the length of a bacterium but extremely thin; used for motility
20. A bacterial sphere
21. Antibiotic that interrupts construction of bacterial cell wall
22. Serves as a buffer between bacterium and external environment
25. Form displayed by typhoid, anthrax, and diphtheria bacilli
27. Alternative name for pilus
29. Bacterium (initials) well-known for its capsule
31. Layer of bacteria and other materials on tooth surface
33. Microscope (abbr) used to visualize ultrasmall bacterial structures
34. Used to stain metachromatic granules in cytoplasm
36. Unlike _____-karyotic cells, bacteria have no nuclei
38. Common site of infection by staphylococci
39. Type of cell (abbr) that does not engulf bacteria if capsule present
42. Pili have _____ function in motility despite their location and appearance

CHAPTER 5

BACTERIAL METABOLISM

ouis Pasteur's discovery of yeast's role in fermentation heralded the beginnings of microbiology because it showed that microscopic organisms could bring about important chemical changes. However, it also opened debate on the chemistry by which yeasts accomplish fermentation. Some scientists offered that grape sugars entered yeast cells to be fermented; others believed that fermentation occurred outside the cells. The question was not resolved until a fortunate accident happened in the late 1890s.

In 1897, two German chemists, Eduard and Hans Buchner, were preparing yeast as a nutritional supplement for medicinal purposes. They ground yeast cells with sand and collected the juice. To preserve the juice, they added a large quantity of sugar (as was commonly done) and set the mixture aside. Several days later they noticed an unusual alcoholic aroma from the mixture. Its taste confirmed their suspicion: the sugar had fermented to alcohol.

The discovery by the Buchner brothers was momentous because it demonstrated that something inside the yeast cells brings about fermentation and that living cells were not required. The substance came to be known as "enzyme," meaning "in-yeast." In 1905, the English chemist Arthur Harden expanded the Buchner study by showing that enzyme was really a multitude of chemical compounds and should better be termed "enzymes." Thus, he added to the belief that fermentation was a chemical process. Soon, many chemists became biochemists, and biochemistry emerged as a new scientific discipline.

Biochemistry, the chemistry of life processes, is one of the most exciting modern sciences. Researchers have uncovered the chemical bases of many biological processes and can understand them well enough to derive some practical value. The products of genetic engineering are typical. Also, numerous principles of biochemistry such as energy utilization underlie the diverse expressions of life, whether bacterial or human. Moreover, the findings in biochemistry are having substantial impact on medicine. For instance, the presence of certain substances in the human blood aids clinical diagnoses, and the mechanism of action of drugs can be explained within biochemical frameworks.

Yeast:
unicellular fungi having the enzymes to convert grape sugars to wine and other products of fermentation.

book'ner

TABLE 5.1

A Comparison of Two Key Aspects of Cellular Metabolism

CATABOLISM	ANABOLISM
Breakdown of large molecules	Buildup of small molecules
Energy is generally released	Energy is generally required
Products are small molecules	Products are large molecules
Glycolysis, Krebs cycle, electron transport, chemiosmosis	Photosynthesis, protein synthesis
Mediated by enzymes	Mediated by enzymes
Reactions converge to major pathways	Reactions diverge from basic pathways

---

Metabolism:
the sum total of all biochemical processes taking place in living cells.

In this chapter we shall focus on the broad aspect of biochemistry known as metabolism. **Metabolism** refers to the sum total of all biochemical processes taking place in living cells. Metabolic processes may be divided into two general categories: **anabolism**, or the synthesis of chemical compounds, and **catabolism**, or the digestion of chemical compounds (Table 5.1). In both processes there are reminders of the origins of biochemistry, because the substances that catalyze anabolism and catabolism are enzymes.

Enzymes and Energy in Metabolism

In order for metabolic processes to take place, living cells such as bacteria must have an adequate supply of enzymes. We shall therefore begin our study of metabolism with a detailed discussion of these substances. In addition, this section will contain some general concepts relating to energy, because the chemical reactions of anabolism generally utilize energy, while those in catabolism often liberate energy. Since the discussions will often mention proteins, carbohydrates, and lipids, you may wish to review Chapter 2 as a refresher.

Enzymes

Enzyme:
a protein molecule that brings about a chemical change while itself remaining unchanged.

Enzymes are a special group of protein molecules that bring about chemical changes while themselves remaining unchanged. They catalyze, or speed up, chemical processes by accomplishing in seconds what otherwise might take hours, days, or longer to happen spontaneously. This is especially true during the bonding of large organic molecules, where the concentration of the molecules is generally low and the reaction areas are often hidden. Random collisions between large molecules are unlikely to occur, and bonding may not otherwise take place. Thus the reaction rate would be very low were it not for the activity of enzymes.

Enzymes are reusable. Once a chemical reaction has occurred, the enzyme is released to participate in another reaction, as illustrated in Figure 5.1. The number of enzymes in a bacterium is therefore quite small compared to the hundreds of thousands of chemical reactions taking place. Enzyme activity is also highly specific, and an enzyme that functions in one type of chemical reaction usually will not participate in another type. The substance acted upon is called the **substrate**, and the products are appropriately termed **end products**. Since the reactions are usually reversible, enzymes may bring about syntheses as well as digestions. This factor is important in metabolism because anabolism often occurs by a reversal of catabolism.

Substrate:
the substance acted upon by an enzyme molecule.

FIGURE 5.1

The Mechanism of Enzyme Action

(a) An enzyme has an active site where the substrate molecules and enzyme can fit together. (b) To accommodate the substrates and fit closely about them, the active site changes shape slightly. This is the "induced fit" phenomenon. The substrates are then converted to end products. (c) Following the reaction, the end products are released from the active site and the latter assumes its original shape, ready to function again.

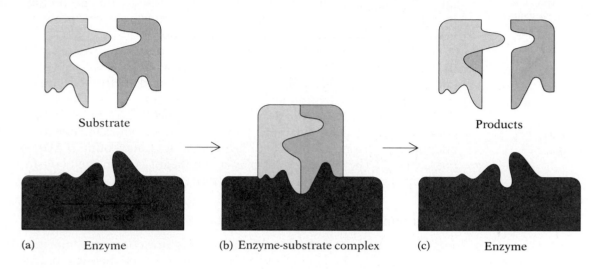

Substrate   Products

Active site

(a)   Enzyme   (b) Enzyme-substrate complex   (c)   Enzyme

Basically, enzymes function by aligning molecules in such a way that a reaction can take place. In a synthesis reaction, for example, the jagged surface of an enzyme molecule holds the substrates in such a way that their electron clouds overlap in the spot where the chemical bond should form. In a digestion reaction, by contrast, the enzyme binds to the substrate and pushes it slightly out of shape so that the bond breaks.

In order for synthesis or digestion to take place, certain critical areas called **active sites** must be present and available on the enzyme molecule. Active sites often contain sulfhydryl groups (—SH), so any substance reacting with sulfhydryl groups will tie up the groups and prevent enzyme activity. A heavy metal such as mercury acts in this fashion and, consequently, it is useful as a disinfectant in such forms as mercuric chloride and Merthiolate.

Another way of inactivating an enzyme is by blocking its active site with a compound closely related to the normal substrate. **Sulfonamide drugs** operate in this way. The drug molecules bind to the bacterial enzyme used to combine substrates during folic acid synthesis. Without folic acid, the bacterial growth is disrupted, and the bacteria soon die. Sulfonamide drugs are used for urinary tract diseases caused by Gram-negative bacteria.

Enzymes were originally named for the chemical reactions they catalyze. For instance, pepsin, from the Greek *pepsis* for digestion, refers to the protein-digesting enzyme of the human gastrointestinal tract. Modern biochemists have adopted the **"-ase" ending** for enzymes, and many enzymes are currently named for the substrate on which they act. Lactase, for example, is the enzyme that digests lactose; sucrase breaks down sucrose; and ribonuclease digests ribonucleic acid.

Some groups of enzymes carry names corresponding to their activity. For example, the **hydrolases** operate in hydrolysis reactions where the products of water are added to the end products. Hydrolases include lipase, which breaks down lipids, and peptidase, which digests peptides to amino acids. Another group, the **oxidases**, catalyze oxidation-reduction reactions. A third group, the **transferases**, operate in transfer reactions. They include kinases, which transfer

sul-fi'dril

Heavy metal:
an element whose atoms have a large atomic weight and are electron donors.

sul-fon'ah-mid

Sulfonamides:
a group of antimicrobial drugs that contain sulfur and amino groups.

Hydrolysis:
a chemical reaction in which a major reactant is split into two products, with the addition of H to one product and OH to the other.

phosphate groups, and transaminase, which exchanges amino groups. Bacterial metabolism utilizes all these enzymes.

Since enzymes are protein molecules, they are sensitive to any physical or chemical agents that injure proteins. Heat can be used to kill bacteria because heat alters the tertiary structure of enzyme proteins. Certain chemicals, such as alcohol and phenol, precipitate enzyme proteins and therefore act as disinfectants. Iodine compounds kill bacteria by combining with certain amino acids in enzyme molecules. And any antibiotic that interferes with protein synthesis automatically interferes with enzyme production.

Some enzymes are made up entirely of protein. An example is **lysozyme**, the enzyme in human tears and saliva that digests the cell walls of Gram-positive bacteria. Lysozyme is composed of 129 amino acids. Other enzymes contain a nonprotein part such as an ion of magnesium, iron, or zinc. Ions such as these are called **cofactors**. When the nonprotein part is an organic molecule, biochemists refer to it as **coenzyme**. Examples of coenzymes include nicotinamide adenine dinucleotide (NAD) and flavin adenine dinucleotide (FAD), as well as coenzyme A. These coenzymes play a significant role in metabolism, and we shall encounter them presently in our study.

Energy and ATP

In bacteria and other living cells, molecules are constantly moving about, a factor that often leads to chemical reactions. Certain chemical reactions yield energy, but in many cases, energy must be supplied for reactions to take place. This is because covalent bonds are forced apart into new combinations. Once a reaction has begun, however, it often gives off enough energy to keep itself going. It may even liberate excess energy in the form of heat or light (MicroFocus: 5.1), or the energy may cause another chemical reaction to occur.

Disinfectant:
a chemical substance used to control microorganisms on a lifeless object.

li′so-zīm

nik″o-tin′ah-mid
di-new′kle-o-tide

Covalent bond:
the force resulting from the sharing of electrons among the atoms in a molecule.

5.1
MICROFOCUS

GLOWING IN THE DARK

If you chanced to trip over a half-rotten log in the woods at night, you might be surprised to see that the exposed parts of the log glow with a strange, eerie light. This fascinating glow is due to a natural emission of light called bioluminescence. One of nature's wondrous beauties, it is caused by a number of plants and animals and, in the case of the log, a bevy of fungi and bacteria.

In 1887 the French physiologist Raphael Dubois studied bioluminescence in sea clams and isolated a substance he named luciferin, after Lucifer, the light-bearer. Dubois demonstrated that luciferin would continue to emit light in cold water for several minutes before fading out. However, the light could be restored by adding a fresh extract of clam. Early in this century, E. Newton Harvey confirmed the report and

offered that bioluminescence was a process requiring not only luciferin but also an enzyme called luciferase. He showed that powder from dried clams, containing both luciferin and luciferase, could be held in the hand and, with a few drops of water, would produce enough light to read a map or letter. In World War II, the Japanese adapted his finding in the field to provide low-intensity light.

It is now known that bioluminescence results from the direct conversion of chemical energy to light energy. Studies of fireflies indicate that the chemical energy is derived from adenosine triphosphate (ATP), as luciferase reacts with luciferin in the presence of oxygen. Studies of bacteria reveal a somewhat different process, in which the role of ATP is still uncertain, but

it is clear that oxygen is a prerequisite for the process.

One dramatic example of luminous bacteria is seen in a group of fish popularly called "flashlight fish." These fish, common in the Gulf of Eilat in the Red Sea, maintain populations of luminescent bacteria in organs under their eyes. The bacteria continuously emit light, but the fish can control the light simply by sliding a darkened lid over the bacteria. Light flashes may be useful in sexual mating rituals, locating food sources, warding off predators, or communicating with other fish. The relationship benefits bacteria, which derive nutrients from their host, and fish, which enjoy the advantages of a light-emitting source that requires no batteries.

Enzymes play a key role in metabolism because they lower the amount of energy required for a reaction to take place. They assist in the destruction of chemical bonds and the creation of new ones by separating or joining atoms in a carefully orchestrated fashion. Thus, by speeding up the reaction, they lower the necessary energy input, the so-called **activation energy**. The reaction could probably occur without enzymes, but far less efficiently.

In the biochemical reactions of metabolism, enzymes are generally supplemented by the chemical energy available in a compound called **adenosine triphosphate**, or simply **ATP** (Figure 5.2). A molecule of ATP acts like a portable battery. It moves to any part of the cell where an energy-consuming reaction is taking place and it provides energy. In a bacterium, ATP supplies energy for binary fission, flagellar motion (Figure 5.2), and spore formation. On a more chemical level, it fuels protein synthesis and carbohydrate breakdown. Scientists believe that a major share of bacterial functions depend on a supply of ATP. When the supply is exhausted, the cell usually dies.

The energy in ATP molecules is released by breaking the high-energy bond holding the last phosphate group onto the molecule, thereby producing adenosine diphosphate (ADP) and a phosphate group. An enzyme called **adenosine triphosphatase (ATPase)** catalyzes this reaction. The energy formerly locked up in the bond is now set free to do work. A single mole of ATP releases 7300 calories of energy when its bonds are broken. (A mole of ATP weighs 507 grams.)

Although ATP molecules are used everywhere in a bacterium to meet energy needs, they are not suitable for storing energy. The molecules are large and bulky, and any significant surplus takes up too much space in a cell. Therefore cells synthesize or obtain small molecules such as glucose or lipids for energy storage. Later, the energy in these molecules can be released in catabolism and used to reform ATP from adenosine diphosphate (ADP) and phosphate. The new ATP is then used to drive the reactions of metabolism and any other activities of the bacterium. This principle is the basis for much of the chemistry in the discussion to follow.

Calorie:
a unit of energy defined as the amount of heat required to raise one kilogram of water 1° C.

Mole:
the quantity of a substance whose weight in grams is numerically equivalent to its molecular weight.

TO THIS POINT
■

We have introduced the concept of metabolism as the sum total of all the biochemical processes taking place in bacteria and other living cells. Metabolism is subdivided into two major categories: anabolism, which implies synthesis of organic compounds, and catabolism, which implies breakdown of these compounds. For anabolism and catabolism to take place, enzymes must be available. We therefore explored these vital proteins in depth with emphasis on their activity, specificity, inhibition, and chemical makeup.

We also discussed the importance of energy as a governing factor in metabolism. Energy is required to assist the construction and destruction of chemical bonds. For the reactions of metabolism to take place, the most important energy source is adenosine triphosphate (ATP). This compound is not stored in the cell, and as it is used up, it must be resynthesized utilizing the energy present in molecules such as glucose and lipid. In the next section we shall follow the chemical events in which glucose energy is released and converted to ATP energy. As we proceed through this involved biochemistry, try to keep in mind that the ultimate goal of the process is to form ATP molecules.

FIGURE 5.2

Energy and Movement in a Bacterium

(1) The structure and activity of adenosine triphosphate (ATP). (a) The ATP molecule is composed of adenine and ribose (adenosine) bonded to three phosphate groups. (b) Energy is released from the molecule during its decomposition to adenosine diphosphate and a phosphate ion.

(a)

High-energy bonds

Adenine — Ribose — Adenosine

Phosphate Phosphate Phosphate — Triphosphate

(b)

$$\text{Adenosine} - \overset{O}{\underset{OH}{P}} - O \sim \overset{O}{\underset{OH}{P}} - O \sim \overset{O}{\underset{OH}{P}} - OH \xrightarrow[\text{ATPase}]{H_2O} \text{Adenosine} - \overset{O}{\underset{OH}{P}} - O \sim \overset{O}{\underset{OH}{P}} - OH + - \overset{O}{\underset{OH}{P}} - OH \quad + 7300 \text{ calories energy per mole}$$

Adenosine triphosphate   Adenosine diphosphate   Phosphate ion

(2) A scanning electron micrograph of the Gram-negative rod *Proteus mirabilis*. This very large cell contains hundreds of flagella, and it requires great quantities of energy to obtain the motion they impart. Much of that energy is supplied by ATP. Bar = 5 μm.

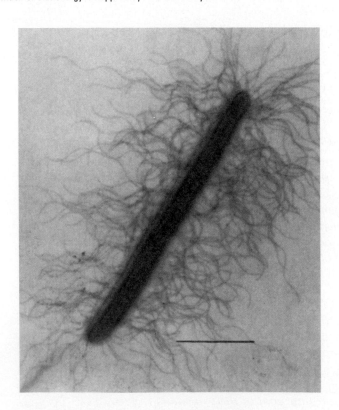

Catabolism of Glucose

One of the most thoroughly studied and best-understood aspects of metabolism is the catabolism of **glucose**. Since the early part of this century, the chemistry of glucose catabolism has been the subject of intense investigation by biochemists because glucose is a principal source of energy for ATP production. Moreover, the process of glucose catabolism is essentially similar in myriad living things, be they bacterium, plant, animal, or human being. We shall therefore give close scrutiny to the process.

A mole of glucose (180 grams) contains about 690,000 calories of energy. This fact can be demonstrated in the laboratory by setting fire to a mole of glucose and measuring the energy released. In a bacterium, however, not all the energy is set free from glucose, nor can the bacterium trap all that is released. Instead, a bacterium traps 277,400 calories of the energy in 38 moles of ATP formed during glucose catabolism (38 moles of ATP $\times$ 7300 cal per mole $=$ 277,400 calories). In the discussions that follow, we shall see how these 38 moles of ATP emerge. You might make note of the number "38," for it is the key number in this paragraph. The process accounts for the transfer of about 40 percent of the glucose energy to ATP energy. To simplify matters we shall use the word "molecule" instead of "mole" and follow the fate of a glucose molecule.

The catabolism of a glucose molecule does not take place in one chemical reaction, nor do 38 molecules of ATP form all at once. Instead, the process involves one or more metabolic pathways. A **metabolic pathway** is a sequence of chemical reactions, usually catalyzed by enzymes, in which the product of one reaction serves as a substrate for the next reaction. Should a reaction in the sequence require energy, then a side reaction often takes place to supply the energy. Also, if a reaction happens to yield excess energy, then a side reaction may take place to trap and preserve the energy. Thus a metabolic pathway involves a sequence of reactions, many of which are coupled with side reactions. You will note an abundance of these instances as we proceed.

Any series of biochemical reactions in which energy is liberated is referred to as **respiration**. Therefore glucose catabolism is a form of respiration. In this text we shall use the terms "respiration" and "catabolism" interchangeably. Respiration may occur in the presence of oxygen, in which case it is **aerobic respiration**. In other instances, it may take place in the absence of oxygen, in which case it is called **anaerobic respiration**. Glucose catabolism is known to take place by both methods.

To begin our study of glucose catabolism, we shall follow the process of aerobic respiration as it occurs in bacteria. There are numerous metabolic pathways for aerobic respiration, but the one we shall discuss is represented by the following chemical formula.

$$C_6H_{12}O_6 + 6\ O_2 + 38\ ADP + 38\ P \longrightarrow 6\ CO_2 + 6\ H_2O + 38\ ATP$$

Glucose   Oxygen   Carbon   Water
dioxide

This straightforward equation summarizes a complex series of metabolic reactions conveniently divided into three processes: glycolysis, the Krebs cycle, and oxidative phosphorylation. We shall examine each in turn.

Glycolysis

The chemical breakdown of glucose is called **glycolysis**, from *glyco-*, referring to glucose, and *lysis*, meaning to break. Several metabolic pathways for this breakdown exist. The best known pathway, and the one we shall follow, is called the **Embden–Meyerhof pathway**, after Gustav Embden and Otto Meyerhof,

Glucose:
a 6-carbon carbohydrate important in energy metabolism.

ATP:
adenosine triphosphate, a high-energy molecule that serves as an immediate energy source for cells.

Metabolic pathway:
a sequence of chemical reactions in which the product of one reaction serves as the substrate for the next reaction.

Respiration:
a series of biochemical reactions in which energy is liberated.

fos"for-ĭ-la'shun

gli-kol'ĭ-sis

two German biochemists who described many of its details in the 1930s. The process occurs in the cytoplasm of bacteria and involves the conversion of glucose to a 3-carbon organic acid called pyruvic acid. Between glucose and pyruvic acid there are nine chemical steps, and even though glycolysis is part of the overall scheme of aerobic respiration, it takes place in the absence of oxygen. Each step in the process is catalyzed by a specific enzyme. Figure 5.3 displays the process. For easy referral, numbers have been inserted in the figure for each step, and it would be helpful to refer to the figure as the discussion proceeds.

In glycolysis one 6-carbon glucose molecule is eventually converted into two 3-carbon **pyruvic acid** molecules. For the conversion to take place, two molecules of ATP must be supplied. Note in Figure 5.3 that one molecule of ATP is used up in reaction (1) at the beginning of glycolysis and that a second ATP molecule is needed for reaction (3). In both cases, the phosphate group from ATP attaches to the substrate molecule. Thus, reaction (1) gives us glucose-6-phosphate, and reaction (3) yields fructose-1-6-diphosphate ("di" for two phosphate molecules).

An important split occurs in reaction (4). The fructose-1-6-diphosphate molecule breaks apart to yield two molecules, each with three carbons. One is dihydroxyacetone phosphate (DHAP); the other is **phosphoglyceraldehyde (PGA)**. Note that an enzyme converts the DHAP molecule to another PGA molecule. This is important because we now have two molecules of PGA.

In the next series of steps, each PGA molecule passes through a series of conversions and ultimately forms pyruvic acid (Figure 5.3). These conversions occur in steps (5) through (9). A significant event takes place in reaction (6). As the enzyme conversion proceeds, enough energy is released to synthesize an ATP molecule from ADP and phosphate in a side reaction. This chemistry happens again in reaction (9). Thus, each time a PGA molecule is broken down to pyruvic acid through the sequence, two ATP molecules are formed. But two PGA molecules are available. Therefore as the second PGA molecule breaks down, two more ATP molecules result. The total is four molecules of ATP. Considering that we "invested" two ATP molecules in reactions (1) and (3) and received back four molecules, the net gain from glycolysis is two molecules of ATP.

Before we proceed, it is well to take note of reaction (5). In this reaction, two high-energy electrons and two protons are released and shuttled to NAD. This event will have great significance shortly. Also of interest is the fact that, as a result of glycolysis, the cell has gained two ATP molecules for use in its metabolism even though no oxygen has been utilized. This has importance in fermentation reactions to be outlined later. For the moment, we shall follow the fate of pyruvic acid.

The Krebs Cycle

The **Krebs cycle** is a series of chemical reactions named for Hans A. Krebs, who won the 1953 Nobel Prize for the discovery of several participating substances. It is referred to as a cycle because the substance formed at the end of the sequence is identical to the substance at the beginning. All the reactions are catalyzed by enzymes and take place along the cell membranes of bacteria. In eukaryotic organisms such as protozoa and humans, the reactions occur in the mitochondria.

The Krebs cycle is somewhat like a wheel constantly turning. Each time the wheel comes back to the starting point, something must be added to spin it for another rotation. That something is a pyruvic acid molecule derived from glycolysis. Figure 5.4 displays the Krebs cycle. Letters are used in the figure to guide us through its reactions.

Pyruvic acid:
a 3-carbon molecule that results from the breakdown of glucose in glycolysis.

fos"fo-glis"er-al'dĕ-hīd

Proton:
the positively charged nucleus of a hydrogen atom.

NAD:
an acronym for the coenzyme nicotinamide adenine dinucleotide.

FIGURE 5.3

Glycolysis by the Embden–Meyerhof Pathway

Carbon atoms are represented by circles. The dark circles represent carbon atoms bonded to phosphate groups. ATP is supplied to glucose in reaction (1) and to fructose-6-phosphate in reaction (3). The splitting of F-1-6-diphosphate in reaction (4) yields PGA and DHAP. DHAP then converts to another PGA molecule. Both PGA molecules proceed through reactions (5) to (9) and yield two molecules of pyruvic acid. Two ATP molecules are generated in reaction (6) and two more in reaction (9). The total ATP produced is four molecules and, since two are used up in glycolysis, the net gain is two molecules of ATP. Also, in reaction (5), high-energy electrons and protons are captured by an NAD molecule. Note that the original six carbon atoms of glucose exist in two pyruvic acid molecules.

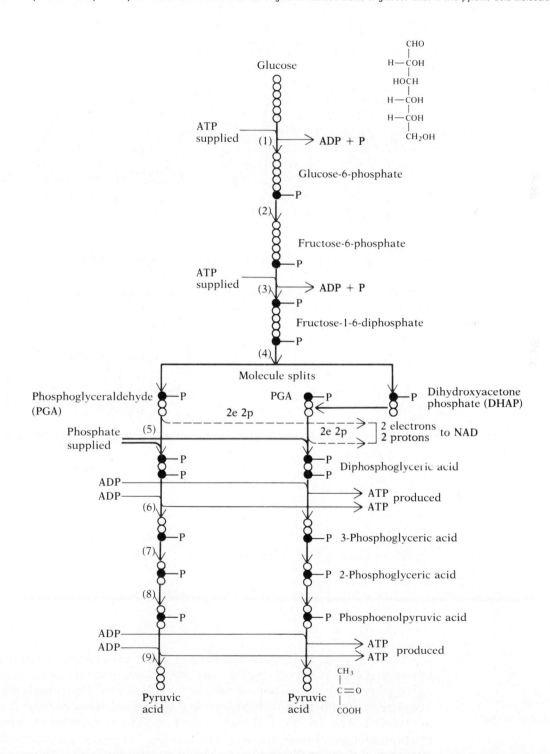

FIGURE 5.4

The Krebs Cycle

Pyruvic acid from glycolysis combines with coenzyme A to form acetyl-CoA in reaction (A). This molecule then condenses with oxaloacetic acid to form citric acid (B). In reactions (C) through (F), citric acid is converted to alphaketoglutaric acid and then to succinic, fumaric, and malic acids in succession. Malic acid regenerates oxaloacetic acid in the last reaction, (G). During the process, the three carbons of pyruvic acid are liberated as three molecules of carbon dioxide. ATP is formed in reaction (D), and high-energy electrons and protons are liberated during certain of the reactions. These are captured by NAD or FAD molecules, and the electrons are utilized for oxidative phosphorylation.

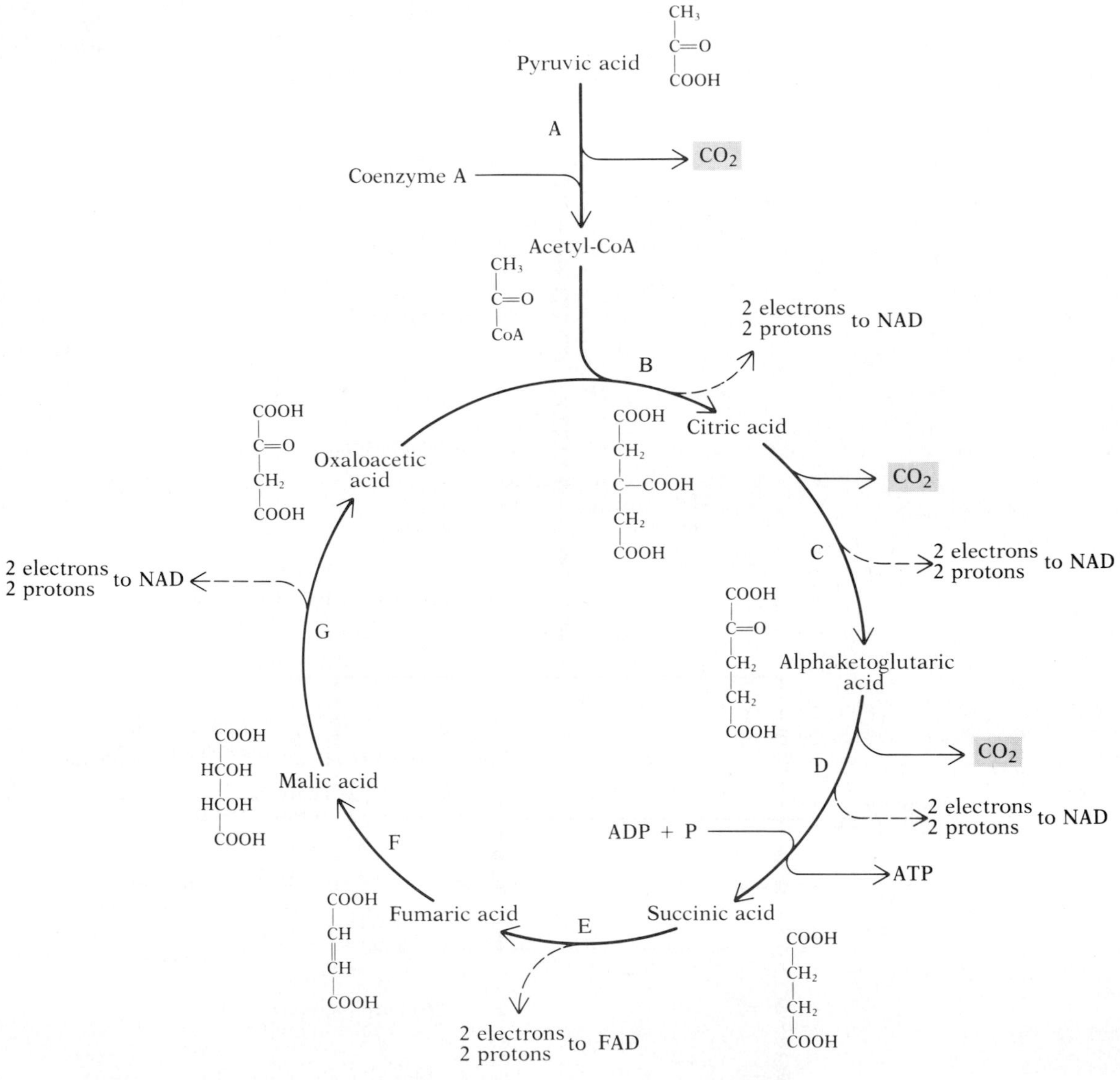

Before pyruvic acid enters the Krebs cycle it undergoes a change indicated in reaction (A). An enzyme removes a carbon atom from the pyruvic acid molecule and releases the carbon as carbon dioxide (CO_2). The remaining two carbons are then combined with coenzyme A to form acetyl-coenzyme A, or simply **acetyl-CoA**. The discovery of coenzyme A by Fritz A. Lipmann provided a key to the understanding of the Krebs cycle. For his work, Lipmann shared the 1953 Nobel Prize in Physiology or Medicine with Krebs.

Coenzyme:
an organic molecule that is the nonprotein part of an enzyme.

The two remaining carbons from pyruvic acid are now ready to enter the Krebs cycle. In reaction (B), acetyl-CoA unites with the 4-carbon substance oxaloacetic acid to form **citric acid**, a 6-carbon organic acid. Citric acid may be familiar as a component of soft drinks. The citric acid molecule changes in reaction (C) to a substance called alphaketoglutaric acid. Note in Figure 5.4 that alphaketoglutaric acid has only five carbon atoms. The sixth atom has been lost as CO_2. In the next step, reaction (D), alphaketoglutaric acid is converted by an enzyme to succinic acid, which has four carbon atoms. The fifth carbon atom is lost as CO_2. Succinic acid then changes to fumaric acid, which then becomes malic acid, and malic acid converts to oxaloacetic acid (reactions E, F, and G). The cycle is now complete, and oxaloacetic acid is ready to unite with another molecule of acetyl-CoA.

ox-ah'lo-acetic

al"fah-ke"to-gloo-tar'ik

suk-sĭ'nik

ox-ah'lo-acetic

Several features of the Krebs cycle merit closer scrutiny. First we shall follow the carbon. Pyruvic acid, with three carbon atoms, entered the cycle, but after one turn, the carbon atoms exist in three molecules of CO_2. There were two molecules of pyruvic acid from glycolysis, so when the second molecule enters the Krebs cycle, its carbon atoms will also form three CO_2 molecules. Remember that we began with a 6-carbon glucose molecule, and now six CO_2 molecules have been produced. This fulfills part of the equation for aerobic respiration:

$$C_6H_{12}O_6 + 6\ O_2 + 38\ ADP + 38\ P \longrightarrow 6\ CO_2 + 6\ H_2O + 38\ ATP$$

The second feature of the Krebs cycle that draws our attention is reaction (D). Here an ATP molecule forms in the conversion to succinic acid because the reaction is an energy-yielding conversion. Since we have two pyruvic acid molecules entering the cycle (per molecule of glucose), a second ATP molecule will form when the second pyruvic acid passes through the cycle. Combining the two ATPs with the net gain of two ATPs from glycolysis, the total gain rises to four ATPs. Impressive as this is, the major gain of ATP is still to come.

Oxidative Phosphorylation

Oxidative phosphorylation refers to a sequence of reactions in which two events happen: pairs of electrons are passed from one chemical substance to another, and the energy released during the passage is used to combine phosphate with ADP to form ATP. The adjective "oxidative" is derived from the term "oxidation," which refers to the loss of electron pairs from chemical molecules in the sequence; the noun "phosphorylation" implies the union of phosphate with ADP (to phosphorylate a molecule is to add phosphate to it). Like the Krebs cycle, oxidative phosphorylation takes place at the cell membrane in bacteria and the mitochondria of eukaryotic cells.

fos"for-ĭ-la'shun

Oxidation:
a chemical reaction in which electrons are lost from the reactant.

Oxidative phosphorylation is the place where the major percentage of ATP molecules forms. Indeed, 34 of the 38 molecules produced during glucose catabolism are manufactured here. The overall sequence begins with glycolysis and the Krebs cycle. In these processes, high-energy electrons and protons are released during several reactions. During **oxidative phosphorylation**, the electrons pass among a series of chemical molecules, and the energy in the electrons is lost. But the energy is not lost in the sense that it is gone forever. Instead, the energy is used to combine ADP molecules with phosphate molecules to yield ATP molecules. Each time a pair of electrons passes through the system, enough energy is released to generate three ATP molecules.

Two important coenzymes that function in oxidative phosphorylation are NAD and FAD. Another important group of molecules are the **cytochromes**. Cytochromes are a set of cellular pigments (cyto-"chrome") containing iron

FAD:
an acronym for the coenzyme flavin adenine dinucleotide.

ions that accept and release electrons during the sequence. Cytochromes are designated by letters, as shown in Figure 5.5. The last link in the chain is oxygen. Because oxidative phosphorylation involves the passage of electrons, the process is often called **electron transport**.

FIGURE 5.5

Oxidative Phosphorylation

Reactions in glycolysis and the Krebs cycle yield two protons and two high-energy electrons. The electrons and one proton are captured by NAD and then passed to FAD. The FAD also acquires a second proton from the cytoplasm. Electrons are then passed to cytochrome Q, and the protons are liberated to the medium. Cytochrome Q passes the electrons to cytochromes B, C, A, and A_3, and the electrons are finally captured by an oxygen atom. The oxygen acquires two protons from the cytoplasm and becomes a water molecule (H_2O). During the process, energy is liberated at various points, and three ADP molecules combine with three phosphate ions to form three molecules of ATP.

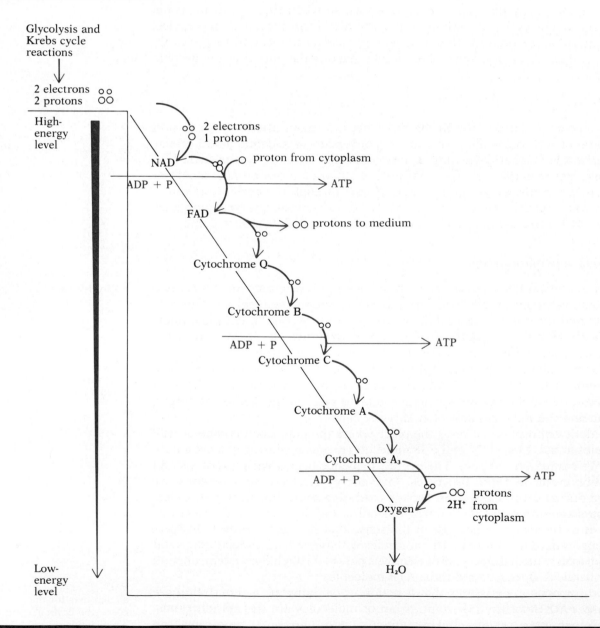

When oxidative phosphorylation is in operation (Figure 5.5), two electrons and one proton are given off to NAD during selected metabolic reactions. The NAD now becomes $NADH^+$. (A second proton is liberated to the cytoplasm.) The electron pair then passes to an FAD molecule, which also takes on two protons to become $FADH_2$. This molecule passes the electron pair to the cytochrome chain and releases the protons to the cytoplasm. The release regenerates FAD, which can then accept more electrons. The cytochromes now pass the electron pair along to one another until the electron pair is finally accepted by an oxygen atom. Now the **oxygen atom** acquires two protons ($2H^+$) from the cytoplasm and becomes water (H_2O). Oxygen's role is of great significance because if oxygen were not present, there would be no way for cytochromes to unload their electrons, and the entire system would soon come to a halt. This role is also reflected in the equation for aerobic respiration:

$$C_6H_{12}O_6 + \mathbf{6\ O_2} + 38\ ADP + 38\ P \longrightarrow 6\ CO_2 + \mathbf{6\ H_2O} + 38\ ATP$$

We have noted that for every transport of an electron pair along the sequence, three ATP molecules are formed. The actual mechanism for ATP formation is called **chemiosmosis** because it involves both chemical and transport ("osmosis") processes. First proposed by Nobel Prize winner Peter Mitchell, chemiosmosis uses the power of proton motion across a membrane to generate energy for ATP synthesis.

What happens in chemiosmosis is this: as electrons move between the coenzymes and cytochromes, significant amounts of energy are released at three transition points (Figure 5.6). The energy powers the pumping of protons from inside the bacterial cytoplasm across the cell membrane to the outside of the membrane (the energy is the so-called "proton motive force"). Soon a large number of protons have built up outside the membrane, and because they cannot easily reenter the cell, they represent a large concentration of potential energy (much like a boulder at the top of a hill). The protons are positively charged, so there is also a buildup of charges outside the membrane.

Suddenly a series of channels open and the proton flow reverses. Each "channel" is apparently a protein pore lined with molecules of a large enzyme complex called ATP synthetase. This enzyme complex has binding sites for ATP and ADP. As the protons rush through the pore, they release their energy, and the energy is used to synthesize ATP molecules from ADP and phosphate ions. Usually three molecules of ATP can be synthesized for each pair of electrons whose transport began the process.

Chemiosmosis occurs only in structurally intact membranes. Indeed, if the membrane is damaged so proton movement cannot take place, the synthesis of ATP ceases even though electron transport through the cytochrome system continues. With the end of ATP production, the organism rapidly dies. This is one reason why damage to the bacterial membrane, such as with antibiotics or detergent antiseptics, is so harmful to a bacterium.

At this point we shall see how the electron pairs are supplied for oxidative phosphorylation and chemiosmosis. Reference to the figures for glycolysis (Figure 5.3) and the Krebs cycle (Figure 5.4) would be helpful to follow the development of thought.

Note in the Krebs cycle that reactions (B) and (C) yield electron pairs and protons to NAD and that reactions (D) and (G) do likewise. In each case, oxidative phosphorylation takes place. At a rate of three ATPs per reaction, the four reactions (B, C, D, and G) give us a total of 12 ATPs. In reaction (E), the electron pair and protons are passed to FAD directly. In this situation, only two ATPs result by oxidative phosphorylation. This brings the total to 14 ATPs per turn of the Krebs cycle.

NAD:
an acronym for the coenzyme nicotinamide adenine dinucleotide.

kem"e-os-mo'sis

Chemiosmosis:
a biochemical process in which energy from electrons powers the movement of protons across the bacterial membrane, a process that leads to ATP formation.

FIGURE 5.6

Chemiosmosis in Bacteria

(a) A coenzyme (e.g., NAD) from the Krebs cycle and/or glycolysis transports electron pairs to cytochromes in the cell membrane. (b) The coenzyme is regenerated for reuse. (c) As cytochromes transport the electron pairs among themselves, the energy released fuels the transport of protons across the cell membrane at three points. (d) Each set of protons then reenters the cytoplasm of the cell through a protein channel lined with ATP synthetase. (e) An ADP molecule is joined with a phosphate each time a set of protons moves through the channel, thereby accounting for three ATP molecules produced for each electron pair. (f) The electrons combine with other protons to form water molecules.

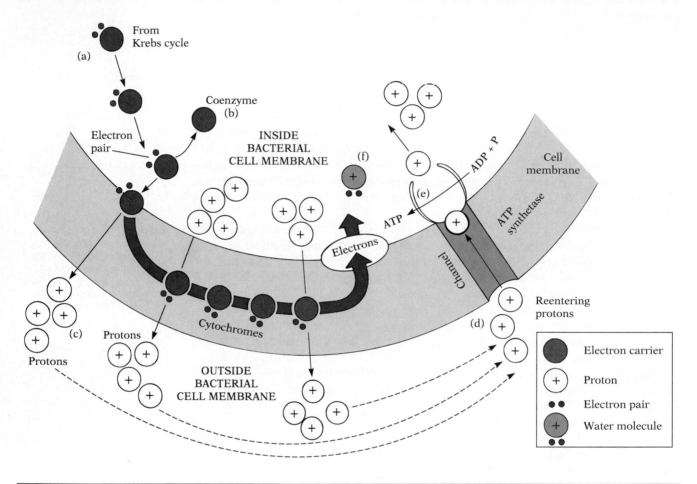

But there are two turns of the cycle because two molecules of pyruvic acid come from glucose in glycolysis. Therefore we gain another 14 ATPs as the second turn of the cycle takes place. We thus arrive at a total of 28 ATPs from the Krebs cycle.

gli-kol'ĭ-sis

Our attention now turns to glycolysis (Figure 5.3). Reaction (5) of this process was highlighted previously. The reaction yields an electron pair and a proton that proceed to NAD. Oxidative phosphorylation takes place and the result is three more ATPs. But reaction (5) takes place two times because two PGA molecules are formed. Therefore with the second reaction, another oxidative phosphorylation occurs and three more ATPs form. This gives us a total of six from the reactions originating in glycolysis.

We may summarize the ATP yield as follows:

2 ATPs from the net gain in glycolysis
2 ATPs from reaction (D) in the Krebs cycle
14 ATPs from the first turn of the Krebs cycle
14 ATPs from the second turn of the Krebs cycle
6 ATPs from reaction (5) in glycolysis

The grand total of ATPs is therefore 38 molecules of ATP. At 7300 calories per mole of ATP, this yields 277,400 calories of energy preserved from the energy in glucose. It also completes the equation for aerobic respiration.

$$C_6H_{12}O_6 + 6\ O_2 + \textbf{38 ADP} + \textbf{38 P} \longrightarrow 6\ CO_2 + 6\ H_2O + \textbf{38 ATP}$$

TO THIS POINT ■

We have discussed the catabolism of glucose by exploring the biochemical reactions of glycolysis, the Krebs cycle, and oxidative phosphorylation. In these processes, glucose is first converted to pyruvic acid, which then feeds into the Krebs cycle. The carbon of glucose is ultimately released as carbon dioxide. In addition, ATP molecules are formed from reactions that accompany the main metabolic pathway. The major percentage of ATP, however, results from oxidative phosphorylation as electrons from glycolysis and Krebs cycle reactions pass among coenzymes and cytochromes and eventually wind up in oxygen atoms. The energy released during the passages is used to form ATP. We saw that a total of 38 ATP molecules can be formed from the energy in a single glucose molecule through the reactions in metabolism.

Glucose is fundamental to the life of a bacterium because it provides the energy for chemical activities. However, glucose is not the only chemical substance used for energy. Modern biochemists have found that numerous other carbohydrates, as well as proteins and fats, contribute to the energy metabolism of bacteria. In the next section we shall see how these compounds are processed. In addition, we shall study mechanisms by which certain microorganisms obtain their energy even though oxygen is not available as an electron acceptor.

Other Aspects of Catabolism

The catabolism of glucose is a complex process and, for the student of microbiology, it is a demanding aspect of chemistry to learn and understand. However, the process is central to the metabolism of bacteria as well as a great many other living things (MicroFocus: 5.2), and it provides a glimpse of how living things obtain energy for life. Moreover, the process of glucose catabolism is a main thoroughfare to which many other biochemical pathways lead and from which numerous pathways extend. In this section we shall examine how cells obtain energy from various other organic substances by incorporating them into the process of glucose catabolism. We shall also see how modifications of the basic scheme account for the utilization of glucose in the absence of oxygen. The use of one major pathway with multiple offshoots lends a sense of economy to the metabolism of a cell.

Catabolism of Other Carbohydrates

Polysaccharide:
a carbohydrate composed of multiple units of monosaccharides.

A wide variety of monosaccharides, disaccharides, and polysaccharides serve as useful energy sources for bacteria and other living cells. All of these carbohydrates undergo a series of preparatory conversions before they are processed in glycolysis, the Krebs cycle, and oxidative phosphorylation.

In preparation for entry into the scheme of metabolism, different carbohydrates utilize different pathways. **Sucrose**, for example, is first digested by the enzyme sucrase into its constituent molecules, glucose and fructose. The glucose molecule enters the glycolysis pathway, but the fructose molecule must first be converted to fructose-1-phosphate. The latter then undergoes further conversions and a molecular split before it enters the scheme as dihydroxyacetone phosphate (DHAP). **Lactose**, another disaccharide, is broken in two by the enzyme lactase to glucose and galactase. Glucose enters the pathway, as before, but galactase undergoes a series of changes before it is ready to enter glycolysis in the form of glucose-6-phosphate.

Galactase:
an isomer of glucose bound to glucose in lactose.

Glycogen:
an animal tissue polysaccharide composed of glucose units.

Polysaccharides also undergo a system of changes before entering the mainstream of glycolysis. Starch and glycogen, for instance, are metabolized by removing one glucose unit at a time and converting the glucose molecule to glucose-1-phosphate. An enzyme converts this compound to glucose-6-phosphate, ready for entry to the glycolysis pathway. We shall not examine the intricate details of these processes, but most biochemistry books explain the conversions in detail. The main point is that carbohydrates other than glucose can be used just as glucose is used for energy metabolism.

Catabolism of Proteins and Fats

The economy of metabolism continues when we consider protein and fat catabolism. Although proteins are generally not considered energy sources, cells utilize them for energy when carbohydrates and fats are in short supply. Fats, by contrast, are extremely valuable energy sources because their chemical bonds contain enormous amounts of chemical energy. There is so much energy, in fact, that when a bacterium has excessive amounts of carbohydrates, its enzymes convert the carbohydrates to fats to store the energy for later use.

Both proteins and fats use the pathway of glucose catabolism as well as many other smaller pathways. Basically, the proteins and fats undergo a series of enzyme-catalyzed conversions and form components normally occurring in carbohydrate metabolism. These components then continue along the scheme as if they originated from carbohydrates. **Proteins**, for example, are broken down to amino acids. Enzymes subsequently convert many of the amino acids to pathway components by removing the amino group and substituting a hydroxyl group. This process is called **deamination**. Alanine, for example, is converted to pyruvic acid, and aspartic acid is converted to oxaloacetic acid. For certain amino acids the process is more complex than a straightforward deamination, but the end result is the same: the amino acids become pathway intermediaries and are metabolized to produce energy in ATP.

de-am″ĭ-na′shun

Fats consist of one or more fatty acids bonded to a glycerol molecule. To be useful for energy purposes, the fatty acids must first be separated from the glycerol by the enzyme lipase. Once this has taken place, the glycerol portion is converted to DHAP. For fatty acids, there is a complex series of conversions called the **fatty acid spiral**. In this process, each long-chain fatty acid is broken by enzymes into 2-carbon units, and enzymes convert each unit to a molecule of acetyl-CoA ready for the Krebs cycle. For each turn of the Krebs cycle, we previously noted that 14 molecules of ATP are derived through oxidative phosphorylation. A quick calculation should illustrate the substantial output of energy obtained from a 16-carbon fatty acid. This is why fats are so important as energy sources.

Fatty acid spiral:
a series of biochemical reactions in which fatty acids are converted to 2-carbon units.

Anaerobic Respiration

Certain bacteria metabolize carbohydrates through **anaerobic respiration**, a process in which oxygen is not used as an electron acceptor. But even without oxygen, the mechanism of electron transport takes place by oxidative phosphorylation. Instead of oxygen, anaerobic bacteria use some inorganic molecule as a final electron acceptor. For example, *Escherichia coli* uses **nitrate ions** (NO_3^-) at the end of the cytochrome chain when it metabolizes anaerobically. Electrons combine with the nitrate ions and convert them into nitrite ions (NO_2^-). Microbiologists take advantage of this chemistry in a laboratory diagnostic test for identifying nitrite producers.

esh″er-i′ke-a

Members of the genus *Desulfovibrio* use **sulfate ions** (SO_4) for anaerobic respiration. The sulfate combines with electrons and changes to hydrogen sulfide (H_2S). This gas gives a rotten egg smell to the environment (such as mud). A final example is exhibited by members of the genera *Methanobacterium* and *Methanococcus*. These bacteria use **carbon dioxide** as an electron acceptor and form large amounts of methane (CH_4). Some scientists believe that because of this chemistry, methane-producing bacteria existed on Earth when the only available gas was carbon dioxide. They postulate that methane may actually have entered the atmosphere for the first time through the activity of these bacteria. Methanogens of the Archaeobacteria (Chapter 4) are typical.

de-sul″fo-vib′re-o

meth-an″o-bak-te′re-um
meth-an″o-kok′us

Fermentation

The chemical process of **fermentation** (Figure 5.7) is a type of anaerobic respiration because it does not use oxygen as a final electron acceptor. Fermentation is a unique process because an organic molecule, usually an

Fermentation:
an anaerobic form of metabolism in which an intermediary in the process acts as an electron acceptor.

FIGURE 5.7

The Relationship of Fermentation to Glycolysis

In reaction (5) of glycolysis, two electrons and two protons are generated. The electrons and one proton are captured by NAD molecules to form NADH. Intermediaries in the process are then used to remove the electrons and proton from the NADH. (a) In yeast cells, the pyruvic acid is converted to acetaldehyde, which then acquires the electrons and proton to become ethyl alcohol. (b) In *Streptococcus* species, the pyruvic acid is converted to lactic acid using the electrons and proton from NADH. In both cases, NAD is regenerated to function again in reaction (5).

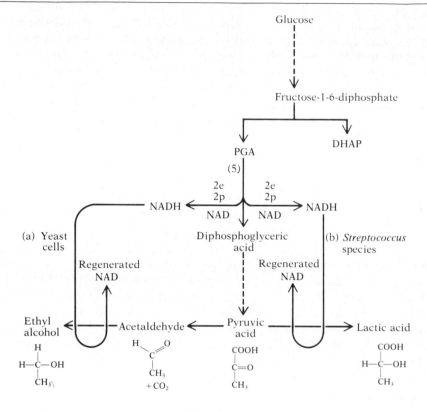

Streptococcus lactis:
a Gram-positive chain of cocci important in many industrial fermentations.

sak"ah-ro-mi′sēz

as"et-al′dĕ-hīd

intermediary of the chemistry, accepts the electrons. For example, in the fermentation of glucose by certain bacteria and yeasts, an intermediary accepts the electrons and proton from NADH formed in reaction (5) of glycolysis. This regenerates the NAD molecules for reuse as electron acceptors. NAD exists in limited supply in the cytoplasm and must be continually regenerated so that glycolysis may proceed. (When oxidative phosphorylation is taking place, the NAD is regenerated by giving up its electrons and proton to FAD.)

The bacterium *Streptococcus lactis* practices fermentation by using pyruvic acid to accept the electrons and proton from NADH. An enzyme reaction converts the pyruvic acid to lactic acid in the process. In a dairy plant, the metabolism is carefully controlled to make buttermilk from fresh milk (Figure 5.8).

The fermentation chemistry in yeasts such as *Saccharomyces* is somewhat different because yeasts contain a different enzyme. In these cells the pyruvic acid is first converted to acetaldehyde, a process in which carbon dioxide evolves. Acetaldehyde then serves as an acceptor for the electrons and proton of NADH, and it changes to ethyl alcohol. The liquor industry uses the ethyl alcohol produced in fermentation to make alcoholic beverages such as beer and wine (Chapter 25). Yeast fermentation of carbohydrates to alcohol may also take place in the human body, as explained in MicroFocus: 5.3.

The energy benefits to *Streptococcus lactis* and yeasts are far less in fermentation than they are in aerobic respiration. The only ATP that evolves are the two molecules resulting from glycolysis. This is in sharp contrast to the 38 molecules evolving in aerobic respiration. It is clear that aerobic respiration is the better

FIGURE 5.8

A Scanning Electron Micrograph of a Species of *Streptococcus*

A species of *Streptococcus* called *S. lactis* is used in industrial processes to produce dairy products by the fermentation of milk.

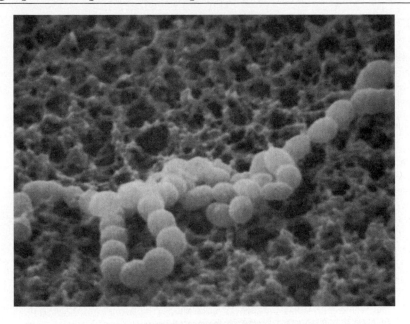

choice for energy production, but under the circumstances of an oxygen-free environment, there is little alternative if life for *Streptococcus lactis* or *Saccharomyces* (yeast) is to continue.

In the food industry, fermentation results in a broad variety of useful products. Vinegar, for instance, is a fermentation product of *Acetobacter* species. Swiss cheese develops its flavor partly from the propionic acid of fermentation and its holes from fermentation gases. Pickles and sauerkraut are sour because bacteria ferment the carbohydrates in cucumbers and cabbage, respectively. Sausage tastes like sausage because bacteria ferment the meat proteins. Thus fermentation is useful not only to the microorganisms but also to consumers who enjoy the products of fermentation.

as″e-to-bak′ter

TO THIS POINT

■

We have surveyed the process of aerobic respiration in glucose, and we saw how numerous carbohydrates, proteins, and fats can be metabolized by bacterial cells through modifications of the scheme. In the latter cases, enzymes first convert the organic molecules to intermediary compounds of the process. The intermediates then proceed along the metabolic pathway and give up their energy to yield ATP. This chemistry illustrates a basic economy in cell metabolism because a centralized process is used, and various organic molecules fit into it.

The discussion then shifted gears to explore how energy may be obtained when oxygen is not available as a final electron acceptor. We described how inorganic molecules are employed as electron acceptors under anaerobic conditions so that the metabolism can continue. Some time was spent with fermentation, a unique process that utilizes an organic intermediary as an electron acceptor. Fermentation has great significance in the food and liquor industries.

5.3

MICROFOCUS

"BUT I HAVEN'T HAD A DROP!"

Charlie Swaart had been a social drinker for years, but in 1945 a nightmare began that would make medical history. While stationed in Tokyo during World War II, Swaart suddenly became drunk for no apparent reason. For years thereafter, the episodes continued—bouts of drunkenness and monumental hangovers without so much as a beer. His doctor warned him not to drink because his liver was becoming damaged. Swaart followed the advice to the letter, and still he got drunk.

In 1964, Swaart learned of a similar case in Japan. For 25 years a man had endured social and professional disgrace because of his drunken behavior. Finally, Japanese doctors found a yeastlike fungus, *Candida albicans*, flourishing in his intestine. Apparently, the fungus was fermenting carbohydrate to alcohol right in his body. An antibiotic killed the fungus and solved the problem; other cases were later found.

With this knowledge in hand, Swaart approached his physician and requested that laboratory tests be performed to learn if he was similarly infected. Sure enough,

massive colonies of *Candida albicans* showed up. Little attention had been given them previously because *Candida* is a common inhabitant of the human intestine. But Swaart's *Candida* was different because it could convert carbohydrate to alcohol. The sugar in a cup of coffee could bring on drunkenness, not to mention the carbohydrate in pasta, cake, and candy. An antifungal drug was prescribed, and the problem abated.

The drug lost its effect in 1970, however, and the symptoms returned, much worse than before. Now the hunt was on for a new drug. Swaart traveled to Japan and underwent hospital tests until a useful drug was finally located. Regular medication continued until 1975, when his *Candida* was eliminated for good.

Were Swaart and his Japanese counterparts the only ones in the world? Many researchers think not. They suggest that the atomic blasts at Hiroshima and Nagasaki may have caused a strain of *Candida* to mutate so that it could ferment carbohydrates to alcohol. If this theory holds, there

may be thousands of people throughout the world with the same type of affliction. For Charlie Swaart, though, the nightmare was over.

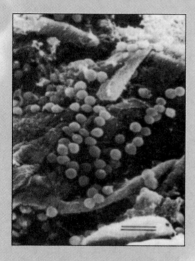

A scanning electron micrograph of the yeastlike fungus *Candida albicans*. Mutant strains of this organism may ferment carbohydrates to ethyl alcohol. Note the typically oval structure of the yeasts. Bar = 10 μm.

> *We shall now turn our attention to anabolism and discuss the methods by which carbohydrates and proteins are synthesized in bacterial cells. Fats will not be considered because they are formed essentially by a reversal of catabolism. The anabolism of carbohydrates and that of proteins are complex processes, and both are essential to an appreciation of bacterial metabolism. Our discussions will be relatively brief because we are more interested in an overview than in specific details. The processes of anabolism, together with those of catabolism, provide a window to the fundamental biochemistry of bacterial life.*

Anabolism of Carbohydrates

Although the anabolism of carbohydrates takes place through various mechanisms in bacteria, the unifying feature is the requirement for energy. In many bacteria, energy comes from sunlight, but in several species, energy is derived from chemical reactions. On these bases, two general patterns of bacterial anabolism exist: photosynthesis and chemosynthesis. In addition, bacteria obtain the carbon for carbohydrate synthesis from either of two different sources, as we shall see presently.

Photosynthesis

Photosynthesis is a process in which light energy is converted to chemical energy, which is then used to synthesize organic compounds from carbon dioxide. The process takes place in organisms having the pigment chlorophyll. Adenosine triphosphate (ATP) is a key intermediary compound in the process, and glucose is a major end product.

Photosynthesis occurs in both eukaryotic and prokaryotic microorganisms. In eukaryotic microorganisms, it takes place in such organisms as diatoms, dinoflagellates, and algae (Figure 5.9). Among the prokaryotes, we find photosynthesis occurring in the cyanobacteria (formerly called blue-green algae), the green sulfur bacteria, and the purple sulfur bacteria. Our discussion will focus on the prokaryotes.

Cyanobacteria practice photosynthesis in much the same manner as eukaryotic microorganisms and green plants. The cyanobacteria absorb light energy in their green pigment **chlorophyll a**. The light excites pigment molecules and each molecule loses one electron. Dislodged electrons are then accepted by an iron-containing compound called **ferredoxin**. Ferredoxin passes electrons to a series of cytochromes, and eventually the electrons are taken up by a new series of chlorophyll molecules. When these molecules are excited by light energy, the electrons are again boosted out of the pigment molecules to another series of cytochrome molecules and finally to a molecule of nicotinamide adenine dinucleotide phosphate (NADP). The latter then receives hydrogen ions from water molecules and becomes **NADPH$_2$**.

During the electron transfer in cytochromes, a proton motive force develops in the cell membrane of the cyanobacterium, and chemiosmosis takes place. As described previously, **ATP** is formed when protons pass back across the membrane and release their energy. Hence, two major products, ATP and NADPH$_2$, result from this phase of photosynthesis. The phase is termed the **light reaction** (or "energy-fixing reaction") because light energy is trapped and

> Photosynthesis: carbohydrate anabolism using light as an energy source.

> Cyanobacteria: prokaryotic aquatic microorganisms formerly called blue-green algae.
>
> fer″ĕ-dok′sin

FIGURE 5.9

Photosynthetic Microorganisms

(a) Four different species of diatoms obtained by dragging a fine-mesh net through seawater off the coast of Rhode Island (× 740). Photosynthetic pigments are located within the cytoplasm of these silica-encased organisms. (b) The dinoflagellate *Perdinium* found in subtropical seawater (× 455). This organism is an amoeba enclosed in two armorlike plates that contain photosynthetic pigments. Certain species of dinoflagellates cause red tides, seawater in which the abundance of red-pigmented organisms makes the water appear blood-red.

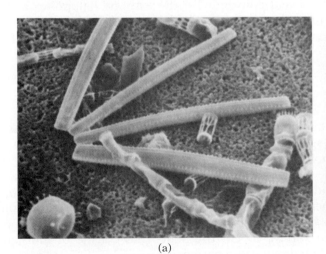

(a)

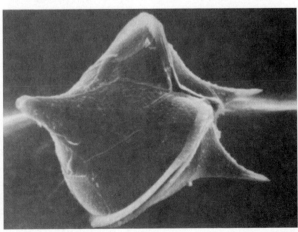

(b)

converted to (or "fixed" as) chemical energy. It is important to note that electrons are replaced in the chlorophyll molecules by water molecules (which also supply hydrogen ions, as noted above). The residual portions of the water molecules recombine with one another and yield oxygen, the oxygen that fills the atmosphere and is used by living creatures in respiration. Organisms that produce oxygen in photosynthesis are said to be **oxygenic**.

In the next step, carbohydrate is formed. The process is known as the **dark reaction** because it can take place without light energy. (It is also known as the "carbon-fixing reaction.") Using the energy stored in ATP from the light reaction, an enzyme bonds carbon dioxide to a 5-carbon organic substance called ribulose-5-phosphate. The resulting 6-carbon molecule then splits to form two molecules of phosphoglyceraldehyde (PGA). This is the substance formed in step (4) of glycolysis. Hydrogen ions for the reaction are obtained from $NADPH_2$, which also supplies electrons to the process. The two molecules of PGA then condense with each other. Now a reversal of glycolysis reactions take place and a molecule of glucose forms. Thus the overall formula for photosynthesis may be expressed as

$$6 \ CO_2 + 6 \ H_2O + ATP \longrightarrow C_6H_{12}O_6 + 6 \ O_2 + ADP + P$$

You may correctly note that this equation is essentially opposite to the equation for aerobic respiration. The essential difference is that aerobic respiration is an energy-yielding process, while photosynthesis is an energy-trapping process.

In addition to the cyanobacteria, several other groups of prokaryotes trap energy by photosynthesis. Two such groups are the **green sulfur bacteria** and **purple sulfur bacteria**, so-named because of the colors imparted by their pigments. These bacteria have pigments known as **bacteriochlorophylls** to distinguish them from other chlorophylls. Bacteriochlorophylls a and b are found in purple sulfur bacteria. In the production of carbohydrate in the dark reaction, the organisms do not use water as a source of hydrogen ions. Consequently, no oxygen is liberated and the bacteria are said to be **anoxygenic**. Instead of water, a series of organic or inorganic substances are utilized for hydrogen ions. Certain species of green sulfur bacteria use hydrogen sulfide (H_2S) as a hydrogen ion source and convert the hydrogen sulfide to sulfur. Species of purple sulfur bacteria use small fatty acids as hydrogen ion donors. The green and purple sulfur bacteria commonly live under anaerobic conditions in environments such as sulfur springs and stagnant ponds.

Another variation of bacterial photosynthesis occurs in the archaeobacteria (Chapter 4). Instead of the usual chlorophylls, the extreme halophiles of the archaeobacteria group contain a pigment called **bacteriorhodopsin**. In the presence of oxygen, extreme halophiles can synthesize ATP with the aid of this pigment.

Chemosynthesis

Bacteria of the type we have discussed are commonly known as **photoautotrophs**, meaning "light-self-feeders," because they utilize light energy to synthesize foods in photosynthesis. Another group of bacteria are the **chemoautotrophs**. These organisms use chemical reactions to obtain energy from inorganic compounds for carbohydrate synthesis. The process of carbohydrate anabolism is therefore called **chemosynthesis** rather than photosynthesis (MicroFocus: 5.4).

Chemoautotrophs can survive in environments that consist of inorganic substances, as long as carbon dioxide and oxygen are available. Certain bacteria obtain energy from ATP formed during reactions taking place in their cytoplasm.

Bacteriochlorophyll:
a type of chlorophyll found in certain species of bacteria.

Halophiles:
bacteria that live in high-salt environments.
bak-te"re-o-ro-dop'sin

fo"to-aw'to-troph

ke"mo-aw'to-troph

Chemosynthesis:
carbohydrate anabolism using chemical reactions as energy sources.

5.4
MICROFOCUS
A VOCABULARY OF BACTERIAL METABOLISM

metabolism: the sum total of all chemical reactions that contribute to the life of a cell and ultimately an organism.

catabolism: the chemical digestion of organic compounds in a cell, usually for the purpose of obtaining energy for life processes or building blocks for chemical synthesis.

anabolism: the chemical synthesis of organic compounds in a cell, usually for the purpose of forming cell structures or energy-containing compounds.

enzyme: any of a series of protein molecules that catalyze a cell's chemical reactions while themselves remaining unchanged.

adenosine triphosphate (ATP): a molecule composed of adenine, ribose, and three phosphate groups that stores energy in its chemical bonds and releases the energy to fuel a cell's chemical reactions.

glycolysis: a series of chemical reactions in which a molecule of glucose is converted to two molecules of pyruvic acid, with the accompanying release of enough energy to form four ATP molecules.

Krebs cycle: a series of chemical reactions in which an acetyl-CoA molecule is metabolized to yield two carbon dioxide molecules and a number of energy-rich compounds for further reactions.

oxidative phosphorylation: a series of chemical reactions in which electrons and energy are released from glycolysis and Krebs cycle products and are used to form ATP molecules.

aerobic respiration: a form of metabolism in which a series of energy-liberating reactions takes place in the presence of oxygen.

anaerobic respiration: a form of metabolism in which a series of energy-liberating

reactions takes place in the absence of oxygen.

fermentation: a type of anaerobic respiration in which an organic compound formed during the reaction series serves as an electron acceptor at the end of the series.

photosynthesis: the reaction series in which a cell forms carbon compounds using light as an energy source.

chemosynthesis: the reaction series in which a cell forms carbon compounds using chemical reactions as an energy source.

transcription: the aspect of protein synthesis in which a nitrogenous base sequence of DNA is used as a template to formulate the nitrogenous base sequence in a messenger RNA molecule.

translation: the aspect of protein synthesis in which the nitrogenous base sequence of messenger RNA is used as a template to formulate the amino acid sequence in a protein.

For example, species of *Nitrosomonas* convert ammonium ions (NH_4^+) into nitrite ions (NO_2^-) under aerobic conditions and in the process obtain ATP. Bacteria of the genus *Nitrobacter* then convert the nitrite ions into nitrate ions (NO_3^-), also as an ATP-generating mechanism. In addition to providing energy to both species of bacteria, these reactions have great significance in the environment because they preserve nitrogen in the soil in the form of nitrate. The nitrate may then be used by green plants to form amino acids in the nitrogen cycle (Chapter 25). Other bacteria living near ocean vents can obtain their energy from hydrogen sulfide (Figure 5.10). Marine bacteria can also use varying electron acceptors (Microfocus: 5.5).

ni-tro"so-mon'as

Carbon Sources

In addition to their classification as photoautotrophs and chemoautotrophs, bacteria fall into either of two other groups. These groups are the **photoheterotrophs** and **chemoheterotrophs**. "Photo-" or "chemo-" refers to whether light or chemical reactions are used for energy. The stem "hetero-" refers to "other" substances, meaning that photoheterotrophs and chemoheterotrophs use organic compounds instead of carbon dioxide to obtain the carbon for carbohydrates. Alcohols, fatty acids, and other organic acids are examples of the organic compounds that may supply the carbon. Photoheterotrophs include certain green nonsulfur and purple nonsulfur bacteria in nature, while chemoheterotrophs include the vast majority of bacteria as well as all fungi, protozoa, and animals. Saprobes and parasites are chemoheterotrophic bacteria because they use organic nutrients as carbon sources for their own carbohydrates. The "chemos" and "photos" can be difficult to remember but Table 5.2 may help.

fo"to-het'er-o-troph
ke"mo-het'er-o-troph

Saprobe:
an organism that feeds on
dead organic matter.

FIGURE 5.10
Chemoautotrophs

A scanning electron micrograph of bacteria and other microorganisms attached to natural surfaces near cracks in the floor of the Pacific Ocean. These samples were retrieved from a depth of 2550 meters (about 1.68 miles) near vents where the temperature was over 400° F. Hydrogen sulfide from the vent is a major source of energy for these marine bacteria. Bar = 5 μm.

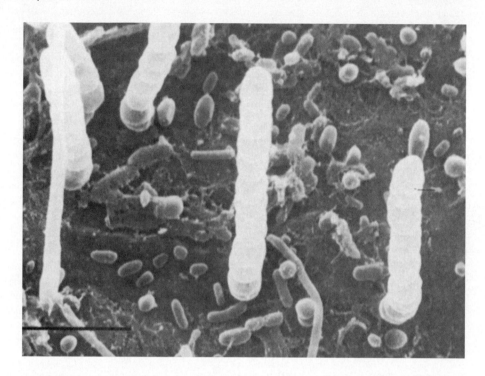

TABLE 5.2
Nutritional Classification of Organisms

NUTRITIONAL TYPE	ENERGY SOURCE	CARBON SOURCE	EXAMPLES
Photoautotroph	Light	Carbon dioxide (CO_2)	Photosynthetic bacteria (green sulfur and purple sulfur bacteria), cyanobacteria, extreme halophiles
Photoheterotroph	Light	Organic compounds	Purple nonsulfur and green nonsulfur bacteria
Chemoautotroph	Chemical reactions	Carbon dioxide (CO_2)	*Nitrosomonas, Nitrobacter*
Chemoheterotroph	Chemical reactions	Organic compounds	Most bacteria, fungi, protozoa, and all animals

Anabolism of Proteins (Protein Synthesis)

The anabolism of proteins stands in stark contrast to the anabolism of carbohydrates, and we must reorder our thinking because there is no glycolysis, Krebs cycle, photosynthesis, or "trophs" in this aspect of metabolism. Instead, protein

5.5

"YUK!"

Ever wonder why fish smells so fishy? Blame it on changes occurring in a compound called trimethylamine-N-oxide (TNO). When a fish is taken out of its natural habitat, it dies quickly, and the natural process of decomposition begins. Marine bacteria proliferate and use TNO as an electron acceptor in their metabolism. They convert the TNO to trimethylamine (tri-methyl-amine). The latter gives rotting fish its dreadful smell.

Is there an up side to all this? Possibly so—at least we know for certain when the fish is bad.

anabolism, or **protein synthesis** as it is more commonly known, is a process in which amino acids are carefully bound together in a sequence determined by the hereditary information in the cell. The compounds resulting from protein synthesis are utilized as cellular enzymes, structural components, toxins, or in other forms.

Before the 1950s, scientists were perplexed by how proteins are assembled from amino acids. An early theory suggested that the amino acid chain of a protein served as a template for the construction of new proteins. However, the transformation experiments of Griffith and the work of Avery and his co-workers (Chapter 6) focused attention on the deoxyribonucleic acid (DNA) of a cell. Finally, in 1953, the epic announcement of the double helix structure of DNA by James D. Watson and Francis H.C. Crick provided a glimpse into how DNA serves as a code for the manufacture of proteins. Their work was based in large measure on the X-ray diffraction studies by Maurice Wilkins and Rosalind Franklin. The work on DNA did not displace anything in science. Rather, it filled a critical gap in understanding how a cell produces protein. Watson, Crick, and Wilkins shared the 1962 Nobel Prize in Physiology or Medicine for their insight.

Today it is universally recognized that protein synthesis requires not only the DNA of the bacterial chromosome but also ribonucleic acid (RNA) and there is evidence that RNA may act independently (MicroFocus: 5.6). Both DNA and RNA are described in Chapter 2. A review of their structure is recommended if their functions are to be fully comprehended. Protein synthesis also utilizes ATP as an energy source, as well as a series of important enzymes. Amino acids must likewise be available. A refresher on their structure and the peptide bonds that hold them together in a protein (Chapter 2) may be of value.

The central theme of protein synthesis holds that segments of DNA on the chromosome, known as **genes**, provide a code for the production of fragments of RNA (Figure 5.11). The genetic code of DNA is expressed in RNA by a process called **transcription**. One type of RNA then functions as a messenger by carrying the code to other areas of the cytoplasm where amino acids are fitted together in a precise sequence to form the protein. This sequencing process, called **translation**, reflects the genetic code in the DNA. The overall process is summarized as follows:

Amino acids: nitrogen-containing compounds that combine to form proteins.

Deoxyribonucleic acid: the nucleic acid that forms the hereditary information of the cell and provides the genetic code for protein synthesis.

Genes: segments of DNA that provide the biochemical code for protein synthesis.

$$\text{DNA} \xrightarrow{\text{Transcription}} \text{RNA} \xrightarrow{\text{Translation}} \text{Protein}$$

The following paragraphs describe transcription and translation in detail.

5.6

MICROFOCUS

OUT WITH THE OLD

Up to the 1980s, one of the bedrock principles of biochemistry was the division of labor in cells: nucleic acids (DNA and RNA) hold the information for directing the biochemical reaction in the cell; proteins serve as the functional molecules (the enzymes) that catalyze the thousands of chemical reactions taking place. But chemistry research in the late 1970s and early 1980s helped overturn this principle. Contemporary scientists believe that RNA acting by itself can trigger chemical reactions.

The seminal research on RNA was performed independently by Thomas R. Cech of the University of Colorado and Sidney Altman of Yale University. In the late 1970s, Altman found an unusual enzyme in bacteria, an enzyme composed of RNA and protein. Initially, he thought the RNA was a contaminant, but when he separated the RNA from the protein, the enzyme could not function. After several years, Altman and his colleagues showed that RNA was the enzyme's key component because under carefully controlled laboratory conditions, it could act alone. At about the same time, Cech discovered that RNA molecules from protozoa could catalyze certain reactions

under laboratory conditions. He went further and showed that a molecule of RNA could cut internal segments out of itself and splice together the remaining segments.

Many biologists met the findings of Cech and Altman with disbelief. The implication of the research was that proteins and nucleic acids are not necessarily interdependent as had been assumed. The research also opened the possibility that RNA could have evolved on Earth without protein and that a self-catalyzing form of RNA could have been the first primitive molecule able to reproduce itself. Perhaps, scientists reasoned, the biochemical machinery for translating the DNA-based genetic code in modern cells evolved much later. In essence, there arose a whole new way of imagining how life might have begun on Earth. The Nobel Prize committee was equally impressed. In 1989, it awarded the Nobel Prize in Chemistry to Cech and Altman.

By 1990, the self-reproducing molecule of RNA had a name—ribozyme. Biochemists at the Massachusetts General Hospital soon modified a ribozyme by removing its internal segments. Then they showed that the new ribozyme could join together sep-

arate short nucleotide segments aligned on specially designed, external templates. The research was a step toward designing a completely self-copying RNA molecule. Would such a ribozyme enclosed in a membrane constitute a primitive cell? If so, the cell would be quite different from what most scientists have imagined the first cells to be. Stay tuned.

Transcription

Hydrogen bonds:
weak bonds between protons and adjacent pairs of electrons.

o-cho'ah

In **transcription**, various types of RNA are produced according to the code of nitrogenous bases in the DNA molecule (Table 5.3). The DNA thus serves as a template for new RNA molecules. The process begins with an uncoupling of the two DNA strands as the hydrogen bonds between opposing bases break down. The DNA double helix then unwinds. This separation or "unzipping" does not begin at the end of the molecule, but rather at certain internal regions.

At this point, RNA nucleotides containing ribose, phosphate groups, and nitrogenous bases align themselves along one strand of DNA. The bases of the RNA are positioned to complement the DNA bases. For example, guanine stands opposite cytosine, cytosine opposes guanine, and adenine opposes thymine. Because RNA contains no thymine, an adenine base on the DNA positions a uracil base opposite itself. The RNA components are then linked together, beginning at a specific site, by an enzyme called **RNA polymerase**. This enzyme was first described by Severo Ochoa, who shared the 1959 Nobel Prize in Physiology or Medicine. The sequence of bases in DNA thus forms a complementary image of itself in the RNA. The language, or genetic code of DNA has been transcribed.

FIGURE 5.11

An Electron Micrograph of a Segment of DNA Isolated from *Escherichia coli* (× 150,000)

The segment is estimated to be 1.4 μm long. Its genes function in the production of an enzyme used in galactose metabolism.

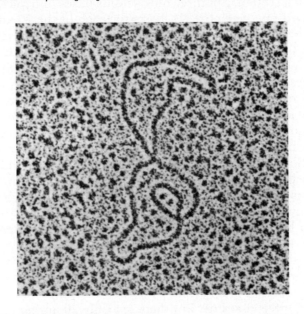

TABLE 5.3

A Comparison of DNA and RNA

DNA (DEOXYRIBONUCLEIC ACID)	RNA (RIBONUCLEIC ACID)
In prokaryotes, found in the nucleoid and plasmids; in eukaryotes, found only in the nucleus	In prokaryotes and eukaryotes, found dissolved in the cytoplasm and at the ribosomes; in eukaryotes, found in the nucleolus
Always associated with chromosome (genes); each chromosome has a fixed amount of DNA	Found mainly in combinations with proteins in ribosomes in the cytoplasm, as messenger RNA, and as transfer RNA
Contains a pentose (5-carbon) sugar called deoxyribose	Contains a pentose (5-carbon) sugar called ribose
Contains bases adenine, guanine, cytosine, thymine	Contains bases adenine, guanine, cytosine, uracil
Contains phosphorus (in phosphate groups) that connects various sugars with one another	Contains phosphorus (in phosphate groups) that connects various sugars with one another
Functions as the molecule of inheritance	Functions in the synthesis of protein

Three types of RNA result from transcription. One type, **ribosomal RNA**, forms from certain regions of the DNA. Together with protein, this RNA serves as the basic framework of submicroscopic particles called ribosomes. Over 30,000 ribosomes are present in each bacterial cell. They are the sites at which amino acids assemble into protein.

Ribosomes: **submicroscopic particles of RNA and protein that function in protein synthesis.**

Transfer RNA:
the RNA molecule that unites with amino acids in the cytoplasm and transports them to the ribosomes for assembly into proteins.

A second type of RNA is **transfer RNA**, or simply **tRNA**. Many of the early discoveries related to this molecule were perfomed by Robert Holley, a corecipient of the 1968 Nobel Prize. Transfer RNA is the smallest RNA molecule. Each molecule has a molecular mass of about 25,000 daltons and contains approximately 75 nucleotides. It is shaped roughly like a cloverleaf, with one point exhibiting a sequence of three nitrogenous bases (a triplet) that functions as a code. At least one type of tRNA exists for each of the 20 amino acids, and a high degree of specificity exists between the tRNA and its amino acid. For example, the amino acid alanine binds only to the tRNA specialized to transport alanine. Transfer RNA molecules function by delivering amino acids to the ribosome for assembly into proteins, as we shall see presently.

Both ribosomal RNA and transfer RNA are produced at irregular intervals and remain for long periods of time in the cytoplasm. The third form of RNA appears each time a protein is manufactured. This RNA is called **messenger RNA (mRNA)** because it carries the genetic message for protein synthesis. Discovered by Sol Spiegelman and his co-workers, mRNA provides the template that indicates which positions the amino acids are to occupy in the protein. Its message consists of a series of three-base codes or **codons**. Each codon specifies an individual amino acid to be slotted into position. Codons were discovered by a number of biochemists, among them Har Gobind Khorana, a 1968 Nobel laureate in Physiology or Medicine.

Codon:
a three-base code on an mRNA molecule that is specific for an amino acid.

One of the startling discoveries of biochemistry is that a three-base code exists for each amino acid. Since there are four nitrogenous bases available to work with, mathematics tells us that 64 possible combinations can be made of the bases, using three at a time. But there are only 20 amino acids for which a code must be supplied. Therefore how do scientists account for the remaining 44 codes? It is now known that 61 of the 64 codes are used for an amino acid and that most amino acids have multiple genetic codes (Table 5.4). For example, GCU, GCC, GCA, and GCG all code for alanine. The three remaining codes act like punctuation marks and stop the addition of amino acids to a growing chain of protein.

Prokaryotes:
simple organisms such as bacteria that lack a nucleus and organelles, and reproduce by fission.

The genetic code of bases is apparently universal for all species, be they prokaryotic or eukaryotic. Thus the GCU code for alanine in bacteria is also

---

TABLE 5.4

The Genetic Codes for Several Amino Acids

UUU	Phenylalanine	CUU		AUU	Isoleucine	GUU	
UUC	(Phe)	CUC	Leucine	AUC	(Ile)	GUC	Valine
UUA	Leucine	CUA	(Leu)	AUA		GUA	(Val)
UUG	(Leu)	CUG		AUG — Methionine (Met)		GUG	
UCU		CCU		ACU		GCU	
UCC	Serine	CCC	Proline	ACC	Threonine	GCC	Alanine
UCA	(Ser)	CCA	(Pro)	ACA	(Thr)	GCA	(Ala)
UCG		CCG		ACG		GCG	
UAU	Tyrosine	CAU	Histidine	AAU	Asparagine	GAU	Aspartic acid
UAC	(Tyr)	CAC	(His)	AAC	(Asn)	GAC	(Asp)
UAA	Chain	CAA	Glutamine	AAA	Lysine	GAA	Glutamic acid
UAG	terminators	CAG	(Gln)	AAG	(Lys)	GAG	(Glu)
UGU	Cysteine	CGU		AGU	Serine	GGU	
UGC	(Cys)	CGC	Arginine	AGC	(Ser)	GGC	Glycine
UGA — Chain terminator		CGA	(Arg)	AGA	Arginine	GGA	(Gly)
UGG — Tryptophan (Trp)		CGG		AGG	(Arg)	GGG	

Boldface codes are chain terminators.

the code for alanine in animal cells. However, the complete set of codes varies in sequence among species, thus providing a constancy as well as a variation in all living things.

There is an important difference between use of the codes in prokaryotes and eukaryotes. In prokaryotes, all the codes are apparently transcribed to an mRNA molecule in protein synthesis, but in eukaryotes some are left out when the final mRNA is produced. In 1977, Philip Sharp and his associates at the Massachusetts Institute of Technology (MIT) found that certain portions of eukaryotic DNA are not transcribed to the final mRNA. Apparently these portions are a type of "genetic gibberish" that must be removed from the mRNA before the molecule is able to function. The nonsense genes were promptly labeled **introns** because they were *intra*genic segments, while the sensible genes were called **exons** because they are *ex*pressed. Prokaryotes, such as bacteria, do not have these intervening sequences; the genes are entirely sensible. Sharp was a Nobel laureate in 1993.

Exons:
segments of DNA that are expressed as a genetic code in protein synthesis.

Translation

In the process of **translation**, the genetic code expressed in mRNA translates into a sequence of amino acids in a molecule of peptide. The process takes place at the ribosome, where the mRNA molecule meets the tRNA molecules that are transporting their amino acids (Figure 5.12). The amino acids have been attached to their individual tRNA molecules using energy from ATP and specific enzymes. Now they are ready to be assembled into a peptide.

Peptide:
a relatively small chain of amino acids.

FIGURE 5.12

The Translation Process in Protein Synthesis

The messenger RNA moves to the ribosome, where it is met by transfer RNA molecules bonded to different amino acids. The tRNA molecules align themselves opposite the mRNA molecule and bring the amino acids into position. A peptide bond forms between adjacent amino acids on the growing protein chain, after which the amino acid leaves the tRNA. The tRNA returns to the cytoplasm to bond with another molecule of the same amino acid.

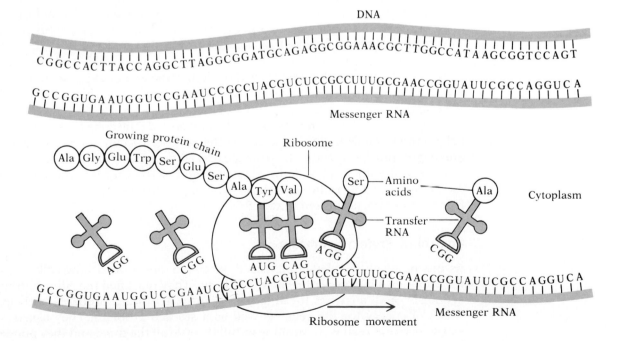

TABLE 5.5
Some Characteristics of Major Pathways of Bacterial Metabolism

FORM	PATHWAY	LOCATIONS	REACTANTS	PRODUCTS
Glycolysis	Embden-Meyerhoff pathway	Cytoplasm	Glucose	2 pyruvic acid, 8 ATP (if oxygen present) 2 ATP (if not present)
Aerobic respiration	Acetyl CoA formation	Bacterial membranes	Pyruvic acid	NADH, CO_2, acetyl CoA
	Krebs cycle (2 turns)	Bacterial membranes	Acetyl CoA	2 ATP, 6 NADH, 4 CO_2, 2 $FADH_2$
	Electron transport and chemiosmosis	Bacterial membranes	10 NADH, 2 $FADH_2$	28 ATP, H_2O
Anaerobic respiration	Electron transport	Bacterial membranes	Nitrate or sulfate or carbon dioxide	Nitrite or H_2S or methane
Fermentation	Alcoholic (yeast)	Cytoplasm	Pyruvic acid	Ethanol, CO_2
	Lactic acid (bacteria)	Cytoplasm	Pyruvic acid	Lactic acid
Photosynthesis (cyanobacteria)	1. Light reaction	Bacterial membranes	$6 CO_2 + 6 H_2O$	Oxygen gas, ATP, NADPH
	2. Dark reaction	Cytoplasm	ATP, NADPH	Glucose, H_2O
Chemosynthesis (green sulfur bacteria)	Energy metabolism	Cytoplasm	H_2S	Elemental sulfur
Protein synthesis	1. Transcription	Nucleoid	DNA	Messenger RNA, transfer RNA, ribosomal RNA
	2. Translation	Ribosomes	Amino acids	Protein

Anticodon:

a three-base code on a tRNA molecule that complements the codon on an mRNA molecule.

Translation takes place as the ribosome moves along the mRNA, with the codons in mRNA exposed. The three-base sequences of the tRNA molecules are called **anticodons**. The anticodons match themselves with the codons, and the tRNAs thus bring their amino acids into a carefully ordered position. For example, a codon of GCC on the mRNA will match with an anticodon of CGG on the tRNA carrying alanine. Alanine will now be slotted into position. Hydrogen bonds between the codon and anticodon bases momentarily hold the tRNA in position while an enzyme forms a peptide bond between alanine and the adjacent amino acid. **Guanosine triphosphate (GTP)** supplies energy for the reaction. The tRNA then breaks free, leaving its alanine molecule on the peptide chain. The next codon attracts a tRNA with its amino acid, and the process continues. When a codon on the mRNA signals the end of chain formation, the process comes to a halt and the chain falls away from the ribosome. This termination code may be UAA, UAG, or UGA.

In practice, a single mRNA molecule may provide the code for several identical peptides at a time. This is because other ribosomes can attach to the mRNA and begin translating the code while the first ribosome is still there. After peptide synthesis, the linear peptide breaks away and twists into its secondary form. As the peptide folds into a tertiary structure, other groups may be added to form the complete protein (Table 5.5).

Control of Protein Synthesis

As biochemical knowledge increased, scientists questioned how cells regulate the complex machinery of protein synthesis. They reasoned that the continuous synthesis of all enzymes for all possible nutrients would probably represent a waste of energy for the cells, as well as a storage problem. They asked what would prevent a cell from running at full throttle all the time, and they pondered how bacteria could economize the synthesis of protein. Experiments indicated that certain enzymes appeared only when their substrates were present and that a delicate and flexible regulation of metabolism appeared to be the rule rather than the exception.

In 1961 two Pasteur Institute scientists, François Jacob and Jacques Monod, proposed a mechanism for controlling protein synthesis. Jacob and Monod suggested that bacterial genes fall into several groups: structural genes, which provide genetic codes for proteins in the process we have discussed; an adjacent operator gene, which stimulates and controls the expression of structural genes; and a distant repressor gene, which controls the operator gene. Jacob and Monod named the entire unit for expressing a particular trait an **operon**. In 1965 they won the Nobel Prize in Physiology or Medicine for their discovery. In recent years scientists have identified a promoter region next to the operator gene.

The operon theory helps explain the control of protein synthesis. When a nutrient is absent from the cytoplasmic environment, the repressor gene codes for mRNA, and a **repressor protein** forms (Figure 5.13). In 1965, Mark Ptashne of Harvard University first isolated a repressor protein. The repressor protein binds to the operator gene and overlaps the adjacent promoter region. This prevents the operator and promoter from stimulating the structural genes, and the cell cannot produce enzyme protein.

zhah-kob'
mo-no'

Operon:
a unit of various genes that control the synthesis of protein.

FIGURE 5.13
One View of the Operon Theory

(a) Under normal conditions, the repressor gene produces a repressor protein that blocks the operator gene and prevents the operator from stimulating the synthesis of protein by the structural genes. (b) When an inducer substance is present, it combines with the repressor protein and prohibits it from functioning. The operator gene now stimulates the structural genes to produce protein.

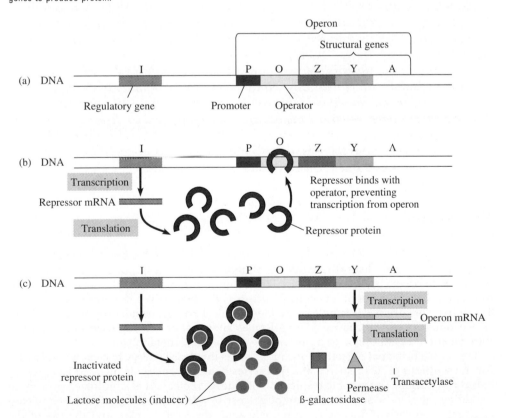

Lactose:
a disaccharide composed of glucose and galactose units.

Tryptophan:
a ring-containing amino acid.

At this point a nutrient enters the environment. This nutrient, known as an **inducer**, binds to the repressor protein and inactivates it. With the repressor protein gone, the operator gene and promoter region can stimulate the structural genes to produce protein. This system was first worked out in *E. coli*. It explains why the lactose-digesting enzyme forms only when lactose (the inducer) is present in the bacterial cytoplasm.

A second type of control mechanism, also traced to the operon, explains how enzyme production can be shut down when products of gene activity accumulate in the environment. For instance, bacteria such as *E. coli* cease their production of a tryptophan-synthesizing enzyme when tryptophan accumulates in the cytoplasm. The tryptophan itself does not cause the enzyme production to stop, nor does a repressor protein shut down the mechanism. Instead, tryptophan molecules unite with the repressor protein, and the tryptophan-repressor complex links to the operator gene. As the operator gene is inhibited, the structural genes cease production of tryptophan-synthesizing enzyme. In this way, tryptophan helps to control its own synthesis.

An understanding of gene activity and protein synthesis had broad ramifications in the 1980s when scientists learned how to control the process and reengineer bacteria to produce useful materials. The research that led to this breakthrough is explored in Chapter 6.

NOTE TO THE STUDENT

This chapter contains some of the most difficult concepts you will encounter in microbiology. The concepts are also among the most fundamental because the biochemistry applies to all life, be it bacterial life, plant life, or human life. I hope you can step back and see the forest as well as the trees. Virtually all living things obtain their energy by glucose digestion; all creatures depend ultimately on the carbohydrate produced in photosynthesis; and all protein in the world is produced essentially as we have described it.

One of the corollary benefits of studying the biochemistry of microorganisms is that you come away with a better understanding of living things in general. Nowhere is this more apparent than in this chapter. Bacteria and other microorganisms are mere tools used to discover the biochemical foundations that govern all living things. Beyond the window dressing of variation, there is an underlying kinship among all forms of life. To understand one form is to understand them all.

Summary

The two major themes of bacterial metabolism are catabolism (the breakdown of organic molecules) and anabolism (the synthesis of organic molecules). For either of these two processes to occur, bacteria and other living things utilize enzymes, a series of protein molecules that bring about a chemical change while themselves remaining unchanged. The activity of enzymes is supplemented by chemical energy, usually supplied by adenosine triphosphate (ATP). Since ATP molecules are constantly being used in living cells, they must be resynthesized to maintain the reactions of metabolism.

One of the principal objectives of catabolism is to release energy in organic molecules for the synthesis of ATP. Such a reaction series by which energy is released is a process called respiration. The aerobic respiration of glucose molecules is a multistep procedure including the step-by-step conversion of glucose to pyruvic acid (gycolysis), the release of energy and carbon dioxide from pyruvic acid (the Krebs cycle), and the use of the energy to form ATP molecules (oxidative phosphorylation). Other carbohydrates as well as proteins and fats also utilize this reaction series. The anaerobic respiration of glucose molecules involves a conversion to pyruvic acid, after which an inorganic molecule of an intermediary of the process is used as an electron acceptor to produce ions, gases, or acids and alcohols (fermentation). Less energy is released in the anaerobic process.

The anabolism of carbohydrates occurs by photosynthesis. In this process, light energy is used to synthesize ATP and the latter is then used to fix atmospheric carbon dioxide into carbohydrate molecules. Various pigments such as chlorophyll and bacteriorhodopsin can be used by various bacteria, and chemical energy can be used instead of light energy. Even the carbon can come from different sources.

The anabolism of proteins takes place by a complex mechanism in which the genetic information in DNA is first transcribed to a genetic message in RNA, then translated to a sequence of amino acids in the protein. Various forms of RNA, including mRNA, tRNA, and ribosomal RNA, function in the process, and a careful ordering of codons and anticodons is essential to successful completion of the process. Various control factors influence the mechanism and provide balance to the overall scheme of metabolism.

Questions for Thought and Discussion

1. Some years ago a magazine cartoon pictured newspapers being carried into a "conversion plant" and beef cattle coming out. The cartoonist envisioned that bacteria in the plant would convert the cellulose of the newspapers into the protein of beef cattle. Can you explain the chemistry behind this look into the future?

2. Citrase is the enzyme that converts citric acid to alphaketoglutaric acid in the Krebs cycle. A chemical company has located a mutant microorganism that cannot produce this enzyme and proposes to use the microorganism to manufacture a particular product. What do you suppose the product is? How might this product be useful?

3. For some mysterious reason, the bacteria normally cultivated in a standard bacteriological medium fail to grow. Questioning reveals that a plumber recently resealed the water pipes with lead. What is the connection between the two events?

4. In order to stimulate fermentation, vintners must seal wine vats tightly and permit the wine to remain still. Why?

5. One of the most important steps in the evolution of life on Earth was the appearance of certain organisms in which photosynthesis takes place. Why was this critical?

6. If ATP is such an important energy source in bacteria, why do you think it is not added routinely to the growth medium for these organisms?

7. One student maintains that organisms use proteins to synthesize enzymes. A second student counters that organisms use enzymes to synthesize proteins. Which student is right? Why?

8. A major pharmaceutical company has developed a new pesticide for use in controlling mosquitoes. However, laboratory tests indicate that the pesticide combines with and alters chlorophyll molecules. What horror story would occur if this chemical were sprayed in the environment?

9. Suppose a bacterium had the choice of producing ATP by respiration or by fermentation. Which would be the better alternative? Why?

10. The formula for a growth medium for a particular bacterium stipulates that riboflavin must be added for the synthesis of a certain chemical compound of the catabolism process. Can you guess which compound? What would be the effect of omitting riboflavin?

11. One reason why a stagnant pond has a putrid odor is that hydrogen sulfide has accumulated in the water. A microbiologist recommends that tons of green sulfur bacteria be added to remove the smell. What chemical process does the microbiologist have in mind? Do you think it will work?

12. Certain antibiotics such as streptomycin are known to bind to messenger RNA molecules. What effect will this have on the metabolism of a bacterium?

13. Suppose glycolysis came to a halt in a bacterial cell. Would this mean that the Krebs cycle would also stop? Why?

14. An essential factor in the growth media for bacteria is phosphate. How many places can you cite where it is needed?

15. When the author was completing his doctorate degree he was required to sit for an 8-hour comprehensive examination on each of three successive Saturdays. The biochemistry portion consisted of a single-line question: "Discuss the interrelationships between anabolism and catabolism." How might you have answered this question?

Review

When you have completed your study of bacterial metabolism, test your knowledge of its important facts and concepts by circling the choices that best complete each of the following statements:

1. The sum total of all a bacterium's biochemical reactions is known as (catabolism, metabolism); it includes all the (synthesis, digestion) reactions called anabolism and all the breakdown reactions known as (inactivation, catabolism).

2. Enzymes are a group of (carbohydrate, protein) molecules that generally (slow down, speed up) a chemical reaction by converting the (substrate, substart) to end-products.

3. The aerobic respiration of glucose begins with the process of (oxidative phosphorylation, glycolysis) and requires that (amino acids, energy) be supplied by (ATP, GTP) molecules.

4. The process of (fermentation, Krebs cycle) takes place in the absence of (oxygen, magnesium) and begins with a molecule of (glucose, protein) and ends with molecules of (amino acid, alcohol).

5. In oxidative phosphorylation, pairs of (neutrons, electrons) are passed among a series of (chromosomes, cytochromes) with the result that (oxygen, energy) is released for (NAD, ATP) synthesis.

6. In the Krebs cycle, (glucose, pyruvic acid) undergoes a series of changes and releases its (carbon, nitrogen) as (nitrous oxide, carbon dioxide) and its electrons to (NAD, TPA).

7. For use as energy compounds, proteins are first digested to (hydrochloric, amino) acids, which then lose their (carboxyl, amino) groups in the process of (fermentation, deamination) and become intermediates of respiration.

8. Ribulose phosphate receives (carbon monoxide, carbon dioxide) molecules in the process of (fermentation, photosynthesis), a process that ultimately results in molecules of (gluconic acid, glucose).

9. Chemoautotrophs use energy from (light, chemical reactions) to synthesize (carbohydrates, proteins) and are typified by species of (*Staphylococcus*, *Nitrosomonas*).

10. The synthesis of proteins begins with the (respiratory, genetic) code in molecules of (DNA, RNA), and that code is transferred to molecules of (rRNA, mRNA) in the process called (transamination, transcription).

11. At the (lysosome, ribosome) of the cell, the codon in an mRNA molecule is used to attract the (anticodon, supercodon) of the tRNA molecule in order to bring (a carbohydrate, an amino acid) molecule into position for protein synthesis.

12. Control of protein synthesis can be exerted by a (repressor, detractor) protein that binds to the (stimulator, operator) gene and prevents this gene from stimulating the (protein, structural) gene.

CHAPTER 6

BACTERIAL GENETICS

During the years that Louis Pasteur was performing his classic work on fermentation, an Augustinian monk named Gregor Mendel was busy developing the principles of genetics. Mendel was a priest and teacher at a monastery in Brunn, a town in Austria. In the 1860s, he performed a series of experiments with pea plants and came to the conclusion that inherited traits were passed from parents to offspring by a series of invisible markers. Mendel's work suggested that two markers exist for each trait and that one marker dominates the other. The markers appeared to separate during the formation of male and female reproductive cells, then come together when these cells unite to form a new plant.

Mendel's experiments were largely ignored until early in the twentieth century when Thomas Hunt Morgan revitalized research in genetics. Morgan worked with fruitflies and showed that the sex of these tiny insects is regulated by the invisible markers Mendel had described. Inheritance research quickly gathered momentum, and by the 1920s scientists had coined a new term, gene, to refer to Mendel's markers. The word was derived from the Greek *gena*, meaning birth or descent, a reference to the concept that inherited traits are derived from these units.

Microbiology began to merge with genetics after the 1928 experiments of Frederick Griffith. As we shall see presently, Griffith showed that traits could be transferred between bacteria, but he could not identify the transferring substance. Another 16 years passed before Oswald Avery and his group pinpointed the substance as deoxyribonucleic acid (DNA). Now the pace of discovery quickened considerably, and biochemists redefined genes as segments of the DNA molccule. With the epic announcement of the structure of DNA by James D. Watson and Francis H.C. Crick in 1953, a theory emerged as to how genes could function in the synthesis of protein. There began a period of intense research in molecular genetics that continues to the present day.

Bacteria stand at the forefront of genetic research for several reasons. They have only a single chromosome and, thus, are far easier to work with than organisms such as Mendel's pea plants or Morgan's fruitflies (where two genes exist for a trait). Also, bacteria can be cultivated conveniently in test tubes, and they produce the next generation much more rapidly than fruitflies, which take several weeks, or pea plants, which require a growing season of several months.

6.1

MICROFOCUS

BIG WINNERS

In the modern age, when time is measured in minutes and seconds, the mind finds it difficult to imagine the colossal 4.5 billion years that the Earth has been in existence. Try, however, to think of Earth's history as a single year. In our "historic" year, the Earth was a lifeless ball of rock until mid-July, when bacteria or something akin to bacteria first appeared. These were the only creatures on Earth until mid-October, when multicellular organisms emerged. Not until the end of November did the first land plants come into being, and not until early December did animals move out of the sea onto the land. The dinosaurs were in existence from December 19th to December 25th, and by December 27th, the earth bore a resemblance to modern Earth. Finally, on December 31st, close to midnight, humans appeared.

We take this trek through geologic time to help us appreciate why bacteria have done so well in evolution. They have been successful primarily because they have been around the longest. Bacteria have been on Earth about 3.5 billion years (versus about 200,000 years for humans). During this time, mutations have been constantly occur-

ring in bacteria, and nature has used the bacteria to test its newest genetic traits. The "bad" mutations have been eliminated together with the bacteria unlucky enough to get them, while the "good" mutations have thrived and have been passed onto the next generation—and to the present day.

And so it is that modern bacteria enjoy the fruits of mutation. Because bacteria have survived so long, the mutations have encouraged bacteria to survive everywhere, whether it be the snows of the Arctic or the boiling hot vents of the deep oceans. Nor can any organism compare to bacteria in sheer numbers—a pinch of rich soil has more bacteria than people living in the United States today—so it stands to reason that mutations must exist in at least some bacteria. Finally, consider a bacterium's multiplication rate—a new generation every half hour—and it is easy to see how a useful mutation (such as drug resistance) can be propagated quickly in an environment (one containing a drug) that is selecting out the fit bacteria (the ones with the mutation).

Any one of the factors—time on Earth, sheer numbers, multiplication rate—would be sufficient to understand how bacteria have evolved drug resistances through mutations. Combined in one type of organism, the factors help us appreciate why bacteria have done very well in the evolutionary lottery; very well, indeed.

We shall begin our study of bacterial genetics by exploring the nature of the bacterial chromosome. We shall then see how it can be altered in two ways: by changing the structure of the DNA in a process called **mutation** (MicroFocus: 6.1); and by acquiring new DNA from another organism via **recombination**. Both of these processes help explain some of the phenomena that take place in nature. We shall then move on to the topic of genetic engineering, in which the genetic material of bacteria (or yeast, or other cells) can be altered for profitable use. Research in this area is among the most elegant of our era.

The Bacterial Chromosome

Most of the genetic information in a bacterial (or prokaryotic) cell is contained within the **chromosome**, a single molecule of DNA arranged as a double helix in a closed loop (Figure 6.1). The chromosome exists freely in the cytoplasm without the surrounding membrane or protein support found in eukaryotic cells. It occupies about 10 percent of the total volume of a bacterial cell, and when extended its full length, it is about 1 millimeter (mm) long. This is approximately 1000 times the length of the bacterium that contains it. The tight packing accounts in part for the explosive release of DNA when the membrane of the

FIGURE 6.1

The Chromosome of *Escherichia coli*

(a) The chromosome is composed of double-stranded DNA wound as a double helix. The chromosome is a closed loop, with sugar molecules bonded to phosphate molecules to form the backbones of the helix. Pairs of nitrogenous bases connect the sugar-phosphate backbones of the molecule. (b) An electron micrograph showing the tangled mass of looping strands of DNA emerging from a disrupted *E. coli* cell (× 6000).

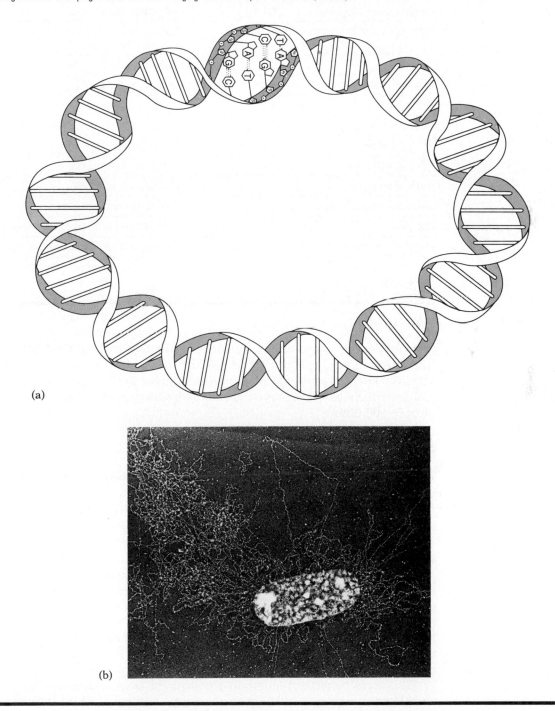

(a)

(b)

cell is broken. The eukaryotic chromosome is compared to the bacterial (prokaryotic) chromosome in Table 6.1.

The chromosome of the intestinal bacterium *Escherichia coli* has probably been studied more thoroughly than any other single chromosome. Distributed

esh″er-ik′e-a

TABLE 6.1

Characteristics of Prokaryotic (Bacterial) and Eukaryotic Chromosomes

PROKARYOTIC (BACTERIAL) CHROMOSOME	EUKARYOTIC CHROMOSOME
Replicates during binary fission	Replicates during mitosis
Single molecule of DNA per genetic trait (haploid)	Two molecules of DNA per genetic trait (diploid); some organisms haploid
Closed loop	Linear
No protein present	Histone protein present
No dominance of recessiveness in genes	Genes are dominant or recessive
Organized at the nucleoid	Organized in the nucleus with a nuclear membrane
No introns present	Introns present
DNA also in plasmids	DNA also in mitochondria
Mutations occur in DNA	Mutations occur in DNA
Genetic recombinations occur	No genetic recombinations occur
4000 genes in *E. coli* chromosome	100,000 genes in total of human chromosomes
About 1 mm in length	Tens or hundreds of millimeters in length
Replicates by a semiconservative method or rolling circle method	Replicates by semiconservative method

---

around the chromosome are individual sites to which genetic activity can be traced. Each site, called a **locus** (pl., loci), consists of several genes for that activity. The chromosome of *E. coli* has about 4000 genes. Some viruses, by contrast, have as few as seven genes, while human chromosomes have a total of over 100,000.

Genes:
segments of DNA that provide the biochemical code for protein synthesis.

Replication of the Chromosome

The bacterial chromosome replicates during the process of binary fission. At the beginning of the replication sequence, the DNA is usually anchored to a particular point on the bacterial cell membrane. The double helix then unwinds, and enzymes synthesize a new strand of DNA for each of the original strands. This is accomplished by joining together nucleotides whose bases complement bases in the original strand. One of the important enzymes, **DNA polymerase**, was discovered by Arthur Kornberg, a co-recipient of the 1959 Nobel Prize in Physiology or Medicine. Each new strand then combines with a parent strand, and the replicated DNA twists to reform the double helix. This combination of new and parent strands was first observed in *E. coli* in 1958 by Matthew J. Meselson and Franklin W. Stahl. It is called the **semiconservative method of replication** because one strand of the parent DNA is conserved in the new DNA molecule and one strand is newly synthesized.

DNA polymerase:
an enzyme used in the synthesis of DNA from a series of nucleotides.

The semiconservative method accounts for DNA replication, but it does not shed light on how a closed loop chromosome replicates. This problem perplexed microbiologists until 1962 when John Cairns and his co-workers clarified some of the details of the process.

Cairns' experiments showed that DNA unwinds at a fixed point, whereupon an enzyme nicks the closed loop at a site called the **origin of replication**. The two strands now separate, or "unzip" to establish a V-shaped replicating fork, as Figure 6.2 displays. Synthesis of DNA then occurs along the sides of the fork.

In later years microbiologists established that along one side of the fork DNA is manufactured by continuous assembly of nucleotides, beginning at the origin. However, along the other side, the synthesis begins at the point of forking and proceeds in a discontinuous fashion (Figure 6.2). Here the DNA is synthesized in a series of segments that later join with the help of an enzyme called DNA ligase. The segments came to be known as **Okazaki fragments**, after Reiji Okazaki who discovered them in 1968. Each of the new DNA strands then combines with a parent strand as Meselson and Stahl postulated.

o-ka-zak'-e

A second type of DNA replication is called the **rolling circle mechanism**. This process takes place in bacteria undergoing the sexual mating process of conjugation. While one strand of DNA remains in a closed loop, an enzyme nicks the other strand. The broken strand then "rolls off" the loop and serves as a template for synthesis of a DNA strand complementary to itself. When the two strands combine, a double helix reforms. Meanwhile the intact loop revolves 360 degrees and serves as a template for a strand of DNA complementary to itself. The loop combines with its new strand to form a second double helix. The bacterium now has two chromosomes, one of which will be used in mating.

Conjugation: a mating process in which DNA passes from one bacterium to another after they have joined together.

Plasmids

Some microorganisms contain genetic material in closed loops of DNA called **plasmids**. Plasmids exist apart from the chromosome as independent units in the cytoplasm. They contain about 2 percent of the total genetic information of the cell and multiply independently of the chromosome.

Plasmid: a closed loop unit of bacterial DNA existing apart from the chromosome.

Plasmids are not essential to the life of the cell, but they may confer selective advantages for those organisms that have them. For example, some plasmids called **R factors** ("resistance" factors) carry genes for antibiotic resistance, while other plasmids allow bacteria to transfer their genetic material to receptive cells in recombination processes. Still other plasmids contain genes for the production of **bacteriocins**, a group of proteins toxic to other bacteria, and other genes code for toxins that affect human cells and processes. Gram-negative bacteria are notable for the presence of plasmids. We shall have much to say about these extrachromosomal units as we proceed in this chapter. For the present, we shall discuss how changes occur in the bacterial chromosome by mutation and recombination.

bak-te"re-o'sinz

Bacterial Mutation

One method by which the information in a bacterial chromosome may be altered is through a permanent change in the DNA called a **mutation**. In most cases a mutation involves a disruption of the nitrogenous base sequence in the DNA molecule or the loss of significant parts of a gene. Often this leads to the production of a miscoded messenger RNA molecule and the insertion of one or more incorrect amino acids into protein molecules during synthesis. Since proteins govern virtually all activities of a bacterium, it follows that a mutation will alter its biochemical functions. Some alterations have little effect, but others may be significant, such as when a bacterium loses its ability to produce a toxin or changes its chemistry (MicroFocus: 6.2).

Mutation: a permanent change for good or ill in the DNA of an organism.

Spontaneous Mutations

Spontaneous mutations are mutations that take place in nature without human intervention or identifiable cause. It has been estimated that one such mutation may occur for every 10^6 to 10^{10} replications of a bacterium. This implies that in a colony of a billion (10^9) bacteria, at least one mutant may be present.

Spontaneous mutation: a mutation that takes place in nature without an identifiable cause.

FIGURE 6.2

Replication of the Closed Loop Chromosome in *E. coli*

(a) The chromosome is a double-stranded DNA molecule. (b) An enzyme nicks the DNA at the origin and separates it into two V-shaped replicating forks. One replicating fork is shown in the inset. Along one side, DNA synthesis proceeds by continuous assembly of nucleotides. Along the other side, Okazaki fragments form. (c) New strands of DNA form along the four sides of the two forks. (d) Synthesis continues and the chromosome is about three-quarters replicated. (e) The inner chromosome moves to a position outside the other chromosome to prepare for separation. (f) Two chromosomes are fully formed, each consisting of one new strand and one old strand of DNA.

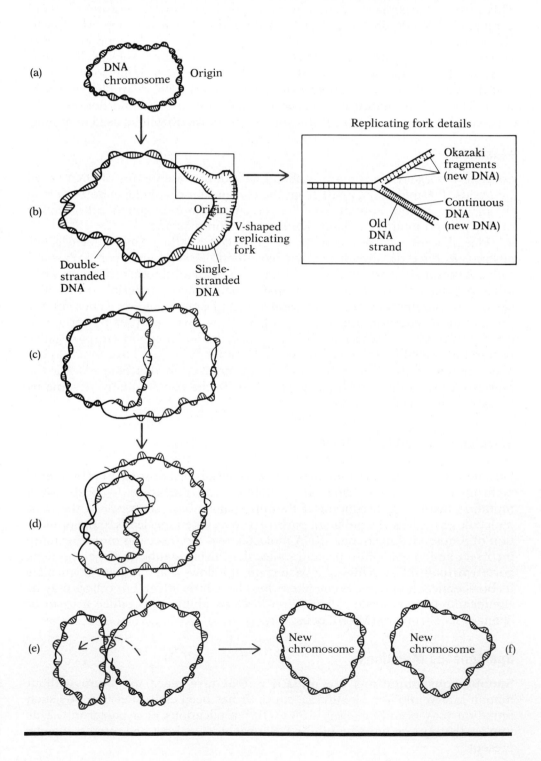

Making cheese is pretty easy—you curdle the milk, then add some ripening micro-organisms, and let them do their thing. If it's going to be Swiss cheese, then some gas-producing *Propionibacterium* is used; if Cheddar is the end-product, then lacto-bacilli are used; and if you are after blue cheese, a blue-green *Penicillium* species is inoculated to the curds.

But occasionally the cheese turns out bad—the holes are too small in the Swiss, the aroma is off in the Cheddar, and the blue cheese is simply too "ripe." The cheese-maker can be blamed in some cases, but other times it's the fault of the micro-organisms, a fault traced to a mutation—when the DNA changes, the enzymes change, and the chemistry changes. Now there are new tastes, new smells, and . . . new adventures. Hmmm. Maybe the new cheeses aren't that bad after all. And maybe the mutations can be put to profitable use. Maybe we've just invented a new cheese. Maybe . . .

In the normal course of events, cells arising from a spontaneous mutation are masked by normal cells. However, should a selective agent be introduced, then the mutant will survive, multiply, and emerge. For example, doctors have used penicillin to treat gonorrhea for many decades. Since 1976, however, a penicillin-resistant strain of *Neisseria gonorrhoeae*, the gonococcus, has been emerging in human populations. Many investigators believe that the resistant strain arose by spontaneous mutation at some unknown time, perhaps centuries ago, and as penicillin gradually eliminated susceptible strains, the resistant strain filled the niche. An equally disturbing problem has recently been encountered with antibiotic-resistant *Salmonella* serotypes (MicroFocus: 6.3).

nĭ-se're-ah

Induced Mutations

Most of the information gathered on mutation has come from induced mutations. **Induced mutations** are those in which the cause can be identified. Occasionally these mutations occur by accident. More often they result from planned experiments in which laboratory personnel subject bacteria to chemical or physical agents. The agents are referred to as **mutagens**. Experiments in molecular genetics reveal that a broad variety of induced mutations may arise, depending on the type of mutagen used.

Induced mutation: **an experimental mutation in which a cause can be identified.**

Mutations arising from treatment with **ultraviolet light** are among the best understood (Figure 6.3). Ultraviolet light is a form of energy not perceived by the human eye. When this energy is absorbed by DNA, it induces adjacent thymine molecules to link together. The DNA thus loses its ability to insert adenine bases in mRNA molecules during protein synthesis. Ultraviolet light is sometimes used for sterilization purposes because it quickly kills bacteria. In the Gram-negative rod *Serratia marcescens*, ultraviolet light induces mutations that prevent the organism from forming its normal red pigment.

sĕ-ra'she-ah
mar-ses'ens

Nitrous acid will also cause an induced mutation. This chemical mutagen converts DNA's adenine molecules to hypoxanthine molecules. Adenine will normally slot thymine into position during DNA replication, but hypoxanthine slots cytosine into place. Later, when protein synthesis takes place, the new DNA codes for guanine in the mRNA instead of the normal adenine (Figure 6.4).

hi"po-zan'thēn

Mutations can also be induced by a series of **base analogs**. These are substances that bear a chemical resemblance to nitrogenous bases. One example, 5-bromouracil, is taken up by cells and incorporated into DNA where thymine should be positioned. The new DNA functions poorly with the analog in place. In the treatment of diseases due to DNA viruses, a base analog is valuable because nucleic acid directs viral replication, and a virus with a functionless

bro"-mo-ur'ah-cil

CONNECTIONS

Antibiotics like penicillin, streptomycin, and tetracycline have revolutionized medicine while providing wonder drugs to agriculture as well. Today, about two-thirds of cattle and virtually all poultry, hogs, and veal calves in the United States are raised on feed that includes antibiotics. According to a recent study, animals consume almost 8 million pounds of drugs annually, nearly 40 percent of all that is produced in the United States. The antibiotics keep the animals healthy in pens and, for reasons not yet clear, appear to speed up growth.

After decades of largely uncritical acceptance, scientists are questioning the use of antibiotics in animal feeds. The trouble stems from the observed increase in the resistance of bacteria to drugs. Researchers believe that drug-resistant mutant strains are emerging and when meat and poultry products transfer mutant strains to humans, they cause disease that defies antibiotic treatment.

Evidence for this belief was largely circumstantial until 1983. Early that year, 18 people in four Midwestern states became infected with *Salmonella newport*. This Gram-negative rod grows in the human intestine and causes extensive diarrhea. Tetracycline and two forms of penicillin proved useless in treatment, and 11 patients were hospitalized and one died.

Scientists from the federal Centers for Disease Control and Prevention (CDC) began an exhaustive search for the origin of the disease. In the ensuing weeks they traced the *Salmonella* back to a beef herd in South Dakota and learned that the farmer regularly added a handful or so of tetracycline to a ton of feed to promote growth. The farmer had won local awards for cleanliness, but he offered that several of his calves died of diarrhea some months before and that he sent samples of one calf to a federal laboratory in Iowa. The investigators called the lab and secured samples of the

calf tissue. Not only did they isolate *Salmonella newport*, they also found plasmids in the *Salmonella* biochemically identical to plasmids from the *Salmonella* in victims.

Now the search intensified. Researchers determined that some of the cattle were slaughtered in Minnesota, processed in Nebraska, and shipped as hamburger meat to certain Minneapolis–St. Paul supermarkets. Coincidentally, eight of the *Salmonella* victims shopped for their meat at these stores, and all had eaten ground beef in the days before their illness. Two other patients purchased hamburger meat at Iowa supermarkets also supplied by the processor.

Though *Salmonella newport* was never found in supermarket beef, the investigators connected enough dots to persuade skeptics that the cattle were the biological factories in which drug-resistant *Salmonella newport* emerged. To many scientists, the findings provided the smoking gun linking the use of antibiotics in animal feed to human illness.

FIGURE 6.3

Two Morphological Forms of *Candida albicans*

Phase-contrast micrographs of the agent of yeast infections in humans. (a) The oval yeast form commonly seen in vaginal infections. (b) The filamentous moldlike form often observed in invaded tissue. Bar = 40 μm. In 1990, investigators demonstrated that the yeast form of *C. albicans* could be converted to the moldlike form by treating the organism with nitrous acid and ultraviolet light, thereby mutating its genetic material.

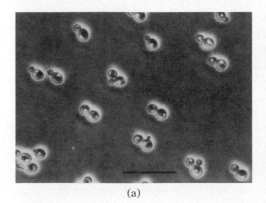

(a)

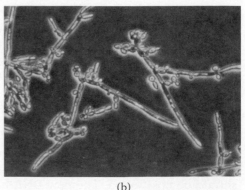

(b)

FIGURE 6.4

Mutations in Bacteria

(a) Nitrous acid causes induced mutations in bacteria. The nitrous acid alters the adenine molecule (asterisk) and changes it to a hypoxanthine molecule. Hypoxanthine slots cytosine into position during DNA replication instead of the normal thymine. The altered daughter DNA propagates the mutation by forming altered and mutated granddaughter DNAs. (b) Base analogs induce mutations by substituting for nitrogenous bases in the synthesis of DNA. The diagrams show two base analogs and their close structural relationship to the normal nitrogenous bases.

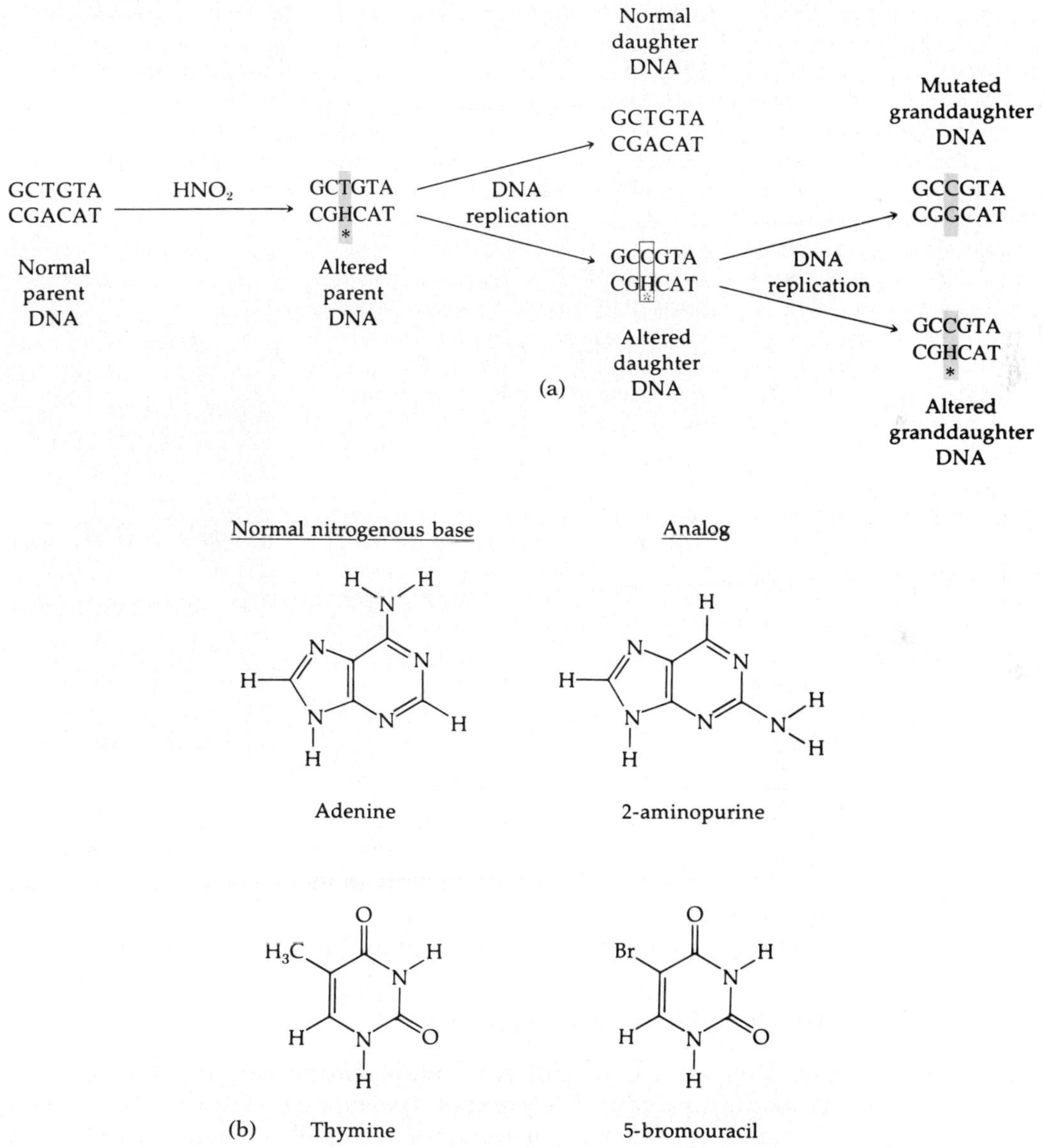

(a)

Normal nitrogenous base   Analog

Adenine   2-aminopurine

(b)   Thymine   5-bromouracil

DNA molecule cannot replicate. The drug acyclovir is a base analog that works against herpes viruses.

Other mutations involve the **deletion** or **insertion** of a nucleotide into the DNA molecule. For example, benzopyrene, present in industrial soot and smoke, and aflatoxin, a fungal toxin found in certain animal products and foods, can induce mutations this way. Substances as benzopyrene and aflatoxin do not themselves become part of the DNA molecule, but they cause the deletion or insertion of extra nucleotides during replication. This may lead to a **frameshift mutation**

Aflatoxin:
a toxin produced by certain pathogenic species of fungi.

6.4

MICROFOCUS

JUMPING GENES

In the early 1950s, scientists assumed that genes were fixed elements, always found in the same position on the same chromosome. But in 1951, Barbara McClintock unveiled her research with corn plants at a symposium at Cold Spring Harbor Laboratory on Long Island, New York. McClintock described genes that seemed to exist in sections and apparently moved from one chromosome to another. The audience listened in respectful silence. There were no questions after her talk, and only three individuals requested copies of her paper.

Like Gregor Mendel a hundred years before, McClintock kept close watch over the changes in color in her plants. Whereas Mendel cultivated peas, however, McClintock grew Indian corn, or maize. In the 1940s she noticed curious patterns of pigmentation on the kernels. Other scientists might have missed the patterns as random variations of nature, but McClintock's record-keeping and careful analysis revealed a method to nature's madness. The pigment genes causing the splotches of color appeared to be switched on or off in particular generations. Still more remarkable, the "switches" seemed to occur in a later generation at different places along the same chromosome. Some switches even showed up in different chromosomes. Such "controlling elements," as McClintock called them, were available whenever needed to turn the genes on or off.

In the modern lexicon of molecular genetics, McClintock's elements are recognized as a two-element system. One is an activator gene; the other is a dissociation gene. The activator gene, for reasons unknown, can direct a dissociation gene to "jump" along the arm of the ninth chromosome in maize plants where color is regulated. When the jumping gene reinserts itself, it turns off the neighboring pigmentation genes, thereby altering the color of the kernel.

The jumping gene is identical to the transposon found in bacteria. It may cause important mutations and gene rearrangements. Many scientists suspect that jumping genes play significant roles in the development of a fertilized egg into a mature organism, while serving as a driving force in evolution.

For Barbara McClintock, recognition came 30 years after the symposium at Cold Spring Harbor. In 1981 (at the age of 79) she received eight awards, among them a $60,000-a-year lifetime grant from the MacArthur Foundation and the $15,000 Lasker prize. In 1983, she was awarded the Nobel Prize in Physiology or Medicine. When informed of the Nobel award, she replied to an interviewer's question that "it seemed unfair to reward a person for having so much pleasure over the years, asking the maize plants to solve specific problems and then watching their response."

because the "reading frame" is modified, and the wrong mRNA forms beyond the point of mutation. A protein with an incorrect amino acid sequence results. Benzopyrene and aflatoxin are believed to be cancer-causing (carcinogenic) substances partly because of the mutations they induce in eukaryotic cells.

Transposable Genetic Elements

Insertion sequence:
a segment of DNA that forms copies that move elsewhere on the chromosome.

Mutations of a totally different nature may be caused by fragments of DNA called **transposable genetic elements**. Two types are known: insertion sequences and transposons. **Insertion sequences** are small segments of DNA with about 1000 base pairs. They are found at one or more sites on the bacterial chromosome and appear to have no specific genetic information other than for the ability to insert onto a chromosome as we shall see during our discussion of conjugation. Insertion sequences form copies of themselves, and the copies move from their normal position into other areas of the chromosome. Here they may interrupt the coding sequence, thereby inducing the wrong protein or no protein to form. Some investigators consider insertion sequences to be a prime force behind spontaneous mutation.

Transposon:
a segment of DNA that carries functional genes from one chromosomal location to another.

Within the past decade, scientists have made rapid advances in understanding the second type of transposable genetic elements called **transposons**. These are the so-called "jumping genes" for which Barbara McClintock won the 1983 Nobel Prize in Physiology or Medicine (MicroFocus: 6.4). First identified and named in 1974 by British microbiologists R.W. Hedges and A.E. Jacobs, transposons are larger than insertion sequences and appear to carry information for protein synthesis. They induce mutations in the same way as insertion sequences, that is, by interrupting the genetic code.

The movement of transposons appears to be nonreciprocal, meaning that an element moves away from its location and nothing takes its place. This contrasts with insertion sequences where copies move. Transposons may move from plasmid to plasmid, from plasmid to chromosome, or from chromosome to plasmid. The presence of inverted repetitive base sequences at the ends of the element (Figure 6.5a) appears to be important in establishing the ability to move.

Of particular significance is the finding that transposons often contain genes for antibiotic resistance. If the plasmid containing the transposon moves from one bacterium to the next, as plasmids are known to do, the transposon will move along with it, thus spreading the genes for antibiotic resistance among bacteria. Moreover, the movement of transposons among plasmids may explain how a single plasmid acquires numerous genes for resistance to different antibiotics. The results of a mutation or series of mutations occurring in a bacterium may also be propagated in the same way.

Plasmid:
a closed loop of DNA in the cytoplasm of a cell.

Antibiotic:
a chemical product of a microorganism that has an inhibitory effect against another microorganism.

FIGURE 6.5
Stimulants of Mutation

(a) Structure of a typical transposon. The transposon illustrated contains one or more active genes bordered by inverted repetitive sequences. Note that the base sequence in A is the reverse and complement of the base sequence in B. Such a situation is called a palindrome. Also, there are numerous sequences such as C-C and G-G. (b) The effect of overlapping genes. The single-stranded DNA provides the genetic code for two amino acid sequences. Below the DNA strand, the base sequence for the third amino acid (Met) includes thymine, a base from the previous base sequence. A mutation at the thymine (shaded) would thus change two base codes.

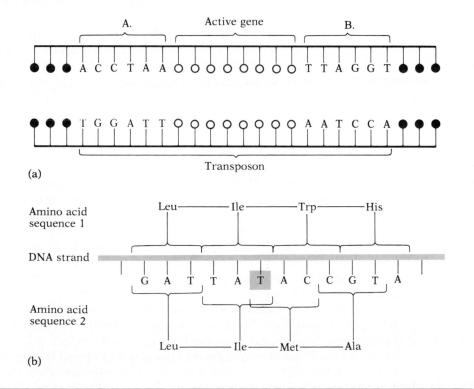

(a)

(b)

Overlapping Genes

For years scientists assumed that a mutation in a gene affected the production of a single protein. However, Frederick Sanger's 1977 work indicated otherwise. Sanger determined the entire nucleotide sequence of a viral DNA molecule and mapped its 5386 bases. Analysis of the DNA by Sanger's co-workers showed that at least four of the genes are overlapping, as shown in Figure 6.5b. This implies that an individual triplet of bases might code for two different amino acids. Thus, a mutation at a base could affect not one amino acid but possibly two, and, in effect, not one protein, but two. In 1958, Sanger had won the Nobel Prize in Chemistry for sequencing the amino acids in insulin. Now, in 1980, he shared a second Nobel Prize in Chemistry for mapping the bases of the viral DNA.

The Ames Test

Cancer:
a condition characterized by the uncontrolled growth and multiplication of cells.

Histidine:
one of 20 amino acids found in proteins.

Some years ago scientists observed that about 90 percent of the agents that cause cancer in humans also cause mutations in bacteria. Working on this premise, Bruce Ames of the University of California developed a procedure to determine whether a chemical can induce a bacterial mutation and thus be a potential agent of cancer. The procedure, called the **Ames test**, is widely used as well as relatively inexpensive, rapid, and accurate.

To perform the Ames test, a technician inoculates a histidine-requiring strain of *Salmonella typhimurium* onto a plate of bacteriological medium lacking histidine. Normally this strain of *Salmonella* will not grow in the medium because the gene allowing the bacterium to synthesize histidine is mutated and hence not active. Now the potential cancer agent is added to the medium, and the plate is incubated. If bacterial colonies appear, then one may conclude that the agent mutated the bacterial gene so it can now code for the enzyme needed for histidine synthesis. Thus, because the agent is a mutagen it is therefore a possible cause of cancer. If bacterial colonies fail to appear, no mutation took place (Figure 6.6).

The Ames test is a sensitive and rapid method for testing numerous compounds. However, the appearance of bacterial colonies does not prove that the agent is cancer-causing, nor does the lack of bacterial colonies indicate that an agent is safe to use. The test works only within specified limits, and these must be understood and appreciated if the proper conclusions are to be made.

TO THIS POINT
■

We have surveyed the bacterial chromosome, a single molecule of DNA arranged as a closed loop, which is free in the cytoplasm without a surrounding membrane or protein support. The chromosome replicates by the semiconservative method when the bacterium is undergoing binary fission, or by the rolling circle method when conjugation is taking place. Small loops of DNA called plasmids exist apart from the DNA and are key players in conjugation and genetic engineering, as we shall see presently.

The discussion then turned to mutation, a process in which the chromosome is altered through a permanent change in the DNA. We compared spontaneous mutations, which take place in nature without human intervention, to induced mutations, which usually occur under laboratory conditions. Various types of induced mutations illustrated how changes in DNA can be brought about by different mutagens. Possible

FIGURE 6.6

The Ames Test for Detecting Chemical Mutagens

A histidine-requiring mutant of *Salmonella typhimurium* is spread on the surface of an agar medium that lacks histidine. A well is cut into the agar and filled with the suspected mutagen. The chemical diffuses from the well and comes in contact with the bacteria. If the chemical is mutagenic, it causes a reversion of the genes of the bacteria so that they can synthesize histidine. Colonies of the histidine-synthesizing bacteria then grow in the zone of diffusion. The greater the mutagenicity of the chemical, the more bacterial colonies develop on the medium.

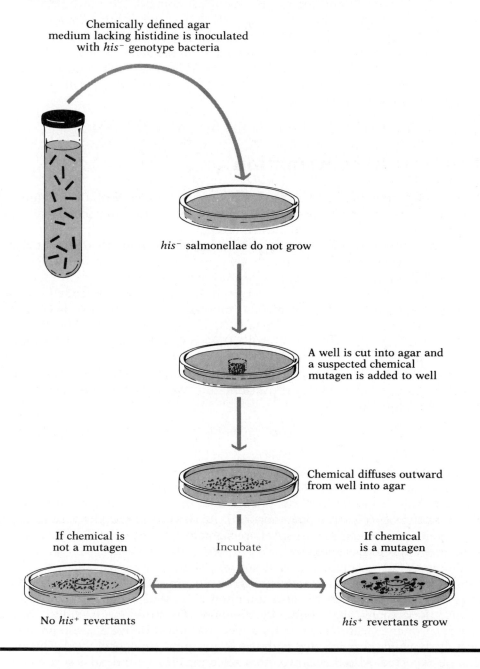

Chemically defined agar medium lacking histidine is inoculated with *his⁻* genotype bacteria

his⁻ salmonellae do not grow

A well is cut into agar and a suspected chemical mutagen is added to well

Chemical diffuses outward from well into agar

If chemical is not a mutagen

Incubate

If chemical is a mutagen

No *his⁺* revertants

his⁺ revertants grow

mutagens include insertion sequences and transposons. These fragments of DNA slot into areas of gene activity, thereby changing the genetic code and inducing a chromosome or plasmid alteration. The concept of overlapping genes was mentioned to show how a mutation may affect more than one

gene, and the Ames test was discussed to illustrate how muta-
tions are used for practical benefit.

We shall now turn our attention to bacterial recombina-
tion. In this remarkable process of bacterial genetics, two orga-
nisms are involved in the transfer of DNA and an alteration of
the genetic material occurs. We shall examine three processes:
transformation, conjugation, and transduction. In transforma-
tion, an organism acquires DNA from its extracellular environ-
ment; in conjugation, two organisms come together and DNA
passes from one bacterium to the other; and in transduction,
a virus transports DNA between bacteria. Research in this
area has many applications to clinical medicine. Moreover,
an understanding of bacterial recombination was to lead to
the process of genetic engineering, as we shall see later in
this chapter.

Bacterial Recombination

The second general method for altering the genetic material of a bacterium is through the process of **recombination**. Although mutation may be considered a form of recombination, we shall use recombination in the more restricted sense to mean the genetic process in which two organisms are involved: a **donor cell** and a **recipient cell**. The donor cell contributes chromosomal DNA or plasmid DNA to the recipient cell. If plasmids are obtained, they exist independently in the recipient's cytoplasm and begin to multiply and encode proteins immediately. If the recipient obtains chromosomal DNA, the new DNA pairs with a complementary region of recipient DNA and replaces it. Thus, there is no change in quantity of the recipient's DNA but there may be a substantial change in its quality.

Microbiologists have identified three methods for bacterial recombination: transformation, conjugation, and transduction. We shall examine each in the following paragraphs.

Transformation

In 1928, an English bacteriologist named Frederick Griffith published the results of an interesting set of experiments with *Streptococcus pneumoniae*. This organism is a major cause of bacterial pneumonia; the bacterium is often referred to as a pneumococcus (pl., pneumococci). At Griffith's time, the results were considered unusual, but in retrospect, microbiologists note that they gave some of the first clues to gene activity.

Pneumococci occur in different strains. There is an encapsulated strain, designated S, because the organisms grow in **smooth** colonies. This strain causes pneumonia. There is also an unencapsulated strain, designated R, because the colonies are **rough**. Organisms in this strain are harmless. In an early experiment, Griffith showed that mice injected with live S strain pneumococci died, while those injected with live R strain pneumococci lived. This was what he expected. Also, he showed that mice injected with dead S strain organisms lived (Figure 6.7). Again, this result was not unusual.

What happened next puzzled Griffith. He mixed dead **S strain** bacteria with live **R strain** bacteria and injected the mixture into mice. The mice died. Griffith questioned how a mixture of live harmless bacteria (R) and dead pathogenic bacteria (S) could kill the mice. His answer came when he autopsied the animals:

Recombination:
a change in an organism's DNA resulting from the acquisition and incorporation of another organism's genes.

Pneumonia:
an infectious disease of the lungs, due to several types of microorganisms, including bacteria.

S strain pneumococci:
those that form smooth colonies and are pathogenic.

R strain pneumococci:
those that form rough colonies and are nonpathogenic.

FIGURE 6.7

The Transformation Experiments of Griffith

(a) When live pathogenic pneumococci (S strain) were injected to mice, the animals died and pathogenic bacteria were isolated from the animals' tissues. (b) When live harmless pneumococci (R strain) were injected, the animals remained healthy. (c) When heat-killed pathogenic bacteria were injected to the animals, they lived and were healthy and no colonies were isolated on the medium. All these results were as expected. (d) However, when live harmless bacteria (R strain) were mixed with heat-killed pathogenic bacteria and the mixture injected to the animals, the animals died. On autopsy, Griffith isolated live harmless bacteria (R strain) as well as live pathogenic bacteria (S strain). Apparently, some of the harmless bacteria had been transformed to pathogenic bacteria.

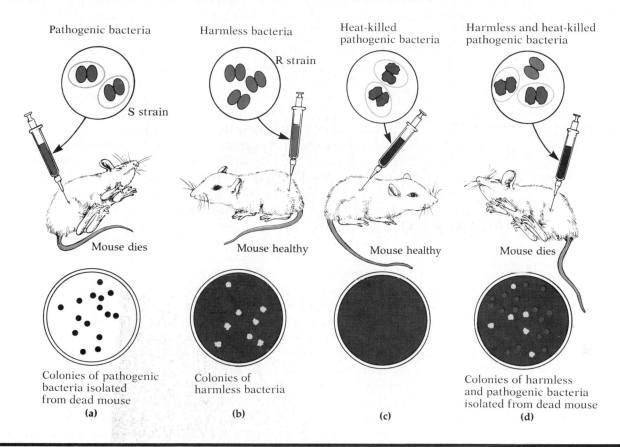

they were full of live S strain pneumococci. Apparently the live R strain bacteria had been transformed to live S strain bacteria.

Five years later (1933), James L. Alloway of the Rockefeller Institute in New York confirmed Griffith's work. Alloway used fragments and debris from the dead S strain cells to transform the R strain cells. However, it was not until 1944 that Oswald T. Avery and his associates Colin M. MacLeod and Maclyn N. McCarty, also of the Rockefeller Institute, purified the transforming substance. These investigators found that the substance was not protein, as had been anticipated, but a then obscure organic compound called deoxyribonucleic acid (DNA).

The majority of scientists were blind to Avery's discovery and were reluctant to accept DNA as a hereditary substance. Geneticists of that period were not trained as chemists, and Avery's experiments were difficult to repeat. Also, many scientists believed that experimental results from bacteria could not necessarily be applied to eukaryotes. Moreover, preoccupation with World War II had restricted the dissemination and flow of scientific knowledge. Not until the 1950s was DNA widely accepted as the molecule of heredity.

Eukaryotes: complex organisms such as plants and animals whose cells have a nucleus and organelles.

Modern scientists regard **transformation** as an important method for recombination even though it takes place in less than 1 percent of a bacterial population. In transformation, a number of donor cells break apart and an explosive release and fragmentation of DNA follows. A segment of double-stranded DNA containing about 10 to 20 genes then passes through the cell wall and membrane of a **recipient cell**, as shown in Figure 6.8. After entry to the recipient cell, an enzyme dissolves one strand of the DNA, leaving the second strand to be incorporated. This second strand then displaces a segment of a DNA strand from the recipient's chromosome. The displaced DNA is dissolved by another enzyme in the cell. Transformation may also take place by the reception of plasmids, in which case no chromosomal DNA is displaced. The plasmid adds genes to the cell and multiplies when the cell multiplies, thereby transforming the bacterium.

The ability of a cell to be transformed depends on its **competence**, defined as the ability of a recipient bacterium to take up DNA from the environment. Competence is an intriguing property that varies among bacteria. For example, in *Streptococcus pneumoniae*, the agent of bacterial pneumonia, competence is displayed in the entire cell population late in the logarithmic phase of growth when a competence-provoking factor is produced. By contrast, *Bacillus* species display competence during the early events leading to spore formation. Competence develops better in certain growth media than in others, and sometimes it does not develop at all.

Factors affecting the cell surface are important to competence, particularly changes in **membrane permeability** or **surface receptors**. The uptake of DNA by *E. coli*, for example, can be induced in the laboratory by chilling bacteria to 4° C in the presence of calcium chloride, then quickly heating them to 42° C together with the DNA. This treatment apparently alters the membrane and encourages passage of DNA strands. Competent streptococci and *Bacillus* cells have approximately 50 binding sites where DNA uptake can occur.

Under natural conditions, transformation takes place in organisms whose DNA is very similar to the DNA being received, which generally implies cells of the same species. The process has been observed in the species mentioned above as well as in *Haemophilus*, *Neisseria*, and *Azotobacter* species. One of the effects of transformation may be to increase an organism's pathogenicity. In pneumococci, the acquisition of genes for capsule formation allows an organism to avoid body defenses and survive and cause disease, as Griffith showed. Microbiologists have also demonstrated that when mildly pathogenic strains of bacteria take up DNA from other mildly pathogenic strains, there is a cumulative effect and the degree of pathogenicity increases. Observations such as these may help explain why especially pathogenic strains of bacteria appear from time to time. Transformed bacteria may also display drug resistance from the acquisition of R factors. Pigment production may be another characteristic derived through transformation.

How significant is transformation as a means of genetic recombination under natural conditions? No one knows for sure, but it appears certain that transformation occurs regularly where bacteria exist in very crowded conditions, such as in the human intestine.

Conjugation

In the recombination process called **conjugation**, two live bacteria come together, and the donor cell transfers genetic material to the recipient cell. This process was first postulated in 1946 by Joshua Lederberg and Edward Tatum in a series of experiments with *E. coli*. Lederberg and Tatum mixed two different

Competence:
the ability of a bacterium to take up DNA from the extracellular environment.

Streptococcus pneumoniae:
a Gram-positive chain of diplococci that causes pneumonia, usually in compromised individuals.

he-mof'ĭ-lus
ah-zo'to-bak"ter

Pathogenic:
able to cause disease.

R factors:
plasmids that carry genes, which confer antibiotic resistance on a bacterium.

FIGURE 6.8

Bacterial Transformation

(a) A segment of donor DNA is released from a disintegrating bacterium and (b) is taken up by a competent living recipient cell. (c) Once the DNA segment passes through the cell wall and membrane of the recipient cell, one strand of the DNA dissolves (d), leaving one strand to be integrated. (e) A strand of recipient cell DNA leaves the chromosome and is replaced by the strand of donor DNA. The cell is now considered transformed. Replication of the cell during binary fission will yield other transformed bacteria.

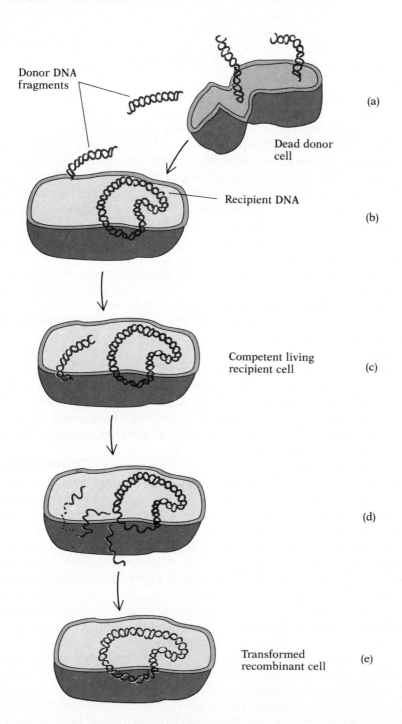

Donor DNA
fragments

Dead donor
cell

(a)

Recipient DNA

(b)

Competent living
recipient cell

(c)

(d)

Transformed
recombinant cell

(e)

zhah-kob′

F factor:
a plasmid that functions in conjugations among bacteria.

F pili:
short hairlike fibers of protein that connect donor and recipient cells during conjugation.

Hfr strain:
a strain of bacteria in which chromosomal material passes to a recipient during conjugation.

Insertion sequence:
a segment of chromosomal DNA that specifies where an acquired DNA segment is to insert.

strains of bacteria and found that genetic traits could be transferred among them as long as contact occurred. The investigators shared the 1958 Nobel Prize in Physiology or Medicine for their work.

Experiments in the 1950s by Francois Jacob and Elie L. Wollman established that conjugating bacteria were of two mating types. Certain "male" types donated their DNA. These were designated **F⁺ cells**. Other "female" types received the DNA and came to be known as **F⁻ cells**. Jacob and Wollman found that F⁻ cells (recipients) became F⁺ cells (donors) when they acquired a small amount of DNA. This eventually led William Hayes to discover the **F (fertility) factors** present in the donor cell. Although the male and female nomenclature is still used today, the process of bacterial conjugation is fundamentally different from sexual mating, since the mechanisms of DNA incorporation are very different.

In contemporary microbiology, the F factors are known to be **plasmids**, the double-stranded loops of DNA existing apart from the chromosome. The factors (plasmids) contain about 20 genes, most of which are associated with conjugation. These genes code for enzymes that replicate and move DNA during conjugation and for enzymes and structural proteins needed to synthesize special pili at the cell surface. Known as **F pili**, these hairlike fibers contact the recipient bacteria, then retract so that the surfaces of the donor and recipient are very close or touching one another. At the area of contact, a channel or conjugation bridge is believed to form (Figure 6.9).

Once contact has been made, the F factor begins replicating by the rolling circle mechanism discussed earlier. A single strand of the factor then passes over or through the channel to the recipient. When it arrives, enzymes synthesize a complementary strand, and a double helix forms once again. The double helix bends to a loop and reforms an F plasmid, thereby completing the conversion of the recipient from F⁻ cell to F⁺ cell. Meanwhile, back in the donor cell, a new strand of DNA forms to complement the leftover strand of the F plasmid.

The transfer of F factors involves no activity of the bacterial chromosome; therefore the recipient does not acquire new genes other than those on the F factor. A type of conjugation that accounts for the passage of chromosomal material does exist in bacteria, however. Strains of bacteria that display the ability to donate chromosomal genes are called **high frequency of recombination**, or **Hfr strains** (Figure 6.9). Such strains were discovered in the 1950s by William Hayes in *E. coli*.

In Hfr strains, the F factor attaches to the chromosome. This integration is a rare event and requires that an **insertion sequence** be present on the chromosome to recognize the F factor. At the point of attachment, the chromosome opens and a copy of one strand of the chromosomal DNA is made by the rolling circle mechanism. A portion of single-stranded DNA then passes via the channel into the recipient cell, where it replaces the complementary region in the recipient's chromosome. The first genes to enter the recipient are F factor genes, but these are not the ones that control the donor state. Instead, the last segment to enter the recipient are the F factor genes for the donor state. These rarely enter the recipient, however, because conjugation is often interrupted by such things as movement that break the attachment. (An estimated 100 minutes is required for a complete transfer of an *E. coli* chromosome.) Thus the F⁻ cell usually remains a recipient, but it has some new genes acquired from the donor.

In certain cases the entire F factor is transferred to the recipient (Figure 6.10). When this happens, the factor usually detaches from the recipient's chromosome, and enzymes synthesize a strand of complementary DNA. The factor now forms a loop to assume an existence as a plasmid, and the recipient becomes a donor (an F⁺ cell).

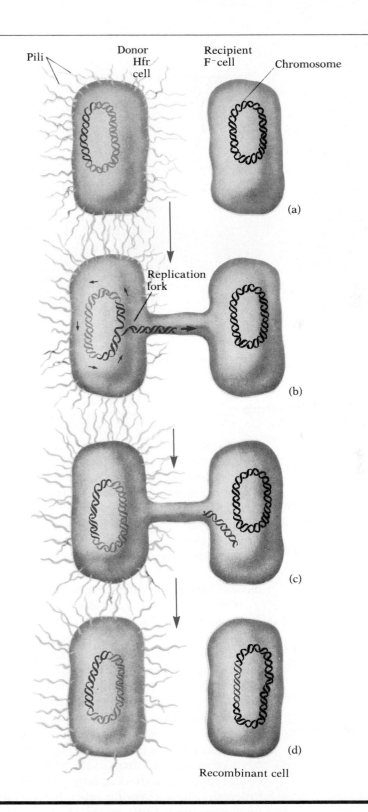

FIGURE 6.9

Conjugation in Bacteria

(a) The Hfr donor cell is distinguished by the presence of pili on its surface, while the recipient cell (F⁻) has no pili. (b) As conjugation begins, the donor cell forms a conjugation tube. The Hfr chromosome opens and a copy of one strand of DNA is made by the rolling circle method. A portion of the new strand then enters the conjugation tube. (c) The new strand enters the recipient cell. The longer the conjugation lasts, the more of the new strand will enter the recipient cell. (d) Once the new strand (or piece of new strand) has entered the recipient cell, it displaces a segment of the recipient's chromosome. The displaced segment disintegrates, and the recipient cell has been recombined.

Pili

Donor Hfr cell

Recipient F⁻ cell

Chromosome

(a)

Replication fork

(b)

(c)

(d)

Recombinant cell

FIGURE 6.10

Two *E. coli* Cells in Contact During Conjugation

The cell with the pili (on the left) is the donor cell.

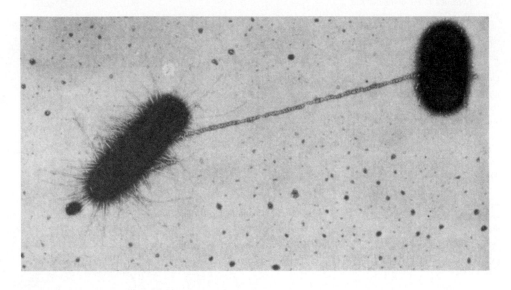

On occasion, the F factor breaks free from the chromosome of an Hfr cell before conjugation takes place and resumes an independent status. The Hfr cell then reverts to an F$^+$ cell. Sometimes when the F factor leaves the chromosome, it takes along a segment of chromosomal DNA. The factor with its extra DNA is now called an **F′ factor** (pronounced F-prime). When the F′ factor is transferred during a subsequent conjugation, the recipient acquires chromosomal genes from the donor, a process known as **sexduction**. Sexduction results in a recipient with its own genes for a particular process as well as additional genes from the donor for that same process. In the genetic sense, the recipient is a partially diploid organism since there are two genes for a given function.

Conjugation has been demonstrated among various genera of bacteria, in contrast to transformation, which appears to occur only among cells of the same species. For example, conjugation is known to occur between such Gram-negative bacteria as *Escherichia* and *Shigella*, *Salmonella* and *Serratia*, and *Escherichia* and *Salmonella*. Intergenic transfer has great significance in the transfer of antibiotic resistance genes carried on plasmids (MicroFocus: 6.5). When the F factors are plasmids, the transfer occurs readily. Moreover, when the genes are attached to transposons, the transposons may "jump" from ordinary plasmids to F factors, after which transfer may occur.

Gram-positive bacteria also appear capable of conjugation. Microbiologists have experimented extensively with *Streptococcus mutans*, a common cause of dental cavities. In this organism, conjugation appears to involve only plasmids, especially those carrying genes for antibiotic resistance. Similar observations have been made in *Bacteroides* and *Clostridium* species. To this point, however, the mechanism of plasmid transfer in Gram-positive bacteria is poorly understood, and chromosomal transfer has not been demonstrated (Table 6.2).

Diploid:
two sets of genes for a
particular trait.

sĕ-ra′she-ah

Transposon:
a movable segment of DNA.

bak″tĕ-roi′dēz
klo-strid′e-um

TABLE 6.2

A Comparison of Transformation, Conjugation, and Transduction

CHARACTERISTIC	TRANSFORMATION	CONJUGATION	TRANSDUCTION
Method of DNA transfer	Movement across wall and membrane of recipient	Through channel after cell-to-cell contact	By an intermediary
Amount of DNA transferred	Few genes	Few genes to entire chromosome	Few genes
Plasmid transferred	Yes	Yes	Not likely
Entire chromosome transferred	No	Sometimes	No
Virus required	No	No	Yes
Live bacteria required	Yes	Yes	Yes
Cell debris required	Yes	No	No
Used to acquire antibiotic resistance	Yes	Yes	Not likely

Transduction

Bacterial recombination by the process of **transduction** was first reported in 1952 by Joshua Lederberg and Norton Zinder. While working with *Salmonella* cells, Lederberg and Zinder observed recombination, but ruled out conjugation and transformation because the cells were separated by a thin membrane and DNA was absent in the extracellular fluid. Eventually, they discovered a virus in the fluid and learned the details of what was taking place.

The virus that participates in transduction is called a **bacteriophage**, or simply **phage** (Figure 6.11). Though invisible at the time, the activity of a bacteriophage (literally "bacteria-eater") was described in 1915 by Frederick Twort and two years later by Felix d'Herelle. Bacteriophages (or phages) were originally assumed to be a type of poison because they dissolve bacteria. Today scientists recognize them as viruses composed of a core of DNA or RNA surrounded by a coat of protein. Not all phages participate in transduction—those that do are called **transducing phages**. All the latter contain DNA.

In the life cycle of a bacteriophage, the phage interacts with bacteria in two ways. In one way the phage invades a bacterium, reproduces within the bacterium, and then destroys the bacterium as new phages are released. This cycle is called the **lytic cycle** of infection because at the cycle's conclusion the invaded bacteria lyse, or rupture. Phages that cause lysis are known as **virulent phages**.

The second way that phages interact with bacteria also involves invasion of the bacterium. In this case, however, the phage DNA codes for a repressor protein that prevents replication. The phage DNA may remain in the cytoplasm as a plasmid or it may form a closed circle, align next to the bacterial chromosome, and integrate into the bacterial chromosome, as the F factor does in Hfr strains. This process of nonreplication is called **lysogeny**. The phage that enters into lysogeny is known as a **temperate phage** and the integrated viral DNA is a **prophage**. Because there are two forms of phage interaction with bacteria, there are two ways in which phages can be involved in the transfer of bacterial genes: generalized transduction and specialized transduction.

bak-te're-o-faj″
Bacteriophage:
a virus that attacks bacteria.

Lytic cycle:
the process in which a phage replicates within a bacterium, thereby destroying the bacterium.

Lysogeny:
the process in which a phage incorporates into bacterium and does not replicate itself or destroy the bacterium.

6.5

MICROFOCUS

TRANSFERABLE DRUG RESISTANCE

In 1968, an extremely serious form of bacterial dysentery broke out in Guatemala. Dysentery is a disease of the human intestine accompanied by loss of large volumes of fluid. The responsible bacterium, *Shigella dysenteriae*, resisted treatment with chloramphenicol, tetracycline, streptomycin, and sulfanilamide, any of which are normally used in therapy. In the three years that the epidemic raged, 100,000 people were infected and 12,000 died.

This particular outbreak of drug-resistant dysentery points up the consequences of what could happen when antibiotic treatment is stifled by resistant bacteria. How *Shigella* may have acquired the resistance was first shown in the 1950s in a remarkable set of experiments by a Japanese team of investigators.

The story began in 1955 when a Japanese woman suffered a case of dysentery caused by bacteria resistant to the same quartet of drugs observed years later in Guatemala. Doctors at Tokyo University, led by Tomoichiro Akiba, investigated the case and found that drug-resistant *Shigella* strains were fairly widespread in Japan. They also noted a surprising coincidence: patients with drug-resistant *Shigella* also had in their intestine a strain of *Escherichia coli* with resistance to the same four drugs. Since simultaneous mutations were highly unlikely, researchers postulated that the resistance had been transferred between organisms.

Akiba and his colleagues devised a series of experiments to test this hypothesis. They mixed liquid suspensions of drug-resistant *E. coli* with drug-sensitive *Shigella dysenteriae*. Then they carefully isolated and tested the *Shigella*. The results were startling: *Shigella* was now resistant to the same drugs as *E. coli*. A transfer had indeed taken place.

In the following years, numerous studies verified transferable drug resistance, and R factor plasmids were identified as the medium of transfer. Epidemics such as that in Guatemala soon broke out as drug-resistant bacteria began appearing in humans. By the 1990s, microbiologists had identified resistance in such diverse organisms as streptococci, gonococci, leprosy and tuberculosis bacilli, and malaria parasites. In the 1980s, an article in *Discover* magazine described a Detroit epidemic due to a drug-resistant strain of *Staphylococcus aureus*. The strain was quickly dubbed "super staph," and the headline was an eye-grabbing "Bugs That Won't Die." Transferable drug resistance had made the popular media.

FIGURE 6.11

Bacteriophages

(a) A transmission electron microscope view of a T2 bacteriophage. Note that each bacteriophage is composed of a head and a tail with a complex endplate. In two cases (arrows), the tail is contracted and a core may be seen protruding. This structure penetrates the cell wall and membrane during viral replication. (b) Bacteriophages adsorbed to receptor sites on a *Bacillus cereus* cell. Note the relative sizes of the viruses and the bacterium.

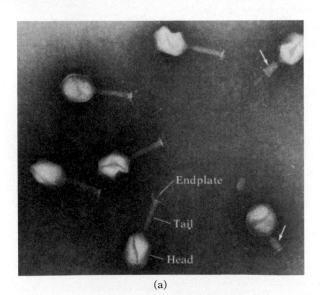

(a)

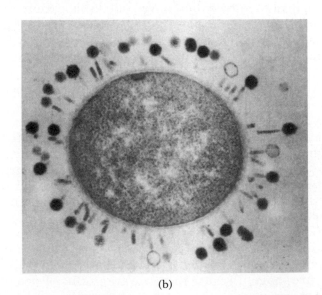

(b)

Generalized transduction is carried out by virulent phages that have a lytic cycle of infection (Figure 6.12). During phage replication a long, linear chain of DNA is first produced. This long DNA is then enzymatically cut into sections that will become phage DNAs. During this process, segments of bacterial DNA may accidentally get caught up in the cutting process, and bacterial segments may end up incorporated to phages where phage sections should be. This event is considered rare (1 in every 100,000 phage progeny may have bacterial DNA). Such a phage is fully formed but carries no phage genes. Even though it carries DNA, it cannot encode its own replication.

Now the transduction takes place. The transducing phage is released and attaches to a new (recipient) bacterium and injects its bacterial DNA. Once released in the recipient, the new genes may pair with a section of the recipient's DNA and replace the section. The recipient has now been transduced (changed) using genes from the donor bacterium and the phage intermediary. Two examples of transducing phages are phage P1 that infects *E. coli* and P22 that infects *Salmonella typhimurium*.

Specialized transduction involves the same genes each time, unlike generalized transduction that can involve any bacterial genes. Specialized transduction occurs as a result of **lysogeny**, the state in which DNA from the phage has incorporated into the DNA of the bacterium. The site of incorporation or insertion site may vary for different phages and is often a region where an insertion sequence is located. When incorporation takes place, phage genes are actually integrating into the sequence of bacterial genes rather than replacing bacterial genes (as in transformation). In the integrated state, the set of phage genes is called a **prophage**, as noted previously, and the bacterium carrying the prophage is said to be lysogenic.

At some time in the future, a mutagen such as UV light or a DNA inhibitor activates an enzyme system that excises the prophage out of the integrated state and causes it to reenter the lytic cycle (Figure 6.13). Most of the time the excision occurs precisely and results in release of an intact set of phage genes (prophage). Sometimes, however, an **imprecise excision** occurs and the excised prophage carries along bacterial genes adjacent to the insertion site while leaving behind some phage genes. At the conclusion of replication, multiple copies of the phage, each with some bacterial genes, are produced.

Because the phage nucleic acid is missing some essential phage genes, the phage resulting from the lytic cycle is somewhat defective. It can infect another bacterium and transfer the defective gene set to a recipient but the genes cannot code for a lytic cycle. Instead, they integrate into the bacterial chromosome carrying the bacterial genes with them. As before, the recipient bacterium has acquired genes from the original bacterium and the recipient is now considered transduced (Table 6.3).

Specialized transduction is an extremely **rare event** in comparison to the generalized form, because genes do not easily break free from the bacterial chromosome. However, the potential for recombination is substantial because lysogeny with prophages has been well established in clinical microbiology. For example, the diphtheria bacillus, *Corynebacterium diphtheriae* harbors a prophage that provides the genetic code for a toxin that destroys organ cells. Other toxins believed to be coded by prophages include staphylococcal enterotoxin in food poisoning, clostridial toxins in some forms of animal botulism, and streptococcal toxins in scarlet fever. Also, *Salmonella* cells carry prophages that code for lipopolysaccharides. These lipopolysaccharides provide the basis for separating *Salmonella* into serological types (serotypes) rather than species.

Certain viruses are known to remain in human body cells for years, expressing themselves at unspecified intervals. The viruses of herpes simplex, infectious

Generalized transduction: transduction in which phages carry random fragments of donor DNA into a recipient cell.

Specialized transduction: transduction in which phages carry selected segments of DNA from the chromosome of a donor cell into a recipient cell.

Prophage: viral genes integrated into a bacterial chromosome.

ko-ri"ne-bac-te're-um
dif-the're-ā

Lipopolysaccharides: polysaccharides with lipid component.

FIGURE 6.12
Generalized Transduction

During the reproduction of progeny bacteriophages (phages) within a bacterium, a rare phage particle may form that contains bacterial genes in place of phage genes. This is a rare, randomly occurring event. Any bacterial gene may be incorporated into the phage. The phage particle so produced cannot replicate itself but remains capable of attaching to another bacterium and inserting its genes. In this way genes from the originally infected bacterium are transferred to a second, recipient bacterium. The second bacterium is transduced.

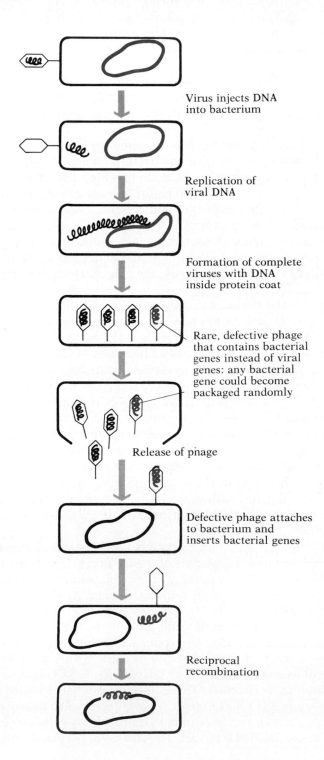

Virus injects DNA
into bacterium

Replication of
viral DNA

Formation of complete
viruses with DNA
inside protein coat

Rare, defective phage
that contains bacterial
genes instead of viral
genes: any bacterial
gene could become
packaged randomly

Release of phage

Defective phage attaches
to bacterium and
inserts bacterial genes

Reciprocal
recombination

FIGURE 6.13

Specialized Transduction

Specialized transduction begins with lysogeny and the integration of a phage DNA into a bacterial chromosome. In the diagram, the viral DNA is inserted between genes A and Z in the bacterium. During the release of the prophage from the integrated state, imprecise excision may occur and genes flanking the prophage may replace phage genes. Three possibilities are shown in the diagram: a phage with bacterial gene A, a phage that is normal, and a phage with bacterial gene Z. The defective phages cannot reproduce themselves but can attach to and invade a second, recipient bacterium. Once inside the bacterium, the defective phage genes (including bacterial genes) may integrate into the recipient's chromosome. In this manner the second, recipient bacterium is transduced, using genes from the originally infected bacterium.

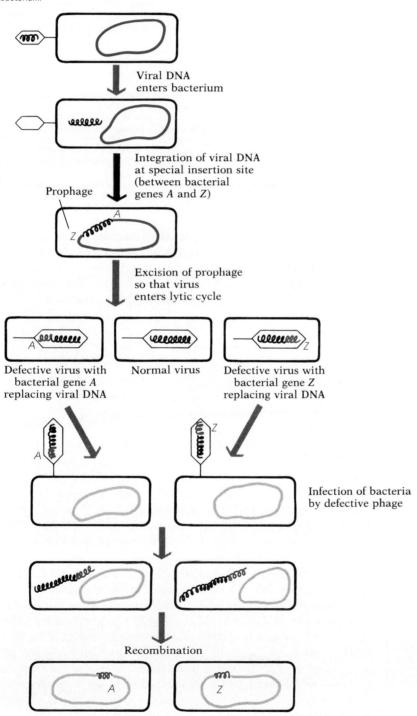

TABLE 6.3

Generalized Transduction Compared to Specialized Transduction

GENERALIZED TRANSDUCTION	SPECIALIZED TRANSDUCTION
1. Viruses (phages) penetrate bacterial cell and enter lytic cycle	1. Viruses (phages) penetrate bacterial cell and enter lysogenic cycle
2. Viral DNA begins to replicate immediately in the bacterial cytoplasm	2. Viral DNA incorporates into the DNA of the bacterium. Replication begins at a later time.
3. During replication, viral DNA accidentally uses some bacterial DNA to make new viruses	3. During release from chromosome in replication, viral DNA accidentally excises some bacterial DNA to make new viruses
4. Any bacterial genes are randomly packaged into new viruses	4. Bacterial genes adjacent to incorporated virus are packaged into new viruses
5. Some new viruses have bacterial DNA and no viral genes	5. Some new viruses have both viral DNA and bacterial DNA
6. Transducing viruses enter recipient bacterium	6. Transducing viruses enter recipient bacterium
7. Old bacterial genes are incorporated into chromosome of new bacterium	7. Old bacterial genes are incorporated into chromosome of new bacterium together with transducing viruses

mononucleosis, and chickenpox are examples. In addition, many tumor viruses are believed to associate with human chromosomes and transform the normal cell to a tumor cell. Such viruses may be considered types of prophages, and the ability to transduce human cells under clinical conditions may be analogous to what is taking place in bacterial recombination (MicroFocus: 6.6).

TO THIS POINT
■

We have described how the genetic material of a bacterium may be altered by mutation and by three forms of recombination processes. Our study of the first form, transformation, began with Griffith's classic experiment that revealed a genetic alteration, and followed with Avery's isolation of DNA. We then moved to the modern era and described the conditions under which competent cells obtain DNA from the local environment and incorporate it into the bacterial chromosome. In the second method, conjugation, we observed how two organisms come together and permit DNA to move from the donor cell to the recipient cell. The importance of the F factor was highlighted, and the Hfr strain was described. Plasmid transfer is common in conjugation, but chromosomal transfer is a rare event.

In the third type of recombination, transduction, a virus functions as an intermediary between cells. We began by describing the lytic cycle of viral replication and contrasted this with lysogeny. We then explained how in generalized transduction, the virus may randomly incorporate segments of bacterial DNA to itself and carry the genes to the next cell when lysogeny is established. This process contrasts with specialized transduction, where bacterial genes are excised from the chromosome with the prophage and replicated along with the virus. Some practical applications of transduction were noted.

6.6

MICROFOCUS

A VOCABULARY OF BACTERIAL GENETICS

chromosome: in bacteria, a single molecule of DNA arranged as a double helix in a closed loop without a surrounding membrane or protein support.

plasmid: a closed loop of DNA existing in the cytoplasm as an independent unit apart from the chromosome.

spontaneous mutation: a permanent change in the DNA of an organism taking place in nature without human intervention.

induced mutation: a permanent change in the DNA of an organism brought about by a planned laboratory experiment.

insertion sequence: a small segment of DNA on a bacterial chromosome that has no apparent genetic information other than denoting the point at which a foreign DNA segment can be inserted to the chromosome.

transposon: a small segment of DNA on a chromosome or plasmid that has genetic information and is able to move to a different location on the chromosome or plasmid.

recombination: a permanent change in the DNA of an organism taking place by the acquisition and incorporation of "foreign" genes by transformation, conjugation, or transduction.

transformation: the recombination process in which a "competent" bacterium acquires a segment of free-existing DNA from its surrounding environment.

conjugation: the recombination process in which a recipient bacterium acquires DNA from the chromosome or plasmid of a donor bacterium during a period of cell-to-cell contact.

transduction: the recombination process in which a recipient bacterium acquires DNA from a donor bacterium by using a virus as an intermediary.

F factor: a plasmid that functions in conjugation.

F$^+$ cell: a donor cell in conjugation.

F$^-$ cell: a recipient cell in conjugation.

Hfr strain: a type of bacteria in which chromosomal DNA passes between donor and recipient cells.

bacteriophage (phage): a DNA-containing virus that functions in transduction.

virulent phage: a bacteriophage that replicates within and destroys a bacterium.

temperate phage: a bacteriophage that exists for a time within a bacterium without replicating in or destroying the bacterium.

prophage: a set of bacteriophage genes inserted to a bacterial chromosome.

genetic engineering: the biochemical process in which "foreign" genes are inserted to existing DNA molecules for some research or practical purpose and are induced to encode their proteins in this unusual environment.

recombinant DNA: DNA that has been altered (or recombined) by the addition or exclusion of a DNA segment.

In the final section, we shall focus on experiments that alter bacterial DNA in the process of genetic engineering. This is where the knowledge from bacterial genetics is applied to the insertion of foreign genes to bacteria. We shall see how the process emerged and how the modern applications of genetic engineering yield products to enhance the quality of life.

Genetic Engineering

Experiments in bacterial recombination entered a new dimension in the late 1970s when it became possible to insert genes into bacterial DNA and thereby establish a cell line that would produce proteins according to the instructions of microbiologists. The use of basic research to solve practical problems had far-reaching ramifications, and an entirely new industry, **genetic engineering**, emerged.

The History of Genetic Engineering

The first glimmer of interest in genetic engineering surfaced in the 1960s with the discovery and isolation of a group of bacterial enzymes called **endonucleases**. These enzymes cleave phosphate-sugar bonds in the backbone of nucleic acids and they could be used to open a bacterial chromosome. Moreover, each endonuclease scanned a DNA chain and cleaved it at a restricted point. For this

Endonucleases: enzymes that cleave sugar-phosphate bonds in nucleic acids.

reason, the endonucleases came to be known as **restriction enzymes**. Their existence was first postulated by Werner Arber when he noted bacterial enzymes scissoring viral DNA at selected spots. Hamilton Smith subsequently isolated a restriction enzyme from Gram-negative rod *Haemophilus influenzae*. In 1971, Daniel Nathans used Smith's enzyme to split the DNA of simian virus 40 (SV40), a cause of tumors in monkeys. In 1978, the Nobel Prize in Physiology or Medicine was awarded to the three scientists. By that time over 100 different restriction enzymes had been isolated and characterized.

Restriction enzymes:
endonucleases that act at
restricted sites on a DNA
molecule.

Among the first scientists to attempt a genetic manipulation was Paul Berg of Stanford University. In 1971, Berg and his co-workers opened the DNA molecule from the SV40 virus and spliced it to a bacterial chromosome. In doing so, they constructed the first **recombinant DNA molecule**. However, the process was extremely tedious because the bacterial and viral DNAs had blunt ends. Berg therefore had to utilize exhaustive enzyme chemistry to form staggered ends that would combine easily. Nevertheless, he achieved a momentous first and was later honored as a co-recipient with Sanger of the 1980 Nobel Prize in Chemistry.

Recombinant DNA:
a DNA molecule resulting
from an alteration of its
structure such as insertion of
a DNA segment.

While Berg was performing his experiments in 1971, an important breakthrough came from Herbert Boyer and his group at the University of California. Boyer isolated a restriction enzyme that would nick a chromosome and leave it with mortiselike staggered ends. The bits of single-stranded DNA extending out from the chromosome easily attached to a new fragment of DNA in recombinant experiments. Scientists quickly dubbed the single-stranded extensions "sticky ends."

During that same period, Stanley Cohen of Stanford was accumulating data on the plasmids of *Escherichia coli* (Figure 6.14). Cohen found that he could isolate **plasmids** from the bacterium and insert them into fresh bacteria by suspending the organisms in calcium chloride and heating them suddenly to achieve a transformation. Once inside *E. coli* cells, the plasmids multiplied independently and produced copies of themselves in succeeding generations. Cohen's data indicated that scientists need not work with the larger, less manageable chromosome.

Transformation:
the uptake of DNA from the
local environment by a
competent recipient
bacterium.

DNA ligase:
an enzyme that combines
fragments of DNA.

The final link to the process was provided by the **DNA ligases**. These enzymes, known since the 1960s, function during the replication of DNA and the repair of broken DNA molecules. Essentially they operate in a manner opposite to the endonucleases; that is, they seal together DNA fragments.

Progress came rapidly. Working together, Boyer and Cohen isolated plasmids from *E. coli* and opened them with restriction enzymes (MicroFocus: 6.7). Next they inserted a segment of foreign DNA using DNA ligase. Then they implanted the plasmids into fresh *E. coli*. By 1973, they had successfully spliced genes from *Staphylococcus aureus* into *E. coli*. The recombined plasmids were termed **chimeras** from the mythical lion-goat-serpent monster of Greek literature.

Genetic engineering experiments intrigued the scientific community because genes from widely divergent species could be spliced together. However, some scientists saw the specter of disaster if certain experiments involving cancer genes were performed. They therefore called for a meeting to discuss the ramifications of genetic engineering experiments, and in 1975, an international group met at Asilomar, California, to map out guidelines to be overseen by the National Institutes of Health (NIH). As the years unfolded, scientists realized that their fears were exaggerated, and most NIH restrictions were relaxed. Only the most risky ones remain under federal regulation.

There is also the legal problem of who owns genetically altered bacteria. In the late 1970s, researchers engineered a strain of *Pseudomonas* to dissolve oil

FIGURE 6.14

The Tools of Genetic Engineering

(a) A transmission electron micrograph of the *E. coli* plasmid. The closed loop form of the plasmid is clearly seen. (b) Opening and closing the plasmid. An original DNA fragment in a plasmid is opened as a restriction enzyme nicks the DNA at two points and leaves it with mortiselike "sticky ends." A segment of foreign DNA can be spliced in at this point. To close the plasmid, the DNA is cooled and DNA ligase is added. The enzyme repairs the nicks in the DNA by reattaching the broken portions of the DNA backbone.

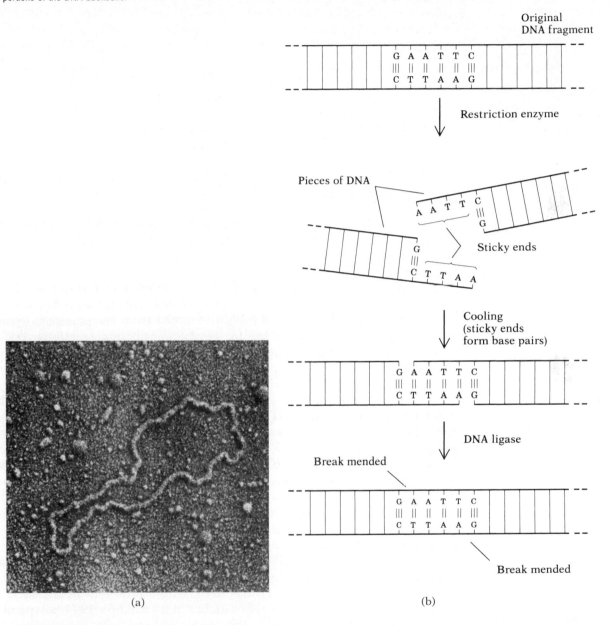

(a)   (b)

rapidly. They hoped to use the new bacteria to clean up oil spills. The question arose as to whether a patent could be obtained for the organism. Testimony was presented in the courts and, finally, in 1980, the U.S. Supreme Court ruled by a 5 to 4 vote that a patent could, indeed, be issued for the new organism.

6.7

MICROFOCUS

OF CORNED BEEF AND PLASMIDS

In 1972, they met at a scientific conference in Hawaii—Stanley Cohen and Herbert Boyer. Cohen was there to talk about his work with plasmids, the submicroscopic loops of DNA in the bacterial cytoplasm. Boyer was delivering a lecture about the restriction enzyme that could cut DNA —any DNA—at a precise point. As he sat in the audience and listened to Cohen, Boyer's mind stirred. Could his enzyme cut Cohen's plasmid and allow a foreign piece of DNA to attach?

Scientific conferences are the last place to talk about science, so Boyer invited Cohen to lunch at a local delicatessen in Waikiki. The corned beef was good that day, and the sandwiches hit the spot. The deli mustard was biting hot, and the beer was ice cold. The time was ripe to talk history —and talk history they did. They would collaborate on a set of genetic engineering experiments, the ones that, in retrospect, revolutionized the science of microbiology. As the afternoon wore on, the ideas flowed and the friendship took root. Only one thing about that historic lunch has remained a mystery: Who picked up the tab?

Modern Applications

By the 1990s, thousands of companies worldwide were working on the industrial applications of genetic engineering and DNA technology. Some were research companies with special units for these studies, while others were established solely to pursue and develop new products by gene-splicing techniques.

The pharmaceutical products of DNA are numerous and diverse. In 1986 the Eli Lilly company began marketing Humulin, a bacteria-produced form of **insulin**. For decades, insulin had been obtained from the pancreas tissues of animals, a factor that accounted for occasional allergic reactions. With the advent of genetic engineering, the insulin could be produced by inserting "insulin genes" to bacterial plasmids and using bacteria such as *E. coli* as chemical factories (Figure 6.15).

Also during the 1980s, a genetically engineered form of **interferon** came into use. This antiviral drug (Chapter 11) is produced by genetically altered *E. coli* cells. Interferon has been approved by the federal Food and Drug Administration (FDA) for use against a form of leukemia as well as for Kaposi's sarcoma (a cancer occurring in AIDS patients), malignant myeloma, and multiple myeloma. Research is continuing on the use of **antisense molecules** as therapeutic agents. Antisense molecules are fragments of nucleic acids that unite with and neutralize mRNA molecules carrying the genetic message for protein synthesis. To treat AIDS, for example, an antisense molecule would block the mRNA used for synthesis of new HIV particles (Chapter 13).

DNA technologists have also used the new biochemical methods to produce **human growth hormone**. In 1986, the genetically engineered form of this hormone became available as Protropin to treat pituitary dwarfism. (One child suffering from this disease grew five inches in a year.) By 1994, a type of synthetic **Factor VIII** was available for use in patients with hemophilia, and a protein called **tissue plasminogen activator (TPA)** was in use to dissolve blood clots. To prevent infectious diseases, there was a genetically engineered **vaccine** for hepatitis B, and research was continuing on an AIDS vaccine using components produced in bacterial cells. Optimism was also high for innovative vaccines for cholera, malaria, and herpes simplex.

Genetic engineering has extended into many realms of science. In **agriculture**, for example, genes for herbicide resistance have been transplanted from bacteria into tobacco plants demonstrating that plants can be engineered to better tolerate the herbicides used for weed control. For tomato growers, a

---

Insulin:
a pancreatic hormone that assists the passage of glucose molecules from the blood into the cells, thereby preventing diabetes.

in″ter-fēr′on

Interferon:
a naturally produced human protein that interferes with viral replication.

Pituitary gland:
A pea-sized, hormone-producing gland at the base of the brain.

Herbicide:
a plant-killing chemical used often to control weeds.

FIGURE 6.15

Developing New Products Using Genetic Engineering

Once a plasmid or other vector has been reengineered with foreign genes, the plasmid can be inserted to a bacterium or other cell, and the cell can be used to make gene products or to increase the number of genes themselves. Myriad applications are possible.

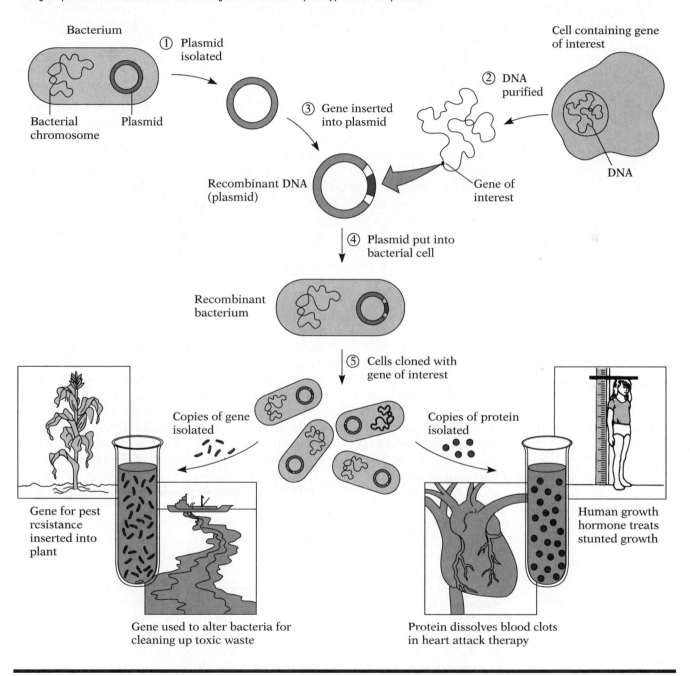

notable advance was made in 1988 when researchers at Washington University spliced genes from a pathogenic virus into tomato plant cells and demonstrated that the cells would produce viral proteins at their surface. The viral proteins then blocked viral encroachment (Chapter 11) and lent resistance to the tomato plants.

ag"ro-bak-te're-um
toom-e-fa'shens

For gene transfer experiments in plants, the mechanism often used for transfer is a plasmid from *Agrobacterium tumefaciens*. This organism causes a plant tumor called crown gall when DNA from the bacterium inserts itself into the plant cell's chromosomes during lysogeny (Figure 6.16). Researchers can therefore splice the desired gene into the plasmid and allow the bacterium to infect the plant. In 1989, scientists successfully removed the tumor-inducing genes from the bacterial plasmid so it can now serve as a gene carrier without causing plant damage.

The **dairy industry** may be the first to feel the dramatic effect of the new DNA technology. In the 1980s, for example, researchers at Cornell University injected dairy cows with bacteria-produced bovine growth hormone and reported a 41 percent increase in milk from the experimental cows. Also being researched is a pig with more meat and less fat, a product of genetically engineered porcine growth hormone. And in 1989, scientists at Auburn University endowed young carp with extra copies of activated growth hormone genes, hoping to enable the fish to grow more efficiently in aquacultural surroundings.

Gene probe:
a segment of single-stranded DNA that binds with a complementary single-stranded segment during diagnostic procedures.

In the **medical laboratory**, diagnosticians are optimistic about the use of **gene probes**, a set of relatively small amounts of DNA that can recognize and bind to certain DNA sites in a host cell's chromosome. The probe hybridizes to its complementary nucleic acid sequence, much like strips of Velcro™ stick together. To make a probe, scientists must first identify the DNA segment (or gene) that will be the object of a probe. Using this segment, they synthetically construct an mRNA fragment, then use reverse transcriptase (Chapter 11) to synthesize a complementary fragment of DNA. This newly formed segment of DNA is the probe. It is a single-stranded molecule that can be stored for later use. Usually it is tagged with a radioactive compound.

When diagnostic information is necessary, cells are obtained and the DNA is isolated. Now the DNA is treated to open all the strands and the radioactive gene probe is added. The probe "searches" among the thousands of genes and ultimately binds to, or hybridizes, the complementary gene. In doing so it concentrates the radioactivity at that site, and when the accumulated radioactivity is detected it can be assumed that a match has been made. The genes for such human diseases as cystic fibrosis, Huntington's chorea, and Alzheimer's disease can now be identified using specific gene probes. Indeed, gene probes are currently used to detect a variety of genetic disorders as well as to locate and identify the bacteria, viruses, and other pathogens of numerous human diseases.

Cystic fibrosis:
a gene-related human disease in which abnormally thick mucus clogs the respiratory passageways.

Detection of a different sort can be accomplished through the law enforcement technology of **DNA fingerprinting**, derived from observations reported by Alec J. Jeffreys in 1985. Jeffreys noted that short, repetitive segments of DNA of unknown function exist between the body's functional genes. He found that the segments appear in all people, but the number of times they are repeated changes per person: the segments may be repeated two times in person A, 20 times in person B, and 200 times in person C. And since the segments often follow one another in a chromosome, the length of a chromosomal fragment varies from person to person.

DNA fingerprint:
the presence of certain identifiable segments of DNA in the chromosomal material of an individual

To make a DNA fingerprint, human cells are obtained, the DNA is extracted, and restriction enzymes scissor the DNA into fragments. The fragments are then separated according to size by a process called electrophoresis. An electric current drives the fragments through a gel in a narrow channel. The shortest fragments meet the least resistance and travel the fastest, moving to the far end of the groove; the longer, slow-moving fragments meet more resistance and remain closer to the starting point (the near end). Radioactive gene probes are then used to seek out the fragments and mark them with radiation. In this way the fragment positions can be revealed as dark bands, looking similar to a bar code. DNA fingerprinting was introduced to the legal system in 1987, but,

FIGURE 6.16

Agrobacterium tumefaciens as a Vector in Plasmid Technology

(a) *A. tumefaciens* induces tumors in plants and causes a disease called crown gall. A lump of tumor tissue forms at the infection site, as the photograph shows. (b) The source of infection is a plasmid carrying DNA which is transferred from the bacterium to the plant cell. Genetic engineers use this ability to their advantage. They isolate the responsible plasmid, designated Ti for tumor-inducing, and insert the fragment of DNA desired. The recombined plasmid is then combined with plant cells in culture. When the Ti plasmid interacts with the plant cell, it transfers the desired DNA, and the plant cell is recombined.

(a)

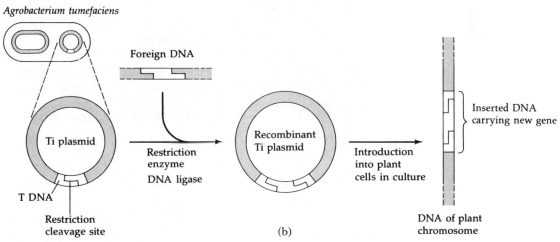

(b)

although accepted in many courts, the technique has been rejected by some judges because results can be inconclusive.

One of the most ambitious projects in the history of molecular genetics, indeed the history of biology, was launched in January 1989. That month a group of biologists, ethicists, industrial scientists, computer experts, and engineers began a monumental effort to map the **human genome**, that is, identify all the nitrogenous bases and their sequence in the 100,000-odd human genes. Their goal: to spell out for the world the entire genetic message hidden in the code of bases in human DNA.

je′nōm

Human genome:
all the nitrogenous bases and
their sequence in the full set
of human genes.

With the National Institutes of Health supporting the project, a staggering $3 billion may be expended during the 15 years estimated for completion. There are about three billion pairs of bases that need to be sequenced, so the project should cost $1 per base pair. Most genes consist of about 10,000 to 150,000 base pairs, and between them are endless stretches of bases that appear to have no meaning, at least in the views of contemporary scientists. The human genome to be deciphered has been equated to a rope two inches in diameter and 32,000 miles long, all neatly arranged in a structure (the nucleus) about the size of a domed stadium.

Determining the nature of the human genome will require a multistep approach beginning with assigning genes to particular chromosomes, continuing with locating genes on a chromosome, and ending with determining the sequence of base pairs in the DNA chain. Each step will require technology not yet invented and the imagination of scientists who are still in grade school. Nevertheless, optimism remains high that the benefits of the project will outweigh the enormous expense. On a broad scale, mapping the human genome will contribute to our understanding of growth, development, and other life processes. James D. Watson, the effort's original leader, has written: "We used to think our fate was in the stars. Now we know that, in large measure, our fate is in our genes."

NOTE TO THE STUDENT

Louis Pasteur once wrote, "There are science and the applications of science, separate yet bound together as the fruit to the tree which bears it." The truth of this statement is particularly apparent in genetic engineering. Genetic engineering represents an apex in microbiology research. It is the equivalent of the engineer's dream of landing on the moon and the physicist's vision of peaceful uses of nuclear power.

The capabilities of genetic engineering have turned microbiology from an analytical science into a synthetic science. By capturing genes from one species and inserting them to another, microbiologists have acquired the ability to make large quantities of proteins previously available only in minute amounts. Triumph has followed triumph. Rarely in scientific history have the discoveries of pure research had such immediate applications and implications upon the society of their time.

I would recommend that you browse through this chapter and try to visualize the growth of molecular genetics beginning with Griffith's puzzling observations in 1928, and continuing with Avery's work in 1944 and the discoveries in bacterial genetics in the 1950s. It was in this period that the idea of gene manipulations arose. Watch for daily announcements in the press of new products that are derived from genetic engineering. It is a historic time for microbiology, and we should all share the excitement. Louis Pasteur would have been proud to see this day.

Summary

Bacterial genetics is concerned with the gene changes that take place in a bacterium's chromosome and plasmids, changes that reflect themselves in the morphology, physiology, and pathology of the organism. The changes can take place by mutation and by recombination.

Mutation is a permanent change in the cellular DNA, occurring spontaneously in nature or by induced methods in the laboratory. Such things as ultraviolet light, chemicals, and base analogs may induce mutations. Deletions or insertions involving the chromosome as well as transposons and insertion sequences may also be mutagenic.

Recombinations imply an exchange of DNA between bacteria and thus, an acquisition of genes. In one form of recombination called transformation, a "competent" bacterium takes up DNA from the local environment; this DNA has been left behind by a disrupted bacterium. The new DNA displaces a segment of equivalent DNA in the recipient cell, and new genetic characteristics are assumed. In another form, conjugation, a live donor cell contributes a portion of its DNA to a recipient cell. Plasmids are commonly transferred, but chromosomal DNA may also be contributed. In the third form of recombination, transduction, a virus enters a bacterium and later replicates within it. During replication, the virus incorporates some bacterial DNA to its protein coat and transports that DNA to a new bacterium. In the specialized form of transduction, the virus first attaches to, then detaches from the bacterial chromosome taking a segment of bacterial DNA with it. Thus, all three forms of recombination are characterized by the introduction of new genes to a recipient bacterium.

Genetic engineering is an outgrowth of studies in bacterial genetics. In one type of this technology, plasmids are isolated from a bacterium, spliced with foreign genes, then inserted to fresh bacteria where the foreign genes are expressed as protein. Bacteria are thus used as the biochemical factories for the synthesis of such proteins as insulin, interferon, human growth hormone, and others. But genetic engineering is only one branch of modern biotechnology. This technology utilizes DNA-based techniques to perform diagnoses, identify individuals in forensic medicine, detect and treat genetic diseases, and spark innovative approaches to agriculture. The practical benefits of research in bacterial genetics have helped revolutionize myriad fields of human endeavor.

Questions for Thought and Discussion

1. Try to put yourself in Griffith's position in 1928. Genetics is poorly understood; DNA is virtually unknown; and bacterial biochemistry has not been clearly defined. How would you explain transformation?

2. In hospitals it is common practice to clear the air bubble from a syringe by expelling a small amount of the syringe contents into the air. One microbiologist estimates that this practice results in the release of up to 30 liters of antibiotic into a typical hospital's environment annually. How might this lead to the appearance of antibiotic-resistant mutants in hospitals?

3. Some geneticists maintain that the movement of transposons in a bacterial cell is a form of recombination, specifically "illegitimate recombination." What arguments can be made for and against calling the movement a recombination, and why do you suppose it is labeled "illegitimate"?

4. Which of the recombination processes (transformation, conjugation, or transduction) would be most likely to occur in the natural environment? What factors would encourage or discourage your choice from taking place?

5. The author of a general biology textbook writes in reference to the development of antibiotic resistance: "The speed at which bacteria reproduce ensures that sooner or later a mutant bacterium will appear that is able to resist the poison." What might be the origin of this mutant bacterium? Do you agree with the statement? Does this bode ill for the future use of antibiotics?

6. How could the process of recombination have a profound effect on the classification system for bacteria?

7. Certain developments such as Koch's cultivation methods for bacteria are pivotal because they open the door to other discoveries and spark research in other areas. Which discoveries in bacterial genetics do you believe to be pivotal?

8. Some scientists suggest that mutation is the single most important event in evolution. Do you agree? Why or why not?

9. In 1976, an outbreak of pulmonary infections among participants at an American Legion convention in Philadelphia led to the identification of a new disease, Legionnaires' disease. The bacterium responsible for the disease had never before been known to be pathogenic. From your knowledge of bacterial genetics, can you postulate how it acquired the ability to cause disease?

10. What factors have made the bacterial chromosome appeal to geneticists and biochemists as a tool in their work?

11. The development of genetic engineering has been hailed as the beginning of a second Industrial Revolution. Do you believe this label is justified? How many products of genetic engineering or applications of the process can you think of?

12. In 1994, the CDC reported that the percentage of antibiotic-resistant isolates of *Haemophilus influenzae* had risen from 4.5 percent to 28 percent over the previous five years. What factors might have accounted for this change?

13. Since the 1950s, the world has been plagued by a broad series of influenza viruses that differ genetically from one another. For example, we have heard of swine flu, Hong Kong flu, Bangkok flu, and Victoria flu. How might the process of transduction help to explain this variability?

14. It is not uncommon for students of microbiology to confuse the terms "reproduction" and "recombination." How do the terms differ?

15. In modern medicine, physicians are strongly urged to prescribe an antibiotic that is specifically geared to the organism causing the present disease rather than a broad-spectrum antibiotic that kills many different bacteria including the present organism. Why?

Review

Use the following syllables to compose the term that answers each of the following clues from bacterial genetics. The number of letters in each term is indicated by the dashes, and the number of syllables used is shown by the number in parentheses. Each syllable is used only once, and the answers are listed in the appendix.

A ASE BAC CHI CHROM CLE COC COM CON CUS DO DUC EN
FER FER FITH GA GASE GE GEN GRIF I I IN JU LA LENT LI LI
LY MER MIDS MO MO MU MU NEL NU NY O O ON PE PHAGE
PI PLAS PNEU PO SAL SEX SO SOME SON TA TA TENCE TER TER
TIL TION TION TION TRANS TY U VIR

1. Closed loops of DNA   (2) __ __ __ __ __ __ __ __

2. Type of conjugation   (3) __ __ __ __ __ __ __ __ __ __

3. Genetic change   (3) __ __ __ __ __ __ __ __

4. Used in Ames test   (4) __ __ __ __ __ __ __ __ __ __

5. Transforming property   (3) __ __ __ __ __ __ __ __ __ __

6. Factor for conjugation   (4) __ __ __ __ __ __ __ __ __

7. Transduction virus   (5) __ __ __ __ __ __ __ __ __ __ __ __

8. Recombinant DNA enzyme   (5) __ __ __ __ __ __ __ __ __ __ __ __

9. Transformed bacterium   (4) __ __ __ __ __ __ __ __ __ __ __

10. Conjugation structures   (2) __ __ __ __ __

11. Viral nonreplication   (4) __ __ __ __ __ __ __ __

12. Recombined plasmid   (3) __ __ __ __ __ __ __

13. DNA linking enzyme   (2) __ __ __ __ __ __ __

14. Gene engineered drug   (4) __ __ __ __ __ __ __ __ __
15. Movable genetic element   (3) __ __ __ __ __ __ __ __ __
16. Discovered transformation (2) __ __ __ __ __ __ __ __
17. Type of recombination   (4) __ __ __ __ __ __ __ __ __ __ __
18. Induces genetic change   (3) __ __ __ __ __ __ __
19. Phage that causes lysis   (3) __ __ __ __ __ __ __ __
20. Only one per bacterium   (3) __ __ __ __ __ __ __ __ __

Bacterial Diseases of Humans

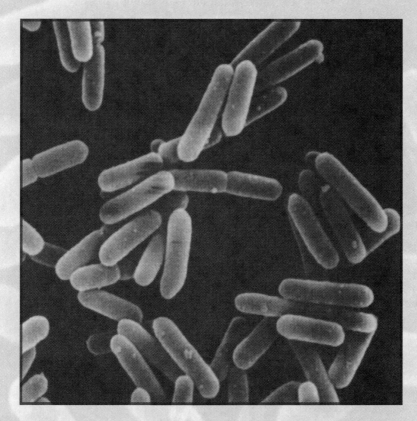

Throughout history, bacterial diseases have posed a formidable challenge to humans. Diseases have often swept through populations virtually unchecked. In the eighteenth century, for example, the first European visitors to the South Pacific found the islanders robust, happy, and well adapted to their environment. But the explorers introduced syphilis, tuberculosis, and pertussis (whooping cough), and soon these diseases spread like wildfire. Hawaii was struck with unusually terrible force. The islands' population was about 300,000 when Captain Cook landed there in 1778; by 1860, it had been reduced to fewer than 37,000 people. ■ With equally devastating results, the Great Plague came to Europe from the Orient, and cholera spread eastward from India. Together with tuberculosis, diphtheria, and dysentery, these bacterial diseases ravaged European populations for centuries and insidiously wove themselves into the pattern of life. Infant mortality was particularly shocking: England's Queen Anne, who reigned in the early 1700s, lost 16 of her 17 babies, primarily to disease; and until the mid-1800s, only half the children born in the United States reached their fifth year. ■ Today, humans are better able to cope with bacterial diseases. The credit is often given to wonder drugs, but the major health gains have resulted from understanding disease and the body's resistance mechanisms, coupled with modern sanitary methods that prevent microorganisms from reaching their targets. Immunization has also played a key role in preventing disease. Indeed, very few people in our society die of the bacterial diseases that once accounted for the majority of all deaths. ■ In Part III of this text, we shall study the bacterial diseases of humans over the course of four chapters. The diseases have been grouped according to their major mode of transmission: airborne diseases are discussed in Chapter 7, foodborne and waterborne diseases in Chapter 8, soilborne and arthropodborne diseases in Chapter 9, and sexually transmitted, contact, and miscellaneous diseases in Chapter 10. Many of the diseases we shall study are of historical interest and are currently under control. But just as the garden always faces new onslaughts of weeds and pests, so too, the human body is continually confronted with newly emerging diseases. In this regard the modern era is no different than Europe or the South Pacific islands of past centuries. On the fundamental level, disease has not changed. Only the pattern of disease has changed.

MICROBIOLOGY PATHWAYS

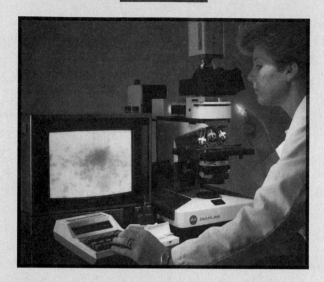

CLINICAL MICROBIOLOGY

One of the most famous books of this century is *Microbe Hunters* by Paul De Kruif. In his book, first published in the 1920s, De Kruif describes the joys and frustrations of Pasteur, Koch, Ehrlich, von Behring, and many of the original microbe hunters, The exploits of these scientists make for fascinating reading and help us understand how the concepts of microbiology were formulated. I would urge you to leaf through the book at your leisure. Perhaps you will decide to set aside a period of time to read it closely.

Microbe hunters did not come to end with Pasteur, Koch, and their contemporaries, nor did the stories of microbe hunters end with the publication of De Kruif's book. Microbe hunters continue to work in the clinical microbiology laboratories. The men and women working in hospital, public, and private laboratories are today's detectives of microbiology. These individuals search for the pathogens of disease. Many travel to far corners of the world studying organisms, but many more remain close to home, identifying the pathogens in samples sent by physicians. Microbiologists even work in dental clinical labs, since many bacteria are involved in tooth decay and periodontal disease.

Clinical microbiology also offers an outlet for the talents of those who prefer to tinker with machinery. New instruments and laboratory procedures are constantly being researched in an effort to shorten the time between detection and identification of microorganisms. Many tests reflect human ingenuity. For example, there is a test that detects bacteria by its interference with the passage of light and ability to scatter it at peculiar angles. Such modern devices as laser beams are used in this kind of instrumentation.

The microbe hunters have not changed materially in the past hundred years. The objectives of the search may be different, but the fundamental principles of the detective work remain the same. The clinical microbiology is today's version of the great masters of a bygone era.

CHAPTER 7

AIRBORNE BACTERIAL DISEASES

T he produce looked good that day. The lettuce was green and crisp, and the carrots just seemed to exude good health. The broccoli was a deep, forest green and the cauliflower a lily white. It was probably due to the ultrasonic humidifier spraying its mist over the vegetable section. That fresh, clean mist made everything look good.

So it was that November day in Bogaloosa, Louisiana, when officials from the local Department of Health walked into the grocery store. They had a disturbing story: 28 people were sick with Legionnaires' disease and 2 had already died of disease-related pneumonia. Most of the patients, it seemed, had been in the grocery store within the previous ten days.

The investigators were friendly but firm: how long, they asked, did the air conditioner run each day? Were the vegetables washed before putting them out and if so, for how long and where? Had any of the employees been sick recently? Who was the grocery store's supplier? And that humidifier—how long had it been there?

Three days later the investigators were back with the answers. It was the humidifier! The lab had found *Legionella pneumophila* in the water in its reservoir. Could the owner please remove it from the store and clean it out with disinfectant before reusing it? Within a day the humidifier was gone. The broccoli didn't look quite so good, and the lettuce leaves seemed to droop. But the air was safer to breathe, and that was important.

Though known only since 1976, Legionnaire's disease is recognized today throughout the world, and outbreaks such as this 1989 incident in Louisiana are reported regularly to the CDC. In the United States alone, microbiologists estimate that between 25,000 and 50,000 cases occur annually, though most go unreported. Legionnaires' disease is one of the contemporary airborne bacterial diseases that we shall discuss in this chapter. Such diseases are generally spread where people crowd together. In many instances, reduced body resistance is a prime motivating factor in the establishment of infection.

Pneumonia:
an often-serious microbial disease of the bronchi and lungs.

le"gion-el'lah

Legionnaires' disease:
a bacterial disease of the respiratory tract transmitted by the air and often accompanied by pneumonia; named for an outbreak at a convention of the American Legion in 1976.

We shall survey the airborne bacterial diseases according to two general categories. The first category will include diseases of the upper respiratory tract such as strep throat, scarlet fever, diphtheria, pertussis, and several forms of meningitis. The second category will include diseases of the lower respiratory tract: tuberculosis, pneumococcal pneumonia, primary atypical pneumonia, Legionnaires' disease, and others. As we proceed, note how the initial focus of infection may be a point from which the bacteria spread to other organs. Also note that antibiotics are available for the treatment of these diseases and that immunizations are used to protect the community at large.

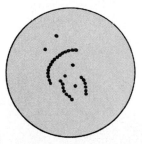

STREPTOCOCCUS PYOGENES

he"mo-lit'ik

C substance:
a specific carbohydrate in the cell walls of streptococci.

pi-oj'ĕ-nez

Pharynx:
the throat region.

Lymph nodes:
bean-shaped pockets of tissue containing white blood cells that respond to disease organisms.

Antibodies:
highly specific protein molecules produced by the immune system in response to microorganisms and chemical substances.

Diseases of the Upper Respiratory Tract

Streptococcal Diseases

Streptococci are a large and diverse group of encapsulated Gram-positive bacteria. They cause streptococcal sore throat (strep throat), scarlet fever, and several other diseases in humans. The bacteria divide in one plane and cling together to form chains of various lengths. This observation is apparent in streptococci grown in liquid media, but it may not be obvious in streptococci from solid media (Figure 7.1).

Microbiologists classify the numerous types of streptococci by several systems of which two are widely accepted. The first system, developed by J.H. Brown in 1919, divides streptococci into hemolytic ("blood-digesting") groups, depending on how they affect red blood cells. For example, **alpha-hemolytic streptococci** turn blood agar an olive-green color as they partially destroy red blood cells in the medium. By contrast, colonies of **beta-hemolytic streptococci** are surrounded by clear, colorless zones due to the complete destruction of red blood cells; and **gamma-hemolytic streptococci** have no effect on red blood cells and thus cause no change in blood agar.

The second classification system, suggested by Rebecca Lancefield in 1935, is based on variants of a specific carbohydrate, the C substance, in cell walls of streptococci. Groups A through O streptococci have been distinguished. Most streptococcal diseases in humans are caused by species of Group A streptococci; *Streptococcus pyogenes* is the most common species. This beta-hemolytic organism is the one generally meant when physicians refer to "Group A beta-hemolytic streptococci."

Streptococcus pyogenes is the cause of streptococcal sore throat, popularly known as **strep throat**. *S. pyogenes* is one of the few bacteria that can invade the normal tissues of the pharynx and cause a **primary disease**, that is, one developing in an otherwise healthy body. The streptococci reach the upper respiratory tract within airborne **droplets**, the tiny particles of mucus expelled during coughing and sneezing. After a short incubation period (one to three days), a strep throat develops. Patients experience a high fever, coughing, swollen lymph nodes and tonsils, and a fiery red "beefy" appearance to pharyngeal tissues owing to tissue erosion. About 500,000 Americans suffer strep throat annually, many cases complicated by infection in the middle ear, **otitis media**.

The pathogenicity of *Streptococcus pyogenes* is enhanced by a substance called the **M protein** (Figure 7.1b). This protein, located in the cell wall and pili, encourages adherence to the pharyngeal tissue and retards phagocytosis. Over 60 specific types of M protein have been identified, and complete immunity to streptococcal disease requires that a person produce antibodies against all 60 types. Since this is improbable, an individual may suffer multiple bouts of strep throat over the course of a lifetime.

FIGURE 7.1

Two Views of Streptococci

Streptococci seen in macroscopic and microscopic views. (a) A scanning electron micrograph of a colony of streptococci on a blood agar medium ($\times$ 4800). Note that the streptococci appear in small clusters rather than in the usual chains. This is characteristic of growth on a solid medium. Some clearing of the medium has occurred in the area immediately about the colony, an indication of hemolysis. (b) An electron micrograph of group A streptococci exhibiting M protein fibrils on the cell surface. The fibrils enhance streptococcal pathogenicity by permitting adherence to the tissue and by discouraging phagocytosis ($\times$ 50,000).

(a)

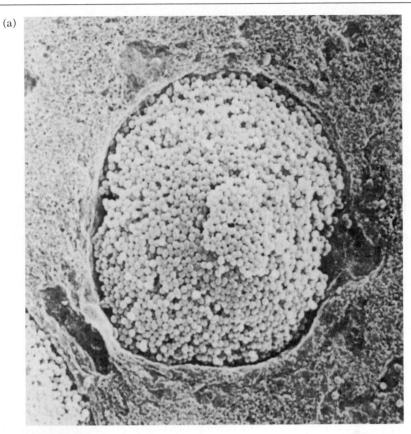

(b)

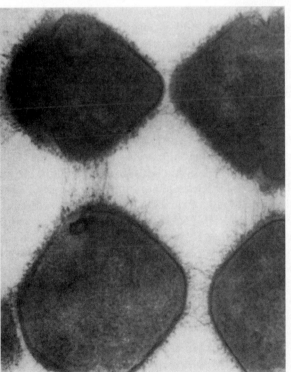

Antitoxins:
highly specific antibodies that react with toxins.

Bacteriophages:
viruses that penetrate bacteria and often remain within them.

Antigen:
a chemical substance that stimulates a response by the immune system.

glo-mer"u-lo-ně-fri'tis

er"ĭ-sip'e-las

pu-er'per-al

dif-the're-ah
bret-o-no'

Lef'ler

ko-ri"ne-bac-te're-um
dif-the're-ā

Exotoxin:
a poisonous chemical substance produced by bacteria and immediately excreted.

Scarlet fever is a strep throat accompanied by a skin rash. The rash develops as a pink-red blush, especially apparent on the neck, chest, and soft-skin areas of the arms. It results from blood leaking through the walls of capillaries damaged by a toxin. This toxin, called an **erythrogenic** ("red-forming") **toxin**, is produced only by certain strains of *S. pyogenes*. Normally an individual experiences a single case of scarlet fever per lifetime because the immune system produces special antibodies, called **antitoxins**, which circulate in the bloodstream and neutralize the toxins in succeeding episodes.

Treatment of streptococcal diseases is generally successful with antibiotics such as penicillin and erythromycin. However, widespread resistance to antibiotics has been noted since the early 1970s, and research indicates that many strains of *S. pyogenes* carry bacteriophages having antibiotic-resistance genes.

An important complication of streptococcal disease is **rheumatic fever** (MicroFocus: 7.1). This condition is characterized by fever and inflammations of the small blood vessels. Joint pain is common, but the most significant long-range effect is permanent scarring and distortion of the heart valves, a condition called **rheumatic heart disease**. The damage arises from a reaction of the body's antibodies with streptococcal antigens bound to the heart tissue. Antibodies may also react with tissue antigens identical to streptococcal antigens. Another manifestation of this reaction may develop in the kidneys, where the condition is called **glomerulonephritis**.

S. pyogenes is commonly present in the human nose and throat, and transmission to many other tissues is possible. For example, streptococci may cause disease in open wounds or skin abrasions, resulting in **erysipelas** (Figure 7.2). If streptococci invade and destroy the underlying fascia over the muscles and beneath the skin, they cause a condition called **necrotizing fasciitis** and an unsightly degeneration of the skin tissues. (Tabloid newspapers and TV programs had a field day with this disease in 1994 as they heralded an outbreak of the "flesh-eating bacteria.") If the tissue of the uterus becomes infected during the birth process the mother may suffer **puerperal sepsis**, also known as **childbed fever**. Moreover, streptococci may infect the lower respiratory tract and cause streptococcal pneumonia as a complication to viral disease. In past generations, many died of **septicemia** or "blood poisoning," a disease now known to be due to several organisms, including streptococci. One of Louis Pasteur's daughters was a victim of this disease.

Diphtheria

Diphtheria was first recognized as a clinical entity in 1826 by French pathologist Pierre F. Bretonneau. Bretonneau named the disease *la diphtherite* from the Greek *diphthera* for membrane, a reference to the "membranes" that appear in the throats of patients. In 1883, Edwin Klebs observed a bacterium in material from a patient's throat, and the next year, Friederich Löeffler successfully cultivated the organism (MicroFocus: 7.2). The so-called Klebs-Löeffler bacillus has the scientific name *Corynebacterium diphtheriae* because it resembles a club ("coryne" is Greek for club) and forms membranes.

Corynebacterium diphtheriae is a Gram-positive bacillus containing numerous **metachromatic granules** that show up with special stains as cytoplasmic dots. The bacteria remain close to one another after multiplying and form a palisade layer, a picket-fence arrangement. The staining and reproductive characteristics are aids to identification.

Diphtheria is acquired by inhaling respiratory droplets. Infection occurs in the upper respiratory tract near the tonsils. The bacteria produce a potent **exotoxin** that interferes with protein synthesis in epithelial cells, the cubelike cells that line the skin and body cavities such as the respiratory tract. As dead

7.1

MICROFOCUS

AN UNHAPPY DANCE

The doctor was baffled. Here was an alert, charming, and seemingly healthy young girl of 10 who had completely lost control of the right side of her body. It had begun a month earlier when her right hand took on a life of its own, writhing and shaking to its own rhythm. The scary movements worsened, and soon her writing was illegible. Then her gait grew wobbly and the right side of her face grimaced with contortions. Even her tongue was twisted. It was then that the child was brought to the hospital's emergency room.

The lab performed diagnostic tests for meningitis, brain tumors, and other neurological impairments, but all were fruitless. With a rapidly diminishing list of possibilities, the doctor ordered blood tests for a series of bacterial diseases, even though no symptoms of infection were present. Success finally came—blood tests revealed antibodies against Group A streptococci.

The child was suffering from rheumatic fever, a disease that can manifest itself as arthritis, endocarditis, skin nodules, or in the bewildering symptoms she displayed. This particular syndrome is called Sydenham's chorea (for Thomas Sydenham who described it in the late 1600s and the Greek word *choreia* for dance, a reference to the jerky movements). The syndrome is also known as Saint Vitus' dance.

Before 1945, rheumatic fever was a fearful cause of death in children, leaving many of the survivors debilitated with damaged hearts. Then, with the linking of streptococci to the disease and with penicillin intervention, the disease waned. Since 1985, however, Group A streptococci have resurfaced in the United States and have exacted a heavy toll (including the life of Muppets creator Jim Henson). Outbreaks of rheumatic fever and other streptococcus-related disorders such as pneumonia, impetigo, and

a form of toxic shock syndrome (called "toxic strep syndrome") have been reported throughout the United States. Whether the new streptococci are mutants of the older forms or a reappearance of the older forms is uncertain. What is interesting is that a large percentage of infections arise from breaks in the skin, not normally the mode of entry for Group A streptococci. In late 1993, the streptococci were also related to Kawasaki disease (Chapter 12).

For those affected, it is reassuring to know that penicillin is effective for treating the Group A streptococci. For the young girl, the infection meant a course of penicillin and a month of physical therapy to help the chorea disappear. The chorea would later resurface on the opposite side of her body but that too would disappear. When last seen, the girl was back at school, blissfully unaware of the unusual memories that she had left with her doctors.

FIGURE 7.2

Erysipelas of the Trunk

A view of erysipelas of the trunk, a disease caused by a species of *Streptococcus*. Note the patches of inflammation on the skin. Erysipelas is one of many streptococcal diseases.

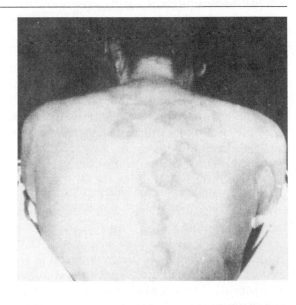

7.2
MICROFOCUS

A PREDICTION FULFILLED

It was the early eighteen-eighties, and diphtheria, which several times each hundred years seems to have violent ups and downs of viciousness—diphtheria was particularly murderous then. The wards of the hospitals for sick children were melancholy with a forlorn wailing; there were gurgling coughs foretelling suffocation; on the sad rows of narrow beds were white pillows framing small faces blue with the strangling grip of an unknown hand.

Paul de Kruif, *Microbe Hunters,*
                                          1925

So it was. Five of every ten cots sent their tenants to the morgue, and there in the morgue, Friederich Löeffler (pronounced "lef'ler") searched for the organisms of diphtheria.

Löeffler, the student of Robert Koch, worked tirelessly. Finally, in 1884, he was sure he had the culprit: a club-shaped rod isolated from the throats of children. But the rods were nowhere else in the body. It was inconceivable that these few organisms, staying in the throat, could kill a human or

animal so huge. Yet it was happening, time and again. Perhaps there was a poison, a toxin that leaked out of the bacilli and spread to other parts of the body.

Four years later, Löeffler's prediction came true in an outrageous experiment performed by Louis Pasteur's young assistant, Emile Roux (pronounced "roo"). Roux was smitten by the toxin theory. Although no one had ever separated a toxin from bacteria, he was determined to try. Carefully he inoculated large quantities of broth with diphtheria bacilli. After four days, he separated the bacilli from the broth using a porcelain filter and injected the broth into animals. Nothing happened. He tried a larger dose, again without success. Still larger doses brought more failure.

Now came the experiment that would change the medical view of diphtheria. It was born of frustration, desperation, and a touch of insight. Roux injected a huge 35 milliliters of the broth into a guinea pig and an equal amount into a rabbit. This was equivalent to injecting a bucketful of fluid

into a human. Privately he scoffed at his own experiment and speculated whether the sheer volume of broth would kill the animals. But they survived the injection, and 48 hours later, both the guinea pig and rabbit showed unmistakable signs of diphtheria. The toxin was there but the amount was infinitesimal.

Next came refinement. Roux found that if he cultivated the bacilli for weeks instead of days, the amount of toxin increased dramatically. Soon he could show that an unbelievably tiny amount of toxin caused diphtheria. The door was now open to an understanding of the disease. Löeffler had been right after all!

In the years thereafter, another Emil, Emil von Behring (the Germans omit the last "e") found a way to cure diphtheria by administering antitoxins. Deaths from diphtheria soon declined, and with the development of the diphtheria vaccine in the 1920s, the incidence of disease dropped dramatically. The strangling grip of diphtheria was finally loosened.

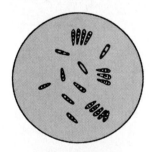

***CORYNEBACTERIUM
DIPHTHERIAE***

Toxoid:
**a preparation of altered
toxin molecules used for
immunization purposes.**

tissue accumulates with mucus, white blood cells, and fibrous material, a leathery **pseudomembrane** forms ("pseudo" because it does not fit the definition of a true membrane). Respiratory blockage and death may follow, especially in children. In adults, the toxin often spreads to the bloodstream, where it causes heart damage and destruction of the fatty sheaths surrounding nerves.

Treatment of diphtheria requires both antibiotics for the bacteria and antitoxins to neutralize the toxins. At present, the number of cases of diphtheria in the United States is less than a dozen annually but the disease remains a health problem in many regions of the world, especially parts of Asia. In 1995, for example, a major outbreak occurred in 13 of the 14 new states of the former Soviet Union and affected almost 50,000 people, with about 2000 deaths.

Immunization against diphtheria may be rendered by an injection of diphtheria **toxoid** contained in the diphtheria-pertussis-tetanus (DPT) vaccine. A toxoid consists of toxin molecules treated with formaldehyde or heat to destroy their toxic qualities. The toxoid induces the immune system to produce antitoxins that circulate in the bloodstream throughout the person's life.

In 1961, V.J. Freeman discovered that a virus attached to the bacterial chromosome contains the genetic code for exotoxin production. The virus, termed a **corynephage**, exists in a lysogenic relationship with the bacterium (Chapter 6). Strains of *C. diphtheriae* lacking the virus are harmless, but they may become pathogenic if the virus infects them.

Pertussis (Whooping Cough)

Pertussis, also known as **whooping cough**, is one of the more dangerous diseases of childhood years. Although the incidence of this disease has declined substantially from the 250,000 annual cases reported during the 1930s, the number of cases in the United States still remains significant. For example, the CDC recorded about 4000 cases in 1995, and the numbers were growing. Almost half occurred in children under six months of age.

Pertussis is caused by ***Bordetella pertussis***, a small Gram-negative rod first isolated by Jules Bordet and Octave Gengou in 1906, and commonly known as the Bordet-Gengou bacillus. The bacillus is spread by droplets and uses its pili to adhere to the cilia of epithelial cells of the upper respiratory tract. There is generally no invasion of tissue but the ciliated cells are destroyed and mucus movement is impaired.

Typical cases of pertussis occur in three stages. The initial stage is marked by general malaise, low-grade fever, and increasingly severe cough. During the next stage disintegrating cells and mucus accumulate in the airways and cause labored breathing. Patients experience multiple paroxysms of rapid-fire **staccato coughs** all in one exhalation, followed by a forced inspiration over a partially closed glottis. The rapid inspiration results in the characteristic "whoop" (hence, whooping cough). Ten to fifteen paroxysms may occur daily, and exhaustion usually follows each. During the third stage, sporadic coughing may continue for several weeks, even after the bacteria have vanished.

Treatment of pertussis is generally successful when penicillin or erythromycin is administered before the respiratory passageways become blocked. Diagnosis of the disease is performed by obtaining swabs from the posterior pharyngeal wall and identifying *B. pertussis* on selective media. A fluorescent antibody test (Chapter 19) is also used. Although several toxins have been isolated from the bacillus, no one toxin is considered the prime factor in the disease.

As with diphtheria, the declining incidence of pertussis stems partly from use of the DPT vaccine (Figure 7.3). The pertussis component consists of *B. pertussis* cells killed with Merthiolate in the manner first described by Pearl Kendrick and Grace Eldering in 1939. Physicians recommend that infants receive the first of these injections at the age of two months, since newborns are highly susceptible to the disease.

Use of the pertussis vaccine is not without risk, however. About one in every 300,000 vaccinated individuals may suffer temporary seizures, and several may experience high fever for a short period of time. Reports such as these have raised public concern about use of the vaccine and as of 1993, public health officials were recommending the newer acellular pertussis vaccine prepared from *Bordetella pertussis* extracts. Combined with diphtheria and tetanus toxoids, the triple vaccine is given the acronym DTaP. At this writing, DTaP is recommended as the fourth and fifth doses of the five-dose regimen given to children. The whole-cell pertussis vaccine in DPT continues to be suggested for the initial three doses.

Meningococcal Meningitis

The term **meningitis** refers to several diseases of the meninges, the three membranous coverings of the brain and spinal cord. Meningitis may be caused by viruses, fungi, protozoa, or bacteria, and different forms of the disease have different mortality rates. In all forms, the meninges become inflamed, causing pressure on the spinal cord and brain. Patients experience headaches, neckaches, and lower backaches.

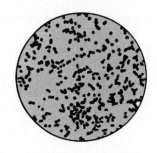

BORDETELLA PERTUSSIS

bor-da'
zhan-goo'

par-oks'ism
Paroxysm:
a severe attack.
Glottis:
the slitlike opening to the respiratory tract.

Erythromycin:
an antibiotic that inhibits protein synthesis in various types of bacteria.

Merthiolate:
a heavy metal compound often used as an antiseptic and preservative.

Mortality rate:
the rate of death in a population from infectious disease.

FIGURE 7.3

The Incidence of Pertussis (Whooping Cough) in the U.S. 1957 to 1993

Note that the number of deaths parallels the number of cases, indicating the seriousness of the disease. The relative rise of pertussis cases during the 1980s is shown in the inset. The recent dropoff probably coincides with increased use of the pertussis vaccine. (Courtesy of Centers for Disease Control and Prevention, *CDC Summary of Notifiable Diseases*, 1993.

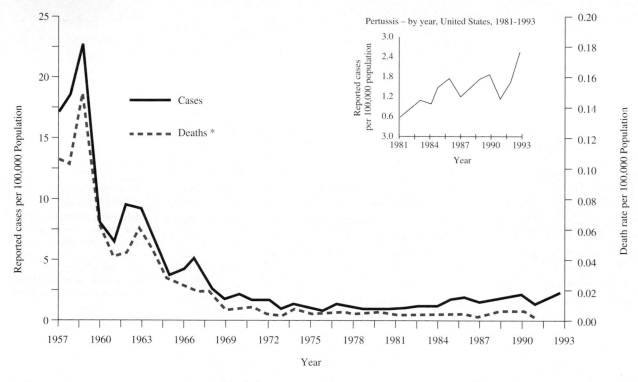

Note: DTP vaccine licensed 1948.

*Data on mortality are not yet available for 1992 and 1993.

FIGURE 7.4

The Spinal Tap Procedure

The spinal needle is inserted between the lumbar vertebrae into the space between the meninges. Fluid is collected in a vial for analysis. Cases of meningitis are diagnosed by searching for bacteria in the fluid.

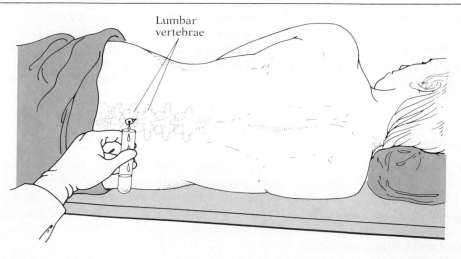

A particularly dangerous form of meningitis is **meningococcal meningitis**. This disease is caused by *Neisseria meningitidis*, a small Gram-negative, encapsulated diplococcus commonly called the **meningococcus**. Groups A, B, C, Y, and W-135 groups of meningococci are known. They enter the body by droplets, often from a reservoir. In most cases, the disease consists of an influenzalike upper respiratory infection. However, the infection sometimes spreads to the bloodstream, where bacterial toxins may overwhelm the body in as little as two hours and cause death. This condition is called **meningococcemia**. In survivors, the meninges become inflamed, and patients experience a characteristic stiff, arched neck and pounding headache. A rash also appears on the skin, beginning as bright red patches, which progress to blue-black spots. Fifty percent of untreated cases may be fatal.

nī-se′re-ah

me-ning″go-kok-se′me-ah

Early diagnosis and treatment of meningococcal meningitis can prevent irreversible nerve damage or death. A principal criterion for diagnosis is the observation of Gram-negative diplococci in samples of spinal fluid obtained by a spinal tap (Figure 7.4). In active cases of disease, *N. meningitidis* may be cultivated from this fluid. Treatment with rifampin, penicillin, or sulfonamide drugs is usually recommended. A vaccine containing capsular polysaccharides from several groups has been available for immunization since 1985 but it is used only under special circumstances.

sul-fon′a-mid

Rifampin:
an antibiotic that interferes with RNA synthesis in various types of bacteria.

Cases of meningococcal meningitis are sometimes complicated by the formation of lesions in the adrenal glands. This condition, called the **Waterhouse-Friderichsen syndrome**, may be a manifestation of a hypersensitivity reaction taking place in the body (Chapter 20). Hormone imbalances usually signal its development.

Adrenal glands:
endocrine glands located atop the kidneys and producing multiple hormones.

Neisseria meningitidis is a fragile organism that does not survive easily in the environment and must be maintained in nature by person-to-person transfer. Meningococcal meningitis is therefore prevalent where people are in close proximity for long periods of time. Grade school classrooms, military camps, and prisons are examples. Though most people suffer nothing worse than a respiratory disease, the Centers for Disease Control and Prevention reported about 2500 cases of *Neisseria*-linked meningitis in 1994 (Table 7.1).

TABLE 7.1

A Comparison of Meningococcal Meningitis and *Haemophilus* Meningitis

CHARACTERISTIC	MENINGOCOCCAL MENINGITIS	HAEMOPHILUS MENINGITIS
Agent	*Neisseria meningitidis*	*Haemophilus influenzae* b
Description	Gram-negative diplococcus	Gram-negative rod
Transmission	Respiratory droplets	Respiratory droplets
Early symptoms	Respiratory disease	Respiratory disease
Blood symptoms	Meningococcemia, serious fever, malaise	Few symptoms
Nervous symptoms	Stiff, arched neck; headache	Stiff, arched neck; headache
Skin rash	Red to blue-black spots	Rare
Age group affected	All	Children primarily
Mortality rate	High	Low
Vaccine available	Not to general public	Available to all
Complications	Waterhouse-Friderichsen syndrome	Few

TABLE 7.2				
A Summary of Airborne Bacterial Diseases of the Upper Respiratory Tract				
DISEASE	**CAUSATIVE AGENT**	**DESCRIPTION OF AGENT**	**ORGANS AFFECTED**	**CHARACTERISTIC SIGNS**
Strep throat Scarlet fever	*Streptococcus pyogenes*	Gram-positive encapsulated streptococcus	Upper resp. tract Blood Skin	Sore throat Skin rash Septicemia
Diphtheria	*Corynebacterium diphtheriae*	Gram-positive rod	Upper resp. tract Heart, nerve fibers	Pseudomembrane
Pertussis (whooping cough)	*Bordetella pertussis*	Gram-negative rod	Upper resp. tract	Mucous plugs Paroxysms of cough with "whoop"
Meningococcal meningitis	*Neisseria meningitidis*	Gram-negative encapsulated diplococcus	Upper resp. tract Blood Meninges	Toxemia Paralysis Skin spots
Haemophilus meningitis	*Haemophilus influenzae* b	Gram-negative encapsulated rod	Upper resp. tract Meninges	Respiratory symptoms Paralysis

Haemophilus Meningitis

fi′fer

he-mof′ĭ-lus

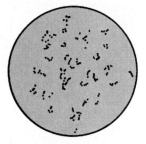

HAEMOPHILUS INFLUENZAE B

rif-am′pin

Polysaccharides:
complex carbohydrate molecules, each consisting of multiple units of a monosaccharide.

In 1892, Richard Pfeiffer isolated a small, Gram-negative encapsulated rod he thought was the cause of influenza. Because of this relationship and due to its attraction to blood ("hemo-philus") the organism was named *Haemophilus influenzae*. However, during the great influenza epidemic of 1918–1919, Pfeiffer's bacillus was identified as a secondary cause of disease, and influenza itself was attributed to a virus rather than a bacterium (Chapter 12).

Haemophilus influenzae is now regarded as an infrequent cause of respiratory tract infections, occurring primarily as a complication to a previous disease. Also, in recent years *H. influenzae* type b has attracted widespread attention as an important cause of meningitis in children between the ages of six months and two years. (The disease is sometimes called "Hib disease.") The organism moves from the respiratory tract to the blood and then to the meninges, where it causes ***Haemophilus* meningitis**. Symptoms of disease include stiff neck, severe headache, and other evidence of neurological involvement such as listlessness, drowsiness, and irritability. Combinations of drugs including rifampin are used for treatment because the many strains are resistant to one or more drugs. The disease has a mortality rate of about 5 percent.

In 1986, about 18,000 cases of *Haemophilus* meningitis were occurring in the United States annually and *H. influenzae* b was named the most common cause of bacterial meningitis in the United States. By that time, however, a vaccine had been licensed by the FDA (Figure 7.5). The vaccine consists of polysaccharides from the organism's capsule and is thus an acellular vaccine. As of 1993, the vaccine was combined with the DPT vaccine for distribution to children as Tetramune, and by 1994, the number of annual cases in the United States was down to 280. Indeed, the CDC had already launched an effort to eliminate *Haemophilus* meningitis by the end of 1996. This disease and others of the upper respiratory tract are summarized in Table 7.2.

TOXIN INVOLVED	TREATMENT ADMINISTERED	IMMUNIZATION AVAILABLE	COMMENT
Erythrogenic toxin	Penicillin Erythromycin	None	Rheumatic fever or glomerulonephritis possible Beta-hemolytic strains
Yes	Penicillin Antitoxin	Toxoid in DPT	Corynephage involved Metachromatic granules
Not established	Penicillin Erythromycin	Vaccine of dead bacteria in DPT; acellular vaccine also available as DTaP	Vaccine may cause side effects 2500 cases annually Mucus movement impaired
Yes	Rifampin Penicillin Sulfonamides	Polysaccharide vaccine	Possibly fatal Adrenal gland involement
Not established	Rifampin	Polysaccharide vaccine to Type b only	Widespread in young children Six bacterial types known

FIGURE 7.5

The Decline of *Haemophilus* Meningitis

The graph illustrates the declining number of cases of *Haemophilus* meningitis in the United States between 1987 and 1994 in children under 5. This decline is set against the increase in the number of states participating in the reporting mechanism. The graph indicates that the decline is due in part to the surveillance mechanism set up in the states. Note that the number of cases in persons over the ages of five (small box) has remained constant in this period. A very small number of cases occur in this group. (Courtesy of Centers for Disease Control and Prevention, *MMWR* 44[10], 1995.)

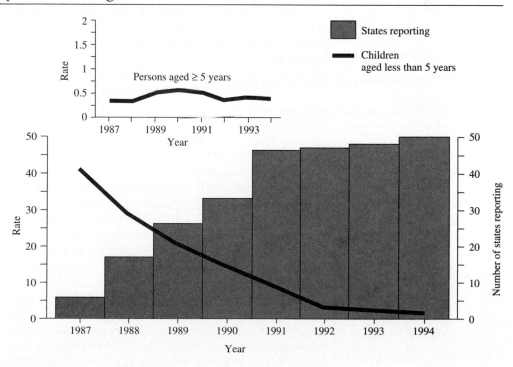

TO THIS POINT ■

We have studied a number of airborne bacterial diseases in which the initial focus of infection is in the upper respiratory tract. Symptoms generally include coughing, inflammation of the throat tissues, and the accumulation of mucus in the respiratory passageways. In diseases such as pertussis this is the extent of disease, but in other diseases, the bacteria invade the blood, and cause serious damage elsewhere. This spread is evident in scarlet fever, diphtheria, and two forms of bacterial meningitis.

The next group of diseases involve the lower respiratory tract. Infection takes place in the lung tissue, and as the breathing capacity is reduced, a life-threatening situation is established. We shall see that lung damage occurs in tuberculosis, pneumococcal pneumonia, Legionnaires' disease, and other diseases surveyed. We shall also examine a number of respiratory diseases that do not spread easily from person to person, but develop from organisms already in the patient's respiratory tract. Our survey will conclude with two diseases of the lower respiratory tract caused by rickettsiae, a group of small bacteria.

Diseases of the Lower Respiratory Tract

Tuberculosis

MYCOBACTERIUM TUBERCULOSIS

At the turn of the century, **tuberculosis** was the world's leading cause of death from all causes, accounting for one fatality in seven (MicroFocus: 7.3). Today's statistics, though improved, are still very high. In underdeveloped countries, public health officials report more deaths from tuberculosis than from any other bacterial disease. In the United States, about 27,000 new cases are identified annually, with almost 3000 fatalities, most among individuals from minority groups (Figure 7.6). The World Health Organization has referred to the global situation as a scandal and estimates that 1.7 billion people are infected.

Tuberculosis is caused by ***Mycobacterium tuberculosis***, the "tubercle" bacillus originally isolated by Robert Koch in 1882. It is a small rod that enters the respiratory tract in droplets, and multiple exposures are generally necessary. People who live in urban ghettoes often contract tuberculosis because malnutrition and a generally poor quality of life contribute to the establishment of disease, and overcrowding increases the concentration of bacilli in the air.

spu'tum

About 10 percent of people who contract tuberculosis become ill within three months. They experience chronic cough, chest pain, and high fever, and they expel **sputum**, the thick matter accumulating in the lower respiratory tract. Often the sputum is rust-colored, signaling that blood has entered the lung cavity. The remaining 90 percent of victims exhibit no readily distinguishable symptoms, except malaise, fever, and weight loss. In these cases the body responds to the disease by forming a wall of white blood cells, calcium salts, and fibrous materials around the organisms. As these materials accumulate in the lung, a hard nodule called a **tubercle** arises (hence the name tuberculosis). This tubercle may be visible in the chest X ray. Despite the body response, the bacilli are not killed, and the tubercle may expand as the lung tissue progressively deteriorates (Figure 7.7).

Tubercle:
a hard nodule consisting of mycobacteria surrounded by white blood cells, salts, and fibrous material.

7.3

MICROFOCUS

"NOT GUILTY!"

Poor Chris Columbus! Historians have accused you of bringing smallpox to the New World—and measles, and whooping cough, and tuberculosis, and almost every other conceivable disease. One can almost imagine that the stately *Santa Maria* was a hospital ship!

Well, rest easy, Chris, for you have been exonerated of bringing at least one disease—tuberculosis. Your defense is based on 1995 reports of the work of Arthur Aufderheide (pronounced OFF-der-hide) from the University of Minnesota. Some years before Aufderheide was studying the remains of a mummified woman from Peru when he noticed in her lung tissues several

lumps reminiscent of tuberculosis. He enlisted the help of a molecular biologist to extract DNA from the lumps and amplify it so there was enough to identify. Sure enough, the DNA turned out to be identical to that of *Mycobacterium tuberculosis*, the tubercle bacillus. Why was that important? Well, Chris, the mummy was a thousand years old—that's right, one thousand years. Apparently both the woman and the tuberculosis were already here hundreds of years before you arrived. It's even possible you might have taken some back with you to Europe—along with syphilis, and . . . . Oops! Sorry!

In many instances the tubercle breaks apart and bacteria spread to other organs such as the bone, liver, kidney, and meninges. The disease is now called **miliary tuberculosis** from the Latin *milium*, for seed. Tubercle bacilli produce no discernible toxins, but growth is so unrelenting that the tissues are literally consumed, a factor that gave tuberculosis its older name, **consumption**. In children, tuberculosis of the spine is known as **Pott's disease**.

Mycobacterium tuberculosis contains a layer of fatty waxy material in its cell wall that greatly enhances resistance to environmental fluctuations. In the laboratory, stain must be accompanied by heat to penetrate this barrier or a lipid-dissolving material must be used. Once stained, however, the organisms resist decolorization, even when subjected to a 5 percent acid-alcohol solution. Thus the bacilli are said to be acid-resistant, or **acid-fast**. The acid-fast test is an important diagnostic screening tool when used with a sample of patient's sputum.

Early detection of tuberculosis is aided by the **tuberculin test**, a procedure that begins with the application of a purified protein derivative (PPD) of *M. tuberculosis* to the skin. One method of application, devised by Charles Mantoux in 1910, utilizes a superficial injection to bring PPD to the outer skin layers (the Mantoux test). If the patient has been exposed to tubercle bacilli, the skin becomes thick and a raised, red welt develops within 48 to 72 hours. A positive test does not necessarily reflect tuberculosis but may indicate a recent immunization, previous tuberculin test, or past exposure to the disease. It does suggest a need for further tests.

Tuberculosis is an extremely stubborn disease. Physicians once recommended fresh air (MicroFocus: 7.4) but now they treat it with such drugs as isoniazid (INH) and rifampin, and to a lesser extent, ethambutol and streptomycin. Alternatives include pyrazinamide, ethionamide, viomycin, and cycloserine,

Pott's disease:
tuberculosis of the spine in children.

Tuberculin test:
a rapid screening procedure for tuberculosis performed by applying PPD to the skin and noting a characteristic reaction.
man-too′

i″so-ni′-ah-zid
pi″ra-zin′-ah-mid
eth″i-on′ah-mid
si″clo-ser′ēn

FIGURE 7.6

Tuberculosis in the United States, 1993

Two analyses of cases of tuberculosis. (a) Distribution of tuberculosis cases by age, race, and ethnicity. Among minority populations, most tuberculosis occurs in young adult to middle-aged individuals, while in non-Hispanic whites, the disease is seen in elderly individuals. Minorities account for a higher number of total cases than whites. (b) Distribution of tuberculosis by race and ethnicity. Over 35% of tuberculosis cases occur in non-Hispanic blacks but only 11.5% of the U.S. population is non-Hispanic black; by contrast, 27.3% of cases occur in whites, but 79.6% of the U.S. population is white. Hispanics make up 6.4% of the U.S. population but 20% of tuberculosis cases occurs in these individuals. Figures such as these point up the disproportionate distribution of the disease. (Courtesy of Centers for Disease Control and Prevention, *CDC Summary of Notifiable Diseases*, 1993. *MMWR* 42[53], 1994.)

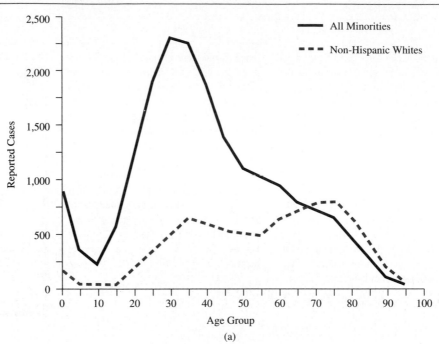

(a)

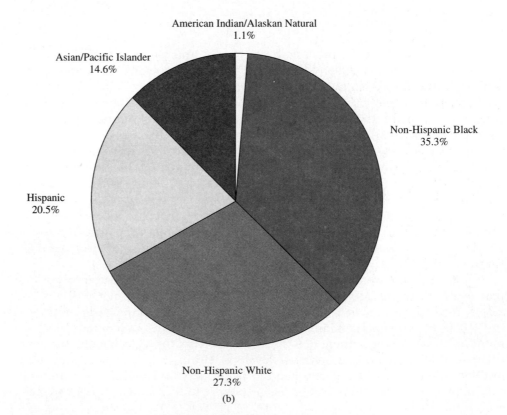

American Indian/Alaskan Natural
1.1%

Asian/Pacific Islander
14.6%

Non-Hispanic Black
35.3%

Hispanic
20.5%

Non-Hispanic White
27.3%

(b)

FIGURE 7.7
The Progress of Primary Tuberculosis

A series of diagrams depicting the development of tuberculosis lesions. (a) *Mycobacterium tuberculosis* is inhaled to the lungs and (b) enters the alveolus. (c) The bacilli multiply within white blood cells called macrophages, and soon the body's immune system dispatches other specialized macrophages and T-lymphocytes to the site. Multinucleated giant cells develop as the cells join together. (d) A wall of cells, calcium salts, and fibrous materials eventually forms around the giant cell. This is the tubercle.

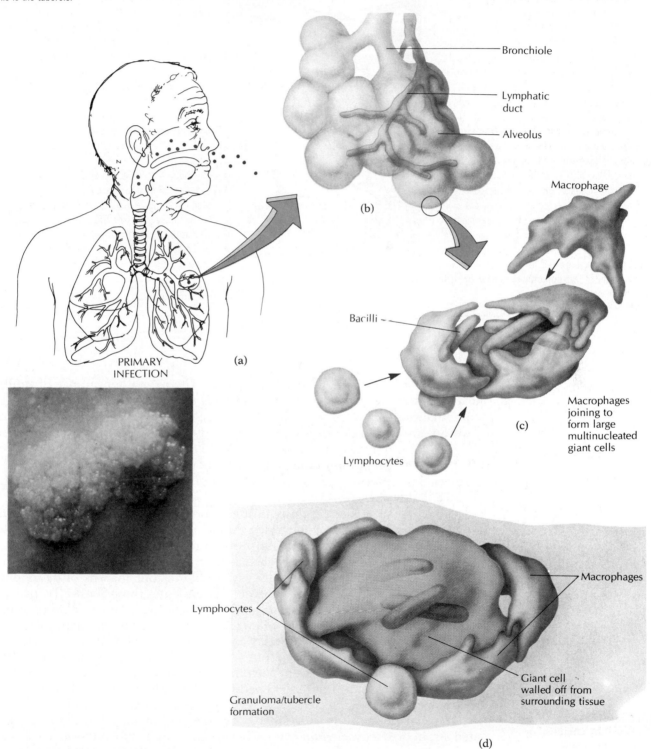

7.4
MICROFOCUS

A BREATH OF FRESH AIR

Edward Livingston Trudeau, an American physician, understood the danger in nursing his brother sick with tuberculosis. It was 1865. Many scientists believed that tuberculosis was passed among individuals, but the cause of the disease remained unknown. In Europe, Louis Pasteur was busy with his swan-necked flask experiments, and Robert Koch was still in medical school.

Several years later, Trudeau's worst fears became reality when he developed tuberculosis. His left lung was almost completely involved, and the disease had begun to appear in his right lung. Trudeau's doctor advised him to go South and exercise; hopefully, he would survive a few months. Instead, Trudeau resolved to spend his last days in the Adirondack Mountains of New York. In 1873 he left for Saranac Lake and a precious last opportunity to hunt and fish in the mountains he loved.

However, death was not yet ready to claim Trudeau. While in the mountains, he found his tuberculosis slowly but steadily going into remission, and within months, his recovery was complete. Trudeau became a strong advocate for the open-air treatment of tuberculosis and in 1884 he built two small cottages and established the sanatarium at Saranac Lake to care for tuberculosis patients. Death did eventually come, but in 1915, a full 42 years after he was told to expect it.

Today the Trudeau Institute at Saranac Lake remains a research facility for the study of human disease. And it continues to be watched over by a Trudeau—the latest director, Francis Trudeau, is the grandson of Edward. (His son, incidentally, is Gary Trudeau, author of "Doonesbury.")

---

but these drugs have more toxic side effects. The recent appearance of **multidrug-resistant *Mycobacterium tuberculosis* (MDR-TB)** has necessitated that combinations of two or more drugs be used to help delay the emergence of resistant strains. In addition, the drug therapy must be extended over a period of six to nine months or more, partly because the organism multiplies at a very slow rate (its generation time is about 18 hours). Early relief, boredom, and forgetfulness often cause the patient to stop taking the medication and the disease flares anew.

Tuberculosis is a particularly insidious problem to those who have **AIDS**. In these patients the T-lymphocytes that normally mount a response to *M. tuberculosis* are being destroyed, and the patient cannot respond to the bacterial infection. HIV-infected patients face a mortality rate from tuberculosis of between 70 and 90 percent, usually within one to four months of developing symptoms. Unlike most other tuberculosis patients, those with HIV usually develop tuberculosis in the lymph nodes, bones, liver, and numerous other organs. Ironically, AIDS patients often test negative for the tuberculin skin test because without T-lymphocytes, they cannot produce the telltale red welt that signals infection. Tuberculosis is often the first disease to occur in the AIDS patient, even before any of the opportunistic diseases appear, and it is generally more intractable than in non-AIDS patients. The WHO estimates that about 4.4 million people are coinfected with HIV and tuberculosis worldwide.

Immunization to tuberculosis may be rendered by injections of an attenuated strain of *Mycobacterium bovis*. This species causes tuberculosis in cows as well as humans. The attenuated strain is called **bacille Calmette Guérin**, or **BCG**, after Albert Calmette and Camille Guérin, the two French investigators who developed it in the 1920s (MicroFocus: 19.1). Though the vaccine is utilized

Attenuated:
weakened by chemical or cultural processes.

bah-cil′ā
kal′met
ga-ran′

FIGURE 7.8

Scrofula and Fashion

Portrait of a Woman by Franz Hals demonstrating the high-collar fashions of Elizabethan Europe of the late 1600s. The prevalence of scrofula is believed to have influenced the popularity of high collars, or ruffs, since they hid the swollen neck glands.

in other parts of the world, many scientists oppose its use in the United States because they point to the success of early detection and treatment and to the vaccine's occasional side effects. Indeed, in 1989 the CDC announced a strategic plan to eliminate tuberculosis from the United States by the year 2010. The plan emphasized more effective use of existing prevention and control methods and the development of new technologies for diagnosis and treatment.

Several other species of *Mycobacterium* bear mentioning. The first, ***Mycobacterium cheloni***, is an acid-fast rod frequently found in soil and water. During the 1980s, microbiologists first recognized this bacillus as a cause of lung diseases, wound infections, arthritis, and skin abscesses. ***Mycobacterium haemophilum*** surfaced as a pathogen in 1991 when 13 cases occurred in immuno-compromised individuals in New York City hospitals. Cutaneous ulcerating lesions and respiratory symptoms were observed in the patients.

ki-lōn′e

Another species is ***Mycobacterium avium-intracellulare***. This acid-fast rod may be the cause of lung disease, especially in persons who have acquired immune deficiency syndrome (AIDS). In AIDS patients, the body's immune system is severely depressed, and the mycobacteria rapidly invade the lungs, often causing death. The disease is commonly called **mycobacteriosis** or **MAI disease**.

The final organism, ***Mycobacterium scrofulaceum***, causes tuberculosis of the neck tissues, a disease commonly known as **scrofula** (from the Latin *scrophula* for "glandular swelling"). Scrofula is manifested by growth of mycobacteria in the lymph nodes of the neck and substantial swelling of the tissues. The disease has been recognized since ancient times. Its prevalence in the Elizabethan era may have influenced people to wear the high-neck collars characteristic of that period (Figure 7.8).

skrof″u-la′ce-um
skrof′u-lah

Pneumococcal Pneumonia

The term **pneumonia** refers to microbial disease of the bronchial tubes and lungs. A wide spectrum of organisms, including many different viruses and bacteria, may cause pneumonia. Over 80 percent of bacterial cases are due to *Streptococcus pneumoniae*, a Gram-positive encapsulated chain of diplococci traditionally known as the **pneumococcus**. The disease is commonly called **pneumococcal pneumonia**.

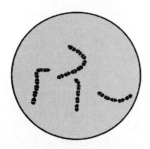

STREPTOCOCCUS PNEUMONIAE

Alveolar sacs:
millions of microscopic air sacs that make up the human lung.

Walking pneumonia:
a colloquial expression for a mild case of pneumonia.

klo"ram-fen'ĭ-kol

Pneumococcal pneumonia exists in all age groups, but the mortality rate is highest among older persons and those with underlying medical conditions. ***Streptococcus pneumoniae*** is usually acquired by droplets or contact, and the pneumococci exist in the respiratory tract of the majority of Americans. However, the natural resistance of the body is high and disease usually will not develop until the defenses are compromised. Malnutrition, smoking, viral infections, and treatment with immune-suppressing drugs typify conditions that may precede pneumonia.

Patients with pneumococcal pneumonia experience high fever, sharp chest pains, difficult breathing, and rust-colored sputum. The color results from blood seeping into the alveolar sacs of the lung as bacteria multiply and cause the tissues to deteriorate. Involvement of an entire lobe of the lung is called **lobar pneumonia**. If both left and right lungs are involved, then the condition is called **double pneumonia**. Scattered patches of infection in the respiratory passageways are referred to as **bronchopneumonia**. The term "walking pneumonia" has no clinical significance but is merely a colloquial expression for a long-lasting and relatively mild case of pneumonia.

The traditional drug of choice for pneumococcal pneumonia has been penicillin, with tetracycline and chloramphenicol used for people who are allergic to penicillin. However, public health microbiologists have noted an increasing incidence of resistant strains of *S. pneumoniae*, and as early as 1981 the first strain resistant to all three antibiotics was reported. In such cases erythromycin is useful.

Microbiologists have identified over 80 strains of *S. pneumoniae* based on the presence of different components in its capsule. Unfortunately, recovery from one strain lends immunity to that strain alone. In 1983, the FDA licensed a vaccine for immunization to 23 strains of the organism. Since these strains are responsible for 87 percent of cases, there is hope that pneumococcal pneumonia may be controlled in high-risk patients. In 1993, the CDC reported that an estimated 40,000 pneumonia deaths occur annually.

Primary Atypical Pneumonia (PAP)

During the 1940s, a significant increase in the number of unusual cases of human pneumonia prompted a search for a new infectious agent. At Harvard University, Monroe Eaton isolated a tiny, viruslike agent from the respiratory tracts of patients and cultivated it on media supplemented with blood (Figure 7.9a). The organism was subsequently named ***Mycoplasma pneumoniae***, the Eaton agent. Its disease came to be known as **primary atypical pneumonia (PAP)**, "primary" because it occurs in previously healthy individuals (pneumococcal pneumonia often is a secondary disease), "atypical" because the organism differs from the typical pneumococcus and because symptoms are unlike those in pneumococcal disease.

Mycoplasma pneumoniae is recognized as one of the smallest bacteria causing human disease. Mycoplasmas measure about 0.2 μm in size and are **pleomorphic**, that is, they assume a variety of shapes (Figure 7.9b). They have no cell wall and thus no Gram reaction or sensitivity to penicillin. *M. pneumoniae* is very fragile and does not survive long periods outside the human or animal host. Therefore it is maintained in nature by passage from host to host, and multiple exposures to droplets are usually necessary to establish disease.

PAP resembles viral pneumonia (Chapter 12) and other respiratory diseases in its symptoms. The patient experiences fever, fatigue, and a characteristic dry, hacking cough. Research indicates that the organisms attach to and destroy the ciliated cells lining the respiratory tract. Blood invasion does not occur, and the disease is rarely fatal. Often it is called **walking pneumonia** or ***Mycoplasma***

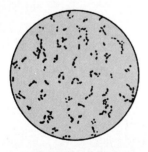

MYCOPLASMA PNEUMONIAE

Pleomorphic:
occurring in a variety of shapes.
ple"o-mor'fic

FIGURE 7.9

Mycoplasma pneumoniae

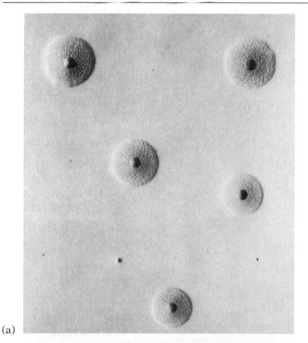

(a)

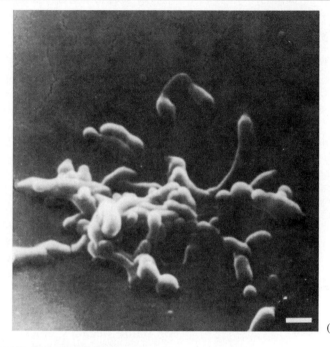

(b)

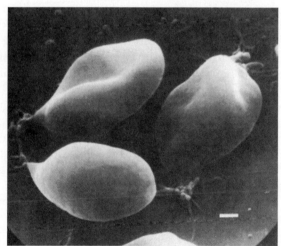

(c)

Three views of *Mycoplasma pneumoniae*, the agent of primary atypical pneumonia. (a) Colonies of *M. pneumoniae* on solid culture medium showing the typical "fried egg" appearance. (b) A scanning electron micrograph of *M. pneumoniae* demonstrating the pleomorphism exhibited by mycoplasmas. Note that the cells appear in multiple shapes, many in filamentous forms. Bar = 1 μm. (c) Three human red blood cells distorted after attachment to mycoplasmas. Bar = 1 μm.

pneumonia. Epidemics are common where crowded conditions exist, such as in college dormitories, military bases, and urban ghettoes. Erythromycin and tetracycline are commonly used as treatments.

Research in the 1940s established that antibodies produced against *Mycoplasma pneumoniae* agglutinate Type O human red blood cells at 4° C but not at 37° C. This observation was used to develop the **cold agglutinin screening test (CAST)** in which a patient's serum is combined with red blood cells at cold temperatures and the red cells are observed for clumping. Diagnosis is also assisted by isolation of the organism on blood agar and observation of a distinctive "fried egg" colony appearance. Experimental vaccines sprayed into the nasal passageways appear to reduce the incidence of disease, but none is licensed for use at this writing.

Agglutinate:
an alternate expression for clump.

Klebsiella Pneumonia

kleb"se-el'lah

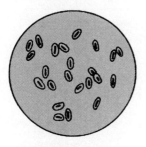

KLEBSIELLA PNEUMONIAE

Secondary disease:
a disease that develops in an already-compromised body.
nos"o-ko'me-al

In 1882, Carl Friedländer isolated ***Klebsiella pneumoniae***, a bacillus he believed was the sole cause of bacterial pneumonia (Figure 7.10). In the years thereafter, *Streptococcus pneumoniae* was found to be the major cause of pneumonia, and Friedländer's bacillus was relegated to a secondary position as the cause of about 5 percent of cases.

Klebsiella pneumoniae is a Gram-negative encapsulated rod. The bacillus is acquired by droplets, and often it occurs naturally in the respiratory tracts of humans. ***Klebsiella*** **pneumonia** may be a primary disease or a secondary disease. As a primary pneumonia, it is characterized by sudden onset and a gelatinous reddish-brown sputum. The organisms grow over the lung surface and rapidly destroy the tissue, often causing death. In its secondary form, *Klebsiella* pneumonia is a **nosocomial**, or hospital-acquired, disease (*nosos* is Greek for disease; *komein* means to care for, as in a hospital). Bacteria are spread to hospitalized patients by such routes as clothing, intravenous solutions, foods, and the hands of health-care personnel. Usually, the patient is suffering from an already established disease such as emphysema or diabetes, hence the "secondary" connotation. The symptoms, while similar to those in the primary disease, are compounded by the patient's other illness.

Serratia Pneumonia

sě-ra'she-ah
mar-ses'ens

For many decades, microbiologists considered ***Serratia marcescens*** a nonpathogenic bacillus and often used it as a test organism in their experiments (MicroFocus: 7.5). Today they view this Gram-negative rod as a cause of nosocomial disease. A patient may be predisposed to infection by such conditions as chronic illness, impaired immunity due to immunosuppressive therapy, radiation, or surgical treatments such as urinary catheterization, lumbar puncture, or blood transfusion.

Serratia **pneumonia** is accompanied by patches of bronchopneumonia and, in some cases, substantial tissue destruction in the lungs. The major clinical problem in treating the disease is resistance to antibiotic therapy, apparently due to R factors (Chapter 6). The literature of the 1990s has also cited *Serratia marcescens* as a cause of eye infections, bone disease, arthritis, and meningitis. In addition, it is an agent of urinary tract diseases.

Legionnaires' Disease (Legionellosis)

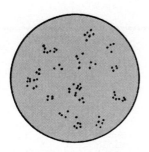

LEGIONELLA PNEUMOPHILA

From July 21 to July 24, 1976, the Bellevue-Stratford hotel in Philadelphia was the site of the fifty-eighth annual convention of Pennsylvania's chapter of the American Legion. Toward the end of the convention, 140 conventioneers and 72 other people in or near the hotel became ill with fever, coughing, and pneumonia. Eventually, 34 individuals died of the disease or its complications.

Initially some scientists suspected that the outbreak was the first wave of a swine flu epidemic predicted for that year, but as the weeks wore on, it became apparent that no known microorganism was responsible. By early December, the mystery had deepened to the point that a writer from the respected journal *Science* described the "investigation that failed." Finally, in January 1977 investigators from the CDC announced the isolation of a previously unknown bacterium from the lung tissue of one of the patients. The organism appeared responsible not only for the Legionnaires' diseases (as the disease had come to be known) but also for a number of other unresolved pneumonialike diseases. The writer from *Science* acknowledged that "the investigation may be successful after all."

FIGURE 7.10

Klebsiella pneumoniae

A transmission electron micrograph of *Klebsiella pneumoniae*, one of the agents of pneumonia. Note that the bacilli have pili, a series of hairlike appendages that assist attachment to the tissue (× 24,400).

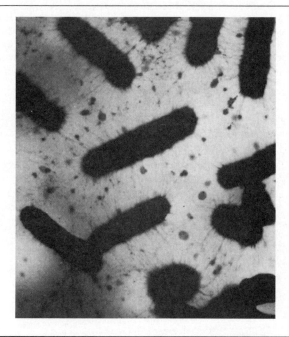

In contemporary microbiology, **Legionnaires' disease** (or **legionellosis**) is known to be caused by the organism isolated in 1977, a Gram-negative rod named ***Legionella pneumophila*** (Figure 7.11). The bacillus exists where water collects, and apparently it becomes airborne in wind gusts and breezes. Cooling towers, industrial air-conditioning units, lakes, stagnant pools, and puddles of water have been identified as sources of bacteria. Humans breathe the contaminated droplets into the respiratory tract, and disease develops a few days later.

nu-mof'ĭ-lah

The symptoms of Legionnaires' disease include fever, a dry cough with little sputum, and some diarrhea and vomiting. In addition, chest X rays show a characteristic pattern of lung involvement. Erythromycin is effective for treatment, and person-to-person transmission is uncommon. In the laboratory, the organisms are difficult to cultivate, but they can be grown on media containing hemoglobin and special amino acid additives.

Hemoglobin:
the red pigment in human red blood cells.

Once *L. pneumophila* was isolated in early 1977, microbiologists found that the organism was responsible for many epidemics of pneumonia in previous years. One such outbreak, called **Pontiac fever**, took place in Michigan in 1968; another was recorded in Washington, D.C., in 1965. In the years after 1977, reports of Legionnaires' disease occurred throughout the world and, invariably, water was involved. For example, 66 cases in Sweden were linked to water collecting on the rooftop of a shopping center, and 23 cases in Italy were traced to well water used for bathing. In South Dakota in 1990, 26 cases and ten deaths were linked to the water in showers at a particular hospital. Investigators isolated *Legionella pneumophila* from shower heads in all the patients' rooms where illness broke out.

Since *Legionella*'s discovery in water, microbiologists have been perplexed as to how such fastidious bacilli could survive in an aquatic environment that is often hostile. An answer was suggested by a 1983 study showing that the bacilli could live and grow within the protective confines of waterborne protozoa. In the South Dakota episode cited above, amebas were found in abundance in the hospital's water supply. The water was treated by heating and by addition of chlorine to help quell the spread of legionellae.

Fastidious:
having special nutritional requirements for growth.

FIGURE 7.11

Legionella pneumophila

Three views of *Legionella pneumophila*, the agent of Legionnaires' disease. (a) A transmission electron micrograph of *Legionella pneumophila*. Both inner (cytoplasmic) and outer membranes can be seen as well as several evaginations (blebs) at the outer membrane. (× 105,000.) (b) A cross-section of the protozoan *Tetrahymena pyriformis* containing *L. pneumophila* (L) within vacuoles (V). The protozoan nucleus (N) is apparent. Studies indicate that *L. pneumophila* may live in the environment within protozoa such as these. Bar = 10 μm. (c) A scanning electron micrograph of *L. pneumophila* about to be attacked and ingested by an ameba.

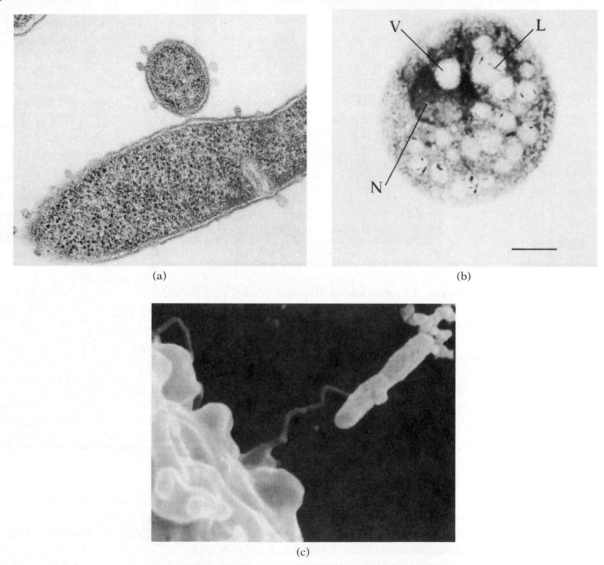

(a)   (b)

(c)

Over 1000 cases of Legionnaires' disease were reported to the CDC in 1994, and some health officials estimate that up to 50,000 cases occur annually in the United States (Figure 7.12). Thus far, six species of *Legionella* have been identified, including *L. pneumophila*. A host of "legionlike organisms" is presently under investigation as causes of human disease.

Q Fever

Q fever is one of several diseases caused by a group of bacteria known as **rickettsiae** (sing., rickettsia). Once regarded as an intermediate type of organ-

rik-et′se-e

FIGURE 7.12

Legionnaire's Disease on a Cruise Ship

1. On June 25, 1994, the cruise ship *Horizon* set sail from New York City with hundreds of passengers bound for Bermuda. All were looking forward to a delightful few days at sea and new adventures at their island destination. They were eager to enjoy all the ship had to offer.

2. Among the popular accommodations were a set of whirlpool baths on deck. Passengers found them to be an exhilarating way of relaxing and passing the hours at sea. They could look out into the water as they let the warm water massage their troubles away. The cruise ended on July 2, 1994.

3. Beginning on July 15, the New Jersey State Department of Health began receiving reports of passengers who had coughing, fever, and pneumonia. Sputum cultures of the patients yielded *Legionella pneumophila*, the agent of Legionnaires' disease. Antibody tests confirmed the diagnosis.

4. The attention of health investigators was drawn to the whirlpool baths since most ill patients had used them. In the sand filters, they located the same strain of *L. pneumophila* as in the patients. Breathing the spray apparently transmitted the bacteria to the ill passengers.

ism between bacteria and viruses, the rickettsiae are now recognized as "small bacteria" and are classified in the first volume of *Bergey's Manual of Systematic Bacteriology*. They have no flagella, pili, or capsules, and they reproduce by binary fission. With a few exceptions, rickettsiae are cultivated in the laboratory only within living tissue cultures such as fertilized eggs or animals. Arthropods are usually involved in their transmission.

7.5

"KEEP IT SHORT, PLEASE!"

Defining, developing, and proving the germ theory of disease was one of the great triumphs of scientists of the late 1800s. Applying the theory to practical problems was another matter, however, because people were reluctant to change their ways. It would take some rather persuasive evidence to move them.

In the summer of 1904, influenza struck with terrible force among members of Britain's House of Commons. Soon the members began wondering aloud whether they should ventilate their crowded chamber. Accordingly, they hired British bacteriologist Mervyn Henry Gordon to determine whether "germs" were being transferred through the air and whether ventilation would help the situation.

Gordon devised an ingenious (and, by today's standards, hazardous) experiment. He selected as his test organism *Serratia marcescens*, a bacterium that forms bright red colonies on Petri dishes of nutrient agar. Gordon prepared a liquid suspension of the bacteria and gargled with it. (*S. marcescens* is now considered a pathogen in some individuals.) Gordon then stood in the chamber and delivered a two-hour oration consisting of selections from Shakespeare's *Julius Caesar* and *Henry V.* His audience was not the august members of Parliament, but rather, hundreds of open Petri dishes containing nutrient agar. The theory was simple: if bacteria were transferred during Gordon's long-winded speeches, then they would land in the plates and form red colonies.

And land they did. After several days, red colonies appeared on agar plates opened right in front of Gordon as well as in distant reaches of the chamber. The members proposed a more constant flow of fresh air to the chamber and shorter speeches. And no one was about to object to either solution, especially the latter.

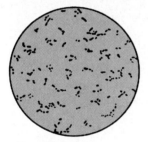

COXIELLA BURNETTI
kok″se-el′lah
bur-net′e-e

Raw milk:
unpasteurized milk.

The term **Q fever** was first used by E.H. Derrick in 1937 to describe an illness that broke out among workers at a meat-packing plant in Australia. The "Q" may have been derived either from "query," meaning unknown, or from Queensland, the province in which the disease occurred. The rickettsia that causes Q fever is named ***Coxiella burnetii*** (Figure 7.13), after H.R. Cox, who isolated it in Montana in 1935, and Frank Macfarlane Burnet, who studied its properties in the late 1930s.

Q fever is prevalent worldwide among livestock, especially dairy cows, goats, and sheep, and outbreaks may occur wherever these animals are raised, housed, or transported. Transmission among livestock and to humans is accomplished primarily by airborne dust particles as well as by respiratory droplets and by ticks. In addition, humans may acquire the disease by consuming raw milk infected with *C. burnetii* or milk that has been improperly pasteurized. Patients experience severe headache, high fever, a dry cough, and occasionally, lesions on the lung surface. The mortality rate is low, and treatment with tetracycline is effective (MicroFocus: 7.6).

Q fever is an occupational hazard of those who work with domestic animals. Q fever can also be contracted from domestic cats. In 1987 in Nova Scotia, for example, twelve members of a poker group came in contact with the homeowner's infected cat and all contracted the disease. One died of disease-related pneumonia. The organism is among the most heat-resistant of all bacteria, and its elimination is a prime object of milk pasteurization (Chapter 25).

sit-ah-ko′sis

Psittacosis and Chlamydial Pneumonia

klah-mid′e-ah
klah-mid′e-e

Psittacosis and chlamydial pneumonia are airborne diseases caused by different species of chlamydia (pl., chlamydiae). Chlamydiae are a subgroup of rickettsiae and are among the smallest bacteria (0.25 μm) related to disease in humans.

7.6

MICROFOCUS

SHEEP, PYTHONS, AND Q FEVER

Since Q fever was first reported in 1937, sporadic outbreaks and epidemics have occurred in over 50 countries on five continents. Though often associated with dairy cattle, the disease may be traced to other animals as the following cases illustrate.

In 1979, Q fever was identified in an American who had recently returned from the Middle East. The traveler had joined a group tour for a camel ride through the countryside in Jordan. When the group paused at an oasis, he remained behind long enough to be exposed to a dust cloud from a Bedouin's flock of goats and sheep. Several days later he was ill with disease.

The second outbreak occurred in 1978 at a Long Island, New York, firm that imported pythons from Ghana. The employees were responsible for removing ticks from the snakes. Apparently, several were either bitten by the ticks or exposed to their excrement. Four individuals soon came down with Q fever.

Another incident took place in Switzerland in 1983 and involved hundreds of cases. The epidemic began shortly after October 9, when 12 flocks of sheep were brought down from their mountain pastures. From October 15 to December 15, more than 300 persons reported Q fever symptoms and 191 cases were diagnosed by laboratory tests. The highest attack rates occurred in people living close to the road on which the sheep traveled. Among the sheep, 166 of 432 tested showed evidence of the disease.

Fortunately Q fever need not be a fatal disease. The Middle East traveler, the python workers, and the Swiss residents were treated with tetracycline. All recovered.

They are cultivated only in living human cells and have a complex life cycle that includes a number of different forms (Figure 7.14). Three species of chlamydiae are recognized: *Chlamydia trachomatis*, the cause of trachoma and two sexually transmitted diseases (Chapter 10); **Chlamydia psittaci**, the cause of psittacosis; and *Chlamydia pneumoniae*, the cause of chlamydial pneumonia.

trah-ko'mah-tis

sit'ah-si

FIGURE 7.13

Coxiella burnetii

An electron micrograph of *Coxiella burnetii*, the agent of Q fever (× 55,000). Note the coccobacillary form of the organism.

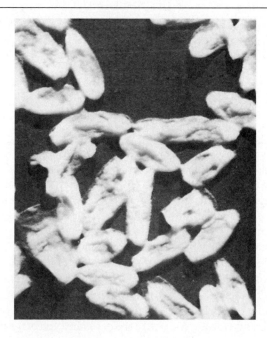

FIGURE 7.14

The Chlamydiae

Various electron microscopic views of chlamydiae. (a) Isolated chlamydiae shown in a series of photographs. In views 1A and 1B, *Chlamydia pneumoniae* is seen as pear-shaped organisms, while in 1C, *C. trachomatis* is more rounded, and in 1D and 1E, *C. psittaci* is also round; "d" refers to electron-dense granules, "s" to the space between membranes, and "e" and "r" to different forms of the chlamydia. Bar = 0.5 μm. (b) *Chlamydia pneumoniae* within host cells. The "I" refers to the inclusion body, the cellular organelle that contains the chlamydiae; "N" and "G" refer to the nucleus and Golgi complex of the cell. Bar = 1 μm.

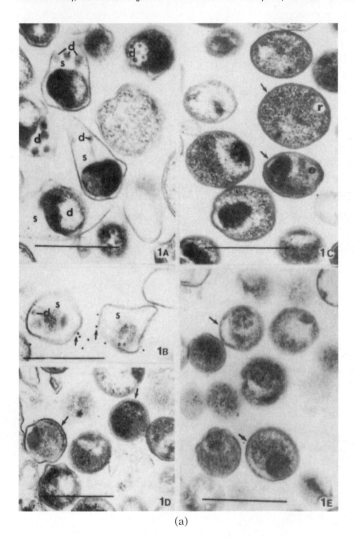

(a)

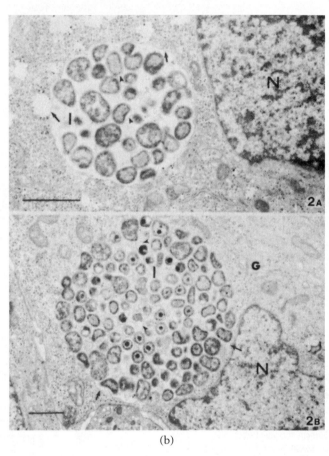

(b)

sit'ah-sin

Psittacosis affects parrots, parakeets, canaries, and other members of the psittacine family of birds (*psittakos* is the Greek word for parrot). The disease also occurs in pigeons, chickens, turkeys, and seagulls, and some microbiologists prefer to call it **ornithosis** to reflect the more widespread occurrence (*orni-* refers to bird). Humans acquire *Chlamydia psittaci* by inhaling airborne dust or dried droppings of infected birds. Sometimes the disease is transmitted by a bite from a bird or via the respiratory droplets from another human. A notable series of cases occurred in 1992 after individuals came in contact with infected parakeets and cockatiels (Figure 7.15).

The symptoms of psittacosis resemble those of influenza, Q fever, or primary atypical pneumonia. Fever is accompanied by headaches, dry cough, and scattered patches of lung infection. Without antibiotics, the mortality rate outbreaks

FIGURE 7.15
An Outbreak of Psittacosis

Two outbreaks of psittacosis traced to birds distributed by a single supplier. All the infected individuals were treated with tetracycline and/or erythromycin and recovered. This incident occurred in 1992. Coincidence linked the disease to the distributor's birds even though there was no evidence of widespread disease among his stock.

1. On February 13, a distributor in Mississippi shipped a supply of parakeets and cockatiels to retail pet stores in Massachusetts and Tennessee. The birds had been supplied to him by domestic breeders.

2. Four days later, on February 17, a man purchased one of the parakeets from the Massachusetts store. The man noted that the bird became very "tired-looking" some days after he brought it home.

3. On February 19, a family from Tennessee purchased a cockatiel sent by the distributor. The bird was taken home and kept in a bird cage where all family members could enjoy it. Some days later, family members observed that the bird was "irritable."

4. On March 1, the man from Massachusetts was hospitalized with pneumonia after two days of fever and sore throat. Two other members of his family were also sick. On the same date, six members of the Tennessee family were experiencing fever, cough, and sore throat. In both families a diagnosis of psittacosis was made.

may reach 20 percent, but tetracycline therapy reduces this number to less than 1 percent. The incidence of psittacosis in the United States is currently very low (about 100 to 250 cases per year) partly because federal law requires a 30-day quarantine for imported psittacine birds. In addition, birds are given chlortetracycline hydrochloride (CTC)–treated water and CTC-impregnated feed. Psittacosis and other diseases of the lower respiratory tract are summarized in Table 7.3.

Chlamydial pneumonia is caused by *Chlamydia pneumoniae*, an organism formerly called **TWAR** because two of the original isolates of the organism were specified TW-183 and AR-39. The chlamydia is transmitted by respiratory

---

TABLE 7.3

A Summary of Airborne Bacterial Diseases of the Lower Respiratory Tract

DISEASE	CAUSATIVE AGENT	DESCRIPTION OF AGENT	ORGANS AFFECTED	CHARACTERISTIC SIGNS
Tuberculosis	*Mycobacterium tuberculosis*	Acid-fast rod	Lungs, bones Other organs	Tubercle
Pneumococcal pneumonia	*Streptococcus pneumoniae*	Gram-positive encapsulated diplococcus in chains	Lungs	Rust-colored sputum
Primary atypical pneumonia	*Mycoplasma pneumoniae*	Mycoplasma No cell wall, 0.2 μm	Lungs	Dry cough
Klebsiella pneumonia	*Klebsiella pneumoniae*	Gram-negative encapsulated rod	Lungs	Pneumonia
Serratia pneumonia	*Serratia marcescens*	Gram-negative rod Red pigment at 25°C	Lungs	Pneumonia
Legionnaires' disease (legionellosis)	*Legionella pneumophila*	Gram-negative rod	Lungs	Pneumonia
Q fever	*Coxiella burnetii*	Rickettsia 0.45 μm	Lungs	Influenzalike symptoms
Psittacosis	*Chlamydia psittaci*	Chlamydia 0.25 μm	Lungs	Influenzalike symptoms
Chlamydial pneumonia	*Chlamydia pneumoniae* (TWAR)	Chlamydia	Lungs	Influenzalike symptoms

---

droplets and causes a mild "walking pneumonia" principally in young adults and college students. The disease is clinically similar to psittacosis and primary atypical pneumonia and is characterized by fever, headache, nonproductive cough, and infection of the lower lobe of the lung. Treatment with tetracycline or erythromycin hastens recovery from the infection. The organism was first observed in Seattle, Washington, in 1983 and is believed to infect many thousands annually in the United States.

NOTE TO THE STUDENT

In this text we shall survey a broad group of diseases caused by an equally broad group of microorganisms. You will note, however, that other microorganisms generally do not induce the life-threatening situations posed by bacterial diseases. To be sure, there are nonbacterial diseases such as malaria, AIDS, and rabies that are deadly, but bacterial diseases are the ones that have ravaged humans for centuries. Tuberculosis, diphtheria, meningococcal meningitis, and pneumococcal pneumonia are examples in this chapter. Typhoid fever, syphilis, cholera, and plague are examples encountered in other chapters.

Why do we not fear these bacterial diseases any more in the United States, and why can we blithely discuss them as if they were occurring on another planet? How often, for example, do newspaper headlines trumpet diphtheria or pertussis epidemics?

TOXIN INVOLVED	TREATMENT ADMINISTERED	IMMUNIZATION AVAILABLE	COMMENT
None	Isoniazid Rifampin Pyrazinamide Ethambutol	Bacille Calmette Guérin (BCG)	Extended treatment necessary Diagnosis by tuberculin test 25,000 new cases annually Related to AIDS
Not established	Penicillin Tetracycline Erythromycin	Polysaccharide vaccine to 23 strains	Over 80 strains identified Natural resistance high Deterioration of alveoli
Not established	Erythromycin	None	Called walking pneumonia Eaton agent involved Diagnosis by CAST
Not established	Various antibiotics	None	Common nosocomial disease Urinary tract infections also
Not established	Various antibiotics	None	Common nosocomial disease Urinary tract infections also
Not established	Erythromycin	None	Associated with airborne water droplets
Not established	Tetracycline	Vaccine for high-risk workers	Occurs in dairy cows Associated with raw milk
Not established	Tetracycline	None	Occurs in parrots and parrotlike birds
Not established	Tetracycline	None	Young adults affected

The fact of the matter is that serious bacterial diseases do break out occasionally, and some of us may be affected by them, but by and large we are insulated from widespread epidemics. This is because we understand bacterial diseases much better than in the past and can deal with them effectively by such means as sanitation measures, good hospital care, and a multitude of antibiotics. Perhaps this is a fitting place to pause and reflect on how much medical research has contributed to the quality of life. We might begin by trying to imagine what it might be like to fear disease as our ancestors once feared it.

Summary

This chapter surveys a number of bacterial diseases of the respiratory tract, beginning with diseases of the upper tract and concluding with diseases of the lower tract and lungs. In many cases, the diseases are not confined to the respiratory organs, and a spread to distant organs takes place.

Among the diseases of the upper tract are streptococcal diseases, diphtheria, pertussis, and several forms of meningitis. Toxins contribute to the development of the first two diseases, and all the diseases are accompanied by fever, cough, and destruction of the local tissues. In strep throat there is tissue erosion; in diphtheria, a pseudomembrane forms; and in pertussis, a narrowing of respiratory passages leads to the characteristic "whoop." Respiratory symptoms tend to be milder for *Neisseria*- and *Haemophilus*-induced meningitis, but serious problems arise after the bacteria pass through the blood and involve the meninges.

The lower tract infections include tuberculosis, various forms of pneumonia, Legionnaires' disease, Q fever, and psittacosis. In these instances, bacteria grow on the lung surface, and the destruction of lung tissue limits the body's ability to exchange gases. Oxygen starvation can develop in the more serious diseases such as tuberculosis and pneumococcal pneumonia, but the infection is usually milder in *Mycoplasma* pneumonia, Q fever, and psittacosis. In virtually all cases, antibiotics are available to control the diseases and help speed recovery.

Questions for Thought and Discussion

1. One of the remarkable public health stories of the ten-year period between 1986 and 1996 was the virtual elimination of *Haemophilus* meningitis as a concern to doctors and parents. Indeed, at the beginning of the period there were 18,000 cases annually in the United States, but in 1994, only 280 cases were reported. What factors probably contributed to the decline of the disease?

2. In the isolation of viruses from lung specimens and fluids, antibiotics are often used to destroy any stray bacteria and reduce the possibility of bacterial contamination. How might this practice have contributed to the failure to recognize *Legionella pneumophila* until 1976?

3. A virus is apparently responsible for the ability of the diphtheria bacillus to produce the toxin that leads to disease. Do you believe that having the virus is advantageous to the bacterium? Why or why not?

4. In New York City in lower Manhattan there stands a building between Water Street and Franklin D. Roosevelt Drive. The building has strange rounded edges that make it appear like a weird huge planter. Constructed in 1901, the building was a hospital for immigrants, especially those with tuberculosis. Can you guess what the shape of the building had to do with the disease?

5. At present there is no licensed vaccine for the prevention of any streptococcal diseases even though these are among the most commonly experienced bacterial diseases in the United States. Can you postulate why a vaccine, especially one composed of killed streptococci, might pose a threat to health?

6. In Germany, a type of spa called a "Kurhaus" exists in the mountains. Victims of lung diseases such as tuberculosis are often encouraged to stay at one of these spas for a time. Why might such a stay be helpful?

7. Bacteria are generally designated Gram-positive or Gram-negative but you may have noted that small bacteria such as rickettsiae and chlamydiae do not have this designation. Why do you think this is so? Also, why do you suppose the designation is lacking for members of the genus *Mycobacterium*?

8. It was February 1987. The patient was admitted to the hospital with high fever and a respiratory infection. Pneumococci and streptococci were eliminated as causes. Penicillin was ineffective. The most unusual sign was a continually dropping count of red blood cells. Can you guess the final diagnosis?

9. "Be sure to dress warmly when you go out or you'll catch pneumonia." This precaution, or something like it, is familiar to almost every child. In what respect is it wrong? Why is it right?

10. Isoniazid, the drug used for tuberculosis therapy, is said to be "cheap as salt." Despite the availability of this drug, however, the number of tuberculosis cases remains high in the United States. Why is this so?

11. Pertussis is particularly dangerous in infants and young children but less of a problem in adults. How many reasons for this observation can you suggest?

12. Now that *Haemophilus influenzae* is no longer considered to be the agent of influenza, would you agree that a name change is in order? If so, what name might you suggest?

13. One of the major world health stories of 1995 was the outbreak of diphtheria in the new independent states of the former Soviet Union. What factors might have contributed to this international public health emergency and what do you think was the plan to help quell the spread of the diphtheria?

14. In this chapter we have encountered organisms that are commonly named for their discoverers. Examples are Friedländer's bacillus, Pfeiffer's bacillus, and the Bordet-Gengou bacillus. However, certain organisms such as *Legionella* and *Chlamydia* are not often referred to by the names of their discoverers. Can you postulate why?

15. How many professions can you name in which workers might be exposed to the organism of Q fever? Why is it likely that many more cases have occurred than have been diagnosed?

Review

A major topic of this chapter has been the respiratory diseases caused by bacteria. To test your recall of these diseases, fill in the following crossword puzzle. The answers to the puzzle are in the appendix.

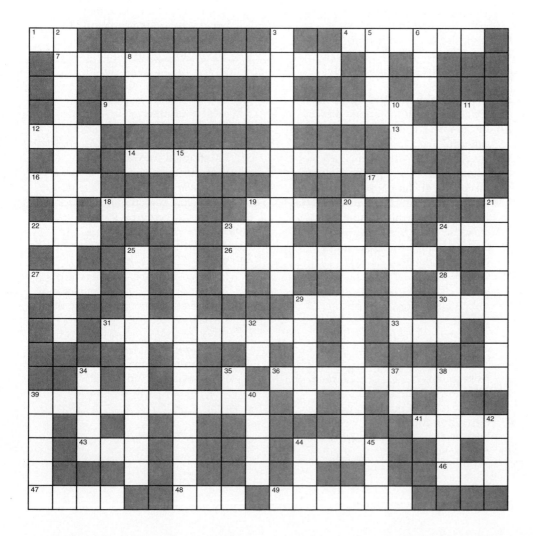

ACROSS

1. Scarlet fever is caused by a species of streptococc_____.
4. Respiratory secretions sometimes containing blood
7. Caused by a species of *Mycobacterium*
9. Genus of acid-fast rods
12. Used for immunization against two respiratory diseases
13. Bacteria can _____ the body by respiratory droplets
14. Not visible with the light microscope
16. Shape of *Legionella* bacterium
17. A valuable drug for treating meningitis is _____-ampin
18. Former name for *Chlamydia pneumoniae*
19. Incidence of diphtheria in the United States
22. Immunization to diphtheria is rendered using a _____-oid
24. The diphtheria toxin is a _____-tein
26. A species causes a form of meningitis
27. Used for immunization against tuberculosis
29. Global health group (abbr) concerned with infectious disease
30. In the CAST test, red blood cells agglutin_____.
31. Species cause psittacosis and pneumonia
33. Streptococci can cause a blood disease called _____-ticemia
36. Transmitted by airborne droplets of water
39. Gram-negative rod that causes whooping cough
41. Certain strains of mycobacteria resist _____-biotics
43. Abbreviation for a mycobacterial disease
44. Accompanies meningococcal meningitis
46. The counterstain in the Gram stain technique is _____ranine.
47. Abrasions of the _____ permit streptococci to enter
48. Spreading tuberculosis is said to be _____-iary
49. *Haemophilus meningitis* is not common in _____

DOWN

2. Alpha, beta, and gamma hemolytic forms
3. Disease of parrots, parakeets, and canaries, as well as humans
5. *S. pneumoniae* is often called the _____-mococcus
6. Psittacosis may be carried by exo_____ birds
8. Cases of legionellosis are treated with _____-thromycin
10. May be due to *Neisseria* or *Haemophilus*
11. *Mycoplasma* species have no _____ wall
15. Causes pseudomembranes to form
20. Small Gram-negative rod involved in meningitis in children
21. Rickettsia that causes Q fever
23. Drug (abbr) used to treat tuberculosis
25. Due to a species of *Corynebacterium*
28. A spinal _____ may be necessary to locate meningitis organisms
29. A symptom of meningitis may be the _____-house-Friderichsen syndrome
32. Q fever occurs in _____-mestic animals
34. The agent of pneumococcal pneumonia is _____-positive
35. Pneumonia may affect the termin_____ air sacs
37. *Klebsiella* may cause a _____socomial infection
38. Organs affected by *Mycobacterium* species
39. Animals in which *C. psittaci* infects
40. An _____-fast rod is involved in cases of tuberculosis
42. A bout of _____luenza may lead to streptococcal disease of the throat
44. Color of the pigment formed by *Serratia marcescens*
45. How a person feels when the fever is present in strep throat.

CHAPTER 8

FOODBORNE AND WATERBORNE BACTERIAL DISEASES

Poultry products and eggs are well known for their connections to *Salmonella* diseases, but the signs that summer did not point to the poultry and eggs; they pointed to the fruit salad.

There were 75 people at the party in New Jersey that June night of 1991. By the next week, 17 were ill with nausea, vomiting, diarrhea, abdominal cramps, and fever. Soon investigators were visiting each one, asking questions and requesting stool samples. All seventeen had eaten the fruit salad—the watermelon, cantaloupe, honeydew melon, strawberries, and grapes. The investigators scratched their heads because fruit salad and *Salmonella* just don't go together. Or do they? And if they do, which was the responsible fruit? They waited for additional cases to surface.

They did not have to wait long. In July, 20 cases of salmonellosis were reported in Minnesota, and several more occurred in Illinois, Pennsylvania, North Dakota, Missouri, and Michigan. Even Canada contributed 72 cases. All told, there were 400-plus cases of salmonellosis in 23 states and Canada that June–July period. And the culprit seemed to be the cantaloupe.

At least the laboratory researchers suspected it was the cantaloupe because even though they could not isolate *Salmonella* from the fruit, they recovered *Salmonella* from enough patients who had eaten cantaloupe and they were able to pinpoint *Salmonella poona* as the epidemic's cause. In addition, they traced all the cantaloupes to a farm in south Texas where the soil was heavily contaminated by *Salmonella* and where *S. poona* had probably infected the rinds. But people do not eat cantaloupe rinds; they eat the fruit. So how did the *Salmonella* get from the rind to the fruit? Any guesses?

Salmonellosis will be one of the major foodborne and waterborne diseases we consider in this chapter. Though many diseases such as salmonellosis are rarely life-threatening, others such as botulism, typhoid fever, and cholera can bring an individual to the brink of death if aggressive therapy is not instituted immediately. Public health officials must be constantly vigilant to prevent the transmission of these diseases in the food we eat and the water we drink.

Sometimes, however, even their strongest precautions are not good enough, and simply slicing through a piece of fruit can bring bacteria from the rind to the edible portion inside, as probably happened hundreds of times during a certain epidemic in the summer of 1991.

We shall categorize the foodborne and waterborne diseases as intoxications or infections. **Intoxications** are diseases in which bacterial toxins, or poisons, are ingested in food and water. Examples are botulism, staphylococcal food poisoning, and clostridial food poisoning. By contrast, **infections** refer to diseases where live bacteria are ingested in food and water and subsequently grow in the body. These infections include typhoid fever, salmonellosis, shigellosis, cholera, diarrheas, and several other bacterial diseases.

Foodborne and Waterborne Intoxications

Botulism

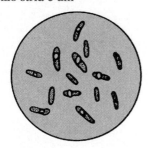

klo-strid'e-um

CLOSTRIDIUM BOTULINUM

Exotoxin:
a poisonous chemical substance produced by bacteria and immediately excreted.

flak'sid
ge-yan'bar-ra'

as"ĕ-til-ko'lēn
Acetylcholine:
the chemical substance that mediates impulse transfer where two nerves or a nerve and muscle come together.

Antitoxins:
antibody molecules that unite specifically with toxin molecules.

Of all the foodborne intoxications in humans, none is more dangerous than **botulism**. The causative agent, *Clostridium botulinum*, produces a toxin so powerful that one pint of the pure material could eliminate the entire population of the world; one ounce would kill all the people in the United States.

Clostridium botulinum is a Gram-positive anaerobic bacillus that forms spores. The spores exist in the intestines of many humans as well as numerous fish, birds, and barnyard animals. They reach the soil in manure, organic fertilizers, and sewage, and often, they cling to harvested products. When spores enter the anaerobic environment of cans or jars, they germinate to bacilli, and the bacilli produce the toxin (MicroFocus: 8.1). The toxin, a protein of high molecular weight (900,000 daltons), is called an **exotoxin** because the bacilli release it to the environment as it is manufactured. Contaminated cans of food usually lack evidence of the toxin. The food tastes and smells normal, and the can is rarely swollen and produces no swish of gas when opened. The bacteria themselves are of little consequence, but the toxin is lethal once absorbed into the bloodstream.

The symptoms of botulism develop within hours. Patients suffer blurred vision, slurred speech, difficulty in swallowing and chewing, and labored breathing. The limbs lose their tone and become flabby, a condition called **flaccid paralysis**. Often the patients are thought to be suffering from a stroke or Guillain-Barré syndrome (Chapter 12) or myasthenia gravis. These symptoms result from a complex process in which the toxin penetrates the end brushes of nerve cells and inhibits the release of the neurotransmitter acetylcholine into the junction of nerves and muscles (Figure 8.1). Without acetylcholine, nerve impulses cannot pass into the muscles, and the muscles do not contract. Failure of the diaphragm and rib muscles to function leads to respiratory paralysis and death within a day or two.

Because botulism is a type of foodborne intoxication (or poisoning), antibiotics are of no value in therapy. Instead, large doses of specific antibodies called **antitoxins** must be administered to neutralize the toxins. Life support systems such as respirators are also utilized. Frequently the nature of the problem is not recognized because the patient has done nothing extraordinary in the previous hours. Identification of the toxin by injecting leftover food samples or patient's serum into laboratory animals often comes too late. Although the annual number of cases in the United States is low, the percentage of deaths is considerable.

FIGURE 8.1

The Botulism Toxin

The mechanism of action of botulism toxin. (a) Under normal circumstances, the neurotransmitter acetylcholine is released in the neuromuscular junction from vesicles within the nerve end brush. The acetylcholine enters the synapse and mediates the transfer of the nerve impulse by bridging the gap between the end brush and receptors on muscle cells. (b) Toxin molecules now enter the area. (c) They penetrate the end brush and (d) bind to the inside surface of the cell membrane. This prevents the vesicles from releasing acetylcholine. Paralysis follows.

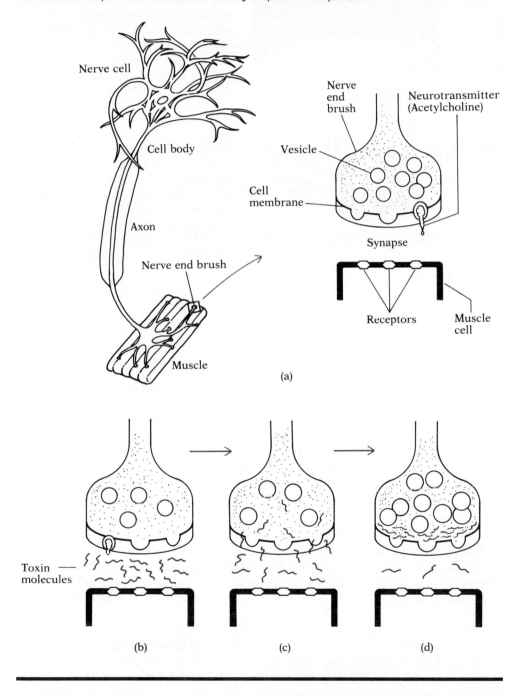

Botulism can be avoided by heating foods before consumption, because the toxin is destroyed on exposure to temperatures of 90° C for ten minutes. However, experience shows that most cases are related to food eaten cold. The largest

FIFTY-FIVE MILLION CANS OF SALMON

History's second largest recall of canned food took place in the spring of 1982 when officials of the Food and Drug Administration (FDA) ordered that grocery shelves be cleared of $7\frac{3}{4}$-ounce cans of Alaska salmon. By the end of April, eight separate recalls had been issued and over 55 million cans were removed from stores.

The problem surfaced in February of that year when a Belgian man died of botulism after eating pâté made from salmon imported from the United States. Shortly thereafter, in March, a woman in Connecticut also became ill with botulism a day after eating salmon. Markings on the cans led the FDA back to Alaskan canneries.

Investigators soon established that the canning process was adequate enough to destroy botulism spores in the fish. How-

ever, several cans contained barely perceptible pinholes. Further study revealed that the machine used for shaping the cans had inadvertently punctured them. Officials theorized that *Clostridium botulinum* entered the cans after processing and that oil or pieces of fish pressing against the sides of the cans sealed the opening.

The ramifications of the recalls were soon evident: sales of Alaska salmon dropped dramatically as warehouses swelled with unwanted cans. With sales of over $300 million annually, the salmon industry was Alaska's largest employer, and many canners forecast economic ruin. The recall also had an effect on fishermen because their catch was now unneeded. As the effects cascaded, local businesspeople began their campaign for recovery.

---

episode recorded in the United States, for example, occurred in Michigan in 1977 when 58 people became ill after eating home-canned peppers at a restaurant. Another notable outbreak, involving canned salmon (MicroFocus: 8.1), took place in 1982, and chopped garlic in oil was apparently the source in a 1985 Canadian incident. Other foods linked to botulism include mushrooms, olives, salami, and sausage. (Indeed, the word "botulism" is derived from the Latin *botulus*, for sausage.)

Scientists have identified seven types of *Clostridium botulinum* depending on variant of toxin produced. The types are lettered A through G. Types A, B, and E cause most human disease; types C and D cause most animal disease, and type F, most fish disease. Knowing which type of *C. botulinum* caused the disease is important because antitoxin therapy must be type-specific. In animals, botulism is manifested as **fodder disease**, acquired when cattle ingest toxin from silage and feed; in fowl, it is called **limberneck**. Massive fish kills also occur periodically from the consumption of botulism toxin in water.

Silage:
animal food made by fermenting plants in tall silos.

Studies in recent decades indicate that *Clostridium botulinum* can grow in the anaerobic tissue of a wound and cause **wound botulism**. Moreover, researchers have implicated the organisms in cases of **infant botulism** and have suggested that the disease may lead to sudden infant death syndrome (SIDS), a condition affecting 8000 infants annually and sometimes called "crib death." In this instance, clostridial spores may germinate in the infant's intestine before other organisms arrive and establish competition. Honey has been implicated as a possible source of the spores. Another species of *Clostridium*, *C. difficile*, may also be the cause.

Before we leave botulism we should note that in recent years, the botulism toxin has been put to use in the service of humanity. Scientists have found that, in minute doses, the toxin can relieve a number of movement disorders (the so-called dystonias) caused by involuntary sustained muscle contractions. For example, "botox," as the toxin is known, can be used to treat strabismus, or misalignment of the eyes commonly known as cross-eye. Also, the toxin may be valuable to relieve stuttering, uncontrolled blinking, musician's cramp (the bane of the violinist), and spastic closure of the anal and urinary sphincter. In addition, the toxin could one day become an important treatment for cerebral palsy and multiple sclerosis.

Staphylococcal Food Poisoning

Years ago it was common for people to complain of ptomaine poisoning shortly after eating contaminated food. Indeed, the recovery from the intestine of **ptomaines** appeared to justify their reputation. Modern microbiologists, however, have exonerated the ptomaines and placed the blame for most food poisonings on the Gram-positive bacterium ***Staphylococcus aureus*** (Figure 8.2). Today, **staphylococcal food poisoning** ranks as the second most reported of all types of foodborne disease, with *Salmonella*-related illnesses ranking first. Because most staphylococcal outbreaks probably go unreported, it may well be that staphylococcal food poisoning is the most common type.

to-mān′

Ptomaines: foul-smelling nitrogen compounds often recovered from the intestine.

FIGURE 8.2

Staphylococcus aureus

A macroscopic and a microscopic view of *Staphylococcus aureus*. This organism is responsible for most cases of common food poisoning. (a) Colonies of staphylococci growing on a nutrient agar plate. (b) A scanning electron microscope view of two cocci (× 57,000).

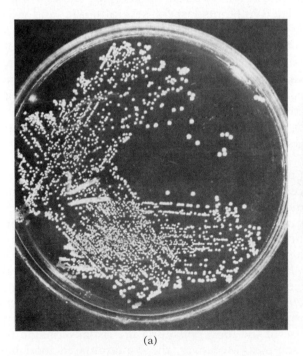

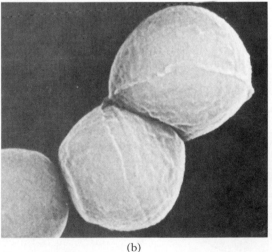

(a)   (b)

8.2

MICROFOCUS

HAPPY EASTER?

On April 2, 1983, in Modesto, California, an estimated 850 children gathered for a grand Easter egg hunt. For hours they searched through fields and hideaways to locate the brilliantly colored eggs, and when there were no more to be found, they returned home with their treasures. Before the next day, however, a number of children developed diarrhea, and parents began calling one another. Surprisingly there were other sick children, and others, and others. Within a week, over 300 children developed diarrhea. All had been at the Easter egg hunt that spring afternoon.

By this time, public health microbiologists were alerted and they promptly located *Staphylococcus aureus* in unconsumed eggs. Next, the search for a source of staphylococci led them to a cook who had boiled and dyed all 3600 eggs. The cook had lesions on his hand that yielded *Staphylococcus aureus* of the same strain as in the eggs. Moreover, he had left the boiled eggs at room temperature for three days before the Easter egg hunt.

The investigation appeared to be complete: staphylococci from the cook probably contaminated the water in which the eggs were dyed, and the three-day period provided an opportunity for growth in the eggs. But, parents asked, how could staphylococci get into hard-boiled eggs? In a 1984 article in the *Journal of the American Medical Association*, G. Alexander Merrill from the University of California provided an answer. Merrill's experiments showed that during boiling and dyeing with vinegar, the heat and acid weakens the shell and membranes of eggs, thereby permitting fluid to pass through. Up to 2 ml of fluid may pass into boiled eggs during a few minutes and if staphylococci are present, they may contaminate the egg. The lesson is simple: cooked eggs should be handled as aseptically as possible and cooled promptly. For Easter eggs, this means a quick transfer from the Easter basket to the refrigerator.

Enterotoxin:
a toxin whose effects are experienced in the intestine.

Like botulism, staphylococcal food poisoning is caused by an exotoxin excreted in foods during bacterial growth. Since the symptoms are restricted to the intestine, the toxin is called an **enterotoxin** ("entero" refers to intestine). Patients experience abdominal cramps, nausea, vomiting, prostration, and diarrhea as the toxin encourages the release of water (the word diarrhea is derived from the Greek stems *dia* meaning through and *rhein* meaning to flow; hence water "flows through" the intestine). Sufferers often report feelings of fever, but few actually have fever. The symptoms last for several hours, and recovery is usually rapid and complete.

Incubation period:
the time that elapses between entry of the disease agent to the host and the appearance of symptoms.

The **incubation period** for staphylococcal food poisoning is a brief one to six hours. Often the individual can think back and pinpoint the source. Examples are spoiled meats and fish, dairy products, cream-filled pastries, and salads such as potato salad and coleslaw. Foods containing *S. aureus* lack an unusual taste, odor, or appearance, and the only clues to possible contamination are factors such as moisture content, low acidity, and improper heating previous to arrival at the table. The staphylococcal enterotoxin is among the most heat-resistant of all exotoxins.

Boil:
a raised pus-filled lesion.

The main reservoir of *S. aureus* in humans is the nose. Thus an errant sneeze may be the source of staphylococci in foods. Studies indicate, however, that the most common **mode of transmission** is from boils or abscesses on the skin that shed staphylococci (MicroFocus: 8.2). Persons who sell, prepare, or serve food should be particularly alert to these situations. The staphylococci grow over the broad temperature range of 8° C to 45° C and since refrigerator temperatures are generally set at about 5° C, refrigeration is not an absolute safeguard. Ham is particularly susceptible because staphylococci tolerate salt.

Staphylococcus aureus normally does not grow in the human intestine because of competition by other organisms. Therefore public health investigators searching for a cause of the food poisoning are usually unable to locate the organisms in stool samples. Moreover, the contaminated food has often been consumed completely. Thus case reports are often based on symptoms,

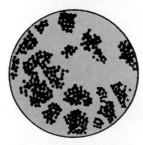

STAPHYLOCOCCUS AUREUS

FIGURE 8.3

Typing of *Staphylococcus aureus* with Bacteriophages

The plate of nutrient medium was seeded with the unknown strain of staphylococci, and numbered bacteriophages were then placed into different areas. The clear areas indicate which of the phages interacted specifically with the bacteria. In this case, the strain of *S. aureus* is one that interacts with phages 6, 47, 53, 81, and 83. Tests like this are important in relating a specific strain of *Staphylococcus aureus* to an outbreak of food poisoning.

patterns of outbreak, and type of food eaten. When investigators locate staphylococci, they can identify the organisms by growth on mannitol salt agar, Gram staining, and testing with bacteriophages to learn the strain involved (Figure 8.3). Methods such as these were used in 1989 to relate a multicity outbreak of staphylococcal food poisoning to canned mushrooms imported from China.

Mannitol salt agar:
a selective and differential medium in which staphylococci usually ferment the mannitol and produce yellow colonies.

Clostridial Food Poisoning

Since its recognition in the late 1960s, **clostridial food poisoning** has risen to prominence as the second most common type of reported food poisoning, after staphylococcal food poisoning. The causative organism, ***Clostridium perfringens***, is widely known as an agent of gas gangrene (Chapter 9). This Gram-positive anaerobic sporeformer commonly contaminates protein-rich foods such as meat, poultry, and beans (Figure 8.4). If the spores survive the cooking process, they germinate to vegetative cells that produce an enterotoxin within oxygen-free portions of the food. Consumption of the toxin leads to illness. Large numbers of vegetative cells may also be consumed into anaerobic pockets in the large intestine where the enterotoxin is produced.

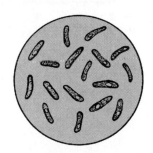

CLOSTRIDIUM PERFRINGENS

The incubation period for clostridial food poisoning is about 8 to 14 hours, a factor that distinguishes it from staphylococcal food poisoning. Moderate to severe cramping, abdominal pain, and watery diarrhea are common symptoms, but nausea and vomiting are rare. The enterotoxin encourages the outward movement of water from epithelial cells lining the intestine. Recovery is rapid, often within 24 hours, and therapy is generally unnecessary.

Research studies have suggested that *C. perfringens* spores are present in virtually every carcass of beef, pork, and lamb, and thus the potential for food poisoning is enormous. Outbreaks usually result from failure to refrigerate cooked meats promptly or adequately.

Epithelial cells:
cubelike cells that line the skin, blood vessels, and body cavities such as the respiratory and gastrointestinal tracts.

FIGURE 8.4

A Textbook Case of *Clostridium perfringens* Food Poisoning

This outbreak occurred in Ojai, California, during an outdoor Mexican-style fiesta.

1. On September 17, 1977, dried pinto beans were boiled, mashed, and stored in a cafeteria refrigerator overnight. *Clostridium perfringens* spores probably contaminated the beans, and they germinated to form vegetative cells.

2. The next day the mashed beans were transported to the outdoor fund-raiser in Ojai, California. Toxin from the cells continued to accumulate in the beans.

3. The beans were reheated in small open containers and were used to make burritos. The temperature in the shade that day was 85°F. The *Clostridium* probably continued to grow.

4. Some noted that the beans were not heated through. This may have permitted the *Clostridium* toxin to remain active.

5. On September 19, 181 persons who had attended the fiesta reported abdominal illness. Over 95% had diarrhea and within 24 hours the symptoms had disappeared.

6. Laboratory findings indicated over 4 million *Clostridium perfringens* per gram of mashed beans.

TO THIS POINT
■

We have surveyed three foodborne diseases in which toxins are responsible for the characteristic symptoms. In botulism the toxin affects the transmission of nerve impulses and causes a life-threatening situation, while in staphylococcal food poisoning and clostridial food poisoning, the toxins affect the intestinal lining and induce a loss of water by diarrhea. The water loss has a less severe impact on the body than nerve interference, but the incidence of cases of food poisoning is much higher than for botulism.

We shall now move on to a series of foodborne and waterborne diseases in which live bacteria enter the body and grow profusely. These diseases are more correctly called food infections than food intoxications. We shall begin with a serious health problem, typhoid fever, then survey other Salmonella-related diseases, and next focus on shigellosis and cholera, where patient dehydration is a prime concern. The discussion will conclude with a series of intestinal disturbances that generally are not life-threatening but have become widely recognized among individuals as detection methods continue to be improved.

Foodborne and Waterborne Infections

Typhoid Fever

Typhoid fever is among the classical diseases that have ravaged humans for generations. The disease captured the attention of microbiologists a century ago and was studied by Karl Eberth and Georg Gaffky, both associates of Robert Koch. Eberth observed the causative organism, *Salmonella typhi*, in spleen tissue in 1880, and Gaffky isolated it in pure culture four years later.

Salmonella typhi, a Gram-negative rod, displays high **resistance** to environmental conditions outside the body. This factor enhances its ability to remain alive for long periods of time in water, sewage, and certain foods. *S. typhi* causes disease only in humans and is transmitted by the five Fs: flies, food, fingers, feces, and fomites (MicroFocus: 8.3).

Salmonella typhi is acid-resistant, and with the buffering effect of food and beverages, it generally survives passage through the stomach. In the small intestine it invades the tissues, causing deep ulcers and bloody stools, but little diarrhea. Blood invasion follows, and after a few days, the patient experiences mounting fever, lethargy, and delirium ("typhoid" is derived from the Greek

SALMONELLA TYPHI

Fomites:
lifeless objects that transmit the agents of disease.

8.3

M I C R O F O C U S

TYPHOID MARY

By 1906, typhoid fever was claiming about 25,000 lives annually in the United States. During the summer of that year a puzzling outbreak occurred in the town of Oyster Bay on Long Island, New York. One girl died, and five others contracted the disease, but local officials ruled out contaminated food or water as sources. Eager to find the cause, they hired George Soper, a well-known sanitary engineer from the New York City Health Department.

Soper's suspicions centered on Mary Mallon, the seemingly healthy family cook. But she had disappeared three weeks after the disease surfaced. Soper was familiar with Robert Koch's theory that infections like typhoid fever could be spread by people who harbor the organisms. Quietly he began to search for the woman who would become known as Typhoid Mary.

Soper's investigations led him back over the ten years' time during which Mary Mallon cooked for several households. Twenty-eight cases of typhoid fever occurred in those households and each time, the cook left soon after the outbreak. One epidemic in 1903 in Ithaca, New York, spread to the community and claimed 1300 lives. Ironi-

cally, Soper had gained his reputation during this episode.

Soper tracked Mary Mallon through a series of leads from domestic agencies and finally came face-to-face with her in March 1907. She had assumed a false name and was now working for a family in which typhoid had broken out. Soper explained his theory that she was a carrier, and pleaded that she be tested for typhoid bacilli. When she refused to cooperate, the police forcibly brought her to a city hospital on an island in the East River off the Bronx shore. Tests showed that her stools teemed with typhoid organisms, but fearing that her life was in danger, she adamantly refused the gall bladder operation that would eliminate them. As news of her imprisonment spread, Mary became a celebrity. Soon public sentiment led to a health department policy deploring the isolation of carriers. She was released in 1910.

But Mary's saga had not ended. In 1915, she turned up again at New York City's Sloane Hospital working as a cook under a new name. Eight people had recently died of typhoid fever, most of them doctors and nurses. Mary was taken back to the island,

this time in handcuffs. Still she refused the operation and vowed never to change her profession. Doctors placed her in isolation in a hospital room while trying to decide what to do. The weeks wore on.

Eventually Mary became less incorrigible and assumed a permanent residence in a cottage on the island. She gradually accepted her lot and began to help out with routine hospital work. However, she was forced to eat in solitude and was allowed few visitors. Mary Mallon died in 1938 at the age of 70 from the effects of a stroke. She was buried without fanfare in a local cemetery.

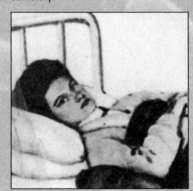

Rose spots:
red-colored, blotchy skin
spots reflecting hemorrhaged
blood.

ve-dahl′

Bacteriophages:
viruses that attack bacteria
and often replicate within
them.

typhos, for smoke or cloud, a reference to the delirium). The abdomen becomes covered with **rose spots**, an indication that blood is hemorrhaging in the skin. Bowel perforation and gall bladder infection are other possible complications.

Diagnostic tests for typhoid fever include the isolation of *S. typhi* from tissue specimens such as blood and stools. Additional data may be obtained from the **Widal test**, named for Georges Fernand I. Widal who devised it in 1896. In this procedure, the patient is tested for the presence of *Salmonella* antibodies by mixing a sample of serum with chemicals derived from *Salmonella* flagella or cells to determine whether a clumping reaction will take place (Chapter 19). A third procedure is to ask the patient to swallow a gelatin capsule tied to a string. After several hours, the capsule is retrieved and tested for *S. typhi*. Typing with bacteriophages can also detect the organism.

About 400 cases of typhoid fever are reported annually to the CDC (Figure 8.5). Treatment is generally successful with the antibiotic chloramphenicol. In the last decade, however, antibiotic resistance due to R factors has been noted (Chapter 6), and other drugs have been substituted. About 5 percent of recoverers continue to harbor and shed the organisms for one year or more. Individuals such as these are termed **carriers**. Because they are possible sources of disease to others, the public health department usually monitors the activities of carriers until they are found free of bacteria. On occasion, gall bladder removal may be recommended since this is often the site in which the bacilli remain. The experiences of one of history's most famous carriers, Typhoid Mary, is recounted in MicroFocus: 8.3.

FIGURE **8.5**

The Incidence of Typhoid Fever

Reported cases of typhoid fever per 100,000 population in the United States by year, 1955–1993. For 1991, 501 cases were reported. In about half the cases the disease was acquired during foreign travel. Note the reduction in the incidence of disease. (Courtesy of Centers for Disease Control and Prevention, *Annual Summary*, 1993.) Portrayed on the graph are the five Fs important in the transmission of typhoid fever: (a) flies, (b) food, (c) fingers, (d) feces, and (e) fomites.

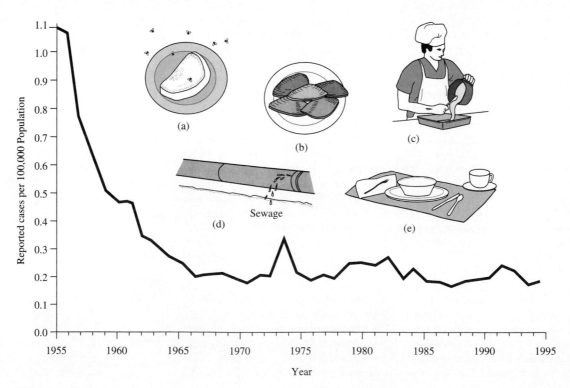

Immunization to typhoid has traditionally been rendered by injecting a vaccine consisting of dead *Salmonella* cells. However, this vaccine has carried an element of risk because of the presence of whole bacterial cells, and a newer oral vaccine has been FDA-licensed since 1990. This second vaccine (the Ty21a vaccine) consists of weakened viruses genetically engineered to carry *Salmonella* antigens for stimulating an antibody response. At this writing there is under consideration a third vaccine, one consisting of capsular polysaccharides from *S. typhi* (and named Typhim Vi). Proponents hope to use it to preclude the adverse reactions possible with the first vaccine and to immunize children under 6, which the second vaccine does not accomplish.

Salmonellosis

Salmonellosis currently ranks as the most reported of all foodborne diseases in the United States, with almost 40,000 cases occurring annually (Table 8.1). It is caused by hundreds of serotypes (serological types) of **Salmonella**. Serotypes are used for *Salmonella* instead of species because of the uncertain relationships existing among the organisms. The most common causes of salmonellosis include *S. typhimurium* (Figure 8.6a), *S. heidelberg*, *S. enteritidis*, and *S. newport*. These organisms are Gram-negative rods with peritrichous flagella (Figure 8.6b) that usually enter the human intestine in foods.

After an incubation period of one to three days, the patient with salmonellosis experiences fever, nausea, vomiting, diarrhea, and abdominal cramps. Ulceration of the intestine is usually mild compared to that in typhoid fever, and blood invasion is uncommon. The symptoms may last a week or more and generally, the disease is allowed to run its course rather than treat the patient with antibiotics that may damage the tissues. Dehydration may occur in some patients and fluid replacement may be necessary. Diagnosis usually consists of isolating the *Salmonella* serotype from stool specimens or rectal swabs using differential media. Often the infection is called **gastroenteritis** or simply, **enteritis**.

sal″mo-nel-lo′sis

Serotypes:
variants of organisms that differ according to the antibodies they elicit from the immune system.

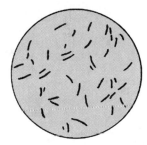

SALMONELLA TYPHIMURIUM

Dehydration:
excessive loss of body fluids.

TABLE 8.1

A Comparison of Staphylococcal Food Poisoning and Salmonellosis

CHARACTERISTIC	STAPHYLOCOCCAL FOOD POISONING	SALMONELLOSIS
Agent	*Staphylococcus aureus*	*Salmonella* serotypes
Description	Gram-positive cluster of cocci	Gram-negative rod
Toxin involved	Yes (enterotoxin)	No
Incubation period	Few hours	One to three days
Incidence in U.S.	Possible millions annually	40,000 reported annually
Symptoms	Cramps, nausea, diarrhea; no ulceration; no fever	Cramps, nausea, diarrhea; some ulceration; fever
Duration of symptoms	Few hours	Few days
Foods involved	All	All, especially poultry/eggs
Animals involved	No	Yes (especially reptiles)
Treatment	None	Antibiotics possible
Source of bacteria	Skin infection, nose	Fecal contamination
Diagnosis	Isolate bacteria from food	Isolate bacteria from feces

FIGURE 8.6
Salmonella Species

Two views of *salmonella* involved in salmonellosis. (a) *S. typhimurium* observed on the collagen fibers of muscle tissue from an infected chicken (× 12,000). (b) A transmission electron micrograph of a *Salmonella* serotype showing peritrichous flagella (× 40,000). Note the long length of the flagella relative to the cell.

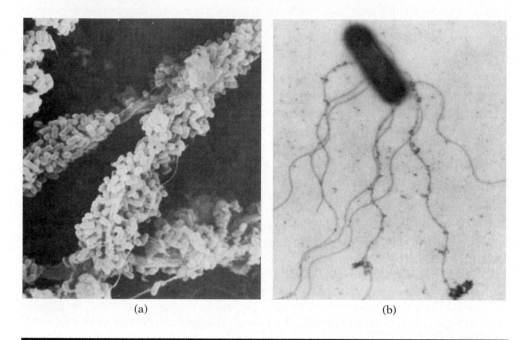

(a)   (b)

With increased awareness and modern methods of detection, salmonellosis has been linked to a broad variety of foods. Pasteurized milk was linked to 5770 cases in the Midwest in 1985 (Chapter 25), and frozen pasta products were the source of dozens of cases in the Northeast in 1986. **Poultry products** are particularly notorious because *Salmonella* serotypes commonly infect chickens and turkeys. The organisms may be consumed directly from the poultry or in pot pies, processed chicken roll or turkey roll, or chicken salads. In 1991, the CDC reported an outbreak of salmonellosis in a Connecticut hospital. The outbreak was traced to improperly cooked turkey and was of particular concern because many of the hospital employees were infected and were spreading the disease to already-ill patients. With over 4 billion chickens and turkeys consumed annually in the United States, the possibilities for *Salmonella* contamination are plentiful.

Eggs are another widespread source of salmonellosis when used in foods such as custard pies, cream cakes, egg nog, ice cream, and mayonnaise (Chapter 24). In 1992 the CDC reported on 38 cases associated with Caesar salad dressing made with raw eggs (Chapter 4). In the past, salmonellosis was associated with cracked or contaminated egg shells, but researchers are now coming to believe that *Salmonella* serotypes infect the ovary of the hen and pass into the egg before the shell forms (Figure 8.7). If this is so, then even the best grade eggs may be contaminated. Consumers should store eggs in the main compartment of the refrigerator, refrigerate leftover egg dishes quickly in small containers (to permit faster cooling), and avoid "runny" or undercooked eggs. The current catchphrase at the CDC is "scramble or gamble." Indeed, in 1992, New Jersey outlawed "eggs over easy." (It later rescinded the ban.)

FIGURE 8.7
Three Incidents of Salmonellosis

All three incidents of salmonellosis occurred in 1989, and all were related to the same organism.

1. On July 1st, 24 persons attended a baby shower in Suffolk County, New York. Twenty individuals became ill with salmonellosis. Homemade baked ziti was served at the get-together.

2. On August 24, 32 persons attended an office party in Carbon County, Pennsylvania. Twelve individuals contracted salmonellosis. All the ill attendees had had a piece of custard pie prepared by a local bakery.

3. On April 8th, 27 patrons of a restaurant in Knox County, Tennessee developed salmonellosis two days after visiting the restaurant. All had been served Hollandaise or Bernaise sauce with their meals.

4. The common threads among all these cases were the bacterium and where health investigators located it — *Salmonella enteritidis* and contaminated eggs.

Live **animals** are also known to transmit salmonellosis, and many states now prohibit the sale of Easter chicks and ducklings for this reason. Moreover, some years ago, the Food and Drug Administration prohibited the distribution of pet turtles less than four inches long because a significant number of salmonellosis outbreaks were attributed to handling these animals. The sale of iguanas is also restricted, and a savannah monitor lizard was apparently the cause of an infant's disease in 1992 (MicroFocus: 8.4).

Shigellosis (Bacterial Dysentery)

Members of the genus *Shigella* were first described by the Japanese investigator Kiyoshi Shiga in 1898, and two years later by the European microbiologist, Simon Flexner. *Shigella* bacilli are small Gram-negative rods found mainly in humans and other primates. They cause digestive disturbances ranging from

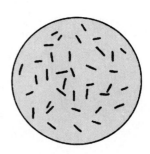

SHIGELLA SONNEI

8.4

"THANKS BUT NO THANKS!"

"It just wasn't a nice pet at all ..." the parents told the health department people, "... but the children really wanted a lizard, so we gave in." There was a pause, then they continued. "It had diarrhea all the time we had it, and the cage was so big we had to climb in to clean it. And oh, those heat rocks in the bottom of the cage—they all had to come out for cleaning. A month ago we got rid of it."

The "it" of the conversation was a 2-foot-long savannah monitor lizard. The lizard was the prime suspect in a case of salmonellosis affecting the eight-week-old infant of the family. A week previously, the infant had been brought to the hospital clinic suffering from bloody diarrhea, severe flatulence, and a temperature of 101° F. The lab isolated *Salmonella poano* from its stools, and health investigators were now trying to track down the source of the *Salmonella*. It was June 1992 in suburban Salt Lake City, Utah.

"Mind if we take some samples?" the visitors asked. In the next few minutes they swabbed the cage, the rocks, the lizard's water dish, and the floor of the cage. Two days later, sure enough—*Salmonella poano* in almost every sample. But how could the infant have come in contact with the animal? It was probably the fault of the parents—cleaning the cage, handling the lizard, washing the rocks in the kitchen sink, forgetting to wash their hands or doing so poorly. Perhaps the bacteria made their way to the infant by feeding utensils.

No additional studies were performed partly because the lizard could not be traced. Moreover, the family had already resolved that no more lizards would set foot in their house. A month before they had traded the lizard back to the pet shop for another pet ... a python.

---

mild diarrhea to a severe and sometimes fatal dysentery. **Dysentery** is a syndrome manifested by waves of intense abdominal cramps and frequent passage of small-volume, bloody mucoid stools. For many years, *Shigella* diseases were known as **bacterial dysentery**. However, watery diarrhea without blood or mucus is a more common symptom than dysentery, and the name shigellosis is currently preferred.

shĭ"gel-lo'sis
sōn'e-i

Shigellosis may be caused by any of four serotypes of *Shigella*, including *Sh. sonnei*, *Sh. dysenteriae*, *Sh. flexneri*, and *Sh. boydii*. Of these, *Sh. sonnei* is the most common. Humans can ingest the organisms in contaminated water as well as in many foods, especially eggs, vegetables, shellfish, and dairy products. *Shigella* usually penetrates the epithelial cells lining the intestine and, after two to three days, it produces sufficient enterotoxins to encourage the cells to release water. Infection of the small intestine produces watery diarrhea, but infection of the large intestine results in the expulsion of bloody mucoid stools characteristic of dysentery. Indeed, the dysentery can be quite indisposing. It is said that at the Battle of Crécy in 1346, the English army was racked with dysentery —when the French attacked, they literally caught the English with their pants down.

Intravenous:
injection directly into the vein.

Most cases of shigellosis subside within one week without complications. However, patients who lose excessive fluids must be given salt tablets or oral solutions or intravenous injections of salt solutions for **rehydration**. This treatment is most often applied to infants and debilitated adults in whom circulatory collapse may lead to death. Antibiotics are sometimes effective, but many strains of *Shigella* are resistant due to R factors, and the shigellosis may be fatal. Recoverers generally become carriers for a month or more and continue to shed the bacilli in their feces. Identification of the causative agent is confirmed by the isolation of *Shigella* from stool specimens. Immunizing agents are not available, but 586 passengers aboard the cruise ship *Viking Serenade* wish they were. They were the unfortunate victims of a shigellosis outbreak in 1994 traced to bacteria in the water supply. Trapped hopelessly aboard ship, they could only wait for landfall and dream of a more pleasant serenade the next time.

Cholera

No diarrhea can compare with the massive diarrhea associated with **cholera** (MicroFocus: 8.5). In the most severe cases, a victim may lose up to one liter of fluid per hour for several hours. The fluid is colorless and watery with characteristic **rice-water stools** reflecting the conversion of the intestinal contents to a thin material like barley soup. The patient's eyes become gray and sink into their orbits. The skin is wrinkled, dry, and cold, and muscular cramps occur in the arms and legs. Despite continuous thirst, sufferers cannot hold fluids. The blood thickens, urine production ceases, and the sluggish blood flow to the brain leads to shock and coma. In untreated cases, the mortality rate may reach 70 percent.

Cholera is caused by *Vibrio cholerae*, a curved Gram-negative rod (Figure 8.8) first isolated by Robert Koch in 1883. The bacilli enter the intestinal tract in contaminated water or food. Vegetables from fields fertilized with nightsoil are a possible source and shellfish such as raw oysters can be involved, as a 1988 case in Colorado illustrated.

Vibrio cholerae is extremely susceptible to stomach acid. However, if sufficient numbers are ingested, enough remain to colonize the intestine. As the bacilli move along the intestinal epithelium, they secrete an enterotoxin that stimulates the unrelenting loss of fluid characteristic of the disease. Antibiotics such as tetracycline may be used to kill the bacteria, but the key treatment is replacement of lost fluids and restoration of the water balance in the body. Often, this entails **intravenous injections** of salt solutions and the consumption of glucose and salts. More commonly, all but severely dehydrated adults and children can be treated with oral rehydration solution (ORS), an oral solution of electrolytes that will restore the normal balances in the body.

kol′er-ah

Rice-water stools: thin, souplike intestinal contents characteristic of cholera.

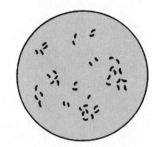

VIBRIO CHOLERAE

Nightsoil: human feces used in some parts of the world as fertilizer.

Tetracycline: an antibiotic that interferes with protein synthesis, especially in Gram-negative bacteria.

F I G U R E **8.8**

The Bacilli of Cholera

Two views of *Vibrio cholerae*, the cholera bacillus. (a) A view of the edge of a colony of *Vibrio cholerae* on a solid culture medium (× 5400). Many cells are seen in various stages of division, and the irregular arrangement of the vibrios is visible. (b) Two free cells, one of which is dividing (× 20,000). Note the commalike shape of the cells.

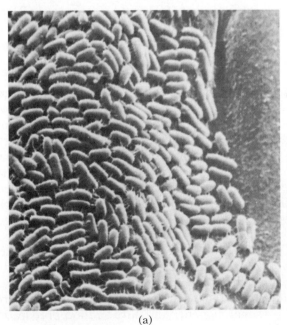

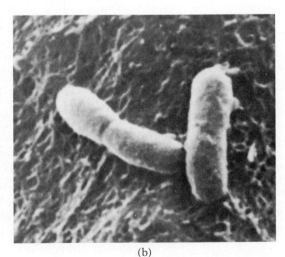

(a)   (b)

8.5

MEMORIES

It was supposed to be a simple evening together at a quiet, local restaurant—six women, dinner, some interesting scuttlebutt, and a pleasant memory or two. The August night was hot that summer in Maryland in 1991. All would go as planned, except for the memory. It would not be pleasant.

Two days later one woman developed vomiting and severe, watery diarrhea so bad she had to be hospitalized. Then two others had acute diarrhea requiring medical attention. When blood samples from the three were analyzed for antibodies, the results raised the eyebrows of health officials—it was cholera.

The first thing that came to mind were the crabs served that night. But Marylanders are proud of their crabs, and health investigators were relieved to learn that crabs from the same batch had been served to others at the restaurant with no apparent effect. What about the Thai-style rice pud-

ding? It came with a topping made from imported coconut milk, and several un-

opened packages of the milk were still in the freezer. Did the sick women have the rice pudding with the topping? Yes. How about the other restaurant patrons that night? They had ordered rice pudding but had passed on the topping. Aha! Could the health officials please have the unopened milk packages for testing?

During the next several days, emergency rooms were checked for additional cases of cholera but none surfaced. Sewage collection points were tested for cholera bacilli but samples turned up negative. Secondary contacts of the women were reached to see if they had any unusual symptoms; none reported any. By now the laboratory results on the coconut milk were ready. The milk was positive for *Vibrio cholerae* and a number of other bacteria. A voluntary product recall was issued by the distributor. Case closed.

Pandemic:
a worldwide epidemic.

    Cholera has been observed for centuries in human populations, and seven pandemics have been documented. The current pandemic began in 1961 in Indonesia and now involves about 35 countries. A major outbreak occurred in Peru and Ecuador early in 1991 and from there spread throughout the region, accounting for 731,000 cases by the end of 1992 and 6300 deaths in 21 countries of the Western Hemisphere (Figure 8.9). In 1992, the United States reported 102 cases of cholera, more than in any year since surveillance began in 1961. Seventy-five additional cases occurred in airline passengers arriving in Los Angeles from South America. The passengers had consumed contaminated seafood aboard the plane. The agent, *V. cholerae* biotype El Tor serotype Inaba, has been identified in estuaries along the U.S. Gulf Coast and in ballast waters of ships from South America. By 1994, the epidemic was spreading throughout Africa at what the World Health Organization called a "catastrophic rate."

    At present, travelers to cholera regions of the world are immunized with preparations of dead *V. cholerae* and receive protection for approximately six months. This may change because public health officials anticipate a genetically engineered cholera vaccine that will give longer-lasting protection without the risk of using dead pathogenic cells. In 1992, researchers at the University of

Enterotoxin:
a toxin affecting the gastrointestinal tract in humans.

Maryland identified the genes for enterotoxin production and successfully removed the genes from experimental *V. cholerae* cells. So treated, the vibrios became nonpathogenic but still alive (i.e., attenuated), and they could be used in a vaccine to stimulate an antibody response. Field trials for the new vaccine are ongoing at this writing. In the meantime, the most important preventive measures are sanitation, personal hygiene, and care in food preparation. Outbreaks in the United States are rare, but the disease is estimated to cause an annual four to five million deaths throughout the world.

FIGURE 8.9

A Spreading Epidemic

An illustration depicting the spread of the Latin American cholera epidemic through December 1992.
(Courtesy of Centers for Disease Control and Prevention, *MMWR* 42 [5]: 89, 1993.)

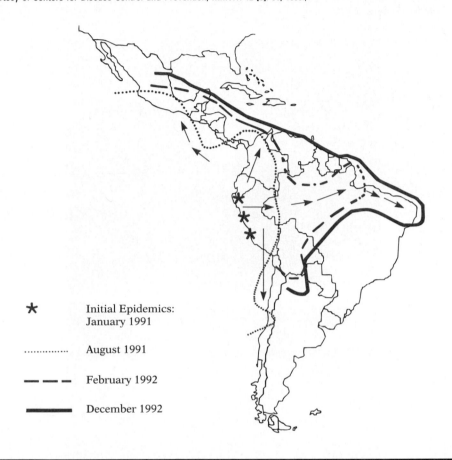

★	Initial Epidemics: January 1991
....	August 1991
– – –	February 1992
▬▬▬	December 1992

E. Coli Diarrheas

One of the major causes of **infantile diarrhea** is the Gram-negative rod *Escherichia coli* (Figure 8.10). This organism may induce diarrhea by either of two mechanisms: certain "enterotoxic" strains produce an enterotoxin similar to that in cholera, while other "enteroinvasive" strains penetrate the intestinal epithelium as in shigellosis. The toxin causes fluid loss in the small intestine, while the penetration occurs primarily in the large intestine. Both mechanisms lead to dehydration and salt imbalance substantial enough to be life-threatening in infants. Vigorous antibiotic therapy and fluid replacement are usually effective. Often the infection is a nosocomial infection.

   Traveler's diarrhea is a term usually applied to a disease in which the victim experiences diarrhea within two weeks of traveling to a tropical location; the diarrhea lasts one to ten days. Sometimes this disease is called **Montezuma's revenge**, a reference to the leader of the Aztec nation decimated by smallpox and other diseases brought from Europe in the 1500s (MicroFocus: 18.1). A number of organisms including several bacteria, viruses, and protozoa may cause traveler's diarrhea, but several recent studies point to *E. coli* as the principal agent. The bacilli adhere to the intestinal lining with pili and produce enterotoxins, which induce water loss. The volume lost is usually low, but occasionally

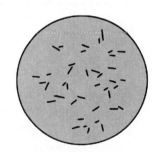

ESCHERICHIA COLI

Nosocomial:
hospital-acquired.

Pili:
short hairlike appendages
that anchor bacteria to
surfaces.

FIGURE 8.10

Escherichia coli

A transmission electron micrograph of *Escherichia coli*, the cause of many forms of traveler's and domestic diarrheas. The hairy fibers extending from the cell surface are pili, short appendages of protein that the bacterium uses to adhere to the infected tissue. (× 40,000).

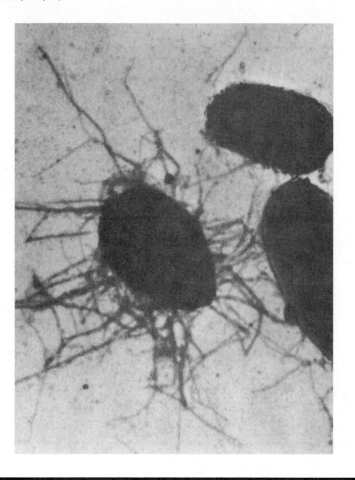

Enterotoxic:
a strain of bacteria able to produce toxins affecting the gastrointestinal tract.

it may be considerable, and dysentery may occur. The possibility of traveler's diarrhea may be reduced by careful hygiene and attention to the food and water consumed during visits to other countries.

In recent years, it has become apparent that one can contract a rather serious *E. coli* diarrhea without traveling. In 1993, for example, ***E. coli* 0157:H7**, an enterotoxic strain, was the cause of over 500 cases of serious illness in patrons of the Jack-In-The-Box fast-food chain (Chapter 3). Also that year, mayonnaise contaminated with *E. coli* was implicated in numerous cases of bloody diarrhea contracted at Sizzler restaurants in Oregon; and some years ago the McDonald's chain had to face an outbreak of *E. coli* infection traced to its hamburgers. In 1991, *E. coli* 0157:H7 was seemingly responsible for 70 cases of illness among attendees at an agricultural threshing show. Statistical analyses showed that contaminated roast beef served at the dinner was the most likely source of bacteria. Slices of the meat were served from warm roasting pans probably of sufficient temperature to encourage bacterial growth.

Epidemiologists at the CDC now estimate that approximately 20,000 cases of *E. coli* 0157:H7 infection occur annually in the United States, with about 250 deaths. Forty-six clusters of infections were recognized in 1994. As of 1995, 32 states required that *E. coli* 0157:H7 isolates be reported to the state health department, and physicians were alerted to watch for cases of bloody diarrhea. A toxin similar to that produced by *Shigella* species has been identified in *E. coli* isolates (it is called the "shigalike toxin"). Fresh-picked carrots in a garden salad were implicated in two incidents in 1993, and dry-cured salami was implicated in two 1994 outbreaks.

Peptic Ulcer Disease

One of the more remarkable discoveries of the modern era is that many cases of peptic (stomach) ulcers are caused by a bacterium. Traditionally, ulcers have resulted from "excess acid" due to factors such as nervous stress, smoking, alcohol consumption, diet, and physiological dysfunction. However, the work of Barry Marshall and his co-workers (MicroFocus: 8.6) has made it clear that the bacterium *Helicobacter pylori* is involved. This Gram-negative curved rod is similar to *Campylobacter* species and is apparently transmitted by contaminated water. Doctors have revolutionized the treatment of ulcers by prescribing antibiotics such as tetracycline and have achieved cure rates of up to 90 percent; and relapses are uncommon.

he"li-co-bak'ter
py-lor'e

How *H. pylori* manages to survive in the intense acidity of the stomach is interesting. Apparently, the bacterium twists its way through the mucous coating of the stomach lining and attaches to the stomach wall. There it secretes the enzyme urease. Urease digests urea in the area and produces ammonia as an end product. The ammonia neutralizes acid in the stomach, and the organism begins its destruction of the tissue supplemented by digestive enzymes normally found in the stomach tissue. Soon a sore appears in the stomach lining of perhaps one-quarter to one-half inch diameter (although some ulcers may be up to an inch in diameter). The pain is severe and is not relieved by food or an antacid (as is a duodenal ulcer).

Urease:
a bacterial enzyme that digests urea to ammonia, an alkaline substance.

The association of peptic ulcers with a bacterium has brought permanent relief to thousands of sufferers worldwide. But the research is not yet done because the modes of transmission must be studied, the treatments await refining, and the organism's relationship to stomach cancer remains to be elucidated. Still, these are historic times for patients and physicians alike.

Campylobacteriosis

Since the early 1970s, **campylobacteriosis** has emerged from being an obscure disease in animals to recognition as a widespread intestinal disease in humans. The causative organism is ***Campylobacter jejuni***, a curved (*campylo-* means "curved"), Gram-negative rod that moves by means of a single polar flagellum (Figure 8.11). Reservoirs for the organism include the intestinal tracts of many animals, including dairy cattle, chickens, and turkeys. Contaminated water is also a source of infection.

kam"pi-lo-bak"te-re-o'sis

kam'pĭ-lo-bak"ter

The clinical symptoms of campylobacteriosis range from mild diarrhea to severe gastrointestinal distress with fever, abdominal pains, and bloody stools. *Campylobacter* colonizes the small or large intestine, causing inflammation and occasional mild ulceration. However, the signs and symptoms of campylobacteriosis are not unique. Most patients recover in less than a week without therapy, but some have high fevers and bloody stools for prolonged periods. Erythromycin therapy hastens recovery.

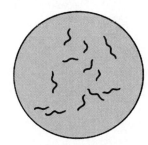

CAMPYLOBACTER JEJUNI

In 1982, eleven states reported a total of 3900 isolations of *Campylobacteria jejuni*, and in two states, the number of isolations exceeded that for *Salmonella*. Raw milk was identified as a source of *Campylobacter* in 18 outbreaks between

8.6

ULCERS ARE TRANSMISSIBLE?

The old cliche, "You're giving me an ulcer!" may be truer than anyone believed if the latest laboratory evidence holds up. Why? The evidence indicates that ulcers are a transmissible disease.

The story began in 1982 when two Australian gastroenterologists, Barry J. Marshall and J. Robin Warren, identified bacteria living in the stomach lining of over 100 patients who had ulcers. The initial discovery was serendipitous because Marshall and Warren could not cultivate the bacterium under normal conditions. Only when they were swamped with work did they leave their culture plates in the incubator too long. And only then did the colonies of bacteria appear. Marshall (photo (a)) and Warren identified the organism as the Gram-negative curved rod *Campylobacter pyloridis*. Since then, the organism's name has been changed to *Helicobacter pylori* (photo (b)).

Marshall's initial speculation that *H. pylori* causes ulcers aroused widespread skepticism, but it also intensified research. It took ten years, but the research has appeared to confirm Marshall's contention. In 1993, a study published in the respected *New England Journal of Medicine* indicated that 48 of 52 peptic ulcer patients could be cured of their ulcers in six weeks if treated with two antibiotics over a 12-day period. Another 52 patients received a placebo and 39 seemed to be cured, but a year later, the ulcers had returned in all 39 patients. By comparison, only four of the patients receiving antibiotics experienced a recurrence.

Marshall now works at the University of Virginia Medical School and supervises an ongoing program to further elucidate the cause of ulcers. The bacterial cause of ulcers conflicts with the conventional wisdom, which says that stress, diet, or other factors trigger excess acid secretion and ulcer formation. (Indeed, two of the top-selling drugs in the United States are the antacids cimetidine [Tagamet] and ranitidine [Zantac], both used to control acid secretion.) Nevertheless, the evidence continues to mount that *H. pylori* is a major factor and in another study at Baylor University, researchers obtained results similar to Marshall's. The data has compelled some researchers to begin thinking of ulcers as a transmissible disease and has encouraged doctors to prescribe tetracycline together with Pepto-Bismol. Two biotechnology firms are even attempting to develop a vaccine against *H. pylori*.

And what of Marshall? It is said that once he drank a culture of *H. pylori* to demonstrate it causes peptic ulcers (his thesis was proven true). Nowadays, he has adopted a more conservative approach to research, leaving the demonstration and thesis-proving to his graduate students.

(a) Barry J. Marshall and (b) *Helicobacter pylori*, the bacterium believed is involved in most cases of peptic ulcers. Bar = 1 μm.

(a)

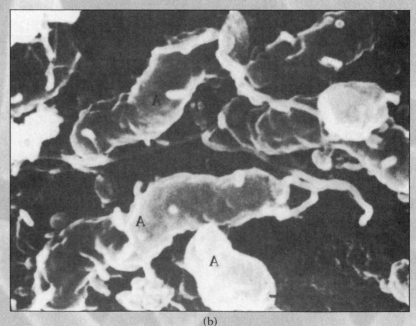

(b)

FIGURE 8.11
Campylobacter jejuni

Two views of *Campylobacter jejuni*, the cause of campylobacteriosis. (a) A scanning electron micrograph of *C. jejuni* taken from a colony of cells. Bar = 1 μm. Note the curved shape of most organisms and the coccus shape of several. (b) A transmission electron micrograph of negatively stained cells showing flagellation of both types of cells. Bar = 0.5 μm.

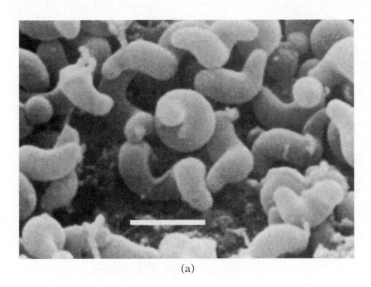

(a)

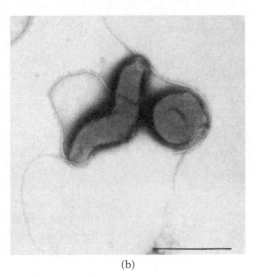

(b)

1981 and 1990, most outbreaks involving school field trips to local dairy farms. A notable incident in New Zealand in 1990 occurred at a camp where contaminated water was apparently the source of the *Campylobacter* (MicroFocus: 8.7). Other species, including *C. coli* and *C. upsaliensis*, have also been identified as human pathogens, and as methods of detection improve, reports of campylobacteriosis will probably continue to rise.

Brucellosis

Brucellosis is an occupational hazard of farmers, veterinarians, dairy and meat plant workers, and others who work with large animals. Transmission can occur by splashing milk into the eye, by accidental passage of fluids through skin abrasions, by contact with animals, and by consumption of milk and other dairy products. In one example of food transmission, the CDC reported 29 cases in Mexican emigrants living in Houston, Texas. All had eaten goat cheese made from unpasteurized goat's milk and sold by street vendors. Human-to-human transmission is virtually unknown.

Among the organisms responsible for brucellosis are ***Brucella abortus*** from cattle, *B. suis* from swine, and *B. mellitensis* from goats and sheep. Another species, *B. canis*, is associated with dogs and may be acquired by contact with pets. All are small Gram-negative rods. In animals, brucellosis manifests itself in several organs, especially those of reproduction since it multiplies in the tissues of the uterus. Sterility is a common complication, and pregnant animals with the disease are known to abort their young. In veterinary literature, the disease is often referred to as **contagious abortion**.

broo"sel-lo'sis

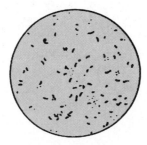

BRUCELLA ABORTUS

Undulant fever: another name for brucellosis based on alternating periods of fever and chills.

The major focus of brucellosis in humans is the blood-rich organs such as the spleen and lymph glands. Patients experience influenzalike weakness as well as backache, joint pain, and a fever that is high with drenching sweats in the daytime and low with chills in the evening. This fever pattern gives the disease its alternate name of **undulant fever** (*undulat-* is Latin for "vary"). The disease seldom causes death in humans, and tetracycline speeds recovery. Pasteuriza-tion of milk and livestock immunization have reduced the incidence of brucello-sis in humans from 6300 cases in 1947 to about 100 cases annually (Figure 8.12).

Much of the early work on brucellosis was performed by David Bruce, after whom the organism is named. Bernard L.F. Bang studied the disease in cows in the 1890s and for a while it was known as **Bang's disease**. Another name, **Malta fever**, was derived from an outbreak among soldiers on the island of Malta in the 1880s. Alice Evans later showed that these various diseases were all related to one another.

Other Foodborne and Waterborne Diseases

A number of other bacterial organisms transmitted by food and water merit consideration in this chapter. Some have been recognized as pathogens only in recent years; others have been known for decades.

par"a-he"mo-lit'ik-us

Among the recent concerns is ***Vibrio parahaemolyticus***. This Gram-negative rod is a major cause of foodborne infections in Japan and other areas of the world where seafood is the main staple of the diet. Patients experience acute abdominal pain, vomiting, diarrhea, and watery stools. Some years ago, an outbreak aboard an American cruise ship was linked to seafood salad, and another incident in Louisiana was traced to unrefrigerated cooked shrimp.

Bacillus cereus is a Gram-positive sporeforming bacillus that causes food poisoning in two distinct forms, both due to enterotoxins. The first form ("diar-rheal") is accompanied by diarrhea and abdominal pain, while the second form ("emetic") is characterized by substantial vomiting, frequently experienced after consuming cooked rice. Neither form involves fever, and most patients recover within two days without therapy. In the 1980s, investigators traced a notable outbreak to a college cafeteria where a macaroni and cheese dish harbored a million bacilli per gram. Heat-resistant spores had apparently survived the cook-ing process. Fried rice prepared at a local restaurant was the cause of a 1993 outbreak at a Virginia day-care center.

FIGURE 8.12

The Incidence of Brucellosis

Reported cases of brucellosis per 100,000 population in the United States by year, 1945–1993. The sharp drop from 1947 to 1965 was due in part to widespread pasteurization of dairy products and the eradication of brucellosis in cows. (Courtesy of Centers for Disease Control and Prevention, *Annual Summary*, 1993.) Portrayed on the graph are some modes of transmission for brucellosis including (a) contact with an animal, (b) consumption of dairy products, (c) consumption of meat, (d) accidental passage from meat through skin, and (e) handling sick animals.

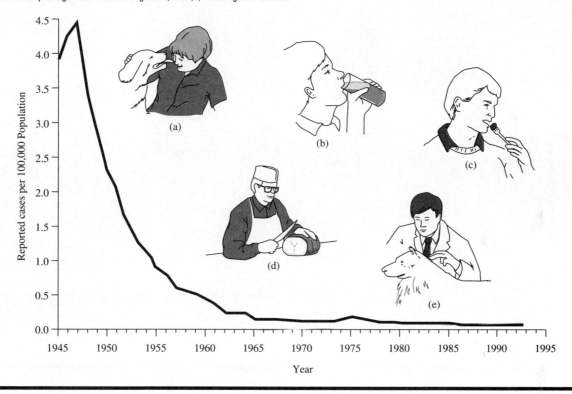

Another emerging cause of foodborne illness is **Yersinia enterocolitica** (Figure 8.13). This Gram-negative rod is widely distributed in animals and in river and lake water. The largest episode of disease thus far recorded in the United States occurred in 1982 and involved 172 patients in Arkansas, Tennessee, and Mississippi. Patients experienced fever, diarrhea, and abdominal pain, and most required hospitalization. Investigators believed that pasteurized milk was the source. A 1990 outbreak in Georgia involved raw chitterlings (pork intestines). In this case, *Y. enterocolitica* was apparently transferred to a group of children by hand-to-hand contact with an individual preparing the food for consumption.   *Yer-sin′i-a en″ter-o-co-lit′ĭ-ka*

Still another cause of intestinal illness is **Plesiomonas shigelloides**, a Gram-negative, facultatively anaerobic rod. *P. shigelloides* is normally limited to the tropical and subtropical climates of Asia and Africa, and it is commonly found in the gut of tropical fish. It can cause intestinal illness in people who eat raw seafood or who travel in the tropics and consume contaminated water or eat contaminated food. A 1990 case, however, appeared to fit neither description. The case occurred in a year-old child who had apparently been infected by water in the bathroom tub. Her parents previously had poured water from their aquarium into the tub. The aquarium was home to a collection of piranhas, a type of tropical fish the family kept as pets.   *ples-e-o-mo′nas shig-el-oi′des*

Clinical reports in the 1980s have documented that *Aeromonas* species, especially **Aeromonas hydrophila**, cause human gastrointestinal disease. The organisms are Gram-negative rods commonly found in soil and water. They appear to be transmitted by food. Both choleralike and dysenterylike diarrheas,   *a″er-o-mo′nas hī-drof′ĭ-lah*

FIGURE 8.13

Yersinia enterocolitica

Transmission electron micrographs of invasive *Yersinia enterocolitica*. (a) A number of bacteria attached to the plasma membrane of infected cells. One bacterium appears to be undergoing cell division and entering the cell at the same time. (b) Several *Y. enterocolitica* cells observed within the cytoplasm of an infected cell. The bacteria are in membrane-bound vacuoles. Researchers postulate that a single gene is responsible for the invasive capacity displayed by this organism.

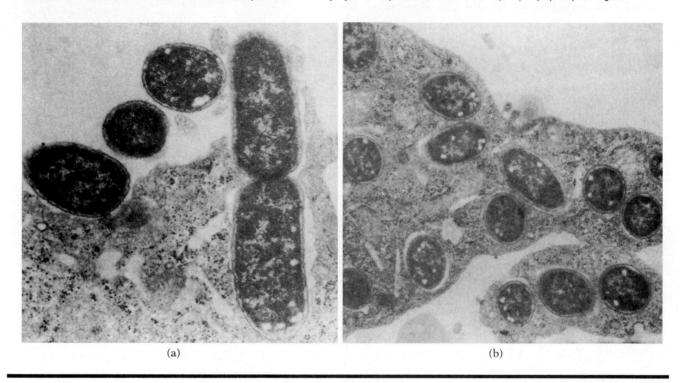

(a)   (b)

ranging from mild to severe, have been reported in patients. An enterotoxin may be responsible for the symptoms. Foodborne and waterborne bacterial diseases are summarized in Table 8.2.

NOTE TO THE STUDENT

For over a half-century Escherichia coli *has been the workhouse of biological research. In the 1940s, it was used as a host organism to determine the life cycle of viruses. Many of the important metabolic pathways, including the renowned Krebs cycle, were worked out first in this organism. Then, in the 1950s, biochemists employed* E. coli *to discover the three forms of microbial recombination. In the 1960s, it was the major research organism for deciphering the genetic code and learning how genes work. In the 1970s,* E. coli *became a valuable indicator of water pollution and an industrial giant for producing enzymes, growth factors, and vitamins. Since the 1980s, biochemists have used it as a living factory to produce an array of genetically engineered pharmaceuticals. An in the 1990s,* E. coli *continues to illustrate how bacteria can be put to work in the interest of science and for the betterment of humanity.*

Then came E. coli *0157:H7. To be sure, this strain has caused much human misery and pain because of its propensity to invade the intestinal tissues, pass to the blood, and cause serious injury to the kidney. It has made us more careful of what we eat and has caused us to think twice about having a rare hamburger. Along the way it has become the "germ of*

TABLE 8.2

A Summary of Foodborne and Waterborne Bacterial Diseases

DISEASE	CAUSATIVE AGENT	DESCRIPTION OF AGENT	ORGANS AFFECTED	CHARACTERISTIC SIGNS	TOXIN INVOLVED	TREATMENT ADMINISTERED	IMMUNIZATION AVAILABLE	COMMENT
Botulism	*Clostridium botulinum*	Gram-positive sporeforming rod	Neuromuscular junction	Paralysis	Yes	Antitoxin	None	Most powerful toxin known Infant and would botulism possible
Staphylococcal food poisoning	*Staphylococcus aureus*	Gram-positive staphylococcus	Intestine	Diarrhea Nausea Vomiting	Yes	None	None	Affects millions of Americans annually Due to an enterotoxin Brief incubation period
Clostridial food poisoning	*Clostridium perfringens*	Gram-positive sporeforming rod	Intestine	Diarrhea Cramping	Yes	None	None	Common in protein-rich foods Spores survive cooking
Typhoid fever	*Salmonella typhi*	Gram-negative rod	Intestine Blood Gall bladder	Ulcers Fever Rose spots	Not established	Chloramphenicol	Vaccine of dead bacteria	Spread from carriers About 400 cases annually in U.S.
Salmonellosis	*Salmonella* serotypes	Gram-negative rods	Intestine	Fever Diarrhea Vomiting	Not established	Not recommended	None	Associated with poultry products Most reported foodborne disease in U.S. About 40,000 cases annually
Shigellosis	*Shigella* serotypes	Gram-negative rods	Intestine	Diarrhea Dysentery	Not established	Rehydration Antibiotics	None	May be accompanied by dysentery R factors limit antibiotic use
Cholera	*Vibrio cholerae*	Gram-negative curved rod	Intestine	Rice-water stools Extreme diarrhea Shock	Yes	Rehydration	Vaccine of dead bacteria	Danger from dehydration Rare in U.S. Raw seafood involved
E. coli diarrheas	*Escherichia coli*	Gram-negative rod	Intestine	Diarrhea	Yes	Antibiotics Rehydration	None	Symptoms due to enterotoxin Enteroinvasive strains observed
Campylobacteriosis	*Campylobacter jejuni*	Gram-negative curved rod	Intestine	Diarrhea Fever	Not established	Erythromycin	None	Associated with raw milk Symptoms mild to serious
Brucellosis	*Brucella* species	Gram-negative rods	Spleen Lymph glands	Undulating fever Joint pain	Not established	Tetracycline	None	Induces abortion in barnyard animals Hazardous to animal workers
Other foodborne and waterborne diseases	*Vibrio parahaemolyticus*	Gram-negative rod	Intestine	Diarrhea	Not established	Various antibiotics	None	Associated with seafood
	Bacillus cereus	Gram-positive sporeforming rod	Intestine	Diarrhea	Yes	None	None	Spores survive in cooked foods
	Yersinia enterocolitica	Gram-negative rod	Intestine	Diarrhea	Not established	Various antibiotics	None	Widely found in animals
	Pleisomonas shigelloides	Gram-negative rod	Intestine	Diarrhea	Not established	Various antibiotics	None	Occurs in tropical climates
	Aeromonas hydrophila	Gram-negative rod	Intestine	Diarrhea	Yes	Various antibiotics	None	Common in soil and water

the week" on "Primetime Live," "Dateline," "20-20," and other television news programs, where it has been portrayed as the chief villain among a world of villains. Unfortunately, it has also made us forget all the good things that E. coli *has done for us. And what a shame that is! Perhaps one bad apple can, indeed, spoil the whole barrel.*

Summary

A recurring theme of this chapter is that foodborne and waterborne bacterial diseases primarily affect the intestines and are of two types: intoxications and infections.

Foodborne intoxications are exemplified by botulism, staphylococcal food poisoning, and clostridial food poisoning. In all three instances, the growing and multiplying bacteria deposit toxins in contaminated food, and the toxins bring on the illness after the food is consumed. In botulism, the toxins interfere with nerve transmission and induce paralysis, while in the remaining two diseases, the toxins induce a mild to moderate bout of diarrhea. The incubation periods tend to be short since bacterial growth in the body is not a prerequisite to disease.

By contrast to the intoxications, the food- and water-related infections have longer incubation periods because bacterial growth must occur before symptoms are experienced. In typhoid fever, there is tissue invasion and severe intestinal ulceration, followed by a life-threatening invasion of the blood. Cases of salmonellosis are less dangerous, most characterized by several days of fever, cramps, and diarrhea. In *Shigella*-related infections, some erosion of the intestinal tissue occurs, and small-volume, bloody stools may occur, a condition called dysentery. Massive often-fatal diarrhea is the key characteristic of cholera, and fluid-electrolyte imbalances are usually severe. Other infections, including traveler's diarrhea and campylobacteriosis, are accompanied by varying intestinal symptoms, but in brucellosis the characteristic sign is a high-low fever pattern called undulating fever. This latter occurs because, although acquired by contaminated food, brucellosis is a blood disease with few evidences of intestinal infection.

Questions for Thought and Discussion

1. The story is told of a turn-of-the-century doctor in New York City who was an expert at diagnosing typhoid fever even before the symptoms of disease appeared. The doctor's forte was the tongue. He would go up and down the rows of hospital beds feeling the tongues of patients and announcing that the patient was in the early stages of typhoid. Sure enough, a few days later the symptoms would surface. What do you think was the secret to his success?

2. It is a good idea when you are traveling to eat a cup of yogurt every so often. Why do you think this is good advice? And another thought: the yogurt should be "made from pasteurized milk"; it should not be "pasteurized yogurt." Why?

3. A two-pound piece of leftover roast beef is warmed up for dinner. Twelve hours later, family members complain of intestinal cramps and diarrhea, but experience no vomiting. What organism might have caused this problem?

4. In July 1988, a local New York newspaper reported that many ducks died of limberneck at a municipal pond. Analysis of the water revealed that botulism toxin was present in the water. What conditions might have contributed to the accumulation of toxin and death of the ducks?

5. At present there is no immunizing agent available to the general public for protection against botulism. Would you support research aimed at developing such a vaccine? Why?

6. Some years ago, the CDC noticed a puzzling trend: reported cases of salmonellosis seemed to soar in the summer months, then drop radically in September. Can you venture a guess why this is so?

7. In March 1984, a local newspaper reported over a hundred cases of "salmonella food poisoning" aboard a British Airways *Concorde* jet. What is wrong with the phrase in quotation marks?

8. A laboratory instructor proposes to demonstrate the growth of soilborne organisms in canned food in the following way: a can is to be punctured with an ice pick and a small pinch of rich soil introduced to the can; the hole will then be sealed with candle wax and the can incubated. A fellow laboratory instructor hears of the experiment, reacts with concern, and advises the first instructor not to perform it. Why?

9. Between February 18 and 22, 1987, eleven cases of botulism were diagnosed in patrons of a restaurant in Vancouver, B.C. The disease was subsequently traced to mushrooms bottled and preserved in the restaurant. What special cultivation practice enhances the possibility that mushrooms will be infected with the spores of the organism of botulism?

10. Clams are sometimes referred to as "typhoid grenades." Can you guess why this term is used?

11. In preparation for a summer barbecue, a man cuts up chickens on a wooden carving board. After running the board under water for a few seconds, he uses it to cut up tomatoes, lettuce, peppers, and other fixings for salad. What sort of trouble is he asking for?

12. When a woman died of botulism in 1980, a public official was quoted in an interview as saying: "This death might have been prevented if any of the doctors involved had recognized and acted early on the symptoms of the disease they were dealing with." Do you believe the doctors were at fault? What might be your reply to the official?

13. During 1992, the North Carolina Department of Health received reports of illness in 18 workers at a local pork processing plant. All the affected employees worked on the "kill floor" of the plant. All had Gram-negative rods in their blood. Their symptoms included fever, chills, fatigue, sweats, and weight loss. Which disease was pinpointed in the workers?

14. Two days ago two hamburgers were purchased at the local market. One was frozen, the other remained chilled in the refrigerator section. Both are now placed on the grill. All other things being equal (size of hamburger, cooking time, source of meat, use of condiments, and so on), which hamburger might be safer to eat if you wish to avoid food poisoning?

15. A newspaper cartoon once showed a man comically slumped over the table, obviously dead. The man had just finished his meal and the waiter was suggesting to a passerby that, in view of what he had been eating, it was definitely a case of suicide. (In the previous weeks the newspaper had carried several stories of foodborne illness.) Imagine you were the cartoon writer. What foods might you have included in the cartoon?

16. In 1982, a Michigan outbreak of foodborne illness caused by *E. coli* was traced to hamburgers fried at a local fast-food restaurant. Most patients who contracted the disease ate hamburgers during peak lunchtime hours. Why would this fact alert the attention of health inspectors?

17. In 1986, a New Rochelle, New York, frozen-food manufacturer recalled thousands of packages of jumbo stuffed shells and cheese lasagna after a local outbreak of salmonellosis. Which parts of the pasta products would attract the attention of inspectors as possible sources of *Salmonella*? Why?

Review

The preceding pages have summarized some of the major bacterial diseases transmitted by food and water. To test your knowledge of the chapter's contents, rearrange the scrambled letters and insert the correct word in each of the missing spaces. The answers are listed in the appendix.

1. To treat patients who have botulism, large doses of ＿＿＿＿＿＿＿＿＿ must be administered.　　　　　Ⓘ Ⓘ Ⓣ Ⓐ Ⓧ Ⓝ Ⓝ Ⓣ Ⓞ

2. One of the most excessive diarrheas observed in an intestinal disease is associated with _____.   Ⓡ Ⓛ Ⓗ Ⓔ Ⓐ Ⓒ Ⓞ

3. The fever pattern in brucellosis gives the disease its alternate name of _____ fever.   Ⓐ Ⓤ Ⓣ Ⓛ Ⓝ Ⓝ Ⓓ

4. Disease associated with *Shigella* species is usually accompanied by a syndrome of cramps and bloody stools called _____.   Ⓝ Ⓢ Ⓓ Ⓔ Ⓨ Ⓨ Ⓣ Ⓡ

5. In cases of typhoid fever, the abdomen is often covered with a series of _____ spots.   Ⓔ Ⓢ Ⓡ Ⓞ

6. _____ food poisoning has a relatively short incubation period and is caused by a toxin deposited in food during aerobic bacterial growth.   Ⓞ Ⓗ Ⓛ Ⓐ Ⓨ Ⓢ Ⓟ Ⓒ Ⓒ Ⓣ Ⓛ Ⓞ Ⓒ Ⓐ

7. Because *Salmonella* species commonly infect _____, any products derived from these animals are potentially infected.   Ⓗ Ⓒ Ⓝ Ⓚ Ⓘ Ⓔ Ⓢ Ⓒ

8. A small percentage of those who recover from typhoid fever remain _____ and can spread the organism in their feces.   Ⓐ Ⓒ Ⓡ Ⓔ Ⓘ Ⓡ Ⓢ Ⓡ

9. *Escherichia coli* is a common Gram-_____ rod that can be a cause of infantile and traveler's diarrhea.   Ⓖ Ⓝ Ⓥ Ⓘ Ⓐ Ⓔ Ⓣ Ⓔ

10. A sporeforming aerobic rod that can cause foodborne illness is *Bacillus* _____.   Ⓡ Ⓒ Ⓔ Ⓢ Ⓤ Ⓔ

11. A curved rod belonging to the genus _____ is known to cause mild to severe diarrhea in humans.   Ⓨ Ⓔ Ⓐ Ⓟ Ⓞ Ⓒ Ⓒ Ⓛ Ⓐ Ⓣ Ⓡ Ⓜ Ⓑ

12. In an animal such as a cow, infection with *Brucella* may result in the _____ of the fetus.   Ⓞ Ⓣ Ⓞ Ⓐ Ⓡ Ⓘ Ⓝ Ⓑ

13. A major symptom in patients experiencing botulism is _____ of the limbs and respiratory muscles.   Ⓐ Ⓛ Ⓢ Ⓟ Ⓐ Ⓨ Ⓢ Ⓡ Ⓘ

14. Many instances of staphylococcal food poisoning originate with staphylococci found in the _____.   Ⓞ Ⓢ Ⓔ Ⓝ

15. *Salmonella typhi* is able to reach the human intestine because it is highly resistant to _____.   Ⓒ Ⓘ Ⓐ Ⓓ

16. One of the dangers of shigellosis is excessive _____ loss in the patient.   Ⓛ Ⓘ Ⓓ Ⓤ Ⓕ

17. Outbreaks of cholera in the United States can best be described as _____.   Ⓐ Ⓔ Ⓡ Ⓡ

18. Consumers of seafood are particularly vulnerable to intestinal infection due to a species of _____.   Ⓞ Ⓘ Ⓑ Ⓡ Ⓥ Ⓘ

19. The most reported foodborne infection in the United States is that caused by types of _____.   Ⓐ Ⓜ Ⓔ Ⓐ Ⓢ Ⓞ Ⓛ Ⓛ Ⓛ Ⓝ

20. To avoid staphylococcal food poisoning, consumers should be sure to _____ all leftover foods before eating them.   Ⓔ Ⓗ Ⓣ Ⓐ

21. The organism *Clostridium perfringens* multiplies in foods only under _____ conditions.   Ⓔ Ⓞ Ⓘ Ⓝ Ⓡ Ⓒ Ⓐ Ⓐ Ⓑ

22. Diagnosis of many of the intestinal diseases is assisted by the isolation of bacteria from _____ specimens.   Ⓞ Ⓞ Ⓣ Ⓢ Ⓛ

23. Diseases such as shigellosis, typhoid fever, and cholera may be contracted by consuming contaminated _____.   Ⓞ Ⓔ Ⓕ Ⓞ Ⓓ Ⓢ Ⓐ

24. In recent years, raw _____ has been implicated in many outbreaks of campylobacteriosis.   Ⓛ Ⓚ Ⓜ Ⓘ

25. Those who work with large _____ may be exposed to the bacterium that causes brucellosis.   Ⓜ Ⓝ Ⓛ Ⓐ Ⓐ Ⓘ Ⓢ

CHAPTER 9

SOILBORNE AND ARTHROPODBORNE BACTERIAL DISEASES

The Black Death (the "plague") was probably the greatest catastrophe ever to strike Europe. It swept back and forth across the continent for almost a decade, each year increasing in ferocity. By 1348, two thirds of the European population was stricken and half of the sick had died. Houses were left empty, towns were abandoned, and a dreadful solitude hung over Europe. The sick died too fast for the living to bury them, and at one point, the Rhone River was consecrated as a graveyard for plague victims. Contemporary historians wrote that posterity would not believe such things could happen, because those who saw them were themselves appalled. To many people, it was the end of the world.

Before the century concluded, the Black Death visited Europe at least five more times in periodic reigns of terror. During one epidemic in Paris, an estimated 800 people died each day; in Siena, the population dropped from 42,000 to 15,000; and in Florence, 75 percent of the citizenry is believed to have perished. Flight was the chief recourse for people who could afford it, but ironically, the escaping travelers spread the disease. Those who remained in the cities were locked in their homes until they succumbed or recovered.

The immediate effect of the plague was a general paralysis in Europe. Trade ceased and wars stopped. Peasants who survived encountered unexpected prosperity because landowners had to pay higher wages to obtain help. Land values declined and class relationships were upset as the system of feudalism gradually crumbled. The authority of the clergy, already in decline, deteriorated further because the Church was helpless in the face of the disaster. With many priests and monks dead, a new order of reformers arose to found Protestantism. Medical practices became increasingly sophisticated, with new standards of sanitation and a 40-day period of detention (a "quarantine") imposed on vessels docking at ports. Even the mechanical clock came into widespread use, reflecting the urgency of life.

The graveyard of plague left fertile earth for the renewal of Europe during the Renaissance. To many historians, the Black Death remains a major turning point in Western civilization.

Arthropods:
animals, such as insects, with jointed appendages, segmented bodies, and outer skeletons of chitin.

Nor did it end there. European populations were also devastated by typhus and relapsing fever. Both of these diseases, like plague, are transmitted by arthropods and both can be interrupted by arthropod control. Neither is a major problem in our society, but other arthropodborne diseases such as Lyme disease, tularemia, and Rocky Mountain spotted fever occasionally are reported in the news media. We shall study each of these diseases in this chapter.

To begin, we shall examine a number of soilborne diseases where organisms enter the body through a cut, wound, or abrasion, or by inhalation. Among these are anthrax, a feared disease in agriculture, and tetanus, a concern to anyone who has stepped on a nail or piece of glass. We shall also study leptospirosis and listeriosis, two diseases that are receiving wider recognition as detection methods improve. These soilborne diseases as well as the arthropodborne diseases are primarily problems of the blood. Thus you will note the general patterns of blood entry and a series of symptoms that reflect involvement of the cardiovascular system.

Soilborne Bacterial Diseases

Anthrax

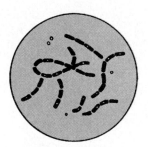

BACILLUS ANTHRACIS

Anthrax is primarily a disease of large animals such as cattle, sheep, and goats. It is caused by ***Bacillus anthracis***, a Gram-positive sporeforming rod pictured in Figure 9.1. Animals ingest the spores from the soil during grazing and are soon overwhelmed with bacteria. Their organs fill with bloody black fluid (*anthrac* is the Greek word for coal; the disease name is thus a reference to the blackening of the blood). About 80 percent of untreated animals die, and since the bacteria remain in the dead body as spores, it is often necessary to cremate the carcass or bury it deeply in lime to prevent soil contamination. Burning the field may also be required.

Humans acquire anthrax in a number of ways. Workers who tan hides, shear sheep, or process wool may inhale the spores and contract pulmonary anthrax, often called **woolsorter's disease**. Consumption of contaminated meat may lead to gastrointestinal anthrax. Contact with spores may lead to anthrax of the skin. Animal products related to outbreaks of anthrax in the past have included violin bows, shaving bristles, goatskin drums, and leather jackets.

Anthrax spores germinate rapidly on contact with human tissues. The thick capsule of the cells impedes phagocytosis, and the organisms produce a number of toxins. Pulmonary anthrax develops into a severe blood infection with extensive hemorrhaging. Skin and intestinal anthrax are accompanied by boillike lesions covered with black crusts (Figure 9.2). Blood invasion follows, and violent dysentery with bloody stools accompanies the intestinal form. Penicillin is used for therapy.

Dysentery:
waves of intestinal cramps and frequent passage of bloody, mucoid stools.

Attenuated:
weakened by chemical or cultural processes.

klo-strid′e-um

Tetanus:
a physiological expression meaning sustained uncontrolled contractions of the muscles.

In the 1980s, the total number of anthrax cases in the United States was less than a dozen, in large measure due to the testing of imported animal products. A vaccine prepared with an attenuated strain of *B. anthracis* has been successful in reducing outbreaks in domestic herds. However, the spores may remain in many soils, and the disease is considered a threat in biological warfare (MicroFocus: 9.1).

Tetanus

Tetanus is among the most dangerous of human diseases. *Clostridium tetani*, the bacillus that causes it, occurs everywhere in the environment, especially the

FIGURE 9.1
Bacillus anthracis

Details of *Bacillus anthracis*, the cause of anthrax. (a) Free spores and vegetative cells of *B. anthracis* visualized with the scanning electron microscope ($\times$ 5400). Note the oval shape of the spores and the typical rod shape of the vegetative cells. (b) Anthrax spores in the process of germinating ($\times$ 50,000). The spore coat of the spore in the center of the photograph has divided and is beginning to separate. A vegetative cell may be seen emerging from the spore at the bottom.

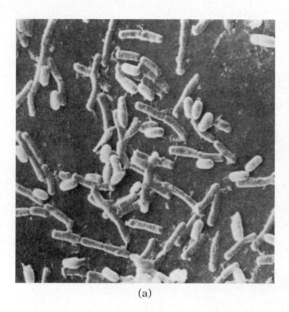
(a)

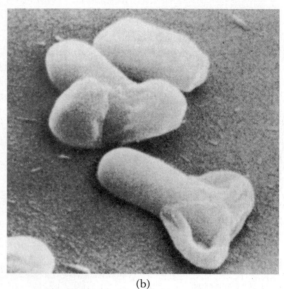
(b)

FIGURE 9.2
An Anthrax Lesion

This cutaneous lesion is a result of anthrax. In this view, the surface of the lesion is covered by a black crust.

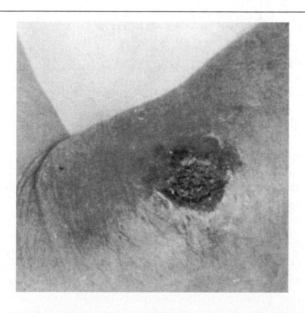

soil. Spores enter a wound in very small numbers and revert to vegetative bacilli that produce the second most powerful toxin known to science (after botulism).

THE LEGACY OF GRUINARD ISLAND

In 1941, the specter of airborne biological warfare hung over Europe. Fearing that the Germans might launch an attack against civilian populations, British authorities performed a series of experiments to test their own biological weapons. The spot chosen was Gruinard Island, a mile-long patch of land off the coast of Scotland. Investigators placed 60 sheep on the island and exploded a bomb containing anthrax spores overhead. Within days, all the sheep were dead.

Warfare with biological weapons never came to reality in World War II, but the contamination of Gruinard Island remained. A series of tests in 1971 showed that anthrax spores were still alive at and below the upper crust of the soil and that they could be spread by earthworms. Officials posted signs warning people not to set foot on the island but did little else.

Then a strange protest occurred in 1981. Activists demanded that the British government decontaminate the island. They backed their demands with packages of soil taken from the island. Notes led government officials to two 10-pound packages of spore-laden soil, and the writers threatened that 280 pounds were hidden elsewhere.

Partly because of the protests, the British government instituted a decontamination of the island in 1986. Using a powerful brushwood killer combined with burning and treatment with formalin in seawater, they managed to rid the soil of anthrax spores. By April 1987, sheep were once again grazing on the island. However, people were somewhat reluctant to return, and Gruinard Island remains a monument of sorts to the effects of biological warfare.

The toxin provokes sustained and uncontrolled contractions of the muscles (tetanus), and spasms occur throughout the body. Patients often experience violent deaths.

Clostridium tetani was first isolated in 1889 by the Japanese bacteriologist Shibasaburo Kitasato. It is a Gram-positive anaerobic sporeforming bacillus found in the intestines of many animals and humans. The bacillus possesses few invasive tendencies and does not cause intestinal disease. However, the spores are excreted in the feces to the soil and later enter the dead oxygen-free tissue of a wound. The wound may result from a fracture, gunshot, dog bite, or puncture by a piece of glass, thorn, or needle. Rusty nails pose a threat because spores cling to the rough edges of the nail and because the nail causes extensive tissue damage as it penetrates. (The rust itself is of no consequence.)

Once inside the tissue, the spores germinate to vegetative cells that produce several toxins. The most important of these toxins appears to be **tetanospasmin**, an exotoxin of high molecular weight. At the synapse, this toxin is believed to inhibit the removal of acetylcholine by interfering with the activity of cholinesterase, the enzyme that normally breaks down acetylcholine. With acetylcholine left in place at the synapse, volleys of spontaneous impulses arise in the nerves and crash into the muscles, causing the muscles to contract (Figure 9.3). Another viable theory is that the toxin blocks the release of transmitter substances, which normally inhibit nerve impulse transfer and serve as controlling devices. With the inhibitor substances not present, nerve impulses enter the muscle fibers and cause uncontrolled contractions.

CLOSTRIDIUM TETANI

Synapse:
the junction of two nerves or a nerve and muscle.

as″ĕ-til-ko′len

Cholinesterase:
the enzyme that breaks down acetylcholine after a nerve impulse has passed the synapse.

FIGURE 9.3

Tetanus

A photograph of a painting by Scottish surgeon Charles Bell showing a soldier dying of tetanus. The soldier was wounded at the battle of Corunna in 1809. Note that the muscles are fully contracting in virtually all parts of the body from head to toes. The soldier's face shows the clenched jaw and fixed ("sardonic") smile characteristic of lockjaw. The artist, Bell, was also the first to describe the paralysis and drooping of the facial muscles known as Bell's palsy.

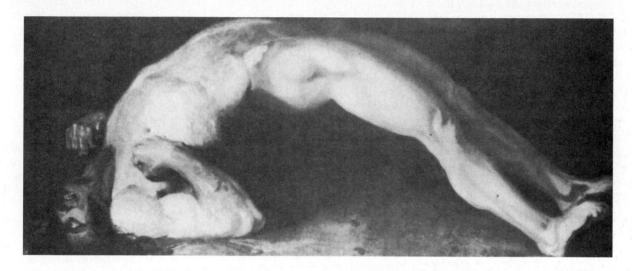

Symptoms of tetanus develop rapidly, often within hours. A patient experiences generalized muscle stiffness, especially in the facial and swallowing muscles. Spasms of the jaw muscles cause the teeth to clench and a condition called **lockjaw**. Severe cases are characterized by a "fixed smile," arching of the back, spasmodic inhalation, and seizures in the diaphragm and rib muscles leading to reduced ventilation and death. Patients are treated with sedatives and muscle relaxants, and are placed in quiet, dark rooms. Physicians prescribe penicillin to destroy the organisms and tetanus antitoxin to neutralize the toxin.

Immunization to tetanus may be rendered by injections of tetanus **toxoid** in the diphtheria-pertussis-tetanus (DPT) vaccine. The toxoid, developed in 1933 by Gaston Ramon, is prepared by treating the toxin with formaldehyde to eliminate its toxic quality. Children usually receive the first of several injections at the age of two months. Booster injections of tetanus toxoid in the Td vaccine ("a tetanus shot") are recommended every ten years to keep the level of immunity high.

The United States has experienced a steady decline in the incidence of tetanus, with approximately 150 cases confirmed annually. Most cases occur in the young, who are more likely to have contact with soil, and in the elderly, in whom antibody levels have dropped or vaccination never took place. In other parts of the world, tetanus remains a major health problem and is often related to traditional customs. In some countries, for example, unsanitary ear-piercing is common, tattooing is widespread, and the umbilical stump of newborns is dressed with soil. In other cases, a simple splinter of wood may be the source of entry (MicroFocus: 9.2).

Diaphragm:
the muscular partition between the thoracic and abdominal cavities; essential in breathing.

Toxoid:
a preparation of altered toxin molecules used for immunization purposes.

Gas Gangrene

To understand gas gangrene as an infectious disease, it is important to understand the physiological term "gangrene." Gangrene is a condition that develops when the blood flow ceases to a part of the body, usually as a result of blockage

9.2

MICROFOCUS

SHOELACES

The untied shoelaces told the story. That and the difficult swallowing and the distorted facial features. She had left the hospital the day before but the symptoms were not gone ... not yet, at least.

It began on Sunday, July 5, 1992. The 4th of July weekend was hot in Rutland, Vermont, and while most of her friends were recovering from the previous day's celebrations she decided to catch up on her gardening. The ground was warm, and she took off her shoes to walk about the garden—nothing like good fertile soil, she must have been thinking.

Then it happened "Ouch!" A splinter entered the base of her right big toe. No matter. Take out the splinter and get on with the gardening. Gone and forgotten.

But it was not forgotten. Three days later the pain on the left side of her face necessitated a visit to her family doctor. Probably a facial infection, she was told. Take the amoxicillin and it should resolve.

It did not resolve—it worsened. Now it was July 12 and her jaw was so tight that she had not eaten for three days. The muscle spasms in her face were intense, and her friends nervously suggested that it looked like lockjaw. When she arrived at the hospital's emergency room, an alert doctor recognized the classic risus sardonicus (the grinning expression caused by spasms of the facial muscles) and the trismus (the lockjaw caused by spasms of the chewing muscles). No question; it was tetanus.

Treatment was swift and aggressive— 3250 units of tetanus antitoxin, intravenous penicillin, and a tetanus booster—and remove the traces of wood still in the wound. She was placed in a quiet room and given muscle relaxants. Fifteen days would pass before she was discharged. She was well on her way to recovery ... except she could still not tie those darn shoelaces.

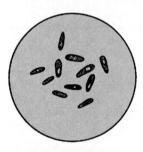

CLOSTRIDIUM PERFRINGENS

by dead tissue. The body part, generally an extremity, becomes dry and shrunken, and the skin color changes to purplish or black (Figure 9.4). The gangrene may spread as enzymes from broken cells destroy other cells, and the tissue may have to be excised ("debrided") or the part amputated. This form of gangrene is called "dry gangrene."

Gas gangrene, or "moist gangrene," occurs when soilborne bacteria invade the dead anaerobic tissue. The organisms responsible are ***Clostridium perfringens*** (formerly called *Clostridium welchii*), as well as several other species of *Clostridium* found in the intestine and obtained from the soil in a wound.

FIGURE 9.4

Dry Gangrene of the Tissues

This photograph shows blackening of the skin color and the dry, shrunken nature of the tissue. The photograph was taken shortly before three digits were amputated. Note that this form of gangrene is usually not due to bacterial infection.

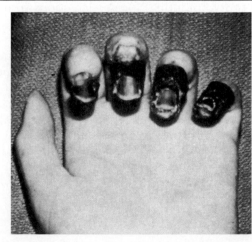

These anaerobic sporeforming Gram-positive rods multiply rapidly, while fermenting the muscle carbohydrates and putrefying the muscle proteins. Large amounts of gas resulting from this metabolism tear the tissue apart. The gas also presses against blood vessels, thereby blocking the flow and forcing cells away from their blood supply. In addition, the organisms secrete **lecithinase**, an enzyme that dissolves cell membranes and releases toxic cellular enzymes. Two other bacterial enzymes, **hyaluronidase** and **hemolysin**, facilitate the passage of bacteria among the cells and destroy red blood cells, respectively. Neurotoxins are also released.

hi′ah-lu-ron′ĭ-das
he-mol′ĭ-sĭn
Neurotoxins:
toxins that affect the nervous system.

The symptoms of gas gangrene include intense pain and swelling at the wound site as well as a foul odor. Initially the site turns dull red, then green, and finally blue-black. Anemia is common, and bacterial toxins may damage the heart and nervous system. Treatment consists of antibiotic therapy as well as debridement, amputation, or exposure in a hyperbaric oxygen chamber. The disease spreads rapidly, and death frequently results. Some microbiologists prefer the name **clostridial myonecrosis** ("muscle-cell-death") to gas gangrene, because gas is not always present in the early stages of disease.

da-brēd′maw
Hyperbaric:
high pressure.

Leptospirosis

Leptospirosis is a typical **zoonosis**, that is, a disease of animals able to spread to humans. The disease affects household pets such as dogs and cats, as well as rats, mice, and barnyard animals. Humans acquire it by contact with these animals or from soil or water contaminated with their urine.

lep″to-spi-ro′sis
Zoonosis:
a disease of animals that may spread to humans.

The agent of leptospirosis is ***Leptospira interrogans***, a small, delicate spirochete usually possessing a hook at one end that resembles a question mark, hence the name "interrogans" (Figure 9.5). The undulating movements of these organisms result from contractions of submicroscopic fibers called **axial filaments**. In animals, the spirochetes colonize the kidney tubules and are excreted in the urine to the soil. They enter the human body through the skin, especially through abrasions and the soft parts of the feet. Following penetration, they infect various organs including the kidney, liver, and meninges. Considerable jaundice may be present as bile seeps from the liver, and the patient may vomit blood from gastric hemorrhages. Despite the numerous tissues involved, the mortality rate from leptospirosis is low, and penicillin is generally used with success.

spi′ro-kēt

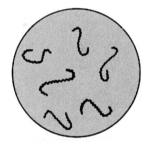

LEPTOSPIRA INTERROGANS

Leptospirosis occurs throughout the United States, and recognition of its symptoms together with modern methods of diagnosis have led to increased reports of its occurrence. The disease is considered an occupational hazard of veterinarians as well as tunnel diggers, sugarcane cutters, dock workers, miners, and others who work in wet areas where rodents are present. Dogs may be immunized with a leptospirosis vaccine usually combined with rabies, parvovirus, and distemper vaccines.

Listeriosis

lis-ter″e-o′sis
mon″o-si-toj′ĕ-nez

Listeriosis is caused by ***Listeria monocytogenes***, a small Gram-positive rod. At one time, this organism was dubbed the "Cinderella" of pathogenic bacteria because its ability to cause disease was not widely recognized. The bacillus is commonly found in the soil and in many animals, including barnyard animals, dairy cattle, and household pets. It is transmitted to humans by contact with the soil as well as by consumption of animal foods. In 1992 the CDC reported on outbreaks of listeriosis over a previous three-year period and indicated that

FIGURE 9.5

Leptospira interrogans

Details of *Leptospira interrogans*, the agent of leptospirosis. (a) A scanning electron micrograph of *L. interrogans* ($\times$ 6000). Note the tightly coiling spirals and the absence of flagella in this spirochete. (b) Details of the terminal hook of one spirochete ($\times$ 15,000).

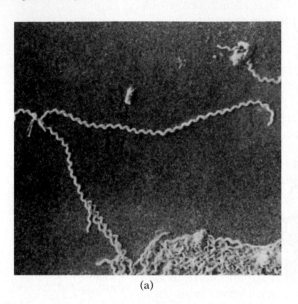

(a)

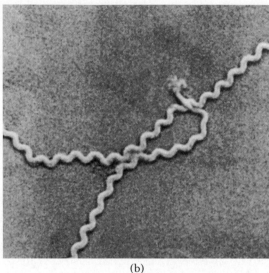

(b)

soft cheeses (e.g., Brie, Camembert, feta, and blue-veined cheeses) as well as delicatessen cold cuts were associated with a significant number of cases.

Listeriosis occurs in many forms. One form, called **listeric meningitis**, is characterized by headaches, stiff neck, delirium, and coma. Another form is a

TABLE 9.1

A Summary of Soilborne Bacterial Diseases

DISEASE	CAUSATIVE AGENT	DESCRIPTION OF AGENT	ORGANS AFFECTED	CHARACTERISTIC SIGNS
Anthrax	*Bacillus anthracis*	Gram-positive sporeforming rod	Blood Lungs Skin	Hemorrhaged blood Boil-like lesions
Tetanus	*Clostridium tetani*	Gram-positive sporeforming anaerobic rod	Nerves at synapse	Spasms Tetanus
Gas gangrene	*Clostridium perfringens*	Gram-positive sporeforming anaerobic rod	Muscles Nerves Blood cells	Gangrene Swollen tissue
Leptospirosis	*Leptospira interrogans*	Spirochete	Kidney Liver Spleen	Jaundice Vomiting
Listeriosis	*Listeria monocytogenes*	Gram-positive rod	Nervous system Blood	High monocyte count
Melioidosis	*Pseudomonas pseudomallei*	Gram-negative rod	Heart Lung	Abscesses

blood disease accompanied by high numbers of white blood cells called mono-cytes (hence the organism's name). A third form is characterized by infection of the uterus with vague influenzalike symptoms. If contracted during pregnancy, the disease may result in miscarriage of the fetus or mental damage in the newborn.

Listeriosis does not appear to be transmissible among humans. Prompt and prolonged treatment with antibiotics such as tetracycline is effective, but relapses are common. CDC epidemiologists estimate that 2000 cases of listeriosis occur in the United States annually and that a quarter result in the patient's death.

Monocyte:
a type of large white blood cell that functions in phagocytosis.

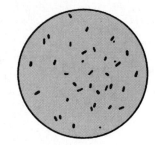

LISTERIA MONOCYTOGENES

Melioidosis

Melioidosis is a disease that, until recently, has been rare in the United States. It is caused by a Gram-negative rod named ***Pseudomonas pseudomallei***. This organism is common in soil and water, especially in Southeast Asia. Humans contract melioidosis by absorbing bacteria through soil-contaminated wounds, by inhalation, or by consuming contaminated food or water. The disease also affects horses, rodents, dogs, and cats.

Melioidosis has been called a "medical time bomb" for its propensity to lie dormant in the body for years. Abscesses may eventually occur in the heart, lungs, liver, or spleen, and symptoms may vary widely, depending on which organs are affected. Because melioidosis can mimic diseases like pneumonia, arthritis, or heart attack, it is not often recognized. Tropical experts have known of melioidosis since the early 1900s, but Americans did not appear to be involved until the Vietnam War, when several cases surfaced. Soilborne diseases are summarized in Table 9.1.

me"le-oi-do'sis
soo"do-mo'nas
soo"do-mal-la'e

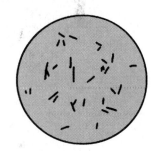

PSEUDOMONAS PSEUDOMALLEI

TOXIN INVOLVED	TREATMENT ADMINISTERED	IMMUNIZATION AVAILABLE	COMMENT
Yes	Penicillin	For animals	High mortality rate Affects many organs Rare disease in humans
Yes	Penicillin Antitoxin	Toxoid in DPT	Second most powerful toxin Inhibits cholinesterase activity
Yes	Penicillin	None	Gas blocks flow of blood Called clostridial myonecrosis
Not established	Penicillin	For animals	Common in animals Typical zoonosis
Not established	Tetracycline	None	Pregnant women at risk Associated with dairy products
Not established	Tetracycline	None	Symptoms develop slowly Occurs in Southeast Asia

TO THIS POINT ∎

We have surveyed several bacterial diseases commonly transmitted from the soil. Diseases such as tetanus and gas gangrene involve contamination of wounds with bacteria in the form of spores. As the bacilli multiply in the dead anaerobic tissue of the wound, the symptoms of disease surface. Tetanus bacilli produce powerful toxins that lead to sustained muscle contractions, while gas gangrene bacilli ferment and putrefy the organic compounds in muscle to produce large amounts of gas that lead to gangrene.

Anthrax, leptospirosis, listeriosis, and melioidosis may also be contracted from the soil, as well as by other means, such as contact with animals or consumption of contaminated food or water. Anthrax occurs in a number of forms, including pulmonary, gastrointestinal, and skin forms, all of which may be fatal. Leptospirosis, often spread by domestic animals and household pets, affects many abdominal organs. Listeriosis is a rare disease that may involve the nervous system, the blood, or the uterus. Melioidosis has contemporary significance because many cases in Americans were contracted from the soil in Southeast Asia during the Vietnam War. You may note that in many of the soilborne diseases there is substantial involvement of the blood since bacteria spread through this tissue from the initial site of invasion.

We shall now turn to a series of bacterial diseases transmitted by arthropods. In these diseases, infected arthropods introduce bacteria into the circulation, and the symptoms of disease commonly occur in the bloodstream. High fever is usually associated with the disease, and a general feeling of illness is common. Our survey will include bubonic plague, one of the most historically important diseases, as well as two diseases that were observed for the first time in the United States.

Arthropodborne Bacterial Diseases

Bubonic Plague

Few diseases have had the rich and terrifying history of **bubonic plague**, nor can any match the array of social, economic, and religious changes wrought by this disease.

Pandemic:
a worldwide epidemic.

The first documented pandemic of plague probably began in northeast Africa during the reign of the Roman emperor Justinian in 542 A.D. It lasted for 60 years, killed millions, and contributed to the downfall of Rome. The second pandemic was known as the **Black Death** because of the purplish-black splotches on victims and the terror it evoked in the 1300s (MicroFocus: 9.3). The Black Death decimated the world and, by some accounts, killed an estimated 40 million people in Europe, almost a third of the population of that continent. When it was over, feudalism was finished, Protestantism was on the rise, and the Renaissance was ready to flower. A deadly epidemic occurred in London in 1665, where 70,000 people succumbed to the disease. Daniel Defoe's *Journal of the Plague Year* recounts how people reacted (Figure 9.6).

In the late 1800s, Asian warfare facilitated the spread of a Burmese focus of plague, and migrations brought infected individuals to China and Hong Kong. During this epidemic in 1894, the causative organism was isolated by Alexandre Yersin and, independently, by Shibasaburo Kitasato. Plague first appeared in

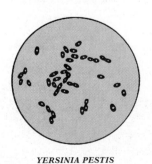

YERSINIA PESTIS

9.3

"WE'RE OUTA' HERE!"

In 1343, a group of merchants from Genoa, Italy, found themselves trapped behind the walls of the far-distant Asian city of Caffa in the Crimea. The dreaded Tartars had laid siege to the city, and things did not look good—indeed the siege would continue for three long years.

Then one day, the Tartars began catapulting corpses over the walls. The corpses were their own men dead of plague. It didn't take long for the disease to sweep through Caffa. The townspeople were terrified—they would either die of the plague inside the walls, or the Tartars would kill them outside the walls. But the Tartars were equally terrified and were withdrawing. Sensing an opportunity to escape, the Genoese merchants ran for their ships and sailed away.

Unfortunately their voyage home would be a voyage of the damned. Many of the merchants died on board, and the survivors managed to spread the plague wherever they stopped along the way. Finally, the remaining few reached home, only to doom the city of Genoa. Before long, all of Europe was caught in the fangs of the Black Death. European life was about to change forever.

the United States in San Francisco in 1900, carried by rats on ships from the Orient. The disease spread to ground squirrels, prairie dogs, and other wild rodents, and it is endemic in the southwestern states today, where it is commonly called **sylvatic plague**.

Bubonic plague is caused by ***Yersinia pestis*** (formerly *Pasteurella pestis*), a Gram-negative rod. This bacillus stains heavily at the poles of the cell, giving it a safety-pin appearance and a characteristic called **bipolar staining**. As first demonstrated by Masaki Ogata in 1897, the bacillus is transmitted by the rat flea *Xenopsylla cheopis*. A living organism such as a flea that transmits disease agents is called a **vector**, from the Latin *vehere*, meaning "to carry." Normally, the fleas infest only rats, but as the rats die, the fleas jump to humans and feed in the skin, thus transferring the bacilli into the bloodstream.

Sylvatic: occurring in nature; wild.

zen"op-sil'ah ke-op'is

FIGURE 9.6

Bubonic Plague

An engraving drawn during the London plague of 1665. Bodies heaped on "dead carts" are being pulled into huge pits by dragging the corpses off the cart with a hook attached to a ten-foot pole. This practice is probably the origin of the saying, "I wouldn't touch it with a ten-foot pole." The pipes being smoked by the men give off noxious fumes to ward off the disease.

Bubo:
a swelling, usually in the lymph nodes.

Bubonic plague is basically a blood disease. The bacteria multiply in the bloodstream and localize in the lymph nodes, especially those of the armpits, neck, and groin. Hemorrhaging in the lymph nodes causes substantial swellings called **buboes** (hence, bubonic plague). Dark, purplish splotches from hemorrhages can also be seen through the skin (the "rosies" in "Ring-a-ring of rosies"). From the buboes, the bacilli spread to the bloodstream, where they cause a form of the disease called **septicemic plague**. Over 50 percent of untreated cases of bubonic or septicemic plague are fatal.

Human-to-human transmission of plague during epidemics has little to do with fleas. Instead, it is dependent upon respiratory droplets because *Y. pestis* localizes in the lungs of many of the first victims. In this form, the disease is called **pneumonic plague** and is highly contagious. Lung symptoms are similar to those in pneumonia, with extensive coughing and sneezing. Hemorrhaging and fluid accumulation are common. Many suffer cardiovascular collapse, and death is common within 48 hours of onset of symptoms. Mortality rates for pneumonic plague approach 100 percent. One 1992 death was linked to a cat and rodents (Figure 9.7) and a notable outbreak appears to have occurred in India in 1994.

FIGURE 9.7
An Episode of Plague

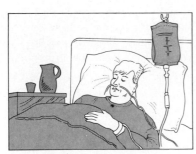

1. On August 19, 1992 a man reached under the crawlspace of a house in Chaffee County, Colorado to help a cat that had become trapped there.

2. The cat was apparently quite sick because it had abscesses under its jaw bone and lesions around its mouth.

3. The man returned home to Tucson, Arizona the next day. Three days later he experienced high fever, abdominal cramps, and a severe cough.

4. Two days later he was hospitalized with septic shock, difficult breathing, and right lobar pneumonia. Despite aggressive antibiotic therapy, he died the next day.

5. When the laboratory tests on the man's tissues were complete, the laboratory director reported the presence of *Yersinia pestis*, the agent of plague.

6. At the house, health investigators saw many dead rodents, and laboratory tests revealed *Yersinia pestis* in their tissues. The cat had died on August 19th, the day it was rescued, with symptoms of plague.

Plague may be treated with tetracycline or streptomycin when cases are detected early. Diagnosis consists of the laboratory isolation of *Y. pestis* together with tests for plague antibodies and typing with bacteriophages. The disease occurs sporadically in Native American populations and in travelers through the Southwest. Small-game hunters, taxidermists, veterinarians, zoologists, and others who have occasion to handle small rodents should be aware of the possibility of contracting plague. About two dozen cases are reported to the CDC annually. A vaccine consisting of dead *Y. pestis* cells is available for high-risk groups.

Tetracycline:
a broad-spectrum antibiotic that inhibits protein synthesis in bacteria.

Tularemia

Tularemia is one of several microbial diseases first recorded in the United States (others include St. Louis encephalitis, Rocky Mountain spotted fever, Lyme disease, and Legionnaires' disease). In 1911, a U.S. Public Health Service investigator named George W. McCoy reported the plaguelike disease in ground squirrels from Tulare County, California, and within a year he isolated the responsible bacillus. By 1920, researchers identified tularemia in humans, and Edward Francis assumed a detailed study of the disease. Over the next quarter century, Francis amassed data on over 10,000 cases, including his own. In the 1960s, when a name was coined for the causative organism, *Francisella tularensis* was selected to honor him.

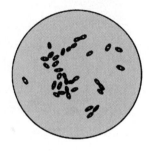

FRANCISELLA TULARENSIS

Francisella tularensis is a small Gram-negative rod that displays bipolar staining. It occurs in a broad variety of wild animals, especially rodents, and it is particularly prevalent in rabbits. Cats and dogs may acquire the bacillus during romps in the woods, and humans are infected by arthropods from the fur of animals. Ticks are important vectors in this regard, as evidenced by an outbreak of 20 tickborne cases in South Dakota in the 1980s. Other methods of transmission include contact with an infected animal (such as skinning an animal), consumption of rabbit meat, splashing bacilli into the eye, and inhaling bacilli, even from laboratory specimens.

Vector:
a living organism that transmits disease agents.

Various forms of tularemia exist, depending on where the bacilli enter the body. An arthropod bite, for example, may lead to a skin ulcer (Figure 9.8) and swollen salivary glands; splashing into the eye may cause an eye lesion; and inhalation may lead to pulmonary symptoms. In all forms, the disease is difficult to recognize because the symptoms are mild and nonspecific. Various patients with tularemia have been mistakenly treated for strep throat, rickettsial disease, lymphoid cancer, and cat scratch fever. Diagnosis depends on tests for tularemia antibodies and isolation of *F. tularensis* from body tissues.

Cat Scratch Fever:
a bacterial disease acquired by a cat bite or scratch and accompanied by swollen lymph nodes, usually on one side of the body.

Tularemia usually resolves on treatment with tetracycline or streptomycin, and few people die of the disease. Epidemics are unknown, and evidence suggests that tularemia may not be communicable among humans despite the many modes of entry to the body. Physicians report about 200 cases annually in the United States, most from the Midwest. MicroFocus: 9.4 describes how the disease unexpectedly entered Massachusetts.

Lyme Disease

One of the major emerging problems of the contemporary era is Lyme disease. Lyme disease is currently the most common tickborne (indeed, arthropodborne) illness in the United States. It has been reported in 47 states, and in 1995, it accounted for over 10,000 cases of infectious disease in the United States.

Lyme disease is named for Old Lyme, Connecticut, the suburban community where a cluster of cases was observed in 1975. That year, researchers led by Allan C. Steere of nearby Yale University traced the disease to ticks of the

FIGURE 9.8
The Lesions of Tularemia

The lesions of tularemia occur where the bacilli enter the body. In (a), the patient acquired the disease by handling infected rabbit meat. In (b), infection was preceded by the bite of an infected arthropod. The craterlike form of the lesion can be seen in both cases.

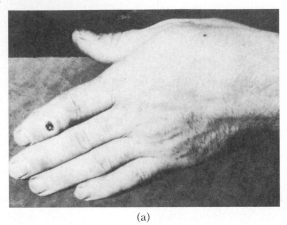

(a)

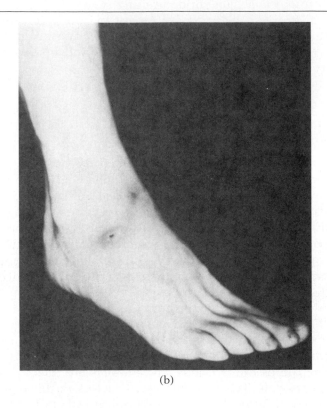

(b)

FIGURE 9.9
Borrelia burgdorferi

Two views of *Borrelia burgdorferi*, the agent of Lyme disease. (a) A scanning electron micrograph displaying an abundance of spirochetes. (b) A transmission electron micrograph in which the plasma membranes of two spirochetes have been disrupted, thereby releasing the so-called endoflagella. Normally these endoflagella run the length of the spirochete under the membrane and contribute to motility. Bar = 0.5 μm.

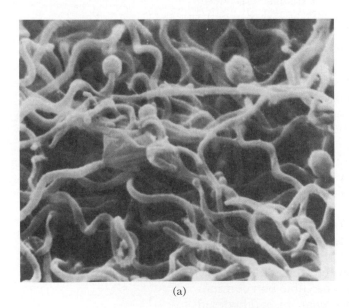

(a)

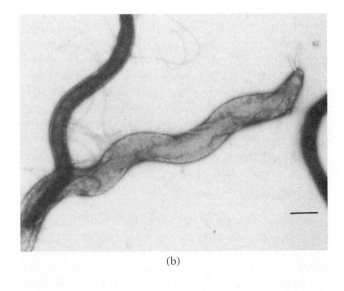

(b)

9.4

"BUT WE'VE NEVER HAD TULAREMIA ON CAPE COD!"

An epidemiologist is one who studies the pattern of disease in a community and attempts to locate its source and predict its spread. In the 1930s, the surprise observation of tularemia on Cape Cod, Massachusetts, tested the best epidemiologists of the day.

The story began when a young girl from Cape Cod was hospitalized with a fever lasting about three weeks. She had few other symptoms of illness, and physicians soon exhausted their list of diagnostic tests. When the girl's symptoms abated, her parents suggested that she be allowed to return home, and her doctors agreed. However, before she left, another round of tests was ordered. A week later, one test showed up positive: the girl apparently had tularemia.

The physicians were perplexed. Cape Cod had plenty of rabbits and ticks, but no case of tularemia had ever been reported. The girl could not recall a tick bite, but she did mention that her puppy had been ill with fever and was eating poorly some weeks before. Doctors tested the puppy and found that, indeed, it had been suffering from tula-

remia. The pieces were starting to fit, but investigators still had no clue on how the disease might have entered the area.

There followed many weeks of judicious questions and possible leads, one of which was especially promising. Epidemiologists established that Cape Cod had experienced a short supply of rabbits that year. Rabbits were a favorite target of sportsmen, and a number of wealthy hunters had pooled their funds to import two freight cars full of rabbits from Kansas. Kansas, coincidentally, was a "homeland" of tularemia, and the ticks from Kansas proved identical to their Cape Cod relatives.

Now the puzzle was complete: the girl with the mysterious fever, the positive test for tularemia, the puppy also sick with tularemia, the imported rabbits, and the tick carriers. It seemed logical that infected rabbits had brought the disease to Cape Cod, and that the puppy was bitten by a rabbit tick during one of its romps in the woods. An affectionate slurp had probably transmitted the disease to the girl.

The story has a bittersweet ending: the young girl recovered without complications; the puppy grew to be a healthy dog; and hunters had plenty of game. But tularemia came to Cape Cod to stay.

genus *Ixodes*. However, they were unable to locate a causative agent. Several years would pass until 1982, when a spirochete was observed in diseased tissue by investigators at the State University of New York at Stony Brook. Still another two years passed before the spirochete was isolated and cultivated (Figure 9.9). Researchers named it **Borrelia burgdorferi** for Willy Burgdorfer, the microbiologist who had studied the spirochete in the gut of an infected tick.

iks-o'dēz

bŏ-rel'e-ah
burg-dorf'er-e

Lyme disease has a variable incubation period that can be as long as six to eight weeks. One of the first signs to appear is a slowly expanding red rash at the site of the tick bite. The rash is called **erythema** (red) **chronicum** (persistent) **migrans** (expanding), or ECM. Beginning as a small flat or raised lesion, the rash increases in diameter in a circular pattern over a period of weeks, sometimes reaching a diameter of 10 to 15 inches. It has an intense red border and a red center, and it resembles a bull's eye (the "bull's eye rash"). It can vary in shape and is usually hot to the touch, but it need not be present in all cases of disease (Figure 9.10). Indeed, about one-third of patients do not develop ECM. The initial tick bite can be distinguished from a mosquito bite because the latter itches, while a tick bite does not. Fever, aches and pains, and flulike symptoms usually accompany the rash.

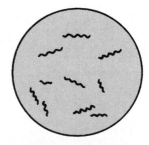

BORRELIA BURGDORFERI
ECM:
the persistent expanding red rash that accompanies Lyme disease.

In the rash (or "early localized") stage of Lyme disease, effective treatment can be rendered with penicillin and tetracycline. Left untreated, some cases resolve spontaneously, but others enter a second stage ("early disseminated") where the disease spreads. Weeks to months later, the patient may experience

FIGURE 9.10
Erythema Chronicum Migrans (ECM)

A view of erythema chronicum migrans (ECM), the rash that accompanies many cases of Lyme disease. Note that the rash consists of a large patch with an intense red border.

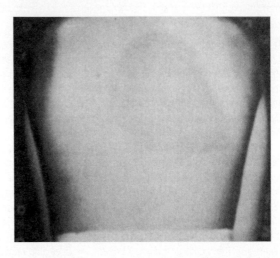

pain, swelling, and **arthritis** in the large joints, especially the knee, shoulder, ankle, and elbow joints. Multiple rash sites may also be present at this stage reflecting spread of the spirochete. Vigorous treatment at this stage may effectively eliminate the spirochete, but the joint damage may be irreversible.

In the third stage ("late chronic"), the arthritis may be complicated by damage to the cardiovascular and nervous systems. Patients often display irregular heartbeats, migraine headaches, hearing and vision abnormalities, and loss of muscle tone especially in the facial muscles (Bell's palsy). Although Lyme disease is not known to have a high mortality rate, the overall damage to the body can be substantial.

The tick that transmits most cases of Lyme disease in the Northeast and Midwest is ***Ixodes scapularis*** (formerly *I. dammini*); in the West the major vector is ***I. pacificus***. In its adult form, the tick is about the size of a pinhead or the period at the end of this sentence. It is smaller than the American dog tick, and it lives and mates in the fur of the white-tailed deer. Eventually it falls into the tall grass, where it waits for an unsuspecting dog, rodent, or human to pass by. The tick then attaches to its new host and penetrates into the skin; for the next 24 to 48 hours, it takes a blood meal and swells to the size of a small pea. While sucking the blood, it also defecates its intestinal cells into the wound and, should the tick be infected, spirochetes within the cells will be transmitted (Figure 9.11). If the tick is observed on the skin, it should be removed with a forceps or tweezers, and the area should be thoroughly cleansed with soap and water and an antiseptic applied.

The diagnosis of Lyme disease is usually based on symptoms, and the physician will often use a pen to mark off the border of the skin rash to see if it expands with time. The patient's recent activities are noted (hiking in the woods, living in a tick-infested area, camping), and a blood sample may also be taken for a test for spirochetal antibodies. The test, however, may not be accurate because a number of weeks are required for the body to produce enough antibodies to show up in the test. A newer test designed to detect spirochetal DNA is now being developed.

ix-o′des
scap-u-lar′is
dam-in′e
pa-cif′i-cus

FIGURE 9.11

Life Cycle of the Tick

The life cycle of *Ixodes scapularis* and *Ixodes pacificus*, the ticks that transmit Lyme disease.

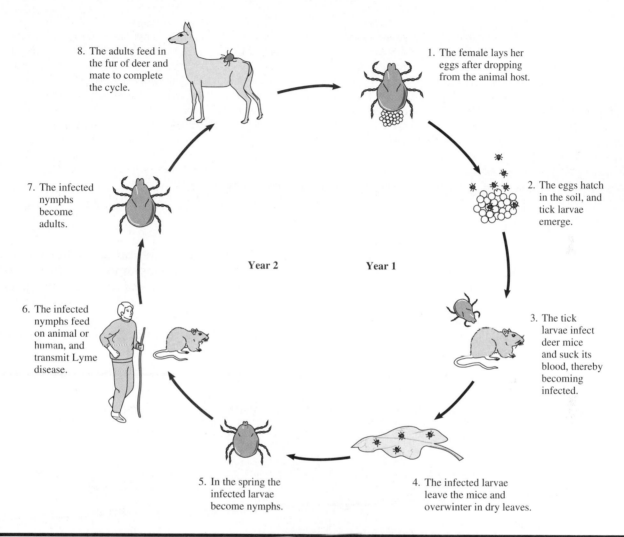

8. The adults feed in the fur of deer and mate to complete the cycle.

1. The female lays her eggs after dropping from the animal host.

7. The infected nymphs become adults.

2. The eggs hatch in the soil, and tick larvae emerge.

Year 2　　**Year 1**

6. The infected nymphs feed on animal or human, and transmit Lyme disease.

3. The tick larvae infect deer mice and suck its blood, thereby becoming infected.

5. In the spring the infected larvae become nymphs.

4. The infected larvae leave the mice and overwinter in dry leaves.

Congenital transfers of *B. burgdorferi* have been documented, and pregnant women are encouraged to seek early diagnosis and treatment for the disease if they suspect they have been infected. At this writing, a vaccine for dogs has been licensed and is in routine use, but a human vaccine is not yet available. Although Lyme disease is considered a "new" disease, there is evidence that it has existed for some decades even though it has escaped recognition (Micro-Focus: 9.5). The disease is also called Lyme borreliosis.

re"cur-ren'tis

Relapsing Fever

Relapsing fever is caused by a long spirochete currently referred to as ***Borrelia recurrentis***. The word "currently" is significant because various names have been given to strains of the spirochete depending on the arthropod vector involved. The CDC, for example, refers to tickborne *Borrelia* species as *B. hermsii* and *B. turicatae*. However, we shall use the traditional name *B. recurrentis* pending the outcome of DNA studies to determine the relationships among different strains.

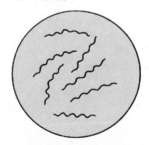

BORRELIA RECURRENTIS

9.5

NOT SO NEW AFTER ALL

Is Lyme disease a totally new disease, or has it been present but unrecognized in human populations for some time?

One of the hotspots for Lyme disease is Long Island, a 120-mile-long island off the coast of New York City and the home of a certain author. At the easternmost tip of Long Island is the town of Montauk, an old fishing village and a resort community. It seems that for decades local doctors have spoken of "Montauk knee," a disease accompanied by swollen and inflamed knee joints and a skin rash. Was Montauk knee what we now call Lyme disease?

Pathologist David Pershing of the Mayo Clinic attempted to find out. He and several colleagues sought out 140 specimens of ticks collected between 1924 and 1951, many from natural history collections. Then they dissected the ticks and removed the intestines. Now they amplified the DNA in the intestinal matter by a technique called the polymerase chain reaction. Next they set out to identify any DNA that might belong to *Borrelia burgdorferi*. The theory was simple: if the DNA of *B. burgdorferi* was present, then the spirochete must also have been present; and if *B. burgdorferi* were present, Lyme disease would predate 1975. And present it was. Pershing's group identified the DNA of *B. burgdorferi* in 13 ticks. All 13 had been collected near Montauk.

bor-rel'e-a

or"ni-tho-dor'us

The borreliae that cause relapsing fever are transmitted by lice and ticks (Figure 9.12). Lice are natural parasites of humans, and they thrive where personal hygiene is poor. Ticks of the genus *Ornithodorus* transmit the borreliae from rodent hosts. The ticks normally inhabit the rodent burrows and nests where the natural infection cycle proceeds without apparent disease in the rodents. Humans are incidental hosts, often bitten briefly and without notice

FIGURE 9.12

Three Arthropods That Transmit Bacterial Disease

(a) The tick (*Ixodes*) releasing numerous eggs. (b) The body louse (*Pediculus*). (c) The flea (*Xenopsylla*).

(a)

(c)

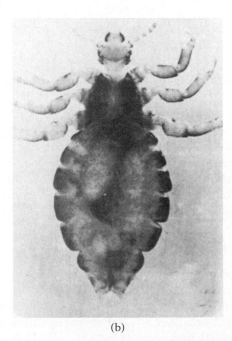

(b)

FIGURE 9.13
A Common Source Outbreak of Tickborne Relapsing Fever

1. In August, 1989, a family visited Big Bear Lake in San Bernardino County, California for a week's vacation.

2. The family was booked into cabin #4. They looked forward to a peaceful week in the country free from the city's hustle and bustle.

3. The next week a second group visited the lake and stayed in the same cabin #4. Several children were in the group. A good time was had by all.

4. After they left, cabin #4 was occupied by a third family. As before, the vacationers anticipated that nothing unusual would happen during their stay at the lake.

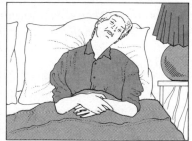

5. By September, however, six persons all from cabin #4 were suffering from relapsing fever. Each of them had recurring bouts of high fever, severe headache, and extreme prostration.

6. On visiting the park, health officials noted that many tick-infested ground squirrels had their burrows underneath cabin #4. They concluded that ticks had probably spread the disease from the animals. The cabin was sprayed with insecticides and the burrows were removed. No further cases occurred.

by the ticks at night. Cabins in wilderness areas are attractive nesting sites for infected rodents and their ticks especially when food is made available by cabin users (Figure 9.13).

Cases of relapsing fever are characterized by substantial fever, shaking chills, headache, prostration, and drenching sweats. The symptoms last for about two days, then disappear only to reappear up to ten times during the following weeks. Microbiologists believe that different serological types of *B. recurrentis* emerge from the organs after each attack. The serological types are the strains noted above. Treatment with tetracycline hastens recovery.

When lice are present, relapsing fever can occur in epidemics. One such epidemic descended upon Napoleon's lice-infested soldiers during the Russian campaign. Together with serious losses from another louseborne disease typhus (to be discussed presently), this disease so decimated the French army that the balance tipped in favor of the Russians. Peter Ilyich Tchaikovsky wrote the 1812 Overture to celebrate the great Russian victory, one that might not have been possible without the intervention of microorganisms.

TO THIS POINT
■

We have studied four arthropodborne diseases of varying significance. Bubonic plague has had substantial impact on the course of history because it is among the most prolific killers of humans. Plague is still present in today's world. Tularemia, by contrast, is a plaguelike disease that has only been recognized in this century. Its symptoms are much milder than plague and it often goes undiagnosed in patients. Lyme disease has a history that is even more recent than tularemia's, having first been recorded in 1975. A distinctive lesion at the site of spirochete entry and developing arthritis are characteristic signs of the disease. In relapsing fever, we see a disease accompanied by recurring periods of fever.

In all these cases, the key element in control is the elimination of the arthropod vector. If the chain of transmission can be broken, the disease will not spread to the body. This is where the pest control operator plays a significant role in our system of public health. It is also where sanitation and personal hygiene contribute to good health.

In the final section of this chapter, we shall continue to examine arthropodborne diseases, emphasizing those diseases caused by rickettsiae, a group of small bacteria. Once again we shall encounter ticks, lice, and other arthropods as vectors, and we shall focus on their place in the disease process. Antibiotics are helpful in treatment of the diseases, but vector control is of prime importance.

Rickettsial Arthropodborne Diseases

Rocky Mountain Spotted Fever

rik-et'se-e

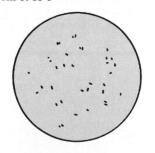

RICKETTSIA RICKETTSII
am''ble-o'mah
der''mah-cent'or

Macules:
pink skin spots.

Chloramphenicol:
an antibiotic that interferes with protein synthesis in a broad variety of bacteria.

In 1909, Howard Taylor Ricketts, a University of Chicago pathologist, described a new organism in the blood of victims of Rocky Mountain spotted fever and showed that ticks transmit the disease. A year later, he located a similar organism in the blood of animals infected with Mexican typhus, and discovered that fleas were the important vectors in this disease. Unfortunately, in the course of his work, Ricketts fell victim to the disease and died. When later research indicated that Ricketts had described a unique group of microorganisms, the name rickettsiae was coined to honor him. Originally, the rickettsiae were set apart from the bacteria, but microbiologists now consider them to be "small bacteria."

The agent of Rocky Mountain spotted fever is **Rickettsia rickettsii**. This organism is transmitted by ticks, especially those of the genera *Amblyomma* and *Dermacentor*. The hallmarks of Rocky Mountain spotted fever are a high fever lasting for many days and a skin rash reflecting damage to the small blood vessels. The rash begins as pink spots called macules, and progresses to pink-red pimplelike spots known as papules. Where the spots fuse, they form a **maculopapular rash**, which becomes dark red and then fades without evidence of scarring. The rash generally begins on the palms of the hands and soles of the feet and then progressively spreads to the body trunk. Mortality rates of untreated cases are variable, with some outbreaks in Montana recording a rate as high as 75 percent. Treatment with tetracycline or chloramphenicol reduces this rate significantly.

Accurate and rapid diagnosis is essential in Rocky Mountain spotted fever. Evidence of a tick bite, progress of the rash, and the recent activities of the patient (camping, backpacking, and other outdoor activity) are taken into account. Traditionally, a procedure called the **Weil-Felix test** has been employed as well. It is performed by mixing a sample of the patient's serum with the bacterium *Proteus OX19*. The bacteria clump together if the serum contains rickettsial antibodies. Absence of clumping means that the disease must be something else, such as measles or scarlet fever. The test works because rickettsiae possess antigens also located in *Proteus* cells and, therefore, both organisms stimulate the immune system to produce the same antibodies. However, the test is nonspecific and relatively insensitive, so antibody-detection tests are advised more strongly.

vīl fe′liks

Antigens: substances that stimulate the immune system, often resulting in antibody production.

Rocky Mountain spotted fever was first observed in early settlers to the American Northwest. About a thousand cases were reported annually in the United States in the early 1980s, but public education about the disease and improved methods of diagnosis and treatment accompanied a drop to about 600 cases by 1994 (Figure 9.14). Contrary to its name, Rocky Mountain spotted fever is not commonly reported from the West any longer, but it remains a problem in many southeastern and Atlantic coast states. Children are its primary victims because of their contact with ticks.

Epidemic Typhus

Epidemic typhus (also called **typhus fever**) is among the most notorious of all bacterial diseases. It is considered one of the most prolific killers of humans, and on several occasions it has altered the course of history such as when it helped decimate the Aztec population in the 1500s. Historians report that Napoleon marched into Russia in 1812 with over 200,000 French soldiers but his forces were hit hard by typhus. Hans Zinsser's classic book *Rats, Lice, and History* details events such as these and provides an engaging look at the effects of several infectious diseases on civilization (MicroFocus: 9.6).

Epidemic typhus is caused by ***Rickettsia prowazekii***. The bacteria are transmitted among individuals by head and body lice of the genus *Pediculus*. This was first noted by Charles Nicolle, winner of the 1928 Nobel Prize in Physiology or Medicine. Lice are natural parasites of humans (MicroFocus: 9.7).

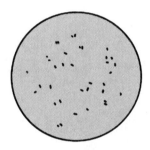

RICKETTSIA PROWAZECKII
pro″vaht-zek′e
pe-dik′u-lus

9.6

MICROFOCUS

RATS, LICE, AND HISTORY

The title of this miniessay is the title of a book written by Hans Zinsser. Zinsser was born in New York City in 1878. He achieved fame for isolating the bacterium of epidemic typhus, but he is equally well-known for his prose. Zinsser wrote some of the most uniquely personal, wise, and witty books of the twentieth century. Indeed, the unabridged title of his most famous book, published in 1935, is (with tongue-in-cheek)

Rats, Lice, and History: Being a Study in Biography, which, after Twelve Preliminary Chapters Indispensable for the Preparation of the Lay Reader, Deals With the Life History of Typhus Fever.

On a more serious note, here is a sample of Zinsser's writing: "Soldiers have rarely won wars. They often mop up after the barrage of epidemics. And typhus, with its brothers and sisters—plague, cholera,

typhoid, dysentery—has decided more campaigns than Caesar, Hannibal, Napoleon, and all the other generals of history. The epidemics get the blame for defeat, the generals the credit for victory. It ought to be the other way 'round ..."

If you think you'd enjoy learning more about disease and its effect on civilization, Zinsser's book would be a worthwhile investment of your time.

FIGURE 9.14
The Decline of Rocky Mountain Spotted Fever

The graph shows the decline of Rocky Mountain spotted fever in the United States since the late 1970s and 1980s. Tick control and antibiotic use have contributed to the decline. The inset shows the typical development of the maculopapular rash associated with the disease. (Courtesy of Centers for Disease Control and Prevention, *Annual Summary*, 1993.)

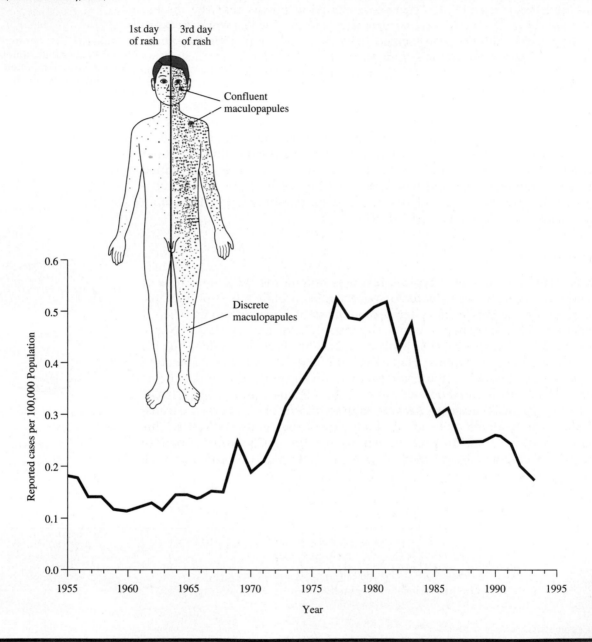

They flourish where sanitation measures are lacking and hygiene is poor. Often these conditions are associated with war, famine, poverty, and a generally poor quality of life.

The characteristic fever and rash of rickettsial disease are particularly evident in epidemic typhus. The rash is maculopapular, but unlike the rash of Rocky Mountain spotted fever, it appears first on the body trunk and then progresses to the extremities. Intense fever, sometimes reaching 104° F, remains for many days as the patient hallucinates and becomes delirious (typhus, like typhoid, takes its name from the Greek word *typhos*, meaning smoke or cloud). Some patients suffer permanent damage to the blood vessels and heart, and over 75 percent of sufferers die in epidemics. Tetracycline and chloramphenicol reduce this figure substantially, and the disease is rare in the United States.

The diagnosis of epidemic typhus depends on observation of symptoms, evidence of lice infestation, and a number of serological tests (Chapter 19). For definitive diagnosis, samples of patient's serum are injected into guinea pigs, where characteristic changes occur in the reproductive organs. Control of the disease requires destruction of lice populations through good hygiene. Insecticide powders on the clothes and body surface are also of value. During World War II, delousing with DDT was common practice. The fear generated by typhus was used to create the deception outlined in MicroFocus: 9.8.

Serum:
the straw-colored fluid remaining after blood cells have been removed from blood.

Endemic Typhus

Endemic typhus is a second form of typhus. This disease occurs sporadically in human populations because the flea that transmits it, *Xenopsylla cheopis*, is not a natural parasite of humans. However, the disease is prevalent in rodent populations where fleas abound (such as rats and squirrels), and thus it is called **murine typhus** from the Latin *murine*, referring to the mouse. Cats and their fleas are also involved and lice may harbor the bacilli.

zen"op-sil'ah
ke-op'is

The agent of endemic typhus is ***Rickettsia typhi***. Both rodent and flea may harbor it for weeks without displaying any effects, but when an infected flea feeds in the human skin, it deposits the organism into the wound. Alternatively, a person may crush the flea while scratching the wound. Endemic typhus is usually characterized by a mild fever, persistent headache, and a maculopapular rash. Often the recovery is spontaneous, without need of drug therapy. However, lice may transport the rickettsiae to other individuals and initiate an epidemic.

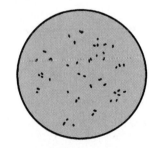

RICKETTSIA TYPHI

In the Southwest, endemic typhus is known as **Mexican typhus**, the disease from which Ricketts died. Since first described in 1570, it has also been called **tabardillo** from *tabardo*, meaning colored cloak, a reference to the mantlelike rash that covers the body. The sharp reduction from 3400 cases in 1941 to about 49 cases in 1987 is mainly due to rodent and flea control.

tab"ar-dēl'yo

Other Rickettsial Diseases

Several other rickettsial diseases have significance in microbiology because they occur sporadically. These diseases are transmitted among humans by arthropod vectors, and most are characterized by fever and rash. The mortality rates are generally low.

---

trom-bik'u-lah
soot"soo-gah-moosh'e

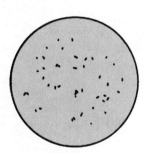

RICKETTSIA AKARI

ro"kah-li-me'ah

boo-ton-ez'

Scrub typhus is so-named because it occurs in scrubland where the soil is sandy or marshy and vegetation is poor. *Trombicula*, the mite that transmits the disease, lives in areas such as these. Scrub typhus is also called **tsutsugamushi** from the Japanese words *tsutsuga* for disease, and *mushi* for mite. The causative agent, ***Rickettsia tsutsugamushi***, enters the skin during mite infestations and soon causes fever and a rash, together with other typhuslike symptoms. Outbreaks may be significant as evidenced by the 7000 U.S. servicemen affected in the Pacific area in World War II. Tetracycline is used in treatment.

Rickettsialpox was first recognized in 1946 in an apartment complex in New York City. Investigators traced the disease to mites in the fur of local mice, and named the disease rickettsialpox because the skin rash was similar to that in chickenpox. Rickettsialpox is now considered a benign disease. It is caused by ***Rickettsia akari*** (*acari* is Greek for mite). Fever and rash are typical symptoms, and fatalities are rare. In many cases the disease runs its course without being recognized. Russia, Korea, and the northeastern United States account for the few cases detected annually.

Except for the great influenza pandemic of 1918–1919 (Chapter 12), **trench fever** was the most widespread disease encountered during World War I. An estimated million soldiers are thought to have been infected. The disease is caused by ***Rochalimaea quintana***, one of the few rickettsiae cultivated outside of living cells. Transmission takes place by head and body lice, and symptoms include a maculopapular rash and a fever occurring at irregular intervals. Trench fever is prevalent where lice abound, and trench warfare provided a near-ideal setting. Mortality rates are generally low. The disease is sometimes called **His-Werner disease** for the two investigators who studied it early in this century.

In various regions of the world, ***Rickettsia conorii*** causes a series of **tickborne fevers**. These diseases are known as boutonneuse fever, Marseilles fever, Nigerian typhus, and South African tick bite fever. All appear to be mild versions of Rocky Mountain spotted fever and all are accompanied by a fever and rash.

In addition, a black spot (*la tache noire*) develops where the tick has fed in the skin. Wild rodents harbor the disease in nature, and outbreaks occur sporadically when they make contact with human populations.

tahsh nwahr

In textbooks from the early part of this century an illness called **Brill-Zinsser disease** was described as a mild form of typhus. The high fever was a distinguishing characteristic, and it was recognized that if a patient was infested with lice, an epidemic of typhus might ensue. Today microbiologists believe that Brill-Zinsser disease is a recurrence of an earlier case of typhus in which *Rickettsia prowazekii* lay dormant in the patient for many years.

Ehrlichiosis, the final disease we shall consider, was first described in humans in 1986. Formerly believed confined to dogs, ehrlichiosis has been recognized in two forms in the United States: **human monocytic ehrlichiosis (HME)** caused by the rickettsia *Ehrlichia chaffeensis* (because the first case was observed at Fort Chaffee, Arkansas); and **human granulocytic ehrlichiosis (HGE)** due to *E. equi*, also a rickettsia. Patients suffer headache, malaise, and fever, but no rash or other distinctive physical findings. There is also a marked reduction of the platelet (thrombocyte) count, a condition known as thrombocytopenia. HME affects the monocytes ("monocytic"), while HGE affects the neutrophils ("granulocytic"). The agents of both diseases are transmitted by ticks, and tetracycline antibiotics are used in therapy. A notable outbreak of HGE affected 68 patients in the New York City area in 1994. All recovered. Table 9.2 summarizes the arthropodborne diseases.

er-lik"e-o'sis

NOTE TO THE STUDENT

Several diseases discussed in this chapter have occurred in broad-scale epidemics in past centuries and have accounted for widespread death. Bubonic plague, relapsing fever, and epidemic typhus are three examples. In today's world, microbiologists can quell the spread of these diseases by controlling arthropod populations.

But there is one disease in this chapter that may be spread by airborne spores. In the human body, its effects are so devastating that whole populations may be wiped out. The disease is anthrax. Its causative agent, Bacillus anthracis, is a potent weapon in biological warfare.

Consider how easily biological warfare can be carried on with B. anthracis spores. Spores released into the atmosphere may settle onto food and into water for human consumption. They may also be inhaled into the lungs or enter the blood through an abrasion in the skin. The terrifying symptoms of anthrax quickly follow any of these exposures, as we have noted in this chapter.

Can warfare of this type come to pass? Apparently so, because for many years several countries have explored its possibilities. To preclude biological warfare, over 100 nations signed an international treaty in 1972 banning the development of biological weapons. Under one of its terms, all stockpiles of anthrax spores were to be destroyed, and all such weapons research was to cease. However, in 1979, observers reported an epidemic of anthrax in the city of Sverdlovsk in the Siberian region of Russia. The epidemic came only a few days after an explosion at a nearby military installation. Soviet officials denied that the two events were related but failed to provide a satisfactory explanation to critics. Then, in 1993, after the breakup of the Soviet Union, examination of autopsy notes and pathological specimens from victims confirmed that lesions in the lungs were characteristic of anthrax.

TABLE 9.2

A Summary of Arthropodborne Bacterial Diseases

DISEASE	CAUSATIVE AGENT	DESCRIPTION OF AGENT	ORGANS AFFECTED	CHARACTERISTIC SIGNS
Bubonic plague	*Yersinia pestis*	Gram-negative bipolar staining rod	Lymph nodes Blood Lungs	Buboes Pneumonia Septicemia
Tularemia	*Francisella tularensis*	Gram-negative bipolar staining rod	Eyes Skin Blood	Eye lesion Skin ulcer Pneumonia
Lyme disease	*Borrelia burgdorferi*	Spirochete	Skin Joints Heart	Large, round red skin lesion Arthritis
Relapsing fever	*Borrelia recurrentis*	Spirochete	Blood Liver	Fever Jaundice
Rocky Mountain spotted fever	*Rickettsia rickettsii*	Rickettsia	Blood Skin	Fever Rash
Epidemic typhus	*Rickettsia prowazekii*	Rickettsia	Blood Skin	Fever Rash
Endemic typhus	*Rickettsia typhi*	Rickettsia	Blood Skin	Fever Rash
Scrub typhus	*Rickettsia tsutsugamushi*	Rickettsia	Blood Skin	Fever Rash
Rickettsialpox	*Rickettsia akari*	Rickettsia	Blood Skin	Fever Rash
Trench fever	*Rochalimaea quintana*	Rickettsia	Blood Skin	Fever Rash
Tickborne fevers	*Rickettsia conorii*	Rickettsia	Blood Skin	Fever Rash
Ehrlichiosis (HME) (HGE)	*Ehrlichia chaffeensis* *E. equi*	Rickettsia	Blood	Fever

> *Many scientists believe that the threat of biological warfare still looms large, and the concern for the enemy's use of bacteriological weapons surfaced during the Persian Gulf War of 1992 (especially when stocks of biological weapons were discovered in 1995). The possibilities of genetically engineered biological weapons have added a new dimension to the issue and compounded the threat.*

Summary

The common themes underlying this chapter are soil and arthropods and their ability to transmit bacterial disease. Soil is the source of sporeforming bacilli that cause anthrax, tetanus, and gas gangrene, and it can also harbor the organisms of listeriosis, leptospirosis, and melioidosis. For all these diseases, there is some level of blood involvement. Anthrax is accompanied by severe blood hemorrhaging, and tetanus and gas gangrene use the blood for transmission of their toxins. In a similar manner, the agents of leptospirosis and listeriosis use the blood to distribute themselves throughout the body.

VECTOR	TOXIN INVOLVED	TREATMENT ADMINISTERED	IMMUNIZATION AVAILABLE	COMMENT
Rat flea	Yes	Tetracycline	Killed bacteria for high risk	Great historic significance Septicemic and pneumonic forms Rodents infected
Flea Tick	Yes	Tetracycline	None	Resembles many other diseases Many modes of transmission No epidemics
Tick	Not established	Penicillin Tetracycline	None	Recognized since 1975 Most prevalent tickborne disease Named for Connecticut town
Louse	Not established	Tetracycline	None	Relapses common Rose spots develop on skin
Tick	Not established	Tetracycline	Killed bacteria for high risk	Most common rickettsial disease Rash first on extremities
Louse	Not established	Tetracycline	None	High mortality rate Rash first on body trunk
Flea	Not established	Tetracycline	None	Mild typhuslike disease Prevalent in rodents
Mite	Not established	Tetracycline	None	Occurs in scrubland Prevalent in Far East
Mite	Not established	Tetracycline	None	Rare disease Resembles chickenpox
Louse	Not established	Tetracycline	None	Common in World War I Cultivation in artificial medium
Tick	Not established	Tetracycline	None	Different forms
Tick	Not established	Tetracycline	None	Recognized since 1986

Arthropods are the important modes of transmission for plague, tularemia, Lyme disease, Rocky Mountain spotted fever, typhus, and a number of other diseases, many due to rickettsiae. Since arthropods inject the bacteria to the blood or deposit them into a wound, the involvement of blood is substantial. In plague there is intense septicemia, tularemia is characterized by malaise and mild fever, and Lyme disease is accompanied by a red skin rash, with fever and headaches that reflect blood involvement. The pattern of skin rash and fever continues for the rickettsial diseases as the blood remains the focus of infection.

Because of the involvement of soil and/or arthropods, the diseases in this chapter are more difficult to transmit than airborne or food- and waterborne diseases. Soilborne diseases do not occur in epidemics and incidences tend to be low, and arthropodborne diseases require that the tick, louse, flea, mite, or other arthropod be present in the environment for epidemics to occur. When the arthropod is available, however, epidemics can be substantial, as the experiences with plague, Lyme disease, Rocky Mountain spotted fever, and epidemic typhus illustrate in this chapter.

Questions for Thought and Discussion

1. In February 1980, a patient was admitted to a Texas hospital complaining of fever, headache, and chills. He also had greatly enlarged lymph nodes in the left armpit. A sample of blood was taken and Gram stained, whereupon Gram-positive diplococci were observed. The patient was treated with cefoxitin, a drug for Gram-positive organisms, but soon thereafter he died. On autopsy, *Yersinia pestis* was found in his blood and tissues. Why was this organism mistakenly thought to be diplococci and what error was made in the laboratory? Why were the symptoms of plague missed?

2. There is a town outside of London known as Gravesend. The town apparently acquired its name during the 1660s in connection with a great medical upheaval. What was the name of that upheaval and how do you suppose the name came about?

3. Although the tetanus toxin is second in potency to the toxin of botulism, many physicians consider tetanus to be a more serious threat than botulism. Would you agree? Why?

4. Some estimates place epidemic typhus among the all-time killers of humans; one listing even has it in third place behind malaria and plague. What conditions in history have contributed to outbreaks of this disease?

5. On February 13, 1976, a newspaper article requested purchasers of a certain brand of Pakistani wool yarn to check with their local health departments because the wool was thought to be contaminated with anthrax spores. The article read, "Spores of anthrax, which is a livestock disease, have been found on the yarn. The disease affects people as a skin ailment and is usually not fatal." Does the article minimize a potential medical emergency? If yes, construct a letter to the editor in reply. If not, explain why not.

6. In 1993, a 57-year-old man died of tetanus in a Kansas hospital. His experience had begun on August 14 with a puncture wound to the foot and ended on September 16. Family members reported that the man had never received a vaccination with tetanus toxoid. Hospital costs for his experience totaled $145,329. Administration of a dose of tetanus vaccine, by comparison, costs $3.30. Setting aside the value of a human life for the moment, what does this disparity of costs tell you?

7. At various times, local governments are inclined to curtail deer hunting. How might this lead to an increase in the incidence of Lyme disease?

8. In 1982, endemic typhus was observed in five members of a Texas household. On investigation, epidemiologists learned that family members had heard rodents in the attic, and two weeks previously had used rat poison on the premises. Investigators concluded that both the rodents and the rat poison were related to the outbreak. Why?

9. Leptospirosis has been contracted by individuals working in such diverse locales as subway tunnels, gold mines, rice paddies, and sewage treatment plants. What precautions might be taken by such workers to protect themselves against the disease?

10. A young woman was hospitalized with excruciating headache, fever, chills, nausea, muscle pains in her back and legs, and a sore throat. Laboratory tests ruled out meningitis, pneumonia, mononucleosis, toxic shock syndrome, and other diseases. On the third day of her hospital stay, a faint pink rash appeared on her arms and ankles. By the next day, the rash had become darker red and began moving from her hands and feet to her arms and legs. Can you guess what the diagnosis eventually was?

11. In Chapter 9 of the Bible, in the Book of Exodus, the sixth plague of Egypt is decribed in this way: "Then the Lord said to Moses and Aaron, 'Take a double handful of soot from a furnace, and in the presence of Pharaoh, let Moses scatter it toward the sky. It will then turn into a fine dust over the whole land of Egypt and cause festering boils on man and cattle throughout the land.'" Which disease in this chapter is probably being described?

12. During the Civil War, 92,000 soldiers died of battle wounds but 190,000 died of battle-related diseases. Which diseases in this chapter probably contributed to the enormous mortality? What conditions encouraged each disease cited?

13. In autumn it is customary for some homeowners in some communities to pile leaves at the curbside for pickup. How might this practice increase the incidence of tularemia, Lyme disease, and Rocky Mountain spotted fever in the community?

14. At various times in past centuries, it was believed that cats were the medium through which witches spoke. Fearing cats, people would try to eliminate these animals from their neighborhood. Bubonic plague would occasionally break out shortly thereafter. Why?

15. Centuries ago, the habit of shaving one's head and wearing a wig probably originated in part as an attempt to reduce lice infestations in the hair. Why would this practice also reduce the possibilities of certain diseases? Which diseases?

Review

The bacterial diseases transmitted by soil and arthropods are the principal focus of this chapter. To test your understanding of the chapter contents, match the statement on the left with the disease on the right by placing the correct letter in the available space. A letter may be used once, more than once, or not at all. The appendix contains the answers.

___ 1. Accompanied by erythema chronicum migrans.

___ 2. Transmitted by lice; caused by *R. prowazekii*.

___ 3. May result in miscarriage if contracted during pregnancy.

___ 4. May be transmitted by contact with dogs.

___ 5. Caused by a sporeforming rod that produces hemolysis and lecithinase.

___ 6. Tickborne disease; caused by *R. rickettsii*.

___ 7. Blood hemorrhaging in large animals such as cattle, sheep, and goats.

___ 8. Also known as tabardillo and Mexican typhus.

___ 9. Treated with antitoxins; caused by an anaerobic sporeformer.

___ 10. Caused by *Borrelia burgdorferi*; transmitted by a tick.

___ 11. Bubonic, septicemic, and pneumonic stages.

___ 12. Maculopapular rash beginning on extremities and progressing to body trunk.

___ 13. Caused by a spirochete that infects kidney tissues in pets and humans.

___ 14. Occurs in small game animals, especially rabbits.

___ 15. Transmitted by mites; also known as scrub typhus.

___ 16. Caused by a Gram-negative rod with bipolar staining; transmitted by rat flea.

___ 17. Long-range complications include arthritis in large joints.

___ 18. Up to ten attacks of substantial fever, joint pains, and skin spots; *Borrelia* involved.

___ 19. Immunization rendered by the DPT vaccine.

A. Plague
B. Epidemic typhus
C. Anthrax
D. Melioidosis
E. Relapsing fever
F. Tularemia
G. Lyme disease
H. Listeriosis
I. Tsutsugamushi
J. Endemic typhus
K. Ehrlichiosis
L. Leptospirosis
M. Tetanus
N. Rocky Mountain spotted fever
O. Gas gangrene

___ 20. Pulmonary, intestinal, and skin forms possible; due to a *Bacillus* species.

___ 21. A typical zoonosis; caused by a spiral bacterium.

___ 22. Caused by a *Francisella* species; has multiple modes of transmission.

___ 23. Epidemics where sanitation is poor; louseborne rickettsial disease.

___ 24. Sustained and uncontrolled contractions of the body's muscles.

___ 25. Most commonly reported tickborne disease in the United States.

CHAPTER 10

SEXUALLY TRANSMITTED, CONTACT, AND MISCELLANEOUS BACTERIAL DISEASES

The Victorian era of the 1800s was notorious for its prudish attitude toward sex. Bulls were called "he cows," and the legs of a piano were modestly covered with pantaloons. Women did not disrobe when visiting a physician but, instead, pointed to a chart to show where they hurt. An 1863 etiquette book stipulated that the works of male and female authors should be separated on bookshelves unless the authors were married. Even the biology books of that era mirrored the taboo on sex by carefully avoiding mention of the human reproductive system. It was as if the human species did not provide for the next generation.

Our era, by contrast, has been marked by broad sexual freedoms, but it has seen an alarming rise in diseases of the reproductive organs. Formerly, these diseases were called venereal diseases (VDs) from Venus, the Roman goddess of love. However, health departments now prefer to call them sexually transmitted diseases (STDs) to focus on the mode of transmission and to avoid the "love" connotation. The STDs discussed in this chapter are of bacterial origin and include familiar names such as syphilis and gonorrhea, as well as emerging problems such as chlamydial urethritis and ureaplasmal urethritis. To underscore the seriousness of the STD problem, the American Social Health Association estimates that for each day in 1995, over 27,000 Americans contracted a sexually transmitted disease.

The upswing in sexually transmitted diseases is but one example of how changing social patterns can affect the incidence of disease. Several other examples are evident in the diseases discussed in this chapter. For instance, the incidence of leprosy in the United States has risen because in the last decade immigrant groups have brought the disease with them. Toxic shock syndrome was first recognized widely in 1980 when a new brand of high-absorbency tampon appeared on the commercial market. Finally, the mortality rate from nocardiosis rose when this disease complicated cases of acquired immune deficiency syndrome (AIDS).

Sexually Transmitted Diseases

Syphilis

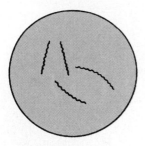

TREPONEMA PALLIDUM

trep"o-ne'mah

Over the centuries, Europeans have had to contend with four pox diseases: chickenpox, cowpox, smallpox, and the Great Pox, a disease now known as syphilis. The first European epidemic was recorded in the late 1400s shortly after the conquest of Naples by the French army (MicroFocus: 10.1). For decades the disease had various names, but by the 1700s, it had come to be called syphilis.

Syphilis is caused by **Treponema pallidum**, literally the "pale spirochete." This spiral bacterium moves by means of axial filaments and spreads by human-to-human contact, usually during sexual intercourse. It penetrates the skin surface through the mucous membranes or via a wound, abrasion, or hair follicle, and it causes a disease that occurs in three stages. The variety of clinical symptoms that accompany the stages and their similarity to other diseases have led some physicians to call syphilis the "Great Imitator."

shang'ker

The incubation period for syphilis varies greatly but averages about three weeks. **Primary syphilis** is the first stage to appear. This stage is characterized by the **chancre**, a painless circular, purplish ulcer with a raised margin and hard edges described as being like cartilage (Figure 10.1). The chancre develops at the site of entry of the spirochetes, often the genital organs. However any area of the skin may be affected, including the pharynx, rectum, or lips. (Centuries ago people discovered that kissing could transmit syphilis and even though they did not know what was causing the dreaded disease, they began to substitute hand-shaking and gentle kisses on the cheek for lip kissing.) The chancre teems with spirochetes. It persists for two to six weeks, and then it disappears spontaneously.

Several weeks later, the patient experiences **secondary syphilis**. Symptoms include fever and a constitutional influenzalike illness, as well as swollen lymph nodes reminiscent of infectious mononucleosis. The skin rash that appears may be mistaken for measles, rubella, or chickenpox (Figure 10.1). Loss of eyebrows often occurs, and the patchy loss of hair causes "moth-eaten" areas commonly seen on the head. Involvement of the liver may result in jaundice and suspicion of hepatitis. In untreated patients, the symptoms may last several weeks and death may occur. Most patients recover, but they bear pitted scars from the lesions and remain "pockmarked." These individuals now enter a latent stage during which they continue to be infectious.

Infectious mononucleosis:
a viral disease accompanied by swollen lymph nodes, sore throat, and mild fever.

Jaundice:
yellow tinting of the eyes and skin due to seepage of bile into the bloodstream.

About one-third of untreated patients eventually develop **tertiary syphilis**. This stage may occur in many forms but most commonly it involves the skin, cardiovascular system, and nervous system. The hallmark of tertiary syphilis is the **gumma**, a soft, "gummy" granular lesion (Figure 10.1). In the cardiovascular system, gummas weaken the major blood vessels, causing them to bulge and burst. In the spinal cord and meninges, gummas lead to degeneration of the tissues and eventual paralysis. In the brain they alter the patient's personality and judgment and cause insanity so intense that for many generations people with tertiary syphilis were confined to mental institutions. It is conceivable that our ancestors failed to equate the chancre of primary syphilis with the horrible symptoms of tertiary syphilis because the stages were so distantly separated in time.

gum'ah

Syphilis is a particular problem in pregnant women because the spirochetes are able to penetrate the placental barrier after the fourth month of pregnancy and cause **congenital syphilis** in the fetus. Syphilitic skin lesions and open sores may be apparent in the newborn, or symptoms may develop weeks after

FIGURE 10.1

The Stages of Syphilis

Views of the skin lesion in the three stages of syphilis. (a) The chancre of primary syphilis as it occurs on the body surface. The chancre is circular with raised margins and is usually painless. It occurs where penetration of syphilis spirochetes has taken place. (b) The flat, wartlike lesions characteristic of secondary syphilis. (c) The gumma that forms in tertiary syphilis. Note the granular, diffuse nature of this lesion compared with the primary chancre shown in (a). (d) A scanning electron micrograph of *Treponema pallidum* in animal tissue. The spirochetal shape is apparent.

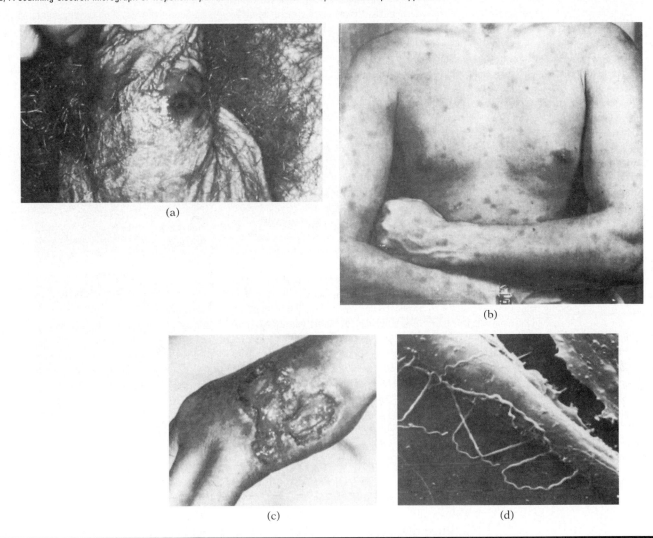

(a)

(b)

(c)

(d)

birth. Affected children often suffer poor bone formation, meningitis, or **Hutch-inson's triad**, a combination of deafness, impaired vision, and notched, peg-shaped teeth. This triad was first described by Joseph Hutchinson of the London Hospital in 1861. The cornerstone of syphilis control is the identification and treatment of sexual contacts of patients. Penicillin is the drug of choice for the primary, secondary, and latent stages of the disease, but antibiotics are ineffective in tertiary syphilis because gummas appear to be an immunological response to the spirochetes. *T. pallidum* multiplies very slowly in the tissues, partly because of its 33-hour generation time. This factor adds to the success of therapy.

Generation time:
the time period that elapses between divisions in bacteria.

10.1
MICROFOCUS

ORIGIN OF A DISEASE

Among the more intriguing questions of medical history are how and why syphilis suddenly emerged in Europe in the late 1400s. Writers of that period tell of an awesome new disease that swept over Europe and on to India, China, and Japan. But where did the disease come from?

One oft-told story is that syphilis existed in the New World and that members of Columbus' crew acquired it during stopovers in the Caribbean islands. Columbus returned to Palos in Northern Italy in 1493, and some of his crew reportedly joined the army of Charles VII of France. In 1494, Charles' army attacked Naples in Italy, but mounting losses from the strange new sickness forced him to withdraw. His army of 30,000 French, German, Swiss, English, Polish, Spanish, and Hungarian troops returned to their native lands, and apparently brought the disease home with them.

A second theory holds that the disease first came to Spain and Portugal with slaves imported from Africa in the mid-1400s. An African disease called yaws is very similar to syphilis in causative organism, transmission, and stages of development. Certain historians believe that some unknown factor caused yaws to flare up in the form of syphilis in the late 1400s. They speculate that

the army of Charles VII provided a highly susceptible population of diverse men who spread the disease wherever they traveled. Indeed, recent studies of Native American burial grounds show no traces of syphilis when the individuals died before the arrival of Columbus. By contrast, remains of those living after Columbus' arrival show signs that the disease was present in the community.

With its devastating effects, syphilis inspired a variety of epithets. The Italians called it the French disease (*morbus Gallicus*), while the French called it the Italian disease (*la maladie Italienne*); to the Japanese it was the Chinese disease. The English impartially termed it the Great Pox. The name "syphilis" derives from the works of Girolamo Fracostoro, a sixteenth-century poet-scientist of Verona. In 1530, Fracostoro wrote a long poem about a shepherd named Syphilus who momentarily left his pastoral responsibilities to commit a sexual indiscretion. The angered gods punished him with the horrible sores of the disease. In a later work on disease transmission, Fracostoro suggested that the illness be called syphilis after the mythical shepherd.

Syphilis was as international in effect as in name, and proved no respector of rank.

Henry VIII of England, Napoleon of France, and Peter the Great of Russia all contracted the disease. Poets like Keats, musicians like Beethoven, and artists like Gauguin also succumbed, as did millions of common people. For many generations epidemics of syphilis swept back and forth across the world. As Lord Byron wrote in one of his poems:

"The smallpox has gone out of late;
Perhaps it may be follow'd by the Great."

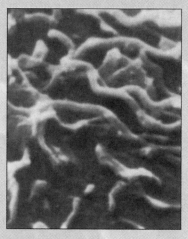

A scanning electron micrograph of *T. pallidum*, the cause of syphilis (× 13,500).

Dark-field microscope:
an instrument that displays live, unstained bacteria on a dark background.

Since first observed in 1905 by Fritz R. Schaudinn and P. Erich Hoffman, exhaustive attempts have been made to cultivate *T. pallidum* on laboratory media, but none have been successful. Diagnosis in the primary stage therefore depends on the observation of spirochetes from the chancre under the dark-field microscope. As the disease progresses, a number of tests to detect syphilis antibodies become useful including the rapid plasma reagin test and the VDRL test. Tests such as these detect **nonspecific treponemal antibodies**, because the antibodies react with numerous chemical substances including those on *T. pallidum*.

A number of **specific treponemal antibody** tests are also available, including the *Treponema pallidum* immobilization (TPI) test. In this test, serum from a patient is combined with spirochetes from a laboratory-infected rabbit to see if antibodies in the serum will immobilize the spirochetes. Another procedure is the fluorescent treponemal antibody absorption (FTA-ABS) test, described in Chapter 19. The complement fixation test devised by Wassermann is rarely used anymore.

FIGURE 10.2

Incidences of Primary and Secondary Cases of Syphilis by Sex in the United States, 1956–1993

Much of the increase is attributed to a rise in syphilis among adolescents and drug users, where sex is often used as payment for drugs. The recent increase in females may reflect increased drug use. (Courtesy of Centers for Disease Control and Prevention, *Annual Summary*, 1993.) The inset shows *Treponema pallidum*, the syphilis spirochete as seen through the dark-field microscope.

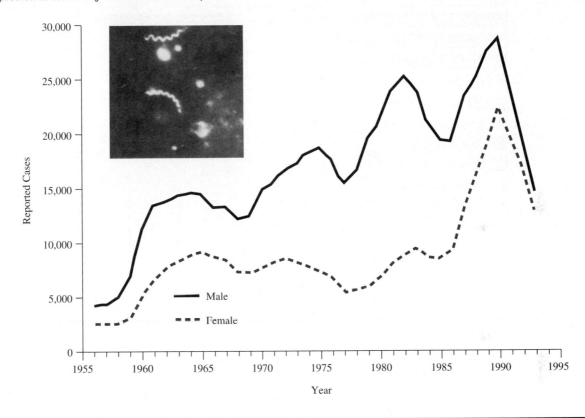

Syphilis is currently the third most reported microbial disease in the United States, after gonorrhea and chickenpox (Figure 10.2). Statistics indicate that about 125,000 people are afflicted with the disease annually, of whom about 43,000 are in the primary or secondary stage. Taken alone, these figures suggest the magnitude of the syphilis epidemic, but some public microbiologists believe that for every case reported, as many as nine cases go unreported. Millions of people may therefore be involved.

Gonorrhea

Gonorrhea has the distinction of being the most frequently reported microbial disease in the United States (Figure 10.3). During the 1960s, the incidence of gonorrhea rose dramatically and since 1975, several hundred thousand cases have been reported annually. Epidemiologists suggest that three or four million cases go undetected or unreported each year. Despite effective antibiotic therapy and national attempts at close surveillance, gonorrhea remains out of control.

The agent of gonorrhea is ***Neisseria gonorrhoeae***, a small Gram-negative diplococcus named for Albert L.S. Neisser, who isolated it in 1879. The organism, commonly known as the **gonococcus**, has a characteristic double-bean

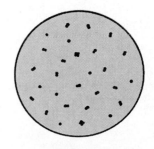

NEISSERIA GONORRHOEAE
ni-se′re-ah gon″o-re′a

FIGURE 10.3

Reported Cases of Microbial Diseases in the United States, 1993

For the year, gonorrhea was three times as prevalent as the next most common reportable disease, chickenpox. AIDS was the third most common. (Courtesy of Centers for Disease Control and Prevention, *Summary of Notifiable Diseases*, 1993.

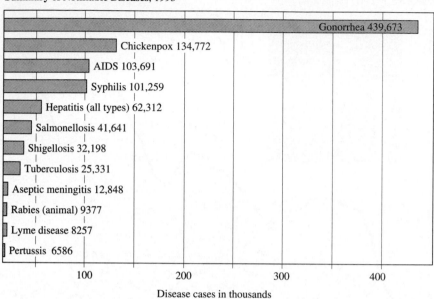

Summary of Notifiable Diseases, 1993

Gonorrhea 439,673
Chickenpox 134,772
AIDS 103,691
Syphilis 101,259
Hepatitis (all types) 62,312
Salmonellosis 41,641
Shigellosis 32,198
Tuberculosis 25,331
Aseptic meningitis 12,848
Rabies (animal) 9377
Lyme disease 8257
Pertussis 6586

100   200   300   400

Disease cases in thousands

Cervix:
the opening to the uterus.

Urethra:
the tube that leads from the bladder to the exterior.

sal"pin-ji'tis

Ectopic pregnancy:
a pregnancy taking place in the Fallopian tube.

shape exhibited by neisseriae. *N. gonorrhoeae* is a very fragile organism susceptible to most antiseptics and disinfectants. It survives only a brief period outside the body and is rarely contracted from a dry surface such as a toilet seat. The great majority of cases of gonorrhea are therefore transmitted in person-to-person contact during sexual intercourse (gonorrhea is sometimes called "the clap" from the French *clappoir* for brothel).

The incubation period for gonorrhea ranges from 2 to 6 days. In females, the gonococci invade the epithelial surfaces of the cervix and the urethra. The cervix may be reddened and a discharge may be expressed by pressure against the pubic area. Patients often report abdominal pain and a burning sensation on urination, and the normal menstrual cycle may be interrupted.

In some females, gonorrhea also spreads to the Fallopian tubes, which extend from the uterus to the ovaries. As these thin passageways become riddled with pouches and adhesions, the passage of egg cells becomes difficult. Complete blockage, or **salpingitis**, may take place. A condition of the pelvic organs such as this is referred to as **pelvic inflammatory disease (PID)**. Sterility may result from scar tissue remaining after the disease has been treated, or a woman may later experience an ectopic pregnancy. It should be noted that symptoms are not universally observed in females, and that an estimated 50 percent of affected women exhibit no symptoms. Such asymptomatic women may spread the disease unknowingly.

In males, gonorrhea occurs primarily in the urethra, the tube from the bladder that passes through the penis. Onset usually is accompanied by a tingling sensation in the penis followed in a few days by pain when urinating. There is also a thin, watery discharge at first and later, a more obvious whitened, thick fluid that resembles semen. Frequency of urination and an urgency to urinate develop as the disease spreads further into the urethra. The lymph glands of the groin may also swell, and sharp pain may be felt in the testicles. Unchecked infection of the epididymis may lead to sterility. Symptoms tend to be more acute in males than in females, and males thus tend to seek diagnosis and treatment more readily.

Gonorrhea does not restrict itself to the urogenital organs. **Gonococcal pharyngitis**, for example, may develop in the pharynx if bacteria are transmitted by oral-genital contact; patients complain of sore throat or difficulty in swallowing. Infection of the rectum, or **gonococcal proctitis**, is also observed, especially in homosexual males. Transmission to the eyes may occur by fingertips or towels, and conjunctivitis or keratitis may develop.

Gonorrhea is particularly dangerous to infants born to infected women. The infant may contract gonococci during passage through the birth canal and develop a disease of the eyes called **gonococcal ophthalmia**. To preclude the blindness that may ensue, most states have laws requiring that the eyes of newborns be treated with drops of 1 percent silver nitrate or antibiotics such as erythromycin or tetracycline.

Traditionally, gonorrhea has been treated with a single large dose of penicillin. In 1976, however, a strain of *N. gonorrhoeae* appeared that resists the drug. These organisms produce an enzyme called penicillinase that converts penicillin to harmless penicilloic acid. The strain was named **penicillinase-producing** *Neisseria gonorrhoeae*, or **PPNG**. In succeeding years, therapy shifted to spectinomycin and then to tetracycline. The latter is still recommended together with ceftriaxone. A study reported in 1992 indicates that good therapeutic results can be obtained with a single oral dose of cefixime. An attack of gonorrhea does not immunize one to future attacks, apparently because the immune system does not respond strongly enough to the first attack. No vaccine is available at this writing, but an antipili vaccine is in the experimental stage.

Gonorrhea can be detected by observing Gram-negative diplococci in the discharge from the urogenital tract as well as in colonies from swab samples cultivated on Thayer-Martin medium (Figure 10.4). In an immunological test called the **Gonozyme test**, physicians take a swab sample and dip the swab into a solution called gonozyme. An immunoassay reaction takes place (Chapter 19) and within a few hours, a color reaction indicates the presence or absence of gonococci. The test allows doctors to detect gonorrhea early so that treatment can start immediately.

Chlamydia (Chlamydial Urethritis)

Chlamydia, also known as chlamydial urethritis, is one of several diseases collectively known as **nongonococcal urethritis**, or **NGU**. Nongonococcal urethritis is a general term for a condition in which persons without gonorrhea have a demonstrable infection of the urethra usually characterized by inflammation, and often accompanied by a discharge. Evidence is convincing that over 50 percent of cases of NGU are actually chlamydia (chlamydial urethritis). Another 25 percent of cases are believed to be ureaplasmal urethritis, and the remaining 25 percent are of unknown cause.

Chlamydia is a gonorrhealike disease transmitted by sexual contact. The causative agent is *Chlamydia trachomatis*, a species of chlamydiae. *Chlamydia trachomatis* is an exceptionally small organism, measuring about 0.25 µm in diameter. It grows only in living tissue such as fertilized chicken eggs and tissue cultures and has a complex reproductive cycle, as illustrated in Figure 10.5. The organism appears to be a specific parasite of humans.

Epididymis:
the thin tube leading from the testicle.

Pharynx:
the throat region.

Keratitis:
disease of the cornea of the eye.

of-thal′me-ah

Penicillinase:
a bacterial enzyme that converts penicillin to penicilloic acid.

cef-tri-ax′ōn

cef-ix′ēm

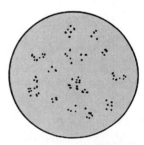

CHLAMYDIA TRACHOMATIS
klah-mid′e-ah tra-ko′mah-tis

FIGURE 10.4

Neisseria gonorrhoeae

Two views of *Neisseria gonorrhoeae*, the agent of gonorrhea. (a) A scanning electron micrograph of the edge of a colony of *Neisseria gonorrhoeae* (× 5400). (b) An enlarged view of the cocci (× 10,000). Note the typical pairing of the cells and the dustlike granules on the cell surfaces. The significance of these granules is not clear.

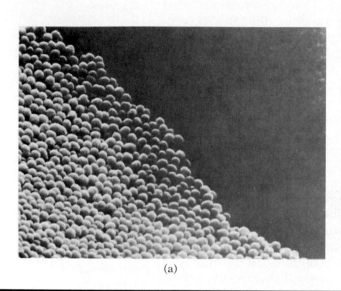

(a)

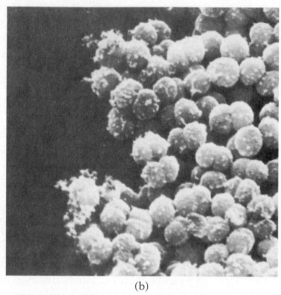

(b)

As early as 1986, a public health estimate suggested that three to five million Americans experience chlamydia annually. The disease has an incubation period of about one to three weeks and the symptoms are remarkably similar to those in gonorrhea, although somewhat milder. Females often note a slight vaginal discharge as well as inflammation of the cervix, since the cervix is an important site of infection. Burning pain is also experienced on urination, reflecting disease in the urethra. In complicated cases, the disease may spread to the Fallopian tubes, causing adhesions that block the passageways (salpingitis). Certain researchers report that **pelvic inflammatory disease (PID)** is a more likely consequence of chlamydia than of gonorrhea. (PID from gonorrhea and chlamydia is believed to affect about 50,000 women in the United States annually.) Often, however, there are few symptoms of disease before the salpingitis manifests itself, thus adding to the danger.

Salpingitis: blockage of the Fallopian tubes.

In males, chlamydia is characterized by painful urination and a discharge that is more watery and less copious than in gonorrhea. The discharge is often observed after urine has been passed for the first time in the morning. Tingling sensations in the penis are generally evident. Inflammation of the epididymis may result in sterility but this complication is uncommon. Chlamydial pharyngitis and proctitis are also possible.

Ophthalmia: severe inflammation of the eyes.

Newborns may contract *C. trachomatis* from an infected mother and develop a disease of the eyes known as **chlamydial ophthalmia**. The silver nitrate used to prevent gonococcal ophthalmia is not effective as a preventative, and erythromycin therapy is required to prevent blindness. Studies in the 1980s also revealed that **chlamydial pneumonia** may develop in newborns from an exposure to *C. trachomatis* during birth. Health officials estimate that 75,000 newborns suffer chlamydial ophthalmia and that 30,000 newborns experience chlamydial pneumonia annually.

FIGURE 10.5
The Chlamydiae

(A) Reproductive cycle of the chlamydiae. (a) The chlamydia enters the cytoplasm of a susceptible cell. (b) It enlarges to form an initial body after about 12 hours. (c) It then undergoes several binary fissions to form small particles called elementary bodies. (d) Elementary bodies accumulate after 30 hours. (e) These emerge as new chlamydiae after 48 hours. (B) An electron micrograph of an ultrathin section through a microcolony of *Chlamydia* in the cytoplasm of a tissue cell after 48 hours of incubation (× 34,000). The various developmental forms are labeled. Note where the vesicular membrane has ruptured and chlamydiae are being released into the cytoplasm.

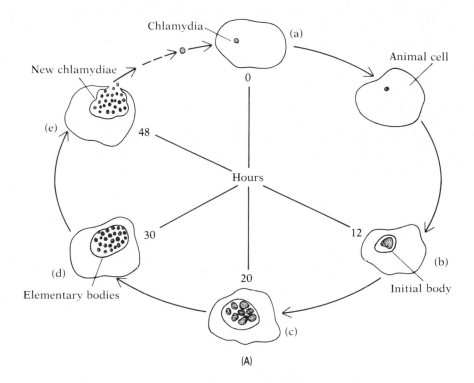

(A)

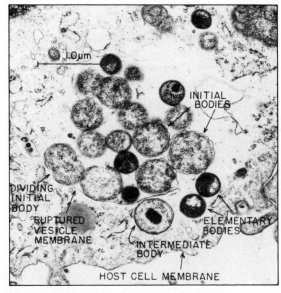

(B)

Chlamydia may be successfully treated with tetracycline. If a woman is pregnant, erythromycin is substituted since tetracycline affects bone formation in newborns. Oftentimes, the disease is so similar in symptoms to gonorrhea that physicians are uncertain which disease is present and, therefore, which antibiotic to prescribe.

Since 1983, two relatively fast and simple laboratory tests have been available to detect *C. trachomatis*. The first test, called the **MicroTrak** test, requires that a physician take a swab sample from the penis or the cervix (as in a pap smear) and place the swab in a vial of fluid for transport to a local laboratory. A fluorescent antibody test using monoclonal antibodies (Chapter 19) is then performed, and within 30 minutes the results are available. The second test, called the **Chlamydiazyme assay**, is an immunoassay test, also performed with a swab sample. It is completed in the doctor's office and is similar to the Gonozyme test for gonorrhea.

Epidemiological studies of chlamydia have been hampered by the fact that physicians are not required to report cases of NGU to their state health departments or the CDC. In Great Britain, where chlamydia is a reportable disease, the magnitude of the epidemic is widely accepted. If the estimate of three to five million annual cases in the United States is correct, chlamydia may be among the most prevalent diseases in our society.

Monoclonal antibodies: antibodies experimentally produced against a single type of cell or substance.

Ureaplasmal Urethritis

u-re"ah-plaz'ma
u-re"ah-lit'ĭ-kum

Ureaplasmal urethritis is another type of nongonococcal urethritis. It is caused by *Ureaplasma urealyticum*, a type of mycoplasma, so-named because of its ability to digest urea in culture media. The organism is often referred to as a **T-mycoplasma** because "tiny" colonies of the organisms develop on laboratory media. *U. urealyticum* is one of the smallest known bacteria that cause human disease, measuring about 0.15 μm in size.

The symptoms of ureaplasmal urethritis are similar to those in gonorrhea. A distinction can be made between the two diseases because in ureaplasmal urethritis the discharge is variable in quantity, and the urethral pain is usually aggravated during urination. Symptoms are often very mild. Transmission is generally by sexual contact.

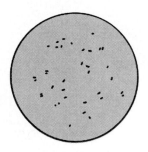

UREAPLASMA UREALYTICUM

Penicillin cannot be used to treat ureaplasmal urethritis because *U. urealyticum* has no cell wall. Tetracycline is currently the drug of choice. Diagnosis often depends on eliminating gonorrhea or other types of NGU as possibilities, and for this reason, cases are not often recognized.

Studies in the 1980s have indicated the serious consequences of ureaplasmal urethritis. Infertility is one consequence because low sperm counts and poor movement of sperm cells have been observed in males. Salpingitis in females has also been described. Moreover, *Ureaplasma* is capable of colonizing the placenta during pregnancy, and reports have linked it to spontaneous abortions and premature births. As previously noted, 25 percent of NGU cases may be ureaplasmal urethritis.

shang'kroid

Chancroid

Chancroid is a sexually transmitted disease believed to be more prevalent worldwide than gonorrhea or syphilis. The disease is endemic in many undeveloped nations and is common where climates are tropical and public health standards are low. Populations of immigrants account for a large percentage of the approximate 3500 cases detected in the United States annually.

The causative agent of chancroid is **Haemophilus ducreyi**, a small Gram-negative rod named for Augusto Ducrey, who observed it in skin lesions in 1889. The earliest sign of chancroid is a tender papule surrounded by a narrow zone of erythema. The papule quickly becomes pus-filled and then breaks down to leave a shallow saucer-shaped ulcer. The ulcer has ragged edges and soft borders, a characteristic that distinguishes it from the primary lesion of syphilis. For this reason the disease is often called **soft chancre**. In addition, the lesion bleeds easily and is extremely painful, unlike the syphilitic chancre.

The lesions in chancroid most often occur on the penis in males and the labia or clitoris in females. Substantial swelling of the inguinal lymph nodes may be observed. However, the disease generally goes no further. The clinical picture in chancroid makes the disease recognizable, but definitive diagnosis depends on isolation of *H. ducreyi* from the lesions.

Transmission of chancroid depends on contact with the lesion, and sexual contact is the usual mode. Tetracycline, erythromycin, and sulfonamide drugs are useful for therapy, but the disease often disappears without treatment.

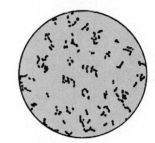

he-mof'i-lus doo-krāy'e

HAEMOPHILUS DUCREYI
Inguinal lymph nodes:
lymph nodes of the groin.

Other Sexually Transmitted Diseases

In addition to the ones discussed, a number of other sexually transmitted diseases merit brief attention in this chapter. None of these diseases normally presents a life-threatening situation to humans.

Lymphogranuloma venereum (LGV) is caused by a variant of **Chlamydia trachomatis**, slightly different from the variant that causes chlamydial urethritis. LGV is more common in males than females, and is accompanied by fever, malaise, and swelling and tenderness in the inguinal lymph nodes. Females may experience infection of the rectum (proctitis), if the chlamydiae pass from the genital orifice to the nearby intestinal orifice. Unless medical treatment is forthcoming, the disease may become chronic and result in rectal blockage. Approximately 500 cases are detected in humans annually.

lim"fo-gran"u-lo'mah
ven-e're-um

Lymphogranuloma venereum is prevalent in Southeast Asia and Central and South America. Sexually active persons returning from these areas may show symptoms of the disease, but treatment with tetracycline leads to rapid resolution.

Granuloma inguinale is a rare disease in Europe and North America, but it remains an endemic problem in tropical and subtropical areas of the world such as Caribbean countries and Africa. It is caused by **Calymmatobacterium granulomatis**, a small Gram-negative encapsulated bacillus. The disease begins with a primary lesion starting as a nodule and progressing to a granular ulcer that bleeds easily. In most cases this ulcer forms in the external genital organs but it may spread to other regions by contaminated fingers. The inguinal lymph nodes may swell, but fever and other body symptoms are usually absent, a factor that distinguishes the disease from LGV. The lesion is less painful and more regular than in chancroid. Tissue samples reveal masses of bacteria called **Donovan bodies** within phagocytes in the lesion. The disease responds to various antibiotics, especially tetracycline.

kah-lim"mah-to-bak-te're-um
gran-u-lo'mah-tis

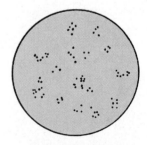

GARDNERELLA VAGINALIS

Vaginitis is a general term for various mild infections of the vagina and sometimes the vulva. One cause of vaginitis is **Gardnerella vaginalis**, formerly called *Haemophilus vaginalis*. This small Gram-negative rod usually lives uneventfully in the vagina, but flare-ups of disease may take place and transmission by sexual contact may occur during such times. A foul-smelling discharge is the most prominent symptom, and tetracycline therapy generally provides relief.

FIGURE 10.6

A Species of *Mycoplasma*

A species of *Mycoplasma* seen with the scanning electron microscope. A closely related species called *M. hominis* causes mycoplasmal urethritis, an STD that is similar to chlamydia. Note the pleomorphic (multishape) appearance of the organism. Bar = 1 μm.

The final organism that we shall consider is ***Mycoplasma hominis*** (Figure 10.6). This mycoplasma causes **mycoplasmal urethritis**, a disease similar to ureaplasmal urethritis. In addition, the organism can colonize the placenta and cause spontaneous abortion and premature birth. *M. hominis* is distinguished from *Ureaplasma* by its inability to digest urea as well as by its larger colony growth on agar, and by its preference for anaerobic conditions during laboratory cultivation. Tetracycline is prescribed for active cases of mycoplasmal urethritis. Apparently the disease can also be caused by ***Mycoplasma genitalium***, as reported first in 1993 by a joint American-British research team. (*M. genitalium* has another distinction in microbiology—it was one of the first organisms of any kind to have its entire genome deciphered, an achievement completed in 1995 by Craig Venter and his associates.) Table 10.1 summarizes the sexually transmitted bacterial diseases.

TO THIS POINT
■

We have studied a series of sexually transmitted diseases and have noted their prevalence in modern society. Syphilis is the only disease in the group that regularly spreads to other organs of the body. The remaining diseases are largely restricted to the urogenital organs. In addition syphilis is the only STD that is life-threatening. The other diseases generally are mild and may resolve even without treatment, but as we have seen in gonorrhea and NGU, there may be long-range complications such as sterility or spontaneous abortions.

We also pointed out some sexually transmitted diseases that are less publicized, but which may occur widely in the United States. Chlamydial urethritis and ureaplasmal urethritis are examples. It is conceivable that the incidence of lymphogranuloma venereum, granuloma inguinale, and Mycoplasma and Gardnerella *infections may be equally high but that detection methods are currently lacking. The true extent of sexually transmitted diseases may never be known.*

We shall now turn our attention to a group of bacterial diseases usually acquired by contact, though generally not sexual contact. Some of these diseases have long histories and, left untreated, they may persist for years. Others were unknown a generation ago and are of short duration. Few are fatal, but all have contributed to our knowledge of the broad spectrum of bacterial diseases.

Contact Bacterial Diseases

Leprosy (Hansen's Disease)

For many centuries, **leprosy** was considered a curse of the damned. It did not kill, but neither did it seem to end. Instead, it lingered for years, causing the tissues to degenerate and deforming the body. In biblical times, victims were required to call out "Unclean! Unclean!" and usually they were ostracized from the community. Among the more heroic stories of medicine is the work of Father Damien de Veuster, the Roman Catholic priest who in 1870 established a hospital for leprosy patients on the island of Molokai in Hawaii. An equally heroic story was written in this century (MicroFocus: 10.2).

10.2
MICROFOCUS

THE "STAR" OF CARVILLE

On December 1, 1894, seven leprosy patients arrived at an old plantation on a crook in the Mississippi River. Soon thereafter, four nuns of the Order of St. Vincent de Paul joined them. Together this small band formed the nucleus of what was to become the National Hansen's Disease Center at Carville, Louisiana.

Change came slowly. In 1921, the federal government acquired the institution, but it remained essentially a prison, patrolled by guards and surrounded by a cyclone fence with barbed wire. Then, in 1931, a leprosy patient named Stanley Stein arrived. (Figure) Stanley Stein was not his real name—he had forsaken that for fear of bringing shame to his family. Soon, Stein instituted a weekly paper to bring a sense of community to the patients. Originally named *The Sixty-Six Star* (Carville was Marine Hospital Number 66), the name was eventually shortened to *The Star.*

As the circulation of *The Star* increased, Stein and others launched a campaign for

change. In 1936, the patients acquired a telephone so they could hear the voices of their families. Three years later, the swamps were drained to reduce the incidence of malaria. Soon there came a better infirmary, a new recreation hall, and removal of the barbed wire. In 1946, the State of Louisiana allowed the patients to vote in local and national elections.

Through all these years, Stein's leprosy worsened. Originally he had tuberculoid leprosy, the form in which the nerves are damaged. Afterward, however, he developed lepromatous leprosy, which causes lesions to form on the face, ears, and eyes. Soon he was totally blind. Without feeling in his fingers, he could not even learn Braille.

But Stein was not finished. He and his newspaper tirelessly fought for a new post office and weekend passes for patients. In 1961, President Kennedy paid tribute to *The Star* on its thirtieth anniversary and singled out its indomitable editor for praise. Stanley

Stein died in 1968. By that time, *The Star* had a circulation of 80,500 in all 50 states and 118 foreign countries.

Stanley Stein at his home "Wit's End" in Carville

TABLE 10.1
A Summary of Sexually Transmitted Bacterial Diseases

DISEASE	CAUSATIVE AGENT	DESCRIPTION OF AGENT	ORGANS AFFECTED	CHARACTERISTIC SIGNS
Syphilis	*Treponema pallidum*	Spirochete	Skin Cardiovascular organs Nervous system	Chancre Gumma
Gonorrhea	*Neisseria gonorrhoeae*	Gram-negative diplococcus	Urethra, cervix Fallopian tubes Epididymis Eyes, pharynx	Pain on urination Discharge Salpingitis
Chlamydia (Chlamydial urethritis)	*Chlamydia trachomatis*	Chlamydia	Urethra, cervix Fallopian tubes Epididymis Eyes, pharynx	Pain on urination Watery discharge Salpingitis
Ureaplasma urethritis	*Ureaplasma urealyticum*	Mycoplasma	Urethra Fallopian tubes Epididymis	Pain on urination Variable discharge Salpingitis
Chancroid (Soft chancre)	*Haemophilus ducreyi*	Gram-negative rod	External genital organs Inguinal lymph nodes	Soft chancre Erythema Swollen inguinal lymph nodes
Lympho-granuloma venereum	*Chlamydia trachomatis*	Chlamydia	Inguinal lymph nodes Rectum	Swollen inguinal lymph nodes Proctitis
Granuloma inguinale	*Calymmato-bacterium granulomatis*	Gram-negative rod	External genital organs	Bleeding ulcer Swollen inguinal lymph nodes
Vaginitis	*Gardnerella vaginalis*	Gram-negative rod	Vagina	Foul-smelling discharge
Mycoplasmal urethritis	*Mycoplasma hominis M. genitalium*	Mycoplasma	Urethra Fallopian tubes Epididymis	Pain on urination Variable discharge Salpingitis

MYCOBACTERIUM LEPRAE

Koch's postulates:
A series of procedures for relating a specific organism to a specific disease.

Peripheral nervous system: the system of nerve fibers that extend from the brain and spinal cord to the body tissues.

The agent of leprosy is *Mycobacterium leprae*, an acid-fast rod related to the tubercle bacillus. This organism, first observed in 1874 by the Norwegian physician Gerhard Armauer Hansen, is referred to as Hansen's bacillus, and leprosy is commonly called **Hansen's disease**. Ironically, *M. leprae* has not yet been cultivated in artificial laboratory medium and thus, Koch's postulates (Chapter 1) have not been fulfilled. In 1960, researchers at the CDC succeeded in cultivating the bacillus in the footpads of mice, and in 1969 scientists at the National Hansen's Disease Center in Carville, Louisiana, found that it would grow in the tissues of the armadillo. Growth in apes was reported in 1982.

Leprosy is spread by multiple skin contacts as well as by droplets from the upper respiratory tracts. The disease has an unusually long incubation period of three to six years, a factor that makes identification very difficult. Because the organisms are heat-sensitive, the symptoms occur in the skin and peripheral nervous system in the cooler parts of the body such as the hands, feet, face, and earlobes. Severe cases also involve the eyes and the respiratory tract.

TOXIN INVOLVED	TREATMENT ADMINISTERED	IMMUNIZATION AVAILABLE	COMMENT
Not established	Penicillin	None	Primary, secondary, and tertiary stages The "Great Imitator" Congenital transmission
Not established	Penicillin Spectinomycin Tetracycline Ceftriaxone	None	Most-reported microbial disease in U.S. Complicated by pelvic inflammatory disease PPNG a possible cause
Not established	Tetracycline Erythromycin	None	Leads to infertility Chlamydiazyme test for diagnosis Estimated 3–5 million cases annually in U.S.
Not established	Tetracycline	None	Leads to infertility Linked to spontaneous abortion
Not established	Tetracycline Erythromycin Sulfonamides	None	Few complications Many immigrant cases
Not established	Tetracycline	None	Prevalent in Central and South America
Not established	Tetracycline	None	Donovan bodies seen in lesions
Not established	Tetracycline	None	Organism commonly found in the vagina
Not established	Tetracycline	None	Similar to ureaplasmal urethritis

Patients with leprosy experience disfigurement of the skin and bones, twisting of the limbs, and curling of the fingers to form the characteristic **claw hand**. Loss of facial features accompanies thickening of the outer ear and collapse of the nose. Tumorlike growths called **lepromas** may form on the skin and in the respiratory tract (lepromatous leprosy), and the optic nerve may deteriorate. However, the largest number of deformities develop from loss of pain sensation due to nerve damage (tuberculoid leprosy). Inattentive patients can pick up a pot of boiling water without flinching, and many accidentally let cigarettes burn down and sear their fingers.

The principal drug for the treatment of leprosy is a sulfur compound known commercially as **dapsone** ("Zap it with dap"). In many instances, such as shown in Figure 10.7, the results may be dramatic. However, studies show that *M. leprae* is becoming increasingly resistant to this drug, and alternative antibiotics such as rifampin and clofazimine are being employed. In 1992, the WHO began a campaign to treat the disease with oflaxacin. Some attempts have been made to immunize populations with BCG, the vaccine used for tuberculosis. Early

Lepromas: **tumorlike growths accompanying leprosy in the tissues.**

clo-faz'i-mēn
of-laks'a-cin

FIGURE 10.7

Treating Leprosy

The child with leprosy is pictured (a) before treatment with dapsone and (b) some months later, after treatment. Note that the lesions of the ear and face and the swellings of the lips and nose have largely disappeared.

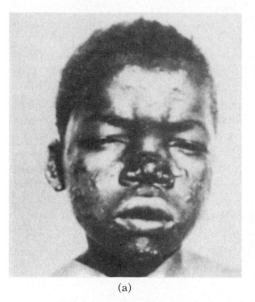

(a)

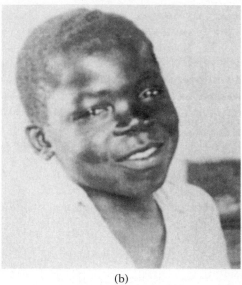

(b)

BCG:
a preparation of modified tubercle bacilli used for immunization to tuberculosis.

diagnosis relies on a procedure called the lepromin test, performed in basically the same way as the tuberculin test (Chapter 7).

World Health Organization reports estimate over ten million victims of leprosy in the world today. Cases in the United States have risen dramatically over the last generation (Figure 10.8), in large measure due to infected immigrants. Approximately 4000 patients are presently under treatment in American hospitals. About 300 come each year to Carville (MicroFocus: 10.2) for two to six weeks of initial diagnosis and treatment, followed by subsequent care at any of nine specially designated medical facilities around the country.

Staphylococcal Skin Diseases

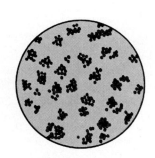

STAPHYLOCOCCUS AUREUS

Abscess:
a circumscribed pus-filled lesion, often due to staphylococci.

Staphylococci are normal inhabitants of the human skin, mouth, nose, and throat. Commonly, they live in these areas without harm, but they can cause extensive disease when they penetrate the skin barrier or the mucous membranes. Penetration is assisted by open wounds, damaged hair follicles, ear-piercing, dental extractions, or irritation of the skin by scratching. *Staphylococcus aureus*, the grapelike cluster of Gram-positive cocci, is the species usually involved in disease.

The hallmark of staphylococcal skin disease is the **abscess**, a circumscribed pus-filled lesion (Figure 10.9). A **boil** is a skin abscess, which often begins as a pimple. Deeper skin abscesses, called **carbuncles**, occur when the staphylococci work their way down to the tissues below the skin. Skin contact with other individuals spreads the disease. Food handlers should be aware that staphylococci from boils and carbuncles can be transmitted to food, where they can cause food poisoning (Chapter 8).

Another skin disease caused by *S. aureus* is the **scalded skin syndrome**, or **Ritter's disease**, occasionally seen in infants. The skin becomes red, wrin-

FIGURE 10.8

Reported Number of Cases of Leprosy in the United States, 1955–1993

The increase in the number of cases in the mid-1980s was due to a rise in the number of foreign-acquired cases. Approximately 150 cases acquired in the United States are reported annually. (Courtesy of Centers for Disease Control and Prevention, *Annual Summary*, 1993.) The inset shows the "claw hand" often found in severe cases of tuberculoid leprosy.

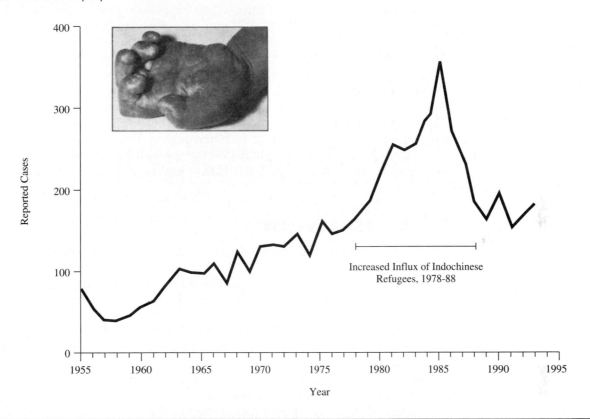

Increased Influx of Indochinese
Refugees, 1978-88

FIGURE 10.9

A Staphylococcal Abscess

A severe abscess on the head of a young boy. Abscesses often begin as trivial skin pimples and boils, but they can become serious as staphylococci penetrate to the deeper tissues.

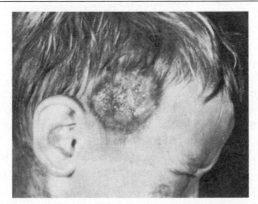

kled, and tender to the touch, with a sandpaper appearance. It may then peel off. Toxins produced by the staphylococci living at a point distant from the skin appear to be responsible for this condition. Mortality in untreated cases may be high.

im-pē-tī′go con-ta″je-o′sum

A more widespread staphylococcal skin disease is **impetigo contagiosum**. Here the infection is more superficial and involves patches of epidermis just below the outer skin layer. Impetigo first appears as thin-walled blisters that ooze a yellowish fluid and form yellowish-brown crusts. Usually the blisters occur on the exposed parts of the body, but they may also occur around the nose and upper lip after a child has had a cold with a runny nose, since the constant irritation provides a mechanism for penetration by the staphylococci. *Streptococcus pyogenes* (Chapter 7) may also cause impetigo.

pi-oj′ĕ-nez

Staphylococcal skin diseases are commonly treated with penicillin, but resistant strains of *S. aureus* are well known, and physicians may need to test a series of alternatives before an effective antibiotic is located. A staphylococcal skin disease can also be treated by vigorously scrubbing the lesions with Betadine applied with two-by-two inch gauze pads under the direction of a physician. The disease should be treated with caution because staphylococci commonly invade the blood and penetrate to other organs. For example, staphylococcal blood poisoning (septicemia) may develop, as well as staphylococcal pneumonia, meningitis, or nephritis. A trivial skin boil is often the source.

Betadine:
an iodine-based antiseptic that kills many bacterial species.

Nephritis:
disease of the kidney.

Toxic Shock Syndrome (TSS)

In 1978, James Todd of Children's Hospital in Denver, Colorado, coined the name **toxic shock syndrome (TSS)** for a blood disorder characterized by sudden fever and circulatory collapse. The name remained in relative obscurity until the fall of 1980, when a major outbreak occurred in menstruating women who used a particular brand of highly absorbent tampons. News of TSS dominated the media for about six months and led to a recall of the tampons. With that episode, toxic shock syndrome assumed a position of significance in modern medicine.

Toxic shock syndrome is caused by a toxin-producing strain of ***Staphylococcus aureus***. The earliest symptoms of disease include a rapidly rising fever accompanied by vomiting and watery diarrhea. Patients then experience a sore throat, severe muscle aches, and a sunburnlike rash with peeling of the skin, especially on the palms of the hands and soles of the feet. A sudden drop in blood pressure also occurs, possibly leading to shock and heart failure. Antibiotics may be used to control the growth of bacteria, but measures such as blood transfusions must be taken to control the shock.

Studies indicate that the staphylococci involved in TSS may exist in various places in the body, but the ones that inhabit the vagina have received the most attention. During the 1980 outbreak, scientists speculated that lacerations or abrasions of the tissue by tampon inserters gave the staphylococci access to the tissues. Others suggested that staphylococci grow in the warm, stagnant fluid during the long period that the tampon is in place. With the recall of the tampons, the incidence of TSS diminished (Figure 10.10). It has continued to be low partly because less absorbent tampons are now used more frequently and because the material makeup of tampons has been changed to create an environment less conducive to bacterial growth.

It appears certain that menstruation and tampon use play a role in toxic shock syndrome, but it is equally certain that other factors are involved. This is because males, prepubertal girls, and postmenopausal women also have been stricken. Indeed, in 1994 one woman contracted TSS from a contaminated needle (MicroFocus: 10.3). Studies indicate that millions of Americans harbor the staphylococcus and pass it along by contact but that the vast majority of individuals are apparently resistant to the toxin. About 500 cases of TSS are reported to the CDC annually.

FIGURE 10.10

Reported Cases of Toxic Shock Syndrome, by Month of Onset, in the United States, 1979–1993

Note that the total number of cases per year is the sum of the numbers of monthly cases. As the chart shows, the disease reached a peak in the fall of 1980 and then, the incidence per month dropped with the recall of the highly absorbent tampons. Nonmenstrual episodes currently account for a significant number of cases. (Courtesy of Centers for Disease Control and Prevention, *Annual Summary*, 1993.) The inset shows the peeling skin often associated with TSS.

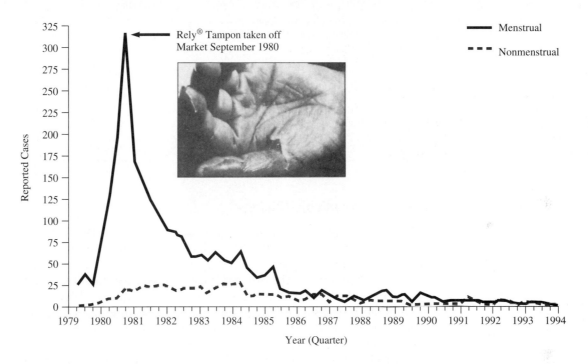

Rely® Tampon taken off Market September 1980

— Menstrual

--- Nonmenstrual

* Includes only cases meeting the CDC case definition (N= 3245) for staphylococcal disease.

Trachoma

Trachoma is a disease of the eyes. It occurs in hot, dry regions of the world and is prevalent in Mediterranean countries, parts of Africa and Asia, and in the southwestern United States among Native American populations. Hundreds of millions throughout the world are believed afflicted by it.

The cause of trachoma is a variant of ***Chlamydia trachomatis***, the organism responsible for chlamydia and lymphogranuloma venereum. Fingers, towels, optical instruments, and face-to-face contact are possible modes of transmission. The chlamydiae grow in the conjunctiva, the thin membrane that covers the cornea and forms the inner eyelid. A series of tiny, pale **nodules** forms on this membrane, giving it a rough appearance ("trachoma" is derived from the Greek *trachi-* for "rough"). In serious cases, the upper eyelid turns in, causing abrasion of the cornea by the eyelashes. Blindness develops from corneal abrasions and lesions.

Tetracycline and erythromycin are helpful in reducing the symptoms of trachoma, but, in many patients, the relief is only temporary because chlamydiae reinfect the tissues. The World Health Organization maintains squads of trachoma nurses whose duty is to travel from village to village identifying trachoma cases in their early stages and treating them before they become serious. Trachoma is believed to be the world's leading cause of preventable blindness.

kla-mid'e-ah
tra-ko'-mah-tis

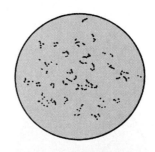

CHLAMYDIA TRACHOMATIS

Conjunctiva:
the thin membrane that covers the cornea and forms the inner eyelid.

10.3

"IT SEEMED LIKE A GOOD IDEA!"

The idea of a tattoo seemed okay. All her friends had them, and a tattoo would add a sense of uniqueness to her personality. After all, she was already 22. It took some pushing from her friends, but she finally made it into the tattoo parlor that day in Fort Worth, Texas.

Two weeks later the pains started—first in her stomach, then all over. Her fever was high, and now a rash was breaking out—it looked like her skin was burned and was

peeling away. One visit to the doctor, then immediately to the emergency room of the local hospital. The gynecologist guessed it was an inflammation of the pelvic organs (pelvic inflammatory disease, they called it), so he gave her an antibiotic and sent her home.

But it got worse the fever, the rash, the peeling, the pains. Back to the emergency room. This time they would admit her

to the hospital, give her blood transfusions and antibiotics intravenously, keep her for 11 days, and discover a severe blood infection with *Staphylococcus aureus*. And there was an unusual diagnosis ... "toxic shock syndrome? Didn't women get that from tampons?" Most do, she was told, but a few get it from staph entering a skin wound—a wound that can be made by a contaminated tattoo needle.

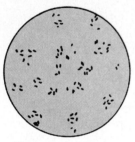

HAEMOPHILUS AEGYPTIUS

Photophobia:
impaired vision in bright light.

Neomycin:
an antibiotic that binds to ribosomes in many Gram-negative bacteria.

per-ten'u

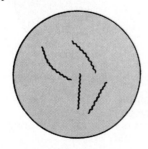

TREPONEMA PERTENUE

Bacterial Conjunctivitis (Pinkeye)

Several microorganisms cause conjunctivitis, among them the bacterium *Haemophilus aegyptius*, the Koch-Weeks bacillus. This organism is also known in the literature as *Haemophilus influenzae* biotype III because of its close relationship to *H. influenzae*. The organism is a small Gram-negative rod that grows in chocolate agar, a rich medium that contains disrupted red blood cells (*Haemophilus* means "blood-loving") The red cells are a source of the X factor (hemin) and the V factor (NAD) that are required for growth of *Haemophilus* species.

Conjunctivitis is a disease of the conjunctiva. When infected, the membrane becomes inflamed, a factor that imparts a brilliant pink color to the white of the eye (hence the name "pinkeye"). A copious discharge runs down the cheek in the waking hours and crusts the eyelids shut during sleep. The eyes are swollen and itch intensely, and vision in bright light is impaired (photophobia).

Conjunctivitis may be transmitted in a number of ways, including face-to-face contact and airborne droplets. Optometric instruments, microscopes, and contaminated towels may also transmit the bacilli. The disease normally runs its course in about two weeks and therapy is usually not required. Neomycin may be administered to hasten recovery. Conjunctivitis is extremely contagious, especially where people congregate.

In recent years, a variant of *H. aegyptius* has also been isolated from the blood of patients suffering from a disease called **Brazilian purpuric fever**. This life-threatening disease is accompanied by nausea, vomiting, fever, and hemorrhagic skin lesions. Many patients display conjunctivitis before the onset of more serious symptoms, an observation that is consistent with the pinkeye noted above.

Yaws

Yaws commonly occurs in tropical countries of Africa, South America, and Southeast Asia. It is caused by *Treponema pertenue*, a spirochete identical in appearance and similar in chemistry to the syphilis spirochete (MicroFocus: 10.4). Yaws is usually acquired by skin contact with a victim. A red, raised lesion called a **mother yaw** develops at the site of entry. Blood associated with the lesion gives it the appearance of a raspberry, and the disease is sometimes termed **frambesia**, from the French *framboise*, for raspberry.

10.4
MICROFOCUS

"YOU WIN A FEW, YOU LOSE A FEW"

Scientists suspected that yaws and syphilis were related—the agents, symptoms, and treatments were extremely similar—but they had no idea how close the relationship was until they tried to eliminate one of them.

Their effort at elimination began in the early 1950s when the World Health Organization mounted a campaign against yaws. Doctors gave massive doses of penicillin to patients and administered preventative doses to their contacts. The results were spectacular—in Brazil, for instance, the incidence of yaws dropped from 16,000 cases in 1965 to 188 cases in 1975. Impressive results were also obtained from other regions of the world.

But now a new problem emerged—syphilis. There had been little syphilis in

Chad, New Guinea, and Thailand, but suddenly an unusual number of cases were reported in all three countries. In Columbia, there had been 4 cases of syphilis for every case of yaws, but by the mid-1970s, there were 1000 syphilis patients for every yaws patient.

What had happened was depressing. People with yaws were immune to syphilis because their antibodies against the yaws organism also protected them against syphilis. Now there was no more yaws ... and no more yaws antibodies. The syphilis organism had a clear path to causing infection. And it was doing so with a vengeance. Success had turned to despair.

In time the lesion disappears, but months later the patient develops numerous other yaws. Left untreated, these also disappear, only to reappear as soft granular lesions. The similarities of organisms and symptoms in yaws and syphilis have led many microbiologists to believe that the two diseases are related. Penicillin is valuable in treatment. **Bejel** and **pinta** are two other diseases very similar to yaws. Table 10.2 summarizes the contact bacterial diseases.

TO THIS POINT
■

We have discussed several bacterial diseases spread by contact by focusing on six examples. In leprosy, multiple contacts may be necessary for passage of the bacteria. Staphylococcal skin diseases are also spread by contact, although many episodes of infection begin by simply irritating the skin and allowing staphylococci to penetrate. Tissue irritation may also be the source of the staphylococci involved in toxic shock syndrome. Indeed, researchers who studied the tampon problem of 1980 concluded that the bacteria were already present in the body. Contact spreads the staphylococci to others.

In trachoma, conjunctivitis, and yaws we discussed three other diseases spread by contact. Trachoma and yaws are prevalent in warm countries where less clothing is worn than in temperate climates. This yields a greater skin surface for contact with an infected individual. Bacterial conjunctivitis spreads where people congregate. You might note that the symptoms of the six diseases studied are largely restricted to the skin, where penetration has occurred. Except for the staphylococcal diseases, deeper organs seldom become involved, and life-threatening situations are uncommon. Also, we see the broad range of bacterial organisms that can cause disease

TABLE 10.2
A Summary of Contact Bacterial Diseases

DISEASE	CAUSATIVE AGENT	DESCRIPTION OF AGENT	ORGANS AFFECTED	CHARACTERISTIC SIGNS
Leprosy (Hansen's disease)	*Mycobacterium leprae*	Acid-fast rod	Skin, bones Peripheral nerves	Tumorlike growths Skin disfigurement "Claw hand"
Staphylococcal skin diseases	*Staphylococcus aureus*	Gram-positive staphylococcus	Skin	Abscess, boil Scalded-skin syndrome Impetigo contagiosum
Toxic shock syndrome	*Staphylococcus aureus*	Gram-positive staphylococcus	Blood	Fever Watery diarrhea Sore throat Sunburnlike rash
Trachoma	*Chlamydia trachomatis*	Chlamydia	Eyes	Nodules on conjunctiva Scarring in eye
Bacterial conjunctivitis	*Haemophilus aegyptius*	Gram-negative rod	Eyes	Pinkeye Photophobia
Yaws	*Treponema pertenue*	Spirochete	Skin	Red, raised lesion

including two rods, a staphylococcus, a chlamydia, and a spirochete.

In the final section of this chapter, we shall consider a number of miscellaneous bacterial diseases grouped into discrete categories. Certain of these diseases (the endogenous diseases) are caused by bacteria already in the body; other diseases (the animal bite diseases) are associated with bite wounds; still other diseases (oral diseases) occur in the mouth; and a final group (nosocomial diseases) are contracted during a stay in the hospital. We shall also make brief mention of urinary tract and burn infections as we complete our survey.

Miscellaneous Bacterial Diseases

Endogenous Bacterial Diseases

Endogenous bacterial diseases are caused by organisms that normally inhabit the body. Natural host resistance generally prevents proliferation of the causative organisms, but when the resistance is suppressed, tissue invasion and disease may follow.

An example of an endogenous disease is **actinomycosis**. This disease is caused by ***Actinomyces israelii***, named for James A. Israel, who described the bacillus in 1878. It is a Gram-positive anaerobic funguslike rod often found in the gastrointestinal and respiratory tracts. When it enters the gum tissues during a dental extraction, *A. israelii* multiplies and grows toward the facial surface, causing a red swelling that is lumpy and hard as wood. The condition, known as **lumpy jaw**, may develop into a skin problem with draining sinuses. Another form of actinomycosis involves **draining sinuses** of the chest wall, while a third form is characterized by abdominal sinuses, often as a complication of ulcers.

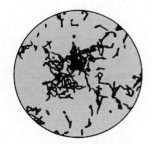

ACTINOMYCES ISRAELII

ak"ti-no-mi-ko'sis

Sinuses:
cavities that often express fluid.

TOXIN INVOLVED	TREATMENT ADMINISTERED	IMMUNIZATION AVAILABLE	COMMENT
None	Dapsone Rifampin Clofazimine	BCG	Lepromin test for diagnosis Long incubation period
Probable	Penicillin	None	Antibiotic resistance in staphylococci Food handlers involved
Yes	Pencillin Blood transfusions	None	Occurs in all groups, especially menstruating women 500 cases annually
Not established	Tetracycline Erythromycin	None	Leading cause of preventable blindness
Not established	Neomycin	None	Extremely contagious Copious discharge
Not established	Penicillin	None	Found in tropical countries Related to bejel and pinta

Actinomycosis of the reproductive tract may result from use of the intrauterine device (IUD) for contraceptive purposes.

Nocardiosis, another example of an endogenous disease, is due to an acid-fast, funguslike rod called ***Nocardia asteroides***. The disease strikes the lungs where multiple abscesses form. Fever, coughing, and bloody sputum accompany the disease, and the symptoms may be mistaken for tuberculosis. Reports of death from nocardiosis have been linked to AIDS, the condition in which the body's immune system is severely suppressed (Chapter 13). *N. asteroides* may also cause abscesses and swelling of the foot, a condition called **Madura foot**, if penetration occurs from the soil via a wound.

spu'tum
Sputum:
thick, expectorated matter.

Species of ***Bacteroides*** are Gram-negative anaerobic rods that inhabit the large intestine and the feces of most individuals (Figure 10.11). These organisms may enter the bloodstream when a person sustains an intestinal injury and cause blood clots that clog the vessels causing oxygen depletion and possible gangrene in the tissues. *Bacteroides fragilis* is the most common pathogenic species of the group.

bak"tĕ-roi'dēz

One of the side effects of excessive antibiotic use in the intestine is the elimination of many species of bacteria that normally keep other species in check. Under the circumstances, an endogenous disease may develop from infection by ***Clostridium difficile***, a Gram-positive anaerobic rod found in the large intestine of many individuals. As other organisms disappear during therapy, the clostridia multiply and produce a series of toxins that induce a condition called **pseudomembranous colitis**. In this condition, a collection of yellowish-green membranous lesions cover the intestinal lining, and patients experience diarrhea with watery stools. Severe dehydration is possible. Infants appear to be particularly susceptible to this condition, especially if a normal population of bacteria has not yet been established. Research in the 1980s also linked *Clostridium difficile* and its toxins to sudden infant death syndrome (SIDS).

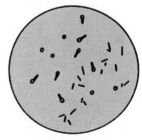

CLOSTRIDIUM DIFFICILE

dif'ē-seel

FIGURE 10.11

Endogenous Microorganisms

A scanning electron micrograph of the lining of the human intestine. The photograph shows various forms of bacteria that normally inhabit the environment. Endogenous disease may develop when bacteria like these invade the tissue during periods of suppressed host resistance. Bar = 1 μm.

Animal Bite Diseases

Public health officials estimate that each year in the United States about 3.5 million people are bitten by animals. Most of these wounds heal without complications but in certain cases, bacterial disease may develop.

mul-toc'i-da

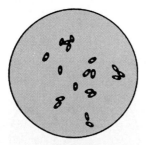

PASTEURELLA MULTOCIDA

Malaise:
a general feeling of illness.

roche"ah-lim-a'ah hens'e-la

an"ge-o-mah-to'sis

An important cause of bite infections is ***Pasteurella multocida***, a Gram-negative rod. This organism is a common inhabitant of the pharynx of the cat and dog where it causes a disease called **pasteurellosis**. In humans, the symptoms of pasteurellosis develop rapidly, with local redness, warmth, swelling, and tenderness at the site of the bite wound. Abscess formation occurs frequently, especially if the wound has been sutured. Arthritis may be encountered in some patients. Despite antibiotic therapy, the disease responds slowly.

Although cats transmit few diseases to humans, a notable problem is **cat scratch disease** (also known as cat scratch fever). The disease affects an estimated 20,000 Americans each year, primarily children, and is transmitted by a scratch, bite, or lick from a cat (or, in some cases, a dog). Symptoms include a papular or pustular lesion at the site of entry, followed by a headache, malaise, and low-grade fever. Swollen lymph glands, generally on the side of the body near the site, accompany the disease. In rare cases, the brain and central nervous system may be involved. Most episodes of disease end after several days or weeks and antibiotics such as rifampin hasten recovery. An agent has not yet been isolated with certainty as of this writing, but a leading candidate is ***Rochalimaea henselae***, a rickettsia. Antibodies against this organism are invariably observed in persons with cat scratch disease, and the rickettsia is usually isolated from cats whose scratch has led to disease in a child. *R. henselae* also appears to cause **bacillary angiomatosis**, a disease characterized by tumors whose cells organize themselves to form blood vessels. Occurring in the body's cardiovascular system, bacillary angiomatosis resembles Kaposi's sarcoma (Chapter 13) and may be a manifestation of cat scratch disease. Another candidate is *Afipia felis*, a Gram-negative rod.

a-fip′e-ah

SPIRILLUM MINOR STREPTOBACILLUS MONILIFORMIS

There are two different bacteria, each of which can cause **rat bite fever**. One bacterium is *Streptobacillus moniliformis*, a Gram-negative rod that occurs in long chains. It is found in the pharynx of rats and other rodents, and may be transmitted by a bite. Patients experience a lesion at the site of the bite, and then a typical triad of fever, arthritislike pain in the large joints, and skin rash. The second organism is *Spirillum minor*, a rigid spiral bacterium with polar flagella. A lesion occurs at the wound site and a maculopapular rash spreads out from this point. In Japan and other parts of Asia, *Spirillum*-related rat bite fever is known as **sodoku**. Antibiotic therapy is generally recommended for both forms of rat bite fever. Table 10.3 summarizes the miscellaneous bacterial diseases that are transmitted by animal bites or are endogenous.

Maculopapular rash: a rash formed by the fusion of pink-red skin spots.

Oral Diseases

The oral cavity is a type of ecosystem, with complex interrelationships between the resident population of microorganisms and the oral environment. The cavity has various ecological niches, each with a different physical property and nutrient supply dictating the number and type of microorganism that can survive. At least 20 different species of bacteria have been isolated from the normal oral environment, among them a variety of streptococci, diphtherialike bacilli, lactobacilli, spirochetes, and filamentous bacteria. A rough estimate is that there are between 50 billion and 100 billion bacteria in the adult mouth at any one time. (To put the number in perspective, consider that about 78 billion people have ever lived on Earth.)

The material that accumulates on the tooth surface is known by many terms, the most common of which is **dental plaque**. Plaque is essentially a deposit of dense gelatinous material consisting of protein, polysaccharide, and an enormous mass of bacteria. By some accounts there are more than a billion bacteria per gram of net weight of plaque. Most bacteria have been cultivated in the laboratory, and two-thirds are either anaerobic or facultative species.

Dental plaque: an oral deposit of dense gelatinous material consisting of organic compounds and bacteria.

Dental caries, or tooth decay, takes its name from the Latin *cariosus*, meaning rotten. In order for dental caries to develop, three elements must be present: a caries-susceptible tooth with a buildup of plaque; dietary carbohydrate, usually in the form of sucrose; and acidogenic (acid-producing) plaque bacteria. The bacteria produce acid that breaks down the calcium phosphate salts in hydroxyapatite, the major compound in the enamel and underlying dentin.

Sucrose: a disaccharide composed of glucose and fructose.

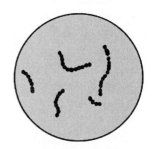

STREPTOCOCCUS MUTANS

One of the primary bacterial causes of caries is acidogenic *Strepococcus mutans*. This Gram-positive coccus has a high affinity for the smooth surfaces, pits, and fissures of a tooth. Its enzymes react with the glucose and fructose in sucrose and convert them to long-chain carbohydrates called glucans and levans. These materials give *S. mutans* its special adherence qualities. The bacilli then ferment dietary carbohydrates to lactic acid, with smaller amounts of acetic acid, formic acid, and butyric acid. The acids dissolve hydroxyapatite, after which protein-digesting enzymes break down any remaining organic materials. Other streptococcal species involved in caries include *S. sanguis*, *S. mitis*, and *S. salivarius*.

The prevention of dental caries relates to three principal areas: protecting the tooth, modifying the diet, and combatting cariogenic bacteria (Figure 10.12). Tooth protection may be accomplished by ingestion and topical application of **fluorides**. These compounds displace hydroxyl ions in hydroxyapatite, thus reducing the solubility of the enamel. Protection may also be rendered by applying polymers to cover pits and fissures in the teeth, thereby preventing bacterial adhesion. Diet modification requires minimizing sucrose in foods.

Fluorides: derivative compounds of the chemical halogen fluorine.

F I G U R E 10.12

Dental Caries

(a) Overlapping circles depicting the interrelationships of the three factors which lead to caries activity. (b) *Streptococcus mutans*, a major cause of dental caries, as visualized by the scanning electron microscope ($\times$ 10,000).

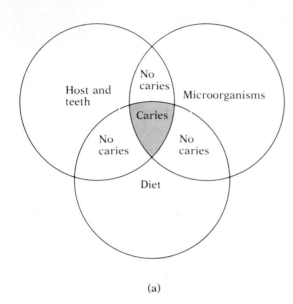

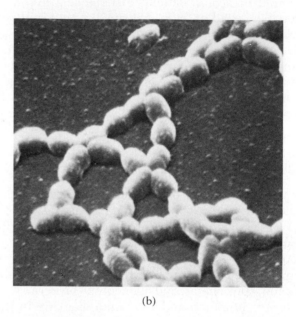

(a)

(b)

MIXED BACTERIA
lep″to-trik′e-ah
i″ken-el′ah

Some success has been observed by substituting **xylitol**, an alcoholic derivative of xylose, for the sugar in candy. Efforts to eliminate cariogenic bacteria focus on preventing the synthesis of dextrans and levans by streptococci, as well as stimulating antibody production against the bacteria. Some dental researchers foresee an eventual vaccine against tooth decay.

Caries is not the only form of dental disease. The teeth are surrounded by tissues that provide the support essential to tooth function. These tissues, called the periodontal tissues, may be the site of a periodontal disease called **acute necrotizing ulcerative gingivitis (ANUG)**. Microbiologists believe that ANUG is due to the invasion of the tissues by several bacteria, among which are *Leptotrichia buccalis*, a long, thin Gram-negative rod; *Treponema vincentii*, a spirochete; and species of *Eikenella*, another Gram-negative rod. Indeed, in 1990 researchers at Mount Sinai Hospital in New York reported that *E. corrodens* can cause severe cellulitis and arthritis of the knee after entering the blood through trauma of the gingival tissues.

ANUG is characterized by punched-out ulcers that appear first along the gingival margin and interdental papillae, and then spread to the soft palate and tonsil areas. Infections in the latter area are sometimes called **Vincent's angina** after Jean Hyacinthe Vincent, who described the spirochete in 1892. A foul odor and bad taste come from gases produced by anaerobic bacteria. As the periodontal tissues decay, the teeth may become loosened and lost (Figure 10.13). Diagnosis depends on the observation of rods and spirochetes from the ulcers. The disease is sometimes called **fusospirochetal disease** because the rods have a long thin fusiform shape and are mixed with spirochetes. Antibiotic washes may be used in therapy, and some physicians suggest painting the area with a traditional remedy of gentian violet.

FIGURE 10.13

A Case of ANUG

The photograph displays infection in the interdental papillae. In cases such as these, the teeth may become loosened and periodontal infection may ensue.

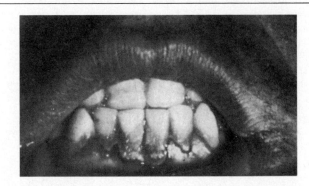

ANUG has long been associated with conditions such as malnutrition, viral infection, excessive smoking, poor oral hygiene, or mental stress. The disease was common among soldiers in World War I and was known at that time as **trench mouth**. It is not considered a transmissible disease, but one that develops from bacteria already in the mouth.

Nosocomial Diseases

Nosocomial diseases are those acquired during hospitalization. The CDC has estimated that up to 10 percent of all hospital patients may develop some nosocomial disease during their stay, with surgical patients particularly susceptible. Certain types of operations such as amputations and intestinal surgery are accompanied by an infection rate approaching 30 percent. Over a million patients may be involved annually and an estimated $6 billion in hospital costs is spent to treat the nosocomial diseases.

A hospital's patient population is a high-density community composed of unusually susceptible individuals. A variety of pathogenic bacteria abound (Figure 10.14), and new ones are continually being introduced as new patients arrive. In addition, the extensive use of antimicrobial agents contributes to the development of resistant strains of microorganisms, and staff members become carriers of these strains. Many of the patients have already experienced some interference with their normal defense mechanisms, such as a breach of the skin barrier in surgery, radiation therapy for cancer, medication to suppress the immune system, or indwelling apparatus such as catheters and intravenous tubes. When all these factors meet, nosocomial diseases break out.

The organisms that cause nosocomial diseases are generally **opportunistic**, that is, they do not cause disease in normal humans but they are dangerous in compromised individuals. Among the most common opportunistic bacteria are the Gram-negative rods such as *Escherichia coli*, *Serratia marcescens*, *Enterobacter aerogenes*, *Enterobacter cloacae*, *Klebsiella pneumoniae*, and *Proteus* species. Often these bacteria are the cause of urinary tract infections. *Staphylococcus aureus* is another important cause of nosocomial disease, as are various types of streptococci. *Pseudomonas aeruginosa*, a Gram-negative rod, is a particular problem in burn victims (Figure 10.15). The organism grows rapidly and produces a sickly sweet odor as well as a green fluorescent pigment that causes the tissue to glow under ultraviolet light.

Opportunistic: **referring to microorganisms that invade the tissues of a compromised individual.**
sĕ-ra′she-ah mar-ses′ens

a″er-jin-o′sa

FIGURE 10.14

A Nosocomial Problem

A scanning electron micrograph of *Serratia marcescens* found in a hospital on the inner surface of a plastic bottle containing the disinfectant chlorhexidine. This photograph was part of a study to determine the source of *S. marcescens* infection in hospitalized patients. It illustrates how bacteria can survive in certain disinfectants and spread to patients when the disinfectant is applied. Bar = 5 μm.

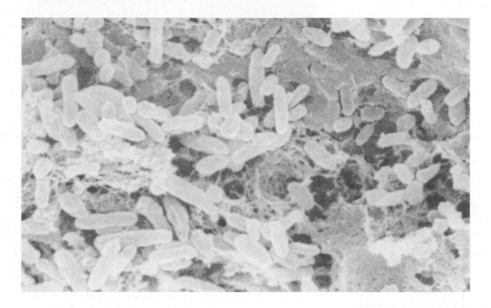

To deal with nosocomial diseases, hospitals designate a specialist, usually a **nurse epidemiologist**, whose primary responsibility is to locate problem areas and report them to an infection-control committee. This practice was established in 1976 on a recommendation by the Joint Committee on Accreditation of Hospitals (JCAH). The local committee consists of nurses, doctors, dieticians, engineers, and housekeeping and laboratory personnel who monitor equipment and procedures to interrupt the disease cycle. All authorities agree that frequent and conscientious hand-washing is the all-important first step.

NOTE TO THE STUDENT

We have surveyed numerous diseases in this chapter, but one group stands out and merits additional comment. I am referring to the sexually transmitted diseases.

Prior to World War II, doctors had to contend with five classical diseases transmitted by sexual contact. These were syphilis, gonorrhea, chancroid, lymphogranuloma venereum, and rarely, granuloma inguinale. With the widespread use of antibiotics after the war, the annual incidence of syphilis and gonorrhea declined and the other three diseases virtually disappeared.

Then came the 1960s and the sexual revolution. It was a time of affluence and defiance of traditional values. Birth control pills, vasectomies,

FIGURE 10.15

Pseudomonas aeruginosa

A scanning electron micrograph of *Pseudomonas aeruginosa*, an important cause of nosocomial disease especially in patients who have suffered burns. Bar = 1 μm.

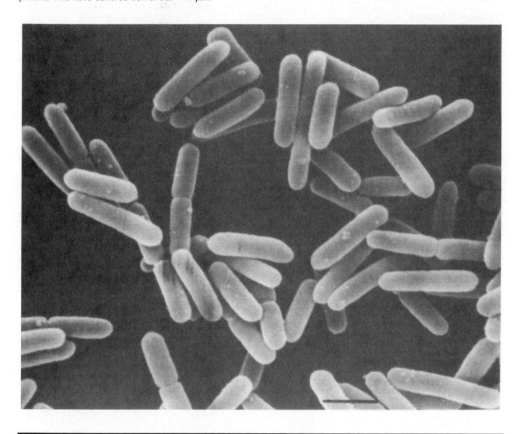

and contraceptive devices offered sexual liberation to go along with social and economic freedoms. Expectedly, the incidence of sexually transmitted diseases soared and today, the United States is in the grip of an STD epidemic of unprecedented proportions. The statistics are awesome: public health officials estimate that 1 in 4 Americans between the ages of 15 and 55 will acquire an STD at some time in his or her life; 10 million Americans visit clinics and doctors' offices annually for treatment; and over $2 billion is spent each year in health-care costs for STDs. Moreover, the list of sexually transmitted diseases continues to expand, and at least 25 diseases are now involved, including the ones noted in this chapter as well as hepatitis B, shigellosis, genital herpes, candidiasis, AIDS, and numerous diseases discussed in other chapters.

Is the epidemic likely to end? Some sociologists contend that fear of an STD may be a motivating factor in limiting promiscuous sex, but a more realistic view is that there has been a shift in attitude toward such things as premarital sex and sex in books and films. Condom use can do much to interrupt transmission of an STD, but while values continue to be sorted out, it appears that the incidence of STDs will remain high, and that an end to the epidemic is still beyond expectation.

TABLE 10.3
A Summary of Miscellaneous Bacterial Diseases

DISEASE	CAUSATIVE AGENT	DESCRIPTION OF AGENT	ORGANS AFFECTED	CHARACTERISTIC SIGNS
ENDOGENOUS DISEASES				
Actinomycosis	*Actinomyces israelii*	Gram-positive anaerobic funguslike rod	Gum tissues Chest wall Abdominal organs	Draining sinus
Nocardiosis	*Nocardia asteroides*	Acid-fast funguslike rod	Lungs	Abscesses Rusty sputum Madura foot
Bacteroides infection	*Bacteroides fragilis*	Gram-negative anaerobic rod	Intestine Blood	Blood clots
Pseudomembranous colitis	*Clostridium difficile*	Gram-positive sporeforming rod	Intestine	Intestinal lesions Diarrhea
ANIMAL BITE DISEASES				
Pasteurellosis	*Pasteurella multocida*	Gram-negative rod	Skin	Abscess at site of bite Arthritis
Cat scratch fever	*Rochalimaea henselae* (?)	Rickettsia	Skin	Lesion at site of bite Swollen lymph glands
Rat bite fever	*Streptobacillus moniliformis*	Gram-negative streptobacillus	Skin	Lesion at site of bite Rash, fever
	Spirillum minor	Spirochete	Skin	Lesion at site of bite Rash, fever

Summary

Two general types of bacterial diseases are discussed in Chapter 10: contact diseases and miscellaneous diseases. The contact diseases are spread among individuals by sexual contact and by skin contact. Diseases transmitted by sexual contact are among the most numerous in society today. They range from the life-threatening disease syphilis to the milder diseases gonorrhea and chlamydia. Untreated, syphilis occurs in three stages, the last stage involving mental and cardiovascular deterioration. Although gonorrhea and chlamydia result in less damage to the body, an important effect of both is the possibility of sterility owing to blockage of the reproductive tubules. Other sexually transmitted diseases such as ureaplasmal urethritis, chancroid, lymphogranuloma, vaginitis, and mycoplasmal urethritis are also relatively mild diseases.

Contact diseases discussed in this chapter include leprosy, staphylococcal skin diseases, trachoma, conjunctivitis, and yaws. Staphylococcal infections can be complicated by blood involvement, but the other diseases rarely go further than the skin, and they can be controlled by drug therapy.

The miscellaneous bacterial diseases fall into identifiable groups that include endogenous diseases and animal bite diseases. Endogenous diseases result from bacteria that normally reside in the body but cause infection when body defenses diminish such as in an immunocompromised host. Animal bite diseases include pasteurellosis from dogs and cats, cat scratch fever, and rat bite fever. Oral diseases and nosocomial diseases are also discussed briefly to provide an overview of their importance.

TOXIN INVOLVED	TREATMENT ADMINISTERED	IMMUNIZATION AVAILABLE	COMMENT
Not established	Various antibiotics	None	Lumpy jaw possible Associated with IUD use Normal lung inhabitant
Not established	Various antibiotics	None	Associated with AIDS Endogenous disease Soil transmission via wound
Not established	Various antibiotics	None	May lead to gangrene Normal intestinal inhab-itant
Possible	Various antibiotics	None	Accompanies excessive antibiotic use
Not established	Various antibiotics	None	Common in cats and dogs
Not established	Various antibiotics	None	Treatment not always necessary
Not established	Various antibiotics	None	Typical fever, arthritis, and rash
Not established	Various antibiotics	None	Common in Japan

Questions for Thought and Discussion

1. Among the bacteria that cause periodontal disease, few are apparently more common than the Gram-negative rod *Porphyromonas gingivalis*. In 1992, researchers from Denmark and the United States announced that they had successfully immunized laboratory rats against *P. gingivalis* by injecting them with proteins from the bacterium's pili. In effect, they had developed a vaccine against periodontal disease. What do you suppose the prospects for using such a vaccine in humans will be?

2. One of the major problems of the current worldwide epidemic of AIDS is the possibility of transferring the human immunodeficiency virus (HIV) among persons who have a sexually transmitted disease. Which diseases in this chapter would make a person particularly susceptible to penetration of HIV into the bloodstream? What explanation can you give for each example?

3. Studies indicate that most cases of *Staphylococcus*-related impetigo occur during the summer months. Why do you think this is the case?

4. Suppose a high incidence of leprosy existed in a particular part of the world. Why is it conceivable that there might be a correspondingly low level of tuberculosis?

5. It has been suggested that women should avoid vaginal douching because the practice can encourage the development of pelvic inflammatory disease (PID) if there is an underlying STD. How can you explain the connection to an inquisitive friend?

6. One day in the late 1980s, there reported to a New York hospital a Senegalese patient with an upper lip swollen to about three times its normal size. Probing with a safety pin at facial points where major nerve endings terminate showed that the area to the left of the nose and above the lip was without feeling. When a biopsy of the tissue was examined it revealed round reservoirs of immune system cells called granulomas within the nerves. On bacteriological analysis, acid-fast rods were observed in the tissue. What disease do all these data suggest?

7. On January 9, 1984, an undergraduate psychology student was bitten by a laboratory rat on the left index finger. Within 12 hours her finger was swollen and throbbing. Soon thereafter she was hospitalized with swollen lymph nodes, a skin rash, fever, and exquisite sensitivity of the finger. Gram-negative branching rods were found in the tissue. What disease was she suffering from?

8. Researchers at the University of Maryland have suggested a "dip and brush" method of controlling oral diseases. The idea is to dip a toothbrush into an antiseptic several times while brushing to reduce the level of plaque bacteria. Do you think this method will reduce dental problems?

9. In some African villages, blindness from trachoma is so common that ropes are strung to help people locate the village well, and bamboo poles are laid to guide farmers planting in the fields. What measures can be taken to relieve such widespread epidemics as this?

10. Certain microscopes have the added feature of a small hollow tube that fits over the eyepiece or eyepieces. Viewers are encouraged to rest their eyes against the tube and thereby block out light from the room. Why is this feature hazardous to health?

11. A quote from the Book of Leviticus reads: "And the leper . . . shall be defiled; he is unclean; he shall dwell alone; without the camp shall his habitation be." Short of editing the Bible, how can this attitude toward leprosy be changed?

12. After a young man suffers an abrasion on the right arm, his affectionate cat licks the wound. Several days later a pustular lesion appears at the site and a low-grade fever develops. He also experiences "swollen glands" on the right side of his neck. What disease has he acquired?

13. A woman suffers two miscarriages, each after the fourth month of pregnancy. She then gives birth to a child, but impaired hearing and vision become apparent as it develops. Also its teeth are shaped like pegs and have notches. What medical problem existed in the mother?

14. In the early 1990s, the CDC was reporting approximately 100,000 instances of ectopic pregnancy in the United States annually, a fivefold increase over the rate in 1970. What could account for this significant increase?

15. One of the diseases discussed in this chapter is probably the most widespread disease in all the world. Students go to special schools and earn a special degree to learn how to deal with it. Then they spend years developing their skill in treating it. What is the disease?

Review

A major topic of this chapter has been the sexually transmitted diseases. To test your recall of these diseases, fill in the following crossword puzzle. The answers to the puzzle are in the appendix.

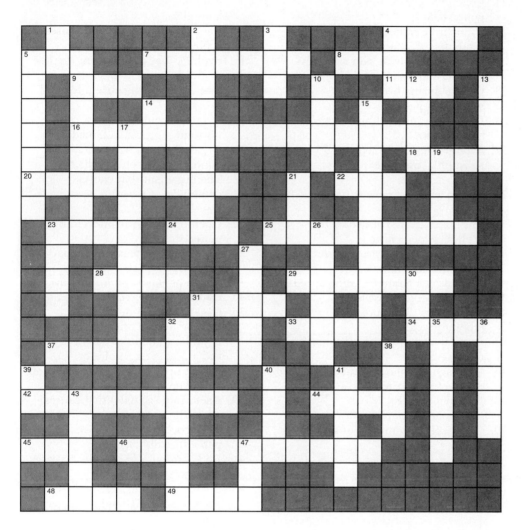

ACROSS

4. The _____ of transmission for STDs is contact.
5. Health organization that charts epidemics of STDs.
7. Occurs during primary syphilis.
8. Number of exposures necessary to establish infection of STD.
9. Living tissue for chlamydia cultivation.
11. Ureaplasmal urethritis is a type of _____ .
16. Accompanied by swelling of inguinal lymph nodes.
18. Gender in which gonorrhea symptoms are more apparent.

DOWN

1. Older name (initials) for sexually transmitted diseases.
2. Tubes invaded by gonococci during time of infection.
3. Gonorrhea is rarely contracted by exposure to a _____ surface.
4. Often display a discharge when gonorrhea present.
5. Sexual _____ is generally required for transmission of syphilis spirochetes.
6. Absent in *Mycoplasma* species.
10. Odor of discharge from vaginitis patient.

ACROSS (*continued*)

20. Caused by *Haemophilus ducreyi*.
22. STD (abbr) caused by *Chlamydia trachomatis*.
23. Colloquial expression for gonorrhea.
24. Number of exposures necessary to establish infection of STD.
25. Syphilis stage with skin rash, loss of hair, and influenzalike illness.
28. Occurs during urination in gonorrhea patient.
29. Urine tube infected by *Neisseria gonorrhoeae*.
31. Syphilis stages are separated by substantial amounts of _____.
33. Container for sample transported to lab.
34. The _____ of choice for syphilis is penicillin.
37. Gonorrhealike disease transmitted by sexual contact.
42. Caused by a Gram (−) diplococcus discovered by Neisser.
44. Digested by T-mycoplasma in laboratory culture.
45. Possible complication of gonorrhea (abbr)
46. Blockage of Fallopian tube as a result of gonorrhea.
48. Continent where lymphogranuloma is prevalent.
49. Can display symptoms of syphilis.

DOWN (*continued*)

12. Stain technique to identify gonococci.
13. Possibly three to _____ million chlamydia cases in U.S. annually.
14. Bacterium (initials) that causes syphilis.
15. Syphilis that passes from mother to child.
17. One species can cause a form of urethritis.
19. Chlamydiae do not grow on nutrient _____ media.
21. Organ that can be infected by gonococci and chlamydiae.
23. The organism that causes chlamydia is a very tiny _____.
26. Opening to uterus infected by gonococci.
27. Soft, granular lesion in tertiary syphilis.
30. Shape of *Haemophilus ducreyi*.
32. Caused by *Treponema pallidum*.
35. Organ infected by gonococci leading to proctitis.
36. Syphilis sometimes called the _____ imitator.
38. Object used to obtain sample from gonorrhea patient.
39. Bacterium (initials) that causes gonorrhea.
40. Gonococci that are resistant to penicillin (abbr).
41. Body region affected during cases of gonorrhea.
43. Swollen _____ nodes accompany LGU and are a major symptom.
47. STDs can also be transmitted by _____-sexual methods.

OTHER MICROORGANISMS

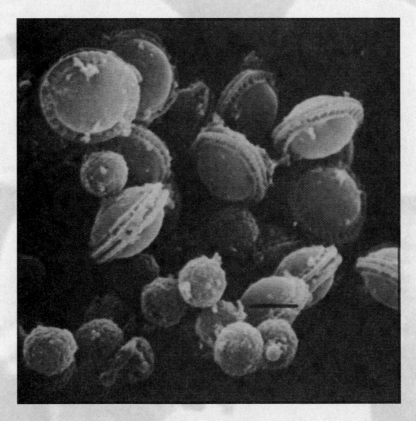

Bacteria are but one of several groups of microorganisms interwoven with the lives of humans. Other prominent groups are the viruses, fungi, protozoa, and multicellular parasites. Knowledge of these microorganisms developed slowly during the early 1900s partly because they were generally more difficult to isolate and cultivate than bacteria. (Indeed, viruses remained unphotographed until the late 1930s.) Another reason is that methods for research into bacteria were more advanced than for other microorganisms, and investigators often chose to build on established knowledge rather than pursue uncharted courses of study. Moreover, the urgency to learn about other microorganisms was not great because other microorganisms did not appear to cause such great epidemics. ■ Much of this changed in the second half of the 1900s. Many bacterial diseases came under control with the advent of vaccines and antibiotics, and the increased funding for biological research allowed attention to shift to other microorganisms. The viruses finally were identified and cultivated, and microbiologists laid the foundations for their study. Fungi gained prominence as tools in biological research, and scientists soon recognized their significance in ecology and industrial product manufacturing. As remote parts of the world opened to trade and travel, public health microbiologists realized the global impact of protozoal disease. Moreover, as concern for the health of the world's peoples increased, observers expressed revulsion at the thought that hundreds of millions of human beings were infected by multicellular parasites. ■ In Part IV we shall study four groups of microorganisms over the course of six chapters. Chapter 11 is devoted to a study of the viruses, and Chapters 12 and 13 outline the multiple diseases caused by these infectious particles. In Chapter 14, the discussion moves to fungi, while in Chapter 15, the area of interest is the protozoa. Throughout these chapters, the emphasis is on human disease. You will note some familiar terms such as malaria, hepatitis, and chickenpox, as well as some less familiar terms such as toxoplasmosis, giardiasis, and dengue fever. The spectrum of diseases continues to unfold as scientists develop new methods for the detection, isolation, and cultivation of microorganisms, even as this text is written.

MICROBIOLOGY PATHWAYS

VIROLOGY

When I was in college, I was part of a group of twelve biology majors. Each of us had a particular area of "pseudoexpertise." I remember John was going to be a surgeon, Jim was interested in marine biology, Walt was a budding dentist, and I was the local virologist. I was fascinated with viruses, the ultramicroscopic bits of matter, and at one time I did a paper summarizing arguments in favor of and against viruses being alive. (At the time, neither side was persuasive, and even my professor gracefully declined to place himself in either camp.)

I never quite made it to being a virologist, but if your fascination with these infectious particles is as keen as mine was (and continues to be), then you might like to consider a career in virology. Virologists investigate dread diseases such as AIDS, polio, and rabies. They also study the beautiful variegations found in certain types of tulips. Some virologists concern themselves with many types of cancer, and others study the chemical interactions of viruses with various tissue culture systems and animal models. Virologists are working to replace agricultural pesticides with viruses that will destroy mosquitoes and other pests. Some virologists are inserting viral genes into plants and are hoping the plants will produce viral proteins to lend resistance to disease. One particularly innovative group is trying to insert genes from hepatitis B viruses into bananas. They hope that one day we can vaccinate ourselves against hepatitis B by having a banana for lunch.

If you wish to consider the study of viruses, there is one very important proviso you should know—virologists are chemists. I realize that to a biologist, "chemistry" ranks with "root canal," but the fact of the matter is that to do virology, you must be able to do chemistry. You must have general, organic, and, if possible, physical chemistry in your undergraduate program. Then, prepare for lots more chemistry in your graduate program as you pursue the dream I was never quite able to attain. (Actually, I must admit it all worked out very well, because I am extremely happy doing what I do.)

CHAPTER 11

THE VIRUSES

Over the span of several generations the word *virus* has had two meanings. A century ago at the time of Pasteur and Koch, "virus" referred to a vague poison associated with disease and death. Physicians would suggest that the air was filled with virus, or that a virus was in the blood. Louis Pasteur and his contemporaries wrote freely about the "cholera virus" and the "rabies virus"; and in the vernacular of Koch's times, a bacterium was the virus of tuberculosis (which makes reading the reports of the 1890s very confusing).

The modern notion of a virus is dramatically different. In today's world viruses are recognized as particles of nucleic acid and protein, often with a covering membrane. They replicate in living cells and cause a number of important diseases such as genital herpes, influenza, hepatitis, and infectious mononucleosis. Viruses vary considerably in size, shape, and chemical composition, and the methods used in their cultivation and detection are completely different than for other microorganisms. The term "rabies virus" is still a common expression in microbiology, but with a vastly different meaning than in Pasteur's time.

In this chapter we shall study the properties of viruses and focus on their unique mechanism for replication. We shall also see how they are classified, how they are inhibited outside the body, and how the body defends against them during time of disease. You will note a simplicity in viruses that has led many microbiologists to question whether they are living organisms or merely fragments of genetic material leading an independent existence. Most of the information in this chapter has only been known since the 1950s, and the current era might be called the Golden Age of Virology. Our survey will begin with a review of some of the events that led to this period.

Foundations of Virology

No one person discovered the viruses. Instead, an understanding of the viruses evolved in the late 1800s with the general understanding of the germ theory of disease. Different diseases had recognizable patterns and, although a bacterium, protozoan, fungus, or other agent could be isolated for most diseases, some diseases had no identifiable agent. Many of these diseases would become the viral diseases.

FIGURE 11.1
Foundations of Virology

(a) Dimitri A. Iwanowski, the Russian pathologist who demonstrated that the agent of tobacco mosaic disease was a "filterable virus," that is, a toxic agent able to pass through a bacterial filter. (b) A tobacco leaf showing the characteristic symptoms of tobacco mosaic disease. Note the patchwork discolorations of the leaf representing the mosaic appearance.

(a)

(b)

Tobacco mosaic disease: a viral disease that causes tobacco leaves to shrivel and assume a mosaic appearance.

bi'jer-ink

A mechanical device, the **filter**, led to the early interest in viruses. In 1884, Charles Chamberland, an associate of Pasteur, devised a porcelain filter to trap the smallest known bacteria, and in 1892, the Russian pathologist Dimitri A. Iwanowski used Chamberland's filter in his studies on tobacco mosaic disease. In **tobacco mosaic disease**, the tobacco leaves shrivel and assume a mosaic (patchwork) appearance before dying (Figure 11.1). Iwanowski filtered the crushed leaves of a diseased plant and found that the clear sap dripping from the filter (rather than the crushed leaves on the filter) contained the infectious agent. Unable to see any microorganism, Iwanowski reported that a "**filterable virus**" was the agent of disease. This merely meant that the unseen agent, whatever its nature, would pass through a bacterial filter. At the time it was a remarkable discovery, because scientists could scarcely comprehend anything smaller than bacteria as disease agents.

Six years later, in 1898, Martinus Beijerinck of Delft (Van Leeuwenhoek's city) repeated Iwanowski's work and added considerably to the knowledge of viruses. Beijerinck found that the tobacco mosaic virus remained active in dried

WHY NOT TRY?

When bacteriophages were identified in 1915, some scientists immediately thought they might be used to cure the dreaded bacterial diseases. After all, these ultramicroscopic viruses could destroy bacteria in test tubes, so why not try them in the human body? Unfortunately, the bacteriophages turned out to be highly specific, attacking only certain strains of bacteria. Indeed, no one knew much about them or had the means to work with them effectively.

Fast forward to the modern era of electron microscopes, sophisticated laboratory analyses, and genetic engineering. Technology had increased substantially, and the Polish microbiologist Stefan Slopek decided to try again. He identified a bacterium in the blood of an ill patient, then searched out a bacteriophage specific for that bacterium. He injected the phages to the patient and held his breath and waited. Over a period of days the infection gradually resolved.

Slopek tried again—this time with 137 patients. Each patient benefited from the treatment, and several were completely cured.

Could bacteriophages be the answer to the growing problem of antibiotic resistance in bacteria? To be sure, some novel approaches to treating infectious disease must be found, and bacteriophages appear to be a logical candidate.

---

leaves and soil. By testing the activity of many dilutions of viral fluid, he was able to demonstrate the fluid's potency and he showed that it was inactivated by boiling. Beijerinck concluded that the disease agent must be a "contagious living fluid" (*contagium vivum fluidum*) rather than a solid object.

Before the excitement of these discoveries subsided, Paul Frosch and Friederich Löeffler reported in 1898 that foot-and-mouth disease was caused by a filterable virus. This finding implied that an invisible agent could be transmitted among animals as well as plants. Three years later, in 1901, Walter Reed and his group in Cuba wrote that yellow fever was also due to a filterable virus, and with this report they established human involvement with viruses. In all these studies, the criterion of filterability remained significant, but scientists soon found that some viruses, such as those of rabies and cowpox, did not easily pass through filters. Hence the agents came to be described simply as "viruses," using the word in a more limited and specific sense than previously.

A unique virus was discovered in 1915 by the English bacteriologist Frederick Twort, and independently in 1917 by the French scientist Felix d'Herrelle. The virus of Twort and d'Herrelle came to be called the **bacteriophage** ("bacteria-eater") for its ability to destroy bacteria (MicroFocus: 11.1). This name stems from experiments in which a drop of virus was placed in a broth culture of bacteria; the bacteria disintegrated within minutes. D'Herrelle believed that bacteriophages (or simply, phages) could be used to kill bacteria in the body, but investigators later found that phages were easily eliminated from the body and were highly specific for the bacterial strain they attacked (MicroFocus: 11.2). Bacteriophages have since become important tools in transduction research (Chapter 6) and in bacterial identification (Chapters 7 and 8).

In any listing of virus-caused diseases, **cancer** remains a puzzle. During the late 1800s, when the germ theory was established, some scientists believed that cancer might prove to be a bacterial disease. But no bacterium could be located. In 1908, a relationship between viruses and cancer was suggested by Wilhelm Ellerman and Olaf Bang. Their investigations showed that leukemia in chickens could be induced by a preparation of "virus." Another researcher, Francis Peyton Rous, demonstrated in 1911 that cancer of the connective tissues could be traced to a virus. The relationship between viruses and cancer is discussed later in this chapter.

lef′ler

Yellow fever:
a mosquito-borne viral disease of the liver and blood.

de-rel′
bak-te′re-o-faj″

Leukemia:
a type of cancer characterized by uncontrolled multiplication of white blood cells.

rows

The Transition Period

Resolving power:
the ability of a lens system to transmit light without variation and permit nearby objects to be clearly distinguished.

By the 1930s it was generally assumed that viruses were a type of invisible microorganism below the resolving power of available microscopes. As the years passed, a long list of viral diseases had developed, and work with plant viruses had been productive because of the ease of viral cultivation. For example, virologists learned that tobacco mosaic viruses were composed exclusively of nucleic acid and protein.

Work with animal viruses, however, was less productive because cultivation in live animals was a difficult procedure. A breakthrough finally came in 1931 when a living analog of Koch's nutrient agar was discovered for viruses. In that year Alice M. Woodruff and Ernest W. Goodpasture of Vanderbilt University published a paper describing the use of **fertilized chicken eggs** as a nutrient for cultivating some viruses (Figure 11.2). The shell of the egg was a type of natural Petri dish for the nutrient medium, and viruses multiplied within the chick embryo tissues.

Antibody:
a protein produced by cells of the immune system in response to a specific chemical substance.

The assumption had been that viruses, though incredibly small, were living things, but a discovery by Wendel M. Stanley of Rockefeller Institute raised doubts. In the early 1930s, three protein enzymes of the human body were crystallized by biochemists, and in 1935, Stanley made the startling announcement that tobacco mosaic viruses could also be crystallized. Crystals are composed of identical molecules, and some scientists now suggested that viruses were as lifeless as any crystalline chemical molecules. Virologists pointed out, however, that the viruses could replicate, cause fevers, and elicit antibody responses, properties not associated with chemical molecules. Stanley's unprecedented discovery was rewarded with the 1946 Nobel Prize in Chemistry. It opened a debate on the living or nonliving nature of viruses that is still unsettled, and it cut across preconceived ideas that only living things could cause disease.

Researchers in the 1940s brought virology into the modern era with two significant developments. One was the electron microscope, which allowed viruses to be visualized and hence, their functions to be understood. The other was the test tube tissue cultivation of viruses, which made vaccines possible for controlling certain viral diseases.

Electron microscope:
a magnifying device that uses an electron beam to increase the resolving power.

Just as one technological device, the filter, led to the early notion of viruses, another technological device, the **electron microscope**, revealed the nature of the viruses. The innovative feature of the electron microscope is the incredibly short wavelength of the electron beam, which dramatically increases the resolving power of the instrument. To understand the principle involved, consider the difference between running and walking on a beach (Figure 11.3). With a running stride, a person misses most grains of sand, just as visible light with its long wavelength passes over the viruses without striking them. When the

FIGURE 11.2
Viral Cultivation

Inoculation of fertilized eggs by a technician in the virology laboratory. Techniques such as these are standard practice in virology research.

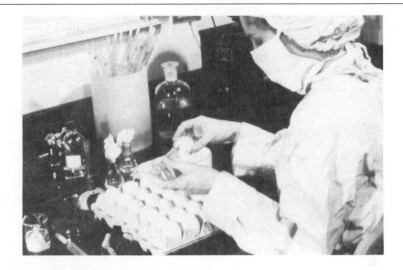

FIGURE 11.3
Wavelength and the Electron Microscope

How the wavelength of the electron beam contributes to the visualization of viruses. (a) When a light microscope is used, the relatively long wavelength of the visible light causes most of the energy to miss the viruses. (b) A person running on a beach misses most grains of sand in similar fashion. (c) Since the electron beam has a short wavelength, the energy strikes many more viruses, bouncing away and creating an image. (d) Similarly, two strollers on the beach step on more grains of sand.

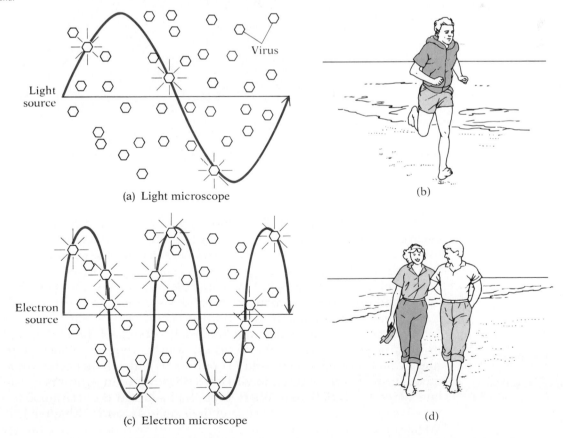

FIGURE 11.4

Cells and Viruses

A transmission electron micrograph of tissue cells and associated viruses. In this view at the margin of the cell, the viruses are the dark, round objects in the space outside the cell. The envelope of the virus is seen as a ring at the viral surface, and the nucleocapsid of the virus is the darker center. These viruses are herpesviruses (× 48,000).

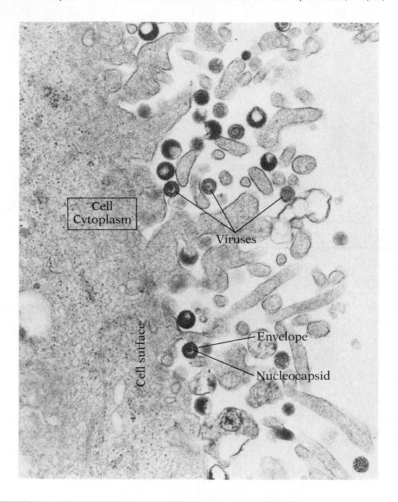

---

stride is shortened, as in walking, more grains of sand are contacted just as the electron beam with its short wavelength contacts more viruses. By 1941, the tobacco mosaic virus and other viruses were visualized (Figure 11.4).

The second discovery of the 1940s was sparked by the national epidemic of poliomyelitis (or simply, polio). Polio is a disease that attacks people indiscriminately and often leads to permanent paralysis. Attempts at vaccine production were stymied by the inability to cultivate polio viruses outside the body, but John Enders, Thomas Weller, and Frederick Robbins of the Children's Hospital in Boston solved that problem. Meticulously, they developed a **test tube medium** of nutrients, salts, and pH buffers in which living cells would remain alive. In the living cells, polio viruses replicated to huge numbers, and by the late 1950s Jonas Salk and Albert Sabin had adapted the technique to produce massive quantities of virus for use in their polio vaccines (Chapter 13). Enders, Weller, and Robbins did not invent tissue cultivation of viruses, but their work

Polio:
a viral disease that may involve the central nervous system and cause paralysis.

pH buffers:
chemical substances that maintain a constant pH in a solution.

showed it had a practical value for vaccine production. In 1954 they shared the Nobel Prize in Physiology or Medicine for their work.

The electron microscope and the test tube cultivation of viruses paved the way for the discoveries in virology that have come forth in our era. We shall explore these discoveries in the following pages.

TO THIS POINT
■

We have outlined some of the major events in the development of virology beginning with the early concept of viruses in the late 1800s and continuing through to the 1940s. Iwanowski was one of the first to note that something smaller than any known bacterium could cause disease. His work was further developed by Beijerinck, and expanded by the discoveries of Löeffler, Frosch, and Reed. In time, viruses were found to affect plants, animals, humans, and bacteria. Viruses were also related to cancer.

A surprising announcement of the 1930s was that viruses could be crystallized. This gave an insight into their simplicity and suggested that they might be nothing more than large chemical molecules. When the electron microscope was used to study viruses in the 1940s, scientists observed that viruses had structural details unlike anything seen in chemical compounds. An understanding of viral functions soon developed, as we shall see presently. Also, great strides were made toward the production of viral vaccines, the advances sparked by the tissue cultivation of polio viruses by Enders, Weller, and Robbins. Vaccines opened the way to prevention of certain viral diseases.

In the next section we shall discuss the characteristics of viruses, beginning with the structure of viruses and continuing with their method of replication. You will note a level of structural simplicity matched by few other disease agents and a replication process not encountered elsewhere in the biological world. These attributes lend uniqueness to viruses.

The Structure and Replication of Viruses

Viruses are among the smallest agents able to cause disease in living things. They range in size from the large 250 nanometers (nm) of poxviruses to the 20 nm of parvoviruses (Figure 11.5). At the upper end of the spectrum, the viruses approximate the size of the smallest bacterial cells, such as the chlamydiae and mycoplasmas; at the lower end, they have about the same diameter as a DNA molecule.

nanometer:
a billionth of a meter.

kla-mid'e-e

Viruses may appear in several shapes. Certain viruses, such as rabies and tobacco mosaic viruses, exist in the form of a **helix** and are said to have helical symmetry. The helix is a tightly wound coil resembling a corkscrew or spring (Figure 11.6). Other viruses, such as herpes simplex and polio viruses, have the shape of an **icosahedron** and hence, icosahedral symmetry. The icosahedron is a polyhedron with 20 triangular faces and 12 corners (MicroFocus: 11.3). Certain viruses have a combination of helical and icosahedral symmetry, a construction described as **complex**. Some bacteriophages, for example, have complex symmetry, with an icosahedral head and a collar and tail assembly in the shape of a helical sheath. Poxviruses, by contrast, are brick-shaped, with submicroscopic filaments or tubes occurring in a swirling pattern at the periphery of the virus.

Helix:
a tightly wound coil.

i-kos"ah-hēd'ron

FIGURE 11.5
Size Relationships Among Microorganisms

(a) The sizes of various viruses related to a eukaryotic cell, a cell nucleus, and the bacterium *E. coli*. The smallpox viruses approximate the chlamydiae and mycoplasmas in size. (b) A scanning electron micrograph of bacterial rods isolated from a water sample on a sieve filter. The surface projections on the rods are bacteriophages. Note their size relative to the host bacterium.

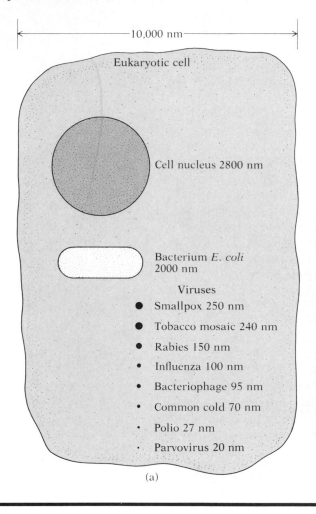

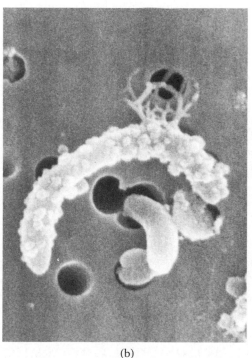

(a)   (b)

---

je′nōm

Capsomeres:
protein subunits in the
capsid of a virus.

All viruses consist of two basic components: a core of nucleic acid called the **genome**, and a surrounding coat of protein known as the **capsid**. The genome contains either DNA or RNA, but not both; and the nucleic acid occurs in double-stranded or single-stranded form. Usually the nucleic acid is unbroken, but in some instances (as in influenza viruses) it exists in segments. The genome is usually folded and condensed in icosahedral viruses, and coiled in helical fashion in helical viruses (Figure 11.7).

The capsid protects the genome. It also gives shape to the virus and is responsible for the helical, icosahedral, or complex symmetry. Generally, the capsid is subdivided into individual protein subunits called **capsomeres**, whose organization yields the symmetry. The number of capsomeres is characteristic for a particular virus. For example, 162 capsomeres make up the capsid in herpesviruses, and the 252 capsomeres compose the capsid in adenoviruses, which cause some common colds.

The capsid provides a protective covering for the genome because the construction of its amino acids resists temperature, pH, and other environmental

FIGURE 11.6
Capsid Symmetries in Viruses

Representatives of the three major types of capsid symmetries in viruses. (a) The helix as illustrated by the capsids of tobacco mosaic and rabies viruses. The helix resembles a tightly wound coil. (b) The icosahedron. This is a polyhedron with 20 triangular faces and 12 corners. (c) Complex viruses showing numerous constructions. The bacteriophage has an icosahedral head with a collar and tail assembly in the form of a helix.

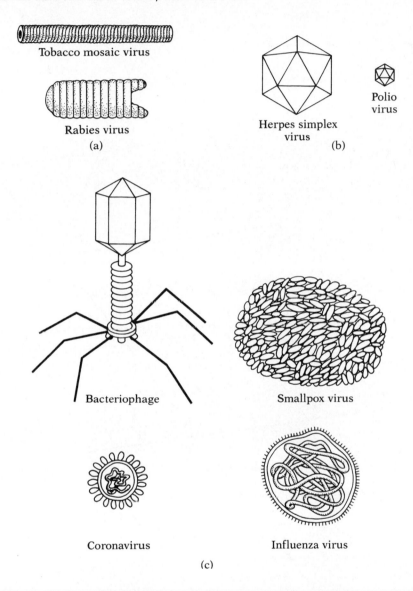

Tobacco mosaic virus

Rabies virus
(a)

Herpes simplex virus
(b)

Polio virus

Bacteriophage

Smallpox virus

Coronavirus

Influenza virus

(c)

fluctuations. In some viruses, capsid proteins are organized into enzymes to assist cell penetration during replication. Also, the capsid is the structure that stimulates an immune response during periods of disease. The capsid plus the genome is called the **nucleocapsid** (though the better term is probably "genocapsid" to maintain the structure-to-structure consistency).

Many viruses are surrounded by a flexible membrane known as an **envelope**. The envelope is composed of lipids and protein and is similar to the host cell membrane, except that it includes viral-specified components. It is acquired

Nucleocapsid:
the capsid plus genome of a virus.

11.3

BUILDING AN ICOSAHEDRON

Saying that a virus has the shape of an icosahedron is easy, but conceptualizing an icosahedron is another matter. You can gain a clearer understanding of this geometric figure if you build your own. Begin by making a copy of the following template on a copy machine, enlarging it if possible. Then trace the template onto some sort of heavy duty paper such as a file folder or manila envelope. Now cut out the template and fold the triangles along the lines. As you fold, bring the sides of the triangles together by matching the numbers (fold down the tabs with the numbers). Use a piece of tape to hold the sides to one another, or you can use glue on the tabs for a more permanent icosahedron. Good luck.

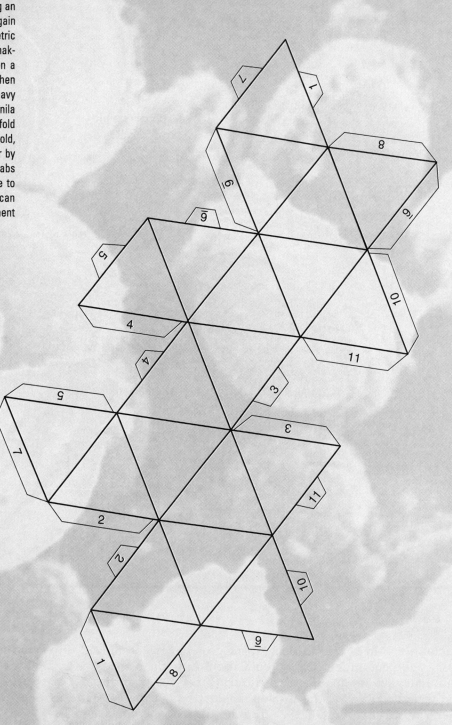

FIGURE 11.7
The Components of Viruses

(a) An icosahedral virus in both naked and enveloped forms. Capsomere units are shown on one face of the capsid. The genome consists of either DNA or RNA and is folded and condensed. (b) A helical virus in both naked and enveloped forms. The genome winds in a helical fashion. The capsomeres are protein subunits that form into the capsid cover.

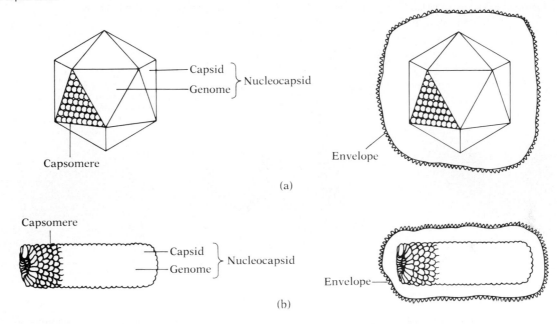

from the cell during replication and is unique to each type of virus. In some viruses, such as influenza and measles viruses, the envelope contains functional projections known as **spikes**. The spikes often contain enzymes to assist the attachment of viruses to host cells. Indeed, enveloped viruses may lose their infectivity when the envelope is destroyed. Also, when the envelope is present, the symmetry of the capsid may not be apparent since the envelope is generally a loose-fitting structure. Hence, in electron micrographs it may not be possible to see the symmetry of enveloped viruses. Indeed, some authors refer to viruses as spherical or cubical because the envelope gives the virus this appearance.

spike:
a projection of the viral envelope that assists attachment to the host cell.

A completely assembled virus outside its host cell is known as a **virion**. (We shall use the terms "virus" and "virion" interchangeably.) Compared to a prokaryote such as a bacterium, a virion is extraordinarily simple (Figure 11.8). As we have seen, it consists essentially of a segment of nucleic acid, a protein coat, and in some cases, an envelope. Virions lack the chemical machinery for generating energy and synthesizing large molecules. Therefore they must rely upon the structures and chemical components of their host cells to replicate themselves. We shall examine how this takes place next.

Virion:
a complete virus outside its host cell.

Replication of Bacteriophages

The process of viral replication is among the most remarkable events of nature. A virion invades a living cell a thousand or more times its size, utilizes the metabolism of the cell, and produces copies of itself, often destroying the cell. The virion cannot replicate independently, but within the cell, the replication takes place with high efficiency.

FIGURE 11.8
Viral Symmetry

Transmission electron micrographs of viruses displaying various forms of nucleocapsid symmetry. (a) The vesicular stomatitis virus, which causes skin sores in bovine animals. This virus has a helical nucleocapsid and appears in the shape of a bullet. It is related to the rabies virus. (Bar = 100 nm). (b) The vaccinia virus, which causes cowpox. This viral nucleocapsid is rectangular with a series of rodlike fibers at its surface. The symmetry is described as complex. (Bar = 100 nm). (c) A bacteriophage displaying an icosahedral head and an extended tail. This symmetry is also designated complex. (Bar = 100 nm).

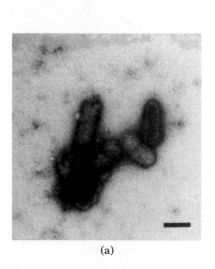

(a)

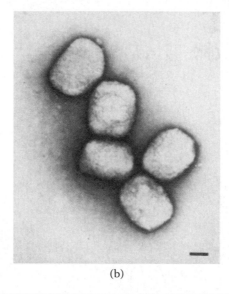

(b)

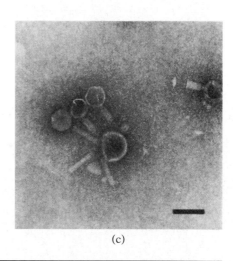

(c)

esh"er-ik'e-a

Receptor site:
the chemical group on a host cell where a virus attaches.

Replication has been studied in a wide range of virions and their host cells. Among the best known processes of replication is that carried on by bacteriophages of the T-even group (T for "type"). Bacteriophages T2, T4, and T6 are in this group. They are large, complex DNA virions with the characteristic head and tail of bacteriophages but without an envelope. We shall use their replication cycle in *Escherichia coli* as a model for the viruses.

It is important to note that the nucleic acid in a virion contains only a few of the many genes needed for viral synthesis and replication. It contains, for example, genes for synthesis of viral structural components, such as capsid proteins, and for a few enzymes used in the synthesis; but it lacks the genes for many other key enzymes, such as those used during nucleic acid production. Its dependence on the host cell for replication is therefore substantial.

The first step in the replication of a bacteriophage is its **attachment** to its host cell (Figure 11.9). There is no long-distance chemical attraction between the two, so the collision is a chance event. For attachment to occur, a site on the phage must match with a complementary **receptor site** on the cell wall of the bacterium. The actual attachment consists of a weak chemical union between virion and receptor site. In some cases, the bacterial flagellum or pilus contains the receptor site.

In the next phase, **penetration**, the tail of the phage releases the enzyme lysozyme to dissolve a portion of the bacterial cell wall. Then the tail sheath contracts, and the tail core drives through the cell wall. As the tip of the core reaches the cell membrane below, the DNA from the phage head passes through the tail core and on through the cell membrane into the bacterial cytoplasm. For most bacteriophages, the capsid remains outside.

Next comes the period of **biosynthesis**. Initially the phage uses the bacterium's nucleotides and enzymes to synthesize multiple copies of phage DNA.

FIGURE 11.9
Viral Attachment

Scanning electron microscope views of the attachment of bacteriophages to their host cells. (a) Numerous bacteriophages are attached to the surface of the cells. The arrow indicates the tail fibers (× 70,000). (b) A remarkable close-up
view of the point of attachment showing the tail assembly (arrow). The magnification is 200,000 ×.

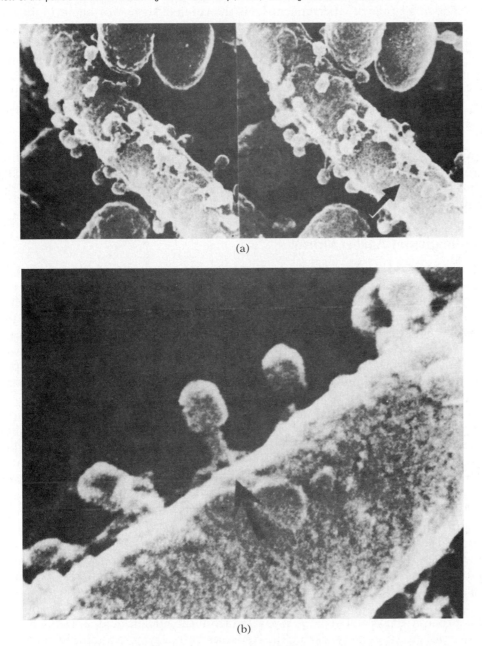

(a)

(b)

Then the DNA is used to encode viral proteins. The RNA appearing in the
bacterial cytoplasm is messenger RNA transcribed from phage DNA (Chapter
6), and biosynthesis of phage enzymes and capsid proteins begins. Ribosomes,
amino acids, and bacterial enzymes are all enlisted for the biosynthesis. Because
viral capsids are repeating units of capsomeres, a relatively simple genetic code
present in just one copy can be used over and over. For a number of minutes,
called the eclipse period, no viral parts are present. Then they begin to appear.

Ribosomes:
**ultramicroscopic particles of
RNA and protein used in
protein synthesis.**

The next period is known as **maturation**. Now the fragments of bacteriophage DNA and the capsids are assembled into complete virions. The products
of certain viral genes guide the assembly in step-by-step fashion. In one
area, phage heads and tails are assembled from protein subunits; in another,
the heads are packaged with DNA; and in a third, the tails are attached
to the heads.

The final stage of viral replication is the **release** stage. For bacteriophages,
this stage is also called the lysis stage because the cell membrane lyses, or breaks
open. For some phages, the important enzyme in this process is lysozyme,
encoded by the bacteriophage genes late in the sequence of events. The enzyme
degrades the bacterial cell wall, and the newly released bacteriophages are set
free to infect other bacteria (Figure 11.10). The progressive disintegration of
bacteria by lysis inspired Twort and d'Herrelle to name the viruses bacteriophages, or "bacteria-eaters."

The time that passes from phage attachment to release of new virions is
commonly referred to as the **burst time**. For bacteriophages the burst time
averages from 20 to 40 minutes. At the conclusion of the process, 50 to 200
new phages emerge from the host cell. This number is commonly called the
burst size.

Replication of Animal Viruses

The method of replication displayed by T-even phages is similar to that in
animal viruses, but with some notable exceptions. One such exception is in the
attachment phase. Like bacteriophages, animal viruses have attachment sites
uniting with complementary receptor sites on animal cell surfaces, but the
attachment sites exist on the cell membrane rather than the cell wall. Furthermore, animal viruses have no tails, so the attachment sites are distributed over
the entire surface of the capsid and the sites themselves vary. For example,
adenoviruses have small fibers at the corners of the icosahedron, while influenza
viruses have spikes on the envelope surface.

An understanding of the attachment phase can have practical consequences
because an animal cell's receptor sites are inherited characteristics. The sites
therefore vary from person to person, which may account for the susceptibility
of different individuals to a particular virus. In addition, a drug aimed at a
virus' attachment site could conceivably bring an infection to an end. Many
pharmaceutical scientists are investigating this approach to antiviral therapy.

Penetration is also different. Phages inject their DNA into the host cell
cytoplasm, but animal viruses are usually taken *in toto* into the cytoplasm. In
some cases, the viral envelope fuses with the cell membrane and releases the
nucleocapsid into the cytoplasm. In other cases, the virion attaches to a small
outfolding of the cell membrane, and the cell then enfolds the virion within a
vesicle and brings it into the cytoplasm like a piece of food during phagocytosis.

Uncoating takes place once the nucleocapsid has entered the cytoplasm. In
this process, the protein coat is separated from the nucleic acid, possibly by the
activity of enzymes derived from the **lysosome**, an enzyme-containing organelle
found in most cells. In a DNA virus, a specific enzyme encoded by the viral
DNA may contribute to uncoating. Cytoplasmic enzymes may also be involved.

Now the process diverges once again because some viruses contain DNA
and some contain RNA. The DNA of a DNA virus supplies the genetic codes for
the enzymes that synthesize parts from available building blocks (such as
enzymes to construct nucleic acids from available nucleotides). A number of
DNA viruses, such as pox viruses, replicate entirely in the host cell cytoplasm.
Other DNA viruses employ a division of labor, with DNA genomes synthesized

FIGURE 11.10

How a Bacterial Virus Reproduces

The replication of a bacteriophage in a host bacterium. (a) The phage attaches to the cell wall of the bacterium by its tail fibers, as shown in the inset. (b) In the penetration phase, the tail sheath contracts and the tail core drives through the cell wall and cell membrane (inset). The nucleic acid enters the bacterial cytoplasm, while the protein capsid remains outside. (c) In the period of biosynthesis, viral components are manufactured using many of the host's enzymes and structures such as ribosomes. (d) The virions are assembled during the maturation period in a step-by-step fashion, as illustrated in the inset. (e) In the final step, the host cell undergoes lysis and releases new phage particles.

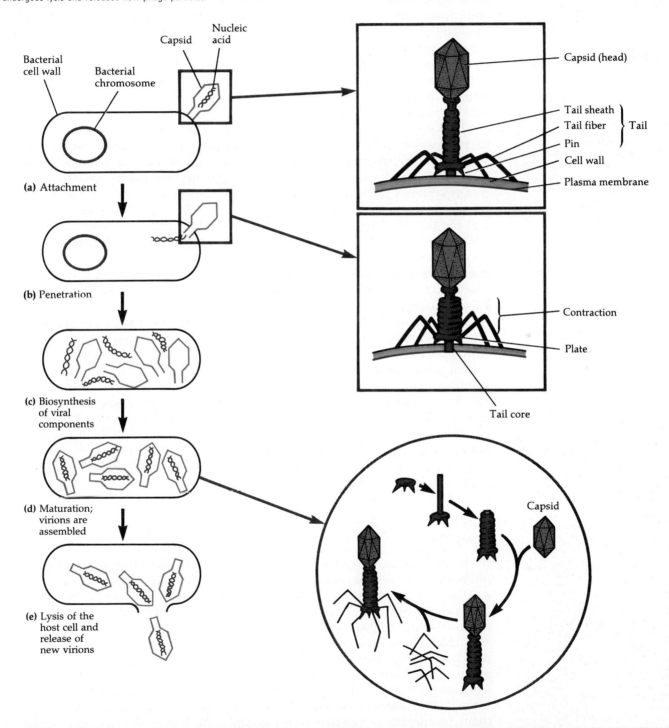

Herpesviruses:
a group of icosahedral DNA viruses involved in herpes simplex, chickenpox, and infectious mononucleosis.

in the host cell nucleus and protein capsids produced in the cytoplasm (Figure 11.11). The proteins then migrate to the nucleus and join with the nucleic acid molecules for assembly. Adenoviruses and herpesviruses follow this pattern.

RNA viruses follow a slightly different pattern. The RNA can act as a messenger RNA molecule (Chapter 6) and immediately begin supplying the codes for protein synthesis using the cell's ribosomes as "workbenches" for the synthesis. Such a virus is said to have "sense"; it is called a positive-stranded RNA virus. In some RNA viruses, however, the RNA is used as a template to formulate a complementary strand of RNA. The latter is then used as a messenger RNA molecule for protein synthesis. The original RNA strand is said to have "antisense"; and the virus is therefore an antisense virus. It is also called a negative-stranded virus. Usually the enzyme RNA polymerase is present in the virus to synthesize the complementary strand for use in protein synthesis. Measles viruses are antisensc (negative-stranded) viruses; while polio viruses are sense (positive-stranded) viruses.

Retrovirus:
a virus whose reverse transcriptase uses RNA as a template to synthesize DNA for incorporation to the cell's nucleic acid.

One RNA virus called the **retrovirus** has a particularly interesting method of replication. Retroviruses carry their own enzyme, called **reverse transcriptase**. The enzyme uses the viral RNA as a template to synthesize single-stranded DNA (the terms reverse transcriptase and retrovirus are derived from this reversal of the usual biochemistry). Once formed, the DNA serves as a template to form a complementary DNA strand. The viral RNA is then destroyed, and the two DNA strands twist with one another to form a double helix. The DNA now migrates to the cell nucleus and integrates to one of the host cell's chromosomes, where it is known as a **provirus**. From this position, the DNA encodes new retroviruses. The process we have described applies to certain leukemia viruses and to the human immunodeficiency virus (HIV) that causes AIDS (Chapter 13).

The final steps of viral replication may include acquisition of an envelope. In this step, envelope proteins are synthesized and incorporated into a membrane within the cytoplasm or at the surface of the cell. Then the virus pushes through the membrane, forcing a portion of the membrane ahead of it and around it, resulting in an envelope. This process, called **budding**, need not necessarily kill the cell during the virus' exit. However, unenveloped viruses leave the cell during rupture of the cell membrane, a process that generally leads to cell death.

Before we leave viral replication, we should note that living cells may not be an absolute necessity for the process to occur. Recent research indicates that cell debris may provide enough essentials for viral replication (Micro-Focus: 11.4).

Lysogeny

li-soj'e-ne

In the replication cycles for bacteriophages and animal viruses, infection need not result in new viral particles or cell lysis. Rather, the virus may incorporate its DNA or its RNA (via DNA) into a chromosome of the cell and achieve a state called **lysogeny** (as described in the previous paragraphs). When bacteriophages are involved, the phage DNA in the lysogenic state is called a **prophage**; when an animal virus, such as a retrovirus, is involved, the viral DNA is known as a **provirus**. In both cases it appears that the viral genome is encoding a repressor protein that prevents activation of the genes necessary for replication.

Lysogeny:
the condition in which viruses and bacteria coexist without damage to each other.

Lysogeny may have several implications. Viruses in the lysogenic state, for example, are immune to body defenses since the body's antibodies cannot reach them, (antibodies do not penetrate into cells). Moreover, the virus is propagated each time the cell's chromosome is reproduced, such as during mitosis in animal

FIGURE 11.11
Replication of a DNA Animal Virus Inside a Host Cell

The virus illustrated here is a herpesvirus such as might cause genital herpes, and the host cell is from human skin. (a) The host cell membrane unites with the viral envelope, thereby permitting entry to the cytoplasm. (b) The viral protein coat is disintegrated by cell enzymes, and the DNA enters the nucleus. (c) New viral DNA forms in the nucleus, while (d) capsid proteins are synthesized in the cytoplasm. (e) Capsid proteins enter the nucleus and combine with viral DNA to form new viruses. (f) The viruses bud through the nuclear envelope or cell membrane to acquire their envelope and (g) leave the cell.

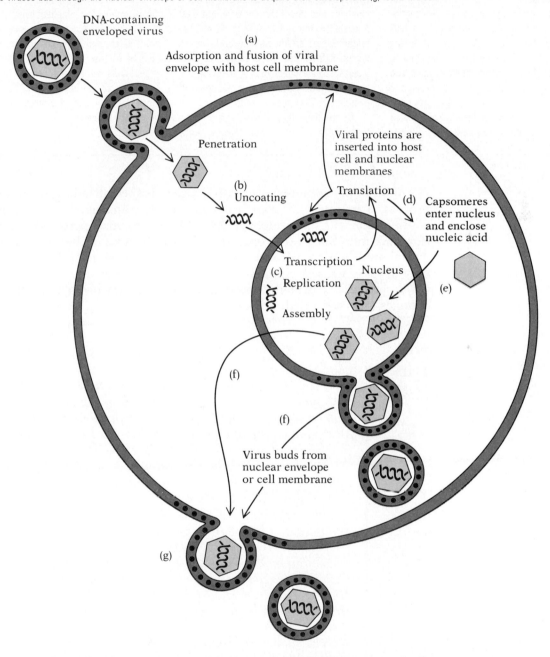

cells. And the prophage or provirus can confer new properties on the infected cell, such as when a toxin-encoding prophage infects a bacterium. A case in point is *Clostridium botulinum*, a bacterium whose lethal toxin is encoded by an indwelling prophage. Another is the bacterium of diphtheria (Chapter 7).

11.4

OUT TO DINNER

A fundamental tenet of microbiology, the fact that viruses replicate only in living cells, crashed to Earth during the winter of 1992 when researchers at the State University of New York at Stony Brook managed to cultivate viruses outside of intact cells. Led by Eckard Wimmer, the research group had for years been trying to replicate polio viruses in the right combination of crushed human cell debris plus salts, ATP, nucleo-tides, and amino acids. Dozens of their combinations had been unsuccessful, but they persisted in the research. And so on that fateful winter afternoon they added fragments of RNA from polio viruses to their latest test tube concoction and set up the myriad controls necessary for a successful experiment. Then they went to dinner.

By the time they returned, history had been made. There in the mixture were whole, intact polio viruses—viruses produced without the benefit of living cells. The scientists gasped in disbelief, congratulated one another, and began to ponder the implications of their discovery. The research implications would be enormous, as would the philosophical implications. At the very least, the definition of a virus would have to change forever.

Another phenomenon traced to lysogeny is specialized transduction. In this process, a fragment of DNA from one cell is transferred to a second cell in combination with bacteriophage DNA (Chapter 6). Also, persons with HIV infection have indwelling proviruses in the T-lymphocytes of their immune systems. A final implication involves cancer. As we shall see later in this chapter, cancer may develop when a virus enters a cell and assumes a lysogenic relationship with that cell. The proteins encoded by the virus may bring about the profound changes associated with this dreaded condition.

TO THIS POINT

■

We have begun a study of viruses by discussing their size, shape, and method of replication. We noted that some viruses have icosahedral symmetry, while others have helical symmetry, and others have a complex symmetry with a variety of shapes. However, all viruses consist of a genome of RNA or DNA, and a protein capsid. Some viruses are surrounded by a flexible membranous envelope. The complete virus is a virion.

We then studied the replication of viruses using the bacteriophage and its bacterial host as a model. The process begins with a union between phage and bacterium, followed by penetration of the viral DNA to the bacterial cytoplasm. Synthesis and release of new virions are the final two stages. Using the model as a basis, we noted how animal viruses differ in their replication mode especially with respect to adsorption and penetration. Differences are also observed in DNA and RNA viruses. The discussion concluded with the concept of lysogeny, where viruses remain in their host cells for long time periods as proviruses or prophages.

The survey of viral characteristics will continue in the final section. We shall see how viruses are classified and how viral diseases are detected. Inhibition of viruses using drugs inside the body will be surveyed, and mention will be made of the types of viral vaccines. We shall also see how viruses can be inactivated outside the body. There will be an extensive treatment of the role of viruses in cancer, and the chapter will conclude with discussions of viroids and prions, two types of infectious subviral particles.

Other Characteristics of Viruses

Like all other microorganisms, viruses have characteristics that set them apart and help virologists understand their activities and deal with them effectively. In this section, we shall examine some of these characteristics.

Nomenclature and Classification

A widely accepted classification system for viruses has not yet been devised, in part because sufficient data is not yet available to determine viral relationships to one another. Viruses therefore lack formal names (a fact that does not seem to bother students). Instead, viruses have acquired their names from several sources. Examples are the measles virus (after the disease), the adenovirus (after the adenoids, where it is commonly located), the Coxsackie virus (after Coxsackie, New York, where it was originally isolated), and the Epstein–Barr virus (after researchers who studied it). The situation is reminiscent of a century ago when different bacteria were called the "tubercle bacillus," or the "diphtheria bacillus," or the "cholera bacillus."

cook-sak'e

This is not to say, however, that no classification system exists for viruses. Indeed, one classification scheme is loosely based on the body tissues affected by the virus. Though inexact, this scheme places viruses into four convenient groups depending on whether they multiply in the respiratory, skin, visceral, or nervous tissue (Table 11.1). We shall employ this scheme in Chapters 12 and 13.

Visceral:
pertaining to the organs in the large cavities of the body, especially the abdominal cavity.

The second classification scheme is currently evolving. Periodically, there is a meeting of the International Committee on Taxonomy of Viruses and a revised and updated viral classification scheme is presented. At this writing, no orders, divisions, and kingdom have been established for viruses, but virologists have prepared a working document in which a viral species is defined as a group of viruses sharing the same genetic information and ecological niche. Names have not yet been established for species, but the viruses have been categorized into genera; each genus name ends with the suffix -*virus*. The genera have then been organized into 73 families, each ending with -*viridae*. A selection of viral families and genera affecting humans together with some of their characteristics is presented in Table 11.2. It should be noted that due to the rapid changes taking place in viral taxonomy, some new names may be in use by the time you

---

TABLE 11.1
Classification of Human Viral Diseases by Tissue Affected

GROUP	TISSUES AFFECTED	IMPORTANT DISEASES
Pneumotropic	Respiratory system	Influenza, respiratory syncytial disease, adenovirus diseases, rhinovirus infection
Dermotropic	Skin and subcutaneous tissues	Chickenpox, herpes simplex, measles, mumps, smallpox, molluscum contagiosum, rubella
Viscerotropic	Blood and visceral organs	Yellow fever, dengue fever, infectious mononucleosis, cytomegalovirus disease, viral fevers, Marburg disease, viral gastroenteritis, hepatitis A, hepatitis B, AIDS
Neurotropic	Central nervous system	Rabies, lymphocytic choriomeningitis, polio, slow virus disease, arboviral encephalitis

TABLE 11.2
Several Viral Families and Characteristics of Their Members

CHARACTERISTICS	VIRAL FAMILY	VIRAL GENUS (WITH REPRESENTATIVE SPECIES) AND UNCLASSIFIED MEMBERS*	DIMENSIONS OF VIRION (DIAMETER IN NM)	CLINICAL OR SPECIAL FEATURES
Single-stranded DNA, nonenveloped	Parvoviridae	*Dependovirus*	18–25	Some depend on coinfection with adenovirus; cause fetal death, gastroenteritis, fifth disease.
Double-stranded DNA, nonenveloped	Adenoviridae	*Mastadenovirus* (adenovirus)	70–90	Various respiratory infections in humans; some cause tumors in animals.
	Papovaviridae	*Papillomavirus* (human wart virus) *Polyomavirus*	40–57	Induce tumors; the human wart virus (papilloma) and certain viruses that produce cancer in animals (polyoma and simian).
Double-stranded DNA, enveloped	Poxviridae	*Orthopoxvirus* (vaccinia and smallpox viruses) *Molluscipoxvirus*	200–350	Smallpox (variola), molluscum contagiosum (wartlike skin lesion), cowpox, and vaccinia.
	Herpesviridae	*Simplexvirus* (herpes simplex viruses 1 and 2) *Varicellavirus* (varicella-zoster virus) *Cytomegalovirus* *Lymphocryptovirus* (Epstein–Barr virus) Human herpes virus 6	150–200	Fever blisters, chickenpox, shingles, and infectious mononucleosis; implicated in a type of human cancer called Burkitt's lymphoma.
	Hepadnaviridae	*Hepadnavirus* (hepatitis B virus)	42	Uses reverse transcriptase to produce its DNA from mRNA; causes hepatitis B and liver tumors.
Single-stranded RNA, nonenveloped + strand	Picornaviridae	*Enterovirus* *Rhinovirus* (common cold virus) Hepatitis A virus	28–30	Polio-, Coxsackie-, and echoviruses; more than 100 rhinoviruses exist and are the most common cause of colds.
Single-stranded RNA, enveloped + strand	Togaviridae	*Alphavirus* *Rubivirus* (rubella virus)	60–70	Transmitted by arthropods (*Alphavirus*); diseases include eastern equine encephalitis (EEE) and western equine encephalitis (WEE). Rubella virus is transmitted by the respiratory route.
	Flaviviridae	*Flavivirus* *Pestivirus* Hepatitis C virus	40–50	Replicate in arthropods; diseases include yellow fever, dengue, St. Louis encephalitis, and Japanese encephalitis.
	Coronaviridae	*Coronavirus*	80–160	Upper respiratory tract infections and the common cold.

* Unclassified viruses have not been assigned to genera; therefore, only their common names are listed.

CHARACTERISTICS	VIRAL FAMILY	VIRAL GENUS (WITH REPRESENTATIVE SPECIES) AND UNCLASSIFIED MEMBERS*	DIMENSIONS OF VIRION (DIAMETER IN NM)	CLINICAL OR SPECIAL FEATURES
(−) strand, one strand of RNA	Rhabdoviridae	*Vesiculovirus* (vesicular stomatitis virus) *Lyssavirus* (rabies virus)	70–180	Rabies and numerous animal diseases.
	Filoviridae	*Filovirus*	80–14,000	Ebola and Marburg viruses.
	Paramyxoviridae	*Paramyxovirus* *Morbillivirus* (measles virus)	150–300	Parainfluenza, mumps, and Newcastle disease in chickens.
(−) strand, multiple strands of RNA	Orthomyxoviridae	*Influenzavirus*	80–200	Influenza.
	Bunyaviridae	*Bunyavirus* (California encephalitis virus) *Hantavirus*	90–120	Hemorrhagic fevers such as Korean hemorrhagic fever associated with rodents.
	Arenaviridae	*Arenavirus*	50–300	Lymphocytic choriomeningitis and hemorrhagic fevers.
	Retroviridae	Oncoviruses *Lentivirus*	100–120	Oncoviruses cause leukemia and tumors in animals; HIV causes AIDS.
Double-stranded RNA, nonenveloped	Reoviridae	*Reovirus* Colorado tick fever virus	60–80	Mild respiratory infections and infantile gastroenteritis; an unclassified species causes Colorado tick fever.

* Unclassified viruses have not been assigned to genera; therefore, only their common names are listed.

FIGURE 11.12
The Cytopathic Effect

Two electron micrographs illustrating the cytopathic effect of viruses. (a) Uninfected cells from a mouse are shown in a single-layer arrangement prior to addition of viruses. (b) The same cells are shown 24 hours after infection of vesicular stomatitis virus (VSV). The cells are visibly damaged and have begun to round up and detach from the glass. Both photographs represent magnifications of 800 ×.

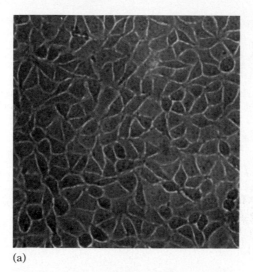
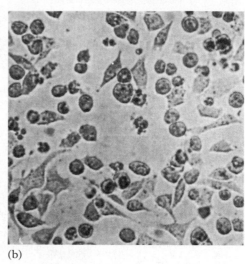

(a)   (b)

Detection of Viruses

The methods used to detect viruses are more involved and considerably more time-consuming than for bacteria and other microorganisms, because viruses require host cells for replication. Plant and animal tissues are difficult and expensive to maintain, and pathogenic viruses often replicate only in human host cells, which causes additional complications. By contrast, bacteriophages are easily cultivated in bacterial cultures, which is why bacteriophages have been used as models to study viral characteristics.

One common method of cultivating viruses is to inoculate them into fertilized eggs. A hole is drilled in the shell of the egg, and a suspension of viral material is introduced. Because different viruses replicate in different membranes or parts of the chick embryo, virologists must anticipate which virus is present. Viral replication is detected by the death of the embryo, or cell damage to the embryo or its membranes, or formation of characteristic lesions at the site of inoculation.

Another method of detecting viruses is to inoculate suspensions of material to **tissue cultures**. To prepare the culture, cells are separated from a tissue with enzymes and suspended in a solution of nutrients, growth factors, pH buffers, and salts. The cells adhere to the wall of the container and reproduce to form a single layer, or monolayer. When viruses replicate in these cells, a noticeable deterioration occurs (Figure 11.12). This is called a **cytopathic effect (CPE)**.

An indirect method for detecting viruses is to search for viral antibodies in a patient's serum. This can be done by combining serum (the blood's fluid portion) with known viruses. In some serological tests, the reaction results in a visible clumping when the viruses are attached to a carrier particle. Certain viruses such as those of influenza, measles, and mumps have the ability to agglutinate (clump) red blood cells. This phenomenon, called **hemagglutination (HA)**, can be used for detection purposes. In addition, it is possible to

Cytopathic effect:
deterioration and destruction of host cells by viruses.

Serological test:
a test that detects the presence of antibodies or antigens.

Hemagglutination:
clumping of red blood cells.

FIGURE 11.13
The Hemagglutination-Inhibition (HAI) Test

(a) Viruses of certain diseases such as measles and mumps are able to agglutinate (clump) red blood cells. (b) In the hemagglutination-inhibition test, known viruses are combined with serum from a patient. If the serum contains antibodies for that virus, the antibodies coat the virus, and when the viruses are then combined with red blood cells, no agglutination will take place. Since the nature of the virus is known, the type of antibody present can be determined.

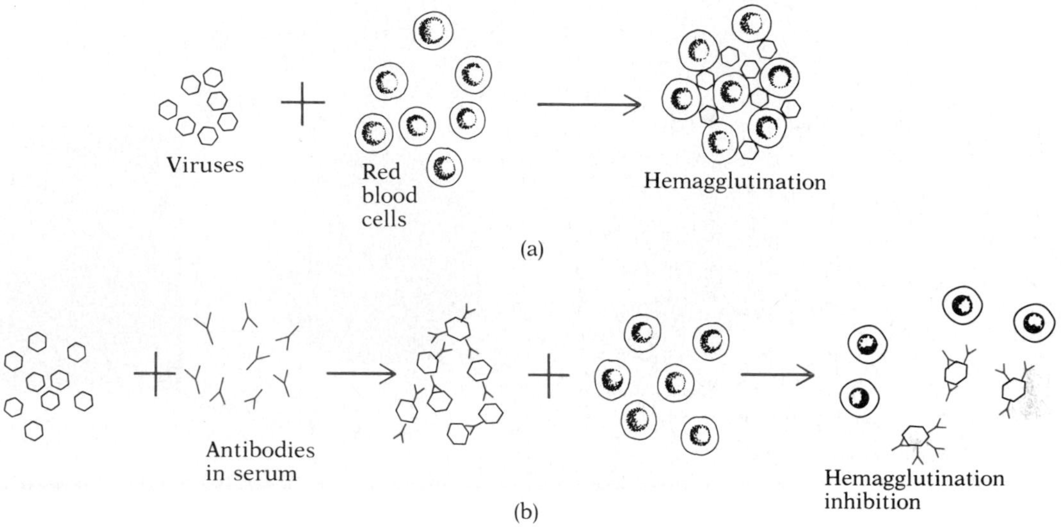

detect antibodies against certain viruses because antibodies react with viruses and tie up the reaction sites, thereby inhibiting hemagglutination. Thus a laboratory test for antibodies can be performed by combining the patient's serum with known viruses and red blood cells. Hemagglutination indicates that antibodies are absent from the serum, but **hemagglutination-inhibition (HAI)** points to the presence of serum antibodies (Figure 11.13). Such a finding implies that the patient has been exposed to the viruses.

In certain instances, viruses leave signs of their presence in infected tissue. For example, the brain cells of a rabid animal contain cytoplasmic granules called **Negri bodies,** and cells from herpes simplex patients have nuclear granules known as **Lipschütz bodies.** Moreover, a series of cellular or tissue modifications may signal the presence of viruses. Infectious mononucleosis, for example, is characterized by large numbers of swollen lymphocytes with foamy, highly vacuolated cytoplasm. Physicians call these cells **Downey cells.** Measles is accompanied by **Koplik spots,** a series of bright red patches with white pimplelike centers on the lateral mouth surfaces. Swollen salivary glands and teardroplike skin lesions are associated with mumps and chickenpox, respectively.

The most obvious method for detecting viruses is by direct observation with the electron microscope. In this procedure, tissue samples may be examined directly or it may be necessary to increase the number of viruses by cultivation before samples are prepared for microscopy. Virologists are often able to identify unknown viruses by comparison to known viruses.

Bacterial viruses, or bacteriophages, may be detected by the formation of plaques. A **plaque** is a clear "moth-eaten" area on a cloudy "lawn" of bacteria where bacteriophages have destroyed the cells. Virologists first cultivate the bacteria on an agar surface, and then add the viruses by spraying or other methods (Figure 11.14). If the viruses are specific for that particular bacterium, they infect and replicate in the cells, thereby destroying them and forming

Lipschütz bodies:
granules in the nucleus of cells infected by herpes simplex viruses.

Lymphocyte:
a type of white blood cell important in specific body defense.

Plaque:
a clear area on a lawn of bacteria where bacteriophages have destroyed the cells.

FIGURE 11.14

The Process Leading to the Formation of Plaques

(a) Susceptible bacteria are inoculated to plates of nutrient medium. (b) Bacterial viruses are then sprayed onto the surface, and the plate is incubated. (c) As viruses replicate in the bacteria, they destroy the cells and leave clear "moth-eaten" areas containing no bacteria. These areas are the plaques shown on the plate in the photograph.

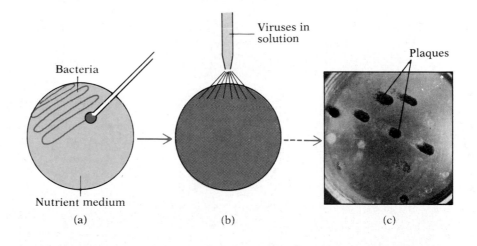

plaques. This method can also be reversed so that a known phage is used to detect an unknown bacterium. Epidemiological surveys of several bacterial diseases such as staphylococcal food poisoning are aided by this procedure, often called **phage typing** (Figure 11.15).

The difficulty in detecting viruses has a bearing on relating a particular virus to a particular disease. In the classical sense, Koch's postulates cannot be applied to a viral disease because the viruses cannot be cultivated in pure culture. To circumvent this problem, Thomas M. Rivers in 1937 expanded Koch's postulates to include viruses by modifying the postulates as follows: Filtrates of the infectious material shown not to contain bacteria or other cultivatable organisms must produce the disease or its counterpart; or the filtrates must produce specific antibodies in appropriate animals. This concept has come to be known as **Rivers' postulate**.

Inhibition of Viruses

During periods of disease, the human body attempts to rid itself of viruses mainly by phagocytosis, neutralization with antibodies, and interactions with T-lymphocytes. **Antibodies** are protein molecules from the immune system that react with the virus stimulating their formation. Usually one antibody molecule unites specifically with a single virion, thereby preventing the virion from uniting with its host cell. Antibodies also clump viruses into large masses efficiently removed by phagocytosis. In addition, viruses activate the complement system, a group of substances that encourage phagocytosis (Chapter 18). Finally, an important source of defense is the body's T-lymphocytes. When activated by viral antigens, T-lymphocytes migrate to virus-infected cells and interact directly with the cells, destroying the cells and the viruses within the cells (Chapter 18).

Normal body defenses to viral disease cannot be supplemented by antibiotics because viruses lack the structures and metabolism with which antibiotics interfere. For example, penicillin is useless for treating viral diseases because viruses have no cell wall. However, several drugs have been found effective

Phage typing:
a method of identifying an unknown bacterium by its reaction with a known bacteriophage.

Phagocytosis:
a defensive measure of the body in which white blood cells engulf and destroy microorganisms.

FIGURE 11.15

A Bacterium and Its Bacteriophage

(a) The bacterium *Caulobacter crescentus*, a Gram-negative rod typically found in freshwater environments. Bar = 0.2 μm. (b) The bacteriophage ϕCR30, the phage that replicates within *C. crescentus*. The icosahedral head and the tail of the phages are visible. The long lines in the photograph are bacterial flagella amid the debris. Bar = 5 μm.

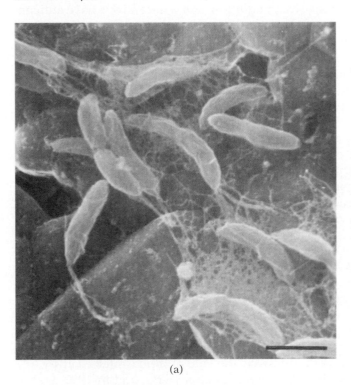

(a)

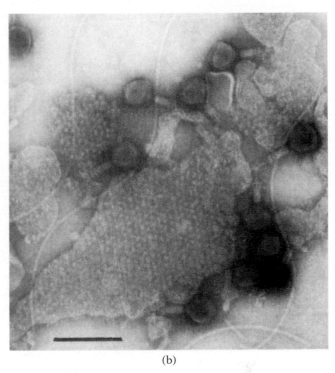

(b)

against viruses. One drug, called **amantadine** (Symmetrel), was licensed by the U.S. Food and Drug Administration (FDA) in 1966 for use against influenza viruses. Virologists believe it prevents attachment of influenza viruses to the host cell surface. Another drug, **vidarabine** (Vira-A), has been in use since 1977 for treating herpes zoster (shingles) and encephalitis (brain disease) due to the herpes simplex virus. The FDA licensed a third drug, **acyclovir** (Zovirax), in 1985 as a topical ointment for genital herpes and more recently for treating chickenpox.

Acyclovir is typical of antimicrobial agents called **base analogs**. These substances resemble nitrogenous bases and are erroneously incorporated into viral DNA. Another base analog, **idoxuridine** (IDU), is taken up by herpesviruses in place of thymine, and the resulting genome cannot replicate itself. IDU has been available since 1964 in ointment form for the treatment of herpes infections of the eye. A newer drug, **trifluridine**, works in a similar way but can also be used in liquid form as eyedrops; the eyedrop is preferable because the ointment blurs vision. The base analog **azidothymidine** (AZT) has been used since 1987 to treat AIDS, and two other analogs called dideoxyinosine (ddI) and dideoxycytidine (ddC) were approved in 1992, also for treating AIDS. Another drug called **foscarnet** has been FDA-approved since 1991 for patients having retinal disease (retinitis) due to the cytomegalovirus (CMV) that often infects AIDS patients. **Ganciclovir** is also used against CMV infection, and a final base analog called **ribavirin** is used to treat respiratory syncytial disease (Chapter 12).

ah-man'tah-dēn

vi-dar'ah-bēn

a-si'klo-vir

i-doks-ur'ĭ-dēn

tri-floor'ĭ-dēn

a-zi"do-thi'-mi-dēn

fos-kar'net

gan-sĭ'klo-vir

ri-ba-vi'rin

in"ter-fēr'on

Interferon represents one of the most optimistic approaches to inhibiting viruses. First identified in 1957 by Alick Isaacs and Jean Lindenmann, interferon is not a single substance but a group of over 20 substances designated alpha, beta, and gamma interferons. Each group has several members, and all appear to be proteins. Interferons are produced by various body cells on stimulation by viruses. They trigger a nonspecific reaction that protects against the stimulating virus, as well as many other viruses. In addition, some interferons have anticancer properties. However, human interferon is the only one that will work in humans. Mouse, chicken, dog, or other animal interferons are ineffective (MicroFocus: 11.5).

Interferons:
a group of cellular proteins that provide protection against viruses.

Unlike antibodies, interferons do not interact directly with viruses, but with the cell they protect. For this reason they have a broader inhibitory effect. Interferons are produced when a virion releases its genome into the cell (Figure 11.16). The viral material (probably a double strand of RNA) induces the cell to synthesize and secrete interferons. These bind to specific receptor sites on the surfaces of adjacent cells and trigger the production of several proteins within those cells. The proteins inhibit viral replication by methods not completely understood, although many virologists believe that at least one protein binds to messenger RNA molecules encoded by the virus. Interferons also appear to mobilize natural killer cells, which attack tumor cells (Chapter 20).

Plasmid:
a closed-loop unit of bacterial DNA existing apart from the chromosome.

The inability to obtain sufficient amounts for research stifled interferon research for many years. One approach was to inoculate huge batches of white blood cells with harmless viruses. However, a breakthrough came in 1980 when Swiss and Japanese scientists deciphered the genetic code for interferon, and spliced *E. coli* plasmids with the DNA code (Chapter 6). Experiments showed that interferon from bacterial factories would reduce hepatitis symptoms, diminish the spread of herpes zoster, and shrink certain cancers. In 1984, a Swiss biotechnology firm began marketing alpha interferon using the trade name Intron. In 1986, the FDA approved the sale of alpha interferon for use against a form of leukemia, in 1988 against genital warts, and in 1992 against chronic hepatitis B.

11.5

MICROFOCUS

A VOCABULARY OF VIROLOGY

genome: the DNA or RNA core of a virus.

capsid: the protein covering enclosing the genome of a virus.

nucleocapsid: the combination of genome and capsid of a virus.

capsomere: subunits of protein that join together to form the capsid.

envelope: the membranous layer enclosing the nucleocapsids of certain viruses but absent in others.

icosahedron: a twenty-sided geometric figure and the shape assumed by the capsids of many viruses such as herpesviruses, polio viruses, and chickenpox viruses.

helix: a tightly wound coil and the shape assumed by the genome and capsids of

many viruses such as rabies viruses and tobacco mosaic viruses.

virion: a complete viral particle outside the host cell; usually synonymous with "virus."

bacteriophage: a virus that replicates within and usually destroys its host bacterium; abbreviated as "phage."

replication: the process wherein a viral genome is released within the cytoplasm of the host cell and thereupon uses host cell resources for the production of new viruses.

lysogeny: the process wherein a viral genome is released within the cytoplasm of a host cell and thereupon integrates itself to the chromosomal material of the host cell to maintain coexistence within the cell.

budding: the process in which a nucleocapsid forces its way through a cell membrane and acquires it as an envelope at the conclusion of the replication sequence.

interferon: a naturally occurring protein that a cell produces on exposure to a virus and that protects adjacent cells against viral penetration.

inactivated virus: a virus chemically or physically treated so it cannot replicate (a component of "dead" virus vaccines).

attenuated virus: a virus replicating at a very low rate and generally incapable of inducing disease (a component of "live" virus vaccines).

FIGURE 11.16

Interferon

The production and activity of interferon. (a) A host cell produces interferon following exposure to the RNA associated with a virus. Viruses replicate in the same cell. (b) The interferon reacts with receptors at the surface of a neighboring cell and induces the cell to produce antiviral proteins. The proteins interfere with viral replication in the cell possibly by binding to messenger RNA molecules.

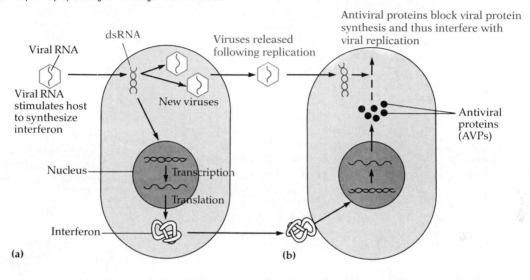

Viral Vaccines

Despite the optimism inspired by the new drugs, the major approach to viral disease continues to be prevention through public education and the use of vaccines. Vaccines stimulate antibody production by the immune system and thereby induce specific defense (MicroFocus: 11.6). They can also stimulate the production of killer T-cells (Chapter 18).

Two types of viral vaccines are currently in use. The first contains **inactivated viruses**, which are viruses treated with a physical agent such as mild heat or a chemical agent such as formaldehyde. This treatment alters the genome of the virus, thereby preventing replication, but it does not severely affect the capsid, which remains intact and stimulates antibody production. Opponents of inactivated viral vaccines point out that treatment may not reach all viruses and that untreated viruses may remain active and cause disease. They also note that the vaccine must be given by injection to ensure an immune response and that the immune response is low relative to that elicited by the second type of vaccine.

The second type of vaccine contains **attenuated viruses**, which are viruses able to continue replicating in body cells but at an extremely low rate. Attenuated viruses stimulate the immune system for a longer period of time than inactivated viruses and thus, the immune response is higher. They are obtained by transferring viruses from culture to culture for a period of months or years until a variant emerges with a greatly reduced replication rate. Opponents of attenuated viral vaccines suggest that the viruses may revert back to their original form and cause disease, and that the viruses may infect cells other than the usual host cells or activate proviruses already in host cells.

Inactivated and attenuated viruses are sometimes called "dead" and "live" viruses, respectively, and the vaccines are therefore known by these terms (e.g., live measles vaccine). The terms "dead" and "alive" are erroneous, however,

Inactivated viruses:
viruses that cannot replicate due to alteration of the genome.

ah-ten'u-a"ted
Attenuated viruses:
viruses that replicate at a very low rate.

11.6
MICROFOCUS

USEFUL INTERFERENCE

In order for a virus to infect a cell it must first dock at the cell's receptor site. After this marriage of sorts has been completed, the viral genome enters the cell by any of a number of means, and replication proceeds. But suppose the cell receptors were hidden from the virus? Would that prevent viral replication from taking place? And would the cell be protected?

Apparently the answer is yes on both accounts. In a canyon near Lake Casitas about 40 miles northwest of Los Angeles there lives a unique population of mice. Somewhere in the ancient past, the cells of these mice acquired the ability to produce proteins normally found in a virus. In an amazing twist of evolution, a virus probably entered the mouse cells, and viral genes that normally encode capsid production became part of the mouse cell chromosomes. So endowed, the mouse cells began producing capsid proteins, and the proteins adhered to viral receptors at the cell surface. With the receptors tied up, the mice became immune to the virus and, as natural selection proceeded, an entire colony of immune mice developed. The key piece of

biochemical evidence was presented in the late 1980s when researchers from the University of California pinpointed the gene for capsid protein inside cells of the Lake Casitas mice. It seems that removing susceptibility to a virus was as easy as adding a gene or two.

Does this finding mean that cells can be protected by incorporating genes for viral proteins? Again, the answer is yes. For several years now, DNA technologists and genetic engineers have been splicing into plant cells the genes for viral proteins and have been developing disease-resistant plants. In 1993, for example, French biotechnologists protected champagne grape vines from disease by incorporating into their cells the capsid genes from grape fan-leaf viruses. Wine from these new transgenic plants is expected to reach market by the year 2000.

Can we protect humans the same way? Perhaps so, but in a slightly modified way. Researchers are currently attempting to develop an AIDS vaccine by using envelope proteins from HIV. These proteins will elicit antibody production in the body. The anti-

bodies would react with and neutralize envelope proteins on HIV, thereby preventing HIV's union with receptor sites on the body's cells. To be sure, this is not the same as the mouse-generated viral proteins, but the principle is similar—interfere with attachment and prevent infection.

---

because they signify the uncertain life status of viruses. The Salk vaccine for polio typifies a vaccine made with inactivated viruses while the Sabin vaccine represents a vaccine made with attenuated viruses. The introduction to Chapter 12 recounts the tortuous development of the measles vaccine.

Inactivation of Viruses

Viruses may be inactivated by many of the physical and chemical agents routinely used for other microorganisms. Among the **physical agents** are heat and ultraviolet light. **Heat** alters the structure of viral proteins and nucleic acids, causing them to unfold and denature. Sterilization temperatures reached in the autoclave (Chapter 21) will destroy all viruses, and boiling water for a few minutes will eliminate most viruses (with the notable exception of the hepatitis A virus, which requires a longer time). **Ultraviolet light** inactivates viruses by stimulating adjacent thymine or cytosine bases on DNA molecules to bind together and form pairs called dimers. The dimers twist the molecule out of shape, and the distorted viral genome cannot replicate. Ultraviolet light is used in the preparation of some vaccines. **X rays** are another type of useful radiation. These radiations cause breaks in the sugar–phosphate backbone of the nucleic acid.

Autoclave:
a device that generates high-temperature steam under pressure for the destruction of microorganisms.

di'mer

Several **chemical agents** can be used outside the body to inactivate viruses, but not inside the body because of their toxic effects on the tissues. Examples are the halogen compounds such as chlorine and iodine derivatives, heavy metal compounds such as mercury and silver derivatives, and phenol derivatives (Chapter 22). These groups of compounds react strongly with protein, thereby altering the viral capsid. **Formaldehyde** is another useful chemical agent because it reacts with free amino groups on adenine, guanine, and cytosine molecules to modify the viral genome and prevent replication. The Salk polio vaccine is prepared with formaldehyde-inactivated viruses. Other valuable chemical agents are **lipid solvents** such as ether, chloroform, and detergents, all of which dissolve the lipid in the envelope of viruses. Enzymes directed against viral proteins and nucleic acids are also useful. Figure 11.17 presents a summary of the activity of chemical and physical agents against viruses.

Halogen:
an element whose atoms have seven electrons in the outer shell, such as chlorine, iodine, and bromine.

Viruses and Cancer

Cancer is indiscriminate. It affects humans and animals, young and old, male and female, rich and poor. In the United States over 450,000 die of cancer annually, making the disease the second most common cause of death after cardiovascular diseases. Worldwide, over two million people die of cancer each year.

Cancer results from the uncontrolled reproduction of cells through the process of mitosis. Cancer cells do not spend less time in mitosis than normal cells; rather, the frequency of mitosis is greater for cancer cells. The cells escape controlling factors and as they continue to multiply, a cluster of cells soon forms. Eventually, the cluster yields an abnormal, functionless mass of cells. This mass is a **tumor**.

Tumor:
an abnormal, functionless mass of cells.

FIGURE 11.17

Six Methods for Inactivating Viruses

(a) Formaldehyde combines with free amino groups on nucleic acid bases. (b) Metals react with the protein of the viral capsid. (c) Lipid solvents dissolve the envelope in enveloped viruses. (d) Phenol reacts with proteins in the capsid. (e) Heat denatures proteins of the capsid. (f) Ultraviolet light binds together thymine molecules in the genome and distorts the nucleic acid.

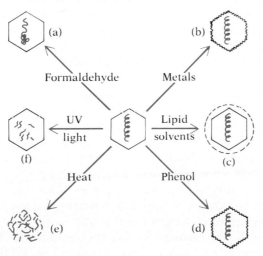

Normally, the body will respond to a tumor by surrounding it with a capsule of connective tissue. Such a tumor is designated **benign**. If, however, the cells multiply too rapidly and break out of the capsule and spread, the tumor is described as **malignant**. The individual now has "cancer," a reference to the radiating spread of cells that resemble a crab ("cancer" is derived from the Greek word *karkinos* meaning "crab"). **Oncology**, the study of cancer, is derived from *onkos*, the Greek word for tumor.

Cancer cells differ from normal cells in three major ways: they grow and undergo mitosis more rapidly than normal cells; they stick together less firmly than normal cells; and they undergo dedifferentiation. **Dedifferentiation** means that normal cells revert to an early stage in their development. For example, when ciliated cells of the bronchi become cancer cells, they lose their cilia and dedifferentiate into formless cells that divide as rapidly as early embryonic cells. Moreover, the cells fail to exhibit **contact inhibition**; that is, they do not adhere tightly to one another as normal cells do. Thus, they overgrow one another to form a tumor. Sometimes, they **metastasize**, or spread, to other body parts where new tumors begin. Also, they invade and grow in a broad variety of body tissues, since they do not stick to tissue cells. There appears to be no boundary limiting the growth of a tumor.

How can such a mass of cells bring illness and misery to the body? Cancer cells by their sheer force of numbers invade and erode local tissues, thereby interrupting normal functions and choking organs to death. For example, a tumor in the kidney prevents kidney cells from performing their excretory function, a brain tumor cripples this organ by compressing the nerves and interfering with nerve impulse transmission, and a tumor in the bone marrow may block blood cell production.

In addition, the tumor cells rob the body of vital nutrients to satisfy their own growth needs. Thus, the cancer patient will commonly experience weight loss even while maintaining a normal diet. Anemia and a general feeling of tiredness may develop when the tumor cells use up essential minerals for red blood cell production. Moreover, some tumor cells are known to produce hormones identical to those normally produced by the body's endocrine glands and thus overload the body with chemical regulators. Some tumors may block air passageways; others interfere with the immune system so that microbial diseases take hold. The tumor cells weaken the body until it fails.

Scientists are uncertain as to what triggers a normal cell to multiply without control. However, they know that certain chemicals are **carcinogens**, that is, cancer-causing substances. The World Health Organization estimates that carcinogens may be associated with 60 to 90 percent of all human cancers. Among the known carcinogens are the hydrocarbons found in cigarette smoke, as well as asbestos, nickel, certain pesticides and dyes, and environmental pollutants in high amounts. Physical agents such as ultraviolet light and X rays are also believed to be carcinogens.

There is considerable evidence that **viruses** are also carcinogens. Experiments with animals indicate that some viruses can induce tumor formation. Federal law, however, prohibits these experiments from being repeated with human volunteers because of the serious consequences. Nevertheless, a number of viruses have been isolated from human cancers, and when these viruses are transferred to animals and tissue cultures, an observable transformation of normal cells to tumor cells takes place. Examples of such viruses are the herpes viruses associated with tumors of the human cervix and the Epstein–Barr virus linked to Burkitt's lymphoma, a tumor of the jaw. As early as 1911, Peyton Rous postulated that a virus was involved in tumors in chickens (MicroFocus: 11.7).

One of the clearest virus-cancer links emerged in 1980 when a research team led by Robert T. Gallo of the National Cancer Institute isolated a virus

Oncology:
the study of cancer.

Metastasize:
to spread to distant locations.

Anemia:
a condition characterized by a below-normal level of red blood cells.

Carcinogens:
cancer-causing substances.

Herpes simplex:
a viral disease of the skin and nervous system, often characterized by blisterlike sores.
rows

11.7

MICROFOCUS

A SOLUTION—60 YEARS LATER

After microbiologists established the existence of viruses at the turn of the century, a search began for a virus that could cause cancer. To many investigators, the search seemed foolhardy because cancer did not appear to be an infectious disease. Nevertheless, one virus did emerge as an apparent cause of a type of cancer.

In 1911, an American physician, Francis Peyton Rous, was studying chickens that had a tumor of the connective tissues called a sarcoma. Rous decided to test the tumor for virus content, and he mashed up a section of tissue and passed it through a bacterial filter. To his astonishment, the clear filtrate caused tumors in healthy chickens. Rous did not refer to the infecting material as a virus, but others gradually did, and for many decades thereafter the "Rous sarcoma virus" remained as a clear-cut example of a cancer-causing virus. The virus soon became an important tool of cancer researchers. In 1966 Rous was awarded the Nobel Prize in Physiology or Medicine, more than 50 years after his discovery.

However, the mystery of the Rous sarcoma virus deepened as knowledge of the transformation of normal cells to cancer cells emerged. Virologists believed that cancer viruses released DNA into the cell cytoplasm, whereupon the DNA inserted itself on the cell's chromosome. But the Rous sarcoma virus was composed of RNA, and RNA had never been found inserted on a chromosome. Moreover, the genetic code for transformation was presumably supplied by DNA, not by RNA.

The mystery was solved by Howard Temin and David Baltimore. In 1970, these investigators announced the discovery of a viral enzyme named reverse transcriptase. This enzyme uses RNA as a template and synthesizes a molecule of DNA complementary to the RNA. The new DNA could then insert on the chromosome and transform the cell. The finding of reverse transcriptase completed the story of Rous' discovery and won for Temin and Baltimore the 1975 Nobel Prize in Physiology or Medicine.

---

that transforms normal T-lymphocytes into the malignant T-lymphocytes found in a rare cancer called **T-cell leukemia**. A year later, the same virus was found responsible for a relatively high rate of both T-cell leukemia and a form of lymphoma in Japan. Gallo identified the virus as an RNA-containing retrovirus and named it HTLV (for *h*uman *T*-cell *l*eukemia *v*irus). His expertise with retroviruses proved valuable in 1984 when his research group set out to isolate the AIDS virus, also an RNA-containing retrovirus. Instead of transforming T-lymphocytes, however, the AIDS virus destroyed them (Chapter 13).

In the 1990s, the evidence has strengthened that HTLV can cause leukemia as well as neurological pain disorder and another pain disorder marked by destruction of the nerve sheaths that surround the nerve fibers. In a study of 2500 patients at Johns Hopkins Hospital, 1 percent of all those admitted to the emergency room displayed evidence of infection with the virus. The virus appears to be bloodborne, and the primary mode of transport is through the use of illegal intravenous drugs. The growing incidence among patients also poses a threat to health-care workers who are exposed to patients' blood. A second virus, named HTLV-II has also been located. Although the virus has not definitely linked to any cancer, it is apparently common in patients suffering from a condition called hairy-cell leukemia, a type of leukemia in which the white blood cells develop long hairlike extensions to their cytoplasm at the surface.

How viruses and other carcinogens transform normal cells into tumor cells remained obscure until the **oncogene theory** was developed in the 1970s. First postulated by Robert Huebner and George Todaro in 1969, this theory suggests that transforming genes, the so-called oncogenes, normally reside in the chromosomal DNA of a cell. In the late 1970s, researchers J. Michael Bishop and Harold Varmus of the University of California at San Francisco located oncogenes in a wide variety of creatures from fruitflies to humans. Bishop and Varmus also made the astonishing discovery that practically the same genes

T-lymphocyte:
an immune system cell that functions in cell-mediated immunity.

Oncogene:
a gene that can transform a normal cell to a cancer cell.

exist in certain viruses, and they hypothesized that the genes could have been captured by the viruses. It appeared that the oncogenes were not viral in origin but part of the genetic dowry of every living cell. Bishop and Varmus received the 1989 Nobel Prize in Physiology or Medicine for their work.

The discovery of oncogenes demonstrated that some forms of cancer had a genetic basis. As research continued, oncology researchers extracted DNA from tumors and used it to turn healthy cells into cancerous ones in the test tube. Moreover, they surmised that the transforming substance was in a small segment of the tumor-cell DNA—probably a single gene. Finally, in 1981, three separate research groups isolated an oncogene residing in a human bladder cancer. At this writing over 60 different oncogenes have been identified.

In more recent years, the theory of oncogene activity has been revised slightly. Researchers now propose that normal genes, called **proto-oncogenes**, are the forerunners of oncogenes. Proto-oncogenes may have important functions as regulators of growth and mitosis. Indeed, research reported in 1985 linked proto-oncogenes to the production of cyclic adenosine monophosphate (AMP), an organic substance central to many physiological processes. That proto-oncogenes exist in diverse forms of life argues for their important role in cell metabolism, perhaps as growth regulators. (As one researcher notes: "They would not have survived through evolution just to make tumors.") It seems certain that proto-oncogenes can be converted to oncogenes by carcinogens, such as viruses, radiation, or chemicals, or by chromosomal breakage and rearrangement, after which tumor formations begin. Oncogenes and proto-oncogenes are so similar that the bladder cancer oncogene differs from its counterpart proto-oncogene by only one nucleotide in 6000.

But how do viruses convert proto-oncogenes to oncogenes? Virologists have observed that when a virus enters a cell, it may enter a lysogenic relationship with the cell and thereby transform it. If the viral genome is composed of DNA, for example, the DNA may attach itself directly to the chromosome (like a prophage attaches to the bacterial chromosome) and become a provirus. If the viral genome is composed of RNA, the enzyme reverse transcriptase synthesizes DNA from RNA. Once the synthesis is complete, the double-stranded DNA inserts into the cell chromosome as a provirus (Figure 11.18). This system applies to the retroviruses studied by Gallo and his associates for human T-cell leukemia and AIDS.

Another mechanism of transformation has been studied with the virus of Burkitt's lymphoma. In this cancer of the lymphoid connective tissues of the jaw, the viral genome appears to insert itself into a chromosome of B-lymphocytes, the white blood cells important in immunity. The insertion triggers certain proto-oncogenes involved in cell growth to move from their position on chromosome 8 to a new position on chromosome 14, far from the influence of their control genes. A segment of DNA from chromosome 14 replaces the proto-oncogenes from chromosome 8. The proto-oncogenes, now oncogenes, appear to produce elevated amounts of their protein product.

Once the proto-oncogene becomes an oncogene, it can influence cellular growth and mitosis in several ways. Oncogenes, for example, may provide the genetic codes for growth factors that stimulate uncontrolled cell development and reproduction. Or they may program the development of protein receptors that receive extracellular messages and transmit them to the nucleus at a higher-than-normal rate. Or the oncogenes may become incapable of producing substances that turn off cell growth, a function the proto-oncogenes once had. In a particularly interesting series of studies at Cold Spring Harbor Laboratories in New York, a team led by Michael Wigler isolated a protein whose production was directed by an oncogene. The protein was injected into normal cells and within hours the normal cells began to multiply rapidly and show unmistakable

---

Proto-oncogenes:
Human genes that can be transformed by carcinogens into oncogenes.

Reverse transcriptase:
an enzyme that synthesizes DNA using the genetic message contained in RNA.

Burkitt's lymphoma:
cancer in the lymphoid connective tissue of the jaw.

FIGURE 11.18

Mechanism by Which an RNA-Containing Retrovirus Becomes a Provirus in a Host Cell

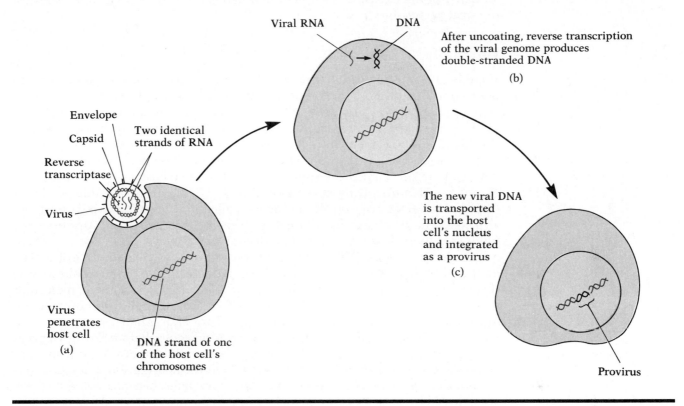

Viral RNA   DNA

After uncoating, reverse transcription of the viral genome produces double-stranded DNA

(b)

Envelope

Two identical strands of RNA

Capsid

Reverse transcriptase

Virus

The new viral DNA is transported into the host cell's nucleus and integrated as a provirus

(c)

Virus penetrates host cell

(a)

DNA strand of one of the host cell's chromosomes

Provirus

signs of conversion to cancer cells. Growth slowed down several hours later, presumably because the supply of protein was exhausted. In a later experiment, they injected antibodies against the protein into cancer cells and observed signs of reversion to normal cells. The experiments strengthened the evidence that abnormal products of oncogenes enhance cell transformation and that, in theory, countermeasures are possible.

One gene that has received considerable attention is a viral oncogene known as *ras*, so named because it induces *ra*t *s*arcoma, a tumor of the connective tissue in rats. When the *ras* gene is inserted in human bladder cells, it causes a human tumor to form. In the spring of 1989, researchers clarified how the *ras* gene gives rise to tumors. An international team of cancer specialists identified a protein programmed by the *ras* proto-oncogene and found that when the protein is activated, it encourages other molecules to induce cell division. Following several cellular divisions (enough, for example, to heal a wound) the protein is normally deactivated. When the *ras* proto-oncogene is mutated to the oncogene, however, *ras* continues to code for the protein, but the protein is in an altered form that cannot be deactivated. The altered protein keeps telling the cell "Divide, divide, divide!" In a short period of time a tumor forms.

Although the advances in cancer research have been remarkable in recent years, investigators are quick to point out that there are no cures in the near future. Nevertheless, several innovative chemical substances loom on the horizon that can influence the course of a cancer. Interferons are one such group of substances, interleukins another, and monoclonal antibodies still another

ras:
a rat sarcoma gene that directs synthesis of a tumor-inducing protein.

Interleukins:
chemical products of white blood cells that enhance the immune process.

(Chapter 20). Oncogene studies are significant because they may aid the early detection of cancers and assist the development of anticancer drugs. The oncogene theory provides an explanation of a unifying mechanism through which all carcinogens may act, regardless of whether they are chemical substances, physical agents, or viruses.

Viroids

Viroids are tiny fragments of nucleic acid known to cause several diseases in plants and thought to be involved in human and animal diseases.

The discovery of viroids resulted from studies conducted in the 1960s at the U.S. Department of Agriculture's Beltsville, Maryland, research center near Washington, D.C. Scientists led by Theodore O. Diener were investigating a suspected viral disease, **potato spindle tuber (PST)**, which results in small, cracked potatoes shaped like spindles. Nothing would destroy the disease agent except an RNA-dissolving enzyme, and in 1971, the group postulated that a fragment of RNA was involved. Diener called the agent a viroid, meaning viruslike. The next year, a team led by Joseph Semancik at the University of California found a similar agent in a disease of citrus trees.

Currently at least a dozen plant diseases have been related to viroids. The largest of these particles is about one-twentieth of the size of the smallest virus (Figure 11.19). The RNA chain of the PST viroid has a known molecular sequence, but it contains so few genetic codes that the replication cycle is not understood. Diener has speculated that the viroids originated as introns, the sections of RNA spliced out of messenger RNA molecules before the messengers are able to function (Chapter 5). The similarity in size between introns and viroids and the ring shape for both have fueled the speculation. Semancik has offered the theory that viroids may be regulatory genes because viroid diseases are characterized by interference with plant growth.

An interesting twist in viroid research occurred in 1981 when it was found that viroids could infect animals. Researchers from Knoxville, Tennessee, reported DNA viroids in hamster colonies afflicted with lymphatic cancer. Currently, some microbiologists believe that viroids may cause the mysterious slow virus diseases of humans (Chapter 13).

Prions

Prions are described as *pro*teinaceous *in*fectious particles thought to cause a number of diseases including the slow virus diseases (Chapter 13). Prions were named by Stanley B. Prusiner, a leading researcher in prion study. Tests indicate that prions can survive the heat, radiation, and chemical treatment that normally inactivate viruses. Moreover, prions appear to be composed only of protein because they are susceptible to protein-digesting enzymes but not nucleases. These factors indicate that prions are not viruses and raise the question of how prions replicate because a central dogma in biology is that inheritance operates through DNA and RNA.

At this writing, the story of prions is far from complete. Some virologists believe that prions are exceptionally small "unconventional" viruses with a protein capsid so dense that it shields the nucleic acid from research probes. Other virologists are awaiting the isolation of the protein since current methods for retrieving it from solution also break it down. It is conceivable, as Prusiner has suggested, that prions use genes in the normal animal and activate them to produce more prions. Perhaps the prions enter the cell and combine with particles already there to initiate the process. Amid the speculation, prions remain one of the more interesting and controversial mysteries of microbiology.

FIGURE 11.19

Viral Relationships

The size relationships of a smallpox virus, polio virus, and viroid. If the viroid were magnified 60,000 times, it would measure only one-eighth of an inch. By contrast, the genome of the smallpox virus would extend almost 14 feet and the genome of the polio virus 4.5 inches. The genome of the potato spindle tuber viroid has 359 nucleotides, while that of the smallpox virus has almost 500,000.

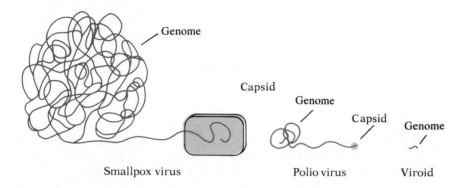

Smallpox virus   Polio virus   Viroid

NOTE TO THE STUDENT

Every science has its borderland where known and visible things merge with the unknown and invisible. Startling discoveries often come from this hazy, uncharted realm of speculation, and certain objects manage to loom large. In the borderland of microbiology, one curious and puzzling object is the virus.

Are viruses alive? At present the tendency of many biologists is to side-step the question. However, I shall put forth a suggestion. Although we have referred to viruses as microorganisms for convenience sake, I have avoided references to "live" or "dead" viruses. Instead I have used the terms "active" for multiplying viruses and "inactive" for viruses unable to multiply. I suggest, therefore, that we consider viruses to be inert chemical molecules with at least one property of living things, the ability to multiply. Thus, viruses are neither totally inert nor totally alive, but somewhere in the threshold between. Perhaps viruses are transitional forms between inert molecules and living organisms. Indeed, a prominent virologist has suggested calling them "organules" or "molechisms," depending on one's preference.

Summary

The fundamental principle underlying this chapter is the anatomy and physiology of viruses. Studying viruses was extremely difficult until the 1940s, when the electron microscope permitted scientists to see viruses and innovative methods allowed scientists to cultivate them. In the following decades it became clear that viruses are extraordinarily small particles composed of nucleic acid surrounded by a protein coat and, in some cases, a membranelike envelope. The nucleic acid core known as the genome can be either DNA or RNA in either a single-stranded or double-stranded form. The protein coat known as the capsid is usually subdivided into smaller units called capsomeres. Capsids can have icosahedral, helical, or complex symmetry. The envelope, when present, is obtained from the host cell during replication and contains viral-specified proteins.

Viral replication occurs only in living cells. The process involves adsorption, penetration, synthesis, assembly, and release phases, and it varies according to the virus. For example, penetration can occur by a number of different methods, and the biochemistry of synthesis varies among DNA and RNA viruses. Viruses have no binomial names, but instead are named according to their disease (e.g., measles virus) or discoverer (e.g., Epstein–Barr virus) or other method. Classification schemes are based on physiological and biochemical characteristics or on the tissue where replication takes place.

Various detection methods for viruses are based on such things as characteristic changes in tissue cultures, antibody responses, or pathological signs in diseased tissue. Certain drugs such as AZT can be used to inhibit viral replication, but the primary public health response to viral disease is through the use of vaccines. Vaccines consisting of inactivated viruses are available for polio and rabies, and vaccines containing weakened (attenuated) viruses are used for measles, mumps, and rubella. Many of the physical and chemical methods routinely used on microorganisms can be employed to inactivate and destroy viruses.

Cancer is a complex condition in which cells multiply without control. Among the many cancer-inducing agents are viruses, and although the mechanisms of cancer induction are still unclear, viruses may bring about cancers by converting preliminary genes into cancer-causing oncogenes. Their relationship to cancers and to infectious diseases has secured important places for viruses in the study of microbiology.

Questions for Thought and Discussion

1. If you were to stop 1000 people on the street and ask if they recognize the term "virus," all would probably respond in the affirmative. If you were then to ask the people to *describe* a virus you might hear answers like "It's very small" or "It's a germ" or a host of other colorful but not very descriptive terms. As a student of microbiology, how would you describe a virus?

2. A textbook author referring to viruses once wrote: "Certain organisms seem to live only to reproduce, and much of their activity and behavior is directed toward the goal of successful reproduction." Would you agree with this concept? Can you think of any creatures other than viruses that fit the description?

3. Oncogenes have been described in the recent literature as "Jekyll and Hyde genes." What factors may have led to this label and what does it imply? In your view is the name justified?

4. Suppose the viroid turned out to be an infectious particle able to cause disease in humans. What difficulties might be presented in dealing with this proteinless nucleic acid fragment both inside and outside the body?

5. Stanley's 1935 announcement that viruses could be crystallized stirred considerable debate about the living nature of viruses. Imagine that one day, a virus was discovered to contain both DNA and RNA. Might this stir an equal amount of controversy on the nature of viruses? Why?

6. Why was the cultivation of viruses in test tubes by Enders' group as significant as the cultivation of bacteria in test tubes by Koch? Why was each achievement critically important to the times? What type of follow-up experiment came after each foundation was established?

7. Researchers studying the bacteria that live in the oceans have long been troubled by the question of why bacteria have not saturated the oceanic environments. What might be a reason?

8. In broad terms, the public health approach to dealing with bacterial diseases is treatment. Can you guess the nature of the general public health approach to viral diseases? What evidence do you have to support your answer?

9. Bacteria can cause disease by using their toxins to interfere with important body processes or by overcoming body defenses such as phagocytosis or by using their enzymes to digest tissue cells or by other similar mechanisms. Viruses, by contrast, have no toxins, cannot overcome body defenses, and produce no digestive enzymes. How, then, do viruses cause disease?

10. An instructor wishes to describe how the immune system's antibodies attack and bind to viruses, and she equates the situation to a dish of spaghetti and meatballs, where the meatballs are surrounded by spaghetti strands. What do you think of the analogy?

11. Some virologists believe that the agent of kuru, scrapie, and other slow virus diseases is a prion. As this chapter notes, the term "prion" is derived from the words *protein-aceous infectious particle*. You will note that if these word parts are put together, the word is proin, not prion. How do you suppose the word got to be prion?

12. A certain highway is crossed by a bridge constructed of heavy stone. On the shoulders of the highway standing in front of the stone abutments are a number of barrels containing sand. The barrels are labeled "impact attenuators." Judging by the name, what do you suppose their purpose is?

13. Many textbooks attempt to simplify biological concepts, and in doing so, often oversimplify a particular thought. For example, it is not unusual to read that the viruses were discovered by Dimitri Iwanowski. How would you react to this statement?

14. How have revelations from studies on viruses, viroids, and prions complicated some of the traditional views on the principles of biology?

15. When discussing the multiplication of viruses, virologists prefer to call the process replication, rather than reproduction. Why do you think this is so? Would you agree with virologists that "replication" is the better term?

Review

Use the following syllables to compose the term that answers each clue from virology. The number of letters in the term is indicated by the dashes, and the number of syllables in the term is shown by the number in parentheses. Each syllable is used only once. The answers are listed in the appendix.

A A A AC AL AN AT AT BAC BO CAP CAP CEP CLO CO CO CY
DE DERS DIES DINE DRON ED EN EN FER FORM GE GEN GENE
HE HE HYDE I I I IN IN LET LEY LIX LY MAN MERES MOR NEG
NOME O O OID ON ON ON ONS OPE PHAGE PRI RE RI SA SID
SO SO STAN TA TED TEN TER TER TI TIV TOR TRA TU U UL
VEL VI VIR VIR VIR Y

1. Viral protein coat   (2) _ _ _ _ _ _
2. Viral shape   (2) _ _ _ _ _
3. Bacterial virus   (5) _ _ _ _ _ _ _ _ _ _ _ _ _
4. Neutralize viruses   (4) _ _ _ _ _ _ _ _ _ _
5. Rabies granules   (2) _ _ _ _ _ _
6. Herpes drug   (4) _ _ _ _ _ _ _ _ _
7. Natural antiviral   (4) _ _ _ _ _ _ _ _ _ _
8. Weakened virus   (5) _ _ _ _ _ _ _ _ _ _
9. Disease RNA fragment   (2) _ _ _ _ _ _
10. Functionless cell mass   (2) _ _ _ _ _ _
11. Virus-inactivating light   (5) _ _ _ _ _ _ _ _ _ _ _
12. Cancer gene   (3) _ _ _ _ _ _ _ _
13. Site where virus attaches (3) _ _ _ _ _ _ _ _
14. Shape of polio virus   (5) _ _ _ _ _ _ _ _ _ _ _
15. Viral core   (2) _ _ _ _ _ _
16. Cultivated polio virus   (2) _ _ _ _ _ _
17. Completely assembled virus (3) _ _ _ _ _ _ _

18. Virus incorporated to cell   (4) _ _ _ _ _ _ _ _
19. Drug for influenza   (4) _ _ _ _ _ _ _ _ _ _
20. Surrounds the nucleocapsid   (3) _ _ _ _ _ _ _ _
21. Crystallized viruses   (2) _ _ _ _ _ _ _
22. Modifies viral genome   (4) _ _ _ _ _ _ _ _ _ _ _ _
23. Protein particles   (2) _ _ _ _ _ _
24. Capsid subunits   (3) _ _ _ _ _ _ _ _ _
25. Vaccine virus   (5) _ _ _ _ _ _ _ _ _ _

THE MICROBES IN COLOR

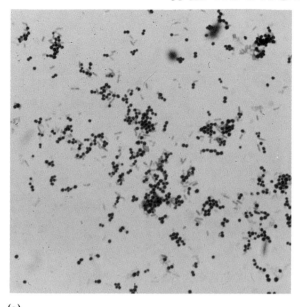

(a)

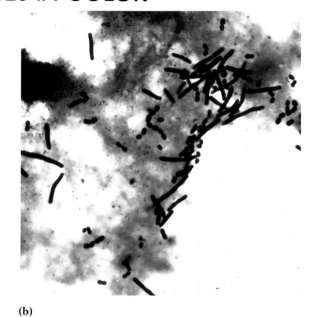

(b)

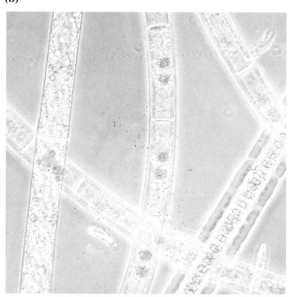

(c)

(d)

PLATE 1
Viewing Microbes via Brightfield and Phase Contrast Microscopy **(a)** A Gram stain of a mixed smear of the Gram-positive coccus *Staphylococcus aureus* (purple) and the Gram-negative rod *Escherichia coli* (pink). **(b)** Two Gram-positive bacteria occur in yogurt—the rods are *Lactobacillus bulgaricus,* and the chains of cocci are *Streptococcus thermophilus.* **(c)** Bacterial endospores within

Bacillus cells as revealed by the Dorner stain method. **(d)** Phase contrast microscopy of the green scum on a pond reveals three different filamentous algae: *Zygnema* (with paired chloroplasts), *Spirogyra* (with spirally coiled chloroplasts), and *Hyalotheca* (a chain of discoidal cells embedded in a gelatinous sheath).

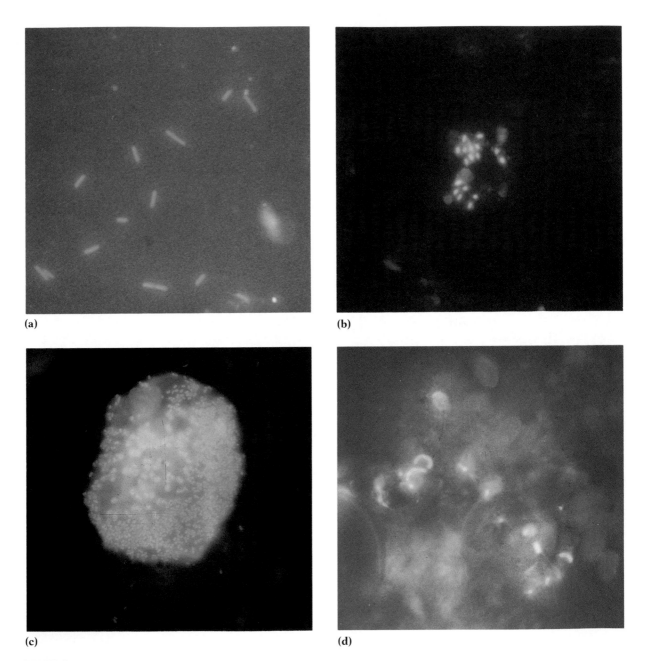

(a)

(b)

(c)

(d)

PLATE 2
Viewing Microbes via Fluorescence Microscopy
(a) Autofluorescence of the rod-shaped archaebacterium *Methanobacterium thermoautotrophicum*.
(b) Staining a sample of subsurface soil with acridine orange reveals a microcolony of bacteria (glowing green). **(c)** Staining with acridine orange reveals a mass of oral streptococci on an epithelial cell. **(d)** Immunofluorescence specifically identifies *Legionella pneumophila* (the bright green rods) on the surface of an aquatic plant—the red oval structures are autofluorescent chloroplasts of diatoms.

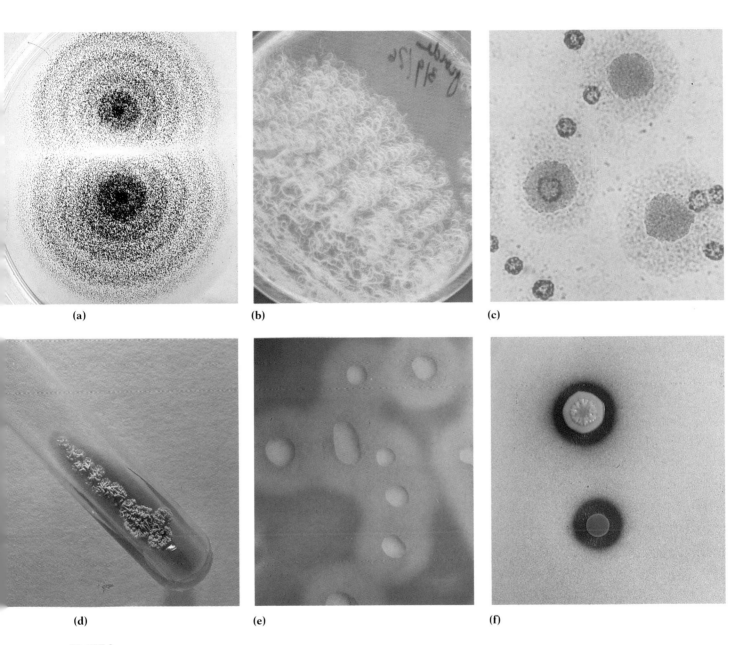

(a)
(b)
(c)
(d)
(e)
(f)

PLATE 3

Growth of Microbes on Agar Masses of microbial growth on agar vary greatly in appearance depending on the particular microbe and the growth conditions. **(a)** A typical fuzzy, black colony of the mold *Aspergillus niger.* **(b)** The bacterium *Bacillus cereus* var. *mycoides* forms white, glistening colonies that whorl in a circular pattern. **(c)** The typical "fried egg" colonies of a *Mycoplasma* species and the tiny colonies of a *Ureaplasma* species. **(d)** The dry, wrinkled growth of a *Streptomyces* species on an agar slant. **(e)** Betahemolysis (complete destruction of red blood cells) around colonies of *Streptococcus equi* on blood agar. **(f)** Colonies of two Gram-negative bacteria that oxidize glucose to gluconic acid—the agar medium contains crystal violet (to inhibit Gram-positive bacteria) and calcium carbonate (which dissolves in acid).

(a)

(b)

(c)

PLATE 4
Microbial Growth in Nature **(a)** A variety of
cheeses produced by the growth of bacteria and
fungi in milk. **(b)** Massive growth of sheathed
bacteria (genus *Sphaerotilus*) in a heavily polluted
stream. **(c)** Pixie cup lichens next to a patch of moss.
(d) The bright orange ferric hydroxide precipitate
of acid mine drainage due to oxidation of ferrous
iron (in pyrites) by the chemolithotroph *Thiobacillus ferrooxidans*.

(d)

CHAPTER 12

PNEUMOTROPIC AND DERMOTROPIC VIRAL DISEASES

How does one go about eradicating a disease from the face of the earth? How indeed, when the disease once killed a third of English newborns; when the disease caused such appalling mortality in Mexico that the highly civilized Aztec nation fell to a few hundred Spanish invaders; and when the disease was so ferocious in New World settlements that it was the major epidemic disease of the American colonies? How does one mount a global campaign against smallpox?

Apparently, the World Health Organization (WHO) was determined to try and in 1966, it received $2.5 million to begin its campaign. The goal was to stamp out smallpox in ten years using global vaccination programs. Under the direction of Donald A. Henderson, vaccine-producing laboratories were established in countries where epidemics were raging. A two-tined needle was developed and donated by Wyeth Laboratories for administering the vaccine in 15 rapid jabs to the arm. By 1971, vaccination programs had been begun in 44 countries.

Then came surveillance containment. Teams were set up to improve the reporting and discovery of outbreaks, and every known contact of victims was vaccinated to break the chain of transmission. Vaccinators used persuasion and, in some cases, coercion to learn the whereabouts of diseased people. A favorite tool was the threat to withhold food ration cards. Rewards were offered for information, and people gradually came forward. By 1970, the number of smallpox countries had dropped to 17, and by 1973, only 6 countries were left.

But the six included India, Pakistan, and Bangladesh, and 700 million people lived there (Figure 12.1). More rigorous techniques were employed, including guards to prevent patients from leaving their homes and vaccinating everyone within five miles of an infected village. By 1976, the job in Asia was done, and only Ethiopia remained. Despite a civil war, famine, several kidnappings, and torrential summer rains, the WHO pressed on. When victory seemed in sight, however, smallpox broke out in neighboring Somalia. Now, all efforts were directed at this African country, and a campaign involving 200,000 health officers, 700 advisors, 55 countries, and untold millions of dollars gradually focused on a hospital cook named Ali Maow Maalin. Maalin developed smallpox on

FIGURE 12.1

Smallpox

A photograph of two young boys infected with smallpox. This photograph was taken during the Bangladesh epidemic in 1973.

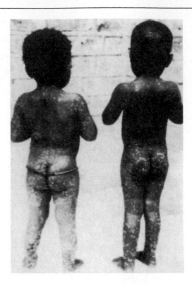

October 26, 1977. He was placed under guard, and vaccination was administered to 161 people previously in contact with him. None developed smallpox. In fact, no one has developed smallpox since that epic date.

In the minds of many scientists (a certain author included), the eradication of smallpox has been "the" medical event of the twentieth century. No claim of eradication has ever been made for a disease, and nowhere have the potential benefits of applied public health been manifested better. Indeed, the eradication of smallpox has probably been public health's finest hour.

So, one might say, that's one less disease to worry about (and one less to learn for the test). Unfortunately not, because we need to know about the classical diseases of history in order to understand the contemporary ones. And that's one reason why we study smallpox and numerous other diseases we probably will not encounter in our lifetimes. We also study viral diseases because, with the control of bacterial diseases, viral diseases have become an important focus of attention. Despite the availability of a vaccine, measles is of continuing concern to public health agencies. In addition, influenza has been an ongoing problem in the twentieth century, and chickenpox continues to be among the most commonly reported diseases of childhood years. Another viral disease, genital herpes, has become so rampant that, by some estimates, 10 to 20 million Americans are currently infected.

But scientists are learning to control many viral diseases even as we study them. For example, mumps, and rubella were part of the fabric of life only a generation ago, but vaccination programs have reduced the annual case reports from hundreds of thousands to mere hundreds, and some officials are bold enough even to whisper the word eradication.

nu"mo-trōp′ik
der"mo-trōp′ik

In this chapter we shall focus on the pneumotropic viral diseases, which affect the respiratory system, and the dermotropic viral diseases, whose symptoms are found in the skin. Each disease is presented as an independent essay so that you can establish an order that best suits your needs. It should be remembered that the pneumotropic and dermotropic divisions represent an artificial classification used simply for grouping convenience. Therefore you may note that the symptoms go beyond the respiratory system or skin, respectively. Indeed, there is some evidence that one dermotropic virus, the herpesvirus, may be involved in clogging of the arteries (MicroFocus: 12.1).

Pneumotropic Viral Diseases

Influenza

Influenza is an acute, contagious disease of the upper respiratory tract transmitted by droplets. The disease is believed to take its name from the Italian word for influence, a reference either to the influence of heavenly bodies, or to the *influenza de freddo* (influence of the cold). Since the first recorded epidemic in 1510, scientists have described 31 pandemics. The most notable pandemic of this century occurred in 1918 (Chapter 3); others took place in 1957 (the Asian flu), and in 1968 (the Hong Kong flu).

The influenza virion belongs to the Orthomyxoviridae family of viruses. It is composed of eight single-stranded segments of RNA, each wound helically and associated with protein to form a nucleocapsid. Additional protein surrounds the core of segments, and an envelope lies outside the protein. The envelope contains a series of projections called **spikes** (Figure 12.2). One type of spike contains the enzyme **hemagglutinin (H)**, a substance that facilitates the attachment of influenza viruses to host cells. The second type contains another enzyme, **neuraminidase (N)**, a compound that assists the entry of the virion into host cells for replication and out of the host cell when replication is complete. Both enzymes are antigens.

Three types of influenza virus are recognized: type A, which causes most pandemics; type B, which is less widespread than type A; and type C, which is rare. Each type is known for its **antigenic variation**, a process in which chemical changes occur periodically in hemagglutinin and neuraminidase, thereby

Droplets:
tiny particles of mucus and saliva expelled from the respiratory tract.
Pandemic:
a worldwide epidemic.

or"tho-mik"so-vir' ĭ-da

hem"ah-gloo'tin-in

nūr-ah-min' ĭ-dās

Antigen:
a chemical substance that stimulates a response by the immune system.

FIGURE 12.2

The Influenza Virus

(a) A schematic diagram of the influenza virus showing its eight segments of RNA and its envelope with spikes. (b) A scanning electron micrograph of a kidney cell infected with influenza virus. The viruses are the tiny knoblike objects at the cell surface. (c) A closer view of the cell surface. The infected cell displays a distorted surface with fingerlike projections of cytoplasm. Influenza viruses are budding from the cell, and they can be seen pushing through the projections.

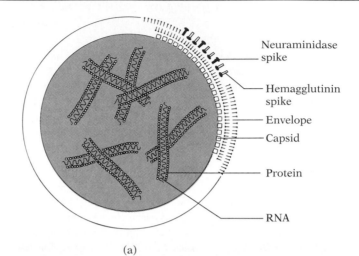

Neuraminidase spike

Hemagglutinin spike

Envelope

Capsid

Protein

RNA

(a)

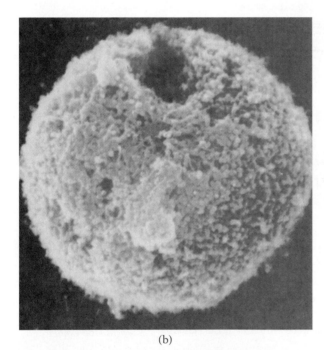

(b)

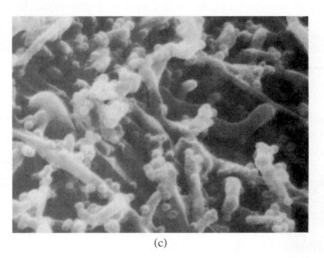

(c)

yielding new strains of virus. These changes have practical consequences because antibodies produced during a previous attack of influenza fail to recognize the new strain, and a person may suffer another attack of disease. Also, antigenic variation precludes the development of a universally effective vaccine, although several have been developed over the years for particular strains. Moreover, the nomenclature for influenza viruses is based on the variant of antigen present. It is not uncommon to see references to strains such as A(H3N2), which predominated in the U.S. in 1995–1996.

The onset of influenza is abrupt, with sudden chills, fatigue, headache, and pain most pronounced in the chest, back, and legs. Over a 24-hour period the fever rises to 104° F, and a severe cough develops. Patients experience an obstructed nose, dry throat, and tight chest, the latter a probable reflection of viral invasion of tissues of the trachea and bronchi. Despite these severe symptoms, influenza is normally short-lived with a favorable prognosis. The disease is self-limiting and usually disappears in a week to ten days. Secondary complications may occur if bacteria such as staphylococci or *Haemophilus influenzae* (Pfeiffer's bacillus) invade the damaged respiratory tissue.

Trachea: the cartilage-lined passageway leading to the bronchi and lungs.

The diagnosis of influenza is based on several factors, including the pattern of spread in the community, observation of disease symptoms, laboratory isolation of viruses, and agglutination of human type O red blood cells. Active cases of type A influenza may be treated with **amantadine** (Symmetrel), a drug believed to interfere with uncoating in the replication cycle. An alternate drug is **rimantadine** approved in 1993. High-risk individuals such as the elderly and very young may be immunized with inactivated influenza viruses of the type and strain predicted for the impending flu season. As of 1996, tests were also ongoing with a new vaccine containing pure hemagglutinin protein. The protein is produced by insect cells biochemically altered by incorporating the genetic code that encodes the protein.

ah-man'tah-dēn
sim-met'-rel
ri-man'tah-dēn

Two serious complications of influenza have surfaced during recent decades. One complication is **Guillain–Barré syndrome (GBS)**, named for Georges Guillain and Jean A. Barré, the two French doctors who described it in 1916. This condition is characterized by nerve damage, poliolike paralysis, and coma. Some nonparalytic cases are characterized by numbness in the arms and legs, and general weakness and shakiness.

ge-yan' bar-ra'

The second complication is **Reye's syndrome**, named for Ramon D.K. Reye, who reported it in 1963. Reye's syndrome (pronounced "ray's" by some and "rye's" by others) usually makes its appearance when a child is recovering from influenza or chickenpox. The child's fever rises, and repeated, protracted vomiting continues for a period of hours. Though initially coherent, the child may become lethargic, sleepy, and glassy-eyed (or "starry-eyed"), as well as disoriented, incoherent, and combative, though weak from vomiting. Reye's syndrome, like GBS, is thought to be due to activity of the immune system because viruses cannot be found in sufficient numbers to explain the symptoms. Studies in the early 1980s suggested a link between aspirin and Reye's syndrome, and parents were advised to give their children aspirin substitutes.

Lethargic: drowsy or indifferent.

Adenovirus Infections

Adenoviruses are a collection of at least 31 types of icosahedral virions having double-stranded DNA and belonging to the family Adenoviridae (Figure 12.3). They multiply in the nuclei of host cells and induce the formation of **inclusions**, a series of bodies composed of numerous virions arranged in a crystalline pattern. The viruses take their name from the adenoid tissue from which they were isolated in 1953.

Inclusions: bodies composed of numerous virions in a crystalline pattern.
Adenoids: patches of lymphoid tissue in the pharynx of humans.

Adenoviruses are one of the many causes of upper respiratory diseases collectively called the **common cold**. Adenoviral colds are distinctive because the fever is substantial, the throat is very sore, and the cough is severe. In addition, the lymph nodes of the neck swell and a whitish-gray material appears over the throat.

One strain of adenovirus is a major cause of **keratoconjunctivitis**, an inflammation of the cornea ("kerato-") and conjunctiva of the eye. Patients experience reduced vision for several weeks, but recovery is usually spontaneous and complete. Transmission may be by respiratory droplets, contact, or ophthalmic instruments. Contaminated water may also transmit the viruses, as illustrated in an early 1970s outbreak where the chlorinator in a swimming pool

ker'ah-to-kon-junk"tĭ-vi'tis

Ophthalmic: referring to the eye.

FIGURE 12.3

A Transmission Electron Micrograph of Adenoviruses

These viruses have no envelope, and the icosahedral symmetry of the nucleocapsid can be observed. The rough surface of the viruses is due to the presence of capsomeres. Adenoviruses are agents of common colds. Bar = 100 nm.

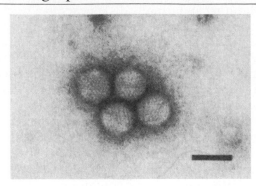

Meninges:
the three membranous coverings of the brain and spinal cord.

sin-sish'al

par"ah-mik"so-vir' ĭ-da

sin-sish'ah
sin-sish'e-um

pi-ko"nah-vir' ĭ-da

malfunctioned resulting in 44 cases of eye infection ("swimming pool conjunctivitis"). Another strain of adenovirus is an agent of **viral meningitis**, an inflammation of the meninges. Viral meningitis is often called **aseptic meningitis** because no visible agent (such as a bacterium or fungus) can be located. Still another type of adenovirus causes tumors in animals.

Respiratory Syncytial Disease

In 1985, CDC microbiologists first reported that **respiratory syncytial (RS) disease** is the most common lower respiratory tract disease in infants and children under two years of age. Infection takes place in the bronchioles and air sacs of the lungs, and the disease is often described as **viral pneumonia**. A life-threatening crisis may arise in children with compromised cardiac and respiratory systems. Maternal antibodies passed from mother to child probably provide protection during the first few months of life, but the risk of infection increases as these antibodies disappear naturally. Extending this theory, researchers have successfully used preparations of antibodies ("immune globulin") to treat established cases of RS disease.

Respiratory syncytial disease takes its name from the virus that causes it. The virus is an enveloped RNA helical virion of the Paramyxoviridae family. When the virus is cultivated in tissue cells, the infected cells tend to fuse and form giant cells called **syncytia** (sing., syncytium). Syncytium formation is also induced by the human immunodeficiency virus (HIV).

RS disease can also occur in adults, usually as an upper respiratory disease with influenzalike symptoms. Outbreaks occur yearly throughout the United States, but most cases are misdiagnosed or unreported. Some virologists believe that up to 95 percent of all children have been exposed to the disease by the age of five, and CDC epidemiologists estimate that 90,000 hospitalizations and 4500 deaths occur in infants and children each year in the United States as a result of RS disease. Ribavirin has also been used successfully for therapy.

Rhinovirus Infections

Rhinoviruses are a broad group of over 100 different RNA viruses with icosahedral symmetry. They belong to the family Picornaviridae (*pico-* means "small"; hence small RNA viruses, pico-rna-viridae). Rhinoviruses take their name from the Greek *rhinos*, meaning nose. The viruses are among the major causes of upper respiratory tract infections, generally called the **common cold**. Many people also refer to the condition as a head cold.

A **head cold** involves a regular sequence of symptoms beginning with headache, chills, and a dry, scratchy throat. A "running nose" and obstructed air passageways are the dominant symptoms. Cough is variable and fever is often absent or slight. Some children suffer **croup**. Provided there are no complications, the symptoms disappear in four to ten days. Antihistamines can sometimes be used to relieve the symptoms because the symptoms are often due to histamines released from damaged host cells.

Croup:
hoarse coughing.

Rhinoviruses thrive in the human nose, where the temperature is a few degrees cooler than in the remainder of the body. This may be one reason why the fumes from hot chicken soup appear to hasten recovery from a cold. Research on the use of **vitamin C** as a preventive has been inconclusive but promising. One study, for example, indicates that this vitamin induces the body to produce interferon, while another suggests that vitamin C encourages the formation of collagen to strengthen the "intercellular cement."

In 1985, scientists reported identification of the receptor sites in nasal tissues where rhinoviruses attach during the replication process. Furthermore, they announced synthesis of an antibody that would bind to these sites and block viral attachment. Supposedly the antibody could be used as an anticold drug. Another group, from the Kimberly-Clark Corporation, conducted tests with a three-ply Kleenex tissue composed of two regular tissues sandwiched around a middle tissue impregnated with acidic compounds (the press dubbed them "killer Kleenexes"). One wipe of the nose was reported to be capable of destroying up to a million rhinoviruses.

Receptor site:
an area of chemical activity on a cell surface where a virus can attach.

Part of the hope for interrupting the spread of colds comes from an understanding of their transmission. Research with rhinoviruses indicates that hand-to-hand contact, sneezing, and coughing are the most important transmission paths. In one study, 16,000 rhinoviruses were counted from a single sneeze. In another study, rhinoviruses were recovered from the hands of over 50 percent of cold sufferers. A third study showed that shaking hands with cold sufferers was more risky than kissing them.

Rhinovirus:
a virus that multiplies in the upper respiratory tract, especially the nose.

The prospects for developing a cold vaccine are not promising, since so many different viruses are involved. In addition to the rhinoviruses, common cold viruses include adenoviruses and respiratory syncytial viruses, as well as several other viral agents such as coronaviruses, Coxsackie viruses, echoviruses, and reoviruses. A new nasal spray of interferon has stimulated interest, but the side effects of this compound need to be understood before commercial products appear on the pharmacy shelf. Pneumotropic viral diseases are summarized in Table 12.1.

in"ter-fēr'on

TABLE 12.1

A Summary of Pneumotropic Viral Diseases

DISEASE	CLASSIFICATION OF VIRUS	TRANSMISSION	ORGANS AFFECTED	VACCINE	SPECIAL FEATURES	COMPLICATIONS
Influenza	Orthomyxoviridae	Droplets	Respiratory tract	Available for high-risk	Antigenic variation Strains A, B, C	Reye's syndrome Guillain–Barré syndrome
Adenovirus infections	Adenoviridae	Droplets Contact	Lungs, meninges Eyes	Not available	Common cold syndrome	Pneumonia Aseptic meningitis
Respiratory syncytial disease	Paramyxoviridae	Droplets	Respiratory tract	Not available	Syncytia of respiratory cells	Pneumonia
Rhinovirus infections	Picornaviridae	Droplets Contact	Upper respiratory tract	Not available	Head-cold syndrome	None

**TO
THIS
POINT**
■

We have surveyed a number of pneumotropic viral diseases affecting the respiratory tract. Our initial emphasis was on influenza, one of the most common diseases in our society. The influenza virus has a unique composition among viruses, with eight segments of RNA in its nucleocapsid. Antigenic variation in the virus accounts for the myriad strains that appear from year to year and that make resistance to influenza and vaccine development very difficult. Secondary infection as well as Reye's and Guillain–Barré syndromes can complicate cases of influenza.

We then concentrated on respiratory viruses that cause the familiar common cold. The adenoviruses were discussed as cold agents, and their role in keratoconjunctivitis, meningitis, and tumors was briefly mentioned. We then turned to the respiratory syncytial virus, one of the most common causes of lower respiratory diseases in children. Finally, we focused on rhinoviruses, a group of RNA viruses that cause the widely encountered head cold. Throughout the discussions, the broad variety of viral strains and types that cause respiratory diseases were noted. This variety will probably preclude the development of a vaccine for many years.

We shall now turn our attention to the dermotropic viral diseases. These are viral diseases of the skin. In this group we shall discuss herpes simplex and chickenpox, two of the most widespread diseases in humans; measles and rubella, two viral diseases whose incidences are declining; and smallpox, a viral disease that is apparently extinct. In this diversity we see the spectrum of viral diseases in the modern era.

Dermotropic Viral Diseases

Herpes Simplex

Lipschütz bodies:
granules in the nucleus of cells infected with herpes simplex or other herpesviruses.
jin′jĭ-vo-sto″mah-ti′tis

Herpes simplex is a collection of viral diseases caused by a large DNA virion. The virion has icosahedral symmetry and an envelope with spikes. It belongs to the family Herpesviridae. The herpes simplex virus is one of the most common viruses in the environment. Some virologists contend that over 90 percent of Americans have been exposed to it by age 18. The virus passes from one cell to another by intercellular bridges and remains in the nerve cells until something triggers it to multiply. Evidence for the viruses in living cells may be ascertained by the presence of granules called **Lipschütz bodies** in the cell nucleus.

There are many manifestations of herpes simplex infection including: **cold sores** (fever blisters), the unsightly lesions that form around the lips or nose; **herpes encephalitis**, a rare but potentially fatal brain disease; **neonatal herpes**, a life-threatening disease transmitted by mothers during childbirth; **gingivostomatitis**, a series of cold sores of the throat usually occurring in children; **herpes keratitis**, a disease of the eye and an important cause of blindness in young adults; and genital herpes, one of the most troublesome sexually transmitted diseases. The word *herpes* is Greek for "creeping," a reference to the spreading nature of herpes infections through the body (MicroFocus: 12.2).

The sores and blisters of herpes simplex have been known for centuries. In ancient Rome, an epidemic was so bad that the Emperor Tiberius banned kissing; Shakespeare, in *Romeo and Juliet*, writes of "blisters o'er ladies lips"; and in the 1700s, genital herpes was so common that French prostitutes considered it a vocational disease. In current times **genital herpes** is estimated to affect

Genital:
referring to the organs of sexual reproduction.

GLADIATORS

The situation appeared normal: a camp for high school wrestlers in Minnesota from July 2 to July 28, 1989. The camp attracted one hundred seventy-five wrestlers from around the United States. As they gathered that first day they looked forward to daily sessions of grunting and wrestling their way to excellence. Three groups would participate: lightweights, middleweights, and heavyweights.

But this would be no ordinary experience. During the final week of the camp, the first case of herpes simplex was observed. Then there was another, and another. Soon, there were too many infected participants

for the camp to continue and, with two days remaining on the schedule, the camp was suspended and the wrestlers sent home.

Subsequent contacts by the federal Centers for Disease Control and Prevention revealed that 60 of the 175 participants had contracted herpes simplex during that two-week period. All experienced symptoms during the camp session or within one week of leaving. Lesions developed on the head or neck (73% of cases), the extremities (42% of cases), and the trunk (28% of cases). Five individuals experienced infection of the eye, and wrestlers in the heavyweight division were most frequently involved.

Herpes simplex in wrestlers and rugby players is called herpes gladiatorum ("herpes-of-the-gladiators," probably a whimsical term at first, but now technically acceptable). First described in the mid-1960s, herpes gladiatorum has broken out several times since then. Transmission occurs primarily by skin contact, and autoinoculation can account for infection at several body sites. The disease illustrates another possible manifestation of herpes-related illness, and its swift passage by contact through a group of susceptible individuals demonstrates the ease of transfer.

between 10 and 20 million Americans yearly, of whom about 500,000 are believed to be new cases. The figures are inexact because genital herpes is not currently a reportable disease. Signs generally appear within a few days of sexual contact, often as itching or throbbing in the genital area. This is followed by reddening and swelling of a small area on which painful blisters erupt. The blisters crust over and the sores disappear, usually within about three weeks. In the majority of cases, however, the symptoms reappear, often in response to stresses such as sunburn, fever, menstruation, or emotional disturbance. Subsequent outbreaks last about five days. Persons with active herpes lesions pass the viruses to others by sexual contact.

In the 1960s, scientists learned that the herpes simplex virus has two different forms: type I and type II. For reasons that are unclear, **type I** virus often inhabits areas above the waist and is the cause of herpes keratitis and most cold sores (Figure 12.4), while **type II** virus appears to be prevalent below the waist. This principle does not always hold true, however. Type I viruses have been isolated from cases of genital herpes, and type II viruses have been recovered from cold sores. This finding implies that the type I viruses causing cold sores may be transferred to the genital region by such means as contaminated fingers or oral–genital contact, and that genital herpes may follow.

Type II herpes simplex virus is especially worrisome because it is the virus associated with **cancer** of the cervix, a disease that strikes over 15,000 women annually in the United States, with a 50 percent mortality rate. Though evidence is inconclusive that the virus actually causes the cancer, studies show that women who have suffered from genital herpes are several times more likely to develop cervical cancer than those who have not had herpes. Papanicolaou (Pap) smears at frequent intervals are recommended for women in this high-risk group.

Herpes encephalitis is a brain disease often accompanied by blindness, convulsions, or a range of neurological disorders including mental retardation and death. The herpesviruses may be acquired by contact with an infected individual and may possibly reach the brain by the olfactory nerves after breathing into the nose. Herpes encephalitis can also occur in a newborn, where it is

Keratitis:
infection of the cornea of the eye.

Cold sores:
herpes-induced blisters occurring on the lips, gums, nose, and adjacent areas.

pap"ah-nic"o-lah'-oo

Olfactory nerves:
nerves used in the sense of smell.

FIGURE 12.4

Ultraviolet Light and Herpes

An experiment showing the effects of ultraviolet light on formation of herpes simplex sores of the lips. This patient usually experienced sores on the left upper lip. (A) The patient was exposed to ultraviolet light from a retail cosmetic sunlamp on the left upper and lower lips in the area designated by the line. The remainder of the face was protected by a sunscreen. (B) Sores formed on the left upper lip. (C) The patient was later exposed a second time to the sunlamp but only on the left upper lip. (D) Sores formed and were larger than the previous ones. The results indicate that herpes sores can be experimentally stimulated by ultraviolet light such as in sunlamps and sunlight.

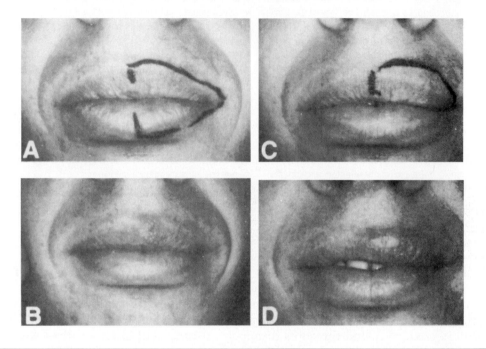

Acronym:
a word formed from the first letters of several words.

i-doks-ur′ĭ-dēn
tri-floor′ĭ-dēn
vi-dār′ah-bēn
a-si′klo-vir

called **neonatal herpes**. In this case, the viruses infect the infant during passage through the birth canal. Indeed, if a woman is diagnosed as having active genital herpes, her obstetrician may recommend birth by cesarean section to prevent the infant's contact with the viruses. Passage of viruses across the placenta may also occur. In recent years the acronym **TORCH** has been coined to focus attention on diseases with congenital significance: T for toxoplasmosis; R for rubella; C for cytomegalovirus; H for herpes; and O for other diseases such as syphilis.

At present, certain drugs have been approved by the Food and Drug Administration (FDA) for the treatment of the various forms of herpes simplex. For example, **idoxuridine (IDU)** and **trifluridine** are used for herpes keratitis, and **vidarabine** is used for eye infections and herpes encephalitis, especially in newborns. Currently, the only drug approved for use in genital herpes is **acyclovir** (Zovirax). Acyclovir is a guanine derivative and a base analog that interferes with viral replication (Figure 12.5).

Before leaving herpes simplex, we shall briefly mention a herpeslike illness called **B virus infection**. B virus infection is a relatively benign and common disease of Old World monkeys. Caused by a herpesvirus related to that of herpes simplex, the human disease is accompanied by serious neurological symptoms such as pain and numbness, with dizziness, local paralyses, and possible respiratory arrest. It is relatively rare in humans, but laboratory researchers who work with monkeys are susceptible to infection.

FIGURE 12.5

The Mode of Action of Acyclovir

How acyclovir interferes with the replication of herpesviruses. (a) The enzyme thymidine kinase (circle, left) functions by combining phosphate groups with sugar–base combinations (nucleosides) to form nucleotides. Acyclovir resembles nucleosides, and the enzyme mistakenly adds phosphate groups to the acyclovir to form a false nucleotide. (b) During viral replication, another enzyme, DNA polymerase (circle, right) attaches the false nucleotide onto a developing DNA molecule. However, the false nucleotide lacks an attachment point for the next nucleotide. The elongation of DNA thus comes to a halt and viral replication stops.

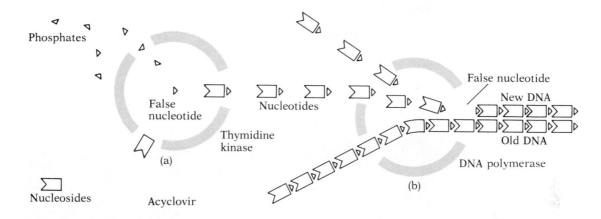

Chickenpox (Varicella)

In the centuries when pox diseases swept across Europe, there was the Great Pox (syphilis), the smallpox, the cowpox, and the chickenpox, a docile disease that made the skin resemble that of a freshly plucked chicken. Today **chickenpox** remains as the second most reported disease in the United States, with about 125,000 cases reported annually (gonorrhea is first). The causative agent is a double-stranded DNA virion with icosahedral symmetry and an envelope. It is a herpesvirus of the family Herpesviridae.

Chickenpox is among the most communicable of all diseases. It is transmitted by respiratory droplets and skin contact, and has an incubation period of two weeks. The disease begins in the respiratory tract, with symptoms of fever, headache, and malaise. The viruses then pass from the respiratory tract to the bloodstream and localize in the peripheral nerves and skin. As they multiply in the cutaneous tissues, they trigger the formation of small, teardrop-shaped, fluid-filled vesicles (Figure 12.6). **Varicella**, the alternate name for chickenpox, is the Latin word for "little vessel."

Malaise:
a general feeling of illness.

The **vesicles** in chickenpox develop over three to four days in a succession of crops. They itch intensely and eventually break open to yield highly infectious virus-laden fluid. Although many refer to the vesicles as pox, the term is more correctly reserved for the pitted scars of smallpox. In chickenpox, the vesicles form crusts that fall off without leaving a scar. Acyclovir has been approved to lessen the symptoms of chickenpox and hasten recovery from the disease.

Pox:
pitted scars remaining in recoverers from smallpox.

The mortality rate from chickenpox is low, and the disease is considered benign. However, **Reye's syndrome** may occur during the recovery period. Public health departments have issued warnings against using aspirin to reduce fever in chickenpox, because aspirin is statistically linked to Reye's syndrome. Other complications of chickenpox include pneumonia, encephalitis (brain inflammation), and bacterial infection of the skin. In pregnant women, the virus has been known to cross the placenta and cause damage to the fetus.

FIGURE 12.6
The Lesions of Chickenpox and Shingles

(a) A child showing a typical case of chickenpox. The lesions may be seen in various stages, with some in the early stage of development and others in the crust stage. (b) Dermal distribution of shingles lesions on the skin of the body trunk. The lesions contain less fluid than in chickenpox and occur in patches as red, raised blotches.

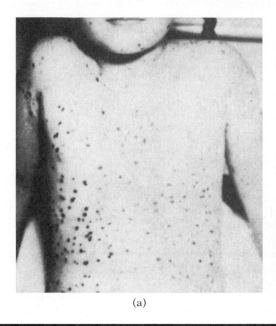
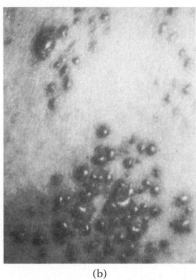

(a)   (b)

ah-ten′u-a″ted
Attenuated virus:
a virus that replicates at a very low rate.

Ganglia:
knots of nerve tissue along the spinal cord through which nerves pass.

In March 1995, the Food and Drug Administration licensed a vaccine for chickenpox. Known as Varivax, the vaccine consists of attenuated viruses administered subcutaneously and is recommended for use in all individuals over 1 year of age. One dose is given to children between ages 1 and 12, and two doses are given to adolescents and adults.

Herpes zoster, or **shingles**, is an adult disease caused by the same virus that causes chickenpox. For this reason, the virus is often referred to as the **varicella-zoster (VZ) virus**. The viruses multiply in the ganglia along the spinal cord, and travel down the nerves to the skin of the body trunk. Here they cause blisters with blotchy patches of red that appear to encircle the trunk (*herpes* is Greek for creeping, and *zoster* is Greek for girdle). Many sufferers also experience a series of headaches as well as facial paralysis and local pain described as among the most debilitating known. The condition can occur repeatedly and is linked to physical and emotional stress, as well as to a suppressed immune system.

Evidence is substantial that herpes zoster is caused by the same virus that caused chickenpox decades before in the individual. Most cases occur in people over 50 years of age, and a person with an active case of herpes zoster can induce chickenpox in another susceptible person. However, a person, regardless of age, who comes in contact with the virus for the first time will normally develop chickenpox, not herpes zoster. AIDS patients may be susceptible to the disease (Figure 12.7). A preparation of purified antibodies called **varicella-zoster immune globulin (VZIG)** is available for the prevention and treatment of chickenpox but not for herpes zoster. For herpes zoster, acyclovir therapy lessens the symptoms (MicroFocus: 12.3).

VZIG:
a preparation of antibodies against the varicella-zoster virus.

FIGURE 12.7

Herpetic Lesions on the Leg of a Patient

This patient suffered from disseminated herpes zoster associated with an immune deficiency due to HIV. The large lesion is a necrotic skin ulcer; the smaller lesions are herpetic pustules. The patient was treated intravenously with acyclovir and the lesions healed. However, she later succumbed to the effects of HIV infection.

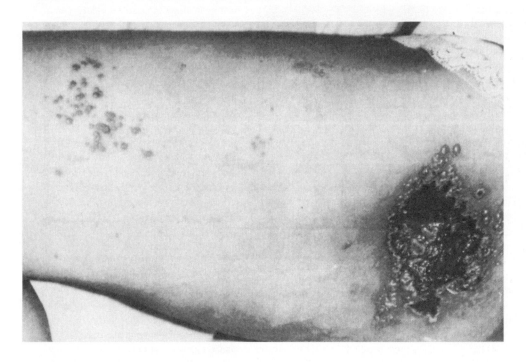

Measles (Rubeola)

Measles is a highly contagious disease due to an RNA helical virion first isolated in tissue culture by John Enders and Thomas Peebles in 1954. The virion is enveloped, with hemagglutinin spikes, and is closely related to the mumps and RS viruses in the Paramyxoviridae family. Transmission usually occurs by respiratory droplets during the early stages of disease.

par"ah-mik"so-vir'ĭ-da

Measles symptoms commonly include hacking cough, sneezing, nasal discharge, eye redness, sensitivity to light, and a high fever that may reach 104° F at times. Red patches with white grainlike centers appear along the gumline in the mouth two to four days after the onset of symptoms. These patches are the **Koplik spots** first described in 1896 by Henry Koplik, a New York pediatrician. They are an important diagnostic sign for measles.

The characteristic red **rash** of measles appears about two days after the first evidence of Koplik spots. Beginning as pink-red pimplelike spots (maculopapules), the rash breaks out at the hairline, then covers the face and spreads to the trunk and extremities (Figure 12.8). **Rubeola**, the alternative name for measles, takes its name from the Latin *rube* for red. Rashes resemble those in scarlet fever, but the severe sore throat of scarlet fever generally does not develop. Within a week, the rash turns brown and fades.

mac"u-lo-pap'ules

Rubeola:
an alternative name for measles.

Measles is usually characterized by complete recovery. In some cases, however, bacterial disease may develop in the damaged respiratory tissue. Another possible problem is **subacute sclerosing panencephalitis (SSPE)**, a rare brain disease characterized by a decrease in intellectual skills and loss of nervous

12.3
MICROFOCUS

THE PREFERRED WAY

Gertrude Belle Elion was getting dressed at 6:30 A.M. on the morning of October 17, 1988. Then the telephone rang. A moment later, Elion was speechless. She had won the Nobel Prize in Physiology or Medicine.

Only a few times in the 90-plus-year history of the Nobel Prize has the award been granted to researchers who developed drugs or worked for drug companies. This was one of those years. Gertrude B. Elion (Figure) shared the award with George H. Hitchings, her former co-worker at Burroughs Wellcome Research Laboratories in North Carolina, and with Sir James Black of King's College Medical School in London. The Nobel Committee named the three scientists for "their discovery of important principles of drug treatment" and for developing an intelligent method for designing new compounds based on an understanding of basic biochemical processes.

For Gertrude Elion, the award culminated a research career that almost wasn't. Even though she had graduated in 1937 from Hunter College with a Bachelor of Science degree in biochemistry, Elion had difficulty obtaining a laboratory position because of her gender. She therefore accepted a job as a chemistry teacher and after World War II came to the Wellcome laboratory, then in New York, as an assistant

to Hitchings. Although she never attained an advanced degree, Elion's technique and expertise were so respected that she soon came to be accepted as a colleague at the laboratory.

In 1944 Elion and Hitchings set out to learn how normal cell growth differs from that of abnormal cells, such as cancer cells, in hopes of finding a way to destroy abnormal cells. They focused on differences in how various species metabolize nucleic acid components, confining their studies mainly to the nitrogenous bases of nucleic acids. In the 1950s, they developed antileukemia drugs called thioguanine and 6-mercaptopurine. Then the biochemical clues led them to a series of other drugs including azathioprine (Imuran), which stalls the rejection mechanism in transplants; allopurinol, which is used to treat gout; and pyrimethamine and trimethoprim for malaria and other diseases. In 1977, they developed acyclovir, now used widely against herpes simplex and more recently against chickenpox. Other colleagues, applying the basic ideas of Elion and Hitchings, synthesized AZT for AIDS patients.

Elion and Hitchings were part of the so-called "fundamentalist" world of chemotherapy. By concentrating on the fundamental physiology and biochemistry of cells, they came to understand the essential metabolic pathways of cells and how to interfere with them. Other researchers, dubbed "screeners," preferred to bypass the cellular biochemistry and devote their efforts to screening a number of compounds, trusting their intuition and luck. It was the more rational approach that the Nobel Committee cited in its award.

FIGURE 12.8

A Young Boy Displaying the Skin Rash of Measles

The rash occurs as a red blush on the body surface and is different from the fluid-filled vesicles occurring in chickenpox.

FIGURE 12.9

Reported Cases of Measles (Rubeola) in the United States by Year, 1950–1991

Note the sharp dropoff in cases after licensure of the vaccine in the mid-1960s. Unfortunately, the immunity from this vaccine was not long-lasting and a new epidemic of measles began in the late 1980s. The inset shows the rise from a low of 1400 cases in 1983 to over 27,000 cases in 1990. The recent reduction in cases is partly due to renewed efforts to revaccinate susceptible individuals. (Courtesy of Centers for Disease Control and Prevention, *Annual Summary*, 1991.)

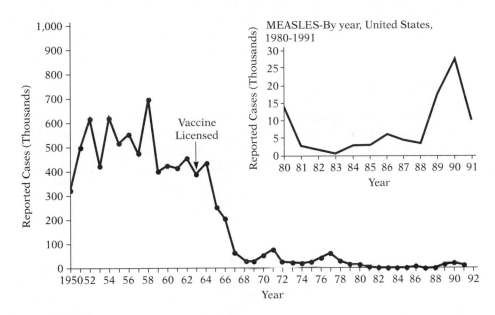

function. Some evidence also exists that the measles virus may be linked to encephalitis, multiple sclerosis, and diabetes.

In 1978, the U.S. Public Health Service launched a campaign to eliminate measles in the United States. The cornerstone of the campaign was immunization of all school-age children with attenuated measles viruses in the **measles-mumps-rubella (MMR)** vaccine. Measles viruses have no known hosts other than humans, a factor that is helpful in containing the disease. In 1983, the total number of reported cases was 1463, a reduction of 99.7 percent from the prevaccine era when over 500,000 cases were reported annually (Figure 12.9). However, the number skyrocketed to over 27,000 cases in 1990, reflecting the problem remaining in unimmunized children and college-age students inoculated with ineffective vaccine. Immunization of children entering grade school is now mandatory in all 50 states, and 42 states require students in all public and private schools below the college level to be immunized. By 1991, the number of cases had dropped to about 9500, and the decline continued in 1992 when about 2500 cases were reported to the CDC. By 1995, there were only 261 cases and the epidemic was over.

Multiple sclerosis: a disease accompanied by patchy destruction of the fatty sheath over nerve fibers.

Rubella (German Measles)

For generations, **rubella** was thought to be a mild form of measles. The distinction was not made until 1829 when Rudolph Wagner, a German physician, suggested that the two diseases were different. Thereafter, the new disease was known as German measles (from Wagner's homeland) to distinguish it from measles. ("German" is also believed to be derived from the Latin *germanus*,

Maculopapular:
pink-red pimplelike spots
that spread.

Mitosis:
the process by which
eukaryotic cells reproduce.

Cataract:
clouding of the lens of the
eye.

Glaucoma:
visual defects due to high
pressures exerted by eye
fluids.

MMR:
a three-component vaccine
providing immunity to
measles, mumps, and
rubella.

er"i-thē'mah
Erythema:
reddening

meaning "akin," in this case, akin to measles.) The name rubella ("small red") was suggested in the 1860s because the disease is accompanied by a slightly red rash.

Rubella is caused by an RNA virus of the Togaviridae family. The virion is icosahedral with an envelope and spikes containing hemagglutinin. Transmission generally occurs by contact or respiratory droplets, and the disease is usually mild. It is accompanied by occasional fever with a variable, pale pink maculopapular rash beginning on the face and spreading to the body trunk and extremities. When present, the rash develops rapidly, often within a day, and fades after another two days. Mild cold symptoms and swollen lymph nodes may accompany the rash. Recovery is usually prompt, but relapses appear to be more common than with other diseases, possibly because the viruses remain active within body cells.

Rubella is a dangerous disease in pregnant women. The viruses localize on the placenta and pass across the umbilical cord, causing damage to the fetus. This condition, called **congenital rubella syndrome**, occurs in about 50 fetuses annually in the U.S. Destruction of the fetal capillaries takes place and blood insufficiency follows. Mitosis is also retarded. The organs most often affected are the **eyes**, **ears**, and **cardiovascular organs**, and children may be born with cataracts, glaucoma, deafness, or heart defects. If rubella is contracted during the first month of pregnancy, the probability of damage to the fetus is about 50 percent. The probability declines sharply thereafter. In the 1965–1966 epidemic of rubella in the United States, over 50,000 instances of stillbirth and fetal deformity were recorded. The R in the TORCH group of diseases stands for rubella.

The **rubella vaccine** has had a dramatic effect on the incidence of rubella since its introduction in 1969. That year, physicians reported 58,000 cases of rubella but by 1995 the number was down to 140. The vaccine consists of attenuated viruses cultivated in human tissue cultures. It is combined with the measles and mumps vaccines (MMR) for subcutaneous inoculation of children. More than 95 percent of children entering grade school now provide evidence of rubella vaccination, but comparable figures do not exist for young adults. As a result, occasional outbreaks of rubella occur in older groups, and the CDC has begun a program to eliminate the disease through immunization, surveillance, and outbreak control. Adult females are advised to avoid pregnancy for three months after immunization as a precaution against contracting rubella from viruses in the vaccine.

Fifth Disease (Erythema Infectiosum)

In the late 1800s, numbers were assigned to diseases accompanied by skin rashes. Disease I was measles, II was scarlet fever, III was rubella, IV was Duke's disease (also known as **roseola** and no longer recognized as a disease but rather, any rose-colored rash), and V was **erythema infectiosum**. This so-called **fifth disease** remained a mystery until the modern era.

Fifth disease affects children primarily. The outstanding characteristic is a fiery red rash on the cheeks and ears that makes it appear as if the child has been slapped. (The disease is sometimes called "**slapped cheek disease**.") The rash may spread to the trunk and extremities, but it fades within several days leaving "lacy" patterns on the skin. Recurrences during ensuing days or weeks are related to bathing, sunlight, exercise, or stress. This characteristic and the "slapped cheek" appearance are important to diagnosis. A diagnosis is also reached by eliminating the possibility of other diseases.

The agent of fifth disease is a strain of **parvovirus** designated **B19** (fifth disease is therefore known as **B19 infection**). The parvovirus is a small DNA virion of the Parvoviridae family, having icosahedral symmetry. Community outbreaks of fifth disease occur worldwide, and transmission appears to be by respiratory droplets. Rubella is often suspected, especially if the child is unimmunized.

Mumps

Mumps takes its name from the English "to mump," meaning to be sullen or to sulk. The characteristic sign is enlarged jaw tissues arising from swollen salivary glands, especially the parotid glands. **Epidemic parotitis** is an alternate name for the disease.

pah-rot′id
pa-ro-ti′tis

The mumps virus is an RNA helical virion of the Paramyxoviridae family. Spikes with hemagglutinin are present in its envelope. The virus was among the first human viruses cultivated in fertilized chicken eggs, an achievement of Claud D. Johnson and Ernest Goodpasture in 1934.

Mumps is generally transmitted by droplets, contact, and fomites, and is considered less contagious than measles or chickenpox. The virus is found in the human blood, urine, and cerebrospinal fluid, even though its effects are observed only in the **parotid glands**. Infection may begin in one gland, but in 75 percent of cases, both glands are involved. Obstruction of the ducts leading from the glands retards the flow of saliva, which causes the characteristic swelling (Figure 12.10). The skin overlying the glands is usually taut and shiny, and patients experience pain when the glands are touched. Complications are rare in children, and attacks generally yield permanent immunity even if only one gland was previously involved. Often there are no symptoms, and children may become immune without realizing they had the disease.

Fomites:
lifeless objects that transmit the agents of disease.

Parotid gland:
the large salivary gland below the ear where the upper and lower jaw bones come together.

FIGURE 12.10

Reported Cases of Mumps in the United States Between 1968 and 1991

Mumps continues to decline in incidence due in large measure to use of the mumps vaccine in the MMR preparation. The inset shows the swollen parotid gland. (Courtesy of Centers for Disease Control and Prevention, *Annual Summary,* 1991.)

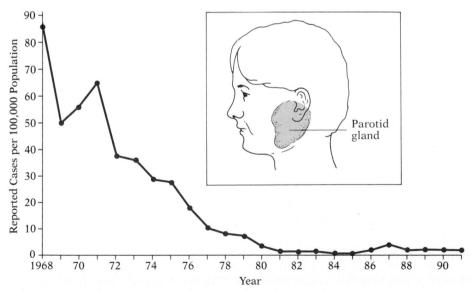

or-ki'tis

Orchitis:
infection of the male genital
organs as a complication to
mumps.

In adults, the mumps virus may pose a threat to the reproductive organs. As long ago as 1790, the Scottish physician Robert Hamilton observed damage to the male genital organs as a complication to mumps. The condition came to be called **orchitis** from the Greek *orchi-*, referring to testis. It is accompanied by swelling of the testes, with some damage to the reproductive tissue. The sperm count may be reduced, but sterility is not common. An estimated 25 percent of mumps cases in postpubic males develop into orchitis.

The vaccine for mumps was developed in 1967 from the Jeryl Lynn strain of the virus. It consists of attenuated viruses and is usually combined with the measles and rubella vaccines (MMR) for administration. Although the campaign against mumps never attained the fame of the campaigns against measles or rubella, the reduction of mumps cases has been equally notable. Almost 200,000 cases were recorded in 1967 but only 2982 cases occurred in 1985, the lowest number ever recorded. That record was broken in 1995, with only 780 cases. Thirty-five states currently require mumps immunizations for various age groups. Humans are the only hosts for the virus.

TO THIS POINT ■

We have surveyed a number of viral diseases whose symptoms occur largely on the skin surface. Herpes simplex infections are manifested as blisterlike lesions in cold sores, genital herpes, and herpes keratitis. With chickenpox, the lesions are more like fluid-filled teardrops. Measles is not accompanied by lesions but rather a blushlike rash that begins on the head and then spreads to the extremities. The rubella rash is similar, but it develops and fades rapidly and often does not occur at all. A fiery red rash on the cheeks and ears is a sign of fifth disease, and swollen parotid glands typify mumps.

None of the dermotropic diseases that we have studied is known to be life-threatening, but the long-ranging effects may be consequential. For example we have seen how SSPE is associated with measles, how orchitis may complicate mumps, how herpes zoster is an adult form of chickenpox, and how genital herpes recurs in a patient for many years. Also a severe congenital problem is associated with rubella. Thus many of the diseases formerly seen as benign are currently viewed in a new light.

In the concluding section of this chapter, we shall study other dermotropic diseases, including smallpox, a viral disease that has been known to be potentially fatal for centuries. The remarkable feature of smallpox is that it has not been observed in humans for about twenty years. This claim cannot be made for any other disease. We shall also give brief mention to the skin warts caused by different viruses, and to Kawasaki disease, a malady not yet related to a virus with certainty.

Other Dermotropic Viral Diseases

Smallpox (Variola)

Smallpox has ravaged people of the world since prebiblical times. It moved swiftly across Europe and Asia, often doubling back on its path, and was apparently brought to the New World in the 1500s by Cortez' troops. There it eventually killed 3.5 million Native Americans and contributed to the collapse of the

F I G U R E   12.11

The Smallpox Virus

(a) A negatively stained electron micrograph of the smallpox virus cultivated in cell culture (× 300,000). Note the bricklike shape of the virus and the characteristic rods at its surface. Smallpox viruses have no envelope. (b) A transmission electron micrograph of a cell infected with smallpox viruses (× 30,000). Rectangular mature virions can be observed (A) as well as immature virions (B). Also visible are the cell nucleus (N), mitochondrion (M), and lysosome (L).

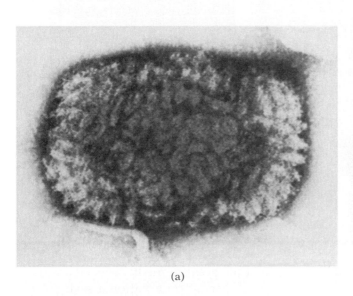

(a)

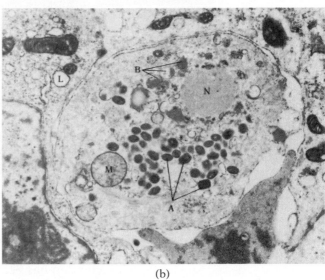

(b)

Inca and Aztec civilizations. Few people escaped the pitted scars that accompanied the disease, and children were not considered part of the family until they had survived smallpox.

Smallpox is caused by a brick-shaped DNA particle of the Poxviridae family (Figure 12.11). It is one of the largest virions, approximately the size of chlamydiae. The nucleocapsid is surrounded by a series of fiberlike rods; the virion has no envelope. Transmission is by contact.

The earliest signs of smallpox are high fever and prostration. Pink-red spots, called macules, soon follow, first on the face and then on the body trunk. (In chickenpox the spots appear randomly in crops.) The spots become pink pimples, or papules, and then, fluid-filled vesicles so substantial that the disease is also called **variola**, from the Latin *varus* for vessel (Figure 12.12). The vesicles soon become deep pustules, which break open and emit pus. Should the person survive, the pustules will leave pitted scars, or **pocks**. These are generally smaller than the lesions of syphilis (the Great Pox) or varicella (chickenpox).

Centuries ago, people discovered that they could survive smallpox if they were fortunate enough to experience it during a mild year. The custom thus arose of "buying the pox"; one would approach a person who had a mild form and offer money to rub skin together. An Oriental custom of injecting oneself with pox fluid eventually spread to Europe and later, North America. In 1721, a hospital was established in Boston for anyone interested in **variolation**, as the process was called.

In 1798, an English physician named Edward Jenner noted that milkmaids contracted a mild form of smallpox named **cowpox**, or **vaccinia** (*vacca* is Latin for cow). Anyone who experienced cowpox apparently did not later contract smallpox. Jenner therefore utilized material from a cowpox lesion for variolation and established the process of **vaccination** (Figure 12.13). His method was so successful that Napoleon ordered his entire army vaccinated in 1806. The effort to vaccinate the American population was led by President Thomas Jefferson.

Nucleocapsid:
the nucleic acid core and protein coat of the virus.

Pustules:
deep pus-containing lesions that reach down to the lower skin layers.

Cowpox:
a pox disease in animals that may also occur in humans.

FIGURE 12.12
The Lesions of Smallpox

(a) The lesions are raised, fluid-filled vesicles similar to those in chickenpox. For this reason, cases of chickenpox have been misdiagnosed as smallpox. Later, the lesions will become pustules (b), and then form pitted scars, the pocks.

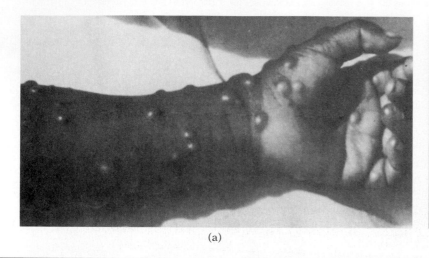

(a)

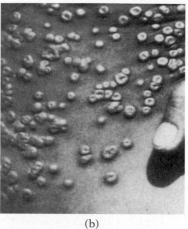

(b)

Vaccination has been hailed as one of the greatest medical-social advances because it was the first attempt to control disease on a national scale. It was also the first effort to protect the community rather than the individual. A century passed before it was understood that the antibodies produced against the mild cowpox virus were equally effective in neutralizing the smallpox virus.

In 1966, the World Health Organization (WHO) received funding to attempt the global eradication of smallpox. **Surveillance containment** methods were used to isolate every known pox victim, and all contacts were vaccinated. The eradication was aided by the fact that smallpox viruses apparently do not exist anywhere in nature except in humans. On October 26, 1977, health workers reported isolation of the last case, and the WHO instituted a two-year waiting period to see if any new cases would appear. Finally, in 1979, the WHO announced that smallpox had been eradicated from the earth, the first such claim made for any disease.

As of this writing, no naturally occurring cases of smallpox have been reported (over 150 suspected cases turned out to be chickenpox). Smallpox viruses still remain in two laboratories (one at the CDC, the other in Moscow), and the 1978 death from smallpox of a laboratory worker in England points up their hazard. The WHO has recommended that stocks of smallpox viruses be destroyed, especially since scientists have deciphered the base sequence of the smallpox virus genome. There was a question, however, as to whether the base sequence of the genome should be published. By the time you read this, the questions should have been resolved and the important decisions made (Micro-Focus: 12.4). For safety's sake, the World Health Organization maintains stocks of smallpox vaccine at depots throughout the world to immunize 300 million people.

Molluscum Contagiosum

Molluscum contagiosum is a viral disease characterized by wartlike skin lesions. The lesions are firm, waxy, and elevated with a depressed center. When pressed, they yield a milky, curdlike substance. Although usually flesh-toned,

mol-lus'kum
kon-ta"je-o'sum

FIGURE 12.13
The First Vaccination

A photograph of a painting by Robert Thom showing Edward Jenner vaccinating a young boy. The woman at the right is holding her wrist at the spot from which cowpox material was taken.

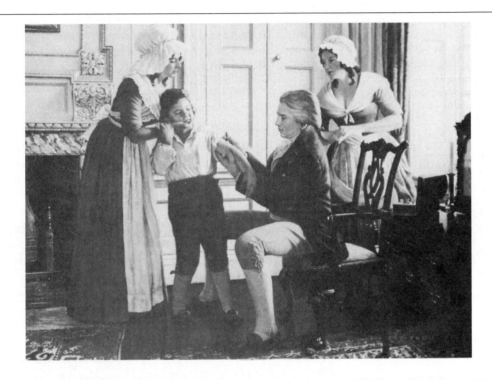

12.4
MICROFOCUS

"SHOULD WE OR SHOULDN'T WE?"

One of the liveliest debates in microbiology is whether the last remaining stocks of smallpox viruses should be destroyed. Herewith some of the arguments.

For Destruction:
- People are no longer vaccinated so if the virus should escape the laboratory, a deadly epidemic could ensue.
- The DNA of the virus has been sequenced and many clones of fragments are available for performing research experiments, so the whole virus is no longer necessary.
- Eradicating the disease means eradicating the remaining stocks of laboratory virus, and the stocks must be destroyed to complete the project.

- If the United States and Russia destroy their smallpox stocks, it will send a message that biological warfare cannot be tolerated.

Against Destruction:
- Future studies of the virus are impossible without the whole virus. Indeed, certain sequences of the viral genome defy deciphering by current laboratory means.
- Studying the genome of the virus without the whole virus will not give insights into how the virus causes disease.
- Mutated viruses could cause smallpoxlike diseases, so continued research on smallpox is necessary in order to be prepared.

- Smallpox viruses may be secretly retained in other labs in the world, so destroying the stocks may create a vulnerability. Smallpox viruses may also remain active in buried corpses.
- Destroying the virus impairs the scientist's right to perform research, and the motivation for destruction is political not scientific.

Now we turn to you. Can you add any insights to either list? And which argument do you prefer? Please let us know when you can.

P.S. June, 1999 could be an important date in medical history.

the lesions may appear white or pink. Possible areas involved include the facial skin and eyelids in children, and the genital organs in adults. The lesions may be removed by cutting (excising) them.

The virus of molluscum contagiosum is a DNA virion of the Poxviridae family. Transmission is generally by contact, such as by sexual contact. A characteristic feature of the disease is the presence of large cytoplasmic bodies called **molluscum bodies** in infected cells from the base of the lesion.

Warts

Warts are small, usually benign skin growths that are commonly due to viruses. Plantar warts are warts that occur on the soles of the feet. Vaginal and genital warts are another form of warts, often transmitted in sexual contact. These warts are sometimes called **condylomata** from the Greek *kondyloma* meaning knob, a reference to the figlike appearance of the warts. They are usually moist and pink.

Among the major causes of warts are the **human papilloma viruses**, a collection of over 25 types of icosahedral DNA virions of the Papovaviridae family. In most cases, the skin warts they cause are a minor problem. However, evidence suggests that certain types of papilloma virus may be associated with cervical cancer. Indeed, in 1994 researchers reported the identification of DNA from 25 types of papilloma viruses in the tumor cells of 95 percent of patients having cervical cancer.

Molluscum bodies: large cytoplasmic bodies in cells infected with molluscum contagiosum viruses.

con"di-lo-mah'tah

TABLE 12.2

A Summary of Dermotropic Viral Diseases

DISEASE	CLASSIFICATION OF VIRUS	TRANS-MISSION	ORGANS AFFECTED
Herpes simplex	Herpesviridae	Contact	Skin Pharynx Genital organs
Chickenpox (varicella)	Herpesviridae	Droplets Contact	Skin Nervous system
Measles (rubeola)	Paramyxoviridae	Droplets Contact	Respiratory tract Skin, blood
Rubella (German measles)	Togaviridae	Droplets Contact	Skin, blood
Fifth disease (B19 infection)	Parvoviridae	Droplets (?)	Skin, blood
Mumps (epidemic parotitis)	Paramyxoviridae	Droplets	Salivary glands Blood
Smallpox (variola)	Poxviridae	Contact Droplets Fomites	Skin Blood
Molluscum contagiosum	Poxviridae	Contact	Skin
Warts	Papovaviridae	Contact	Skin
Kawasaki syndrome	(Unknown)	(Unknown)	Skin, blood

Genital warts due to papilloma viruses may be transferred during sexual intercourse, and some virologists suggest that these warts may be more prevalent than genital herpes. Malignancies of the cervix, vagina, penis, and anus have also yielded papilloma viruses. Moreover, a pregnant woman with genital warts may transmit the viruses during the birth process, and studies show that in the newborn the viruses may lodge in the larynx, trachea, and the lungs. Table 12.2 presents a summary of dermotropic viral diseases.

pap-i-lo′mah

Kawasaki Disease

Although Kawasaki disease has not yet been identified as a viral disorder, the course of the disease suggests that an infection is involved. We shall therefore consider it here.

Kawasaki disease takes its name from the Japanese pediatrician Tomisaku Kawasaki, who described its symptoms in 1967. Children one to two years of age account for most of its victims. The illness includes a high fever and sore throat, and red spots appear, first on the extremities and then the body trunk. As the disease progresses, there is a characteristic peeling of the skin about the fingers, a phenomenon called **desquamation**. Complications of Kawasaki disease usually involve the cardiovascular system, and internal bleeding may occur. Treatment is generally directed at minimizing these possibilities.

des″kwa-ma′shun

Kawasaki disease was first detected in the United States in 1971. Since then it has been reported with increasing frequency and about 2500 cases were

VACCINE	SPECIAL FEATURES	COMPLICATIONS
Not available	Characteristic lesions Lipschütz bodies Acyclovir treatment	Encephalitis Congenital infections Neonatal herpes
Attenuated viruses	Characteristic lesions Crops Acyclovir treatment	Herpes zoster (shingles) Reye's syndrome Pneumonia
Attenuated viruses	Koplik spots Progress of rash Hemagglutination inhibition	SSPE Pneumonia Encephalitis
Attenuated viruses	Skin rash Mild cold symptoms	Congenital rubella syndrome
Not available	"Slapped cheek" appearance Lacy skin patterns	None established
Attenuated viruses	Swollen glands Hemagglutination inhibition	Orchitis Encephalitis Meningitis
Cowpox viruses	Characteristic lesions Macules, papules, vesicles, pox	Permanent scarring
Not available	Characteristic lesions	None
Not available	Characteristic lesions	None
Not available	Skin rash Desquamation	Heart involvement

Syndrome:
a collection of symptoms.

reported through the 1980s. A notable outbreak in 1985 in the Denver, Colorado, area involved 61 people. Why the disease occurs in clusters is not yet explained, but there is some laboratory evidence that a staphylococcus or a streptococcus may be involved, as reported in 1993.

The possibility also exists that Kawasaki disease is an immune-related syndrome complicating some unknown viral disease in the same manner that Reye's syndrome follows influenza and chickenpox. Some physicians therefore refer to the illness as **Kawasaki syndrome** rather than Kawasaki disease. Indeed, the current trend appears to favor the outlook that neither Kawasaki disease nor Kawasaki syndrome is a clinical entity. The CDC has made no mention of the disease in its reports since 1989, and most other textbooks of microbiology do not discuss the disease. Perhaps the disease will also disappear from this textbook one day, but for the time being it is important to be aware of it.

NOTE TO THE STUDENT

In this chapter we have had an opportunity to note the changing complexion of microbiology. We have seen, for example, how measles viruses are increasingly associated with neurological problems; how Reye's syndrome has become linked to influenza and chickenpox; how an association is growing between cervical tumors and herpes simplex viruses; and how chickenpox and herpes zoster have become so closely related that a single virus causes both. Moreover, we note that despite modern detection devices, there is still no identifiable agent for Kawasaki disease.

The point is that microbiology is a dynamic and ever-changing science. This dynamism implies that microbiologists and physicians must continually adjust to new truths as they emerge. A friend once told me that if you dip a tennis ball into the ocean, the water dripping from the ball represents all that is known; the ocean represents all that is waiting to be discovered.

Summary

Within broad limits, viral diseases usually occur in specific parts of the body. Accordingly, the diseases are classified into four categories, two of which are pneumotropic viral diseases and dermotropic viral diseases. Pneumotropic diseases occur in the respiratory tract, while dermotropic diseases display their symptoms on the skin or close to the skin surface.

Among the important pneumotropic diseases are influenza, adenovirus infections, respiratory syncytial disease, and rhinovirus infections. Varying degrees of fever and respiratory distress accompany all these diseases, and life-threatening situations are rare except if secondary infection takes place, as is possible with influenza. Various sites of infection within the respiratory tract are seen, ranging from the air sacs for respiratory syncytial disease to the nose for rhinoviral infection. Adenovirus infections can occur outside the respiratory tract as well.

The dermotropic viral diseases include mild infections such as chickenpox and serious infections such as smallpox. Infection can begin with skin contact, such as in herpes simplex, smallpox, or warts, or it may be initiated by airborne viruses. In the latter case, the viruses infect the respiratory tract causing mild respiratory symptoms, then invade the blood and localize near the skin where they induce manifestations of disease. The skin rashes of measles, rubella, and fifth disease are typical. Mumps also follows this pattern but a skin rash is not present; rather, there is a salivary gland swelling and a painfully tight skin. Kawasaki disease is included in this chapter due to its skin symptoms, even though a virus has not yet been identified.

Questions for Thought and Discussion

1. In the mid-1980s, the nation's colleges for the deaf reported an unprecedented demand for admission. For example, at the National Technical Institute for the Deaf at Rochester Institute of Technology, the student body swelled from 750 students to 1250 students. How was this related to the events of a previous generation involving rubella?

2. In February 1992, the CDC reported an outbreak of measles at an international gymnastics competition in Indianapolis, Indiana. A total of 700 athletes and numerous coaches and managers from 51 countries were involved. Although the potential for a disastrous international epidemic was high, it never materialized. What steps do you think the local health agencies took to quell the spread of the disease?

3. A little girl experiences frequent and severe vomiting for a period of hours. She soon becomes sleepy and glassy-eyed. When disturbed, she quickly becomes irritated and combative. One week before, she recovered from chickenpox. What is she experiencing and what course of action must be taken?

4. Thomas Sydenham, the "English Hippocrates," was a London physician of the seventeenth century. In 1661, he differentiated measles from scarlet fever, smallpox, and other fevers, and set down the foundations for studying these diseases. How would a modern Thomas Sydenham go about distinguishing the variety of lookalike skin diseases in this chapter?

5. In the United Kingdom, the approach to rubella control is to concentrate vaccination programs on young girls just before they enter the childbearing years. In the United States, the approach is to immunize all children at the age of 15 months. Which approach do you believe preferable? Why?

6. From February to May 1991 an outbreak of rubella occurred among Amish people living near Lancaster, Pennsylvania. Members of the Amish community traditionally refuse vaccination, and the great majority of those affected had no history of rubella immunization. Once the epidemic had subsided, health officials instituted a campaign to convince the Amish community of the benefits of immunization. How do you think they proceeded with their campaign?

7. In 1994 the fitness file of a local newspaper carried a story on "The New Herpes." The reference was to human papilloma viruses and genital warts as the herpes simplex of the 90s. How many similarities can you find between these two diseases? Would you agree with the comparison?

8. What is so unique about influenza viruses that it probably accounts for the continuing emergence of new strains? Why should you be hesitant about accepting a flu shot if you are allergic to eggs?

9. Most physicians agree that there would be great demand for a genital herpes vaccine. However, there is much opposition to marketing a vaccine that contains attenuated herpes simplex viruses. Why is this so? What alternatives are there for a useful vaccine against genital herpes?

10. It is not uncommon for a person with respiratory disease to visit the doctor and be told: "Don't worry about it. It's just a touch of the flu. I'll give you an injection of penicillin before you leave." Suppose the person wanted to know if it really was influenza. What diagnostic tests would have to be performed? Also, why is the doctor inclined to give a penicillin injection?

11. A child experiences "red bumps" on her face, scalp, and back. Within 24 hours, they have turned to tiny blisters and become cloudy, some developing to sores. Finally all become brown scabs. New "bumps" keep appearing for several days, and her fever reaches 102° F by the fourth day. Then the blisters stop coming and the fever drops. What disease has she had?

12. The great seventeenth-century physician William Harvey, who discovered how blood circulates, was a great fan of garlic therapy for disease. In one of his writings, Harvey recommended putting a clove of garlic inside your shoe when you have a respiratory illness. What do you think of Harvey's recommendation?

13. One day in March 1977, a Boeing 737 bound for Kodiak, Alaska, developed engine trouble and was forced to land. While the company rounded up another aircraft, the passengers sat waiting for four hours in the unventilated cabin. One passenger, it seemed, was in the early stages of influenza and was coughing heavily. By the week's end, 38 of the 54 passengers on the plane had developed influenza. What lesson does this incident provide?

14. Although smallpox viruses are considered to be gone from the environment, there remains a closely related virus that causes monkeypox in nature. By what genetic mechanisms could this virus conceivably become a smallpox virus?

15. A man experiences an attack of shingles and is warned by his doctor to stay away from children as much as possible. Why is this advice given? Is it justified?

Review

On completing your study of pneumotropic and dermotropic viral diseases, test your comprehension of the chapter contents by circling the choices that best complete each of the following statements:

1. Rhinoviruses are a collection of (RNA, DNA) viruses having (helical, icosahedral) symmetry and the ability to infect the (air sacs, nose) causing (mild, serious) respiratory symptoms.

2. Herpes simplex is a viral disease that can be transmitted by (breathing contaminated air, contact) and is characterized by thin-walled (blisters, ulcers) that often appear during periods of (emotional stress, exercising) but can be treated with a drug called (deoxycyclovir, acyclovir).

3. In children, the skin lesions of chickenpox occur (all at once, in crops) and resemble (teardrops, pitted scars), but in adults the lesions are known as (shingles, erythemas) and resemble blotchy patches of (blue, red) that are very (itchy, painful).

4. For generations, rubella was thought to be a mild form of (chickenpox, measles) because it was also accompanied by (a skin rash, brain lesions) and was transmitted by (contaminated water, airborne droplets).

5. The complications of influenza include (Reye's, Koplik's) syndrome, while for mumps the complication is a disease of the (testes, pancreas) called (colitis, orchitis) and for measles it is a disease of the (brain, liver) known as (SSPE, GBS).

6. One of the early signs of (smallpox, measles) is a series of (Koplik spots, Lipschütz bodies) occurring in the (lungs, mouth) and signaling that a (red rash, blue-green rash) is forthcoming.

7. Although the agent of (fifth, sixth) disease has not been identified with certainty, a leading candidate is the (B29, B19) strain of (picornavirus, parvovirus), a small (DNA, RNA) virus.

8. After transmission by (mosquitoes, airborne droplets), the virus of (mumps, Kawasaki disease) spreads by the blood to the (salivary, sweat) glands where it interferes with fluid secretion.

9. Although now eradicated, (mumps, smallpox) can be prevented by immunizations with (fowlpox, cowpox) virus in a method first devised in 1798 by Edward (Jennings, Jenner).

10. Antigenic variation among (mumps, influenza) viruses seriously hampers the development of a highly effective (vaccine, treatment), and a life-threatening situation can occur if secondary infection due to (fungi, bacteria) complicates the primary infection.

11. Respiratory syncytial disease is caused by a (DNA, RNA) virus that infects the (lungs, intestines) of (adults, children) and induces cells to (clump together, move apart) and form giant cells called (syncytia, tumors).

12. Warts are small, benign skin growths caused by human (parvoviruses, papilloma viruses), transmitted by (sexual contact, contaminated milk), and somewhat similar to the growths associated with (chickenpox, molluscum contagiosum).

13. Adenoviruses include a collection of (DNA, RNA) viruses that induce the formation of (granules, inclusions) and are responsible for (yellow fever, common colds), as well as infections of the (eye, ear) and (kidneys, meninges).

14. Genital herpes is caused by a (helical, icosahedral) virus that is believed to affect 10 to 20 (thousand, million) Americans each year, causing blisters with (thick, thin) walls that disappear in about three (days, weeks) only to reappear when (stress, physical injury) occurs.

15. The TORCH diseases are a set of (infectious, physiological) diseases transmitted by (airborne droplets, transplacental passage), occurring in (the elderly, newborns), and including (rubeola, rubella) and (herpes simplex, humoral disease).

16. The MMR vaccine contains (inactivated, attenuated) viruses and is used primarily in (children, older adults) to provide (long-term, short-term) immunity to such diseases as (measles, molluscum contagiosum), (mononucleosis, mumps), and (German measles, influenza).

CHAPTER 13

VISCEROTROPIC AND NEUROTROPIC VIRAL DISEASES

Yellow fever was probably the most dramatic disease ever to strike the United States. During the 1700s historians chronicled 35 separate outbreaks as the disease ravaged the country (MicroFocus: 13.1). Nowhere did the pestilence strike harder than in Philadelphia.

In 1793, Philadelphia was the capital of the United States and its largest city, with a population of 40,000. When yellow fever broke out, the panic rivaled that in Europe during the plague years. In sick patients, the eyes glazed, the flesh yellowed, and delirium set in. People died, not here and there, but in clusters and in alarming patterns. Friends recoiled from one another. If they met by chance, they did not shake hands but nodded distantly and hurried on. The very air felt diseased, and people dodged to the windward of those they passed. The deaths went on, great ugly scythings of humanity.

At the height of the epidemic, thousands of people fled Philadelphia, and officials posted warning notices on all remaining houses where people were infected. Those who could not get away, including most of the city's poor, sought protection by breathing through cloth masks soaked in garlic juice, vinegar, or camphor. Benjamin Rush, a noted American physician (and signer of the Declaration of Independence), prescribed a frightening course of purges, bloodlettings, vomiting, and immersion in icewater to reduce fever. Nearly all the 24,000 people who remained in Philadelphia were afflicted. Almost 5000 died.

Yellow fever will be the first disease discussed in this chapter. Microbiologists now know that the disease is caused by a virus, but in 1793 only a few doctors were bold enough to suggest the involvement of a contagious agent. The majority of physicians, Rush included, believed that yellow fever was primarily a malfunction of the person's own bloodstream and, therefore, not transmissible. Over a century elapsed before mosquitoes were identified as agents of transmission, and still another 50 years passed before the responsible virus was photographed. Currently, yellow fever is under control in the United States.

The diseases we shall study in this chapter fall into two general categories. Some illnesses, like yellow fever, are regarded as viscerotropic diseases because they affect the blood and organs of the viscera. Other viscerotropic diseases

vis″er-o-trōp′ik

Viscera:
the abdominal cavity.

nur"o-trōp'ik

include hepatitis, infectious mononucleosis, and a series of gastrointestinal illnesses. The second category of illnesses are the neurotropic diseases, such as polio and rabies, which affect the central nervous system. As in Chapter 12, each disease is presented as a separate essay, and you may select the order of study most suitable to your needs.

Viscerotropic Viral Diseases

Yellow Fever

Yellow fever occurs primarily in tropical and subtropical countries, where it is referred to as "yellow jack." The disease takes its name from the lemon-yellow tint seen in the skin of victims with severe cases.

The earliest known outbreak of yellow fever in the western hemisphere took place in Central America in 1596. The disease spread rapidly, and soon made large regions of the Caribbean and tropical Americas almost uninhabitable. In time, natives developed immunity or suffered only mild cases, but the mortality rate in outsiders remained high. Finally, in 1901, a group led by Walter Reed identified mosquitoes as the agents of transmission (MicroFocus: 13.2), and with widespread vector control, the incidence of the disease gradually declined.

Yellow fever was the first human disease associated with a virus. The causative agent is an RNA virion of the Flaviviridae family with icosahedral symmetry and an envelope. It is one of the smallest known viruses and is often referred to as an **arbovirus** because it is *arthropod-borne*.

hem"ah-go'gus
a-e'dez a-gip'ti

Yellow fever occurs naturally in monkeys and other jungle animals, where the virus is transmitted by various **mosquitoes**, including species of *Haemogogus*. In the cities, a different mosquito, *Aedes aegypti*, transmits the virus among humans. (The tiger mosquito *Aedes albopictus* has also been implicated in transmission.) *Aedes aegypti* is common in the Caribbean region and in the southern and eastern United States, and yellow fever was a problem in these areas for many generations. In 1803, for example, Napoleon sent troops to quell an uprising in Haiti, but yellow fever killed thousands of his men. Soon thereafter, Napoleon came to think of the Americas as a fever-ridden land, and when President Thomas Jefferson sent emissaries to France to negotiate the purchase of the New Orleans region, Napoleon offered the entire Louisiana territory at a bargain price.

13.2

"FOR THE CAUSE OF HUMANITY. . ."

During the Spanish American War, the U.S. Government became disease-conscious because more soldiers were dying from disease than from bullet wounds. Yellow fever was particularly bad in Cuba, where it exacted a heavy toll from those who lived there. When the war was over, Cuba remained under U.S. control, and the surgeon general sent a commission of four men to study the disease. Led by Major Walter Reed, the group included three assistant surgeons: James Carroll, Jesse W. Lazear, and Aristides Agaramonte. On June 25, 1900, the four men assembled in Cuba and began their work.

At first the commission devoted its energy to isolating a bacterium, but none could be found. As the weeks passed, the investigators were impressed with the peculiar way the disease jumped from house to house, even when there was no contact with infected persons or contaminated objects. They also visited Carlos J. Finlay, a physician from Havana who insisted that mosquitoes were involved in transmission. If his theory was true then the disease could be interrupted by simply killing the mosquito.

By now it was August, and Reed had been called back to Washington. Carroll, Lazear, and Agaramonte pushed forward and bred mosquitoes from eggs given them by Finlay. They allowed the mosquitoes to feed on patients with established cases of yellow fever, and then applied the insects to the skin of volunteers, including themselves. The results were inconclusive: some volunteers got yellow fever, but others did not. Two accidents that saved the research then followed. One was fortunate, the other tragic.

One day in late August, Carroll decided to feed an "old" mosquito some of his own blood, lest it die. Three days later, Carroll was ill with the fever. Lazear's notebook recorded that the insect had fed "*twelve*

days before on a yellow fever patient, who was then in his *second day* of disease." This, they discovered, was the proper combination of two factors necessary for a successful transmission. If properly adhered to, the disease could be reproduced over and over again.

Then came tragedy. Lazear was working at the bedside of a yellow fever patient when a stray mosquito settled on his wrist. For reasons not clear, Lazear let the insect drink its fill. Five days later he developed yellow fever; on the seventh day of his illness, he died. Lazear had been bitten previously, but apparently by uninfected mosquitoes. This time, the mosquito was infected.

Reed returned to Cuba in October, and a new set of experiments was planned to prove once and for all that clothing and

other objects could not transmit yellow fever. The experiments were gruesome: some volunteers slept in the blood-soaked and vomit-stained garments of disease victims; others allowed themselves to be bitten mercilessly by mosquitoes; still others came forward to be injected with blood from yellow fever victims. By late 1900 there was no doubt that mosquitoes were the carriers of yellow fever.

It has been said that the experiments performed in Cuba are among the noblest in the history of medicine. Two volunteers, Private John R. Kissinger (Figure) and clerk John J. Moran, were asked why they were agreeing to such life-threatening experiments. "We volunteer," they replied, "solely for the cause of humanity and in the interest of science."

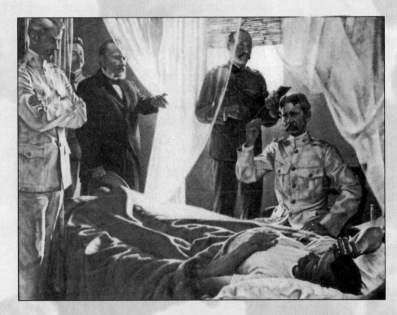

Photograph of a painting by Robert Thom depicting members of the yellow fever commission at the bedside of Private John Kissinger after he was bitten by infected mosquitoes. From left to right, Major W.C. Gorgas (Havana sanitation officer), Aristides Agaramonte, Carlos J. Finlay, James Carroll, and Walter Reed. The experiments were conducted at Camp Lazear, named for Jesse Lazear, who had died of yellow fever the previous summer.

Bile:
the yellow-brown mixture of acids, salts, pigments, and other substances produced by the liver and stored in the gall bladder; assists fat digestion.

ti'ler

deng'e

Breakbone fever:
an alternate name for dengue fever, accompanied by bone-breaking sensations.

B-lymphocyte:
a type of mononuclear white blood cell that functions in the immune system.

Yellow fever can be a fatal disease. Mosquitoes inject the viruses to the bloodstream, and fever mounts within days. Infection of the liver causes an overflow of bile pigments to the blood, a condition called **jaundice**, and the complexion becomes yellow. Successive waves of fever accompany viral invasions of the bloodstream from the liver. The gums bleed, the stools turn bloody, and the delirious patient often vomits blood. Mortality rates approach 50 percent in some epidemics.

Except for supportive therapy, no treatment exists for yellow fever. However, the disease can be prevented by immunization with either of two **vaccines**. The more widely used vaccine contains the 17D strain of yellow fever virus cultivated in chicken eggs. Max Theiler, a South African physician, won the 1951 Nobel Prize in Physiology or Medicine for the development of this vaccine. The second vaccine contains viruses from mouse brain tissue. Both vaccines provide immunity for about ten years.

Dengue Fever

Dengue fever has been known since David Bylon, a physician in the Dutch East Indies, described an outbreak in 1779. The disease takes its name from the Swahili word *dinga*, meaning cramplike attack, a reference to the symptoms. Dengue fever is caused by an RNA icosahedral virion of the Flaviviridae family. The virus is closely related to the yellow fever virus, except that four types of dengue fever virus are known to exist. Transmission is by the *Aedes aegypti* **mosquito** and by the tiger mosquito *Aedes albopictus* (Figure 13.1).

High fever and prostration are early signs of dengue fever. These are followed by sharp pain in the muscles and joints, and patients often report sensations that their bones are breaking. The disease is therefore called **breakbone fever**. Another name, **saddleback fever**, is used because of up-and-down temperature fluctuations. After about a week the symptoms fade. Death is uncommon, but if one of the other dengue viruses enters the body, a condition called **dengue hemorrhage fever** may occur. In this condition, a rash appears on the face and extremities, and severe vomiting and shock ensue.

Dengue fever has traditionally been a problem in Southeast Asia. However, in 1963, the disease broke out in Central America, and it has occurred sporadically in the Americas since then. In 1994, public health officials identified 46 cases of various types in 17 states of the United States, mostly from travelers to Caribbean regions.

Infectious Mononucleosis

The name **infectious mononucleosis** (or "mono" in the vernacular) is familiar to young adults because the disease is common in this age group. It is sometimes called the "kissing disease" because it is spread by mouth-to-mouth contact. Droplets and fomites, such as table utensils and drinking glasses, may also carry the virus. Studies show that the "infectious" nature of infectious mononucleosis may be overly emphasized because the disease is not highly contagious.

Infectious mononucleosis is a blood disease, especially of cells of the lymph nodes and spleen. Enlargement of the lymph nodes ("swollen glands") is accompanied by a sore throat, fever, and a high count of **B-lymphocytes**, a type of mononuclear white blood cells (hence, mononucleosis). The virus infects the lymphocytes and causes them to proliferate. Mononucleosis usually runs its course in three to four weeks. However, cases may be complicated and the effects of disease may remain for several months. Among the most dangerous

FIGURE 13.1

The Dengue Fever Virus and Its Vector

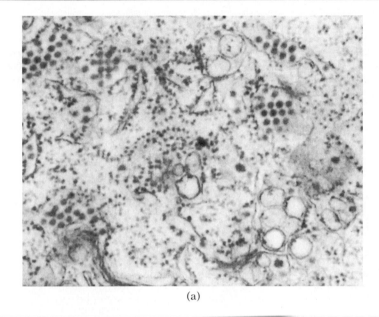

(a)

(b)

(a) A transmission electron micrograph of dengue fever viruses in the cytoplasm of host tissue cells. The viruses are the darkly stained bodies occurring in clusters ($\times$ 51,200).
(b) The *Aedes aegypti* mosquito that transmits dengue fever as well as yellow fever.

complications are heart defects, facial paralysis, and rupture of the spleen. The liver may be involved and jaundice may occur, a condition some physicians mistake for hepatitis. Those who recover usually become carriers for several months and shed the viruses into their saliva.

The diagnostic procedures for mononucleosis include detection of an elevated lymphocyte count and the observation of **Downey cells**, a type of B-lymphocyte with vacuolated and granulated cytoplasm. The patient also experiences an elevation of **heterophile antibodies**, that is, antibodies reacting with antigens from unrelated species. Such antibodies can be detected by the **Paul–Bunnell test**, a procedure first proposed in 1932 by John R. Paul and W. Willard Bunnell. The test is performed by mixing samples of the patient's serum with sheep or horse erythrocytes and observing the cells for agglutination. Figure 13.2 describes an adaptation of the test.

The virus of infectious mononucleosis is a DNA herpesvirus having icosahedral symmetry and an envelope. A substantial body of evidence indicates that it is identical to the **Epstein–Barr virus**. This virus has been detected in patients who have Burkitt's lymphoma, a cancer of the connective tissues of the jaw that is prevalent in areas of Africa. First isolated in the early 1960s by British virologists M. Anthony Epstein and Yvonne M. Barr, the Epstein–Barr virus was a surprising revelation and an important breakthrough in medicine because it demonstrated the link between viruses and cancer. In 1966, Gertrude and Werner Henle located the Epstein–Barr virus in mononucleosis patients.

Contemporary virologists continue to search for reasons why the Epstein–Barr virus is associated with tumors on one continent and infectious mononucleosis on another. Some cancer specialists theorize that the malaria parasite, common in Africa, acts as an irritant of the lymph gland tissue, thereby stimulating tumor development. Another possibility is that there are really two viruses, a mononucleosis virus and an Epstein–Barr virus, and that one triggers the other to function.

het'er-o-fil"
Antigens:
substances that stimulate the immune system, often resulting in antibodies.

Burkitt's lymphoma:
a type of cancer occurring in connective tissues of the jaw.

Malaria:
a serious protozoal disease of the red blood cells, transmitted by mosquitoes.

FIGURE 13.2

The Monospot Slide Test for Infectious Mononucleosis

The test is based on an agglutination reaction between horse erythrocytes and infectious mononucleosis antibodies. (a) Blood is taken from the patient. (b) The serum is separated from the cells. (c) A drop of serum is placed on a slide containing guinea pig tissue extract and the two are mixed. A reaction occurs that adsorbs any closely related antibodies in the serum. (d) Horse erythrocytes are then added and the components are mixed. (e) If infectious mononucleosis antibodies are present in the serum, the erythrocytes will agglutinate and give a positive reaction. (f) Absence of agglutination indicates absence of mononucleosis antibodies and therefore a negative reaction. Suitable controls not shown in the figure must also be included.

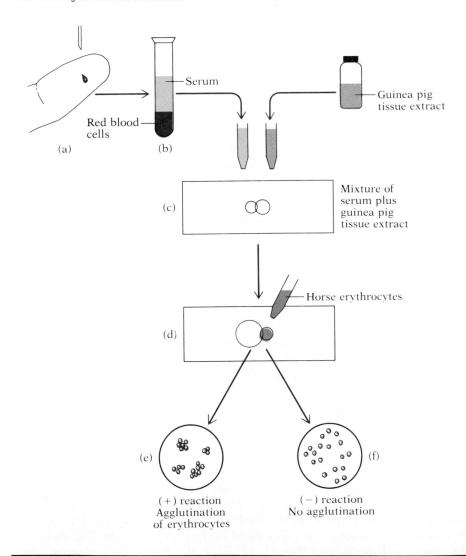

Herpesviruses:
a group of enveloped DNA viruses that can remain latent in cells.

The mononucleosis virus resembles other herpesviruses in its ability to remain in the body for years. Some virologists suggest that the virus enters the body when the person is very young, then replicates and remains inside the lymphocytes before emerging in the young adult. A vaccine is unavailable, mainly because the virus has not yet been isolated with certainty. However, interesting work is being done with proteins from sheep blood cells as immunizing agents. Infectious mononucleosis strikes about 100,000 persons annually in the United States, but this figure is only an estimate because physicians are not required to report cases to the CDC.

The Epstein–Barr virus has also been linked to **chronic fatigue syndrome**. The symptoms of chronic fatigue syndrome include sore throat; aching muscles; swollen lymph nodes; and prolonged, overwhelming fatigue (patients say they feel as limp as Raggedy Ann dolls). Because the syndrome often occurs in young, urban professionals, it has been dubbed the "Yuppie Flu." Evidence of EB virus involvement is based on the high levels of EB viruses and their antibodies in affected individuals. The relationship is not conclusive, however. A 1988 study, for example, showed that treatment specifically aimed at Epstein–Barr virus did little to relieve the symptoms. There is also a body of evidence dating from 1991 that points to a retrovirus or retroviruslike agent called a **spumavirus** as the possible cause. Even this evidence is questionable, however, because research reported in 1993 suggests that there is little correlation between positive laboratory tests for the retrovirus and patients having the syndrome. The tendency at present is to diagnose chronic fatigue syndrome by identifying specific symptoms reported by the patient and by excluding other potential causes of prolonged fatigue.

Hepatitis A

Some years ago, the members of a university football team were drinking water taken from a local well. But this was no ordinary water. During the previous week, the water had been contaminated by viruses seeping into the well from a cesspool high above. Within days, all who had drunk the water began to feel ill, and soon the unmistakable signs of hepatitis appeared.

Hepatitis is an acute inflammatory disease of the liver caused by several viruses. **Hepatitis A (infectious hepatitis)** is the form that is most commonly transmitted by the fecal–oral route through contamination of food or water by the feces of an infected individual. An infected food handler is often involved (MicroFocus: 13.3), but outbreaks have also been traced to day-care centers where contact may take place with contaminated feces. In addition, the disease may be transmitted by the consumption of raw shellfish such as clams and oysters, since these animals filter and concentrate the viruses from contaminated seawater. Saliva contact, sexual contact, and arthropods have also been implicated in transmission.

Hepatitis A is caused by a small RNA virion considered by most virologists to belong to the Picornaviridae family and referred to as a heparnavirus (*hepa*titis-*RNA*-*virus*). The virion lacks an envelope and appears to have cubic symmetry (resembling a cube), but this factor is not fully documented (Figure 13.3). Hepatitis A viruses are very resistant to chemical and physical agents, and several minutes of exposure to boiling water may be necessary to inactivate them.

pi-kor"nah-vir'ĭ-da

The incubation period for hepatitis A is usually between two and four weeks. Therefore hepatitis A is sometimes called **short-incubation hepatitis** relative to hepatitis B, or long-incubation hepatitis. The disease is generally regarded as an illness of limited duration, but with considerable variability depending on such factors as dose of the virus ingested, activity of the patient's immune system, virulence of the viral strain, and age. Children generally develop milder disease than adults. Initial symptoms include anorexia, nausea, vomiting, and low-grade fever. Discomfort in the upper right quadrant of the abdomen follows as the liver tissue becomes enlarged. Patients sometimes experience appendicitis-like pain and dark urine. Considerable jaundice usually follows the onset of symptoms by one to two weeks, but many cases are without jaundice. The symptoms may last for several weeks, and relapses are common. A long period of convalescence is generally required, during which alcohol and other liver irritants are excluded from the diet.

Incubation period:
the time that elapses between entry of the microorganism to the host and the appearance of symptoms.

Anorexia:
loss of appetite.

Jaundice:
a yellowing of the complexion and eyes caused by overflow of bile pigments to the bloodstream.

13.3
MICROFOCUS

THE UNLUCKY 32

For some, the number 13 is unlucky, but for Peter's Creek, Alaska, the unlucky number was 32. It was May 1988, and the weather was unusually hot for that time of year. Between May 23 and June 10, 32 unfortunate people contracted hepatitis A, and things went downhill fast.

The outbreak of hepatitis was traced to a local convenience market, and the culprit was the ice slush that so many people enjoy on a hot day. The slush was contaminated, possibly by a certain store employee. Although the employee refused to be tested, his sister had had hepatitis A recently and he had looked somewhat jaundiced at the time. He was one of two store employees responsible for preparing the slush each day. (Later, it was learned that he used water from the bathroom sink to make the slush.)

For the unlucky 32, there were many days of abdominal pain, fever, jaundice, and a serious liver disease. They would have to avoid fats and oils (no fried foods, mayonnaise, or oily salad dressings), and they could not have alcohol of any type. For many there was the added burden of knowing they had infected others, because 23 additional cases soon developed. It was not a summer to remember fondly.

Diagnostic procedures for hepatitis A are based on liver function tests, observation of characteristic symptoms, and the demonstration of hepatitis A antibodies in the serum. Virus isolation tests are generally unproductive because viruses are not present in considerable amounts during the acute stage. On the contrary, the virus is excreted in large numbers in the stools about two weeks before symptoms appear. In one recent incident, for example, a restaurant worker in New Jersey fell ill on May 9 but showed no unusual symptoms even though he was shedding hepatitis viruses. Hepatitis symptoms developed by the end of May, and during the first three weeks of June, 56 cases of hepatitis broke out among patrons of the restaurant.

There is no treatment for hepatitis A except for prolonged rest. In people who have been exposed to the virus, it is possible to prevent development of the disease by administering injections of **hepatitis A immune globulin** within two weeks of infection. This preparation consists of hepatitis antibodies obtained from blood donations at blood banks. Blood is routinely screened for hepatitis antibodies and if large amounts are found, the blood is used for immune globulin. In the New Jersey outbreak cited previously, 1430 persons were given injections during a two-day clinic held June 19 and 20.

Maintenance of high standards of personal and environmental hygiene, and identification and removal of the source of contamination are essential to

Immune globulin:
an antibody preparation used as a preventative or a treatment for disease.

FIGURE 13.3
Hepatitis A Viruses

An electron micrograph of hepatitis A viruses (× 241,000). The particles were coated with antibodies to assist staining. This accounts for the halo around the viruses.

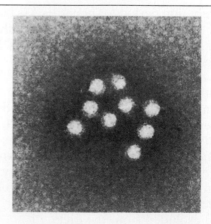

TABLE 13.1
A Comparison of Three Types of Hepatitis

CHARACTERISTIC	HEPATITIS A	HEPATITIS B	HEPATITIS C
Alternate names	Infectious hepatitis	Serum hepatitis	Posttransfusion hepatitis NANB hepatitis
Virus	RNA virus Picornaviridae	DNA virus Dane particle Hepadnaviridae	RNA virus Flaviviridae
Incubation period	2 to 4 weeks	4 weeks to 6 months	2 weeks to 6 months
Major transmission	Food and water Saliva contact Sexual contact	Body fluids Blood Sexual contact	Blood
Symptoms	Jaundice Abdominal pain	Jaundice Abdominal pain	Jaundice Abdominal pain
Illness severity	Moderate	High	High
Diagnosis	Liver function tests Symptoms Antibodies in serum	Liver function tests Symptoms HBsAg in serum	Liver function tests Symptoms Antibodies in serum
Carrier state	Rarely develops	Develops	Develops
Nosocomial problem	No	Yes	Yes
Prevention	Vaccine	Vaccine	None
Liver cancer	Not likely	Possible	Not established

interrupting the spread of hepatitis. In 1995, the Food and Drug Administration licensed a vaccine composed of inactivated viruses. Known commercially as Havrix, the vaccine is administered into the deltoid muscle in two doses to those between ages 2 and 18 (pediatric formulation) and in three doses to those over 18 years (adult formulation). Over 20,000 Americans contract hepatitis A each year, although unreported cases may make the actual figures much higher. Table 13.1 compares hepatitis A with hepatitis B, the other major type of hepatitis.

Deltoid muscle: the shoulder muscle.

Hepatitis B

Hepatitis B (serum hepatitis) is the second major form of hepatitis. It is caused by a DNA virus known as a hepadnavirus (*hepa*titis-*DNA*-*virus*) of the family Hepadnaviridae. The hepadnavirus can appear in three forms. In its most frequently observed form, the virus is seen as spherical particles measuring about 22 nm in diameter. These small particles appear to be composed exclusively of an antigenic protein substance called **hepatitis B surface antigen**, or **HBsAg**. A second form are elongated particles also composed of HBsAg in tubular or filamentous rods up to 200 nm long. The third form is the more traditional virus, a virion containing HBsAg as an outer surface or envelope surrounding an inner nucleocapsid of double-stranded DNA enclosed in a core antigen called hepatitis B core antigen, or HBcAg. This third form has been called the **Dane particle** since first reported by D. S. Dane in 1970. Apparently, the overproduction of the surface HBsAg during viral replication leads to the presence of this protein in large amounts. The antigen was previously known as the **Australia antigen**, and it has remained an important diagnostic sign for hepatitis B since reported by Baruch Blumberg in Australian aborigines in the early 1970s. Blumberg won the Nobel Prize for his work.

hep-ad′na-virus

Australia antigen: an antigen located on the surface of the hepatitis B virus.

Fiber optic endoscope:
a fiberlike strand inserted into tissue to visually observe the tissue.

Renal dialysis:
the process of cleansing the blood in an artificial kidney machine.

Transmission of hepatitis B usually involves direct or indirect contact with an infected body fluid such as blood, urine, saliva, or semen. For example, transmission may occur by contact with blood-contaminated needles used in hypodermic syringes or for tattooing, acupuncture, or ear-piercing. Blood-contaminated objects such as fiber optic endoscopes, instruments, and renal dialysis tubing are also implicated. Moreover, transmission may take place by contact with saliva, including contact made in the dental office. Hepatitis B is also an important sexually transmitted disease, particularly when anal intercourse takes place. This is because bleeding often occurs during anal intercourse, and viruses are able to enter the bloodstream of the receptive partner from the semen of the infected individual. A similar situation holds for AIDS.

The clinical course of hepatitis B is basically the same as for hepatitis A, but more severe illness is generally associated with hepatitis B. The disease has an incubation period of four weeks to six months and is therefore known as **long-incubation hepatitis**. Among adults, the common findings during the early stage are fatigue, anorexia, and taste changes. Smokers experience a notable distaste for cigarettes. Dark urine and clay-colored stools are present several days before jaundice appears. Other symptoms diminish as the jaundice intensifies, but an uncomfortable sense of fullness and tenderness is present in the upper right quadrant of the abdomen. Recovery usually requires about three to four months after the onset of jaundice, with about 10 percent of patients remaining carriers for several months. In rare cases, extensive liver damage may occur.

Prophylactic:
protective.

No specific treatment except injections of alpha interferon (Intron A) can influence the course of hepatitis B. However, prophylactic therapy may be given with injections of **hepatitis B immune globulin**. This preparation consists of antibodies concentrated from the serum of donated blood. In addition, the disease can be prevented by immunization with the hepatitis B vaccine. Since 1987 this **vaccine** has consisted of hepatitis B surface antigens (HBsAg) produced by genetically engineered yeast cells. The vaccine, known commercially as Recombivax HB or Engerix-B (depending on the company that produces it) involves no blood. The older vaccine, by comparison, was made from purified HBsAg from blood donors. The vaccine is recommended for all age groups and is particularly valuable for health-care workers who might be exposed to blood from patients. Indeed, in 1991 the CDC and the American Pediatric Association also issued a recommendation that the vaccine be administered to children as early as 4 months of age. Some hospitals administer it to infants just before they leave the nursery for home.

Nosocomial:
hospital-acquired.

Hepatitis B may occur in over 100,000 persons annually in the United States, but only about 20,000 cases are reported to the CDC. Virologists believe that cancer of the liver (hepatocarcinoma) may be a long-term effect of hepatitis B because, statistically, recoverers are at high risk for cancer. Also, there is evidence that hepatitis B is a major nosocomial disease. An estimated million people in the U.S. are believed to carry the virus.

Hepatitis C and Delta Hepatitis

Public health officials estimate that over 100,000 persons in the United States contract hepatitis each year from blood transfusions and blood products. Some virologists refer to this hepatitis as **posttransfusion hepatitis**. Often the disease is hepatitis B, but in a large number of instances the disease is identified as **hepatitis C** (in the older literature, known as non-A non-B [NANB] hepatitis). Up to 40 percent of all cases of hepatitis, including the majority of cases of posttransfusion hepatitis, may be hepatitis C. The disease is generally less severe than hepatitis A or hepatitis B, but hepatitis C often leads to chronic liver disease.

In the fall of 1984, researchers made a tentative identification of the virus of hepatitis C and reported that it was a retrovirus. Retroviruses are RNA viruses that use the enzyme reverse transcriptase to form DNA from the base code in RNA and thus reverse the normal process for replication (Chapter 11). By 1989, a new test was available to detect the virus in drawn blood and the incidence of hepatitis C began to decline dramatically. There was also a new virus for hepatitis C. More recent evidence was pointing to a **flavivirus** as the agent of the disease. Flaviviruses are enveloped RNA viruses of the type that cause yellow fever and dengue fever. The flavivirus theory holds to this writing.

Retrovirus:
an RNA virus that uses reverse transcriptase to form DNA from the base code in RNA.

Delta hepatitis is a recently recognized form of hepatitis. The disease appears to be caused by two viruses, the hepatitis B virus and the delta virus ("delta" is the fourth letter of the Greek alphabet). Delta viruses were discovered in 1977 by Mario Rizzetto in Turin, Italy. They are composed of a protein antigen called the **delta antigen** and a small fragment of RNA. Although transmissible as an independent agent, delta viruses can only infect and cause illness when hepatitis B viruses are present. In 1984, virologists presented evidence that the hepatitis B virus assists the replication of the delta virus (the press labeled delta viruses "piggyback viruses").

Antigen:
a chemical substance that stimulates the immune system.

Delta hepatitis is apparently common in southern Italy, the Middle East, and parts of Africa and South America. In late 1984, an outbreak of 75 cases occurred in Massachusetts among illicit drug users. The disease is similar to hepatitis B, although detailed studies are still lacking.

It is now apparent that the hepatitis C and delta viruses are part of a large group of **non-A non-B hepatitis viruses**. A virus for hepatitis E has been identified (and related to a group of nonenveloped RNA icosahedral viruses called calciviruses), and a number of other viruses are known to be involved in the typical hepatitis syndrome. Most of these latter cases are identified by exclusion, that is, if none of the A to E viruses can be located then the disease is simply called non-A non-B hepatitis.

TO THIS POINT
■

We have begun a study of the viscerotropic viral diseases by focusing on several diseases of the blood and internal organs of the viscera. The first two diseases, yellow fever and dengue fever, are caused by viruses injected directly into the bloodstream by mosquitoes. Yellow fever has great historical interest because epidemics were widespread in past generations, but vaccines and arthropod control limit its spread in many parts of the world. Dengue fever remains a threat in many countries. High fever reflects viral invasion of the blood in both cases.

We then turned to infectious mononucleosis and hepatitis. Research into these diseases continues to reveal new information while questions continue to surface. For example, the role of the Epstein–Barr virus in infectious mononucleosis is still uncertain, and the relationship of the disease to Burkitt's lymphoma is a debatable issue. In hepatitis, we saw how numerous different forms have emerged since the 1970s, each caused by a different virus. Infection of the liver is the common element in all types of hepatitis, and convalescence is generally long because damage to the liver is not easily repaired. Liver disease is also a consequence of yellow fever, and often it occurs in infectious mononucleosis.

We shall now give attention to a series of viral diseases of the digestive tract as we study viral gastroenteritis and a number of viral fevers where the agents affect the bloodstream. In addition, acquired immune deficiency syndrome (AIDS) will be considered. Many of the viscerotropic diseases in this section occur sporadically and are not well known. However, the potential for epidemics is great.

Other Viscerotropic Viral Diseases

Viral Gastroenteritis

gas"tro-en-ter-i'tis

Viral gastroenteritis is a general name for a common illness occurring in both epidemic and endemic forms. It affects all age groups worldwide and may include some of the frequently encountered travelers' diarrheas. Public health officials believe that the disease is second in frequency to the common cold among infectious illnesses affecting people in the United States. (In developing nations, gastroenteritis is estimated to be the second leading killer of children under the age of 5, accounting for 23 percent of all deaths in this age group.) Clinically, the disease varies but usually it has an explosive onset with varying combinations of diarrhea, nausea, vomiting, low-grade fever, cramps, headache, and malaise. It can be severe in infants, the elderly, and patients compromised by other illnesses. Some people mistakenly call it "stomach flu."

Rotavirus:
A circular RNA virus that can cause gastroenteritis.

One cause of viral gastroenteritis is the human **rotavirus**, a virus first described in 1973 in Australia. The virion contains segmented, double-stranded RNA as well as inner and outer capsids. It is a member of the Reoviridae family and is so named because of its circular appearance (*rota* is Latin for wheel). Figure 13.4 displays the virus.

The CDC considers rotaviruses the single most important cause of diarrhea in infants and young children admitted to hospitals. Many cases involve severe dehydration and death. In 1984, researchers in Finland announced the development of a vaccine containing a strain of attenuated rotaviruses. The vaccine is not yet licensed in the United States.

A second cause of viral gastroenteritis is the **Norwalk virus**, named for Norwalk, Ohio, where it caused a notable outbreak of intestinal disease in 1968. The virion is a nonenveloped particle whose nucleic acid was determined to be RNA as recently as 1992. It is transmitted by contaminated water and food, especially seafood (Figure 13.5). Outbreaks of disease can be detected by the symptoms and rise in antibodies to the virus. In 1987, over 300 participants at a South Dakota outing were infected by Norwalk virus from well water. The well was adjacent to a septic dump station.

Viral gastroenteritis may also be caused by either of two **enteroviruses**. Enteroviruses are small, icosahedral RNA virions of the Picornaviridae family. They are currently considered a less frequent cause of gastroenteritis than other viruses and are notable for the variety of infections they may cause.

cook-sak'e

One enterovirus is the **Coxsackie virus**, first isolated in 1948 by Gilbert Dalldorf and Grace Mary Sickles from the stool of a patient residing in Coxsackie, New York. The viruses occur in many strains within two groups, A and B. Strains B4 and B5 Coxsackie viruses are most commonly associated with gastroenteritis. Group B viruses are also implicated in **pleurodynia** (or Bornholm's disease), a painful disease of the rib muscles, and **myocarditis**, a serious disease of the heart muscle and valves sometimes resulting in a heart attack or need for a heart transplant. In addition, group B Coxsackie viruses are among the most frequent cause of **aseptic meningitis**. Group A viruses have been

ploor"o-din'e-ah

Aseptic meningitis:
meningitis in which no microbial agent is observed.

FIGURE 13.4

Two Viruses That Cause Gastroenteritis

(a) Rotaviruses observed in the diarrheal stool of an infant with gastroenteritis. Bar = 100 nm. Note the wheel-like circular appearance from which the rotavirus takes its name. (b) Norwalk viruses from stool specimens of a patient with gastroenteritis. Bar = 100 nm. The viruses appear to have icosahedral symmetry.

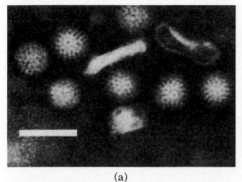

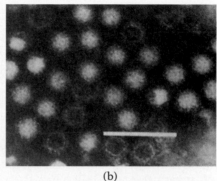

(a)   (b)

isolated from respiratory infections and from cases of **herpangina**, a disease of children with abrupt fever onset and punched-out vesicles on the soft palate, tongue, tonsils, and hands. Some virologists believe that Coxsackie viruses are the so-called "24-hour viruses." And an intriguing 1994 report in the respected *New England Journal of Medicine* indicated that Coxsackie viruses might trigger the development of insulin-dependent diabetes in genetically susceptible individuals. According to the report, the viruses stimulate the immune system to produce antibodies that attack the pancreas cells responsible for producing insulin.

The second major enterovirus is the **echovirus**. These viruses were discovered in the early 1950s. They take their name from the acronym ECHO, which stands for *e*nteric (intestinal), *c*ytopathogenic (pathogenic to cells), *h*uman (human location), and *o*rphan (a virus without a famous disease). Echoviruses occur in many strains and cause gastroenteritis as well as aseptic meningitis. The meningitis, however, is usually less severe than bacterial meningitis. Echoviruses are also a cause of respiratory infections and maculopapular skin rashes called **exanthems**. A Massachusetts outbreak called the Boston exanthem attracted attention in 1954.

Viral Fevers

A series of viruses may cause human illnesses characterized by high fever. These diseases occur sporadically and are usually rare in the United States.

An example of a viral fever is **Colorado tick fever**. This disease is caused by an RNA virus of the Reoviridae family. It is transmitted by the tick *Dermacentor andersonii* and accounts for about 200 cases of disease per year, chiefly in the state of Colorado. Characteristic symptoms include alternating periods of fever and relief (saddleback fever), with pain in the muscles, joints, and eyes. Leukopenia and the presence of viruses inside the red blood cells are other factors that mark the disease.

Sandfly fever is another example of a viral fever. The disease is prevalent where sandflies of the genus *Phlebotomus* abound. It occasionally breaks out in Mediterranean regions, Southeast Asia, and parts of Central America. Victims suffer recurrent high fever and joint and bone pains resembling those in dengue fever. Fatalities are rare, however. The responsible virus is an RNA virion of the Bunyaviridae family.

Diabetes:
a disease in which the passage of glucose into body cells is interrupted.

eg-zan'them
Exanthem:
a skin rash that develops rapidly.

der-ma-cen'tor

Leukopenia:
a reduced number of white blood cells.
flĕ-bot'o-mus

FIGURE 13.5

A Textbook Case of Norwalk Virus Gastroenteritis

This outbreak occurred in New Jersey during December 1979. It was one of the first reports of Norwalk virus disease linked to any food other than shellfish.

1. In early December 1979, a New Jersey restaurant and catering facility received a shipment of lettuce from a produce market in Philadelphia. The lettuce was to be served at the restaurant later that day and in the days following.

2. Some of the lettuce was washed in a kitchen sink generally used for preparing meat and seafood. In this particular case, shrimp was washed in the sink previous to the lettuce, and the same table was used for preparing both shrimp and lettuce.

3. On December 6, 1979, a group attended a luncheon banquet at the restaurant and was served green salad made with the lettuce. A second group received cole slaw instead of salad. The luncheon was successful and a good time was had by all.

4. About 30 hours later, 63 of the 87 people who ate green salad developed gastroenteritis. None of those who had cole slaw became ill. Health department microbiologists identified Norwalk virus as the agent and theorized that it entered the lettuce from contaminated shrimp.

Rift Valley fever is named for a region in eastern Africa called the Rift Valley. This is an immense earthquake-prone area where major plates of the earth come together. In addition to affecting humans, the disease affects animals and causes massive losses of sheep and cattle. Indeed, the virus is so dangerous to animals that federal law prohibits its transport to the mainland of the United States. (Research on the virus is performed on offshore islands.) Its potential for use in biological warfare prompted vaccine development in 1978. Transmission is by several genera of mosquitoes, and the disease is not considered to be exceptionally serious in humans. Denguelike pain in the bones and joints accompanies the fever, which lasts for about a week. The virion is an RNA virus of the Bunyaviridae family.

bun″ya-vir′i-dā

FIGURE 13.6

An "Emerging Virus"

A suggested method for the emergence of Lassa fever from nature. (a) A staple of the diet of certain African natives is the bush rat. The traditional rat hunt begins with a fire in the grasslands that drives the rats into the open, where they are clubbed. (b) On occasion the rats will run into local houses for shelter. Lassa fever viruses apparently are transmitted by the arthropods in the fur of the rat (c) or by the dust of the rat (d) to susceptible individuals not normally residents of the area.

Certain viral fevers are accompanied by severe hemorrhagic lesions of the tissues. These diseases are classified as **viral hemorrhagic fevers**. One example, **Lassa fever**, is so named because it was first reported in the town of Lassa, Nigeria, in 1969 (Figure 13.6). The disease is caused by an RNA virus of the Arenaviridae family (an arenavirus). Infection is accompanied by severe fever, prostration, and patchy blood-filled hemorrhagic lesions of the throat. The fever persists for weeks, the pharyngeal lesions bleed freely, and profuse internal hemorrhaging is common. At least four epidemics have been identified in Africa since 1969, and the disease is occasionally seen in the United States. John G. Fuller has written a lucid and vivid account of the discovery of Lassa fever in his book *Fever! The Hunt for a New Killer Virus*. The book should be read by anyone who believes there are no remaining frontiers in medicine.

a-re″na-vir′i-dā

Hemorrhagic:
referring to bleeding and accumulating blood.

404 VISCEROTROPIC AND NEUROTROPIC VIRAL DISEASES

Saddleback fever:
alternating periods of fever
and relief.

Another viral hemorrhagic fever is **Marburg disease**, named for Marburg, West Germany, where an outbreak occurred in 1967. Virologists identified the virus in tissues of green vervet monkeys imported from Africa. Saddleback fever, bleeding from the gums and throat, and gastrointestinal hemorrhaging are characteristic features, and mortality rates during epidemics tend to be high. The viral agent is a filovirus, a long threadlike virus (*filum* is Latin for thread) that often takes the shape of a fishhook or U. The genome is composed of RNA.

e-bol'ah

Another filovirus, the **Ebola virus**, captured headlines in 1995. During that summer, an outbreak of **Ebola hemorrhagic fever** was reported in Zaire and Sudan, and a wide-eyed world watched as newspaper reports spoke of hemorrhaging blood spouting from patients' orifices, organs turning to liquid, and a horror story similar to the one recounted in *The Hot Zone* by Richard Preston. (The difference was that the Ebola viruses in Preston's book were let loose near Washington, D.C.) To make matters worse, theaters had recently screened *Outbreak* and a nervous public watched as Dustin Hoffman and Rene Russo battled an Ebolalike epidemic in Africa. Now the real thing was happening. When it was over, hundreds had died and the media were alerting readers to an unknown microbial world lurking in the wilderness. The term "emerging viruses" was being mentioned with increased frequency on the late news.

Other viral hemorrhagic fevers, all caused by Bunyaviridae or Arenaviridae, are Congo–Crimea hemorrhagic fever, which occurs worldwide; Oropouche fever, which affects regions of Brazil; and Junin and Machupo, the hemorrhagic fevers of Argentina and Bolivia, respectively. Still another hemorrhagic fever is Korean hemorrhagic fever caused by the **Hantaan virus**, named for the Hantaan River in Korea. American servicemen experienced the disease during the Korean War, and an outbreak occurred among Marines in Korea in 1986. Then, in the summer of 1993, a brief epidemic occurred among Native Americans living in the southwest United States (the "four corners disease" for the place where four states come together—MicroFocus: 13.4). Symptoms of the disease included an influenzalike prodrome with blood hemorrhaging and acute respiratory failure. The disease was given the technical name **hantavirus pulmonary syndrome** and the strain of virus was named Sin Nombre virus (another strain isolated later was called Muerto Canyon virus). CDC investigators identified the deer mouse as the vector for the virus. By the year's end, 91 cases were confirmed in 20 states. Forty-eight patients died.

Cytomegalovirus Disease

si'to-meg"ah-lo-vi'rus

Epithelium:
the tissue that lines blood
vessels and numerous body
cavities.

The cytomegalovirus is an icosahedral DNA virion of the herpesvirus group. The virus takes its name from the enlarged cells ("cyto-megalo") found in infected tissues. Often these are cells of the salivary glands, epithelium, or liver.

Cytomegalovirus disease may be among the most common diseases in American communities. Fever, malaise, and possibly, an enlarged spleen develop but few other signs of disease are observed. Most patients recover uneventfully. However, if a woman is pregnant, a serious congenital disease may ensue as the viruses pass into the fetal bloodstream and damage the fetal tissues. Mental retardation of the young is sometimes observed. For example, in a study at a Rochester, New York, hospital, virologists discovered the virus in the tissues of 52 of 8600 newborns. When examined years later, the 52 children had lower IQs than a comparable noninfected group and many had hearing defects. The "C" in the **TORCH** group of diseases refers to cytomegalovirus disease. The other letters stand for toxoplasmosis (T), rubella (R), and herpes simplex (H). The "O" is for other diseases such as syphilis.

TORCH:
an acronym for diseases
transmitted from a pregnant
woman to her unborn child.

The cytomegalovirus has also demonstrated its invasive tendency in patients who have acquired immune deficiency syndrome (AIDS). AIDS patients often

13.4

MICROFOCUS

"WE CAN'T FIND IT BUT WE KNOW IT'S THERE!"

A microbial identification of a unusual sort was made in 1994. That year, a strange epidemic of viral disease broke out in the southwestern United States at the border area of four states (Arizona, Colorado, New Mexico, and Utah). The disease came to be known as the "four corners disease." It spread quickly among Native Americans causing severe hemorrhaging, pneumonia, and kidney disease, and it resulted in 35 deaths among 52 patients. Scientists performed a battery of tests, hunting for myriad viruses and their antibodies, but they came up empty each time.

Then they hit on a novel idea. Instead of looking for the virus, why not look for the viral genome—the DNA of the virus. Scientists obtained DNA from numerous different viruses in culture collections and replicated the DNA fragments to produce a variety of "gene probes." A gene probe is a small, single-stranded molecule of DNA that will hunt down and unite with its complementary single-stranded DNA molecule in a morass of DNA material. Acting like a right hand searching for a left hand, the gene probe emits a radioactive signal when the union has taken place (if the complementary strand cannot be found, no signal is sent).

Now the scientists were ready to give their theory a try. They began by extracting DNA from the tissues of disease patients. Then they combined the DNA with gene probes from a constellation of viruses and held their breath to see which would send a signal. Their answer came a few short minutes later when a sooty band of radioactivity appeared on their instruments—the DNA from the patient was uniting with the gene probe from the hantavirus, a very rare virus named for the Hantaan River in Korea. The infecting virus must be a hantavirus.

Not only was the mystery resolved, but scientists now had a useful tool for diagnosing and tracking the disease. And they could work to interrupt its spread because they knew how to locate the disease and what was causing it. They could also point out the face of the enemy—well, not really, because they had not seen the virus, only its footprint.

---

experience **CMV-induced retinitis**, a serious infection of the retina that can lead to blindness. Ganciclovir is often used to treat the disease, with foscarnet as an alternative drug. In immunocompromised individuals, cytomegaloviruses can also infect the lungs, liver, brain, and kidneys and cause death. Patients undergoing cancer therapy or receiving organ transplants may be susceptible to CMV disease because immune-suppressing drugs are often administered to these patients.

gan-ci′klo-vir
fos-car′net

HIV Infection and Acquired Immune Deficiency Syndrome (AIDS)

In the early 1980s physicians described a syndrome involving a deficiency of the immune system in young homosexual men. This clinical entity involved the development of opportunistic diseases, including an unusual type of skin cancer called Kaposi's sarcoma. Initially, it was believed that many factors were responsible for the immune deficiency. The most plausible factor was a virus, since it would account for the state of immune deficiency in the host and the risk for developing an opportunistic disease. During the first months of the epidemic, the syndrome was termed gay-related immunodeficiency, or GRID.

Kap′o-sē sar-ko′mah

It soon became apparent that male homosexuals were not the only persons at risk for the disease. The disease appeared in blood transfusion recipients, heterosexuals, and intravenous drug users, and it became clear that a broader definition for the syndrome was needed; therefore the syndrome was termed **acquired immune deficiency syndrome**. Initially the cause was elusive and multiple factors were implicated, including viruses, recreational drugs, sperm antibodies, and antigenic factors. However, the agent was identified in 1984 almost simultaneously by research teams from the Pasteur Institute (Paris) and the National Cancer Institute (United States). The French group, headed by **Luc Montagnier**, named the virus lymphadenopathy-associated virus (LAV), while the American group, led by Robert Gallo, called it the human T-cell lymphotropic virus type III (HTLV-III). In 1986 an international commission recommended the single name of **human immunodeficiency virus (HIV)**. By 1993, Montagnier was acknowledged as the discoverer.

mon′tan-yā
lim-fad″ĕ-nop′ah-the
lim-fo-trop′ik

Homeostasis:
referring to the stability in the internal environment of a complete organism achieved by interacting control mechanisms.

Cell-mediated immunity:
immunity derived from the activity of T-lymphocytes that encompasses an interaction between stimulated T-lymphocytes and antigen-bearing cells.

Lymph nodes:
pockets of white blood cells located in the neck, armpits, groin, and other body regions; site of the cells of the immune system.

Mammalian:
pertains to warm-blooded vertebrate animals that possess hair and nurse their young.

Helper T-lymphocytes:
a T-lymphocyte that functions in humoral immunity and cell-mediated immunity and helps regulate both processes; one of the host cells for HIV.

Normal Immunologic Response The immunologic response is an integral part of the homeostasis of any mammalian system. This response must be adaptable to all types of environments and capable of recognizing the diversity of pathogens as outlined in Chapter 18. The goal of the immune system is to neutralize the pathogen prior to the development of disease.

The main cells of immunity are the lymphocytes, one type of white blood cell. Lymphocytes have two broad immunologic tasks. The first is to participate in **antibody-mediated (humoral) immunity** and produce antibodies that will bind to a specific antigen. Lymphocytes that do this are B-lymphocytes (or B-cells), each of which responds to one antigen (Chapter 18). The second is to participate in **cell-mediated immunity**, a task performed by T-lymphocytes (or T-cells). The T-lymphocytes differentiate into mature cells by passing through the thymus. T-lymphocytes recognize antigens and collaborate with macrophages through secretion of humoral factors. The result is a proliferation of B-lymphocytes, phagocytosis of the antigen, and self-renewal of the T-lymphocytes to form cytotoxic T-lymphocytes (Chapter 18). B-lymphocytes also have self-renewal properties to respond to antigens throughout the lifetime of the individual.

Special lymphoid tissues are present in the lymph nodes, spleen, intestines, pharynx, and bone marrow. These are locations where the B-lymphocytes, T-lymphocytes, and macrophages interact with antigens to bring about the immune response. Deficiencies of the immune response, acquired through disease, medication, or congenital deficiency, will lead to susceptibility to pathogens and cancers (neoplasms). Infection by HIV results in a failure of the immune response and places the host at risk for pathogens and cancers similar to those in patients with other acquired or congenital immune deficiency states.

The Human Immunodeficiency Virus The human immunodeficiency virus (HIV) is a unique type of virus that utilizes its RNA for replication, rather than the DNA in mammalian genetic material (Figure 13.7). For most physiologic functions in the mammalian system, the genetic message in DNA is transferred to RNA and then to an amino acid sequence during protein production (Chapter 5). HIV, by contrast, has its genetic message in RNA. It transfers the message to DNA utilizing the enzyme reverse transcriptase to make a DNA copy of itself. The viral DNA, now termed a **provirus**, integrates into the host's DNA where it may lay dormant for an unknown period of time. Once activated, the proviral DNA may be transcribed into viral RNA and the viral RNA—along with viral proteins—can be combined to form whole viruses. The whole virus can then "bud" from the host cell and infect other cells (Figure 13.8).

The major cell infected by HIV is the **helper T-lymphocyte** that bears the CD4 receptor site (the helper T-lymphocyte is also called a CD4+ cell because it has the receptor site). Once HIV has attached to the CD4 receptor site, HIV is incorporated into the T-lymphocyte cytoplasm, the viral genome is released, and, as with other retroviruses, a proviral state may occur. Strictly speaking, the person who is infected with HIV is probably infected forever. Other cells that the HIV may infect include certain cells in the central nervous system, megakaryocytes, and the macrophage/monocyte cell. These cells also possess the CD4 receptor sites. The macrophage/monocyte is thought to be an important reservoir for HIV.

In the HIV-infected person, an estimated 1 out of 500 to 1000 circulating helper T-lymphocytes are infected with virus. For a period of time, the body can replace the T-lymphocytes destroyed, but eventually the infected individual suffers a gradual decline of helper T-lymphocytes. The T-lymphocyte acts as an important central communicating lymphocyte. It also secretes soluble factors

FIGURE 13.7

A Schematic Diagram of the Human Immunodeficiency Virus (HIV)

The virus consists of two molecules of RNA and molecules of reverse transcriptase. A protein capsid surrounds the genome and an envelope with **spikes of protein** lies outside the capsid.

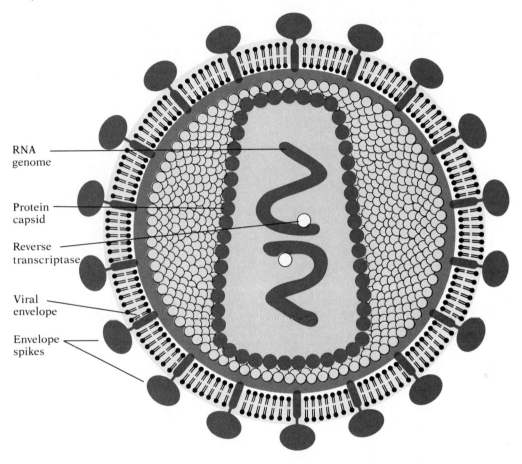

RNA genome

Protein capsid

Reverse transcriptase

Viral envelope

Envelope spikes

that result in the replication of other T-lymphocytes as well as B-lymphocytes. As the HIV infection continues in the host, the decrease of helper T-lymphocytes results in failure of the immune system involving both the antibody-mediated and cell-mediated branches (Figure 13.9). Therefore, the failure of the immune system places the infected individual at risk for common pathogens and certain types of neoplasms. It is important to realize that specific social preferences of the host have nothing to do with the immunodeficiency, and that anyone who is infected with HIV will be susceptible to very similar infections and neoplasms.

HIV is one of many retroviruses that is thought to cause disease in susceptible mammalian systems. After the discovery of HIV type I (HIV-1), a similar but slightly different virus termed **HIV-2** was discovered in patients in West Africa who experienced a related human immune deficiency illness. A virus found in monkeys, called the simian immunodeficiency virus (SIV), is similar to HIV-2. It is important to note, however, that the most common virus currently associated with AIDS is HIV-1. Studies have shown that the pathogenicity and clinical immune deficiency of HIV-2 may be different than for HIV-1. Further studies are required to clarify this point.

HIV-2:
a strain of HIV related to HIV-1, the AIDS virus.

FIGURE 13.8

FIGURE 13.8
The Replication Cycle of the Human Immunodeficiency Virus (HIV)

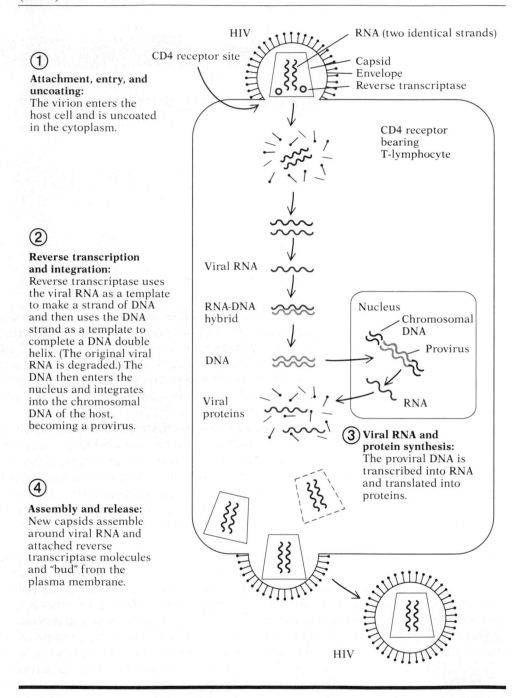

① **Attachment, entry, and uncoating:** The virion enters the host cell and is uncoated in the cytoplasm.

② **Reverse transcription and integration:** Reverse transcriptase uses the viral RNA as a template to make a strand of DNA and then uses the DNA strand as a template to complete a DNA double helix. (The original viral RNA is degraded.) The DNA then enters the nucleus and integrates into the chromosomal DNA of the host, becoming a provirus.

④ **Assembly and release:** New capsids assemble around viral RNA and attached reverse transcriptase molecules and "bud" from the plasma membrane.

HIV
CD4 receptor site
RNA (two identical strands)
Capsid
Envelope
Reverse transcriptase

CD4 receptor bearing T-lymphocyte

Viral RNA

RNA-DNA hybrid

DNA

Viral proteins

Nucleus
Chromosomal DNA
Provirus
RNA

③ **Viral RNA and protein synthesis:** The proviral DNA is transcribed into RNA and translated into proteins.

HIV

Diagnosis Once the host is infected by HIV, a detectable antibody response occurs in most cases within six to eight weeks. These antibodies can be detected by the **ELISA test** (enzyme-linked immunosorbent assay) discussed in Chapter 19. The ELISA test, when performed properly, is close to 100 percent sensitive

FIGURE 13.9

Transmission Electron Micrographs of the AIDS Virus and Its Host Cell, the T-lymphocyte

The large photograph shows a cross-section of the T-lymphocyte with viruses present at the lower right surface (arrows). The inset magnifies the viral particles. Photographs A, B, C, and D show viruses in various stages of "budding out" from the host cell. Bar = 100 nm.

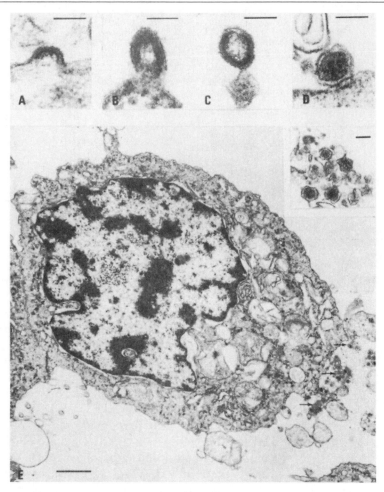

and specific (a positive result indicates HIV infection, while a negative result indicates lack of infection). A positive ELISA result automatically is repeated, and if again positive, should be confirmed using the **western blot test**. The western blot test detects individual antibodies that have been produced against individual HIV antigens. By using these two tests, almost 100 percent of people infected with HIV are detected. If both tests are negative, the person has less than a one percent chance of HIV infection. These assays, however, are only indirect tests and do not detect the virus itself. More direct tests using gene probes and amplification methods—such as the polymerase chain reaction (PCR)—detect HIV genetic material (RNA or DNA) or viral fragments (Chapter 19). A gene probe test should be performed when HIV infection is suspected even though the ELISA and western blot tests are negative.

Gene probe: a segment of radioactive DNA used in diagnostic tests to seek out and combine with a complementary segment of DNA.

Transmission The transmission of HIV is well understood at the current time. Casual contact does not result in transmission of HIV. However, intimate sexual contact and contact with blood are common modes of transmission. Since HIV

TABLE 13.2

Number of Cases of AIDS Reported to the CDC by June 30, 1995

A)	Homosexual or bisexual contact	244,235
B)	Homosexual IVDU	31,024
C)	IVDU*	118,694
D)	Transfusion recipients	7,128
E)	Heterosexual contact	35,683
F)	Hemophilia or other coagulant disorder	3,872
G)	Unknown	29,652
H)	Pediatrics	
	1) Hemophilia	226
	2) Parent at risk	5,925
	3) Transfusion	359
	4) Unknown	101
	TOTAL	476,899

* Intravenous drug user

Condom:
a sheath of rubber, latex, or natural "skin" that is placed over the penis during sexual intercourse.

exists in blood, semen, and vaginal secretions, any behaviors that transmit these fluids put a person at risk for HIV infection. **High-risk sexual activities** include unprotected rectal intercourse, vaginal intercourse, and oral sex. There may be differences in sexual activities in terms of the risk for HIV infection, but all are associated with HIV transmission and all should be considered unsafe sexual practices. The use of condoms has been shown to decrease the transmission of HIV, but condom use may not be 100 percent effective. The sharing of blood-contaminated needles either for recreational use or medical purposes also has been shown to transmit HIV. Needle exchange programs and rinsing contaminated needles in a 10 percent solution of bleach (a concentration that neutralizes HIV) may result in a decreased risk of HIV transmission. Patients receiving blood products for treatment of medical illness, including packed red blood cells, whole blood, platelets, and protein factor concentrates, may also be at risk for HIV infection, especially if such products were transfused prior to April of 1985, when testing of blood products with the ELISA was instituted. Another mode of transmission is possible in the perinatal period when either direct placental transfer of the virus or transvaginal infection may occur in the newborn. The present risk for perinatal transmission to the newborn is believed to be approximately 33 percent. Transmission of HIV from the mother to the newborn may occur during breast feeding, but this is rare.

Transvaginal infection:
infection acquired by the newborn during passage through the vagina.

Needle stick:
injury to the skin tissues arising from unintentional injection with a needle.

Health-care workers also risk acquiring HIV during professional activities. This risk is estimated to be approximately 0.4 percent when a significant exposure to HIV occurs. The most common mode of HIV exposure to health-care workers is through accidental needle sticks. Transmission of HIV rarely has been documented in health-care workers exposed to contaminated blood through mucosal membranes or skin. Healthcare workers must therefore practice infection-control procedures that have been established for the workplace.

The Spectrum of HIV Illness   By the end of June 1995 a total of 476,899 cases of AIDS have been reported to the Centers for Disease Control and Prevention, with 295,473 deaths. Table 13.2 describes the major risk groups and number of individual cases in each group. It is important to understand that AIDS is the end result of HIV infection and therefore only represents a minor group

---

TABLE 13.3

CDC Classification of HIV Infections

NEW CLASSIFICATION		COMMON NAME
Group I	Acute HIV infection (mononucleosis-like, aseptic meningitis)	Acute infection
Group II	Asymptomatic	Healthy carrier
Group III	Persistent generalized lymphadenopathy	ARC
Group IV	Other disease	
	A. Constitutional disease (fever, weight loss, diarrhea)	ARC
	B. Neurological disease (dementia, myelopathy, peripheral neuropathy)	ARC
	C. Secondary infections	
	1. CDC-defined AIDS-associated[a]	AIDS
	2. Other specified infections[b]	ARC
	D. Secondary cancers (CDC-defined AIDS-associated)[c]	AIDS
	E. Other conditions attributed to HIV infection or immunosuppression	ARC

CDC, Centers for Disease Control and Prevention; HIV, human immunodeficiency virus; ARC, AIDS-related complex.

[a] *Pneumocystis carinii* pneumonia, toxoplasmosis, cryptococcosis, chronic cryptosporidiosis, extra-intestinal strongyloidiasis, isosporiasis, candidiasis (esophageal, bronchial, or pulmonary), histoplasmosis, mycobacterial infection with *Mycobacterium avium-intracellulare* complex or *M. kansasii*, cytomegalovirus infection, chronic mucocutaneous or disseminated herpes simplex infection, and progressive multifocal leukoencephalopathy.

[b] Multidermatomal herpes zoster, oral hairy leukoplakia, nocardiosis, tuberculosis, recurrent *Salmonella* bacteremia, oral candidiasis.

[c] Kaposi's sarcoma, non-Hodgkin's high grade lymphoma, primary central nervous system lymphoma.

of the total HIV-infected group. Approximately 1.0 to 1.5 million Americans are believed to have HIV infection. Estimates continue to be made to better understand the financial and social impacts of this disease on the world's resources. Epidemiologists estimate that approximately 10 to 15 million people worldwide are infected with HIV.

Because patients may have HIV infection but not manifest the actual syndrome of AIDS, a reclassification of **HIV infection** into four stages is used by the Centers for Disease Control and Prevention (Table 13.3). The first stage is acute infection by HIV. Patients with acute disease may experience fever, diarrhea, rash, lymphadenopathy, night-sweats, and a general fatigue similar to that in acute infectious mononucleosis. These symptoms may last a few weeks or up to 6 months.

After the acute infection, most patients will be asymptomatic. The asymptomatic state is designated **stage II disease**. At this time, the patient may have few or no symptomatic complaints. There is antibody evidence of HIV infection, but it may be difficult to culture the virus from T-lymphocytes. The patient may also have minor laboratory abnormalities but immune deficiency related infections or cancers are not present.

The third stage of disease is termed **persistent generalized lymphadenopathy** (PGL). Patients in this stage have generalized lymphadenopathy for greater than 6 months in two or more areas of their bodies (excluding the inguinal area) and have antibody evidence for HIV (positive ELISA and western blot tests). Physicians once believed that these patients would not develop AIDS, but as time passed, these patients did develop symptomatic HIV infection; now it is clear that PGL is not an end-state of HIV infection but an early manifestation

Lymphadenopathy:
disease of the lymph nodes generally accompanied by obvious swelling.

a"simp-to-mat'ik
Asymptomatic:
without symptoms.

Inguinal:
referring to the groin.

Symptomatic:
having symptoms.

sin-sish'al

mi"e-lop'ah-the
Myelopathy:
functional disturbances or
pathological changes in the
spinal cord.

car-in'e-e

of AIDS. Studies reveal that there is probably a 6 to 10 percent chance per year that patients with stage II and stage III disease will develop symptomatic HIV infection. There are estimates that 100 percent of the patients with HIV infection will eventually develop stage IV disease.

Stage IV disease can also be viewed as "**symptomatic HIV infection.**" At this stage, patients begin experiencing symptoms from their immunodeficiency due to gradual loss of helper T-lymphocytes. The actual reason for loss of the T-lymphocytes is under speculation. Replication of HIV in the cells may cause their destruction, or because the T-lymphocyte now has the virus infection, the immune system may recognize the infected cell as being abnormal and therefore attack this cell. In the laboratory, HIV-infected T-lymphocytes form a large syncytial mass (a collection of cells into one mass) that would be eliminated by the immune system if this happened in the body.

Stage IV is further subdivided into five different sections in relation to the specific symptoms of immunodeficiency. Patients in stage IVA exhibit constitutional disease, fever, weight loss, or unexplained diarrhea. Stage IVA was formerly termed **AIDS-related complex (ARC)**. Patients in stage IVB display symptomatic neurologic disease including dementia, myelopathy, or peripheral neuropathy. The secondary infections that have been classically associated with AIDS are in stage IVC, along with other infections having a higher incidence in patients with HIV infection, including multidermatomal herpes zoster, oral hairy leukoplakia, and oral candidiasis. Stage IVD includes the patients with secondary cancers, Kaposi's sarcoma (KS), non-Hodgkin's lymphoma, and primary central nervous system lymphoma. Other conditions attributable to HIV infection or immunosuppression by HIV comprise stage IVE. With this expanded classification of HIV infection, it is quite evident that AIDS is but one manifestation of HIV infection. Many physicians believe that a classification incorporating the symptomatic illnesses along with specific immunologic markers would be helpful. This type of classification has been proposed under the Walter Reed classification described in Table 13.4.

On January 1, 1993, the CDC revised the surveillance case definition. The latest definition takes into account the number of CD4 T-lymphocytes present in the patient and adds three new conditions to the previous 23 indicators of AIDS. According to the definition, a person infected with HIV has progressed to AIDS if the count of CD4 T-lymphocytes falls below 200 per microliter of blood or if the person has one or more of 26 specified conditions, including the three new conditions of invasive cervical carcinoma, pulmonary tuberculosis, and recurrent bacterial pneumonia. The new definition reflects increased use of T-lymphocyte testing, while circumventing classification problems arising when drugs alter the course of clinical symptoms and remove the condition that designated an AIDS case. It also includes more HIV-infected women in the case definition since early cervical cancer is often seen in association with HIV. Moreover, the CD4 count per microliter will be used to classify a person's HIV infection into three categories: a count above 500 is considered least severe disease; between 200 and 499 reflects intermediate severity; and below 200 is the severest form of disease.

Opportunistic Infections and Neoplasms   As described previously, patients with HIV infection suffer a gradual decline of helper T-lymphocytes together with an increase in viruses. This combination allows for continued immune deficiency and the development of opportunistic diseases and cancers.

Pneumocystis carinii **pneumonia (PCP)** is the most common opportunistic disease that patients with HIV infection develop (Chapter 15). With

---

TABLE 13.4

TABLE 13.4
A Walter Reed Classification for HIV Infection

STAGE	HIV Ab, Ag, OR CULTURE	PGL	CD4 CELLS /mm^3	SKIN TESTS	THRUSH	*OIs
WR 0	−	possible or probable exposure				
WR 1	+	0	>400	normal	0	0
WR 2	+	+	>400	normal	0	0
WR 3	+	±	<400	normal	0	0
WR 4	+	±	<400	partial loss	0	0
WR 5	+	±	<400	complete	+	0
WR 6	+	±	<400	partial or complete	±	+

* OIs = opportunistic infections

---

prophylaxis and early intervention, the survival of patients with PCP has dramatically improved in recent years. Patients with symptomatic HIV infection can also experience toxoplasmosis, especially of the brain (Chapter 15), oral candidiasis (Chapter 14), and cryptosporidiosis (Chapter 15). Recently, an association with *Mycobacterium tuberculosis* has been elucidated, and an increased incidence of significant mycobacterial disease in patients with HIV infection has been established. Other mycobacterial species that commonly cause disease associated with HIV infection include *Mycobacterium avium-intracellulare* complex and *Mycobacterium kansasii* (Chapter 7). Patients may also contract systemic or central nervous system cryptococcosis, histoplasmosis (Chapter 14), cytomegalovirus, or disseminated herpes simplex infection.

krip″to-spoor-id′e-o′sis

krip″-to-kok-o′-sis

The presence of **Kaposi's sarcoma (KS)** in patients at risk for HIV infection was noted early in the 1980s. The classic KS usually was seen in patients who were older men, usually of Mediterranean descent. In most of these patients, the cancer was benign or responded to local radiation. In patients infected with HIV, on the other hand, KS often takes a very aggressive course. Patients with KS may present initially with limited cutaneous involvement and later develop widespread visceral disease. Approximately 20 to 25 percent of patients with HIV infection eventually develop KS. The tumor itself is thought to be dependent on growth factors, which may be elaborated by HIV-infected T-lymphocytes. Eventually, the endothelial proliferation may become independent of the growth factors and thus, a true neoplasm (cancer). Patients who have lung, GI, or liver involvement may suffer severe organ failure and death.

Endothelial:
pertaining to the cell layer that lines the heart cavities, the blood vessels, and certain body cavities.

Lymphoma:
a general term applied to cancer of the tissues of the lymphoid systems.

An epidemic of high-grade B-lymphocyte lymphomas also has been seen in patients at risk for HIV infection. Patients with non-Hodgkin's lymphoma (NHL) and HIV infection often present with disseminated disease, which responds rapidly to therapy. Patients usually have a shortened survival because of the development of resistant lymphomas and of opportunistic disease, but long-term survivors are common.

Most patients with end-stage HIV infection have **neurologic abnormalities**. A primary HIV-related dementia has been described but rarely at the beginning of the disease. The patients may manifest early mental dysfunction but have no other opportunistic diseases or AIDS-defining neoplasms. Most patients by the end of their HIV infection have evidence of central nervous system problems, which may manifest as mental confusion, nerve or muscular dysfunction, memory loss, seizures, or depression.

a-zi"do-thi'-mi-dēn

Treatment   Because HIV infection is probably a life-long disease of the host, treatment strategies must address the biology of HIV. The life cycle of the HIV can be interrupted in different steps. These steps include decreasing the binding of the HIV to susceptible T-lymphocytes with medications like soluble CD4 (sCD4). Synthesis of the proviral nucleic material (DNA) could be inhibited by reverse transcriptase inhibitors. One such drug which now has been licensed by the Food and Drug Administration (FDA) is zidovudine, originally called **azidothymidine** and commonly known as AZT. AZT is similar to an important building block (deoxythymidine) used in the synthesis of viral DNA. High concentrations of AZT, therefore, act as a chain terminator and inhibit DNA synthesis. Studies have shown that AZT not only decreases viral load in patients, but also increases survival in patients with symptomatic HIV infection. Unfortunately, AZT does not cure patients but only inhibits viral replication temporarily.

di"de-ox"e-in'o-sēn
di"de-ox"e-si'ti-dēn

Other anti-HIV agents that have been approved in recent years include **dideoxyinosine (ddI)** and **dideoxycytidine (ddC)**, which are similar to AZT but have fewer side effects. Other agents that have been utilized to interrupt HIV infection are segments of RNA or DNA, which attach to their complementary strand of RNA or DNA but do not allow synthesis of the normal viral nucleic acids. Other potential anti-HIV agents include foscarnet and ampligen.

A team of industrial biochemists has also studied the therapeutic value of a sulfur-containing analog of deoxycytidine known as 2'-deoxy-3'-thiacytidine, or 3TC (also known as **lamivudine**). The target of 3TC is reverse transcriptase, and combinations with AZT have proven effective in reducing the HIV burden of test patients. Indeed, in 1995, researchers reported on four separate studies (two in the United States and two in Europe) and pointed to tenfold reduction of virus as well as a significant increase in the CD4 T-lymphocyte count in infected persons. In late 1995, the U.S. Food and Drug Administration approved 3TC (marketed as Epivir) for use in combination with AZT as an AIDS treatment.

Still another group of anit-HIV drugs are the **protease inhibitors**. These drugs interfere with the final processing steps of the protein used in the HIV capsid. They inhibit the enzyme protease, which is responsible for sectioning molecules of large protein for final use in capsid protein. By 1996, the FDA had approved three protease inhibitors: **saquinavir** (Invirase), **indinavir** (Crixivan), and **ritonavir**.

Attempts to decrease the transmission of HIV through extensive educational programs are probably the most important step in curbing the rise of HIV infection. A vaccine for HIV infection is being sought with great anticipation, but at this writing a vaccine that protects humans from HIV has not been developed (MicroFocus: 19.2). There is evidence that inactivated whole virus may protect susceptible animals. Other vaccines utilizing synthetic viral fragments have recently shown promise in producing protective antibody responses in laboratory animals. It is believed that a combination of antibody production and cell-mediated response will be important for the final vaccine when and if that becomes available. As development of this vaccine continues, emphasis must be placed on prevention of HIV infection because it may be many years before a trusted and safe vaccine is developed. For further information on this and other issues regarding AIDS, you might enjoy reading *AIDS: The Biological Basis* (1997, W.C. Brown; Dubuque, Iowa). The viscerotropic viral diseases are summarized in Table 13.5.

TABLE 13.5

A Summary of Viscerotropic Viral Diseases

DISEASE	CLASSIFICATION OF VIRUS	TRANSMISSION	ORGANS AFFECTED	VACCINE	SPECIAL FEATURES	COMPLICATIONS
Yellow fever	Flaviviridae	*Aedes aegypti* (mosquito)	Liver Blood	Inactivated viruses	Jaundice	Hemorrhaging
Dengue fever	Flaviviridae	*Aedes aegypti* (mosquito)	Blood Muscles	Not available	Breakbone fever	Hemorrhagic fever
Infectious mononucleosis	Herpesviridae	Saliva Contact Droplets	Blood Lymph nodes Spleen	Not available	Downey cells Paul–Bunnell test	Splenic rupture Jaundice
Hepatitis A	Picornaviridae	Food Water Contact	Liver	Inactivated virus	Jaundice	Liver damage
Hepatitis B	(Unclassified) Hepadnaviridae	Contact with body fluids	Liver	Synthetic proteins	Jaundice	Liver cancer
Hepatitis C	Flaviviridae	Contact with body fluids	Liver	Not available	Jaundice	Liver damage
Viral gastroenteritis	Many viruses	Food Water	Intestine	Not available	Diarrhea	Dehydration Meningitis
Viral fevers	Many viruses	Contact Arthropods Animals	Blood	Not available	Fever Joint pain	Hemorrhaging
Cytomegalovirus disease	Herpesviridae	Contact Congenital transfer	Blood Lung	Not available	Enlarged cells	Fetal damage
Acquired immune deficiency syndrome	Retroviridae	Contact with body fluids and blood products	T-lympho- cytes Brain	Not available	Immune deficiency	Opportunistic diseases

TO THIS POINT

■

We have discussed additional viscerotropic viral diseases, beginning with a series of diseases of the gastrointestinal tract. Many viruses are involved in this disorder, including the rotavirus, the Norwalk virus, and two types of enteroviruses. We noted that infections are accompanied by varying combinations of diarrhea and malaise, and that the diseases can be severe in certain individuals.

We then surveyed a number of viral fevers, three of which are transmitted by arthropods. The viral fevers are rare in the United States, but they commonly occur in other parts of the world such as South America and Africa and represent a potential source of epidemics if they break out in susceptible populations. We also saw how cytomegalovirus is widespread in Americans but of little consequence except in pregnant women and those with suppressed immune systems such as people infected with HIV. The section closed with a discussion of HIV infection and AIDS, a disease that is currently considered a major health problem in the United States. The disease is caused by human immunodeficiency virus (HIV). It involves *the immune system and paves the way for opportunistic microorganisms.*

In the final section of this chapter we shall survey several neurotropic viral diseases. These diseases have substantial importance because they affect the nervous system and often result in death or permanent paralysis. Two of the diseases, rabies and polio, can be prevented with immunization. Another neurotropic disease, called slow virus disease, represents an area of intense investigation in microbiology.

Neurotropic Viral Diseases

Rabies

Rabies is notable for having the highest mortality rate of any human disease once the symptoms have fully materialized. Few people in history have survived rabies and in those who did, it is uncertain whether the symptoms were due to the disease or the therapy.

Rabies can occur in most warm-blooded animals in nature, including dogs and cats, horses and rats, and skunks and bats. The disease has been identified in Alaskan caribou, Russian wolves, and Western prairie dogs. Statistics released in the 1980s indicate that rabies is more common in cats than dogs. Public health microbiologists have also voiced concerns for the raccoon epizootic that is currently working its way up the eastern seaboard toward New England (MicroFocus: 13.5). Raccoons are well-adapted to urban dwelling (one writer calls them "garbage can gourmets"), and the symbiotic relationship with humans may encourage the spread of rabies.

Epizootic:
an epidemic in animals.

rab″do-vir′ĭ-da

The rabies virus is an RNA virion of the Rhabdoviridae family. It is rounded on one end, flattened on the other, and looks like a bullet (Figure 13.10). The virus enters the tissue through a skin wound contaminated with the saliva, urine, blood, or other fluid from an infected animal. The air in a cave inhabited by diseased bats can also transmit the virus.

The incubation period for rabies varies according to the amount of virus entering the tissue and the wound's proximity to the central nervous system. As few as six days or as long as a year may elapse before symptoms appear. A bite from a rabid animal does not ensure transmission, however, because experience shows that only 5 to 15 percent of inoculated individuals develop the disease.

Early signs of rabies are abnormal sensations such as tingling, burning, or coldness at the site of the bite. Fever, headache, and increased muscle tension develop, and the patient becomes alert and aggressive. Soon there is paralysis, especially in the swallowing muscles of the pharynx, and saliva drips from the mouth. Brain degeneration together with the inability to swallow increases the violent reaction to the sight, sound, or thought of water (the word "rabies" comes from the Latin *rabere* for rage). The disease has therefore been called **hydrophobia**, meaning fear of water. Death usually comes within days from respiratory paralysis (MicroFocus: 13.6).

Hydrophobia:
an alternate name for rabies based on extreme sensitivity to water.

Carnivore:
a meat-eating animal with canine teeth.

A person who suffers an animal bite, particularly by a wild carnivore, should be treated as if the animal were rabid. Before 1980, this meant up to two dozen injections of duck embryo vaccine given at a 45 degree angle in the abdominal fat. Since 1980, however, physicians have used a vaccine composed of inactivated viruses cultivated in human embryonic lung cells. Because human tissue

13.5

BEWARE THE GARBAGE CAN GOURMET

Human cases of rabies are very rare (10 confirmed cases in the 1980 to 1992 period), but animal rabies continues to be common. In 1991, for instance, 6972 cases of rabies among animals were reported to the CDC. Almost half the cases (3072) were reported in raccoons.

Rabid raccoons were probably introduced to the mid-Atlantic region of the United States in the mid-1970s, when hunters brought raccoons up from the Southeast to replenish hunting stocks. While raccoons remained in their original territories, rabies was reasonably under control, but in a new rabies-free area, the disease spread rapidly among the raccoon population. The first cases occurred in West Virginia (1977). Then the disease was detected in raccoons from Virginia (1978), Maryland (1981), Pennsylvania (1982), New Jersey (1989), New York (1990), Connecticut (1991), and New Hampshire (1992). The disease also spread east to North Carolina (1991) and north to Ohio (1992).

Given that raccoons live in close proximity with humans in urban, suburban, and rural areas, the possibility of contracting rabies from a raccoon is real. Pet immunizations may interrupt the chain of transmission to humans, but stray raccoons remain a threat for direct transmission to humans via a bite. And should that occur, it is somewhat sobering to know that death is a virtual certainty once rabies symptoms have developed. Indeed, a CDC writer recently stated: "...there is no evidence that any pharmocologic intervention is effective for the treatment of human rabies" (*MMWR*, 1992, 41:663).

To interrupt the animal epidemic, researchers are testing oral rabies vaccines distributed by means of raccoon baits. On Parramore Island, a barrier island off the coast of Virginia, thousands of doses of a new vaccine have been distributed inside baited tubes to attract the raccoons living on the strip of land. The island must be used because the vaccine has been genetically engineered by grafting genes from the rabies virus to the harmless vaccinia (cowpox) virus formerly used against smallpox. Such genetically engineered vaccines cannot be released without careful supervision, and the island provides the opportunity to control the spread of the virus. Over the months, raccoons will be trapped to determine whether the vaccine is eliciting an immune response. The hope is that one day the vaccine can be routinely used to interrupt the raccoon epidemic.

While the vaccine work is continuing public education has remained the chief method for intervention. Health departments warn: minimize exposures to wild raccoons; keep pet immunizations current; seek prompt medical attention if bitten by or exposed to an animal; and consider pre-exposure immunization if you must work where there is an animal population.

FIGURE 13.10

The Rabies Virus

An accumulation of rabies viruses in the salivary gland tissue of a canine (× 58,000). The viruses are elongated particles, some having the tapered shape of bullets. This configuration is characteristic of the rhabdoviruses.

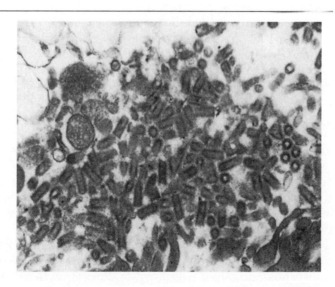

is used, allergic responses are much reduced. For people suffering from animal bites, five injections are given in the arm. These injections are preceded by thorough cleansing and one dose of rabies immune globulin to provide immediate antibodies at the site of the bite. A **postexposure immunization** is usually accompanied by a tetanus booster. For high-risk individuals such as veterinarians, trappers, and zoo workers, a preventive immunization of three injections may be given (Figure 13.11). This **preexposure immunization** currently costs about $125.

Rabies has historically been a major threat to animals. One form, called **furious rabies**, is accompanied by violent symptoms as the animal becomes wide eyed, drools, and attacks anything in sight. In the second form, **dumb rabies**, the animal becomes docile and lethargic, with few other symptoms. In 1959, the number of wildlife cases first exceeded the number of domestic cases, and in 1991, almost 7000 wildlife cases were reported throughout the United States. However, the number of human cases is usually less than 5 per year. This is due in part to postexposure immunizations, of which over 25,000 are given annually.

Polio

The term **polio** is an abbreviation of **poliomyelitis**, a word derived from the Greek terms *polios* for gray and *myelon* for matter. The "gray matter" is the nerve tissue of the spinal cord and brain, which are affected in the disease. Viruses that cause polio are among the smallest virions, measuring 27 nm in diameter. They are composed of RNA and are icosahedral virions of the Picornaviridae family (Figure 13.12).

Polio viruses usually enter the body by contaminated water and food. They multiply first in the tonsils and then in lymphoid tissues of the gastrointestinal tract, causing nausea, vomiting, and cramps. In many cases, this is the extent of the problem. Sometimes, however, the viruses pass through the bloodstream

Tonsils:
patches of lymphoid tissue in the pharynx of humans.

FIGURE 13.11
FIGURE 13.11
An Unusual Case of Rabies

1. A wedding reception takes place out-side a rural farmhouse. The happy couple is in the background. In the foreground, a guest is petting a young calf.

2. Some days later, the newly-married couple is watching helplessly as the calf gets sick. It is lying on the ground trying hard to breathe. A vet is nearby. The calf dies.

3. The calf's head has been sent to the Health Department lab where a micro-biologist examines the brain tissues for signs of rabies lesions.

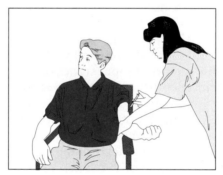

4. The guests at the reception and the family members come to the doctor's office where they are given rabies immunizations in the arm.

and localize on the meninges, where they cause **meningitis**. Paralysis of the arms, legs, and body trunk may result. In the most severe form of polio, the viruses infect the medulla of the brain, causing **bulbar polio** (the medulla is bulblike). Nerves servicing the upper body torso are affected. Swallowing is difficult and paralysis develops in the tongue, facial muscles, and neck. Paralysis of the diaphragm muscle causes labored breathing and possibly, death.

Virologists have identified three types of polio virus: type I, the **Brunhilde** strain, causes a major number of epidemics and is sometimes a cause of paraly-sis; type II, the **Lansing** strain, occurs sporadically but invariably causes paraly-sis; and type III, the **Leon** strain, usually remains in the intestine. Once the method of laboratory cultivation of polio viruses was established by Enders, Weller, and Robbins, a team led by Jonas Salk grew large quantities of the viruses and inactivated them with formaldehyde to produce the first polio vac-cine in 1955 (MicroFocus: 13.7). Albert Sabin's group subsequently developed a vaccine containing attenuated (weakened) polio viruses. This vaccine was in

Medulla:
the stem of the brain.

FIGURE 13.12
Polio Viruses

An electron micrograph of the viruses that cause polio (× 286,000). Although the particles appear to be circular, their symmetry has been found to be icosahedral. These viruses are among the smallest that cause human disease and have a diameter of about 27 nm.

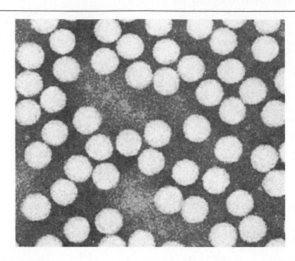

Trivalent vaccine: one that contains three strains of virus.

widespread use by 1961 and could be taken orally, as compared with Salk's vaccine, which had to be injected. Both vaccines are referred to as **trivalent** because they contain all three strains of virus. The vaccines have contributed substantially to the reduction of polio.

The gradual disappearance of polio from media headlines has brought some complacency regarding the need for immunization, but polio still breaks out occasionally. For example, a limited outbreak occurred in 1978 in an unimmunized population of Amish people in southern Pennsylvania. The disease also broke out in Finland in 1985, and a 1993 epidemic affected close to 100 individuals in The Netherlands who had chosen to refuse immunization. Recent events in the Americas have been more positive, however. The attempt at polio eradication by the Pan American Health Organization has apparently been successful and no case of polio due to a "wild" virus (not from a vaccine or lab) has occurred since August 23, 1991. This does not necessarily mean that polio viruses have been eradicated from individuals. Indeed, when CDC researchers undertook a screening of Canadian residents related to the affected individuals from The Netherlands (above), they isolated the identical polio virus that had affected their European relatives.

A discouraging legacy of the polio epidemics is the **postpolio syndrome**, a series of ailments that include fatigue, muscle weakness and atrophy, and some instances of difficult breathing. A 1994 report indicated that these symptoms are believed to affect about 500,000 Americans, nearly a third of the 1.6 million survivors of past polio epidemics. Researchers have found fragments of polio viruses in the tissues of these individuals, leading to the suggestion that the infection is reemerging. Other researchers, however, offer that viral fragments do not indicate reinfection, and at this writing, no whole viruses have been found. Another suggestion is that Coxsackie B viruses may be involved.

Slow Virus Disease

Certain viruses appear to cause slow-developing degenerative diseases in which many years elapse between the original infection and any detectable symptoms. The viruses, known as **slow viruses**, have not yet been visualized with the electron microscope because of their minute size. Moreover, no nucleic acid

"THANKS, DR. SALK!"

In the spring of 1954, I was a lad of thirteen growing up in the Bronx and looking forward to a carefree summer. But I could feel the tension in my parents' voices as they anticipated the days ahead, for summer was the dreaded polio season.

And sure enough, when July came the tension had turned to outright fear. I was told to avoid the public pool and the lusciously cool air-conditioned movie house. I had to report any cough or stiff neck promptly. "Stick with your old friends," my father told me, "you've already got their germs." Most of the time I was indoors, and the only baseball I got to play was in my imagination as I listened to the Yankees every afternoon on my portable radio (they lost the pennant to the Indians that year). And there would be no circus that summer—mixing with strange people was just too dangerous.

Our family was one of the lucky ones to have a television, and each night we watched the row upon row of iron lungs, and we saw the faces of the kids inside. (Iron lungs, I was told, help you breathe when paralysis affects the respiratory muscles.) We heard and read about the daily toll from polio, where the victims lived, how many kids had died, and how many would be captured forever inside their iron prisons.

But there was hope. My mom and her friends were out collecting dimes to fight polio (they called it the "Mothers' March Against Polio"), and the National Foundation of Infantile Paralysis said it had 75 million dimes to help fund the tests of a new vaccine—Dr. Salk's vaccine. Two million children would be getting shots. Maybe next year would be different.

Boy, was it ever different. On April 12, 1955, at a televised news conference, Dr. Jonas Salk declared: "The vaccine works!" The celebration was wild. Our school closed for the day, and the church bells rang even though it was only Thursday. I could tell my mother was relieved—we had steak for dinner that night.

Two months later, things were back to normal. I was now fourteen and eager to show off my baseball skills to any girl who cared to watch. Down at the neighborhood pool I was learning how to dive (when no one was watching). And for a quarter, I got to cheer for the cavalry at the Saturday movie. Summer was back. Thanks, Dr. Salk.

has been detected in material that transmits the disease. Though many scientists believe that viruses will eventually be seen and that nucleic acid will be identified, nevertheless, some investigators hold to the theory that slow viruses are really prions (Chapter 11).

pre'onz

Virologists have related three diseases to slow viruses. The first such disease to be identified was **kuru**, a fatal disorder found in villages of the Fore people who live in the remote highlands of New Guinea. Victims suffer a gradual loss of limb control and eventually all bodily control because of the destruction of their brain cells. Kuru was considered a genetic disease until D. Carleton Gajdusek of the National Institutes of Health researched it and helped prove that it was due to an infectious agent. For his work, Gajdusek shared the 1976 Nobel Prize. The time between injection of the infectious material and the first signs of disease is measured in months or years.

koo'roo

gad'u-sek

The second disease in this group is **scrapie**, a disease of sheep and goats. Known for centuries, scrapie interferes with nervous coordination in animals so they cannot walk or stand. Tortured by itching, the animals scrape constantly against rocks and tree trunks, hence the disease's name. In the late 1930s, over 1500 sheep in Scotland died from a scrapie-contaminated vaccine. A somewhat similar disease in cows has recently emerged. Known technically as **bovine spongiform encephalopathy (BSE)**, the disease is euphemistically called **mad cow disease** (MicroFocus: 13.8).

Scrapie:
a disease of sheep and goats thought to be due to a slow virus.

The last disease is a disorder called **Creutzfeldt–Jakob disease**. This disease, much like kuru in effects and deadliness, is found throughout the world. Virologists are unsure how Creutzfeldt–Jakob disease is transmitted, but at least one case has been contracted from a corneal transplant. Another was contracted from contaminated needle electrodes inserted for diagnostic purposes into the brains of individuals suspected of having epilepsy. And in

kroits'felt
yak'ob

13.8

MICROFOCUS

THE DAY THE COWS WENT MAD

It appeared in England in 1986. First one cow became apprehensive and twitchy, over-reacting to a sound or touch; then another and another showed the same symptoms. Soon the whole herd developed a peculiar, high-stepping, swaying gait with an unsteady lurch. Some cows became overly aggressive, and soon the local residents were talking about mad cow disease.

By 1990, cows were becoming ill at the rate of hundreds per week, and cartoon writers were having a field day with mad cows descending on farm houses and milk factories. To health officials, however, the situation was more foreboding than mirthful. Were the milk and meat supplies contaminated, and would humans be next? Indeed,

there were reports of mad cats, and speculation arose that cat food made from bovine parts was to blame.

Through all the months, virologists had been drawing a parallel between the unknown disease and scrapie, the slow virus disease in which sheep develop neurological symptoms. Under the microscope, scrapie-infected tissue looks spongy with tiny fluid-filled holes, and so did the cows' brain tissue. But how was the agent transferred to cows? Veterinary researchers soon discovered that young calves are often fed a protein-rich feed made from the carcasses of sheep.

Today the term "mad cow disease" is used only by the more frivolous researchers.

The proper name for the disease is bovine (cow) spongiform (spongy appearance) encephalopathy (brain disease), or BSE. Certain researchers believe BSE is a slow virus disease, while others consider it a prion disease. Most investigators think of it in the same terms as kuru and scrapie.

In 1996, ten cases of BSE were identified in Great Britain and at least 20 countries (including the United States) instituted or continued their embargoes on English beef. Even the people at McDonald's were concerned—until they could obtain foreign beef, veggie burgers would replace hamburgers at their restaurants in London.

1988, a case was linked to transplanted dura mater, a covering membrane of the brain.

altz'hi-mer

a-mi"o-trof'ik

Work on slow viruses has substantial importance with respect to other diseases. For example, there is evidence that slow viruses may be involved in Alzheimer's disease, a condition that involves deterioration of physical health and the mind. Other possibilities are that slow viruses may cause Parkinson's disease and amyotrophic lateral sclerosis (Lou Gehrig's disease) as well as Gerstmann-Straussler-Scheinker (GSS) syndrome, a spongiform encephalopathy that affects humans as BSE affects cows. In addition, research may shed light on the relationship between childhood diseases and the slow-developing, mind-destroying illnesses that afflict the elderly. Measles viruses, for instance, are thought to be involved in multiple sclerosis, a disease characterized by degeneration of the myelin sheaths in nerves.

Arboviral Encephalitis

Encephalitis:
acute inflammation of the brain.
en-cef"a-li'tis

The term **encephalitis** refers to an acute inflammation of the brain. Used in the general sense, encephalitis may refer to any brain disorder, much as pneumonia refers to a lung disorder. In this section we shall use encephalitis to mean a number of viral disorders that are *ar*thropod*bo*rne (hence "arboviral").

Arboviral encephalitis may be caused by a series of RNA viruses, usually of the Togaviridae or Bunyaviridae families. In humans, viral encephalitis is characterized by sudden, very high fever and a severe headache. Normally, the patient experiences pain and stiffness in the neck, with general disorientation. Victims become drowsy and stuporous, and may experience a number of convulsions before lapsing into a coma. Paralysis and mental disease may occur in recoverers. Mortality rates are generally high.

There are many forms of arboviral encephalitis and various vectors of the disease. One form is **St. Louis encephalitis (SLE)**, named for the city of St. Louis, where it was first identified in 1933. This disease, transmitted by mosqui-

FIGURE 13.13
An Outbreak of Eastern Equine Encephalitis

1. During the summer of 1991, a group of horses in northern Florida were observed displaying the erratic behavior characteristic of Eastern equine encephalitis (EEE). The horses were unsteady on their feet, walked in circles, and held their heads low, as if asleep.

2. Shortly thereafter, a number of trainers, jockeys, and other individuals associated with the horses developed piercing headaches, occasional numbness in the arms and legs, and stiff necks. Their symptoms indicated Eastern equine encephalitis.

3. As part of the investigation, researchers from the local and federal health departments collected *Aedes albopictus* mosquitoes from a tire depot in central Florida.

4. The virus of Eastern equine encephalitis was isolated from the mosquitoes, the first such isolation of the EEE virus from *A. albopictus*. Measures were instituted to control the mosquito population to interrupt the epidemic.

toes, resurfaced in 1975 with 1300 cases nationwide. Other forms are California encephalitis, La Crosse encephalitis, and Japanese B encephalitis, all transmitted by mosquitoes. Russian encephalitis and Louping ill encephalitis are forms transmitted by ticks. In 1994, 100 cases of arboviral encephalitis were reported in the United States.

Arboviral encephalitis is also a serious problem in **horses**, causing erratic behavior, loss of coordination, and fever. In addition to the economic loss sustained by the death of the horse, the disease is transmissible to humans by ticks, mosquitoes, and other arthropods (Figure 13.13). Important forms are eastern equine encephalitis (EEE), western equine encephalitis (WEE), and Venezuelan eastern equine encephalitis (VEEE). The diseases occur in many animals in nature and are a particular problem in birds because birds are natural reservoirs of the viruses and spread them during annual migrations.

TABLE 13.6

A Summary of Neurotropic Viral Diseases

DISEASE	CLASSIFICATION OF VIRUS	TRANSMISSION	ORGANS AFFECTED	VACCINE	SPECIAL FEATURES	COMPLICATIONS
Rabies	Rhabdoviridae	Contact with body fluids	Brain Spinal cord	Inactivated viruses	Paralysis Hydrophobia	Death
Polio	Picornaviridae	Food Water Contact with human feces	Intestine Spinal cord Brain	Inactivated viruses Attenuated viruses	Infection of medulla	Permanent paralysis
Slow virus disease	(Unknown)	Not established	Brain	Not available	Brain degeneration	Paralysis Mind destruction
Arboviral encephalitis	Many viruses	Arthropods	Brain	Not available	Encephalitis	Coma Seizures
Lymphocytic choriomeningitis	Arenaviridae	Dust Contact	Brain Meninges	Not available	Lymphocytes in brain	Paralysis

Lymphocytic Choriomeningitis

lim″fo-sit′ĭk
ko″re-o-men″in-ji′tis

Lymphocytic choriomeningitis (LCM) is usually found in mice, hamsters, and other **rodents**, where it is transmitted by feces and dustborne viruses from urine. Several outbreaks in humans have been recorded since 1960, most related to pet hamsters or hamster colonies in laboratories. In 1984, a case was traced to mice in the patient's home.

LCM is a mild disorder, often with influenzalike symptoms. Fever and malaise precede headache, drowsiness, and stupor, and the meninges of the brain are infiltrated with large numbers of lymphocytes (thus, the disease's name). The symptoms subside within a week and the mortality is very low. An RNA virus of the Arenaviridae family appears to be the cause. Aseptic meningitis may be the only symptom of disease. Table 13.6 summarizes the neurotropic viral diseases.

NOTE TO THE STUDENT

In the remote tropics of South American and Africa there lurks a coterie of viruses that infect animals, but seldom bother humans. Occasionally, however, humans stumble into their paths, with results horrifying enough to mark the annals of medicine. The hemorrhagic fever viruses, for example, make the internal organs bleed and rot, and patients ooze contagious, virus-laden blood from their eyes, ears, nose, and other orifices. Marburg viruses appeared in 1967, then Lassa fever viruses in the 1970s, followed by Ebola viruses and hantaviruses in the years thereafter. In 1994, another hemorrhagic fever virus, the Sabia virus, escaped from a high-security laboratory at Yale University and roamed the streets of New Haven and Boston before felling its human host with symptoms (the patient recovered and none of his 80 contacts developed the disease.)

Where have these viruses come from and what do they portend? The strange new pathogens erupting on American soil are believed to be ancient organisms that lacked the opportunity to attack until humans blundered into their habitat. Lassa virus, Marburg virus, Ebola virus, hantavirus—

all were probably confined for millennia to isolated human groups or animals; then civilization's encroachment on the forest let them broaden their range. Now it was up to nature to trigger an expansion of the host organism. Previous to the hantavirus epidemic, for instance, heavy rains had disturbed the desert and mountain ecology leading to an abundance of pinon nuts and grasshoppers. Suddenly, the deer mouse population exploded, and hantaviruses had a ready mode of transport. In other cases, humans luck out before the viruses can find an alternate host: Ebola virus kills its victims so quickly that it does not have time to spread. And it does not have an animal host—not yet, at least.

As the world shrinks to a village, a great biological soup is emerging. It is a soup into which dangerous microbes of long-isolated ecologies are being mixed. Air travel provides an excellent opportunity for an ill person to carry viruses across the ocean in a matter of hours. Military bases in far corners of the globe and increased trade with new nations promote the exposure of foreigners to exotic viruses. Europeans once encountered dangerous new microorganisms in the New World. Now their descendants are confronting novel pathogens in their own New Worlds.

To many scientists these are exciting times because their search for disease agents is reminiscent of searches made by Pasteur and Koch a hundred years ago. To others, the times are foreboding because they must deal with immediate threats to life. And, they ask, what else is out there? What, for instance, is the agent of Kawasaki disease? Or kuru? Or slow virus disease?

Contrary to what some people believe, science does not have an answer for everything. If you find yourself becoming complacent about disease and its control, I urge you to read Fuller's book about Lassa fever (Fever) or Preston's book about Ebola fever (The Hot Zone). Better yet, read Laurie Garrett's The Coming Plague. *It will help you realize that the wave of emerging diseases is far from spent.*

Summary

The major thrusts of this chapter are the viscerotropic and neurotropic viral diseases. Viscerotropic diseases are those that affect organs of the visceral cavity, while neurotropic diseases occur in the spinal cord and brain.

Two important viscerotropic diseases are yellow fever and dengue fever, both of which occur in the tropics where the agent of transmission, the mosquito, is common. Viral invasion of the blood, or viremia, characterizes both diseases and high fever is a characteristic symptom. Several forms of hepatitis, including hepatitis A, B, and C are also considered viscerotropic diseases since the major organ affected is the liver. Hepatitis A is caused by a resistant RNA virus that can remain active outside the body and be transmitted by a fecal-oral route. Hepatitis B, by comparison, is caused by a fragile DNA virus that must be transferred directly from person to person by blood or semen contact to remain active. Infectious mononucleosis, another viral disease in this category, is also transferred from person to person, often by saliva contact.

The viscerotropic diseases also include several forms of viral gastroenteritis that occur in the intestine and are accompanied by diarrhea. A number of serious viral fevers including Lassa fever, Ebola fever, and sandfly and Colorado tick fevers are in the group, as are cytomegalovirus and AIDS. AIDS is due to RNA retrovirus called the human immunodeficiency virus (HIV). The virus attacks and destroys the body's lymphocytes, thus weakening the immune system and encouraging disease by opportunistic organisms. HIV is transmitted by blood-to-blood contact, or semen to blood, and HIV disease is the first of several syndromes culminating in AIDS.

Among the neurotropic viral diseases, rabies is notable for its high mortality rate. The disease affects the brain tissue of most animals in nature and is generally transmitted by inoculation with their saliva. Another neurotropic disease, polio, is currently under control due in large measure to mass immunization programs using Salk and Sabin polio vaccines. Also in the group are a collection of slow-developing brain diseases such as kuru, scrapie, and Creutzfeldt–Jakob syndrome. Several arthropodborne CNS diseases collectively called encephalitis are likewise considered neurotropic diseases.

Questions for Thought and Discussion

1. In 1995 the respected New York newspaper *Newsday* reported that during the previous year, 1700 cases of rabies had occurred in New York State. Public health officials report, to the contrary, that rabies cases in the entire United States rarely exceed single digits. What might have been the source of *Newsday*'s error?

2. During an outbreak of infectious mononucleosis in Massachusetts in 1969, a state public health official urged students to reduce kissing to a minimum or, if necessary, to adopt the European custom of kissing on the cheeks. Was this good advice? Why?

3. A *New York Times* crossword puzzle once contained a space in the Across column for a term containing 14 letters. The only clue given was "dengue." By using the Down column, it was learned that the second last letter was an e and that the fifth last letter was an f. What was the answer?

4. In 1985, the Pan American Health Organization established the goal of eliminating poliomyelitis from the Western Hemisphere by 1990. The organization came close to success, and the last confirmed case of paralytic polio due to a "wild" virus occurred in 1991 in Peru. In 1988, the World Health Assembly set out to achieve global eradication of polio by the year 2000. What do you think will be the cornerstones of this ongoing campaign?

5. In 1984, a group of Texas teenagers, wishing to imitate a certain rock singer, bit a series of wild bats. When local health officials found out about the escapade, they insisted that the teenagers be immunized against a certain viral disease. Which disease do you suspect? Why?

6. The intense effort to learn more about HIV and to develop treatments and vaccines for AIDS has necessitated a regular traffic of primates and other animals to the United States and European countries from remote parts of the world. Why has this practice increased the possibility that previously unknown viruses may emerge in American and other populations?

7. Since 1965 the vaccine of choice for polio has been the Sabin oral vaccine composed of attenuated viruses. In 1995, federal health officials recommended a switchover to the Salk vaccine composed of inactivated viruses. This vaccine requires injection to the muscle and is more difficult to administer. What, do you suppose, prompted the recommendation to change?

8. Written on some blood donor cards are the words "CMV(+)." What do you think the letters mean, and why are they placed there?

9. A student in a biology laboratory uses a sterile lancet from a package to pierce the skin and obtain blood for a blood typing exercise. The student places the lancet down on the desktop, whereupon a nearby student picks it up and uses it again to pierce the skin. What is the danger?

10. Sometimes the use of acronyms becomes absurd. A recent headline in the medical section of a newspaper blared: "CDC blames HPS on MCV." Would you care to take a stab at what this gibberish means?

11. At a college campus some years ago, a group of students were passing around a wine bottle. Several days later, one member of the group developed hepatitis A and the other members requested that the college infirmary distribute immune globulin shots. When the infirmary refused, the students demonstrated, causing a stir. A reporter from a local newspaper wrote about the incident and recounted it accurately

except for the last line of the article, which read: "Hepatitis A is normally transmitted by contaminated syringes." What is microbiologically wrong with this statement and what are its implications?

12. Since 1990, the makers of one of the recombinant hepatitis B vaccines (Engerix-B) have offered free immunizations to emergency medical technicians (EMTs) throughout the United States. If you were an EMT would you accept the offer?

13. The control of yellow fever in Central America was a principal factor in the construction of the Panama Canal. The work began in 1904 and was completed in 1914. It is said that during construction, a group of workers made up the following phrase: A MAN A PLAN A CANAL—PANAMA. What two things are unique about this phrase?

14. Sicilian barbers are renowned for their skill and dexterity with razors (and sometimes their voices). In 1995, French researchers studied a group of 37 Sicilian barbers and found that 14 had antibodies to hepatitis C despite never having been sick with the disease. By comparison, when a random group of 50 blood donors was studied, none had the antibodies. What might account for the high incidence of exposure to hepatitis C among the barbers?

15. In many diseases, the immune system overcomes the infectious agent and the person recovers. In several other diseases, the infectious agent overcomes the immune system and death follows. Compare this broad overview of disease and resistance to what is taking place in AIDS and explain why AIDS is probably unlike any other disease encountered in medicine.

Review

On completing your study of these pages, test your understanding of their contents by deciding whether the following statements are true or false. If the statement is true, add the word "True" in the space. If false, substitute a word for the underlined word to make the statement true.

1. _____ Both yellow fever and dengue fever are caused by a DNA virus transmitted by the mosquito.

2. _____ One of the most common causes of death in AIDS patients is pneumonia due to Toxoplasma gondii.

3. _____ The Coxsackie virus is a well-known cause of gastroenteritis in humans.

4. _____ Downey cells are a characteristic sign of the viral disease infectious mononucleosis.

5. _____ The echovirus is known to cause disease of the spleen in humans who acquire the virus.

6. _____ Polio may be caused by any of three strains of polio virus.

7. _____ The term hydrophobia means fear of water and is commonly associated with patients who have encephalitis.

8. _____ Hepatitis is primarily a disease of the liver.

9. _____ The cell most often affected by the AIDS virus is the human monocyte.

10. _____ Because of the characteristic symptoms, dengue fever is sometimes called breakbone fever.

11. _____ A small RNA virus is regarded as the cause of hepatitis A.

12. _____ Norwalk virus and rotavirus are both considered to be agents of viral encephalitis.

13. _____ The cytomegalovirus can cause serious disease of the lungs in AIDS patients.

14. _____ The Epstein–Barr virus is most probably the cause of infectious mononucleosis.

15. _____   A vaccine is available to prevent <u>yellow fever</u> but not to prevent AIDS.

16. _____   Slow virus diseases include <u>kuru</u> and scrapie, but not encephalitis.

17. _____   The Salk and Sabin vaccines are used for immunizations against <u>hepatitis</u>.

18. _____   HIV is a <u>reovirus</u> in which the RNA of the genome is used as a template to synthesize DNA.

19. _____   Pleurodynia and myocarditis are both related to infection by <u>Norwalk</u> virus.

20. _____   Hepatitis B is most commonly transmitted by contact with infected semen or infected <u>blood</u>.

21. _____   One of the organs that suffer damage during infections with yellow fever virus is the <u>kidney</u>.

22. _____   The vaccine currently in use to protect against hepatitis A consists of <u>genetically engineered proteins</u>.

23. _____   One of the most important causes of diarrhea in infants and young children admitted to hospitals is the <u>Epstein–Barr virus</u>.

24. _____   Filoviruses are long, threadlike viruses that cause hemorrhagic fevers and include the <u>Marburg virus</u>.

25. _____   Both *Aedes aegypti* and the *Aedes albopictus* may be capable of transmitting the virus of <u>dengue fever</u>.

26. _____   Enlargement of the lymph nodes, sore throat, mild fever, and a high count of B-lymphocytes are characteristic symptoms in persons who have <u>polio</u>.

27. _____   The resistance of hepatitis A viruses to chemical and physical changes in the environment is generally considered to be <u>low</u>.

28. _____   The "four corners disease" that broke out in the United States in 1993 was eventually related to <u>Lassa fever viruses</u>, which were possibly transmitted among individuals by the deer mouse.

29. _____   The C in the TORCH group of diseases stands for the <u>cephalovirus</u>, which is transmitted from the pregnant woman to her <u>unborn child</u>.

30. _____   The current protocol for rabies immunizations is to give the injections into the <u>stomach</u>.

CHAPTER 14

THE FUNGI

I reland of the 1840s was an economically depressed country of eight million people. Most were tenant farmers paying rent to landlords who were responsible, in turn, to the English owners of the property. The sole crop of Irish farmers was potatoes, grown season after season on small tracts of land. What little corn was available was usually fed to the cows and pigs.

Early in the decade, heavy rains and dampness portended calamity. Then, on August 23, 1845, *The Gardener's Chronicle and Agricultural Gazette* reported: "A fatal malady has broken out amongst the potato crop. On all sides we hear of the destruction. In Belgium, the fields are said to have been completely desolated."

The potatoes had suffered before. There had been scab, drought, "curl," and too much rain, but nothing was quite so destructive as this new disease. It struck down the plants like frost in the summer. Beginning as black spots, it decayed the leaves and stems, and left the potatoes a rotten, mushy mass with a peculiar and offensive odor. Even the harvested potatoes rotted.

The winter of 1845–1846 was a disaster for Ireland. Farmers left the rotten potatoes in the fields, and the disease spread. First the farmers ate the animal feed and then the animals. They also devoured the seed potatoes, leaving nothing for spring planting. As starvation spread, the English government attempted to help by importing corn and establishing relief centers. In England, the potato disease had few repercussions because English agriculture included various grains, but in Ireland famine spread quickly.

After two years, the potato rot seemed to slacken, but in 1847 it was back with a vengeance. Despite relief efforts by the English, over two million Irish suffered death from starvation. Eventually, about 900,000 survivors set off for Canada and the United States. Those who stayed had to deal with economic and political upheaval as well as misery and death.

The potato blight faded in 1848, but did not vanish. Instead, it emerged again during wet seasons and blossomed anew. In the end, hundreds of thousands of Irish left the land and moved to cities or foreign countries. During the 1860s, great waves of Irish immigrants came to the United States. Many Americans are descended from those starving, demoralized farmers.

14.1

MICROFOCUS

A MOLDY SOLUTION

Two reports in 1989 gave insight to the industrial use of fungi as biological insecticides. The first report came from the northeast United States where ravenous multitudes of gypsy moths were devastating trees from New Hampshire to Pennsylvania. In the middle of June, however, the caterpillars of the gypsy moths began to die off *en masse*. It was then that mycologists from the Connecticut Agricultural Experiment Station in New Haven began a systematic study to determine the cause of the die-off. Their report: a fungus named *Entomophaga* ("insect-eater") was killing the caterpillars. The fungus had not been identified in nature in the United States before 1989. Eventually it was located in seven states.

Almost simultaneously, mycologists from North Dakota were reporting that *Entomophaga grylii* can be used to kill crop-destroying grasshoppers. Apparently the fungus produces enzymes that help it penetrate the tough outer skeleton of the insect. Then it circulates in the insect's blood, attacking its body tissues and fat reserves. The grasshopper dies within a week, but the fungus lives on by producing spores that attack other grasshoppers. This particular strain of *E. grylii* is known to exist in Australia, and it may be identical to the fungus observed in the gypsy moth caterpillars. It is apparently superior to strains found in the United States with respect to its resistance to environmental fluctuations and spectrum of insects killed. Scientists anticipate that laboratory safety tests will be followed by the release of infected grasshoppers to crop-plagued areas to test the efficiency of the fungus as an insecticide. Then, on to the consumer ... maybe.

Such are the historic, political, economic, and sociological effects of one species of fungus. Other fungal diseases of fruits, grains, and vegetables can be equally devastating, and we shall see numerous examples in this chapter as we survey the fungi. In addition, we shall take note of several widespread human and animal diseases that are due to fungi and we shall encounter many beneficial fungi such as those used to make antibiotics, bread, and foods or used as insecticides (MicroFocus: 14.1). Our study will begin with a focus on the structures, growth patterns, and life cycles of fungi.

fun′jī

Characteristics of Fungi

Eukaryote:
a complex organism whose cells have organelles and a nucleus; eukaryotes reproduce by mitosis.

ki′tin

The fungi (sing., fungus) are a diverse group of eukaryotic microorganisms, with over 80,000 identifiable species. For many decades, fungi were classified as plants, but laboratory studies have revealed a set of four properties that distinguish fungi from plants: fungi lack chlorophyll, while plants have this pigment; the cell walls of fungal cells contain a carbohydrate called chitin not found in plant cell walls; though generally filamentous, fungi are not truly multicellular like plants, because the cytoplasm of one fungal cell mingles through pores with the cytoplasm of adjacent cells; and fungi are heterotrophic eukaryotes, while plants are autotrophic eukaryotes. Mainly for these reasons, fungi are placed in their own kingdom **Fungi**, in the Whittaker classification of organisms.

Saprobes:
organisms that feed on dead organic matter.

Fungi generally are saprobes with complex life cycles usually involving spore formation. A major subdivision of fungi, the **molds**, grow as long, tangled strands of cells that give rise to visible colonies (Figure 14.1). Another subdivision, the **yeasts**, are unicellular organisms whose colonies resemble bacterial colonies.

FIGURE 14.1

A Typical Fungus

A scanning electron micrograph of *Cladosporium cladosporioides*, one of the most common fungi isolated from air samples. Outdoors, this fungus is commonly found on decaying vegetation. Indoors, it may be isolated from refrigerator moldings, tile grout in showers, and vinyl shower curtains. The conidiophores and conidia of the fungus can be seen. (× 2970)

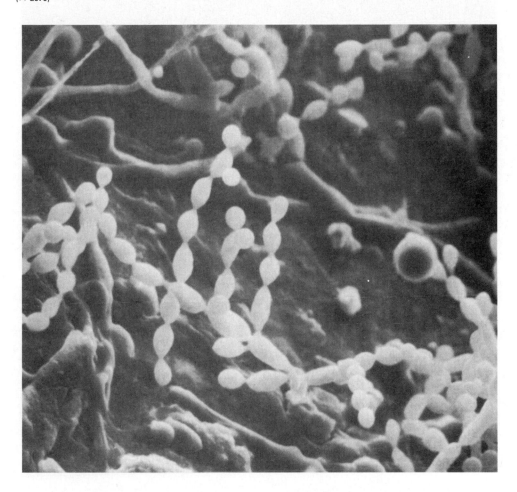

Structure of the Fungi

With the notable exception of yeasts, the fungi consist of masses of intertwined filaments of cells called **hyphae** (sing., hypha). Each cell of the hypha is eukaryotic, with a distinct nucleus surrounded by a nuclear membrane and other eukaryotic organelles. The cell wall is composed of small amounts of cellulose and large amounts of chitin. **Cellulose** is a polysaccharide composed of glucose units linked together in such a way that most organisms cannot digest it. **Chitin** is a polymer of acetylglucosamine units, that is, glucose molecules containing amino and acetyl groups. Chitin gives the cell wall rigidity and strength, a function it also performs in the exoskeletons of arthropods.

hi′fā
hi′fah

as″e-til-gloo-kos′ah-mēn

Fungal cells lack chlorophyll, and photosynthesis is therefore impossible. Since they consume preformed organic matter, fungi are described as **heterotrophic** organisms. They are **saprobic**, except for the parasitic fungi, which cause disease. Together with the bacteria, fungi decompose vast quantities of dead organic matter that would otherwise accumulate and make the earth uninhabitable.

FIGURE 14.2
The Hypha

A phase-contrast photomicrograph of *Candida albicans* in its hyphal form. In the human body, phagocytes have a difficult time coping with these filamentous structures, a factor that may contribute to the pathogenicity of the fungus. Bar = 40 μm.

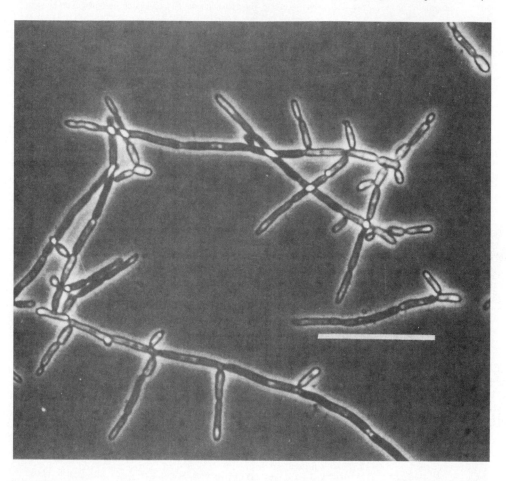

se-no-sit′ik
ri-zo′pus

mi-se′le-um

Mycology:
the study of fungi.

In many species of fungi the individual cells are separated by cross walls, or **septa** (sing., septum). The septa are not complete, however, and pores allow a mixing of adjacent cytoplasms. In other fungal species, the cells have no septa, and the cytoplasms and organelles of neighboring cells mingle freely. These fungi are said to be **coenocytic**. The common bread mold *Rhizopus stolonifer* is coenocytic, while the blue-green mold that produces penicillin, *Penicillium notatum*, has septa.

The hypha is the morphological unit of the fungus and is seen only with the aid of a microscope (Figure 14.2). Hyphae have a broad diversity of forms, as photographs in this chapter illustrate, and many are highly branched with reproductive structures called **fruiting bodies**. A thick mass of hyphae is called a **mycelium** (pl., mycelia). This mass is usually large enough to be seen with the unaided eye (MicroFocus: 14.2), and generally it has a rough, cottony texture. The study of fungi is called **mycology**; and the individual who studies the fungi is a mycologist. Invariably, the prefix "myco-" will be part of a word referring to fungi, since *mycote* is Greek for fungus.

14.2

MICROFOCUS

BIG AS THE BLUE WHALE

The most obvious thing to call it is a "humongous fungus," for indeed, it is gigantic. The fungus weighs over 100 tons; it occupies 37 acres of forest ground; and it is about 1500 years old. Located in the Upper Peninsula of Michigan, it was found in 1992 by James B. Anderson, Myron Smith, and their colleagues from the University of Toronto.

The enormous underground fungus is called *Armillaria bulbosa*. It consists of innumerable hyphae that intertwine to unimagined lengths and push their fruiting bodies up to the surface as edible mushrooms called the honey fungus. To prove their case that the massive structure is a single fungus, Anderson and Smith gathered 20 samples of the fungus and performed DNA analyses on 16 fragments from each sample. The genetic material was identical in every specimen.

Can the behemoth stand with the redwood trees and other domineering giants of our times? Possibly so. But then again, is a field of grass with all its blades coming from a single root system a single organism? It depends on how you interpret the rules. Nevertheless, it is becoming clear that a fungus is certainly not a lesser organism in the scheme of things—certainly not in terms of size.

Growth of the Fungi

In nature, the fungi are important links in ecological cycles because they rapidly digest animal and vegetable matter. In doing so, they release carbon and minerals back to the environment and make them available for recycling in plants. However, fungi may be a bane to industries because they also contaminate leather, hair products, lumber, wax, cork, and polyvinyl plastics.

Many fungi live in a harmonious relationship with other plants in nature, a condition called **mutualism**. In the southwestern Rocky Mountains, for instance, a fungus of the genus *Acremonium* thrives on the blades of a species of grass called *Stipa robusta* ("robust grass"). The fungus produces a powerful poison that can put an animal such as a horse to sleep for about a week (the grass is called "sleepy grass" by the locals). Thus the grass survives where others are nibbled to the ground, reflecting the mutually beneficial interaction between plant and fungus.

ac-re-mo′ne-um

Other fungi called **mycorrhizal fungi** also live harmoniously with plants. The hyphae of these fungi invade the roots of plants (and sometimes their stems) and plunge into their cells. Though poised to suck the plants dry, the fungi are in fact gentle neighbors. Mycorrhizal fungi consume some of the carbohydrates produced by the plants, but in return they contribute certain minerals and fluids to the plant's metabolism. Mycorrhizal fungi have been found in plants from salt marshes, deserts, and pine forests. Indeed, in 1995, researchers from the University of Dayton reported that over 50 percent of the plants growing in the large watershed area of southwestern Ohio contain mycorrhizal fungi.

Most fungi grow best at approximately 25° C, a temperature close to normal room temperature (about 75° F). The notable exceptions are the pathogenic fungi, which thrive at 37° C, body temperature. Usually these fungi also grow on nutrient media at 25° C. Such fungi are described as **biphasic** (two phases) or **dimorphic** (two forms). Many have a yeastlike phase at 37° C and a moldlike phase at 25° C. Certain fungi grow at still lower temperatures, such as the 5° C found in the normal refrigerator.

dī-mor′fik

Many fungi thrive under acidic conditions at a pH from 5 to 6. Acidic soil may therefore favor fungal turf diseases, and lime should be used to neutralize the soil. Mold contamination is also common in acidic foods such as sour cream, applesauce, citrus fruits, yogurt, and most vegetables. Moreover, the acidity in

Lime: calcium carbonate.

breads and cheese encourages fungal growth. Blue cheese, for example, consists of milk curds in which the mold *Penicillium roqueforti* is growing.

Fungi are aerobic organisms, with the notable exception of the fermentation yeasts that multiply in the presence or absence of oxygen. Normally, a high concentration of sugar is conducive to growth, and laboratory media for fungi usually contain extra glucose in addition to an acidic environment. Examples of such media are **Sabouraud dextrose agar** and **potato dextrose agar** (Micro-Focus: 14.3).

sab'oo-rō

Reproduction in Fungi

Reproduction in fungi may take place by asexual processes as well as by a sexual process. The principal structure of asexual reproduction is the **fruiting body**. This structure usually contains thousands of **spores**, all resulting from the mitotic divisions of a single cell and all genetically identical. Each spore has the capability of germinating to reproduce a new hypha that will become a mycelium (Figure 14.3).

Fruiting body:
a spore-containing fungal structure used in asexual reproduction.

Certain spores develop within a sac called a sporangium. Appropriately, these spores are called **sporangiospores**. Other spores develop on supportive structures called conidiophores (Figure 14.4). These spores are known as **conidia** (sing., conidium), from the Greek *conidios*, meaning dust. The bread mold *Rhizopus* produces sporangiospores, while the blue-green mold *Penicillium* produces conidia. Fungal spores are extremely light and are blown about in huge numbers by wind currents. Many people suffer allergic reactions when they inhale spores, and communities therefore report the mold spore count to alert sufferers.

Conidia:
unprotected asexually produced fungal spores formed on a supportive structure.

Some asexual modes of reproduction do not involve a fruiting body. For example, spores may form by fragmentation of the hypha. This process yields **arthrospores**, from the Greek stem *arthro-* for joint. The fungi that cause ath-

FIGURE 14.3

A Germinating Spore

A scanning electron micrograph of a spore of the fungus *Cephalosporium* germinating to form a hypha (× 6000). Note the septa between cells of the hypha.

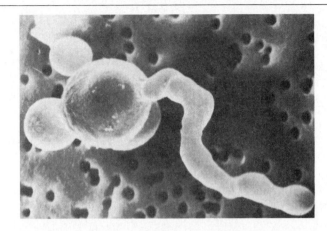

FIGURE 14.4

The Fungus *Aspergillus niger*

(a) A scanning electron micrograph showing the moldlike phase of *Aspergillus niger* (× 600). Many conidiophores are present within the mycelium. These conidiophores contain conidia, the unprotected spores of certain fungi. This photograph demonstrates the three-dimensional image possible with the scanning electron microscope. (b) A close-up view of one conidiophore with a mass of conidia.

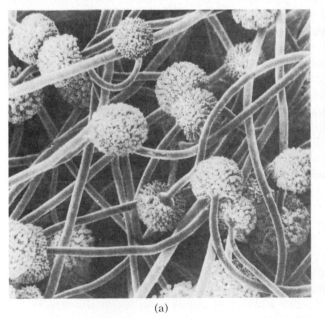 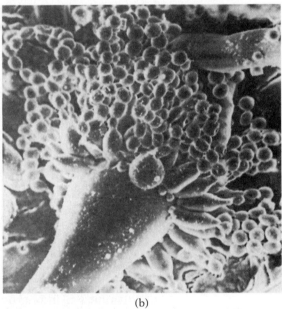

(a)   (b)

lete's foot multiply in this manner. Another asexual process is called **budding**. Here, the cell becomes swollen at one edge, and a new cell called a **blastospore**, or **bud** develops from the parent cell and breaks free to live independently. Yeasts multiply in this way. **Chlamydospores** and **oidia** are other forms of spores produced without a fruiting body. Chlamydospores are thick-walled spores formed along the margin of the hypha, while oidia form at the tip of the hypha.

Budding:
an asexual reproductive process in which new cells form at the periphery of parent cells.
kla-mid'o-spōr
o-id'e-ah

Many fungi also produce spores by a sexual process of reproduction. In this process, the cells of opposite mating types of fungi come together and fuse. A fusion of nuclei follows and the mixing of chromosomes temporarily forms a double set of chromosomes, a condition called **diploid** (from the Greek *diploos* for twofold). Eventually the chromosome number is halved, and the cell returns to the condition where it has a single set of chromosomes, the so-called **haploid** condition (from the Greek *haploos* for single). Spores develop from cells in the haploid condition.

Sexual reproduction is advantageous because it provides an opportunity for the evolution of new genetic forms better adapted to the environment than the parent forms. For example, a fungus may become resistant to fungicides as a result of chromosomal changes during sexual reproduction. Separate mycelia of the same fungus may be involved in sexual reproduction, or the process may take place between separate hyphae of the same mycelium. The process is essentially similar to that taking place in complex animals and plants.

Classification of Fungi

miks"o-mi-ko'tah
li'ken
mī"ko-fi"ko-mi-ko'ta

Variations in the sexual process of reproduction provide important criteria for the classification of fungi. The true fungi, such as we are discussing, belong to the division **Eumycota** in the kingdom Fungi, as set forth by Whittaker (Table 14.1). Other fungi in this kingdom are the slime molds in the division Myxomycota, and the lichens in the division Mycophycomycota. **Slime molds** are complex organisms having an amebalike motile stage and a funguslike spore-producing stage. **Lichens** consist of a fungal mycelium containing a number of unicellular algae or cyanobacteria that perform photosynthesis.

Members of the division Eumycota (true fungi) are divided into five classes, based mainly on the type of sexual spore produced. Generally the distinctions among the fungi are made on the basis of structural differences or physiological or biochemical patterns. However, DNA analyses are becoming an important tool for drawing relationships among the fungi. Indeed, the first-place winner of the 1993 Westinghouse Science Talent Search was an Illinois student named Elizabeth M. Pine who showed that two structurally related mushrooms should probably be reclassified on the basis of their DNA content. We shall briefly examine each of the five classes in the next paragraphs.

Oomycetes

o"o-mi-se'tēz

o'o-spōr

o"o-mi'sēt
zo'o-spōr

Fungi of the class **Oomycetes** are commonly called "water molds," a reference to the fact that most species are aquatic fungi. During sexual reproduction, the members of this group form clusters of egglike bodies at the tips of hyphae. Other nearby hyphae grow toward the bodies and fuse with them. Nuclear fusion leads to the formation of sexual spores called **oospores**, which germinate to produce new hyphae.

A notable feature of an oomycete is the **zoospore**, a flagellated spore formed in the asexual process of reproduction. No other fungi produce motile cells. Also, fungi of the Oomycetes class have diploid cells during most of their life cycle, whereas most other fungal species have haploid cells. Moreover, the cell walls lack chitin. Because of these characteristics, some mycologists postulate that oomycetes may be the product of an evolutionary development entirely separate from other fungi.

TABLE 14.1	
Classification in the Kingdom Fungi	
Division I	Eumycota: true fungi
	Class 1. Oomycetes
	Class 2. Zygomycetes
	Class 3. Ascomycetes
	Class 4. Basidiomycetes
	Class 5. Deuteromycetes
Division II	Myxomycota: slime molds having a mobile ameboid stage in the life history
Division III	Myxophycomycota: lichens consisting of a fungus and an alga growing together

Aquatic oomycetes are familiar as the molds that plague fish in an aquarium. Some terrestrial oomycetes are parasites of insects and plants, and certain ones cause downy mildew in grapes, white rust disease of cabbage, and the infamous late blight of potatoes. The effects of this disease, caused by *Phytophthora infestans*, are described in the introduction to this chapter.

fi-tof'tho-rah

Zygomycetes

The second class of Eumycota is **Zygomycetes**, a group of terrestrial fungi with coenocytic hyphae. Sexual reproduction in these organisms results in **zygospores** from the mating of hyphae (Figure 14.5). Both sexually and asexually produced spores are dispersed on air currents.

zi"go-mi-se'tēz
Coenocytic: lacking cross walls between adjacent cells.

The well-known member of the class Zygomycetes is the common bread mold, ***Rhizopus stolonifer***. The hyphae of this fungus form a white or gray mycelium on bread, with upright sporangiophores each bearing globular sporangia. Thousands of sporangiospores are formed in each sporangium. Occasional contamination of bread is compensated by the beneficial roles *Rhizopus* plays in industry. One species, for example, ferments rice to sake, the rice wine of Japan; another species is used in the production of cortisone, a drug that reduces inflammation in body tissues. These processes are explored further in Chapter 25.

ri-zo'pus

Ascomycetes

Members of the class **Ascomycetes** are very diverse, varying from unicellular yeasts to powdery mildews, cottony molds, and large and complex "cup fungi." The latter form a cup-shaped structure composed of hyphae tightly packed together. The hyphae of an ascomycete are septate, with large pores allowing a continuous flow of cytoplasm.

as"ko-mi-se'tēz

as"ko-mi'sēt

Though their mycelia vary considerably, all ascomycetes form a reproductive structure called an **ascus** during sexual reproduction. An ascus is a sac within which up to eight haploid **ascospores** form (Figure 14.6). Most of the ascomycetes also reproduce asexually by means of conidia, produced in chains at the end of a conidiophore.

Certain members of the Ascomycetes class are extremely beneficial. One example is the yeast ***Saccharomyces***, used in brewing and baking. Another example is ***Aspergillus***, which produces such products as citric acid, soy sauce, and vinegar and is used in genetics research (Figure 14.7). A third is ***Penicillium***, various species of which produce the antibiotic penicillin as well as such cheeses as Roquefort and Camembert (Chapter 24).

sak"ah-ro-mi'sēz
as"per-jil'us

FIGURE 14.5

Reproduction in Fungi

A sequence of scanning electron micrographs showing zygospore formation in the mold *Rhizopus*. (a) Sexually opposite hyphae fuse and form a fusion septum (FS). (b) Cells at the septum begin to swell and show early signs of zygospore formation. (c) The outer primary wall begins to rupture. (d) The rupturing continues and (e) the zygospore is revealed. The zygospore continues to mature as the primary wall separates away. (f) A magnified view of the zygospore showing its surface characteristics and the remnants of the primary wall.

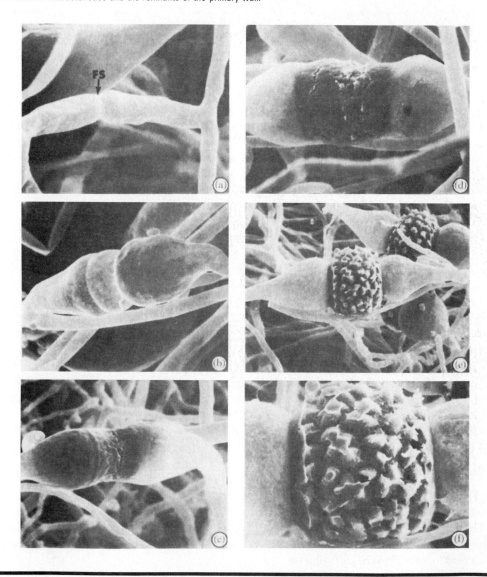

On the deficit side, some ascomycetes attack valuable plants. For instance, one member of the class parasitizes crops and ornamental plants, causing powdery mildew. Another species has almost entirely eliminated the chestnut tree from the American landscape. Still another ascomycete is presently attacking elm trees in the United States (Dutch elm disease) and is threatening extinction of this plant. Two other ascomycete pathogens are ***Claviceps purpurea***, which causes ergot disease of rye plants, and ***Aspergillus flavus***, which attacks a variety of foods and grains (Chapter 24).

klav'ĭ-seps
pur-pur'e-a

FIGURE 14.6

Two Types of Fungal Spores

Ascospores (A) and conidia (B) of the fungus *Aspergillus quadrilineatus*. This fungus was cultivated from the nasal sinuses of an ill patient who had recently received a bone marrow transplant. The sexually produced ascospores display a series of so-called "equatorial crests" at their midlines. The asexually produced conidia are rounder, with tightly folded ("rugous") surfaces. Bar = 1μm.

Basidiomycetes

Members of the class **Basidiomycetes** are commonly called "club fungi." They include the common mushroom (MicroFocus: 14.4), as well as the shelf fungi, puffball, and other fleshy fungi, plus the parasitic rust and smut fungi. The name basidiomycete refers to the reproductive structure on which sexual spores are produced. The structure, resembling a club, is called a **basidium**, the Latin term for "small pedestal." Its spores are known as **basidiospores**.

bah-sid′e-o-mi-se′tēz

ba-sid″e-o-mi′sēt
bah-sid′e-um

Perhaps the most familiar member of the class is the edible **mushroom**. Indeed, the Italian "fungi" means mushroom. Its mycelium forms below the ground and after sexual fusion has taken place, the tightly compacted hyphae force their way to the surface and grow into the mushroom cap (Figure 14.8). Basidia develop on the underside of the cap along the gills, and each basidium may have up to eight basidiospores. Edible mushrooms belong to the genus **Agaricus**, but one of the most potent toxins known to science is produced by another species of a visually similar genus, **Amanita**. Sixteen outbreaks of mushroom poisoning, most related to this genus, were reported to the CDC in recent years. Another mushroom, the huge puffball, caused serious respiratory illness in eight persons when the spores were inhaled in an incident in Wisconsin in 1994.

a-gar′i-cus
am″ah-ni′tah

FIGURE 14.7
Normal and Mutant Fungi

Scanning electron micrographs of the conidiophores of *Aspergillus nidulans*. (a) The normal or "wild type" of the fungus is depicted. At the tip of the hypha, the conidiophore contains hundreds of asexually produced spores (conidia) any of which can germinate to reproduce the fungus. (b) A mutated form of *A. nidulans*. This organism was produced by mutating the regulatory genes of the fungus. As a result of the molecular manipulations, distinctive structural variations have occurred in the fungus and the production of spores has been interrupted.

(a)

(b)

Agricultural losses due to rust and smut diseases are considerable. **Rust diseases** are so named because of the orange-red color of the infected plant. The diseases strike wheat, oats, and rye, as well as trees used for lumber, such as white pines. Many rust fungi require alternate hosts to complete their life cycles, and local laws often prohibit the cultivation of certain crops near rust-sensitive plants. For example, it may be illegal to raise gooseberries near white pine trees. **Smut diseases** give a black, sooty appearance to plants. They affect corn, blackberries, and a number of grains, and cause untold millions of dollars of damage yearly.

Deuteromycetes

doo"ter-o-mi-se'tēz

Certain fungi lack a known sexual cycle of reproduction and consequently are labeled with the botanical term "imperfect." These imperfect fungi are placed in the fifth class, **Deuteromycetes**, where reproduction is only by an asexual method (Table 14.2). It should be noted that a sexual cycle probably exists for these fungi, but it has thus far eluded mycologists.

14.4

A GUIDE FOR THE MUSHROOM HUNTER

In ancient Rome, mushrooms were the food of the gods, and only the emperors were permitted to partake of their delights. Today, exotic mushrooms enjoy an equally high reputation among gourmets of the world. Some experts know how to spot them in the wild, but for amateurs the key word is caution because in mushroom hunting, ignorance is disaster.

Mushrooms come in a huge variety of shapes, colors, and sizes. Among the interesting wild mushrooms are the Jack-O-Lantern fungus, known for its luminous gills; the Beefsteak fungus, whose cap resembles a piece of raw beef; and the Bird's Nest fungus, in which the fruiting body and its spores look like a bird's nest with eggs. On the debit side, about 100 of the 2000 known species can cause mushroom poisoning and death. High on the list of dangerous organisms are *Amanita verna*, the Destroying Angel, and *Amanita phalloides*, the Deathcap. Mortality rates of 50 percent have been observed in people who consume these mushrooms.

Botanists urge that mushrooms be hunted with a camera rather than a fork and plate. They point out that the colors and settings encourage prize-winning photography, and they urge that mushroom consumption be limited to those species cultivated for use as food. After all, they reason, birdwatchers do not eat birds, so why should mushroom-watchers eat mushrooms?

For those who insist on stalking wild mushrooms, mycologists recommend joining a society, reading extensively, and treading lightly into this hobby. As the sage writes:

There are old mushroom hunters,
And there are bold mushroom hunters,
But there are no old, bold mushroom
   hunters.

When the sexual cycle is discovered, the deuteromycete is reclassified into one of the other four classes. A case in point is the fungus known as ***Histoplasma capsulatum***. This fungus causes histoplasmosis, a disease of the human lungs and other internal organs. When the organism was found to produce ascospores, it was reclassified with the Ascomycetes and given the new name *Emmonsiella capsulata*. However, some traditions die slowly, and certain mycologists insisted on retaining the old name because it was familiar in clinical medicine. Thus, mycologists decided to use two names for the fungus: the new name, *Emmonsiella capsulata*, for the sexual stage, and the old name, *Histoplasma capsulatum*, for the asexual stage.

Many fungi pathogenic for humans are classified as Deuteromycetes. These fungi usually reproduce by budding or fragmentation, and segments of hyphae

doo″ter-o-mi′sēt
his″to-plaz′mah
cap-su-lat′um

ĕ″mon-si-el′ah

TABLE 14.2

Comparison of the Classes of Fungi

CLASS NAME	COMMON NAME	CROSS WALLS	SEXUAL STRUCTURE	SEXUAL SPORE	REPRESENTATIVE GENERA
Oomycetes	Water molds	−	None	Oospore	*Phytophthora* *Saprolegnia*
Zygomycetes	Terrestrial molds	−	None	Zygospore	*Rhizopus*
Ascomycetes	Sac fungi	+	Ascus	Ascospore	*Penicillium* *Aspergillus* *Saccharomyces*
Basidiomycetes	Club fungi	+	Basidium	Basidiospore	*Agaricus* *Amanita*
Deuteromycetes	Imperfect fungi	+	Unknown	Unknown	*Candida* *Epidermophyton*

FIGURE 14.8

Development of the Mushroom

(a) Sexual fusion of hyphae takes place below the ground. (b) The hyphae compact together tightly and form a fleshy mass, the button. (c) The button pushes through the ground and grows into a fruiting body, the mushroom (d). On the underside of the mushroom cap basidiospores form on basidia along the gills (e).

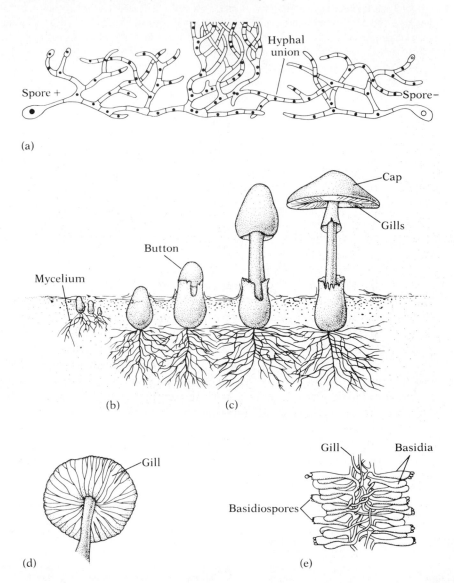

are commonly blown about by dust or deposited on environmental surfaces. For example, fragments of the athlete's foot fungus are sometimes left on towels and the shower room floor. Recently discovered fungi are also placed here until more is known about them (MicroFocus: 14.5).

The Yeasts

The word "yeast" refers to a large variety of unicellular fungi (as well as the single cell stage of any fungus). Included in the group are nonsporeforming yeasts of the class Deuteromycetes, as well as certain yeasts that form basidiospores or ascospores and thus belong to the Basidiomycetes or Ascomycetes

14.5

NOT ALL FUNGI ARE BAD

In 1989, researchers at Johns Hopkins University discovered that taxol, a chemical derived from yew trees, could greatly reduce the size of tumors in women suffering from ovarian cancer. Two years later, in January 1993, the Food and Drug Administration approved taxol for ovarian cancer, while noting that the drug might be useful for breast, head, and neck tumors. Unfortunately, the exhilaration attending approval of the new treatment was counterbalanced by the cost of the drug (about $1000 per treatment cycle) and the fear that the yew tree might be overfarmed to provide bark for the drug.

Then, in 1993, a new twist was added to the taxol story. In April, Montana researchers discovered growing within the bark of the yew a fungus that produces taxol on its own. Plant pathologist Gary Strobel and chemist Andrea Stierle both from Montana State University led the research (Figure). Under Strobel's intuitive direction, Stierle searched the Montana woods for local yews (*Taxus pacifica*) that would yield taxol. Finding one such yew, they went a step further and isolated a fungus from within the folds of the yew's bark. The fungus continued to produce taxol even after removal from its host plant. They named the fungus *Taxomyces andreanae* (Andrea's taxus-fungus). Although *T. andreanae*'s yield of taxol is low, the potential for increasing the yield is great. For example, enormous fermentation tanks can be used to produce enormous amounts of the fungus and much larger amounts of the drug. Moreover, genetic engineering techniques can be used to pinpoint and clone the taxol genes, then transfer them to high-yield vector organisms such as bacteria. Apparently the drug companies believe that these and other approaches can work —months before their scientific paper appeared in print, the Montana researchers had secured a patent on the fungal production of taxol and were being courted by numerous drug companies. The fungus' future appears bright.

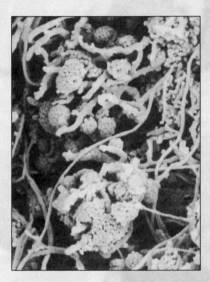 

A fungus and its discoverers. (a) The fungus *Taxomyces andreanae* showing its hyphal strands and fruiting bodies with spores ($\times$ 2300). (b) Strobel and Stierle with the yews from which the fungus was obtained.

classes. The yeasts we shall consider here are the species of *Saccharomyces* used extensively in brewing, baking, and as a food supplement. Pathogenic yeasts will be discussed presently.

sak"ah-ro-mi'sēz

Saccharomyces translates literally to "sugar-fungus," a reference to the ability of the organism to ferment sugars. The most commonly used species of *Saccharomyces* are *S. cerevisiae* and *S. ellipsoideus*, the former used for bread baking and alcohol production, the latter for alcohol production. Yeast cells are about 8 μm long and about 5 μm in diameter. They reproduce chiefly by budding (Figure 14.9), but a sexual cycle also exists in which cells fuse and form an enlarged cell (an ascus) containing smaller cells (ascospores). The organism is therefore an ascomycete.

ser"e-vis'e-a
e-lip-soid'e-us

The cytoplasm of *Saccharomyces* is rich in B vitamins, a factor that makes yeast tablets valuable **nutritional supplements**. One pharmaceutical company adds iron to the yeast and markets its product as Ironized Yeast, recommended for people with iron-poor blood.

The **baking** industry relies heavily upon *S. cerevisiae* to supply the texture in breads. Flour, sugar, and other ingredients are mixed with yeast, and the

FIGURE 14.9

Yeasts

Two views of *Saccharomyces*, the common baking and brewing yeast. (a) A photomicrograph of yeast cells used in beer fermentation (× 1000). (b) A scanning electron micrograph of *Saccharomyces cerevisiae*. A bud (A) is present on one of the cells, and a bud scar (B) can be seen at the lower right.

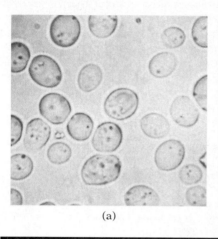

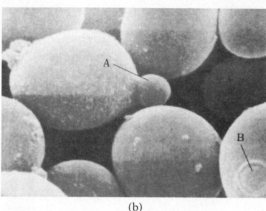

(a)   (b)

Lactobacillus:
a genus of Gram-positive rods that ferment lactose to lactic acid.

Glycolysis:
the multistep enzyme process in which glucose is converted to pyruvic acid.

dough is set aside to rise. During this time, the yeasts break down glucose and other carbohydrates, and produce carbon dioxide through the chemistry of glycolysis and the Krebs cycle (Chapter 5). The carbon dioxide expands the dough, causing it to rise. Protein-digesting enzymes, also from the yeast, partially digest the gluten protein of the flour to give bread its spongy texture (MicroFocus: 14.6). To make bagels, the dough is boiled before baking; for sour dough bread, *Lactobacillus* species are added to give an acidic flavor to the bread; for rye bread, rye flour is substituted. In all these modifications, yeast remains an essential ingredient.

Yeasts are plentiful where there are orchards or fruits (the haze on an apple is a layer of yeasts). In natural alcohol **fermentations**, wild yeasts of various *Saccharomyces* species are crushed with the fruit; in controlled fermentations, *S. ellipsoideus* is added to the prepared fruit juice. Now the chemistry is identical with that in dough: the fruit juice bubbles profusely as carbon dioxide evolves through the reactions of glycolysis and the Krebs cycle. When the oxygen is depleted, the yeast metabolism shifts to fermentation, and the pyruvic acid from glycolysis changes to consumable ethyl alcohol (Chapter 5).

The products of yeast fermentation depend on the starting material. For example, when yeasts ferment barley grains, the product is **beer**; if grape juice is fermented, the product is **wine** (Figure 14.10). Sweet wines contain leftover sugar, but dry wines have little sugar. Sparkling wines such as champagne continue to ferment in thick bottles as yeast metabolism produces additional carbon dioxide. For **spirits** such as whiskey, rye, or scotch, some type of grain is fermented and the alcohol is distilled off. Liqueurs are made when yeasts ferment fruits such as oranges, cherries, or melons. Virtually anything that contains simple carbohydrates can be fermented by *Saccharomyces*. The huge share of the U.S. economy taken up by the wine and spirits industry is testament to the significance of the fermentation yeasts. A fuller discussion of fermentation processes is presented in Chapter 25.

FIGURE 14.10

Aging Wine

Wooden casks of wine aging in a cool cellar at a California winery. Wines remain in casks such as these for months, and sometimes, for years, before bottling.

TO THIS POINT

■

We have discussed aspects of the structure, growth, and reproductive patterns in fungi. We began by exploring some details of the hypha and mycelium, and then we examined the temperature, pH, and oxygen requirements for growth. Most fungi grow at temperatures close to room temperature and under conditions that are acidic.

The discussion then turned to the two processes of reproduction that take place in the fungi. All fungi reproduce by an asexual process, and most reproduce by a sexual process in which sexually opposite hyphae fuse to form sexual spores. Genetic variation is an advantage of such a process. The sexual cycle is also the basis for the classification of Eumycota into five classes. Members of the class Oomycetes form oospores. Members of the class Zygomycetes form single free zygospores. In fungi of the Ascomycetes class, a sac forms containing up to eight ascospores. Members of the class Basidiomycetes form a clublike supportive structure, the basidium, on which basidiospores develop. Deuteromycetes have no known sexual cycle. The section closed with a discussion of the economically important baking and fermentation yeasts.

We shall now focus our attention on the fungal diseases of humans. These diseases generally do not occur in widespread epidemics and thus, their names may be unfamiliar. However, some diseases may endanger human life, especially if the immune system has been suppressed. For this reason fungal diseases are commonly found as complications of other diseases or in situations where a patient is undergoing therapy for an unrelated problem.

Fungal Diseases of Humans

Cryptococcosis

krip"to-kok-o'sis

Cryptococcosis is considered the most dangerous fungal disease in humans. It affects the lungs and the meninges, the coverings of the brain and spinal cord, and is estimated to account for over 25 percent of all deaths from fungal disease.

krip"to-kok'us
nē-o-form'anz

Cryptococcosis is caused by a yeast known as ***Cryptococcus neoformans***. The organism is found in the soil of urban environments and grows actively in the droppings of pigeons but not within the pigeon itself. Cryptococci may become airborne with gusts of wind and subsequently enter the respiratory passageways of humans. Air conditioner filters represent a hazard because they trap large numbers of cryptococci.

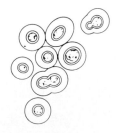

CRYPTOCOCCUS NEOFORMANS

Cryptococcus neoformans cells have a diameter of about 5 to 6 μm and are embedded in a gelatinous capsule that provides resistance to phagocytosis (Figure 14.11). The cells penetrate to the air sacs of the lungs, but symptoms of infection are generally rare. However, if the cryptococci pass to the bloodstream and localize in the meninges and brain, the patient may experience piercing headaches, stiffness in the neck, and paralysis. Diagnosis is aided by the observation of encapsulated yeasts in respiratory secretions or cerebrospinal fluid obtained by a spinal tap.

am"fo-ter'ĭ-sin

Untreated cryptococcosis may be fatal. However treatment with the antifungal drug **amphotericin B** is usually successful, even in severe cases. Because this drug has toxic side effects such as kidney damage and anemia, the patient should be continually monitored.

Resistance to cryptococcosis appears to depend upon proper functioning of a branch of the immune system governed by the T-lymphocytes (or T-cells). When these cells are lacking, the immune system is severely suppressed and cryptococci invade the tissues as opportunists. In patients with acquired immune deficiency syndrome (AIDS), one of the causes of death is cryptococcosis.

FIGURE 14.11
The Agent of *Cryptococcus*

A negatively stained photomicrograph of *Cryptococcus neoformans.* A distinct capsule surrounds each oval, yeastlike cell. This capsule provides resistance to phagocytosis and enhances the pathogenic tendency of the fungus.

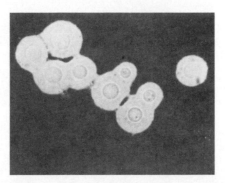

During the early 1980s, mycologists located a sexual stage for *C. neoformans.* The stage is related to the smut fungi of the class Basidiomycetes, and is called *Filobasidiella neoformans.*

fi"lo-bah"sid-ē-el'ah

Candidiasis

Candida albicans is often present in the skin, mouth, vagina, and intestinal tract of healthy persons and animals, where it lives without causing disease (Figure 14.12). The organism is a small yeast that forms filaments called pseudohyphae when cultivated in laboratory media. As body defenses are compromised or when changes occur in the microbial population, *C. albicans* flourishes and causes any number of forms of **candidiasis**. Older texts refer to the condition as **moniliasis** because the organism was once called *Monilia albicans.*

kan-di-di'ah-sis

One form of candidiasis occurs in the vagina and is often referred to as **vulvovaginitis** or the "**yeast disease**." Symptoms include itching sensations (pruritis), burning internal pain, and a white "cheesy" discharge. Reddening (erythema) and swelling of the vaginal tissues also occur. Diagnosis is performed by observing *C. albicans* in a sample of vaginal discharge or vaginal smear, and by cultivating the organisms on laboratory media. Treatment is usually successful with **nystatin** (Mycostatin) applied as a topical ointment or suppository. **Miconazole**, **clotrimazole**, and **ketoconazole** are useful alternatives, with ketoconazole taken orally.

nis'tah-tin
mĭ-kon'ah-zōl
kē"to-kon'ah-zōl

Vulvovaginitis is considered a sexually transmitted disease, but the disease is usually much milder in men than in women. Studies in women have shown that birth control pills may change the pH of the vaginal environment and encourage loss of the rod-shaped lactobacilli that are normally present. Without lactobacilli as competitors, *C. albicans* flourishes. Antibiotic therapy has also been shown to eliminate lactobacilli. Other predisposing factors are the intra-uterine device, corticosteroid treatment, pregnancy, diabetes, and tight-fitting garments, which increase the local temperature and humidity.

CANDIDA ALBICANS

Oral candidiasis is known as **thrush**. This disease is accompanied by small, white flecks that appear on the mucous membranes of the oral cavity and then grow together to form soft, crumbly, milklike curds. When scraped off, a red, inflamed base is revealed. Oral suspensions of nystatin ("swish and swallow")

FIGURE 14.12

The Agent of Candidiasis

A scanning electron micrograph of *Candida albicans* associated with the tissues of an animal. The oval structure of the cells and the tendency to form hyphae are apparent. Bar = 10 μm.

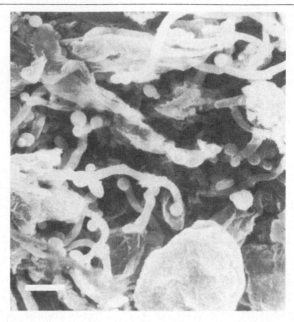

and gentian violet are effective for therapy. The disease is common in newborns, who acquire it during passage through the vagina of infected women. Children may also contract thrush from nursery utensils, toys, or the handles of shopping carts.

Candidiasis in the **intestine** is closely tied to the use of antibiotics, which destroy bacteria normally found in this organ and allow *C. albicans* to flourish. In the 1950s, yogurt became popular as a way of replacing the bacteria. Today when intestinal surgery is anticipated, the physician often uses nystatin to curb *Candida* overgrowth. Candidiasis may also be related to a suppressed immune system. Indeed, in 1984 it was first pointed out that thrush may be an early sign of AIDS in a patient. Moreover, people whose hands are in constant contact with water may develop a hardening, browning, and distortion of the fingernails called **onychia**, also caused by *C. albicans*.

o-nik′e-ah

Dermatomycosis

der-mah″to-mi-ko′sis

Dermatomycosis is a general name for a fungal disease of the hair, skin, and nails caused by a wide variety of fungi. The diseases are commonly known as **tinea infections**, from the Latin *tinea* for "worm" because in ancient times worms were thought to be the cause. The tinea diseases include tinea pedis (**athlete's foot**), tinea capitis (**ringworm of the head**), tinea corporis (ringworm of the body), tinea cruris (ringworm of the groin or "jock itch"), tinea unguium (ringworm of the nails), and tinea favosa (ringworm of the scalp, or favus).

tri-kof′ĭ-ton
ar″thro-derm′ah
na-niz′e-ah
ep″e-der-mof′ĭ-ton

The causes of dermatomycosis are a series of fungi called **dermatophytes**. One example is species of ***Trichophyton***, an ascomycete whose sexual stage is named *Arthroderma*; another example is certain species of ***Microsporum***, also an ascomycete, whose sexual stage is named *Nannizzia*; a third is species of ***Epidermophyton***, currently considered a deuteromycete.

FIGURE 14.13

Two Views of Dermatomycosis

(a) Ringworm of the scalp due to *Trichophyton mentagrophytes*. The lesions have crusted to form scaly blisters in this view. (b) Athlete's foot due to *Trichophyton rubrum*. The scaly blisters can be seen on the soles of the feet and in the webs between the toes.

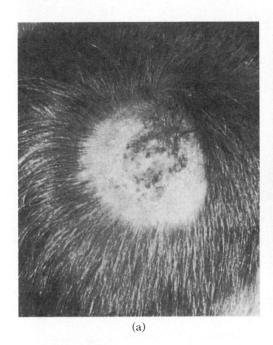

(a)

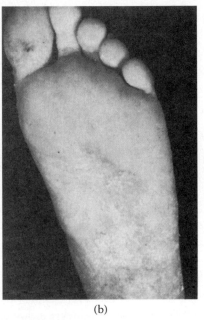

(b)

Dermatomycosis is commonly accompanied by blisterlike lesions appearing on the skin, along the nail plate, or in the webs of the toes or fingers. Often a thin, fluid discharge exudes when the blisters are scratched or irritated. As the blisters dry, they leave a scaly ring (Figure 14.13). Centuries ago, worms were thought to inhabit the scaly ring; hence, the term "ringworm." The symptoms of dermatomycosis vary considerably and may include loss of hair, change of hair color, and local inflammatory reaction.

The majority of dermatophytes grow readily on Sabouraud dextrose agar, and trained mycologists can usually diagnose the disease by observing the type of hypha and arthrospore present. Moreover, infected hairs and fungal cultures fluoresce in ultraviolet light. If protected from dryness, the dermatophytes live for weeks on wooden floors of shower rooms or on mats. People transmit the fungi by contact (Figure 14.14) and on towels, combs, hats, and numerous other types of fomites (inanimate objects). They also acquire the fungi by contact with household pets, because tinea diseases are known to affect cats and dogs.

Treatment of dermatomycosis is often directed at changing the conditions of the skin environment. Commercial powders dry the diseased area, while ointments change the pH to make the area inhospitable for the organism. Certain acids such as undecylenic acid (Desenex) and mixtures of acetic acid and benzoic acid (Whitfield's ointment) are active against the fungi. Also, tolnaftate (Tinactin) and miconazole (Micatin) are useful as topical agents for infections not involving the nails and hair. **Griseofulvin**, administered orally, is a highly effective chemotherapeutic agent for severe dermatomycoses. This drug causes shriveling of the hyphae, possibly by interfering with nucleic acid synthesis.

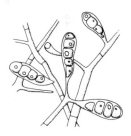

TRICHOPHYTON

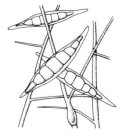

MICROSPORUM

un″dec-ĕ-len′ik
tahl-naf′tate

gris″e-o-ful′vin

FIGURE 14.14

An Outbreak of Ringworm

This outbreak occurred among participants at an international wrestling meet. The incident happened in Schaumberg, Illinois, in 1992. It was believed to be one of the first epidemics of transmissible ringworm to be reported in the United States.

1. In the fall of 1992, a number of wrestlers from the United States and abroad attended a multi-day meet in Schaumberg, Illinois.

2. The wrestlers competed in several divisions until the divisional champions were decided. Skin contacts were routine during the bouts.

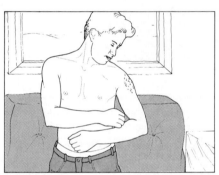

3. On returning home a number of participants noticed scaly, pink blotches on their shoulders, neck, or face.

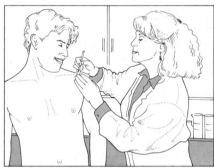

4. Mycologists took skin samples and scrapings from those affected and cultivated *Trichophyton tonsurans*. Their diagnosis was ringworm (tinea corporis).

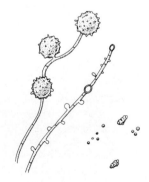

HISTOPLASMA CAPSULATUM

his"to-plaz-mo'sis
his"to-plaz'mah
cap-su-lat'um

Histoplasmosis

On January 4, 1988, 17 students from an American university entered a cave in a national park in Costa Rica to observe the 500 bats whose droppings covered the floor. Within three weeks, 15 students developed fever, headache, cough, and severe chest pains. Twelve patients gave a positive test for *Histoplasma capsulatum*. All were treated by their physicians for histoplasmosis.

Histoplasmosis is a lung disease prevalent in the Ohio River valley and the Mississippi River valley. The causative agent is ***Histoplasma capsulatum***, an ascomycete whose sexual phase is called *Emmonsiella capsulata*. Infection usually occurs from the inhalation of spores in dry, dusty soil, and the disease is often called "summer flu." Most people recover without treatment. However, a small percentage of the population develops a disseminated form of histoplasmosis with tuberculosislike lesions of the lungs and other visceral organs. AIDS patients are vulnerable to this condition. Amphotericin B or ketoconazole may be used in treatment.

The fungus of histoplasmosis is often found in the air of **chicken coops** and **bat caves**. *Histoplasma* does not affect birds or bats but it grows in the droppings of these animals, as the outbreak in Costa Rica illustrates. Prolonged exposure to the air may therefore be hazardous. The disease is sometimes called **Darling's disease** after Samuel Darling, who described the cause in 1915.

Blastomycosis

Blastomycosis occurs principally in Canada, the Great Lakes region, and areas of the United States from the Mississippi River to the Carolinas. It is due to *Blastomyces dermatitidis*, a member of the Ascomycetes class whose sexual phase is named *Ajellomyces dermatitidis*. The fungus is dimorphic, appearing in the human as a yeast with a figure-8 appearance.

ah"jĕ-lo-mi'sēz

Blastomycosis is associated with dusty soil and **bird droppings**, particularly in and near barns and sheds. Entry to the body may occur through cuts and abrasions, and raised wartlike lesions are often observed on the face, hands, and legs. Inhalation leads to lung lesions with persistent cough and chest pains. Healing is generally spontaneous.

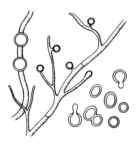

The progressive form of blastomycosis may involve many internal organs and may prove fatal. Amphotericin B used in therapy is thought to change the permeability of the fungal cell membrane and induce the leakage of cytoplasm. Blastomycosis is also referred to as **Gilchrist's disease** for Thomas C. Gilchrist, the American dermatologist who described it in 1896.

BLASTOMYCES DERMATITIDIS

Other Fungal Diseases

A number of other fungal diseases bear brief mention because they are important in certain parts of the United States or affect individuals in certain professions. Generally the diseases are mild, although complications may lead to serious tissue damage.

Travelers to the San Joaquin Valley of California and dry regions of the southwestern United States may be exposed to a fungal disease known as **coccidioidomycosis**, or "valley fever." Its cause is *Coccidioides immitis*, a protozoanlike fungus of the Deuteromycetes class. The organism produces arthrospores by a unique process of endospore and **spherule** formation shown in Figure 14.15. When inhaled to the human lungs, *C. immitis* induces an influenzalike disease, with dry, hacking cough, chest pains, and high fever. For most of the 1980s, there were about 450 cases of coccidioidomycosis reported per year to the CDC. In 1991, however, that number jumped to over 1200 cases and in 1992, the number of reports exceeded 4500, most from southern California. Investigators from the CDC postulated that the outbreak might have been associated with the recent, protracted drought followed by heavy rains. An alternative theory was the movement of large numbers of previously unexposed persons into the area.

wah-kēn'
kok-sid"e-oi"do-mi-ko'sis
kok-sid"e-oi'dēz

COCCIDIOIDES IMMITIS

Aspergillosis is a unique disease because the fungus enters the body as conidia, then grows as a mycelium. Disease usually occurs in a compromised host or where an overwhelming number of conidia has entered the tissue. The most common cause is *Aspergillus fumigatus*, an ascomycete. Infection of the lung may yield a round ball of mycelium called an **aspergilloma**, requiring surgery for removal. Conidia in the earwax lead to a painful ear disease known as **otomycosis**. Disseminated *Aspergillus* causes blockage of blood vessels, inflammation of the inner lining of the heart, or clots in the heart vessels. Amphotericin B therapy is usually necessary.

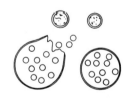

ASPERGILLUS FUMIGATUS

FIGURE 14.15

The Cycle of Development of *Coccidioides immitis*

Outside the body, the organism exists as a septate mycelium. It segments to form airborne arthrospores, which are inhaled. In the respiratory tract, the arthrospores swell to yield a large body, the spherule, that segments and breaks down to release endospores. When released to the environment, the endospores form germ tubes and, then, the mycelium.

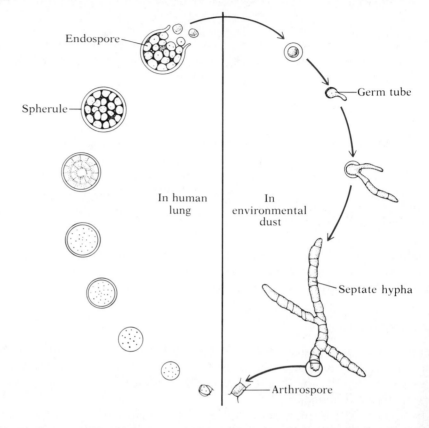

A closely related fungus ***Aspergillus flavus***, produces toxic compounds called **aflatoxins**. The mold is found primarily in warm, humid climates, where it contaminates agricultural products such as peanuts, grains, cereals, sweet potatoes, corn, rice, and animal feed. Aflatoxins are deposited in these foods and ingested by humans where they are thought to be carcinogenic, especially in the liver. Contaminated meat and dairy products are also thought to be sources of the toxins. Fungal toxins are called **mycotoxins**.

Another fungus that produces a powerful toxin is ***Claviceps purpurea*** (MicroFocus: 14.7). This member of the Ascomycetes class grows as hyphae on kernels of rye, wheat, and barley. As hyphae penetrate the plant, the fungal cells gradually consume the substance of the grain, and the dense tissue hardens into a purple body called a **sclerotium**. A group of peptide derivatives called alkaloids are produced by the sclerotium and deposited in the grain as a substance called **ergot**. Products such as bread made from rye grain may cause ergot rye disease, or **ergotism**. Symptoms may include numbness, hot and cold sensations, convulsions with epileptic-type seizures, and paralysis of the nerve endings. Lysergic acid diethylamide (LSD) is a derivative of an alkaloid in ergot. Commercial derivatives of these alkaloids are used to cause contractions of smooth muscles such as to induce labor or relieve migraine headaches.

Mycotoxins:
fungal toxins.

Sclerotium

FIGURE 14.16
Sporothrix schenkii

(a) A transmission electron micrograph of conidia of *Sporothrix schenkii* formed at the tip of a conidiophore. Bar = 10 μm. (b) A patient showing the lesions of sporotrichosis on an infected arm. Characteristic "knots" can be felt under the skin.

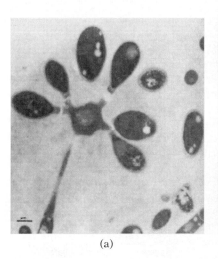

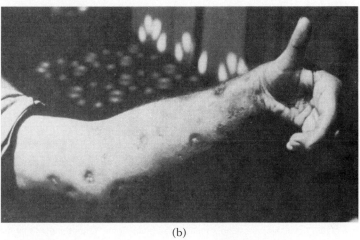

(a)   (b)

Sporotrichosis is an occupational hazard of those who work with wood, wood products, or the soil. The disease can be contracted by handling sphagnum (peat) moss used to pack tree seedlings. It is also transmitted by punctures with rose thorns and is often referred to as **rose thorn disease**. The causative agent is ***Sporothrix schenkii*** (sometimes called *Sporotrichum schenkii*), a dimorphic fungus (Figure 14.16). Pus-filled purplish lesions form at the site of entry and "knots" may be felt under the skin. Dissemination, though rare, may occur to the bloodstream, where blockages may cause swelling of the tissues (edema). In 1988, 84 cases of cutaneous sporotrichosis occurred in persons who handled conifer seedlings packed with sphagnum moss from Pennsylvania (Figure 14.17). Cutaneous infections are controlled with potassium iodide, but systemic infections require amphotericin B therapy. Table 14.3 summarizes this and other fungal diseases of humans.

SPOROTHRIX SHENKII

NOTE TO THE STUDENT

During the 1980s, an epidemiologist at the University of Virginia initiated a study of infectious diseases acquired in the hospital, fully expecting most diseases to be caused by viruses and bacteria. To his surprise, almost 40 percent of the diseases were due to fungi.

The study was one of the first pieces of evidence pointing to an emerging threat of fungal disease. In most cases, an impaired immune system is involved—a system overwhelmed with human immunodeficiency virus; or a system depressed by cancer chemotherapy or antirejection drugs following a transplant. But there is also the problem of a meager arsenal of antifungal drugs; and the fungal pathogens are developing resistances to drugs of long-standing use. Moreover, clinical research in mycology has not kept up with that in clinical virology and bacteriology, in part due to the lack of established cultivation methods for fungal pathogens.

14.7

A FUNGUS AND THE FRENCH REVOLUTION

In the early summer of 1789, a great wave of panic spread over France. Rumors circulated that brigands were everywhere, and many townsfolk fled to the woods to hide. Peasants stockpiled arms, and soon turned their hostility on landowners, burning homes, and destroying records of their debts.

The incident came to be called the Great Fear (*la Grand Peur*). After it subsided, the rich remained apprehensive. They gradually realized that the peasants had enough power to seize property and commit acts of violence. The momentum for reform soon built to a fever pitch and on the night of August 4, 1789, the French Assembly met and voted to abolish many ancient rights of the nobility. The French Revolution was under way.

Historians have often wondered what roused the peasants and precipitated the events of 1789. Essentially the fears were groundless: no more brigands than usual were about; and no evidence of a conspiracy existed among the peasants because the panic broke out in widely scattered communities, some separated by mountains. The episode seemed to be one of sheer wildness and not necessarily a manifestation of resentment. Nor did the timing seem to fit any political, economic, or sociological pattern.

In 1984, Mary Kilbourne Matossian, from the University of Maryland, proposed a solution to the mystery of the Great Fear. She blamed the episode on ergot rye disease. Her studies of provincial records revealed deterioration of public health in sections of France in mid-1789, and instances of nervous attacks and manic behavior. "Bad flour" was thought to be the cause, a factor that would tie in with the ergot rye theory. Another important clue had surfaced in 1974, when a historian reported that the rye crop of the late 1700s was "prodigiously" affected by ergot. He reported evidence of *Claviceps purpurea* in one-twelfth of all the rye (today if one three-hundredth of the rye is infected, it cannot be sold).

But why would so many peasants eat bad bread in 1789? Apparently it was a bad year for rye, and the cold winter and wet spring contributed to widespread ergot disease. Coincidentally, the Great Fear broke out just after the rye harvest. Thus the timing of the panic and behavior of the peasants appear to go hand-in-hand.

In retrospect, it is clear that the peasants resembled victims of ergot poisoning. Hallucinations are common and delirium often sets in. There are seizures, jaundice, numbness, and a belief that ants are crawling under the skin. Tremors, loss of speech, and a sense of suffocation occur. During the Middle Ages, the disease was called the "holy fire."

It would be simplistic to suggest that a fungus precipitated the French Revolution. Nevertheless, the evidence is substantial that ergot disease was a contributing cause. Certainly, the wild displays of the peasants must have been a terrifying sight to landowners. No doubt, the fever pitch of the panic and the far-reaching consequences are better explained by the political and cultural climate of the times. Still, if the ergot disease had not happened . . . .

*The list of fungal pathogens also appears to be expanding: In addition to the well-known pathogens (*Candida, Histoplasma, Cryptococcus, and others*), mycologists are proposing that up to 150 species of fungi may be pathogenic.* Aspergillus *species, for example, have increasingly been related to pneumonia; the plant pathogen* Fusarium *has been isolated from human blood disease; the mushroom* Coprinus *riddled a prosthetic heart valve in a case of endocarditis; and even the yeast* Saccharomyces *was found to cause infection in one person's burnt tissue.*

At this writing, the CDC requests physicians to report 49 infectious diseases, but not one is of fungal origin. But this policy may change in the years ahead because more and more fungi are being recognized for their pathogenic potential. Once considered a nuisance, fungi are now coming out of the shadows as serious threats to human health.

FIGURE 14.17

An Outbreak of Sporotrichosis

In this outbreak, *Sporothrix schenkii* was isolated from the packing moss used at the Pennsylvania nursery, which supplied the trees and seedlings. Eighty-four cases of sporotrichosis in 15 states were identified in the outbreak.

1. In May, 1988 a man visited an Illinois physician complaining of a swollen right hand and forearm and almond-sized "knots" under the skin of his right arm. The physician diagnosed sporotrichosis and placed the man on potassium iodide therapy.

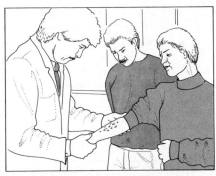

2. On the next visit, the man brought along his neighbor who had similar symptoms. Once again, the physician diagnosed sporotrichosis. The neighbor offered that he and the first man raised and sold Christmas trees as a side-job.

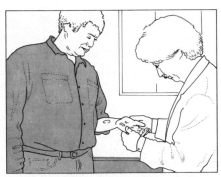

3. In July, the physician examined a third patient. On questioning, the patient explained that he had recently participated in a sale of Colorado blue spruce seedlings to children. The source of the seedlings was the same Pennsylvania nursery that supplied Christmas trees to the first two men.

4. The physician reported his observations to the Illinois State Health Department. He learned that his patients were part of a growing list of cases of sporotrichosis being reported nationally. All the cases were related to fungus-contaminated moss used to pack trees and seedlings at the same nursery in Pennsylvania.

Summary

Fungi are a group of eukaryotic microorganisms distinguished from plants by their lack of chlorophyll, by differences in their cell walls, and by the fact that fungi are not truly multicellular. Moreover, fungi are heterotrophic (they utilize organic matter for food), while plants are autotrophic (they synthesize their own food).

TABLE 14.3
A Summary of Fungal Diseases of Humans

ORGANISM	CLASS	DISEASE	TRANSMISSION
Cryptococcus neoformans	Basidiomycetes	Cryptococcosis	Airborne cells
Candida albicans	Deuteromycetes	Candidiasis Vaginitis Thrush, onychia	Airborne Sexual contact Skin contact
Trichophyton species *Microsporum* species *Epidermophyton* species	Ascomycetes Deuteromycetes	Tinea pedis Tinea capitis Tinea corporis Tinea barbae	Contact with hyphal fragments
Histoplasma capsulatum	Ascomycetes	Histoplasmosis	Airborne spores
Blastomyces dermatitidis	Ascomycetes	Blastomycosis	Airborne spores Open wound
Coccidioides immitis	Deuteromycetes	Coccidioidomycosis	Airborne spores
Aspergillus fumigatus	Ascomycetes	Aspergillosis Otomycosis	Airborne spores
Sporothrix schenkii	Deuteromycetes	Sporotrichosis	Spores Puncture wound

Fungi generally consist of masses of intertwined filaments of cells called hyphae. Cross walls separate the cells of the hypha in some fungal species but not in others. Reproductive structures referred to as fruiting bodies generally occur at the tips of hyphae. Masses of asexually produced spores within or at the tip of the fruiting body provide the mechanisms for propagating the fungi. Spores can also be produced by a sexual mode, in which case the format of the reproductive process provides a basis for separating fungi into four distinctive groups. In a fifth group, no sexual reproduction structure is formed and reproduction occurs solely by an asexual mode. Most fungi grow best at room temperature and prefer conditions that are acidic.

Fungi are very diverse microorganisms capable of causing a broad variety of plant diseases. Many species have industrial significance as producers of valuable products. Yeasts, for example, are nonfilamentous fungi that ferment carbohydrates into alcohol; they also produce large amounts of carbon dioxide that encourages bread to rise, and they are useful vitamin supplements.

Among the many fungal diseases of humans are cryptococcosis and candidiasis. Both diseases occur in immunosuppressed individuals such as AIDS patients and both are opportunistic diseases. Cryptococcosis affects the lungs and spinal cord, while candidiasis can occur in numerous organs such as the skin, intestine, vaginal tract, and oral cavity. Various antibiotics are available to alleviate the symptoms of these diseases.

Other important fungal diseases include the ringworm infections of the skin, a series of airborne lung diseases such as histoplasmosis and blastomycosis, and toxin-induced ergot disease of rye and other grains. A final disease, sporotrichosis, occurs on the skin and within blood vessels following a puncture wound.

ORGAN AFFECTED	DIAGNOSIS	TREATMENT	COMMENT

Lungs
Spinal cord
Meninges
	Examination of
	  spinal fluid
		Amphotericin B	Associated with
			  pigeon
			  droppings

Intestine
Vagina
Skin, mouth
	Urine examination
	Vaginal smears
	Lab cultivation
		Nystatin
		Miconazole
		Ketoconazole
			Normally in
			  human
			  intestine

Skin	Lab cultivation
	Tissue examination
		Undecylenic acid
		Griseofulvin
			Widely
			  encountered
			  skin diseases

Lungs Various organs	Lab cultivation Tissue examination	Amphotericin B	Associated with birds and bats
Lungs Various organs	Lab cultivation Tissue examination	Amphotericin B	Associated with bird droppings

Lungs	Lab cultivation
	Tissue examination
		Amphotericin B	Common in
			  southwestern
			  United States

Lungs Ears	Lab cultivation Tissue examination	Amphotericin B	Hyphae grow in body
Skin Lymph vessels	Lab cultivation Tissue examination	Amphotericin B Potassium iodide	Associated with rotten wood

Questions for Thought and Discussion

1. In the 1980s in a suburban community, a group of residents obtained a court order preventing another resident from feeding the flocks of pigeons that regularly visited the area. Microbiologically, was this action justified? Why?

2. A homemaker decides to make bread. She lets the dough rise overnight in a warm corner of the room. The next morning she notices a distinct beerlike aroma in the air. What is she smelling, and where did the aroma come from?

3. Fungi are extremely prevalent in the soil and yet we rarely contract fungal disease by consuming fruits and vegetables. Why do you think this is so?

4. In 1991, the federal Food and Drug Administration approved for over-the-counter sales a number of antifungal drugs such as clotrimazole and miconazole (Gyne-Lotrimin and Monistat 7, respectively). It thus became possible for a woman to diagnose and treat herself for a vaginal yeast infection. Should she do it?

5. Why is it a good idea to occasionally empty a package of yeast into the drain leading to a cesspool or a septic tank? Why are yeasts accused of having "metabolic schizophrenia"?

6. A woman has a continuing problem of ringworm, especially of the lower legs in the area around the shins. Questioning reveals that she has five very affectionate cats at home. Is there any connection between these facts?

7. A student of microbiology proposes a scheme to develop a strain of bacteria that could be used as a fungicide. Her idea is to collect the chitin-containing shells of lobsters and shrimp, grind them up, and add them to the soil. This, she suggests, will build up the level of chitin-digesting bacteria. The bacteria would then be isolated and used to kill fungi by digesting the chitin in fungal cell walls. Do you think her scheme will work? Why?

8. A certain restaurant advertises on its menu "Pizza con Funghi." Suppose you ordered this dish. What would you receive?

9. Mr. A and Mr. B live in an area of town where the soil is acidic. Oak trees are common, and azaleas and rhododendrons thrive in the soil. In the spring Mr. A spreads lime on his lawn, but Mr. B prefers to save the money. Both use fertilizer, and both have magnificent lawns. Come June, however, Mr. B notices that mushrooms are popping up in his lawn and that brown spots are beginning to appear. By July, his lawn has virtually disappeared. What is happening in Mr. B's lawn and what can Mr. B learn from Mr. A?

10. A mushroom walks into a bar and orders a beer. "Sorry," says the bartender, "We don't serve mushrooms." The mushroom thinks for a moment and replies "But I'm a fun-guy." When you have recovered from this dreadful attempt at humor you might like to try your hand at another "fun-guy" joke; or a "fun-gus" joke.

11. On June 27, 1995, a crew of five workers began a partial demolition of an abandoned city hall building in a Kentucky community. Three weeks later all five required treatment for acute respiratory illness, and three were hospitalized. Cells obtained from the patients by lung biopsy revealed oval bodies. When the construction site was inspected, epidemiologists found an accumulation of bat droppings, and neighbors said they had seen bats in the area in recent weeks. From the information can you surmise the nature of the disease in the demolition crew?

12. The baking or fermentation yeast *Saccharomyces* can be pronounced at least three ways: sa-KAR-o-myces, sak-a-ro-MY-ces, and sak-a-ROM-a-ces. Which pronunciation does your instructor prefer and how do you suppose the three pronunciations evolved?

13. In 1992, residents of a New York community, unhappy about the smells from a nearby composting facility and concerned about the health hazard posed by such a facility, had the air at a local school tested for the presence of fungal spores. Investigators from the testing laboratory found abnormally high levels of *Aspergillus* spores on many inside building surfaces. Is there any connection between the high spore count and the composting facility? Is there any health hazard involved?

14. On January 17, 1994 a serious earthquake struck the Northridge section of Los Angeles County in California. From that date through March 15, 170 cases of coccidioidomycosis were identified in adjacent Ventura County. This number was almost four times the previous year's number of cases. Can you guess the connection between the two events?

15. The clublike instrument that a priest or minister uses to sprinkle holy water on the congregation is called an aspergillum. Why might a mycologist's attention be aroused at learning this piece of information?

Review

The significance of the fungi is broad and diverse, as this chapter has demonstrated. To test your knowledge of the important fungi, match the statement on the left to the organism on the right by placing the correct letter in the available space. A letter may be used once, more than once, or not at all.

_____ 1. Causes late blight of potatoes.

_____ 2. Produces a widely used antibiotic.

_____ 3. Converts carbohydrates to alcohol in beer.

_____ 4. Growth can be interrupted with griseofulvin.

_____ 5. Causes "valley fever" in SW United States.

_____ 6. Common white or gray bread mold.

_____ 7. Poisonous mushroom.

_____ 8. Sexual phase known as *Emmonsiella*.

_____ 9. Agent of rose thorn disease.

_____ 10. Edible mushroom.

_____ 11. Can overgrow the intestine when antibiotic consumed.

_____ 12. Known to cause Darling's disease.

_____ 13. A coenocytic mold.

_____ 14. Produces citric acid, soy sauce, and vinegar.

_____ 15. Associated with the droppings of pigeons.

_____ 16. Blue-green mold that has septa.

_____ 17. Agent of ergot disease in rye plants.

_____ 18. Cause of the "yeast disease" in women.

_____ 19. Used to produce wine from grape juice.

_____ 20. One of the causes of athlete's foot.

_____ 21. Often found in chicken coops and bat caves.

_____ 22. Produces a toxic aflatoxin.

_____ 23. Agent of dermatomycosis.

_____ 24. Reproduction includes a spherule.

_____ 25. Nystatin and miconazole to inhibit.

A. *Amanita phalloides*
B. *Claviceps purpurea*
C. *Sporothrix schenkii*
D. *Blastomyces dermatitidis*
E. *Agaricus* species
F. *Mucor racemosus*
G. *Aspergillus* species
H. *Saccharomyces ellipsoideus*
I. *Phytophthora infestans*
J. *Myzeloblastanon krausi*
K. *Coccidioides immitis*
L. *Mucor mellitensis*
M. *Epidermophyton floccosum*
N. *Uncinocarpus reesii*
O. *Cryptococcus neoformans*
P. *Candida albicans*
Q. *Penicillium notatum*
R. *Acrotheca pedrosoi*
S. *Histoplasma capsulatum*
T. *Wangiella dermititidis*
U. *Rhizopus nigricans*
V. *Volutella graphii*
W. *Saccharomyces cerevisiae*
X. *Drechslera rostrata*
Y *Streptothrix bovis*
Z. *Aspergillus flavus*

CHAPTER 15

THE PROTOZOA

A pril 12, 1993 should have been a festive day in Milwaukee, Wisconsin. The baseball home opener was scheduled for that day, and fans were eager to see the Brewers play the California Angels. But the scoreboard contained an ominous message: "For your safety, no city of Milwaukee water is being used in any concession item." The city was in the throes of an epidemic, and a protozoan was to blame.

The protozoan was *Cryptosporidium coccidi*, an intestinal parasite that causes mild to serious diarrhea, especially in infants and the elderly. As the protozoa attach themselves to the intestinal lining, they mature, reproduce, and encourage the body to release large volumes of fluid. The infection is accompanied by abdominal cramps, extensive water loss, and in many cases, vomiting and fever.

krip"to-spor-id'e-um
kok-sid'e-e

Even as the first ball was being thrown out at the stadium, health inspectors were checking Milwaukee's two water purification plants to see how a protozoan could be pumped into the city's water supply. *Cryptosporidium* is a waterborne parasite commonly found in the intestines of cows and other animals. Perhaps, they guessed, the heavy rain and spring thaw had washed the protozoan from farm pastures and barns into the Milwaukee River. The river might have brought *Cryptosporidium* into Lake Michigan from which the city drew its water. Indeed, the mouth of the river was unusually close to the intake pipe from the lake. Moreover, they added, *Cryptosporidium* can resist the chlorine treatment used to control bacteria in water; and the tests to detect bacterial contamination do not detect protozoa, such as *Cryptosporidium*.

As researchers worked to unravel the mystery, the game went on. Soda was available, but only from bottles. Drinking fountains were turned off. Two huge U.S. Army water tanks stood by to provide a reserve for the 50,000 fans in attendance. And in the city, tens of thousands of Milwaukeeans made the mildly embarrassing trip to the drug store to stock up on toilet paper and antidiarrheal medications. (A large window sign at a local Walgreen's proudly proclaimed: "We have Imodium A-D.") Back at the ballgame, things were not going much better—the Brewers lost to the Angels 12 to 5.

FIGURE 15.1

A Waterborne Protozoan

A scanning electron micrograph of *Cryptosporidium coccidi* at the surface of intestinal tissue. The parasites appear as globular saclike bodies 2 to 6 μm in diameter. Within the globes are numerous long, thin forms of the parasite called merozoites. After release from the globes, the merozoites will infect nearby cells. One parasite (A) has lost its membrane and the merozoites can be seen crowded together. The craterlike structures (B) are parasites from which the merozoites have already been released.

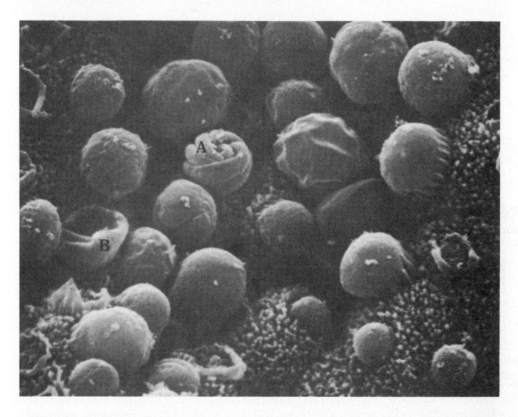

Cryptosporidium coccidi (Figure 15.1) will be one of the protozoa we study in this chapter. We shall encounter other protozoa that infect the human intestine, as well as several protozoa that live primarily in the blood and other organs of the body. Many of the diseases we encounter (for example, malaria) will have familiar names, but others, such as *Cryptosporidium* infections, are emerging diseases in our society (indeed, *Cryptosporidium* was not known to infect humans before 1976). Our study will begin with a focus on the characteristics of protozoa.

Characteristics of Protozoa

Protozoa are a group of about 30,000 species of single-celled organisms. They take their name from the Greek words *protos* and *zoon*, which translate literally to "first animal." This name refers to the position many biologists believe protozoa occupy in the evolution of living things. Though often studied by zoologists, protozoa are also of interest to microbiologists because they are unicellular, have a microscopic size, and are involved in disease. The discipline of **parasitology** is generally concerned with the medically related protozoa and multicellular parasites (Chapter 16).

Parasitology:
the discipline of biology concerned with pathogenic protozoa and multicellular parasites.

Structure and Growth of Protozoa

Protozoa are among the largest organisms encountered in microbiology, some forms reaching the size of the period at the end of this sentence. With only a few exceptions, protozoa have no chlorophyll in their cytoplasm and thus cannot produce carbohydrates by photosynthesis. Although each protozoan is composed of a single cell, the functions of that cell bear a resemblance to the functions of multicellular animals rather than to an isolated cell from that animal.

Most protozoa are free-living and thrive where there is water. They may be located in damp soil and mud, in drainage ditches and puddles, and in ponds, rivers, and oceans. Some species of protozoa remain attached to aquatic plants or rocks, while other species swim about. The film of water on an ordinary dirt particle often contains protozoa. Figure 15.2 illustrates some of the diversity that exists within the group.

Protozoal cells are surrounded only by a membrane. However, outside the membrane, some species of protozoa possess a rigid structure called a **pellicle**. The cytoplasm contains eukaryotic features, each cell having a nucleus and nuclear membrane. In addition, freshwater protozoa continually take in water by the process of osmosis and eliminate it via organelles called **contractile vacuoles**. These vacuoles expand with water drawn from the cytoplasm and then appear to "contract" as they release water through a temporary opening in the cell membrane. Many protozoa also contain locomotor organelles, which permit independent motion.

Protozoa obtain their nutrients by engulfing food particles by phagocytosis or through special organs of ingestion (Figure 15.3). A membrane then encloses the particles to form an organelle called a **food vacuole**. The vacuole joins with another organelle known as the lysosome, and digestive enzymes from the lysosome proceed to break down the particles. Nutrients are absorbed from the vacuole, and the remaining material is eliminated from the cell.

Nutrition in protozoa is primarily heterotrophic, since chlorophyll pigments are generally lacking. Except for the parasitic organisms of disease and the species that feed on bacteria, protozoa are saprobic. All protozoa are aerobic, obtaining their oxygen by diffusion through the cell membrane. The feeding form of a protozoan is commonly known as the **trophozoite** (*troph-* is the Greek stem for food). Another form, the **cyst**, is a dormant, highly resistant stage that develops in some protozoa when the organism secretes a thick case around itself during times of environmental stress.

Reproduction in protozoa usually occurs by the asexual process of mitosis, although many protozoa also have a sexual stage. Whittaker's classification scheme places the organisms together with certain algae in the kingdom **Protista**. Within the kingdom, further classification tends to be controversial, although three phyla are generally recognized: Sarcomastigophora (the ameboid and flagellated protozoa), Ciliophora (the ciliated protozoa), and Apicomplexa (the sporozoan protozoa). Within these three phyla are four major groups, which we shall refer to as "classes" (although general agreement may be lacking on this concept). The classes are distinguished from one another by how the members of the class move. We shall survey the four classes next.

Classification of Protozoa

Sarcodina

Sarcodina is the class of ameboid protozoa. These organisms move as their cell contents flow into temporary formless projections called **pseudopodia** ("false-feet"). The ameba is the classic example of the group, and thus the motion is

Chlorophyll:
a green plant pigment essential to photosynthesis.

Pellicle:
the thick, rigid structure outside the cell membrane of some protozoa.

Osmosis:
the movement of water through a membrane from a region of low concentration of a chemical substance to one of higher concentration.

Lysosome:
a vacuolelike cell organelle that contains digestive enzymes.

Saprobic:
living on dead organic matter.
trof"o-zo'it

sar"co-mas"ti-gof-o-ra
sil"e-of'o-rah
a"pē-com-plex'ah

sar"ko-di'nah

Pseudopodia:
temporary projections of an ameba used for motion and phagocytosis.

FIGURE 15.2

An Illustration of the Diversity Among Protozoa

The four classes are represented, with the organs of motion shown for members of each class.

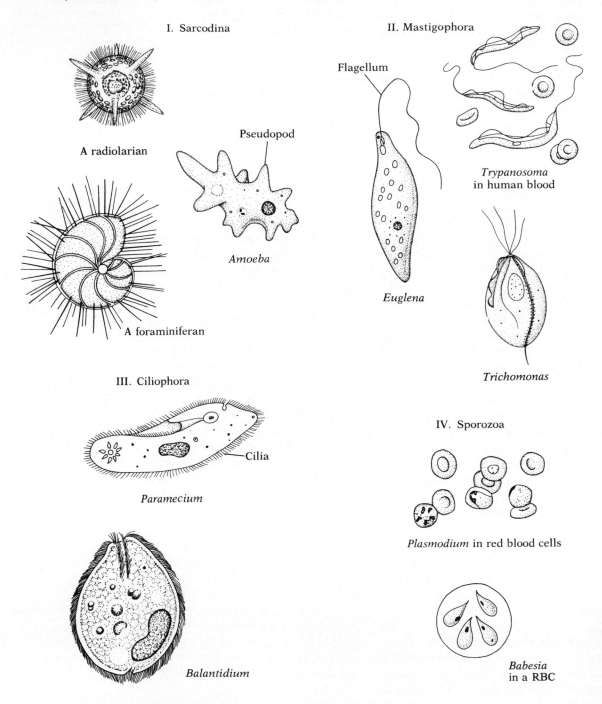

I. Sarcodina

A radiolarian

Pseudopod

Amoeba

A foraminiferan

II. Mastigophora

Flagellum

Trypanosoma
in human blood

Euglena

Trichomonas

III. Ciliophora

Cilia

Paramecium

IV. Sporozoa

Plasmodium in red blood cells

Balantidium

Babesia
in a RBC

called **ameboid motion**. Pseudopodia also capture small algae and other protozoa in the process of phagocytosis.

An ameba may be as large as a millimeter in diameter. It usually lives in fresh water and reproduces by binary fission as shown in the sequence in Figure

FIGURE 15.3

Feeding Behavior in Protozoa

The scanning electron photograph shows three amebas attacking and beginning to devour a fourth, presumably dead, ameba. The amebas are *Naegleria fowleri*, a cause of meningoencephalitis in humans. Each ameba possesses a series of suckerlike structures that apparently stick to the prey and serve as portals for ingestion. The researchers who first described the structures in 1984 recommend that they be called amebastomes. Bar = 10 μm.

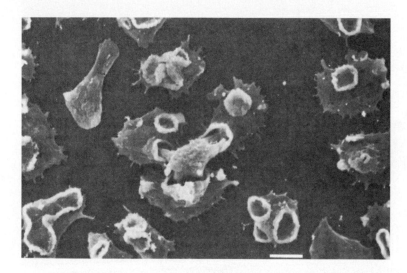

15.4. Amebas may be found in home humidifiers, where they have been known to cause an allergic reaction called **humidifier fever**. Far more serious are the parasitic amebas that cause amebiasis and a form of encephalitis.

Two large groups of marine amebas are included in the class Sarcodina. The first group, the **radiolaria**, are abundant in the Indian and Pacific oceans. These amebas have spherical shells with highly sculptured glassy skeletons, reminiscent of vintage Christmas ornaments. When the protozoa die, their skeletal remains litter the ocean floor with deposits called radiolarian ooze. The second group, the **foraminifera**, have chalky skeletons, often in the shape of snail shells with openings between sections ("foraminifera" means "little window"). Foraminifera flourished during the Palcozoic era about 225 million years ago. Their shells in ocean sediments therefore serve as depth markers for oil drilling rigs and as estimates of the age of the rock. Geologic upthrust has brought the sediments to the surface in several places around the world such as the White Cliffs of Dover (Figure 15.5).

ra"di-o-lar'i-ah

fo-ram"ĭ-nif'er-ah

Reports first published in 1987 indicated that amebas in the genus ***Acanthamoeba*** (e.g., *A. castellani*) can cause corneal infection in persons who wear contact lenses. Wearers were reminded to adhere to recommended lens wear and care procedures, and ophthalmologists and optometrists were advised to increase patient education. Recent research indicates that bacterial infection can be a cofactor in *Acanthamoeba* infection of the eye (Figure 15.6).

a-can"a-me'bah

Mastigophora

Protozoa of the class **Mastigophora** often have the shape of a vase. All members move by means of one or more whiplike, undulating **flagella** (*mastig-* is Greek for whip). The flagellum can either push or pull the organism, depending on the species. Flagella occur singly, in pairs, or in large numbers. Each flagellum

mast"tĭ-gof'o-rah

Flagellum:
the whiplike organ of motion in certain species of protozoa.

F I G U R E 15.4
Binary Fission in *Amoeba proteus*

In this sequence of photomicrographs, the cytoplasm is seen separating to form two new individuals. Few visible changes are apparent for the first 15 minutes but once the cell division begins, separation takes place rapidly.

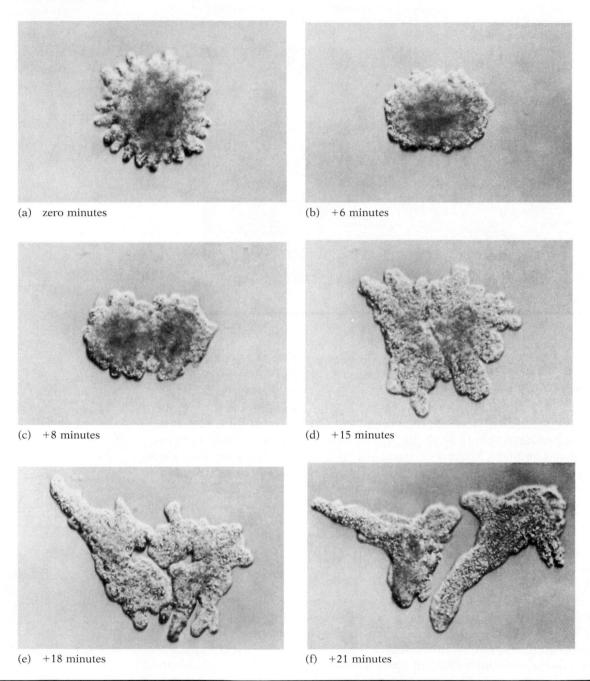

(a)   zero minutes

(b)   +6 minutes

(c)   +8 minutes

(d)   +15 minutes

(e)   +18 minutes

(f)   +21 minutes

has the characteristic 9 + 2 arrangement of microtubules found in all eukaryotic flagella. Undulations sweep down the flagella to the tip, and the lashing motion forces water outward to provide locomotion. The movement resembles the activity of a fish sculling in water. Flagella also occur in bacteria but their structure, size, and type of movement differ.

FIGURE 15.5
The White Cliffs of Dover, England

These cliffs are thought to be composed of the remains of foraminifera that thrived in the oceans millions of years ago.

FIGURE 15.6
Acanthamoeba

A photomicrograph of an *Acanthamoeba* trophozoite in a field of bacteria of the genus *Xanthomonas*. Several bacteria are also contained within vesicles in the ameba. The researchers who conducted this study postulated that contaminating bacteria in contact lens cleaning solutions support the growth of *Acanthamoeba* and thereby encourage it to multiply and adhere to the lens. Later, when the contact lens is inserted the amebas infect the cornea of the eye.

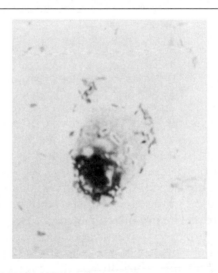

Almost half the known species of protozoa are classified as Mastigophora. An example is the green flagellate **Euglena** often found in freshwater ponds. This organism is unique because it is one of the few types of protozoa that contain chloroplasts with chlorophyll, and is thus capable of photosynthesis. Some botanists claim it to be a plant, but zoologists point to its ability to move and suggest that it is more animal-like. Still other biologists point out that it may be the basic stock of evolution from which animal and plant forms once arose.

Euglena: a flagellated protozoan with plantlike properties.

Some species of flagellated protozoa are free-living, but most live together with plants or animals. Several species, for example, are found in the gut of the termite, where they participate in a symbiotic relationship (MicroFocus: 15.1). Other species are parasitic in humans and cause disease of the nervous, urogenital, or gastrointestinal systems.

Symbiotic relationship: one in which two populations coexist.

15.1

MICROFOCUS

A LAXATIVE FOR TERMITES

Americans spend millions of dollars annually to protect their homes against termites or to repair the damage caused by these insects. However, the notorious reputation acquired by termites may not be completely justified because their voracious appetites for wood are linked by a quirk of nature to protozoa.

The termite's intestine is the dwelling place for a species of protozoa belonging to the genus *Trichonympha*. These multiflagellated organisms are among the few species of living things that can produce the enzyme cellulase. The cellulase is released into the intestinal cavity of the termite, where it breaks down cellulose, the principal component of wood. The protozoan thus

lives in the stable environment of the termite's intestine and returns the favor by digesting the termite's next meal. To the ecologist, this symbiotic relationship is called mutualism. To the homeowner, it spells disaster.

But there may be hope in the future. In the tropics, a method of termite extermination has proven so successful that American companies are examining its feasibility. The method involves placing a paste of ground-up plant material, known to be a termite laxative, into termite tunnels. As termites eat the paste, they develop diarrhea and excrete their protozoal inhabitants. Thereafter, the termites are unable to digest cellulose, and they starve to death. With

insecticides coming under fire for their hazard to the environment, the research into termite laxatives holds substantial promise for the future.

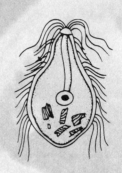

Trichonympha

Ciliophora

Members of the class **Ciliophora** are among the most complex cells on Earth. They range in size from a microscopic 10 μm to a huge 3 mm (about the same relative difference between a football and a football field). All members are covered with hairlike **cilia** (sing., cilium) in longitudinal or spiral rows. The movement of the cilia is coordinated by a network of fibers running beneath the surface of the cell. Cilia beat in a synchronized pattern much like a field of wheat bending in the breeze or the teeth on a comb when the thumb is passed down the row. The organized rowing action that results speeds the ciliate along in one direction. By contrast, flagellar motion tends to be jerky and much slower.

The complexity of ciliates is illustrated by the slipper-shaped ***Paramecium***. This organism has a primitive gullet, as well as a "mouth" into which food particles are swept, a single large macronucleus, and one or more micronuclei.

Micronucleus:
one of several smaller nuclei found in the cytoplasm of certain ciliates.

During **sexual conjugations**, two cells make contact and a cytoplasmic bridge forms between them. A micronucleus from each cell undergoes two divisions to form four micronuclei, of which one remains alive and undergoes division. Now a "swapping" of micronuclei takes place followed by a union to re-form the normal micronucleus (Figure 15.7). This genetic recombination is somewhat analogous to that occurring in bacteria. It is observed during periods of environmental stress, a factor that suggests the formation of a genetically different and perhaps, better adapted organism. Reproduction at other times is by mitosis.

Contractile vacuole:
a clear, circular, cellular organelle used to remove water from the cytoplasm in certain protozoa.

Another feature of *Paramecium* is the **kappa factors**. These nucleic acid particles appear responsible for the synthesis of toxins that destroy ciliates lacking the factors. *Paramecium* species also possess **trichocysts**, organelles that discharge filaments to trap prey. A third feature is the **contractile vacuole** used to "bail out" excess water from the cytoplasm. These organelles are present in freshwater ciliates but not in saltwater species because little excess water exists in the cells.

FIGURE 15.7
Sexual Recombination in *Paramecium*

(a) Two cells make contact and a cytoplasmic bridge forms between them. (b) A micronucleus in each undergoes two divisions to form four micronuclei per cell. (c) Three micronuclei disintegrate and the one remaining alive undergoes division by mitosis. (d) An exchange of micronuclei takes place. (e) The cells separate, the micronuclei fuse, and the macronuclei disintegrate. (f) A new macronucleus forms in each cell from the dividing micronucleus.

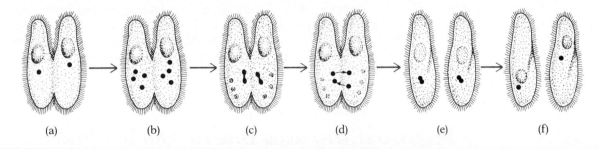

(a)   (b)   (c)   (d)   (e)   (f)

Ciliates have been the subject of biological investigation for many decades. They are readily found in almost any pond or gutter water; they have a variety of shapes; they exist in several colors, including light blue and pink; they exhibit elaborate and controlled behavior patterns; and they have simple nutritional requirements, which makes cultivation easy.

Sporozoa

The class **Sporozoa** includes a number of parasitic protozoa with complex life cycles that include alternating sexual and asexual reproductive phases. These life cycles include intermediary forms that resemble bacterial or fungal spores, a factor from which the class takes its name. However, the spores lack the resistance of other spores, and so the name sporozoa is probably a misnomer.

Sporozoa are notable for the absence of locomotor organelles in the adult form. Two members of the class, the organisms of malaria and toxoplasmosis, are of special significance, the first because it is one of the most prolific killers of humans, the second because of its association with the disease AIDS. Other notable members of the class include *Isospora belli*, a cause of the human intestinal disease coccidiosis, and *Sarcocystis* species, which live in the intestine as well as the muscle tissue of humans and animals.

TO THIS POINT

We have introduced the protozoa and have noted that they are a group of heterotrophic, unicellular, microscopic organisms often involved in human disease. Most protozoa are found in aquatic environments, and the majority are free-living. Nutrients are commonly obtained by phagocytosis by the trophozoite form of the organism. A cyst may be formed under conditions of environmental stress.

We then outlined the salient features of the four classes of protozoa. Members of the Sarcodina class move by means of ameboid motion. Amebas, radiolaria, and foraminifera are important members of this group. The Mastigophora class includes flagellated protozoa, and Euglena is an often-studied, photosynthetic member of the group. Ciliates are classified in the class Ciliophora. These organisms are extremely complex, with many features of multicellular animals. The genetic recombination mechanism exhibited by Paramecium illustrates the complexity. The final class, Sporozoa, contains protozoa whose life cycles are complex. Motion is not observed in the adult forms of these organisms.

We shall now examine several human diseases caused by protozoal parasites and studied by parasitologists. The survey will be organized according to the classes of protozoa, beginning with amebas and concluding with sporozoa. Protozoal diseases occur worldwide, and public health agencies consider them to be a global health problem. The diseases are particularly prevalent in tropical and subtropical regions.

Protozoal Diseases Due to Amebas and Flagellates

Amebiasis

am″e-bi′ah-sis

Amebiasis occurs throughout all areas of the world from tropical to subpolar. It is a disease that primarily affects people who are undernourished and living under unsanitary conditions. The disease is basically an intestinal illness, but it can spread to various organ systems.

en″tah-me′bah his″to-lit′ĭ-ka

The causative agent of amebiasis, ***Entamoeba histolytica***, is a member of the class Sarcodina. In nature, the organism exists in the cyst form. It enters the body by food or water contaminated with feces or by contact with feces. Contact with soiled diapers, such as in a day-care center, may thus be hazardous. The organisms pass through the stomach as cysts, and the trophozoite amebas emerge in the distant portion of the small intestine and in the large intestine (Figure 15.8).

Trophozoite:
the feeding form of a protozoan.

Entamoeba histolytica has the ability to destroy tissue (*histolytica* means "tissue-breaking"). Using their protein-digesting enzymes, the amebas penetrate the wall of the large intestine, causing lesions and deep ulcers. Patients experience appendicitislike sharp pain but relatively little diarrhea or dysentery because the ulcers are separated and do not drastically affect water absorption (the older term, **amebic dysentery**, has thus been replaced by "amebiasis"). In severe cases, tissue invasion will extend to blood vessels of the intestinal wall and bloody stools will follow. Ulcer perforation into the peritoneum is a possibility. The amebas also invade the blood and may spread to the liver or lung, where fatal abscesses may develop.

Peritoneum:
the cavity outside the visceral organs.

me″tro-ni′dah-zōl

par′o-mo-mi″sin

Metronidazole and paromomycin are commonly used to treat amebiasis, but the drugs do not affect the cysts and repeated attacks of amebiasis may occur for months or years. The patient often continues to shed **cysts** in the feces to infect other people. Amebiasis was originally a disease of the tropics, but soldiers returning after World War II brought it to the United States, and about 5 percent of Americans are believed infected. In recent years, waves of immigrants from Mexico and Caribbean nations have added to the incidence of the disease. Barnyard animals are known to harbor cysts, and their manure used for fertilization purposes increases the chance for spread of the disease.

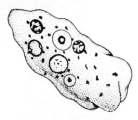

ENTAMOEBA HISTOLYTICA

Primary Amebic Meningoencephalitis

In the summer of 1980, the Centers for Disease Control and Prevention noted an unusual cluster of seven cases of **primary amebic meningoencephalitis (PAM)**. All the patients had been in contact with fresh water in a tropical setting.

FIGURE 15.8

The Course of Amebiasis

Entamoeba histolytica enters the gastrointestinal tract in the cyst form and (a) passes through the stomach. (b) The amebas emerge in the small and large intestines and (c) form deep ulcers. (d) In complicated cases, the organisms reach the blood vessels and travel to the liver and lungs, where additional abscesses form. (e) If the ulcers become perforated, infection of the peritoneal cavity follows. (f) Cysts form as the amebas enter the latter part of the large intestine. (g) Elimination is in this form.

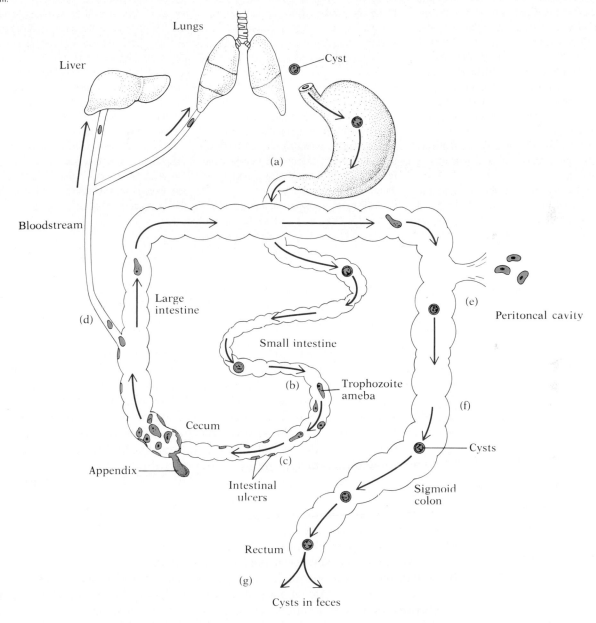

Since the first description of the disease in 1965, fewer than three dozen cases had been reported. In the 1980 cluster all seven victims died.

PAM may be caused by several species of ameba in the genus *Naegleria*, especially ***Naegleria fowleri***. The amebas appear to enter the body through the mucous membranes of the nose and then follow the olfactory tracts to the brain. Nasal congestion precedes piercing headaches, fever, delirium, neck rigidity, and occasional seizures. The symptoms resemble those in other forms of encephalitis and meningitis. *Naegleria* in the spinal fluid is a sign of PAM.

na-gle′rĭ-ah

NAEGLERIA FOWLERI

15.2

MICROFOCUS

HAWAII TO MINNESOTA

On December 13, 1979, the Minnesota Department of Health learned that employees of a local school were being treated for *Giardia lamblia* infections and that 50 percent of the staff was involved. Health officials subsequently distributed a questionnaire requesting information on symptoms, travel, type of water consumed, and contact with other sick people. Those with symptoms were asked for stool samples to test for *Salmonella, Shigella, Campylobacter*, and *Giardia lamblia*. Giardiasis was confirmed.

In the ensuing days, the Department of Health examined the eating, drinking, and plumbing facilities at the school and found all sanitary. Analysis of the questionnaire,

however, provided a promising lead: most of the affected individuals had eaten home-canned salmon. Officials narrowed their search in this direction.

Two months previously, an employee of the school had gone fishing on Lake Michigan and caught a bevy of salmon. He marinated the fish in vinegar and canned it. Wishing to share his bounty, the employee had recently brought several jars to school. Microbiologists were able to obtain four unopened jars of salmon and sample them for *Giardia*. To their dismay, all samples were negative.

Investigators now turned their attention to how the salmon was prepared for serving in the staff lunchroom. They established that

several days before the outbreak, a female employee had opened the jars and tipped them over to drain the juice. In doing so, she held back the fish with her fingertips. When the woman was questioned, she admitted that before coming to work, she had changed her grandson's diaper and had soiled her hands with his loose stools. At the time the boy was visiting from Hawaii, but he had since gone home.

Health officials now contacted their counterparts in Hawaii and requested that the child be tested for *Giardia lamblia*. A few days later, a telephone call confirmed their suspicions: the test proved positive.

Giardiasis

ji"ar-di'ah-sis

je-ar'de-ah

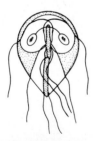

GIARDIA LAMBLIA

Flatulence:
gas expelled from the intestinal tract through the anus.

Since the 1970s, **giardiasis** has emerged to become the most commonly detected protozoal disease of the intestine in the United States. Though not reportable to the CDC, the disease is estimated to attack thousands of Americans annually. The causative agent is a member of the Mastigophora class named ***Giardia lamblia***. This organism is distinguished by four pairs of anterior flagella and two nuclei that stain darkly to give the appearance of eyes on a face. The protozoan can be divided equally along its longitudinal axis and is therefore said to display **bilateral symmetry**. Some microbiologists believe that *Giardia lamblia* was described as early as 1681 by Anton van Leeuwenhoek in samples of his stool.

Giardiasis is commonly transmitted by **water** that contains *Giardia* cysts stemming from cross-contamination of drinking water with sewage. In the 1970s, for example, 38 cases in Aspen, Colorado, broke out after a sewer line was obstructed and sewage leaked into the town's water supply. Recent years have also witnessed outbreaks in day-care centers and schools resulting from contact with feces (MicroFocus: 15.2). In addition, the disease has spread to wild **animals**, where it is now very common and from which it can be obtained via contaminated water. In 1989, a notable cluster of 22 cases in Albuquerque, New Mexico, was related to contamination of taco ingredients by water used to wash the lettuce.

Giardia lamblia passes through the stomach as cysts, and the trophozoites emerge as flagellates in the **duodenum**. They adhere to the intestinal lining using sucker devices and multiply rapidly. The patient feels nauseous, experiences gastric cramps and flatulence, and emits a foul-smelling watery **diarrhea** that may last for weeks. The absence of blood, mucus, and pus from the diarrhea is notable. Some microbiologists maintain that the diarrhea arises from overgrowth of the intestinal wall with parasites, while others suggest injury to the tissue (Figure 15.9).

FIGURE 15.9

Giardia lamblia Infection of the Intestine

(a) A mass of trophozoites adhering to the base of a villus in the intestine of an animal (× 1085). (b) A view of the *Giardia* population at the base (× 1958). Note the flat shape of the cells with numerous flagella and a disklike sucker device for attaching to the tissue. (c) A single *Giardia* trophozoite wedged into the wall of the villus. Three pairs of flagella protrude posteriorly (× 5285).

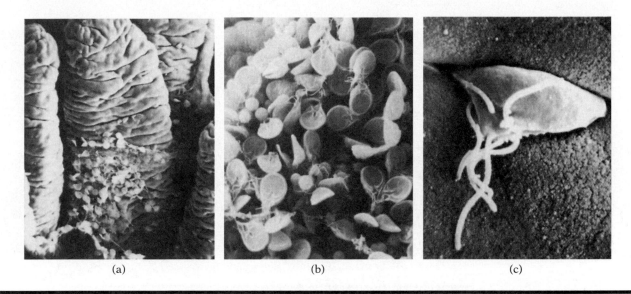

| (a) | (b) | (c) |

Diagnosis of giardiasis depends on the microscopic identification of tropho-zoites or cysts in freshly passed fecal material. A reliable alternative to stool examination is the use of the **Enterotest capsule**. In this procedure the patient swallows a weighted gelatin capsule attached to a string. The free end of the string is then taped to the mouth. After four hours the capsule is withdrawn and the bile-stained mucus is scraped from the capsule and examined for tropho-zoites. The organisms show an erratic turning motion similar to a falling leaf.

Treatment of giardiasis may be administered with drugs such as quinacrine (Atabrine), furazolidine (Furoxone), and metronidazole. However, these drugs have side effects that the physician may wish to avoid by letting the disease run its course without treatment. Those who recover are known to become **carriers** and to excrete the cysts for years. Giardiasis is sometimes mistaken for viral gastroenteritis and is often considered a type of traveler's diarrhea. Latin America, Russia, and the Far East are considered prime areas for contracting the disease.

kwin'ah-krin
fu"rah-zol'ĭ-dēn
me"tro-ni'dah-zōl

Trichomoniasis

Trichomoniasis is among the most common diseases in the United States, with an estimated 2.5 million people affected annually. The disease is transmitted primarily by sexual contact and is considered a **sexually transmitted disease**. It sometimes occurs simultaneously with candidiasis or gonorrhea. Fomites (inanimate objects) such as towels and clothing have also been implicated in transmission.

trik"o-mo-ni'ah-sis
trik"o-mo'nas

Trichomonas vaginalis, the causative agent, is a pear-shaped protozoan with two pairs of anterior flagella and one posterior flagellum. It thrives in the slightly acidic environment of the human vagina. Establishment may be encouraged by physical and chemical trauma including poor hygiene, drug therapy, diabetes, or mechanical contraceptive devices such as the intrauterine device (IUD). The organism has no cyst stage.

TRICHOMONAS VAGINALIS

FIGURE 15.10
Two Views of Trypanosomes

(a) A photomicrograph of a species of *Trypanosoma*. The nucleus (N), the flagellum (F), and the undulating membrane (UM) are visible. The area about the trypanosome contains red blood cells. (b) A scanning electron micrograph of *T. brucei* among red blood cells. Internal details of the trypanosome are not visible but the undulating membrane and short flagellum are clear.

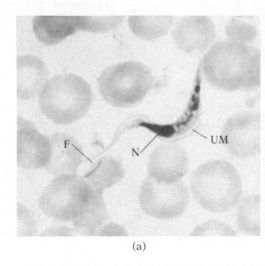
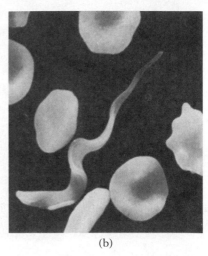

(a)   (b)

Pruritis:
itching.

Asymptomatic:
without symptoms.

ti-nid'ah-zōl
mǐ-kon'ah-zōl

tri-pan"o-so-mi'ah-sis
tri"pan-o-so'mah

In females, trichomoniasis is accompanied by intense itching (pruritis), and burning pain during urination. Usually, a creamy white, frothy discharge is also present. The **symptoms** are frequently worse during menstruation, and erosion of the cervix may occur. In males, the disease occurs primarily in the urethra, with pain on urination and a thin, mucoid discharge. Many males are asymptomatic, however, and they may spread the disease unknowingly. Where sperm counts are marginal, trichomoniasis may reduce them to the level of sterility.

Direct microscopic examination of clinical specimens is the most rapid and least expensive technique for identifying *T. vaginalis*. This is accomplished by making a wet mount preparation of the discharge and observing the quick, jerky motion of the protozoa. The drug of choice for treatment is orally administered metronidazole (Flagyl). Tinidazole and miconazole have also been used with success. Both patient and sexual partner should be treated concurrently to prevent transmission or reinfection.

Trypanosomiasis

Trypanosomiasis is a general name for two diseases caused by species of *Trypanosoma*. Protozoa in this genus are elongated flagellates having a characteristic undulating membrane that waves as the organism moves (Figure 15.10). The two diseases caused by trypanosomes are traditionally known as African sleeping sickness and South American sleeping sickness.

African sleeping sickness is transmitted by the **tsetse fly** *Glossina palpalis*. The insect bites a disease victim, and the trypanosomes localize in the insect's salivary gland. Transmission occurs during the next bite. The point of entry becomes painful and swollen in several days and a chancre similar to that in syphilis is observed. Invasion of the bloodstream follows soon thereafter.

15.3

MICROFOCUS

CLOSE, BUT NO CIGAR

In the late 1800s, new trade routes opened in Africa and travel increased substantially into the hitherto unknown equatorial belt. Among the most adventurous explorers were the British, but they regularly fell ill with the "sleeping sickness." Therefore in 1894 the British government sent a medical team headed by David Bruce to investigate the disease and find its cause. In several types of animals, the researchers located a new parasite (later named *Trypanosoma brucei* for the team leader), but they could not find it in human victims. Interestingly, they also found the parasite in tsetse flies.

As it turned out, the British were not the only ones interested in sleeping sickness.

The Italians also had a medical team in Africa, and the group headed by Aldo Castellani found Bruce's parasite in the nervous system of infected humans. Bruce read Castellani's report and began a search for parasites in human blood, for if they could reach the nervous system, the blood would be the logical route. And sure enough they were in the blood—in scores of thousands. But how did they get there? The tsetse flies, of course! All the pieces of the puzzle seemed to fit: the sickness is transmitted among animals by tsetse flies, which also bite humans and inject the parasites into the blood. Passage to the brain follows.

The solution was obvious: stop the

tsetse flies and stop the sleeping sickness. After all, that's what Gorgas and Reed were doing with mosquitoes and yellow fever in the Caribbean islands and Central America. But it was not to be. Yellow fever is primarily a human disease, but sleeping sickness affects numerous wild animals, and the animals could not be kept away from human populations. Moreover, tsetse flies breed everywhere from waterlogged river banks to arid deserts to savannas and grasslands (mosquitoes breed only where water collects). The effort to stop the disease was valiant, but it was doomed to failure. To this day, sleeping sickness remains a threat to life in Africa.

Two types of African sleeping sickness exist. One form, common in Central Africa, is caused by *Trypanosoma brucei* variety *gambiense*. It is accompanied by chronic bouts of fever, as well as severe headaches, paralysis, and a general wasting away. As the trypanosomes invade the brain, the patient slips into a coma (hence, the name, "sleeping sickness") and dies. The second form, common in East Africa, is due to *Trypanosoma brucei* variety *rhodesiense*. This disease is more acute, with higher fever and rapid coma and death.

African sleeping sickness exists wherever the tsetse fly is found, and Africa provides the right combination of temperature and moisture for this insect. In 1898, David Bruce identified the trypanosomes in tsetse flies and recommended insect control (MicroFocus: 15.3). Bruce's discovery opened the door to the British colonization of Africa. Today the disease is checked by clearing brushlands and treating areas where the insects breed. Patients are treated with a drug called suramin sodium (or Bayer 205), a sulfonic acid derivative administered intravenously. An alternative drug known as pentamidine isethionate is also used.

South American sleeping sickness is caused by *Trypanosoma cruzi*, a trypanosome discovered by Carlos Juan Chagas. In his honor the disease is often called **Chagas' disease**. **Triatomid bugs** of the genera *Triatoma* and *Rhodnius* are essential to transmission of the trypanosomes. The insects are found in the cracked walls of mud and adobe houses. They feed at night and bite where the skin is thin, such as on the lips, face, or forearms. For this reason they are called "kissing bugs."

In humans, South American sleeping sickness is characterized by fever and widespread tissue damage, especially in the **heart** (MicroFocus: 15.4). The trypanosomes destroy the cardiac nerves so thoroughly that the victim experiences sudden heart failure. Organisms may also reach the brain where they induce coma and death. A recent estimate put the number of cases in South and Central America at 12 million. Although there is no effective treatment for the disease, some relief is experienced with nifurtimox, also known as Bayer 2502.

bru′cē-i

TRYPANOSOMA BRUCEI

cru′zi

tri-at′o-mah
rod′ne-us

ni-fur′ti-max

15.4
MICROFOCUS

CHAGAS AND DARWIN

In 1909 Carlos Chagas was a young Brazilian doctor of 29 when he arrived to fight malaria in a small town north of Rio de Janeiro. But there was another problem that caught his attention—many of the local people were suffering from lethargy, shortness of breath, and irregular heartbeat. And no one knew what was the cause.

Chagas set aside his interest in malaria and began a search for the parasite of this strange new disease. He quickly tracked down the vector, a cricketlike triatomid bug that lived in the walls of thatched houses and sucked the blood of sleeping inhabitants. From the bug he extracted a whiplike protozoan similar to the trypanosome of African sleeping sickness. In rapid succession, Chagas proved that the trypanosome could infect monkeys; he found it in a cat in a bug-infested house; and he isolated it from the blood of a young girl displaying the symptoms of the disease (now recognized as Chagas' disease).

To be sure, the disease had not been described previously, but neither was it a

new disease. Unbeknown to Chagas, it had probably claimed the life of Charles Darwin many years before. Historians record that Darwin's health declined perceptibly on his

return to England from South America (from his famous voyage aboard the HMS *Beagle*). Some writers maintain that the illness was psychosomatic—Darwin took a public battering on publication of his theory of evolution—but Saul Adler, a tropical medicine researcher, believes otherwise. Adler believes that Darwin suffered from Chagas' disease. Indeed, while in Argentina, Darwin wrote: "... at night I experienced an attack (for it deserves no less a name) of *Reduvius*, the giant black bug of the Pampas. It is most disgusting to feel soft, wingless insects about an inch long, crawling over one's body. Before sucking they are quite thin, but afterwards they become round and bloated with blood."

The bug that Darwin describes is *Triatoma infestans*, the vector that Chagas would identify generations later. Furthermore, Darwin's symptoms matched those in Chagas' patients. The disease would linger in Darwin's tissues for 40 years and reduce the vigorous adventurer to a shell of his former self.

An interesting theory emerged in the 1980s to explain why victims of trypanosomiasis suffer waves of blood invasion of parasites and accompanying waves of fever. Researchers found that proteins on the membrane surface of the trypanosome were different as each new blood invasion occurred. Thus the antibodies formed against the preceding parasites were ineffective against the new variants. The change in the trypanosome's surface protein is apparently due to chromosomal **insertion elements** that produce copies of themselves and move along the DNA changing the genetic code for membrane protein (Chapter 6). This important finding indicates how insertion elements may affect the progress of a disease.

Insertion elements: segments of chromosomal DNA that produce copies, which insert at other places on the chromosomes.

Leishmaniasis

lēsh"mah-ni′ah-sis

Leishmaniasis is a rare disease in the United States, but it occurs worldwide in large-scale epidemics. (An estimated 12 million individuals in the tropics and subtropics suffer the disease annually.) The responsible protozoa belong to the class Mastigophora and include several species of *Leishmania* such as ***Leishmania donovani*** and ***L. tropica*** (Figure 15.11). Transmission is by the **sandfly** of the genus *Phlebotomus*.

lēsh-ma′ne-ah

flĕ-bot′o-mus

One form of leishmaniasis is a visceral disease called **kala-azar**, meaning "black fever." This disease is characterized by infection of the body's **white blood cells** and is accompanied by fever, progressive anemia, and emaciation. Secondary infection by bacteria or viruses is common. A second form of leish-

FIGURE 15.11

A Photomicrograph of *Leishmania tropica*

L. tropica is shown in the rosette pattern wherein the long, thin flagellates are clustered so as to give the appearance of petals on a flower. The rosette pattern is found in older culture media that poorly support the growth of the organism. Some researchers believe that the organisms are in a feeding frenzy at this point, but evidence to justify this theory is lacking.

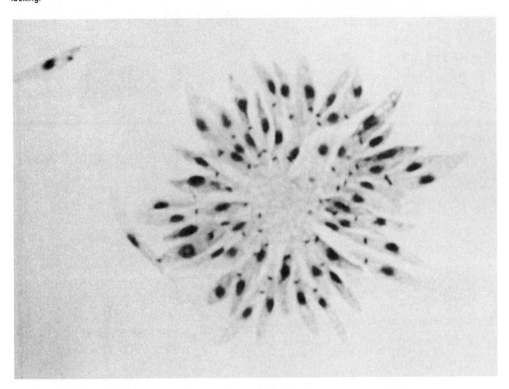

FIGURE 15.12

The Leishmaniasis Sore

Photograph of a cutaneous sore on the wrist of a man with leishmaniasis. The shallow, circular nature of the ulcer is visible.

maniasis is a cutaneous disease, less severe because the parasites remain in the **skin** (Figure 15.12). This disease is often referred to as **Oriental sore** ("Oriental" because epidemics have been documented in China). It is also known in the Middle East as the "rose of Jericho" disease from the flowery appearance of the skin lesion (in some Israeli communities it is called "shoske" from the Hebrew *shoshana*, for rose).

Control of the sandfly remains the most important method for preventing outbreaks of leishmaniasis. Several drugs, including antimony compounds, are

LEISHMANIA DONOVANI

FIGURE 15.13

An Outbreak of Leishmaniasis

This outbreak occurred among military personnel who fought in Operation Desert Storm during 1991.

1. During 1991, approximately 500,000 military personnel took part in Operation Desert Storm in Saudi Arabia, Kuwait, and other countries of the Persian Gulf region. Many personnel were involved in the troop movements and fighting that resulted in the Allied victory.

2. While stationed in the Middle East during and after the fighting, many individuals were subjected to the bites of sandflies, the arthropods that transfer the protozoan *Leishmania tropica*. Exposure to the insects took place in urban as well as in desert areas.

3. On returning to the United States after several months' duty, seven men displayed the symptoms of leishmaniasis, including high fever, chills, malaise, liver and spleen involvement, and gastrointestinal distress. Several had low-volume watery stools and abdominal pain.

4. When bone marrow samples from the seven patients were examined, the tissue yielded evidence of *L. tropica*, the agent of leishmaniasis. The men were treated over a period of weeks with an antimony compound called sodium stibogluconate. All recovered.

available for treating established cases. At least seven cases of visceral leishmaniasis and 16 cases of cutaneous leishmaniasis occurred in military personnel associated with Operation Desert Storm, in 1991 (Figure 15.13). In 1993, visceral leishmaniasis broke out in epidemic proportions in civil-war-torn regions of Sudan.

TO THIS POINT
■

We have considered seven protozoal diseases caused by amebas or flagellates. Amebiasis is a disease of the intestine that has become fairly widespread in the United States since it was introduced from tropical areas. The cyst form may be contracted from contaminated food or water or by contact with infected individuals. A similar situation holds for giardiasis. Sewage leaks are another community problem that may lead to giardiasis.

We also studied trichomoniasis, a sexually transmitted disease that is among the most common illnesses in Americans. Women are particularly susceptible to this disease, but men can be carriers. The discussion then moved to arthropod-borne protozoal diseases, where species of Trypanosoma *and* Leishmania *rely on insects to complete their life cycles. The arthropod represents the "weak link" in the transmission of these diseases, and control can be effected by eliminating the arthropod. This is one reason why leishmaniasis and sleeping sickness are not prevalent where sanitation methods are established. The diseases are uncommon in the United States also because the necessary arthropods are not usually found here.*

We shall now focus our attention on a disease caused by a ciliate and several diseases due to sporozoa. In this group we shall find rapidly emerging problems in medical microbiology as well as one illness, malaria, that is the most widespread human disease recognized today.

Protozoal Diseases Due to Ciliates and Sporozoa

Balantidiasis

Balantidiasis is an intestinal disease caused by ***Balantidium coli***. The protozoan is among the largest organisms to infect humans. It measures up to 100 μm in length by 70 μm in width, and it has a large kidney-shaped nucleus as well as a small micronucleus. Trophozoite and cyst stages exist. The organism is one of the few members of the class Ciliophora to cause disease.

bal″an-tĭ-di′ah-sis
bal″an-tid′e-um

BALANTIDIUM COLI

Balantidiasis is a rare disease in temperate climates. It is spread by contaminated water or food, especially pork. Cysts pass through the stomach, and the ciliated trophozoites emerge in the intestine, where they cause mild ulceration. Profuse **diarrhea**, nausea, and rapid weight loss are characteristic signs of disease. Often the disease is chronic, since cysts remain in the intestinal wall. Metronidazole or paromomycin may be used for treatment. Though cases are uncommon in the United States, some concern has been voiced about symptomless carriers returning from tropical regions of the world.

Toxoplasmosis

Toxoplasmosis was first recognized as a clinical disease in 1909 by Charles Nicolle (the French investigator who associated lice with epidemic typhus). Nicolle assumed that the disease was due to toxins and named it accordingly, but no evidence exists for the involvement of toxins. Instead, the symptoms arise from tissue damage caused by the extensive growth of protozoa.

The cause of toxoplasmosis is ***Toxoplasma gondii***, a member of the Sporozoa class. *T. gondii* exists in three forms: the trophozoite, the cyst, and the oocyst. Trophozoites are crescent-shaped or oval organisms without evidence of locomotor organelles (Figure 15.14a). Located in tissue during the acute stage of disease, they are able to invade all mammalian cells (Figure 15.14b), with the notable exception of erythrocytes. Cysts develop from the trophozoites within host cells and may be the source of repeated infections. Muscle and

toks″o-plaz′mah gon′de-e

o′o-sist

FIGURE 15.14

Toxoplasma gondii, the Cause of Toxoplasmosis

(a) A scanning electron micrograph of numerous parasites in the crescent-shaped trophozoite stage ($\times$ 31,000). (b) A white blood cell simultaneously attacked by several trophozoites ($\times$ 13,500). There are at least three sites where invasion of the white blood cells is taking place.

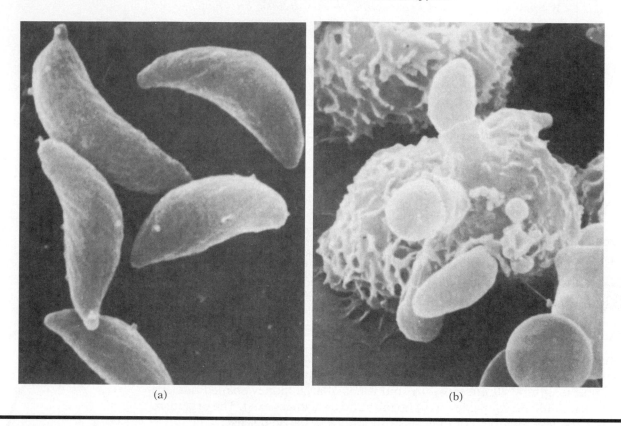

(a)   (b)

TOXOPLASMA GONDII

Trophozoite:
the feeding form of a
protozoan.

Retina:
the layer of light-sensitive
nerve cells along the rear
periphery of the eye.

nerve tissue are common sites of cysts. Oocysts are oval bodies that develop from the cysts by a complex series of asexual and sexual reproductive processes.

Toxoplasma gondii exists in nature in the cyst and oocyst forms. Grazing animals acquire these forms from the soil and pass them to humans via **contaminated beef**, **pork**, **or lamb**. Rare hamburger meat is a possible source. **Domestic housecats** acquire the cysts from the soil or from infected birds or rodents. Oocysts then form in the cat. Humans are exposed to the oocysts when they forget to wash their hands after contacting cat feces while changing the cat litter or working in the garden. Touching the cat can also bring oocysts to the hands, and contaminated utensils, towels, or clothing can contact the mouth and transfer oocysts.

Toxoplasmosis develops after trophozoites are released from the cysts or oocysts in the host's gastrointestinal tract. *T. gondii* invades the intestinal lining and spreads throughout the body via the blood. Patients develop fever, malaise, sore throat, and swelling of the spleen, liver, and lymph nodes. In these respects the disease resembles infectious mononucleosis. Lesions may also occur on the retina, and virtually any organ may be involved. Complications, however, are rare. Diagnosis may be made by isolating trophozoites from the blood or other body fluids. Sulfonamide drugs are commonly used in therapy.

Pregnant women are at risk from toxoplasmosis because the protozoa may cross the placenta and infect the fetal tissues (Figure 15.15). Neurological dam-

FIGURE 15.15

The Cycle of Toxoplasmosis in Nature

The cat acquires the parasites by consuming a contaminated bird (a) or rodent (b) or by contact with contaminated soil or water. It may transmit the parasite by its urine to a sandbox (c) or cat litter (d), and then to an individual. A pregnant woman may pass the parasite to the fetus (e). Otherwise, the contaminated soil may be ingested by cattle (f), from which the organism is transmitted to humans in rare beef (g). Birds and rodents may also acquire the organism from contaminated soil.

age, lesions of the fetal visceral organs, or spontaneous abortion may result. **Congenital infection** is least common during the first trimester, but damage may be substantial. By contrast, congenital infection is more common if the woman is infected in the third trimester, but fetal damage is less severe. Lesions of the retina are the most widely documented complication in congenital infections. The "T" in the **TORCH** group of diseases refers to toxoplasmosis (the others are rubella, cytomegalovirus, and herpes simplex; the "O" is for other diseases such as syphilis).

Toxoplasma gondii is also known to cause severe disease in immune-suppressed individuals. For example, in patients with **AIDS**, the normal defensive mechanisms preventing spread have been destroyed, and *T. gondii* attacks the brain tissue. The infected tissue attracts immune system components, and the resulting inflammation and swelling often result in cerebral lesions, seizures,

AIDS:
a serious viral disease accompanied by destruction of cells of the immune system.

and death. Often the disease stems from opportunistic parasites already in the body. Patients receiving immune-suppressing therapy for cancer or to prevent organ graft rejection are similarly at risk.

Toxoplasma gondii is regarded as a universal parasite. Some researchers suggest that it is the most common parasite of humans and other vertebrates. It is important to ranchers, dairy product producers, pet breeders, and anyone who comes in contact with a domestic cat. Indeed, evidence indicates that where there are no cats, toxoplasmosis is rare. Hundreds of millions of humans worldwide are estimated to be infected.

Vertebrates:
animals with backbones.

Malaria

a'gu

Though the disease malaria has been known to exist since at least 1000 B.C., the word "malaria" has been used only since the 1700s. Before that time, the disease was known as "ague," from the French *aigu*, meaning sharp (a reference to the sharp fever that accompanies the disease). During the 1700s, however, Europeans began using the Italian word for bad air (*mal-aria*), reflecting the theory that the disease was somehow related to some unknown atmospheric influence called miasma. Wave after wave of malaria swept over the world during that century, and few parts were left untouched. American pioneers settling in the Mississippi and Ohio valleys suffered great losses from the disease.

More than 250 million of the world's population now suffer the chills, fever, and life-threatening effects of **malaria**. The disease takes its greatest toll in Africa, where the WHO estimates that over a million children under the age of five die from malaria annually. No infectious disease of contemporary times can claim such a dubious distinction. Though progress has been made in the control of malaria, the figures remain appallingly high. (Even the United States is involved in the malaria pandemic—over 1000 cases are diagnosed annually.)

PLASMODIUM MALARIAE

ah-nof'e-lēz

spor-o-zo'īt
mer-o-zo'īt
gam-e'to-cīt

Malaria is caused by four species of ***Plasmodium***: *P. vivax*, *P. ovale*, *P. malariae*, and *P. falciparum*. All are transmitted by the female *Anopheles mosquito*, which consumes human blood to provide chemical components for her eggs. The life cycle of the parasites has three important stages: the sporozoite, the merozoite, and the gametocyte. Each is a factor in malaria.

The mosquito sucks human blood and acquires the gametocyte stage of the parasite in the red blood cells (Figure 15.16). Within the insect a transition to sporozoites takes place, and the sporozoites then migrate to the salivary gland. When the mosquito bites another human being, several hundred sporozoites enter the person's bloodstream and quickly migrate to the liver. After several hours, the transformation of one sporozoite to 25,000 merozoites has been completed, and the merozoites emerge from the liver to invade the **red blood cells**. While in the red blood cells, the merozoites synthesize about 150 proteins that attach to the RBC membrane and anchor the RBC to the blood vessels. By constantly switching between 150 genes (for 150 proteins), the malarial parasite avoids detection by the body's immune system (Chapter 18).

Merozoite:
an intermediate stage in the life cycle of malaria parasites.

Within the human red blood cells, the merozoites undergo another series of transformations that result in several gametocytes and thousands of new merozoites. In response to a biochemical signal, thousands of red blood cells rupture simultaneously, thereby releasing the parasites and their toxins. Now the excruciating **malaria attack** begins. First, there is intense cold, with shivers and chattering teeth. The temperature then rises rapidly to 104° F, and the victim develops intense fever, headache, and delirium. After two or three hours, massive perspiration ends the hot stage, and the patient often falls asleep, exhausted.

FIGURE 15.16

The Malaria Cycle

(a) *Anopheles* mosquitoes inject malaria parasites into the human bloodstream in the sporozoite form. (b) Liver invasion is followed by transformation into a second form, the merozoite. (c) The merozoites invade the red blood cells and undergo transformation to a ring form (d), then an ameboid form, and finally (e) back to the merozoite form. Fever and chills accompany the disruption of red blood cells. If a mosquito ingests blood at this point, the malaria parasites enter its stomach. (f) Male sperm cells in the mosquito fertilize egg cells to (g) produce a zygote. (h) Nuclei of the zygote eventually form sporozoites, which accumulate in the mosquito's salivary gland and are injected with the next bite.

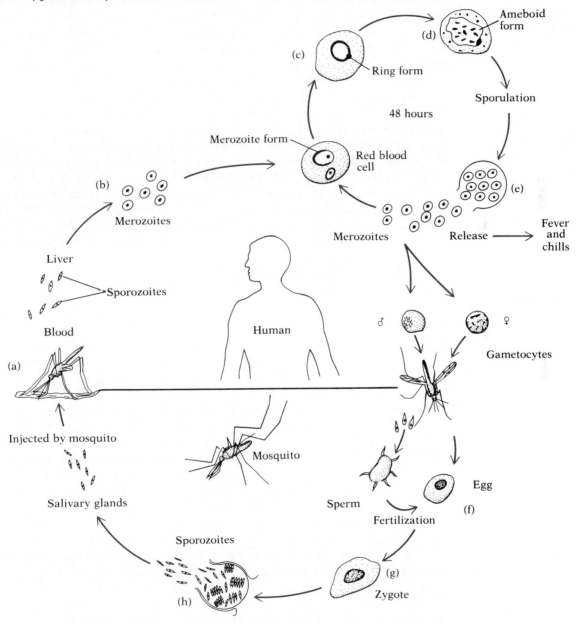

During this quiet period, the merozoites enter a new set of red blood cells and repeat the cycle of transformations. *P. vivax* and *P. ovale* spend about 48 hours in the red blood cells, so that 48 hours pass between malaria attacks. This is **tertian malaria** ("tertian" comes from the Latin for "three-day," based

BABESIA MICROTI

on the Roman custom of calling the day the event happened the first day of the cycle). For *P. malariae*, the cycle takes 72 hours, and the disease is called **quartan malaria** ("four-day" malaria). The cycle of *P. falciparum* is not defined, and attacks may occur at widely scattered intervals. This type of malaria, the most lethal, is known as **estivo-autumnal malaria**, referring to the summer–fall periods when mosquitoes breed heavily.

Death from malaria may be due to a number of factors related to the loss of red blood cells. Substantial anemia develops, and the hemoglobin from ruptured blood cells enters the urine (malaria is sometimes called blackwater fever). Cell fragments accumulate in the small vessels of the brain, kidneys, heart, liver, and other vital organs and cause clots to form. Heart attacks, cerebral hemorrhages, and kidney failure are common. The intense fever leads to convulsions.

Since its discovery about 1640, **quinine** has been the mainstay for the treatment of malaria (MicroFocus: 23.1). When the trees used as a source of this drug were captured by the Japanese in World War II, American researchers developed chloroquine for the active stage of malaria and primaquine for the dormant stage. Chloroquine remained an important mode of therapy until recent years when drug resistance began emerging in *Plasmodium* species. Since 1989, an alternative drug called **mefloquine** has been recommended for individuals entering malaria regions of the world. Another drug called **proguanil** is recommended if the person cannot tolerate mefloquine. Since 1991, the CDC has also recommended using **quinidine gluconate**, a quinine derivative, for treating complicated *P. falciparum* infections. The drug is administered by injection, usually by an intravenous route. Prospects for a malaria vaccine have emerged in the 1990s (MicroFocus: 15.5).

Babesiosis

Babesiosis is a malarialike disease caused by ***Babesia microti***, a member of the class Sporozoa. The protozoa live in **ticks** of the genus *Ixodes*, and are transmitted when these arthropods feed in human skin. Areas of coastal Massachusetts and Long Island, New York, have experienced outbreaks in recent years, and six cases were identified in residents of Connecticut in 1989.

Babesia microti penetrates human **red blood cells**. As the cells disintegrate, a mild anemia develops. Piercing headaches accompany the disease and, occasionally, meningitis occurs. A suppressed immune system appears to favor establishment of the disease. However, babesiosis is rarely fatal and drug therapy is not recommended. Carrier conditions may develop in recoverers, and spread by blood transfusion is possible. Travelers returning from areas of high incidence are therefore advised to wait several weeks before donating blood to blood banks. Tick control is considered the best method of prevention.

Babesia has a significant place in the history of American microbiology because in the late 1800s, Theobald Smith located *B. bigemina* in the blood of cattle suffering from Texas fever (Chapter 1). His report was one of the first linking protozoa to disease and, in part, it necessitated that the "bacterial" theory of disease be modified to the "germ" theory of disease.

Cryptosporidiosis

Before 1976, **cryptosporidiosis** was recognized as a cause of diarrhea in animals but was unknown in humans. Since that year, however, microbiologists have observed the disease in humans, and the number of cases has risen markedly.

15.5

EUPHORIA TO DESPONDENCY AND BACK AGAIN

Early in the 1960s, international health authorities thought they had malaria licked. Drugs such as quinine and chloroquine and insecticides such as DDT had reduced the incidence dramatically, and some officials were bold enough to speak of eradication.

But then reality dawned: *Plasmodium falciparum*, the most lethal of the malaria species, showed signs of resistance to chloroquine; mosquitoes with DDT resistance began emerging; Third World countries became complacent about malaria and relaxed their vigilance; and the campaign to control the disease in remote parts of the globe failed. Between 1972 and 1976, the worldwide number of malaria cases doubled and by various estimates, the incidence is now over 300 million cases per year, with an estimated 2 million deaths.

New hope exists, however, that a vaccine for malaria may be in the near future. The vaccine's story began in 1967 when Ruth Nussenzweig and her co-workers at New York University showed that irradiated sporozoites provoke an immune response in mice. A vaccine with whole sporozoites was not feasible, though, because sporozoites were impossible to obtain in quantity. The advent of genetic engineering in the 1970s solved this problem. Nussenzweig's

group first isolated large amounts of the antibodies that react with sporozoites. Then they used the antibodies to track down the immune-stimulating antigen on the sporozoite surface. By 1984, they located the antigen. The next step was to pinpoint the gene that codes for the antigen; by 1985, this also was accomplished.

Now the scene shifts to Colombia and a research group led by Manuel Elkin Patarroyo. In the late 1980s, Patarroyo's group developed a vaccine containing four different and synthetic protein antigens derived from *P. falciparum*. One antigen is from the sporozoite stage (similar to Nussenzweig's antigen), and the other three are from the merozoite stage, the stage in which plasmodia actually infect red blood cells. In March 1993, the Colombian researchers announced in *Lancet* the results of a clinical trial in which 738 volunteers received the vaccine and displayed a 39 percent reduction of malaria cases. Later that year, a second trial in 468 Ecuadorian volunteers resulted in 68 percent fewer cases among vaccine recipients. The scientific community noted that 39 and 68 percent reductions are not extraordinary for a vaccine; nevertheless, when 2 million people die of malaria annually, these reductions constitute sig-

nificant numbers. The vaccine has been named SPf66.

Research into a malaria vaccine continues today (with 2-year field trials ongoing in Thailand and Tanzania), and Patarroyo's vaccine stands as the prototype for other vaccines. When he began his research Patarroyo pledged that if his vaccine proved effective, he would donate his patent rights to the United Nations. In May 1995 he made good on the pledge, signing over all vaccine rights to the U.N. and its public health agency, the World Health Organization.

There are clinkers in the agreement, however. The agreement specifies that the WHO must select a nonprofit manufacturer suitable to both sides (a leading contender is a firm in Bogotá, Colombia); the field trials have yielded mixed results, and some pessimism regarding the vaccine's efficacy has surfaced; and still to be settled is who determines when the vaccine is ready for large-scale production. There is also the Nussenzweig factor—Ruth Nussenzweig and her collaborators have continued their research into vaccine development, and in 1995 they announced a peptide vaccine that is highly effective in mice. Stay tuned . . . .

Cryptosporidiosis is caused by ***Cryptosporidium parvum*** and other species of *Cryptosporidium* such as ***C. coccidi***. All are members of the Sporozoa class of protozoa. The organism is similar to *Toxoplasma gondii* in that it has a complex life cycle involving trophozoite, sexual, and oocyst stages (Figure 15.17). Patients with competent immune systems appear to suffer limited diarrhea that lasts one to two weeks and does not require hospitalization. However, individuals with suppressed immune systems, as in **AIDS**, experience cholera-like, profuse **diarrhea** that is severe and often irreversible. These patients undergo dehydration and emaciation, and often die of the disease.

Cryptosporidiosis has an incubation period of about one week, which means that few cases are diagnosed properly because most people relate their nausea and diarrhea to something they ate a day or two before. In healthy adults, the infection usually lasts about 10 days, but the diarrhea may remain for up to two months. A fecal-oral route is the major mode of transmission (as noted in

krip″to-spor-id′e-um
kok-sid′e-e

FIGURE 15.17
Cryptosporidium coccidi

A transmission electron micrograph of the intestinal lining of an animal infected with *Cryptosporidium* (× 7500). Many aspects to the complex life cycle are visible including the trophozoite stage (T), and the schizont stage (S) releasing eight merozoites. The macrogametes (M) are large cells that fuse in the sexual reproductive cycle of the organism. Dense dotlike areas within the macrogametes are polysaccharide granules.

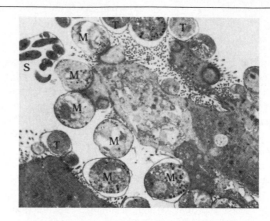

this chapter's opening), but physical contact can also transmit *Cryptosporidium* and children in day-care centers are at risk. Experiments indicate that the dose level to establish infection is very low (as low as 30 organisms), and no drug is widely accepted to treat the disease at this writing, although paromomycin and azithromycin have showed promise. High-level infections can be detected by performing the acid-fast test (Chapter 3) on stool specimens and noting the presence of acid-fast oocysts.

pa-ro'mo-my"sin
a-zith'ro-my"sin

In 1995, researchers reported that a *Cryptosporidium*-like parasite was also involved in outbreaks of diarrhea. The newly recognized parasite is a coccidian protozoan named *Cyclospora cayetanensis*. Occurring in tropical climates of the world (e.g., Peru, Central America, Nepal), the parasite causes long-lasting diarrheal illness but is apparently susceptible to treatment with Bactrim. The incidence of the disease is not certain at this writing.

si-clo-spor'ah
cay"e-tan-en'sis

Pneumocystosis

Pneumocystosis is currently the most common cause of nonbacterial pneumonia in Americans with suppressed immune systems. The causative organism, ***Pneumocystis carinii***, was first observed in 1910 by John Carini in studies with rats. It remained in relative obscurity until the 1980s, when it was recognized as the cause of death in over 50 percent of persons dying from the effects of AIDS.

nu"mo-sis-to'sis

nu"mo-sis'tis car-in'e-e

Pneumocystis carinii has a complex life cycle that takes place entirely in the alveoli of the lung. The trophozoite stage swells to become a precyst stage in which up to eight sporozoites develop. When the cyst is mature, it opens and liberates the sporozoites, which enlarge and undergo further reproduction and maturation to trophozoites. It should be noted, that although this life cycle is reminiscent of a protozoan, there is a recent body of evidence pointing to the possibility that *P. carinii* is a **fungus**. Analysis of its ribosomal RNA, for example, shows a closer relationship to fungal RNA than to protozoal RNA. However, *P. carinii* has historically been considered a protozoan and we shall consider it here until the evidence for reclassification becomes more persuasive.

Alveoli:
the air sacs of the lungs.

Present evidence indicates that *P. carinii* is transmitted by droplets from the respiratory tract. A wide cross-section of individuals harbor the organism without symptoms, mainly due to the control imposed by T-lymphocytes. However, when the immune system is suppressed as in AIDS patients, *Pneumocystis* trophozoites and cysts fill the alveoli and occupy all the air spaces. A nonproductive cough develops, with high fever and difficult breathing. Progressive deterioration leads to consolidation of the lungs and eventually, death. The disease is commonly referred to as ***Pneumocystis carinii* pneumonia (PCP)**.

The current treatment of choice for PCP is pentamidine isethionate, approved for use by the Food and Drug Administration in October 1984 and now commercially available. Another drug, trimetrexate, was approved in 1988. These drugs, however, have limited value in immune-suppressed individuals, as evidenced by the high death rate in AIDS patients. Table 15.1 summarizes the protozoal diseases of humans.

pen-tam′ĭ-dēn i-se-thī′o-nāt

tri-meh-trex′ate

NOTE TO THE STUDENT

On an October day in 1983, a local medical laboratory notified the Los Angeles Department of Health Services of a large number of stool samples containing Entamoeba histolytica, the protozoan of intestinal amebiasis. The laboratory had also reported 38 cases of amebiasis in the previous two months. Department officials were mystified because there had been no increase in the number of specimens the laboratory examined, no clustering of cases of amebiasis, few patients in high-risk categories (tourists, immigrants, or institutionalized patients), and no instances of increased reporting from other laboratories. When Health Department investigators reexamined 71 slides of fecal material from the 38 cases, they found only four with Entamoeba histolytica. Officials concluded that the laboratory was in error, and that the bodies thought to be amebas were in fact white blood cells.

Diagnostic methods for protozoal diseases have not changed fundamentally for many generations, and the sophisticated devices used for bacterial and viral detection have not yet reached the parasitology laboratory. Cultivation methods for pathogenic protozoa are difficult, and a correct diagnosis often depends on direct observation of tissue specimens together with a sense of intuition and understanding of the patterns of disease. In the case cited in Los Angeles, these were apparently lacking.

With the increasing prevalence of protozoal diseases in recent years, parasitologists have strengthened their role in the healthcare delivery system. Giardiasis has become a well-known intestinal disease, and trichomoniasis remains among the most common sexually transmitted diseases. Diseases such as cryptosporidiosis, toxoplasmosis, and pneumocystosis are relatively new to the 1990s, and increasing coverage in media reports is testimony to their importance. Even malaria is still encountered in travelers and immigrant groups.

It takes years of training and experience to be a good parasitologist. A keen and discriminating eye is essential, and the right questions must be asked to distinguish between organisms and tissue debris. You might like to consider parasitology as a career if your future goals are not yet established.

TABLE 15.1

A Summary of Protozoal Diseases in Humans

ORGANISM	CLASS	ORGANS OF MOTION	DISEASE	TRANSMISSION
Entamoeba histolytica	Sarcodina	Pseudopodia	Amebiasis	Water Food
Naegleria fowleri	Sarcodina	Pseudopodia	Primary amebic meningoencephalitis	Water
Giardia lamblia	Mastigophora	Flagella	Giardiasis	Water Contact
Trichomonas vaginalis	Mastigophora	Flagella	Trichomoniasis	Sexual contact
Trypanosoma brucei	Mastigophora	Flagella	African sleeping sickness	Tsetse fly (*Glossina*)
Trypanosoma cruzi	Mastigophora	Flagella	South American sleeping sickness	Triatomid bug (*Triatoma*)
Leishmania donovani	Mastigophora	Flagella	Leishmaniasis (Kala-azar)	Sandfly (*Phlebotomus*)
Balantidium coli	Ciliophora	Cilia	Balantidiasis	Food Water
Toxoplasma gondii	Sporozoa	None in adult	Toxoplasmosis	Domestic cats Food
Plasmodium species	Sporozoa	None in adult	Malaria	Mosquito (*Anopheles*)
Babesia microti	Sporozoa	None in adult	Babesiosis	Tick (*Ixodes*)
Cryptosporidium parvum	Sporozoa	None in adult	Cryptosporidiosis	Water, food
Pneumocystis carinii	Sporozoa	None in adult	Pneumocystosis	Droplets

Summary

Protozoa are of interest to microbiologists because protozoa are unicellular, have a microscopic size, and cause infectious disease. They are usually very large organisms, and the functions of their cells bear a resemblance to the functions of multicellular organisms. The majority of protozoa are heterotrophic, and some exist in the trophozoite and cyst forms, the latter a very resistant form.

Protozoa are divided into four classes according to the method of locomotion they display. Members of the Sarcodina class move by means of pseudopodia and include the amebas; flagellated protozoa are grouped into the Mastigophora class; ciliated protozoa make up the Ciliophora class; and the nonmotile protozoa are classified as Sporozoa. Chlorophyll-containing protozoa, species that undergo sexual conjugations, and various other interesting forms are found within the classes.

The most serious Sarcodina-related disease is amebiasis due to *Entamoeba histolytica*. Intestinal ulcers and sharp, appendicitislike pain accompany the disease. Giardiasis is also an intestinal disease but the protozoa do not penetrate the tissue. The agent, a

ORGAN AFFECTED	DIAGNOSIS	TREATMENT	COMMENT
Intestine Liver Lungs	Stool examination	Paromomycin Metronidazole	Deep intestinal ulcers
Brain	Spinal fluid examination	None effective	Uncommon in U.S.
Intestine	Stool examination Gelatin capsule	Quinacrine Furazolidine	Incidence increasing in U.S.
Urogenital organs	Urine or swab examination	Metronidazole Tinidazole	May result in sterility
Blood Brain	Blood smear	Various drugs	Two types, depending on region
Blood Brain Heart	Blood smear	Various drugs	Common in South America
White blood cells Skin Intestine	Tissue examination	Antimony	Ulcers yield skin disfiguration
Intestine	Stool examination	Paromomycin Metronidazole	Symptomless carriers common
Blood Eyes Tissue cells	Blood examination	Sulfonamide drugs	Congenital damage possible Associated with AIDS
Liver Red blood cells	Blood examination	Quinine Mefloquine Proguanil	World's most urgent public health problem
Red blood cells	Blood examination	None recommended	Carrier state possible
Intestine	Stool examination	None effective	Associated with AIDS
Lungs	Lung examination	Pentamidine isethionate	Associated with AIDS

flagellate named *Giardia lamblia*, is acquired from contaminated food and water. Trichomoniasis, a sexually transmitted disease due to a flagellate, is among the most common diseases in the United States.

Arthropod vectors can play a role in protozoal disease. Sleeping sickness (trypanosomiasis) and leishmaniasis are transmitted by tsetse flies and sandflies, respectively. Blood invasion and tissue involvement characterize both diseases. Among the ciliates, *Balantidium coli* is known as an intestinal parasite acquired by a fecal-oral route. A sporozoan, *Toxoplasma gondii*, causes serious disease in AIDS patients, as do other sporozoans named *Cryptosporidium coccidi* and *Pneumocystis carinii*. *Toxoplasma* affects various organs especially those of the nervous system. *Cryptosporidium* causes choleralike diarrhea, and *Pneumocystis* is an agent of pneumonia. Another sporozoan, *Plasmodium*, is the cause of malaria, one of the most serious global health problems. Transmitted by mosquitoes, *Plasmodium* species destroy the red blood cells and often bring on death. A similar disease called babesiosis is tickborne and more localized in occurrence.

Questions for Thought and Discussion

1. *Giardia lamblia* has been imaginatively described as a "cross-eyed tennis racket." From the marginal diagram and electron micrograph of this organism, can you think of any other such descriptions? Considering the other protozoa in this chapter, what innovative descriptions can you give of them?

2. A newspaper article written in the 1980s asserted that parasitology is a "subject of low priority in medical schools, largely because the diseases are considered as exotic infections that occur in remote areas of the world." Do you agree with the contention that they are "exotic diseases that occur in remote parts of the world"? Would you favor increased attention to protozoal diseases and more study of the general field of parasitology in the medical school curriculum? Why?

3. In the early part of this century, quinine syrup was taken to prevent malaria during visits to tropical regions of the world. To most people, the taste of quinine is very bitter, but the British found a way to make it less objectionable by mixing it with a type of liquor. What do you suppose was that liquor, and how did the drink come to be known?

4. On returning from the Persian Gulf War, American servicemen were not accepted as blood donors because there was concern about their having been exposed to *Leishmania* species. What is the connection?

5. The incidence of malaria is said to be skyrocketing in Africa in part because of the antidrought measures being employed in certain regions and the improvements in agriculture being developed in other areas. Why is the malaria increase related to these measures?

6. The term "irritable bowel syndrome" is often used when a physician cannot explain why a patient is suffering extended diarrhea, abdominal pain, and fatigue. (Such a diagnosis is euphemistically called a wastebasket diagnosis.) What candidates for this syndrome might be found in this chapter, and what is the description of each organism sought out in the cases you mention?

7. Recent outbreaks of waterborne protozoal disorders of the intestine have raised the possibility that municipalities may have to begin filtering their water instead of just chlorinating it. Cost estimates for the filtering technology could reach into the millions or billions of dollars. Do you think the average taxpayer will consider the money well-spent?

8. An anonymous CDC epidemiologist once said in an interview: "Day-care centers are the open sewers of the twentieth century." Would you agree? Why or why not?

9. You and a friend who is three months pregnant stop at a hamburger stand for lunch. Based on your knowledge of toxoplasmosis, what helpful advice can you give your friend? On returning home, you notice that she has two cats. What additional information might you be inclined to share with her?

10. Malaria is widely accepted as a contributing factor to the downfall of Rome. Over the decades, the incidence of malaria seemed to increase with expansion of the Roman empire—one that stretched from the Sahara desert to the borders of Scotland, and from the Persian Gulf to the western shores of Portugal. How do you suspect the disease and the expansion are connected?

11. The flagellated protozoan *Giardia lamblia* is named for Alfred Giard, French biologist of the late 1800s, and Vilem Dusan Lambl, Bohemian physician of the same period. Unfortunately, this information does not tell us much about the organism (except for who did much of the descriptive work). Quite to the contrary, the names of other protozoa in this chapter tell us much. What are some examples of the more informative names?

12. It has been said that until recent times, many victims of a particular protozoal disease were buried alive because their life processes had slowed to the point where they could not be detected with the primitive technology available. Which disease was probably present?

13. In 1989, giardiasis was the most reported disease in the state of Vermont. Assuming that Vermont residents have reasonably good standards of hygiene, how is it that a disease like giardiasis can top the list of reported diseases in that state?

14. The outbreak of infections due to *Cryptosporidium* has prompted a reevaluation of drinking water purification systems in the United States. Of particular concern is the observation that *Cryptosporidium* species are unaffected by chlorine, the major disinfectant used in drinking water. Indeed, in a 1995 interview, the head of a research team studying the chlorine resistance of *Cryptosporidium* said, "You can wash these things in Clorox and they will smile right back at you." And an earlier CDC study showed that *Cryptosporidium* could actually grow on Clorox powder. In view of these findings, what alternatives might be available for safeguarding the nation from *Cryptosporidium*-contaminated water?

15. African herdsmen know that to prevent trypanosomiasis, cattle should be driven across the "fly belt" from one grazing ground to another only at night. What does this bit of folk knowledge tell you about tsetse flies?

Review

A major theme of this chapter has been the protozoal diseases that affect humans. To review the chapter contents, solve the following crossword puzzle. The answers to the puzzle are contained in the appendix.

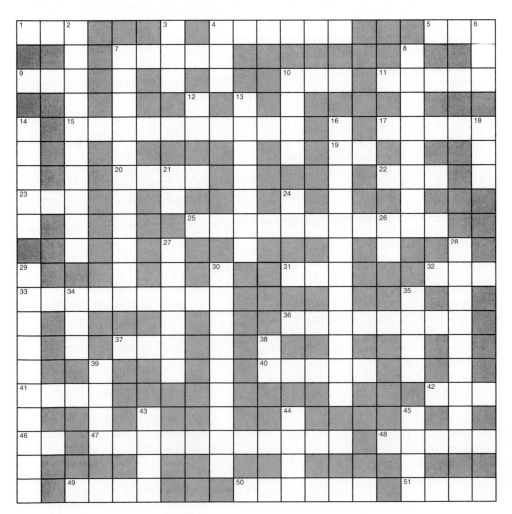

ACROSS

1. Pet that is a reservoir for *Toxoplasma gondii*.
4. One of the species that causes malaria (abbr).
5. *Trichomonas* has _____ posterior flagellum.
7. Discovered the agent of African sleeping sickness.
9. Pregnant women should not clean an animal's litter _____.
10. Agency that summarizes protozoal disease statistics.
11. Specimen examined to detect *Trichomonas* infection.
15. Genus of sporozoans involved in malaria.
17. Possible mode of transmission of *Giardia lamblia*.
19. Ameba (initials) that may cause nervous system illness.
20. Arthropod that transmits *Babesia microti*.
22. Climate where malaria is commonly found.
23. Emitted from intestine during cases of giardiasis.
25. Sexually transmitted flagellated protozoan.
31. Type of cell (abbr) infected by *Toxoplasma gondii*.
32. Number of malaria attacks possible before death.
33. Induces intestinal ulcers and sharp pains after water transmission.
36. Arthropod that transmits *Plasmodium* species.
37. System (abbr) affected during cases of trypanosomiasis.
40. Mode of locomotion of Ciliophora.
41. How a protozoan is able to _____ determines its class designation.
42. Once believed to transmit malaria.
46. There are _____ flagella in *Balantidium coli*.
47. Genus of protozoa involved in sleeping sickness.
48. Extraordinarily high during cases of malaria.
49. Resistant form present in *Entamoeba* and *Giardia*.
50. Insect fluid that contains malarial parasites.
51. Number of classes of protozoa.

DOWN

2. Causes a disease that resembles infectious mononucleosis.
3. *Triatoma* species live in the _____ walls of South American huts.
4. Can be infected by *Toxoplasma*.
6. Body organ infected by *Acanthamoeba* species.
7. Ciliated protozoan that causes intestinal illness.
8. Genus of arthropod that transmits *T. cruzi*.
10. Possible complication of trypanosomiasis.
12. Sporozoa display _____ evidence of locomotion in the adult form.
13. Mastigophora species with two nuclei and eight flagella.
14. Symptom present in persons with *Pneumocystis* infection.
16. Causes a lung disease that is common in AIDS patients.
18. Cell (abbr) in which *Plasmodium* undergoes its life cycle.
21. Sporozoan (intls) that induces choleralike diarrhea.
24. Trichomoniasis is associated with an elevated vaginal _____.
26. State (abbr) where babesiosis is known to occur.
27. Fly that transmits protozoan that causes trypanosomiasis.
28. Stage in the life cycle of *Plasmodium*.
29. Flagellated protozoan that causes kala-azar and Oriental sore.
30. Class to which amebas belong.
34. Number of recognized forms of trypanosomiasis.
35. Serious human disease complicated by protozoal infection.
38. Protozoan (intls) that causes a life-threatening lung disease.
39. Rare hamburger _____ can be a mode of transmission for *Toxoplasma*.
43. Resistant form present in *Entamoeba* and *Giardia*.
44. Environment from which amebic cysts can be obtained.
45. The motion of *Giardia* resembles that of a falling _____.

CHAPTER 16

THE MULTICELLULAR PARASITES

O n February 13, 1984, eighteen students from the United States arrived in Kenya for various purposes of study. The weather was hot and humid, so the students decided to improvise a small swimming pool. Carefully they placed rocks and branches across a small stream and within hours, they had their pool. The water was cool and refreshing, and they congratulated themselves on their ingenuity.

But soon a problem developed. Several students broke out with an itchy rash shortly after emerging from the water, and subsequently, 14 of the group became acutely ill with fever, diarrhea, malaise, and weight loss. Two of the students developed paralysis of the lower extremities and had to return home for treatment. Stool examinations revealed the eggs of *Schistosoma mansoni*, a parasitic worm. Apparently the water had been contaminated. What had begun as a pleasant day eventually became a nightmare.

shis-to-so'mah

Schistosomiasis is one of the diseases caused by the multicellular parasites studied in this chapter. Multicellular parasites include the flatworms and roundworms that probably infect more people on a global basis than any other group of organisms. They range in size from the tiny flukes, which must be studied with a microscope, to the tapeworms, which sometimes reach 20 feet in length. In the strict sense, flatworms and roundworms are animals, but they are also studied in microbiology because of their small size and relationship to disease. Together with the protozoa discussed in Chapter 15, they are the subject of study of the biological discipline known as parasitology.

shis"to-so-mi'a-sis

Our review of the multicellular parasites will be a brief one. We shall include descriptions of the parasites, their life cycles, and the types of organisms and tissues they infect. You may note that the diseases in this chapter have few unique symptoms and are characterized by an abundance of parasites in a particular area of the body. Often the body will tolerate the parasites until the worm burden becomes immense. At that point, interference with an organ's function develops, and disease symptoms follow.

Flatworms

plat″e-hel-min′thēz

Flatworms belong to the animal phylum Platyhelminthes, a name derived from the Greek terms *platy-* for flat and *helminth* for worm. All the parasites in this phylum have flattened bodies that are slender and broadly leaflike, or long and ribbonlike (Figure 16.1). The animals exhibit **bilateral symmetry**, meaning that when cut in the longitudinal plane, the division yields identical halves.

As multicellular animals, flatworms have tissues functioning as organs in organ systems. Many species have a gut consisting of a sac with a single opening. Complex reproductive systems are found in many animals of the group, and a large number of species have both male and female reproductive organs. These organisms are termed **hermaphroditic**.

her-maf′ro-dit″ik
tur″be-la′re-ah

trem″ah-to′dah

Flatworms are divided into three classes. The first class, Turbellaria, includes free-living flatworms that do not cause disease and consequently, will not receive our attention. The common planarian studied in general biology programs is a typical member of the class. The second class, Trematoda, includes the flukes. We shall begin our coverage with this group of parasites. The third class, Cestoda, consists of tapeworms. These parasites will be studied presently.

General Description of Flukes

Fluke:
a leaflike parasite of the class Trematoda.

Flukes are leaflike parasitic worms of the class **Trematoda**. Generally flukes have complex life cycles that may include encysted egg stages and temporary larval forms. Sucker devices are commonly present to allow the parasite to hold fast to its host. In many cases more than one host exists: an **intermediate host**, which harbors the larval form, and a **definitive host**, or final host, which harbors the sexually mature adult form. In this chapter, we shall be concerned with parasites whose definitive host is the human being.

Miracidia:
tiny, ciliated larvae of flukes.

Cercariae:
tadpolelike stages in the life cycles of flukes.

Flukes may inhabit various parts of the human body including the blood, lung, liver, and intestine. The life cycle of a fluke often contains several phases. In the human host, the parasite produces fertilized eggs generally released in the feces. When the eggs reach water, they hatch and develop into tiny ciliated larvae called **miracidia** (sing., miracidium). The miracidia penetrate **snails** (the intermediate host) and go through a series of asexual reproductive stages, often including **sporocyst** and **redia** stages. Rediae become tadpolelike **cercariae**, which are released to the water. Now the cercariae develop to encysted forms called **metacercariae**, which make their way back to humans. We shall see variations on this basic life cycle in the discussions of fluke diseases to follow.

Blood Fluke Disease

shis-to-so′mah
jah-pon′ĭ-kum
hēm″ah-tōb′e-um

bil″har-zi′ah-sis

Three important species of flukes are able to invade the bloodstream in humans: ***Schistosoma mansoni***, **S. *japonicum***, and **S. *haematobium***. The first is distributed throughout Africa and South America, the second in the Far East, and the third mainly in Africa. The WHO estimates that 250 million people worldwide are infected with *Schistosoma*, including about 400,000 individuals in the United States. The disease is called **schistosomiasis**. In some regions, the term **bilharziasis** is still used from the older name for the genus, *Bilharzia*.

Species of *Schistosoma* measure about 10 mm in length. Male and female species mate in the human liver and produce eggs that are released in the feces (Figure 16.2). The eggs hatch to miracidia in water, and the miracidia make their way to snails, where conversions to sporocysts and cercariae take place.

FIGURE 16.1

Examples of Two Flatworms

(a) A scanning electron micrograph of the fluke *Fasciola hepatica*. Note the flattened, broad, and leaflike shape of this parasite. *F. hepatica* thrives in the liver of humans. The ventral and oral suckers used for attachment to the tissue can be seen. (b) An unmagnified view of the tapeworm *Taenia saginata*. This flatworm is long and ribbonlike, and consists of hundreds of visible segments. *T. saginata* infects the human intestine.

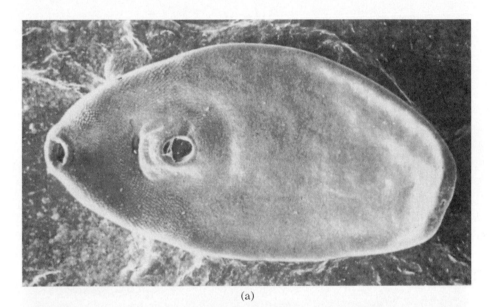

(a)

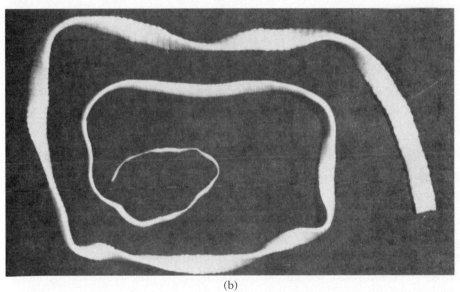

(b)

The cercariae escape from the snails and attach themselves to the bare skin of humans. Cercariae become young schistosomes, which infect the bloodstream and cause fever and chills. The major effects of disease are due to eggs formed in the liver. Liver damage may be substantial. Eggs may also gather in the intestinal wall causing ulceration, diarrhea, and abdominal pain. Bladder infection may be signaled by bloody urine and pain on urination. Outbreaks may have substantial consequences (MicroFocus: 16.1).

SCHISTOSOMA MANSONI

FIGURE 16.2

Life Cycle of the Blood Fluke, *Schistosoma mansoni*

(a) Eggs of the parasite are deposited in water in human feces. (b) Miracidia escape from the eggs and swim to snails.
(c) Here they develop to cercariae. (d) The cercariae penetrate the skin of an individual who walks in the water and
(e) pass into the bloodstream, where infection is established.

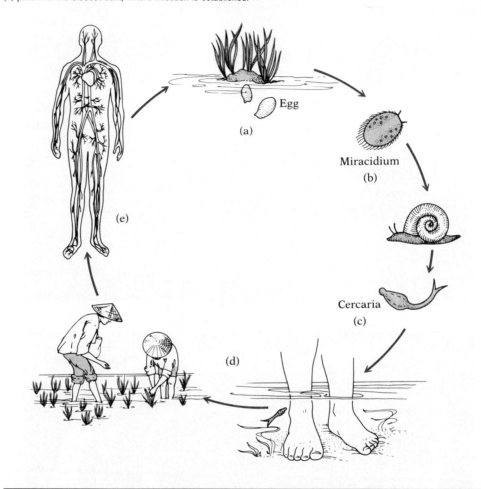

Certain species of *Schistosoma* penetrate no farther than the skin because the definitive hosts are birds instead of humans. The cercariae of these schistosomes cause dermatitis in the skin and a condition commonly called **swimmer's itch**. After penetration, the cercariae are attacked and destroyed by elements of the body's immune system, but they release allergenic substances that cause the itching and body rash (Figure 16.3). The condition, not a serious threat to health, is common in northern lakes in the United States.

Swimmer's itch:
a skin condition caused by
allergic substances released
by *Schistosoma* species.

Chinese Liver Fluke Disease

The **Chinese liver fluke** is so-named because it infects the liver and is common in many regions of the Orient, especially China, southern Asia, and Japan. The organism is named ***Clonorchis sinensis***. It has oral and ventral sucker devices (Figure 16.4) and is hermaphroditic, with a complex reproductive system.

klo-nor′kis si-nen′sis

FIGURE 16.3
An Outbreak of Schistosomiasis in Delaware

1. On October 19, 1991, 37 students from a local high school biology class visited a shellfishing area at Cape Henlopen State Park in Delaware. The tide was low, the water was calm, and the weather was sunny and unseasonably warm.

2. The students and their teacher spent about two hours wading in the water collecting specimens for examination at school. There were many clams, oysters, and snails in the water, as well as other marine specimens of interest. Several ducks and geese were nearby.

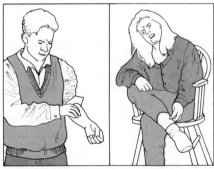

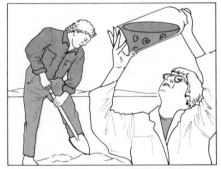

3. Within ten days after the field trip, 29 of the 37 students developed pruritic dermatitis, an itching skin condition. Eleven of the affected students visited their doctors for treatment. Eventually, 36 of the 37 individuals developed pruritis; the only unaffected individual was one who did not enter the water.

4. Public health investigators examined snails from the sea water for signs of schistosomes. Schistosomes normally associated with ducks, geese, and other birds were found in a small percentage of snails examined. However, it is known that thousands of schistosomes can be released from a single snail, and the release is favored by weather conditions seen that day.

Eggs of *C. sinensis* contain complete miracidia, which emerge after the eggs enter water. The miracidia penetrate snails and change into sporocysts, which then produce a generation of elongated rediae. These rediae become cercariae, which escape from the snail and bore into the muscles of fish, the second intermediary host. Now the cercariae develop into encysted metacercariae, and humans acquire the metacercariae by consuming raw or poorly cooked fish. Public health officials recommend that fish be heated to a minimum of 50° C for 15 minutes to destroy the cysts. In humans, the metacercariae become adults and migrate from the intestine up the bile duct to the gall bladder and liver, where infection takes place.

CLONORCHIS SINENSIS

FIGURE 16.4

The Chinese Liver Fluke

Scanning electron microscope views of *clonorchis sinensis*, the Chinese liver fluke. (a) A whole mount showing the anterior aspect of the parasite (× 2000). (b) The oral sucker device (× 6000). (c) The ventral sucker device (× 4000).

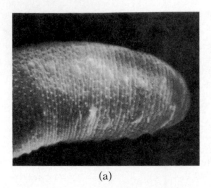

(a)

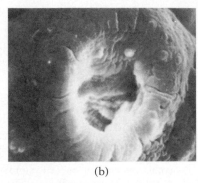

(b)

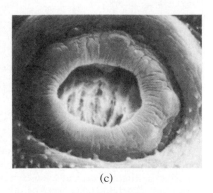

(c)

Gall bladder:
the pouch on the underside
of the liver that stores bile.

The effect of *C. sinensis* on humans depends on the extent of infection. Substantial infection in the gall bladder may lead to duct blockage and poor digestion of fats. There may also be damage to the liver as eggs accumulate in the tissues of this organ (Figure 16.5). Often the patient is without symptoms due to the low number of parasites. Cats, dogs, and pigs may also carry the metacercaria cysts.

Other Fluke Diseases

fas"e-o-lop'sis boo'ski

FASCIOLOPSIS BUSKI

par"ah-gon'ĭ-mus
wes'ter-man-i

PARAGONIMUS WESTERMANI

fah-si-o'lah

FASCIOLA HEPATICA

The **intestinal fluke** of humans is known as ***Fasciolopsis buski***. This is the largest fluke that infects humans, with a length reaching 8 cm. The parasite lives in the duodenum, where it causes diarrhea and intestinal blockage. From the feces, eggs enter water in such places as rice paddies and drainage ditches, and miracidia emerge. Snails are the next host for the miracidia, and cercariae escape from the snail and swim to blades of grass and vegetation where metacercariae form. Such plants as water chestnuts and water bamboo are likely to be contaminated. Consumption of these vegetables raw or poorly cooked leads to human infection.

The human **lung fluke** is ***Paragonimus westermani***. This parasite is common in the Orient and South Pacific. Its eggs are coughed up from the lung, swallowed, and then excreted in the feces. Cercariae develop in snails and metacercariae eventually form in crabs. Infection follows when people eat the poorly cooked crab meat, and the flukes pass from the intestine to the blood to the lungs. Difficult breathing and chronic cough develop as the parasites accumulate in the lung. Fatalities are possible.

The human **liver fluke** is ***Fasciola hepatica***, a leaflike flatworm common in sheep and cattle. Eggs from the animal reach the soil in feces, and if snails are present, the conversions to cercariae and encysted metacercariae follow. Parasite cysts gather on vegetation such as watercress, and ingestion by humans takes place. The parasites penetrate the intestinal wall and migrate to the liver, where tissue damage may be substantial, especially if fluke numbers are high.

FIGURE 16.5

Life Cycle of the Chinese Liver Fluke, *Clonorchis sinensis*

(a) Eggs are deposited in water in human feces. (b) Miracidia emerge from egg cases and penetrate snails. (c) Within the snail the miracidia change into sporocysts, then rediae, and finally cercariae. (d) Cercariae escape from the snail and bore into the muscles of fish, (e) where they form encysted metacercariae. (f) When fish is eaten raw or poorly cooked, the metacercariae enter the human intestine, hatch to adults, and (g) invade the liver, gall bladder, and bile duct. The center photograph shows eggs of *C. sinensis* in human tissue. The thick protective covering of the eggs is evident.

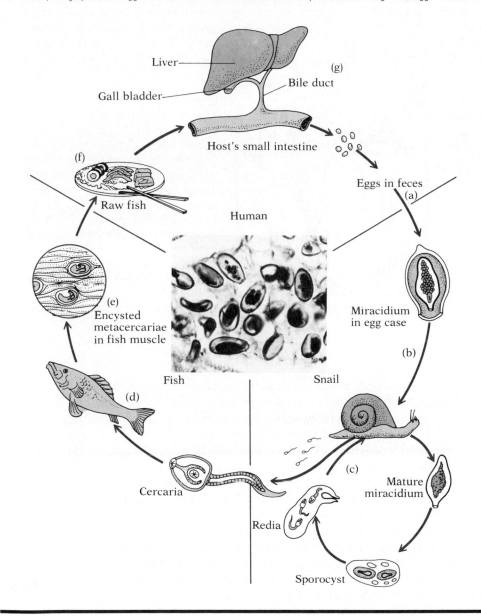

General Description of Tapeworms

Tapeworms belong to the third class of flatworms, **Cestoda**. These worms have long, flat bodies consisting of a head region and a ribbonlike series of segments called **proglottids**. The head region, called the **scolex**, contains hooks or suckerlike devices to allow the worm to hold fast to infected tissue. Behind the

Scolex:
the head region of a
tapeworm.

Gravid proglottids:
segments filled with
fertilized eggs at the distant
end of a tapeworm.

scolex there is a neck region in which new proglottids are formed. These new proglottids constantly push the more mature proglottids to the rear. The most distant ones, called **gravid proglottids**, are filled with fertilized eggs. As they break free, they spread the eggs of the tapeworm.

Tapeworms generally live in the intestine of a host organism (MicroFocus: 16.2). In this environment they are constantly bathed by nutrient-rich fluid from which they absorb their food. Tapeworms have adapted to a parasitic existence and have lost their intestines, but they still retain well-developed muscular, excretory, and nervous systems.

Tapeworms are widespread parasites that infect practically all mammals as well as many other vertebrates. Since they are more dependent on their hosts than flukes, tapeworms have precarious life cycles. Tapeworms have a limited range of hosts, and the chances for completing the cycle are often slim. With rare exceptions, tapeworms require at least two hosts, as we shall see in the following examples of tapeworm diseases.

Beef and Pork Tapeworm Diseases

tā′ne-ah
saj-in-ah′tah
so′le-um

Humans are the definitive hosts for both the **beef tapeworm** *Taenia saginata* and the **pork tapeworm** *Taenia solium*. When people are infected with one of these tapeworms, they expel numerous gravid proglottids daily. The proglottids accumulate in the soil and are consumed by cattle or pigs, respectively. Embryos from the eggs travel to the animal's muscle, where they encyst. Humans then acquire the cysts in poorly cooked beef or pork.

The beef tapeworm may reach 25 feet in length, while the pork tapeworm averages 20 feet long. Each tapeworm may have up to 2000 proglottids. Attachment via the scolex occurs in the intestine, and obstruction of this organ may be experienced. In most cases, however, there are few symptoms other than mild diarrhea, and a mutual tolerance may develop between parasite and host. The notion that a tapeworm causes severe emaciation is largely unfounded.

TAENIA SAGINATA

Fish Tapeworm Disease

The **fish tapeworm** is the longest parasite to infect humans, some species measuring 60 feet in length. The parasite is named ***Diphyllobothrium latum***. Its life cycle is complex and includes two intermediate hosts: a small shrimplike crustacean called a copepod and a fish. The copepod acquires tapeworm embryos from proglottids excreted into water by humans. Next the copepod is eaten by fish such as minnows, and the embryo finds its way to the fish muscle. Minnows may be eaten by larger fish such as trout, perch, and pike, and the embryos are passed along (Figure 16.6). Consumption of raw or poorly cooked fish by humans completes the cycle.

di-fil″o-both′re-um

Copepod:
a small, shrimplike animal common in marine environments.

Infections with fish tapeworms occur throughout the world. In the United States, infections are most common in the Great Lakes region. Obstruction of the small intestine may occur, and anemia may develop in the patient, possibly due to the parasite's consumption of vitamin B_{12} needed for red blood cell formation. Cooking at 50° C for 15 minutes is generally sufficient to destroy the cysts, but the recent popularity of raw fish dishes has contributed to a rise in incidence of the disease.

DIPHYLLOBOTHRIUM LATUM

Other Tapeworm Diseases

The **dwarf tapeworm** *Hymenolepis nana* is so named because it is only 25 mm in length. The most common tapeworm in humans worldwide, *H. nana* lives in the intestine, where it holds fast with four sucker devices and a row of small hooks. Eggs released in the feces spread among humans by contaminated

hi″mĕ-nol′ĕ-pis

HYMENOLEPIS NANA

FIGURE 16.6

Life Cycle of the Fish Tapeworm, *Diphyllobothrium latum*

(a) Gravid proglottids are released from the human intestine into water. (b) Eggs emerge from the proglottid, and (c) a ciliated embryo of the tapeworm develops from each egg. (d) The embryo is ingested by copepods, (e) which in turn are ingested by fish. (f) Consumption of the fish returns the parasite to the human intestine. The center photographs shows *D. latum* isolated from a human patient.

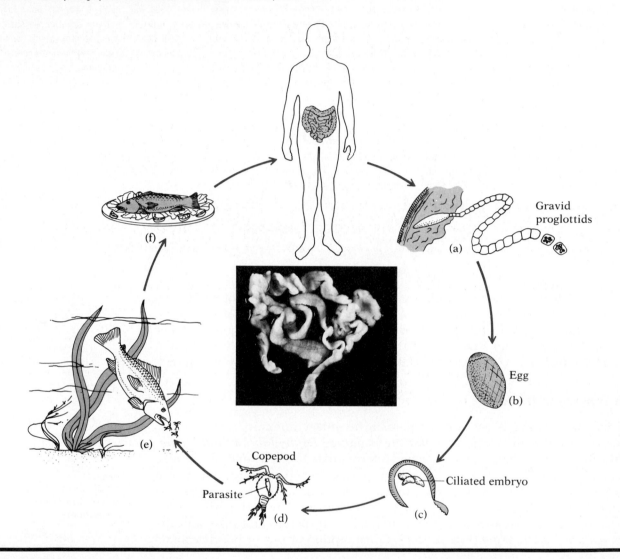

food or contact with objects or an infected individual. People living in the southeastern United States often experience infections.

Dogs and other canines such as wolves, foxes, and coyotes are the definitive hosts for the **dog tapeworm** called ***Echinococcus granulosus*** (Figure 16.7). Eggs reach the soil in feces and spread to numerous intermediary hosts, one of which is humans. Contact with a dog may also account for transmission. In humans, the parasites travel by the blood to the liver, where they form thick-walled **hydatid cysts.** Surgery may be necessary for their removal. Completion of the life cycle takes place when a canine consumes infected animal liver, such as after a kill. Dog food is another possible source of cysts. The adult tapeworm then emerges to parasitize the dog. Flatworm diseases of humans are summarized in Table 16.1.

e-ki″no-kok′us gran-u-lo′sis

ECHINOCOCCUS GRANULOSUS

Hydatid cyst:
a thick-walled liver cyst formed by dog tapeworms.

FIGURE 16.7

The Dog Tapeworm

Two views of the dog tapeworm *Echinococcus granulosus* as seen by phase-contrast microscopy. (a) The head region, or scolex of *E. granulosus*. The hooks at the end of the scolex are used to attach to the infected tissue; the indentations at the base of the scolex emphasize its presence. (b) The hydatid cyst of *E. granulosus* isolated from a lung resection of a 31-year-old woman immigrant from the Sudan. Both magnifications × 580.

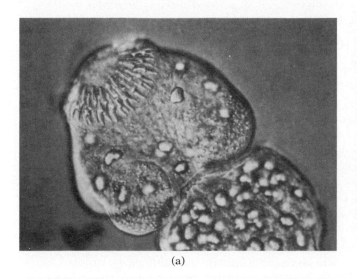

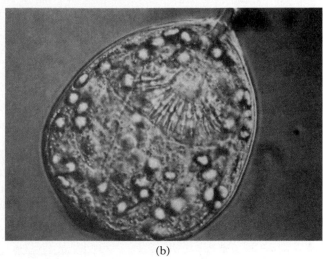

(a)   (b)

TABLE 16.1

A Summary of Flatworm Parasite Diseases in Humans

ORGANISM	DISEASE	TRANSMISSION	ORGANS AFFECTED	CHARACTERISTIC SIGN	ANIMAL HOST
Schistosoma mansoni *S. japonicum* *S. haematobium*	Blood fluke disease	Water contact	Skin Liver Blood	Rash Liver damage Fever	Snail
Clonorchis sinensis	Chinese liver fluke disease	Fish consumption	Gall bladder Liver	Liver damage Poor fat digestion	Snail Fish
Fasciolopsis buski	Intestinal fluke disease	Consumption of water plants	Intestine	Diarrhea	Snail
Paragonimus westermani	Lung fluke disease	Consumption of crabs	Lung	Cough Poor breathing	Snail Crab
Fasciola hepatica	Liver fluke disease	Consumption of water plants	Liver	Liver damage	Snail Cattle
Taenia saginata	Beef tapeworm disease	Beef consumption	Intestine	Diarrhea	Cattle
Taenia solium	Pork tapeworm disease	Pork consumption	Intestine	Diarrhea	Pig
Diphyllobothrium latum	Fish tapeworm disease	Fish consumption	Intestine	Diarrhea Anemia	Copepod Fish
Hymenolepis nana	Dwarf tapeworm disease	Food Contact	Intestine	Diarrhea	None significant
Echinococcus granulosus	Dog tapeworm disease	Contact	Liver	Liver damage	Dog, other canines

TO THIS POINT
■

We have surveyed many of the flatworm parasites that infect human beings. The flatworms include flukes, which are slender and broadly leaflike, and tapeworms, which are long and ribbonlike. Flukes belong to the class Trematoda. They have complex life cycles, with snails serving as an intermediate host. Stages in the life cycle include the miracidium, cercaria, and metacercaria, among others. The blood fluke is acquired by contact with water, while the Chinese liver fluke is ingested in raw or poorly cooked fish. Intestinal flukes and lung flukes are also acquired in foods.

We then discussed the tapeworms of the class Cestoda. These parasites generally live in the human intestine. Proglottids carry tapeworm eggs to the soil or water, from which they are consumed by animals or fish. Human infection is reestablished when contaminated meat or fish is eaten. Beef, pork, and fish tapeworm diseases are passed along in this way. Contaminated food or objects may be the source of dwarf tapeworm disease, and dogs may pass dog tapeworms to humans. In dog tapeworm disease, humans are an intermediary host.

We shall now focus on the roundworms. Roundworms are anatomically more complex than flatworms and are classified in a completely different phylum of animals. We shall see how the life cycles of roundworms are considerably more simple than the cycles of flatworms, and how infection is spread much more easily. Among the roundworms are those that cause pinworm disease and trichinosis, two diseases that are common in the United States.

Roundworms

Roundworms occupy every imaginable habitat on Earth. They live in the sea, in fresh water, and in soil from polar regions to the tropics. Good topsoil, for example, may contain billions of roundworms per acre. They parasitize every conceivable type of animal and plant, causing both economic damage and serious disease.

ask"hel-min'thēz

The roundworms are a subgroup of the phylum **Aschelminthes**, from the Greek stems *asc-* for sac and *helminth* for worm. The sac refers to a digestive tract set apart from the internal muscles in a saclike or pouchlike arrangement, a feature not found in flatworms. In reality, the sac is a tubular intestine open at the mouth and anus. Food can thus move in one direction, a substantial improvement over the blind sac arrangement in Platyhelminthes. This is one reason why the Aschelminthes are considered to be more evolutionarily advanced than the Platyhelminthes.

plat"e-hel-min'thēz

Roundworms have separate sexes. Following fertilization of the female by the male, the eggs hatch to larvae that resemble miniature adults. Growth then occurs by cellular enlargement and mitosis. Damage in hosts is generally caused by large worm burdens in the intestine, blood vessels, or lymphatic vessels (Figure 16.8). Also, the infestation may result in nutritional deficiency or damage to the muscles.

Nematode:
an alternate expression for a roundworm.

Roundworms have been traditionally known as **nematodes** because they are threadlike (*nema* is Latin for thread). Indeed, in some texts the phylum

FIGURE 16.8
A Typical Roundworm

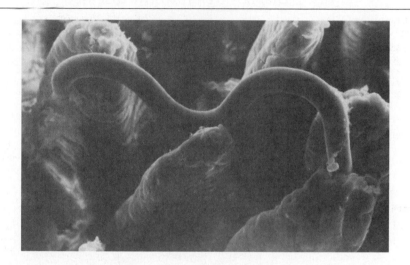

A scanning electron micrograph of the roundworm *Trichinella spiralis* in human intestinal tissue. This parasite is the cause of trichinosis. In the photograph, the worm is emerging from one intestinal villus and entering another villus.

of roundworms is called Nematoda, and in other books it is referred to as Nemathelminthes.

Pinworm Disease

The widely encountered **pinworm** is a roundworm called ***Enterobius vermicularis***. This worm is considered the most common helminthic parasite in the United States, with an estimated 30 percent of children and 16 percent of adults serving as hosts. The male and female worms live in the distant part of the small intestine and in the large intestine, where the symptoms of infection include diarrhea and itching in the anal region. Females are about 10 mm in length, and the male about half that size.

en"ter-o'be-us ver"mik-u-la'ris

The life cycle of the pinworm is relatively simple. Females migrate to the anal region at night and lay a considerable number of eggs. The area itches intensely, and scratching contaminates the hands and bed linens with eggs. Reinfection may then take place if the hands are brought to the mouth or if eggs are deposited in foods by the hands. The eggs are swallowed, whereupon they hatch in the duodenum and mature in the regions beyond.

ENTEROBIUS VERMICULARIS

Diagnosis of pinworm disease may be accurately made by applying the sticky side of cellophane tape to the area about the anus and examining the tape microscopically for pinworm eggs (Figure 16.9). Several drugs are effective for controlling the disease, and all members of an infected person's family should be treated because transfer of the parasite has probably taken place. Even without medication, however, the worms will die in a few weeks, and the infection will disappear so long as reinfection is prevented.

Whipworm Disease

The **whipworm *Trichuris trichiura*** acquired its name from the observation that its anterior end is long and slender like a buggy whip. Infection takes place in the human intestine, especially near the junction of the small and large

trik-u'ris trik-e-u'rah

FIGURE 16.9

Diagnosing Pinworm Disease

The transparent tape technique used in the diagnosis of pinworm disease. (a) Clear plastic tape is pulled back over the end of the slide to expose the gummed surface. (b) The tape, still attached to the slide, is looped over a wooden stick. (c) Now the gummed surface of the tape is touched several times to the anal region. (d) The tape is then replaced on the slide, and (e) smoothed down with cotton or gauze. The slide is placed under the microscope and examined for pinworm eggs.

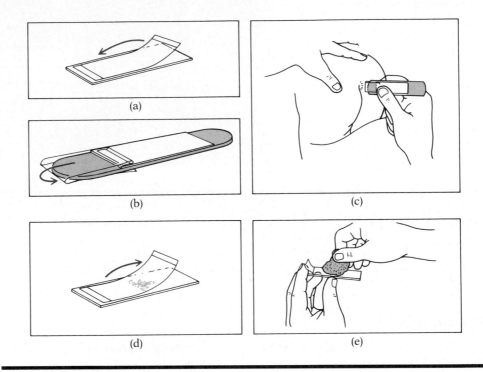

(a)

(b)

(c)

(d)

(e)

TRICHURIS TRICHIURA

as'kah-ris
lum"brĭ-koid'ēz

intestines as shown in Figure 16.10. Damage to the intestinal lining may be severe, and appendicitislike pain is sometimes experienced, with some anemia resulting from ingestion of blood by the parasite.

The female whipworm is approximately 40 mm long; the male is shorter with a characteristically curled tail. Eggs eliminated in the human feces hatch to larvae after about two weeks in the soil. Transmission occurs by soil-contaminated food and water as well as by contact with soiled hands. Whipworm disease is generally encountered where the environment is hot and moist, such as in the tropics, and where poor sanitary facilities exist. Patients often have concurrent infections with other parasites. Diagnosis generally depends on the identification of eggs in the feces.

Roundworm Disease

Infection with "roundworms" usually implies infection with **Ascaris lumbricoides**. One of the largest of the intestinal nematodes, the female *A. lumbricoides* may be up to a foot long, and the male, 8 inches long. The parasite resembles an earthworm and is the most wormlike of the helminthic parasites.

A female *Ascaris* is a prolific producer of eggs, sometimes generating over 200,000 per day. The eggs are fertilized and passed to the soil in the feces, where they hatch to larvae. The larvae then attach to plants and are ingested. In many parts of the world, human feces, or nightsoil, is used as fertilizer for crops.

FIGURE 16.10

The Whipworm

(a) A view of whipworm, *Trichuris trichiura*, in human intestinal tissue. The worms have a relatively small size compared to other roundworms discussed in this chapter. Whipworms have a long slender form resembling a buggy whip (arrows). A close-up view is shown in part (b).

(a)   (b)

This adds to the spread of the parasite. Contact with contaminated fingers and consumption of water containing soil run-off are other possible modes of transmission.

After the larvae have been consumed, the tiny worms grow in the intestine. Abdominal symptoms develop as the worms reach maturity in about two months. Intestinal blockage may be a consequence when tightly compacted masses of worms accumulate, and perforation of the intestine is possible. In addition, roundworm larvae may pass to the blood and infect the lungs, causing pneumonia. If the larvae are coughed up and then swallowed, reinfection of the intestine occurs.

Except for pinworms, *A. lumbricoides* is the most prevalent multicellular parasite in the United States. The WHO estimates that hundreds of millions are infected worldwide. Tropical and subtropical regions are the primary foci of disease, but areas of the southwestern United States are heavily infested because eggs remain viable in the moist clay soil of this region.

Pneumonia:
disease of the lung tissues.

Trichinosis

Trichinosis is a term familiar to anyone who enjoys pork and pork products, because packages of pork usually contain warnings to cook the meat thoroughly to avoid this disease. Ironically, trichinosis is common where living standards are high enough for pork to be eaten routinely, but the disease is rare where pork is a luxury.

Trichinosis is caused by the small roundworm, ***Trichinella spiralis***. The worm lives in the intestines of pigs and several other mammals. Larvae of the worm migrate through the blood and penetrate the pig's skeletal muscles where they remain in cysts (Figure 16.11). When raw or **poorly cooked pork** is consumed, the cysts pass into the human intestine and the worms emerge. Intestinal pain, vomiting, nausea, and constipation are common symptoms.

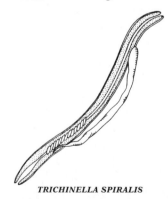

TRICHINELLA SPIRALIS

trik"ĭ-nel'ah spir-al'is

FIGURE 16.11
Life Cycle of *Trichinella spiralis*

(a) Parasite cysts enter the environment in human feces or waste. (b) They are acquired by pigs that feed in the soil. (c) The larvae emerge from the cysts in the pig's intestine and (d) migrate to the muscle where they form cysts. (e) Consumption of poorly cooked pork results in human infection, initially in the intestine and later in the muscles (f).

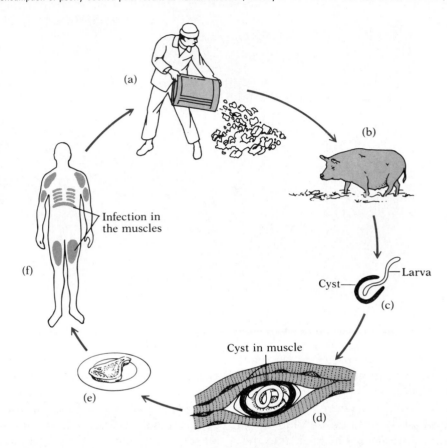

ASCARIS LUMBRICOIDES

Cyst:
a dormant, highly resistant form of an organism such as a protozoan or multicellular parasite.

Complications of trichinosis occur when *T. spiralis* larvae migrate to the muscles and form cysts. The patient commonly experiences pain in the breathing muscles of the ribs, loss of eye movement due to cyst formation in the eye muscles, swelling of the face, and hemorrhaging in various body tissues. Some victims also develop cough and skin lesions, and the larvae may invade the brain, where they cause paralysis. Paralysis and death occurred in a man in New York in 1985 after he contracted trichinosis by consuming pork inadvertently ground with beef. The man had a habit of nibbling the meat while preparing hamburgers.

The cycle of trichinosis is completed as cysts are transmitted back to nature in the human feces. Consumption of human waste and garbage then brings the cysts to the pig. Modern methods of agriculture provide standardized feed for pigs, but many pigs are still exposed to the cysts. The consumer should be aware that poorly cooked pork is the principal source of the approximate 150 cases of trichinosis that are reported in the United States annually. Pickling, smoking, and heavy seasoning are not adequate substitutes for thorough heating, but freezing greatly reduces larval viability. Routine inspection of pigs for *T. spiralis* cysts is not common practice in slaughterhouses, and thus the burden for prevention of trichinosis falls to the consumer. In 1986, the U.S. government approved the use of low-dose irradiation of fresh pork to combat trichinosis.

Hookworm Disease

Hookworms are roundworms having a set of hooks or sucker devices for firm attachment to tissues of the host. Two hookworms, both about 10 mm in length, may be involved in human disease: the first is the **Old World hookworm**, *Ancylostoma duodenale*, which is found in Europe, Asia, and the United States; the second is the **New World hookworm**, *Necator americanus*, which is prevalent in the Caribbean islands (where it may have been brought by slaves from Africa).

an″kĭ-los′to-mah
du-od-in-al′e
ne-ka′tor a-mer-i-ca′nus

Hundreds of millions of people around the globe are believed infected by hookworms. These parasites live in the human intestine, where they suck blood from the tissues. Hookworm disease is therefore accompanied by blood loss and is generally manifested by anemia. Cysts may also become lodged in the intestinal wall, and ulcerlike symptoms may be experienced.

The life cycle of a hookworm involves only a single host, the human. Hookworm eggs are excreted to the soil, where the larvae emerge as long, rodlike **rhabditiform** larvae. These later become threadlike **filariform** larvae that attach themselves to vegetation in the soil. When contact with bare feet is made, the filariform larvae penetrate the skin layers and enter the bloodstream (Figure 16.12). Soon they localize in the lungs and are carried up to the pharynx in secretions, then swallowed to the intestine.

NECATOR AMERICANUS

rab-dit′ĭ-form

Hookworms are common where the soil is warm, wet, and contaminated with human feces. The disease is prevalent where people wear no shoes or protective foot covering. Drugs may be used to reduce the worm burden, and the diet may be supplemented with iron to replace that lost in the loss of blood. It should be noted that dogs and cats also harbor hookworm eggs and pass them in the feces.

Strongyloidiasis

Strongyloidiasis is caused by *Strongyloides stercoralis*, a parasite that resembles hookworms in appearance, distribution, and life cycle. Adult worms inhabit the small intestine, especially the duodenum, where they cause abdominal pains, nausea, vomiting, and diarrhea alternating with constipation. Pulmonary symptoms mimic the pneumonia induced by hookworms. A drug called thiabendazole provides effective therapy.

stron″jĭ-loi-di′ah-sis
stron″ji-loi′dez ster-ko-ral′is

Strongyloidiasis is of importance to Americans because many Vietnam veterans were exposed to the parasites in Southeast Asia. In a 1981 study, for example, doctors tested 530 veterans of Pacific wars for the parasite and found that 43 harbored it in their stools. In some cases this was almost 40 years after the initial infection. Many reported regular five-day episodes of itchy skin rash, another sign of the disease.

STRONGYLOIDES STERCORALIS

Filariasis

Filariasis is a parasitic disease caused by a roundworm named *Wuchereria bancrofti*. The worm breeds in the tissues of the human lymphatic system and causes extensive inflammation and damage to the lymphatic vessels and lymph glands. After years of infestation, the arms, legs, and scrotum swell enormously and become distorted with fluid. This condition is known as **elephantiasis** because of the gross deformity of tissues and the resemblance of the skin to elephant hide (Figure 16.13).

fil″ah-ri′ah-sis
voo″ker-e′re-ah ban-krof′ti

The female form of *Wuchereria bancrofti* is about 100 mm long. Its fertilized eggs give rise to tiny eel-like microfilariae, which enter the human bloodstream. The microfilariae are then ingested by **mosquitoes** during a blood meal, where they develop into infective larvae passed along to another human during the

WUCHERERIA BANCROFTI

FIGURE 16.12

Hookworms

Life cycle of the hookworms *Ancylostoma duodenale* and *Necator americanus*. (a) Eggs pass from the human intestine into the soil, (b) where larvae emerge as rodlike rhabditiform larvae. (c) They revert to threadlike filariform larvae that (d) contact the skin when one walks in the soil and pass into the tissues. (e) The larvae move through the bloodstream to the lungs, from which they are coughed up and swallowed. (f) Infection then occurs in the intestine. The inset shows an egg lodged in intestinal tissue.

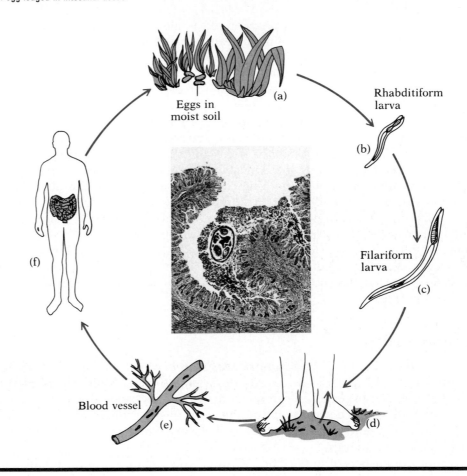

next blood meal. The larvae subsequently grow to adults and infect the lymphatic system. Infection is limited to the number of larvae injected by the mosquito, since microfilariae cannot grow to adulthood without first passing through the mosquito.

Filariasis is prevalent where mosquitoes are plentiful, such as in the hot, humid climates of Central and South America and the Caribbean islands. Missionaries, emigrants, and visitors from these areas often carry the worms in their tissues.

Guinea Worm Disease

The **Guinea worm** is a roundworm, thought to have been introduced to Africa, India, and the Middle East by navigators who plied the sealanes between these areas and South Pacific islands such as New Guinea. The worm's scientific name is ***Dracunculus medinensis***.

drah-kung'ku-lus
med-i-nen'sis

FIGURE 16.13

The Effect of Filariasis

Elephantiasis of the leg caused by the parasite *Wuchereria bancrofti*. The worm breeds in the tissues of the lymphatic vessels and causes damage to the vessels. As fluid accumulates, the legs swell and become distorted.

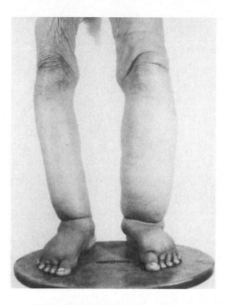

The male Guinea worm is small, but the female may be up to 80 cm in length. Often the worm lies just below the skin of humans and causes swellings that resemble varicose veins. It causes a skin ulcer through which larvae are discharged into water. A small shrimplike crustacean called a **copepod** picks up the larvae and transmits them to another human when the copepod is consumed in water. Alternately, the copepod is eaten by fish, which then transmit the larvae (MicroFocus: 16.3). Once in humans, the larvae migrate to the human skin and grow to adults.

The Guinea worm exposes itself through the skin ulcer and thus can be removed by careful winding on a stick. This primitive but time-honored method must be performed cautiously and slowly, and it may take several weeks. The site of the ulcer burns intensely, and the fever and local burning are often described as "fiery." Partly because of this perception, it is believed that the "fiery serpents" mentioned in the Bible's Book of Numbers may have been Guinea worms. Also the stick with worms wound around it may be the source of the serpent on a staff that is the symbol of healing used by the medical profession.

As of 1990, Guinea worm disease affected an estimated 5 million persons in 17 African countries and parts of India and Pakistan. By that time, eradication programs had been established in Ghana and Nigeria to interrupt the spread of the disease by providing safe sources of drinking water, by teaching populations at risk to boil or filter contaminated water, and by treating drinking water with chemicals. By 1993, four years of decline were reported in both countries, and prospects were raised for complete eradication in the near future.

DRACUNCULUS MEDINENSIS

Eyeworm Disease

The **eyeworm** is a type of roundworm called ***Loa loa***. This parasite, native to West and Central Africa, is about 60 mm long. It is often found in insects such as deerflies and horseflies and is ingested into the subcutaneous tissues

LOA LOA

WELCOME TO NEW YORK CITY!

The Guinea worm had long been a parasite among Scandinavian fishermen and their families. Individuals acquired the worm by drinking water that contained infected crustaceans or by eating fish that had previously eaten the crustaceans. For the Guinea worm, New York City appeared to be a long way off, but a somewhat circuitous route made the transition possible in the 1800s.

Scandinavians immigrated to the United States in substantial numbers during the nineteenth century. They settled along the shores of the Great Lakes and, as their feces made their way into the lakes, the Guinea worm followed along. Eventually, fish from the Great Lakes became infected with larvae of the worm.

As the years passed, substantial commerce in fish developed between the Midwest and New York City. Jewish homemakers were particularly fond of pike, pickerel, and carp from the Great Lakes, and they used the fish to make a delicacy called gefilte fish. Carefully they pressed minced fish, eggs, and seasonings into balls and boiled the mixtures. During cooking, how-ever, they often tasted the gefilte fish to see if it was done. While sampling the uncooked fish, they unwittingly acquired roundworm larvae and became new hosts for the Guinea worms. The transition was complete.

Guinea worm infection is now rare in New York City or elsewhere in the United States. Sanitary practices, fish inspection, and commercial production of gefilte fish have limited spread of the worm. Gefilte fish is still popular among Jewish people, but the unwanted hitchhikers have been largely eliminated.

of humans during an **insect bite**. The parasite may remain in the connective tissues for many months.

Loa loa is called the eyeworm because it is often attracted to the surface of the eye by warm temperatures. Commonly it appears on the cornea (Figure 16.14). Here it causes conjunctivitis and painful irritation of the eye muscles. Swellings of the extremities are also observed. The parasite returns to the subcutaneous tissues as the skin temperature cools. Table 16.2 summarizes the human diseases caused by roundworms.

FIGURE 16.14

Two Views of the Human Eye Showing Infection with *Loa loa*, the Eyeworm

(a) The worm is seen below the conjuctiva along the white of the eye. (b) The worm is being removed from the eye.

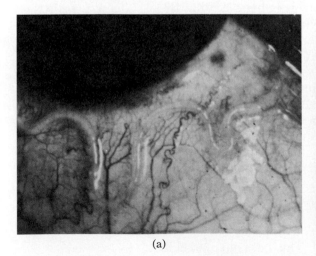

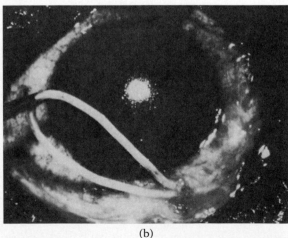

(a)   (b)

TABLE 16.2

A Summary of Roundworm Parasite Diseases in Humans

ORGANISM	DISEASE	TRANSMISSION	ORGANS AFFECTED	CHARACTERISTIC SIGN	ANIMAL HOST
Enterobius vermicularis	Pinworm disease	Contact Food, clothing	Intestine	Anal itching	None significant
Trichuris trichiura	Whipworm disease	Food Water	Intestine	Abdominal pain	None significant
Ascaris lumbricoides	Roundworm disease	Food, water Contact	Intestine Lungs	Emaciation Pneumonia	None significant
Trichinella spiralis	Trichinosis	Pork consumption	Intestine Muscles Eyes	Diarrhea Muscle pain Loss of eye movement	Pig
Ancylostoma duodenale *Necator americanus*	Hookworm disease	Contact with moist vegetation	Intestine Lungs Lymph	Anemia Abdominal pain	None significant
Strongyloides stercoralis	Strongyloidiasis	Contact with moist vegetation	Intestine Lung	Anemia Abdominal pain	None significant
Wuchereria bancrofti	Filariasis	Mosquito	Lymph vessels	Edema Elephantiasis	Mosquito
Dracunculus medinensis	Guinea worm disease	Food Water	Skin	Skin ulcer	Copepod Fish
Loa loa	Eyeworm disease	Deerflies Horseflies	Eye Connective tissues	Conjunctivitis	Insects

NOTE TO THE STUDENT

When we think of commensalism we often conjure thoughts of microorganisms living in our intestines, mouths, and other organs while causing no apparent harm. We sometimes discover that microorganisms such as certain bacteria actually benefit us.

Distasteful as it may be, we must now broaden our definition of commensalism to include certain multicellular parasites, for, indeed, many of us harbor these worms in our organ menageries. As we have seen in this chapter, infection by multicellular parasites need not be a fatal experience. Moreover, we should realize how widespread the parasites are. If estimates by the World Health Organization are correct, more people are infected with multicellular parasites than anyone previously imagined. With the emergence of Third World countries and travels by Americans to remote parts of the world, the parasites will continue to enter our society in ever-increasing numbers. They may represent an area of public health concern in the years ahead.

Summary

Multicellular parasites are worms of two general types: flatworms and roundworms. Flatworms are further subdivided as flukes and tapeworms. Flukes are leaflike worms belonging to the class Trematoda in the animal phylum Platyhelminthes. Their life cycles include several phases including miracidium, sporocyst, redia, cercaria, and metacercaria phases. Snails are intermediate hosts for flukes. The major fluke parasites of humans include the blood fluke (*Schistosoma*), the Chinese liver fluke (*Clonorchis*), the intestinal fluke (*Fasciolopsis*), the lung fluke (*Paragonimus*), and others.

Tapeworms are long, segmented worms belonging to the class Cestoda in the animal phylum Platyhelminthes. The worms consist of segments called proglottids and a head region called the scolex, often with hooks or sucker devices for attachment to the tissue. Tapeworms infect all mammals and are generally transmitted to humans in foods. Examples of foodborne tapeworms include the beef tapeworm (*Taenia*) and the fish tapeworm (*Diphyllobothrium*). Other tapeworms are acquired from the soil.

Roundworms belong to the animal phylum Aschelminthes. The life cycles of roundworms are relatively simple (compared to those of the flatworms), and include separate male and female sexes. Roundworm eggs from the soil are a common method of infection. The pinworm (*Enterobius*) is acquired this way, as are the whipworm (*Trichuris*) and the "roundworm" (*Ascaris*). An important human parasite is the pork roundworm (*Trichinella*) acquired in undercooked pork. Hookworms (*Necator* and *Ancylostoma*) are soilborne, and the filarial worm (*Wuchereria*) that causes elephantiasis is mosquitoborne. Arthropods also transmit the eyeworm (*Loa loa*).

Most of the damage due to roundworms arises from the worm burden in the tissues. Symptoms of disease generally emerge from the worm's interference with body functions, and the diseases due to multicellular parasites are worldwide in scope.

Questions for Thought and Discussion

1. A diplomat visits a foreign country with which relations have recently been established. She observes that the people eat raw fish, wear no shoes, consume snails as a regular part of their diet, enjoy watercress and water chestnuts in their salads, and do not believe in pesticides. What report might she make on her return to the United States?

2. As of 1991, the World Health Organization reported that malaria was the most prevalent tropical disease (300 million cases per year). The next two were schistosomiasis and filariasis (200 million and 90 million cases per year, respectively). How do you believe the incidence of these diseases can be reduced on a global scale?

3. Because tapeworms have no intestines, they must obtain their nutrients by absorbing organic matter from the external environment in the human or animal intestine. How does this observation dispel the notion that evolution always yields animals more complex than their predecessors?

4. Federal law now stipulates that food scraps fed to pigs must be cooked to kill any parasites present. It is also known that feedlots for swine are generally more sanitary than they have been in the past. As a result of these and other measures, the incidence of trichinosis in the United States has declined and the acceptance of "pink pork" has increased. Do you think this is a dangerous situation? Why?

5. In the 1960s, the Aswan High Dam was built in Egypt to retain the water of the Nile River for irrigation. The agricultural productivity of the region increased significantly but was accompanied by a dramatic increase in the snail population in the water. Before long, the number of cases of schistosomiasis in farmers had doubled. How are all these events related?

6. The Greek stems *di-* and *phyllo-* mean "double-thin" (filo dough is thin dough used in Greek pastry). *Bothros* is also a Greek word, meaning "dirty water." *Latum* is the Latin word for broad or wide. How does this knowledge of classical languages help in remembering the description of the parasite *Diphyllobothrium latum*?

7. In Japan, eating raw fish is a traditional and honored custom. Are there any hazards involved in this practice?

8. Why are some veterinarians inclined to recommend that dog and cat owners avoid buying pet food that contains liver?

9. A person finds that he has difficulty digesting fats. Mayonnaise makes him ill, fried foods cause diarrhea, and he cannot eat salads made with oily salad dressings. However, he does enjoy raw fish dishes such as sushi and sashimi. What might be his problem?

10. A group of campers was sitting around a fire on a chilly summer evening when one noticed a threadlike body in the eye of another camper. What might the body have been and what conditions might have led to its appearance at that time?

11. A newspaper report in the 1980s indicated that for every case of cancer, the National Institutes of Health spends $209 annually on research. By contrast, for every case of blood fluke disease (schistosomiasis), the figure is $0.04. What factors might account for this sharp distinction in figures? Would you support increased expenditures for schistosomiasis research?

12. Trichinosis may occur in Italian, Polish, and German communities, but it is extremely rare in Jewish communities. Why is this so? How can the local butcher play a key role in preventing the spread of trichinosis among his customers?

13. It has been suggested that with fewer and fewer of the world's peoples walking barefoot, the possibility of infection by skin-penetrating parasites will continue to decline. How many other instances can you name where a change in living habits will bring about the decline of parasitic diseases? Now consider the reverse. How many instances can you name where changes in peoples' habits lead to an increasing incidence of parasitic disease?

14. Certain restaurants offer a menu item called steak tartare. Aficionados know that the beef in this dish is served raw. What hazard might this meal present to the restaurant patron?

15. A biologist, writing in a 1984 publication, stated, "Perhaps the most important reason for discussing parasitical diseases is that they highlight just how enmeshed we are in the web of life. . . ." How many examples can you find in this chapter to support this concept?

Review

Consider the characteristic on the left and the three possible choices on the right. Select the disease(s) or parasite name(s) that best apply to the characteristic and place the letter(s) next to the characteristic.

1. __ Transmitted by an arthropod
    a. filariasis
    b. trichinosis
    c. hookworm disease

2. __ Animal tapeworm
    a. *Taenia piscium*
    b. *Taenia latum*
    c. *Taenia saginata*

3. __ Type of fluke
    a. *Schistosoma*
    b. *Necator*
    c. *Fasciola*

4. __ Eggs the mode of transmission
    a. dog tapeworm disease
    b. lung fluke disease
    c. whipworm disease

5. __ Infects the human intestine
    a. *Trichinella spiralis*
    b. *Ascaris lumbricoides*
    c. *Fasciolopsis buski*

6. __ Type of tapeworm

    a. *Paragonimus*
    b. *Hymenolepis*
    c. *Fasciola*

7. __ Snail the intermediate host

    a. blood fluke
    b. dog tapeworm
    c. intestinal fluke

8. __ Attaches to tissue by hooks

    a. *Necator*
    b. *Diphyllobothrium*
    c. *Loa*

9. __ Affects pigs as well as humans

    a. *Loa loa*
    b. *Trichinella spiralis*
    c. *Clonorchis sinensis*

10. __ Life cycle includes miracidium and cercaria

    a. *Schistosoma*
    b. *Fasciola*
    c. *Paragonimus*

11. __ Acquired by consuming contaminated meat

    a. *Taenia soleum*
    b. *Loa loa*
    c. *Necator americanus*

12. __ Classified in the phylum Aschelminthes

    a. *Dracunculus*
    b. *Wuchereria*
    c. *Trichinella*

13. __ Infects tissues of the human eye

    a. *Schistosoma*
    b. *Clonorchis*
    c. *Loa*

14. __ Male and female forms exist

    a. *Hymenolepis*
    b. *Ascaris*
    c. *Trichuris*

15. __ Forms hydatid cysts

    a. fish tapeworm
    b. Guinea worm
    c. dog tapeworm

PART 5

DISEASE AND RESISTANCE

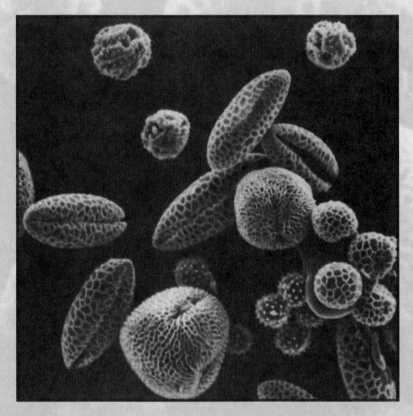

In past centuries, the spread of disease appeared to be willfully erratic. Illnesses would attack some members of a population while leaving others untouched. A disease that for many generations had taken small, steady tolls would suddenly flare up in epidemic proportions. And strange, horrifying plagues descended unexpectedly on whole nations. ■ Scientists now know that humans live in a precarious equilibrium with the microorganisms that surround them. Generally the relationship is harmonious, because humans can come in contact with most microorganisms and develop resistance to them. However, when the natural resistance is unable to overcome the aggressiveness of microorganisms, disease sets in. In other instances, the resistance is diminished by a pattern of human life that gives microorganisms the edge. For example, during the Industrial Revolution of the 1800s, many thousands of Europeans moved from rural areas to the cities. They sought new jobs, adventure, and prosperity. Instead, they found endless labor, unventilated factories, and wretched living conditions; and they found disease. ■ In Part V of this text we shall explore the disease process and the mechanisms by which the body responds to disease. Chapter 17 will open with an overview of the host–parasite relationship and the factors that contribute to the establishment of disease. In Chapter 18, the discussion turns to nonspecific and specific methods by which body resistance develops, with emphasis on the immune system. Various types of immunity are explored in Chapter 19, together with a survey of laboratory methods that utilize the immune reaction in the diagnosis of disease. In Chapter 20 the discussion centers on immune disorders that lead to serious problems in humans. In these chapters we shall ferret out the roots of disease and resistance and come to understand them at the fundamental level.

MICROBIOLOGY PATHWAYS

HEALTH AND MEDICINE

One of the givens of studying health and allied health is the importance of a course in microbiology. It should not be difficult to convince someone that the study of infectious diseases is central to the study of health and medicine. To have an appreciation of AIDS, tuberculosis, influenza. malaria, and other notable diseases is to have an appreciation of contemporary life itself.

But contemporary microbiology is more than infectious diseases. Over the last 20 years or so, the discipline of immunology has ingrained itself into modern health care. For instance, dealing with allergy is dealing with an immunological problem. Diseases such as lupus and rheumatoid arthritis are the domain of immunology (as is Rh disease of the newborn). Geriatric medicine is related to immunology because the immune systems of the elderly function less efficiently and therefore leave individuals more susceptible to infectious diseases.

Many microbiological tests performed in today's laboratories are based in immunology. Where older tests attempted to detect the presence of microorganisms, the newer tests detect antibodies produced in response to these organisms. Antibody tests are also being used to detect hormones in blood and urine (such as the hormone test for pregnancy). Before a transplant is performed, immunological tests are required to determine how compatible the recipient and donor tissues are. Antibody tests are even used in forensic medicine to identify blood types and other secretions.

As a subdiscipline of microbiology, immunology has emerged to become an important facet of the health care system. Many health science curricula offer separate courses in immunology, but for many undergraduates, the exposure to immunology begins and ends with the microbiology course. For this reason, it is important to understand not only the philosophical significance of immunology, but also the practical benefit of studying the immune system.

CHAPTER 17

INFECTION AND DISEASE

C **holera** broke out in Europe in the 1840s and reached London in June 1849. One of the worst-affected areas was the district around Golden Square, where in a ten-day period in August over 500 people died of the disease.

A physician named **John Snow** lived close to Golden Square. Snow had long been interested in cholera, and the outbreak provided an opportunity to continue and concentrate his study. He was of the opinion that bad air and direct contact played negligible roles in the spread of the disease, and his beliefs would be strengthened by his observations.

Snow noted that most of the cholera victims in the Golden Square district drew their water from a well on Broad Street. The water was obtained by a hand-operated pump accessible to all. Snow found that the well was contaminated by the cesspool overflow from a tenement in which a cholera patient lived, and he concluded that water was the source of the disease. On September 7, 1849, he presented his findings to the local community council. Snow's study impressed the council members, and they inquired how he intended to stop the epidemic. Snow thought for a moment and replied with the now classic solution: "Take the handle off the Broad Street pump." By the following day, the handle was gone, and shortly thereafter the epidemic subsided.

Unfortunately, not all epidemics are quite so easy to interrupt. In this chapter we shall discuss the complex mechanisms that underlie the spread and development of disease. Individual diseases are considered in many other chapters of this text, and our purpose here is to bring together many concepts of disease and synthesize an overview of the host–parasite relationship. We shall summarize much of the important terminology used in medical microbiology and focus on the methods used by microorganisms to establish themselves in the tissues. The study will prepare us for a detailed discussion of resistance mechanisms in the following chapters.

The Host–Parasite Relationship

The word **infection** is derived from Latin origins meaning to mix with or corrupt. The term refers to the relationship between two organisms, the host and the parasite, and the competition for supremacy that takes place between them. A host whose resistance is strong remains healthy, and the parasite is either driven from the host or assumes a benign relationship with the host. By contrast, if the host loses the competition, disease develops. The term **disease** appears to have originated from Latin stems that mean living apart, a reference to the separation of ill individuals from the general population. Disease may be conceptualized as any change from the general state of good health. It is important to note that disease and infection are not synonymous; a person may be infected without becoming diseased.

The Normal Flora

The concept of infection in the host–parasite relationship is expressed in the body's normal flora. The **normal flora** is a population of microorganisms that infect the body without causing disease. Some organisms in the population establish a permanent relationship with the body, while others are present for limited periods of time. In the large intestine of humans, for example, *Escherichia coli* is almost always found, but streptococci are transient.

The relationship between the body and its normal flora is an example of a **symbiosis**. In some cases the symbiosis is beneficial to both the body and the microorganisms. This relationship is called **mutualism**. For example, species of *Lactobacillus* live in the human vagina and derive nutrients from the environment while producing acid to prevent the overgrowth of other organisms. In other cases, the symbiosis is beneficial only to the microorganisms, in which case the symbiosis is called **commensalism**. *Escherichia coli* is generally presumed to be a commensal in the human intestine, although some evidence exists for mutualism because the bacteria produce certain amounts of vitamins B and K.

A normal flora may be found in several body tissues. On the **skin**, for instance, there are various forms of viruses, fungi, and bacteria, particularly staphylococci and *Propionibacterium acnes*. The **oral cavity** commonly contains members of the genera *Neisseria*, *Leptotrichia*, and *Bacteroides*, as well as many diphtherialike bacilli (diphtheroids), fungal spores, and streptococci. The upper **respiratory tract** is the site of all these organisms, as well as pneumococci and species of *Haemophilus* and *Mycoplasma*. These organisms may cause respiratory disease if the body defenses are compromised.

The stomach in humans is generally without a normal flora mainly due to the low pH of its contents. However, the latter part of the **small intestine** and the **large intestine** abound with microorganisms. *Bacteroides* species are numerous, together with *Clostridium* spores, various streptococci, and a number of Gram-negative rods including species of *Enterobacter*, *Klebsiella*, *Proteus*, and *Pseudomonas*. *Escherichia coli* is a well-known resident of the intestine, as is *Candida albicans*, the yeast (Figure 17.1). In females, *Lactobacillus* is a notable component of the **vagina**; other organisms may be located near the urogenital orifices in both males and females. The blood and urine are usually sterile unless disease is in progress.

Organisms of the normal flora are introduced when the child passes through the birth canal. Additional organisms enter when breathing begins and upon first feeding. Within two to three days most organisms of the flora have

FIGURE 17.1

The Normal Flora

A scanning electron micrograph of an intestinal membrane showing *Candida albicans* attached to the surface. The yeast form of the cells is apparent.

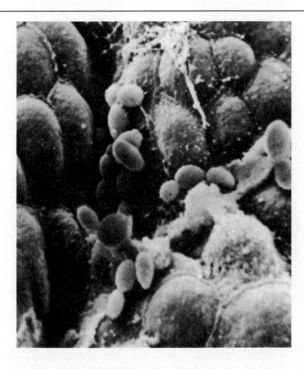

appeared. During the next few weeks, contact with the mother and other individuals will expose the child to additional microorganisms. The normal flora remains throughout life, undergoing changes in response to the internal environment of the individual.

Pathogenicity

Pathogenicity refers to the ability of a parasite to gain entry to the host's tissues and bring about a physiological or anatomical change resulting in a change of health and thus disease. The word "pathogenicity" is derived from the Greek term *pathos*, meaning suffering. The term **pathogen** has the same root and refers to an organism having pathogenicity. The symbiotic relationship between host and parasite is called **parasitism**.

Parasites vary greatly in their pathogenicity. Certain parasites such as the cholera, plague, and typhoid bacilli are well known for their ability to cause serious human disease (Figure 17.2). Other organisms, such as common cold viruses, are considered less pathogenic because they induce milder illnesses. Still other organisms are **opportunistic**. These organisms may be commensals in the body until the normal defenses are suppressed, at which time the commensals seize the "opportunity" to invade the tissues and express their pathogenicity. An example is observed in individuals with **acquired immune deficiency syndrome (AIDS)**. These patients are highly susceptible to opportunistic organisms such as *Pneumocystis carinii*, *Toxoplasma gondii*, and species of *Cryptosporidium*. None of these protozoa (Chapter 15) was considered a serious pathogen before the emergence of AIDS.

Pathogenicity:
the ability of a parasite to gain entry to host tissues and bring about disease.

Opportunistic organisms:
those that invade the tissues when body defenses are suppressed.

nu"-mo-sis'tis car-in'e-e
krip"to-spor-id'e-um

FIGURE 17.2

Salmonella Penetration of Intestinal Cells

Researchers have discovered that microvilli of the intestinal epithelium undergo dramatic changes when *Salmonella* cells come into close contact. The microvilli themselves disappear and tiny membrane blebs (or "blisters") spring up in their place, engulfing the salmonellae. Two hours later the microvilli reappear but the bacteria are now within the cells. In this scanning electron micrograph, the membrane blebs (A) can be seen as puffy structures. A bacterium (B) is within the cavity of one bleb and another is adjacent to it.

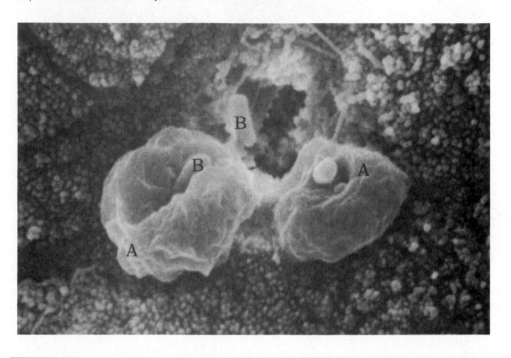

Candida albicans:
a yeast fungus that can cause infection of the intestine, vagina, or oral cavity.

Diseases caused by opportunistic organisms illustrate how a shift in the body's delicate balance of controls may convert "infection" to "disease." Another example happened in the 1950s when antibiotics came into widespread use. As bacteria in the intestine's normal flora disappeared, the biological controls exerted on **Candida albicans** vanished and candidiasis became common. Today, a number of new organisms are entering the United States from foreign lands and are assuming commensal relationships with Americans. A chance upset in resistance mechanisms or control shifts may enhance the ability of these newcomers to establish disease (MicroFocus: 17.1).

A sobering concept has emerged in recent years. Microbiologists once believed that organisms were either pathogenic or nonpathogenic. This distinction has been blurred by the realization that normally benign organisms may be pathogenic when body defenses weaken or fail. The AIDS epidemic points up this concept well. Transplant and cancer patients treated with radiation and immune-suppressing drugs are also affected. It is now recognized that pathogenicity is a function of the aggressive nature of the parasite as well as the level of resistance in the host.

Virulence:
the degree of pathogenicity of a parasite.

The word **virulence** is used to express the degree of pathogenicity of a parasite. This term is derived from the Latin *virulentus*, meaning full of poison. An organism such as the typhoid bacillus that invariably causes disease is said to be "highly virulent." By comparison, an organism such as *Candida albicans* that sometimes causes disease is labeled "moderately virulent." Certain organisms, described as **avirulent**, are not regarded as disease agents. The lactobacilli and streptococci found in yogurt are examples. However, it should be

Avirulent:
lacking the ability to cause disease.

17.1

MICROFOCUS

UPSETTING THE BALANCE OF NATURE

An incident that took place on the South Pacific island of Borneo illustrates how an unthinking approach to the eradication of disease may have dire consequences.

In 1955, Borneo was in the throes of a malaria epidemic. With the specter of widespread death looming, the government issued an appeal for assistance to the World Health Organization. The WHO obliged by sending teams of technicians to spray the natives' huts with DDT and dieldrin, two powerful insecticides. The spraying successfully reduced the mosquito population and halted the spread of malaria. However, it also touched off a bizarre series of events.

It seemed that the insecticides also killed massive numbers of houseflies. The houseflies were eagerly consumed by tiny lizards called geckos, and as the insecticides concentrated in the geckos, they too died. Geckos, in turn, were eaten by the island's population of housecats, which likewise perished. Only then did ecologists realize that housecats had been keeping the rat population under control; within weeks rats were everywhere. Now a new problem emerged because rats carry the fleas that transmit bubonic plague. Before long, plague was consuming the island.

But the World Health Organization was not about to admit defeat. It rounded up thousands of housecats, placed them in boxes, and parachuted the boxes into the island's remote villages. The project was labeled "Operation Cat Drop." Eventually the cats established themselves on Borneo and brought the rat population (and the plague) under control. Looking back on the operation, one observer wryly noted: "That was the day it rained cats."

noted that any microorganism has the ability to change genetically and become virulent. *E. coli*, for example, was long considered an avirulent commensal of humans, but toxin-producing strains (Chapter 8) are now isolated during outbreaks of human gastroenteritis. Virulence is determined in part by invasiveness and toxigenicity, as we shall see presently.

The Progress of Disease

Disease is a dynamic series of events expressing the competition between parasite and host (although certain diseases such as botulism follow an ingestion of toxins). In most instances, there is a recognizable pattern in the progress of the disease following the entry of the parasite. Certain periods may be distinguished as follows.

The episode of disease begins with a **period of incubation** reflecting the time that elapses between the entry of the parasite to the host and the appearance of symptoms (Figure 17.3). Incubation periods may be a short one to three days as in cholera, a moderate two weeks as in chickenpox, or a long three to six years as in leprosy. Such factors as the number of parasites, their generation time and virulence, and the level of host resistance determine the period's length. Also the location of entry may be a determining factor. For instance, the incubation period for rabies may be as short as several days or as long as a year depending on how close to the central nervous system the viruses enter the body.

The next phase in disease is the **period of prodromal symptoms**. This period is characterized by general symptoms such as nausea, fever, headache, and malaise, which indicate that the competition for supremacy has begun. The **period of acme** follows. This is the acute stage of the disease, when specific symptoms appear. Examples are the skin rash in scarlet fever, jaundice in hepatitis, teardrop-shaped vesicles in chickenpox, and swollen lymph nodes in infectious mononucleosis. Often, patients suffer high fever and chills, the latter reflecting differences in temperature between the superficial and deep areas of the body. Dry skin and pale expression may result from constriction of the skin's blood vessels to conserve heat.

Period of incubation: the time that elapses between entry of the parasite to the host and the appearance of symptoms.

pro-drōm'al

Period of acme: the acute stage of the disease.

FIGURE 17.3

The Course of Disease as Typified by Measles

(a) A child is exposed to measles viruses in respiratory droplets from another child, and the period of incubation begins. (b) At the end of this period, the child experiences fever, respiratory distress, and general weakness as the period of prodromal symptoms ensues. (c) The period of acme begins with the appearance of specific measles symptoms such as the body rash. (d) The period reaches a peak as the rash covers the body. (e) The rash fades first from the face and then the body trunk as the period of decline takes place. (e) With the period of convalescence, the body returns to normal. (f) The child later returns to school.

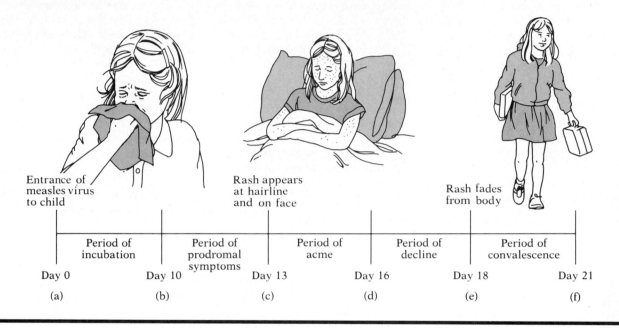

def"er-ves'ens

Period of convalescence: **the period during which the body systems return to normal.**

Subclinical disease: **a disease accompanied by few or no symptoms.**

Droplets: **tiny particles of mucus expelled from the respiratory tract in a cough or sneeze.**

As the symptoms subside, the **period of decline**, or defervescence, sets in. This period may be preceded by a crisis period after which recovery is often rapid. In other cases the period may last a long time. Sweating is common as the body releases excessive amounts of heat, and the normal skin color soon returns as the blood vessels dilate. The sequence comes to a conclusion during the **period of convalescence**. During this time, the body's systems return to normal.

Diseases may be described as clinical or subclinical. A **clinical** disease is one in which the symptoms are apparent, while a **subclinical** disease is accompanied by few obvious symptoms. Many people, for example, have experienced subclinical cases of mumps or infectious mononucleosis and have developed immunity to future attacks. By contrast, certain diseases are invariably accompanied by clearly recognized clinical symptoms (MicroFocus: 17.2). Measles and malaria are examples.

Transmission of Disease

The agents of disease may be transmitted by a broad variety of methods conveniently divided into two general categories: direct methods and indirect methods, as displayed in Figure 17.4.

Direct methods of transmission imply close or personal contact with one who has the disease. Hand-shaking, kissing, sexual intercourse, and contact with feces are examples. Such diseases as gonorrhea and genital herpes are spread by direct contact. Direct contact may also mean exposure to **droplets**, the tiny particles of mucus expelled from the respiratory tract during a cough

FIGURE 17.4

How Diseases Can Be Communicated

(a) By respiratory droplets. (b) By dust. (c) By arthropod bites. (d) By contact with animals. (e) By contaminated food. (f) By contaminated water. (g) By contact with contaminated objects and instruments. (h) By injection of contaminated soil.

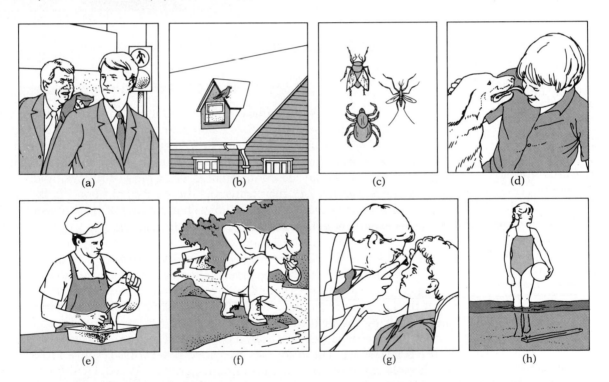

(a) (b) (c) (d)

(e) (f) (g) (h)

or sneeze. Diseases spread by this method include influenza, measles, pertussis (whooping cough), and streptococcal sore throat. For some diseases, direct contact with an animal is necessary. Rabies, leptospirosis, and toxoplasmosis are typical.

Indirect methods of disease transmission include consumption of contaminated food or water (MicroFocus: 17.3) as well as contact with fomites. Foods are contaminated during processing or handling, or they may be dangerous

fo′mītz

17.3

MICROFOCUS

DINNER AT LA CASA CUCURACHA

"...We're having a wonderful time in this tropical wonderland. Last night, we went to a great restaurant. We walked down Cabeza de Vaca Boulevard, just as they told us, and there on the corner of Ponce de Leon square, we saw it—La Casa Cucuracha.

The shrimp were so icy cold, we couldn't eat them fast enough (Uh oh. Should never have cold seafood in that part of the world). The salad was crisp and delicious, especially the lettuce (Shouldn't have salad —the fixins' were probably washed in the local water). And the waiter was so helpful—he peeled the orange for me right there at the table (Another no no; his hands probably touched the peeled fruit).

Dinner was superb. We had crab meat *au gratin* just warm enough to bring out the flavor (It should have been steaming hot). The local vegetables were good enough to eat raw (Another mistake); but the potato was a run-of-the-mill baked potato (Probably the only safe thing you ate). For dessert we had local fruits on a lovely bed of crushed ice (The water, again).

How am I feeling today? OK, I guess. On second thought, I do have this little pain in my stomach...."

when made from diseased animals. Poultry products, for example, are often a source of salmonellosis because *Salmonella* species frequently infect chickens; and pork may spread trichinosis because *Trichinella* parasites live in muscles of the pig. **Fomites** are inanimate objects that carry disease organisms. For instance, bed linens may be contaminated with pinworm eggs; and contaminated syringes and needles may transport the viruses of hepatitis B.

Arthropods represent another indirect method for transmission. Living organisms, such as arthropods, that carry disease agents from one host to another are called **vectors**. In some cases the arthropod may be a **mechanical vector** of disease because it transports microorganisms on its legs and other body parts. In other cases, the arthropod itself is diseased and serves as a **biological vector**. In malaria and yellow fever, for instance, disease organisms infect the arthropod and accumulate in its salivary gland, from which they are injected during the next bite.

For disease to perpetuate itself a continuing source of disease organisms in nature is necessary. These sources are called **reservoirs** (Figure 17.5). In smallpox the sole reservoir of viruses is humans, and the World Health Organization was able to limit the spread of the virus and eradicate smallpox from the earth by locating all human victims (Chapter 12). A **carrier** is a special type of reservoir. Generally a carrier is one who has recovered from the disease but continues to shed the disease agents. For instance, people who have recovered from typhoid fever or amebiasis are usually carriers for many weeks after the symptoms of disease have left. Their feces may spread the disease to others via contaminated food or water.

Animals may also be reservoirs of disease. Domestic housecats usually show no symptoms of toxoplasmosis but are able to transmit *Toxoplasma gondii* to

Fomites:
inanimate objects that carry disease organisms.

Vectors:
living organisms that transmit disease organisms.

Reservoir:
organisms that harbor disease agents but show no sign of disease.

Carrier:
one who has recovered from a disease but continues to shed the disease agents.

Toxoplasmosis:
a protozoal disease of the blood accompanied by mononucleosislike symptoms.

FIGURE 17.5

How Disease May Be Introduced to a Population by a Reservoir

Measles broke out in a Chicago neighborhood late in August 1983. A two-year-old Gypsy boy contracted the virus and, during the period of incubation, he returned home to Billings, Montana. On September 4, he broke out in a rash and infected boy Number 2. The second boy developed a rash on September 19 while at a wedding in Spokane, Washington. The wedding was attended by about 375 people from Gypsy communities in Idaho, Montana, Oregon, and Washington. Seventeen children at the wedding were infected. They eventually spread measles to numerous individuals in non-Gypsy populations, as well as to healthcare workers and families of healthcare workers, as displayed in the diagram. In all, 42 persons contracted the disease from the initial reservoir. (Courtesy of Centers for Disease Control and Prevention, *MMWR*, December 23, 1983.)

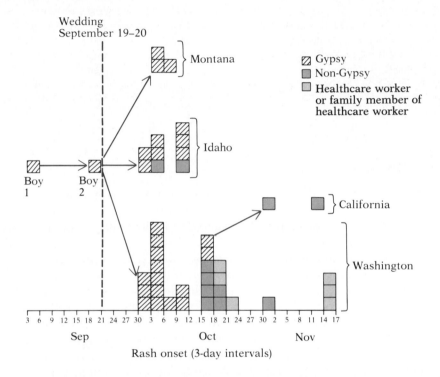

humans, where the disease manifests itself. Water and soil may likewise be considered reservoirs since they are often contaminated with disease agents.

Most diseases studied in this text are **communicable** diseases; that is, they are transmissible among hosts. Certain communicable diseases are further described as **contagious**, because they pass with particular ease among hosts. Chickenpox, measles, and genital herpes fall into this category. One reason for easy transmission is that the focus of disease is on or close to the body surface. **Noncommunicable** diseases are singular events where the agent is acquired directly from the environment and is not easily transmitted to the next host. In tetanus, for example, penetration of *Clostridium tetani* spores to the dead anaerobic tissue of a wound must occur before this disease develops.

Diseases may also be described as endemic diseases, epidemic diseases, or pandemic diseases. An **endemic** disease is one that occurs at a low level in a certain geographic area. By comparison, **epidemic** diseases (or "epidemics") break out in explosive proportions within a population, and **pandemic** diseases ("pandemics") occur worldwide. MicroFocus: 17.4 indicates how an endemic disease may become epidemic.

Contagious disease:
a disease that passes with particular ease among hosts.

Epidemic disease:
a disease that breaks out in explosive proportions in a population.

17.4

MICROFOCUS

ENDEMIC TO EPIDEMIC

For centuries, nomadic Mongol tribesmen of the Siberian steppes observed well-established customs when dealing with marmots. These large, burrowing rodents could be shot but never trapped. A sluggishly moving animal was taboo. The lump of tissue under the animal's arm was not to be eaten because legend said it contained the soul of a dead hunter. And, if a marmot colony was deemed sick, custom required that the tribesmen strike their tents and

move away. Contemporary microbiologists know that prescriptions as these reduced the possibility of acquiring plague, a disease often carried by marmots. So long as the laws were obeyed, the disease remained endemic.

But in 1910 furriers suddenly increased their demand for marmot skins. The skins were an ideal, low-cost substitute for sable used in women's coats. Thousands of greedy fur trappers invaded the steppes, led

by inexperienced Chinese guides. Unaware of the local Mongol traditions, the hunters captured numerous sick marmots among the healthy ones. The results were predictable: plague soon broke out among the hunters, and they brought the disease back to China. In a seven-month period during 1910–1911, an estimated 60,000 Chinese people died as the disease assumed epidemic proportions. Historians note that this was the last epidemic of plague ever recorded.

Types of Diseases

The dictionary of medical microbiology contains numerous terms that describe types of disease. For example, such words as "communicable" or "noncommunicable" are commonly applied to a disease to denote its ability to be transmitted, and "endemic" and "epidemic" indicate how widespread the disease is. In these paragraphs we shall briefly summarize some other types of disease.

The words "acute" and "chronic" are applied to diseases as relative measurements of their severity. An **acute** disease develops rapidly, is usually accompanied by severe symptoms, comes to a climax, and then fades rather quickly. Cholera, epidemic typhus, and yellow fever are examples of acute diseases. **Chronic** diseases, by contrast, often linger for long periods of time. The symptoms are slower to develop, a climax is rarely reached, and convalescence may continue for several months. Hepatitis A, trichomoniasis, and infectious mononucleosis illustrate chronic diseases. Sometimes an acute disease may become chronic when the body is unable to rid itself completely of the parasite. For example, one who has giardiasis or amebiasis may experience sporadic symptoms for many years.

Infections may develop as primary or secondary diseases. A **primary** disease occurs in an otherwise healthy body while a **secondary** disease develops in a weakened individual. In the 1918–1919 influenza pandemic, hundreds of millions of individuals contracted influenza as a primary disease and then developed pneumonia as a secondary disease. Most of the deaths in the pandemic were due to pneumonia's complications.

As the names imply, **local** diseases are restricted to a single area of the body, while **systemic** diseases are those disseminating to the deeper organs and systems. Thus a staphylococcal skin boil beginning as a localized skin lesion may become more serious when staphylococci spread and cause systemic disease of the bones, meninges, or heart tissue. The word commonly used for dissemination of bacteria through the bloodstream is **bacteremia**. Another term, **septicemia**, means an infection spreading in the blood and is often used as a synonym. The "septicemia" or "blood poisoning" appearing in older textbooks is now regarded as streptococcal or staphylococcal blood disease. **Fungemia** refers to the spread of fungi, **viremia** to spread of viruses, and **parasitemia** to spread of protozoa and multicellular worms to the blood.

Chronic disease:
one that lingers for a long time, rarely reaches a climax, and disappears slowly.

Secondary disease:
one that develops in a weakened host.

Systemic disease:
a disease that disseminates to the deep organs and tissues.

Viremia:
dissemination of viruses to the bloodstream.

A word of caution might be appropriate at this juncture. We have used the word "disease" loosely to define a clinical condition initiated by an infectious microorganism. It should not be inferred, however, that this description is true of all diseases. For example, physiological diseases such as diabetes mellitus are due to a malfunction of a body organ or system; nutritional diseases such as scurvy and beri-beri are due to a dietary insufficiency; and genetic diseases such as sickle cell anemia and phenylketonuria (PKU) are traced to defects in human genes. In these cases, infectious agents or parasites are not involved. Therefore as we use the word "disease," it is well to remember that we are referring to "infectious disease."

fen"il-ke"to-nu're-ah

TO THIS POINT
■

We have begun our study of infection and disease by studying the host–parasite relationship. Infection refers to the relationship between the host and parasite, while disease results from the change in the state of the host's health brought about by damage caused by the parasite. We noted that the body is inhabited by a normal flora of organisms, some of which are commensals and opportunists. The term pathogenicity was explored in the general sense, and we noted how a shift in the delicate balance of body control can lead to disease. Virulence refers to the degree of pathogenicity of a parasite.

In the next paragraphs we surveyed the periods during the progress of disease and used many examples from various diseases to illustrate the basic concepts. Clinical and subclinical diseases vary on the basis of clinical symptoms, and the transmission of disease occurs by direct and indirect methods, with particular reference to reservoirs and carriers. We defined communicable and noncommunicable diseases, acute and chronic diseases, primary and secondary diseases, and local and systemic diseases.

We shall now focus on the factors contributing to the establishment of disease. Portal of entry, dose, and tissue penetration are examples of these factors. We shall then survey some of the enzymes and toxins used by the parasite to overcome body defenses and interfere with vital metabolic processes. The section will end with a discussion of the blood, since the blood is often the site of infection and spread of parasites, and it is the organ where a major aspect of body resistance takes place.

The Establishment of Disease

A **parasite** must possess unusual abilities if it is to overcome host defenses and bring about the anatomical or physiological changes leading to disease. Before it can manifest these abilities, however, the parasite must first gain entry to the host in sufficient numbers to establish a population (Figure 17.6). Next, it should be able to penetrate the tissues and grow at that location. Disease is therefore a complex series of interactions between parasite and host. In this section we shall examine some of the factors that determine whether disease can occur, with a focus on the parasite.

FIGURE 17.6

FIGURE 17.6

A Population of Intestinal Bacteria

Two scanning electron micrographs of the villi of the intestine showing filamentous bacteria. (a) The bacteria appear to emerge from clefts in the tissue and extend out onto the surface of the villus in this view ($\times$ 425). (b) When magnified 2000 times, the attachment in the cleft of tissue can be seen more clearly. Note the three-dimensional image afforded by the scanning electron microscope.

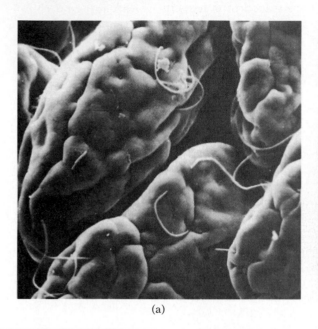

(a)

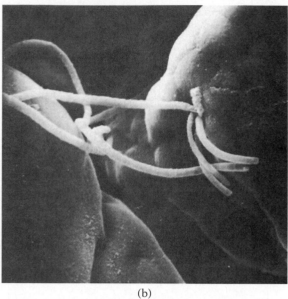

(b)

Portal of Entry

Portal of entry:
site where microorganisms enter the body of the host in anticipation of disease.

Portal of entry refers to the site at which the parasite enters the host. It varies considerably for different organisms and is a key factor in the establishment of disease. For example, **tetanus** may occur if *Clostridium tetani* spores are introduced from the soil to the anaerobic tissue of a wound but tetanus will not develop if spores are consumed with food because the spores do not germinate in the human intestine. This is why one can eat a freshly picked radish without fear of tetanus. One reason offered for specific portals of entry is the presence of adhesive factors on the surface of parasites. For example, gonococci attach by means of pili to specific receptor sites on tissues of the urogenital system.

Pili:
short hairlike appendages used by bacteria for attachment to the tissues.

Q fever:
a rickettsial disease due to *Coxiella burnetii* characterized by flulike symptoms.

Certain parasites have multiple portals of entry. The **tubercle bacillus**, for instance, may enter the body together with respiratory droplets, contaminated food and milk, and skin wounds. The **Q fever** organism enters by all these methods as well as by an arthropod bite. The **tularemia bacillus** may enter the eye by contact, the skin by an abrasion, the respiratory tract by droplets, the intestine by contaminated meat, or the blood by an arthropod bite. Indeed, arthropodborne diseases will break out or disappear depending on the extent of the arthropod population (MicroFocus: 17.5).

Dose

Dose refers to the number of parasites that must be taken into the body for disease to be established. Experiments indicate, for example, that the consumption of a mere million **typhoid bacilli** in contaminated water will probably lead to disease. By contrast, many times that number of **cholera bacilli** must be

17.5

AT HOME IN A TIRE

Each year Americans discard about 200 million tires, most of which are buried in landfills or left to rot in vacant lots. In these settings, discarded tires make perfect homes for mosquitoes. The tires fill with water and garbage, and the mosquitoes grow larger, live longer, and complete their life cycles sooner than their relatives from swamps and drainage ditches.

Since 1963, the state of Ohio has recorded over 600 cases of La Crosse encephalitis. First observed in La Crosse, Wisconsin in 1960, La Crosse encephalitis is endemic in a crescent-shaped area of the Great Lakes region extending from southeastern Minnesota to New York. A serious viral disease of the brain, La Crosse encephalitis is transmitted by mosquitoes of the genus *Aedes*. When the Ohio Department of Health investigated 69 cases of the disease in the 1980s, it found that discarded tires were connected in some way with almost three-quarters of the cases. It also located *Aedes* mosquitoes in tires associated with half the cases. A notable episode of La Crosse encephalitis occurred in 1983 in the daughter of a rural Ohio businessman. His business was recycling tires.

Some years later, a case of La Crosse encephalitis was observed in a Milwaukee suburb, well outside the normal area for the disease. While pursuing leads, health officials noted that a neighbor of the patient had previously lived near La Crosse. They questioned the neighbor and learned that when he moved, he brought along one very significant item: his collection of old tires.

ingested if cholera is to be established. One explanation offered is the high resistance of typhoid bacilli to the acidic conditions in the stomach in contrast to the low resistance of cholera bacilli. Also, it may be safe to eat fish when the water contains hepatitis A viruses, but eating raw clams from that same water can be dangerous because clams are filter feeders, and the concentration (or dose) of hepatitis A viruses is much higher in these animals.

Often the host is exposed to low doses of a parasite and as a result, develops immunity. For instance, many people can tolerate low numbers of mumps viruses without exhibiting disease. They may be surprised to find that they are immune to **mumps** when it breaks out in their family at some later date. Other individuals have developed immunity to fungal pathogens after several low-grade exposures. Cases of **histoplasmosis**, for example, are regarded as a mild "summer flu" in the Ohio and Mississippi valleys, but histoplasmosis can be serious in an individual not previously exposed to the disease. It should be noted that immune people remain healthy because they enjoy strong resistance, again pointing to the notion that resistance and disease are fundamentally inseparable.

Mumps:
a viral disease accompanied by swollen salivary glands.

his"to-plaz-mo'sis

Tissue Penetration

Most knowledge of tissue penetration (Figure 17.7) has been developed from studies on histological preparations, tissue cultures, and animals, and an understanding of the penetration process is generally incomplete. Experiments reported by Stanley Falkow and his co-workers at Stanford University indicate that genes for cell penetration exist on the chromosome of certain bacteria. These genes appear to code for surface proteins that assist penetration. In the late 1980s, Falkow's group successfully isolated the penetration genes from

FIGURE 17.7
Tissue Invasion

A series of transmission electron micrographs displaying the interaction between invasive *Salmonella typhimurium* and epithelial cells of the human intestine. (a) 30 minutes postinfection, bacteria are seen adhering to the tips of intestinal microvilli. (b) A bacterium is being engulfed by an epithelial cell. (c) One hour postinfection, salmonellae can be observed within vacuoles in the cells. (d) 12 hours postinfection, a number of vacuoles containing bacteria unite with one another, and bacteria multiply within this large vacuole. (e) 24 hours postinfection, the epithelial cell is filled with salmonellae and is breaking down to release the bacteria. Bar = 1 μm.

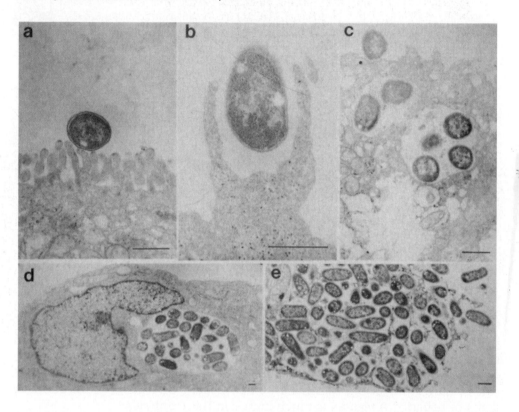

Yersinia:
a genus of Gram-negative rods that display bipolar staining.

Pertussis:
a bacterial disease accompanied by mucus accumulation in the respiratory tract.

Yersinia pseudotuberculosis and inserted them into *E. coli*, which then displayed penetration (Figure 17.8). Other researchers have reported genes for penetration on the plasmids of *Shigella flexneri*.

The ability of a parasite to penetrate tissues and cause structural damage is a virulence factor called **invasiveness**. The bacilli of typhoid fever and the protozoa that cause amebiasis are well known for their invasiveness. By penetrating the tissue of the gastrointestinal tract, these organisms cause ulcers and sharp, appendicitislike pain characteristic of the respective diseases. Another organism noted for its invasiveness is ***Escherichia coli* 0157:H7**. Since 1993, this organism has caused notable outbreaks of bloody diarrhea when consumed in contaminated meat (e.g., the Jack-In-The-Box incident—Chapter 3). The presence of blood in the stools of patients confirmed that this strain of *E. coli* is able to invade the intestinal tissue.

In some cases, penetration may not be critical to disease. The **pertussis bacillus**, for example, remains on the surface layers of the respiratory tract while producing the toxins that lead to disease. The **cholera bacillus** does likewise in the intestine.

FIGURE 17.8

Two Examples of Tissue Penetration by Parasites

(a) The fungus *Dactylaria* attacking the roundworm *Panagrellus* in two places. The fungus penetrates the outer membranes of the worm and parasitizes its tissues. (b) An *Escherichia coli* cell experimentally reengineered to produce a surface protein that permits invasion of tissue cells. In this photograph, *E. coli* is invading a human lung cell. The fingerlike protrusions are part of the normal outer membrane of the lung cell.

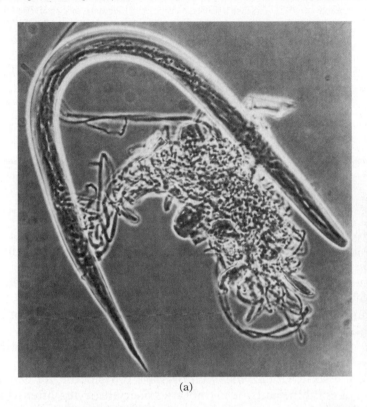

(a)

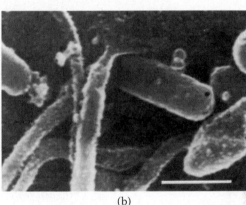

(b)

Enzymes

The virulence of a parasite depends to some degree on its ability to produce a series of enzymes directed at host defenses. The enzymes act on host cells and interfere with certain functions meant to retard invasion. Table 17.1 summarizes the activities of these enzymes.

An example of a bacterial enzyme is the **coagulase** produced by virulent staphylococci. Coagulase catalyzes the formation of a blood clot from fibrinogen proteins in the human blood. The clot sticks to **staphylococci**, protecting them from phagocytosis. Part of the walling-off process observed in a staphylococcal skin boil is due to the clot formation. Coagulase-positive staphylococci may be identified in the laboratory by combining staphylococci with human or rabbit plasma. Formation of a clot in the plasma indicates coagulase activity.

Many streptococci have the ability to produce the enzyme **streptokinase**. This substance dissolves fibrin clots used by the body to restrict and isolate an infected area. Streptokinase thus overcomes an important host defense and permits further tissue invasion by the parasites.

Hyaluronidase is sometimes called the spreading factor because it enhances penetration of a parasite through the tissues. The enzyme digests hyaluronic acid, a polysaccharide that binds cells together in a tissue. The term "tissue cement" is occasionally applied to this polysaccharide. Hyaluronidase

Coagulase:
a bacterial enzyme that catalyzes the formation of a blood clot.

Streptokinase:
a bacterial enzyme that dissolves fibrin clots.

hi″ah-lu-ron′ĭ-dās

TABLE 17.1			
A Summary of Enzymes That Add Virulence			
ENZYME	SOURCE	ACTION	EFFECT
Coagulase	Staphylococci	Forms a fibrin clot	Allows resistance to phagocytosis
Streptokinase	Streptococci	Dissolves a fibrin clot	Prevents isolation of infection
Hyaluronidase	Pneumococci Streptococci Staphylococci	Digests hyaluronic acid	Permits tissue penetration
Leukocidin	Staphylococci Streptococci Certain rods	Disintegrates phagocytes	Limits phagocytosis
Hemolysins	Clostridia Staphylococci	Dissolves red blood cells	Induces anemia and limits oxygen delivery

is an important virulence factor in pneumonococci and certain species of streptococci and staphylococci. In addition, gas gangrene bacilli use it to facilitate spread through the muscle tissues.

loo"ko-si'din

he-mol'ĭ-sin

Leukocidin and hemolysins are enzymes that destroy blood cells. **Leukocidin** is a product of staphylococci, streptococci, and certain bacterial rods. The enzyme disintegrates circulating neutrophils and tissue macrophages, both of which are active phagocytes. Usually disintegration occurs before phagocytosis has taken place, but on some occasions it occurs after the phagocyte has engulfed the parasite.

Hemolysins:
a series of enzymes that dissolve red blood cells.

Hemolysins are a series of enzymes that dissolve red blood cells. Studies show that hemolysins combine with the membranes of erythrocytes, after which lysis takes place. In gas gangrene, hemolysins lead to substantial anemia. Staphylococci and streptococci are also known to produce these virulence factors. In the laboratory, hemolysin producers can be detected by hemolysis, a destruction of blood cells in a blood agar medium.

Toxins

Toxins are microbial poisons that profoundly affect the establishment and course of disease because a single toxin can make an organism virulent. Two types of toxins are recognized: exotoxins and endotoxins (Table 17.2).

Exotoxins:
microbial poisons released to the tissues as they are produced.

Exotoxins are produced chiefly by Gram-positive bacteria. They are protein molecules, manufactured during the metabolism of bacteria. Exotoxins are released into the surrounding environment of the tissue as they are produced (*exo-* is the Greek stem for outside). They dissolve in the blood fluid and circulate to their site of activity. The symptoms of disease soon develop.

The exotoxin produced by the botulism bacillus *Clostridium botulinum* is among the most lethal toxins known. One pint of the pure toxin is believed sufficient to destroy the world's population. In humans, the toxin inhibits the release of acetylcholine at the synaptic junction, a process that leads to the paralysis in botulism (Chapter 8). Another exotoxin is produced by tetanus bacilli. In this case, the exotoxin inhibits the removal of acetylcholine in the synapse, thereby permitting volleys of spontaneous nerve impulses and uncontrolled muscle contractions.

as"e-til-ko'lēn

Acetylcholine:
a neurotransmitter released in the junction of two nerves or a nerve and muscle.

TABLE 17.2
A Comparison of Exotoxins and Endotoxins

CHARACTERISTIC	EXOTOXINS	ENDOTOXINS
Usual source:	Mainly Gram-positive bacteria	Mainly Gram-negative bacteria
Location in parasite:	Cytoplasm	Cell wall
Chemical composition:	Protein	Lipid–polysaccharide–peptide
Antibodies elicited:	Yes	No
Conversion to toxoid:	Possible	Not possible
Liberation of toxin:	On production by the parasite	On disintegration of the parasite
Representative effects:	Interfere with synapsis activity Interrupt protein synthesis Increase capillary permeability Increase water elimination	Increase body temperature Increase hemorrhaging Increase swelling in tissues Induce vomiting, diarrhea

A third toxin is produced by *Corynebacterium diphtheriae*, the diphtheria bacillus. The exotoxin interferes with the assembly of proteins in the cytoplasm of epithelial cells of the upper respiratory tract. Disintegrated cells then accumulate with mucus, bacteria, fibrous material, and white blood cells to cause life-threatening respiratory blockages. Other exotoxins are formed by the bacteria that cause scarlet fever, staphylococcal food poisoning, pertussis, and cholera.

When toxins function in a particular organ system, they are given more clearly defined terms. For example, the botulism toxin is called a **neurotoxin** because of its activity in the nervous system, while the staphylococcal toxin is called an **enterotoxin** since it functions in the gastrointestinal tract.

The body responds to exotoxins by producing special antibodies called **antitoxins**. When toxin and antitoxin molecules combine with each other, the toxin is neutralized. This process represents an important defensive measure in the body. Therapy in people who have botulism, tetanus, or diphtheria often includes injections of antitoxins to neutralize the toxins.

Because exotoxins are proteins, they are susceptible to the heat and chemicals that normally react with proteins. A chemical such as formaldehyde may be used to alter the toxin and destroy its toxicity without hindering its ability to elicit an immune response in the body. The result is a **toxoid**. When the toxoid is injected to the body, the immune system responds with antitoxins that circulate and provide a measure of defense to disease. Toxoids are used for diphtheria and tetanus immunizations in the diphtheria-pertussis-tetanus (DPT) vaccine.

Endotoxins are part of the cell wall of bacteria and as such, they are released only upon disintegration of the parasite. They are present in many Gram-negative bacilli and are composed of lipid–polysaccharide–protein complexes. Endotoxins do not appear to stimulate an immune response in the body, nor can they be altered to prepare toxoids.

Endotoxins manifest their presence by certain signs and symptoms. Usually an individual experiences an increase in body temperature, substantial body weakness and aches, and general malaise. Damage to the circulatory system and shock may also occur. In this case, the permeability of the blood vessels changes and blood leaks into the intercellular spaces, where it is useless. The tissues swell, the blood pressure drops, and the patient may lapse into a coma. This condition, commonly called **endotoxin shock**, may accompany antibiotic treatment of diseases due to Gram-negative bacilli because endotoxins are released as the bacilli disintegrate.

ko-ri"ne-bac-te're-um dif-the're-ā

Antitoxins: antibodies that neutralize toxin molecules.

Toxoid: an altered toxin used for immunization purposes.

Endotoxins: microbial poisons released upon disintegration of the parasite cell.

Endotoxin shock: a condition arising from release of endotoxins and characterized by low blood pressure, swollen tissues, and shock.

Endotoxins usually play a contributing rather than a primary role in the disease process. Certain endotoxins reduce platelet counts in the host and thereby hinder clot formation. Other endotoxins are known to increase hemorrhaging. Like exotoxins, the endotoxins add to the virulence of a parasite and enhance its ability to establish disease.

The Human Circulatory System

The circulatory system of the human host serves as the principal vehicle for the dissemination of parasites and their toxins. Many important factors in the host defense systems also operate within the circulation. These two concepts add to the importance of a basic understanding of the circulatory system and its capabilities. Our outline of the system will also serve as a bridge between the disease process discussed in this chapter and the resistance process surveyed in the next chapter. The emphasis will be on the blood and its components.

Blood Components

Serum:
the fluid portion of the blood minus the clotting agents.

Plasma:
serum plus the clotting agents in blood.

There are three major components of the blood: the fluid, the clotting agents, and the cells. The fluid portion, called **serum**, is an aqueous solution of minerals, salts, proteins, and other organic substances. When clotting agents such as fibrinogen and prothrombin are present, the fluid is referred to as **plasma**. The pH of arterial blood is about 7.35 to 7.45.

Three types of cells circulate in the blood: the red blood cells, or erythrocytes; the white blood cells, or leukocytes; and the platelets, or thrombocytes. **Erythrocytes** arise in the bone marrow and carry oxygen to the tissues loosely bound to the red pigment hemoglobin. A normal adult has about 5 million erythrocytes per cubic millimeter (mm^3) of blood. After circulating for about 120 days, erythrocytes disintegrate in the spleen, liver, and bone marrow. The hemoglobin is then converted to bilirubin, a pigment that gives bile a deep yellow color. Normally, bilirubin is carried to the liver for degradation in a protein-bound form. However, if the liver is damaged by a disease, or if too many erythrocytes disintegrate, excess bilirubin and bile pigments may enter the bloodstream. This condition, called **jaundice**, is responsible for the yellow color of the complexion during cases of hepatitis and yellow fever.

Bilirubin:
a pigment derived from hemoglobin that gives bile a deep yellow color.

The white blood cells, or **leukocytes**, have no pigment in their cytoplasm and therefore they appear gray when unstained. These cells are also produced in the bone marrow. They number about 5000 to 9000 per mm^3 of blood and have different life spans depending on the type of cell. Table 17.3 displays the different types of white blood cells.

pol″e-mor″fo-nu′kle-ar

One type of leukocyte has a multilobed nucleus and is therefore referred to as a **polymorphonuclear cell**, or **PMN**. In its cytoplasm it has many lysosomes that contain digestive enzymes. The cell is also called a neutrophil, because its granules stain with neutral dyes. **Neutrophils** (Figure 17.9) function chiefly as phagocytes. They pass out of the circulation through pores in the vessels and squeeze into narrow passageways among the cells to engulf particles. Approximately 55 to 60 percent of the leukocytes are neutrophils. Their life span is about 12 hours.

ba′so-fil

Basophils are types of leukocytes whose granules stain with basic dyes such as hematoxylin. The basophils number about 50 to 90 per mm^3 of blood and represent about 1 percent of the total number of leukocytes. They function in

---

TABLE 17.3

The Cellular Composition of Human Blood

TYPES OF CELLS			NUMBER OF CELLS PER CUBIC MILLILITER	
Erythrocytes		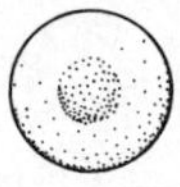	Men: 4,900,000–5,500,000 Women: 4,400,000–5,000,000	
				Approximate percentage
Leukocytes			5000–9000	
Polymorphonuclear cells (granulocytes)	Neutrophil	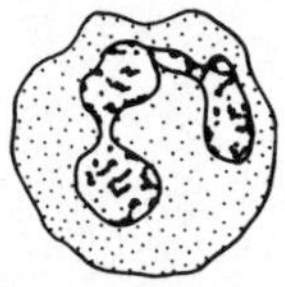	2875–5175	55–60
	Eosinophil	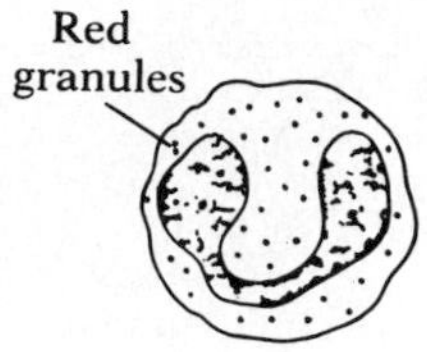 Red granules	100–180	2
	Basophil	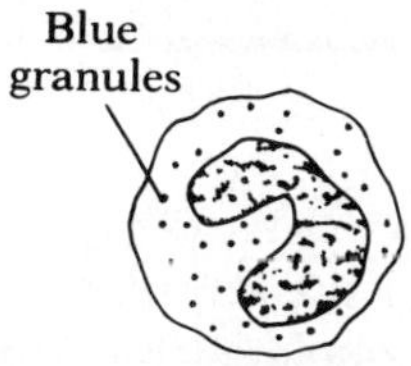 Blue granules	50–90	1
Mononuclear cells (agranulocytes)	Monocyte	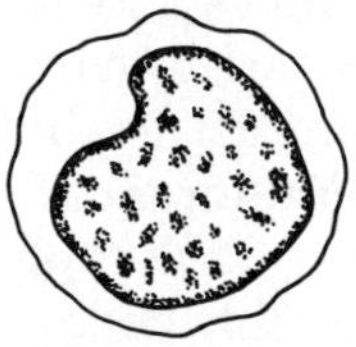	525–607	5–8
	Lymphocyte	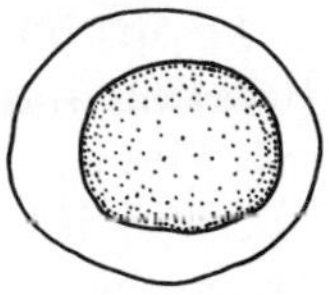	1625–2925	30–35
Platelets			250,000–400,000	

---

allergic reactions as their granules release physiologically active substances such as histamine. This process is discussed in detail in Chapter 20.

Eosinophils are the third type of leukocyte, representing about 2 percent of the total number. These cells exhibit red cytoplasmic granules when an acidic dye such as eosin is applied. Substances in the granules are thought to neutralize the active chemicals in basophil granules. (Eosinophils, basophils, and neutrophils are often called granulocytes.) Eosinophils are postulated to have phagocytic activity.

Histamine:
a physiologically active substance that contributes to the contraction of smooth muscles.

e″o-sin′o-fil

e′o-sin

FIGURE 17.9

A View of Neutrophils

A scanning electron micrograph of normal neutrophils. Note the irregular shapes of the cells, demonstrating the amebalike projections used in phagocytosis and mobility.
Bar = 10 μm.

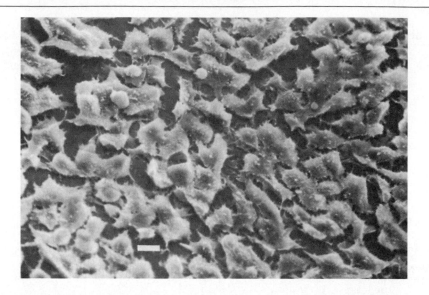

Macrophage:
a type of large monocyte that functions as a phagocyte in the tissues.

Lymphocyte:
a type of leukocyte associated with the body's immune system.

Another major phagocyte of the circulatory system is the **monocyte**. This cell has a single, bean-shaped nucleus that takes up most of the area of the cytoplasm. Monocytes lack granules and account for approximately 5 to 8 percent of the leukocytes. In the tissues, monocytes mature into a type of phagocyte called the **macrophage**. In contrast to the two-week life span of monocytes, macrophages may live for several months.

Another type of leukocyte is the **lymphocyte**. This cell arises in the bone marrow and migrates to the lymph nodes after modification. It has a single, large nucleus and no granules. (Monocytes and lymphocytes are often called agranulocytes.) Lymphocytes make up about 30 to 35 percent of the white blood cells in the human body. They function in the immune system as B-lymphocytes and T-lymphocytes (Chapter 18). Their numbers increase dramatically during the course of certain diseases, such as infectious mononucleosis.

Blood platelets, or **thrombocytes**, represent the third type of cell in the circulatory system. They are small disk-shaped ("plate-let") cells that originate from cells in the bone marrow. Platelets have no nucleus and function chiefly in the blood clotting mechanism.

The Lymphatic System

Lymph:
the cell-free fluid that surrounds tissue cells and fills the intercellular spaces.

The fluid that surrounds the tissue cells and fills the intercellular spaces is called tissue fluid, or **lymph**. Lymph is similar to serum except that lymph has fewer proteins. Lymph bathes the body cells, supplying oxygen and nutrients while collecting wastes. It is pumped along in tiny vessels by the contractions of skeletal muscle cells. Eventually the tiny lymph vessels unite to form large vessels that compose a lymphatic system. On the right side, the system empties into a large vein just before the heart. Figure 17.10 illustrates how the lymphatic system interrelates with the circulatory system in humans.

FIGURE 17.10

How the Lymphatic System Interrelates with the Circulatory System

Blood fluid passes out of the arteries in the upper and lower parts of the body. It enters a system of lymphatic ducts that arise in the tissues. The fluid, called lymph, passes through lymph nodes and on the right side makes its way back to the general circulation via a duct called the thoracic duct. The thoracic duct enters a main vein just before the vein enters the heart. A similar system exists on the left side.

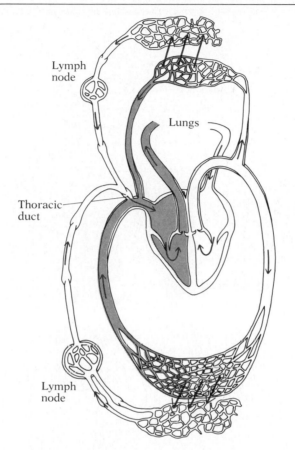

Lymph node

Lungs

Thoracic duct

Lymph node

Pockets of lymphatic tissue located along the lymph vessels are known as **lymph nodes**. Lymph nodes are prevalent in the neck, armpits, and groin. They are bean-shaped organs containing phagocytes that engulf particles in the lymph and lymphocytes that respond specifically to substances in the circulation. Since resistance mechanisms are closely associated with the lymph nodes, it is not surprising that they become enlarged during time of disease (sometimes they are called "swollen glands"). The tonsils, adenoids, spleen, Peyer's patches of the intestine, and appendix are specialized types of lymph nodes. Their locations are noted in Figure 17.11.

Optimal functioning of the circulatory system is an essential prerequisite to good health. Parasites possess a wealth of virulence factors that add to pathogenicity, but hosts have an equally formidable array of defensive capabilities. For example, in the 1300s an estimated one-third of the population of Europe fell to bubonic plague. However, two-thirds survived, and without the benefit of antibiotics, antibody injections, or other treatments now used for disease. The nature and function of the defensive mechanisms that led to this survival and the resistance modes that operate in all humans are the major topics of Chapter 18.

Lymph nodes:
pockets of lymphocytes, phagocytes, and lymphatic tissue along the lymph vessels.

FIGURE 17.11

The Human Lymphatic System

The human lymphatic system consists of lymphocytes, lymphatic organs, lymph vessels, and lymph nodes located along the vessels. The lymphatic organs are illustrated and the preponderance of lymph nodes in the neck, axilla, and groin is apparent.

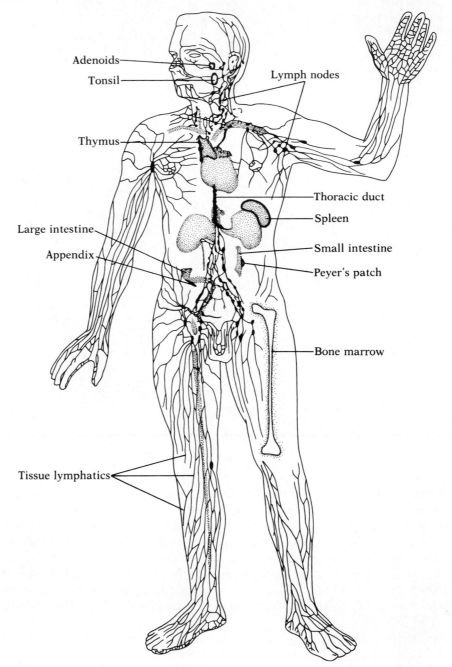

NOTE TO THE STUDENT

You may have noted that infection, disease, and host death are widely separated phenomena with many intermediary steps. Consider, for example, what happens when cholera strikes. The bacillus, Vibrio cholerae, *multiplies along the walls of the intestine and produces toxins. The toxins cause such massive diarrhea that up to six quarts of fluid may be lost over the next several hours. The blood thickens, urine production ceases, the skin becomes dry, wrinkled, and cold, and the sluggish flow of blood to the brain leads to shock, coma, and death.*

Now consider where this leaves the cholera bacilli. After the host dies, the bacilli are pushed to the side by the enormous number and variety of other microorganisms normally found in the intestine. For the cholera bacilli, their source of nourishment is gone and the environment is now filled with hostile predators. The situation has gotten thoroughly out of hand and soon, the cholera bacilli will perish just as the host perished.

It is clear that, in the long run, host death is rarely beneficial to the parasite. Is it possible that death of the host is, in fact, a biological accident?

Summary

A review of certain concepts of infections and disease and a discussion of the factors leading to the establishment of disease are the two principal thrusts of this chapter. Infection refers to the relationship between two organisms, the host and the parasite, and the competition taking place between them. Disease may be considered a change from the general condition of good health arising from the parasite's victory in the competition. Pathogenicity, the ability of the parasite to cause disease, is fundamentally inseparable from the host's ability to resist disease. Thus, disease and resistance go hand-in-hand. Virulence refers to the degree of pathogenicity that a parasite displays.

The progress of disease follows a regular pattern that includes periods of incubation, prodrome, acme, decline, and convalescence. Disease may be transmitted by direct as well as indirect methods and may be communicable such as measles, or noncommunicable such as tetanus. Disease may also be classified as acute or chronic, primary or secondary, local or systemic.

A parasite must have unusual abilities to bring about disease, and certain conditions must exist if disease is to be established. For example, the correct portal of entry and dose must be fulfilled. The possibility of disease is enhanced if a parasite has the ability to penetrate tissue or produce enzymes to overcome body defenses or synthesize toxins to interfere with body processes. Toxins can be classified as exotoxins or endotoxins depending on their physiology and activity. Consideration of factors as these help us appreciate why a parasite is pathogenic. An understanding of the blood and circulation is important because the disease process often occurs here and because resistance arises from factors in the circulatory system.

Questions for Thought and Discussion

1. In 1840 Great Britain introduced penny postage and issued the first adhesive stamps. However, politicians did not like the idea because it deprived them of the free postage they were used to. Soon, a rumor campaign was started saying that these "gummed" labels could spread disease among the population. Can you see any wisdom in their contention? Would their concern "apply" today?

2. A woman takes an antibiotic to relieve a urinary tract infection caused by *Escherichia coli*. The infection resolves, but in two weeks she develops a *Candida albicans* infection of the vaginal tract. What conditions may have caused this to happen? What nonantibiotic course of treatment might a doctor prescribe to solve this problem?

3. How might you describe the characteristics of a chronic, subclinical, systemic, non-communicable, secondary disease?

4. The transparent windows placed over salad bars are commonly called "sneeze bars" because they help prevent nasal droplets from reaching the salad items. What other suggestions might you make to prevent disease transmission via the salad bar?

5. In 1892, a critic of the germ theory of disease named Max von Pettenkofer sought to discredit Robert Koch's work by drinking a culture of cholera bacilli diluted in water. Von Pettenkofer suffered nothing more than mild diarrhea. What factors may have contributed to the failure of the bacilli to cause cholera in his body?

6. It is generally conceded that bad breath is due to the gases produced by the oral bacteria that digest particles of food and other organic materials in the mouth. With this in mind, can you guess why "morning breath" tends to be worse than breath odor at other times of the day?

7. While slicing a piece of garden hose, a certain textbook author cut himself with a sharp knife. The wound was deep but it closed quickly. Shortly thereafter, he reported to the emergency room of the community hospital, where he received a tetanus shot. What did the tetanus shot contain and why was it necessary?

8. Suppose there were no infectious diseases to worry about. Would that be beneficial to the human race?

9. It has been estimated that in a sneeze, 4600 droplets are shot forth with a muzzle velocity of 152 feet per second. Scientists believe that the droplets from a single sneeze hang suspended in the air for up to 30 minutes, and collectively may contain over 35 million viruses. How many diseases can you name that are transmitted by droplets such as these?

10. In October 1995, an article entitled "Dr. Darwin" appeared in *Discover* magazine. The thrust of the article was that disease symptoms (such as fever and diarrhea) are a product of the evolutionary development of humans and are meant to be defensive mechanisms helping us to cope with disease. Presumably, the symptomatic responses to infection would make the body inhospitable to the invading parasites. What do you think of this theory? How might buying into this theory alter your approach the next time you become infected?

11. In the mid-1980s, epidemiologists reported that shigellosis broke out in two different day-care centers in Seattle, Washington. Health officials noted that one difficulty experienced in controlling outbreaks of shigellosis was the low dose levels of the bacilli. What does this mean?

12. A man takes a roll of dollar bills out of his pocket and "peels" off a few to pay the restaurant tab. Each time he peels, he wets his thumb with saliva. What is the hazard involved?

13. In your opinion, would an epidemic disease or an endemic disease pose a greater threat to public health in the community? Why? Given the choice, would you rather experience a chronic disease or an acute disease?

14. While touring a day-care center, a visitor notices that a nail brush is next to each sink. When queried as to the reason, the center director replies that all personnel are requested to use the brush after they wash their hands, especially after changing a child's diaper. Why is this a good idea?

15. When Ebola fever broke out in Africa in 1995, public health epidemiologists noted how quickly the responsible virus killed its victims and guessed that the epidemic would end shortly. Sure enough, within three weeks it was over. What was the basis for their prediction? What other conditions had to apply for them to be accurate in their guesswork?

Review

When you have finished this chapter test your knowledge of its contents by determining whether the following statements are true or false. If the statement is true, write the word "true" in the space. If false, substitute a word for the underlined word to make the statement true.

1. _____ An <u>epidemic</u> disease is one that occurs at a low level in a certain geographic area.

2. _____ A parasite that is <u>invasive</u> has the ability to penetrate tissues and cause structural damage.

3. _____ Among the microbial enzymes that are able to destroy blood cells are <u>hemolysins</u> and leukocidins.

4. _____ The term <u>disease</u> refers to a living together of two organisms and the competition that takes place between them for supremacy.

5. _____ Organs of the human body that do not have a normal flora include the blood and the <u>small intestine</u>.

6. _____ A <u>commensalism</u> is a form of symbiosis in which only one organism benefits but no damage occurs to the other.

7. _____ An organism with <u>high</u> virulence is generally unable to cause disease in the body.

8. _____ The period of <u>decline</u> follows the period of prodromal symptoms in the progress of a disease.

9. _____ A <u>biological</u> vector is an arthropod that carries pathogenic micro-organisms on its feet and body parts.

10. _____ Those organisms that cause disease when the immune system is depressed are known as <u>opportunistic</u> organisms.

11. _____ A <u>reservoir</u> is one who has recovered from a disease but continues to shed the disease agents.

12. _____ The human body responds to the presence of toxins by producing <u>endotoxins</u>.

13. _____ The term <u>bacteremia</u> refers to the spread of bacteria through the bloodstream.

14. _____ A pathogenic staphylococcus is able to form a fibrin clot through its production of <u>hyaluronidase</u>.

15. _____ A <u>toxoid</u> is an immunizing agent prepared from an exotoxin.

16. _____ Few symptoms are displayed by an individual who has a <u>subclinical disease</u>.

17. _____ Among the <u>indirect</u> methods of disease transmission are kissing, hand-shaking, and contact with feces.

18. _____ Certain parasites such as the bacterium that causes <u>tuberculosis</u> have multiple portals of entry.

19. _____ A <u>chronic</u> disease is one that develops rapidly, is usually accompanied by severe symptoms, and comes to a climax.

20. _____ The period of <u>acme</u> is the time between the entry of the parasite to the host and the appearance of symptoms.

CHAPTER 18

RESISTANCE AND THE IMMUNE SYSTEM

T hink of the human body as a doughnut. Just as the hole passes through the center of the doughnut, so too the gastrointestinal tract passes through the center of the human body. The body secretes enzymes out into the hole, digests the food we eat, absorbs what it wants, and lets the remainder pass out of the hole. There is no natural opening to the internal tissues in the gastrointestinal tract, and the fact that something is in the GI tract does not mean that it is in the body. The same holds true for the respiratory and urinary tracts: the respiratory tract leads to dead-end pouches of the alveoli, while the urinary tract leads first to the urinary bladder and then to the kidneys, where it terminates at the cuplike structures called Bowman's capsules. In both instances, a layer of cells shields the blood system and the body's internal organs from the outside environment. Close examination of other parts of the body reveals similar dead ends, and it becomes clear that the body, like the doughnut, is a closed container. Only after the walls of this container are penetrated can most diseases be established.

The skin and its extensions into the gastrointestinal, respiratory, and urinary tracts represent a major form of resistance to infection and disease. This form is **nonspecific**, because it exists in all humans and is present from the earliest time of life. Also, it protects against all parasites. Other forms of resistance are **specific**, because they come about in response to a particular parasite and are directed solely at that parasite (Figure 18.1). The major thrust of this chapter will be to examine both forms of resistance and to show how good health depends upon their proper functioning. The discussion is related to the previous chapter's study of disease, except that where the emphasis was formerly on the parasite, it now turns to the host.

Nonspecific Resistance

Nonspecific resistance to disease involves a broad series of factors, many of which are still not defined. It depends upon the general well-being of the individual and proper functioning of the body's systems. Accordingly, it takes into

FIGURE 18.1

The Relationship Between Host Resistance and Disease

(a) Host resistance to microorganisms and numerous other agents depends upon multiple factors that must function well in the individual. In the diagram, the factors to the left of the umbrella stem are nonspecific factors, while those to the right are specific factors. (b) Resistance may begin to break down when one or more factors are inoperable. In the diagram, phagocytosis fails to take place and some infectious agents penetrate the umbrella of defense. (c) Disease develops when many host defenses are compromised. Under these conditions, the body is unable to defend itself. Even when defenses are not compromised, the aggressiveness and toxicity of the pathogen may lead to infection.

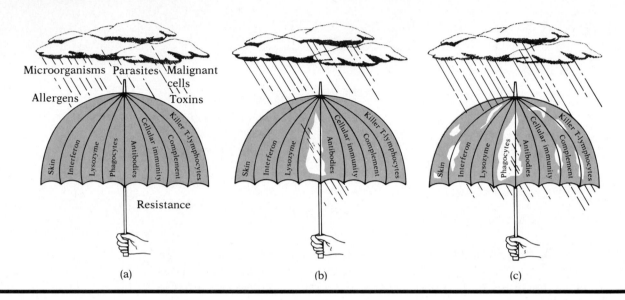

(a)　　　　　(b)　　　　　(c)

account such determinants as nutrition, fatigue, age, sex, and climate. Specific examples of these factors are highlighted in discussions of individual diseases, and therefore we shall not pause to delineate them here.

One form of nonspecific resistance is called **species immunity**. This implies that diseases affecting one species will not affect another. For example, humans do not contract hog cholera, while hogs do not contract AIDS (their cells lack the necessary receptor sites for HIV). Similarly, cattle plague is unknown in humans, while gonorrhea does not occur in cattle. Immunities such as these are probably based on physiological, anatomical, and biochemical differences. In chickens, a physiological difference lends resistance to anthrax. The normal body temperature of chickens is 45° C, and anthrax bacilli do not grow well at this elevated temperature. However, when the temperature is lowered to 37° C, susceptibility increases. This phenomenon, first demonstrated by Pasteur in 1878, reduces resistance to a simple factor.

Behavioral immunities exist among various races and peoples of the world. Many of these immunities are due to nonspecific factors related to a people's way of life. For example, in the 1700s, Tamil laborers were brought from southern India to work on plantations in Malaya. The laborers continued the custom of bringing water into their houses only once a day and not storing it between times. This deprived mosquitoes of indoor breeding places and reduced the incidence of malaria among the laborers.

Racial immunities reflect the evolution of resistant humans. For instance, black Africans affected by the genetic disease sickle cell anemia do not suffer **malaria**, presumably because the parasite cannot penetrate distorted red blood cells. Some investigators hold to the theory that population immunities exist

Species immunity:
immunity existing in one species of organisms but not other species.

Anthrax:
a serious disease of the blood caused by *Bacillus anthracis* and accompanied by severe hemorrhaging and lesion formation.

Racial immunity:
immunity existing in one race of humans but not other races.

Sickle cell anemia:
an inherited disease in which a defect in the genetic code leads to distorted red blood cells.

18.1

MICROFOCUS

CONQUEST BY DISEASE

History books teach that Spain's conquest of the Aztec nation of Central America was due to horses, gunpowder, and the superior force of Spanish arms. But conquest meant overcoming millions of people and overturning long-standing traditions of religion and culture; and there were only 800 men with Hernando Cortez the day he landed in Mexico in 1518. Cortez and the Spanish eventually toppled the Aztec nation, but their strongest ally was not gunpowder and arms; it was disease.

Mexico was totally unprepared for smallpox. The disease was new to the country, and the Aztec population was without any trace of immunity. The Spanish, on the other hand, had contended with smallpox for generations, and they were relatively immune. Little thought was given to the consequences when sick slaves arrived with the Spaniards. Soon the Aztec community was infected.

By April 1521, Cortez had established a colony on Mexican soil and marched inland to attack Mexico City, the stronghold of the Aztec nation. The siege continued for four months, until the city finally fell on August 13, 1521. Expecting to plunder the city, the Spanish rushed in, but were shocked to find the houses filled with dead people. A smallpox epidemic was raging. Half the population had succumbed.

Nor did it end here. Eleven years later, Spanish invaders introduced another epidemic of disease, now believed to be measles. Thousands of Aztecs died. In 1545, disease broke out again. Contemporary writings describing the symptoms indicate that it was either epidemic typhus or typhoid fever. In one province 150,000 people are estimated to have died. Influenza raged in 1558 and 1559, and mumps broke out with fatal consequences in 1576.

By the end of the century, the Aztec nation was battered into submission. Native authority figures crumbled and the surviving Aztecs dutifully obeyed the commands of Spanish landowners, tax collectors, and missionaries. To both the conqueror and conquered, the divine and natural orders

had spoken out loudly against the native beliefs; the Aztecs would offer no further resistance. By one historian's account, almost 19 million of the original Aztec population of 25 million was destroyed by 1595.

because parasites have adapted to the body's environment. Americans, for example, generally view **measles** as a mild disorder, but when the disease was introduced to Greenland in the early 1960s, it exacted a heavy toll of lives in a population that had no previous exposure to the virus. A similar event took place when the Spanish conquistadors arrived in the New World in the 1500s (MicroFocus: 18.1).

In addition to species and other immunities, the body possesses a number of identifiable processes for nonspecific resistance. Like the immunities explored previously, they are present in the body from birth, but they differ because they operate against all parasites. We shall review a number of these processes next.

Mechanical and Chemical Barriers

As noted in the introduction to this chapter, the intact **skin** and the **mucous membranes** that extend into the body cavities are among the most important resistance factors. Toxins notwithstanding, unless penetration of these barriers occurs, disease is rare.

But penetration of the skin barrier is a fact of everyday life. A cut or abrasion, for example, allows staphylococci to enter the blood, and an arthropod bite acts as a hypodermic needle permitting many different organisms to enter. Yellow fever viruses, malaria parasites, certain rickettsiae, and plague bacilli are but a few examples. Other means of penetrating the barrier include splinters, tooth extractions, burns, shaving nicks, war wounds, and injections.

Certain features of the mucous membranes provide resistance to parasites. For instance, cells of the mucous membranes along the lining of the respiratory passageways secrete **mucus**, which traps heavy particles and microorganisms. The **cilia** of other cells then move the particles along the membranes up to the throat, where they are swallowed. Stomach acid now destroys any microorganisms.

Resistance in the vaginal tract is enhanced by the low pH. This develops when *Lactobacillus* **species** in the normal flora break down glycogen to various acids. Many researchers believe that the disappearance of these bacilli during antibiotic treatment encourages diseases such as candidiasis and trichomoniasis to emerge. In the urinary tract, the slightly acidic pH of the urine promotes resistance to parasites, and the flow of urine flushes microorganisms away.

A natural barrier to the gastrointestinal tract is provided by **stomach acid**, which has a pH of approximately 2.0. (A cotton handkerchief placed in stomach acid would dissolve in a few short moments.) Most organisms are destroyed in this environment. Notable exceptions include typhoid and tubercle bacilli, protozoal cysts, and polio and hepatitis A viruses. **Bile** from the gall bladder enters the system at the duodenum and serves as an inhibitory substance. In addition, duodenal enzymes digest the proteins, carbohydrates, fats, and other large molecules of microorganisms.

A chemical inhibitor of a nonspecific nature is the enzyme **lysozyme**. This protein was described in the early 1920s by Alexander Fleming, who later gained recognition for the discovery of penicillin. Lysozyme is found in human tears and saliva. It disrupts the cell walls of Gram-positive bacteria by digesting peptidoglycan. Another inhibitor is **interferon**. Interferon is actually a group of substances (interferons) produced by body cells in response to invasion by viruses. Interferons trigger the production of inhibitory substances that "interfere" with viral reproduction. A thorough account of the interferons is presented in Chapter 11. Table 18.1 summarizes many of the nonspecific resistance mechanisms.

Mucus:
a thick fluid secreted by membranes that line body cavities opening to the exterior.

kan″dĭ-di′ah-sis
trik″o-mo-ni′ah-sis

Bile:
the yellow-brown mixture of acids, salts, pigments, and other substances produced by the liver and stored in the gall bladder for fat digestion.
li′so-zīm

in″ter-fēr′on

TABLE 18.1

Nonspecific Mechanical and Chemical Barriers to Disease

RESISTANCE MECHANISM	ACTIVITY
Skin layers	Provide a protective covering to all body tissues
Mucous membranes of body cavities	Trap airborne particles in mucus Sweep particles along by cilia
Acidity in the vagina and stomach	Acidic pH toxic to microorganisms
Bile	Inhibitory to most microorganisms
Duodenal enzymes	Digest structural and metabolic chemical components of microorganisms
Lysozyme in tears, saliva, secretions	Digests cell walls of Gram-positive bacteria
Interferons	Inhibit replication of viruses

FIGURE 18.2
Phagocytosis

A transmission electron micrograph of a dividing bacillus being engulfed by a polymorphonuclear cell. The extensions of the PMN surround the bacilli, and membranes of the phagocytic vesicle follow the contours of the organism engulfed.

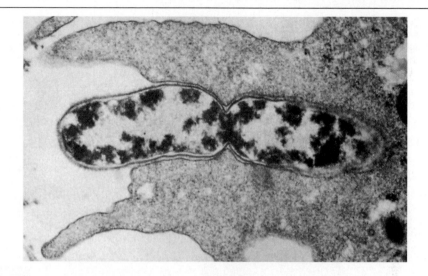

Phagocytosis

Shortly after the germ theory of disease was verified by Koch, a native of the Ukraine named **Elie Metchnikoff** made a chance discovery that clarified how living cells could protect themselves against microorganisms. Metchnikoff noted that motile cells in the larva of a starfish gathered around a wooden splinter placed within the cell mass. He suggested that the cells actively sought out and engulfed foreign particles in the environment to provide resistance. Metchnikoff's theory of **phagocytosis**, published in 1884, was received with skepticism, because it appeared to conflict with the antitoxin theory. Many investigators believed that antitoxins produced by the body against toxins were the sole basis for resistance, but in succeeding years they came to appreciate phagocytosis as equally important. Metchnikoff later became an associate of Pasteur and was a co-recipient of the 1908 Nobel Prize in Physiology or Medicine.

In contemporary microbiology, phagocytosis is viewed as a major form of nonspecific defense in the body. The cells involved are called **phagocytes** (Figure 18.2). They are polymorphonuclear cells (PMNs) and monocytes of the circulatory system (Chapter 17), as well as cells of the **reticuloendothelial system (RES)**. This system, also called the **mononuclear phagocyte system**, is a collection of monocyte-derived cells that leave the circulation and undergo modification in the tissues. They include Kupffer cells of the liver and **macrophages** of the spleen, bone marrow, lymph nodes, brain, and connective tissues. RES cells are larger than monocytes and have more lysosomes and a longer life span. Some phagocytes are termed **resting cells** because they are stationary, while other phagocytes are **wandering cells** because they are actively motile.

Phagocytosis begins with an invagination and pinching of the cell membrane to form a phagocytic vesicle, or **phagosome** (Figure 18.3). The phagosome then fuses with a **lysosome**, an organelle that contributes digestive enzymes, lysozyme, and an acidic pH to the digestion process. Lysosomal substances also increase the permeability of capillaries, which brings more phagocytes

Metch′ni-koff

fag″o-sī-to′sis

Antitoxins:
antibodies produced specifically against toxins.

Monocytes:
large, phagocytic white blood cells having a single, bean-shaped nucleus.

Lysosome:
a saclike organelle in eukaryotic cells that contains digestive enzymes and other factors to aid digestion.

FIGURE 18.3
The Mechanism of Phagocytosis

In this process, (a) the macrophage attaches to the bacterium assisted by antibodies called opsonins. (b) Pseudopodia then surround the bacterium and (c) bring it into the cytoplasm of the macrophage where union with lysosomes takes place. (d) Within the newly formed phagolysosome, the bacterium disintegrates through the activity of lysosomal enzymes. (e) The process concludes with the elimination of bacterial debris during egestion. (f) A scanning electron micrograph showing a phagocyte engulfing a yeast cell.

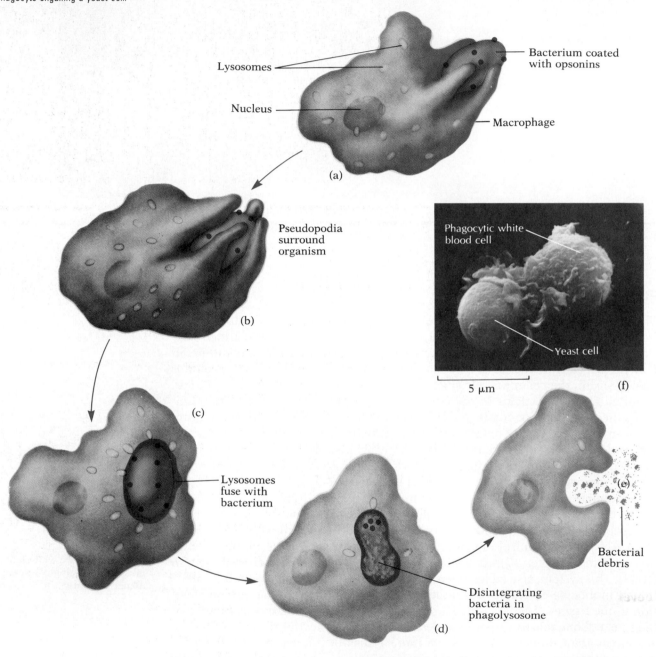

to the area. The process is completed as waste materials are egested from the phagocyte.

A chemical attraction called a **chemotaxis** exists between parasite and phagocyte. This attraction is mediated by an unidentified substance released by the parasite. The interaction between parasite and phagocyte is also enhanced by the presence of antibodies. These protein molecules attach to parasites and thereby increase adherence to the phagocytic cell at specific receptor sites. In other situations, components of the complement system (to be discussed presently) bind the parasite to the phagocyte. Enhanced phagocytosis is called **opsonization**, and the antibodies or complement components that encourage it are termed **opsonins**, from Latin stems meaning "to prepare for." Opsonins were first described in 1903 by Almroth Wright, who imagined them as serum proteins that increase the susceptibility of parasites to phagocytosis. At the time it was a way of unifying Metchnikoff's phagocytosis theory with the anti-toxin theory.

Complement:
a series of 11 proteins that function in disease resistance.

Inflammation

Inflammation is a nonspecific defensive response by the body to an injury in the tissue. It develops after a mechanical injury such as an injury or blow to the skin, or from exposure to a chemical agent such as acid or bee venom. It may also be due to a physical agent such as heat or ultraviolet radiation, or to a living organism such as a parasite.

The irritant sets into motion a process that limits the extent of the injury (Figure 18.4). Dilation of blood vessels leads to increased capillary permeability, followed by a flow of plasma into the tissue and fluid accumulation at the site of irritation. Neutrophils adhere to the vessels close to the injury and migrate through the wall to begin phagocytosis of the irritant. Soon, macrophages replace the neutrophils. The area shows the four characteristic signs of inflammation: a red color from blood accumulation (**rubor**); warmth from the heat of the blood (**calor**); swelling from the accumulation of fluid (**tumor**); and pain from injury to the local nerves (**dolor**).

Neutrophils:
phagocytic white blood cells having a multilobed nucleus.

A product of phagocytosis during inflammation is the mixture of plasma, dead tissue cells, leukocytes, and dead bacteria known as **pus**. When this material becomes enclosed in a wall of fibrin through activation of the clotting mechanism, a sac may form. This sac is the **abscess**, or **boil**. When several abscesses accumulate, an enlarged structure called a **carbuncle** results.

Abscess:
a fibrin-enclosed sac of pus.

Inflammation and phagocytosis are thus related to each other. The object is to confine the irritant to the site of entry and to repair or replace the tissue that has been injured.

Fever

Fever is an abnormally high body temperature that may provide a nonspecific mechanism of defense to disease. Scientists believe that bacteria, viruses, and other microorganisms affect a region at the base of the brain called the **hypothalamus** and stimulate it to raise the body temperature several degrees. As this takes place, cell metabolism increases and blood vessels constrict, thus denying blood to the skin and keeping its heat within the body. Patients thus experience cold skin and chills along with the fever.

Fever may be beneficial because it appears to inhibit the growth of certain organisms. The increased metabolism in cells also encourages rapid tissue repair and increases the level of phagocytosis. However, if the temperature rises above 45° C, convulsions and death may occur.

FIGURE 18.4

The Process of Inflammation

(a) A nail pierces the skin causing a mechanical injury and bringing bacteria into the tissue. (b) Capillary walls open (dilate) and plasma flows to the site of injury making it red, swollen, and warm. (c) Neutrophils arrive and begin phagocytosis, and macrophages follow to continue the process. (d) A fibrin wall accumulates around the mixture of plasma, leukocytes, bacteria, and tissue cells, collectively called pus. An abscess becomes apparent. (e) Lancing of the abscess releases the pus.

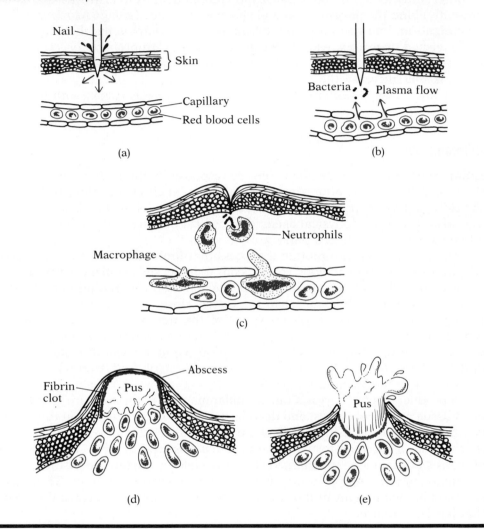

Natural Killer Cells

Natural killer (NK) cells are a unique group of defensive cells that roam the body in blood and lymph and kill cancer cells and virus-infected cells before the immune system is enlisted. NK cells were originally believed to be lymphocytes that help mediate the immune response, but this outlook has proven untrue, because NK cells act spontaneously and without any involvement of the immune system. The name "natural" killer cells reflects the nonspecific nature of the killing activity.

**TO
THIS
POINT**
■

We have begun the study of resistance to disease by outlining some nonspecific mechanisms that defend against all parasites from the time of birth. Species and racial immunities are examples. Other examples are mechanical and chemical barriers, including the barriers presented by the skin and mucous membranes, the acidity in the stomach and vagina, and the antimicrobial substances lysozyme and interferon.

We examined phagocytosis in depth since this process is a major form of nonspecific defense. The process may involve particles as well as dissolved materials. Chemotaxis and antibodies increase the efficiency of phagocytosis. We also noted the value of inflammation, fever, and natural killer cells in nonspecific resistance.

We shall now focus on specific resistance as reflected by the immune system. After a brief discussion of the beginnings of immunology, we shall survey antigens, the substances that stimulate the immune response. From there, we shall explore the origin of the immune system and see how it is established in the body. The system is actually dual in nature, although the processes are interlocking. The discussion will therefore diverge, first to study the system of cellular immunity and later, to examine antibody-mediated immunity. In a sense this is the most important section in the entire textbook because it outlines the processes by which specific resistance to specific parasites takes place. How well the resistance operates dictates whether a person will remain free of infectious disease or will recover from disease.

Specific Resistance and the Immune System

In the late 1800s, the mechanisms of specific resistance to disease were largely obscure because no one was really sure how the body responded to disease. However, medical bacteriologists were aware that certain proteins of the blood unite specifically with chemical compounds of microorganisms. The blood proteins were named **Bence Jones proteins** after Henry Bence Jones, a British physician who identified them in the urine of a patient in 1847. In 1922, researchers from Johns Hopkins University showed that Bence Jones proteins (later found to be parts of antibodies) were unlike normal serum proteins. Despite this finding, the nature of the proteins was still a matter of conjecture a generation later.

Until the 1950s, specific resistance to disease was virtually synonymous with immunity. By that time, vaccines were available for numerous diseases, and immunologists saw themselves as specialists in disease prevention. But the explosion of interest in the biological sciences after World War II spilled over to immunology, and soon it became apparent that **specific resistance** is a phenomenon with broader implications. For example, organ transplantation is associated with immunology because the rejection mechanism is a type of specific resistance. Researchers also found that allergic reactions and resistance to cancer could be explained in immunological terms. In addition, both diagnostic laboratories and vaccine research were reaping benefits from advances in immunology; and the groundwork was being laid for deciphering the nature and function of the Bence Jones proteins. This work would lead to the elucidation of the antibody structure in the 1960s, and the maturing of immunology to one of the key scientific disciplines of our times.

In this section, we shall study the immune system as it relates to specific resistance to disease. But it should be noted that the immune system also responds to myriad chemical substances in the natural environment and, in some cases, to a person's own chemical compounds. Therefore our study of specific resistance and the immune system will take into account the infectious disease patient as well as the cancer patient, transplant recipient, laboratory diagnostician, and person who has hay fever. Some of these topics are explored in detail in Chapters 19 and 20. Research in immunology provides an important window to the disease process as well as many other processes of life. Our study will begin with a survey of the substances that stimulate the immune response.

Antigens

Antigens:
chemical substances that elicit a response by the body's immune system.

Antigens are chemical substances capable of mobilizing the immune system and provoking an immune response. Most antigens are large, complex molecules ("macromolecules"), which are not normally found in the body and are consequently referred to as "nonself." Antigens exhibit two important properties: **immunogenicity**, the ability to stimulate cells of the immune system; and **reactivity**, the ability to react with products of the immune system cells or the cells themselves.

The list of antigens is enormously diverse. It includes milk proteins, substances in bee venom, hemoglobin molecules, bacterial toxins, and chemical substances found in bacterial flagella, pili, and capsules (Figure 18.5). The most common antigens are proteins, polysaccharides, and the chemical complexes formed between these substances and lipids or nucleic acids. Proteins are the most potent antigens because their amino acids have the greatest array of building blocks, an array leading to a huge variety of combinations and hence, diversity in three-dimensional structures. Among the polysaccharide antigens are the capsular polysaccharides of pneumococci and the blood group antigens A and B. Polysaccharides are less potent antigens than proteins because they lack chemical diversity and rapidly break down in the body environment.

Dalton:
a unit of weight equal to the weight of one hydrogen atom.

Antigens usually have a molecular weight of over 10,000 daltons. Because of this large size, the antigen molecule is easily phagocytized by macrophages, the necessary first step in the immune process. Antigens do not stimulate the immune system directly. Rather, the stimulation is accomplished by a small area of activity on the molecule called the **antigenic determinant**, or **epitope**. An antigenic determinant (epitope) contains about six to eight amino acid molecules or monosaccharide units. Each antigenic determinant has a characteristic three-dimensional shape and a molecular weight of about 350 daltons. An antigen may have numerous antigenic determinants, and a structure such as a bacterial flagellum may have hundreds of these molecules. The antigenic determinants are unique microbial fingerprints to which the immune system responds.

Hapten:
a small molecule that complexes with a protein or a polysaccharide to form an antigen.

Small nucleotides, hormones, peptides, and other molecules are not capable of stimulating the immune system by themselves. However, when they link to proteins in the body, the immune system may recognize the combination as foreign and respond to it. Allergy reactions (Chapter 20) are examples of immune responses to these combinations. The small molecule in the combination provides the key functional unit and is known as a **hapten** (*haptein* is Greek for grasp, reference to the interactions between the hapten molecule and the immune system's antibody). Examples of haptens include penicillin molecules, molecules in poison ivy plants, and molecules in certain cosmetics and dyes.

FIGURE 18.5

A Display of the Various Antigens Possible on a Bacterial Cell

Each different antigen is capable of stimulating the immune system. In this idealized bacterium, antigens are located in the cell membrane (CM), cell wall (CW), capsule (C), flagellum (F), pilus (P), and cytoplasm (Cy), Exotoxins (Ex) are also antigenic.

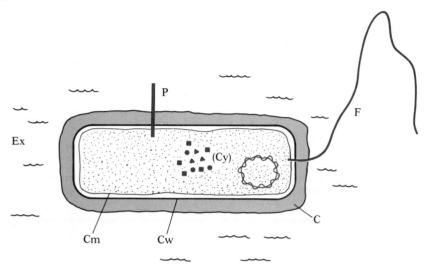

Under normal circumstances, one's own chemical substances do not stimulate an immune response. This failure to stimulate the immune system occurs because the substances are interpreted as "self." Immunologists believe that prior to birth, the body's proteins and polysaccharides inactivate the immune system cells that otherwise might respond to them (Figure 18.6). In the fetal stage, these responsive cells are easily paralyzed. The individual thereby develops a tolerance of "self" and remains able to respond only to "nonself." This theory, known as **specific immunologic tolerance**, was advanced in the late 1950s by Frank MacFarlane Burnet (a 1960 Nobel laureate) and David Talmadge. Cells responding to self are also destroyed in the thymus, as we shall see presently.

Antigens enter the body through a variety of portals such as the mucous membranes of the respiratory tract, a wound, an arthropod bite in the skin, or an injury to the gastrointestinal tract. **Autoantigens** are a person's own chemical substances that stimulate an immune response when self-tolerance breaks down (as in lupus erythematosus, Chapter 20). **Alloantigens** are antigens existing in certain but not all members of a species. The A, B, and Rh antigens of humans are typical alloantigens. **Heterophile** antigens are antigens found in unrelated species. For instance, the erythrocytes of horses and the viruses that cause infectious mononucleosis have certain identical antigens. Antigens of these and other types are responsible for eliciting the immune response, as we shall see next.

Autoantigens:
a person's own chemical substances that elicit an immune response.

het′er-o-fīl″

Heterophile antigens:
identical antigens found in apparently unrelated species.

Origin of the Immune System

The **immune system** is a general term for complex series of cells, factors, and processes providing an adaptive and specific response to antigens associated with microorganisms or with potentially harmful molecules such as microbial toxins. As such, the system lends specific resistance to infection and disease.

FIGURE 18.6
A Hypothesis for the Induction of Immunologic Tolerance

In the fetal stage, the tissue cells produce protein or polysaccharide antigens that contact and inactivate those cells that might later respond to them. The inactivated cells are paralyzed, and the ones left behind will respond to nonself antigens. In the figure, tissue cells 36, 64, and 77 produce antigens that inactivate responsive cells 36, 64, and 77.

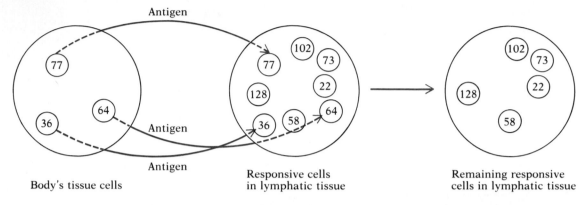

ĕ-rith″ro-poi-et′ik

lim-fo″poi-et′ik

Thymus:
a flat bilobed organ in the upper thorax of humans where T-lymphocytes are formed.

The cornerstones of the immune system are a set of body cells known as **lymphocytes**. These cells are distributed throughout the body, where they comprise the lymphoid system (Chapter 17). Lymphocytes are small cells, about 10 to 20 μm in diameter, each with a large nucleus taking up almost the entire space of the cytoplasm. Under the microscope, all lymphocytes look similar. However, two types of lymphocytes can be distinguished on the basis of developmental history, cellular function, and unique biochemical differences. The two types are B-lymphocytes and T-lymphocytes. **B-lymphocytes** are responsible for antibody-mediated immunity (AMI), while **T-lymphocytes** are responsible for cell-mediated immunity (CMI). In AMI, antibodies provide resistance to disease, while in CMI, a direct cell-to-cell contact is involved.

The immune system arises in the fetus approximately two months after conception (Figure 18.7). At this time, lymphocytes originate from primitive cells in the bone marrow known as **stem cells**. Stem cells differentiate into two types of cells: **erythropoietic cells** (myeloid cells), which become red blood cells; and **lymphopoietic cells**, which become lymphocytes of the immune system. (The Greek word *poien* means "to make," thus lymphopoietic cells are "lymphocyte-making cells.") We shall now follow the fate of the lymphopoietic cells.

Lymphopoietic cells take either of two courses. Some of the cells proceed to an organ of the thoracic cavity called the **thymus**. This flat, bilobed organ lies below the thyroid gland near the top of the heart. The thymus is large in size at birth and increases in size until the age of puberty, when it begins to shrink. Within the thymus, the lymphopoietic cells mature over a two-or three-day period and are modified by the addition of surface receptor proteins. They emerge from the organ as T-lymphocytes ("T" for thymus). Mature T-lymphocytes, or **T-cells**, are ready to engage in cell-mediated immunity and are said to be **immunocompetent**. The cells colonize the lymph nodes, spleen, tonsils, and other lymphoid organs, and they become a major portion of the lymphoid system.

It should be noted at this juncture that a large percentage of lymphopoietic cells are destroyed in the thymus. These cells would normally react with self antigens. Therefore, the mature T-lymphocytes emerging are the cells able to interact with nonself antigens. In this way, the thymus helps select out cells that will later mount an immune response.

FIGURE 18.7

Origin of the Immune System

Stem cells in the bone marrow differentiate into erythropoietic cells that become erythrocytes, or lymphopoietic cells that become lymphocytes. The lymphopoietic cells take either of two courses. Some pass through the bursa of Fabricius in embryonic chicks, or liver or bone marrow in humans, to form B-lymphocytes that colonize the lymphoid tissues. Others pass through the thymus to form T-lymphocytes that also colonize the lymphoid tissues. The location of lymphoid tissues, and hence the immune system, is shown at the bottom of the figure.

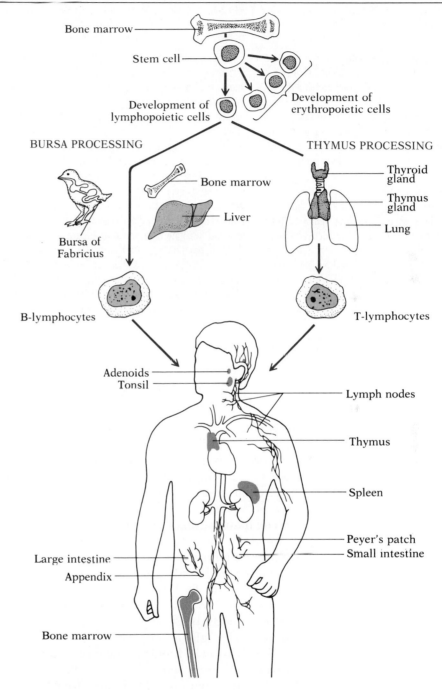

The B-lymphocytes mature and become immunocompetent in a site that has not been determined with certainty in humans. Much evidence suggests that the bone marrow is the maturation site, but some immunologists favor the liver, spleen, or gut-associated tissue as the site. In the embryonic chick, the maturation site has been identified as the bursa of Fabricius (MicroFocus: 18.2).

Bursa of Fabricius:
a gland in the gastrointestinal tract of chicks where B-lymphocytes are formed.

18.2

"YOU GAVE ME BAD CHICKENS!"

For a young graduate student studying poultry science at Ohio State University, it was a rather embarrassing end to a routine laboratory experiment.

Timothy S. Chang had been assigned to show some younger students how chickens develop immunity when injected with *Salmonella* bacilli. For the demonstration, he borrowed a dozen healthy chickens from Bruce Glick, a fellow graduate student. Glick had been using the chickens to study the functions of a mysterious gland in their intestines. Chang and his students carefully inoculated the chickens with *Salmonella*, waited a week, and then drew blood samples to test for evidence of *Salmonella* antibodies. To Chang's surprise and chagrin, 10 of the 12 chickens failed to show any sign of antibodies.

Chang went down the hall to Glick's lab and asked if he were pulling some kind of joke. Puzzled at his friend's results, Glick checked his lab records and found he had removed the mysterious gland from the 10 "bad" chickens. The two animals that developed antibodies still had the gland intact.

The year was 1954. The gland was the bursa of Fabricius, named for Hieronymus Fabricius, a seventeenth century anatomist who discovered it. Glick had developed an interest in the obscure gland after observing it in the intestine of a goose and learning that its function was unknown. He had removed the gland in newborn chicks and found there was no effect on growth. His research was at a dead end until Chang's experiment gave him a clue to the true function of the gland. Glick and Chang repeated the experiment and published their results in 1956 in *Poultry Science*, a specialty journal.

The accidental discovery might have gone unnoticed except that zoologists at the University of Wisconsin spotted it and confirmed the findings. As word spread to immunologists, Robert A. Good at the University of Minnesota assigned a team of colleagues to develop the theory of "bursa-derived immunity." In 1965, Good's team delivered its report before the American Academy of Pediatrics. Several immunologists cautioned against drawing conclusions about humans from work with chickens, but many saw the discovery as an important first step to understanding the origin of antibodies.

For this reason, the lymphocyte is known historically as the bursa-derived or B-lymphocyte (although the letter "B" can be related to bone marrow). Like the T-lymphocytes, B-lymphocytes mature with receptor proteins on their surfaces. Once they are immunocompetent, the B-lymphocytes, or **B-cells**, move through the circulation to colonize organs of the lymphoid system, where they join the T-lymphocytes.

The maturation of both T-lymphocytes and B-lymphocytes is accompanied, as noted, by the insertion of **surface receptor proteins**. The receptor protein enables the lymphocyte to recognize a specific antigen and bind to it. After this recognition and binding has occurred, the lymphocyte is said to be **committed** to that antigen. In T-lymphocytes, the receptor protein at the cell surface is composed of two chains of glycoproteins linked to one another by sulfur-sulfur chemical bonds. In B-lymphocytes, the receptor protein is an antibody molecule believed to be IgD (as we shall see presently).

Glycoprotein:
a protein molecule having a number of attached carbohydrate molecules.

The location of highly specific receptor proteins on the lymphocyte surface is one of the more remarkable discoveries of contemporary immunology. The implication is that even before an antigen enters the body, a recognition site is already waiting for it. Moreover, the genetic code for synthesizing the receptor is present in an individual even though that individual has not ever been exposed to an antigen. We may never experience malaria, for example, yet we already have receptor sites for recognizing and binding to the antigens of malaria parasites. This concept is engaging and thought-provoking.

MHC molecules:
molecules on the surface of body cells that define the uniqueness of an individual.

The recognition between antigenic determinants and lymphocytes depends on receptor proteins plus another set of glycoprotein molecules called the **major histocompatibility complex molecules**, or **MHC molecules**. (These proteins are sometimes called "MHC antigens" because they stimulate an antibody response in other individuals.) The MHC molecules are embedded in the membranes of all cells of the body. At least 20 different genes encode MHC molecules, and at least 50 different forms of the genes exist. Thus, the variety of MHC molecules existing in the human population is enormous, and the chance of two individuals having the same MHC molecules is incredibly small. (The notable exception is identical twins.) The MHC molecules define the uniqueness of the individual and play a role in the immune response.

There are two important classes of MHC molecules, and both help define the individual as self. Class I MHC molecules are found on virtually all cells of the body, but class II MHC molecules can be found only on B-lymphocytes and macrophages. These class II molecules function in the immune response. There is also evidence that certain human diseases are associated with the MHC molecules (Chapter 20). Certain researchers postulate a set of immune response (IR) genes within the MHC genes, and they believe that IR genes influence the T-lymphocyte's receptor sites.

The actual immune response originates with the entry of antigens into the body and the passage to the lymphatic or cardiovascular system. Here the antigens are phagocytized by macrophages and other phagocytic cells, and the antigens are broken down to release the antigenic determinants (Figure 18.8). Certain macrophages transport the antigenic determinants (epitopes) to the lymphoid organs where the T-lymphocytes and B-lymphocytes are waiting. The phagocytosis and transport are extremely important because research evidence indicates that unprocessed antigens stimulate the immune system poorly.

Within the tissues of the lymphoid organs (spleen, lymph nodes, tonsils, and so on), the T-lymphocytes and B-lymphocytes are waiting to implement either of the two arms of the immune system: antibody-mediated immunity (previously known as **humoral immunity**) will result in antibodies that react with microorganisms, soluble antigens, and nonself cells within the body's environment; cell-mediated immunity will result in activated T-lymphocytes for direct interaction with antigen-marked cells (such as virus-infected and transplanted cells). In the following paragraphs, we shall discuss both types of immunity, beginning with cell-mediated immunity.

Cell-Mediated Immunity (CMI)

The body's defense against microorganisms existing within its cells is centered in cell-mediated immunity (CMI). Cell-mediated immunity responds to cells that have been infected with pathogens such as viruses, rickettsiae, and certain bacteria such as *Mycobacterium tuberculosis*. The pathogens announce their presence by altering the infected cells and changing the molecular architecture of the cell's surface. The infected cells therefore act as signposts and become the objective of an attack by cells of CMI. Protozoa and fungi also stimulate CMI, as do cancer cells and the cells of transplanted tissue (MicroFocus: 18.3).

Tuberculosis:
an airborne bacterial disease of the lung tissue accompanied by severe cough and lesion formation.

The T-lymphocytes participating in CMI are of four major types in two categories. One category contains the **effector T-lymphocytes**, so-named because they bring about or "effect" CMI. The effector T-lymphocytes include cytotoxic T-lymphocytes and delayed hypersensitivity T-lymphocytes (Chapter 20). The other category of T-lymphocytes are **regulator T-lymphocytes**. These cells oversee the immune process. They include helper T-lymphocytes and suppressor T-lymphocytes.

Regulator T-lymphocyte:
the helper and suppressor T-lymphocytes that oversee the immune process.

When a macrophage or other phagocyte enters the lymphoid organ, it displays on its surface the antigenic determinants derived from the microorganism (Figure 18.9). At this point, a binding occurs between the antigenic determinant at the surface of the macrophage and the receptor protein at the surface of the

FIGURE 18.8

The Phagocyte and Phagocytosis

(a) A scanning electron micrograph of a macrophage displaying the highly irregular fluid nature of the cell surface of the cell. (b) A phagocytic macrophage in the process of engulfing a protozoan of the genus *Trypanosoma*. Cytoplasmic extensions have begun to wrap themselves around the protozoan. (c) A higher magnification of the same photograph. Approximately half of the protozoan has already been engulfed by the macrophage. The immune process generally begins with phagocytosis of an antigen-bearing organism or other particle.

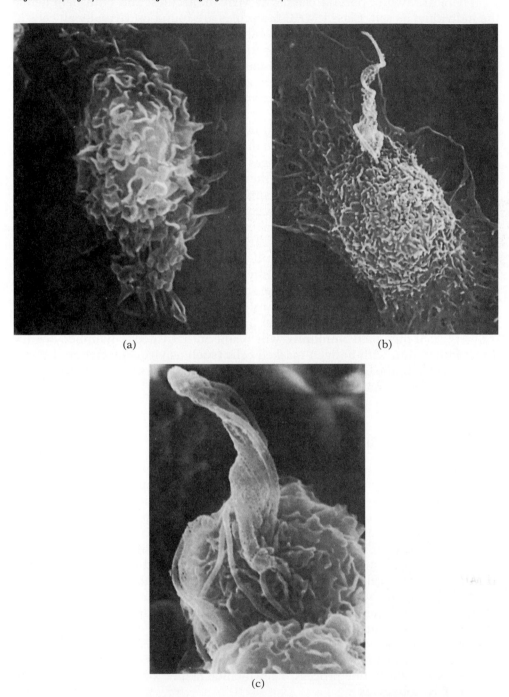

(a)   (b)

(c)

18.3

MICROFOCUS

A VOCABULARY OF IMMUNOLOGY

antigen: a substance, usually a protein or polysaccharide, that is interpreted as nonself and is capable of stimulating the immune system.

antigenic determinant: a small segment of an antigen that is the actual stimulant of the immune system; also called an epitope.

self: a person's own biochemicals that normally do not elicit an immune response.

nonself: any "foreign" biochemicals that normally elicit an immune response in a person.

hapten: a small molecule that must complex to a large protein in the body in order to be antigenic.

B-lymphocyte: a type of white blood cell that is stimulated by an antigen and reverts to a plasma cell for antibody production.

T-lymphocyte: a type of white blood cell that is stimulated by an antigen and reverts to specialized T-lymphocytes, which attack and destroy antigen-bearing cells.

antibody-mediated immunity (AMI): immunity based on the activity of antibodies

derived from plasma cells; also called humoral immunity.

cell-mediated immunity (CMI): immunity based on the activity of T-lymphocytes that attack antigen-bearing cells.

lymphokines: a series of small glycoproteins produced by T-lymphocytes that enhance the activity of T-lymphocytes and macrophages in specific resistance.

antibodies: a series of proteins produced by plasma cells and occurring in five different types; antibodies unite with antigens and bring about specific resistance.

immunoglobulins: another expression for antibodies.

IgG: a monomeric antibody molecule that predominates in the blood and is a major factor in the primary and secondary responses to microbial infection; also called gamma globulin.

IgM: a pentameric antibody molecule that predominates in the blood and is a major factor in the primary response to microbial infection.

IgA: a dimeric antibody molecule that predominates in the body secretions and is a major factor in immunity at the body surfaces.

IgD: a monomeric antibody molecule that is apparently the receptor molecule at the surface of a B-lymphocyte.

IgE: a monomeric antibody molecule that functions in allergic reaction.

antitoxins: antibody molecules that unite with and neutralize toxins.

opsonins: antibody molecules that unite with microbial surfaces and encourage phagocytosis.

complement system: a complex system of numerous molecules that are stimulated into action by an antigen–antibody reaction and that bring about destruction of microorganisms.

T-lymphocyte. Before the two cells can combine, however, they must locate one another among the myriad different types of T-lymphocytes, a process that requires considerable searching.

Once they have found each other, the T-lymphocyte and macrophage will bind, but not before another recognition is accomplished. This second recognition happens between the class II MHC molecules on the macrophage and the MHC receptor site on the T-lymphocyte. Indeed, the T-lymphocyte is able to recognize the antigenic determinant only within the spatial context of the class II MHC molecules. It is as if the surface receptor of the T-lymphocyte will bind to the antigenic determinant only if the T-lymphocyte first ascertains that the macrophage is from the same body. Two mechanisms have been postulated to explain the reaction, as Figure 18.10 explains. Thus, for proper binding, the antigenic determinants must be nestled within class II MHC molecules.

Once the recognitions and bindings have taken place (i.e., the square pegs have fit into the square holes and the round pegs into the round holes), the T-lymphocytes enlarge and divide rapidly to form a colony (or clone) of activated T-lymphocytes. The activated cells are **cytotoxic T-lymphocytes** if microbial infection is involved. The cytotoxic T-lymphocytes enter the lymph and blood vessels and circulate until they come upon cells displaying the telltale antigenic

Cytotoxic T-lymphocyte: a T-lymphocyte that contacts and lyses cells containing certain antigens.

FIGURE 18.9

T-Lymphocyte Activation in Cell-Mediated Immunity

The antigens (actually, antigenic determinants) of a particle such as a virus are preserved by the macrophage that performed phagocytosis. The antigens are presented to a committed T-lymphocyte within the spatial context of the class II MHC molecules. The receptor of the T-lymphocyte reacts with the antigen/class II MHC combination and is activated to form a cytotoxic T-lymphocyte. When the cytotoxic T-lymphocyte encounters cells infected by the same virus, it docks on the surface of the infected cell at the class I MHC molecules. Then it discharges the protein perforin, which assists the lysis of the infected cell.

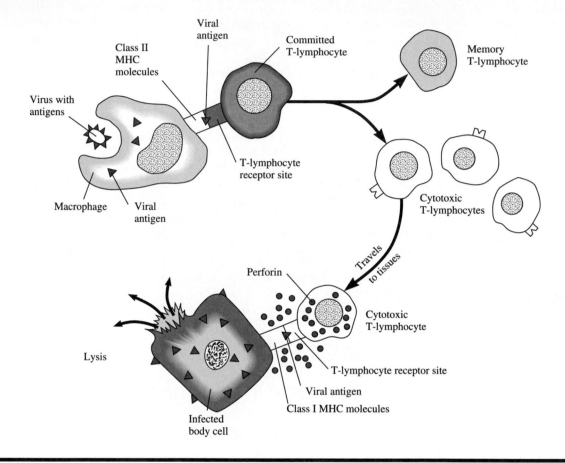

determinants on their surface. These cells are virus-infected or tuberculosis-infected cells, for example. The cytotoxic cell interacts with the infected cell after it has recognized its class I MHC molecules and its nonself antigens. It then lyses (destroys) the cell (Figure 18.11), which is why the cytotoxic cell is also called a **killer T-lymphocyte**. The interaction is called a "lethal hit" and appears to be mediated by a toxic chemical called **perforin**. Perforin is released from granules of the T-lymphocyte, whereupon it inserts into the plasma membrane of the infected cell and dissolves the membrane, thereby bringing about lysis.

Cytotoxic T-lymphocytes are also the source of **lymphokines**, a series of low molecular weight glycoproteins used to enhance the defensive capabilities of the body. Immunologists often refer to lymphokines as **cytokines**, because they are attracted to and act on cells (*cyto-* refers to cell). One lymphokine (cytokine) is the **macrophage activating factor (MAF)**. MAF attracts macrophages to the infection site and activates its enzymes to encourage cellular digestion. Another lymphokine, known as **migration inhibitor factor (MIF)**,

lim'fo-kīn″

Lymphokine:
a protein produced by lymphoblasts that increases the efficiency of phagocytosis.

FIGURE 18.10

A Comparison of Two Mechanisms for the Union of an Immunocompetent T-Lymphocyte with an Infected Cell

(a) In both the dual recognition and altered self mechanisms, the normal body cell is altered by infection with viruses. In the dual recognition mechanism, the virus-associated antigen is displayed on the cell surface alongside the MHC molecule. In the altered self mechanism, the MHC molecule has been altered by the viral antigen and a single receptor site now exists. (b) When the T-lymphocyte unites with the infected cell, the union may involve two separate components for recognition (dual recognition), or a single recognition site with the MHC molecule modified by the foreign viral antigen (altered self).

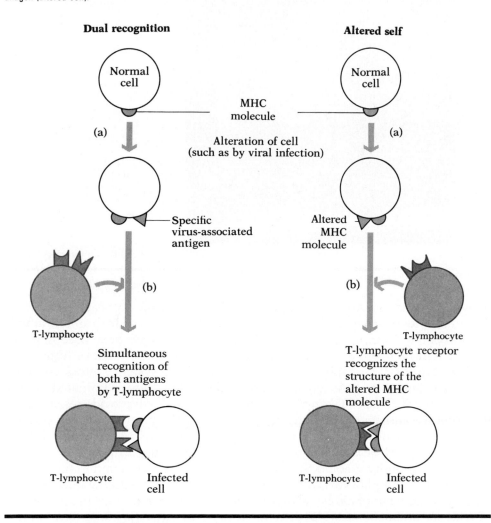

prevents macrophages from leaving the infection sites. A third, called **transfer factor (TF)**, mobilizes other T-lymphocytes in the area and encourages their conversion to cytotoxic cells. The latter continue the destruction of infected cells and augment the process of immunity.

The lymphokines described above are but a few of the broad series of lymphokines described since the 1970s. The term **interleukin** has been used interchangeably with lymphokine since the substances are produced by white blood cells (*-leuko*) and affect other white cells (*inter-*). One lymphokine, called **interleukin-1**, is believed to stimulate the maturation of T-lymphocytes after

Interleukins: substances produced by white blood cells that influence the activity of other white blood cells.

FIGURE 18.11
A "Lethal Hit"

A scanning electron micrograph of a cytotoxic T-lymphocyte attacking a budding *Cryptococcus neoformans* cell. *Cryptococcus neoformans* is an important fungal pathogen that may cause meningitis. The activity of T-lymphocytes represents a key defensive mechanism by the immune system. Bar = 1 μm.

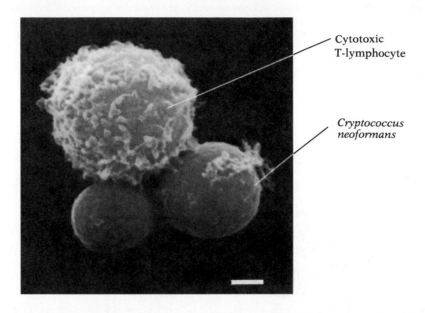

Cytotoxic T-lymphocyte

Cryptococcus neoformans

Clone:
a colony of identical cells.

phagocyte binding. Another lymphokine, **interleukin-2**, appears to activate T-lymphocytes to grow rapidly and divide, a process resulting in dramatic amplification of the T-lymphocyte number. Both substances have been used in medicine to activate killer T-lymphocytes, especially in the treatment of tumors.

The cytotoxic T-lymphocytes and other immune system factors continue the destruction of the infected cells until the latter have been eliminated. At this point, a special clone of T-lymphocytes forms, one that will provide resistance in the event the antigen reenters the body anytime in the future. The lymphocytes so-formed are called **memory T-lymphocytes**. They distribute themselves to virtually all parts of the body and remain in the tissues to provide a type of long-term immunity. Should the antigens be detected once again in the tissues, the memory T-lymphocytes will multiply rapidly, interact with the infected cells quickly, secrete lymphokines without delay, and set into motion the process to provide instantaneous CMI. This is one reason why we enjoy long-term immunity to disease after having suffered a bout.

Before leaving CMI, we shall make brief reference to the other three types of T-lymphocytes. **Delayed hypersensitivity T-lymphocytes** are effector cells functioning in allergic reactions and long-term inflammations. They secrete a deluge of lymphokines to bring on the symptoms of contact dermatitis (Chapter 20). **Helper T-lymphocytes**, a type of regulator cells, do not participate directly in cell killing, but they interact with antigenic determinants and stimulate the proliferation of both T-lymphocytes and B-lymphocytes, as we shall see presently. **Supressor T-lymphocytes** dampen the activity of T-lymphocytes and B-lymphocytes and temper the immune response as the antigenic stimulus wanes. Furthermore, they also appear to be important in controlling an immunological reaction to oneself.

TO THIS POINT
■

We have delved into specific resistance by describing the antigens that stimulate the immune system. The list of possible antigens is enormously diverse but most are polysaccharides and proteins with a series of antigenic determinants. Haptens may function as antigenic determinants, but first they must complex with protein or polysaccharide. We saw why the immune system will not usually respond to the body's substances in the theory of specific immunologic tolerance. Among the identifiable antigens are autoantigens, alloantigens, and heterophile antigens.

The discussion then moved on to the immune system. We explored the origin of the system in terms of the B-lymphocytes and T-lymphocytes, and we noted how each colonizes the lymphoid tissues to form the immune system. The process of immunity generally begins with the phagocytosis of antigens and delivery of antigenic determinants to the lymphoid tissues. The immune process then diverges to either of two systems. The focus next turned to cell-mediated immunity, where T-lymphocytes migrate to the antigen site and interact with cells, while stimulating phagocytosis through a series of lymphokines. The roles of other forms of T-lymphocytes were also mentioned briefly.

We shall now explore the second form of immunity, called antibody-mediated immunity. Here the B-lymphocyte plays a central role, and antibodies function in the immune response. We shall examine the structures and types of antibodies, how they are related to the genes of B-lymphocytes, and the processes by which they interact with antigens and provide specific resistance. Together with cellular immunity, this process of immunity represents the key to continued good health.

Specific Resistance and Antibodies

Cell-mediated immunity (CMI) is provided by attacking cells, but antibody-mediated immunity (AMI) is centered in antibodies, a series of protein molecules circulating in the body's fluids. (A body fluid is commonly known as a "humor," and antibody-mediated immunity has traditionally been called **humoral immunity**, but we shall use the more contemporary name.) In AMI, the antibodies react with such things as toxin molecules in the bloodstream, as well as with antigens on microbial surfaces or microbial structures (e.g., flagella, pili, capsule), and with viruses in the extracellular fluids. The reaction between antibody and antigen leads to elimination of the antigen by various means, as we shall discuss in this section.

Antibody-mediated immunity, the second arm of the immune system, is better known than CMI because its origins trace back to the 1890s. In that period Emil von Behring and Shibasaburo Kitasato determined that animals surviving a serious bacterial infection have "protective factors" in their blood to increase resistance to future attacks of the same infection. Today, these protective factors are known to be antibodies. Antibody formation and activity is the major topic of this section.

Antibody-Mediated (Humoral) Immunity

Like cell-mediated immunity, antibody-mediated immunity (AMI) begins with an encounter between macrophages bearing antigenic determinants and cells of the immune system. In this case, however, the major participating cell is the B-lymphoctye. The antigenic determinants stimulating the B-lymphocytes are usually derived from antigens free in the bloodstream, such as those associated with bacteria, viruses, and certain organic substances. Capsular polysaccharides and bacterial toxins also serve as antigens to stimulate B-lymphocytes (Figure 18.12).

Antigenic determinant:
the small part of the antigen that stimulates the immune system.

Activation of the B-lymphocytes begins when macrophages find their way to the lymph tissue and bring the antigenic determinants close to the appropriate B-lymphocytes that can respond. As with CMI, there is a considerable search

FIGURE 18.12

Clonal Selection and Antibody-Mediated Immunity

Antigen receptors on the surfaces of B-lymphocytes unite with antigens (actually, antigenic determinants) delivered by macrophages. Note the various shapes of antigen receptors of the six B-lymphocytes pictured. Only one B-lymphocyte will react with and be "selected out" by the antigen. That B-lymphocyte will reproduce and give rise to a population of plasma cells that will secrete antibodies specific for that antigen. The antibodies enter the circulation to provide antibody-mediated immunity. Memory cells for the antigen are also produced for long-term immunity.

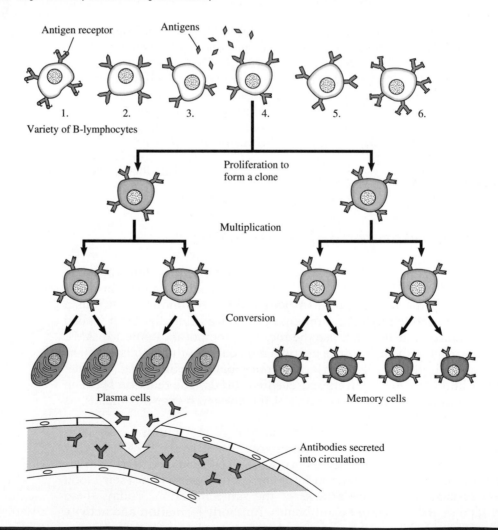

process involved as the antigenic determinants attempt to match with surface receptor proteins on the B-lymphocytes. The surface proteins on B-lymphocytes are antibodies of the type IgD. Activation of the B-lymphocytes requires that multiple antigenic determinants bind simultaneously with multiple antibody molecules and pull the antibodies into a continuous cluster, a process called **capping** (Figure 18.13). Small haptens may be poor stimulators of B-lymphocytes because of their inability to perform capping.

The selection of a single cluster of B-lymphocytes by a particular antigenic determinant is the underlying basis for the **theory of clonal selection**. First postulated in the 1950s by Frank MacFarlane Burnet and Peter Medowar, the theory of clonal selection explains how antigenic determinants "select out" which B-lymphocytes will function in the immune process from the enormous variety of possible types available in the lymphoid tissue. This is what the theory suggests:

Once the antigenic determinant has bound with its corresponding surface receptor, and after capping has occurred, the combination of receptor protein and antigenic determinant is taken into the cytoplasm of the B-lymphocyte. Now the B-lymphocyte displays the antigenic determinants on its surface. While these processes have been going on, macrophages bearing the antigenic determinants have contacted helper T-lymphocytes, and the macrophages have been recognized as belonging to the human body (self). This recognition has occurred because the antigenic determinants are cradled within the correct class II MHC molecules on the macrophage surface. Now the helper T-lymphocyte is activated, and it forms a clone of T-lymphocytes keyed to the particular antigenic determinant. These activated helper T-lymphocytes then recognize the same antigenic determinant/MHC molecule complex on the B-lymphocyte surface, and they bind to the B-lymphocytes (Figure 18.14). This immunologic cooperation between the macrophage, the B-lymphocyte, and T-lymphocyte continues the immune response.

Once activated by helper T-lymphocytes, the B-lymphocytes multiply and give rise to a clone of **plasma cells**. Plasma cells are large, complex cells having no surface protein receptors. Their sole purpose is to produce **antibodies**. A plasma cell lives about four to five days, during which time it produces an incredible 2000 antibody molecules per second. Each antibody molecule has the same antigen-binding property as the receptor on the B-lymphocyte, and all the antibody molecules eventually reach the circulation. In a matter of several hours the body becomes saturated with antibodies. Antibody molecules fill the blood, lymph, saliva, sweat, and all other body secretions. By sheer force of numbers they come upon the original site of antigens, bind to the antigens, and mark them for destruction. (In many cases, however, they are too late to prevent the initial surge of infection and symptoms may already be present when they arrive to provide their specific defense.)

Plasma cells:
antibody-producing cells
derived from B-lymphocytes.

At any one time, antibody molecules represent about 17 percent of the total protein in a person's blood fluid. Antibodies are extremely diverse, and the cells of a single species of bacterium may elicit the formation of hundreds of different kinds of antibodies corresponding to hundreds of different kinds of bacterial antigens. Each unique antibody molecule is then capable of uniting with the unique antigen molecule to effect a neutralization. The term **immunoglobulin (Ig)** is used interchangeably with "antibody" because antibodies exhibit the properties of globulin proteins and are used in the immune response. In this text, we shall use the terms antibody and immunoglobulin interchangeably.

Immunoglobulin:
an alternate expression for
an antibody.

We shall study the interaction of antibodies and antigens shortly, but at this point it is important to note that certain B-lymphocytes do not become plasma cells. Instead, they become a clone of **memory B-lymphocytes**. Memory

Memory B-lymphocytes:
stimulated B-lymphocytes
that remain in the lymphoid
tissues anticipating the
reappearance of a certain
antigen.

FIGURE 18.13

Capping of B-Lymphocytes

In this process a B-lymphocyte is activated when antigenic determinants bind to the surface receptors of B-lymphocytes. Multiple antigenic determinants appear to pull the surface receptors into a continuous cluster. Once capping has occurred, the antigen-receptor complex is engulfed into the cytoplasm of the B-lymphocyte.

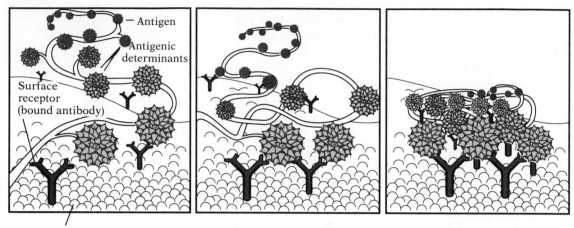

Surface membrane of B-lymphocyte

B-lymphocytes remain in the lymphoid tissues for many years, in some cases for a person's lifetime. Should the bacterium and its antigens reenter the body, the memory cells will immediately revert to plasma cells and produce antibodies without delay. The antibodies rapidly flood the bloodstream and bring about immediate neutralization of the antigens. The symptoms of infection rarely occur this time.

Structure and Types of Antibodies

Details about the nature of antibodies began to emerge in the late 1950s and early 1960s, with the union of immunology and molecular biology. Research focused on the Bence Jones proteins, identified a century earlier by Henry Bence Jones. A substantial body of information and numerous theories existed about the functions and origins of these proteins, but little was known about their detailed chemical composition. Investigators expected that an understanding of the structure of the proteins (now known to be antibodies) would provide an understanding of their unique specificity for antigens. The answers were provided by **Gerald M. Edelman** and **Rodney M. Porter**, who described the chemical composition and structure of the proteins and received the 1972 Nobel Prize in Physiology or Medicine for their work.

Polypeptide:
a chain of amino acids.

The basic antibody molecule consists of four polypeptide chains: two identical "heavy" (H) chains and two identical "light" (L) chains. These chains are joined together by sulfur-to-sulfur (disulfide) linkages to form a Y-shaped structure illustrated in Figure 18.15. Each heavy chain consists of about 400 amino acids, while each light chain has about 200 amino acids. The antibody molecule formed is called a **monomer**. It has two identical halves, each half consisting of a heavy and a light chain.

FIGURE 18.14

B-lymphocyte Activation in Antibody-Mediated (Humoral) Immunity

(a) Clonal selection has occurred as shown in Figure 18.12, and capping has taken place as in Figure 18.13. The antigenic determinant–receptor complex has been taken into the B-lymphocyte and the antigenic determinant is now displayed on the B-lymphocyte's surface within the spatial context of the class II MHC molecules. (b) Meanwhile, macrophages bearing the antigenic determinants have contacted special types of regulator T-lymphocytes called helper T-lymphocytes. The helper T-lymphocyte recognizes the macrophage as being "self" by recognizing the class II MHC molecules. (c) Now, the helper T-lymphocyte is activated to form a clone of T-lymphocytes unique for that antigenic determinant. (d) An activated helper T-lymphocyte then locates a stimulated B-lymphocyte and binds to the antigenic determinant–MHC molecule at its surface. (e) This activation spurs the B-lymphocyte to convert to a clone of plasma cells that secrete antibodies for antibody-mediated immunity.

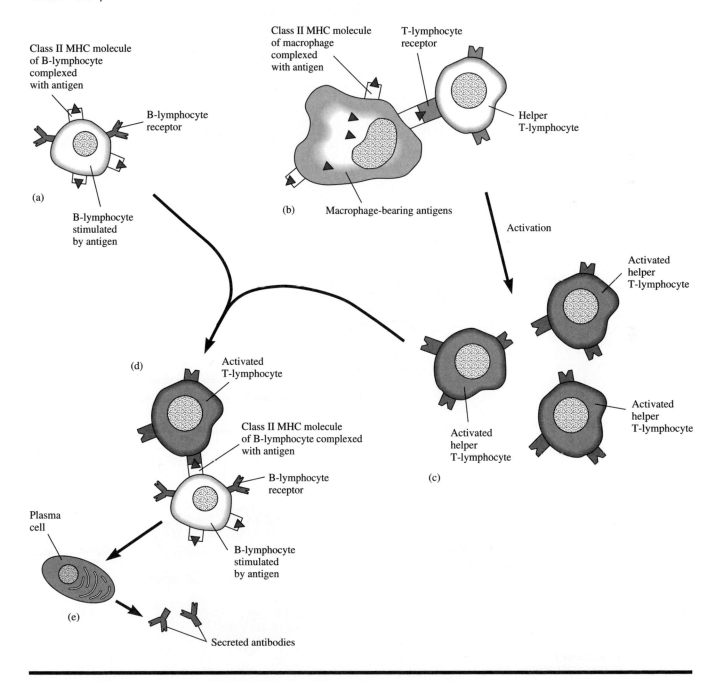

FIGURE 18.15
Details of the Antibody Molecule

(a) The antibody molecule consists of four chains of protein: two light chains and two heavy chains connected by disulfide (S-S) linkages. The heavy chains bend at a hinge point. The variable region is where the amino acid compositions of various antibodies differ. In the constant region, the amino acid compositions are similar among different antibodies. On treatment with papain enzyme, cleavage occurs at the hinge point and three fragments result: two Fab fragments and one Fc fragment. (b) A more detailed view of the protein chains showing where disulfide linkages are found and where loops exist within the chain to form domains. All light chains have a single variable domain (V_L) and a single constant domain (C_L). Heavy chains contain a variable domain (V_H) and three or four constant domains. (c) The reaction between antibody molecules and the antigenic determinants of an antigen showing the specificity that occurs. In this case the antigens are on the surface of a microorganism. Note that the antibody molecules unite with the triangular antigenic determinants but not with the circular ones.

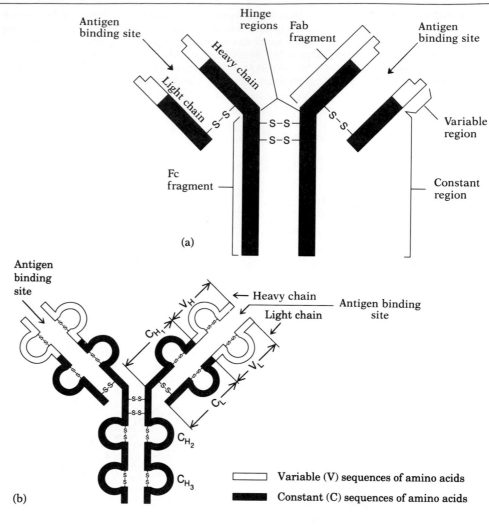

(a)

(b)

☐ Variable (V) sequences of amino acids

■ Constant (C) sequences of amino acids

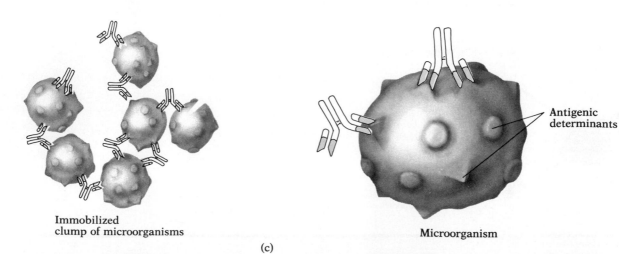

Immobilized
clump of microorganisms

Microorganism

Antigenic
determinants

(c)

FIGURE 18.16

The Structures of Five Types of Antibodies

Note the complex structures of IgM (a pentamer) and IgA (a dimer). IgG, IgE, and IgD consist of monomers, each composed of two heavy chains and two light chains of amino acids.

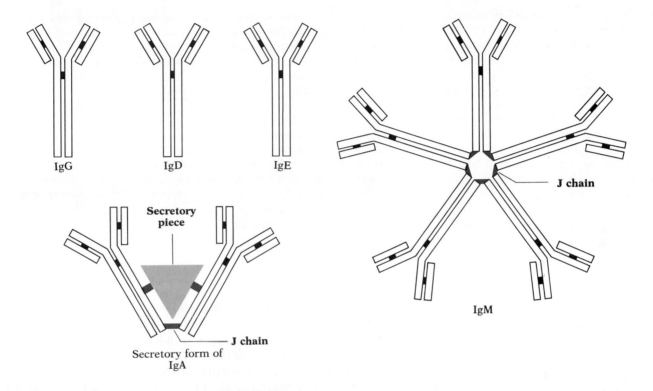

Constant and variable regions exist within each polypeptide chain. The amino acids in the **constant regions** of both light and heavy chains are virtually identical among antibodies. However, the amino acids of the **variable region** vary among the hundreds of thousands of different antibodies. Thus the variable regions of a light and heavy chain combine to form a highly specific, three-dimensional structure somewhat analogous to the active site of an enzyme. This portion of the antibody molecule is uniquely shaped to "fit" a specific antigen or antigenic determinant. Moreover, the "arms" of the antibody are identical so that a single antibody molecule may combine with two antigen molecules. These combinations may lead to a complex of antibody and antigen molecules.

When an antibody molecule is sectioned with papain, an enzyme from the papaya fruit, the molecule separates at the hinge joint, and two functionally different segments are isolated (Figure 18.15). One segment, called the **Fab fragment**, for "*f*ragment-*a*ntigen-*b*inding," is the portion that will combine with the antigenic determinant. The second segment, called the **Fc fragment**, for "*f*ragment able to be *c*rystallized," serves various functions: it is the part of the antibody molecule that combines with phagocytes in opsonization; it appears to neutralize viral receptor sites; it attaches to certain cells in allergic reactions; and it activates the complement system in resistance mechanisms.

To the present time, five types of antibodies have been identified, based on differences in the heavy (H) chains in the constant region. Using the abbreviation Ig (immunoglobulin), the five classes are designated IgM, IgG, IgA, IgE, and IgD. The structures of these antibodies are illustrated in Figure 18.16. IgM

Variable region:
the region of an antibody molecule where the amino acid composition varies among antibodies.

Fab fragment:
the portion of an antibody molecule that combines with the antigenic determinant.

18.4

MICROFOCUS

THE OTHER ARMY

Most people are familiar with the body's army headquartered downtown—the antibodies and T-lymphocytes that constitute the blood-centered system of immunity. Few, however, know much about the army of the suburbs—the specialized cells and antibodies that protect the body at its vulnerable entryways in locations such as its respiratory, gastrointestinal, and urinary tracts. The cells and antibodies in the fragile mucous membranes lining these systems constitute a network now known as the mucosal immune system.

The extent of communication between the blood and mucosal systems is uncertain

at this writing. Injected vaccines, for example, rarely evoke IgA production for mucosal immunity. However, stimulating one region of mucosal tissue to elicit IgA appears to produce an immune response at a distant mucosal surface. Thus, a vaccine in a nasal spray can elicit an antibody response along the gastrointestinal tract. This interrelatedness has led immunologists to think in terms of a single mucosal system.

Studies in mucosal immunity have lagged behind those of bloodborne immunity in part because tissue samples from the mucosa are difficult to work with. But studies are being encouraged because stopping

infection at its point of entry is highly desirable. Also, vaccine research would benefit considerably because oral vaccines are preferred to injectable types, and developing an oral vaccine requires a basic understanding of mucosal immunity. Indeed, the output of the mucosal system appears to outdistance that of the bloodborne system. It has been estimated, for instance, that over 3 grams of IgA are secreted as compared to only 1 gram of IgA released into the bloodstream. On this basis alone, the "suburban army" merits a greater share of attention.

J chain:
a glycoprotein in IgM and IgA molecules that connects the five subunits of the antibody.

Primary antibody response:
the first antibody reaction of the immune system to the presence of antigens.

Serum:
the straw-colored fluid portion of the blood minus the clotting agents.

Maternal antibodies:
antibodies from a pregnant woman that cross the placenta and provide protection to the fetus.

consists of a pentamer (five monomers), whose tail segments are connected by a glycoprotein called a joining, or J, chain. IgA is a dimer (two monomers), with the segments also connected by a J chain. In addition, there is a secretory component of IgA that enables the molecule to leave secretory epithelial cells (MicroFocus: 18.4). The remaining three antibodies are monomers.

IgM is the first antibody to appear in the circulation after stimulation of B-lymphocytes. It is the principal component of the primary antibody response (Figure 18.17) and the largest antibody molecule ("M" stands for macroglobulin). Because of its size, most IgM remains in the circulation. Research indicates that IgM is formed during fetal infections with rubella or toxoplasmosis, indicating that a certain immunological competence exists in the fetus. About 5 to 10 percent of the antibody components of normal serum consists of this antibody. IgM is one of the antibodies bound to the B-lymphocyte surface as a receptor protein.

IgG is the classical gamma globulin. This antibody is the major circulating antibody, comprising about 80 percent of the total antibody content in normal serum. IgG appears about 24 to 48 hours after antigenic stimulation and continues the antigen–antibody interaction begun by IgM. It is thus the antibody of the secondary antibody response. In addition, it provides long-term resistance to disease as a product of the memory B-lymphocytes. Booster injections of a vaccine raise the level of this antibody considerably in the serum. IgG is also the **maternal antibody** that crosses the placenta and renders immunity to the fetus until the child is able to make antibodies at about six months of age.

Approximately 10 percent of the total antibody in normal serum is **IgA**. One form of this antibody, called **serum IgA**, exists in the serum and is similar to IgG. A second form accumulates in body secretions and is referred to as **secretory IgA**. This antibody provides resistance in the respiratory and gastrointestinal tracts, possibly by inhibiting the attachment of parasites to the tissues. It is also located in tears and saliva, and in the colostrum, the first milk secreted by a nursing mother. When consumed by a child, the antibodies provide resistance to gastrointestinal disorders, as MicroFocus: 18.5 indicates. The secretory

18.5

THE BENEFITS OF BREAST FEEDING

Ironically, the literature from a major baby formula company says it best: "Breast feeding is the most natural and satisfying conclusion to the normal cycle of pregnancy and birth." Years of research have bolstered the truth of this statement.

The immune system of a child is primitive at birth and its intestinal wall is poorly developed. These factors make the infant easy prey for the bacteria that cause diarrhea. Breast feeding may help to solve this problem because the milk contains IgA, which accumulates in the intestine and provides surveillance against intestinal pathogens. Indeed, studies show that epidemics of diarrhea can be limited by feeding infants with colostrum, the antibody-rich "pre-

milk" of nursing mothers. Moreover, up to 80 percent of the cells in colostrum are macrophages, the leukocytes that perform phagocytosis.

Nursing may also limit allergies later in the child's life. Immunologists postulate that the IgA in mother's milk blocks the entry into the circulation of allergenic materials that would otherwise pass through the immature intestinal wall. Freedom from allergenic substances in the newborn may portend freedom from allergy in the years ahead.

It has also been suggested that a nursing mother's immune system may be programmed to provide specific antibodies for the child. During contact with her child, the mother is exposed to antigens such as dis-

ease organisms. She may then produce antibodies that would return to the child via the milk. Obviously these antibodies could not be provided by cow's milk. Some immunologists even foresee the day when women may be immunized against diseases that commonly infect newborns. After giving birth, they could serve as personalized immunological factories for their offspring.

Certain manufacturers have researched the possibility of adding antibodies to artificial formulas to simulate mother's milk. However, data gathered for decades indicate that bottle feeding is a pale substitute for a process that has withstood the test of centuries. As the sage puts it, "Human milk is for humans; cow's milk is for calves."

FIGURE 18.17

The Primary and Secondary Antibody Responses

After the initial antigenic stimulation, IgM is the first antibody to appear in the circulation. It is the principal component of the primary antibody response. Later, IgM is supplemented by IgG. On second exposure to the same antigen, the production of IgG is more rapid, and the concentration in the serum reaches a higher level than previously. Thus the IgG antibody is more concentrated in the secondary antibody response.

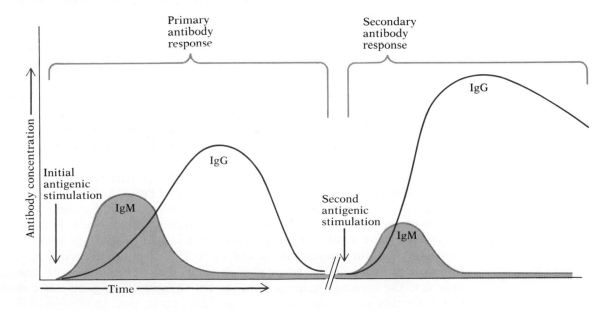

TABLE 18.2
Properties of Five Types of Antibodies

DESIGNATION	PERCENTAGE OF ANTIBODY IN SERUM	LOCATION IN BODY	MOLECULAR WEIGHT (DALTONS)	NUMBER OF FOUR-CHAIN UNITS	CROSSES PLACENTA	CHARACTERISTICS
IgM	5–10	Blood Lymph	900,000	5	No	Principal component of primary response
IgG	80	Blood Lymph	150,000	1	Yes	Principal component of secondary response
IgA	10	Secretions Body cavities	400,000	2	No	Protection in body cavities
IgE	<1	Blood Lymph	200,000	1	No	Role in allergic reactions
IgD	0.05	Blood Lymph	180,000	1	No	Possible receptor sites on B-lymphocytes

component comes from epithelial cells and helps move the antibody into the secretions.

IgE plays a major role in allergic reactions by sensitizing cells to certain antigens. This process is discussed in Chapter 20. Both the functions and significance of **IgD** are presently unclear, but evidence indicates that the antibody is a cell surface receptor on the B-lymphocytes together with IgM. Table 18.2 summarizes the characteristics of the five types of antibodies.

Diversity of Antibodies

For decades, immunologists were puzzled as to how an enormous variety of antibodies (perhaps a million or more different types) could be encoded by the limited number of genes associated with the immune system. Because antibodies, like all proteins, are specified by genes, it would be reasonable to assume that an individual must have a million or more genes for antibodies. But geneticists point out that human cells have only about 100,000 genes for all their functions.

The solution to antibody diversity has been apparently resolved in recent years by showing that the genes for each antibody are not present in embryonic cells. Rather than harboring a set of "antibody genes," embryonic cells contain about 300 genetic segments that can be shuffled like transposons and combined in each B-lymphocyte as it matures. The process, known as **somatic recombination**, is a random mixing and matching of gene segments to form unique antibody genes. Information encoded by these genes is then expressed in the surface receptors of B-lymphocytes and in the antibodies later expressed by the stimulated clone of plasma cells.

The process of somatic recombination was first postulated in the late 1970s by **Susumu Tonegawa**, winner of the 1987 Nobel Prize in Medicine or Physiology. According to the process, the gene segments coding for light and heavy chains of an antibody are located on different chromosomes. The light and heavy chains are synthesized separately, then joined to form the antibody. One constant gene, four joiner genes, and up to 150 variable genes can be used to form a light chain. One variable gene is selected and combined with one joiner gene and the constant gene to form the active light chain gene. After deletion of intervening genes, the new gene can function in protein synthesis. For the

Transposons: segments of DNA that carry functional genes from one chromosomal location to another.

heavy chain of the antibody, the process is similar but even more complex because there is a greater variety of variable genes and more possible combinations.

Tonegawa's discovery was revolutionary because it questioned two dogmas of biology: that the DNA for a protein needs to be one continuous piece (for antibody synthesis the gene segments are separated); and that every body cell has exactly the same DNA (the antibody genes for different B-lymphocytes differ). Current evidence suggests that no more than 600 different antibody gene segments exist per cell. From this relatively small collection, however, the cell can generate over 200 million different antibody molecules through somatic recombination.

Antigen–Antibody Interactions

In order for specific resistance to develop, antibodies must interact with antigens in such a way that the antigen is altered. The alteration may result in death to the microorganism that possesses the antigen, inactivation of the antigen, or increased susceptibility of the antigen to other body defenses.

Certain antibodies, called **neutralizing antibodies**, react with viral capsids and prevent viruses from entering their host cells. Influenza viruses are inhibited by neuraminidase antibodies in this way (Figure 18.18). Neutralizing antibodies also bind viruses together in clumps, thereby encouraging phagocytosis. Moreover, neutralizing antibodies represent a vital defensive mechanism to toxins. The antibodies, known as **antitoxins**, alter the toxin molecules near their active sites and mask their toxicity. Neutralization of toxin molecules also increases their size, thus encouraging phagocytosis while lessening their ability to diffuse through the tissues.

Antibodies called **agglutinins** react with antigens on the surface of organisms such as bacteria. This action causes clumping, or agglutination, of the organisms and enhances phagocytosis. Movement is inhibited if antibodies react with antigens on the flagella of microorganisms, and the organisms may even be clumped together by their flagella. The reaction of antibodies with pilus antigens prohibits attachment of an organism to the tissues while agglutinating them and increasing their susceptibility to phagocytosis.

Another form of antibodies is the **precipitins**. These antibodies react with dissolved antigens and convert them to solid precipitates. In this form, antigens are usually inactive and more easily phagocytized.

Opsonins are antibodies that stimulate phagocytosis by direct intervention. The antigen attaches to the Fab fragment of the antibody while the Fc portion inserts to a receptor site on the phagocyte. An example occurs in the inactivation of *Streptococcus pneumoniae*, the pneumococcus. Normally the organism resists phagocytosis but when antibodies react with the M protein in the bacterial capsule, phagocytosis takes place quickly.

A final example of antigen–antibody activity involves the **complement system**. Originally described in 1895 by Jules Bordet at the Pasteur Institute, the complement system is a series of at least 11 proteins that function in a cascading series of reactions. The system exists in all normal sera and is activated by IgM or IgG. Complement works with antibodies to cause opsonization, chemotaxis, and lysis of bacterial cells.

The pathway for complement activation is set into motion by the interaction of antigen and antibody molecules as shown in Figure 18.19. Usually the interaction takes place on the surface of a cell such as a bacterium. The cascade results in a series of substances, one of which is an **attack complex** (C5b, C6, C7, C8,

Antitoxins: antibodies that combine with and neutralize toxin molecules.

ah-gloo′ti̇̆-nin

Precipitins: antibodies that react with dissolved antigens to yield precipitates.

bor-da′

FIGURE 18.18

Five Different Mechanisms by Which Antibodies Interact with Antigens and Alter the Antigens

(a) Antibodies react with molecules at viral surface and prevent viral attachment to cells. (b) Antitoxins combine specifically with toxins and neutralize the toxicity site of the toxin. (c) Agglutinins combine with antigens on the cell surface and bind the cells together or restrict movement by binding with flagellar antigens. (d) Precipitins combine with dissolved antigens and form latticelike arrangements that precipitate out of solution. (e) Opsonins encourage phagocytosis by forming a bridge between parasites and receptor sites on the phagocyte.

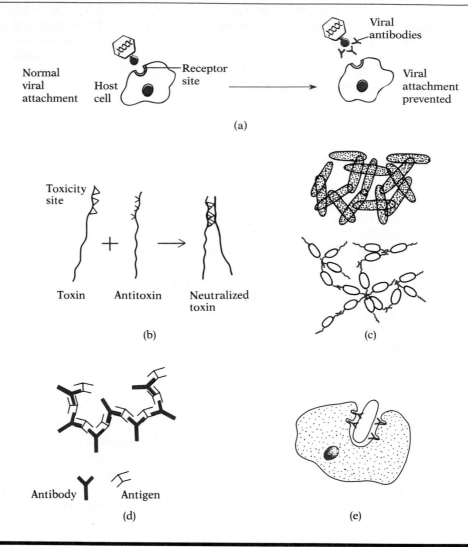

C9). This complex increases cell membrane permeability and induces the cell to undergo lysis through leakage of its cytoplasm. Another substance in the cascade (C5a) attracts phagocytes through chemotaxis, while another substance (C2a, C3b, C4b) facilitates phagocytosis by binding the cell to the phagocyte. Still other fragments (C3 and C5a) are **anaphylatoxins**. These substances react with tissue cells called mast cells (Chapter 20), and induce the release of histamine, which contracts smooth muscles and thereby increases the movement of phagocytes out of blood vessels to the infection site.

Complement activity may also take place through an **alternative pathway**. In this system several complement components are bypassed, and an antigen–antibody reaction is not required for stimulation. A serum protein named **properdin** functions in the process. The alternative pathway is stimulated by endotoxins as well as by capsular polysaccharides in streptococci and by certain fungi.

an"ah-fi"lah-tok'sin

Anaphylatoxins:
substances that react with mast cells and cause the release of histamine, which contracts smooth muscles.

FIGURE 18.19
The Pathway of Complement Activation

(a) The complement components as they exist just prior to antigen–antibody reaction on the surface of a cell such as a bacterium. (b) The results of the complement cascade showing the fate of the components after the cascade has been stimulated by antigen–antibody activity. (c) Some of the effects of complement activity: C3a and C5a attract phagocytes and induce histamine release to bring in more phagocytes; C2a, C3b, C4b increases adherence to phagocytes; C5b, C6, C7, C8, C9 induces lysis at the cell surface.

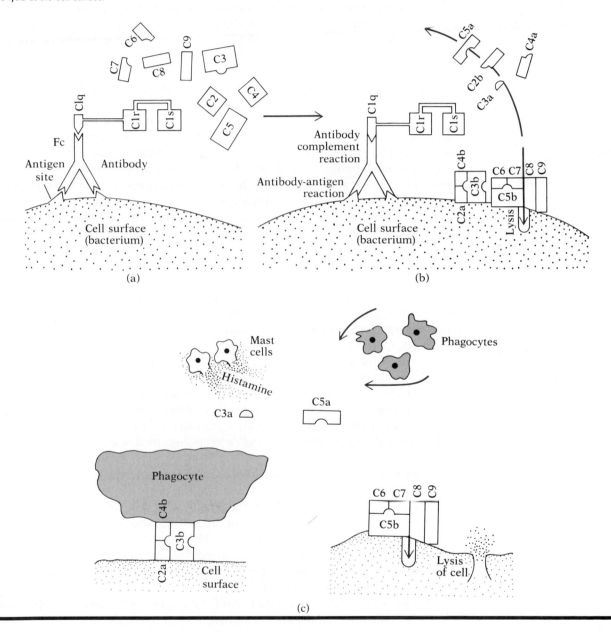

The complement system is particularly useful against Gram-negative bacteria because it alters their cell walls and makes them susceptible to lysozyme. Moreover, studies show that a complement component encourages the release of lysozyme from local macrophages. In Gram-positive bacteria, the peptidoglycan of the cell wall resists the attack complex, but lysozyme from macrophages degrades the cell wall, thereby leading to cell disintegration. Moreover, enhanced phagocytosis of all bacteria occurs, regardless of their type.

Lysozyme:
an enzyme that degrades the cell walls of certain bacteria.

Thinking Well

The idea that mental states can influence the body's susceptibility to and recovery from disease has a long history. The Greek physician Galen asserted that cancer struck more frequently in melancholy women than in cheerful women. During the last quarter century, the concept of mental state and disease has been researched more thoroughly, and a firm foundation has been established linking the nervous system to the immune system (MicroFocus: 18.6).

One such link occurs between the hypothalamus and the T-lymphocytes. The **hypothalamus** is a portion of the brain located beneath the cerebrum. It produces a chemical-releasing factor that induces the pituitary gland, positioned just below the hypothalamus, to secrete the hormone ACTH that targets the adrenal glands. The adrenal glands, in turn, secrete steroid hormones (glucocorticoids) that influence the activity of T-lymphocytes in the thymus gland.

Another link is established by branches of the **autonomic nervous system** that extend into the lymph node and spleen tissues. The autonomic (or "automatic") nervous system typically operates on its own to regulate the involuntary action of such organs as the heart, stomach, and lungs through myriad nerve fibers. The direct anatomical link between the immune and nervous systems permits a direct two-way communication.

A third link between the nervous and immune systems starts with **thymosins**, a family of substances originating in the thymus gland. When experimentally injected into brain tissue, thymosins stimulate the pituitary gland via the hypothalamus to release hormones, including the one that stimulates the adrenal gland. Although the precise functions of thymosins are still to be determined, it appears that they serve as specific molecular signals between the thymus and the pituitary gland. A circuit is apparently present that stimulates the brain to adjust immune responses and the immune system to alter nerve cell activity.

The outcome of these discoveries is the emergence of a strong correlation between a patient's mental attitude and the progress of disease. Rigorously controlled studies conducted in recent years have shown that the aggressive determination to conquer a disease can increase one's lifespan. Therapies can consist of relaxation techniques as well as mental imagery that disease organisms are being crushed by the body's stalwart defenses. Behavioral therapies of this nature can amplify the body's response to disease and accelerate the mobilization of its defenses.

Few reputable practitioners of behavioral therapies believe that such therapies should replace drug therapy. However, the psychological devastation associated with many diseases such as AIDS cannot be denied, and it is this intense stress that the "thinking well" movement attempts to address. Very often, for instance, a person learning of a positive HIV test goes into severe depression, and because depression can adversely affect the immune system, a double dose of immune suppression ensues. Perhaps by relieving the psychological trauma, the remaining body defenses can adequately handle the virus.

As with any emerging treatment method, there are numerous opponents of behavioral therapies. Some opponents argue that naive patients might abandon conventional therapy; another argument is that therapists might cause enormous guilt to develop in patients whose will to live cannot overcome failing health. Proponents counter with the growing body of evidence showing that patients with strong commitments and a willingness to face challenges—signs of psychological hardiness—have relatively greater numbers of T-lymphocytes than passive, nonexpressive patients. To date, no study has proven that mood or personality has a life-prolonging effect on immunity. Still, doctors and patients are generally inspired by the possibility that one can use his or her mind to help stave off the effects of infectious disease. Though unsure of what it is, they generally agree that *something* is going on.

18.6

SICK OVER EXAMS

At the end of every semester, medical students at Ohio State University suffer through the Day of the Big Bleed. First, they "bleed" over their final exams; then they bleed, quite literally, for psychologist Janice Kiecolt-Glaser and immunologist Ronald Glaser, her husband. The Glasers are attempting to learn whether the stress of final exams diminishes the activity of the immune system.

The "thinking well" phenomenon has been known for years—a positive attitude can assist the immune system's activities, while stress can place a burden on the system. Providing proof of this phenomenon has been difficult, but the Glasers are among those determined to show that a correlation exists between stress and reduced immune activity. Already, they have demon-strated reduced activity in natural killer cells in blood taken from students during exam week. Also, they have shown that in herpes-infected students, the virus is more active when exams are going on (a reflection of reduced body defense).

In their latest study reported in 1993, the Glasers gave the Ohio State students hepatitis B vaccinations and then tested them for antibody responses. Not surpris-ingly, the more stressed-out and anxious students consistently responded with lower antibody levels. Perhaps these students might have contracted hepatitis B more readily if the virus were present but that would be difficult to test. The lessons appear straightforward: prepare thoroughly for exams; think of clear mountain streams; and stay healthy.

NOTE TO THE STUDENT

It may have occurred to you that an episode of disease is much like a war. First, the invading microorganisms must penetrate the natural barriers of the body; then they must escape the phagocytes and other chemicals that constantly patrol the body's waterways and tissues; and finally, they must elude the antibodies or T-lymphocytes that the body sends out to combat them. How well we do in this battle will determine whether we survive the disease.

Obviously, we have done very well, because most of us are in good health today. As it turns out, most infectious organisms are stopped at their point of entry to the body. Cold viruses, for example, get no farther than the upper respiratory tract. Many parasites come into the body by food or water, but they rarely penetrate beyond the intestine. The staphylo-cocci in a wound may cause inflammation at the infection site, but this is usually the extent of the problem.

To be sure, drugs and medicines help in those cases where diseases pose life-threatening situations. In addition, sanitation practices, vector control, care in the preparation of food, and other public health measures prevent microorganisms from reaching the body in the first place. However, in the final analysis, body defense represents the bottom line in protection against disease and, as history has shown, body defenses work very well. Indeed, as Lewis Thomas has suggested in Lives of a Cell, *a microorganism that catches a human is in considerably more danger than a human who has caught a microorganism.*

Summary

The body's resistance to disease takes two forms: nonspecific resistance and specific resistance. Nonspecific resistance exists in all humans and protects against all parasites. It involves species and population immunities and such mechanical and chemical barriers as the skin surface, mucus secretions, stomach acid, lysozyme, and interferon. Phagocytosis is a nonspecific mechanism in which macrophages and other phagocytes engulf and destroy microorganisms. Fever and inflammation are other forms of this resistance.

Specific resistance develops from the response by the body's immune system to substances called antigens. Antigens are large, complex molecules interpreted as nonself. Proteins, polysaccharides, and an enormous list of substances containing these molecules are antigenic. A small part of the antigen called the antigenic determinant performs the actual stimulation of the immune system. A person's own chemical substances are nonantigenic because they are interpreted as self.

The immune system originates with bone marrow cells that undergo differentiation to form B-lymphocytes and T-lymphocytes. These cells comprise the tissue of the spleen, lymph nodes, and other lymphoid organs, and they are the major underpinnings of the immune system. When T-lymphocytes are stimulated by antigenic determinants, they leave the immune system as cytotoxic cells and travel to the infection site. Here they kill the infecting organisms in a process called cell-mediated immunity. Memory T-lymphocytes remain in the tissue to provide long-lasting protection.

The second process is antibody-mediated (humoral) immunity. In this case, B-lymphocytes are stimulated to form antibody-producing cells called plasma cells. Antibodies are formed in the lymph nodes and are protein molecules composed of light and heavy chains of amino acids. The antibodies enter the circulation and reach the infection site, where they react with and neutralize microorganisms by various mechanisms. Five types of antibodies are recognized, each with its own function and structure. Together with cytotoxic T-lymphocytes, the antibodies impart specific resistance during times of disease and they remain in the body for long-lasting resistance.

Questions for Thought and Discussion

1. The opening paragraphs of this chapter suggest that for many diseases, a penetration of the walls surrounding the human body must take place. Can you think of any diseases where penetration is not a prerequisite to illness?

2. It has been written that ". . . no other system in the human body depends and relies on signals as greatly as the immune system." What evidence can you offer to support or reject this concept?

3. Lysozyme is a valuable natural enzyme for the control and destruction of Gram-positive bacteria. Can you postulate why this substance is not a commercially available pharmaceutical product?

4. The ancestors of modern humans lived in a sparsely settled world where communicable diseases were probably very rare. Suppose that by using some magical scientific invention, one of those individuals was thrust into the contemporary world. How do you suppose he or she would fare in relation to infectious disease? What is the immunological basis for your answer?

5. In the book and movie *Fantastic Voyage*, a group of scientists is miniaturized in a submarine (the "Proteus") and sent into the human body to dissolve a blood clot. The odyssey begins when the miniature submarine carrying the scientists is injected into the bloodstream. Can you think of any natural opening to the body interior they could have used instead of the injection?

6. One of the most intriguing aspects of dental research is that a vaccine may eventually be produced to protect against tooth decay. A study in the 1980s, for example, showed that low caries activity was associated with high levels of salivary IgA specific for *Streptococcus mutans*. What problems do you foresee in the development of such a vaccine and what advantages and disadvantages might there be in its use?

7. A University of Cincinnati immunologist has demonstrated that cockroaches injected with small doses of honeybee venom can develop resistance to future injections of venom that would ordinarily be lethal. Does this finding imply that cockroaches have an immune system? Which might be the next steps that the research might take? What does this research tell you about the cockroach's ability to survive for about three to four years, far longer than most other insects?

8. Cattle are known to have the alcoholic carbohydrate erythritol in their placentas. When brucellosis occurs in pregnant cattle, the causative agent *Brucella abortus* metabolizes the erythritol and grows in the placenta causing abortion of the calf. Humans, by contrast, have no erythritol in the placenta and therefore do not suffer similar effects when they contract brucellosis. What type of immunity does this exemplify?

9. The botulism toxin is the most powerful toxin known to science; the tetanus toxin is the second most powerful. Despite this, the botulism toxin may pose a lesser danger than the tetanus toxin. Why is this so?

10. Why is it a good idea to cry when something gets into your eye?

11. Phagocytes have been described as "bloodhounds searching for a scent" as they browse along the walls of blood vessels. The scent they usually seek is a chemotactic factor, a peptide released by a bacterium. Does it strike you as unusual that a bacterium should release a substance to attract the "bloodhound" that will eventually lead to the bacterium's demise?

12. One of the earliest symptoms of disease in the body is swollen lymph nodes (erroneously referred to as "swollen glands"). Why?

13. A microbiology professor has suggested that an antigenic determinant arriving in the lymphoid tissue is like a parent searching for the face of a lost child in a crowd of a million children. Would you agree with this analogy? Why or why not?

14. A high school biology text published in 1989 by a well-known book company contains the following statement: "T cells do not produce circulating antibodies. They carry cellular antibodies on their surface." What is fundamentally incorrect about this statement, and what would you say in your letter to the publisher after reading the statement?

15. Some years ago, a novel entitled *Through the Alimentary Canal with Gun and Camera* appeared in bookstores. The book described a fictitious account of travels through the human body. What perils, microbiologically speaking, would you encounter if you were to take such an adventure?

Review

A major theme of this chapter has been antigens and antibodies, two key aspects of the immune system. To test your knowledge of antigens and antibodies, solve the following puzzle. The answers are presented in the appendix.

ACROSS

1. Does not induce an immune response
4. Type of immunity (abbr) where T-lymphocytes function
5. Antigenic on rare occasions
6. Immunity related to antibodies is also said to be hum-_____.
10. Small molecules may act as antigens called _____-tens.
12. Alternate name for antibody
14. Protein group; functions in antigen–antibody reaction
16. End of antibody molecule where antigen reacts

DOWN

1. Medium for transport of antibodies
2. Type of acid in antibody molecule
3. Antigens generally have a _____ molecular weight.
7. Numbers of chains in a monomeric antibody molecule
8. Nobel Prize for study of antibody diversity
9. Structure of IgA
11. Sole organic material in antibody molecule
12. Body system depending on antibody activity

ACROSS (*continued*)

19. Antigenic materials are classified
    as _____-self
20. Cells that produce antibody
    molecules
21. Number of monomeric units in
    IgM
23. Antibody response in which IgM is
    the principal component
26. Largest antibody
29. Number of heavy chains in a
    monomeric antibody molecule
30. A person's own chemical substance
    is an _____-antigen
33. Alternate name for an antigenic
    determinant
34. Reaction in which IgE functions
35. Cell that phagocytizes
    microorganisms and begins
    immune response
36. Number of amino acid molecules
    possible in an antigenic
    determinant
37. White blood cells involved in
    immune response
38. In the embryonic chick,
    B-lymphocytes mature in
    the _____ of Fabricius

DOWN (*continued*)

13. Secretory antibody
14. Region of identity among antibody
    molecules
15. Fragment of antibody that
    combines with antigenic
    determinant
17. Sensitizes cells to certain antigens
18. Proposed theory of clonal selection
22. Major antibody of the secondary
    antibody response
24. Antibody molecule with two
    identical halves
25. Type of blood cell (abbr)
    containing alloantigens
26. Antibody at the surface of
    B-lymphocytes
27. An antibody molecule reacts with
    an epi-_____.
28. Type of immunity (abbr) based on
    activity of B-lymphocytes
31. Nodes where immune system cells
    gather
32. Molecules functioning in
    recognition reactions between cells
34. Nucleic _____ usually are
    poor antigens

CHAPTER 19

IMMUNITY AND SEROLOGY

Tradition tells us that as early as the eleventh century, Chinese doctors ground up smallpox scabs and blew the powder into the noses of healthy people to protect them against the ravages of smallpox. Modern vaccines work less haphazardly. They are composed of chemically altered toxins or treated microorganisms. Such vaccines work by exploiting the immune system's ability to recognize antigens and respond with antibodies that remain in the body and later attack and destroy invading organisms.

But no vaccine now available is completely safe. For example, the Sabin oral polio vaccine causes 5 to 10 cases of paralysis in the United States each year, an average of 1 case for every 3 million doses. The whole-cell pertussis vaccine, routinely used for the past 50 years, has been related to brain damage in 1 child per 300,000 immunized.

The future bodes well, however. Using the techniques of genetic engineering, researchers are now striving to eliminate all but the absolutely necessary material required to stimulate the immune system. Developers identify the antigen that triggers the immune response, clone the genes that encode its production, and insert the genes into organisms such as yeasts or *Escherichia coli*. These esh-er-ik'e-a
biological factories promptly churn out multiple copies of the antigen. The antigens are then extracted from the culture media, purified, and concentrated to form the new vaccine. A synthetic vaccine for hepatitis B is now available, and laboratories are working on synthetic vaccines for genital herpes, rabies, typhoid fever, diphtheria, malaria, and cholera. In the view of many scientists, the 1990s has witnessed a golden age of vaccines and a new opportunity to immunize against disease.

In this chapter we shall study the role of vaccines in the immune response while examining the four major mechanisms by which immunity comes about. Antibodies occupy a central position in each of these mechanisms. We shall also study how antibodies may be detected in a disease patient by a variety of laboratory tests. These diagnostic procedures help the physician understand the disease and prescribe a course of treatment. Immune mechanisms generally protect against disease, but when they fail, the laboratory tests provide a clue about what is taking place in the patient.

Immunity to Disease

The word "immune" is derived from the Latin stem *immuno*, meaning safe, or free from. In its most general sense, the term implies a condition under which an individual is protected from disease. This does not mean, however, that one is immune to all diseases, but rather to a specific disease or group of diseases.

Two general types of immunity are recognized: innate immunity and acquired immunity. **Innate immunity** is an inborn capacity for resisting disease. It begins at birth and depends on genetic factors expressed as physiological, anatomical, and biochemical differences among living things. Examples of innate immunity are the lysozyme found in tears, saliva, and other body secretions, acidic pH of the gastrointestinal and vaginal tracts, and interferon produced by body cells to protect against viruses. **Acquired immunity**, by contrast, begins after birth. It depends on the presence of antibodies and other factors originating from the immune system. Four types of acquired resistance are generally recognized, as the following paragraphs will explore. Although the emphasis will be on antibodies and antibody-mediated immunity, it should be remembered that cellular immunity is also an important consideration in the total spectrum of resistance.

Naturally Acquired Active Immunity

Active immunity develops after antigens enter the body and the individual's immune system responds with antibodies. The exposure to antigens may be unintentional or intentional. When it is unintentional, the immunity that develops is called naturally acquired active immunity.

Naturally acquired active immunity usually follows a bout of illness (Figure 19.1) and occurs in the "natural" scheme of events. However, this need not always be the case because subclinical diseases may also bring on the immunity. For example, many individuals have acquired immunity from subclinical cases of mumps or from subclinical fungal diseases such as cryptococcosis.

Memory cells residing in the lymphoid tissues are responsible for the production of antibodies that yield naturally acquired active immunity. The cells remain active for many years and produce IgG immediately upon later entry of the parasite to the host. Such an antibody response is sometimes called the **secondary anamnestic response**, from the Greek *anamnesis*, for recollection.

Artificially Acquired Active Immunity

Artificially acquired active immunity develops after the immune system produces antibodies following an intentional exposure to antigens. The antigens are usually contained in an immunizing agent such as a vaccine or toxoid and the exposure to antigens is "artificial."

Viral vaccines consist of either inactivated viruses incapable of multiplying in the body or attenuated viruses, which multiply at low rates in the body but fail to cause symptoms of disease. The Salk polio vaccine typifies the former, while the Sabin oral polio vaccine represents the latter. Bacterial vaccines fall into similar categories: the older whooping cough (pertussis) vaccine consists of dead cells, while the tuberculosis vaccine is composed of attenuated bacteria (MicroFocus: 19.1). Whole microorganism viral and bacterial vaccines are commonly called **first-generation vaccines**.

One advantage of vaccines made with attenuated organisms is that organisms multiply for a period of time within the body, thus increasing the dose of antigen administered. This higher dose results in a higher level of immune

---

Innate immunity:
an inborn capacity to resist disease.

Acquired immunity:
immunity that develops after the interaction of antigens with the immune system.

Active immunity:
immunity derived from an exposure to antigens and subsequent production of antibodies.

krip″to-kok-o′sis

an″am-nes′tik

Artificial active immunity:
immunity that develops after an intentional exposure to antigens followed by an immune response.

First-generation vaccine:
one that contains whole microorganisms.

FIGURE 19.1
The Four Types of Immunity

(a) Naturally acquired active immunity arises from an exposure to antigens and often follows a disease. (b) Artificially acquired active immunity results from an inoculation of toxoid or vaccine. (c) Naturally acquired passive immunity stems from the passage of IgG across the placenta from the maternal to the fetal circulation. (d) Artificially acquired passive immunity is induced by an injection of antibodies taken from the circulation of an animal or another individual.

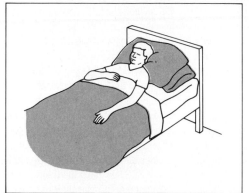

(a) Naturally acquired active immunity

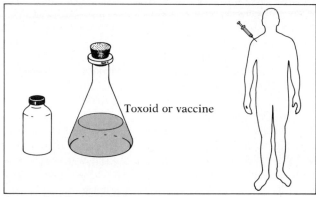

(b) Artificially acquired active immunity

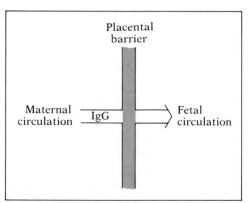

(c) Naturally acquired passive immunity

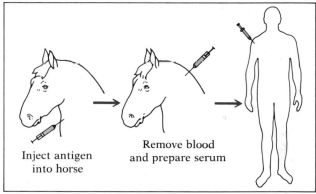

(d) Artificially acquired passive immunity

response than that obtained with the single dose of inactivated organisms. Also, **attenuated organisms** can spread to other individuals and reimmunize them or immunize them for the first time. However, attenuated organisms may be hazardous to health because of this same ability to continue multiplying. In 1984, for example, a recently immunized soldier spread vaccinia (cowpox) viruses to his daughter. She, in turn, infected seven young friends at a slumber party.

With only one notable exception, there are no widely used bacterial vaccines made with whole organisms and used for long-term protection. The exception is the older **pertussis** vaccine, now in the process of being replaced by the acellular pertussis vaccine composed of *Bordetella pertussis* extracts. Other bacterial vaccines made with organisms are used for temporary protection. For instance, when health officials suspect that water contains typhoid bacilli, they may administer a vaccine for typhoid fever. Bubonic plague or cholera vaccines are also available to limit an epidemic. In these cases, the immunity lasts only for several months because the material in the vaccine is weakly antigenic.

Attenuated: weakened.

19.1

MICROFOCUS

TRAGEDY AT LUBECK

Albert Calmette and Camille Guérin were excited. Since 1908 they had been transferring a strain of tubercle bacillus to fresh media every two weeks and now, after 13 long years, they found that the strain had lost most of its virulence. When inoculated to test animals, the bacillus failed to cause tuberculosis. Moreover, it apparently immunized animals to the disease and could be used as a vaccine. Their report, published in 1921, described the vaccine and gave it the name BCG, for bacille Calmette–Guérin.

By 1929, physicians had immunized thousands of children throughout the world and established the value of BCG. Statistical data added to the confidence doctors had in the vaccine, even as reports began to filter in from the Public Health Laboratories at Lubeck.

Lubeck was a major city on the Baltic Sea in what is now western Germany. It had a rich history dating back to the medieval period and was once the capital of the Hanseatic League. Doctors in the bacteriology laboratory at the city hospital had prepared the vaccine as directed by Calmette, and administered it to 252 children. However, to their shock and dismay, 71 of the recipients soon died of tuberculosis. Subsequent studies showed that a contaminating tubercle bacillus had caused the deaths, but the damage was already done. World opinion soon turned against BCG.

In the following years, doctors continued to use BCG, but with caution and skepticism. Some confidence was restored by a 1937 Chicago study showing that in 20,000 vaccinated infants and schoolchildren, there was an 80 percent reduction of expected cases of tuberculosis. By 1964, 200 million people worldwide had been immunized, and the safety of BCG once again was established. However, the United States remained then and continues today as a "blind spot." Many American doctors argue that BCG would destroy the effectiveness of the tuberculin test and that most cases of tuberculosis can be controlled with drugs. Nevertheless, tuberculosis still affects over 25,000 Americans today, and it is plausible that physicians may rethink the use of BCG. No doubt the tragedy at Lubeck will continue to color their decision.

Weakly antigenic vaccines are also available for laboratory workers who deal with rickettsial diseases such as Rocky Mountain spotted fever, Q fever, and typhus. The danger in these vaccines is that the residual egg protein in the cultivation medium for rickettsiae may cause allergic reactions in recipients. Table 19.1 presents a summary of currently available bacterial and viral vaccines.

Toxoid:
an immunizing agent consisting of chemically altered toxins; provides protection to toxins.

Immunizing agents that stimulate immunity to toxins are known as **toxoids**. These agents are currently available for protection against diphtheria and tetanus, two diseases whose major effects are due to toxins. The toxoids are prepared by incubating toxins with a chemical such as formaldehyde until the toxicity is lost.

To avoid multiple injections of immunizing agents, it is advantageous to combine vaccines into a single dose. Experience has shown this possible for the diphtheria-pertussis-tetanus vaccine (DPT), the newer diphtheria-tetanus-acellular pertussis vaccine (DTaP), the measles-mumps-rubella vaccine (MMR), and the trivalent oral polio vaccine (TOP). There is even a vaccine that will immunize against four diseases simultaneously: in 1993, the FDA approved a combined vaccine which includes diphtheria and tetanus toxoids, whole-cell pertussis vaccine, and *Haemophilus influenzae* b (Hib) vaccine. Marketed as Tetramune, the quadruple vaccine is used in children aged 2 months to 5 years to protect against the DPT diseases as well as *Haemophilus* meningitis. Table 19.2 outlines the recommendations for these vaccines in normal infants and children. For other vaccines, however, a combination may not be valuable because the antibody response is lower for the combination than for each vaccine taken separately. Immunologists believe that poor phagocytosis by macrophages is one reason. Activation of suppressor T-lymphocytes may be another reason.

TABLE 19.1

Principal Bacterial and Viral Vaccines Currently in Use*

DISEASE	VACCINE COMPOSITION	USE
BACTERIAL		
Cholera	Killed *Vibrio cholerae*	For those in areas endemic for the disease
Haemophilus meningitis	Purified polysaccharide from *H. influenzae* b	For infants and young children
Meningococcal meningitis	Purified polysaccharide from types A and C	Primarily for high-risk groups
Pertussis (whooping cough)	Killed *Bordetella pertussis* or acellular polysaccharides	Begun at 3 months of age, followed by boosters
Plague	Killed *Yersinia pestis*	For workers in areas endemic for the disease
Pneumonia caused by *Streptococcus pneumoniae*	Mixture of 23 different capsular polysaccharide types	For older adults compromised by other illnesses
Rocky Mountain spotted fever	Killed *R. rickettsii*	For laboratory workers
Tuberculosis	Attenuated *Mycobacterium tuberculosis* (BCG)	Used primarily in Europe
Typhoid fever	Killed *S. typhi*	For travelers to area where endemic
VIRAL		
Hepatitis A	Inactivated virus	For individuals over 2 years of age
Chickenpox	Attenuated virus	For all individuals over 1 year of age
Hepatitis B	Synthetic viral fragments	For all groups
Influenza	Inactivated virus or attenuated virus	Inactivated virus is used in United States; attenuated virus is used in Europe
Measles	Attenuated virus	For infants and college students
Mumps	Attenuated virus	For infants 15 to 19 months of age
Polio	Attenuated (Sabin) or inactivated (Salk) virus	Oral attenuated virus is most widely used in U.S.; begun at 2 to 3 months of age plus boosters
Rabies	Inactivated virus	For workers where rabies is endemic in wildlife and after animal bite
Rubella	Attenuated virus	For infants 15 to 19 months of age and women of childbearing age who are not pregnant
Smallpox	Attenuated virus	Disease eradicated and vaccine no longer recommended
Yellow fever	Attenuated virus	For travelers to areas endemic for the disease

* Toxoid vaccines not included.

Modern immunologists foresee the day when preparations called **subunit vaccines**, or **second-generation vaccines**, will completely replace whole organism vaccines. For example, pili from bacteria may be extracted and purified for use in a vaccine to stimulate antipili antibodies. These would inhibit the

Second-generation vaccine: one that contains parts or subunits of microorganisms.

TABLE 19.2
Approximate Recommended Schedule for Immunizations

AGE	IMMUNIZATION
2–3 months	Diphtheria toxoid ⎫ Pertussis vaccine ⎬ DPT (combined preparation) Tetanus toxoid ⎭ Hepatitis B vaccine *Haemophilus influenzae* b (Hib) vaccine Trivalent oral polio vaccine (TOPV)
3–4 months	DPT OPV (Type III or trivalent preparation)
6 months	DPT TOPV (optional)
15 months	Measles, mumps, and rubella vaccine. These vaccines can be used in combined form (MMR) or as individual preparations.[a] Tuberculin testing also is included at this point.[b] Hib vaccine booster.
16–18 months	DPT or DTaP (preferred if available) TOPV
4–7 years	DPT or DTaP (preferred if available) TOPV
14–16 years	Td (adult) booster[c]
Thereafter every 10 years at 10-year intervals (except in case of injury)	Td (adult) booster[c]

[a] Children between the ages of 12 months and puberty may be vaccinated either with combined vaccines or with single-vaccine preparations.

[b] Initial testing is recommended at 1 year of age. The frequency of testing thereafter depends on the risk of exposure and the prevalence of tuberculosis in the community.

[c] Adult form of tetanus and diphtheria toxoids. The diphtheria toxoid dosage is reduced because of possible undesirable reactions in individuals with several previous inoculations.

attachment of bacteria to tissues and facilitate phagocytosis. Another example is the vaccine for pneumococcal pneumonia, licensed for use in 1983. The vaccine contains 23 different polysaccharides from the capsules of 23 strains of *Streptococcus pneumoniae*. Still another example is the vaccine against *Haemophilus influenzae* b, the agent of *Haemophilus* meningitis (Chapter 7). Also composed of capsular polysaccharides, the so-called Hib vaccine has been available since 1988 and has been a critical factor in reducing the incidence of *Haemophilus* meningitis from 18,000 cases annually (1986) to a few dozen cases in current years (1995).

Another form of vaccine is the **synthetic vaccine**, or **third-generation vaccine**. This preparation represents a sophisticated and practical application of recombinant DNA technology. To produce the vaccine, three major technical problems must be solved: the immune-stimulating antigen must be identified; living cells must be reengineered to produce the antigens; and the size of the antigens must be increased to promote phagocytosis and the immune response. Thus far, the process has been successful for a vaccine for foot-and-mouth disease licensed in 1981.

The genetic engineering process has also worked for a synthetic vaccine for **hepatitis B**. The vaccine is marketed by different companies as Recombivax and Engerix-B. Because the vaccine is not made from blood fragments (as the previous hepatitis B vaccine was), it relieves the fear of contracting human

Third-generation vaccine: one that contains microbial fragments produced by genetic engineering.

19.2

MICROFOCUS

AN AIDS VACCINE—WHY THE WAIT?

Producing an AIDS vaccine might appear rather straightforward: cultivate a huge batch of human immunodeficiency virus (HIV); inactivate it with chemicals; purify it; and prepare it for marketing. Unfortunately, things are not quite so simple when HIV is involved. For example, the effects of a bad batch of vaccine would be catastrophic; and people are generally reluctant to be immunized with whole HIV particles, no matter how inactive they are.

Nevertheless, developing an AIDS vaccine is a major priority of modern scientists because about 20 million people are candidates to receive it. The cohort of candidates include anyone practicing high-risk behaviors (e.g., unprotected sexual intercourse) or using drugs intravenously or having contact with blood. The latter group includes doctors, surgeons, dentists, nurses, medical technologists, morticians, emergency medical technicians, police officers, and firefighters. All would welcome the vaccine when it becomes available.

And the operative word is *when*, for an AIDS vaccine will be available one day. Already, biotechnologists have cloned the genes for the gp120 and gp41 molecules in the HIV envelope, and they are producing these subunit molecules in vast quantities. Others are researching gp160, the precursor molecule to gp120 and gp41, for use as a vaccine subunit; still others are utilizing p17, an HIV core protein, in their candidate vac-

cine. There is a potential vaccine containing HIV minus its envelope, and another composed of simian immunodeficiency virus (SIV). Proponents of the latter vaccine hope to use it as cowpox viruses were once used against smallpox.

Before a vaccine reaches the human population, however, a number of problems must be resolved. For instance, HIV tends to mutate much as influenza viruses do, and a vaccine would have to take HIV variants into account. Then too, HIV remains within T-lymphocytes as a provirus, and antibodies

elicited by a vaccine could not reach it here. There is also the problem of locating an animal model for testing purposes. (Chimpanzees are the only animals that display AIDS symptoms.)

When it comes to field trials, still other problems must be confronted. For example, volunteers can only be used once, and there are many candidate vaccines to test. Also, volunteers will test positive for HIV antibodies after participating in a trial and, should they contract HIV, the diagnostic test for antibodies would not work. Moreover, they might suffer discrimination (for example, in obtaining insurance) when they test positive during and after the vaccine trial. There is also the problem of counseling: when a person volunteers for a vaccine trial, the physician is ethically obliged to counsel the person on methods of avoiding HIV. It would then be difficult to assess whether the person remained free of HIV because of the counseling or because of the vaccine. And, finally, once a person has developed antibodies from the vaccine, it is ethically unsound to inject that person with HIV to see if the antibodies are protective.

To be sure, the problems are daunting. However, the project is equally massive in scope and effort. No vaccine has appeared in the marketplace less than ten years after the inception of development. For AIDS, no vaccine predates 1987, so if we add ten years....

immunodeficiency virus (HIV) from contaminated blood. Many immunologists believe that the synthetic agents will usher in a Renaissance of vaccines. In 1993, for example, biotechnologists announced the development of a cholera vaccine containing *Vibrio cholerae* whose genes for toxin production were experimentally removed. An AIDS vaccine also looms on the horizon (Micro-Focus: 19.2).

Immunizations may be administered by injection, oral consumption, or nasal spray, as currently used for some respiratory viral diseases. **Booster immunizations** commonly follow as a way of raising the antibody level by stimulating the memory cells to induce the secondary anamnestic response. This is why a "tetanus booster" is given to anyone who sustains a deep puncture wound by a soil-contaminated object if they have not had a tetanus immunization in the previous ten years.

TABLE 19.3
Characteristics of Four Types of Immunity

TYPE OF IMMUNITY	IMMUNIZING AGENT	EXPOSURE TO IMMUNIZING AGENT	EFFECTIVE DOSE REQUIRED	RELATIVE TIME UNTIL IMMUNITY	RELATIVE DURATION OF IMMUNITY
Naturally acquired active	Antigens	Unintentional	Small	Long	Long (lifetime)
Artificially acquired active	Antigens	Intentional	Small	Long	Long (months to years)
Naturally acquired passive	Antibodies	Unintentional	Large	Short	Short (4–6 months)
Artificially acquired passive	Antibodies	Intentional	Large	Short	Short (up to 6 weeks)

ad'ju-vant

Adjuvant:
a substance that increases the immunizing potential of a vaccine or toxoid by boosting the availability of the antigen.

Substances called **adjuvants** increase the efficiency of a vaccine or toxoid by increasing the availability of the antigen in the lymphatic system. Common adjuvants include aluminum sulfate ("alum") and aluminum hydroxide in toxoid preparations, as well as mineral oil or peanut oil in viral vaccines. The particles of adjuvant linked to antigen are taken up by macrophages and presented to lymphocytes more efficiently than dissolved antigens. Experiments also suggest that adjuvants may stimulate the macrophage to produce a lymphocyte-activating factor and thereby reduce the necessity for helper T-lymphocyte activity. Moreover, adjuvants provide slow release of the antigen from the site of entry and provoke a more sustained immune response. A high priority in the development of synthetic vaccines is the production of suitable adjuvants.

Naturally Acquired Passive Immunity

Passive immunity:
immunity derived from an infusion of antibodies from an outside source.

Passive immunity develops when antibodies enter the body from an outside source (as compared to active immunity in which individuals synthesize their own antibodies). The infusion of antibodies may be unintentional or intentional, and thus, natural or artificial. When unintentional, the immunity that develops is called naturally acquired passive immunity. Table 19.3 compares this immunity with the other three types.

Naturally acquired passive immunity, also called **congenital immunity**, develops when antibodies pass into the fetal circulation from the mother's bloodstream via the placenta and umbilical cord. These antibodies, called **maternal antibodies**, remain with the child for approximately 3 to 6 months after birth and fade as the child's immune system becomes fully functional. Certain antibodies, such as measles antibodies, remain for 12 to 15 months. The process occurs in the "natural" scheme of events.

Maternal antibodies play an important role during the first few months of life by providing resistance to diseases such as pertussis, staphylococcal infections, and viral respiratory diseases. Because the antibodies are of human origin and are contained in human serum, they will be accepted without problem. The only antibody in the serum is IgG.

kŏ-los'trum

Maternal antibodies also pass to the newborn through the first milk, or **colostrum**, of a nursing mother as well as during future breast feedings. In this instance, IgA is the predominant antibody, although IgG and IgM have also been found in the milk. The antibodies accumulate in the respiratory and gastrointestinal tracts of the child and apparently lend increased resistance to diseases (MicroFocus: 18.5).

USUAL ROUTE OF INTRODUCTION	SOURCE OF ANTIBODIES	FUNCTION	EFFEC-TIVENESS IN NEWBORN	EFFEC-TIVENESS IN ADULT	ORIGIN
Various tissues	Self	Therapeutic Prophylactic	Low	High	Clinical or subclinical disease
Intramuscular or intradermal	Self	Prophylactic	Low	High	Toxoid or vaccine
Intravenous	Other than self	Prophylactic	High	Low	Transplacental passage of antibodies
Intravenous	Other than self	Prophylactic Therapeutic	High	Moderate	Serum that contains antibodies

Artificially Acquired Passive Immunity

Artificially acquired passive immunity arises from the intentional injection of antibody-rich serum into the circulation. The exposure to antibodies is thus "artificial." In the decades before the development of antibiotics, such an injection was an important therapeutic device for the treatment of disease. The practice is still used for viral diseases such as Lassa fever, hepatitis, and arthropodborne encephalitis, and for bacterial diseases where a toxin is involved. For example, established cases of botulism, diphtheria, and tetanus are treated with serum containing the respective antitoxins.

Various terms are used for the serum that renders artificially acquired passive immunity. **Antiserum** is one such term. Another is **hyperimmune serum**, which indicates that the serum has a higher-than-normal level of a particular antibody. If the serum is used to protect against a disease such as hepatitis A, it is called **prophylactic serum**. When the serum is used in the therapy of an established disease, is it called **therapeutic serum**. Should the serum be taken from the blood of a convalescing patient, physicians refer to it as **convalescent serum**. Another common term, **gamma globulin**, takes its name from the fraction of blood protein in which most antibodies are found. Gamma globulin usually consists of a pool of sera from different human donors, and thus it contains a mixture of antibodies including those for the disease to be treated.

Passive immunity must be used with caution because in many individuals, the immune system recognizes foreign serum proteins as antigens and forms antibodies against them in an allergic reaction. When antibodies interact with the proteins, a series of chemical molecules called immune complexes may form (Chapter 20), and with the activation of complement, the person develops a disease called **serum sickness**. This is often characterized by a hivelike rash at the injection site, accompanied by labored breathing and swollen joints. To avoid the disease, it is imperative that the patient be tested for allergy before serum therapy is instituted. If an allergy exists, minuscule doses should be given to eliminate the allergic state, and then a large therapeutic dose can be administered.

Artificially acquired passive immunity provides substantial and immediate protection to disease, but it is only a temporary measure. The immunity that develops from antibody-rich serum usually wears off within days or weeks. Among the serum preparations currently in use are those for hepatitis A and chickenpox. Both are made from the serum of blood donors routinely screened for hepatitis A and chickenpox.

Artificial passive immunity:
immunity that develops after an intentional exposure to antibodies.

Antiserum:
serum that is rich in a particular antibody.

Gamma globulin:
an alternate name for antiserum based on the fraction of blood protein in which antibodies are located.

Serum sickness:
a type of allergic reaction in which the immune system forms antibodies against proteins in antiserum.

TO THIS POINT
■

We have discussed four different types of acquired immunity with examples of each. Naturally acquired active immunity develops after a bout of illness or following a subclinical disease. The exposure to antigens is unintentional, and the immune system produces antibodies. By contrast, artificially acquired active immunity requires an intentional exposure to antigens, such as when one receives a vaccine or toxoid. Subunit vaccines and synthetic vaccines contain no organisms and thus are an improvement over whole organism vaccines. As before, immunity comes about when the immune system produces antibodies.

The third type of immunity is naturally acquired passive immunity. This develops from the passage of antibodies from mother to child across the placenta within the uterus. The antibodies remain active for several months after birth, a time during which the immune system is not fully functional. An injection of antibodies brings about the fourth type of immunity, called artificially acquired passive immunity. Allergic reactions in the recipient limit the use of this form of immunity.

In the next section we shall shift our attention to the activity of antibodies in the laboratory. Our focus will center on serological reactions where antigen–antibody interactions are used for diagnostic purposes. We shall survey multiple forms of these tests and note the advances in technology that have made the serology laboratory a highly sophisticated environment for conducting practical immunology.

Serological Reactions

Serological reaction:
a laboratory reaction in which serum is involved.

Antigen–antibody reactions studied under laboratory conditions are known as **serological reactions** because they commonly involve serum from a patient. In the late 1800s, serological reactions were first adapted to laboratory tests used in the diagnosis of disease. The principle was simple and straightforward: if the patient had an abnormal level of a specific antibody in the serum, a suspected disease agent was probably present. Today, serological reactions have diagnostic significance as well as more broad-ranging applications. For example, they are used to confirm identifications made by other procedures and detect organisms in body tissues. In addition, they help the physician follow the course of disease and determine the immune status; and they assist the determination of groupings below the species level.

Characteristics of Serological Reactions

Serum:
the clear cell-free fluid portion of the blood that contains no clotting agents.

The reactants in a serological reaction generally consist of an antigen and a serum sample. The nature of either must be known; the object is to determine the nature of the other. In some cases the unknown can be determined merely by placing the reactants on a slide and observing the presence or absence of a reaction. However, serological tests are not always quite so direct. For example, an antigen may have only one antigenic determinant and the combination with an antibody molecule on a one-to-one basis may be invisible. To solve this dilemma, a second-stage reaction using an indicator system may be required or a contact signal may be necessary. We shall see how this works in several tests.

Another possible drawback to a successful serological reaction is that antigen or antibody solutions may require considerable dilution to reach a concentration at which reaction is most favorable. This may be used to the physician's advantage, however, because the dilution series is a valuable way of determining the titer of antibodies, as shown in Figure 19.2. The **titer** is the most dilute concentration of serum antibody that yields a detectable reaction with its specific antigen. This number is expressed as a ratio of antibody to total fluid (for example, 1:50), and is used to indicate the amount of antibodies in a patient's serum. For instance, the titer of **influenza** antibodies may rise from 1:20 to 1:320 as an episode of influenza progresses and then continue upward, stabilizing at 1:1280 as the disease reaches its peak. A rise in the titer of a certain antibody also indicates that an individual has a case of that disease, an important factor to diagnosis.

Haptens may pose a problem in a serological reaction because, as partial antigens, they usually have only a single antigenic determinant. Modern technologists have circumvented this problem by conjugating the haptens to carrier particles such as polystyrene beads. When the hapten unites with an antibody, the entire bead is involved in the complex and a visible reaction develops.

Serology has become a highly sophisticated and often automated branch of immunology. As the following tests illustrate, the serological reactions have direct application to the clinical laboratory not only in microbiology but in other fields as well.

tī'ter

Titer:
the most dilute concentration of serum antibody that yields a detectable reaction with a specific antigen.

Hapten:
a partial antigen that complexes to carrier proteins or polysaccharides to form a complete antigen.

Neutralization

Neutralization is a serological reaction in which antigens and antibodies neutralize each other. The reaction is used to identify toxins and antitoxins, as well as viruses and viral antibodies. Normally, little or no visible evidence of a

FIGURE 19.2

Determination of Titer

A sample of antibody-containing serum was diluted in saline solution to yield the dilutions shown. An equal amount of antigen was then added to each tube, and the tubes were incubated. An antigen–antibody reaction may be seen in tubes (a) through (d), but not in tubes (e) or (f) or the control tube, (g). The titer of antibody is the highest dilution of serum antibody in which a reaction is visible, in this case, tube (d). The titer is therefore expressed as 1:160.

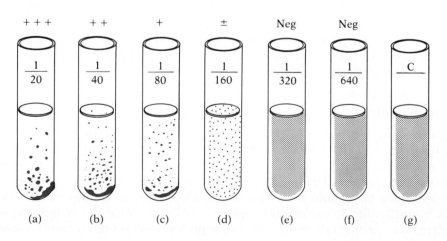

neutralization reaction is present, and the test mixture must therefore be injected into a laboratory animal to determine whether neutralization has taken place.

An example of a neutralization test is the one used to detect **botulism toxin** in food. Normally the botulism toxin is lethal to a laboratory animal, and if a sample of the food contains the toxin, the animal will succumb after an injection. However, if the food is first mixed with botulism antitoxins, the antitoxin molecules neutralize the toxin molecules, and the mixture has no effect on the animal. Conversely, if the toxin was produced by some other organism, no neutralization will occur, and the mixture will still be lethal to the animal.

Precipitation

Precipitation reactions are serological reactions involving thousands of antigen and antibody molecules cross-linked at multiple determinant sites to form a structure called a **lattice**. The lattices are so huge that particles of precipitate form, and the reaction product is visually observed.

Precipitation tests are performed either in fluid media or gels. In **fluid precipitation**, the antibody and antigen solutions are layered over each other in a thin tube. The molecules then diffuse through the fluid until they reach a zone of equivalence, the ideal concentration for precipitation. A visible mass of particles now forms at the interface or at the bottom of the tube. Fluid precipitation is frequently used in forensic medicine to learn the origin of blood stains. The technique works because the albumin proteins in bloods of various animals precipitate only with their respective antibodies.

In **gel precipitation**, the diffusion of antigens and antibodies takes place through a semisolid gel such as agarose. The Oudin tube technique, described in 1946 by Jacques Oudin, is typical. A plug of gel is placed between solutions of antigen and antibody in a thin tube. As the molecules diffuse through the gel, they eventually reach the zone of equivalence where they interact and form a visible ring of precipitate. This type of technique is called a double diffusion process because both reactants diffuse. Another application is in the Ouchterlony plate technique. In this technique, antigen and antibody solutions are placed in wells cut into agarose in Petri dishes. The plates are incubated and precipitation lines form at the zone of equivalence (Figure 19.3).

In the procedure known as **immunoelectrophoresis**, the techniques of electrophoresis and diffusion are combined for the detection of antigens. A mixture of antigens is placed in a reservoir on an agarose slide and an electrical field is applied to the ends of the slide. The different antigens then move through the agarose at different rates of speed depending on the electrical charges they possess. This process is **electrophoresis**. A trough is then cut into the agarose along the same axis, and a known antibody solution is added. During incubation, antigens and antibodies diffuse toward each other and precipitation lines form as in the Ouchterlony technique.

Agglutination

Agglutination is a serological reaction in which antibodies react with antigens on the surface of particular objects and cause the objects to clump together, or agglutinate. Agglutination techniques were among the earliest serological reactions adapted to the diagnostic laboratory. For example, until the 1960s, the diagnosis of typhoid fever was based on the agglutination of *Salmonella* cells by antibodies in the patient's serum. This test, called the **Widal test** after its developer Georges Fernand I. Widal, is now supplemented by more sophisticated procedures.

Botulism: a foodborne bacterial disease in which toxins lead to muscular paralysis.

Zone of equivalence: the ideal concentration of reactants at which precipitation occurs.

oo-deh′

ouk″ter-lōn′e

im″mu-no-e-lek″tro-fo-re′sis

Agglutination: an antigen–antibody reaction accompanied by clumping.

ve-dahl′

FIGURE 19.3

A Precipitation Test

Wells are cut into a plate of purified agar. Different known antibodies are then placed into the two upper wells and a mixture of unknown antigens is placed into the lower well. During incubation, the reactants diffuse outward from the wells, and cloudy lines of precipitate form where the reactants happen to meet. The lines cross one another because each antigen has reacted only with its complementary antibody.

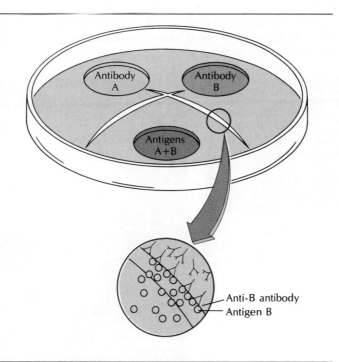

Anti-B antibody
Antigen B

Agglutination procedures are performed on slides or in tubes. For example, emulsions of unknown bacteria are added to drops of known antibodies on a slide, and the mixture is observed for clumping. If none occurs, different antibodies are tried until the correct one is found (Figure 19.4). This process is essentially a trial-and-error method, although the chances for success may be enhanced by using a **polyvalent serum**, that is, one containing a mixture of antibodies. Tube agglutinations may be performed with serum to determine the titer of antibodies present.

Polyvalent serum: serum that contains a mixture of antibodies.

Passive agglutination represents a modern approach to agglutination procedures. In this process, antigens are adsorbed onto the surface of latex spheres, polystyrene particles, red blood cells, bacteria, or other carriers. Serum antibodies are then detected by observing agglutination of the carrier particle. **Hemagglutination** refers to the agglutination of red blood cells. This process is particularly important in the determination of blood types prior to transfusion procedures (Chapter 20). In addition, certain viruses such as measles and mumps viruses agglutinate red blood cells. Antibodies for these viruses may be detected by a procedure in which the serum is first combined with laboratory-cultivated viruses and then added to the red blood cells. If serum antibodies neutralize the viruses, agglutination fails to occur. This test, called the **hemagglutination inhibition test**, is discussed in Chapter 11.

hem"ah-gloo"tĭ-na'shun

Flocculation

The **flocculation** test combines the principles of precipitation and agglutination. The antigen exists in a noncellular particulate form that reacts with antibodies to form large, visible aggregates.

flok"u-la'shun

FIGURE 19.4

Two Views of Agglutination

(a) A slide agglutination procedure. *Salmonella* cells were mixed with *Salmonella* antibodies on the right side of the slide and with antibody-free saline solution on the left side. Visible clumps of cells are evident on the right side. (b) A scanning electron micrograph of human red blood cells agglutinated by a microcolony of *Mycoplasma pneumoniae*, the cause of primary atypical pneumonia. Bar = 1 μm.

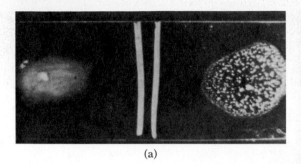

(a)

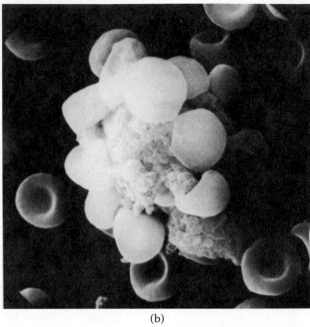

(b)

Cardiolipin:
an alcoholic extract of beef heart used in the flocculation test for syphilis.

VDRL test:
a flocculation test used for the rapid identification of syphilis antibodies.

An example of the flocculation test is the Venereal Disease Research Laboratory (VDRL) test used for the rapid screening of patients to detect **syphilis**. The antigen consists of an alcoholic extract of beef heart called cardiolipin. When diluted with buffer solution, the cardiolipin forms a milky-white precipitate. Serum from a patient is then added. If the serum contains syphilis antibodies, the particles of precipitate react with antibodies and cling together, yielding aggregates. Observation under the low-power objective of the microscope reveals the extent of aggregate formation and gives a clue on the amount of antibodies present. The **VDRL test** has been in use for many decades and is part of the "blood test" that couples may be required to take before obtaining a marriage license.

TO THIS POINT ■

We have surveyed a number of serological reactions used in the diagnostic laboratory to detect the reaction between antigens and antibodies. In each case, the reaction is fairly simple and straightforward, and the laboratory technician can usually determine whether a reaction has occurred. For example, particles clump in agglutination, and precipitates form in precipitation. Many of the basic reactions have been adapted by modern technologists to improve on the fundamental theme of the process.

In the final section of this chapter we shall explore another series of serological reactions and tests. Most of the tests involve a multistep procedure, and the visible manifestation of the reaction usually requires the participation of accessory factors, indicator systems, and specialized equipment. These diagnostic tests are more complicated to perform and require skilled technicians, but their development has ensured the position of laboratory immunology as a key link in the health-care delivery system.

Other Serological Reactions

Complement Fixation

The **complement fixation test** was devised by Jules Bordet and Octave Gengou in 1901. It was later adapted for syphilis by August von Wassermann in 1906, and for many decades it remained a mainstay for syphilis diagnosis. Modern technologists now use it for detection of antibodies against a variety of viruses, fungi, and bacteria.

bor-da'
zhaw-goo'

The test is performed in two parts. The first part, the **test system**, utilizes the patient's serum, a preparation of antigen and complement derived from guinea pigs. The second part, the indicator system, requires sheep red blood cells and hemolysin (antibodies against sheep red cells). Hemolysins cause lysis of red blood cells in the presence of complement.

he-mol'i-sin

The first step in the test is to heat the patient's serum to 56° C for 30 minutes. This destroys any complement present in the serum and allows the laboratory technician to regulate its amount. Next, carefully measured amounts of antigen and guinea pig complement are added to the serum (Figure 19.5). This test system is then incubated at 37° C for 90 minutes. During this period, if antibodies specific for the antigen are present in the serum, an antibody–antigen reaction takes place and the complement is used up, or fixed. However, there is no visible sign of whether a reaction has occurred.

After the incubation period, the **indicator system** (sheep red blood cells and hemolysin) is added to the tube, and the tube is reincubated at 37° C for an additional two hours. If the complement was previously fixed, none would be available to the hemolysin, and lysis of the sheep red blood cells could not take place. The blood cells would therefore remain intact and, when the tube is centrifuged, the technician observes clear fluid with a "button" of blood cells at the bottom. Conclusion: the serum contained antibodies that reacted with the antigen and fixed the complement.

Hemolysin:
antibodies that will lyse sheep red blood cells in the presence of complement.

If the complement was not fixed in the test system, it would still be available to the hemolysin and the hemolysin-complement mixture would lyse the sheep red blood cells. When the tube is centrifuged, the technician sees red fluid, colored by the hemoglobin of the broken blood cells, and no evidence of blood cells at the bottom of the tube. Conclusion: the serum lacked antibodies for the antigen tested.

Hemoglobin:
the red oxygen-carrying pigment in erythrocytes.

The complement fixation procedure is valuable because it may be adapted by varying the antigen. In this way, tests may be conducted for such diverse diseases as encephalitis, Rocky Mountain spotted fever, meningococcal meningitis, and histoplasmosis. The versatility of the test together with its sensitivity and relative accuracy have secured its continuing role in diagnostic medicine.

FIGURE 19.5

The Complement Fixation Test

(a) Specific antigen, the patient's serum, and complement are added to a test tube and incubated. (b) The indicator system is then added. This consists of sheep red blood cells (RBC) and antibodies that will lyse these cells if complement is available. These antibodies are called hemolysins. The test is evaluated as follows: (c) If specific antibodies are present in the patient's serum, a reaction will take place between the antibodies, antigen, and complement. Because the complement has been used up, no lysis of the sheep red blood cells will occur when the indicator system is added. (d) If specific antibodies are not present in the serum, no reaction occurs during the first incubation, and the complement is left free to unite with the sheep red blood cells and hemolysins. Lysis of the red blood cells results.

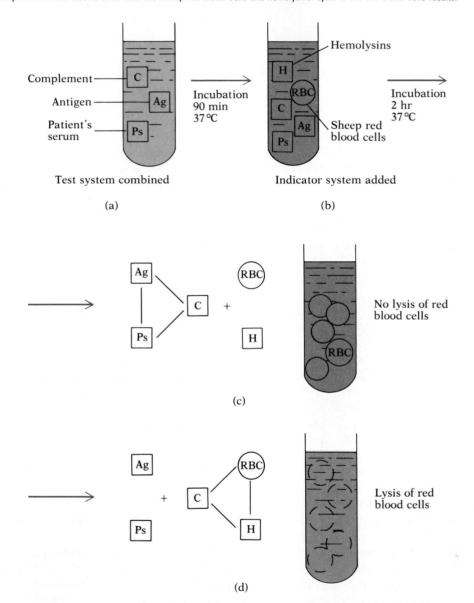

Fluorescent Antibody Techniques

The **fluorescent antibody technique** is a slide test performed by combining particles containing antigens with antibodies and a fluorescent dye. When the three components react, the dye causes the complex to glow on illumination

FIGURE 19.6

The Indirect Fluorescent Antibody Technique for Syphilis

(a) Syphilis spirochetes are fixed to a slide, and (b) serum from the patient is added. If the serum contains syphilis antibodies, they will adhere to the spirochetes (Ag•Ab). (c) Now fluorescein-labeled antiglobulin antibodies (Ab∗) are added. (d) If the spirochetes are coated with syphilis antibodies, the labeled antibodies will react with them (Ag•Ab•Ab∗), and the spirochetes will be coated with dye and fluoresce. (e) If the serum has no syphilis antibodies, no coating takes place, the labeled antibodies remain in the fluid, and the spirochetes fail to fluoresce.

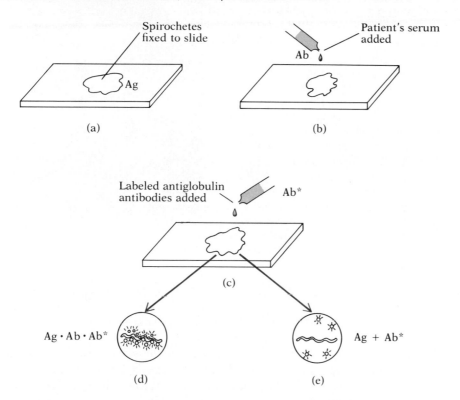

with ultraviolet light under a fluorescent microscope (Chapter 3). Two commonly used dyes are fluorescein, which emits an apple-green glow, and rhodamine, which gives off orange-red light.

Fluorescent antibody techniques may be direct or indirect. In the **direct method**, the fluorescent dye is linked to known antibody molecules. The antibodies are then combined with particles such as bacteria that may contain complementary antigens. If the presumption is correct, the tagged antibodies accumulate on the particle surface and the particle glows under the microscope. In this way, an unknown antigen or unknown organism can be identified.

The **indirect method** is illustrated by the FTA-ABS diagnostic procedure used for detecting **syphilis** antibodies in the blood of a patient (Figure 19.6). A sample of commercially available syphilis spirochetes is placed on a slide, and the slide is then flooded with patient's serum. Next, a sample of fluorescein-labeled **antiglobulin antibodies** is added. Antiglobulin antibodies are antibodies that unite with human antibodies. They are produced by an animal injected with human antibodies. The slide is then observed under the fluorescent microscope.

The test is interpreted as follows. If the patient's serum contains syphilis antibodies, the antibodies bind to the surfaces of spirochetes and the labeled antiglobulin antibodies are attracted to them. The spirochetes then glow from

Fluorescent microscope: a microscope that uses ultraviolet light to illuminate dye-coated particles.

FTA-ABS: fluorescent treponemal antibody absorption test; a diagnostic procedure for syphilis.

Antiglobulin antibodies: antibodies that react with human antibodies in diagnostic tests; produced in animals after injection with human antibodies.

19.3

MICROFOCUS

SOMETHING SPECIAL FROM A SPECIAL SOMEONE

When the Nobel Prize in Physiology or Medicine was announced on October 14, 1977, the scientific community applauded a special person and a special technique: one of the recipients was Rosalyn Sussman Yalow, a developer of the radioimmuno-assay technique. This technique is one of the most highly regarded immunological procedures.

The radioimmunoassay technique has made possible the detection of incredibly small amounts of chemical substances in body fluids. One offshoot was the discovery of certain hormones not previously known to exist in the body. Another was the revelation that individuals receiving insulin injections produce antibodies against insulin. This finding put in serious doubt the contention that the insulin molecules were too small to be antigenic.

The radioimmunoassay technique also permits detection of tumor viruses in the body before the appearance of a tumor. Moreover, it may be utilized to screen for hepatitis B viruses in blood used for transfusions. Radioimmunoassay is said to be sensitive enough to detect trillionths of a gram of a substance. This is equivalent to detecting a lump of sugar in Lake Erie.

Like the process she developed, Rosalyn Yalow also was special. She was educated during a time when opportunities for women were limited, and her work was done in restricted surroundings at the Veterans Hospital in the Bronx, New York. She was the sixth woman honored by the Nobel Committee and only the second in Physiology or Medicine. She was also a product of the same neighborhood of New York City as a certain textbook author. Rosalyn Yalow remains one of the eminent immunologists of our time.

---

the dye. However, if no antibodies are present in the serum, nothing accumulates on the spirochete's surface, and labeled antiglobulin antibodies also fail to gather on the surface. The labeled antibodies remain in the fluid and the spirochetes do not glow.

Fluorescent antibody techniques are adaptable to a broad variety of antigens and antibodies and are widely used in serology. Antigens may be detected in bacterial smears, cell smears, and viruses fixed to carrier particles. The value of the techniques is enhanced because the materials are sold in kits and are readily available to small laboratories.

Radioimmunoassay (RIA)

Radioimmunoassay:
a sensitive serological procedure in which radioactive antigens compete with nonlabeled antigens for reactive sites on antibody molecules.

Radioimmunoassay (RIA) is an extremely sensitive serological procedure used to measure the concentration of low molecular weight antigens such as haptens. Since its development in the 1960s, the process has been adapted for quantitating hepatitis antigens, as well as reproductive hormones, insulin, and certain drugs. One of its major advantages is that it can detect trillionths of a gram of substances (MicroFocus: 19.3).

The radioimmunoassay procedure is based upon the competition between radioactive-labeled antigens and unlabeled antigens for the reactive sites on antibody molecules. A known amount of the radioactive (labeled) antigens, is mixed with a known amount of specific antibodies, and an unknown amount of unlabeled antigens. The antigen–antibody complexes that form during incubation are then separated out, and their radioactivity is determined. By measuring the radioactivity of free antigens remaining in the leftover fluid (unbound labeled antigen), one can calculate the percentage of labeled antigen bound to the antibody. This percentage is equivalent to the percentage of unlabeled anti-

FIGURE 19.7

The Radioimmunoassay Technique

Radioactive antigen molecules of known amount (a) are mixed with unlabeled antigens (b) whose concentration is to be determined. (c) The two are combined with antibodies specific for that antigen. (d) Because each antibody molecule has two reactive sites, some of the sites will be taken up by labeled antigens and some by unlabeled antigens. The number of sites taken up depends on the concentration of antibodies. (e), (f) There will also be a certain amount of each type of antigen left over. (g) It is now possible to determine amounts of bound antigen (Ag*Ab) and leftover, or free, antigen (Ag*) by radioactive measurements. The amounts expressed as bound-to-free (B-F) known antigen will be the same as the percentage for bound-to-free unknown antigen. (h) By consulting a standard curve of B-F ratios, the amount of unknown antigen may be ascertained. In the example cited, the percent of B-F antigen is 52 percent. This means that the ratio of B-F unknown must also be 52 percent and by comparison to the standard curve, this equals 16 μg of antigen.

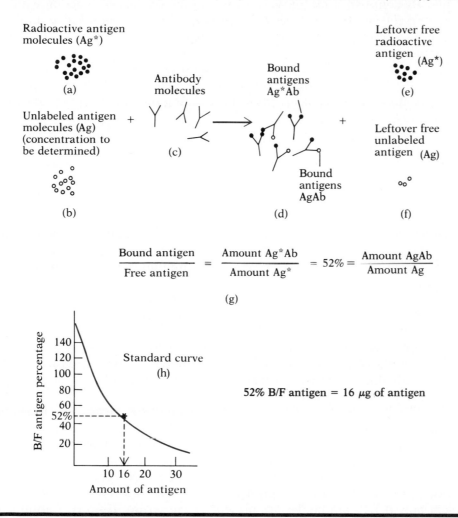

$$\frac{\text{Bound antigen}}{\text{Free antigen}} = \frac{\text{Amount Ag}^*\text{Ab}}{\text{Amount Ag}^*} = 52\% = \frac{\text{Amount AgAb}}{\text{Amount Ag}}$$

(g)

52% B/F antigen = 16 μg of antigen

gen bound to the antibody because the same proportion of both antigens will find spots on antibody molecules. The concentration of unknown unlabeled antigen can then be determined by reference to a standard curve. The curve is constructed from data obtained by allowing varying amounts of unlabeled known antigen to compete with antibody and determining the percentage of bound to free antigens (Figure 19.7).

Radioimmunoassay procedures require substantial investment in sophisticated equipment and carry a certain amount of risk because radioactive isotopes are used. For these reasons, the procedure is not widely used in routine serological laboratories. However, immunologists with access to radioimmunoassay have discovered a wealth of information.

Radioallergosorbent Test (RAST)

The **radioallergosorbent test** (RAST) is an extension of the radioimmunoassay. Another sophisticated procedure, it may be used to detect IgE, other antibodies, or a variety of small antigens.

To detect **IgE** against penicillin, penicillin antigens are attached to a suitable particle. Serum that may contain penicillin IgE is then added. If the antibody is present, it will combine with the penicillin antigens on the surface of the particle. Now another antibody, one that will react with human antibodies, is added. This antiglobulin antibody carries a radioactive label. The entire complex will therefore become radioactive if the antiglobulin antibody combines with the IgE. By contrast, if no IgE was present in the serum, no reaction with the antigen on the particle surface will take place, and the radioactive antibody will not be attracted to the particle. When tested, the particles will not show radioactivity.

The RAST is commonly known as a **"sandwich" technique**. There is no competition for an active site as in RIA, and the type of unknown antibody as well as its amount may be learned by determining the amount of radioactivity deposited.

Antiglobulin antibody: antibodies produced in an animal after injection with human antibodies; react with human antibodies.

Enzyme Linked Immunosorbent Assay (ELISA)

Enzyme linked immunosorbent assay (ELISA) has virtually the same sensitivity as radioimmunoassay and RAST, but does not require expensive equipment or radioactivity. The procedure involves attaching antibodies or antigens to a solid surface and combining (immunosorbing) the coated surfaces with the test material. An enzyme system is then linked to the complex, the remaining enzyme is washed away, and the extent of enzyme activity is measured. This gives an indication that antigens or antibodies are present in the test material.

An application of ELISA is found in the hour-long laboratory test used to detect antibodies against **human immunodeficiency virus (HIV)**. A serum sample is obtained from the patient and mixed with a solution of plastic or polystyrene beads coated with antigens from HIV. Antibodies present in the serum will adhere to the antigens on the surface of the beads. The beads are then washed and incubated with antiglobulin antibodies chemically tagged with molecules of the enzyme horseradish peroxidase. The preparation is washed and a solution of substrate molecules for the peroxidase enzyme is added. Initially the solution is clear, but if enzyme molecules react with the substrate, the solution will become yellow-orange in color. The enzyme molecules will be present only if HIV antibodies are present in the serum. If no HIV antibodies are in the serum, no enzyme molecules could concentrate on the bead surface, no change in the substrate molecules could occur, and no color change would be observed (Figure 19.8).

ELISA procedures may be varied depending on whether one wishes to detect antigens or antibodies. The solid phase may consist of beads, paper disks, or other suitable supporting mechanisms, and alternate enzyme systems such as the alkaline phosphatase system may be used. In addition, the results of the test may be quantitated by noting the degree of enzyme-substrate reactions as a measure of the amount of antigen or antibody in the test sample. Availability of the ELISA tests as inexpensive kits has brought the procedure into the doctor's office and routine serological laboratory, and broad applications of the test are expected in the future.

FIGURE 19.8

ELISA

Enzyme linked immunosorbent assay (ELISA) as used in the HIV antibody test to detect HIV antibodies. (a) Plastic beads coated with HIV antigens are combined with a serum sample from the patient. If the serum contains HIV antibodies, the antibodies combine with antigens on the bead surface. (b) The beads are then incubated with antiglobulin antibodies linked to the enzyme horseradish peroxidase. The antibodies combine with antibodies on the bead surface and the enzyme accumulates. (c) Next a substrate for the enzyme is added and again the mixture is incubated. (d) A yellow-orange color develops in the mixture as the enzyme changes the substrate to a colored compound. This constitutes a positive test for HIV antibodies. If no color develops, the test is negative and implies that no antibodies were present to accumulate on the bead surface.

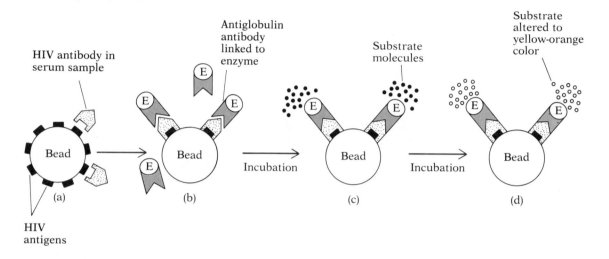

Monoclonal Antibodies

An antibody-secreting cell, like any other cell, can become cancerous. Unchecked, the cell proliferates to become a mass of cells called a **myeloma**. Because a myeloma begins as a single cell, all of its progeny constitute a clone with identical genetic characteristics. The remarkable feature of this clone is that the cells produce only a single type of antibody. Thus the serum of an animal with a myeloma contains substantial amounts of one antibody; and a tissue culture of a myeloma produces only one antibody.

Myeloma:
a mass of cells reproducing at an uncontrolled rate.

In 1975, **Georges J.F. Köhler** of West Germany and **Cesar Milstein** of Argentina developed a method for the laboratory production of antibodies from a single clone of myeloma cells. Antibodies from this clone were named **monoclonal antibodies**. Their method, now in widespread use, begins with the injection of antigens to mice, followed by the extraction of plasma cells from the spleen of the mice (Figure 19.9). The plasma cells are then fused with unstimulated myeloma cells from another mouse to form a clone of hybrid cells. This fusion results in a **hybridoma** (a hybrid myeloma), which is immortal and is programmed to produce a single antibody against the original antigen. The plasma cells supply the program for the antibody, the myeloma cells supply the immortality. In 1984, Köhler and Milstein shared the Nobel Prize in Physiology or Medicine for the development of the monoclonal antibody technique.

ka′ler

hi″brĭ-do′mah
Hybridoma:
a clone of cells produced by the union of antibody-producing cells with myeloma cells.

Monoclonal antibodies and the hybridoma technique have been hailed as one of our era's most important methodological advances in biomedicine. The antibodies differ from ordinary antibodies because they are far more pure and uniform, and exquisitely sensitized to probe for their antigenic targets. Monoclonal antibodies, for example, have been used to pinpoint the antigens on the surfaces of parasites and thus allow researchers to zero in on these antigens for vaccine production. In this regard they are excellent research tools.

FIGURE 19.9
Production of Monoclonal Antibodies

(a) Antigens are injected to mice. The mouse's immune system is stimulated to form antibody-producing plasma cells.
(b) Plasma cells are removed from the mouse, and (c) fused with unstimulated myeloma cells from another mouse.
(d) This fusion results in a clone of hybridoma cells, which are immortal and which produce a single type of antibody, the monoclonal antibody. (e) Monoclonal antibodies are extracted from the cell culture.

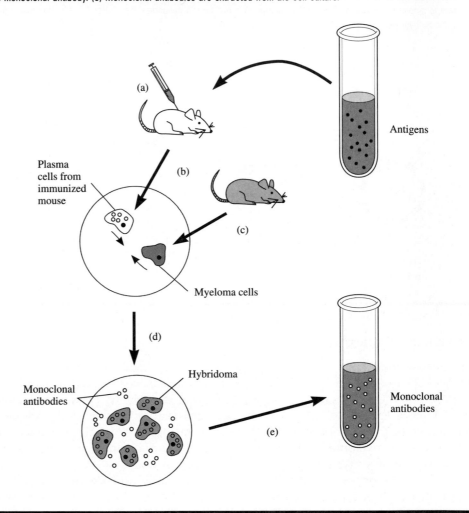

Monoclonal antibodies may also hold the key to the treatment of tumors. Scientists have developed a technique in which tumor cells are removed from a patient and injected into a mouse, whereupon the mouse's spleen begins production of tumor antibodies. Spleen cells are then fused with myeloma cells to produce a hybridoma that produces antibodies for that specific tumor. When the antibodies are injected back into the patient, they react specifically with the tumor cells without destroying other tissue cells. Some researchers foresee the day when monoclonal antibodies can also be used to carry drugs to the tumor and destroy its cells.

Another medical breakthrough was reported in 1983 when investigators reported the development of two **diagnostic tests** using monoclonal antibodies. The tests for gonorrhea and chlamydia are called the Gonozyme and Chlamydiazyme tests, respectively (Chapter 10). They reduce from days to minutes the time required for diagnosis by using monoclonal antibodies to detect minute quantities of disease antigens.

Monoclonal antibodies are reproducible because of how they are manufactured and, therefore, laboratories throughout the world can use identical antibodies. This allows hitherto impossible comparison of tests and research results. Monoclonal antibodies are also used for cleansing bone marrow prior to transplantation, in treating disorders of the immune system, and for an assortment of basic studies and practical approaches to medicine. They represent one of the most elegant expressions of modern biotechnology.

Gene Probes

Antibody tests are a valuable resource to the clinical laboratory, but a new series of diagnostic tests permit the identification of an organism (MicroFocus: 19.4) and its antigens. These tests are based on the use of a DNA fragment called the gene probe and a procedure known as the polymerase chain reaction (PCR). A **gene probe** is a relatively small, single-stranded DNA segment that can hunt for a complementary fragment of DNA within a morass of cellular material, much like a right hand searching for a left hand. When the probe locates its complementary fragment, it emits a signal such as a pulse of radioactivity. If the complementary fragment cannot be found, then no signal is sent. The procedure is remarkable for its accuracy. For example, if we were to line up the forty-six human chromosomes as a two-lane highway, the highway would stretch around the earth 300 times. A gene probe can locate and unite with a few-mile stretch of this highway.

Gene probe: a small DNA fragment having a single strand and uniting with its complementary DNA fragment.

To use a gene probe effectively, it is valuable to increase the amount of DNA to be searched. The **polymerase chain reaction (PCR)** accomplishes this task. The procedure takes a segment of DNA and reproduces it to a billion copies in a few short hours. Target DNA is mixed with polymerase (the enzyme that synthesizes DNA), short strands of primer DNA, and a mixture of nucleotides. The mixture is then alternately heated and cooled during which time the double-stranded DNA unravels, is duplicated, then reforms the double helix. The process is repeated over and over again in a highly automated PCR machine, which is the biochemist's equivalent of an office copier. Each cycle takes about two minutes, and each new DNA segment serves as a model for producing many additional copies, which in turn serve as models for producing more copies. Instead of looking for a needle in a haystack, the gene probe now has a huge number of needles.

Polymerase chain reaction: a biochemical procedure in which a small number of DNA molecules is amplified to a huge number.

One place where gene probes and PCR have been useful is in the detection of **human immunodeficiency virus (HIV)**. T-lymphocytes are obtained from the patient and disrupted to secure the celluar DNA. The DNA is then amplified by PCR and the gene probe is added. The probe is a segment of DNA that complements the DNA in the provirus synthesized from the genome of HIV (Chapter 13). If the person is infected with HIV, the probe will locate the proviral DNA, bind to it, and emit radioactivity. An accumulation of radioactivity thus constitutes a positive test. Because the test identifies viral DNA rather than viral antibodies, the physician can be more confident of the patient's health status. At this writing, the test is not licensed for general use, but it is available to physicians requesting it.

A gene probe test is also available for detecting **human papilloma virus**. This virus is known to cause genital warts. The test utilizes a gene probe to detect viral DNA in a sample of tissue obtained from a woman's cervix. Because certain forms of human papilloma virus have been linked to tumors of the cervix, the test has won acceptance as an important preventative technique, and it has been licensed by the FDA. It is commercially available as the ViraPap test.

19.4
MICROFOCUS

CAUGHT IN THE SPOTLIGHT

Current diagnostic tests for tuberculosis can take several weeks to complete because the tubercle bacillus *Mycobacterium tuberculosis* multiplies very slowly, a binary fission taking place every 24 hours or so. While waiting for a definitive diagnosis, physicians must make treatment decisions based on very limited information. It is therefore possible that ineffective drugs may be prescribed in this interval and that the patient's illness will worsen; also the patient may transmit the disease to others as the wait goes on.

With help from the firefly, researchers have developed an innovative and imaginative diagnostic test for tuberculosis that could shorten the time interval for detection considerably. Only a few days may be required and the test could help determine whether that particular strain of *M. tuberculosis* is drug-resistant. The new approach relies on the firefly enzyme luciferase to produce a flash of light in living *M. tuberculosis*. The process works this way: a bacteriophage (a bacterial virus) specific for *M. tuberculosis* is genetically engineered to carry the gene for luciferase. A sample of phage is then mixed with a culture of bacteria. If the culture contains *M. tuberculosis*, the phage penetrates the bacterium and inserts itself into the bacterial chromosome, carrying the luciferase gene along. The bacterium promptly begins producing luciferase. Now luciferin, a compound attacked by luciferase, is added to the culture together with the high-energy molecule ATP (Chapter 5). If luciferase is present, the enzyme breaks down luciferin and the reaction results in a flash of light. A sensitive instrument detects the light flash and the culture is confirmed to contain *M. tuberculosis*. The report is made to the physician and the diagnosis is complete.

To determine drug susceptibility or resistance, the same procedure is used except a drug is added to the culture. If the bacteria are sensitive to the drug the bacteria die and, quite literally, their "lights go out." If they are resistant, they continue to live, and they produce luciferase; and they give off light.

In the spring of 1993, scientists from New York's Albert Einstein College of Medicine and the University of Pittsburgh reported the test's development in *Science* magazine. As expected, headline writers from numerous publications had a field day as word of the successful test filtered through various journals and newspapers. The well-worn cliche is particularly appropriate in this instance—the future of tuberculosis diagnosis "appears bright."

A similar technique can be used to conduct **water quality tests** based on the detection of coliform bacteria such as *Escherichia coli* (Chapter 25). Traditionally, the latter had to be cultivated in the laboratory and identified biochemically. With gene probe technology a sample of water can be filtered, and the bacteria trapped on the filter can be broken open to release their DNA for PCR and probe analysis. Not only is the process time-saving (many days by the older method, a few short hours by the newer method) but it is extremely sensitive: a single *E. coli* cell can be detected in a 100-ml sample of water. Moreover, the pathogens transmitted by water, rather than the "indicator" *E. coli*, can be detected by DNA analysis. Thus, the identification of *Salmonella*, *Shigella*, and *Vibrio* species will become more feasible in the future as probe analyses become more widely accepted.

Gene probe assays are widely available in kit forms for a variety of bacteria (e.g., streptococci, *Haemophilus*, *Listeria*, *Mycobacterium*, and *Neisseria*) as well as for fungi (e.g., *Blastomyces*, *Coccidioides*, and *Histoplasma*). In many cases, the tests are described as "exquisitely accurate" with a high degree of discrimination and a reliability as high as older identification methods. Since first introduced in the 1970s, the gene probe tests have encountered periods of unbridled enthusiasm counterbalanced by periods of disappointment. The future will depend in part on the development of ways to minimize false-positive reactions due to contamination, on methods of increasing the sensitivity of tests, and on mechanisms for enhancing the signal from bound probes.

NOTE TO THE STUDENT

Ever consider that we are born too soon? Is nine months in the womb enough? Or would eighteen months be preferable?

Absurd you say? Why, then, is a baby born immunologically "unfinished," that is, why is its immune system not fully functional until it is roughly six months old? And why is it so dependent on its parents that it probably could not lead an independent existence until it is perhaps nine months old? Still not convinced? Then consider a newborn colt or a newly hatched chick. Each is able to walk about and gather food within hours of its birth. Certainly the colt and chick will survive better than a newborn human.

If we are willing to buy into the concept that we are born too soon, then the next question is "why does this happen?" The answer, according to Stephen Jay Gould and other evolutionary biologists, is the size of our brain. In proportion to the remainder of our body, our brain is larger than any other animal's brain. To have this large brain, we must have a large head. After nine months, our heads are able to fit through the birth canal, but they could not fit if we stayed inside much longer. So it becomes a matter of give-and-take. Evolution has given us a large brain (and head), but it has also decreed that we must complete our development outside the comfortable confines of our mothers. That development includes immunological development as well as physical development. It also increases our dependence on our parents, and perhaps that's not all so bad—it certainly helps us to appreciate the ones who care for us. Thanks guys.

Summary

Antibodies are the key element in the long-term resistance that the body exhibits to infectious disease. They are also important elements of diagnostic tests used to detect diseases. These two concepts are the major topics of this chapter.

Antibodies render immunity to the body. If the body's immune system produces the antibodies, the immunity is said to be active. By contrast, if antibodies come from some source outside the body, then the immunity is passive. Both active and passive immunities may be natural or artificial. Natural immunity happens in the "natural" course of events such as when a person becomes ill with disease (natural active immunity) or a fetus acquires antibodies from its mother (natural passive immunity). Artificial immunity occurs in an unnatural way such as when a person receives an injection of vaccine (artificial active immunity) or an injection of antibodies (artificial passive immunity).

Laboratory reactions in which antibodies are the focus of attention are called serological reactions because serum is generally involved. A serological reaction may involve direct observation of an antigen–antibody reaction, such as in precipitation or agglutination reactions; or the reaction may involve indirect observation of an antigen–antibody reaction, such as in complement fixation or fluorescent antibody techniques. Several of the indirect reactions have become quite sophisticated and often employ radioactive markers or complex enzyme reactions to denote whether a reaction has occurred. However, all share the property of detecting antibodies in the patient sample as a way of knowing whether a particular disease is present.

Questions for Thought and Discussion

1. It is estimated that when at least half the individuals in a given population have been immunized to a disease, the chances of an epidemic occurring are very slight. The population is said to exhibit herd immunity because members of the population (or herd) unknowingly transfer the immunizing agent to other members and eventually immunize the entire population. What are some ways by which the immunizing agent can be transferred?

2. The tendency of women in the present generation is to marry at an older age than in past generations. How might this present an immunological problem for the newborn?

3. A man is found murdered on the front seat of his automobile, and the police observe blood stains on the floor. It is important to know whether this was the victim's blood, the murderer's blood, or the blood of the victim's dog, which was always with him. However, it could also be fish blood, since the man was an avid fisherman, or blood from the poorly wrapped chicken the man was bringing home from the supermarket. How might the medical examiner proceed?

4. In 1991, scientists first reported success in vaccinating women late in pregnancy to protect their newborns from *Haemophilus* meningitis. The researchers found that levels of meningitis antibodies in the newborns were far above those normally present. Would you favor this approach to protecting newborns? Why or why not?

5. When a child is born in Great Britain, he or she is assigned a doctor. Two weeks later, a social services worker visits the home, enrolls the child on a national computer registry for immunization, and explains immunization to the parents. When a child is due for an immunization, a notice is automatically sent to the home, and if the child is not brought to the doctor, the nurse goes to the home to learn why. Do you believe a method similar to this can work in the United States to achieve uniform national immunization?

6. For passive immunity, serum containing type G immunoglobulins is routinely used. Why do you suppose type M immunoglobulins are not used, especially since they are the important components of the primary antibody response? Do you believe that research in this direction would be fruitful?

7. Given a choice, which of the four general types of immunity would it be safest to obtain? Why? Ultimately, which would be the most helpful?

8. One of the hot research items of 1993 was that naked DNA molecules could conceivably be used to immunize an individual. The theory was that DNA from a virus could be made to penetrate to cells such as muscle cells, which would then display that virus' proteins on their surface. What do you suppose would happen next?

9. A complement fixation test is performed with serum from a patient with an active case of syphilis. In the process, however, the technician neglects to add the syphilis antigen to the tube. Would lysis of the sheep red blood cells occur at the test's conclusion? Why?

10. From 1980 to 1989, the incidence of pertussis increased in the United States, and the greatest incidence was found to be in adolescents and adults. Can you think of any reason why adolescents and adults should have been the targets of the bacillus, especially since these individuals were usually considered immune to the disease? Would you be in favor of using the new acellular pertussis vaccine to reimmunize these populations?

11. Suppose the titer of mumps antibodies from your blood was higher than that for your fellow student. What are some of the possible reasons that could have contributed to this? Try to be imaginative on this one.

12. Children between the ages of 5 and 15 are said to pass through the "golden age of resistance" because their resistance to disease is much higher than that of infants and adults. What factors may contribute to this resistance?

13. In 1985, a vaccine for meningococcal meningitis was licensed for use. The vaccine consists of capsular polysaccharides from four different strains of *Neisseria meningitidis*. What form of vaccine does this vaccine represent, and why is it safer to use than older vaccines made from whole meningococci?

14. The ability to keep the human body alive artificially, though brain dead, has stimulated the idea of keeping the organs functioning to produce vaccines for disease treatment. What arguments can be presented for and against this proposition?

15. In 1985, New York State dropped its requirement for a VDRL test prior to obtaining a marriage license. Why might you support this action? Can you think of any reasons to oppose it?

Review

On completing the section on immunity to disease, test your comprehension of the section's contents by filling in the following blanks with two terms that answer the description best. The appendix contains the answers.

1. Two general forms of immunity:

   _____ and _____.

2. Two types of natural immunity:

   _____ and _____.

3. Two diseases that MMR is used against:

   _____ and _____.

4. Two diseases that DPT is used against:

   _____ and _____.

5. Two types of passive immunity:

   _____ and _____.

6. Two adjuvants used in vaccines:

   _____ and _____.

7. Two names for antibody-containing serum:

   _____ and _____.

8. Two antibody types formed on antigen stimulation:

   _____ and _____.

9. Two ways newborns have acquired maternal antibodies:

   _____ and _____.

10. Two characteristics of serum sickness:

    _____ and _____.

11. Two materials used in second-generation vaccines:

    _____ and _____.

12. Two diseases for which synthetic vaccines are available:

    _____ and _____.

13. Two factors that can determine innate immunity:

   _____ and _____ .

14. Two types of viruses in viral vaccines:

   _____ and _____ .

15. Two bacterial diseases for which toxoids are used:

   _____ and _____ .

16. Two methods for administering vaccines:

   _____ and _____ .

17. Two tracts in which IgA accumulates:

   _____ and _____ .

18. Two viral diseases where passive immunity is used:

   _____ and _____ .

19. Two functions of antibodies in antiserum:

   _____ and _____ .

20. Two bacterial diseases where passive immunity is used:

   _____ and _____ .

CHAPTER 20

IMMUNE DISORDERS

During the Golden Age of Microbiology, investigations were carried out on certain diseases that were not as dangerous as microbial diseases but were a source of great discomfort and inconvenience nevertheless. One such disease was hay fever. In the 1870s, British scientist Charles Harrison Blackley noted that crude pollen placed into the eyes of hay fever sufferers caused swelling of the membranes. Blackley also observed that pollen grains rubbed into a skin scratch produced a local reaction. Some of his critics suggested that the reaction was due to the mechanical injury inflicted by pollen grains, but in 1903, another British investigator, William Philipps Dunbar, supported Blackley's work by showing that saline extracts of pollen grains would cause the same reaction.

Research on hay fever has come a long way since the experiments of Blackley and Dunbar and, as we shall see in this chapter, the disease is now regarded as a disorder of the immune system. About 35 million Americans currently suffer from hay fever, and thousands more are allergic to foods, cosmetics, leather, or metals. Many people cannot keep pets because of severe sensitivities, and between 2000 and 4000 Americans die of asthma annually. All told, an estimated 40 to 50 million people in the United States have some type of allergy.

The common denominator among these problems is a state of increased reactivity known as **hypersensitivity**. First reported in the early 1900s, hypersensitivity stems from activity of the immune system and involves both antibody-mediated and cell-mediated aspects of immunity. It represents a major topic of this chapter and is currently a subject of intense research in immunology.

Also included in the broad category of immune disorders are the autoimmune diseases and various immune deficiency diseases, as well as the principles of transplantation research and tumor immunology. We shall survey each in the pages to follow.

Hypersensitivity

Hypersensitivity is a multistep phenomenon requiring an exposure to an antigen; a dormant (latent) stage during which an individual becomes sensitized; and a reaction following a subsequent exposure to the antigen. The process may involve elements of antibody-mediated immunity or cell-mediated immunity, or sometimes both. The antibody response to the second dose of antigens often occurs within minutes, whereas the cell-mediated response develops over two to three days. For this reason, the terms **immediate hypersensitivity** and **delayed hypersensitivity** have traditionally been used to differentiate two types.

In the early 1970s, **P. G. H. Gell** and **R. A. Coombs** proposed another method for classifying hypersensitivities into four types. They classified immediate hypersensitivity into three types: **type I anaphylactic hypersensitivity**, a process involving IgE, mast cells, basophils, and mediators that induce smooth muscle contraction; **type II cytotoxic hypersensitivity**, which involves IgG, IgM, complement, and the destruction of host cells; and **type III immune complex hypersensitivity**, which involves IgG, IgM, complement, and the formation of antigen–antibody aggregates in the tissues. Gell and Coombs defined delayed hypersensitivity as **type IV cellular hypersensitivity**, in which lymphokines and T-lymphocytes function. Table 20.1 compares these four types.

Type I Anaphylactic Hypersensitivity

an″ah-fĭ-lak′sis

Anaphylactic hypersensitivity can be life-threatening. It is accompanied by **anaphylaxis**, a series of events in which mediators induce vigorous contractions of the body's smooth muscles. The term "anaphylaxis" is derived from Latin stems that mean "against-protection," a reference to the dangerous nature of the condition.

TABLE 20.1

The Gell and Coombs Classification of Hypersensitivity Reactions

HYPERSENSITIVITY TYPE	ORIGIN OF HYPERSENSITIVITY	ANTIBODY INVOLVED	CELLS INVOLVED	MEDIATORS INVOLVED
Type I Anaphylactic	B-lymphocytes	IgE	Mast cells Basophils	Histamine Serotonin Bradykinin SRS-A, etc.
Type II Cytotoxic	B-lymphocytes	IgG IgM	RBC WBC Platelets	Complement
Type III Immune complex	B-lymphocytes	IgG IgM	Host tissue cells	Complement
Type IV Cellular	T-lymphocytes	None	Host tissue cells	Lymphokines

Type I anaphylactic hypersensitivity begins with the entry of an antigenic substance into the body. This antigen, referred to as an **allergen**, may be any of a wide variety of materials such as bee venom, serum proteins, or drugs such as penicillin. In the case of penicillin, the drug molecule itself is the allergen, but the molecule does not stimulate the immune system until after it has combined with tissue proteins to form an allergenic complex. Doses of antigen as low as 0.001 mg have been known to sensitize a person. Allergists refer to such a dose of antigen as the **sensitizing dose**.

The immune system responds to the allergen, and B-lymphocytes produce IgE (MicroFocus: 20.1). This antibody, formerly known as reagin, enters the circulation and fixes itself to the surface of mast cells and basophils. **Mast cells** are connective tissue cells numerous in the respiratory and gastrointestinal tracts and near the blood vessels. They measure about 10 μm to 15 μm in diameter and are filled with granules containing histamine and other physiologically active substances. **Basophils** are circulating leukocytes, also rich in granules. They represent about 1 percent of the total leukocyte count in the circulation and measure about 15 μm in diameter. As IgE accumulates on mast cells and basophils, the individual becomes sensitized.

Sensitization usually requires a minimum of one week, during which time millions of molecules of IgE attach to thousands of mast cells and basophils. The attachment occurs at the Fc end of the antibody leaving the Fab ends pointing outward from the cell. Multiple stimuli by allergen molecules may be required to sensitize a person fully. This is why penicillin is often taken several times before a penicillin allergy manifests itself (Figure 20.1).

The symptoms of anaphylaxis occur rapidly on subsequent exposure to the allergen. Allergen molecules unite with IgE on the surfaces of sensitized cells and appear to form a bridge between adjoining combining sites, as illustrated

Allergen:
an antigenic substance that induces an allergic reaction.

ba'so-fil

Basophils:
granulated, circulating white blood cells that participate in anaphylactic reactions.

IgE:
a monomeric antibody produced by stimulated plasma cells.

TRANSFER OF SENSITIVITY	EVIDENCE OF HYPERSENSITIVITY	SKIN REACTION	EXAMPLES
By serum	30 minutes or less	Urticaria	Anaphylaxis Atopic disease
By serum	Hours to days	Usually none	Transfusion reactions Hemolytic disease of newborns Thrombocytopenia Agranulocytosis Goodpasture's syndrome
By serum	Hours to days	Usually none	Serum sickness Arthus phenomenon SLE Rheumatic fever LCM Organ rejection
By lymphoid cells	Days	Induration Tissue death	Contact dermatitis Infection allergy Skin graft rejection

FIGURE 20.1
Type I Anaphylactic Hypersensitivity

(a) An allergen enters a person's circulation and (b) stimulates the immune system to produce IgE. (c) The IgE fixes itself to mast cells and basophils, thereby sensitizing the person. (d) On subsequent exposure to allergens, the allergens unite with IgE on the surface of the sensitized cells, causing (e) the release of granules and the liberation of mediators that induce the contraction of smooth muscles in the body.

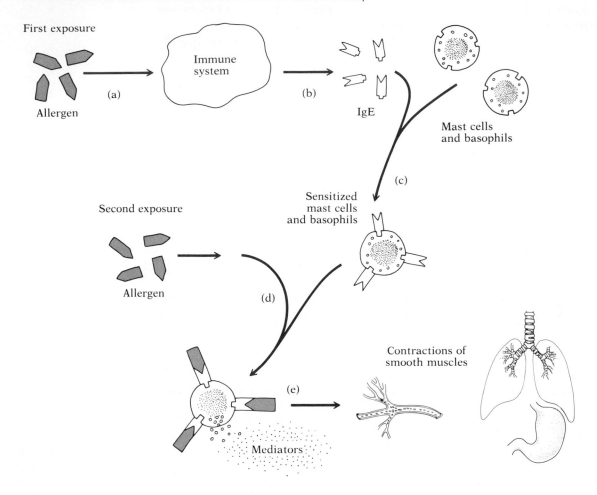

Adenyl cyclase:
an enzyme that digests adenosine triphosphate to adenosine monophosphate and phosphate ions.

his'tah-mēn
ser"o-to'nin
brad"e-ki'nin

in Figure 20.2. This union causes inhibition of the enzyme **adenyl cyclase**, which leads to a reduction of cyclic adenosine monophosphate (cAMP). As this takes place, the cells swell and cellular granules move to the cell surface, where they are released. An understanding of this biochemistry is important because anaphylaxis inhibitors such as epinephrine increase the activity of adenyl cyclase. This leads to an increase in cAMP, and the cAMP inhibits the further release of granules.

As granules flow into the extracellular fluid, they emit a number of mediators having substantial pharmacologic activity. These mediators include **histamine**, a simple compound derived from the amino acid histidine, and **serotonin**, a derivative of tryptophan. Another mediator is **bradykinin**, a peptide consisting of nine amino acids. Still another is an acidic lipopeptide called the **slow reacting substance of anaphylaxis (SRS-A)**. This substance appears shortly after the antigen–antibody reaction and functions for prolonged periods, as the

FIGURE 20.2

The Involvement of Mast Cells and Basophils in Type I Anaphylactic Hypersensitivity

(a) In the sensitized condition, the granules are dispersed throughout the cytoplasm, the level of cyclic AMP is normal, and IgE is bound to the cell's outer surface at the IgE receptors. (b) The cell is activated when allergen molecules with multiple determinant sites react with IgE and form a bridge between adjacent molecules. The level of AMP is reduced, and granules move toward the cell surface. (c) Degranulation occurs as the granules flow into the extracellular fluid and release their mediators. (d) A scanning electron micrograph of a normal mast cell ($\times$ 6500). (e) A degranulated mast cell at the same magnification. In this cell, granules have been released to the extracellular fluid and the irregular surface of the cell reflects the granule loss ($\times$ 6500).

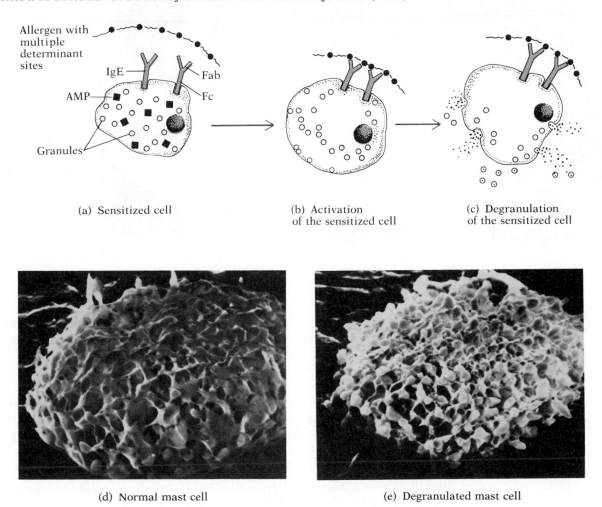

(a) Sensitized cell

(b) Activation of the sensitized cell

(c) Degranulation of the sensitized cell

(d) Normal mast cell

(e) Degranulated mast cell

name implies. It belongs to a class of substances called leukotrienes. Eosinophil- and platelet-activating factors also appear; the former are thought to attract eosinophils that neutralize the effects of the mediators. Prostaglandins are also known to be released.

The principal activity of the mediators is to contract **smooth muscles** in the body. One effect is constriction of the small veins and the expansion of capillary pores forcing fluid out and into the tissues. The skin becomes swollen about the eyes, wrists, and ankles, a condition called **edema**. The edema is accompanied by a hivelike rash and burning and itching in the skin as the sensory nerves are excited. Contractions also occur in the gastrointestinal tract and bronchial muscles, leading to sharp cramps and shortness of breath, respectively. The individual inhales rapidly without exhaling and traps carbon dioxide

loo'ko-trēnz
e"o-sin'o-fil

Edema:
swelling of the body tissues.

20.1

MICROFOCUS

ITCHES AND ANTIBODIES

In the early 1960s, scientists knew that pollen and other allergens caused mast cells and basophils to release histamine. What they did not know was how allergens induced the cells to spill their contents. The immune system appeared to be involved, but researchers were ignorant of the nature or function of the immune mechanism.

At a Denver, Colorado, hospital, two Japanese doctors, Teruko Ishizaka and her husband Kimishige, set out to find an antibody that would stimulate the allergic reaction. Neither scientist had any observable allergies. Therefore they decided to use themselves as guinea pigs. When an extract of ragweed pollen was injected under their skin, no reaction would take place. But if they first injected serum from an allergy

patient then followed it with an injection of pollen extract, a raised itchy welt appeared. It was apparent that something in the patient's serum was responsible for the allergy.

The Ishizakas went to the next step. They separated the serum into as many different components as possible and repeated the skin test with each component. When a particular component caused welts, they purified it further and reinjected themselves. After four years of experiments (and lots of itching), they finally isolated their elusive substance. The substance was an antibody. The Ishizakas named it immunoglobulin E (or IgE) because the antibody was directed against antigen E of ragweed pollen. Then they breathed a sigh of relief that the investigation was over.

in the lungs, an ironic situation where the lungs are fully inflated but lack oxygen. Death may occur in 10 to 15 minutes as a result of asphyxiation if prompt action is not forthcoming (hence, "immediate hypersensitivity"). MicroFocus: 20.2 describes the emergency treatment that may be given in such a case.

A person sensitized to an allergen may undergo **desensitization** therapy to reduce the possibility of anaphylaxis. This procedure involves injections of tiny but increasing amounts of allergen over a period of hours, to effect a gradual reduction of granules in sensitized mast cells and basophils. Such treatment prevents a massive degranulation later. One who is sensitive to immune serum used in disease therapy may need to undergo desensitization before the serum is used in large therapeutic doses.

Another approach to desensitization is to give a series of injections of allergens over a period of weeks. Allergists believe that these exposures cause the immune system to produce IgG, which circulates and later neutralizes the allergens before they contact sensitized cells. The **blocking antibodies**, as they are called, appear to be an effective device for individuals sensitized to bee stings. A promising alternative is to inject Fc fragments of IgE to fill the receptor sites on mast cells and basophils, thereby making the sites unavailable to the person's IgE.

Desensitization:
a process in which antigens are injected to relieve sensitivity to that antigen.

Blocking antibodies:
IgG antibodies that neutralize allergens before they can induce hypersensitivity.

Atopic Diseases

Type I hypersensitivity reactions need not result in the whole body involvement that accompanies anaphylaxis. Indeed, the vast majority of hypersensitivity reactions are accompanied by limited production of IgE and the sensitization of mast cells in localized areas of the body. The result is an **atopic disease**, or **common allergy**.

a-top'ik

WHEN ANAPHYLAXIS STRIKES

Anaphylaxis is a terrifying experience. The skin itches intensely and breaks into hives, the eyes and joints become red and puffy, and the individual doubles over with abdominal pains. Breathing becomes difficult, then belabored, and finally, is reduced to life-sucking gasps. The symptoms develop within minutes, and the individual usually faints and quickly lapses into a coma.

The key to survival is swift action. Epinephrine (adrenalin) is the highest priority drug. Within minutes of injection, it stabilizes basophils and mast cells to prevent further mediator release. It also dilates the bronchioles to reopen the air passageways, and constricts the capillaries to keep fluid in the circulation.

A smooth-muscle relaxant such as aminophylline may also be used. This will help dilate the bronchial airways and pulmonary blood vessels. An antihistamine such as diphenhydramine (Benadryl) may be valuable. This drug competes with histamine for the active sites on smooth-muscle receptors and thus inhibits the action of histamine. Hydrocortisone may be used to reduce swelling in the tissue, and an expectorant may help clear laryngeal edema.

If the patient does not respond rapidly to the drugs, it may be necessary to insert a tube into the respiratory passageway or perform a tracheostomy. Either must be done quickly because life is now reduced to a scant few minutes.

An example of an atopic disease is **hay fever**, technically referred to as allergic rhinitis. This condition develops from springtime inhalations of tree and grass pollens and summer and fall exposures to grass and weed pollens (Figure 20.3). Fall exposures coincide with the haying season, from which the disease first acquired its name (although there is no fever associated with the condition). Immune stimulation by pollen antigens leads to IgE production, and a sensitization of mast cells follows in the eyes, nose, and upper respiratory tract. Subsequent exposures bring on sneezing, tearing, swollen mucous membranes, and other well-known symptoms. In addition, hay fever symptoms may be caused by house dust, mold spores, detergent enzymes, and the particles of animal skin and hair called dander. (Dander itself is not the allergen. Rather, the actual allergens are proteins deposited in the dander from the animal's saliva when it preens itself.)

Allergic reactions also are responsible for triggering the great majority of asthmatic attacks. **Asthma** is characterized by wheezing and stressed breathing, and appears due to the same allergens associated with hay fever. However, little information currently exists on asthma's exact causes or mechanisms, although leukotrienes seem to play a significant role. About 15 million Americans suffer from asthma, including about 3.7 million children and adolescents. Bronchodilators have been traditionally used to widen the bronchioles by relaxing the surrounding muscles, but newer evidence indicates that asthma is due to inflammation of the airways, so many physicians now prescribe anti-inflammatory agents such as inhaled steroids or cromolyn sodium.

az'mah

Food allergies are accompanied by symptoms in the gastrointestinal tract, including swollen lips, abdominal cramps, nausea, and diarrhea. The skin may break out in a rash containing **hives**, each hive consisting of a central puffiness, called a wheal, surrounded by a zone of redness known as a flare. Such a rash is called **urticaria**, from the Latin *urtica* for "stinging needle." Allergenic foods include chocolate, codfish, strawberries, oranges, cow's milk, and fish (MicroFocus: 20.3). A dry food such as flour may also cause respiratory allergy.

ur"tĭ-ka're-ah

According to public health estimates, almost 20 percent of Americans have some type of atopic disease. An interesting avenue of research was opened when it was discovered that the lymphocytes responsible for IgE and IgA production

FIGURE 20.3

Scanning Electron Micrographs of Two Types of Pollen Grains

(a) Pollen grains of *Sphaeralcea munroana*, a desert plant (× 1000). (b) Pollen grains from *Penstemon pruinosus*, a flowering plant that grows wild in the mountains of Southern California (× 2000). Antigens in pollen grains such as these stimulate allergic reactions.

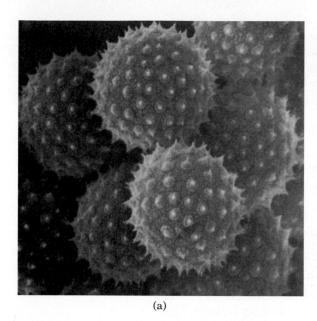

(a)

(b)

20.3

MICROFOCUS

NOT THE REAL THING

"**B**ut I'm not allergic to crabs!" exclaimed the woman. "I've eaten shellfish all my life —clams, oysters, crabs—and I've never had a reaction. Why this time?"

"Let's go over it again," said the allergist. "You had the crabmeat egg rolls and within minutes you were wheezing."

"Yes."

"You're sure it was real crabmeat?"

"Well, the menu said *egg rolls made with crabmeat.*"

"Could it have been surimi, you know, that processed fish used to make imitation crabmeat? They sell it in the supermarkets. It's artificially colored to look like crabmeat, but it's really processed pollack or some other inexpensive fish."

"I guess it could have been...," her voice trailed off.

"Are you allergic to fish?"

"Yes, as a matter of fact I am."

"I think we have the answer."

lie close to one another in the lymphoid tissue, and that the IgA level and its corresponding lymphocytes are greatly reduced in atopic individuals. Immunologists have suggested that in the nonallergic individual, IgA lymphocytes shield IgE lymphocytes from antigenic stimulation but that atopic persons may lack sufficient IgA lymphocytes to block the antigens.

Another theory of atopic disease maintains that allergy results from a breakdown of feedback mechanisms in the immune system. Research findings indicate that B-lymphocytes, which synthesize IgE, are controlled by suppressor T-lymphocytes, regulated in turn by IgE. Under normal conditions, IgE may limit its own production by stimulating suppressor T-lymphocyte activity. However, in atopic individuals the mechanism malfunctions, possibly because the T-lymphocytes are defective, and IgE is produced in massive quantities. Allergic individuals are known to possess almost 100 times the IgE level of normal individuals. Radioimmunoassay techniques and radioallergosorbent tests (Chapter 19) are used to detect the nature and quantity of IgE in an individual.

Some researchers postulate a **positive role** for the allergic response. They suggest, for example, that sneezing expels respiratory pathogens, and that contractions of the gastrointestinal tract force parasites out of the body. It has been further theorized that allergy was once a survival mechanism and that atopic individuals are the modern generation of people who developed this ability to resist pathogens and passed the trait along.

Suppressor T-lymphocytes: **T-lymphocytes that suppress the immune reaction to antigens.**

TO THIS POINT ■	*We have spent considerable time discussing the process of type I anaphylactic hypersensitivity because it has substantial importance to the millions of people who have allergies. We noted how IgE plays a central role in the process, how mast cells and basophils are involved, and how smooth muscle contractions account for the symptoms in anaphylaxis and common allergy. Two forms of desensitization were outlined, and the discussion described two theories on why allergies develop. One theory has to do with the blockage of allergens by IgA lymphocytes; the second involves a defect in the suppressor T-lymphocytes. We also explored some positive roles for the allergic reaction.*
	We shall continue the discussion of hypersensitivity by studying the salient features of the remaining three types of hypersensitivity. Type II involves a destruction of cells, type III leads to the formation of aggregates called immune complexes, and type IV depends upon the exaggeration of T-lymphocyte function. As the reading progresses, you may note how phenomena of the immune system are helping to enlighten scientists on several disease conditions that were poorly understood in past decades.

Other Types of Hypersensitivity

Type II Cytotoxic Hypersensitivity

A **cytotoxic hypersensitivity** is a cell-damaging reaction that occurs when IgG reacts with antigens on the surfaces of cells. Complement is often activated and IgM may be involved, but IgE does not participate, nor is there any degranulation of mast cells. The cells affected in cytotoxic hypersensitivity are known as **target cells.**

Target cells: **cells against which cytotoxic hypersensitivity is directed.**

Agglutination:
a clumping reaction.

A well-known example of cytotoxic hypersensitivity is the **transfusion reaction** arising from the mixing of incompatible blood types. Four major human blood types are known: A, B, AB, and O. Each type is distinguished by unique antigens on the surface of erythrocytes and certain antibodies in the plasma that are directed against antigens not present in the individual's cells (Table 20.2). Before a transfusion is attempted, the laboratory technician must determine the **blood types** of participants so that incompatible types are not mixed. For example, if a person with type A blood donates to a recipient whose type is O, the A antigens on the donor's erythrocytes will react with *a* **antibodies** in the recipient's plasma, and the cytotoxic effect will be expressed in the form of agglutination of donor erythrocytes and activation of complement in the recipient's circulatory system. If the conditions are reversed, the donor's *a* antibodies will react with the recipient's A antigens, although to a lesser degree, because dilution in the recipient's plasma takes place. Most blood banks crossmatch the donor's erythrocytes with the recipient's serum, as well as the reverse, to ensure compatibility (MicroFocus: 20.4).

Rh factor:
a group of antigenic substances on the red blood cells of Rh-positive individuals.

Another expression of cytotoxic hypersensitivity is **hemolytic disease of the newborn**, or **Rh disease**. This problem arises from the fact that erythrocytes of approximately 85 percent of Caucasian Americans contain a surface antigen, first described in rhesus monkeys and therefore known as the **Rh antigen**. Such individuals are said to be Rh-positive. The 15 percent who lack the antigen are considered Rh-negative. For African Americans, the figures are 90 percent and 10 percent, respectively. Evidence indicates that the antigen is really a group of antigens that vary among Rh-positive individuals, but we shall consider the group as a single factor for the purposes of discussion.

The ability to produce the Rh antigen is a genetically inherited trait. When an **Rh-negative woman** marries an **Rh-positive man**, there is a 3 to 1 chance (or 75 percent probability) that the trait will be passed to the child, resulting in an Rh-positive child. Normally, a woman's circulatory system is exposed to her child's blood during the birth process, and if the child is Rh-positive, the Rh antigens enter her blood and stimulate her immune system to produce rh antibodies (Figure 20.4). If a succeeding pregnancy results in another Rh-positive child, these antibodies will cross the placenta (along with other antibodies) and enter the fetal circulation. There they will react with Rh antigens on the fetal erythrocytes and cause complement-mediated lysis of the cells. The fetal circulatory system rapidly releases immature erythroblasts to replace the lysed blood cells, but these cells are also destroyed. From this observation, the disease acquired its older name, **erythroblastosis fetalis**. The result may be stillbirth or, in a less extreme form, a baby with jaundice.

ĕ-rĭth″ro-blas-to′sis

Modern treatment for hemolytic disease of the newborn consists of an injection of rh antibodies (**RhoGAM**). The injection is given within 72 hours of delivery of an Rh-positive child (no injection is necessary if the child is Rh-negative). Antibodies in the preparation interact with Rh antigens and remove them from the circulation, thereby preventing them from stimulating the woman's immune system. The success of this procedure has virtually eliminated concern for disease among expectant parents. It should be noted, however, that an Rh-negative woman may produce rh antibodies as a result of miscarriage or abortion of an Rh-positive fetus, or after a transfusion with Rh-positive blood.

throm″bo-si″to-pe′ne-ah

Thrombocytopenia:
a cytotoxic hypersensitivity in which antidrug antibodies attack thrombocytes.

Other examples of cytotoxic hypersensitivity are less familiar. One example, called **thrombocytopenia**, results from antibodies produced against certain drugs such as aspirin, antibiotics, or antihistamines. The antibodies combine with antigens and drug molecules adhering to the surface of thrombocytes (blood platelets), and as complement is activated, the thrombocytes undergo lysis. The effect is impaired blood clotting, and hemorrhages may appear on

TABLE 20.2

Some Characteristics of Major Blood Groups

	TYPE A	TYPE B	TYPE AB	TYPE O
Red blood cells	Antigen A	Antigen B	Antigens A and B	Neither A nor B antigens
Serum	b antibody	a antibody	Neither a nor b antibody	a and b antibodies
Approximate Incidence U.S. Caucasian Population	40%	10%	5%	45%

FIGURE 20.4

Hemolytic Disease of the Newborn

(a) Hemolytic disease of the newborn can develop when an Rh (+) man has a child by an Rh (−) woman. (b) When an Rh (−) woman gives birth to an Rh (+) child, Rh antigens from the child's blood enter the woman's blood. (c) The antigens stimulate her immune system to produce Rh antibodies that circulate in her blood, but since the child has already been born, there is no effect on the child. (d) In a succeeding pregnancy, if the child is Rh (+), the rh antibodies will pass across the placenta and enter its blood. The rh antibodies attack the red blood cells by uniting with Rh antigens on their surface, and they damage the cells, leading to severe anemia and hemolytic disease.

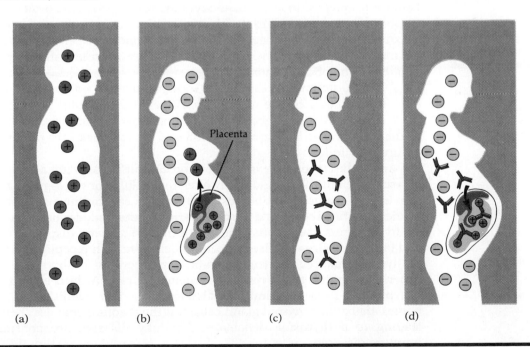

(a) (b) (c) (d)

Since earliest times, people believed that blood contained mysterious powers of rejuvenation. For example, the Romans would rush into the gladiatorial arena to drink the blood of dying gladiators because they thought that blood would restore youth. Transfusions of blood came into popular use after 1667 when Jean Baptiste Denis, physician to Louis XIV of France, temporarily restored a dying boy by transfusing lamb's blood into his veins. Transfusions, however, were a mixed blessing: sometimes they worked, but often they proved fatal. Why this happened perplexed doctors until Karl Landsteiner provided an answer in 1900.

Landsteiner was an 1891 graduate of the medical school at the University of Vienna. After graduation he spent five years in Wurtzburg, Germany, working as a chemist with Emil Fischer, the 1902 Nobel laureate for the synthesis of sugars. Landsteiner was more interested in proteins, and he concluded that protein differences could be fruitfully revealed by studying interactions between serum proteins and the components of living cells. In 1897 he became assistant at the Hygienic Institute in Vienna, and in 1900 he applied his method of protein analysis to blood cells.

Landsteiner observed that when erythrocytes from one person were mixed with the serum from another person, the cells sometimes clumped together. In other cases there was no clumping. It was as if a person's blood was being mixed with its own serum. By meticulous cross-comparisons, Landsteiner concluded that two markers, now called antigens, exist on the surface of the red blood cells. He labeled them with the first two letters of the alphabet, A and B. Landsteiner also surmised that a person's plasma contained antibodies against another person's antigens. These findings formed the basis of the familiar ABO blood groups and explained why some transfusions were successful and others fatal. It now became possible to work out the details for matching up blood groups for safe transfusions.

In World War I, physicians finally recognized the immense value of Landsteiner's work. Over 21 million men were wounded in the war and for many, a blood transfusion was life saving. More than 3 million Americans now receive safe transfusions annually during surgery, childbirth, or in the treatment of disease. In 1930, Landsteiner was the recipient of the Nobel Prize in Physiology or Medicine. Ten years later, while working with New York physician Alexander S. Weiner, he also discovered the Rh factor.

---

a-gran"u-lo-si-to'sis

Autoimmune disease:
one in which the body
produces antibodies against
its own cells.

Glomeruli:
coils of capillaries in the
kidney through which blood
fluid passes into the kidney
capsules.

Thyroxine:
a thyroid hormone that
regulates body metabolism.

the skin and mouth. The symptoms subside as the drug is withdrawn. Another condition, referred to as **agranulocytosis**, results from the destruction of neutrophils by antibodies. This problem, also stimulated by drugs, is manifested as a reduced capacity for phagocytosis. In both conditions, antibodies are directed toward the individual's own cells. The term **autoimmune disease** is therefore applied to the phenomenon.

A third autoimmune disease is **Goodpasture's syndrome**. In this rare disease, antibodies combine with antigens on the glomerular membranes in kidneys. Antibody binding activates the complement system and, as the integrity of the membranes is destroyed, blood and protein leak into the urine. Kidney failure may follow.

Antibodies reacting with antigens on cell surfaces do not always lead to cell destruction, but the reaction may alter the cellular physiology. In **myasthenia gravis**, for example, antibodies react with acetylcholine receptors on membranes covering the muscle fibers. This interaction reduces nerve impulse transfer to the fibers and results in a loss of muscle activity, displayed as weakness and fatigue. In **Graves' disease**, antibodies unite with receptors on the surfaces of thyroid gland cells, causing overabundant secretion of thyroxine. The patient experiences goiter and a rise in the metabolic rate. A third example of altered cell physiology is **Hashimoto's disease**. This is a condition in which antibodies also attack thyroid gland cells, but the reaction changes their chemistry leading to a thyroxine deficiency. All three diseases are considered autoimmune diseases.

Although cytotoxic hypersensitivity is generally cast in a negative role with deleterious effects on the body, the cytotoxic activity may contribute to the body's resistance to disease. For example, the antigen–antibody interaction occurring on the surface of a parasite leads to destruction of the parasite. The interaction also may encourage chemotaxis or histamine release through the activity of C3a and C5a components of the complement system. Increased phagocytosis and membrane damage from the complement attack complex are other by-products of complement activation. These activities probably account for resistance to many disorders.

Chemotaxis:
a chemical attraction.

Type III Immune Complex Hypersensitivity

Immune complex hypersensitivity develops when antibodies combine with antigens and form aggregates that accumulate in blood vessels or on tissue surfaces (Figure 20.5). As complement is activated, the C3a and C5a components increase vascular permeability and exert a chemotactic effect on phagocytic neutrophils, drawing them to the target site. Here the neutrophils release lysosomal enzymes, which cause tissue damage. Local inflammation is common, and fibrin clots may complicate the problem. The antibodies are predominantly IgG, with IgM also found in certain cases.

Neutrophils:
multilobed circulating white blood cells that function as phagocytes.

Serum sickness is a common manifestation of immune complex hypersensitivity. It develops when the immune system produces IgG against residual proteins in serum preparations. The IgG then reacts with the proteins, and immune complexes gather in the kidney over a period of days. The problem is compounded when IgE, also from the immune system, attaches to mast cells and basophils, thereby inducing a type I anaphylactic hypersensitivity. The sum total of these events is kidney damage together with hives and swelling in the face, neck, and joints (Table 20.3).

Serum sickness:
a hypersensitivity reaction that follows injection of serum used for therapy.

Another form of immune complex hypersensitivity is the **Arthus phenomenon**, named for Nicolas Maurice Arthus, the French physiologist who described it in 1903. In this process, excessively large amounts of IgG form complexes with antigens either in the blood vessels or near the site of antigen entry into the body. Antigens in dust from moldy hay and in dried pigeon feces are known to cause this phenomenon. The names **"farmer's lung"** and **"pigeon fancier's disease"** are applied to the conditions, respectively. Thromboses in the blood vessels may lead to oxygen starvation and cell death.

Arthus phenomenon:
a type of hypersensitivity in which immune complexes form in blood vessels near the site of antigen entry.

Systemic lupus erythematosus (SLE) is another example of a type III hypersensitivity (MicroFocus: 20.5). It is an autoimmune disease in which B-lymphocytes produce IgG upon stimulation by nuclear components of disintegrating white blood cells. The terms "autoantigens" and "autoantibodies" apply because the antibodies are formed against the body's own molecules. When the autoantigens and autoantibodies react, immune complexes accumulate in the skin and body organs, and complement is activated. The patient experiences the **butterfly rash**, a facial skin condition across the nose and cheeks (Figure 20.6). Lesions also form in the heart, kidney, and blood vessels. In **rheumatoid arthritis**, another autoimmune disease, immune complexes form in the joints.

loo′pus er″i-them″ah-to′sis

Several microbial diseases are also complicated by immune complex formation. For example, the glomerulonephritis and rheumatic fever that follow **streptococcal diseases** (Chapter 7) appear to be consequences of immune complex formation in the kidney and heart, respectively. In these cases, the deposit of complexes relates to common antigens in streptococci and the tissues. Other immune complex diseases include the hemorrhagic shock that may accompany

glo-mer″u-lo-nĕ-fri′tis

FIGURE 20.5
Type III Immune Complex Hypersensitivity

(a) Antigens stimulate the immune system to produce antibodies. (b) These antibodies react with excessive amounts of antigens or (c) fresh infusions of antigens to form aggregates called immune complexes (d) in the blood vessels or on tissue surfaces. (e) The immune complexes activate complement, and phagocytes are drawn to the area by C3a and C5a. (f) The phagocytes release lysosomal enzymes that cause tissue damage.

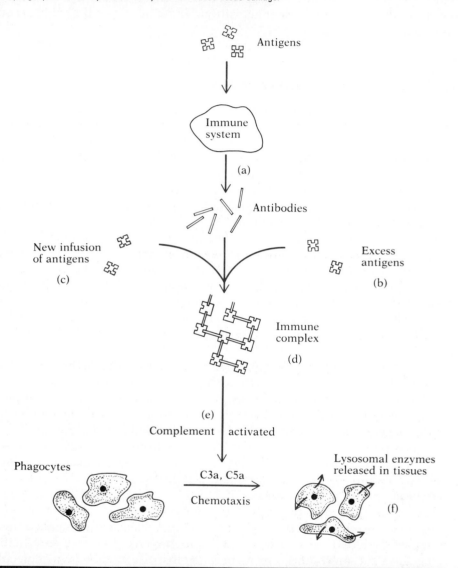

dengue fever, the subacute sclerosing panencephalitis (SSPE) following cases of measles, and the slow-forming kidney deposits associated with lymphocytic choriomeningitis (LCM). Research evidence also suggests that Reye's syndrome and Guillain–Barré syndrome may be related to immune complex formation.

ko"re-o-men"in-ji'tis
ge-yan' bar-ra'

Type IV Cellular Hypersensitivity

Cellular hypersensitivity is an exaggeration of the process of cellular immunity discussed in Chapter 18. The adjective "cellular" was originally applied because

TABLE 20.3

A Summary of Some Immune Disorders

DISORDER	TYPE OF HYPERSENSITIVITY	TARGET TISSUE	ANTIGEN	EFFECT
Thrombocytopenia	Cytotoxic	Thrombocytes (Blood platelets)	Aspirin Antibiotics Antihistamine	Impaired blood clotting Hemorrhages
Agranulocytosis	Cytotoxic	Neutrophils	Drugs	Reduced phagocytosis
Goodpasture's syndrome	Cytotoxic	Kidneys	Own antigens (?)	Kidney failure
Myasthenia gravis	Cytotoxic	Membranes of muscle fibers	Not established	Loss of muscle activity
Graves' disease	Cytotoxic	Thyroid gland	Not established	Abundant thyroxine High metabolic rate
Hashimoto's disease	Cytotoxic	Thyroid gland	Not established	Thyroxine deficiency Low metabolic rate
Serum sickness	Immune complex	Kidney	Serum proteins	Kidney failure
Arthus phenomenon	Immune complex	Blood vessels Site of antigen entry	Environmental antigens	Thromboses in blood vessels
Systemic lupus erythematosus	Immune complex	Skin, heart, kidney, blood vessels	Own nucleo-proteins	Butterfly rash Heart, kidney failure
Rheumatoid arthritis	Immune complex	Joints	Own nucleo-proteins (?)	Swollen joints Arthritis

FIGURE 20.6

Systemic Lupus Erythematosus (SLE)

Two patients showing some of the symptoms of systemic lupus erythematosus (SLE). (a) The butterfly rash on the face. Note how the rash extends out to the cheeks like the wings of a butterfly. (b) A young girl displaying the loss of hair that accompanies some cases of SLE.

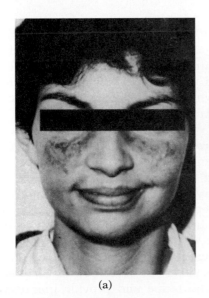

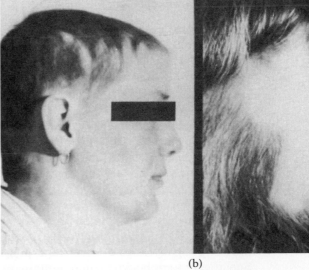

(a)

(b)

A VOCABULARY OF HYPERSENSITIVITY

allergen: an antigen that stimulates a hypersensitivity reaction.

allergy: a localized type I hypersensitivity reaction in which mediator release and smooth muscle contraction lead to discomfort in one body area.

anaphylaxis: a whole body type I hypersensitivity reaction in which mediator release and smooth muscle contraction lead to a life-threatening situation involving multiple body systems.

mast cells: connective tissue cells that release their granules and mediators during type I hypersensitivity reactions.

basophils: circulating white blood cells that release their granules and mediators during type I hypersensitivity reactions.

IgE: the antibody that triggers the type I hypersensitivity reaction by interacting with the allergen, which elicited its production.

histamine: a mediator derived from histidine that stimulates smooth muscle contractions in type I hypersensitivity reactions.

serotonin: a mediator derived from tryptophan that stimulates smooth muscle contractions in type I hypersensitivity reactions.

atopic disease: an alternative term for allergy.

cytotoxic hypersensitivity: a type II hypersensitivity reaction occurring when antibodies attack body cells and incite a process that leads to destruction of the cells.

immune complex hypersensitivity: a type III hypersensitivity reaction occurring when antibodies unite with antigens and form granular masses that cling to body tissues and incite a process that leads to destruction of the tissues.

thrombocytopenia: a cytotoxic hypersensitivity in which antibodies produced against certain drugs unite with antigens on the surface of blood platelets and set off a reaction in which the platelets are destroyed.

agranulocytosis: a cytotoxic hypersensitivity in which antibodies produced against certain drugs unite with antigens on the surface of neutrophils and set off a reaction in which the neutrophils are destroyed.

autoimmune disease: a disease in which a person's immune system produces antibodies that react with and destroy cells or tissues within that same person.

systemic lupus erythematosus: a type III hypersensitivity reaction occurring when antibodies produced against body substances unite with those substances and form immune complexes that fix themselves in body tissues and damage those tissues.

rheumatoid arthritis: a type III hypersensitivity reaction occurring when antibodies produced against body subtances unite with those substances and form immune complexes that fix themselves in the body's joints and damage the joint tissues.

cellular hypersensitivity: a type IV hypersensitivity reaction occurring when an exaggerated T-lymphocyte response to an allergen leads to local tissue destruction.

Lymphokines:
a series of lymphocyte-derived proteins that increases the efficiency of phagocytosis of antigens.

Induration:
thickening and drying of skin tissues in cellular hypersensitivity.

contact between T-lymphocytes and antigens was thought to be necessary for the reaction. However, the later identification of lymphokines as mediators of the process challenged this assumption. The hypersensitivity cannot be transferred to a normal individual by serum lymphokines because their concentration is too low in transfused serum. Transfer is accomplished only with T-lymphocytes.

Type IV hypersensitivity is a **delayed reaction** whose maximal effect is not seen until 24 to 72 hours have elapsed (hence, "delayed hypersensitivity"). It is characterized by a thickening and drying of the skin tissue, a process called **induration**, and a surrounding zone of erythema (redness). Two major forms of this hypersensitivity are recognized: infection allergy and contact dermatitis.

Infection allergy develops when the immune system responds to certain microbial agents. Effector cells known as delayed hypersensitivity T-lymphocytes migrate to the antigen site and release lymphokines. The lymphokines attract phagocytes and encourage phagocytosis, as described in Chapter 18. Sensitized lymphocytes then remain in the tissue and provide immunity to successive episodes of infection. Among the microbial agents that stimulate this type of immunity are the bacterial agents of tuberculosis, leprosy,

FIGURE 20.7

A Positive Tuberculin Test

The raised induration and zone of inflammation indicate that antigens have reacted with T-lymphocytes possibly sensitized by a previous exposure to tubercle bacilli.

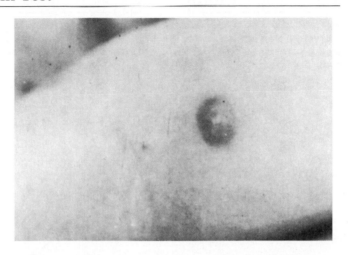

and brucellosis, the fungi involved in blastomycosis, histoplasmosis, and candidiasis, the viruses of smallpox and mumps, and the chlamydiae of lymphogranuloma.

his"to-plaz-mo'sis
kan-dĭ-di'ah-sis

Infection allergy is demonstrated by injecting into the skin an extract of the microbial agent. As the immune response takes place, the area develops induration and erythema, and fibrin is deposited by activation of the clotting system. An important application of infection allergy is the **tuberculin skin test** for tuberculosis (Figure 20.7). A purified protein derivative (PPD) of *Mycobacterium tuberculosis* is applied to the skin by intradermal injection or multiple punctures (tines). Individuals sensitized by a previous exposure to *Mycobacterium* species develop a vesicle, erythema, and induration.

Tuberculosis:
a bacterial disease of the lungs and other tissues accompanied by tubercle formation.

Skin tests based on infection allergy are available for many diseases. Immunologists caution, however, that a positive result does not constitute a final diagnosis. They note that sensitivity may have developed from a subclinical exposure to the organisms, from clinical disease years before, or from a former screening test in which the test antigens elicited a T-lymphocyte response.

Contact dermatitis develops after exposure to a broad variety of antigens such as the allergens in clothing, jewelry, insecticides, coins, cosmetics, and furs. The offending substances may also include formaldehyde, copper, dyes, bacterial enzymes, and protein fibers. In **poison ivy**, the allergen has been identified as **urushiol**, a low molecular weight chemical on the surface of the leaf. In the body, urushiol complexes with tissue proteins to form allergenic compounds. A poison ivy rash consists of itchy pinhead-size blisters usually occurring in a straight row (Figure 20.8).

Contact dermatitis:
a cellular hypersensitivity of the skin that follows repeated exposure to certain antigens.

u-roo'she-ol

The course of contact dermatitis is typical of the type IV reaction. Repeated exposures cause a drying of the skin, with erythema and scaling. Examples are seen on the scalp when allergenic shampoo is used, on the hands when contact is made with detergent enzymes, and on the wrists when an allergy to watchband chemicals exists. Contact dermatitis also occurs on the face where contact is made with cosmetics, the skin where chemicals in permanent press fabrics have accumulated, or the feet when there is sensitivity to dyes in leather shoes. Factory workers exposed to photographic materials, hair dyes, or sewing materials may experience allergies. The list of possibilities is limitless. Applying a

FIGURE 20.8

The Poison Ivy Rash

When a sensitized individual touches a poison ivy plant, a substance called urushiol stimulates T-lymphocytes in the skin and, within 24 hours, a type IV reaction takes place. The reaction is characterized by pinhead-sized blisters that usually occur in a straight row. Poison ivy is easily spread by touching urushiol to other parts of the body via clothing, fingers, or any similar means.

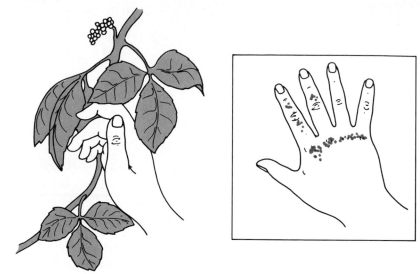

sample of the suspected substance to the skin (a patch test) and leaving it in place for 24 to 48 hours will help pinpoint the source of the allergy. Relief generally consists of avoiding the inciting agents.

TO THIS POINT
■

We have completed a survey of the four types of hypersensitivity, and we have noted the salient differences among them. In type II cytotoxic hypersensitivity we observe an antigen–antibody reaction on the surface of cells that leads to cell destruction. Transfusion reactions and hemolytic disease of the newborn are symbolic of this cytotoxicity. Several autoimmune diseases such as thrombocytopenia, myasthenia gravis, and Graves' disease also result from the cytotoxic reaction. In type III hypersensitivity, immune complex formation takes place as antigens react with antibodies to form masses of granular material. With the activation of complement, a local destruction of body tissue ensues. The effects of serum sickness, the Arthus phenomenon, and systemic lupus erythematosus illustrate how tissue may be affected.

Type IV cellular hypersensitivity is substantially different from the other three types because it involves sensitized T-lymphocytes. An exaggeration of cellular immunity is basic to type IV hypersensitivity. Infection allergy and contact dermatitis are two examples of this hypersensitivity.

In the final section of this chapter, we shall examine immune deficiency diseases, transplantation immunology, and tumor immunology. The immune deficiency diseases rob the body of its natural defenses to infection and demonstrate the vital role that the immune system plays in defense. In transplantation and tumor immunology we shall encounter some of the most recent findings in immunology, and see how scientists are beginning to understand certain functions of the immune system unrelated to infectious disease. Studies in these areas represent major thrusts of modern immunological research.

Immune Deficiency Diseases

The spectrum of immune deficiency diseases ranges from major abnormalities that are life-threatening to relatively minor deficiency states. The latter may be serious in populations where there is malnutrition and frequent contact with pathogenic organisms. Tests to determine immune deficiency diseases include measurements of antibody types, detection tests for B-lymphocyte function, and enumeration of T-lymphocytes. Assays of complement activity and phagocytosis are also of value.

An example of an immune deficiency disease is **Bruton's agammaglobulinemia**, first described by Ogden C. Bruton in 1952 (MicroFocus: 20.6). In this disease, B-lymphocytes fail to develop from pre-B-lymphocyte cells in the bone marrow. The lymphoid tissues lack plasma cells; levels of all five classes of antibodies are low or absent in the patient; and antibody responses to infectious disease are undetectable. Infections from staphylococci, pneumococci, and streptococci are common between the ages of six months and two years. Bruton's agammaglobulinemia is apparently a sex-linked inherited trait much more frequently observed in males than in females. Artificially acquired passive immunity is used to treat infectious disease in patients with this form of immune deficiency.

DiGeorge's syndrome is an immune deficiency disease in which T-lymphocytes fail to develop. The deficiency is linked to failure of the thymus gland to mature in the embryo. Cellular immunity is defective in such individuals, and susceptibility is high to many fungal and protozoal diseases and certain viral diseases. Correction of the defect by grafts of thymus tissue has been reported. Partial thymus insufficiencies exist in some individuals.

Perhaps the most dangerous immune deficiency disease is **severe combined immunodeficiency (SCID)**, a disease accompanied by defects in both antibody and cellular immunity. The B-lymphocyte and T-lymphocyte areas of the lymph nodes are depleted of lymphocytes, and all immune function is suppressed. For many years immunologists believed that the syndrome resulted from a defect in stem cells of the bone marrow, but recent evidence indicates a failure in the normal development of both thymus and bursa equivalent.

The longest surviving victim of SCID was a boy known only as David to protect his anonymity. David lived for 12 years inside a sterile, plastic bubble at the Baylor College of Medicine in Houston (Figure 20.9). On October 21, 1983, he received a bone marrow transplant from his sister as doctors attempted to establish an immune system within his body. Three months later, David was removed from the bubble, but on February 22, 1984, he died, a victim of blood cancer traced to a virus apparently brought into his body by the transplant.

Other immune deficiency diseases are linked to the polymorphonuclear (PMN) cells. In these diseases phagocytes fail to engulf and kill microorganisms

a-gam"ah-glob"u-lĭ-ne′me-ah
Bruton's agammaglobulinemia: an immune deficiency disease characterized by lack of B-lymphocytes.

DiGeorge's syndrome: an immune deficiency disease characterized by lack of T-lymphocytes.

FIGURE 20.9

A Photograph of the Boy in the Plastic Bubble, Known Only As David

David suffered from severe combined immunodeficiency and lived for 12 years within the sterile confines of a plastic bubble. In 1983 he received a bone marrow transplant from his sister and left the bubble. Unfortunately, he developed a blood cancer and died some months thereafter.

Job's syndrome:
an immune deficiency disease in which defective chemotaxis leads to poor phagocytosis.

because of defects in chemotaxis, ingestion, and/or intracellular digestion. For example, in the rare disease known as **Chédiak–Higashi syndrome**, there is delayed killing of phagocytized microorganisms traced to the inability of lysosomes to release their contents. Another disease, called **Job's syndrome**, is accompanied by defective chemotaxis. Scientists sometimes call this disease the "lazy leukocyte syndrome."

Deficiencies in the **complement system** may be life-threatening. As noted in Chapter 18, complement is a series of proteins any of which the body may fail to produce. For reasons not currently understood, many patients with complement component deficiencies suffer systemic lupus erythematosus (SLE) or an SLE-like syndrome. Meningococcal and pneumococcal diseases are often observed in patients who lack C3, probably as a result of poor opsonization.

Although **acquired immune deficiency disease (AIDS)** is considered a viral disease (Chapter 13), we shall also mention it here. In patients with AIDS, cellular immunity is severely depressed, while antibody-mediated immunity may also be altered. The number of T-lymphocytes is sparse, and a striking imbalance is observed between two subgroups of T-lymphocytes—the helper T-lymphocytes and the suppressor T-lymphocytes. Usually, the number of helper cells is twice that of suppressor cells, but in AIDS patients, the number of helper cells dramatically decreases. In scientific writing, the helper/suppressor ratio is expressed as the CD4/CD8 ratio because monoclonal antibody OkT4 identifies helper cells, and monoclonal antibody OkT8 identifies suppressor cells. The

20.6
MICROFOCUS

SOMETHING FROM NOTHING

In the 1940s and early 1950s, scientists frequently debated whether antibodies were essential to immunity. Amid this controversy, a pediatrician at Washington D.C.'s Walter Reed Hospital made a momentous discovery.

The story began when an air force officer's son was admitted to the hospital with acute streptococcal disease. The child's pediatrician, Ogden C. Bruton, used penicillin to control the infection, but two weeks later, the child was back, sick again. Wishing to determine the level of antibodies in the boy's system, Bruton sent a sample of blood for testing on a new machine recently acquired by the hospital.

The next day, the laboratory called Bruton to report that something must be wrong with the machine because it could not detect any antibodies in the blood. Bruton responded by sending over a second sample, but the results were the same: no antibodies. It occurred to Bruton that maybe the trouble was not in the machine but in the blood. Perhaps the blood had no antibodies. Conceivably, this could account for the recurring infections. Bruton began giving the boy monthly injections of antibodies, with outstanding success.

In 1952, Bruton reported his remarkable findings. His report was a medical bombshell because it established the concept of immune deficiency disease, while helping to solidify the role played by antibodies in resistance to infection.

But the story was not finished because certain patients with immune deficiency disease still produced the lymphocytes that immunize against skin grafts. This observation led to the notion that the immune system was actually a dual system: one branch centered in antibodies, a second branch centered in lymphocytes. The sources of antibodies and lymphocytes would be answered almost simultaneously 13 years later.

In 1965, at the meeting of the American Academy of Pediatrics, Robert A. Good's team was presenting evidence on the importance of the bursa of Fabricius to antibody production. The audience was skeptical, but the evidence appeared substantial.

When the speaker finished, a Philadelphia pediatrician named Angelo DiGeorge stepped to the microphone to add a footnote. DiGeorge described how four children at his hospital were struck repeatedly with severe infections. The children had antibodies in their blood but, curiously, each lacked a thymus. DiGeorge's observation was lost in the shuffle, but a question arose in his mind: were the children susceptible to skin grafts or could they possibly be immune?

DiGeorge hurried back to the hospital and devised a set of experiments to determine whether the children could successfully receive skin grafts. After weeks of study he found they could. The thymus was apparently the key to production of lymphocytes and hence to graft immunity. Bruton's patient lacked the bursa type of immunity but DiGeorge patients lacked the thymus type of immunity. The duality of the immune system was thus strengthened and a second immune deficiency disease, DiGeorge's syndrome, entered the dictionary of microbiology.

CD4/CD8 ratio in normal individuals is about 2:1, but in AIDS patients, it is permanently reversed.

Transplantation Immunology

Modern techniques for the transplantation of tissues and organs trace their origins to Jacques Reverdin, who in 1870 successfully grafted bits of skin to wounded tissues. Enthusiasm for the technique rose after his reports, but waned when doctors found that most transplants were rapidly rejected by the body. Then in 1954, a kidney was transplanted between identical twins and again, interest heightened. The graft survived for years until ultimately destroyed by a recurrence of the recipient's original kidney disease. Attempts to transplant kidneys between unrelated individuals were less successful.

Transplantation technology improved considerably during the next few decades, and today, four types of transplantations, or grafts, are recognized, depending upon the genetic relationship between donor and recipient. A graft taken from one part of the body and transplanted to another part of the same body is called an **autograft**. This graft is never rejected since it is the person's own tissue. A tissue taken from an identical twin and grafted to the other twin is an **isograft**. This, too, is not rejected because the genetic constitutions of identical twins are the same.

Autograft:
tissue grafted from one part of the body to another.

Allograft:
tissue grafted between
members of the same species
zen′o-graft

Rejection mechanisms become more vigorous as the genetic constitutions of donor and recipient cells become more varied (MicroFocus: 20.7). For instance, grafts between brothers and sisters, or between fraternal twins, may lead only to mild rejection because many of their genes are similar. Grafts between cousins may be rejected more rapidly, and as the relationship becomes distant, the vigor of rejection increases proportionally. **Allografts**, or grafts between random members of the same species such as two humans, have variable degrees of success, while **xenografts**, or grafts between members of different species such as a monkey and a human, are rarely successful.

Transplanted tissue is rejected by the body if the immune system interprets the tissue to be nonself. The rejection mechanisms may take either of two forms. In the first mechanism, cytotoxic T-lymphocytes aided by helper T-lymphocytes attack and destroy transplanted cells. The process is stimulated by the recognition of foreign MHC molecules (to be discussed next) on the surface of the graft cells. The class II MHC molecules stimulate helper T-lymphocytes, while the class I MHC molecules are the sites that the cytotoxic cells recognize during their attack.

Necrosis:
cell death.

A second rejection mechanism involves helper T-lymphocytes alone. These lymphocytes are stimulated by the class II MHC molecules, after which the lymphocytes release lymphokines (cytokines). The lymphokines stimulate phagocytes to enter the graft tissue. The phagocytes secrete lysosomal enzymes, which digest the tissues, leading to a dryness and thickening as in type IV hypersensitivity. Cell death, or necrosis, follows.

20.7

MICROFOCUS

ACCEPTANCE

Ordinarily a woman will reject a foreign organ, such as a kidney or heart, but she will accept the fetus growing within her womb. This acceptance exists even though half the fetus' genetic information has come from a "foreigner," namely the father. Has her immune system failed?

Apparently not. It seems that the sperm carries an antigenic signal, which induces the woman's immune system to produce a series of so-called blocking antibodies. The blocking antibodies form a type of protective screen that sheaths the fetus and prevents its antigens from stimulating the production of rejection antibodies by the mother. So protected, the fetus grows to term and "escapes" before any immunological damage can be done to its tissues.

But sometimes a rejection in the form of a miscarriage occurs. A number of physicians now believe that at least certain miscarriages have an immunological basis. Their research indicates that the level of

blocking antibodies in some pregnant women is too low to protect the fetus and that the miscarriage is due to the woman's antibodies. Ironically, they have discovered the blocking antibody level may be low because the father's tissue is very similar to the mother's. In such a case, the sperm's antigens elicit a weak antibody response, too low to protect the fetus.

With this knowledge in hand, physicians are now attempting to boost the level of blocking antibodies as a way of preventing miscarriage. They inject white blood cells from the father into the mother, thereby stimulating her immune system to produce antibodies to the cells. These antibodies exhibit the blocking effect. In other experiments, injections of blocking antibodies are administered to augment the woman's normal supply. Both approaches have been successful in trial experiments, and continuing research has given cause for optimism that fetal rejection will give way to acceptance.

A rejection mechanism of a completely different sort is sometimes observed in bone marrow transplants. In this case, the transplanted marrow may contain immune system cells that form immune products against the host after the host's immune system has been suppressed during transplant therapy. Essentially, the graft is rejecting the host. This phenomenon is called **graft versus host reaction** (GVHR). It can sometimes lead to fatal consequences in the host body.

The Major Histocompatibility Complex

During the 1970s, immunologists determined that the acceptance or rejection of a graft depends largely on a relatively small number of genes called the major histocompatibility complex (MHC) genes. These genes encode a series of cell-surface glycoproteins called the **major histocompatibility complex (MHC) molecules**, also known as **human leukocyte antigens (HLAs)**. The 1980 Nobel Prize in Physiology or Medicine was awarded to George Snell, Jean Dausset, and Baraj Benacerraf for their discoveries regarding the MHC genes and molecules.

his"to-kom-pat"ĭ-bil'i-te

The MHC genes are believed to exist on chromosome 6 in humans. The actual gene complex consists of four clusters of genes, each gene having multiple versions, or alleles. Two individuals chosen at random (a husband and wife, for example) are not expected to have many MHC genes in common. Identical twins, by contrast, have identical MHC genes. MHC molecules are of two types: class I and class II. Class II MHC molecules are important in the recognition of nonself antigens when T-lymphocytes combine with macrophages in cell-mediated immunity (Chapter 18). Class I MHC molecules are present on every cell in the human body and help define the uniqueness of a person's tissue. For this reason they are important subjects of transplantation immunology.

Alleles:
different versions of the same gene.

The nature of MHC molecules is a key element in transplant acceptance or rejection: the closer the match between donor and recipient MHC molecules, the greater the chance of a successful transplant. Matching of donor and recipient is performed by **tissue typing** (Figure 20.10). In this procedure, the laboratory uses standardized MHC antibodies for particular MHC molecules. Lymphocytes from the donor are incubated with a selected type of MHC antibodies. Complement and a dye such as trypan blue are then added. If the selected MHC antibodies react with the MHC molecules of the lymphocytes, the cell becomes permeable and dye enters the cells (living cells are not normally invaded by dye). Similar tests are then performed with recipient lymphocytes to determine which MHC molecules are present and how closely the tissues match one another. The blood types of donor and recipient must also be identical.

Tissue typing:
a process used to determine how closely two tissues match genetically.

Studies of the MHC molecules also give clues to the development of certain immune diseases. In **Graves' disease**, for example, antibodies appear to unite with MHC molecules on the surface of thyroid gland cells and overstimulate the gland. Another example is seen in **ankylosing spondylitis**, a spinal disease in which adjacent vertebrae fuse and cause fixation and stiffness of the spine. The relative risk for this disease is far greater when individuals possess particular MHC molecules. A final possibility is that certain viruses, because of their similarity to histocompatibility molecules, induce antibodies that attack body cells. This may be the source of Guillain–Barré syndrome and Reye's syndrome that follow cases of influenza.

ang"ki-lo'sing spon"di-li'tis

Antirejection Mechanisms

Tissue typing and histocompatibility screening help reduce the rejection mechanism in allograft transfers, but they do not eliminate the mechanism completely. To arrest rejection, it is necessary to suppress activity of the immune system.

FIGURE 20.10

Tissue Typing for MHC Molecules

Lymphocytes from an individual are incubated with selected MHC antibodies for a particular MHC molecule. The lymphocytes are then incubated with complement and a dye such as trypan blue. If the antibodies react with the MHC molecules, complement opens pores in the cells and allows the dye to enter. A positive result therefore consists of a stained cell, which indicates that a particular MHC molecule is present on the cell surface.

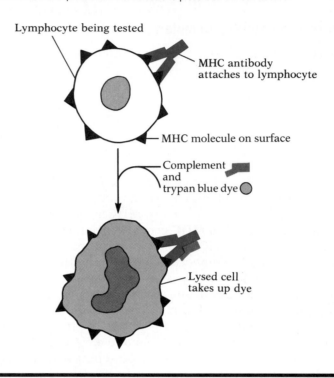

Lymphocyte being tested

MHC antibody attaches to lymphocyte

MHC molecule on surface

Complement and trypan blue dye

Lysed cell takes up dye

a″zah-thi′o-prēn

Cyclosporin A: an antirejection drug that appears to suppress cellular immunity.

pred′nĭ-sōn

One method utilizes **antimitotic drugs**, which prevent multiplication of lymphocytes in the lymph nodes. The drug **azathioprine** is a nucleic acid antagonist used for this purpose. Another drug is **cyclosporin A**. Originally isolated from strains of fungi as an antimicrobial substance, this drug is now produced synthetically. It appears to suppress cell-mediated immunity without killing T-lymphocytes or interfering with antibody formation to pathogens. Immunologists believe that cyclosporin A prevents division of helper T-lymphocytes by blocking formation of the growth and division factor **interleukin 2**. Suppressor T-lymphocytes appear resistant to the drug.

Another means of diminishing the rejection mechanism is to introduce antibodies against the lymphocytes. To obtain these antibodies, lymphocytes from the transplant patient are injected into animals and later, the animals are bled to obtain the antibody-rich serum. When introduced to the transplant recipient (serum sickness notwithstanding), the antibodies interact with the lymphocytes, thereby slowing the rejection process and increasing the survival time of the transplant.

Additional methods of reducing rejection include the use of prednisone, a steroid hormone that suppresses the inflammatory response, and the bombardment of lymphocyte centers with radiations such as X rays. In another experimental treatment, the donor's bone marrow cells are injected into the recipient's bone marrow to induce lymphocytes that will recognize the transplant as self. It should be noted that virtually all treatments leave the recipient in an immune-suppressed condition and increase the susceptibility to a variety of infectious diseases. Immune suppression is, at best, a poor expedient and is viewed as only a stopgap measure until the rejection mechanism can be understood and exploited.

Tumor Immunology

The alteration from a normal cell to a cell with tumor potential is accompanied by many physiological changes, including change to an immature form, loss of contact inhibition, and increased rate of mitosis (Chapter 11). In addition, tumor cells contain antigens not found in surrounding cells since they differentiate and revert to embryological cells. Since many of these antigens localize on the cell surface, they might be expected to induce an immune response and make the cancer cells vulnerable to destruction. However, the cells escape the resistance mechanisms of the body and proliferate to form the tumor. As the tumor breaks apart and spreads, a metastasizing cancer develops (Figure 20.11).

At least four different types of antigens have been identified in tumor cells. One type called the **oncofetal antigen** occurs in certain tumors and fetal tissues. An example is alpha fetoprotein (AFP), an antigen found in liver cancers. Another is carcinoembryonic antigen (CEA) found in cancer of the colon. Tests for these antigens are used in the diagnosis of cancers. The second type of antigen is found in tissues transformed by chemical carcinogens. The third and fourth types of antigens are associated with DNA viruses and RNA viruses, respectively. Viral-induced antigens are unique for each virus and are expressed in all tumor cells.

Host resistance to tumors depends on immune responses directed against the tumor antigens. This process is termed immune surveillance. It is considered an ongoing function of **cytotoxic T-lymphocytes**, which recognize the antigens

Oncofetal antigen:
an antigen found in tumors and fetal tissues.

Cytotoxic T-lymphocyte:
a T-lymphocyte that destroys cells containing certain antigens.

FIGURE 20.11

A Patient Being Treated with Radiation to Reduce a Tumor

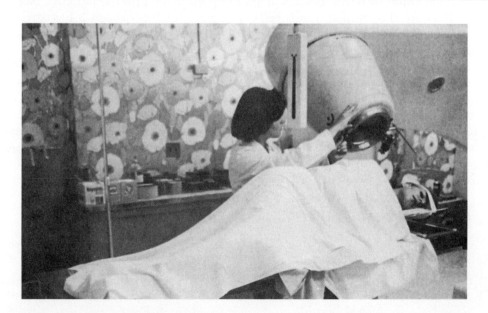

as foreign and destroy the cells containing them in a manner analogous to graft rejection. Another cell thought to function in immune surveillance is the **natural killer (NK) cell** which may be a form of T-lymphocyte. Natural killer cells exert a nonspecific, cytotoxic effect against other cells, including tumor cells. They kill cells in the absence of a prior sensitization and without the involvement of antibody. Interferon appears to regulate their activity and can increase it.

Immunologic Escape and Immunotherapy

Despite immune surveillance, tumors continue to occur in otherwise normal individuals as tumor cells escape destruction. Various theories are offered to account for this **immunologic escape**. One possibility is that individuals learn to tolerate tumor cells before immune competence has developed, and then express the tumors later in life. Another theory is that the majority of cancer cells are destroyed by killer cells, but that rapidly growing variants emerge and outpace the immune system. It is also possible that tumor antigens elicit antibodies that protect against the killer cells by combining with antigens to form a complex to block killer cell activity. A final theory points to the release by tumor cells of substances that suppress the immune system, such as happens in **Hodgkin's disease** of the lymph nodes. Although experimental evidence is offered for each theory, no one is universally accepted.

Immunotherapy for the management of cancer is developing along several lines. It has been suggested, for example, that killer lymphocytes specific for the tumor be injected to the patient or that monoclonal antibodies for tumor antigens be utilized (Figure 20.12). **Interferon** has also been investigated as a way of enhancing the activity of natural killer cells. In 1986, the FDA approved its use against hairy cell leukemia, a cancer of the white blood cells named for the hairlike appendages on malignant cells.

In the 1980s, interest grew in therapy with **interleukin 2**, a T-lymphocyte protein (a lymphokine) that stimulates the rapid multiplication of helper T-lymphocytes. In clinical trials, about 10 billion lymphocytes are removed from the cancer patient's blood and cultured for 3 days with interleukin 2. This process converts the resting lymphocytes to functional helper T-lymphocytes known as **lymphokine-activated killer (LAK)** cells. The LAK cells, along with additional amounts of interleukin 2, are then reinfused into the patient's bloodstream where they attack tumor cells. The gene for interleukin 2 has already been cloned, and biotechnology firms are now producing the lymphokine in volume by recombinant DNA techniques.

Another anti-cancer agent that has drawn attention is the **tumor necrosis factor (TNF)**. TNF, a protein product of macrophages, was discovered in 1975 by researchers at Memorial Sloan-Kettering Cancer Center in New York. The protein is currently produced by genetic engineering techniques and has been shown effective against more than 20 kinds of cancer cells grown in cultures. Combination with interferon appears to improve its anti-tumor abilities. At present, tumor immunology and treatment contains more questions than answers, but the prognosis remains positive.

NOTE TO THE STUDENT

The body's immunologic responses to transplants and tumors represent an ironic dilemma. Against transplants, the immune system responds vigorously (immunologists attempt to suppress the response); against tumors, the immune system responds weakly (immunologists attempt to enhance the response). Would that the situations could be reversed.

FIGURE 20.12

Monoclonal Antibodies in Tumor Therapy

Specific monoclonal antibodies are receiving clinical testing in new approaches to tumor therapy. (a) In a typical procedure, tumor cells, obtained by biopsy, immunize a mouse against the tumor antigens. The mouse is killed and its spleen B-lymphocytes used to produce hybridoma cells. Hybridoma cells that produce highly specific antitumor antibodies are selected and cloned, and their antibodies are purified. (b) Next, the antibodies are chemically coupled to a bacterial exotoxin or a chemotherapeutic drug with the potential to kill human cells. (c) The modified antitumor antibody is now ready for administration to the patient from whom the original biopsy was obtained. The antibodies bind and deliver their lethal drug specifically to the tumor, while normal cells (to which the antibody cannot bind) are unharmed.

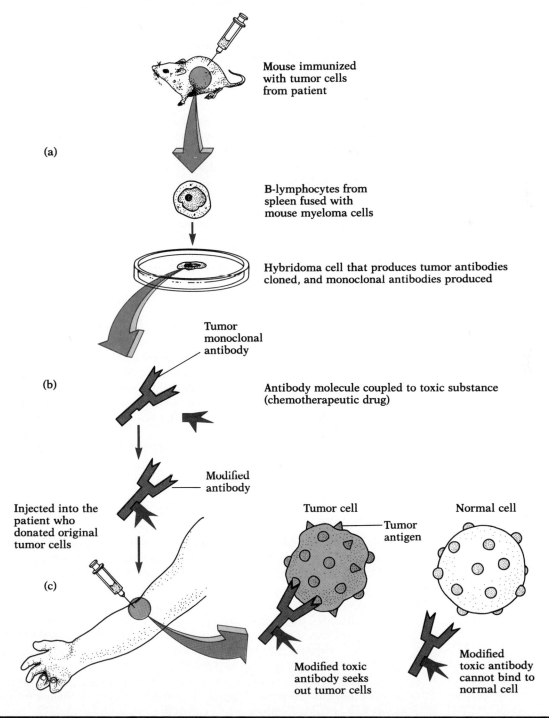

(a) Mouse immunized with tumor cells from patient

B-lymphocytes from spleen fused with mouse myeloma cells

Hybridoma cell that produces tumor antibodies cloned, and monoclonal antibodies produced

Tumor monoclonal antibody

(b) Antibody molecule coupled to toxic substance (chemotherapeutic drug)

Modified antibody

Injected into the patient who donated original tumor cells

(c)

Tumor cell

Tumor antigen

Normal cell

Modified toxic antibody seeks out tumor cells

Modified toxic antibody cannot bind to normal cell

Summary

Among the immune disorders that affect humans are a series of hypersensitivity reactions categorized in four types. Type I hypersensitivity is accompanied by anaphylaxis, a whole body reaction in which a series of mediators induce vigorous and life-threatening contractions by the smooth muscles of the body. The predominant antibody in anaphylaxis is IgE, produced in response to certain antigens and able to fix itself to the surfaces of mast cells and basophils. These cells release the mediators on subsequent exposure to the antigens. A localized anaphylaxis is a common allergy such as hay fever and food allergy.

The second type of hypersensitivity, type II, is called cytotoxic hypersensitivity. In this process, the immune system produces IgG and IgM, both of which react with the body's cells and often destroy the latter. The destruction of platelets in thrombocytopenia and neutrophils in agranulocytosis is typical of the immune disorder. No cells are involved in type III hypersensitivity. Rather, the body's IgG and IgM interact with dissolved antigen molecules to form visible masses of matter called immune complexes. Accumulation of immune complexes in various organs leads to local tissue destructions in such illnesses as serum sickness, Arthus phenomenon, and systemic lupus erythematosus. The final type of hypersensitivity involves no antibodies but is an exaggeration of the process of cellular immunity based in T-lymphocytes. Contact dermatitis may be a manifestation of this hypersensitivity.

Immune disorders also include immune deficiency diseases in which important cells of the immune system, such as B- or T-lymphocytes are not formed. Deficiencies in phagocytic cells or complement components may also be observed. The activity of the immune system is a key factor in the acceptance or rejection of transplanted tissue, and an understanding of immune system functions enhances our understanding of the development of tumors and how to deal with them.

Questions for Thought and Discussion

1. In December 1967, people throughout the world were startled to learn that Christiaan A. Barnard, a South African physician, had made the first successful heart transplant. The patient, Louis Washkansky, survived the rejection mechanism for two weeks before dying of pneumonia. Many scientists consider Barnard's surgery to signal the beginning of modern immunology. What evidence do you think might be offered to support this contention?

2. During war and under emergency conditions a soldier whose blood type is O donates blood to save the life of a fellow soldier of type B. The soldier lives and after the war becomes a police officer. One day he is called to donate blood to a brother officer who has been wounded and finds that it is his old friend from the war. He gladly rolls up his sleeve and prepares for the transfusion. Should he be allowed to proceed? Why?

3. Coming from the anatomy lab, you notice that your hands are red and raw and have begun peeling in several spots. This was your third period of dissection. What is happening to your hands and what could be causing the condition? How will you solve the problem?

4. In a 1993 article in *Discover* magazine, anaphylaxis was described this way: "It is so overwhelming that it can leave virtually every body system in a state of collapse, and so ferocious that a patient can be dead in minutes despite the best medical treatment." Suppose you were the emergency room nurse when a patient in the midst of anaphylaxis was brought in. What would you do?

5. Some weeks ago, a person had several blood transfusions during heart surgery. Now it is autumn and the person displays an allergic reaction to pollen for the first time in his life. Is there a connection between the two events?

6. As part of an experiment, one animal is fed a raw egg, while a second animal is injected intravenously with a raw egg. Which animal is in greater danger? Why?

7. "He had a history of nasal congestion, swelling of his eyes and difficulty breathing through his nose. He gave a history of blowing his nose frequently, and the congestion was so severe during the spring he had difficulty running. . . ." The "he" in this description is a certain President of the United States, and the writer is an allergist from Little Rock, Arkansas. What condition (technically known as allergic rhinitis) is probably being described?

8. When the skin of the feet becomes itchy, red, and scaly, many individuals assume they have athlete's foot, a fungal disease. However, allergists now know that the irritation could be a contact dermatitis resulting from chemicals in the insoles of sneakers. How might the distinction be made by physical examination of the feet and by laboratory testing?

9. A woman is taking the fifth in a weekly series of hay fever shots. Shortly after leaving the allergist's office, she develops a flush on her face, itching sensations of the skin, and shortness of breath. She becomes dizzy, then faints. What is taking place in her body and why has it not happened after the first four injections?

10. You may have noted that brothers and sisters are allowed to be organ donors for one another, but that a person cannot always donate to his or her spouse. Many people feel bad about being unable to help a loved one in time of need. How might you explain to someone in such a situation the basis for becoming an organ donor and why it may be impossible to serve as one?

11. Some years ago there was a commercial for a life insurance company that included the jingle: "There's nobody else exactly like you; nobody else like you." Why is this concept immunologically correct?

12. The story is told that in 1552, the distinguished physician Jerome Cardan of Pavia was called to England to advise treatment for Archbishop John Hamilton. Hamilton had suffered asthma for ten years. Cardan prescribed a carefully controlled diet, plenty of exercise and sleep, and the removal of feathers from the archbishop's mattress. Soon the asthma slackened. Which of Cardan's recommendations was the key to success? Why?

13. Paternity suits are often settled by determining the blood types of parents and child, and then concluding whether the man could be the father of the child. How could determining the major histocompatibility complex be used to replace the blood typing procedure in paternity suits of the future?

14. In order to prevent hemolytic disease of the newborn, Rh antibodies must be injected into an Rh-negative woman shortly after birth of an Rh-positive child. Where do you suppose these antibodies were obtained in the past and what might be their source in the future?

15. The immune system is commonly regarded as one that provides protection against disease. This chapter, however, seems to indicate that the immune system is responsible for numerous afflictions. Even the title says Immune Disorders. Does this mean that the immune system should be given a new name? On the other hand, is it possible that all these afflictions are actually the result of the body's attempts to protect itself? And finally, why can "immune disorder" be considered an oxymoron?

Review

The preceding pages have summarized some of the disorders associated with the immune system. To gauge your understanding of the chapter contents, rearrange the scrambled letters to form the correct word for each of the spaces in the statements. The correct answers are in the appendix.

1. The simple compound _____ is one of the major mediators released during allergy reactions.   Ⓘ Ⓣ Ⓜ Ⓔ Ⓗ Ⓘ Ⓐ Ⓢ Ⓝ

2. An immune deficiency called _____ syndrome is characterized by the failure of T-lymphocytes to develop.   Ⓔ Ⓓ Ⓞ Ⓖ Ⓢ Ⓖ Ⓔ Ⓡ Ⓔ

3. In type IV hypersensitivity a drying and thickening of the skin known as

　　_____ is an observable symptom.　　Ⓐ Ⓝ Ⓤ Ⓘ Ⓡ Ⓞ Ⓝ Ⓘ Ⓓ Ⓣ

4. Cases of rheumatoid arthritis are accompanied by immune complex formation in

　　the body's _____.　　Ⓝ Ⓘ Ⓢ Ⓣ Ⓙ Ⓞ

5. In cases of _____ disease, antibody molecules unite with receptors on
the surface of thyroid gland cells.　　Ⓥ Ⓢ Ⓡ Ⓔ Ⓐ Ⓖ

6. In a _____ hypersensitivity, antibodies unite with cells and
trigger a reaction that results in cell destruction.　　Ⓧ Ⓨ Ⓞ Ⓣ Ⓒ Ⓒ Ⓘ Ⓣ Ⓞ

7. Hay fever is an example of an _____ disease, one in which a local allergy
takes place.　　Ⓞ Ⓒ Ⓣ Ⓘ Ⓐ Ⓟ

8. Immune complex hypersensitivities develop when antibody molecules interact

　　with _____ molecules and form aggregates in the tissues.
　　　　　　Ⓔ Ⓝ Ⓘ Ⓖ Ⓝ Ⓐ Ⓣ

9. In persons suffering from Bruton's agammaglobulinemia, the lymph nodes are

　　noticeably deficient in _____ cells.　　Ⓢ Ⓛ Ⓜ Ⓐ Ⓐ Ⓟ

10. The skin test for _____ relies on a response by T-
lymphocytes to PPD placed in the skin tissues.
　　　　　　Ⓤ Ⓤ Ⓘ Ⓒ Ⓣ Ⓡ Ⓛ Ⓢ Ⓑ Ⓔ Ⓞ Ⓢ

11. Mast cells and _____ are the two principal cells that function
in anaphylactic responses.　　Ⓢ Ⓢ Ⓑ Ⓘ Ⓞ Ⓗ Ⓟ Ⓐ Ⓛ

12. During cases of thrombocytopenia, antibodies react with platelets and cause them

　　to undergo _____.　　Ⓨ Ⓢ Ⓢ Ⓛ Ⓘ

13. Rh disease can develop in a fetus if the father's blood type was Rh-

　　_____ and the mother's type was Rh-negative.
　　　　　　Ⓞ Ⓢ Ⓘ Ⓟ Ⓘ Ⓔ Ⓣ Ⓥ

14. The glomerulonephritis that accompanies streptococcal disease may be due to

　　immune complex formation in the _____.　　Ⓝ Ⓘ Ⓨ Ⓔ Ⓚ Ⓓ

15. A key element in transplant acceptance or rejection is a set of molecules abbre-

　　viated as _____ molecules.　　Ⓗ Ⓒ Ⓜ

16. Urticaria is a form of skin _____ that occurs in an individual undergoing an
allergic reaction.　　Ⓐ Ⓗ Ⓢ Ⓡ

17. An _____ disease occurs when a person's own antibodies are directed against a person's own cells.   Ⓔ Ⓤ Ⓤ Ⓘ Ⓜ Ⓐ Ⓜ Ⓣ Ⓞ Ⓝ

18. The process of _____ hypersensitivity is an exaggeration of the process of cellular immunity in which T-lymphocytes are the key cell.   Ⓡ Ⓛ Ⓔ Ⓤ Ⓛ Ⓒ Ⓛ Ⓐ

19. A butterfly rash occurring on the _____ is an important symptom of systemic lupus erythematosus.   Ⓔ Ⓐ Ⓕ Ⓒ

20. The contractions of _____ muscles are an essential feature in the life-threatening process of anaphylaxis.   Ⓗ Ⓞ Ⓢ Ⓞ Ⓣ Ⓜ

CONTROL OF MICROORGANISMS

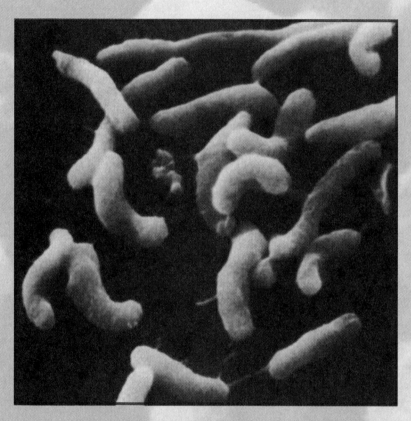

In the late 1800s and the early 1900s, growing support for the germ theory of disease led to a dramatic reduction in the frequency of epidemics. Scientists reasoned that if microorganisms cause disease, then it was possible to control disease by controlling the microorganisms. The idea had been proposed by Pasteur decades before, but not until the turn of the century did it gather momentum and achieve a firm footing in the scientific community. ■ Two types of control gradually evolved: physical control and chemical control. To achieve physical control, scientists employed heat, radiations, filters, and other physical agents to remove microorganisms from instruments, equipment, and fluids. Pasteurization of dairy products was in wide use by 1895 and food preservation methods were gradually updated. To achieve chemical control, doctors utilized antiseptics and disinfectants in medicine, surgery, and wound treatment. Moreover, municipalities began adding chlorine to their water supplies to protect citizens from waterborne diseases. As control methods became a way of life in public health, the chains of microbial transmission broke down and disease outbreaks declined. ■ But little could be done for the patient who already was ill. Here, another type of control was necessary, one that supplemented the body's own natural defenses. That control device would emerge in the 1940s with the discovery and development of chemotherapeutic agents and antibiotics. Now, physicians could do something about diseases they could not hope to treat before. The new drugs and medicines ushered in a second dramatic reduction in the incidence of infectious disease. ■ The control of microorganisms is an essential factor to maintaining good health. In Part VI we shall consider a variety of control methods and examine their uses and modes of action. Our survey will begin in Chapter 21 with an exploration of physical methods of control used for objects outside the body. It will continue in Chapter 22 where we examine chemical methods employed on objects that come in contact with the body (e.g., instruments) and on the body surface itself. Finally, in Chapter 23 we shall move inside the body environment to discuss chemotherapeutic agents and antibiotics. Applied on a broad scale, these control methods remain a major deterrent to infection and disease.

MICROBIOLOGY PATHWAYS

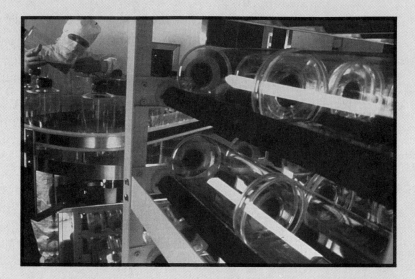

SALES AND RESEARCH

Not all sales people sell vacuum cleaners or brushes. And not all microbiologists wear white coats and work in ultraclean laboratories. It is quite commonplace for a person trained in microbiology to find a comfortable and enjoyable future with a company that sells and distributes instruments, scientific chemicals, and pharmaceuticals. For example, it is valuable for a sterilizer salesperson to understand the microbiological basis for sterilizing such things as instruments, microbial media, and patient materials. Also, it makes sense for a disinfectant salesperson to realize why microorganisms must be eliminated from a particular surface. Finally, it goes without question that an individual selling new antibiotics should have a strong familiarity with the microorganisms that the antibiotic is intended to eliminate. The bottom line is: "Know your product, and know what to use it for."

Before a product is available for sale, however, it must be developed, and once again, the microbiologist is a member of the team. The microbiologist can appreciate why new methods must be developed for preserving dairy products and for storing foods. The microbiologist will be able to set the direction for developing new sterilizing instruments and can have significant input into the development of pharmaceuticals for treating infectious disease. The diagnostic lab will depend heavily on instruments produced under the guidance of microbiologists because the lab's objective is to detect microbes.

The contemporary health industry depends heavily on the talents of microbiologists for the development and sales of innovative and novel approaches to diagnosis and treatment. As you might suspect, a healthy dose of chemistry, physics, and mathematics is valuable, depending upon which road one chooses to follow. The ability to tinker with instruments is important to research and development, while strong interpersonal relationships are important for the sales phase. What may seem like services and sales is really microbiology at its core.

MICROBES AND BIOTECHNOLOGY

(a)

(b)

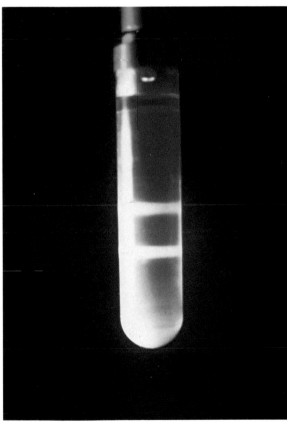

(c)

PLATE 1
Key Components of Genetic Engineering **(a)** The massive industrial-scale production of genetically engineered *E. coli*. Centrifugation harvests *E. coli* as a thick paste—here being scooped up with a spatula. **(b)** Extraction of large quantities of DNA— here DNA that has been precipitated from solution is spooled onto a glass rod. **(c)** Plasmid DNA purified in an ultracentrifuge by cesium chloride density gradient. The upper band is chromosomal DNA and the lower band is plasmid DNA. The DNA fluoresces under ultraviolet light because it is complexed with the fluorescent dye ethidium bromide.

PLATE 2
DNA Analysis   DNA fragments can be separated
on an electrophoresis gel. The bands are made
visible by reacting the DNA with a fluorescent dye
and illuminating with ultraviolet light.

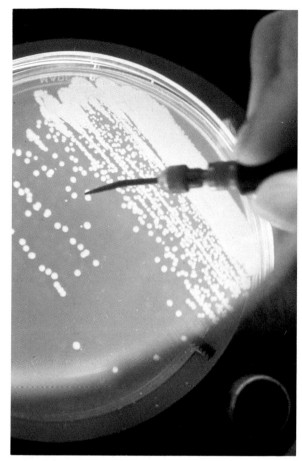

(a)

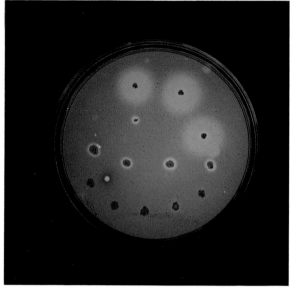

(b)

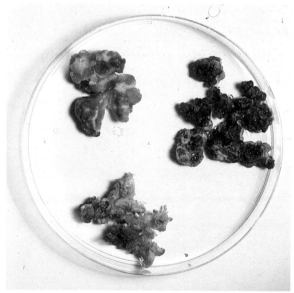

(c)

PLATE 3
Screening and Identification of Genetically Engineered Organisms   (a) Bacteria may be isolated as colonies on streak plates and then evaluated for metabolic capabilities. (b) On a differential growth medium, colonies of bacteria genetically engineered to digest cellulose are surrounded by a light yellow zone. Some colonies have a greater digestive capacity than others. (c) Tobacco tissues genetically engineered to express a bacterial gene. The blue tissue [right] expresses at high level the bacterial enzyme beta-galactosidase—its activity releases a blue dye (indigo) that stains the plant tissue. Other tissues without the bacterial gene stain lightly blue because of a low level of plant beta-galactosidase in normal tobacco cells.

(a)

(b)

(c)

PLATE 4
The Production Phase of Genetic Engineering
(a) Modern 100-gallon fermentation vessels which have been specially modified so genetically engineered microbes cannot escape. (b) Final production phase of interferon synthesized via genetic engineering uses stainless steel manufacturing tanks. Particular attention is paid to maintaining homogeneity of the bulk solution. (c) Yeast cells immobilized on Manville Biocatalyst Carrier R-630, 10,000 ×. With the proper supporting material, microbes such as yeasts can be absorbed, or immobilized, so as to retain their metabolic activity for long periods of time. For example, immobilized yeasts could be used for continuous production of alcohol.

CHAPTER 21

PHYSICAL CONTROL OF MICROORGANISMS

By the best estimates of health officials, the problem began during the first week of July in 1988. One day that week a person unintentionally contaminated the water while swimming in an indoor pool at a school in Los Angeles, California. The same week the pool was used by a water polo team for its match against a local opponent. Shortly thereafter, a class of SCUBA divers used the pool to perfect their skills. A group of elementary school children also enjoyed the pool during a camp field trip.

Beginning in early August, doctors began receiving reports of watery diarrhea, abdominal cramps, and fever. Two persons from the SCUBA class were hospitalized. Eleven patients were tested for stool microorganisms and in seven cases, the protozoan *Cryptosporidium* was located. Indeed, the attack rate was highest for those who were exposed to the water the longest. By the second week of August, 44 persons were affected from a total of 60 swimmers during that fateful July.

When CDC investigators arrived at the pool they inquired about the water purification methods and were told that three filters were used to sterilize the water and chlorine was added to maintain a low bacterial count. Their inspection revealed that the chlorination was adequate, but the filtration left much to be desired. One of the three filters was inoperative, and the flow rate of water through the remaining two filters was only 70 percent of the expected rate. The water filtration had not worked.

The episode in Los Angeles illustrates how microorganisms can spread when proper methods of water purification are lacking. In many cases, these methods are designed to achieve sterilization. **Sterilization** implies the destruction or removal of all life forms. It is an absolute term that cannot be qualified. Thus one cannot assume "partial sterilization" or "incomplete sterilization," but must consider a material "contaminated" until sterilized. It should be pointed out, however, that material may remain harmful even though it is sterile. For example, a solution of bacterial toxin may contain no living forms but still cause physiological damage in the body.

Sterilization:
the destruction or removal of all forms of life.

Toxin:
a bacterial poison able to inflict damage on body tissues.

FIGURE 21.1

A Bank of Three Steam Sterilizer Units Known As Autoclaves

Modern units such as these combine traditional principles of sterilization with computer circuitry to enhance flexibility and productivity. For the destruction of bacterial spores, it is often necessary to employ these machines.

Although chemicals may sometimes be used to sterilize objects, the principal methods for achieving sterilization employ physical agents such as heat (Figure 21.1), as well as radiations, and filtration. These methods are not products of the modern era. They were used by Pasteur, Koch, and microbiologists of a century ago to prevent contamination of their materials and to ensure the accuracy of their work.

Microbiologists recognize that bacterial **spores** are the most resistant forms of life. Anthrax spores, for example, have been found viable after 60 years on dry silk threads. The spores are not metabolically inert. Rather, they carry on life processes at minimum rates and possess the necessary enzymes to transform from a dormant state to actively metabolizing vegetative cells. The extraordinary survival of spores is often attributed to their low water content, which yields a heat-resistant gel-like spore core. Another possible reason is the presence of **dipicolinic acid**, an organic compound that encourages heat resistance by linking to the spore proteins. Destruction of the bacterial spore is the principal aim of sterilization methods, especially those involving heat.

di″-pik-o-lin′ik

Physical Control with Heat

The Citadel is a novel by A.J. Cronin that follows the life of a young British physician, beginning in the 1920s. Early in the story the physician, Andrew Manson, begins his practice in a small coal-mining town in Wales. Almost immediately, he encounters an epidemic of **typhoid fever**. When his first patient dies of the disease, Manson becomes terribly distraught. However, he realizes that the epidemic can be halted, and in the next scene, he is tossing all of the patient's bedsheets, clothing, and personal effects into a huge bonfire.

Typhoid fever:
a serious waterborne and foodborne bacterial disease accompanied by intestinal ulcers and high fever.

The killing effect of **heat** on microorganisms has long been known. Heat is fast, reliable, and relatively inexpensive, and it does not introduce chemicals to a substance, as disinfectants sometimes do. Above maximum growth temperatures, biochemical changes in the cell's organic molecules result in its death. These changes arise from alterations in enzyme molecules or chemical breakdowns of structural molecules, especially in the cell membranes. Heat also drives off water, and since all organisms depend on water, this loss may be lethal.

The killing rate of heat may be expressed as a function of time and temperature. For example, **tubercle bacilli** are destroyed in 30 minutes at 58° C, but in only 2 minutes at 65° C, and in a few seconds at 72° C. Each microbial species has a **thermal death time**, the time necessary for killing it at a given temperature. Each species also has a **thermal death point**, the temperature at which it dies in a given time. These measurements are particularly important in the food industry, where heat is used for preservation (Chapter 24).

Thermal death time: **the time required to kill a population of microorganisms at a given temperature.**

In determining the time and temperature for microbial destruction with heat, certain factors bear consideration. One factor is the type of organism to be killed. For example, if materials are to be sterilized, the physical method must be directed at bacterial spores. Milk, however, need not be sterile for consumption, and heat is therefore aimed at the most resistant vegetative cells of pathogens (Figure 21.2).

Another factor is the type of material to be treated. Powder is subjected to dry heat rather than moist heat, because moist heat will leave it soggy. Saline solutions, by contrast, can be sterilized with moist heat but are not easily treated with dry heat. Other factors are the presence of organic matter and the acidic or basic nature of the material. Organic matter may prevent heat from reaching microorganisms, while acidity or alkalinity may encourage the lethal action of heat.

FIGURE 21.2

Use of the Direct Flame As a Sterilizing Agent

(a) A photograph of the original Bunsen burner invented by Robert W. Bunsen in 1855. Note the holder on the right for objects heated in the flame. (b) Laboratory use of the Bunsen burner. A few seconds in the flame is usually sufficient to effect sterilization.

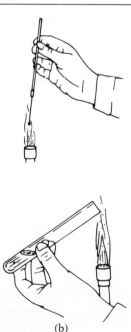

(a)   (b)

Direct Flame

Perhaps the most rapid sterilization method is the **direct flame** method used in the process of incineration. The flame of the Bunsen burner is employed to sterilize the bacteriological loop before removing a sample from a culture tube and after preparing a smear (Figure 21.2). Flaming the tip of the tube also destroys organisms that happen to contact the tip, while burning away lint and dust.

In general, objects must be disposable if a flame is used for sterilization. Disposable hospital gowns and certain plastic apparatus are examples of materials that may be incinerated. In past centuries, the bodies of disease victims were burned to prevent spread of the pestilence (Figure 21.3). It is still common practice to incinerate the carcasses of cattle that have died of **anthrax** and to put the contaminated field to the torch because anthrax spores cannot adequately be destroyed by other means. British law even stipulates that anthrax-contaminated animals may not be autopsied before burning.

Hot-Air Oven

The **hot-air oven** utilizes radiating dry heat for sterilization. This type of energy does not penetrate materials easily and thus, long periods of exposure to high temperatures are necessary. For example, at a temperature of 160° C (320° F), a period of two hours is required for the destruction of bacterial spores. Higher temperatures are not recommended because the wrapping paper used for equipment tends to char at 180° C. The hot-air method is useful for sterilizing dry powders and water-free oily substances, as well as many types of glassware such as pipettes, flasks, and syringes. Dry heat does not corrode sharp instruments as steam often does, nor does it erode the ground glass surfaces of nondisposable syringes.

The effect of **dry heat** on microorganisms is equivalent to that of baking. The heat changes microbial proteins by oxidation reactions and creates an arid internal environment, thereby burning microorganisms slowly. It is essential that organic matter such as oil or grease films be removed from the materials,

Anthrax:
a bacterial disease of the blood and other organs caused by a sporeforming bacillus.

Oxidation:
a chemical reaction involving addition of oxygen atoms or loss of electrons.

FIGURE 21.3

Incineration

A newspaper photograph from the Spanish-American War showing troops burning a yellow fever hospital in Cuba. Such methods were a drastic but accepted way of preventing the spread of yellow fever in that period.

FIGURE 21.4

Various Temperature Environments Used in the Physical Control of Microorganisms

The temperature of heat used in sterilization varies considerably according to the method used for the destruction of organisms.

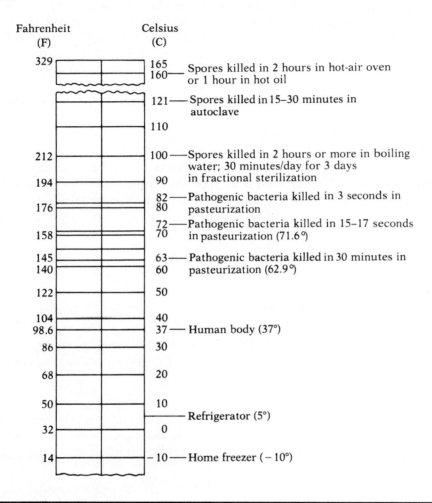

because organic matter insulates against dry heat. Moreover, the time required for heat to reach sterilizing temperatures varies among materials. Thus this factor must be considered in determining the total exposure time.

Boiling Water

Immersion in **boiling water** is the first of several moist-heat methods that we shall consider. **Moist heat** penetrates materials much more rapidly than dry heat because water molecules conduct heat better than air. Lower temperatures and less time of exposure are therefore required than for dry heat (Figure 21.4).

Moist heat kills microorganisms by denaturing their proteins. **Denaturation** involves changes in the chemical or physical properties of proteins. It includes structural alterations due to destruction of the chemical bonds holding proteins in a three-dimensional form. As proteins revert to a two-dimensional structure, they coagulate (denature) and become nonfunctional. Egg protein undergoes a

de-na"chur-a'shun

similar transformation when it is boiled. You might find a review of the chemical structure of proteins in Chapter 2 helpful to your understanding of this process. The coagulation and denaturing of proteins require less energy than oxidation, and therefore, less heat need be applied.

Boiling water is not considered a sterilizing agent because destruction of bacterial spores and inactivation of viruses cannot always be assured. Under ordinary circumstances, with microorganisms at concentrations of less than one million per milliliter, most species of microorganisms can be killed within 10 minutes. Indeed, a few seconds may be all that is necessary. However, fungal spores, protozoal cysts, and large concentrations of hepatitis A viruses require up to 30 minutes' exposure. Bacterial spores often require two hours or more. Because inadequate information exists on the heat tolerance of many microorganisms, boiling water is not reliable for sterilization purposes (MicroFocus: 21.1).

If it is imperative that boiling water be used to destroy microorganisms, materials must be thoroughly cleaned to remove traces of organic matter such as blood or feces. The minimum exposure period should be 30 minutes, except at high altitudes, where it should be increased to compensate for the lower boiling point of water. All materials should be well covered. Washing soda

Hepatitis A:
a viral disease of the liver transmitted by contaminated food and water.

21.1

MICROFOCUS

A HEATED CONTROVERSY

Among the last defenders of spontaneous generation was the British physician Harry Carleton Bastian. Louis Pasteur had stated that boiled urine failed to support bacterial growth, but in 1876, Bastian claimed that if the urine were alkaline, microorganisms would occasionally appear. Pasteur repeated Bastian's work and found it correct. This led Pasteur to conclude that certain microorganisms could resist death by boiling. The spores of *Bacillus subtilis*, discovered coincidentally in 1876 by Ferdinand Cohn, were an example.

Pasteur soon realized that he would have to heat his broths at a temperature higher than 100° C to achieve sterilization. He therefore put his pupil and collaborator Charles Chamberland in charge of developing a new sterilizer. Chamberland responded by constructing a pressure steam apparatus patterned after a steam "digester" invented in 1680 by the French physician Denys Papin (Figure). The sterilizer resembled a modern pressure cooker. It attained temperatures of 120° C and higher, and established the basis for the modern autoclave. Chamberland would also achieve fame in later years for his work with porcelain filters.

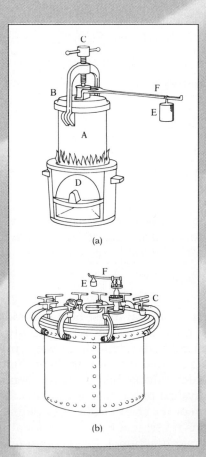

(a)

(b)

But Chamberland's invention was not universally accepted. A German group of investigators, led by Robert Koch, criticized the pressurized steam sterilizer because they believed its higher temperatures might destroy critical laboratory media. Instead they preferred an unpressurized steam sterilizer. In 1881, the German group developed a free-flowing steam sterilizer of the type used in tyndallization. In time, however, they came to appreciate the benefits of pressurized steam as a sterilizing agent, so much so that they modified Chamberland's device to an upright model. Ironically, the instrument became known as the Koch autoclave.

(a)
Origins of the autoclave. Denis Papin's steam digester designed in 1680. The digester consisted of a vessel (A) into which food was placed. The lid (B) and screw (C) sealed the vessel. A furnace (D) raised the temperature, and a weight (E) and safety lever (F) controlled the pressure of steam in the vessel.
(b)
Chamberland's autoclave built in 1880 according to the principles of Papin's digester. The lid is held in place by screws (C) and a weight (E) and safety lever (F) are used to control pressure. This autoclave is basically similar to a home pressure cooker.

may be added at a 2 percent concentration to increase the efficiency of the process.

Autoclave

Moist heat in the form of pressurized steam is regarded as the most dependable method for the destruction of all forms of life, including bacterial spores (Micro-Focus: 21.1). This method is incorporated into a device called the **autoclave**. Over a hundred years ago, French and German microbiologists developed the autoclave as an essential component of their laboratories.

A basic tenet of chemistry is that when the pressure of a gas increases, the temperature of the gas increases proportionally. Because steam is a gas, increasing its pressure in a closed system increases its temperature. As the water molecules in steam become more energized, their penetration increases substantially. This principle is used to reduce cooking time in the home pressure cooker and to reduce sterilizing time in the autoclave. It is important to note that the sterilizing agent is the moist heat, not the pressure.

Most autoclaves contain a **sterilizing chamber** into which articles are placed and a steam jacket where steam is maintained, as shown in Figure 21.5. As steam flows from the steam jacket into the sterilizing chamber, cool air is forced out and a special valve increases the pressure to **15 pounds/square inch (lb/in^2)** above normal atmospheric pressure. The temperature rises to **121.5° C** and the superheated water molecules rapidly conduct heat into microorganisms. The time for destruction of the most resistant bacterial spore is now reduced to about **15 minutes**. For denser objects, up to 30 minutes of exposure may be required.

The autoclave is used to control microorganisms in both hospitals and laboratories. It is employed for blankets, bedding, utensils, instruments, intravenous solutions, and a broad variety of other objects. The laboratory technician uses it to sterilize bacteriological media and destroy pathogenic cultures. The autoclave is equally valuable for glassware and metalware, and is among the first instruments ordered when a microbiology laboratory is established.

aw'to-klāv

Autoclave:
a pressurized steam device used for sterilization purposes.

FIGURE 21.5

Operation of the Autoclave

Steam enters through the port (A) and passes into the jacket (B). After the air has been exhausted through the vent, a valve (C) opens to admit pressurized steam (D) that circulates among and through the materials, thus sterilizing them. At the conclusion of the cycle, steam is exhausted through the steam exhaust valve (E).

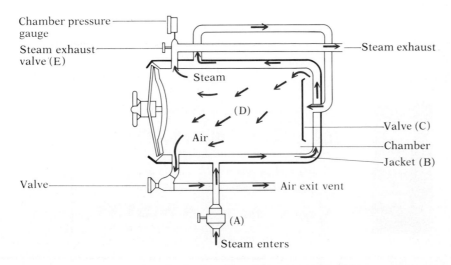

Bacillus:
a genus of sporeforming Gram-positive bacterial rods.

The autoclave also has certain **limitations**. For example, some plasticware melts in the high heat, and sharp instruments often become dull. Moreover, many chemicals break down during the sterilization process, and oily substances cannot be treated since they do not mix with water. To gauge the success of sterilization, a strip containing spores of a *Bacillus* species is included with the objects treated (Figure 21.6). At the conclusion of the cycle, the strip is placed in nutrient broth medium and incubated. If the sterilization process has been successful, no growth will occur, but growth indicates failure.

In recent years a new form of autoclave, called the **prevacuum autoclave**, has been developed for sterilization procedures. This machine draws air out of the sterilizing chamber at the beginning of the cycle. Saturated steam is then used at a temperature of 132° C to 134° C at a pressure of 28 to 30 lb/in². The time for sterilization is now reduced to as little as 4 minutes. A vacuum pump operates at the end of the cycle to remove the steam and dry the load. The major advantages of the prevacuum autoclave are the minimal exposure time for sterilization and the reduced time to complete the cycle.

Fractional Sterilization

Fractional sterilization:
sterilization by exposure to free-flowing steam for 30 minutes on each of three successive days.

tyn″dal-ĭ-za′shun

In the years before development of the autoclave, liquids and other objects were sterilized by exposure to free-flowing steam at 100° C for 30 minutes on each of three successive days. The method was called **fractional sterilization** because a fraction was accomplished on each day. It was also called **tyndallization** after its developer, John Tyndall (MicroFocus: 21.2), and **intermittent sterilization** because it was a stop-and-start operation.

Sterilization by the fractional method is achieved by an interesting series of events. During the first day's exposure, steam kills virtually all organisms except bacterial spores, and it stimulates spores to germinate to vegetative cells. During overnight incubation the cells multiply and are killed on the second day. Again, the material is cooled and the few remaining spores germinate, only to be killed on the third day. Although the method usually results in sterilization, occasions arise when several spores fail to germinate. The method also requires that spores be in a suitable medium for germination, such as a broth.

Fractional sterilization has assumed importance in modern microbiology with the development of high-technology instrumentation and new chemical substances. Often, these materials cannot be sterilized at autoclave temperatures, or by long periods of boiling or baking, or with chemicals. An instrument that generates free-flowing steam, such as the **Arnold sterilizer**, is used in these instances.

Arnold sterilizer:
an instrument that generates free-flowing steam.

FIGURE 21.6
Testing Effective Sterilization

A bacterial test strip containing *Bacillus stearothermophilus* and *Bacillus subtilis* (*globigii*). The strip inside the package is placed into broth medium at the conclusion of the sterilization cycle. If growth fails to appear on incubation, it may be assumed that the sterilization was successful. If growth occurs in the broth, then sterilization may not be assured.

BACTERIAL TEST STRIP
FOR STERILIZATION PROCEDURES
Gas, Dry Heat or Steam (250 or 270 F.)
Bacillus stearothermophilus
Bacillus subtilis (*globigii*)
Mfr. Lot No. 180 BGS   **AMSCO**

21.2
MICROFOCUS

TEDIOUS BUT WORTHWHILE

While Pasteur and Koch were setting down the foundations of microbiology in Europe, a British physicist named John Tyndall was developing a process for killing the most resistant forms of bacteria.

Tyndall believed that airborne microorganisms were associated with dust particles. In the early 1870s, he devised a wooden chamber with a glass front and glass side windows, and passed a beam of light through the chamber to visualize the dust (Figure). A beam of sunlight through a window displays dust particles the same way. Tyndall managed to prepare a sample of dust-free air and showed that it was free of microorganisms, thereby adding credence to Pasteur's theory that microorganisms were present in the air.

In 1876, Tyndall concluded that certain forms of bacteria were more resistant than other forms. His theory was based on observations that samples of old, dried hay were more difficult to sterilize than samples of fresh hay. Though unaware of the existence of spores, he resolved to develop a method to kill the "heat-resistant" bacteria.

Tyndall soon found that extended heating did not work well, but a stop-and-start method seemed useful. He heated hay samples to boiling on five consecutive occasions and allowed the samples to cool to room temperature between heatings. Intervals of 10 to 12 hours between heatings were found most effective in the sterilization process. Tyndall's sterilization method preceded the development of the autoclave by several years. It showed that even the most heat-resistant forms of life could be eliminated, and it made possible the use of sterile broths for verifying the germ theory of disease. Today the method is known as tyndallization, for its developer.

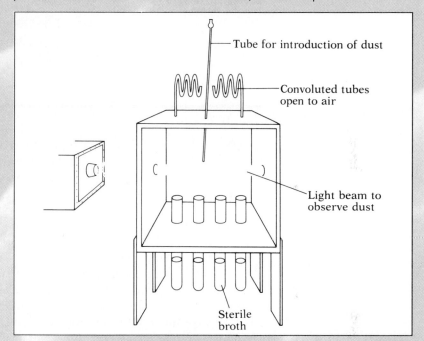

Tyndall's apparatus for observing dust particles and determining the presence of microorganisms. Dust-free air contained few microorganisms, and thus the sterile broths open to the air showed little evidence of growth on incubation. When dust was introduced through the tube, however, most broth tubes became cloudy on incubation, indicating that microorganisms were present in the dust.

Pasteurization

Pasteurization is not the same as sterilization. Its purpose is to reduce the bacterial population of a liquid such as milk and to destroy organisms that may cause spoilage and human disease. Spores are not affected by pasteurization (Figure 21.7).

One method for milk pasteurization, called the **holding method**, involves heating at 62.9° C for 30 minutes. Although thermophilic bacteria thrive at this temperature, they are of little consequence because they cannot grow at body temperature. For decades, pasteurization has been aimed at destroying ***Mycobacterium tuberculosis***, long considered the most heat-resistant bacterium. More recently, however, attention has shifted to destruction of ***Coxiella burnetii***, the agent of Q fever, because this organism has a higher resistance to heat. Since both organisms are eliminated by pasteurization, dairy microbiologists assume that other pathogenic bacteria are also destroyed.

Pasteurization:
a method of reducing microbial numbers in liquids by using heat.

kok″se-el′lah
bur-net′e

FIGURE 21.7

Bacillus subtilis

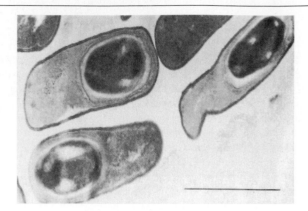

A transmission electron micrograph of *Bacillus subtilis* during the latter stage of spore formation. The spores and their multiple coats are visible within the cellular cytoplasm. Spores such as these are extremely resistant to destruction. The heat employed in pasteurization has little effect on them. Bar = 1 μm.

Two other methods of pasteurization are the **flash pasteurization** method at 71.6° C for 15 seconds, and the **ultrapasteurization** method at 82° C for 3 seconds. These methods are discussed in detail in Chapter 24 on the microbiology of milk and dairy products.

Hot Oil

Some dentists and physicians use **hot oil** at 160° C for the sterilization of instruments. A time period of 1 hour is usually recommended. Hot oil does not rust metals, and minimal corrosion takes place. However, once sterilization is complete, the instruments must be cleaned and dried for storage, and this step may reintroduce contamination. Silicone is sometimes used as an alternative to oil.

TO THIS POINT
■

We have surveyed a number of physical methods for controlling microorganisms that involve heat. These methods are generally aimed at sterilization, a term that denotes the destruction or removal of all life forms including bacterial spores. Incineration with the direct flame is the most rapid heating method but materials must be disposable. Dry heat is used in the hot-air oven where an exposure at 160° C for 2 hours achieves sterilization. A lower temperature of 100° C is used in boiling water, but sterilization cannot always be assured. The moist heat in boiling water is superior to the dry heat in the hot-air oven because moist heat penetrates better.

The autoclave is regarded as the most dependable instrument for sterilization. Moist heat in the form of steam under pressure is used in this instrument, and sterilization may be achieved in 15 to 30 minutes. The prevacuum autoclave utilizes a vacuum to draw out air at the beginning of the cycle and steam at the end, thereby reducing the cycle time. Higher pressures and temperatures are also used. We also noted how fractional sterilization is a useful tool under certain circumstances and how pasteurization is not the same as sterilization. Hot oil was briefly mentioned.

> *We shall now direct our attention to physical methods that do not employ heat. These methods include filtration and the use of ultraviolet light, other radiations, and ultrasonic vibrations. Each method is of value under certain circumstances. For example, ultraviolet light is useful for sterilizing the air in a room. However, the destruction of all organisms may not be as thorough as with heating methods. The chapter will close with a review of certain preservation methods used in foods. Here, too, sterilization may not be achieved, but the number of microorganisms can be substantially reduced.*

Physical Control by Other Methods

Filtration

Heat is a valuable physical agent for controlling microorganisms but sometimes it is impractical to use. For example, no one would suggest removing the microbial population from a tabletop by using a Bunsen burner, nor can heat-sensitive solutions be subjected to an autoclave. In instances such as these and numerous others, a heat-free method must be used. This section describes some examples.

Filters came into prominent use in microbiology as interest in viruses grew in the 1890s. Previous to that time, filters had been utilized to trap airborne organisms and sterilize bacteriological media, but now they became essential for separating viruses from other microorganisms. Among the early pioneers of filter technology was Charles Chamberland, an associate of Pasteur. His porcelain filter was important to early virus research, as noted in Chapter 11. Another pioneer was **Julius Petri** (inventor of the Petri dish), who developed a sand filter to separate bacteria from the air.

Cham'ber-land

The **filter** is a mechanical device for removing microorganisms from a solution. As fluid passes through the filter, organisms are trapped in the pores of the filtering material. The solution that drips into the receiving container is decontaminated or, in some cases, sterilized. Filters are used to purify such things as intravenous solutions, bacteriological media, toxoids, many pharmaceuticals, and beverages.

Toxoid:
a chemically treated toxin used for immunization purposes.

sītz

Several types of filters are available for use in the microbiology laboratory. **Inorganic filters** are typified by the Seitz filter, which consists of a pad of porcelain or ground glass mounted in a filter flask. **Organic filters** are advantageous because the organic molecules of the filter attract organic components in microorganisms. One example, the Berkefeld filter, utilizes a substance called **diatomaceous earth**. This material contains the remains of marine algae known as diatoms. **Diatoms** are unicellular algae that abound in the oceans and provide important foundations for the world's food chains. Their remains accumulate on the shoreline and are gathered for use in swimming pool and aquarium filters, as well as for microbiological filters used in laboratories.

di"ah-to-ma'shus

The **membrane filter** is a third type of filter that has received broad acceptance. It consists of a pad of organic compounds such as cellulose acetate or polycarbonate, mounted in a holding device. This filter is particularly valuable because bacteria multiply and form colonies on the filter pad when the pad is placed on a plate of culture medium (Figure 21.8). Microbiologists can then count the colonies to determine the number of bacteria originally present. For example, if a 100-ml sample of liquid were filtered and 59 colonies appeared on the pad after incubation, it could be assumed that 59 bacteria were in the sample.

Membrane filter:
a filter composed of cellulose acetate on which bacterial colonies may form for enumeration.

FIGURE 21.8

The Membrane Filter Technique

(a) The membrane filter consists of a pad of cellulose acetate or similar material mounted in a holding device. (b) The holding device is secured by a clamp, and a measured amount of fluid is filtered by pouring it into the cup. The solution runs through to a flask beneath and bacteria are trapped in the filter material. (c) The filter pad is placed onto a plate of nutritious medium and the plate is incubated. (d) After incubation, colonies appear on the surface of the filter pad. The colony count reflects the original number of bacteria in the fluid sample.

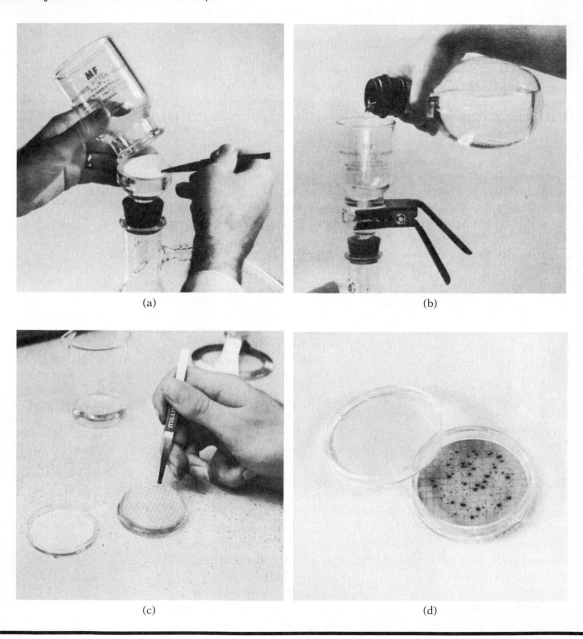

(a)

(b)

(c)

(d)

Air can also be filtered to remove microorganisms. The filter generally used is a **high-efficiency particulate air (HEPA) filter**, which can remove over 99 percent of all particles, including microorganisms having a diameter of over 0.3 micrometer. The air entering surgical units and specialized treatment facilities such as burn units is filtered to exclude microorganisms. In some hospital wards, such as for respiratory diseases, and in certain pharmaceutical filling rooms, the air is recirculated through HEPA filters to ensure its purity.

21.3
MICROFOCUS

OUT OF HARM'S WAY

It has long been recognized that bacterial spores can resist the effects of ultraviolet light and survive where vegetative bacterial cells quickly die. The fact that spores remain alive for decades in soil exposed to sunlight is but one manifestation of the spore's ability to withstand UV light.

In the 1980s, scientists caught a glimpse of how this resistance works. Sporeformers, it seems, have the ability to produce a certain protein during the early stages of sporulation. The protein, referred to as a small, acid-soluble spore protein (SASP), appears to protect the spore from UV light. In virtually all other bacteria, UV light affects adja-

cent thymine molecules in the cell's DNA and binds them together to form gnarled, double-looped structures. The disfigured DNA cannot replicate or be repaired. But in spores with SASP there is no effect on the thymine molecules.

Then, in 1991, new light was shed on the process. Biochemists at the University of Connecticut and Boston University discovered that SASP can bind to DNA and untwist it ever so slightly. This untwisting and change in DNA's geometry apparently makes the DNA resistant to the effects of UV light.

Earthshattering news? Perhaps not. But that's how science works. We might expect scientific experiments to have profound effects on our lives, but the general rule is that scientific endeavors rarely have immediate impact. The more usual occurrence is that experimental findings (like those above) elicit an "Aha!" from the scientist and from colleagues in the scientific community. Then the scientist goes back to work.

Ultraviolet Light

Visible light is a type of radiant energy detected by the sensitive cells of the eye. The wavelength of this energy is between 400 and 800 nanometers (nm). Other types of radiations have wavelengths longer or shorter than that of visible light and therefore, they cannot be detected by the human eye.

One type of radiant energy, **ultraviolet light**, is useful for controlling microorganisms. Ultraviolet light has a wavelength between 100 and 400 nm, with the energy at about 265 nm most destructive to bacteria. When microorganisms are subjected to ultraviolet light, cellular DNA absorbs the energy, and adjacent **thymine molecules** link together. Linked thymine molecules are unable to position adenine on messenger RNA molecules during the process of protein synthesis. Moreover, replication of the chromosome in binary fission is impaired. The damaged organism can no longer produce critical proteins or reproduce, and it dies quickly.

Ultraviolet light effectively reduces the microbial population where direct exposure takes place. It is used to limit airborne or surface contamination in a hospital room, morgue, pharmacy, toilet facility, or food service operation. It is noteworthy that ultraviolet light from the sun may be an important factor in controlling microorganisms in the air and upper layers of the soil, but it may not be effective against all bacterial spores (MicroFocus: 21.3). Ultraviolet light does not penetrate liquids or solids, and it may cause damage in human skin cells.

Ultraviolet light:
a form of electromagnetic energy whose wavelength is between 100 and 400 nm.

Thymine:
one of the four different nitrogenous bases of DNA.

Other Types of Radiation

The spectrum of energies (Figure 21.9) includes two other forms of radiation useful for destroying microorganisms. These are **X rays** and **gamma rays**. Both have wavelengths shorter than the wavelength of ultraviolet light. As X rays and

FIGURE 21.9

The Ionizing and Electromagnetic Spectrum of Energies

A complete spectrum is presented at the bottom of the graph, and the ultraviolet and visible sections are expanded at the top. Note how the bactericidal energies overlap with the UV portion of sunlight. This may account for the destruction of microorganisms in the air and in upper layers of soil.

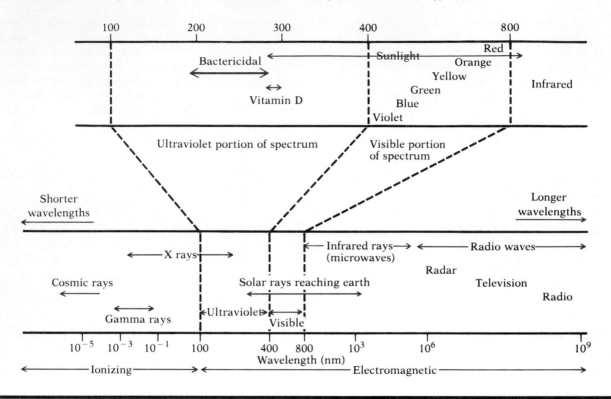

Ionizing radiations:
radiations that cause atoms to change to ions.

gamma rays pass through microbial molecules, they force electrons out of their shells, thereby creating ions. For this reason, the radiations are called **ionizing radiations**. The ions quickly combine with and destroy proteins and nucleic acids such as DNA, causing death. Gram-positive bacteria are more sensitive to ionizing radiations than Gram-negative bacteria. Ionizing radiations are currently used to sterilize such heat-sensitive pharmaceuticals as vitamins, hormones, and antibiotics as well as certain plastics and suture materials.

Another form of energy, the **microwave**, has a wavelength longer than that of ultraviolet light. In a microwave oven, microwaves are absorbed by water molecules. The molecules are set into high-speed motion, and the heat of friction is transferred to foods, which become hot rapidly. Other than the heat generated, there is no specific activity against microorganisms (MicroFocus: 21.4).

Laser beam:
a beam of light energy concentrated to considerably increase its energy.

A final form of radiation that we shall consider is light energy. When concentrated by sophisticated devices, light energy forms a **laser beam**. The term "laser" stands for *l*ight *a*mplification by *s*timulated *e*mission of *r*adiation. Recent experiments indicate that laser beams can be used to sterilize instruments and the air in operating rooms, as well as for a wound surface. Microorganisms are destroyed in a fraction of a second, but the laser beam must reach all parts of the material to effect sterilization.

MICROWAVES: MYTH AND FACT

1. **Myth:** Microwaves cook food.

    Fact: It's true that microwaves cook food but in a somewhat indirect manner. Food is composed of a loosely woven nest of large proteins, fats, carbohydrates, and other organic materials. Water molecules float freely among the materials and, like submicroscopic magnets, they have positive and negative ends. Alternating microwave fields (2.4 billion alternations per second) spin the water molecules about at implausibly high rates, thus creating friction. The heat of friction is transferred to the surrounding food molecules as heat, and the food cooks.

2. **Myth:** Microwaves cook from the inside out.

    Fact: Not really. Microwaves excite all water molecules throughout the food simultaneously. However, some heat at the food surface is lost to the surrounding air, so the outside of the food cools more quickly than the inside, and the perception is that the inside has been heated more thoroughly.

3. **Myth:** There is no solution to soggy and limp pizza heated in the microwave oven.

    Fact: Sure, there is, but first you have to understand why the pizza gets soggy. In a regular oven, hot water molecules from the dough come to the surface where they meet the hot oven air and evaporate, leaving the pizza crust crisp and the inside moist. In the microwave oven, the surrounding air is cool, so when the hot water molecules reach the surface, they condense to liquid and stay there or soak back into the pizza, creating a soggy mess. The answer to this problem is a metallic film mounted on a piece of cardboard. The metal heats up when exposed to microwaves, and the hot water molecules vaporize when they hit it, thereby forming a dry surface and a crisp pizza.

4. **Myth:** Microwaves are useless for sterilization.

    Fact: Microwaves have potential as sterilizing agents. One company is experimenting with a process that shreds infectious waste, sprays it with water, and exposes it to microwaves until the temperature reaches near-boiling levels. Another company is testing a procedure that sterilizes instruments after sealing them within a vacuum in a glass container. Both methods show promise.

Ultrasonic Vibrations

Ultrasonic vibrations are high-frequency sound waves beyond the range of the human ear. When directed against environmental surfaces, they have little value because air particles deflect and disperse the vibrations. However, when propagated in fluids, ultrasonic vibrations cause the formation of microscopic bubbles, or cavities, and the water appears to boil. Some observers call this "cold boiling." The cavities rapidly collapse, and send out shock waves. Microorganisms in the fluid are quickly disintegrated by the external pressures. The formation and implosion of the cavities is known as **cavitation**. Figure 21.10 illustrates this process.

Ultrasonic vibrations are valuable in research for breaking open tissue cells and obtaining their parts for study. A device called the **cavitron** is used by dentists to clean teeth, and ultrasonic machines are available for cleaning dental plates, jewelry, and coins. A major appliance company has also experimented with an ultrasonic washing machine.

As a sterilizing agent, ultrasonic vibrations have received minimal attention because liquid is required and other methods are more efficient. However, many research laboratories use ultrasonic probes for cell disruption and hospitals use ultrasonic devices to clean their instruments. When used with an effective germicide, an ultrasonic device may achieve sterilization, but the current trend

Ultrasonic vibrations: high-frequency sound waves.

kav″ĭ-ta′shun

FIGURE 21.10

How Ultrasonic Vibrations Kill Microorganisms

(a) High-frequency sound waves cause the formation of microscopic bubbles in fluid. (b) As the bubbles collapse, shock waves are created in the fluid, and alternating high- and low-pressure areas impinge upon microorganisms and (c) cause their destruction.

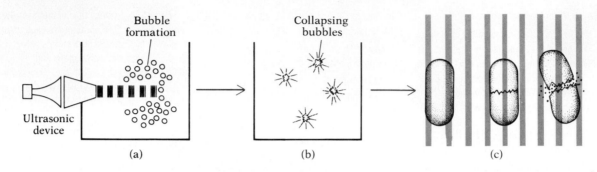

is to use ultrasonic vibrations as a cleaning agent and follow the process by sterilization in an autoclave. Table 21.1 summarizes the physical agents used to control microorganisms.

Preservation Methods

Over the course of many centuries, various physical methods have evolved for controlling microorganisms in food. Though valuable for preventing the spread of infectious agents, these procedures are used principally to retard spoilage and prolong the shelf life of foods, rather than for sterilization.

Drying is useful in the preservation of various meats, fish, cereals, and other foods. Since water is a necessary requisite for life, it follows that where there is no water, there is virtually no life. Many of the foods in the kitchen pantry typify this principle. One example is discussed in MicroFocus: 21.5.

Preservation by **salting** is based upon the principle of osmotic pressure. When food is salted, water diffuses out of microorganisms to the higher salt concentration and lower water concentration in the surrounding environment. This flow of water, called **osmosis**, leaves microorganisms to shrivel and die. The same phenomenon occurs in highly sugared foods such as syrups, jams, and jellies. However, fungal contamination may remain at the surface because aerobic molds tolerate high sugar concentrations.

Low temperatures found in the refrigerator and freezer retard spoilage by reducing the rate of metabolism in microorganisms and, consequently, reducing their rate of growth. Spoilage is not totally eliminated in cold foods, however, and many microorganisms remain alive, even at freezer temperatures. These organisms multiply rapidly when food thaws, which is why prompt cooking is recommended.

Note in these examples that there are significant differences between killing microorganisms, holding them in check, and reducing their numbers. The preservation methods are described as bacteriostatic because they prevent the further multiplication of bacteria. A fuller discussion of food preservation as it relates to public health is presented in Chapter 24.

oz-mo'sis

Osmosis:
the diffusion of water molecules from a region of high water concentration to a region of low water concentration.

TABLE 21.1

A Summary of Physical Agents Used to Control Microorganisms

PHYSICAL METHOD	CONDITIONS	INSTRUMENT	OBJECT OF TREATMENT	EXAMPLES OF USES	COMMENT
Direct flame	A few seconds	Flame	All micro-organisms	Laboratory instruments	Object must be disposable or heat-resistant
Hot air	160° C for 2 hr	Oven	Bacterial spores	Glassware Powders Oily substances	Not useful for fluid materials
Boiling water	100° C for 10 min 100° C for 2 hr+	— —	Vegetating micro-organisms Bacterial spores	Wide variety of objects	Total immersion and precleaning necessary
Pressurized steam	121° C for 15 min at 15 lb/in^2	Autoclave	Bacterial spores	Instruments Surgical materials Solutions and media	Broad application in microbiology
Fractional sterilization	30 min/day for 3 days	Arnold sterilizer	Bacterial spores	Materials not sterilized by other methods	Long process
Pasteurization	62.9° C for 30 min 71.6° C for 15 sec	Pasteurizer	Pathogenic micro-organisms	Dairy products	Sterilization not achieved
Hot oil	160° C for 1 hr	—	Bacterial spores	Instruments	Rinsing necessary
Filtration	Entrapment in pores	Berkefeld filter Membrane filter	All micro-organisms	Fluids	Many adaptations
Ultraviolet light	265 nm energy	UV light	All micro-organisms	Surface and air sterilization	Not useful in fluids
X rays Gamma rays	Short-wave length energy	Generator	All micro-organisms	Heat-sensitive materials	Formation of toxic chemicals
Ultrasonic vibrations	High-frequency sound waves	Sonicator	All micro-organisms	Fluids	Few practical applications

NOTE TO THE STUDENT

As the science of microbiology evolved in the late 1800s, physicians were quick to grasp the notion that contaminated instruments, clothing, and similar articles were important to disease transmission. Also, researchers soon realized that contaminated growth media and materials could ruin their experiments and cast doubt on their findings. The physical methods for controlling microorganisms therefore attained acceptance rapidly, and steam and high-pressure sterilizers became commonplace in medical microbiology.

21.5
MICROFOCUS

A DRY AND TWISTED TALE

The kitchen pantry usually contains many foods that resist microbial contamination simply because they are too dry to support life. In many pantries, one of these foods is the pretzel.

According to a major manufacturer, the origin of the pretzel dates back to the early 600s and a monk in southern France or northern Italy. Legend has it that the monk took leftover strips of bread and fashioned them into twisted loops, similar to the arms folded across the chest in prayer. He baked the loops and distributed them to children who performed good deeds, such as learning their prayers well. Each child received a "pretiola" or "little reward." The name was later shortened to pretzel.

It seems unfortunate that anyone would take issue with such an appealing tale, but Tom Burnam has done just that in *The Dic-*

tionary of Misinformation. Burnam maintains that the word "pretzel" is derived from the German word for "branch," and is not Latin for "little reward." He points out that a pretzel looks like the intertwined branches of a tree, and he attempts to punch holes in the legend by noting that folding the arms across the chest is the conventional coffin arrangement, not the position for prayer. Some traditionalists might dispute this.

Whatever the true story, the pretzel was brought to America by European immigrants, especially the Pennsylvania Dutch (really "Deutsch," for German), who settled in the Northeast. Today, major pretzel factories are located in Lancaster and Lititz, Pennsylvania. Their products are as dry, twisted, and free of microbial contamination as the original pretzels of centuries ago.

A hundred years later, things have not changed substantially. The autoclave still occupies a prominent position in the clinical and research laboratory, and most physical methods for sterilization are essentially as they were a century ago. Even tyndallization, tedious as it is, has reemerged for use in decontaminating the products of modern biotechnology. To be sure, there are more blinking lights in today's sophisticated equipment, but it is reassuring to note that yesterday's principles are still valid. Perhaps this is one example of what is implied by the saying, "The more things change, the more they remain the same."

Summary

The physical methods for controlling microorganisms are generally intended to achieve sterilization. Sterilization implies the destruction or removal of all life forms, with particular reference to the bacterial spore.

Microorganisms can be controlled by various methods. The direct flame, for example, achieves sterilization in a few seconds, while the hot-air oven requires exposure to hot air at 160° C for two hours. Exposure to boiling water at 100° C for two hours may also result in sterilization, but spore destruction cannot always be assured. By contrast the autoclave uses pressurized steam at 121° C to sterilize objects in about 15 minutes. A prevacuum sterilizer shortens this time still further. Other heat methods for sterilization include fractional sterilization (30 minutes exposure to steam on each of three successive days) and hot oil (160° C for one hour). Pasteurization reduces the microbial population in a liquid and is not intended to be a sterilization method.

Certain nonheat methods are also used to control microbial populations. Filters, for instance, use various materials to trap microorganisms within the pores of filtering material. Inorganic, organic, and membrane filters are used. Ultraviolet light is an effective way of killing microorganisms on a dry surface and in the air. Other radiations include X rays and gamma rays, two forms of ionizing radiations used to sterilize heat-sensitive objects. Laser beams and ultrasonic vibrations could also be useful for sterilization, but instruments that utilize these energy forms are not currently available.

For food preservation, drying, salting, and low temperatures can be used to control microorganisms, but sterilization is a virtual impossibility. However, for instruments, pharmaceuticals, medical apparatus, microbial media, and numerous other products, the physical methods noted are effective and widely used means for achieving sterilization.

Questions for Thought and Discussion

1. A concerned parent "sterilizes" the baby's formula according to directions accompanying the sterilizing machine. The bottles are then placed in the refrigerator until used. Is the refrigeration necessary? Why?

2. When the local drinking water is believed contaminated, area residents are advised to boil their water before drinking. Often, however, they are not told how long to boil it. As a student of microbiology, what might be your recommendation?

3. The label on the container of a product in the dairy case proudly proclaims, "This dairy product is sterilized for your protection." However, a statement in small letters below reads: "Use within 30 days of purchase." Should this statement arouse your suspicion about the sterility of the product? Why?

4. Several days ago, you bought a steak and stored it in the refrigerator. Now you find that the surface of the steak has spots of bacterial contamination. You are certain that the bacteria have not penetrated deeply into the meat. Should you broil the steak as you had planned or discard it? Why?

5. A patient is asked over the phone to bring a urine sample to the laboratory for analysis. The immediate thought is to sterilize a small jar by placing it in boiling water. In practical terms how much time might be recommended?

6. Early in the century, a prehistoric woolly mammoth was discovered in the tundra of Siberia. The meat of the animal was so well preserved that it was fed to hunters' dogs. What factors contributed to the meat's preservation?

7. Suppose a liquid needed to be sterilized. What methods could be developed using only the materials found in a normal household?

8. Old metropolitan buildings often had hallway chutes in which garbage could be dumped. The garbage would then drop into an incinerator and burn. How might this be of value to a budding microbiologist who happened to live in the building?

9. Why might ultraviolet light be expected to have a greater killing effect on DNA viruses than on RNA viruses?

10. The Bunsen burner was invented in 1855 by the German chemist Robert W. Bunsen. In how many different ways can it be used to sterilize objects and materials under laboratory conditions?

11. Other than the last few spores not germinating during treatment, what drawbacks are there to the process of fractional sterilization?

12. In view of all the sterilization methods we have reviewed in this chapter, why do you think none has been widely adapted to the sterilization of milk? Which, in your opinion, holds the most promise?

13. Why may sunlight be referred to as nature's great sterilizing agent?

14. The word autoclave is derived from stems that mean "self-closing." This is a reference to the fact that the chamber closes itself by the pressure of the steam. Would it be correct to equate the words autoclave and sterilizer? Why?

15. A liquid that has been sterilized may be considered pasteurized, but one that has been pasteurized may not be considered sterilized. Why not?

Review

Use the following syllables to form the term answering the clue pertaining to sterilization. The number of letters in the term is indicated by the dashes, and the number of syllables in the term is shown by the number in parentheses. Each syllable is used only once. The answers are listed in Appendix F.

A   A   AU   BA   BER   BRANE   BUN   CIL   CLAVE   CLEAN   CRO   CU   DA   DALL
DE   DER   DI   DRY   GAM   HOLD   I   IC   IDS   IN   ING   ING   ING   INS   LET
LO   LUS   MA   MEM   MENTS   MI   MO   NA   O   OX   OS   PLAS   POW   PRES
SEN   SIS   SIS   SOL   SON   SPORE   STRU   SURE   THIR   TIC   TION   TION   TO   TOMS   TOX
TRA   TRA   TU   TUR   TY   TYN   UL   UL   VI   WAVES

1. Instrument for sterilization   (3) _ _ _ _ _ _ _ _ _ _

2. Type of filter   (2) _ _ _ _ _ _ _ _ _

3. Sterilization in an oven   (2) _ _ _ _ _ _ _

4. Occurs in boiling water   (5) _ _ _ _ _ _ _ _ _ _ _ _

5. Developed fractional method   (2) _ _ _ _ _ _ _ _

6. Preserves meat, fish   (2) _ _ _ _ _ _ _

7. High-frequency vibrations   (4) _ _ _ _ _ _ _ _ _ _

8. Short-wavelength rays   (2) _ _ _ _ _ _

9. Most resistant life form   (1) _ _ _ _ _

10. Occurs in dry heat   (4) _ _ _ _ _ _ _ _ _ _

11. Raised in the autoclave   (2) _ _ _ _ _ _ _ _

12. Minutes for tyndallization   (2) _ _ _ _ _ _ _

13. Method of pasteurization   (2) _ _ _ _ _ _ _ _

14. Sterilized with hot oil   (3) _ _ _ _ _ _ _ _ _ _ _

15. Source of an organic filter   (3) _ _ _ _ _ _ _ _ _

16. Light for air sterilization   (5) _ _ _ _ _ _ _ _ _ _ _ _

17. Melts in the autoclave   (2) _ _ _ _ _ _ _ _

18. May remain after filtration   (2) _ _ _ _ _ _ _

19. Not penetrated by UV light   (2) _ _ _ _ _ _ _

20. Water flow from salting   (3) _ _ _ _ _ _ _ _

21. Used to heat water molecules   (3) _ _ _ _ _ _ _ _ _ _

22. Genus of sporeformers   (3) _ _ _ _ _ _ _ _ _

23. Direct flame burner   (2) _ _ _ _ _ _ _

24. Essential pretreatment   (2) _ _ _ _ _ _ _ _

25. Prevented by pasteurization   (5) _ _ _ _ _ _ _ _ _ _ _ _ _

CHAPTER 22

CHEMICAL CONTROL OF MICROORGANISMS

Before the 1900s, hospitals rarely had running water, and what water they had was usually contaminated. Garbage, human waste, and other hospital refuse were usually dumped into a pit in the courtyard; surgeons wiped their hands and instruments on their hospital jackets and trousers; bedclothes were rarely changed; and infection was rampant. Up to one-third of women giving birth died of puerperal fever, a blood disease often due to a species of *Streptococcus*.

As late as the mid-1800s, only a few visionaries recognized the relationship between filth and disease. One such person was the American physician, poet, and jurist, Oliver Wendell Holmes, author of an 1843 paper on the contagious nature of puerperal fever. Another was the Hungarian physician **Ignaz Semmelweis**.

In 1847, Semmelweis was working at a Vienna hospital's obstetrics clinic when he made a remarkable observation: an unusually high incidence of puerperal fever occurred in maternity wards tended by physicians fresh from dissecting cadavers. However, in the ward tended by midwives, the incidence was much lower. Semmelweis reasoned that disease was transmitted by infected hands, and he ordered all the attendants in his ward to wash their hands in chlorine water before ministering to patients. Soon the death rate among his patients fell significantly.

Semmelweis summarized his findings and presented his evidence for the transmission of puerperal fever to hospital administrators. However, they rejected his conclusion that doctors were spreading disease because it cast doctors in a negative light. The debate became vehement, and Semmelweis insisted that rejecting his methods was tantamount to committing murder. Shortly thereafter, an anguished and frustrated Semmelweis left Austria for his native Hungary, where he died in 1865 (Figure 22.1).

Ironically, the death of Semmelweis paralleled Pasteur's landmark experiments in microbiology. As the germ theory of disease gained a foothold, the realization dawned that infectious microorganisms could indeed be transmitted by clothing, utensils, instruments, and other objects. To interrupt the spread of

pu'er-per-al

sem'el-vīs

Puerperal fever: a blood disease accompanied by high fever and often transmitted during childbirth; also called childbed fever; due to a streptococcus or other bacterium.

FIGURE 22.1
Ignaz Semmelweis

A statue of Ignaz Semmelweis erected to honor one of the first physicians to urge the use of disinfectants in disease control. The monument stands in Budapest, Hungary, and refers to Semmelweis as the "savior of mothers."

organisms, doctors began using chemical antiseptics and disinfectants, and before long they witnessed a substantial decline in the incidence of disease.

In this chapter we shall examine a variety of chemical methods used for controlling the spread of microorganisms. Our study begins by outlining some general principles and terminology of disinfection practices, and then proceeds to a discussion of the spectrum of antiseptics and disinfectants. Whether for wounds, swimming pools, industrial machinery, or pharmaceutical products, antiseptics and disinfectants are fundamental to public health practices that ensure continued good health.

General Principles of Chemical Control

The notions of sanitation and disinfection are not unique to the modern era. **The Bible** refers often to cleanliness and prescribes certain dietary laws to prevent consumption of what was believed to be contaminated food. Egyptians used resins and aromatics for embalming even before they had a written language, and ancient peoples burned sulfur for deodorizing and sanitary purposes. Over the centuries, necessity demanded chemicals for food preservation, and spices were used as preservatives as well as masks for foul odors. Indeed, Marco Polo's trips to the Orient for new spices were made out of necessity as well as adventure.

fahr"mah-ko-pe'ah

Medicinal chemicals came into widespread use in the 1800s. As early as 1830, for example, the *U.S. Pharmacopoeia* listed tincture of iodine as a valuable antiseptic, and soldiers in the Civil War used it in plentiful amounts. In the first decades of that century, people found copper sulfate useful for preventing fungal

22.1

MICROFOCUS

SURGERY WITHOUT INFECTION

While Louis Pasteur was speaking and writing about the germ theory of disease, the British physician Joseph Lister was doing his best to reform the practice of surgery.

Lister was an innovative and imaginative individual. He was, for example, among the earliest physicians to put the newly discovered anesthetics to use. By using carefully measured quantities of ether and air, Lister found he could perform operations without torture. But it distressed him that even though an operation could be painless and successful, the patient might still die of infection. Throughout Great Britain roughly one of every two amputations ended in death from "hospital gangrene," "blood poisoning," or other affliction.

Toward the end of 1864, Lister read Pasteur's reports on fermentation and successfully repeated many of Pasteur's experiments. The work convinced Lister that airborne microorganisms were responsible for postsurgical diseases. He decided to test Pasteur's suspicions that microorganisms cause disease.

While searching for an antimicrobial compound, Lister's attention was drawn to a newspaper account describing the use of carbolic acid (phenol) for the treatment of sewage in a town near Glasgow. After exploring the capabilities of this compound in laboratory cultures and finding it effec-

tive, Lister was ready to proceed with human experiments.

The first recorded use of chemical disinfection in a surgical procedure occurred in March 1865 at the Glasgow Royal Infirmary. During surgery to repair a compound fracture, Lister sprayed the air with a fine mist of carbolic acid and soaked his instruments and ligatures in carbolic acid solution. Although the patient subsequently died of infection, Lister remained optimistic. He repeated the experiments, improved the procedures, and finally met with success. In 1867, he reported his results in an article

in *The Lancet*, a British medical journal. Lister wrote that his antiseptic methods reduced the mortality rate in postoperative surgery from 45 percent to 9 percent.

Lister's work was accomplished without a clear knowledge or understanding of pathogenic microorganisms, and in this regard, his achievements merit special note. Though hospital equipment and methods have changed, the principles he established are as valid today as they were a century ago. Before his death in 1912, he was knighted by the British Crown and is remembered as Sir Joseph Lister.

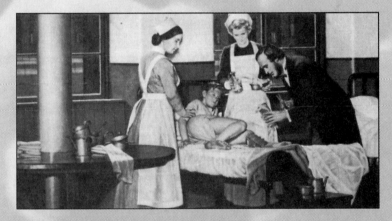

A photograph of a painting by Robert Thom depicting Joseph Lister using antiseptic methods in the surgical treatment of a leg wound.

disease in plants, but unfortunately, this chemical was not well-known during the great potato blight of Ireland (Chapter 14). Mercury was sometimes used for treating syphilis, as first suggested by Arabian physicians centuries before. Moviegoers have probably noted that American cowboys practiced disinfection by pouring whiskey onto gunshot wounds between drinks.

In the 1860s, a physician at the University of Glasgow named **Joseph Lister** established the principles of aseptic surgery. Pasteur's references to airborne microorganisms convinced Lister that microorganisms were the cause of wound infections (MicroFocus: 22.1). Lister experimented with several chemicals to kill microorganisms and finally settled on carbolic acid (phenol). He applied the chemical to wounds and instruments and sprayed it in the air near the operating table. Lister achieved brilliant success and was accorded many accolades as a pioneer microbiologist of his time, although he advocated the same principles recommended by Semmelweis years before.

a-sep'tik

Terminology of Disinfection

Sterilization:
the destruction or removal of
all forms of life.

Disinfection:
the destruction or removal of
pathogenic microorganisms.

The chemical control of microorganisms extends into such diverse areas as the hospital environment, food processing plant, and everyday household. These broad fields have yielded an equally broad terminology that should be explored before we undertake a discussion of individual chemical agents.

The physical agents for controlling microorganisms (Chapter 21) are generally intended to achieve **sterilization**. This implies the destruction of all forms of life, especially bacterial spores. Chemical agents, by contrast, rarely achieve sterilization. Instead, they are expected only to destroy the pathogenic organisms on or in an object. The process of destroying pathogens is called **disinfection**; the object is said to be disinfected. If the object is lifeless, such as a table top, the chemical agent is known as a **disinfectant**. However, if the object is living, such as a tissue of the human body, the chemical is an **antiseptic**. Figure 22.2 displays the fundamental difference between disinfectants and antiseptics. It is important to note that even though a particular chemical may be used in a disinfectant as well as an antiseptic (iodine, for example), the precise formulations are so different that the ability to kill microorganisms differs substantially in the two products.

bak-ter"i-si'dal
bak-te"re-o-stat'ik

Antiseptics and disinfectants are usually bactericidal, but occasionally they may be bacteriostatic. A **bactericidal agent** kills microorganisms, while a **bacteriostatic agent** temporarily prevents their further multiplication without necessarily killing them. For example, a bactericidal chemical may inactivate the enzymes of an organism and interfere with its metabolism so that it dies. A bacteriostatic chemical, by contrast, disrupts a minor chemical reaction and slows the metabolism, resulting in a longer time between divisions. Although a delicate difference sometimes exists between the bactericidal and bacteriostatic nature of a chemical, the terms indicate effectiveness of the chemical agent in a particular situation.

The word **sepsis** is derived from the Greek *seps*, meaning putrid. It refers to contamination of an object by microorganisms and forms the stems for "septicemia," meaning microbial infection of the blood, and "antiseptic," which translates to "against infection." It is also the origin of the term **aseptic**, meaning "free of contaminating microorganisms."

Aseptic:
free of contaminating
microorganisms.

Sanitize:
to reduce the microbial
population to a safe level as
determined by local public
health standards.

Degerm:
to remove microorganisms
from a surface.

Other expressions associated with chemical control are sanitize and degerm. To **sanitize** an object is to reduce the microbial population to a safe level as determined by local public health standards. For example, in dairy and food processing plants, the equipment is usually sanitized. To **degerm** something is merely to remove organisms from a surface. Washing with soap and water degerms the skin surface but has little effect on microorganisms deep in the skin pores.

A final group of terms that bears mention is the -cidal agents. These include the fungicidal agents, which kill fungi; the virucidal agents, for viruses; the sporicidal agents, for bacterial spores; and the germicidal agents, for various types of microorganisms.

Selection of Antiseptics and Disinfectants

To be useful as an antiseptic or disinfectant, a chemical agent must have certain properties, some of which are more desirable than others. The first prerequisite is that it must be able to kill microorganisms. It should also be nontoxic to animals or humans, especially if it is used as an antiseptic. It should be soluble in water and have a substantial shelf life during which its activity is retained. The agent should be useful in very diluted form and perform its job in a relatively short time. Both factors tend to reduce toxic side effects.

FIGURE 22.2
Sample Uses of Antiseptics and Disinfectants

(a) Antiseptics are used on body tissues, such as on a wound or before piercing the skin to take blood. (b) A disinfectant is used on inanimate objects such as a table top or equipment used in an industrial process.

(a) Antiseptic

(b) Disinfectant

Other characteristics will also contribute to the value of a chemical agent: it should not separate on standing; it should penetrate well; and it should not corrode instruments. The chemical will have a distinct advantage if it does not combine with organic matter such as blood or feces, because microorganisms are often found in this type of material. Of course, the chemical should be easy to obtain and relatively inexpensive.

Since disinfection is essentially a chemical process, the **parameters of chemistry** should be considered when selecting an antiseptic or disinfectant. For example, the temperature at which the disinfection is to take place may be important because a chemical reaction occurring at 37° C (body temperature) may not occur at 25° C (room temperature). Also, a particular chemical may be effective at a certain pH but not another. Moreover, the chemical reaction may be very rapid with one agent and slower with another. Thus if long-term disinfection is desired, the second agent may be preferable.

Two other considerations are the type of microorganism to be eliminated and the surface treated. For instance, the removal of bacterial spores requires more vigorous treatment than the removal of vegetative cells. Also, a chemical applied to a laboratory bench is considerably different from one used on a wound. It is therefore imperative to distinguish the antiseptic or disinfectant nature of a chemical before proceeding with use. Indeed, chemical agents formulated as disinfectants are regulated and registered by the federal Environmental Protection Agency, while chemicals formulated as antiseptics are regulated by the federal Food and Drug Administration.

Evaluation of Antiseptics and Disinfectants

Phenol coefficient:
a measure of the effectiveness of an antiseptic or disinfectant as compared to phenol.

At the current time, the United States Environmental Protection Agency lists over 8000 disinfectants for hospital use and thousands more for general use. Evaluating these chemical agents is a tedious process because of the broad diversity of conditions under which they are used.

A standard of effectiveness used for the chemical agents is the **phenol coefficient (PC)**. This is a number that indicates the disinfecting ability of an antiseptic or disinfectant in comparison to phenol under identical conditions (Table 22.1). A PC higher than 1 indicates that the chemical is more effective than phenol; a number less than 1 indicates poorer disinfecting ability than phenol. For example, antiseptic A may have a PC of 78.5, while antiseptic B has a PC of 0.28. These numbers are used relative to each other rather than to phenol because phenol is allergenic and irritating to tissues and thus is rarely used.

The phenol coefficient is determined by a laboratory procedure in which dilutions of phenol and the test chemical are mixed with standardized bacteria such as *Staphylococcus aureus* and *Salmonella typhi* or other species (Figure 22.3). The laboratory technician then determines which dilutions have killed the organisms after a 10-minute exposure but not after a 5-minute exposure. The test has many **drawbacks**, especially since it is performed in the laboratory rather than in a real-life situation. Nor does it take into account such factors as tissue toxicity, activity in the presence of organic matter, or variations of temperature.

A more practical way of determining the value of a chemical agent is by an **in-use test**. For example, swab samples of a floor may be taken before and after application of a disinfectant to determine the level of kill. Another method is to dry standardized cultures of bacteria within small stainless steel cylinders and then expose the cylinders to the test chemical. After an established time, the bacteria are assayed for survival to the chemical. These methods of standardization have value under certain circumstances, but it is conceivable that a universal test may never be developed in view of the huge variety of chemical agents available and the numerous conditions under which they are employed.

FIGURE 22.3

A Scanning Electron Micrograph of *Salmonella typhimurium* (× 40,000)

This Gram-negative bacillus caused thousands of cases of milkborne salmonellosis in the midwestern United States in the spring of 1985.

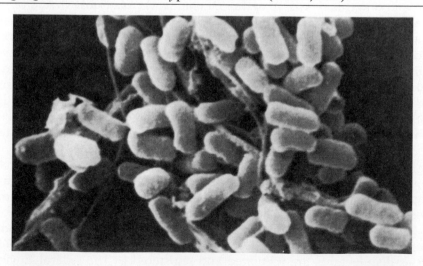

TABLE 22.1

Phenol Coefficients of Some Common Antiseptics and Disinfectants

CHEMICAL AGENT	STAPHYLOCOCCUS AUREUS	SALMONELLA TYPHI
Phenol	1.0	1.0
Chloramine	133.0	100.0
Tincture of iodine	6.3	5.8
Lysol	5.0	3.2
Mercury chloride	100.0	143.0
Ethyl alcohol	6.3	6.3
Formalin	0.3	0.7
Hydrogen peroxide	—	0.01

TO THIS POINT

■

We have examined some of the history and origins of disinfection practices with emphasis on the works of Ignaz Semmelweis and Joseph Lister. We saw how chemicals have been used since ancient times and how increased use occurred in the 1800s as the germ theory of disease was widely propagated. We then outlined some of the important terms applied to chemical agents and distinguished disinfectants used on lifeless objects from antiseptics applied to the skin surface. Sterilization and disinfection were also compared. We explored certain specifications for selecting antiseptics and disinfectants, including characteristics of the chemical agent, the conditions under which it is used, and the type of microorganism it is intended to kill.

The discussion then turned to methods for evaluating the effectiveness of antiseptics and disinfectants with emphasis on the phenol coefficient method. This method is limited because it is a laboratory test and does not necessarily reflect the conditions under which the chemical is to be used. Some in-use tests were briefly mentioned.

We shall now survey the broad spectrum of chemical agents used for disinfection. You will note an equally broad set of applications for these agents. We shall focus on the materials for which the chemical agents are employed and the mechanisms by which they kill microorganisms. Few agents in the first group to be studied are sterilizing agents.

Important Chemical Agents

The chemical agents currently in use for controlling microorganisms range from the very simple substances such as halogen ions to the very complex compounds typified by the detergents. Many of these agents in nature have been employed for generations (MicroFocus: 22.2), while others represent the latest developments. In this section we shall survey several groups of chemical agents and indicate how they are best applied in the chemical control of microorganisms.

22.2

MICROFOCUS

A CLOVE FOR ALL REASONS

Some people love it; some people loathe it. But nearly everyone has an opinion about the merits of garlic. Moreover, their opinions are often as strong as garlic itself.

The garlic controversy has been going on almost as long as garlic has been known to exist. The Egyptians had a love–hate relationship with the bulb. It was fed to slaves to build up their energy during construction of the pyramids, but people with garlic on their breath were forbidden to enter Egyptian temples because garlic was considered unclean. The Greeks perpetuated this attitude by feeding garlic to athletes, but they forbade anyone who ate it to enter holy places.

Garlic was also part of the pharmacopoeia of the Romans. Physicians applied it to wounds as an antiseptic and prescribed it for people who had respiratory problems, high blood pressure, or parasites. Moreover, Roman soldiers consumed it before battle to instill courage.

In recent years, a new group of garlic users has emerged. One scientist, for example, has suggested that joggers eat garlic to offset the pollutants in automobile exhaust fumes. His studies show that garlic binds up lead, mercury, and cadmium, and allows these minerals to pass in the feces. A West German doctor has written that in blood vessels garlic helps break up the cholesterol that might lead to atherosclerosis. In China, researchers reported that 16 cases of meningitis due to *Cryptococcus neoformans* responded to treatment with garlic alone. While the percentage of recoveries was lower than with drugs, nevertheless, the results of garlic treatment were said to be promising.

A central European superstition with blurred origins decrees that vampires cannot rise from their coffins if their mouths are crammed with garlic. For centuries, Asian peoples have eaten garlic to purify their complexions and strengthen their intellects, and in many cultures throughout the world garlic is believed to have aphrodisiac powers. In addition, people in Balkan countries such as Bulgaria and Yugoslavia consume bulbs of garlic as a regular part of their diet, and their reputation for longevity is well known.

Garlic lovers should note that only fresh garlic achieves the desired medicinal effect. Garlic powder and garlic salt are pale substitutes. In Russia, a garlic extract named allicin is used as a type of antibiotic. The *Merck Index* lists allicin as an "antibacterial principle of garlic." Its odor quickly reveals its origin.

Halogens

Halogen:
a highly reactive element whose atoms have seven electrons in the outer shell.

The **halogens** are a group of highly reactive elements whose atoms have seven electrons in the outer shell. Two halogens, chlorine and iodine, are commonly used for disinfection.

Chlorine is available in a gaseous form and as organic and inorganic compounds. It is widely used in municipal water supplies, where it keeps bacterial populations at low levels. Chlorine combines readily with numerous ions in water and therefore, enough chlorine must be added to ensure that a residue remains for antibacterial activity. In municipal water, the residue of chlorine is usually about 0.2 to 1.0 parts per million (ppm) of free chlorine. One ppm is equivalent to 0.0001 percent, an extremely small amount.

Chlorine is also available as **sodium hypochlorite** (NaOCl) or as **calcium hypochlorite** ($Ca(OCl)_2$). The latter, also known as chlorinated lime, was used by Semmelweis in his studies in Vienna. Hypochlorite compounds release free chlorine in solution. They are typified by the 0.5 percent sodium hypochlorite solution of H.D. Dakin used extensively for wounds sustained in World Wars I and II. Dakin's solution remains popular in Europe, where it is used to treat athlete's foot.

Athlete's foot:
a fungal disease accompanied by thin-walled blisters on the skin surface, usually of the feet.

FIGURE 22.4

Some Practical Applications of Disinfection with Chlorine Compounds

Sodium hypochlorite is used as a bleaching agent in the textile industry; commercially available **bleach** contains about 5 percent of this compound. To disinfect **clear water**, the Centers for Disease Control and Prevention recommends a half-teaspoon of household chlorine bleach in 2 gallons of water, with 30 minutes contact time before consumption. Hypochlorites are also useful in very dilute solutions for disinfecting swimming pools and sanitizing factory equipment (Figure 22.4).

The **chloramines** such as chloramine-T are organic compounds that contain chlorine. These compounds release free chlorine more slowly than hypochlorite solutions and are more stable. They are valuable for general wound antisepsis and root canal therapy.

Chlorine is effective against a broad variety of organisms including most Gram-positive and Gram-negative bacteria, and many viruses, fungi, and protozoa. However, it is not sporicidal. In microorganisms, the halogen is believed to cause the release of atomic oxygen, which then combines with and inactivates certain cytoplasmic proteins such as enzymes. Another theory is that chlorine changes the structure of cell membranes, thus leading to leakage.

The **iodine** atom is slightly larger than the chlorine atom and is more reactive and more germicidal. It is widely found in nature in such plants as marine seaweeds, where it is bound to chemical compounds. Iodine acts by halogenating tyrosine portions of protein molecules.

Tincture of iodine, a commonly used antiseptic for wounds, consists of 2 percent iodine and sodium iodide dissolved in ethyl alcohol. For the disinfection of **clear water**, the CDC recommends 5 drops of tincture of iodine in 1 quart of water, with 30 minutes contact time before consumption. Iodine compounds in different forms are also valuable sanitizers for restaurant eating utensils and equipment.

klo′rah-měn
Chloramine:
a disinfectant compound containing chlorine and amino groups.

Tincture of iodine:
a 2% iodine solution in ethyl alcohol.

FIGURE 22.5

When Disinfectants Fail to Work

Scanning electron micrograph of rod-shaped and coccobacillary cells embedded in the interior surface of a pipe located in a manufacturing plant where iodine disinfectants were produced. Investigators located the bacteria in the pipe after a series of nosocomial diseases were traced to a contaminated iodine product made at the plant. Bar = 2 μm.

i-o'do-for

po'vĭ-dōn

soo"do-mo'nas
se-pa'she-ah

Iodophors are iodine-detergent complexes that release iodine over a long period of time and have the added advantage of not staining tissues or fabrics. The detergent portion of the complex loosens the organisms from the surface and the halogen kills them. Some examples of iodophors are Wescodyne, used in preoperative skin preparations; Ioprep, for presurgical scrubbing; Iosan, for dairy sanitation; and Betadine, for local wounds. Iodophors may also be combined with nondetergent carrier molecules. The best known carrier is povidone, which stabilizes the iodine and releases it slowly. However, compounds like these are not self-sterilizing (Figure 22.5). In 1989, for example, four cases of peritoneal *Pseudomonas cepacia* infection were related to a contaminated povidone-iodine product.

Phenol and Phenolic Compounds

Phenol and phenolic compounds (phenolics) have played a key role in disinfection practices since Joseph Lister used them in the 1860s. Phenol remains the standard against which other antiseptics and disinfectants are evaluated in the phenol coefficient test. It is active against Gram-positive bacteria, but its activity is reduced in the presence of organic matter. Biochemists believe that phenol and its derivatives act by coagulating proteins, especially in the cell membrane.

Phenol is expensive, has a pungent odor, and is caustic to the skin, and therefore the role of phenol as an antiseptic has diminished. However, phenol

Cresols:
phenol derivatives containing methyl groups.

derivatives called **cresols** have greater germicidal activity and lower toxicity than the parent compound. Mixtures of ortho-, meta-, and para-cresol (creosote)

FIGURE 22.6
Phenol Derivatives

The chemical structures of some important derivatives of phenol used in disinfection and antisepsis.

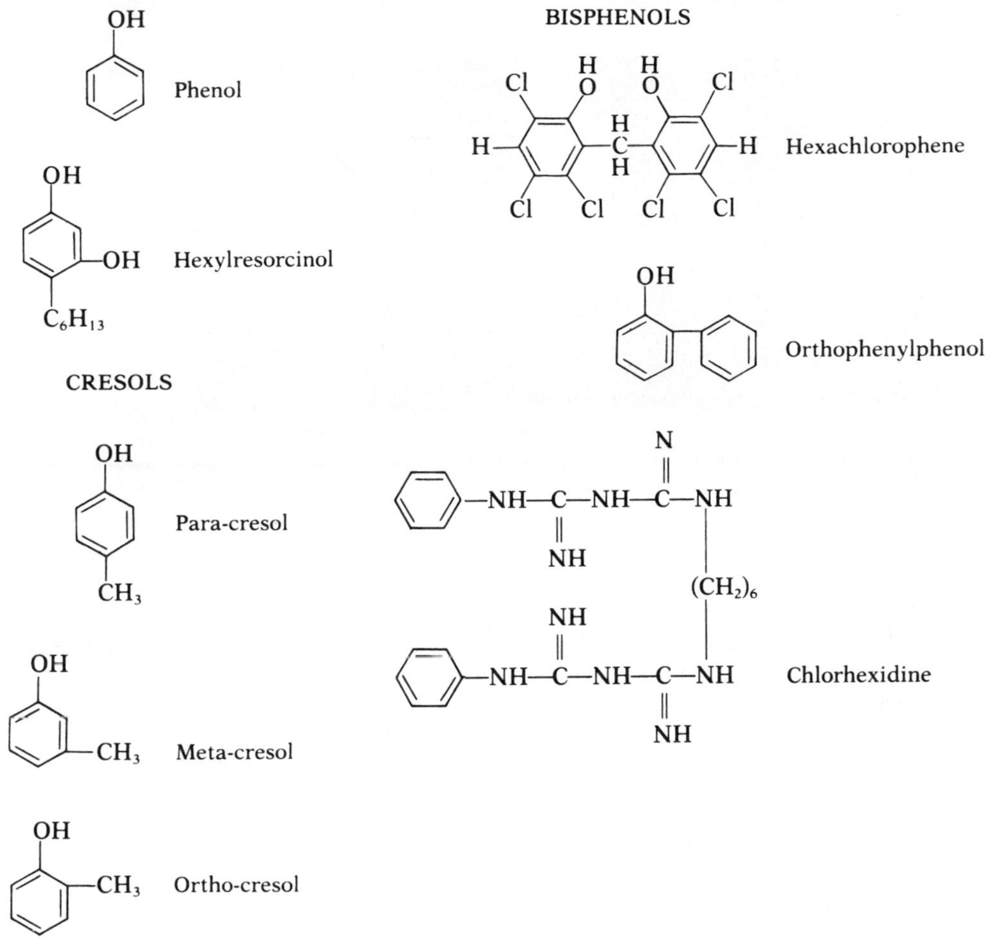

are used commercially as wood preservatives for railroad ties, fence posts, and telephone poles.

Combinations of two phenol molecules called **bisphenols** are prominent in modern disinfection and antisepsis. Orthophenylphenol, for example, is used in Lysol, Osyl, Staphene, and Amphyl. Another bisphenol, **hexachlorophene**, was used extensively in the 1950s and 1960s in such products as toothpaste (Ipana), underarm deodorant (Mum), and bath soap (Dial). One product, **pHiso-Hex**, combined hexachlorophene with a pH-balanced detergent cream. Pediatricians recommended it to retard staphylococcal infections of the scalp and umbilical stump, and for general cleansing of the newborn. However, a late 1960s study indicated that excessive amounts could be absorbed through the skin and cause neurological damage in newborn rats and monkeys, and hexachlorophene was subsequently removed from over-the-counter products. PHiso-Hex is still available, but only by prescription.

An important bisphenol relative is **chlorhexidine** (Figure 22.6). This compound was approved in 1976 by the Food and Drug Administration (FDA) for use as a surgical scrub, hand wash, and superficial skin wound cleanser. A 4

or"tho-fos"fo-fē'nol
hek"sah-klo'ro-fēn

klor-heks'i-dēn

FIGURE 22.7

Contamination

A scanning electron micrograph of the inner surface of a plastic chlorhexidine bottle contaminated with *Serratia marcescens*. Epidemiologists located these bacilli following an outbreak of *S. marcescens* infections in a hospital. The study indicates that *S. marcescens* survives in chlorhexidine and may be transferred in the disinfectant. Bar = 5 μm.

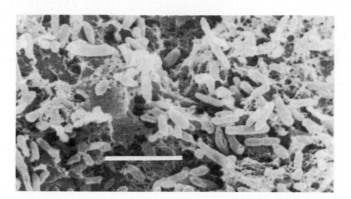

percent chlorhexidine solution in isopropyl alcohol is commercially available as **Hibiclens**. The chemical is believed to act on the cell membrane of Gram-positive and Gram-negative bacteria. Chlorhexidine in a concentration of 0.2 percent is also the most extensively tested and most effective antiplaque and antigingivitis agent. However, there is evidence that bacteria may grow within it (Figure 22.7).

hek″sil-rĕ-sor′sĭ-nol

Another phenol derivative, **hexylresorcinol**, is used in a mouthwash and topical antiseptic (ST37) and in throat lozenges (Sucrets). It has the added advantage of reducing surface tension, thereby loosening bacteria from the tissue and allowing greater penetration of the germicidal agent.

Heavy Metals

The term oligodynamic action, meaning "small power," expresses the activity of heavy metals such as mercury, silver, and copper on microorganisms. The elements are called **heavy metals** because of their large atomic weights and complex electron configurations.

Heavy metal:
an electron-donating element whose atoms are large, with complex electron arrangements.

Mercury is one of the older antiseptics, having been used as mercuric chloride ($HgCl_2$) by the Greeks and Romans for treating skin diseases. However, mercury is very toxic to the host, and the antimicrobial activity of mercury is reduced when other organic matter is present. In certain products such as Mercurochrome, Merthiolate, and Metaphen, mercury is combined with carrier compounds and is less toxic when applied to the skin, especially after surgical incisions.

Copper is active against chlorophyll-containing organisms and is a potent inhibitor of algae. As **copper sulfate ($CuSO_4$)**, it is incorporated into algicides and is used in swimming pools and municipal water supplies. Copper sulfate is also mixed with lime to form the bluish-white Bordeaux mixture used since 1882 to control the growth of fungi (MicroFocus: 22.3).

Silver in the form of **silver nitrate ($AgNo_3$)** is useful as an antiseptic and disinfectant. For example, one drop of a 1 percent silver nitrate solution may be placed in the eyes of newborns to protect against infection by *Neisseria*

FROM OUT OF THE BLUE

One October day in 1882, Professor Alexis Millardet of the University of Bordeaux was strolling through a vineyard in Medoc, France. Downy mildew of grapes, a fungal disease, had been widespread that year, and Millardet was surprised to note that the grapevines beside his path were rather healthy. By contrast, those most everywhere else were ill with disease. He paused to examine the leaves and found them covered with a bluish-white deposit.

Millardet inquired of the owner what the deposit might be. The owner responded that it was customary to spray the vines with a mixture of copper sulfate (blue) and lime (white) to make them look poisonous. This would quickly deter any would-be thieves from helping themselves to the grapes.

Millardet's curiosity was aroused. It was clear that the pathway vines were greener and healthier than their counterparts farther off in the field. He therefore devised a set of experiments to test the antifungal properties of the copper sulfate and lime mixture. Within three years he concluded that the combination of chemicals was an effective deterrent to mildew, and the now famous "Bordeaux mixture" came into being. Today the mixture is one of the most widely used fungicides in all the world.

The story has an ironic twist. While mildew was ravaging French vines, other Mediterranean countries hurriedly planted their own vines anticipating the collapse of the French wine industry. But the collapse never came. The Bordeaux mixture saved French vines and left the neophyte grape growers with lots of grapes but nowhere to sell them.

---

gonorrhoeae. This Gram-negative diplococcus can cause blindness if contracted by a newborn during passage through the birth canal. In 1884, Karl S.F. Credé first used silver nitrate to prevent gonococcal eye disease; many states now require that Credé's method be followed (although erythromycin or tetracycline ointments are often substituted since silver nitrate can cause irritation). Silver compounds are also used to treat suturing threads.

kra-dā'

Heavy metals are very reactive with proteins, particularly at the protein's sulfhydryl groups (—SH), and they are believed to bind protein molecules together by forming bridges between the groups. Because many of the proteins involved are enzymes, the cellular metabolism is disrupted, and the microorganism dies. However, the heavy metals are not sporicidal.

sul-fi'dril

Alcohols

Alcohols are effective skin antiseptics and are valuable disinfectants for medical instruments. For practical use, the preferred alcohol is ethyl alcohol. **Ethyl alcohol** is active against vegetative bacterial cells, including the tubercle bacillus, but it has no effect on spores. It denatures proteins and dissolves lipids, an action that may lead to cell-membrane disintegration. Ethyl alcohol also is a strong dehydrating agent.

Ethyl alcohol: a 2-carbon consumable alcohol used as an antiseptic and disinfectant.

Because ethyl alcohol reacts readily with any organic matter, medical instruments and thermometers must be thoroughly cleaned before exposure. Usually, a 50 to 80 percent alcohol solution is recommended since water prevents rapid evaporation and assists penetration into the tissues. A 10-minute immersion in 70 percent ethyl alcohol is generally sufficient to disinfect a thermometer or delicate instrument.

Alcohol is used to preserve cosmetics, and to treat skin before a venipuncture or injection. It mechanically removes bacteria from the skin and dissolves lipids to clean the skin. **Isopropyl alcohol**, or rubbing alcohol, has high bactericidal activity in concentrations as high as 99 percent. Methyl alcohol is toxic to the tissues and is used infrequently.

i"so-pro'pil

TO THIS POINT
■

We have discussed four major groups of chemical agents used for disinfectant and antiseptic purposes. Halogen compounds, which include chlorine and iodine, are effective against most microorganisms. The halogens can be used either in elemental forms or as derivatives such as chloramines and iodophors. Phenol is rarely used in modern disinfection practices but the phenolic compounds such as bisphenols are valuable skin cleansers. The heavy metals include mercury, copper, and silver. Silver nitrate is often used to prevent the transmission of gonorrhea to newborns. Alcohols are valuable for the disinfection of instruments, especially as 70 percent ethyl alcohol.

Through the discussions of these chemical agents you may have noted certain general considerations that apply to all. For example, the chemical agent will usually combine with any organic matter present, so that cleanliness is an important prerequisite to disinfection. Also, the chemical agents react with a wide spectrum of organisms, not just one type. Moreover, they are generally useless against bacterial spores. Virtually no chemical agent is useful within the human body, but most are employed to interrupt the spread of organisms outside the body.

In the final section of this chapter we shall discuss three chemical agents used for sterilization purposes. These agents are sporicidal if sufficient time is given for them to act and proper conditions are established. We shall also mention a group of other agents that are used on the skin surface and in wounds.

Other Chemical Agents

The chemical agents we have discussed in previous sections are best recognized as disinfectants and antiseptics. Augmenting these are some chemicals that can be used for sterilization purposes, especially for modern high-technology equipment (as well as the rather mundane Petri dish). Three such agents are considered next.

Formaldehyde

Formalin:
a 37% solution of formaldehyde in water.

Formaldehyde is a gas at high temperatures and a solid at room temperatures. When 37 grams of the solid are suspended in 100 ml of water, a solution called **formalin** results. For over a century formalin has been used in embalming fluid for anatomical specimens and by morticians as well as for disinfecting purposes (MicroFocus: 22.4). In microbiology, formalin is utilized for inactivating viruses in certain vaccines and producing toxoids from toxins.

In the gaseous form, formaldehyde is expelled into a closed chamber where it is a sterilant for surgical equipment, hospital gowns, and medical instruments. However, penetration is poor, and the surface must be exposed to the gas for up to 12 hours for effective sterilization. Instruments can be sterilized by placing them in a 20 percent solution of formaldehyde in 70 percent alcohol for 18 hours. Formaldehyde, however, leaves a residue, and instruments must be rinsed before use. Many allergic individuals develop a contact dermatitis to this compound (Chapter 20).

22.4

MICROFOCUS

FOR PEOPLE WHO WEAR SHOES

Here's some good news for people who can't bear to part with those extraordinarily comfortable but o' so smelly shoes: You can get rid of the smell and enjoy many more years with them. But before you do anything, you should understand what's going on.

The first thing you should know is that shoes become smelly when odors accumulate in the material of the shoe. The odors come from gases that bacteria produce while growing in the material. It seems that airborne cocci, recently identified as members of the genus *Micrococcus*, thrive in the sweat produced by the feet (about a gallon per week in some people). This sweat is absorbed by the shoe's material. The bacteria break down the sweat's organic components and produce sulfur compounds not unlike the hydrogen sulfide in a swamp or landfill. These sulfur compounds gather in the material and produce the odor we turn up our nose at.

So what to do? First, try to rotate your shoes as often as possible and let them air out as long as feasible between wearings. (Some of the gas will dissipate.) Try to wear cotton, silk, or other natural fabric socks rather than synthetic materials because bacteria thrive better in synthetic materials. This is because synthetics retain more heat, increase sweating, and limit evaporation. And use a foot powder to absorb sweat to make life difficult for the micrococci.

Now, for those old shoes. Buy some formaldehyde at a local pharmacy and try wiping the insides of the shoes with it. Be sure to follow all the precautions that come with the formaldehyde because it can be poisonous. You can also roll up a rag, stuff it inside the shoe, and soak it with the formaldehyde. Place the shoe in a closed environment such as a box and leave it for a few days in an airy environment where the fumes can disperse. Be sure to dry out the

shoes before wearing them again. And if it works, score another one for the disinfectants.

Formaldehyde is an **alkylating agent**. It reacts with amino and hydroxyl groups of nucleic acids and proteins, and with carboxyl and sulfhydryl groups in proteins by inserting between them a small carbon fragment (an alkyl group) and forming bridges (Figure 22.8). This insertion changes the structures of the molecules and interferes with an organism's chemistry, thereby leading to death. We shall discuss two other alkylating agents, ethylene oxide and glutaraldehyde, in the next paragraphs.

Sulfhydryl group: a chemical group consisting of a sulfur and a hydrogen atom.

Ethylene Oxide

The development of plastics for use in microbiology required a suitable method for sterilizing these heat-sensitive materials. In the 1950s, research scientists discovered the antimicrobial abilities of **ethylene oxide (EtO)** and essentially made the plastic Petri dish and plastic syringe possible.

Ethylene oxide is a small molecule with excellent penetration capacity and sporicidal ability. However, it is both toxic and highly explosive. Its explosiveness is reduced by mixture with Freon gas in Cryoxide or carbon dioxide gas in Carboxide, but its toxicity remains a problem for those who work with it. The gas is released into a tightly sealed chamber where it circulates for up to 4 hours with carefully controlled humidity. The chamber must then be flushed with inert gas for 8 to 12 hours to ensure that all traces of EtO are removed, otherwise the chemical will cause "cold burns" on contact with the skin.

FIGURE 22.8
Alkylating Agents

The chemistry that takes place between alkylating agents and other molecules. (a) Formaldehyde reacts with amino groups on protein molecules and forms bridges between adjacent groups. (b) Ethylene oxide reacts with nucleic acid molecules and forms bridges between adjacent groups. The effect in both cases is to alter the structure of the molecules and change their chemistry.

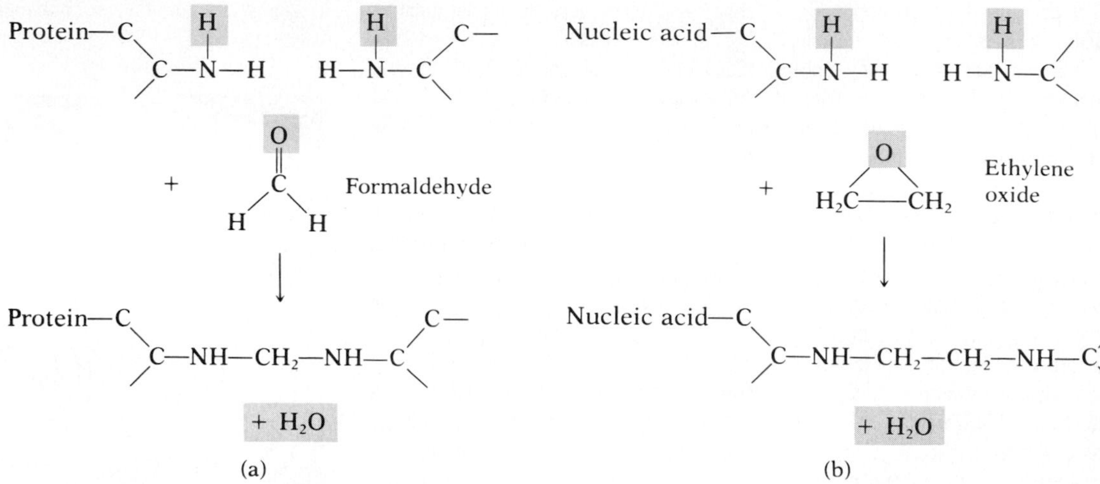

(a)   (b)

Ethylene oxide has been employed to sterilize paper, leather, wood, metal, and rubber products, as well as plastics. In the hospital it is used to sterilize catheters, artificial heart valves, heart–lung machine components, and optical equipment. The National Aeronautics and Space Administration (NASA) uses the gas for sterilization of interplanetary space capsules. Ethylene oxide chambers have become chemical counterparts of autoclaves for sterilization procedures. Often they are called gas autoclaves.

pro"pe-o-lak'tōn

A closely related gas, **beta-propiolactone (BPL)**, is less explosive than ethylene oxide, but its penetrating capacity is more limited. As a liquid it is used to sterilize vaccines, sera, and surgical ligatures. However, scientists have shown it to be carcinogenic, and BPL is therefore used only under restricted circumstances.

Carcinogenic:
able to cause cancer.

Glutaraldehyde

gloo"tah-ral'dĕ-hīd

Glutaraldehyde has become one of the most effective chemical liquids for sterilization purposes. This small molecule destroys vegetative cells within 10 to 30 minutes and spores in 10 hours. Glutaraldehyde is an alkylating agent, usually employed as a 2 percent solution. To use it for sterilization, materials have to be precleaned, immersed for 10 hours, rinsed thoroughly with sterile water, dried in a special cabinet with sterile air, and stored in a sterile container to ensure that the material remains sterile. If any of these parameters is altered, the materials may be disinfected but may not be considered "sterile."

Because the activity of glutaraldehyde is not greatly reduced by organic matter, the chemical is recommended for use on surgical instruments where residual blood may be present. In addition, glutaraldehyde does not damage delicate objects, and therefore it can be used to sterilize optical equipment such as the optic fiber endoscopes used for arthroscopic surgery. It gives off irritating fumes, however, and instruments must be rinsed thoroughly in sterile water.

At pH 7.5, glutaraldehyde kills *S. aureus* in 5 minutes and *M. tuberculosis* in 10 minutes.

Hydrogen Peroxide

Hydrogen peroxide (H_2O_2) is used as a rinse in wounds, scrapes, and abrasions. The area foams and effervesces as **catalase** in the tissue breaks down hydrogen peroxide to oxygen and water. The furious bubbling removes microorganisms mechanically. Hydrogen peroxide decomposition also results in a reactive form of oxygen—the superoxide radical—highly toxic to microorganisms. Anaerobic bacteria are sensitive to hydrogen peroxide because the sudden release of oxygen gas inhibits their growth.

New forms of H_2O_2 such as Super D Hydrogen Peroxide are more stable than traditional forms and therefore do not decompose spontaneously. Such inanimate materials as soft contact lenses, utensils, heat-sensitive plastics, and food processing equipment can be disinfected within 30 minutes. Sterilization can also be achieved after 6 hours of exposure to a 6 percent solution. Microbiologists are currently researching the use of hydrogen peroxide for destroying bacteria in milk and milk products.

> Hydrogen peroxide:
> a simple chemical compound digested by catalase to water and oxygen.

Soaps and Detergents

A **soap** is a chemical compound of fatty acids combined with potassium or sodium hydroxide. The pH of the compound is usually about 8.0, and some microbial destruction is therefore due to the alkaline conditions established on the skin. However, the major activity of soap is as a degerming agent for the mechanical removal of microorganisms from the skin surface.

> Soap:
> a compound of fatty acids and potassium or sodium hydroxide.

Soaps are **wetting agents**, that is, they emulsify and solubilize particles clinging to a surface. The surface tension is also reduced by the soap. In addition, soaps remove skin oils, further reducing the surface tension and increasing the cleaning action.

> Surface tension:
> the attraction between a surface and a particle lying on it.

Detergents are synthetic chemicals acting as strong wetting agents and surface tension reducers. Since they are actively attracted to the phosphate groups of cellular membranes, they also alter the membranes and encourage leakage from the cytoplasm.

Certain **anionic detergents** yield negatively charged ions in solution. The detergents are somewhat active against Gram-positive bacteria, but the negative charges of bacteria usually repel them and limit their use to common laundry products. Anionic detergents such as Triton W-30 and Duponol find value as iodophors.

> Anionic:
> yielding negatively charged ions in solution.

Other detergents are **cationic**. These derivatives of ammonium chloride contain four organic radicals in place of the four hydrogens, and at least one radical is a long-chain alkyl group (Figure 22.9). The positively charged ammonium group is counterbalanced by a negatively charged chloride ion. Such compounds are often called **quaternary ammonium compounds** or simply, quats.

> Cationic:
> yielding positively charged ions in solution.

The compounds in cationic detergents have rather long, complex names, such as benzalkonium chloride in Zephiran and cetylpyridinium chloride in Ceepryn. Other detergents are used in Phemerol and Diaparene. Cationic detergents are bacteriostatic on a broad range of bacteria, especially Gram-positive bacteria, and are relatively stable, with little odor. They are used as sanitizing agents for industrial equipment and food utensils, as skin antiseptics, in mouthwashes and storage solutions for contact lenses, and for disinfecting hospital walls and floors. Mixing with soap, however, reduces their activity, and certain Gram-negative bacteria such as *Pseudomonas aeruginosa* can grow in them.

> ben″zal-ko′ne-um
> se″til-pi″rĭ-din′e-um

> soo″do-mo′nas a″er-u-jin-o′sa

FIGURE 22.9

Cationic Detergents

The chemical structures of some important cationic detergents used in disinfection and antisepsis. Note that a long chain of carbon atoms, called an alkyl group, is included in each molecule, and that nitrogen is bonded to four radicals.

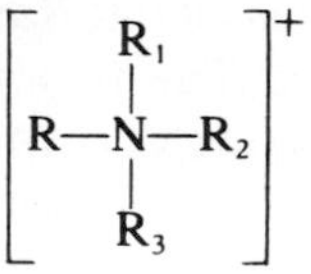

Basic structure of cationic detergents

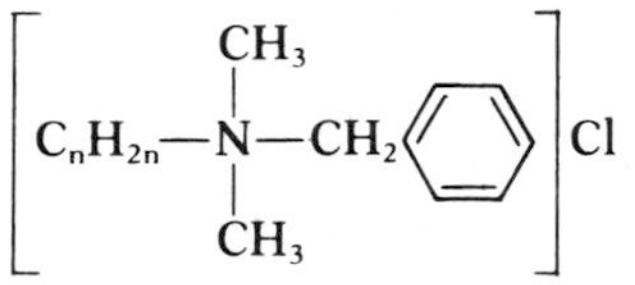

Benzalkonium chloride (Zephiran)

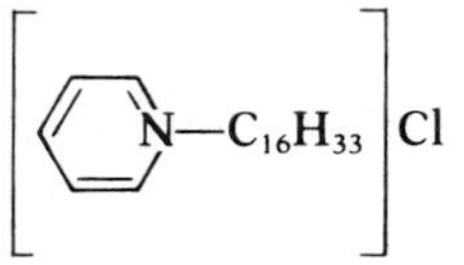

Cetylpyridinium chloride (Ceepryn chloride)

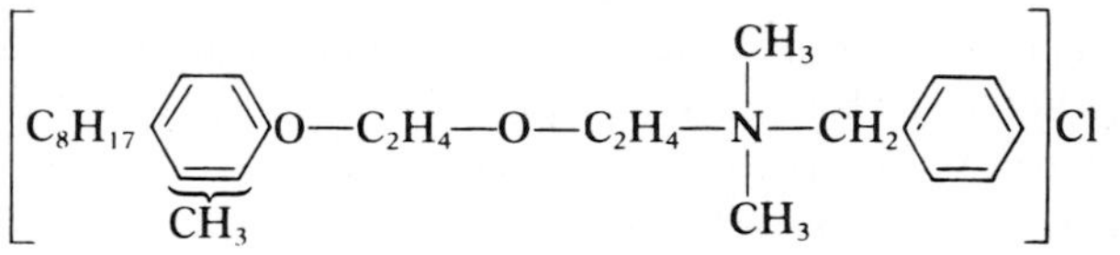

Methylbenzethonium chloride (Diaparene chloride)

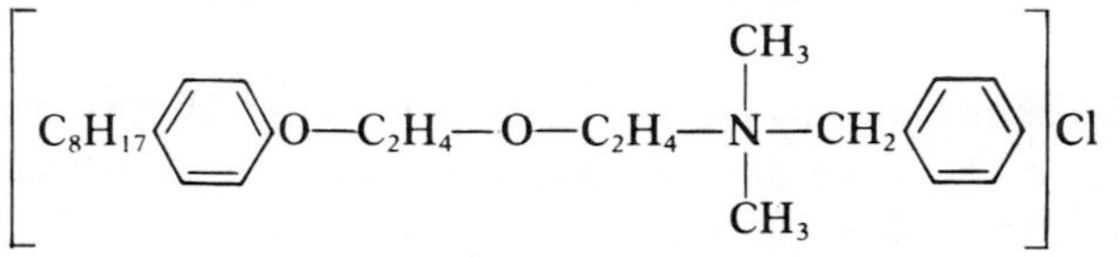

Benzethonium chloride (Phemerol chloride)

Dyes

tri-fen″il-meth′ān

Dyes are useful in microbiology as staining reagents and in laboratory media, where they help select out certain organisms from a mixture. A group of dyes called **triphenylmethane dyes** is also useful as antiseptics against species of *Bacillus* and *Staphylococcus*. The group includes malachite green and crystal violet. Crystal violet has also been used traditionally as **gentian violet** for trench mouth (Chapter 10) and for *Candida albicans* infections such as thrush. Interference with cell wall construction appears to be the mode of activity. The dye is bactericidal at very weak dilutions of less than 1:10,000.

ak″ri-fla′vin

A second group of dyes, the **acridine dyes**, includes acriflavine and proflavine. Both dyes are used as antiseptics for staphylococcal infections in wounds. They apparently act by combining directly with DNA, thereby halting RNA synthesis.

Acids

un″dec-ĕ-len′ik

Certain acids are useful as antiseptics or disinfectants. The popular ones include benzoic, salicylic, and undecylenic acids for tinea infections of the skin. These infections are caused by various species of fungi, as noted in Chapter 14.

Organic acids are particularly valuable as food preservatives. Lactic and acetic acids, for example, are important preservatives in sour foods such as cheeses, sauerkraut, and pickled products. Propionic acid is added to bakery products to keep microbial populations low. Chapter 24 discusses these preservatives in more detail.

In general, acid enhances the effects of disinfectants and antiseptics and makes them more soluble. Also, heat is a more potent sterilizing agent if acid conditions are present, because hydrolysis is increased. Acid is therefore a valuable adjunct to disinfection, albeit in an indirect way. Table 22.2 summarizes the chemical agents.

TABLE 22.2

A Summary of Chemical Agents Used to Control Microorganisms

CHEMICAL AGENT	ANTISEPTIC OR DISINFECTANT	MECHANISM OF ACTIVITY	APPLICATIONS	LIMITATIONS	ANTIMICROBIAL SPECTRUM
Chlorine	Chlorine gas Sodium hypochlorite Chloramines	Protein oxidation Membrane leakage	Water treatment Skin antisepsis Equipment spraying Food processing	Inactivated by organic matter Objectionable taste, odor	Broad variety of bacteria, fungi, protozoa, viruses
Iodine	Tincture of iodine Iodophors	Halogenates tyrosine in proteins	Skin antisepsis Food processing Preoperative preparation	Inactivated by organic matter Objectionable taste, odor	Broad variety of bacteria, fungi, protozoa, viruses
Phenols	Cresols Hexachlorophene Hexylresorcinol Chlorhexidine	Coagulates proteins Disrupts cell membranes	General preservatives Skin antisepsis with detergent	Toxic to tissues Disagreeable odor	Gram-positive bacteria Some fungi
Mercury	Mercuric chloride Merthiolate Metaphen	Combines with —SH groups in proteins	Skin antiseptics Disinfectants	Inactivated by organic matter Toxic to tissues Slow acting	Broad variety of bacteria, fungi, protozoa, viruses
Copper	Copper sulfate	Combines with proteins	Algicide in swimming pools Municipal water supplies	Inactivated by organic matter	Algae Some fungi
Silver	Silver nitrate	Binds proteins	Skin antiseptic Eyes of newborns	Skin irritation	Organisms in burned tissue Gonococci
Alcohol	70% ethyl alcohol	Denatures proteins Dissolves lipids Dehydrating agent	Instrument disinfectant Skin antiseptic	Precleaning necessary Skin irritation	Vegetative bacterial cells, fungi, protozoa, viruses
Formaldehyde	Formaldehyde gas Formalin	Reacts with functional groups in proteins and nucleic acids	Embalming Vaccine production Gaseous sterilant	Poor penetration Allergenic Toxic to tissues Neutralized by organic matter	Broad variety of bacteria, fungi, protozoa, viruses
Ethylene oxide	Ethylene oxide gas	Reacts with functional groups in proteins and nucleic acids	Sterilization of instruments, equipment, heat-sensitive objects	Explosive Toxic to skin Requires constant humidity	All microorganisms, including spores
Glutaraldehyde	Glutaraldehyde	Reacts with functional groups in proteins and nucleic acids	Sterilization of surgical supplies	Unstable Toxic to skin	All microorganisms, including spores

(continued)

TABLE 22.2 (*continued*)

A Summary of Chemical Agents Used to Control Microorganisms

CHEMICAL AGENT	ANTISEPTIC OR DISINFECTANT	MECHANISM OF ACTIVITY	APPLICATIONS	LIMITATIONS	ANTIMICROBIAL SPECTRUM
Hydrogen peroxide	Hydrogen peroxide	Creates aerobic environment Oxidizes protein groups	Wound treatment	Limited use	Anaerobic bacteria
Cationic detergents	Commercial detergents	Dissolve lipids in cell membranes	Industrial sanitization Skin antiseptic Disinfectant	Neutralized by soap	Broad variety of microorganisms
Triphenyl-methane dyes	Malachite green Crystal violet	React with cytoplasmic components	Wounds Skin infection	Residual stain	Staphylococci Some fungi Gram-positive bacteria
Acridine dyes	Acriflavine Proflavine	React with cytoplasmic components	Skin infection	Residual stain	Staphylococci Gram-positive bacteria
Acids	Benzoic acid Salicylic acid Undecylinic acid Lactic and propionic acids	Alter pH	Skin infections Food preservatives	Skin irritation	Many bacteria and fungi

NOTE TO THE STUDENT

In this chapter, we have focused on a broad variety of antiseptics and disinfectants but few, you will note, are of significant value on the body surface and virtually none are used to treat internal disease. Most chemical agents are used for equipment, instruments, materials, and other related purposes. The major reason so few are available as antiseptics is simple: chemicals will often do more harm to the human body than to the microorganisms they are intended to kill.

Witness the spectacular rise and fall of hexachlorophene during the 1960s and 1970s. This compound's antimicrobial capabilities, residual activity, and lack of side effects appeared too good to be true. Eventually, though, it was found to be hazardous to the tissues. Similarly, many thousands and thousands of possible antiseptics have fallen by the wayside. If you have a background in chemistry, then you know of the dazzling array of available chemicals. Yet how many are useful as antiseptics?

Summary

Chemical agents are effectively used to control the growth of microorganisms even though they do not achieve sterilization. Instead, they generally are able to destroy pathogenic microorganisms, a process called disinfection. A chemical agent used on a living object such as the body surface is an antiseptic; one used on a nonliving object such as a tabletop is a disinfectant. Both antiseptics and disinfectants are selected according to certain criteria and are evaluated by the phenol coefficient method.

Among the important chemical agents for disinfection are the halogens and phenols. Halogens such as chlorine and iodine are useful for water disinfection, wound antisepsis, and for various forms of sanitation. Phenol derivatives such as hexachlorophene and chlorhexidine are valuable skin antiseptics and presurgical scrubs. In both cases, the chemical agent reacts with most types of organic matter (including microorganisms), so precleaning is a necessary prerequisite to disinfection. Other useful chemical agents are heavy metals, such as silver in silver nitrate and copper in copper sulfate, and alcohol, which is commonly used in 70 percent solution of ethyl alcohol. Precleaning and complete immersion are required for alcohol use.

Chemical agents can also be used as sterilizing agents so long as a closed chamber is employed for exposure. Formaldehyde, ethylene oxide, and glutaraldehyde are examples of small molecules that unite with amino and hydroxyl groups in proteins and nucleic acids to interrupt the biochemistry of microorganisms. Hydrogen peroxide acts by releasing oxygen to cause an effervescing cleansing action, and detergents have a profound effect on microbial membranes. Other useful chemical agents include dyes and acids, the latter used as preservatives in foods. The broad variety of chemical agents usually ensures that an agent is available for each situation.

Questions for Thought and Discussion

1. Suppose you were in charge of a clinical laboratory where instruments are routinely disinfected and equipment is sanitized. A salesperson from a disinfectant company stops in to spur your interest in a new chemical agent. What questions might you ask the salesperson about the product?

2. In March 1982 six cases of *Pseudomonas aeruginosa* infection occurred in members of a North Carolina coed fraternity after bathing in a redwood hot tub. The infection was accompanied by skin rashes in various measures of severity. What precautions might be taken by hot-tub users to prevent a recurrence of this episode?

3. While on a camping trip, you find that a luxury hotel has been built near the stream where you once swam and from which you drank freely. Fearing contamination of the stream, you decide that before drinking the water some form of disinfection would be wise. The nearest town has only a grocery store, pharmacy, and post office. What might you purchase? Why?

4. European manufacturers have included chlorhexidine in their toothpastes and mouthwashes for many years, but there has been resistance to this practice in the United States. Would you favor or oppose such a move? Why?

5. A study reported in 1992 in the *New England Journal of Medicine* recounted the reluctance of patients to ask their physician a vitally important question before beginning any sort of treatment. Can you guess what that question is? (It is not "how much do you charge?")

6. A portable room humidifier can incubate and disseminate infectious microorganisms. If a friend asked your recommendations on disinfecting the humidifier, what might you suggest?

7. With over 11 million children currently attending day-care centers in the United States, the possibilities for disease transmission among children has mounted considerably. Under what circumstances may antiseptics and disinfectants be used to preclude the spread of microorganisms?

8. Ethylene oxide, glutaraldehyde, and formaldehyde are commonly called high-level germicides. By contrast, most antiseptics are low-level germicides. Why is the distinction made?

9. In 1912, J.W. Churchman coined the word "bacteriostasis" to indicate that certain dyes were inhibitory rather than destructive to bacteria. What does the word mean today and how is it applied to the germicides?

10. Before the advent of antibiotics, a certain dye was used to treat gum infections then known as trench mouth. The dye caused the tissues to become purple, and bacteriologists would occasionally make reference to "kissing a Gram-positive mouth." What was the dye? What has replaced it?

11. A student has finished his work in the laboratory and is preparing to leave. He remembers the instructor's precautions to wash and disinfect his hands before leaving. However, he cannot remember whether to wash first then disinfect, or to disinfect then wash. What advice might you give?

12. Before taking a blood sample from the finger, the blood bank technician commonly rubs the skin with a pad soaked in alcohol. Many people think that this procedure sterilizes the skin. Are they correct? Why?

13. A Clorox advertisement published in 1993 carries the following message: "Raw foods like chicken can carry germs and bacteria that cause salmonella sickness. It's important to kill the bacteria on any surface raw foods touch with a little Clorox. Soap and water won't do the trick." Immediately above the statement was a photograph of a raw chicken, a green pepper, three carrots, and a red Bermuda onion. At the bottom of the page a box describes "a little Clorox" as a solution made by mixing a sink full of water with 1/8 cup of Regular Clorox Liquid Bleach. How many things can you find wrong with this advertisement? Suppose you were the company microbiologist. What would you say in your version of the advertisement?

14. Suppose you had just removed the thermometer from the mouth of your sick child and confirmed your suspicion of fever. Before checking the temperature of the next child, how would you treat the thermometer to disinfect it?

15. Why is it a good idea to discard old bottles of tincture of iodine and hydrogen peroxide from the medicine cabinet?

Review

The chemical agents are a broad and diverse group as this chapter has demonstrated. To test your knowledge of the chapter contents, match the chemical agent on the right to the statement on the left by placing the correct letter in the available space. A letter may be used once, more than once, or not at all.

___ 1. The halogen in bleach.

___ 2. Sterilizes heat-sensitive materials

___ 3. Used to prevent gonococcal eye disease

___ 4. Part of chlorhexidine molecule

___ 5. Oxygen retards anaerobic bacteria

___ 6. Seventy percent concentration recommended

___ 7. Active ingredient in Betadine

___ 8. Quaternary compounds or "quats"

___ 9. Can induce a contact dermatitis

___ 10. Valuable food preservative

___ 11. Often used as a tincture

___ 12. Rinse for wounds and scrapes

___ 13. Example of a heavy metal

___ 14. Enhances antimicrobial activity of heat

___ 15. Two molecules in hexachlorophene

___ 16. Assists mechanical removal of microbes

___ 17. Exerts an oligodynamic action

___ 18. Used by Joseph Lister

A. iodine
B. ethylene oxide
C. hydrogen peroxide
D. ethyl alcohol
E. acid
F. chlorine
G. glutaraldehyde
H. soap
I. phenol
J. silver
K. cationic detergent
L. dye
M. anionic detergent
N. formaldehyde

___ 19. Used for plastic Petri dishes

___ 20. Benzoic and salicylic for tinea

___ 21. Found in Zephiran and Diaparene

___ 22. Used to purify waters

___ 23. Derivatives of ammonium compounds

___ 24. Active ingredient in Dakin's solution

___ 25. Broken down by catalase

CHAPTER 23

CHEMOTHERAPEUTIC AGENTS AND ANTIBIOTICS

For centuries, physicians believed that heroic measures were necessary to save patients from the ravages of infectious disease. They prescribed frightening courses of purges and bloodlettings, enormous doses of strange chemical concoctions, ice water baths, starvation, and other drastic remedies. These treatments probably complicated an already bad situation by reducing the natural body defenses to the point of exhaustion.

But a revolution in medicine took place about 1825 when a group of doctors in Boston and London experimented to see what would happen if such treatments were withheld from diseased patients. Surprisingly, they found the survival rate essentially the same and, in some cases, better. Over the next few decades, the lessons from their experiments spread, and as the worst features of heroic therapy disappeared, doctors adopted a conservative, nonmeddling approach to disease. It became the doctor's job to diagnose the illness, explain it to the family, predict what would happen in the next several days, and then stand by and care for the patient within the limits of what was known.

When the germ theory of disease emerged in the late 1800s, the information about microorganisms added considerably to the understanding of disease and increased the storehouse of knowledge available to the doctor. However, it did not change the fact that little, if anything, could be done for the infected patient (Figure 23.1). Tuberculosis continued to kill one of every seven people dead from all causes; and streptococcal disease was a fatal experience, as were pneumococcal pneumonia and meningococcal meningitis.

Then, in the 1940s, the chemotherapeutic agents and antibiotics burst on the scene, and another revolution in medicine began. Doctors were astonished to learn that they could kill bacteria in the body without doing substantial harm to the body itself. Medicine experienced a period of powerful, decisive therapy for infectious disease, and doctors found they could successfully alter the course of disease. The chemotherapeutic agents and antibiotics effected a radical change in medicine and charted a new course that has followed through to the present day.

FIGURE 23.1
The Doctor

A painting depicting the physician of the 1800s caring for his patient. Physicians were expected to recognize an illness, distinguish it from other illnesses, and minister to the patient as best they could. Without antibiotics, their ability to treat infectious diseases was severely limited. This 1891 painting by Luke Fildes is entitled *The Doctor.*

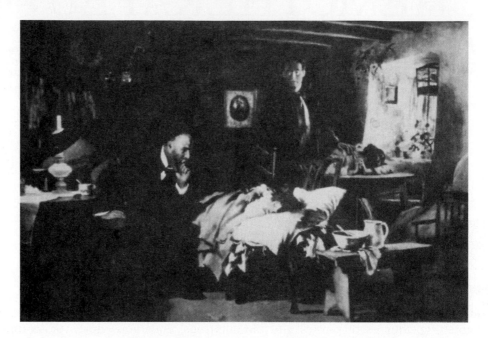

In this chapter we shall discuss the antimicrobial drugs that have become mainstays of our health-care delivery system. We shall explore their discovery and examine their uses, while noting the important side effects attributed to many of them. When Pasteur performed his experiments a hundred years ago, he implied that microorganisms could be destroyed and that some day, a way would be found to successfully treat many diseases. Only since the 1940s has Pasteur's prophecy become reality.

Chemotherapeutic Agents

Chemotherapeutic agent:
a chemical agent used in the body for therapeutic purposes.

Chemotherapeutic agents are chemical substances used within the body for therapeutic purposes. The term generally implies a chemical that has been synthesized by chemists or produced by a modification of a preexisting chemical. By contrast, an **antibiotic** is a product of the metabolism of a microorganism. We shall maintain the distinction in these pages, although many antibiotics are currently produced by synthetic or semisynthetic means and are more correctly "chemotherapeutic agents." Our discussion of chemotherapeutic agents will begin with a brief review of their development.

A Short History of Chemotherapy

In the drive to control and cure disease, the efforts of early microbiologists were primarily directed toward enhancing the body's natural defenses. Sera containing antibodies lessened the impact of diphtheria, typhoid fever, and

FIGURE 23.2
Paul Ehrlich

A painting by Robert Thom depicting Paul Ehrlich and Sahachiro Hata, the two investigators who developed arsphenamine for the treatment of syphilis. Ehrlich is shown writing a work order with the stubby colored pencil he habitually used. He and Hata conducted their experiments at the Institute of Experimental Therapy in Frankfurt, Germany.

tetanus; and effective vaccines for smallpox and rabies (and later, tuberculosis, diphtheria, and tetanus) reduced the incidence of these diseases.

Among the leaders in the effort to control disease was an imaginative investigator named **Paul Ehrlich**. Ehrlich envisioned antibody molecules as "magic bullets" that seek out and destroy disease organisms in the tissues without harming the tissues. His experiments in stain technology indicated that certain dyes also had antimicrobial qualities, and by the early 1900s, his attention had turned to magic bullets of a purely chemical nature.

One of Ehrlich's collaborators was the Japanese investigator **Sahachiro Hata**. Hata wished to perform research on the chemical control of the syphilis spirochete *Treponema pallidum*, and Ehrlich was happy to oblige. Previously, Ehrlich and his staff had synthesized hundreds of arsenic–phenol compounds, and Hata set to work testing them for antimicrobial qualities. After months of painstaking study, Hata's attention focused on **arsphenamine**, compound #606 in the series. Hata and Ehrlich (Figure 23.2) successfully tested arsphenamine against syphilis in animals and human subjects, and in 1910, they made a derivative of the drug available to doctors for use against the disease. Arsphenamine, the first modern chemotherapeutic agent, was given the common name **Salvarsan** because it offered salvation from syphilis and contained arsenic.

ars-fen′ah-min

Salvarsan met with mixed success during the ensuing years. Its value against syphilis was without question, but local reactions at the injection site and indiscriminate use by some physicians brought adverse publicity. Moreover, some church officials used the threat of syphilis as a deterrent to immoral behavior, and they were less than enthusiastic about Salvarsan's therapeutic effect. Ehrlich's death in 1915 together with the general ignorance of organic chemistry and the emerging world war further eroded enthusiasm for chemotherapy. Instead, interest strengthened in serum and vaccine therapy for war-related wounds and diseases.

Significant advances in chemotherapy would not be made for another 20 years. During this interval, German chemists continued to synthesize and manufacture dyes for fabrics and other industries, and they routinely tested their new products for antimicrobial qualities. Among these products was a red dye, **prontosil**, synthesized in 1932. Prontosil had no apparent effect on bacteria in culture.

pron′to-sil

do'mak

But things were different in animals. When the German chemist **Gerhard Domagk** tested prontosil in animals he found a pronounced inhibitory effect on staphylococci, streptococci, and other Gram-positive bacteria. In February 1935, Domagk injected the dye to his daughter Hildegarde, who was gravely ill with septicemia. She had pricked her finger with a needle, and blood infection had followed rapidly. Hildegarde's condition gradually improved, and, to many historians, her recovery set into motion the age of modern chemotherapy. For his discovery, Gerhard Domagk was awarded the 1939 Nobel Prize in Physiology or Medicine (*in absentia*, however, because Hitler forbade him to accept it).

tref'oo-el
sul"fah-nil'ah-mīd

Later in 1935, a group at the Pasteur Institute headed by **Jacques and Therese Tréfouël** isolated the active principle in prontosil. They found it to be **sulfanilamide**, a substance first synthesized by Paul Gelmo in 1908. Sulfanilamide quickly became a mainstay for the treatment of wound-related infections sustained during World War II.

Sulfanilamide and Other Sulfonamides

sul-fon'ah-mīdz

Sulfanilamide was the first of a group of chemotherapeutic agents known as **sulfonamides**. In 1940, the British investigators **D.D. Woods** and **E.M. Fildes** proposed a mechanism of action for sulfanilamide and other sulfonamides, and gave insight on how they interfere with the metabolism of bacteria without damaging body tissues. The mechanism came to be known as **competitive inhibition** (Figure 23.3).

Folic acid:
a chemical substance used by many types of living things in the synthesis of nucleic acids.

According to the mechanism of competitive inhibition, certain bacteria synthesize an important molecule called **folic acid** for use in nucleic acid production. Humans cannot synthesize folic acid and must consume it in foods or vitamin capsules. However, bacteria possess the necessary enzyme to manufacture folic acid and are incapable of absorbing folic acid from the surrounding environment.

par"ah-am"ĭ-no-ben-zo'ik

In the production of folic acid, the bacterial enzyme joins together three important components, one of which is **para-aminobenzoic acid (PABA)**. This molecule is similar to sulfanilamide in chemical structure (Figure 23.3). Therefore if the environment contains large amounts of sulfanilamide, the enzyme selects the sulfanilamide molecule instead of the PABA molecule for use in folic acid production. Once combined with the enzyme, the molecule binds tightly, and effectively inhibits the enzyme, thus making it unavailable for folic acid synthesis. As the production of folic acid is reduced, nucleic acid synthesis ceases, and the bacteria die.

sul"fah-meth-oks'ah-zōl

tri-meth'o-prim

sulf"is-oks'ah-zōl

Modern sulfonamides are typified by **sulfamethoxazole**. Doctors prescribe this drug for urinary tract infections due to Gram-negative rods, and for meningococcal meningitis. Frequently the drug is combined with **trimethoprim**, a drug that inhibits another step in folic acid synthesis. Commercially the drug combination is known as Bactrim. It is frequently used to treat *Pneumoncystis* pneumonia. Another common sulfonamide, **sulfisoxazole**, is marketed as Gantrisin cream for vaginal infections due to Gram-negative bacteria. In some patients, a drug allergy to sulfonamides develops, with a skin rash, gastrointestinal distress, or type II cytotoxic hypersensitivity.

Other Chemotherapeutic Agents

i"so-ni'ah-zid

The discovery and development of sulfanilamide led to the development of numerous other chemotherapeutic agents, many of which are currently in wide use. One example is the antituberculosis drug **isoniazid** (isonicotinic acid hydrazide, INH). Biochemists believe that isoniazid interferes with cell-wall

FIGURE 23.3

Inhibition of Folic Acid Synthesis by Competitive Inhibition

(a) The chemical structures of para-aminobenzoic acid (PABA) and sulfanilamide (SFA) are very similar. (b) Folic acid is made up of three components: pteridine, PABA, and glutamic acid. (c) In the normal synthesis of folic acid, a bacterial enzyme joins the three components to form folic acid. However, in competitive inhibition, the enzyme takes up SFA because it is abundant. The SFA assumes the position normally reserved for PABA, and folic acid cannot form. Without folic acid, nucleic acid metabolism is interrupted and the bacterium dies.

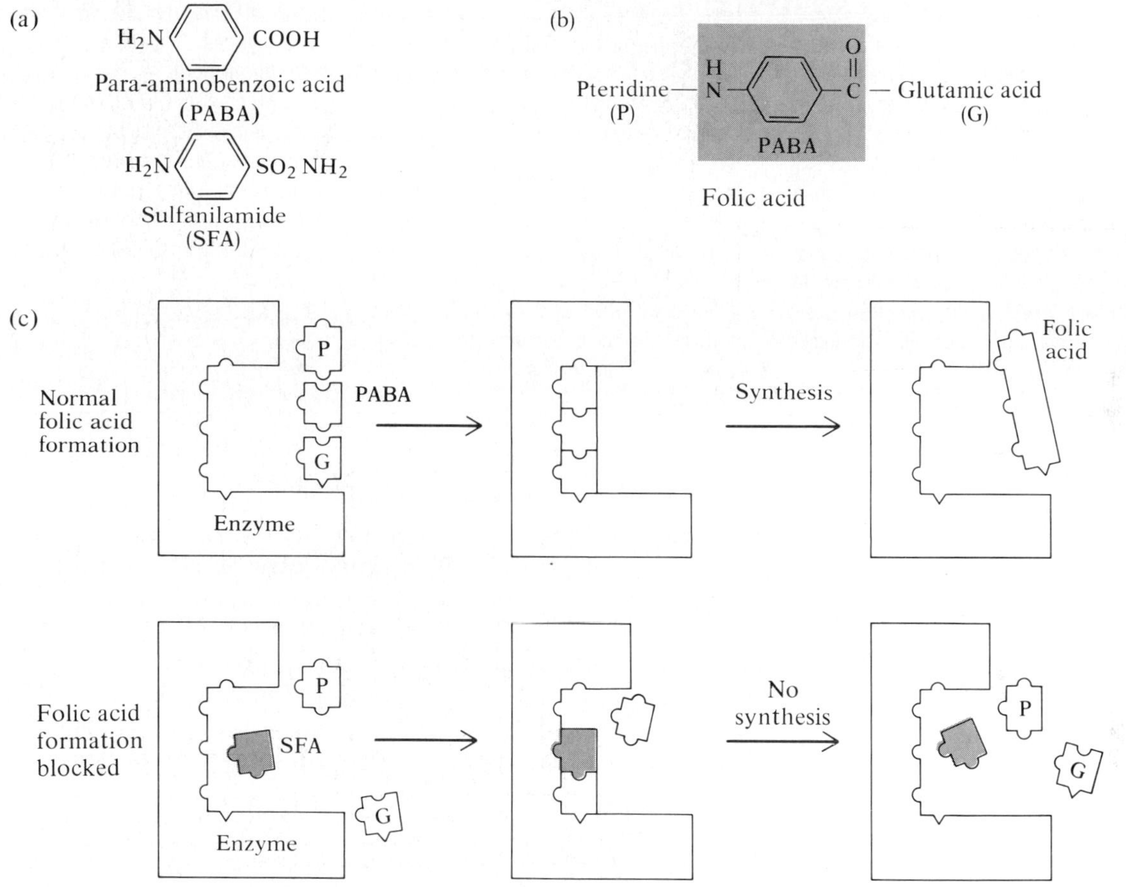

(a)

H_2N ⬡ COOH
Para-aminobenzoic acid
(PABA)

H_2N ⬡ $SO_2 NH_2$
Sulfanilamide
(SFA)

(b)

Pteridine—(P) — N—H ⬡ C=O — Glutamic acid (G)
PABA
Folic acid

(c)

Normal folic acid formation — P, PABA, G — Enzyme → → Synthesis → Folic acid

Folic acid formation blocked — P, SFA, G — Enzyme → → No synthesis → P, SFA, G

synthesis in *Mycobacterium* species by inhibiting the production of mycolic acid, a component of the wall. Isoniazid is often combined in therapy with such drugs as rifampin and ethambutol. **Ethambutol** is a synthetic, well-absorbed drug that is tuberculocidal. Visual disturbances limit its use to treatment of tuberculosis.

eth-am′bu-tol

Another chemotherapeutic agent, a quinolone called **nalidixic acid**, blocks DNA synthesis in certain Gram-negative bacteria that cause urinary tract infections. Synthetic derivatives of nalidixic acid called **fluoroquinolones** are also used in urinary tract infections as well as for gonorrhea and chlamydia and for intestinal tract infections due to Gram-negative bacteria. Examples of the fluoroquinolone drugs are ciprofloxacin (Cipro), enoxacin, and norfloxacin.

nal-ĭ-diks′ik

flu″ro-quin′o-lones
cip″ro-flox′a-cin
e-nox′a-cin
nor-flox′a-cin

Nitrofurantoin is a drug actively excreted in the urine for urogenital infections. **Metronidazole** (Flagyl) has been used for decades against trichomoniasis, amebiasis, and giardiasis. However, evidence that the drug causes tumors in mice has prompted physicians to prescribe it with caution. The treatment of malaria has long depended upon the consumption of **quinine**

ni″tro-fu-ran′to-in
me″tro-ni′dah-zōl
trik″o-mo-ni′ah-sis
ji″ar-di′ah-sis

23.1

MICROFOCUS

THE FEVER TREE

Rarely had a tree caused such a stir in Europe. In the 1500s, Spaniards returning from the New World told of its magical powers in victims of malaria, and before long, the tree was dubbed "the fever tree." The tall evergreen grew only on the eastern slopes of the Andes Mountains but was known throughout the countryside. According to legend, the Countess of Chinchón, wife of the Spanish ambassador to Peru, developed malaria in 1638 and agreed to be treated with its bark. When she recovered, she spread news of the tree throughout Europe, and a century later, Linnaeus named it *Cinchona* after her.

For the next two centuries, cinchona bark remained a staple for malaria treatment. Peruvian Indians called the bark *quina-quina* (bark of bark), and the term quinine gradually evolved. In 1820, two French chemists, Pierre Pelletier and Joseph Caventou, extracted pure quinine from the bark and increased its availability still further. The ensuing rush to stockpile the chemical led to a rapid decline in the supply of cinchona trees from Peru, but Dutch farmers made new plantings in Indonesia, where the climate was similar. The island of Java eventually became the primary source of quinine for the world.

During World War II, Southeast Asia came under Japanese domination, and the supply of quinine to the West was drastically reduced. Scientists synthesized quinine shortly thereafter, but production costs were prohibitive. Finally, two useful substitutes were synthesized in chloroquine and primaquine. Today, as resistance to these drugs is increasingly observed in malarial parasites, scientists are once again looking to the fever tree to help control malaria.

klor'o-kwin

prim'a-kwin

par"ah-am"ĭ-no-sal-ĭ-cil'ik

di-am'ĭ-no-di-fen"il-sul'fōn

(MicroFocus: 23.1). When the tree bark used in its production became unavailable during World War II, researchers quickly set to work to develop two alternatives: chloroquine and primaquine. **Chloroquine** is effective for terminating malaria attacks; **primaquine** destroys the malaria parasites outside red blood cells.

Two other chemotherapeutic drugs, both inhibitory to *Mycobacterium* species, bear brief mention. The first is **para-aminosalicylic acid (PAS)**, a drug that closely resembles sulfonamides and is used for tuberculosis. The second agent is diaminodiphenylsulfone, or **dapsone**, used to treat leprosy.

TO THIS POINT
■

We have explored some of the events that led to the discovery and development of chemotherapeutic agents, first by Paul Ehrlich in the early 1900s and then by Gerhard Domagk in the 1930s. The landmark work of these investigators established the principle that chemical agents could be effective in the therapy of established disease and encouraged other researchers to synthesize new compounds.

The focus next shifted to sulfanilamide. We described the mechanism of action of this compound and illustrated how a chemotherapeutic agent could interfere with an important metabolic process within a microorganism. Though sulfanilamide is rarely used any more, a number of modern derivatives such as sulfamethoxazole are often prescribed. We also mentioned several other chemotherapeutic agents and their uses in order to see the spectrum of these drugs. A variety of Gram-positive, Gram-negative, and acid-fast bacteria as well as several protozoa and fungi can be controlled with chemotherapeutic agents.

We shall now turn our attention to the antibiotics. These are naturally occurring products of the metabolism of microorganisms. As in the previous section, we shall examine the experiments leading to the discovery of antibiotics and then

Many of the names in this section should be familiar since antibiotics are a key to the successful recovery from disease. Antiviral agents are discussed in Chapter 11 and will not be considered here.

Antibiotics

The word antibiotic is derived from Greek stems that mean "against life." In 1889, the French researcher **Paul Vuillemin** coined "antibiotic" to describe a substance he isolated some years earlier from *Pseudomonas aeruginosa*. The substance, called pyocyanin, inhibited the growth of other bacteria in test tubes but was too toxic to be useful in disease therapy. Vuillemin's term has survived to the current era. **Antibiotics** are now considered to be chemical products or derivatives of certain organisms that are inhibitory to other organisms.

soo"do-mo′nas a″er-u-jin-o′sa

Antibiotic:
an antimicrobial agent that is a naturally occurring product of the metabolism of microorganisms.

Scientists are uncertain as to how the ability to produce antibiotics arose in living things, but it is conceivable that random genetic mutations were responsible. Clearly, the ability to produce an antibiotic conferred an extraordinary evolutionary advantage on the possessor in the struggle for survival. In this section we shall discuss the sources of antibiotics, their modes of action, and side effects, and how they are used by physicians to control microorganisms. Our study will begin with Fleming's discovery of penicillin and the events that followed.

The Discovery of Antibiotics

One of the first to postulate the existence and value of antibiotics was the British investigator, **Alexander Fleming** (Figure 23.4). Fleming was a student

FIGURE 23.4

Penicillin and Its Discoverer

(a) Alexander Fleming, the British microbiologist who reported the existence of penicillin in 1928 but was unable to purify it for use as a therapeutic agent. (b) The actual photograph of Fleming's culture plate originally published in the *British Journal of Experimental Pathology* in 1929. The photograph, taken by Fleming, shows how staphylococci in the region of the *Penicillium* colony have been killed (they are "undergoing lysis") by some unknown substance produced by the mold. Fleming called the substance penicillin. Ten years later, penicillin would be rediscovered and developed as the first modern antibiotic.

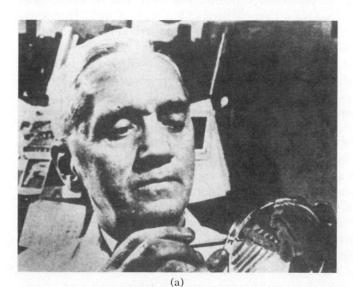

(a)

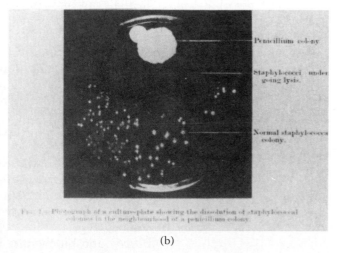

(b)

23.2

COINCIDENCE OR FATE?

In August 1961, the *Bacteriological News*, a publication of the American Society for Microbiology, carried the following story:

In the 1870s, the son of a British nobleman became mired in a bog and was in danger of losing his life when a Scotsman happened along. The Scotsman waded into the bog and pulled the boy free.

When the nobleman learned of the deed, he offered money to the Scotsman, but the Scotsman politely refused. Instead, the Scotsman pointed out that he also had a son and perhaps the nobleman would be willing to educate the boy. The nobleman agreed, and the bargain was struck.

The Scotsman's son was Alexander Fleming. After some years, Fleming attended St. Mary's Hospital School of Medicine and in the course of his research, he discovered penicillin. Meanwhile the nobleman's son was also rising to a prominent position in British politics. During World War II he was stricken with pneumonia but was treated with penicillin and survived. His name was Winston Churchill.

Penicillium:
a genus of molds belonging to the Ascomycetes class of fungi.

doo-bo'

of Almroth Wright, the discoverer of opsonins. During his early years, Fleming experienced the excitement of the Golden Age of Microbiology and spoke up for the therapeutic value of Salvarsan. In a series of experiments in 1921, he described lysozyme, the nonspecific enzyme that breaks down cell walls in Gram-positive bacteria. MicroFocus: 23.2 describes an ironic incident in his life.

The discovery of antibiotics is an elegant expression of Pasteur's dictum: "Chance favors the prepared mind." In 1928, Fleming was performing research on staphylococci at St. Mary's Hospital in London. Before going on vacation, he had spread staphylococci on plates of nutrient agar, and now, on his return, he noted that one plate was contaminated by a green mold. His interest was piqued by the failure of staphylococci to grow near the mold. Fleming isolated the mold, identified it as a species of *Penicillium*, and found that it produced a substance that killed Gram-positive organisms. Though he failed to isolate the substance, he named it **penicillin**.

Fleming was not the first to take note of the antibacterial qualities of *Penicillium* species. Joseph Lister observed a similar phenomenon in 1871, John Tyndall did likewise in 1876, and a French medical student Ernest Duchesne wrote a research paper on the subject in 1896. Whether they were observing the effects of penicillin or some other inhibitor is not known. However, we know that Fleming proposed using penicillin to eliminate Gram-positive bacteria from mixed cultures, and that he unsuccessfully tried the filtered broth on infected wound tissue. At the time, vaccines and sera were viewed as essential to disease therapy, and Fleming's request for financial support went unheeded. Moreover, biochemistry was not sufficiently advanced to make complex separations possible, and funds for research were limited since the Depression had begun. Fleming's discovery was soon forgotten.

In 1935, Gerhard Domagk's dramatic announcement of the chemotherapeutic effects of prontosil fueled speculation that chemicals could be used to fight disease in the body. Then, in 1939, Rene Dubos of the Rockefeller Institute in New York City reported that soil bacteria could produce antibacterial substances. By that time, a group at Oxford University led by pathologist **Howard Florey** and biochemist **Ernst Boris Chain** (Figure 23.5) had reisolated Fleming's penicillin and were conducting trials with highly purified samples. An article in *The Lancet* in 1940 detailed their success. But England was deeply

FIGURE 23.5

The Discoverer and Developers of Penicillin

In 1940, Alexander Fleming (left, standing) learned that Howard Florey and Ernst Boris Chain (center and right, standing) had reisolated penicillin and were conducting tests on laboratory mice. He decided to visit them at the Sir William Dunn School of Pathology, Oxford. This painting by Robert Thom recalls the meeting of the three scientists who would share the Nobel Prize five years later. Norman Heatley, another member of the research group, is at the lower right.

involved in World War II, so a group of American companies developed the techniques for large-scale production of penicillin and made the drug available for commercial use (MicroFocus: 23.3). Fleming, Florey, and Chain shared the 1945 Nobel Prize in Physiology or Medicine for the discovery and development of penicillin.

Penicillin

Since the 1940s, penicillin has remained the most widely used antibiotic because of its low cost and thousands of derivatives. **Penicillin G**, or benzylpenicillin, is currently the most popular penicillin antibiotic and is usually intended when the physician prescribes "penicillin." Other types are penicillin F and penicillin V, all with the same basic structure of a **beta-lactam nucleus** and several attached groups (Figure 23.6).

Beta-lactam nucleus: a distinctive chemical group central to the penicillin molecule.

The penicillins are active against a variety of Gram-positive bacteria, including staphylococci, streptococci, clostridia, and pneumococci. In higher concentrations, they are also inhibitory to the Gram-negative diplococci that cause gonorrhea and meningitis, and are useful against syphilis spirochetes. Penicillin functions during the synthesis of the bacterial cell wall. It blocks cross-linking of carbohydrates in the peptidoglycan layer during wall formation, resulting in such a weak wall that internal pressure causes the cell to swell and burst (Figure 23.7). Penicillin is therefore bactericidal in rapidly multiplying bacteria (as in an infection). Where bacteria are multiplying slowly or are dormant, the drug may have only a bacteriostatic effect or no effect at all.

pep"ti-do-gli'kan

FIGURE 23.6
Some Members of the Penicillin Group of Antibiotics

The beta-lactam nucleus is common to all the penicillins. Different penicillins are formed by varying the side group on the molecule.

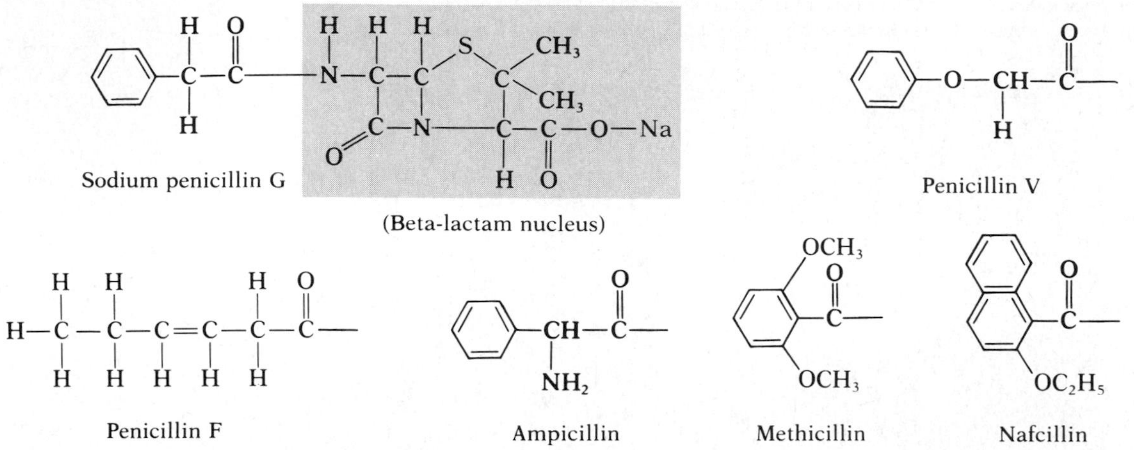

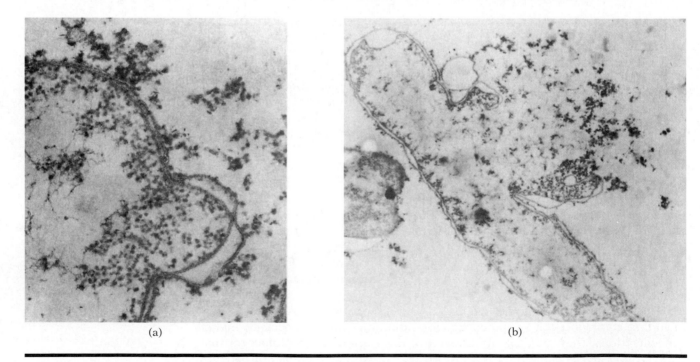

FIGURE 23.7
The Action of Penicillin on Bacteria

Penicillin is known to block synthesis of bacterial cell materials. (a) Twenty minutes after treatment with penicillin, a bacterial rod experiences bulging in its cell wall as the wall weakens from the internal pressures (TEM, × 84,800). (b) Another rod displays a ruptured cell wall and leakage of its cytoplasm (TEM, × 20,640).

(a)   (b)

Anaphylactic reaction:
a whole-body allergic
reaction accompanied by
severe contractions of the
smooth muscles.

Over the years, two major drawbacks to the use of penicillin have surfaced. The first is the **anaphylactic reaction** occurring in allergic individuals (Chapter 20). This allergy applies to all compounds related to penicillin. Swelling about the eyes or wrists, flushed or itchy skin, shortness of breath, and a series of hives are signals that sensitivity exists and that penicillin therapy should cease immediately.

TRANSPORTING A TREASURE

Their timing could not have been worse. Howard Florey, Ernest B. Chain, Norman Heatley, and others of the team had rediscovered penicillin, refined it, and proven it useful in infected patients. But it was 1939, and German bombs were falling on London. This was no time for research into new drugs and medicines.

There was hope, however. Researchers in the United States were willing to attempt the industrial production of penicillin and so, the British workers would move their lab across the ocean. There were many problems, to be sure, but one was particularly interesting—how to transport the vital *Penicillium* cultures. If the molds were to fall into enemy hands or if the enemy were to learn the secret of penicillin, all their work

would be wasted. Then Heatley made a suggestion—they would rub the mold on the inside linings of their coats, deposit the mold spores there, and transport the *Penicillium* cultures across the ocean that way.

And so they did. On arrival in the United States they set to work to reisolate the mold from their coat linings, and they began the laborious task of manufacturing penicillin. One of the great ironies of medicine is that virtually all the world's penicillin-producing mold has been derived from those few spores in the linings of the British coats. Few coats in history have yielded so noble a bounty.

The second disadvantage is the evolution of penicillin-resistant bacteria. These organisms produce **penicillinase** (also called beta-lactamase), an enzyme that converts penicillin into harmless penicilloic acid (Figure 23.8). It is probable that the ability to produce penicillinase has always existed in certain bacterial mutants, but that the ability manifests itself when the organisms are confronted with the drug. Thus a natural selection takes place, and the rapid multiplication of penicillinase-producing bacteria yields organisms against which penicillin is useless. Recent years, for example, have witnessed an increase in penicillinase-producing *Neisseria gonorrhoeae* (PPNG), with the result that penicillin is now less useful for gonorrhea treatment.

pen″ĭ-sil-o′ik

nī-se′re-ah

Semisynthetic Penicillins

In the late 1950s, the beta-lactam nucleus of the penicillin molecule was identified and synthesized, and it became possible to attach various groups to this nucleus and create new penicillins. In the following years, thousands of penicillins emerged from this semisynthetic process.

Ampicillin exemplifies a semisynthetic penicillin. It is less active against Gram-positive cocci than penicillin G, but is valuable against several Gram-negative rods as well as gonococci and meningococci. The drug resists stomach acid and is absorbed from the intestine after oral consumption. **Amoxicillin**, a chemical relative of ampicillin, is also acid-stable and has the added advantage of not binding to food as many antibiotics do. Because ampicillin and amoxicillin are excreted into the urine, they are used to treat urinary tract infections.

ah-moks″ĭ-sil′in

Another semisynthetic penicillin, **carbenicillin**, is used primarily for infections of the urinary tract. Other semisynthetic penicillins include methicillin, nafcillin, piperacillin, and oxacillin. Still another is **ticarcillin**, a penicillin derivative often combined with **clavulanic acid** (the combination is called Timentin) for use against penicillin-resistant organisms. The clavulanic acid inactivates

ti-car′cil-lin
clav″u-lan′ik

FIGURE 23.8

The Action of Penicillinase on Sodium Penicillin G

The enzyme converts penicillin to harmless penicilloic acid by opening the beta-lactam ring (arrow) and inserting a hydroxyl group to the carbon and a hydrogen to the nitrogen.

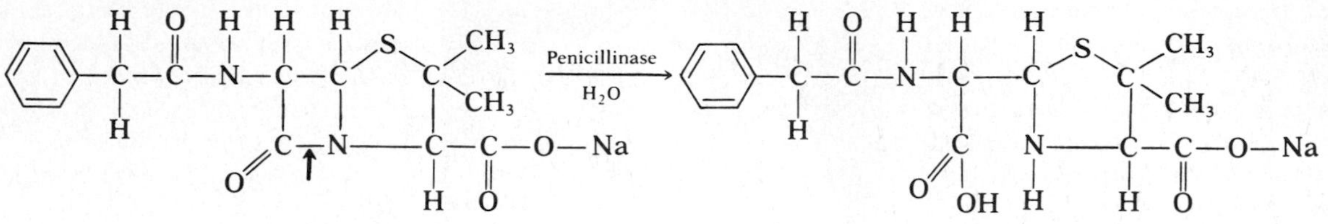

Sodium penicillin G   Sodium penicilloic acid

penicillinase and thus overcomes the resistance. None of these drugs may be prescribed where allergy to the parent drug exists, and many have been implicated in gastrointestinal disturbances and kidney and liver damage.

Cephalosporins

sef"ah-lo-spōr'e-um
ak-re-mōn'e-um
sef"ah-lo-spōr'in

While evaluating sea water samples along the coast of Sardinia in 1945, an Italian microbiologist named Giuseppe Brotzu observed a striking difference in the amount of *E. coli* in two adjoining areas. Subsequently he discovered that a fungus, *Cephalosporium acremonium*, was producing an antibacterial substance in the water. The substance, named cephalosporin C, was later isolated and characterized by scientists, and eventually it formed the basis for a family of antibiotics known as **cephalosporins**.

Cephalosporins are generally arrranged in three groups or "generations." **First-generation** cephalosporins are variably absorbed from the gut and are useful against Gram-positive cocci and certain Gram-negative rods (Figure 23.9). They include cephalexin (Keflex) and cephalothin (Keflin). **Second-generation**

sef'ah-clor
sef-ox'i-tin
sef"u-rox'ime
sef"o-tax'ime
sef"tri-ax'on
sef-taz'i-dime

drugs are active against Gram-positive cocci as well as Gram-negative rods (e.g., *Haemophilus influenzae*) and include cefaclor, cefoxitin, and cefuroxime (Zinacef). The **third-generation** cephalosporins are used primarily against Gram-negative rods (e.g., *Pseudomonas aeruginosa*) and for treating diseases of the central nervous system. Cefotaxime (Claforan), ceftriaxone (Rocephin), and ceftazidime (Fortaz) are in the group.

Cephalosporins resemble penicillins in chemical structure except that the beta-lactam nucleus has a slightly different composition. The antibiotics are used as alternatives to penicillin where resistance is encountered or in cases where penicillin allergy exists. Side effects appear to be minimal but allergic reactions have been reported and thrombophlebitis can occur. The drugs function by interfering with cell wall synthesis in bacteria.

Aminoglycosides

am"i-no-gli'ko-sīd

The **aminoglycosides** are a group of antibiotic compounds in which amino groups are bonded to carbohydrate molecules (glycosides) bonded to other carbohydrate molecules. All aminoglycosides attach irreversibly to bacterial ribosomes, thereby blocking the reading of the genetic code on messenger RNA molecules. Since oral absorption is negligible, the antibiotics must be administered by injection. Their use has declined in recent years with the introduction

FIGURE 23.9
The Effects of Cephalexin on *Vibrio cholerae*

(a) A scanning electron micrograph of control cells grown in a medium free of antibiotic. The cells exhibit the typical vibrio shape with a short curve and an incomplete spiral. Bar = 1 μm. (b) Experimental cells treated with 3.13 μg of cephalexin per ml. The cells have elongated and formed right-handed spirals that are complete. Bar = 5μm.

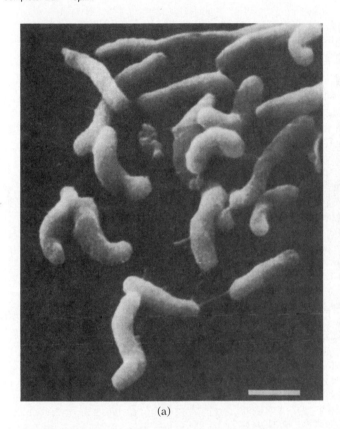

(a)

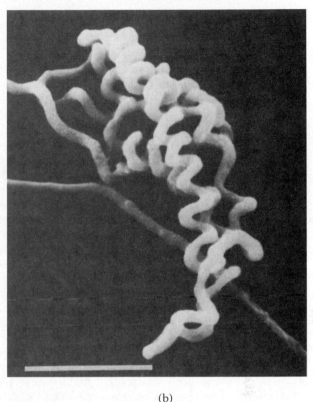

(b)

of second- and third-generation cephalosporins and with the introduction and development of quinolone drugs such as the fluoroquinolones.

In 1943, the first aminoglycoside was discovered by researchers led by **Selman A. Waksman**, a soil microbiologist at Rutgers University. Waksman's group isolated an antibacterial substance from a moldlike bacterium named *Streptomyces griseus* and named the substance **streptomycin** (MicroFocus: 23.4). At the time, the discovery was sensational because streptomycin was useful against tuberculosis and numerous diseases caused by Gram-negative bacteria. Since then it has been largely replaced by safer drugs, but streptomycin is still prescribed on occasion for such diseases as tuberculosis. The major side effect of therapy is damage to the auditory branch of the nerve extending from the inner ear. Deafness may result.

strep-to-mī′cez

Gentamicin, a widely used aminoglycoside, is administered for serious infections caused by Gram-negative bacteria, especially those in urinary tract infections. The antibiotic is produced by species of *Micromonospora*, a bacterium related to *Streptomyces*. Damage to the kidneys and the hearing mechanism reflects its toxicity.

jen″tah-mi′sen

mi″kro-mo-no-spōr′ah

Neomycin and kanamycin are two older antibiotics of the aminoglycoside group, having been isolated from *Streptomyces* species in 1949 and 1957, respectively. **Neomycin** is sometimes used to control infections of the intestine

23.4

MICROFOCUS

SERENDIPITY

They met by chance on a ship from France to the United States: Rene Dubos (doo-bo'), a 23-year-old French student interested in soil science, and Selman A. Waksman, a professor of soil microbiology at Rutgers University in New Jersey. The year was 1924. For the next two decades, their lives would intertwine as each carved out a niche in modern microbiology.

As they chatted on board ship, Waksman suggested that Dubos come to Rutgers for a doctorate in microbiology. Dubos took the advice and by 1927, he had his Ph.D. and a job at Rockefeller Institute in New York City. Here he discovered a bacterial enzyme that destroys the capsules of pneumococci and hastens their death.

But it was 1931, and Dubos' work stalled because biochemistry was still in a state of infancy. Soon, Domagk's work on prontosil burst on the scene and Dubos began searching for ways to destroy whole organisms, not just capsules. He isolated a soil bacterium, *Bacillus brevis*, and in 1939 he extracted from it an antibiotic called tyrothricin. Further extractions yielded a second antibiotic, gramicidin. Both substances killed a variety of Gram-positive bacteria, but both were too toxic for use in the body.

Nevertheless, Dubos' discoveries encouraged Florey and Chain to continue their work with penicillin and showed that antimicrobial substances could be obtained from soil bacteria.

The lesson also had an impact on Selman Waksman. Over the years he had followed his former pupil's research and in 1939 he began testing soil bacteria for antimicrobial compounds. The work was long, systematic, and plodding. In 1940, Waksman's group isolated the toxic antibiotic actinomycin, and in 1942, they found another antibiotic, streptothricin. But before this second drug could be thoroughly evaluated, streptomycin emerged.

The saga of streptomycin began in August 1943. Working with Albert Shatz and Elizabeth Bugie, Waksman isolated the moldlike bacillus *Streptomyces griseus* from the throat of a chicken. The bacillus produced streptomycin, an antibiotic with extraordinary capabilities. Preliminary studies showed its effectiveness against tubercle bacilli, and exhaustive tests at the Mayo Clinic in 1944 confirmed the results. Merck and Company soon began industrial production of the antibiotic, and within a decade, 26 companies throughout the world were

manufacturing it. In 1952, Waksman was awarded the Nobel Prize in Physiology or Medicine for his accomplishment.

Serendipity is a word derived from Horace Walpole's fairy tale *The Three Princes of Serendip*. In the tale, desirable things happen by accident or chance. Many antibiotics are the products of serendipity, but even more fundamental are the serendipitous meetings of two people such as once happened on a ship from France to the United States.

A scanning electron micrograph of *Streptomyces griseus*, the organism isolated by Waksman and one of the first species of bacteria to yield an antibiotic. Bar = 5 μm.

---

pol"e-mik'sin
bas"ĭ-tra'sin

par'o-mo-mi"sin

because it is poorly absorbed, and it is prescribed as an ointment for bacterial conjunctivitis. Commercially, it is available in combination with polymyxin B and bacitracin as Neosporin. **Kanamycin** is used primarily against Gram-negative bacteria in wounded tissue. A derivative of kanamycin called **amikacin** is used for controlling numerous nosocomial diseases and for intestinal diseases. Physicians use **tobramycin** against *Pseudomonas* species, and **paromomycin** against *Entamoeba histolytica*. Both antibiotics are derived from *Streptomyces* species.

Chloramphenicol

klo"ram-fen'ĭ-kol

Broad-spectrum antibiotic: one that is effective against a wide range of bacteria, rickettsiae, chlamydiae, and fungi.

Chloramphenicol was the first broad-spectrum antibiotic discovered. Its isolation in 1947 by John Ehrlich, Paul Burkholder, and David Gotlieb was hailed as a milestone in microbiology because the drug was capable of inhibiting a wide variety of Gram-positive and Gram-negative bacteria, as well as several species of rickettsiae and fungi. During the next 30 years, however, physicians tempered their enthusiasm for chloramphenicol as side effects became apparent and new drugs appeared. Nevertheless, chloramphenicol still retains its importance in the treatment of many diseases.

FIGURE 23.10

The Effect of Chloramphenicol on Bacteria

Escherichia coli was treated with chloramphenicol and photographed with the scanning electron microscope 20 minutes after the exposure. Note the formation of "minicells" and their threadlike attachments to the parent cells. Phenomena such as these are not usually found during normal cell division. They reflect the interference of the antibiotic with the metabolism of the organisms. Bar = 0.5 μm.

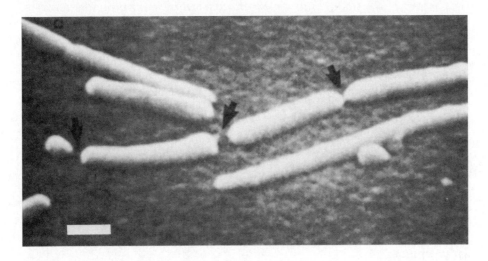

Chloramphenicol is a small molecule that passes into the tissues, where it interferes with protein synthesis in microorganisms (Figure 23.10). It diffuses into the nervous system and is thus useful for meningitis. It also remains the drug of choice in the treatment of typhoid fever and is an alternative to tetracycline for typhus fever and Rocky Mountain spotted fever. Originally isolated from the waste products of *Streptomyces venezuelae*, chloramphenicol became the first synthetic antibiotic when scientists at the Parke-Davis Company manufactured it from raw materials (its trade name is Chloromycetin). The drug has two major side effects: In the bone marrow it prevents hemoglobin incorporation into the red blood cells and induces a condition called **aplastic anemia**; and it accumulates in the blood of newborns causing a toxic reaction and sudden breakdown of the cardiovascular system known as the **gray syndrome**. For these reasons, chloramphenicol is not used in trivial infections.

Aplastic anemia:
a condition in which red
blood cells lack hemoglobin.

TO THIS POINT
■

We have surveyed four major groups of antibiotics and have noted their sources, modes of activity, and uses. We mentioned the side effects associated with each since these are a major consideration in the determination of which drug to prescribe. Penicillin and penicillin derivatives are primarily used for Gram-positive bacteria and have their activity at the cell wall of microorganisms. Penicillinase production and patient allergy limit their use. Cephalosporin antibiotics also function at the cell wall. The aminoglycoside group consists of numerous antibiotics that interfere with protein synthesis and that physicians prescribe for diseases caused by Gram-negative bacteria. Some are used on the skin, while others must be injected. Chloramphenicol is a broad-spectrum antibiotic with numerous uses but potentially lethal side effects.

As the discussion progressed, you may have noted how the antibiotics are products of microorganisms, especially Penicillium, Cephalosporium, *and* Streptomyces *species. Many of the original products are then modified to form semisynthetic antibiotics. In the case of chloramphenicol, the antibiotic is produced totally by synthetic means. This is why antibiotics are often placed under the umbrella of chemotherapeutic agents.*

We shall continue our discussion by examining the tetracycline antibiotics and a miscellaneous group of other drugs including several that are used against fungi. Our survey will conclude with a description of the laboratory tests used to determine the effectiveness of antimicrobial agents under experimental conditions. Some discussion of antibiotic resistance and abuse will also be presented.

Other Antibiotics

Tetracyclines

In 1948, scientists at Lederle Laboratories discovered chlortetracycline, the first of the tetracycline antibiotics. This finding completed the initial quartet of "wonder drugs": penicillin, streptomycin, chloramphenicol, and tetracycline.

Modern **tetracyclines** are a group of broad-spectrum antibiotics with a range of activity similar to chloramphenicols. They include naturally occurring chlortetracycline and oxytetracycline isolated from species of *Streptomyces* (Figure 23.11), and semisynthetic tetracycline, doxycycline, methacycline, and minocycline. All have four benzenelike rings in their chemical structure, as Figure 23.11 illustrates. All interfere with protein synthesis in microorganisms by binding to ribosomes.

dox′e-cy-cline
min′o-cy-cline

Tetracycline antibiotics may be taken orally, a factor that led to their indiscriminate use in the 1950s and 1960s. The antibiotics were consumed in huge quantities by tens of millions, and in some people, the normal bacterial flora of the intestine was destroyed. With these natural controls eliminated, fungi such as *Candida albicans* flourished. Patients then had to take an antifungal antibiotic such as nystatin, but because this drug was sometimes toxic, the preferred course was to replace the intestinal bacteria by consuming large quantities of bacteria-laden yogurt. Tetracyclines also cause a yellow-gray-brown discoloration of teeth and stunted bones in children. These problems are minimized by restricting use of the antibiotic in pregnant women and children through the teen years.

nis′tah-tin

Despite these side effects, tetracyclines remain the drugs of choice for most rickettsial and chlamydial diseases, including the STD chlamydia (Chapter 10). They are used against a wide spectrum of Gram-negative bacteria, and they are valuable for treating primary atypical pneumonia, syphilis, gonorrhea, pneumococcal pneumonia, and certain protozoal diseases as well as acne. Although resistances have occurred, newer tetracyclines such as minocycline (Minocin) and doxycycline (Vibramycin) appear to circumvent these. There is evidence that tetracycline may have been present in the food of ancient people, as Micro-Focus: 23.5 points out.

FIGURE 23.11
Oxytetracycline

(a) A colony of *Streptomyces rimosus*, the organism used in the production of oxytetracycline (Terramycin). (b) The chemical structure of oxytetracycline. Note the presence of four benzene rings that characterizes all the tetracycline antibiotics.

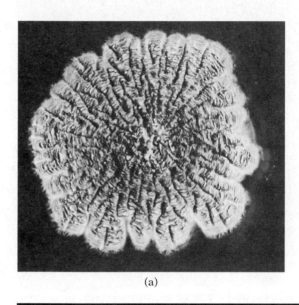

Oxytetracycline

(a)　　　　(b)

Miscellaneous Antibiotics

A miscellaneous group of antibiotics merits brief discussion because the drugs in this group are commonly used in modern therapy.

Erythromycin is a clinically important antibiotic in the group of substances called the macrolides. **Macrolides** consist of large carbon rings attached to unusual carbohydrate molecules. In the 1970s, researchers discovered that erythromycin was effective for treating primary atypical pneumonia and Legionnaires' disease. The antibiotic is a *Streptomyces* product and a protein synthesis inhibitor. It is recommended for use against Gram-positive bacteria in patients with penicillin allergies and against both *Neisseria* and *Chlamydia* species that can affect the eyes of newborns. It has few side effects. *ĕ-rith″ro-mi′sin*

Another macrolide antibiotic is **clarithromycin**, a semisynthetic drug. Clarithromycin (Biaxin) acts by binding to ribosomes to inhibit protein synthesis in Gram-negative bacteria and the same variety of Gram-positive bacteria as erythromycin. Another macrolide called **azithromycin** (Zithromax) has a similar mode of action and spectrum of activity. Both antibiotics are believed dangerous to fetal tissue and should not be taken by pregnant women. *clar-ith″ro-mi′sin* *a-zith″ro-mi′sin*

Vancomycin, a cell-wall inhibitor, is a waste product of a *Streptomyces* species. It is administered by intravenous injection against diseases due to Gram-positive bacteria, especially severe staphylococcal diseases where penicillin allergy or bacterial resistance is found. Its major side effects are damage to the ear and kidney; the drug is not routinely prescribed for trivial situations.

Rifampin is a semisynthetic drug prescribed (in combinations with isoniazid and ethambutol) for tuberculosis and leprosy patients. It is also administered to carriers of *Neisseria* and *Haemophilus* species that cause meningitis and as *rif-am′pin*

soo″do-mem′brah-nus

a prophylactic when exposure has occurred. It acts by interfering with RNA synthesis in bacteria. Rifampin therapy may cause the urine, feces, tears, and other body secretions to assume an orange-red color and may cause liver damage. The drug is administered orally and is well absorbed.

Clindamycin and its parent drug, **lincomycin**, are alternatives where penicillin resistance is encountered. Both are active against Gram-positive bacteria, including several anaerobic species (e.g., *Bacteroides* species). Use of the antibiotics is limited to serious infections, however, because the drugs eliminate competing organisms from the intestine and permit *Clostridium difficile* to overgrow the area. The clostridial toxins may then induce a condition called **pseudomembranous colitis**, in which membranous lesions cover the intestinal wall.

bas″ĭ-tra′sin
pol″e-mik′sin

Both bacitracin and polymyxin B are polypeptide antibiotics produced by *Bacillus* species. These antibiotics are generally restricted to use on the skin because internally they may cause kidney damage and they are poorly absorbed from the intestine. **Bacitracin** is available in pharmaceutical skin ointments and is effective against Gram-positive bacteria such as staphylococci. **Polymyxin B** is valuable against *Pseudomonas aeruginosa* and other Gram-negative bacilli, particularly those that cause superficial infections in wounds, abrasions, and burns. The two antibiotics are often combined with neomycin in Neosporin. Bacitracin inhibits cell wall synthesis, while polymyxin B injures bacterial membranes (Figure 23.12).

spek-tin″o-mī′sin

Spectinomycin is a *Streptomyces* product that shares chemical properties with the aminoglycosides. The antibiotic came into prominence in the 1970s for use against gonorrhea caused by penicillin-resistant gonococci. It is given by intramuscular injection and appears to interfere with protein synthesis.

kro-mo-bak′ter
vi″o-la′she-um
moks-ah-lak′tam

Monobactams are a group of antibiotics first synthesized by researchers at Squibb Laboratories in the early 1980s. The core of monobactam antibiotics is a beta-lactam nucleus isolated from *Chromobacter violaceum*, a purple-pigmented bacterium. One antibiotic, **moxalactam**, is active against a broad variety of Gram-negative bacteria, especially those involved in nosocomial diseases and bacterial meningitis. Resistance to penicillinase appears to be high

FIGURE 23.12

The Sites of Activity in a Bacterial Cell for Various Antibiotics

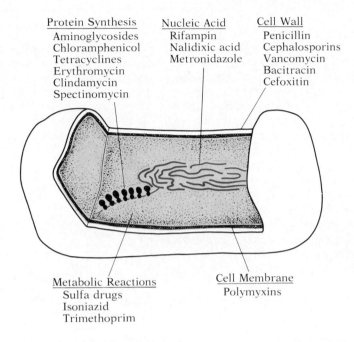

Protein Synthesis
Aminoglycosides
Chloramphenicol
Tetracyclines
Erythromycin
Clindamycin
Spectinomycin

Nucleic Acid
Rifampin
Nalidixic acid
Metronidazole

Cell Wall
Penicillin
Cephalosporins
Vancomycin
Bacitracin
Cefoxitin

Metabolic Reactions
Sulfa drugs
Isoniazid
Trimethoprim

Cell Membrane
Polymyxins

but interference with platelet function and severe bleeding have limited its use. Another set of beta-lactam drugs are the **carbapenems**. The representative of this group called **imipenem** (Primaxin) has good activity against a variety of Gram-positive bacteria and Gram-negative rods as well as anaerobes (e.g., *Bacteroides fragilis*). It is often prescribed where resistances occur, and it appears to have minimal side effects.

car-ba-pen'ems
im-i-pen'em

Antifungal Antibiotics

Fungal diseases pose a special problem for the medical mycologist because there are few drugs available for treatment. For infections of the intestine, vagina, or oral cavity due to *Candida albicans*, the recommended drug is usually **nystatin** (MicroFocus: 23.6). This product of *Streptomyces* is commercially available as Mycostatin or Achrostatin and is sold in ointment, cream, or suppository form. It acts by changing the permeability of the cell membrane by combining with fungal sterols. Often it is combined with antibacterial antibiotics to retard *Candida* overgrowth of the intestines during the treatment of bacterial diseases. Another antifungal product of *Streptomyces*, candicidin, has similar uses but is not as widely used.

nis'tah-tin

Griseofulvin is an antibiotic used for fungal infections of the skin, hair, and nails, such as ringworm and athlete's foot. Griseofulvin interferes with mitosis and causes the tips of molds to curl. It is a product of a *Penicillium* species and is taken orally.

gris"e-o-ful'vin

For serious systemic fungal infections, the drug of choice is **amphotericin B**. This antibiotic degrades the cell membranes of fungal cells, and is effective for treating serious diseases discussed in Chaper 14. However, it causes a wide variety of side effects and therefore is used only in progressive and potentially fatal cases.

am"fo-tĕr'i-sin

23.6
MICROFOCUS
PERSISTENCE

The discovery of nystatin stems from the research of two scientists—Rachael F. Brown and Elizabeth Hazen. Their work was accomplished through persistent effort and intelligent use of basic scientific principles. Sound research led them to a naturally occurring substance that would prove valuable in fighting fungal disease.

Rachael Brown was an organic chemist with a Ph.D. from the University of Chicago. Elizabeth Hazen was a mycologist. In the mid-1940s, the two scientists were employed at the New York State Department of Health where their interest was piqued by the increasing reports concerning antibiotics. Penicillin was in widespread use by that time, and streptomycin had been discovered by Waksman in 1942. Both antibiotics were useful against bacteria, but none had yet been developed for fungal disease. Brown and Hazen would try to fill that gap.

In the 1940s, scientists knew that soilborne bacteria of the genus *Streptomyces* were potential sources of antibiotics. Hazen therefore collected soil samples from various places and tested the waste products of soilborne microorganisms to determine whether they could inhibit fungal growth. A

bacterium from soil on a farm owned by a certain Henry Nourse was especially promising. The organism was apparently unknown before Hazen isolated it, and she therefore named it *Streptomyces noursei* after the farmer. Now it was Brown's turn. Using her skills in chemistry, she isolated, purified, and characterized the active ingredient in the waste product. With Hazen, she demonstrated that minuscule amounts of the active principle were extraordinarily inhibitory to fungi. Hazen and Brown named the ingredient nystatin, for New York State.

Nystatin was introduced to the scientific community at the 1949 meeting of the National Academy of Sciences. Two years later, a patent was issued for production, and E.R. Squibb received exclusive license to manufacture the antibiotic. Before long, nystatin became a key treatment for various forms of candidiasis, and when the Arno River flooded Florence, Italy, in the 1970s, nystatin was used to combat fungi attacking the art treasures. Nystatin also has commercial value for preventing spoilage in foods, especially bananas, and it is used in surgery to preclude fungal infection.

For Brown and Hazen, the accolades

were many, including several honorary degrees and awards. Students at Mount Holyoke College currently vie for the Rachael Brown Fellowship, and students at Mississippi University for Women are eligible for the Elizabeth Hazen Scholarship, both named for alumnae of the respective colleges.

flu-si'to-sēn

klo-tri'mah-zōl
mi-kon'ah-zōl
it"ra-kon'ah-zōl
ke"to-kon'ah-zōl

Other antifungal antibiotics are synthetic compounds. One example, **flucytosine**, is converted in fungal cells to an inhibitor that interrupts nucleic acid synthesis. The drug is used primarily with amphotericin B in systemic diseases. Another example, the **imidazoles**, include clotrimazole, miconazole, itraconazole, and ketoconazole. These compounds interfere with sterol synthesis in fungal cell membranes. Clotrimazole (Gyne-Lotrimin) is used topically for *Candida* skin infections, while the other drugs are used topically as well as internally for systemic diseases. Side effects are uncommon. Miconazole is commercially available in Micatin for athlete's foot and Monistat 7 for yeast disease. Itraconazole is sold as Sporonox for athlete's foot.

Antibiotic Assays and Resistance

Antibiotic Sensitivity Assays

Antibiotic sensitivity assays are used to study the inhibition of a test organism by one or more antibiotics or chemotherapeutic agents. Two general methods are in common use: the tube dilution method, and the agar diffusion method.

The **tube dilution method** determines the smallest amount of antibiotic necessary to inhibit a test organism. This amount is known as the **minimum inhibitory concentration (MIC)**. To determine it, the microbiologist prepares a set of tubes with different concentrations of a particular antibiotic. The tubes are then inoculated with the test organism, incubated, and examined for the growth of bacteria. The extent of growth diminishes as the concentration of antibiotic increases, and eventually an antibiotic concentration is observed at which growth fails to occur. This is the MIC.

The second method, the **agar diffusion method**, operates on the principle that antibiotics will diffuse from a paper disk or small cylinder into an agar medium containing test organisms. Inhibition is observed as a failure of the organism to grow in the region of the antibiotic. A common application of the agar diffusion method is the **Kirby–Bauer test** named after **W.M. Kirby** and **A.W. Bauer**, who developed it in the 1960s (Figure 23.13). This procedure determines the sensitivity of an organism to a series of antibiotics and is performed according to standards established by the Food and Drug Administration (FDA).

The test is set up as illustrated in Figure 23.13. An agar medium such as Mueller–Hinton agar is poured into the plate and inoculated with the organism. Paper disks containing known concentrations of antibiotics are applied to the surface, and the plate is incubated. The appearance of a zone of inhibition surrounding the disk indicates sensitivity. By comparing the diameter of the zones to a standard table, one can determine whether the test organism is susceptible or resistant to the antibiotic. If the organism is susceptible, it will be killed in the patient's bloodstream if the experimental concentration of antibiotic is reached. Resistance indicates that the antibiotic will not be effective at that concentration in the circulation.

Antibiotic Resistance and Abuse

During the past 25 years, an alarming number of bacterial species have evolved with resistance to chemotherapeutic agents and antibiotics. Public health microbiologists note that resistant organisms are increasingly responsible for human diseases of the intestinal tract, lungs, skin, and urinary tract. Those in intensive care units and burn wards are particularly vulnerable, as are children, the elderly, and the infirm. Common diseases like bacterial pneumonia, streptococcal sore throat, and gonorrhea that a few years ago succumbed to a single dose of antibiotics are now among the most difficult to treat.

Microorganisms may **acquire resistance** to antibiotics in a number of ways. In some cases resistance arises from the microorganism's ability to destroy the antibiotic. The production of penicillinase by penicillin-resistant gonococci is an example. Other resistances are traced to changes in the permeability of the microbial cell wall and membrane, thus prohibiting passage of the antibiotic to the interior. In addition, resistance to the drug's activity may develop. An example of the latter takes place when sulfa drugs fail to unite with enzymes that synthesize folic acid because the enzyme's structure has changed. Moreover, drug resistance may be due to an altered metabolic pathway in the microorganism, a pathway that bypasses the reaction normally inhibited by the drug. An altered structural target for the drug may also evolve. For example, the structure of a pathogen's ribosome may change and make a drug that unites with the ribosome useless.

Resistance may develop in bacteria during the normal course of events, but antibiotic abuse encourages the emergence of resistant forms. For example, drug companies promote antibiotics heavily, patients pressure doctors for quick

FIGURE 23.13

The Kirby–Bauer Test for Determining Antibiotic Susceptibility of an Organism by the Agar Diffusion Method

(a) A suitable medium such as Mueller–Hinton agar is poured into a Petri dish. (b) Bacteria from a culture are taken with a swab, and (c) streaked onto the medium. (d) Paper disks containing various antibiotics are placed onto the surface and the plate is incubated. (e) Zones of inhibition around the disks indicate susceptibility of the organism to one or more antibiotics.

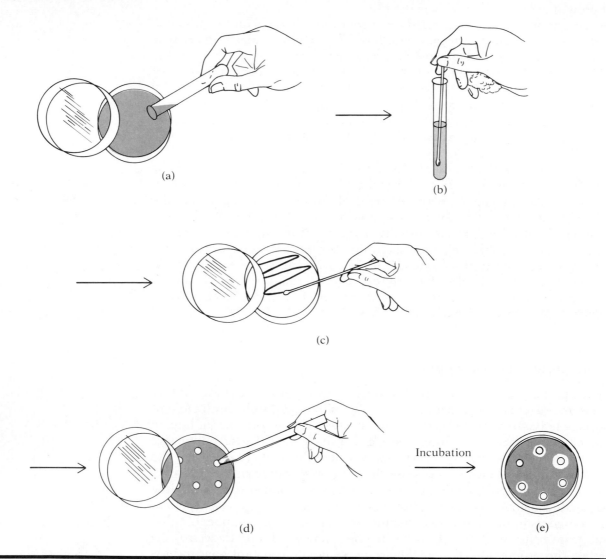

(a)

(b)

(c)

(d)   Incubation   (e)

cures, and physicians sometimes write prescriptions without ordering costly tests to pinpoint the patient's illness. In addition, people may diagnose their own illness and take leftover antibiotics from their medicine chests for ailments where antibiotics are useless.

Hospitals are another forcing ground for the emergence of resistant bacteria. In many cases, physicians use unnecessarily large doses of antibiotics to prevent infection during and following surgery. This increases the possibility that resistant strains will overgrow susceptible strains and subsequently spread to other patients, thereby causing nosocomial disease. Antibiotic-resistant *Escherichia coli*, *Pseudomonas aeruginosa*, *Serratia marcescens*, and *Proteus* species are now widely encountered causes of illness in hospital settings (Figure 23.14).

FIGURE 23.14

Contamination

A scanning electron micrograph of *Pseudomonas aeruginosa* adhering to material used in urinary catheters in hospital settings. Experiments conducted in 1985 indicated that large numbers of these cells remained alive after exposure to high concentrations of tobramycin, an antibiotic to which the bacillus is normally susceptible. The researchers concluded that growth within the material conferred a measure of antibiotic resistance on the cells, although the source of the resistance was not determined. Bar = 5 μm.

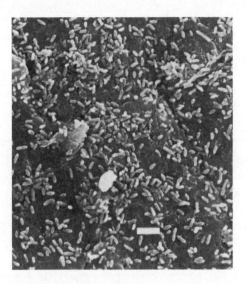

Antibiotics are also abused in **Third World countries** where they are often available without prescription, even though they have toxic side effects. Countries such as Mexico, Brazil, and Guatemala permit some of the most potent antibiotics to be sold over the counter, and large doses encourage resistance to develop. Between 1968 and 1971, 12,000 people died in Guatemala from shigellosis attributed to antibiotic-resistant *Shigella dysenteriae* (Chapter 6).

Moreover, the problem of antibiotic abuse is widespread in **livestock feeds**. An astonishing 40 percent of all the antibiotics produced in the United States finds its way into animal feeds to check disease and promote growth. By killing off less hardy bacteria, chronic low doses of antibiotics create ideal growth environments for resistant strains. Transferred to humans through meat, these resistant organisms may cause intractable illnesses. An outbreak of salmonellosis in 1984 is illustrative (Chapter 6).

Allied to the problem of antibiotic resistance is the concern for transfer of the resistance. Researchers have amply demonstrated that plasmids and transposons account for the movement of antibiotic-resistant genes among bacteria (Chapter 6). Thus the resistance in a relatively harmless bacterium may be passed to a pathogenic bacterium where the potential for disease is supplemented by resistance to standardized treatment.

Antibiotics have traditionally been known as miracle drugs, but there is a growing body of evidence that they are becoming overworked miracles. Some researchers suggest that antibiotics should be controlled as strictly as narcotics. The antibiotic roulette that is currently taking place should be a matter of discussion to all individuals concerned about infectious disease, be they scientist or student. Table 23.1 summarizes the antimicrobial agents currently in use.

TABLE 23.1

A Summary of Major Chemotherapeutic Agents and Antibiotics

CHEMOTHERAPEUTIC AGENT OR ANTIBIOTIC	SOURCE	ANTIMICROBIAL SPECTRUM	ACTIVITY IMPEDED	SIDE EFFECTS
Sulfonamides 　Sulfanilamide 　Sulfamethoxazole	Synthetic	Broad spectrum	Folic acid metabolism	Kidney and liver damage Allergic reactions
Isoniazid	Synthetic	Tubercle bacilli	Cell wall synthesis	Liver damage
Metronidazole	Synthetic	*Trichomonas vaginalis*	Cell metabolism	Tumors in mice
Chloroquine and primaquine	Synthetic	*Plasmodium* species	Cell metabolism	Eye damage
Fluoroquinolones 　Ciprofloxacin 　Enoxacin	Synthetic	Broad spectrum	DNA synthesis	Few reported
Penicillins 　Penicillin G 　Ampicillin 　Amoxicillin 　Nafcillin 　Oxacillin	*Penicillium notatum* and *Penicillium chrysogenum* Some semisynthetic	Broad spectrum, especially Gram-positive bacteria	Cell wall synthesis	Allergic reactions Selection of penicillinase-producing strains
Cephalosporins 　Cephalothin 　Cephalexin	*Cephalosporium* species	Gram-positive bacteria Broad spectrum	Cell wall synthesis	Occasional allergic reactions
Aminoglycosides Gentamicin 　Neomycin 　Amikacin 　Streptomycin	*Micromonospora* species *Streptomyces* species	Broad spectrum, especially Gram-negative bacteria	Protein synthesis	Hearing defects Kidney damage
Chloramphenicol	*Streptomyces venezuelae*	Broad spectrum, especially typhoid bacilli	Protein synthesis	Aplastic anemia Gray syndrome
Tetracyclines 　Chlortetracycline 　Oxytetracycline 　Doxycycline 　Minocycline	*Streptomyces* species Some semisynthetic	Broad spectrum, rickettsiae, chlamydiae, Gram-negative bacteria	Protein synthesis	Destruction of natural flora Discoloration of teeth Stunted bones
Erythromycin	*Streptomyces erythraeus*	Gram-positive bacteria, *Mycoplasma*	Protein synthesis	Gastrointestinal distress
Vancomycin	*Streptomyces orientalis*	Gram-positive bacteria, especially staphylococci	Cell wall synthesis	Ear and kidney damage
Rifampin	*Streptomyces mediterranei*	Tubercle bacilli Gram-negative bacteria	RNA synthesis	Liver damage
Clindamycin Lincomycin	*Streptomyces lincolnesis*	Gram-positive bacteria	Protein synthesis	Pseudomembranous colitis
Bacitracin	*Bacillus subtilis*	Gram-positive bacteria, especially staphylococci	Cell wall synthesis	Kidney damage

CHEMOTHERAPEUTIC AGENT OR ANTIBIOTIC	SOURCE	ANTIMICROBIAL SPECTRUM	ACTIVITY IMPEDED	SIDE EFFECTS
Polymyxin	Bacillus polymyxa	Gram-negative bacteria, especially in wounds	Cell membrane function	Kidney damage
Spectinomycin	Streptomyces spectabilis	Gonococci	Protein synthesis	Few reported
Moxalactam	Chromobacter violaceum	Gram-negative bacteria	Protein synthesis (?)	Few reported
Nystatin	Streptomyces noursei	Fungi, especially Candida albicans	Cell membrane function	Few reported
Griseofulvin	Penicillium janczewski	Fungi, especially in superficial infections	Nucleic acid synthesis	Occasional allergic reactions
Amphotericin B	Streptomyces nodosus	Fungi, especially in systemic infections	Cell membrane function	Fever Gastrointestinal distress
Imidazoles Clotrimazole Ketoconazole	Synthetic	Fungi, especially in superficial infections	Inhibit sterol synthesis	Few reported

NOTE TO THE STUDENT

In the last hundred years, there were two periods in which the incidence of disease declined sharply. The first period took place in the early 1900s. At this time an understanding of the disease process led to numerous social measures such as water purification, careful food production, insect control, milk pasteurization, and patient isolation. Sanitary practices such as these made it possible to prevent virulent microorganisms from reaching their human targets.

The second period began in the 1940s with the development of antibiotics, and blossomed in the years thereafter when physicians found they could treat established cases of disease. Major health gains were made as serious illnesses came under control.

An outgrowth of these events has been the belief by many people that science can cure any infectious disease. A shot of this, a tablet of that, and then perfect health. Right? Unfortunately not.

Scientists may show us how to avoid infectious microorganisms and doctors may be able to control certain diseases with antibiotics, but the ultimate body defense depends upon the immune system and other natural measures of resistance. Used correctly, the antibiotics provide that extra something needed by natural defenses to overcome pathogenic microorganisms. The antibiotics supplement natural defenses; they do not replace them.

The great advances in chemotherapy should be viewed with caution. Antibiotics have doubtlessly relieved much misery and suffering, but they are not the cure-all some people perceive them to be. In the end, it is well to remember that good health comes from within, not from without.

Summary

Chemotherapeutic agents and antibiotics work with the body's natural defenses to stop the growth of bacteria and other microorganisms in the body. Chemotherapeutic agents are drugs produced by synthetic means in the laboratory. One group of such agents, the sulfonamides, interfere with the production of folic acid in bacteria. Other agents have various uses and various chemical compositions.

Although antibiotics are largely produced by synthetic means, they were originally derived from microorganisms. One example, penicillin, was first obtained from the green mold *Penicillium*. The antibiotic interferes with cell wall synthesis in Gram-positive bacteria and is available in numerous synthetic and semisynthetic derivatives such as ampicillin and amoxicillin. Where penicillin allergy or resistance is encountered, a cephalosporin antibiotic can be utilized. Certain cephalosporin drugs are first-choice antibiotics, and a wide variety of these drugs is currently in use.

For Gram-negative bacteria, the aminoglycoside antibiotics can be employed. Gentamicin, neomycin, and kanamycin are useful members of the aminoglycoside group. Chloramphenicol is a broad-spectrum antibiotic and is valuable against both Gram-positive and Gram-negative bacteria, but serious side effects limit its use. Less severe side effects accompany tetracycline use, and this antibiotic is recommended against Gram-negative bacteria as well as against rickettsiae and chlamydiae. These and other antibiotics interfere with protein synthesis in bacteria. It is also possible that an antibiotic will interrupt nucleic acid or cell membrane metabolism. Certain antibiotics such as amphotericin B and the imidazoles are valuable against fungal infections.

A major problem attending antibiotic use is the development of antibiotic-resistant strains of microorganisms. Arising from any of several sources such as changes in microbial biochemistry, antibiotic resistance threatens to put an end to the cures of infectious disease that have come to be expected in contemporary medicine.

Questions for Thought and Discussion

1. In 1877, Pasteur and his assistant Joubert observed that anthrax bacilli grew vigorously in sterile urine but failed to grow when the urine was contaminated with other bacilli. What was happening?

2. Historians report that 2500 years ago, the Chinese learned to treat superficial infections such as boils by applying moldy soybean curds to the skin. Can you suggest what this implies?

3. During World War II, American soldiers were required to carry a full canteen of water. If they found it necessary to sprinkle a sulfa drug on a wound, they were instructed to drink the complete contents of the canteen. Why do you think this was necessary?

4. One of the novel approaches to treating gum disease is to impregnate tiny vinyl bands with antibiotic, stretch them across the teeth, and push them beneath the gumline. Presumably, the antibiotic would kill bacteria that form pockets of infection in the gums. What might be the advantages and disadvantages of this therapeutic device?

5. In May 1953, Edmund Hillary and Tenzing Norgay reached the summit of Mount Everest, the world's highest mountain. Since that time over 150 other mountaineers have reached the summit and groups have come to Nepal from all over the world on expeditions. The arrival of "civilization" has brought a drastic change to the lifestyle of Nepal's Sherpa mountain people. For example, half of all Sherpas used to die before the age of 20, but with antibiotics available for disease, the population has grown from 9 million to 15 million in three decades. Medical enthusiasts are proud of this increase in the life expectancy, but population ecologists see a bleaker side. What do you suspect they foresee and what does this tell you about the impact that antibiotics have on a culture?

6. It was a crude remark, but during a discussion on the side effects of antibiotics, a student blurted out: "Better red than dead!" What antibiotic do you think was being discussed?

7. Most naturally occurring antibiotics appear to be products of the soil bacteria belonging to the genus *Streptomyces*. Can you draw any connection between the habitat of these organisms and their ability to produce antibiotics?

8. Bacitracin derives part of its name from Margaret Tracy, a patient from whom doctors isolated the *Bacillus subtilis* used in antibiotic production. Lincomycin is so named because it was first obtained from a bacterium isolated in Lincoln, Nebraska. From your knowledge of prefixes and suffixes, can you guess how other antibiotics in this chapter got their names?

9. Why would a synthetically produced antibiotic be more advantageous than a naturally occurring antibiotic? Why would it be less advantageous?

10. When the social security system was introduced in 1935, the age for collecting benefits was set at 65. How have the antibiotics helped put the system in jeopardy?

11. Of the thousands and thousands of types of organisms screened for antibiotics since 1940, only five genera appear capable of producing these chemicals. Does this strike you as unusual? What factors might eliminate potentially useful antibiotics?

12. In this chapter, we have encountered several winners of the Nobel Prize including Ehrlich, Domagk, Waksman, Fleming, Florey, and Chain. Can you see any patterns in their accomplishments that might explain some rationale for awarding the prize?

13. Ecologists tell us that by upsetting the balances in nature, a group of organisms may emerge with unusual characteristics. For instance, the continual use of rodenticides in the 1960s and 1970s permitted a variety of pesticide-resistant "super rat" to emerge in certain American cities. How does this principle relate to the appearance of PPNG?

14. Is an antibiotic that cannot be absorbed from the gastrointestinal tract necessarily useless? How about one that is rapidly expelled from the blood into the urine? Why?

15. The antibiotic issue can be argued from two perspectives. Some people contend that because of side effects and microbial resistance, the antibiotics will eventually be abandoned in medicine. Others see the future development of a super antibiotic, a type of "miracle drug." What arguments can you offer for either view? Which direction in medicine would you support?

Review

On completing this chapter, you are invited to test your knowledge of its contents by completing the following crossword puzzle. The solution is in Appendix F.

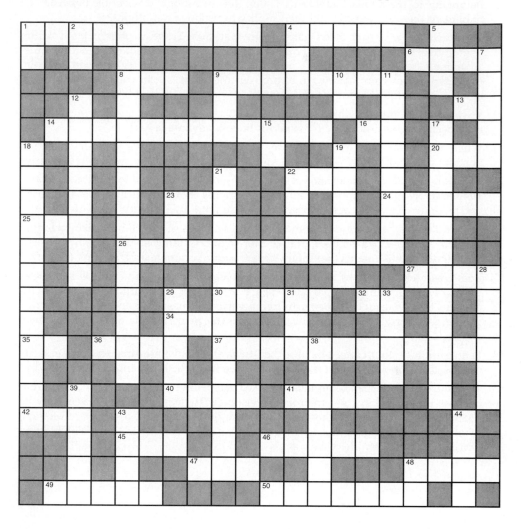

ACROSS

1. Has a beta-lactam nucleus; most familiar antibiotic.
4. Four carbon _____ exist in certain antibiotic molecules.
6. Penicillinase is also known as _____-lactamase.
8. Erythromycin has relatively _____ activity against Gram-negatives.
9. Causes body secretions to assume a red-orange color.
13. An important antifungal drug is _____-conazole.
14. Drug of choice for rickettsial and chlamydial infections.

DOWN

1. Acidity level (abbr) that can influence antibiotic activity.
2. Penicillin has _____ effect on viruses.
3. Broad-spectrum antibiotic that interferes with protein synthesis.
4. Nucleic acid whose synthesis is inhibited by rifampin.
5. Many _____ cephalosporin antibiotics are available today.
7. Type of chemical group in gentamicin and neomycin.
9. Cell (abbr) having no hemoglobin following chloramphenicol use.

ACROSS (*continued*)

16. Country (intls) where penicillin developed after British discovery.
20. Number of major drawbacks to penicillin use.
22. An example of an aminoglycoside antibiotic is _____-tamicin.
23. Organ treated with neomycin to relieve infection.
24. Age group where tetracycline used for acne.
25. Unit (abbr) to establish useful dose of antibiotic.
26. Macrolide antibiotic used as a penicillin substitute.
27. An antibiotic and its receptor are similar to a _____ and key.
30. Body organ possibly damaged by antibiotic use.
32. Bacterium (intls) from which bacitracin isolated.
34. The drug recommended for tuberculosis is _____-niazid.
35. Disease (intls) treated with INH and rifampin.
36. Number of benzene rings in tetracycline molecule.
37. Antibiotic whose major side effect is damage to the auditory mechanism.
40. A desirable trait for antibiotics is _____ excretion to the urine.
41. Type of microorganism that produces certain antibiotics.
42. Organ affected by excessive streptomycin use.
45. A _____-day period of antibiotic use may be required.
46. Method for applying a topical antibiotic.
47. Number of beta-lactam rings in a penicillin molecule.
48. Body part where fungus infection may require antibiotic.
49. Bacterium (abbr) treated with penicillin or vancomycin.
50. Genus of fungus susceptible to nystatin therapy.

DOWN (*continued*)

10. Fungus (intls) from which penicillin first isolated.
11. Antifungal drug to treat yeast infection.
12. Used in the eye for conjunctivitis.
15. After reaction with penicillinase, penicillin becomes _____-active.
17. Genus of antibiotic-producing bacteria.
18. Type of bacteria treated with aminoglycosides.
19. Method of introducing cephalosporins to body.
21. Used against penicillin-resistant bacteria.
22. Bacterial characteristic that often determines antibiotic use.
23. Where penicillin allergy exists, _____-thromycin is useful.
28. Function can be affected by antibiotic use.
29. Microorganism unaffected by many antibiotics.
31. Swollen during penicillin allergy.
33. Ampicillin and amoxicillin are _____-synthetic drugs.
38. Organic compound whose synthesis inhibited by certain antibiotics.
39. Urinary _____ infections are treated with certain antibiotics.
43. Antibiotic synthesis is a multi-_____ process.
44. Part of body on which polymyxin is used.
48. Salt (abbr) of antibiotic that is often used.

MICROBIOLOGY AND PUBLIC HEALTH

Public health has many facets. Several such as mental health, nutritional health, and environmental health lie outside the realm of this text. Other facets, however, are concerned with communicable disease and therefore fall within the scope of microbiology. Immunization, for example, is a public health measure because it creates a barrier within the individual that lessens susceptibility to infection. Case finding and antibiotic treatment are also under the umbrella of public health because they are used to limit the spread of communicable disease. ■ In Part VII, we shall be concerned with sanitation, a facet of public health designed to prevent microorganisms from reaching the body. The word "sanitation" is derived from the Latin *sanitas*, meaning health. Sanitation came into full flower in the mid-1800s and is a relatively recent phenomenon. Before that time, living conditions in some Western European and American cities were almost indescribably grim: garbage and dead animals littered the streets, human feces and sewage stagnated in open sewers, rivers were used for washing, drinking, and excreting, and filth was rampant. ■ The belief that filth was a catalyst to disease fueled the sanitary movement of the mid-1800s. As the Industrial Revolution sparked great population increases in the cities, health problems mounted, and sanitary reformers spoke up for effective sewage treatment, water purification, and food preservation. The germ theory of disease, developed in the 1870s, strengthened the movement and justified its actions because communicable disease could now be blamed on microorganisms in food and water. As sanitary reformers merged with bacteriologists, they issued loud calls for government support of public health and stimulated the development of sophisticated methods for dealing with food and water that we shall study in Chapters 24 and 25. ■ But not all microorganisms are bad. Indeed some contribute mightily to public health. For example, certain microorganisms are used to produce a broad assortment of foods and dairy products that make up a regular part of our diet. In addition, microorganisms play a dominant role in numerous cycles of elements in the environment and forge a link between what is useless and what is useful to other living things. Moreover, scientists employ microorganisms on industrial scales to synthesize a variety of products that we could not obtain otherwise. We shall see examples of these contributions in Chapters 24 and 25.

MICROBIOLOGY PATHWAYS

PUBLIC HEALTH MICROBIOLOGY

If you enjoy the adventure and excitement of travel, then a career in public health microbiology may be for you. Public health deals with the effects of the environment on community health. It involves strategic planning, services delivery, organization, marketing, and economics. Public health practitioners vary from individuals with two-year college degrees to those with M.D. and Ph.D. degrees. Indeed, there is even a special degree for specialists in public health, the D.P.H. (Doctor of Public Health).

The public health microbiologist is often called an epidemiologist, that is, one who studies epidemics and the means to interrupt them. This individual is concerned with serious foodborne and waterborne diseases of the community such as cholera and typhoid fever and with diseases such as AIDS. Even the less serious diseases such as chickenpox and measles are studied because large-scale outbreaks can exact heavy human tolls. Public health also takes into account the new and emerging epidemics occurring in various parts of the world. (Public health microbiologists were the doctors and researchers who arrived to study recent outbreaks of Ebola fever in Africa.)

Public health microbiologists are also involved with the sanitary treatment of water supplies, and they are constantly on the lookout to ensure conformity with public health standards. They oversee the proper disposal of sewage and industrial wastes and work to ensure the safety of milk, dairy products, and other foods we consume. The spread of disease by insects, rodents, and wildlife is also in the domain of the public health microbiologist working locally and at distant parts of the globe. Public health microbiologists attempt to prevent and control the spread of contagious diseases in the urban environment as well as in the forest, open savanna, mountains, and deserts. There is virtually no place on Earth that the public health microbiologist will not put his or her talents to use.

CHAPTER 24

MICROBIOLOGY OF FOODS

The idea was appealing and the price was right: a patty melt sandwich and a soft drink for lunch. The rye bread was toasted, the hamburgers were stacked and waiting to be cooked, the American cheese slices stood next to the grill, and the aroma from the sauteed onions was irresistible. It was October 1983, at a restaurant at the Northwoods Mall in Peoria, Illinois. The stage was set for the third largest recorded outbreak of **botulism** in U.S. history.

Between October 14 and 16, numerous people stopped at the now-defunct Skewer Inn in Peoria and enjoyed patty melt sandwiches. Soon, however, 36 individuals began experiencing the paralyzing signs of botulism. They suffered blurred vision, difficulty in swallowing and chewing, and labored breathing. One by one they called their doctors, and within a week, all were hospitalized. Twelve patients had to be placed on respirators, but after many anxious hours, all but one recovered (MicroFocus: 24.1).

Investigators from the Centers for Disease Control arrived in Peoria shortly thereafter. They obtained detailed food histories from victims and from others who ate at the restaurant during the same three-day period. First they identified patty melt sandwiches as the probable cause (24 of 28 victims interviewed specifically recalled eating the sandwiches); then they began a search to pinpoint the sandwich item that might be responsible. The data pointed to the onions. Investigators isolated *Clostridium botulinum* spores from the skins of fresh onions at the restaurant, and learned that once sauteed, the onions were left uncovered on the warm stove for hours. Furthermore, the onions were not reheated before serving. Spores had probably germinated within anaerobic mounds of warm onions and deposited their deadly toxins.

Incidents like this one point up how most foods, even cooked foods, provide excellent conditions for the growth of microorganisms and the deposit of their toxins. The organic matter in food is plentiful, the water content is usually sufficient, and the pH is either neutral or only slightly acidic. To the food manufacturer or restaurant owner, the growth of microorganisms may spell economic loss or a reputation for bad business. To the consumer, it may mean illness or a waste of money.

The primary thrust of this chapter will be to examine the types of microorganisms that contaminate various foods and to point out the consequences of contamination. We shall also focus on food spoilage and the preservation methods used to prevent spoilage. As we shall see, some forms of microbial growth in food are actually desirable because they lead to numerous fermented foods we consume regularly.

Food Spoilage

Food spoilage has been a continuing problem since humans first discovered they could produce more food than could be consumed in a single meal. School children are taught that Marco Polo traveled to China in the thirteenth century to obtain spices and explore new trade routes. What many fail to realize is that spices were more than just a luxury at that time. They were essential for improving the smell and taste of spoiled food. Refrigeration was virtually unknown, and canning was yet to be invented.

Food is considered **spoiled** when it has been altered from the expected form. Usually, the food has an unpleasant appearance, aroma, and taste. Sometimes, however, these signs may be difficult to detect such as when staphylococci deposit enterotoxins in food or when too few bacteria grow to cause a perceptible change. Consumption of toxins or microorganisms may cause a number of food poisonings or infections, including those noted in Chapter 8.

Contaminating microorganisms enter foods from a **variety of sources**. Airborne organisms, for example, fall onto fruits and vegetables, then penetrate the product through an abrasion of the skin or rind. Crops carry soilborne bacteria to the processing plant. Shellfish concentrate organisms by straining contaminated water and catching the organisms in their filtering apparatus. And rodents and arthropods transport microorganisms on their feet and body parts as they move about among foods.

Human handling of foods also provides a source of contamination. For example, bacteria from an animal's intestine contaminate meat handled carelessly by a butcher. Of even more concern are raw vegetables such as those

Enterotoxins:
toxins that affect the gastrointestinal tract.

FIGURE 24.1

Food Spoilage

How chemical and physical properties affect the type of microorganism that grows in foods. The figure shows the conditions under which foods are likely to spoil quickly or resist spoilage.

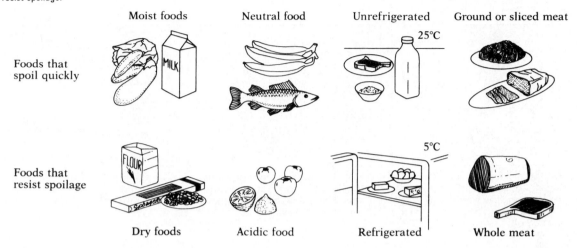

obtained at salad bars. In March 1983, for instance, 107 Maryland residents contracted shigellosis after eating from the salad bar in the cafeteria at a military hospital. Earlier that year, 123 cases of hepatitis A were diagnosed in patients who dined at a salad bar restaurant in Lubbock, Texas. In both cases, investigators believed that food handlers were the source of pathogenic microorganisms.

Shigellosis:
a bacterial disease of the intestine, caused by a Gram-negative rod and accompanied by extensive diarrhea.

The Conditions of Spoilage

Since food is basically a culture medium for microorganisms, the chemical and physical properties of food have a significant bearing on the type of microorganisms growing on or in it (Figure 24.1). **Water**, for instance, is a prerequisite for life, and therefore the food must be moist, with a minimum water content of 18 to 20 percent. Microorganisms do not grow in foods such as dried beans, rice, and flour because of their low water content.

Another important fact is **pH**. Most foods fall into the slightly acidic range on the pH scale, and numerous species of bacteria multiply under these conditions. In foods with a pH of 5.0 or below, acid-loving molds often replace the bacteria. Citrus fruits, for example, generally escape bacterial spoilage but yield to mold contamination.

pH:
a measure of the acidity or alkalinity of a substance.

A third property of the food is its **physical structure**. A steak, for example, is not likely to spoil quickly because microorganisms cannot penetrate the meat easily. However, an uncooked hamburger rapidly deteriorates since microorganisms exist within the loosely packed ground meat as well as on the surface.

The **chemical composition** of the food may be another determining factor in the type of spoilage encountered. Fruits, for instance, support organisms that metabolize carbohydrates, whereas meats attract protein decomposers. Starch-utilizing bacteria and molds are often found on potatoes, corn, and rice products. The presence of certain vitamins encourages particular microorganisms to proliferate, while the absence of other vitamins provides natural resistance to decay.

FIGURE 24.2
A Textbook Case of *Vibrio parahaemolyticus* Food Poisoning

This outbreak occurred in Port Allen, Louisiana, in June 1978. *Vibrio parahaemolyticus* was subsequently found in the leftover shrimp as well as in a major number of stool specimens from patients.

1. On the morning of June 21, shrimp was cooked for a dinner by bringing the water to a "rolling boil". The gas was then turned off.

2. The shrimp was repackaged in the boxes in which it came and was covered with aluminum foil to keep it warm.

3. The shrimp was transported 40 miles in an unrefrigerated truck to the site of the dinner.

4. It was held unrefrigerated until 7:30 P.M. serving time, a total of almost 8 hours since preparation.

5. Seventeen hundred persons attended the dinner that evening. The shrimp was served to all without incident.

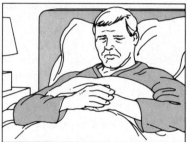

6. Approximately 16 hours later, 1100 of the 1700 guests reported diarrhea, cramps, nausea, and vomiting. The diagnosis was food poisoning.

Anaerobes:
microorganisms that live in the absence of oxygen.

Oxygen and **temperature** are other considerations. Vacuum-sealed cans of food do not support the growth of aerobic bacteria, nor do vegetables or most bakery products support anaerobes. Similarly, the refrigerator is usually too cold for the growth of human pathogens, but the warm hold of a ship or a humid, hot warehouse storeroom is an environment conducive to the growth of these pathogens. It is common knowledge that contamination is more likely in cooked food at warm temperatures than in refrigerated cooked food (Figure 24.2).

The food industry recognizes three groups of foods loosely defined on the basis of their chemical and physical properties. The **highly perishable** foods are those that spoil rapidly. They include poultry, eggs, meats, most vegetables and fruits, and dairy products. Other foods such as nutmeats, potatoes, and some apples are considered **semiperishable**, because they spoil less quickly. The **nonperishable** foods are often stored in the kitchen pantry. Included in this group are cereals, rice, dried beans, macaroni products, flour, and sugar. Figure 24.3 summarizes the three groups.

FIGURE 24.3
Types of Foods

Examples of highly perishable, semiperishable, and nonperishable foods. The physical and chemical properties of these foods are reliable indicators of their rate of perishability.

Highly perishable foods   Semiperishable foods   Nonperishable foods

The Chemistry of Spoilage

Spoilage in foods is often due to the naturally occurring chemistry of contaminating microorganisms. Yeasts, for instance, live in apple juice and convert the carbohydrate into **ethyl alcohol**, a product that gives spoiled juice an alcoholic taste. Certain bacteria convert food proteins into amino acids, then break down the amino acids into foul-smelling end-products. Cysteine digestion, for example, yields hydrogen sulfide, which imparts a rotten egg smell to food. Tryptophan digestion yields **indole** and **skatole**, which give food a fecal odor.

sis-te'in

Indole:
a foul-smelling product of the breakdown of tryptophan.

Two other possible products of the microbial metabolism of carbohydrates are **acid**, which causes food to become sour, and **gas**, which causes sealed cans to swell. Moreover, when fats break down to fatty acids as in spoiled butter, a rancid odor or taste may evolve. **Capsule** production by bacteria causes food to become slimy, and **pigment** production imparts color. In numerous historical incidents, the red pigment from *Serratia marcescens* in bread has been interpreted as a sign of blood (MicroFocus: 24.2).

sĕ-ra'-she-ah mar-ses'ens

Some foods resist spoilage naturally because they contain antimicrobial chemicals. The white of an egg has **lysozyme**, an enzyme that digests the cell wall of Gram-positive bacteria. Garlic contains certain compounds that inhibit many bacteria (MicroFocus: 22.2). Oil of cloves appears to have healing tendencies, while radish and onion extracts both retard bacterial growth.

li'so-zime

Meats and Fish

Meats and fish are originally free of contamination because the muscle tissues of living animals are normally sterile. Spoilage organisms enter during handling, processing, packaging, and storage. For example, if a piece of meat is ground for **hamburger**, microorganisms from the surface accumulate in the teeth of the grinder along with other dustborne organisms. Bacteria from preparers' hands or from a sneeze compound the problem. The grinder may be cleaned well and refrigerated, but rarely is it sterilized. The importance of foodborne infection in ground meat is pointed up by the 1993 outbreak of hemolytic diarrhea traced to hamburger meat contaminated by *Escherichia coli* 0157:H7. Over 500 patrons of Jack-In-The-Box restaurants were involved (Chapter 8).

Clostridium botulinum:
a Gram-positive sporeforming rod that causes botulism.

lac″to-ba-cil′lus
loo″ko-nos′tok

There is also the problem of the so-called "choke points," or places where animals come together and epidemics spread. In the United States, for instance, about 9000 farms produce calves for beef; the animals are then sent to about 46,000 feedlots for developing, and then, to about 80 plants for slaughter. Microorganisms can spread at any of these points. Moreover, animal feeds often contain the entrails of poultry as added protein sources, and another opportunity for bacteria to spread is presented.

Processed meats such as luncheon meats, sausages, and frankfurters represent special hazards because they are handled often. Also, natural sausage and salami casings made from animal intestines may contain residual bacteria, especially botulism spores. When preparers pack such casings tightly with meat, ***Clostridium botulinum*** may multiply and produce its powerful toxins. As early as the 1820s, people recognized the symptoms of "sausage poisoning" and coined the name botulism, which translates loosely to sausage.

Organ meats such as livers, kidneys, and sweetbreads (thymus and pancreas) are less compact tissue than muscle and thus spoil more quickly. Moreover, because the organs contain many natural filtering tissues, bacteria tend to be trapped here. Foods like these should therefore be cooked as soon as possible after purchase.

The extent of contamination in meats often consists of a harmless "greening" seen on the surface of a steak. This discoloration is commonly due to the Gram-positive rod ***Lactobacillus***, or the Gram-positive coccus ***Leuconostoc***. The green color results from pigment alteration in the meat and represents no hazard to the consumer. Slime formation on the outer casings and souring in frankfurters, bologna, or processed meats may also result from these organisms as well as from the *Streptococcus* species.

Microbiologists can often trace spoilage in **fish** to the water from which it is taken or held. Fish tissues deteriorate rapidly and the filleting of fish on a blood-stained block encourages contamination. It is also interesting to note that bacteria in fish are naturally adapted to the cold environment that fish live in and thus, cooling will not affect them so thoroughly; freezing is preferred.

FIGURE 24.4
A Textbook Case of *Staphylococcus aureus* Food Poisoning

This incident took place in Sussex County, Delaware, between March 8 and 10, 1979. Analysis showed that *S. aureus* phage type 95 was present in the nasal swab from the person who ground the chicken as well as in the meat grinder and in the leftover chicken salad.

1. On March 8, 1979, chicken was cooked for chicken salad for a wedding reception to be held 2 days later.

2. The chicken was deboned and refrigerated overnight in a large washtub.

3. On March 9, the chicken was ground in a meat grinder by a person who had toxigenic *Staphylococcus aureus* in his nose.

4. The chicken was then mixed with celery, onions, and mayonnaise and refrigerated overnight in the same washtub.

5. On March 10, the chicken salad was delivered, unrefrigerated, to the wedding reception where it was placed on a table. A total of 7 hours passed until it was consumed.

6. Some hours later, 64 of the 107 guests at the wedding experienced nausea, diarrhea, intestinal cramps, and other symptoms of food poisoning.

Shellfish are of particular concern because they commonly obtain their food by filtering particles from the water. Clams, oysters, and mussels therefore concentrate such pathogens as hepatitis A viruses, typhoid bacilli, cholera vibrios, or amebic cysts. In 1989, six cholera cases were related to raw oysters.

Poultry and Eggs

Contamination in poultry and eggs may reflect human contamination (Figure 24.4), but it usually stems from microorganisms that have infected the bird. Members of the genus *Salmonella* may cause diseases in **chickens** and **turkeys**, then pass to consumers via poultry and egg products. For example, a 1986 outbreak of **salmonellosis** in New York was traced to contaminated eggs used to make cheese lasagna and stuffed shells. Processed foods such as chicken pot pies, whole egg custard, mayonnaise, egg nog, and egg salad may also be sources of salmonellosis. **Psittacosis** is another problem because *Chlamydia psittaci* infects poultry. In July 1981, for instance, 27 employees in a turkey-processing plant contracted psittacosis while preparing consumer products from diseased

Salmonella:
a genus of Gram-negative nonsporeforming rods.

sit-ah-ko′sis

kla-mid′e-ah sit′ah-sī

turkeys. A third problem is **listeriosis**. In 1989 in Oklahoma, a serious outbreak of listeriosis was related to turkey franks. Contamination from the turkey's visceral organs was suspected.

Eggs are normally sterile when laid, but the outer waxy membrane as well as the shell and inner shell membrane may be penetrated by bacteria after several hours. *Proteus* species cause **black rot** in eggs as hydrogen sulfide gas accumulates from the breakdown of cysteine. This gives eggs a rotten odor. Other spoilages in eggs include **green rot** from *Pseudomonas* species and **red rot** from the growth of *Serratia marcescens*. Green rot results from a fluorescent pigment that causes the egg to glow when placed in front of an ultraviolet light. Red rot gives the yolk a blood-red appearance.

The primary focus of contamination in an egg is the yolk rather than the white. This is due to the nutritious quality of the yolk and because the white has a pH of approximately 9.0. Also, the lysozyme in egg white is inhibitory to Gram-positive bacteria.

It should be noted that even though eggs are hard-boiled, contamination is not always prevented. This became apparent in 1983 when over 300 children developed staphylococcal food poisoning after eating hard-boiled eggs following an Easter-egg hunt in Modesto, California.

Breads and Bakery Products

In the production of bakery products, ingredients such as flour, eggs, and sugar are generally the sources of spoilage organisms. Although most contaminants are killed during baking, some bacterial and mold spores survive because **bread** is heated internally to 100° C for only about nine minutes. Members of the sporeforming genus ***Bacillus*** commonly survive, and as they proliferate, their capsular material accumulates, giving the bread a soft cheesy texture with long, stringy threads. The bread is said to be **ropy**.

Cream fillings and toppings in **bakery products** provide excellent chemical and physical conditions for bacterial growth. For example, custards made with whole eggs may be contaminated with *Salmonella* species, and whipped cream may contain dairy organisms such as species of *Lactobacillus* and *Streptococcus*. The acid produced by these bacteria results in a sour taste. High sugar environments of chocolate toppings and sweet icings support the growth of fungi. Most bakery products should be refrigerated during warm summer months.

Grains

Two types of grain spoilage are important in public health microbiology. The first type of spoilage is caused by the ascomycete ***Aspergillus flavus***. This mold produces **aflatoxins**, a series of toxins that accumulate in stored grains such as wheat as well as peanuts, soybeans, and corn. Scientists have implicated aflatoxins in liver and colon cancers in humans. The toxins are consumed in grain products as well as meat from animals that feed on contaminated grain.

The second type of grain spoilage is caused by ***Claviceps purpurea***, the cause of **ergot poisoning** (ergotism). Rye plants are particularly susceptible to this type of spoilage (MicroFocus: 24.3), but wheat and barley grains may also be affected. The toxins deposited by *C. purpurea* may induce convulsions and hallucinations when consumed. The drug LSD is derived from the toxin.

Milk and Dairy Products

Milk is an extremely nutritious food. It is an aqueous solution of proteins, fats, and carbohydrates that contains numerous vitamins and minerals. Milk has a pH of about 7.0 and is an excellent growth medium for humans and animals, as well as microorganisms.

---

sis-te′in

soo″do-mo′nas

Lysozyme:
an enzyme that breaks down the cell walls of Gram-positive bacteria.

Ropy bread:
bread that is soft and stringy due to the presence of capsular material.

Lactobacillus:
a genus of Gram-positive rods known for its production of acid in foods.

klav′ĭ-seps pur-pu′re-ah

24.3

MICROFOCUS

24.3 MICROFOCUS

OF PILGRIMS AND WITCHES

The first descriptions of ergot poisoning appeared in the Middle Ages, although it is probable that the disease was prevalent long before that time. Ergot poisoning was accompanied by burning pain in the extremities, and in the 1100s the disease was often called the Holy Fire (*Ignis Sacer*). For some strange reason, though, it would disappear if people made a pilgrimage to the hospital of St. Anthony, founded in 1039 near Vienne, France. Thus the disease was also called St. Anthony's Fire. Microbiologists now believe that ergot-contaminated rye was the cause of disease, and that people became ill from poisoned rye bread. The cure of St. Anthony probably resulted from a change of diet en route to the hospital because the pilgrims ate uncontaminated bread. Also, the monks at the hospital served wheat bread, not rye bread.

An equally serious story related to ergot poisoning began in December 1691. That month, eight girls from the town of Salem, Massachusetts, developed disorderly speech, odd postures and gestures, and convulsive fits. Observers suggested that perhaps the girls were bewitched, since no other cause for their symptoms could be found. The girls were given a "witch cake" made from rye flour to determine if witchcraft was involved. Now the symptoms worsened: the girls experienced burning pain in the extremities, buzzing in the ears, and sensations of flying through the air "out of the body." Diaries from that time record that the winter of 1690–1691 was very cold, and that rye grains flourished where other crops failed. In retrospect, it is conceivable that ergot-contaminated grain was in the witch cakes and since children and teenagers ingest more food per unit of body weight, the poison may have affected them the most.

The infamous Salem witch trials began on June 2, 1692, and lasted through May of the following year. Nineteen young people were executed for witchcraft. The role of ergot poisoning is minimized by some historians, but it is possible that a mild form of poisoning may have initiated the incident, and that the social and psychological climate of the day made an already bad situation even worse.

About 87 percent of the substance of milk is water. Another 2.5 percent is a protein called **casein**, actually a mixture of three long chains of amino acids suspended in fluid. A second protein in milk is **lactalbumin**. This protein forms the surface skin when milk is heated to boiling. Lactalbumin is a whey protein, one that remains in the clear fluid (the whey) after the casein curdles during milk spoilage or fermentation. Carbohydrates make up about 5 percent of the milk. The major carbohydrate is **lactose**, sometimes referred to as milk sugar (*lactus* is Latin for milk). Rarely found elsewhere, lactose is a disaccharide that can be digested by relatively few bacteria, and these are usually harmless. The last major component of milk is **butterfat**, a mixture of fats that can be churned into butter. Butterfat comprises about 4 percent of the milk and is removed in the preparation of skim milk or low-fat milk. When bacterial enzymes digest fats into fatty acids, the milk develops a sour taste and becomes unfit to drink.

A common type of milk spoilage often takes place in the kitchen refrigerator or dairy case at the supermarket. Here, ***Lactobacillus*** or ***Streptococcus*** species multiply slowly and ferment the lactose in milk. Soon, large quantities of lactic and acetic acids accumulate. Enough acid may develop to change the structure of the protein and cause it to solidify as a curd. Dairy microbiologists refer to such an acidic curd as a **sour curd**. The lactobacilli and streptococci that cause it have usually survived the pasteurization process.

Sweet curdling in milk may result when enzymes from species of *Bacillus*, *Proteus*, *Micrococcus*, or certain other bacteria attack the casein. As weak hydrogen bonds break, casein loses its three-dimensional structure and curdles. The reaction is said to be sweet because little acid production occurs. It is an essential step in the production of cheese. The clear liquid is **whey**, an aqueous solution of lactose, minerals, vitamins, lactalbumin, and other milk components. Whey is used to make processed cheeses and "cheese foods."

Milk may also be contaminated by Gram-negative rods of the coliform group of bacteria, including *Escherichia coli* and *Enterobacter aerogenes*. These bacteria

Whey:
the clear liquid portion of milk remaining after the protein curd has been removed.

Lactose:
a disaccharide composed of a glucose molecule and a galactose molecule covalently linked.

Streptococcus:
a genus of Gram-positive cocci in chain formation.

Hydrogen bonds:
weak bonds resulting from attractions between oppositely charged poles of adjacent molecules.

24.4

MICROFOCUS

GOOD AND NOT-SO-GOOD

In parts of Great Britain, milkmen still visit homes regularly and deliver bottles of fresh milk. That's good. Magpies and crows arrive shortly thereafter and use their strong beaks to peck through the foil caps of the bottles. Then they take a drink. That's bad.

The birds use the milk to feed their young broods in the nest. That's good. But while taking a drink, they transmit *Campylobacter* species to the milk. That's bad.

By having their milk delivered, British families save a trip to the market. That's

good. Unfortunately, they also spend extra time in the bathroom suffering the misery of diarrhea. And that's bad.

The moral of the story? "Bewildered Brits better beware bacteria-bearing birds."

---

produce acid and gas from lactose. The acid curdles the protein, and the gas forces the curds apart, sometimes so violently that they explode out of the container. The result is a **stormy fermentation**. *Clostridium* species also cause this reaction.

Ropiness in milk is similar to that in bread. It develops from the capsule-producing organisms such as *Alcaligenes*, *Klebsiella*, and *Enterobacter*. These Gram-negative rods multiply in milk, even at low temperatures, and deposit gummy material that appears as stringy threads and slime.

Another form of spoilage is caused by species of *Pseudomonas* and *Achromobacter* that produce the enzyme **lipase**. This enzyme attacks butterfats in milk and digests them into glycerol and fatty acids, giving milk a sour taste and a putrid smell. A similar problem may develop in butter.

Additional types of milk spoilage result from the red pigment deposited in dairy products by ***Serratia marcescens***, the blue or green pigment of ***Pseudomonas*** species, and the gray rot caused by certain ***Clostridium*** species. In gray rot, hydrogen sulfide (H_2S) from cysteine imparts a rotten-egg smell to milk, and the H_2S reacts with minerals to yield a gray or black sulfide compound. Spoilage due to wild yeasts is usually characterized by a pink, orange, or yellow coloration in the milk. Acid conditions stimulate mold decay in cheese products.

Milk is normally sterile in the udder of the cow, but contamination occurs as it enters the ducts leading from the udder as well as from other unlikely sources (MicroFocus 24.4). The colostrum, or first milk, is laden with organisms. Species of soilborne *Lactobacillus* and *Streptococcus* are acquired during the passage, together with various coliform bacteria from dust, manure, and polluted water. The milk may derive other organisms from dairy plant equipment and unsanitary handling of dairy products by plant employees.

al″kah-lij′e-nēz
kleb″se-el′ah

soo″do-mo′nas
ah-kro″mo-bak′ter

sĕ-ra′she-ah
mar-ses′ens

TO THIS POINT

■

We have discussed the problem of food spoilage and have indicated how various conditions may contribute to the type of spoilage observed. Among these conditions are the amount of water present, the pH of the food product, the physical structure of the food and its chemical composition, and the oxygen content and temperature of the environment in which the food is found. We also noted how the chemistry of spoilage leads to the deposit of certain end-products in foods. For example, hydrogen sulfide production from cysteine gives a rotten smell to eggs, and pigment production imparts color to foods.

The discussion then turned to various foods and the

spoilage that takes place within them. We examined the sources and types of spoilage within meats and fish, poultry and eggs, breads and bakery products, grains, and milk and dairy products. In each case, some examples of the microbial flora were noted.

We shall now move on to the topic of food preservation. This section will outline various methods utilized to prevent microorganisms from reaching food and causing spoilage. Food preservation is also an essential prerequisite to sanitation and public health because preservation methods halt the spread of infectious microorganisms. Several traditional methods of preservation will be explored together with a number of contemporary methods utilized in food technology.

Food Preservation

Centuries ago, humans battled the elements to keep a steady supply of food at hand. Sometimes there was a short growing season; other times locusts descended on their crops; still other times, they underestimated their needs and had to cope with scarcity. However, experience taught humans they could overcome these difficult times by preserving foods. Among the earliest methods was drying vegetables and strips of meat and fish in the sun. Foods could also be preserved by salting, smoking, and fermenting. Individuals could now trek far from their native habitat, and soon they took to the sea and moved overland to explore new lands.

The next great advance did not come until the mid-1700s. In 1767, **Lazaro Spallanzani** attempted to disprove spontaneous generation by showing that beef broth would remain unspoiled after being subjected to heat. **Nicholas Appert** applied this principle to a variety of foods (MicroFocus: 24.5). Neither Appert nor his contemporaries were quite sure why food was being preserved, but it was clear that the spoilage could be retarded by prolonged heating. The significance of microorganisms as agents of spoilage awaited Pasteur's classic experiments with wine several generations later.

ah-pehr′

Through the centuries, preservation methods have had a common objective: to reduce the microbial population and maintain it at a low level until food could be consumed. Modern preservation methods are mere extensions of these principles. Though today's methods are sophisticated and technologically dynamic, advances in preservation processes are counterbalanced by the great volumes of food that must be preserved and the complexity of food products. Thus the problems that early humans faced do not differ fundamentally from those confronting modern food technologists. In this section, we shall examine the methods of food preservation currently in use.

Heat

Heat kills microorganisms by changing the physical and chemical properties of their proteins. In a moist heat environment, proteins are denatured and lose their three-dimensional structure, reverting to a different three-dimensional form or a two-dimensional form. As structural proteins and enzymes undergo this change, the organisms die. Chapter 21 explores various forms of moist heat and their applications. You might find a brief review helpful in relating heat to food preservation.

Denaturation: a process in which a protein loses its three-dimensional structure and forms a solid.

24.5

MICROFOCUS

TO FEED AN ARMY

Part of Napoleon's genius was understanding the finer points of warfare, including how to feed an army. He realized that thousands of men-at-arms were a glut on the countryside and so he broke up his army into smaller units that foraged on their own as they moved. When the time for battle neared, he reassembled his forces and engaged the enemy.

The shortcomings of this system became painfully clear when Napoleon crossed into Northern Italy in 1800 and engaged the Austrians at the Battle of Marengo. Aware that the French army was scattered about, the Austrians charged

before Napoleon could bring his forces together. Disaster was averted only when units arrived on the flanks to repel the Austrians. Napoleon had learned an important lesson: the next time he went to war, food would go with him.

Included in Napoleon's plan to resurrect France was a ministry that encouraged industry by offering prizes for imaginative inventions. A winemaker named Nicholas Appert attracted attention with his process of preserving food. Appert placed fruits, vegetables, soups, and stews in thick bottles, then boiled the bottles for several hours. He used wax and cork to seal the

bottles and wine cages to prevent inadvertent opening of the bottle. By 1805, Appert had set up a bottling industry outside Paris and had a thriving business.

The Ministry of Industry encouraged Appert to publish his methods and submit samples of bottled foods for government testing. The French Navy took numerous bottles on long voyages and reported excellent food preservation. In 1810, Appert was awarded 12,000 francs for his invention. Two years later, Napoleon assembled hundreds of cannons, thousands of men, and countless bottles of food, and marched off to war with Russia.

The most useful application of heat is in the process of **canning**. Shortly after Appert established the use of heat in preservation (MicroFocus: 24.4), an English engineer named Bryan Donkin substituted iron cans coated with tin for Appert's bottles. Soon he was supplying canned meat to the British navy. In the United States, the tin can was virtually ignored until the Civil War period. In the years thereafter, mass production began and soon the tin can became the symbol of prepackaged convenience.

Blanching:
treatment with heat for a short period of time to destroy cellular enzymes.

Modern canning processes are complex (Figure 24.5). Machines wash, sort, and grade the food product, then subject it to steam heat for three to five minutes. This last process, called **blanching**, destroys many enzymes in the food product and prevents any further cellular metabolism from taking place. The food is then peeled and cored, and its diseased sections are removed. Canning comes next, after which the containers are evacuated and placed in a pressured steam sterilizer similar to an autoclave at a temperature of 121° C or lower depending on the pH, density, and heat penetration rate.

Clostridium:
a genus of Gram-positive anaerobic sporeforming rods.

The sterilizing process is designed to eliminate the most resistant bacterial spores, especially those of the genera *Bacillus* and *Clostridium*. It is considered **commercial sterilization**, however, which is not as rigorous as true sterilization, and some spores may survive. Moreover, should a machine error lead to improper heating temperatures, a small hole allow airborne bacteria to enter, or a proper seal not form, contamination may result.

Coliform bacteria:
a group of Gram-negative rods commonly found in the human and animal intestine.

Contamination of canned food is commonly due to facultative or anaerobic bacteria that produce gas and cause the ends of the can to bulge. Food microbiologists call a can a **flipper** if the bulge can be flattened easily. It is a **springer** if pushing the bulge pushes out the opposite end of the can. A **soft swell** occurs when both ends bulge. If neither end can be pushed in because of the large amount of gas, a **hard swell** is present. The organisms often responsible for gas production are *Clostridium* species as well as coliform bacteria, a group of Gram-negative nonsporeforming rods that ferment lactose to acid and gas. Contamination is usually obvious since the spoiled product generally has a putrid odor.

FIGURE 24.5

Two of the Many Steps in an Industrial Canning Process

(a) Initial inspection of green beans is made after cutting and sieve sizing. The beans are washed in the apparatus in the foreground and then conveyed through sanitary glass piping (at the left) to the blanching machine. (b) A continuous cycle orbital cooker used by modern vegetable processors to reduce cooking time of the product.

(a)   (b)

Growth of acid-producing bacteria presents a different problem because spoilage cannot be discerned from the can's shape. Food has a **flat-sour** taste from the acid and has probably been contaminated by a *Bacillus* species, a coliform, or another acid-producing bacterium that survived the heating.

The process of **pasteurization** was developed by Louis Pasteur in the 1850s to eliminate bacteria in wines. His method was first applied to milk in Denmark about 1870, and by 1895 the process was widely employed. Although the primary object of pasteurization is to eliminate pathogenic bacteria from milk, the process also lowers the total number of bacteria and thereby reduces the chance for spoilage. The more traditional method involves heating the milk in a large bulk tank at 62.9° C (145° F) for 30 minutes. This is the **holding method**, also known as the **LTLT** method for "low temperature, long time." Machines stir the milk constantly during the pasteurization to ensure uniform heating, and cool it quickly when the heating is completed. More concentrated products such as cream or ice cream are often heated at the higher temperature of 69.5° C (155° F) to be certain of successful pasteurization.

The more modern method of pasteurization is called the **flash method**. In this process, raw milk is first warmed using the heat of previously pasteurized milk. Machines then pass the milk through a hot cylinder at 71.6° C (161° F) for a period of 15 to 17 seconds. Next, the milk is cooled rapidly, in part by transferring its heat to the incoming milk. This is the **HTST** method for "high temperature, short time." It is useful for high-quality raw milk, in which the bacterial count is consistently low. A new method called **ultrapasteurization** is used in some dairy plants. In this process, milk and milk products are subjected to heat at 82° C (180° F) for 3 seconds.

Holding method:
a method of pasteurization that employs heat at 62.9° C for 30 minutes.

Flash method:
a method of pasteurization that employs heat at 71.6° C to 15 to 17 seconds.

FIGURE 24.6

Important Temperature Considerations in Food Microbiology

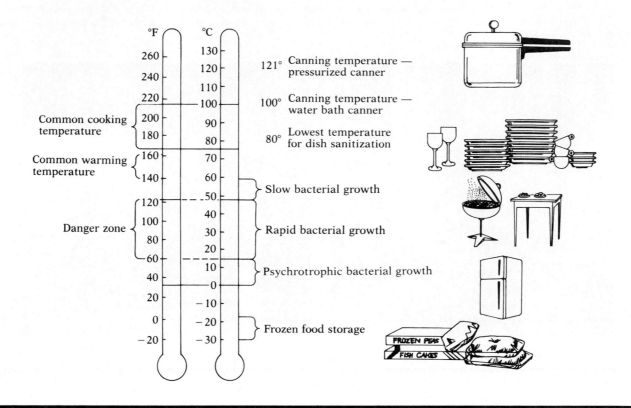

The bacteria that survive pasteurization may be involved in spoilage. *Streptococcus lactis*, for instance, grows slowly in refrigerated milk, and when its numbers reach 20 million per milliliter (ml), enough lactic acid has been produced to make the milk sour. Organisms that survive the heat of pasteurization are described as **thermoduric**. Pasteurization is virtually useless against **thermophilic bacteria**, since they grow naturally at 60° C to 70° C. These organisms do not grow at refrigerator temperatures or cause human disease, however, since conditions are too cool. Pasteurization is also useless against spores.

Although milk in the United States is normally pasteurized and refrigerated, exposure to steam at 140° C for 3 seconds can sterilize it. This **ultra-high temperature (UHT)** results in milk (e.g., Parmalat) with an indefinite shelf life as long as the container remains sealed. Small containers of coffee cream are often prepared this way.

Thermoduric:
heat-enduring.
Thermophilic:
heat-loving.

Low Temperatures

By lowering the environmental temperature one can reduce the rate of enzyme activity in a microorganism and thus lower the rate of growth and reproduction. This principle underlies the process of refrigeration and freezing. Although the organisms are not killed, their numbers are kept low and spoilage is minimized. Ironically, the food is preserved by preserving the microorganisms.

Well before contemporary humans developed refrigerators, the Greeks and Romans had solved the problem of keeping things cold. They simply dug a snow cellar in the basements of their homes, lined the cellar with logs, insulated it

FIGURE 24.7
Food Contamination

A scanning electron micrograph of unidentified
flagellated bacteria growing on the skin of a chicken
carcass. Bacteria such as these contaminate frozen food
and grow to large numbers when the food is thawed and
held for long periods before cooking. Bar = 2 μm.

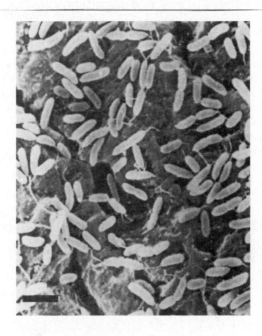

with heavy layers of straw, and packed it densely with snow delivered from far-off mountaintops ("The iceman cometh!"). The compressed snow turned to a block of ice, and foods would remain unspoiled for long periods when left in this makeshift refrigerator. The modern **refrigerator** at 5° C (41° F) provides a suitable environment for preserving food without destroying its appearance, taste, or cellular integrity (Figure 24.6). However, psychrotrophic microorganisms survive and cause green meat surfaces, rotten eggs, moldy fruits, and sour milk. Pathogens such as *Listeria monocytogenes* and *Yersinia enterocolitica* also grow at low temperatures.

si"kro-trof'ik

When food is placed in the **freezer** at −5° C (23° F), ice crystals form rapidly. These crystals tear and shred microorganisms and kill a significant number. However, many microorganisms survive, and the ice crystals are equally destructive to food cells. Therefore, when the food thaws, bacteria multiply quickly. Organisms such as staphylococci produce substantial amounts of enterotoxins, and *Salmonella* serotypes and streptococci and other bacteria grow to large numbers (Figure 24.7). Rapid thawing and cooking are thus recommended. Moreover, food should not be refrozen, because during thawing and refreezing bacteria deposit sufficient enterotoxin to cause food poisoning the next time the food is thawed. Microwave cooking, which requires minimal thawing, may eliminate some of these problems.

Deep freezing at −60° C results in smaller ice crystals, and although the physical damage to microorganisms is less severe, their biochemical activity is reduced considerably. The small ice crystals do not damage food cells as severely as do the larger crystals formed at higher temperatures. Some food producers blanch their product before deep freezing, a process that further reduces the number of microorganisms.

A major problem in freezing is **freezer burn**, which may occur over long periods of time as food dries out from moisture evaporation. Another disadvantage is that the energy cost for freezing is considerable. Nevertheless, freezing

Freezer burn:
the drying of foods subjected to long periods of freezing.

They say it began in 1878 at Moon's Lake House Restaurant in fashionable Saratoga Springs, New York. Cornelius Vanderbilt, the wealthy entrepreneur was visiting the resort, and one evening he took issue with the fried potatoes being served. They were too thick, it seemed, and the chef, a certain George Crum should have known better. But Vanderbilt was magnanimous—Crum would have another chance to make a more acceptable potato dish.

Crum did not take the criticism well at all, and he plotted revenge. If Vanderbilt wanted thinner potatoes, then thinner potatoes he would get. Apparently trying for overkill, Crum sliced the potatoes superthin,

dropped them into boiling oil, and fried them to a crisp. Then he sent the potatoes out to his picky patron, fully expecting to be working elsewhere the next day. You can imagine his surprise when he heard sounds of delight from the dining room—Vanderbilt loved the crunchy potatoes. Soon everyone was coming from miles around to sample the Saratoga chips, as they were called. The chips later became famous as potato chips. Unfortunately, few people knew about their inventor George Crum, ... until now, that is.

has been a mainstay of preservation since Clarence Birdseye first offered frozen foods for retail purchase in the 1920s. Approximately 33 percent of all preserved food in the United States is frozen.

Drying

The advantage of drying foods is best expressed by the phrase "where there is no water, there is no natural life." Indeed, dry foods cannot support microbial growth (MicroFocus: 24.6). In past centuries, people used the sun for drying, but modern technologists have developed sophisticated machinery for this purpose. For example, the **spray dryer** expels a fine mist of liquid such as coffee into a barrel cylinder containing hot air. The water evaporates quickly and the coffee powder falls to the bottom of the cylinder.

Another machine for drying is the **heated drum**. Machines pour liquids such as soup on the drum and the water evaporates rapidly, leaving dried soup to be scraped off the drum. A third machine utilizes a **belt heater** that exposes liquids such as milk to a stream of hot air. The air evaporates any water and leaves dried milk solids. Unfortunately, sporeforming and capsule-producing bacteria are problems because they resist drying.

During the past 20 years, freeze-drying, or **lyophilization**, has emerged as a valuable preservation method. In this process, food is deep frozen, and then a vacuum pump draws off the water in a machine like the one pictured in Figure 24.8. The dry product is sealed in foil and easily reconstituted with water. Hikers and campers find considerable value in freeze-dried food because of its light weight and durability, and freeze-dried products such as coffee are often found on the grocery shelf. However, there is a disquieting note: lyophilization is also a useful method for storing, transporting, and preserving bacterial cultures.

Spray dryer:
a device that dries liquid foods by spraying a fine mist from which the water is evaporated.

li-of″ĭ-li-za′shun

Lyophilization:
the process of freeze-drying.

FIGURE 24.8
An Industrial Model Lyophilizer

This model removes 500 pounds of product moisture in 24 hours of freeze drying. Using vacuum and heat, water is drawn off from ice without passing through the liquid phase to produce the freeze-dried product.

Osmotic Pressure

When living cells are immersed in large quantities of a compound such as salt or sugar, water diffuses out of cells through cell membranes and into the surrounding environment, where it dilutes the high concentration of the compound. The flow of water is called **osmosis** and the force that drives the water is termed **osmotic pressure**.

Osmotic pressure can be used to preserve foods because water flows out of microorganisms as well as food cells. For example, in highly salted or sugared foods, microorganisms dehydrate, shrink, and die. Jams, jellies, fruits, maple syrups, honey, and similar products typify foods preserved by high sugar concentrations. **Salted foods** include ham, cod, bacon, and beef, as well as vegetables such as sauerkraut, which has the added benefit of large quantities of acid. It should be noted, however, that staphylococci tolerate salt and may survive to cause staphylococcal food poisoning.

Osmotic pressure:
the force that drives water through a membrane in osmosis.

Staphylococci:
Gram-positive irregular clusters of aerobic cocci.

Chemical Preservatives

In order for a chemical preservative to be useful in foods, it must be inhibitory to microorganisms while easily broken down and eliminated by the body without side effects. Other requirements are listed in Table 24.1. These requirements have limited the number of available chemicals to a select few.

A major group of chemical preservatives are the organic acids such as sorbic acid, benzoic acid, and propionic acid. Microbiologists believe that these compounds damage microbial membranes and interfere with the uptake of certain essential organic substances such as amino acids. Chemicals are used primarily against molds and yeasts, but their acidity is also a deterrent to bacterial growth.

Sorbic acid, which came into use in 1955, is used in syrups, salad dressings, jellies, and certain cakes. **Benzoic acid**, the first chemical (1908) to be approved by the Food and Drug Administration, protects beverages, catsup, margarine, and apple cider. **Propionic acid** is incorporated in wrappings for butter and

Benzoic acid:
a phenyl-containing organic acid used as a food preservative.

TABLE 24.1
Requirements for Use of a Chemical as a Preservative

Must be economical
Must extend the storage life of the food
Must be safe at levels needed
Must be readily identified by chemical analysis
Must exhibit antimicrobial abilities at pH of the food
Must be readily soluble
Must not lower the quality of the food
Must be easily controlled in the food
Must exhibit activity against spoilage organisms expected
Must be uniformly distributed in the food
Must not retard the action of digestive enzymes
Must be the last alternative after other preservation methods have been exhausted

---

cheese, and is added to breads and bakery products, where it inhibits the ropiness commonly due to *Bacillus* species and prevents the growth of fungi. Other natural acids in food add flavor while serving as preservatives. Examples are **lactic acid** in sauerkraut and yogurt, and **acetic acid** in vinegar.

The process of **smoking** with hickory or other woods accomplishes the dual purposes of drying food and depositing chemical preservatives. By-products of smoke such as aldehydes, acids, and certain phenol compounds effectively inhibit microbial growth for long periods of time. Smoked fish and meats have been staples of the diet for many centuries.

Sulfur dioxide:
a chemical used as a preservative in foods.

Sulfur dioxide has gained popularity as a preservative for dried fruits, molasses, and juice concentrates. Used in either gas or liquid form, the chemical retards color changes on the fruit surface and adds to the aesthetic quality of the product. Another gas, **ethylene oxide**, is employed for the preservation of spices, nuts, and dried fruits, especially those packaged in cellophane bags. This same gas is used for the chemical sterilization of packaged Petri dishes and other plastic devices.

Some foods contain their own natural preservatives. Examples are the antimicrobial substances in garlic, the lysozyme in egg white, and the benzoic acid of cranberries.

Radiation

Though the public is apprehensive about foods exposed to radiation, various forms of radiation are used to sterilize foods. Taste and appearance have been preserved by freezing food in liquid nitrogen and exhausting oxygen from the package before irradiation. Meat storage facilities use **ultraviolet light** to reduce surface contamination while natural enzymes increase meat's tenderness, and water can be treated with UV light when chlorine is not useful.

Ultraviolet light:
a form of energy whose wavelength is shorter than that of visible light.

Gamma rays are utilized to extend the shelf life of fruits, vegetables, fish, and poultry from several days to several weeks. This form of radiation also increases the distances fresh food can be transported and lengthens significantly the storage time for food in the home. Gamma rays are high-frequency forms of electromagnetic energy emitted by a radioactive isotope called **cobalt-60**. The radiations are not radioactive and cannot cause food to become radioactive. They kill microorganisms by reacting with and destroying microbial DNA. Opponents to their use point out, however, that the radiations also break chemical bonds in foods and cause new ones to form, thereby raising the possibility of new and toxic chemical compounds.

FOWL PLAY

It was the annual company picnic and the softball game was finally over. Now the serious eating could begin. There were salads, stuffed eggs, barbecued chickens, lots of desserts, and a picnic table overflowing with goodies. Unfortunately, there was also an unwelcome guest at the picnic: a species of *Salmonella*. During the next three days, over half the attendees would suffer abdominal cramps, diarrhea, headaches, and fever. The common thread was the barbecued chicken—all the sick attendees had eaten it.

Public health inspectors questioned the woman in charge of the chickens. She led inspectors back to a local supermarket, where the barbecued chickens were sold "ready to eat." It seemed that store employees cooked the birds, then put them back in the original trays where the uncooked birds had been stored. Investigators found evidence of *Salmonella* in the trays. Further questioning revealed that the woman had bought the chickens in the morning and stored them for the next seven hours in the trunk of her car, believing they would remain cool. They did not. In fact, they reached incubator temperatures quickly and by dinner time at 7:00 P.M. they were teeming with *Salmonella*. The lessons are clear: keep foods cold; store them in clean, fresh trays after cooking; and play softball *after* dinner, not before.

Interest in gamma radiation for food preservation grew during the 1950s under President Eisenhower's "Atoms for Peace" program. For the next quarter century, the Food and Drug Administration conducted extensive tests to determine whether the process was safe. In March 1981, the FDA permitted radiated foods such as spices, condiments, fruits, and vegetables to be sold to American consumers. Gamma radiation of pork to prevent trichinosis won approval in 1985 and irradiated strawberries made it to market in 1992.

Trichinosis:
a disease of the muscles, caused by a roundworm and transmitted by pork.

Preventing Foodborne Disease

Food may be a mechanical vector for infectious microorganisms or a culture medium for growth. People are then affected by the organism or the toxin it has produced in the food. In the former case, a food infection is established; in the latter, a food poisoning or intoxication occurs.

Food infections are typified by typhoid fever, salmonellosis, cholera, and shigellosis, all of which are of bacterial origin. Among the protozoal infections amebiasis, balantidiasis, and giardiasis represent foodborne diseases. Viral infections are exemplified by hepatitis A. **Food intoxications** include botulism, staphylococcal food poisoning, and clostridial food poisoning. Since full discussions of these diseases are presented elsewhere in this text, we shall not pause to examine them individually.

am″e-bi-′ah-sis
bal″an-tĭ-di′ah-sis
ji″ar-di′ah-sis

In the United States, public health microbiologists estimate that between two and ten million people are affected by foodborne disease annually. Many episodes require medical attention, but the vast majority of patients recover rapidly without serious complications. In many cases the incident might have been avoided by taking some basic precautions. For example, unrefrigerated foods are a prime source of staphylococci and *Salmonella* serotypes (Micro-Focus: 24.7), and perishable groceries such as meats and dairy products should not be allowed to warm up while other errands are performed. Also, a thermometer should be used to ensure that the refrigerator temperature is below 40° F at all times.

Another way to avoid foodborne disease is to cover any skin boils while working with foods since boils are a common source of staphylococci. The hands should always be washed thoroughly after handling raw vegetables or salad fixings to avoid cross-contamination of other foods. It is well to cook meat

FIGURE 24.9
A Case of Botulism Due to Leftover Food

This incident happened in California in August 1984. Because the stew was discarded, it could not be tested for *Clostridium botulinum*. However, the 16-hour interval during which the stew remained at room temperature was sufficient for germination of clostridial spores. The first man was unaffected because he ate the stew immediately after cooking. (Courtesy of Centers for Disease Control and Prevention, *MMWR*, March 22, 1985.)

1. In August, 1984, a man prepared stew from fresh ingredients including meat, unpeeled potatoes, and carrots.

2. After simmering it for 45 minutes, the man ate some of the stew and left the remainder on the stove overnight.

3. Sixteen hours later the man's roommate tasted the stew previous to heating it and noted a sour taste. He decided not to eat it and subsequently threw it away.

4. Forty hours after tasting the stew, the man developed signs of botulism. He was hospitalized, and Type A botulism toxin was detected in his serum. After extensive treatment with antitoxin, he recovered.

from a frozen or partly frozen state; if this is impossible, the meat should be thawed in the refrigerator. Cutting boards should be cleaned with hot, soapy water after use, and old cutting boards with cracks and pits should be discarded.

Studies indicate that leftovers are implicated in most outbreaks of foodborne disease (Figure 24.9). It is therefore important to refrigerate leftovers promptly and keep them no more than a few days. Thorough reheating of leftovers, preferably to boiling, also reduces the possibility of illness.

Many instances of foodborne disease occur during summer months when foods are taken on picnics where they cannot be refrigerated. In general, dairy foods such as custards, cream pies, pastries, and salads should be excluded from the picnic menu. For outdoor barbecues, one dish should be used for carrying hamburgers to the grill and another dish for serving them. Many of these principles apply equally well to fall and winter tailgate parties.

Over 90 percent of **botulism** outbreaks reported to the CDC are traced to home-canned food. To prevent this sometimes fatal foodborne disease, health officials urge that homemakers use the pressure method to can foods. Reliable canning instructions should be obtained and followed stringently. Foods suspected of contamination should not be tasted to confirm the suspicion, but should be discarded immediately. If doubtful, it is well to boil the food for a minimum of 10 minutes and thoroughly wash the utensil used to stir the food. Bulging or leaking cans must be discarded in a way that will not endanger other people or animals.

Botulism:
a foodborne disease of the nervous system accompanied by paralysis.

Milk is an unusually good vehicle for the transmission of pathogenic microorganisms because its fat content protects organisms from stomach acid and, being a fluid, it remains in the stomach a relatively short period of time. The diseases of cows transmitted by milk include bovine tuberculosis, brucellosis, and Q fever (Chapter 8). Since the 1980s, several outbreaks of milkborne disease have been linked to *Salmonella* serotypes. In the spring of 1984, for example, 16 cases of **salmonellosis** due to *Salmonella typhimurium* occurred in nuns at a convent in Kentucky. A failure in milk pasteurization accounted for the episode. And in 1985, federal officials associated *Salmonella typhimurium* with a milkborne epidemic of salmonellosis involving almost 6000 persons in the midwestern United States, as noted in this chapter's introduction.

Brucellosis:
a bacterial disease caused by a Gram-neative rod and accompanied by recurring fever.

Another milkborne organism of significance is ***Campylobacter jejuni***, the cause of **campylobacteriosis**. In 1984, this Gram-negative curved rod was isolated from nine kindergarten children and adults who sampled raw milk at a bottling plant in Southern California while on a school field trip. Scientists estimate that *Campylobacter jejuni* is present in the intestinal tracts of about 40 percent of dairy cattle. Milk products may also be a source of disease. In 1985, a notable epidemic of **listeriosis** was traced to Mexican-style cheese made in California. Manufacturers had to destroy more than 100 tons of the cheese to prevent the spread of *Listeria monocytogenes*, the responsible bacillus.

kam'pǐ-lo-bak"ter jě-joo'ne
kam"pǐ-lo-bak"ter-e-o'sis

lis-ter'e-ah
mon"o-si-toj'ě-nez

Public health microbiologists seek to limit milkborne disease by inspecting dairy plants regularly and making recommendations on improved sanitary practices. In the field, sick animals are treated with antibiotics and immunized. The success of mass brucellosis immunizations in the 1970s showed their effectiveness as a public health measure.

TO THIS POINT ∎

We have focused on the topic of food preservation and have outlined the various ways in which preservation can be achieved. One of the most widespread methods of preservation is by using heat in the canning process. Another method is by employing low temperatures in the refrigerator and freezer. A third is by drying food through various industrial modes, and a fourth is by drawing fluid out of microorganisms by osmotic pressure.

Two additional methods of preservation include chemical preservatives and radiations. Chemicals currently utilized include sorbic, benzoic, and propionic acid, as well as sulfur dioxide and various chemicals in wood smoke. Gamma rays are a type of radiation shown to be an excellent food preservative under experimental conditions. We pointed out some advantages to the use of gamma rays, but extensive testing still needs to be completed and consumer acceptance remains an obstacle to acceptance. Prevention of illness from food and dairy products can be effected by consumers, however, and we noted several methods.

> *The discussion up to this point has cast microorganisms in a negative role as spoilers of food and objects of preservation methods. But microorganisms may also play a positive role in the food and dairy product industries because their chemical activities result in numerous food products. We shall briefly discuss some of these products in the chapter's final section.*

Foods from Microorganisms

Over the centuries, social customs and traditions have brought acceptance of a variety of foods produced by microorganisms. Some individuals regard these foods as "spoiled," but to many people, the food is "fermented" (Figure 24.10).

Fermented foods have three things in common: they are less vulnerable to extensive spoilage than unfermented foods; they are less likely to be vectors of foodborne illness than unfermented foods; and they have not only been accepted by the cultures in which they were developed, but in some cases, they are considered delicacies. It is conceivable that ancient peoples were first attracted to the preservative qualities of fermented foods and coincidentally learned to appreciate their tastes.

Sauerkraut

Sauerkraut (German for "sour cabbage") is not only a well-preserved and tasty form of cabbage but also nutritionally sound. For instance, the vitamin C content of sauerkraut is equivalent to that of citrus fruits, and sauerkraut was often taken on British sea voyages to prevent scurvy because citrus fruits were too expensive.

Sauerkraut:
fermented cabbage.
loo″ko-nos′tok

Sauerkraut is prepared commercially by adding salt to shredded cabbage and packing the cabbage tightly to stimulate anaerobic conditions. The first organisms to multiply are species of **Leuconostoc**, a Gram-positive coccus found naturally in the cabbage. These bacteria ferment carbohydrates in the plant cells and produce acetic and lactic acids. After some days, the acids lower the pH of the cabbage to about 3.5. Species of **Lactobacillus** then proliferate, and the additional lactic acid they produce by fermentation further reduces the pH to about 2.0. The salt helps retard mold contamination while drawing juices out of the plant cells. A compound called diacetyl (the flavoring agent in butter) is produced by *Leuconostoc*, adding aroma and flavor.

Diacetyl:
a compound that gives flavor to butter.

Pickles

In the United States, "pickle" is practically synonymous with "pickled cucumber." Over 37 types of dill, sour, and sweet **pickles** have been categorized, but essentially the fermentations are similar. Cucumbers are placed in a salt solution of 8 percent or higher, at which point the cucumber changes color from bright green to olive green. Next comes curing.

a″er-oj′en-ēz

Three groups of microorganisms are important to the fermentation and curing of cucumbers. *Enterobacter aerogenes*, a Gram-negative rod, produces large amounts of CO_2, which takes up all the air space and establishes anaerobic conditions. *Lactobacillus* and *Leuconostoc* species then dominate and form abundant amounts of acid that softens the tissues and sours the cucumbers (Figure 24.11). Finally, certain yeasts grow and establish many flavors associated

FIGURE 24.10

An Array of Foods and Beverages Produced by Microorganisms

FIGURE 24.11

A Scanning Electron Micrograph of a Cucumber and Its Bacterial Flora

(a) *Lactobacillus plantarum* on the surface of the cucumber. The openings are stomata through which gases pass for the cucumber's metabolism. Note the accumulation of lactobacilli at these stomata. Bar = 10 μm. (b) A longitudinal section through the vascular tissue of a brined cucumber showing *Leuconostoc* species along the tubular walls. Bar = 10 μm.

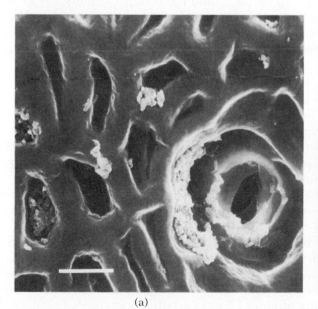

(a)   (b)

FIGURE 24.12

Soy Sauce

Roasted soybeans inoculated with *Aspergillus oryzae* to produce the koji used for soy sauce. Fungi cover the beans and begin the fermentation later completed by bacteria during aging.

with ripe pickles. Dill, garlic, and other herbs and spices are added to finish the product. Most pickles are heat-pasteurized or further acidified to increase their keeping quality, but "kosher-style" pickles are given no further treatment.

Vinegar

Vinegar is a fermented food that traditionally has been made by the spontaneous souring of wine. Indeed, the word "vinegar" is derived from the French *vinaigre*, which means sour wine.

A widely used method for industrial vinegar production follows a procedure first devised in Germany in the early 1800s. Yeasts ferment the fruit juice until the alcohol concentration is about 10 to 20 percent. Machines then spray the juice into a tank containing the bacterium ***Acetobacter aceti*** growing on the surface of wood shavings, gravel, or other substrate. As alcohol percolates through, bacterial enzymes convert it to acetaldehyde and then acetic acid. The vinegar recirculates several times before collection at the bottom of the tank. Residual alcohol evaporates in the heat, and the product usually has an acetic acid content of about 3 to 5 percent. The flavor of vinegar is determined by oils, sugars, and other compounds produced by the bacteria plus the residue of organic compounds in the wine from which it was made.

ah-se'to-bak'ter

as"et-al'dĕ-hīd

Other Fermented Foods

Several other fermented foods are also worthy of note. One example, **soy sauce**, is made from roasted soybeans and wheat inoculated with the fungus ***Aspergillus oryzae*** and allowed to stand for three days (Figure 24.12). The fungus-covered product, called koji, is then added to a solution of salt and microorganisms, and aged for about a year. During this time, lactobacilli produce acid, and yeasts produce small amounts of alcohol. Together with the fungal products, the acid and alcohol determine the flavor. The liquid pressed from the mixture is soy sauce.

as"per-gil'lus o-ri'zā

ko'je

Another example is **fermented sausages**. These are generally produced as dry or semidry products and include pepperoni from Italy, thuringer from Germany, and polsa from Sweden. Curing and seasoning agents are first added to ground meats, followed by stuffing into casings and incubation at warm temperatures. Mixed acids produced from carbohydrates in the meat give the sausage its unique flavor and aroma.

Cocoa and coffee also owe their flavor in part to microorganisms. **Cocoa** derives some of its taste from the microbial fermentation that helps remove cocoa beans from the pulp covering them in the pod. Likewise, **coffee** is believed to obtain some of its flavor from the fermentation of coffee berries when the beans are soaked in water to loosen the berry skins before roasting.

The influence of microorganisms on animal nutrition is evident in the animal food called **silage**. Silage is made in huge, cylindrical silos that commonly stand adjacent to barns. The farmer packs the silo with corn, grain stalks, grass, potatoes, and virtually anything that can be fermented. During storage, innumerable bacteria ferment the plant carbohydrates in a process similar to that for sauerkraut and pickles. Fermentation yields a broad mixture of acids, aldehydes, and ketones as well as diacetyl, which is relished by cattle. After several weeks, the well-preserved plant material is sweet smelling, succulent, and nutritious to animals. It is also inexpensive to the farmer.

Sour Milk Products

The sour milks are typical examples of fermented milk products. **Buttermilk**, for instance, is made by adding starter cultures of *Streptococcus cremoris* and *Leuconostoc citrovorum* to vats of skim milk. The *Streptococcus* ferments lactose to lactic and acetic acids, and the *Leuconostoc* continues the fermentation to yield various aldehydes and ketones, and a compound called diacetyl. These substances, especially diacetyl, give buttermilk its flavor, aroma, and acidity. For **sour cream**, the same process is used except pasteurized light cream is the starting point.

loo"ko-nos'tok sit"ro-vor'um

Acidophilus milk is produced in much the same way as buttermilk, except that the skim milk is inoculated with *Lactobacillus acidophilus*. This bacterium

as"ĭ-dof'ĭ-lus

24.8

MICROFOCUS

MAKING YOUR OWN YOGURT

The popularity of yogurt as a nutritious low-calorie food has prompted many to try their hand at making it at home. Here is a recipe that works well.

Heat one quart of milk to about 170° F (77° C), stirring often and using a thermometer to check the temperature. This heating will evaporate some of the liquid and reduce the bacterial population. Let the milk cool to about 130° F (about 55° C), then add one cup of powdered milk and one-third cup of unflavored commercial yogurt. Mix thoroughly and pour into small containers with lids. Styrofoam coffee cups may be used.

For the incubation step you will need a small cooler of the type used for picnics. Fill the cooler with several inches of water at 130° F (55° C). Now place the containers in the cooler, close the lid tightly, and let the containers stand for about six to eight hours. (A large pan of hot water in the oven also works well or the cups can be wrapped with hot towels.) During this time the bacteria will multiply and the yogurt will thicken. Refrigerate, add fresh or frozen fruit, and enjoy.

is a normal member of the human intestinal flora. Many health practitioners believe that the lactobacilli in acidophilus milk augment naturally occurring lactobacilli and help the digestion process, while keeping molds in check. A newer type of acidophilus milk, **sweet acidophilus milk**, lacks the sour taste. It is prepared by adding *Lactobacillus acidophilus* to pasteurized milk and packaging without fermentation.

Yogurt:
a thick, sour milk product produced by bacterial action on lactose.

    Yogurt is a form of sour milk (MicroFocus: 24.8), made by adding dry-milk solids to boiled milk to achieve a custardlike consistency. The two starter cultures are ***Streptococcus thermophilus*** and ***Lactobacillus bulgaricus***. Yogurt is sometimes called Bulgarian milk because it was popular among peasants in Bulgaria and other Balkan countries. After World War I, many Americans became health-conscious, and Elie Metchnikoff (discoverer of phagocytosis) took note of the longevity of Bulgarian peasants and attributed it to the yogurt they drank. Metchnikoff believed that the streptococci and lactobacilli in yogurt assume residence in the intestine and replace organisms that contribute to aging. Although the aging theory has largely been discounted, many microbiologists hold that bacteria in yogurt assist good health much as the bacteria in acidophilus milk.

Cheese

Rennin:
an animal enzyme that accelerates curdling in milk.

Cheese production begins when the casein curdles out of milk. Usually this accompanies a souring of the milk by streptococci, but the process may be accelerated by adding **rennin**, an enzyme obtained from the stomach lining of a calf. The milk curd is essentially an unripened cheese. It may be marketed as **cottage cheese**, or pot cheese. **Cream cheese** is also unripened cheese with a butterfat content of up to 20 percent.

FIGURE 24.13

An Array of Cheeses Produced by Microorganisms

To prepare ripened cheese, the milk curds are washed, pressed, sometimes cooked, and cut to the desired shape. Often the curds are salted to add flavor, control moisture, and prevent contamination by molds. If **Swiss cheese** is to be made, two types of bacteria grow within the cheese: *Lactobacillus* species, which ferment the lactose to lactic acid, and *Propionibacterium* species, which produce organic compounds and carbon dioxide, which seeks out weak spots in the curd and accumulates as holes, or eyes (Figure 24.13). **Cheddar cheese** is scalded at a lower temperature than Swiss. Cheddar and Swiss are examples of cheese ripened internally by bacteria. Provolone, Edam, and Gouda are others.

pro"pe-on"e-bak-te're-um

Another group of cheeses are somewhat softer in texture, a characteristic deriving from the partial breakdown of the protein curds by microbial enzymes. Growth takes place primarily at the surface of these cheeses, and the products tend to be pungent. Within the group of **soft cheeses** are Muenster, Port du Salut, and Limberger. Yeasts and species of the Gram-positive rod *Brevibacterium* are among the surface flora. The rind of the cheese is derived from microbial pigments.

brev"e-bak-te're-um

The **mold-ripened** cheeses are represented by Camembert and Roquefort. **Camembert cheese** is made by dipping salted curds into *Penicillium camemberti* spores to stimulate a surface growth. The fungus grows on the outside of the curd and digests the proteins, thus softening the curd. **Roquefort** is a blue-veined cheese produced by *Penicillium roqueforti*. The mold penetrates cracks within the curd, creating the distinctive veins within the cheese. Most people, however, would rather remain blissfully ignorant of this fact.

NOTE TO THE STUDENT

Since 1925, only five deaths from botulism have been attributed to commercially canned food in the United States. During this period, almost 100 billion cans of food were produced for sale to consumers.

I believe that these figures are a testament to the high standards achieved by the canning industry. They represent an achievement of which we consumers can be justifiably proud. I say "we consumers" because we are the ones who understand that foods can be a vehicle for disease and we refuse to tolerate a manufacturer's ignorance. Working through our representatives in government agencies, we exact heavy penalties from companies whose products are tainted. Witness the 55 million cans of Alaska salmon recalled in 1982 (Chapter 8) and the 36,000 pounds of ham recalled in 1983 (this chapter). Both incidents were due to the possibility of microbial contamination.

The next time you shop at the supermarket, stop and take note of the broad variety of foods we consume and consider that we buy and eat these foods with full confidence that none will make us ill. It is a confidence that is not shared by peoples in other parts of the world.

Summary

The microbiology of foods is concerned with spoilage that occurs in foods, with methods for preserving foods, and with activities of microorganisms in the formation of certain foods.

Food spoilage has been a continuing problem since ancient times. Certain conditions such as water content, pH, physical structure, and chemical composition determine the extent of food spoilage since they influence the growth of microorganisms. In meats and fish, spoilage organisms enter during processing, while in fish the water is often the source. In poultry, the spoilage may reflect human contamination, but often it is due to members of the genus *Salmonella* that infect the bird. Bakery ingredients generally bring contaminants to bread, and grains may be spoiled by toxin-producing fungi. Dairy products are spoiled by microorganisms surviving pasteurization such as curd, capsule, and pigment producers.

To preserve food from spoilage, a number of methods are used including heat, low temperatures, drying, chemical preservatives, and radiation. The most useful application of heat is in the process of canning, while low temperatures are achieved in the refrigerator and freezer. Pasteurization is used for milk and dairy products. Drying is useful because water is an absolute necessity for life. Dried foods are therefore unable to support microbial life. Various chemicals such as propionic, sorbic, and benzoic acids are used to preserve foods, and ultraviolet light and gamma rays typify the radiations used in processing certain foods.

Certain foods "spoiled" by microorganisms have come to be accepted as the norm. Among these microbial products are sauerkraut, vinegar, and fermented sausages. The spoilage in these foods causes no harm to consumers and the foods reflect the helpful activities that microorganisms perform to add to the quality of our lives. Numerous dairy products such as cheeses are also products of microorganisms.

Questions for Thought and Discussion

1. A writer in a food technology magazine once suggested that refrigerators be fitted with ultraviolet lights to reduce the level of microbial contamination in foods. Would you support this idea?

2. In October 1984, salmonellosis was diagnosed in 124 students at an eating facility on a university campus in New Jersey. Contaminated eggs were the source. Investigators learned that eggs for breakfast were broken into a large vat and ladled onto the grill when students requested an omelet or other egg dish. The unused portion was refrigerated until the next day when more eggs were added to the vat to keep it full. What recommendation do you suppose health officials made?

3. Chicken and salad are two items on the dinner menu at home, and you are put in charge of preparing both. You have a cutting board and knife for slicing up the salad items and cutting the chicken into pieces. Which task should you perform first? Why? What other precautions might you take to ensure that the meal is not remembered for the wrong reason?

4. A novel suggestion for preventing *Salmonella* infection in poultry is to feed newborn chicks capsules of bacteria normally found in the intestine of adult chickens. What is the theory behind this approach to prevention, and do you believe it will be useful?

5. On January 12, 1996, the author opened a container of sour cream that had become lost in the back of the refrigerator some nine months before (the expiration date listed on the bottom was April 27, 1995). The sour cream appeared satisfactory and there was no unusual smell. The author proceeded to spoon it onto a baked potato and dig in. What factors might have contributed to the sour cream's preservation so long after the expiration date?

6. It is a hot Saturday morning in July. You get into your car at 9:00 A.M. with the following list of chores: pick up the custard eclairs for tonight's dinner party, drop off clothes at the cleaners, buy the ground beef for tomorrow's barbecue, deliver the kids to the Little League baseball game, pick up a broiler at the poultry farm. Microbiologically speaking, what sequence should you follow?

7. Suppose you had the choice of purchasing "yogurt made with pasteurized milk" or "pasteurized yogurt." Which would you choose? Why? What are the "active cultures" in a cup of yogurt?

8. How would you answer a child who asks, "Who puts the holes in Swiss cheese?" Why might blue (Roquefort) cheese pose a possible threat to someone who has an acute allergy to penicillin?

9. It is 5:30 P.M. and you arrive on campus for your evening college class. You stop off at the cafeteria for a bite to eat. Which foods might you be inclined to avoid purchasing?

10. To avoid *Salmonella* infection when preparing eggs for breakfast, the operative phrase is "scramble or gamble." How many foods can you name that use uncooked or undercooked eggs and that can pose a hazard to health?

11. At 10:00 A.M. a typhoid carrier managed to contaminate a sample of food with 100 typhoid bacilli. The food was left in a warm environment until 2:30 P.M., when it was served. Assuming a generation time of 30 minutes, how many bacilli did the food contain at serving time?

12. Which principles of preservation ensure that each of the following remains uncontaminated on the pantry shelf: vinegar, olive oil, brown sugar, tea bags, spaghetti, hot cocoa mix, pancake syrup, soy sauce, rice?

13. On Saturday, a man buys a steak and a pound of calves' liver and places them in the refrigerator. On Monday a decision must be made on which to cook for dinner. Microbiologically, which is the better choice? Why?

14. One food mentioned in this chapter is a fruit that is easily contaminated by molds when its skin is broken. No other word in the English language rhymes with the name of this fruit. What is its name?

15. One day in 1985, the students in a microbiology class presented the instructor with a basket of "microbial cheer" in recognition of his efforts on their behalf. From your knowledge of this and other chapters, can you guess some of the things that the basket contained?

Review

On completing this chapter on food microbiology, test your knowledge of its contents by using the following syllables to compose the term that answers the clue. Each term is a genus of microorganism important in food microbiology. The number of letters in the genus is indicated by the dashes, and the number of syllables in the genus is shown by the number in parentheses. Each syllable is used only once, and the answers are listed in Appendix F.

A A A AS AS BA BA BAC BAC CE CEPS CHLA CIL CIL CLAV CLO CLO CO COC CUS DI DO EN GIL I I I LA LAC LEU LUS LUS LUS MO MON MY NEL NOS O PER PRO PSEU RA SAL SER STREP STRID STRID TE TER TER TER TI TO TO TO TOC UM UM US

1. Common poultry contaminant (4) _ _ _ _ _ _ _ _ _ _
2. Discolors meat surface (5) _ _ _ _ _ _ _ _ _ _ _ _
3. Destroyed in canning (4) _ _ _ _ _ _ _ _ _ _
4. Red pigment in bread (4) _ _ _ _ _ _ _ _
5. Causes psittacosis (4) _ _ _ _ _ _ _ _ _
6. Black rot in eggs (3) _ _ _ _ _ _ _
7. Sours dairy products (4) _ _ _ _ _ _ _ _ _ _ _
8. Sporeforming contaminant (3) _ _ _ _ _ _ _ _
9. Used to make vinegar (5) _ _ _ _ _ _ _ _ _ _ _
10. Sauerkraut producer (4) _ _ _ _ _ _ _ _ _ _ _
11. Causes ergot poisoning (3) _ _ _ _ _ _ _ _ _
12. Produces potent exotoxins (4) _ _ _ _ _ _ _ _ _ _ _
13. Green rot in foods (4) _ _ _ _ _ _ _ _ _ _
14. Used to cure cucumbers (5) _ _ _ _ _ _ _ _ _ _ _ _
15. Aflatoxins in grains (4) _ _ _ _ _ _ _ _ _ _ _

CHAPTER 25

ENVIRONMENTAL AND INDUSTRIAL MICROBIOLOGY

I n the early 1800s, the steam engine and its product, the Industrial Revolution, brought crowds of rural inhabitants to European cities. To accommodate the rising tide, row houses and apartment blocks were hastily erected, and owners of existing houses took in tenants. Not surprisingly, the bills of mortality from typhoid fever, cholera, tuberculosis, dysentery, and other diseases mounted in alarming proportions.

As the death rates rose, a few activists spoke up for reform. Among them was an English lawyer and journalist named **Edwin Chadwick**. Chadwick subscribed to the then novel idea that humans could shape their environment and could eliminate diseases of filth by doing away with filth. In 1842, he published a landmark report indicating that poverty-stricken laborers suffered a far higher incidence of disease than people from middle or upper classes. Chadwick attributed the difference to the abominable living conditions of workers, and he declared that most of their diseases were preventable. His report laid the basis for the Great Sanitary Movement, a wave of reform that began in Europe and spread to developed countries.

Chadwick was not a medical man, but his ideas captured the imagination of both scientists and social reformers. He proposed that sewers be constructed using smooth ceramic pipes, and that enough water be flushed through the system to carry waste to some distant depository. In order to work, the system required installation of new water and sewer pipes, development of powerful pumps to bring water into homes, and elimination of older sewage systems. He foresaw intrusions upon private property to permit water mains and extensive construction to allow straight sewer pipes. The cost would be formidable.

Chadwick's vision eventually came to reality, but it might have taken decades longer without the intervention of **cholera**. In 1849, a cholera epidemic broke out in London and terrified so many people that public opinion began to form in favor of Chadwick's proposal. Another epidemic occurred in 1853, during which John Snow proved that water was involved in transmission of the disease (Chapter 17). In both outbreaks, the disease reached the affluent as well as the poor, and the mortality rate exceeded 50 percent. Construction of the sewer system began shortly thereafter, with John Simon, London's first Medical Officer of Health, in command of the project.

The proverbial "icing on the cake" came in 1892 when a devastating epidemic of cholera erupted in Hamburg, Germany. For the most part, Hamburg drew its water directly from the polluted Elbe River. Adjacent to Hamburg lay Altona, a city where the German government had previously installed a water filtration plant. Altona remained free of cholera. The contrast was further sharpened by a street that divided Hamburg and Altona. On the Hamburg side of the street multiple cases of cholera broke out; across the street none occurred. Chadwick and his fellow sanitarians could not have imagined a more clear-cut demonstration of the importance of water purification and sewage treatment.

Dealing with water pollution and treating sewage are but two of the myriad activities where microbiologists occupy important places in public health. Water pollution and sewage treatment concerns are environmental in scope, but other forms of industrial microbiology are more oriented to the developmental laboratory where products are manufactured for improving the quality of life. Thus, industrial microbiology encompasses food production as well as the synthesis of organic compounds, antibiotics, and insecticides. The thrust in environmental microbiology is to study microorganisms as they affect the natural environment, while the emphasis in industrial microbiology is to study microorganisms for the beneficial uses to which they can be placed. We shall see examples of both these thoughts in the chapter ahead.

Water Pollution

For purposes of simplification, scientists classify water into two major types: ground water and surface water. **Ground water** originates from deep wells and subterranean springs and, because of the filtering action of soil, deep sand, and rock, it is virtually free of microorganisms. As the water flows up along channels, contaminants may enter it and alter its quality. **Surface water** is found in lakes, streams, and shallow wells. Its microbial population may reflect the air through which rain has passed, the meat-packing plant by which a stream flows, or the sewage treatment facility located along a river's banks.

Certain terms are significant in water microbiology. For example, water is considered **contaminated** when it contains a chemical or biological poison, or infectious agent. In water that is **polluted**, the same conditions apply, but the poison or agent is obvious. Polluted water carries an unpleasant taste, smell, or appearance. **Potability**, by contrast, refers to the drinkability of water. Potable water is fit for consumption, while unpotable water is unfit.

Various Water Environments

A body of **unpolluted** water such as a mountain lake or stream is usually low in organic nutrients, and thus, only a limited number of bacteria are present, perhaps a few thousand per milliliter. Most bacteria are soil organisms that have run off into the water during a rainfall. An example are the **actinomycetes**, a group of moldlike bacteria that give a musty odor to soil. Other inhabitants include yeasts and bacterial and mold spores. **Cellulose digesters** such as members of the genus *Cellulomonas* are also found. These bacteria digest cellulose in plant cell walls. **Autotrophic bacteria** are also common, and free-living protozoa such as *Paramecium*, *Euglena*, *Tetrahymena*, and *Amoeba* abound (Figure 25.1).

Cholera:
a bacterial disease of the intestine characterized by unrelenting diarrhea.

Surface water:
rainwater accumulating in lakes, streams, and shallow wells.

po'ta-bl

ak"tĭ-no-mi-se'tēz

Autotrophic:
able to synthesize its own food from simple inorganic compounds.

FIGURE 25.1

Electron Micrographs of Microorganisms Commonly Found in Water Environments

(a) An SEM of the protozoan *Tetrahymena pyriformis* (× 1790). This large organism is approximately 500 μm long, with numerous cilia covering the entire body. *Tetrahymena* inhabits unpolluted water and reproduces asexually by binary fission and sexually by mating. The photograph shows three pairs of mating cells. (b) A TEM of bacteriophages of *Vibrio parahaemolyticus* (× 150,000). Bacteriophages like these are commonly found in sewage-contaminated polluted water.

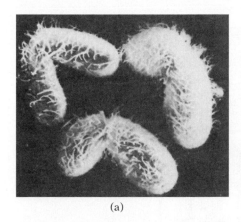
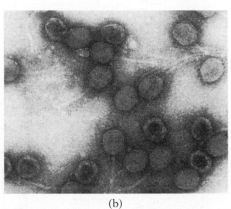

(a)   (b)

A **polluted** body of water such as a polluted lake or river presents a totally different picture. The water contains large amounts of organic matter from sewage, feces, and industrial sources, and the population of microorganisms is usually **heterotrophic**. A major type of bacteria in polluted water is **coliform bacteria**, a group of Gram-negative nonsporeforming bacilli usually found in the human intestine. Coliform bacteria ferment lactose to acid and gas. Included in this group are *Escherichia coli* and species of *Enterobacter*. Noncoliform bacteria also common in polluted water include *Streptococcus*, *Proteus*, and *Pseudomonas* species.

> Heterotrophic: obtains its food from preformed organic compounds.
>
> Coliform bacteria: Gram-negative intestinal rods that ferment lactose to acid and gas.

In polluted water, microorganisms contribute to a chain of events that drastically alters the ecology of the environment. When phosphates accumulate in the water, algae may bloom. The algae supply nutrients to microorganisms, which multiply rapidly (Figure 25.2) and use up the available oxygen. Soon, other protozoa, small fish, crustaceans, and plants die and accumulate on the bottom. Anaerobic species of bacteria such as *Desulfovibrio* and *Clostridium* then thrive in the mud and produce gases that give the water a stench.

> de-sul″fo-vib′re-o

The aquatic environment of the oceans illustrates another view of microbial populations in water. In the high salt concentration of ocean water, **halophilic**, or salt-loving, microorganisms survive. In addition, the organisms must be **psychrophilic** since it is very cold below the surface. Those at the bottom must also withstand great pressure and are therefore **barophilic**, or pressure-loving. The organisms in these environments pose no threat to humans because they cannot grow in the body tissues.

> Halophilic: salt-loving

Marine microorganisms are vital to ecological cycles because they form the foundations of many food chains. For example, marine algae such as **diatoms** (Figure 25.3) and organisms such as the **dinoflagellates** capture the sun's energy and, using carbon dioxide, convert the energy to chemical energy in carbohydrates. The microorganisms are then consumed by other animals in the food pyramid. Dinoflagellates have made the news in recent years because certain species of *Gonyaulax* and *Gymnodinium* are responsible for the red tide.

> di′ah-tomz
>
> gon″e-aw′laks
> jim″no-din′e-um

FIGURE 25.2

Death of a River

(a) Nutrients enter the river from such sources as sewage treatment facilities, and the river suddenly develops a high nutrient content. (b) Algae bloom rapidly. (c) The algae die and settle to the bottom as sediment. (d) Microorganisms from the sewage multiply furiously and decompose the sediment. (e) This process quickly uses up the available oxygen in the water. (f) Fish and other small animals and plants then die for lack of oxygen.

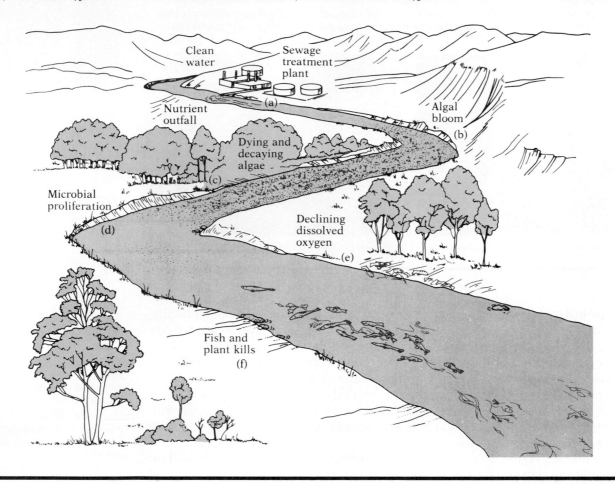

Most marine microorganisms are found along the shoreline, or **littoral zone**, because this is where nutrients are plentiful. Certain unusual types of microorganisms have also been found on the ocean floor in the **benthic zone** and even at the bottom of 6-mile-deep trenches, the **abyssal zone**. Two types of protozoa, the foraminiferans and radiolarians, are of special interest to oil companies because these protozoa were dominant species during formation of the oil fields, and their fossils serve as markers for oil-bearing layers of rock.

a-bis'al

Types of Water Pollution

Water is vital to such industries as food processing, meat packing, and paper manufacturing. It is also used extensively in pharmaceutical plants and mines, and for cooling purposes in power-generating units. It irrigates agricultural lands and provides the focus for many recreational facilities. Uses such as these, however, commonly add to contamination and pollution of water.

FIGURE 25.3

A Scanning Electron Micrograph of a Collection of Diatoms ($\times$ 90)

Note the broad variety of shapes and sizes of microorganisms in this group. Diatoms trap the sun's energy in photosynthesis and use it to form carbohydrates that are passed on to other marine organisms as food.

Physical pollution of water occurs when particulate matter such as sand or soil makes the water cloudy or when cyanobacteria bloom during midsummer and give water the consistency of pea soup. **Chemical pollution** results from the introduction of inorganic and organic waste to the water. For example, water passing out of a mine contains large amounts of copper or iron. Other chemical pollutants in water include phosphates and nitrates from laundry detergents as well as acids such as sulfuric acid.

Cyanobacteria: **prokaryotes with photosynthetic capabilities.**

The third type of pollution, **biological pollution**, is the main concern of our discussion. This type of pollution develops from microorganisms that enter water from human waste, food processing and meat-packing plants, medical facilities, and similar sources (Figure 25.4). Normally, water can handle biological material because heterotrophic microorganisms digest organic matter to carbon dioxide, water, and useful ions (phosphates, nitrates, and sulfates). With the rapid movement of the water, aeration is constant and waste is soon diluted and eliminated. However, when water stagnates or is overloaded with waste, it cannot deal with biological material and becomes polluted.

A critical measurement in polluted water is the **biochemical oxygen demand**, or **BOD**. This refers to the amount of oxygen that microorganisms require to decompose the organic matter in water. As the number of microorganisms increases, the demand for oxygen increases proportionally. In the laboratory, the BOD is determined by measuring the dissolved oxygen content of water immediately after collection and then after incubation at 20° C for five days. The difference in oxygen content represents the amount used up by microorganisms in the water sample. Results are generally expressed as parts per million (ppm), with a BOD of several hundred ppm usually considered high.

BOD: **the amount of oxygen required by metabolizing microorganisms during a five-day period of incubation.**

Diseases Transmitted by Water

Water may be the vehicle for transfer of a broad variety of microbial diseases, including **bacterial diseases** such as typhoid fever, cholera, and shigellosis. Waterborne epidemics of these diseases, however, are rare due to continual

FIGURE 25.4

A Coliform Bacterium

A scanning electron micrograph of *E. coli* on the microvilli of an animal's small intestine ($\times$ 3900). *E. coli* is commonly found in water that is biologically polluted. The bacillus represents the coliform group of bacteria and is often used as an indicator of bacterial pollution of water.

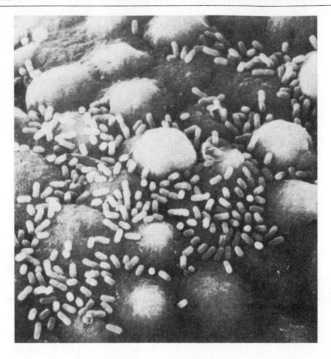

kam″pĭ-lo-bak′ter

surveillance. Many waterborne illnesses are due to less familiar bacteria such as species of *Yersinia* and *Campylobacter*, and toxin-producing strains of *Escherichia coli*. An emerging pathogen associated with contaminated water is ***Vibrio vulnificus***, a Gram-negative bacterium that can cause serious illness in persons with preexisting liver disease or compromised immune systems. In the ten-year period preceding 1993, 125 persons became infected with *V. vulnificus* and 44 died. Raw oyster consumption was implicated in the majority of deaths.

cook-sak′e

je-ar′de-ah

Viral diseases transmitted by water include hepatitis A, gastroenteritis due to Coxsackie or Norwalk virus, and in rare instances, polio. These diseases are generally related to fecal contamination of water. Many **protozoa** form cysts that survive for long periods in water. For this reason water may be a vehicle for the transfer of *Entamoeba histolytica* and *Giardia lamblia*. A notable outbreak of *Cryptosporidium* infection occurred in 1993 when the municipal water supply of Milwaukee, Wisconsin, became contaminated (Chapter 15).

gon″e-aw′laks

gam″be-er-dis′cus

Two **dinoflagellates** bear mention because of their involvement in human poisonings. The first, *Gonyaulax catanella*, produces a toxin that may cause muscular paralysis and death from asphyxiation. The toxin is ingested from shellfish that feed on the dinoflagellate. The second dinoflagellate is *Gambierdiscus toxicus*. This marine microorganism is consumed by small fish that concentrate the toxin and pass it to larger fish such as sea bass and red snapper. Human consumption of the fish leads to neurological and muscular intoxication and a condition called **ciguatera fish poisoning** (from *cigua* for poisonous snail, originally thought to be the cause).

se″gwah-ta′rah

2 5 . 1

MICROFOCUS

JELLO PREFERRED

Jello-wrestling may not be quite as chic as mud-wrestling but it is presumably safer. At least that is the opinion of physicians confronted with students suffering from skin rashes after a mud-wrestling contest.

The "contest" occurred in the spring of 1992 on Seattle's University of Washington campus. Seven students began experiencing tiny red and pus-filled bumps on their skin within hours after participating. The rash was heaviest on their arms and legs. Cultures of pus from the pustules yielded species of *Enterobacter*, a common soil bacterium associated with fecal matter and manure. Other intestinal bacteria were also found. One student was treated with antibiotics, and no serious aftereffects accompanied any of the cases.

For the record, microbiologists gave the new disease a name—*dermatitis* (skin) *palaestrae* (the Greek arena for wrestling events) *limosae* ("pertaining to mud") or, literally, "dermatitis-of-mud-wrestling." They also recommended that if Jello were not available for future "events," the students might consider investing in sterilized potting soil.

Treatment of Water and Sewage

Some years ago, health workers in Africa asked villagers to name their single greatest need. The answer was almost unanimous: "Water." A startling survey by the World Health Organization indicates that three of every four humans alive today do not have enough water to drink or, if water is available, the supplies are contaminated.

Water is unfit to drink when it contains human sewage, animal waste, or other pollutants. However, the situation can be reversed through the proper management of water resources. Water purification procedures prevent pathogenic microorganisms from reaching the body, while sewage treatment processes remove pathogens from body waste products (MicroFocus: 25.1). In this section we shall examine how these are accomplished.

Water Purification

Three basic steps are included in the preparation of water for drinking: sedimentation, filtration, and chlorination (Figure 25.5). In the **sedimentation** step, leaves, particles of sand and gravel, and other materials from the soil are removed in large reservoirs or settling tanks. Chemicals such as aluminum sulfate (alum) or iron sulfate are dropped as a powder onto water and they form jellylike masses of coagulated material called **flocs**. The flocs fall through the water and cling to organic particles and microorganisms, dragging a major portion to the bottom sediment in the process of **flocculation**.

The **filtration** step is next. Although different types of filtering material are available, most filters utilize a layer of sand and gravel to trap microorganisms. A **slow sand filter** using fine particles of sand several feet deep is efficient for smaller scale operations. Within the sand a layer of microorganisms acts as a supplementary filter. This layer is called a **schmutzdecke**, or dirty layer. A slow sand filter may purify over 3 million gallons of water per acre per day. To clean the filter, the top layer is removed and replaced with fresh sand.

Flocs:
jellylike masses of coagulated material.

Flocculation:
the formation of jellylike masses of coagulated material.

shmoots′dek-ĕ

Steps in the Purification of Municipal Water Supplies

(a) In the reservoir, large objects are removed. (b) The water is then sprayed in the air to increase its oxygen content. (c) Next the water is piped to a mixing chamber where flocculating agents are added. (d) The flocculating agents are churned in the water and large jellylike masses, or flocs, form. (e) The flocs settle to the bottom of the sedimentation tank. (f) The water is then filtered, and (g) chlorinated before being piped off to storage tanks.

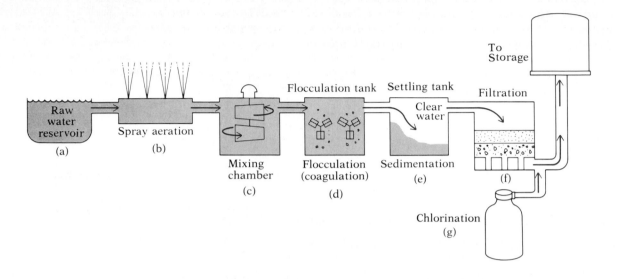

Schmutzdecke:
a layer of microorganisms that forms in a slow sand filter.

A **rapid sand filter** contains coarser particles of gravel. A schmutzdecke does not develop in this filter, but the rate of filtration is much higher, with over 200 million gallons purified per acre per day. This type of filter is commonly used in municipal water systems. It must be cleaned more often than the slow sand filter, a process accomplished by forcing water back through the filter by mechanical pressure. Both slow and rapid sand filters remove approximately 99 percent of the microorganisms from water.

The final step is **chlorination**, in which chlorine gas is added to the water. Chlorine is an active oxidizing agent that reacts with any organic matter in water. It is important, therefore, to continue adding chlorine until a residue is present. A residue of 0.2 to 1.0 parts of chlorine per million (ppm) of water is often the standard used. Under these conditions, most remaining microorganisms die within 30 minutes.

At this point, some communities **"soften"** water by removing magnesium, calcium, and other salts. Softened water mixes more easily with soap, and soap curds do not form. Water may also be fluoridated to help prevent tooth decay. Scientists believe that fluoride strengthens tooth enamel and makes it more resistant to the acid produced by anaerobic bacteria in the mouth.

Sewage Treatment

Systems for the treatment of human waste range from the primitive outhouse, which is nothing more than a hole in the ground, to the sophisticated sewage treatment facilities used by many large cities. All operate under the same basic principle: water is separated from the waste, and the solid matter is broken down by microorganisms to simple compounds for return to the soil and water.

Cesspool:
a concrete cylindrical ring with pores; used for waste disposal.

In many homes, human waste is emptied into underground **cesspools**. These are concrete cylindrical rings with pores in the wall. Water passes into the soil through the bottom and pores of the cesspool, while solid waste accumulates on the bottom. Microorganisms, especially anaerobic bacteria, digest the solid matter into soluble products that enter the soil and enrich it. Some hardware stores sell enzymes and dried bacteria, usually *Bacillus subtilis* spores, to accelerate sludge digestion.

Some homes use a **septic tank**, an enclosed concrete box that receives waste from the house. Organic matter accumulates on the bottom of the tank while water rises to the outlet pipe and flows to a distribution box. The water is then separated into pipes that empty into the surrounding soil. Since digested organic matter is not absorbed into the earth, the septic tank must be pumped out regularly.

Sewers are at least as old as the Cloaca Maxima of Roman times. Until the mid-1800s, however, sewers were simply elongated cesspools with overflow pipes at one end. They collected filth and had to be pumped out regularly. Finally, in 1842, Edwin Chadwick's report raised the possibility that sewage spreads disease, and soon thereafter a movement (fueled by a cholera outbreak) sprang up to sanitize European cities.

Small towns collect sewage into large ponds called **oxidation lagoons**. Here, the sewage is left undisturbed for up to three months. During that time, aerobic bacteria digest organic matter in the water, while anaerobic organisms break down sedimented material. Under controlled conditions, the waste may be totally converted to simple salts such as carbonates, nitrates, phosphates, and sulfates. At the cycle's conclusion, the bacteria die naturally, the water clarifies, and the pond may be emptied into a nearby river or stream.

Large municipalities rely on a mechanized sewage treatment facility to handle the massive amounts of waste and garbage generated daily (Figure 25.6). The first step in the process, **pretreatment**, involves grit and insoluble waste removal. Next comes **primary treatment**, in which raw sewage is piped into huge open tanks for organic waste removal. This waste, called **sludge**, is passed into sludge tanks for further treatment. Flocculating materials such as aluminum and iron sulfate are then added to the raw sewage to drag microorganisms and debris to the bottom.

The **secondary treatment** of sewage has two phases, a liquid phase and a solid phase. The **liquid phase** involves aeration of the water portion to encourage aerobic growth of microorganisms. As they grow, microorganisms digest proteins into simple amino acids, carbohydrates into simple sugars, and fats into fatty acids and glycerol. Acids and alcohols are also produced, and carbon dioxide evolves. The water then is passed through a clarifier and filter to remove the microorganisms and remaining organic matter, after which it flows into a stream or river.

The **solid phase** of secondary treatment is carried on in a **sludge tank**. Within the tank, microbial growth is encouraged either by aerobic or anaerobic processes. In the aerobic process, compressed air is forced into the sludge, and the suspended particles form tiny gelatinous masses swarming with microorganisms, which thrive on the organic matter. ***Zoogloea ramigera***, a Gram-negative rod, produces the slime to which other microorganisms attach and congregate. The **activated sludge**, as it is termed, gathers to itself much of the microorganisms, organic material, color, and smell of the sewage. The activated sludge is drawn off and dried.

In the anaerobic method of sludge digestion, sewage is held in the tank for up to 30 days while the sludge ferments. Gases such as methane, carbon dioxide, and nitrogen are derived from this process. The methane may be captured and used to run the machinery of the sewage facility. Other gases such as ammonia and hydrogen sulfide are of value to chemical industries. The digested sludge, together with the dried activated sludge, may be used as agricultural fertilizer

Septic tank:
an enclosed concrete box with fluid outlet pipes; used for waste disposal.

Oxidation lagoon:
a large pond into which sewage is piped for natural digestion of organic matter.

Sludge tank:
a tank in which bacteria digest sedimented organic matter from primary sewage treatment.

zo"o-gle'ah ra-me'jer-ah

FIGURE 25.6
A Schematic View of a Waste Treatment Facility

(a) Sewage is initially pretreated with a bar screen to remove grit. (b) The sewage is then piped to a settling tank where organic waste passes out to a sludge tank. This is primary treatment. (c) In the liquid phase of secondary treatment, microorganisms digest the soluble organic matter as the water percolates through a filter. (d) The water is separated from the microorganisms and passes out, while the sedimented material flows to the sludge tank. (e) In the solid phase of secondary treatment, sludge is treated in an activated sludge tank with thorough aeration. (f) This is followed by separation and flow of sedimented material to the anaerobic sludge tank. (g) In the anaerobic sludge tank, sludge from three processes (b, d, and f) is held for several weeks, during which anaerobic bacteria break down the sludge to usable end-products. (h) The water from the settling tanks may be further processed in tertiary treatment.

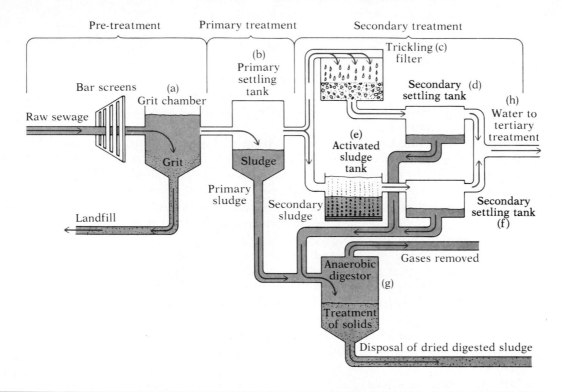

since it contains many valuable salts, or the sludge may be carted to landfill sites or offshore dumping grounds.

The separation of solid sludge leaves a certain amount of water that may be further processed in **tertiary treatment** by purifying the water. Sedimentation is followed by filtration and chlorination, after which the water is placed back into circulation and made available to the consumer. In many municipalities it is also important to remove salts such as phosphates from the water because these salts may spark blooms of algae.

Tertiary treatment: purification of the water remaining after the sludge has been removed.

Bacteriological Analysis of Water

Many methods are available for the detection of bacterial contamination of water, and various ones are selected according to the resources of the testing laboratory. Since it is impossible to test for all pathogenic microorganisms, water quality bacteriologists have adopted the practice of testing for certain indicator bacteria normally found in the human intestinal tract. If these bacteria are present, fecal contamination has probably taken place.

Among the most frequently used indicator organisms are the coliform bacteria. **Coliform bacteria** are normally found in the intestinal tracts of humans

Coliform bacteria: Gram-negative intestinal rods that ferment lactose to acid and gas.

FIGURE 25.7
Use of the Membrane Filter Technique for Water Analysis

(a) The technician collects water from the area to be tested. (b) The sample is poured into the cup and passes through the filter where bacteria are trapped. (c) After incubation, bacterial colonies form on the surface of the membrane filter. (d) Waterborne *Escherichia coli* cells trapped on a membrane filter are seen with the scanning electron microscope.

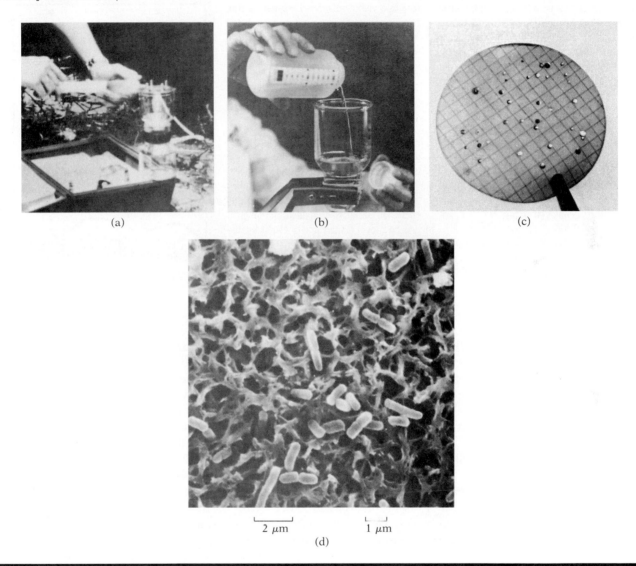

(a)   (b)   (c)

2 µm   1 µm

(d)

and many warm-blooded animals. They are able to survive for extensive periods of time in the environment, and they are relatively easy to cultivate in the laboratory. *Escherichia coli* is the most important indicator organism within the group.

The **membrane filter technique** is a popular laboratory test in water microbiology because it is straightforward and can be used in the field. A technician holds a specially designed collecting bottle against the current and takes a 100-ml sample (Figure 25.7). The water is then passed through a membrane filter, and the filter pad is transferred to a plate of bacteriological medium, as outlined in Chapter 21. Bacteria trapped in the filter will form colonies, and by counting the colonies, the technician may determine the original number of bacteria in the sample.

25.2
MICROFOCUS

THE HORSESHOE CRAB IN MICROBIOLOGY

The horseshoe crab, *Limulus polyphenus*, has existed essentially unchanged for about 200 million years. Many biologists refer to it as a living fossil, because it is apparently unrelated to anything presently in the sea. Despite its name, zoologists point out that it is not even a crab but more closely related to a waterborne spider.

Since 1968 the horseshoe crab has assumed a position of importance in the microbiology laboratory. That year scientists discovered that a solution of broken *Limulus* blood cells quickly forms a gel in the presence of endotoxins. Soon thereafter they developed a test for detecting microorganisms.

The test, called the *Limulus* endotoxin assay, detects minute amounts of endotoxins in fluids. Polluted water, for example,

is mixed with a solution of horseshoe crab blood cells, and the preparation is incubated at 37° C for 60 minutes. If the mixture shows an increase in viscosity or if a gel forms, the test is considered positive for endotoxins. Inasmuch as endotoxins often signal the presence of Gram-negative bacteria, microbiologists have encouraged use of the test as a presumptive sign for these organisms.

Because the test is highly sensitive, endotoxins may also be detected in such specimens as urine, cerebrospinal fluid, and serum, and many diagnostic laboratories are now incorporating the *Limulus* assay into their standard procedures. Though it has no known relative and is misnamed, the lowly horseshoe crab has apparently found a future in microbiology.

Standard plate count:
a technique for determining the total number of bacteria per ml of fluid.

Another method for testing water is the **standard plate count technique (SPC)**. Samples of water are diluted in sterile buffer solution and carefully measured amounts are pipetted into Petri dishes. Agar medium is added, and the plates are set aside at incubation temperatures. A count of the colonies multiplied by the reciprocal of the dilution (the dilution factor) yields the total number of bacteria per ml of the original sample.

MPN test:
a statistical evaluation of the number of coliform bacteria in a water sample.

A third test is a statistical evaluation called the **most probable number (MPN)**. In this procedure, a technician inoculates water in 10-ml, 1-ml, and 0.1-ml amounts into lactose broth tubes. The tubes are incubated and coliform organisms are identified by their production of gas from lactose. Referring to an MPN table, a statistical range of the number of coliform bacteria is determined by observing how many broth tubes showed gas.

A more extensive bacteriological analysis involves specific tests for coliform bacteria and gene probe analyses. MicroFocus: 25.2 explores another method for detecting Gram-negative organisms.

TO THIS POINT
■

We have given close scrutiny to water pollution and the methods used in water purification and sewage treatment. We noted the three basic steps in the preparation of water for drinking: (1) sedimentation to remove bulky objects; (2) filtration to eliminate the vast majority of microorganisms; and (3) chlorination to destroy the last remnants of microbial life. We also saw how sewage treatment may involve something as simple as the outhouse, cesspool, or septic tank, or a highly sophisticated operation such as used in metropolitan centers. Primary treatment involves screening the sewage, while secondary treatment is more concerned with the microbial decomposition of organic waste. Sludge tanks are used in the secondary treat-ment. Tertiary treatment is also practiced in some communities.

The section then surveyed some of the tests used by the bac-teriologist to determine the fitness of water for consumption. Emphasis was placed on coliform bacteria as indicators of water pollution. The membrane filter technique, standard plate count, and most probable number (MPN) procedure were out-lined to illustrate the range of available tests.

In the next section of this chapter, we shall shift gears and examine the important positions occupied by microorganisms in the cycles of elements. We shall use three examples, the car-bon, sulfur, and nitrogen cycles, to show how microorganisms contribute to the quality of life. Working in their countless numbers in water and soil, microorganisms effect a series of chemical changes fundamental to all life processes on Earth.

The Cycles of Elements in Water and Soil

The thought of microorganisms usually conjures a negative reaction because of their disease implication, and the pages of this chapter have doubtlessly added to that notion. It would be unwise, however, to neglect the positive role of the microorganisms in water and soil, because it is a substantial one. We have alluded to their role in the decay of organic matter, and we shall expand the idea by briefly examining the place of microorganisms in three vital cycles of nature: the carbon, sulfur, and nitrogen cycles.

The Carbon Cycle

The earth is composed of numerous elements, among which is a defined amount of **carbon** that must constantly be recycled to allow the formation of organic compounds of which all living things are made. **Photosynthetic organisms** take carbon in the form of carbon dioxide and convert it into carbohydrates using the sun's energy and chlorophyll pigments. The vast jungles of the world, the grassy plains of the temperate zones, and the plants of the oceans show the results of this process. Photosynthetic organisms, in turn, are consumed by grazing animals, fish, and humans who use some of the carbohydrates for energy and convert the remainder to cell parts. To be sure, some carbon dioxide is released back in respiration, but a major portion of the carbon is returned to the earth when the animal or plant dies (Figure 25.8).

It is here that the microorganisms exert their influence, for they are the **primary decomposers** of dead organic matter. Working in their countless bil-lions in the water and soil, bacteria, fungi, and other microorganisms consume the organic substances and release carbon dioxide for reuse by the plants. This activity results from the concerted action of a huge variety of microorganisms, each with its own nutritional pattern of protein, carbohydrate, or lipid digestion (Figure 25.9). Without the microorganisms, the earth would be a veritable gar-bage dump of animal waste, dead plants, and organic debris accumulating in implausible amounts.

But there is more. Microorganisms also break down the carbon-based chem-icals produced by **industrial processes** including herbicides, pesticides, and plastics. In addition, they produce methane, or natural gas, from organic matter and are probably responsible for the conversion of plants to petroleum and coal deep within the recesses of the earth. Moreover, many microorganisms trap

FIGURE 25.8
A Simplified Carbon Cycle

Photosynthesis represents the major method for incorporating carbon dioxide to organic matter and respiration accounts for its return to the atmosphere. Microorganisms are crucial to all decay in soil and ocean sediments.

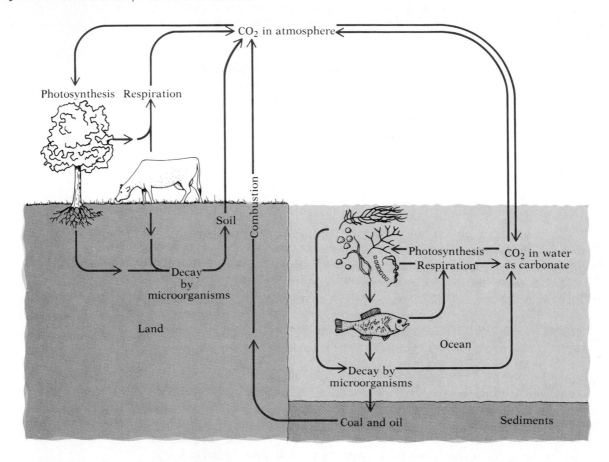

CO_2 from the atmosphere and form carbohydrates to supplement the results of photosynthesis. In these activities, the microorganisms represent a fundamental underpinning of organic creation.

The Sulfur Cycle

The sulfur cycle may be defined in more specific terms than the carbon cycle. **Sulfur** is a key constituent of such amino acids as cystine, cysteine, and methionine, all of which are important components of proteins. Proteins are deposited in water and soil as living things die, and bacteria decompose the proteins and break down the sulfur-containing amino acids to yield various compounds including hydrogen sulfide. Sulfur may also be released in the form of sulfate molecules commonly found in organic matter. Anaerobic bacteria such as those of the genus ***Desulfovibrio*** subsequently convert sulfate molecules to hydrogen sulfide.

The next set of conversions involves several genera of bacteria, including members of the genera ***Thiobacillus***, ***Beggiatoa***, and ***Thiothrix***. These bacteria

Cystine:
a sulfur-containing amino acid found in many proteins.

de-sul"fo-vib're-o
thi"o-bah-sil'us
bei"je-ah-to'ah
thi'o-thricks

FIGURE 25.9

A Soil Bacterium

Low voltage scanning electron micrographs of the soil bacterium *Myxococcus xanthus*. This bacterium belongs to a group that displays gliding motility, complex social interactions, and under certain circumstances, a resistant spore called the myxospore. In this series of electron micrographs of increasing magnification, the fibrils connecting the cells at the surface can be seen. The fibrils branch and form an interconnecting network among the cells. Note also the rough texture of the cell surface.

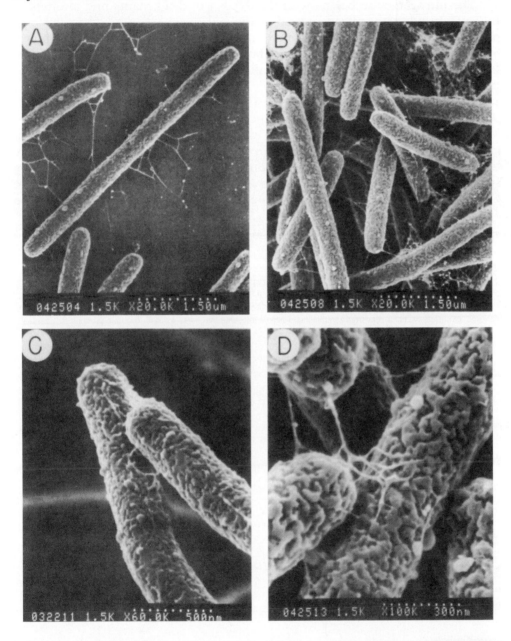

release sulfur from hydrogen sulfide during their metabolism and convert it into sulfate. The sulfate is now available to plants where it is incorporated into the sulfur-containing amino acids. Consumption by animals and humans completes the cycle.

The Nitrogen Cycle

The cyclic transformation of nitrogen is of paramount importance to life on Earth. **Nitrogen** is an essential element in nucleic acids and amino acids. Although it is the most common gas in the atmosphere (about 80 percent of air), animals cannot use nitrogen in its gaseous form, nor can any but a few species of plants. The animals and plants thus require the assistance of microorganisms to trap the nitrogen.

Urea:
the nitrogen-containing product of amino acid decomposition that comprises the main component of urine.

The nitrogen cycle begins with the deposit of dead plants and animals in the soil. In addition, nitrogen reaches the soil in urea contained in urine. A process of digestion and putrefaction by soil bacteria and other microorganisms follows, thus yielding a mixture of amino acids (Figure 25.10). Amino acids are further broken down in microbial metabolism, and the ammonia that accumulates may be used directly by plants.

Next, **mineralization** takes place. In this process, complex organic compounds are finally converted to inorganic compounds and additional ammonia. Much of the ammonia is converted to nitrite ions by **Nitrosomonas** species, a group of aerobic Gram-negative rods. In the process, the bacteria obtain energy for their metabolic needs. The nitrite ions are then converted to nitrate ions by species of **Nitrobacter**, another group of aerobic Gram-negative rods, which obtain energy from the process. Nitrate is a crossroads compound: it can be

---

FIGURE 25.10

A Simplified Nitrogen Cycle

Plant and animal protein and metabolic wastes are decomposed by bacteria into ammonia. The ammonia may be utilized by plants or it may be converted by *Nitrosomonas* and *Nitrobacter* species to nitrate, which is also used by plants. Some nitrate is broken down to atmospheric nitrogen. This nitrogen is returned to the leguminous plants by nitrogen-fixing microorganisms as nitrate, which is converted to ammonia. Animals consume the plants as food to obtain proteins that contain the nitrogen.

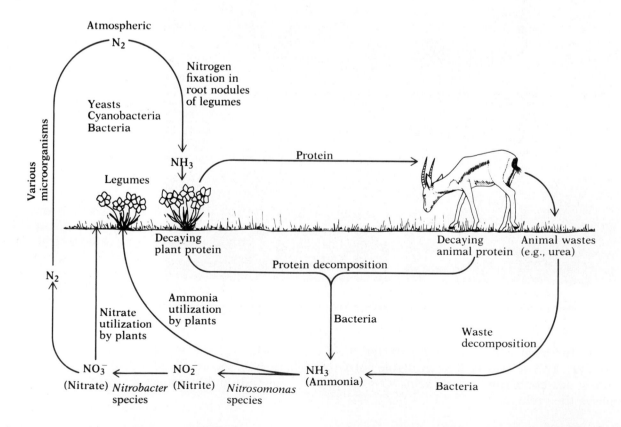

used by plants for their nutritional needs, or it can be liberated as atmospheric nitrogen by certain microorganisms.

For the nitrogen released to the atmosphere, a reverse trip back to living things is an absolute necessity for life to continue as we know it. The process is called **nitrogen fixation**. Once again microorganisms in water and soil play a key role because they possess the enzyme systems that trap atmospheric nitrogen and convert it to compounds useful to plants. In nitrogen fixation, gaseous nitrogen is incorporated to ammonia that fertilizes plants.

Two general types of microorganisms are involved in nitrogen fixation: free-living species and symbiotic species. **Free-living species** include bacteria of the genera *Bacillus, Clostridium, Pseudomonas, Spirillum,* and *Azotobacter,* as well as types of cyanobacteria and certain yeasts. Generally, the free-living species fix nitrogen during their growth cycles. The nitrogen-fixing ability of these species cannot be overemphasized.

Symbiotic species of nitrogen-fixing microorganisms live in association with plants that bear their seeds in pods. These plants, known as **legumes**, include peas, beans, soybeans, alfalfa, peanuts, and clover. Species of Gram-negative rods known as ***Rhizobium*** infect the roots of the plants and live within swellings, or nodules, in the roots (Figure 25.11). Although complex factors are involved, the central theme of the relationship is that *Rhizobium* fixes nitrogen and makes nitrogen compounds available to the plant while taking energy-rich carbon compounds in return. The bulk of the nitrogen compounds accumulates when *Rhizobium* cells die. Legumes then use the compounds to construct amino acids and, ultimately, protein. Animals consume the soybeans, alfalfa, and other legumes and convert plant protein to animal protein, thereby completing the cycle.

Humans have long recognized that soil fertility can be maintained by rotating crops and including a legume. The explanation lies in the ability of rhizobia to fix nitrogen within the nodules of legumes. So much nitrogen is captured, in fact, that the net amount of nitrogen in the soil actually increases after a crop of legumes has been grown. When cultivating legumes, there is no need to add nitrogen fertilizer to the soil. In addition, when crops such as clover or alfalfa are plowed under, they markedly enrich the soil's nitrogen content. Thus,

Nitrogen fixation:
the chemical process in which atmospheric nitrogen is incorporated to organic compounds.

ah-zo"to-bak'ter

Legume:
a plant that bears its seeds in pods.
ri-zo'be-um

FIGURE 25.11

Nodules on the Roots of a Cowpea, a Legume Plant

Species of *Rhizobium* live within the nodules and fix nitrogen to nitrogen-containing compounds. When the bacteria die, the compounds are utilized by the legume to synthesize amino acids.

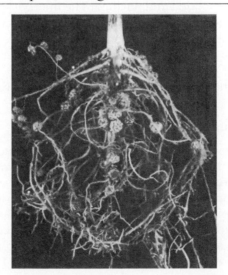

humans are indebted to microorganisms for such edible plants as peas and beans, as well as for the indirect products of nitrogen fixation, namely, steaks, hamburgers, and milk.

TO THIS POINT ■	*In the previous section, we have explored the essential places that microorganisms occupy in the recycling of elements in the environment. In the carbon, nitrogen, and sulfur cycles, micro-organisms convert the basic elements of life into usable forms, and they replenish the environment to nourish all living things. By far, the great majority of microorganisms are engaged in constructive, cooperative, healthy, and wholesome activities. And, it should be noted, these activities happen after death, not before.*
	It is now time to shift gears once again as we move from the outside environment to the inside environment within the laboratory. Here we shall see how microorganisms are used in industrial processes to enhance the quality of our lives with numerous products. It is appropriate that this should be the last section of this book because it points up microorganisms in a positive light. In so many chapters we have portrayed microorganisms as sordid threats to our health, but the opposite side of the coin is that microorganisms add to our lives significantly. The following pages should show you how.

Microorganisms in Industry

Certain properties make microorganisms well suited for industrial processes. Microorganisms not only possess a broad variety of enzymes to make an array of chemical conversions possible, but they also have a relatively high metabolic activity that permits conversions to take place rapidly. In addition, they have a large surface area for quick absorption of nutrients, and release of end-products. Moreover, they usually multiply at a high rate, as evidenced by the 20-minute generation time for *Escherichia coli* under ideal conditions.

In the industrial process, microorganisms act like chemical factories. To be effective, they should liberate a large amount of a single product that can be efficiently isolated and purified. The microorganisms should be easy to maintain and cultivate, and should have genetic stability with infrequent mutations. Their value is enhanced if they can grow on an inexpensive, readily available medium that is a by-product of other industrial processes. For example, a large amount of whey is produced in cheese manufacturing, and microorganisms that convert whey components to lactic acid add to the overall profit of the cheese industry.

Whey:
the clear fluid remaining after protein curdles from milk.

Production of Organic Compounds

Microorganisms are used in industry to produce a variety of organic compounds including acids, growth stimulants, and enzymes. In some cases the production results from an apparent accident in nature where an organism manufactures many thousands of times the amount necessary for its own metabolism.

One of the first organic acids to be made in bulk by microorganisms was **citric acid**. Manufacturers use this organic compound in soft drinks, candies, inks, engraving materials, and in a variety of pharmaceuticals such as anticoagulants and effervescent tablets (Alka-Seltzer). The organism most widely used in citric acid production is the mold ***Aspergillus niger***. Microbiologists inoculate

as″per-jil′us ni′jer

FIGURE 25.12

Chemistry of Citric Acid Production

Aspergillus niger is grown in a mixture of nutrients, where it digests glucose into pyruvic acid. The pyruvic acid is then converted to acetyl-CoA, which condenses with oxaloacetic acid in the Krebs cycle to yield citric acid. However, the chemistry goes no further, because the next enzyme in the cycle is lacking. Citric acid therefore accumulates and is isolated for use in various products as shown.

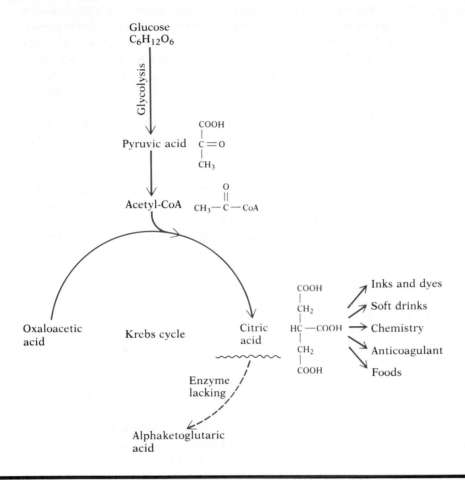

the mold to a medium of corn meal, molasses, salts, and inorganic nitrogen in huge shallow pans or fermentation tanks. The absence of a Krebs cycle enzyme in the mold prevents the metabolism of citric acid into the next component of the cycle, and the citric acid accumulates in the medium. Figure 25.12 outlines this chemistry.

Another important microbial product is **lactic acid**, a compound employed to preserve foods, finish fabrics, prepare hides for leather, and dissolve lacquers. Lactic acid is commonly produced by bacterial activity on the whey portion of milk. ***Lactobacillus bulgaricus*** is widely used in the fermentation because it produces only lactic acid from lactose.

Gluconic acid, another valuable organic acid, is useful in medicine as a carrier for calcium, because gluconic acid is easily metabolized in the body leaving a store of calcium for distribution. This acid is produced from carbohydrates by *A. niger* and species of the bacterium ***Gluconobacter*** cultivated in fermentation tanks. Calcium gluconate is also added to the feed of laying hens to provide calcium that strengthens the eggshells.

gloo"ko-no-bak′ter

IT SMELLED BAD, BUT IT WORKED

In past centuries, industrial pectinase was not available for retting flax, nor was protease for bating hides. Nevertheless, the processes were carried on efficiently and successfully.

The retting process began by bundling flax plants and drying them in stacks. The stacks were then placed in a long trench several feet deep, covered with water, and weighted down with stones to exclude as much air as possible. After a few days, the water turned black, and an unmistakable stench signaled that retting was taking place. Two weeks later the flax was so soft and pliable that the fibers could be easily removed by pounding with wooden blocks. Today's microbiologists point out that *Clostridium* species were probably producing pectinase in the trenches.

The method for treating hides was equally messy. Skins were mixed with dog or fowl manure and set aside to cure. Fragments of tissue and hair gradually dissolved in the muck, and soon, the hide became soft and pliable. Apparently, the proteases from fecal bacteria were responsible for the digestion. Bating hides was another smelly process, to be sure, but like the method for retting, it was usually reliable.

---

ar″thro-bak′ter
brev″e-bak-te′re-um

When the amount of amino acid produced by a microorganism exceeds the need, the remainder is excreted into the environment. Such is the case with **glutamic acid** produced by certain species of *Micrococcus*, *Arthrobacter*, and *Brevibacterium*. Glutamic acid is a valuable food supplement for humans and animals, and its sodium salt, monosodium glutamate, is utilized in food preparations.

di-am″ĭ-no-pim-el′ik

In the production of **lysine**, another amino acid, two organisms are involved. *E. coli* is first cultivated in a medium of glycerol, corn steep liquor, and other ingredients, and the compound diaminopimelic acid (DAP) accumulates. Several days later, *Enterobacter aerogenes* is added to the mixture. This organism produces an enzyme that removes the carboxyl group from DAP to produce the lysine used in breads, breakfast cereals, and other foods.

ash′be-ah go-sip′e-e
si″ah-no-ko-bal′ah-min
pro″-pe-on″e-bak-te′re-um

Two important vitamins, riboflavin (vitamin B$_2$) and cyanocobalamin (vitamin B$_{12}$), are also products of microbial growth. **Riboflavin** is a product of ***Ashbya gossypii***, a mold that produces 20,000 times the amount it needs for its metabolism. **Cyanocobalamin** is produced by selected species of *Pseudomonas*, *Propionibacterium*, and *Streptomyces* grown in a cobalt-supplemented medium. The vitamin prevents pernicious anemia in humans and is used in bread, flour, cereal products, and animal feeds.

Enzymes and Other Products

The production of microbial enzymes for commercial exploitation has been an important industry since the emergence of industrial microbiology. Currently, over two dozen types of microbial enzymes are in use, and several others are in the research or developmental stage.

o-ri′za

The important microbial enzymes include amylase, pectinase, and several proteases. **Amylase** is produced by the mold ***Aspergillus oryzae***. It is used as a spot remover in laundry presoaks, as an adhesive, and in baking, where it digests starch to glucose. **Pectinase**, a product of a ***Clostridium*** species, is employed to ret flax for linen. In this process, manufacturers mix the flax plant with pectinase to decompose the pectin "cement" that holds cellulose fibers together. The cellulose fibers are then spun into linen. MicroFocus: 25.3 describes a more traditional process for retting. Pectinase is also used to clarify fruit juices.

Retting:
the process by which enzymes remove the pectin that holds cellulose fibers together in flax plants.

Proteases are a group of protein-digesting enzymes produced by *Bacillus subtilis*, *Aspergillus oryzae*, and other microorganisms. Certain proteases are used for bating hides in leather manufacturing, a process in which organic tissue is removed from the skin to yield a finer texture and grain. Other proteases find value as liquid glues, laundry presoaks, meat tenderizers, drain openers, and spot removers.

Bating: the process by which organic material is removed from hides in leather manufacturing.

One of the most appreciated but lesser known uses of a microbial enzyme is in making soft-centered chocolates. **Invertase**, an enzyme from yeast, is mixed with flavoring agents and solid sucrose, and then covered with chocolate. The enzyme converts some of the sucrose to liquid glucose and fructose, forming the soft center of the chocolate. In medical microbiology, doctors use another microbial enzyme, **streptokinase**, to break down blood clots formed during a heart attack. Still another enzyme, **hyaluronidase**, is used to facilitate the absorption of fluids injected under the skin.

hi″ah-lu-ron′ĭ-dās

Gibberellins are a series of plant hormones that promote growth by stimulating cell elongation in the stem. Botanists use the hormones to hasten seed germination and flowering, and agriculturalists find them valuable for setting blooms in the plant. This increases the yield of fruit and in the case of grapes, enhances their size. Gibberellins are produced during the metabolism of the fungus **Gibberella fujikuroi** and may be extracted from these organisms for commercial use.

gib-ber-el′in

gi-ber-el′ah fu″ji-kur′oi

In addition to the major products that we have surveyed, microorganisms provide a number of specialized materials. Typical of the miscellaneous microbial products is **alginate**, a sticky substance used as a thickener in ice cream, soups, and other foods. Another product of microbial origin is **perfume**. Musk oil, for example, is prepared from ustilagic acid, a product of the mold **Ustilago zeae**, which, ironically, causes smut disease (Chapter 14). Production by purely chemical means is prohibitively expensive. Moreover, there are numerous pharmaceutical products derived from the ergot poisons of the mold **Claviceps purpurea**. These derivatives are prescribed to induce labor, treat menstrual disorders, and control migraine headaches.

us″ti-lah′jĭk
us″tĭ-la′go ze′a

Alcoholic Beverages

The fermentation of beer, wine, and other alcoholic beverages is among the most venerable and universal of human domestic activities (Figure 25.13). The origin of beer fermentation, for example, has been traced as far back as 4000 B.C. when legend tells us that Osiris, the god of agriculture, taught Egyptians the art of brewing. Wine production apparently has an equally long history because archaeologists have discovered evidence of grape cultivation in the Nile Valley during the same period. Thus it is conceivable that the Egyptians, Sumerians, Assyrians, and other Near East peoples were among the earliest consumers, if not connoisseurs, of alcoholic beverages.

Beer

As early as 3400 B.C., a tax was placed on beer in the ancient Egyptian city of Memphis on the Nile. The Greeks later brought the art of brewing to Western Europe, and the Romans refined it. Indeed, the main drink of Caesar's legions was beer. During the Middle Ages, monasteries were the centers of brewing, and by the 1200s, breweries and taverns were commonplace in Great Britain. However, centuries passed before beer made its appearance in cans. That auspicious event took place in the United States in Newton, New Jersey, in 1935. The six-pack was a logical successor.

FIGURE 25.13
Ancient Wine Making

Photograph of a Greek vase from the sixth century B.C. showing grapes being crushed for wine by two satyrs. In mythology, satyrs were part man, part goat creatures (note the horns, tails, and feet), who attended Dionysus, the Greek god of wine. Bacchus was the Roman counterpart of Dionysus.

Malting:
the enzymatic process in which starch in barley grains converts to smaller carbohydrates such as maltose.

hu′mu-lus loop′u-lus

sak″ah-ro-mi′sēz
ser″e-vis′e-ā

karls″berg-en′sis

lah′ger-ing

Sake:
an Oriental beer derived from rice starch.

The word **beer** is derived from the Anglo-Saxon *baere*, meaning barley. Thus beer is traditionally a product of yeast fermentations of barley grains. However, yeasts are unable to digest barley starch, and therefore it must be predigested for them (Figure 25.14). This is accomplished in the process of **malting**, where barley grains are steeped in water while naturally occurring enzymes digest the starch to simpler carbohydrates, principally maltose (malt sugar).

At the brewery, the malt is ground with water to achieve further digestion of the starch. This process, called **mashing**, often includes corn as a starch supplement. Brewers then remove the liquid portion, or **wort**, and boil it to inactivate the enzymes. Dried petals of the vine *Humulus lupulus*, called **hops**, are now added to the wort, giving it flavor, color, and stability. Hops also prevent contamination of the wort, because the leaves contain at least two antimicrobial substances. At this point, the fluid is filtered, and yeast is added in large quantities.

The yeast usually employed in beer fermentation is one of two species of **Saccharomyces** developed for centuries by brewers. One species, *S. cerevisiae*, gives a uniform dark cloudiness to beer and is carried to the top of the fermentation vat by foaming carbon dioxide. This yeast is therefore called a **top yeast**. It is used primarily in English-type brews such as ale and stout. The second species, *S. carlsbergensis*, ferments the malt more slowly and produces a lighter, clearer beer, having less alcohol. This yeast sediments and is thus called a **bottom yeast**. Its product is pilsener or lager beer. Almost three quarters of the world's beer is lager beer.

A normal fermentation requires approximately seven days in a fermentation tank. The young beer is then transferred to vats for secondary aging, or **lagering**, which may take an additional six months. If the beer is intended for canning or bottling, it may be pasteurized at 140° F for 55 minutes to kill the yeasts, or filtered through a membrane filter. Some yeast is used to seed new wort, and the remainder may be dried for animal feed or pressed to tablets for human consumption. The alcoholic content of beer is approximately 4 percent.

Although commonly referred to as rice wine, the Oriental beverage **sake** is more like a rice beer. It is produced by allowing *Aspergillus oryzae* to convert rice starch to fermentable sugar. *Saccharomyces* species then ferment the sugar until the alcohol level is approximately 14 percent. The sake is now ready for consumption.

FIGURE 25.14

FIGURE 25.14

A Generalized Process for Producing Beer

Barley grains are held in malting tanks while the seeds germinate to yield fermentable sugars. The digested grain, or malt, is then mashed in a mashing tank and the fluid portion, the wort, is removed. Hops are added to the wort in the next step, followed by the yeast growth and alcohol production in fermentation. The young beer is aged in primary and secondary aging tanks. When it is ready for consumption, it is transferred to kegs, bottles, or cans.

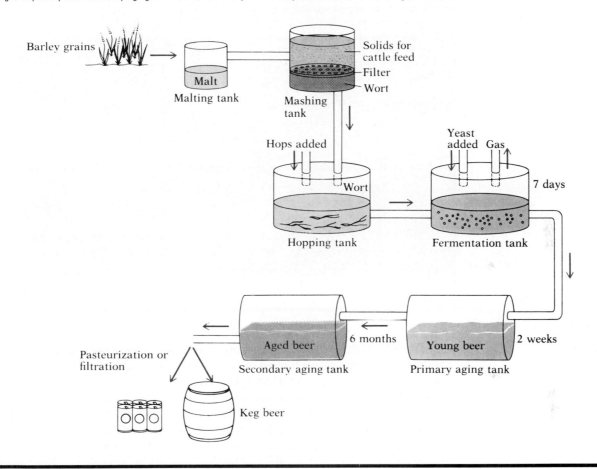

Wine

Essentially, all **wines** are derived from the natural conversion of grape or other fruit sugars to ethyl alcohol by the enzymes in *Saccharomyces*. Wild yeasts naturally occurring on the grapes may be used, but in the United States, the general trend is to use controlled cultures of yeast, usually *S. cerevisiae* variety *ellipsoideus*.

e″lip-soid′e-us

Wine may be made from fruit, fruit juice, or plant extracts such as dandelions. Among the grapes, the species *Vitis vinifera* is recognized as the highest-quality fruit. The wine-making process begins with crushing to produce the juice, or **must**. For **red wine**, black grapes are used, including skins and sometimes, the stems. **White wine**, by contrast, is made from black or white grapes without their skins or stems. Yeasts begin to multiply immediately and initiate the fermentations. Anaerobic conditions are soon established as carbon dioxide evolves and takes up all the air space. The mixture froths with CO_2 and "ferments" in the original sense of the word.

Must:
fruit juice used to make wine.

FIGURE 25.15
Two Steps in the Large-scale Production of Wine

(a) A view of industrial model wine presses. The press on the right is open for loading with crushed grapes. Once loaded, a rubber bag is inflated in the center of the press to gently squeeze the grapes against the inside walls of the press. This extracts as much juice as possible from crushed grapes without breaking the seeds. (b) Bottles move along a conveyor belt and are mechanically filled with wine.

(a)   (b)

Alcohol production requires only a few days, but the **aging process** in wooden casks may go on for weeks or months. During this time wine develops its unique flavor, aroma, and bouquet. These result from the array of alcohols, acids, aldehydes, and other organic compounds produced by the yeast during aging. Soil and climate conditions also contribute to the wine because they determine what organic compounds are present in the grape. The type of yeast and nature of wood derivatives from the fermentation casks are other determining factors. Thus there are "vintage years" and "poor years." (Figure 25.15).

The broad variety of available wines result from modifications of the basic fermentation process. In **dry wines**, for example, most or all of the sugar is metabolized, while in **sweet wines**, fermentation is stopped while there is residual sugar. **Sparkling wines**, including champagne, sparkle because of a second fermentation taking place inside the bottle. For a sweet **sauterne**, vintners enhance the sugar content of grapes by a controlled infection with the mold *Botrytis cinera*. The mold literally sucks water out of the grapes, thereby increasing the sugar concentration.

bo-tri′tis sin-er′a

The strongest natural wines measure about 15 percent alcohol because yeasts cannot tolerate alcohol levels above this. Most table wines average about 10 to 12 percent alcohol, with **fortified wines** reaching 22 percent alcohol. In fortified wines, brandy or other spirits are added to produce such wines as Port, Sherry, and Madeira. For mass production, wine is pasteurized to increase its shelf life, filtered, and bottled.

Distilled Spirits

Proof number: twice the percentage of alcohol in a fermented product such as distilled spirits.

Distilled spirits contain considerably more alcohol than beer or wine. Each is designated with a **proof number**, which is twice the percentage of alcohol. For example, a 90 proof product contains 45 percent alcohol.

The production of distilled spirits begins like a wine fermentation. A raw product is fermented by *Saccharomyces* species, then aged, and finally matured in casks. At this point the process diverges as manufacturers concentrate the alcohol by a distillation apparatus using heat and vacuum. Next they mature the product in wooden casks to introduce unique flavors from various chemicals such as aldehydes and volatile acids. Finally the alcohol is standardized by diluting it with water before bottling.

Four basic types of distilled spirits are produced: brandy, whiskey, rum, and neutral spirits. **Brandy** is made from fruit or fruit juice, while **rum** is produced from molasses. **Whiskey** is a product of various malted cereal grains, such as scotch from barley, rye from rye grain, and bourbon from corn. The final type, **neutral spirits**, includes vodka, which is made from potato starch and left unflavored, and gin, which is flavored with the oils of juniper berries.

Other Microbial Products

In addition to the products we have discussed, microorganisms are the sources of antibiotics and a number of valuable insecticides. Moreover, they are the biological factories for the genetic engineering technology that has revolutionized industrial microbiology. In the final section of this text, we shall study the methods for antibiotic and insecticide production and discuss some details of the genetic engineering process.

Antibiotics

Penicillin was the first antibiotic to be produced on an industrial scale. In 1941, Robert H. Coghill of the Fermentation Division of the USDA made the suggestion that the deep-tank method used to produce vitamins might be applied to penicillin. In the ensuing months he offered several modifications to stimulate the growth of **Penicillium notatum** and increase the penicillin yield. For example, corn steep liquor in the culture medium increased the output 20 times, and substitution of lactose for glucose made penicillin production still more efficient. Moreover, the search for a higher-yielding producer of the drug led researchers to **Penicillium chrysogenum**, a mold isolated from a rotten canteloupe from a Peoria, Illinois, supermarket (Figure 25.16). Treatment with ultraviolet light resulted in a mutant with still higher penicillin yields. By 1943, the United States was producing enough penicillin for the allied forces, and by 1945 sufficient amounts were available for the civilian population.

To the present time, over 5000 antibiotic substances have been described and approximately 100 such drugs are available to the medical practitioner. Although most antibiotics are produced by species of *Streptomyces*, a significant number are products of *Penicillium* or *Bacillus* species. The worldwide production of antibiotics exceeded 25,000 tons in 1990, and two-thirds were penicillins.

Antibiotic production is carried on in huge, aerated tanks of stainless steel similar to those used in brewing. A typical tank may hold 30,000 gallons of medium. Older methods employed enormous mats of fungus or actinomycetes on the surface of the tank. Newer technology, however, employs small fragments of submerged hyphae or cells, rotated and agitated in the medium with a constant stream of oxygen. After several weeks of growth, the microorganisms are removed, and the antibiotic is extracted from the medium for further conversion to the desired product. The remaining brown mash of microorganisms may be dried and sold as an animal feed additive. Another alternative is to process it for use as human food.

Penicillin:
an antibiotic that prevents cell wall synthesis primarily in Gram-positive bacteria.

krĭ-soj′en-um

FIGURE 25.16

An Antibiotic Producer

A symmetrical colony of *Penicillium chrysogenum*, the mold isolated from a rotten canteloupe in Peoria, Illinois. This mold produces a large proportion of the world's supply of penicillin.

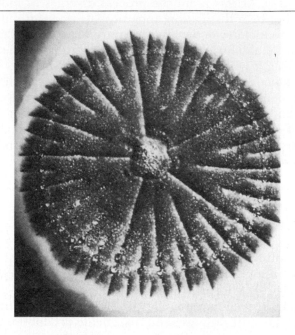

Insecticides

To be useful as an insecticide, a microorganism should be relatively specific for an insect pest and should act rapidly. It should be stable in the environment and easily dispensed, as well as inexpensive to produce.

thur"in-jen'sis

In the early part of this century, a scientist named G.S. Berliner found that sporulating cells of a *Bacillus* species were inhibitory to moth larvae. Berliner named the organism **Bacillus thuringiensis** after the European province Thuringia where he lived. The bacillus remained in relative obscurity until recent years when scientists learned that *B. thuringiensis* produces **toxic crystals** in older cells during the process of sporulation (Figure 25.17). The toxic substance, an alkaline protein, is deposited on leaves and ingested by caterpillars, the larvel forms of butterflies, moths, and related insects. In the caterpillar gut, the insecticide dissolves substances that cement the cells of the gut wall together. As gut liquid diffuses between the cells, the larvae experience paralysis, and bacterial invasion soon follows.

Bacillus thuringiensis (Bt) appears to be harmless to plants and other animals. It is produced by harvesting bacteria at the onset of sporulation and drying them into a commercially available dusting powder. The product is useful on tomato horn worms and gypsy moth caterpillars. Its success led to the discovery of a new strain called *B. thuringiensis israelensis* (Bti), first isolated in Israel.

pop-il'e-a

In 1941, **Bacillus popilliae** was introduced as a control measure for Japanese beetles. The bacillus infects beetle larvae and causes **milky spore disease**, so named because the blood of the larvae becomes milky white. The larvae eventually die and the bacillus spores remain in the soil to infect other larvae.

Viruses also show promise as pest control devices partly because they are more selective in their activity than bacteria. Once released in the field, the viruses spread naturally. It is also possible to harvest early disease victims, grind

FIGURE 25.17

Bacillus thuringiensis

A transmission electron micrograph of *Bacillus thuringiensis* ($\times$ 60,300) showing its endospore (ES) and toxic crystal (TC). The toxic crystal contains an alkaline protein that functions as an insecticide by dissolving the cementing substance around the gut cells of insect larvae. Also worthy of note in this micrograph are the multiple coverings of the spore, and the separate cell wall (CW) and cell membrane (CM).

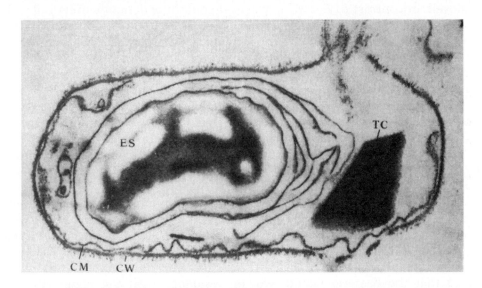

them up, and use them to disseminate the virus to new locations. Among the insects successfully controlled with viruses are the cotton boll worm, cabbage looper, and alfalfa caterpillar.

Industrial Genetic Engineering

In 1973, **Herbert Boyer** of the University of California at San Francisco and **Stanley Cohen** of Stanford University performed the first practical experiment in genetic engineering. Working with *Escherichia coli*, they removed the genes for kanamycin resistance and spliced them to a plasmid that already carried genes for tetracycline resistance. The plasmids were then mixed with *E. coli* cells not resistant to either antibiotic. Finally, the cells were streaked on plates of culture medium containing both kanamycin and tetracycline. The bacteria that grew were ones that took up the plasmids and were now resistant to both antibiotics. Feats as these launched the modern era of biotechnology. As one observer later noted: "Biotechnology used to be BBC (before Boyer–Cohen). Now, it is ABC (after Boyer–Cohen)."

Genetic engineering based in plasmid technology has been hailed as the beginning of modern industrial microbiology. **Plasmids** are ultramicroscopic ringlets of double-stranded DNA that exist apart from the chromosome. A single plasmid may contain between 2 and 250 genes. The significance of **plasmid technology** lies in the fact that plasmids can be spliced with fragments of DNA from unrelated organisms and inserted into host organisms. The host organisms then produce the protein whose genetic message is carried by the foreign DNA. Mechanics of this reengineering process are explored in Chapter 6. Some contemporary products already obtained by plasmid technology include interferon, insulin, a vaccine for hoof-and-mouth disease, human growth hormone, and urokinase.

Plasmids:
loops of DNA apart from the chromosome that contain several genes and are used in genetic engineering experiments.

When microbiologists first developed the art of plasmid technology in the 1970s, the likely choice for the prototype "bacterial factory" was **E. coli**. Nonpathogenic strains had long been used as test organisms in the laboratory, and the genetics of *E. coli* was well understood. In the 1980s, however, attention shifted to **Bacillus subtilis** and yeasts as host organisms. Advocates of *B. subtilis* point out that this Gram-positive bacillus normally secretes the proteins it makes, while *E. coli* retains them. Also *B. subtilis* is not regarded as a human pathogen, in contrast to *E. coli*, nor does it contain endotoxins in its cell wall. **Yeast** supporters point to the traditional role of their organism in fermentation processes and, therefore, the public acceptance of an organism without disease potential. One industrial firm has reengineered a *Saccharomyces* species to produce a synthetic vaccine for hepatitis B, while a second firm has used yeast to obtain rennin, the enzyme used to make cheese.

The implications of genetic engineering are far-reaching and thought-provoking. One group of biotechnologists has focused on the antibiotic-producing genes of plasmids and is attempting to remove the genes from *Streptomyces* species and insert them to more rapidly growing organisms. An allied group is trying to amplify the number of plasmids in antibiotic producers so as to obtain a higher yield of product.

In 1985, scientists at the Monsanto Company announced the development of a new **microbial insecticide** for corn plants. Toxin-producing genes were extracted from *Bacillus thuringiensis* and inserted to *Pseudomonas fluorescens*, a common Gram-negative rod that colonizes the roots of corn plants. The Monsanto group sprayed corn plants with reengineered *Pseudomonas* and determined that the bacteria would take up residence with the plants, thereby providing a built-in insecticide.

F I G U R E 25.18

An Artist's Fanciful Conception of a Superplant of the Future Produced by Genetic Engineering

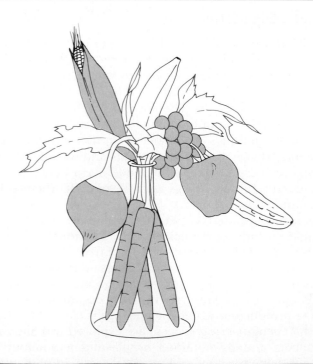

In the 1960s, a "Green Revolution" took place in which high-yielding supergrains were exported to poor countries throughout the world to encourage self-sufficiency. However, the program stalled when stocks of essential petroleum fertilizers fell to high oil prices. Agriculturalists now hope for a second Green Revolution sparked by genetic engineering (Figure 25.18). They foresee the day when the genes for **nitrogen fixation** can be extracted from bacteria such as *Rhizobium* and inserted into grain plants such as wheat, rye, and barley. The most optimistic planners look to the future and foresee fertilizers becoming obsolete, plants using microbial toxins to drive off insects, grains growing in salty water, and crops living for weeks without water.

Other research projects in genetic engineering have a more contemporary application. Biochemists at breweries are attempting to insert the starch-digesting enzyme amylase into yeasts to eliminate the malting process in beer production. At chemical companies, scientists are seeking to incorporate genes for cellulase activity into microorganisms so as to make cellulose a useful source of glucose. For the 20,000 hemophiliacs in the United States, new hope dawned when a genetic engineering company announced that it had isolated and cloned the genes for Factor VIII, an essential blood-clotting protein missing in hemophilia. The next step was to splice the genes to bacteria and hope for an inexpensive, safe product.

Nor does it end here. Research in genetic engineering holds promise for new strategies for cancer prevention, new diagnostic and screening procedures for microbial diseases and genetic abnormalities, new methods to correct genetic disorders, new hormones, antibiotics, and vaccines, and a generally improved quality of life. The discoveries and insights made possible by genetic engineering have been described as breathtaking. In the future we can look to startling developments in medicine, agriculture, and the pharmaceutical and chemical industries. Indeed, it is an exciting time to be a microbiologist.

NOTE TO THE STUDENT

It should be clear from this chapter that the microorganisms make a substantial contribution to the quality of life. Rather than gush with enthusiasm and rhetoric on the positive roles they play, I would prefer to paraphrase several concepts of applied microbiology set down by the late industrial microbiologist David Perlman. In a 1980 publication, Perlman wrote:

1. *The microorganism is always right, your friend, and a sensitive partner;*

2. *There are no stupid microorganisms;*

3. *Microorganisms can and will do anything;*

4. *Microorganisms are smarter, wiser, and more energetic than chemists, engineers, and others; and*

5. *If you take care of your microbial friends, they will take care of your future.*

Summary

Environmental microbiology is concerned in large measure with the pollution brought on by microorganisms. A polluted environment contains a huge variety of heterotrophic organisms from sewage, feces, and industrial sources. Coliform bacteria, the

Gram-negative rods of the human and animal intestine, abound in polluted water. The biochemical oxygen demand (BOD) is a measure of the amount of biological pollution, and concern exists for a variety of bacterial, viral, and protozoal diseases transmitted by water.

To prepare the water for drinking purposes, municipalities employ various levels of water purification including sedimentation, filtration, and chlorination. Sewage can also be treated by different steps according to the needs of the municipalities. Cesspools and septic tanks are used for local treatment, and variations of oxidation lagoons and secondary and tertiary treatments are used on larger scales. To test the effectiveness of purification procedures several bacteriological tests are available, including the membrane filter technique, the standard plate count, and the most probable number test.

In the biosphere of the water and soil, microorganisms are positive factors in the cycles of carbon, sulfur, and nitrogen. In the carbon cycle, bacteria and fungi are essential to the breakdown of organic matter and the release of carbon back to the atmosphere for recycling. Anaerobic bacteria fill a similar niche in the sulfur cycle. In the nitrogen cycle, many microorganisms release nitrogen from urea, amino acids, and nitrogenous organic matter. Numerous nitrogen-based conversions are performed by microorganisms, and bacteria are essential in the steps of nitrogen fixation where nitrogen is brought back into the cycle. Indeed, the processes performed by nitrogen-fixing bacteria are so essential that life as we know it would probably not exist without bacterial intervention.

Microorganisms occupy an important place in industry as the producers of many important products such as organic compounds, alcoholic beverages, antibiotics, and insecticides. Among the organic compounds synthesized on an industrial scale are organic acids (such as citric, lactic, and gluconic acids), various amino acids and vitamins, and a series of enzymes such as amylase, pectinase, and proteases. In addition, microorganisms produce plant hormones known as gibberellins. Beer, wine, and spirits are among the products for public consumption that yeast cells produce.

Though antibiotics were originally a microbial product, most of these drugs are synthetically produced today. By contrast, the spores of *Bacillus thuringiensis* and *B. popilliae* are used in the live form as insecticides for plants. The future of industrial microbiology will center on genetic engineering and plasmid technology. By inserting foreign genes to vector organisms such as bacteria and yeasts, biotechnologists can produce rare proteins on an industrial scale and provide treatments for such diseases as diabetes, hemophilia, and cancer. Medicine, agriculture, and industry eagerly anticipate the fruits of modern biotechnology.

Questions for Thought and Discussion

1. When sewers were constructed in New York City in the early 1900s, engineers decided to join storm sewers carrying water from the streets together with sanitary sewers bringing waste from the homes. The result was one gigantic sewer system. In retrospect, was this a good idea? Why?

2. In the 1970s, a popular bumper sticker read: "Have you thanked a green plant today?" The reference was to photosynthesis taking place in plants. Suppose you saw a bumper sticker reading: "Have you thanked a microorganism today?" What might the owner of the car have in mind?

3. Certain beer companies have developed strains of yeasts that break down more of the carbohydrate in barley malt than traditional yeasts. What do you think their product is called?

4. A student notes in her microbiology class that a particular species of bacteria actively dissolves fats, greases, and oils. Her mind stirs and she wonders whether such an organism could be used to unclog the cesspool that receives waste from her house. What do you think she is considering? Will it work?

5. The water in a particular bay is relatively free of bacteria in the wintertime but generally polluted with bacteria in the summer. One reason is that summer boaters illegally empty their holding tanks into the bay before docking. Another is that fishermen clean their catch along the docks and dump the refuse into the bay. How many other reasons can you suggest for the summertime pollution?

6. The victims of disease, including animals and people, are buried underground and yet the soil is generally free of pathogenic organisms. Why?

7. A product called Dipel contains *Bacillus thuringiensis*. It is used widely in the Northeast for destruction of the gypsy moth caterpillar. What dangers might result from extensive application of this product?

8. What information might you offer to dispute the following four adages common among campers and hikers? (1) Water in streams is safe to drink if there are no humans or large animals upstream. (2) Melted ice and snow is safer than running water. (3) Water gurgling directly out of the ground or running out from behind rocks is safe to drink. (4) Rapidly moving water is germ-free.

9. Certain bacteria produce many thousands of times their requirement of specific vitamins. Some biologists suggest that this makes little sense since the excess is wasted. Can you suggest a reason for this apparent overproduction in nature?

10. Why is it unlikely that wine, beer, or distilled spirits will contain any microorganisms that might cause disease in the human body?

11. The editors of *The Economist*, a British magazine, have referred to genetic engineering as "one of the biggest industrial opportunities of the late 20th century." What evidence can you offer to support this contention and why might you be inclined to reject it?

12. How many times in the last 24 hours have you had the opportunity to use or consume the industrial product of a microorganism?

13. The author of a biology textbook writes: "Because the microorganisms are not observed as easily as the plants and animals, we tend to forget about them, or to think only of the harmful ones . . . and thus overlook the others, many of which are indispensable to our continued existence." How do the carbon, sulfur, and nitrogen cycles support this outlook?

14. A park in a local community has two swimming pools: an Olympic-sized pool for swimmers and a small wading pool for toddlers. In which pool does the greater potential for disease transmission exist, and what precautions may be taken to limit the transmission of microorganisms?

15. The tale on the inside back cover of this book is adapted from a Middle Eastern story. It has been placed there for a particular reason. Can you guess why?

Review

Using your knowledge of environmental microbiology, consider the characteristic on the left and the three possible choices on the right. In the space, place the letter or letters of the most appropriate choice(s).

1. ___ Genus (genera) of coliform bacteria
    a. *Escherichia*
    b. *Staphylococcus*
    c. *Enterobacter*

2. ___ Where halophilic bacteria live
    a. lake
    b. ocean
    c. mountain stream

3. ___ Used as markers for oil drilling
    a. bacteriophages
    b. radiolaria
    c. foraminifera

4. ___ Waterborne microbial disease(s)

    a. hepatitis A
    b. amebiasis
    c. hepatitis B

5. ___ Step(s) in water purification

    a. filtration
    b. chlorination
    c. sedimentation

6. ___ Found in polluted water

    a. *Proteus* species
    b. *E. coli*
    c. AIDS virus

7. ___ Test(s) for oxygen consumption in water

    a. SPC
    b. BOD
    c. MPN

8. ___ Type(s) of pollution when microorganisms are present

    a. biological
    b. physical
    c. chemical

9. ___ Toxin-producing dinoflagellate(s)

    a. *Entamoeba*
    b. *Gambierdiscus*
    c. *Gonyaulax*

10. ___ Produce(s) flocs in water

    a. iron sulfate
    b. copper sulfate
    c. aluminum sulfate

11. ___ Needed to perform the standard plate count

    a. Petri dishes
    b. pipettes
    c. agar medium

12. ___ Possible cause(s) of red tide

    a. *Gonyaulax*
    b. *Streptococcus*
    c. *Gymnodinium*

13. ___ Found in anaerobic mud at lake bottom

    a. *Giardia*
    b. diatoms
    c. *Clostridium*

14. ___ Release(s) carbon in carbon cycle

    a. viruses
    b. fungi
    c. bacteria

15. ___ Function(s) in nitrogen cycle

    a. *Thiobacillus*
    b. *Nitrobacter*
    c. *Thiothrix*

Appendix A

Established Incubation Periods for Selected Microbial Diseases

DISEASE	USUAL INCUBATION PERIOD	RANGE
anthrax	2–7 da	
Bacillus cereus food poisoning	1–16 hr	
bacterial conjunctivitis	3–5 da	
botulism	12–48 hr	6 hr–8 da
brucellosis	5–21 da	
bubonic plague	2–6 da	
campylobacteriosis	5–8 da	
chancroid	6–7 da	
chlamydia	1–3 wk	
cholera	1–3 da	few hours–5 da
clostridial food poisoning	10–12 hr	8–22 hr
diphtheria	2–6 da	
endemic typhus	5–10 da	
epidemic typhus	1–2 wk	
gas gangrene	1–5 da	7 hr–6 wk
gonorrhea	3–5 da	1–14 da
Haemophilus meningitis	3–5 da	
Legionnaires' disease	1–10 da	
leprosy	3 yr	3 mo–20 yr
leptospirosis	2–20 da	
Lyme disease	3–32 da	
lymphogranuloma venereum	5–21 da	
meningococcal meningitis	1–7 da	
Mycoplasma pneumonia	1–3 wk	
pertussis	2 wk	
pneumococcal pneumonia	variable	
pneumonic plague	2–4 da	
psittacosis	4–15 da	
Q fever	14–26 da	
relapsing fever	2–10 da	
rickettsialpox	3–7 da	
Rocky Mountain spotted fever	4–8 da	3–12 da
salmonellosis	12 hr	6–72 hr
scarlet fever	2–5 da	
shigellosis	1–2 da	
Spirillum rat bite fever	10–30 da	
staphylococcal food poisoning	1–7 hr	
Streptobacillus rat bite fever	3–10 da	
streptococcal sore throat	2–5 da	
syphilis	21 da	10–90 da
tetanus	8 da	3 da–3 wk
toxic shock syndrome	2–4 da	
trachoma	5–14 da	
tuberculosis	2–10 wk	
tularemia	3 da	1–10 da
typhoid fever	7–12 da	
Vibrio parahaemolyticus infection	7–48 hr	
whooping cough	5–21 da	
yaws	3–4 wk	
Yersinia enterocolitica infection	12–36 hr	
adenovirus infection	5–6 da	
California encephalitis	5–15 da	
chickenpox	14 da	10–21 da

Appendix A *(continued)*

Established Incubation Periods for Selected Microbial Diseases

DISEASE	USUAL INCUBATION PERIOD	RANGE
Colorado tick fever	3–6 da	
Coxsackie virus diseases	3–5 da	2 da–2 wk
cytomegalovirus disease	3–5 wk	
dengue fever	3–12 da	
echovirus disease	3–5 da	
hepatitis A	25 da	15–40 da
hepatitis B	6 wk–6 mo	
herpes simplex	4 da	2–12 da
infectious mononucleosis	2–8 wk	
influenza	1–3 da	
Lassa fever	10–14 da	
lymphocytic choriomeningitis	7–21 da	
measles	10–12 da	
molluscum contagiosum	2–7 wk	
mumps	18 da	14–21 da
polio	7–14 da	5–35 da
rabies	3–8 wk	
respiratory syncytial disease	3–7 da	
rhinovirus infection	several days	
rubella	16–18 da	14–21 da
St. Louis encephalitis	4–21 da	
smallpox	12 da	7–17 da
yellow fever	3–6 da	
coccidioidomycosis	1–3 da	
histoplasmosis	5–18 da	
tinea capitis	2–3 wk	
tinea pedis	2–3 wk	
American trypanosomiasis	5–15 da	
amebiasis	2 wk	variable
balantidiasis	few days	variable
giardiasis	1–2 wk	variable
malaria	depends on species	8–37 da
primary amebic meningoencephalitis	1 wk	
hookworm disease	6 wk	
pinworm disease	2–6 wk	
roundworm disease	8–10 wk	
trichinosis	2–28 da	
whipworm disease	90 da	

Appendix B
Metric Measurement

	SYMBOL	FUNDAMENTAL UNIT	QUANTITY	NUMERICAL UNIT	AMERICAN MEASUREMENT
Length		meter (m)			39.37 inches
	km		kilometer	1,000 m	0.62137 miles
	cm		centimeter	0.01 m	0.3937 inches
	mm		millimeter	0.001 m	
	μm		micrometer	0.000001 m	
	nm		nanometer	0.000000001 m	
	Å		angstrom	0.0000000001 m	
Volume (liquids)		liter (l)			1.06 quarts
	ml		milliliter	0.001 l	
	μl		microliter	0.000001 l	
Mass		gram (g)			0.035 ounces
	kg		kilogram	1,000 g	2.2 pounds
	mg		milligram	0.001 g	
	μg		microgram	0.000001 g	

Appendix C

Temperature Conversion Chart

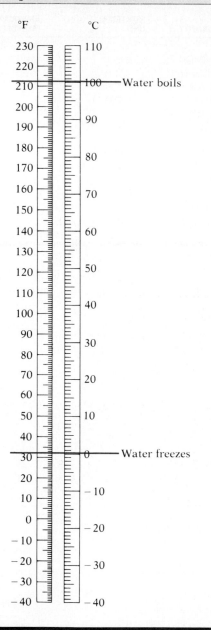

°F   °C

230 ─ 110
220
210 ─── 100 ── Water boils
200 ─ 90
190
180 ─ 80
170
160 ─ 70
150
140 ─ 60
130
120 ─ 50
110
100 ─ 40
90
80 ─ 30
70 ─ 20
60
50 ─ 10
40
30 ─── 0 ── Water freezes
20
10 ─ − 10
0
− 10 ─ − 20
− 20 ─ − 30
− 30
− 40 ─ − 40

To convert Fahrenheit to Celsius, use this formula:

$$°C = \tfrac{5}{9}(°F - 32)$$

To convert Celsius to Fahrenheit, use this formula:

$$°F = \tfrac{9}{5}°C + 32$$

Appendix D

The Classification of Bacteria According to *Bergey's Manual of Systematic Bacteriology*, 1984

Kingdom Procaryotae
  Division I: Gracilicutes
  Division II: Firmicutes
  Division III: Tenericutes
  Division IV: Mendosicutes

VOLUME 1

Section 1: The Spirochetes
  Order I: Spirochaetales
    Family I: Spirochaetaceae
      Genus I: *Spirochaeta*
      Genus II: *Cristispira*
      Genus III: *Treponema*
      Genus IV: *Borrellia*
    Family II: Leptospiraceae
      Genus I: *Leptospira*
    OTHER ORGANISMS: Hindgut spirochetes of termites and *Cryptocercus punctulatus*

Section 2: Aerobic/microaerophilic, motile, helical/vibrioid Gram-negative bacteria
    Genus: *Aquaspirillum*
    Genus: *Spirillum*
    Genus: *Azospirillum*
    Genus: *Oceanospirillum*
    Genus: *Campylobacter*
    Genus: *Bdellovibrio*
    Genus: *Vampirovibrio*

Section 3: Nonmotile (or rarely motile), Gram-negative curved bacteria
  Family I: Spirosomaceae
    Genus I: *Spirosoma*
    Genus II: *Runella*
    Genus III: *Flectobacillus*
  OTHER GENERA
    Genus: *Microcyclus*
    Genus: *Meniscus*
    Genus: *Brachyarcus*
    Genus: *Pelosigma*

Section 4: Gram-negative aerobic rods and cocci
  Family I: Pseudomonadaceae
    Genus I: *Pseudomonas*
    Genus II: *Xanthomonas*
    Genus III: *Frateuria*
    Genus IV: *Zoogloea*
  Family II: Azotobacteraceae
    Genus I: *Azotobacter*
    Genus II: *Azomonas*
  Family III: Rhizobiaceae
    Genus I: *Rhizobium*
    Genus II: *Bradyrhizobium*
    Genus III: *Agrobacterium*
    Genus IV: *Phyllobacterium*
  Family IV: Methylococcaceae
    Genus I: *Methylococcus*
    Genus II: *Methylomonas*

Family V: Halobacteriaceae
  Genus I: *Halobacterium*
  Genus II: *Halococcus*
Family VI: Acetobacteraceae
  Genus I: *Acetobacter*
  Genus II: *Gluconobacter*
Family VII: Legionellaceae
  Genus I: *Legionella*
Family VIII: Neisseriaceae
  Genus I: *Neisseria*
  Genus II: *Moraxella*
  Genus III: *Acinetobacter*
  Genus IV: *Kingella*
OTHER GENERA
  Genus: *Beijerinckia*
  Genus: *Derxia*
  Genus: *Xanthobacter*
  Genus: *Thermus*
  Genus: *Thermomicrobium*
  Genus: *Halomonas*
  Genus: *Alteromonas*
  Genus: *Flavobacterium*
  Genus: *Alcaligenes*
  Genus: *Serpens*
  Genus: *Janthinobacterium*
  Genus: *Brucella*
  Genus: *Bordetella*
  Genus: *Franciscella*
  Genus: *Paracoccus*
  Genus: *Lampropedia*

Section 5: Facultatively anaerobic Gram-negative rods
  Family I: Enterobacteriaceae
    Genus I: *Escherichia*
    Genus II: *Shigella*
    Genus III: *Salmonella*
    Genus IV: *Citrobacter*
    Genus V: *Klebsiella*
    Genus VI: *Enterobacter*
    Genus VII: *Erwinia*
    Genus VIII: *Serratia*
    Genus IX: *Hafnia*
    Genus X: *Edwarsiella*
    Genus XI: *Proteus*
    Genus XII: *Providencia*
    Genus XIII: *Morganella*
    Genus XIV: *Yersinia*
  OTHER GENERA OF THE FAMILY ENTEROBACTERIACEAE:
    Genus: *Obesumbacterium*
    Genus: *Xenorhabdus*
    Genus: *Kluyvera*
    Genus: *Rahnella*
    Genus: *Cedecea*
    Genus: *Tatumella*
  Family II: Vibrinoaceae
    Genus I: *Vibrio*
    Genus II: *Photobacterium*

    Genus III: *Aeromonas*
    Genus IV: *Plesiomonas*
  OTHER GENERA
    Genus: *Zymomonas*
    Genus: *Chromobacterium*
    Genus: *Cardiobacterium*
    Genus: *Calymmatobacterium*
    Genus: *Gardnerella*
    Genus: *Eikenella*
    Genus: *Streptobacillus*

Section 6: Anaerobic Gram-negative straight, curved, and helical rods
  Family I: Bacteroidaceae
    Genus I: *Bacteroides*
    Genus II: *Fusobacterium*
    Genus III: *Leptotrichia*
    Genus IV: *Butyrivibrio*
    Genus V: *Succinomonas*
    Genus VI: *Succinivibrio*
    Genus VII: *Anaerobiospirillum*
    Genus VIII: *Wolinella*
    Genus IX: *Selenomonas*
    Genus X: *Anaerovibrio*
    Genus XI: *Pectinatus*
    Genus XII: *Acetovibrio*
    Genus XIII: *Lachnospira*

Section 7: Dissimilatory sulfate- or sulfur-reducing bacteria
    Genus: *Desulfuromonas*
    Genus: *Desulfovibrio*
    Genus: *Desulfomonas*
    Genus: *Desulfococcus*
    Genus: *Desulfobacter*
    Genus: *Desulfobulbus*
    Genus: *Desulfosarcina*

Section 8: Anaerobic Gram-negative cocci
  Family I: Veillonellaceae
    Genus I: *Veillonella*
    Genus II: *Acidaminococcus*
    Genus III: *Megasphaera*

Section 9: The rickettsias and chlamydias
  Order I: Rickettsiales
    Family I: Rickettsiaceae
      Genus I: *Rickettsia*
      Genus II: *Rochalimaea*
      Genus III: *Coxiella*
    Family II: Ehrlichieae
      Genus I: *Ehrlichia*
      Genus II: *Cowdria*
      Genus III: *Neorickettsia*
    Family III: Wolbachieae
      Genus I: *Wolbachia*
      Genus II: *Rickettsiella*
    Family IV: Bartonellaceae
      Genus I: *Bartonella*
      Genus II: *Grahamella*

Appendix D *(continued)*

The Classification of Bacteria According to *Bergey's Manual of Systematic Bacteriology*, 1984

Family V: Anaplasmataceae
  Genus I: *Anaplasma*
  Genus II: *Aegyptianella*
  Genus III: *Haemobartonella*
  Genus IV: *Eperythrozoon*
Order II: Chlamydiales
  Family I: Chlamydiaceae
    Genus I: *Chlamydia*

Section 10: The mycoplasmas
Order I: Mycoplasmatales
  Family I: Mycoplasmataceae
    Genus I: *Mycoplasma*
    Genus II: *Ureaplasma*
  Family II: Acholeplasmataceae
    Genus I: *Acholeplasma*
  Family III: Spiraplasmataceae
    Genus I: *Spiroplasma*
  OTHER GENERA
    Genus: *Anaeroplasma*
    Genus: *Thermoplasma*

Section 11: Endosymbionts
  A. Endosymbionts of Protozoa
    Genus I: *Holospora*
    Genus II: *Caedibacter*
    Genus III: *Pseudocaedibacter*
    Genus IV: *Lyticum*
    Genus V: *Tectibacter*
  B. Endosymbionts of Insects
    Genus: *Blattabacterium*
  C. Endosymbionts of Fungi and Invertebrates
  Other than Arthropods

VOLUME 2

Section 12: Gram-positive cocci
  Family I: Micrococcaceae
    Genus I: *Micrococcus*
    Genus II: *Stomatococcus*
    Genus III: *Planococcus*
    Genus IV: *Staphylococcus*
  Family II: Deinococcaceae
    Genus I: *Deinococcus*
  OTHER ORGANISMS
    "Pyogenic" streptococci
    "Oral" streptococci
    "Lactic" streptococci and enterococci
    Genus: *Leuconostoc*
    Genus: *Pediococcus*
    Genus: *Aerococcus*
    Genus: *Gemella*
    Genus: *Peptococcus*
    Genus: *Peptostreptococcus*
    Genus: *Ruminococcus*
    Genus: *Coprococcus*
    Genus: *Sarcina*

Section 13: Endospore-forming Gram-positive rods and cocci
  Genus: *Bacillus*

Genus: *Sporolactobacillus*
Genus: *Clostridium*
Genus: *Desulfotomaculum*
Genus: *Sporosarcina*
Genus: *Oscillospora*

Section 14: Regular, non-sporing, Gram-positive rods
  Genus: *Lactobacillus*
  Genus: *Listeria*
  Genus: *Erysipelothrix*
  Genus: *Brochothrix*
  Genus: *Renibacterium*
  Genus: *Kurthia*
  Genus: *Caryophanon*

Section 15: Irregular, non-sporing, Gram-positive rods
  Animal and Saprophytic Corynebacteria
  (*Corynebacterium*)
  Plant Corynebacteria (*Corynebacterium*)
  OTHER GENERA
    Genus: *Gardnerella*
    Genus: *Arcanobacterium*
    Genus: *Arthrobacter*
    Genus: *Brevibacterium*
    Genus: *Curtobacterium*
    Genus: *Caseobacter*
    Genus: *Microbacterium*
    Genus: *Aureobacterium*
    Genus: *Cellulomonas*
    Genus: *Agromyces*
    Genus: *Arachnia*
    Genus: *Rothia*
    Genus: *Propionibacterium*
    Genus: *Eubacterium*
    Genus: *Acetobacterium*
    Genus: *Lachnospira*
    Genus: *Butyrivibrio*
    Genus: *Thermoanaerobacter*
    Genus: *Actinomyces*
    Genus: *Bifidobacterium*

Section 16: Mycobacteria
  Family I: Mycobacteriaceae
    Genus I: *Mycobacterium*

Section 17: Nocardiforms
  Genus: *Nocardia*
  Genus: *Rhodococcus*
  Genus: *Nocardioides*
  Genus: *Pseudonocardia*
  Genus: *Oerskovia*
  Genus: *Saccharopolyspora*
  Genus: *Micropolyspora*
  Genus: *Promicromonospora*
  Genus: *Intrasporangium*

VOLUME 3

Section 18: Gliding, non-fruiting bacteria
  Order I: Cytophagales
    Family I: Cytophagaceae
      Genus I: *Cytophaga*

Genus II: *Sporocytophaga*
Genus III: *Capnocytophaga*
Genus IV: *Flexithrix*
Genus V: *Flexibacter*
Genus VI: *Microscilla*
Genus VII: *Saprospira*
Genus VIII: *Herpetosiphon*
Order II: Lysobacterales
  Family I: Lysobacteraceae
    Genus I: *Lysobacter*
Order III: Beggiatoales
  Family I: Beggiatoaceae
    Genus I: *Beggiatoa*
    Genus II: *Thioploca*
    Genus III: *Thiospirillopsis*
    Genus IV: *Thiothrix*
    Genus V: *Achromatium*
  Family II: Simonsiellaceae
    Genus I: *Simonsiella*
    Genus II: *Alysiella*
  Family III: Leucotrichaceae
    Genus I: *Leucothrix*
  Family IV: Pelonemataceae
    Genus I: *Pelonema*
    Genus II: *Achroonema*
    Genus III: *Peloploca*
    Genus IV: *Desmanthus*
  Families and genera *incertae sedis*
    Genus: *Toxothrix*
    Genus: *Vitreoscilla*
    Genus: *Chitinophagen*
    Genus: *Desulfonema*

Section 19: Anoxygenic phototrophic bacteria
  Purple bacteria
  Family I: Chromatiaceae
    Genus I: *Chromatium*
    Genus II: *Thiocystis*
    Genus III: *Thiospirillum*
    Genus IV: *Thiocapsa*
    Genus V: *Amoebobacter*
    Genus VI: *Lamprobacter*
    Genus VII: *Lamprocystis*
    Genus VIII: *Thiodictyon*
    Genus IX: *Thiopedia*
  Family II: Ectothiorhodospiraceae
    Genus I: *Ectothiorhodospira*
  Purple nonsulfur bacteria
    Genus: *Rhodospirillum*
    Genus: *Rhodopseudomonas*
    Genus: *Rhodobacter*
    Genus: *Rhodomicrobium*
    Genus: *Rhodopila*
    Genus: *Rhodocyclus*
  Green bacteria
  Green sulfur bacteria
    Genus: *Chlorobium*
    Genus: *Prosthecochloris*
    Genus: *Ancalochloris*

Appendix D *(continued)*

The Classification of Bacteria According to *Bergey's Manual of Systematic Bacteriology*, 1984

Genus: *Pelodyction*
Genus: *Chloroherpeton*
Genus: Symbiotic consortia
Multicellular filamentous green bacteria
Genus: *Chloroflexus*
Genus: *Heliothrix*
Genus: *Oscillochloris*
Genus: *Chloronema*
Genera *incertae sedis*
Genus: *Heliobacterium*
Genus: *Erythrobacter*

Section 20: Budding and/or appendaged bacteria
Prosthecate bacteria
Budding bacteria
Genus: *Hyphomicrobium*
Genus: *Hyphomonas*
Genus: *Pedomicrobium*
Genus: *"Filomicrobium"*
Genus: *"Dicotomicrobium"*
Genus: *"Tetramicrobium"*
Genus: *Stella*
Genus: *Ancalomicrobium*
Genus: *Prosthecomicrobium*
Non-budding bacteria
Genus: *Caulobacter*
Genus: *Asticcacaulis*
Genus: *Prosthecobacter*
Genus: *Thiodendron*
Non-prosthecate bacteria
Genus: *Planctomyces*
Genus: *Pasteuria*
Genus: *Blastobacter*
Genus: *Angulomicrobium*
Genus: *Gemmiger*
Genus: *Ensifer*
Genus: *Isophaera*
Non-budding stalked bacteria
Genus: *Gallionella*
Genus: *Nevskia*
Morphologically unusual budding bacteria involved in iron and manganese deposition
Genus: *Seliberia*
Genus: *Metallogenium*
Genus: *Caulococcus*
Genus: *Kuznezovia*
OTHERS
Spinate bacteria

Section 21: Archaeobacteria
Methanogenic bacteria
Genus: *Methanobacterium*
Genus: *Methanobrevibacter*
Genus: *Methanococcus*
Genus: *Methanomicrobium*
Genus: *Methanospirillum*
Genus: *Methanosarcina*
Genus: *Methanococcoides*
Genus: *Methanothermus*

Genus: *Methanolobus*
Genus: *Methanoplanus*
Genus: *Methanogenium*
Genus: *Methanothrix*
Extreme halophilic bacteria
Genus: *Halobacterium*
Genus: *Halococcus*
Extreme thermophilic bacteria
Genus: *Thermoplasma*
Genus: *Sulfolobus*
Genus: *Thermoproteus*
Genus: *Thermococcus*
Genus: *Desulfurococcus*
Genus: *Thermodiscus*
Genus: *Pyrodictium*

Section 22: Sheathed bacteria
Genus: *Sphaerotilus*
Genus: *Leptothrix*
Genus: *Haliscominobacter*
Genus: *Lieskeella*
Genus: *Phargmidiothrix*
Genus: *Crenothrix*
Genus: *Clonothrix*

Section 23: Gliding, fruiting bacteria
Order I: Myxobacteriales
Family I: Myxococcaceae
Genus I: *Myxococcus*
Family II: Archangiaceae
Genus I: *Archangium*
Family III: Cystobacteraceae
Genus I: *Cystobacter*
Genus II: *Melittangium*
Genus III: *Stigmatella*
Family IV: Polyangiaceae
Genus I: *Polyangium*
Genus II: *Nannocystis*
Genus III: *Chondromyces*
Genus *incerta sedis*
Genus: *Angiococcus*

Section 24: Chemolithotrophic bacteria
Nitrifiers
Family I: Nitrobacteraceae
Genus I: *Nitrobacter*
Genus II: *Nitrospina*
Genus III: *Nitrococcus*
Genus IV: *Nitrosomonas*
Genus V: *Nitrosopira*
Genus VI: *Nitrosococcus*
Genus VII: *Nitrosolobus*
Sulfur oxidizers
Genus: *Thiobacillus*
Genus: *Thiomicrospira*
Genus: *Thiobacterium*
Genus: *Thiospira*
Genus: *Micromonas*
Obligate hydrogen oxidizers
Genus: *Hydrogenbacter*

Metal oxidizers and depositers
Family I: Siderocapsaceae
Genus I: *Siderocapsa*
Genus II: *Naumaniella*
Genus III: *Ochrobium*
Genus IV: *Siderococcus*
OTHER MAGNETOTACTIC BACTERIA

Section 25: Cyanobacteria

Section 26: Others
Order I: Prochlorales
Family I: Prochloraceae
Genus I: *Prochloron*

VOLUME 4

Section 27: Actinomycetes that divide in more than one plane
Genus: *Geodermatophilus*
Genus: *Dermatophilus*
Genus: *Frankia*
Genus: *Tonsilophilus*

Section 28: Sporangiate actinomycetes
Genus: *Actinoplanes* (including *Amorphosporangium*)
Genus: *Streptosporangium*
Genus: *Ampullariella*
Genus: *Spirillospora*
Genus: *Pilimelia*
Genus: *Dactylosporangium*
Genus: *Planomonospora*
Genus: *Planobispora*

Section 29: Streptomycetes and their allies
Genus: *Streptomyces*
Genus: *Streptoverticillium*
Genus: *Actinopycnidium*
Genus: *Actinosporangium*
Genus: *Chainia*
Genus: *Elytrosporangium*
Genus: *Microellobosporia*

Section 30: Other condidiate genera
Genus: *Actinopolyspora*
Genus: *Actinosynnema*
Genus: *Kineospora*
Genus: *Kitasatosporia*
Genus: *Microbispora*
Genus: *Micromonospora*
Genus: *Microtetraspora*
Genus: *Saccharomonospora*
Genus: *Sporichthya*
Genus: *Streptoalloteichus*
Genus: *Thermomonospora*
Genus: *Actinomadura*
Genus: *Nocardiopsis*
Genus: *Exellospora*
Genus: *Thermoactinomyces*

Appendix E
Answers to Questions for Thought and Discussion

CHAPTER 1

1. Rhazes believed that the rotting of meat was somehow related to the rotting of human flesh during time of disease, and that something in the air was responsible. Therefore he picked the spot where the elusive something was least abundant, as evidenced by the slow rotting of meat.

2. Pasteur tackled more philosophical and fundamental questions than Koch. For example, he successfully discredited spontaneous generation and proposed the germ theory of disease. Koch's work was of a more practical nature. Also, as the son of a tanner, he displayed an "upwardly mobile" life, and his works were of a broader scope than Koch's.

3. Since measles is caused by a virus and since the measles viruses cannot be seen or cultivated by methods that work for bacteria, your work would probably result in a series of dead-ends and your frustrations would mount quickly.

4. Believing in spontaneous generation meant that people could create living things and therefore they could have dominion over them. To have dominion over something means to be able to understand it and control it. It followed that the laws of nature and living things were accessible to humans, much as physical and chemical laws were coming under scrutiny in that period.

5. Some candidates for the key event in the development of microbiology might be Pasteur's fermentation work, Koch's cultivation methods, the spontaneous generation controversy, Pasteur's discovery of immunity, Koch's proof of the germ theory of disease, and so on.

6. Even though cures were many years off, the realization evolved that infectious diseases were due to specific organisms. Once this concept and the methods of transmission were understood, it became possible to interrupt the spread of the microorganisms by such things as insect control, milk pasteurization, water purification, instrument sterilization, and patient isolation. You might like to draw the parallel between this outlook and the discovery of the AIDS virus in the spring of 1984, three long years after the disease was first recognized as a clinical entity.

7. In 1911, scientists were so smitten by the germ theory of disease that they assumed all human diseases were due to microorganisms. In an extraordinary departure from beliefs of an earlier generation, they simply could not conceive of anything other than a "germ" being the cause of illness. Hence, they advised Funk to drop his ill-conceived theory and get on the germ bandwagon. Fortunately, Funk persevered, the mystery of beriberi was solved, and the dietary deficiency remedied.

8. The doctrine of spontaneous generation focused attention on microorganisms because Needham maintained they originated from lifeless matter, while Spallanzani denied it. A hundred years later, Pasteur was forced to devise experiments to refute the doctrine in order to salvage his germ theory of disease. Once more, questions on the origin of life entered the picture. To reject this concept, a case can be made that the science of microbiology evolved from the riddle of what caused disease and how to prevent its spread.

9. As I researched the material for Chapter 1, the year 1884 kept popping up and it soon became obvious that this was a remarkable year. Among the key events: cultivation of the typhoid bacillus by Gaffky, isolation of the diphtheria bacillus by Löeffler, description of phagocytosis by Metchnikoff, and announcement of the use of agar by Koch.

10. It is possible that in 1879 Sedillot did not think of microorganisms as "organisms." The word microbe may have been his way of expressing some entity intermediary between an organism and a microscopic being of a nonliving nature.

11. Proving that spontaneous generation did not occur in the hamburger meat can be a tricky proposition because a way must be devised to sterilize equal portions of meat and then expose one to the air while keeping the other portion sterile. You may wish to explore how this can be done and what the results will indicate.

12. The improvements in microscopy that paralleled the interest in bacteria in the 1880s were probably not a coincidence because one probably stimulated the other. For example, the interest in bacteria may have demanded that the technology be improved for the developing science. Conversely, the improved microscope may have stimulated interest in microorganisms because knowledge of microorganisms became more accessible.

13. To understand why microbiology did not develop as a science after van Leeuwenhoek's time, it would be well to understand the political, social, and economic climate of those years in Europe. Lack of interest in science, preoccupation with warfare and conquest, unavailability of funds for experimentation, lack of a learned population augment the facts that technology was not developed and people simply failed to understand the relationship between microorganisms and disease.

14. In 1798, science was barely a topic of conversation in European society, much less a well-organized body of knowledge. There were no laboratories, no technology, and precious few scientists. It would not be another hundred years before the great laboratories of Europe were founded to explore scientific principles. I would love to take a course in persuasion from Edward Jenner. Jenner convinced people to be inoculated with cowpox material even though they cringed with fear at the mere mention of the "pox." This man must have put up an extremely convincing argument. He could probably sell refrigerators to Eskimos.

15. Pasteur was something of a revolutionary in the way he thought and attacked the medical establishment. His germ theory of disease was severely at odds with accepted outlooks of his time and his persistent hammering away at contemporary dogmas (the nature of fermentation; cause and transmission of infectious disease; the ability to interrupt epidemics; and others) initiated a revolution that has carried through to the present day. In 1910, the French honored their "rebel" in connection with Mexican "rebels." Pasteur was also a great patriot, as were the Mexicans.

CHAPTER 2

1. It is unlikely that "good chemistry" will occur when calcium and magnesium ions are brought together since both have electrons to give up. Both are donor atoms, and both require a receiver atom to be reactive.

2. Lactose intolerance occurs when an individual's digestive cells are unable to synthesize lactase, the enzyme that digests lactose to its constituents, glucose and galactose.

3. Detergents act on the lipid portions of bacterial membranes. They dissolve the membranes and cause leakage from the cytoplasm and subsequent death to the cell.

4. Organic molecules invariably contain carbon atoms, and carbon atoms form covalent bonds with other atoms or radicals at four places. Soon these evolve a Tinkertoy-like arrangement of atoms and a rather large molecule, especially if multiple carbon atoms are present.

5. A billion seconds is equivalent to about 30 years. Ten billion seconds would equate to about 300 years. Therefore, your counting would not be complete for 300 years.

6. Wöhler showed that organic molecules could be synthesized, in effect demonstrating that living things could be synthetically produced. This finding stood in stark contrast to the pre-1800s belief that living things were something special and not within the ability of scientists to create. It also brought chemistry into the realm of biology, and made biologists begin thinking as chemists.

7. Soap is a combination of fat and sodium or potassium hydroxide (lye). These latter compounds are basic, and therefore, soap has a pH close to 8. Bacteria cannot live well under these conditions.

8. Nothing. It is empty space.

Appendix E *(continued)*

Answers to Questions for Thought and Discussion

9. The organic matter of an individual consists of proteins, lipids, carbohydrates, and nucleic acids. Oxygen is a key component of each type of these groups of organic compounds. Also, there are large volumes of water in the body. It is not surprising, therefore, that a 120-pound individual contains 78 pounds of oxygen.

10. Different arguments can be presented for destroying different groups of chemical substances in bacteria as a way of preventing their spread. For example, destroying proteins would eliminate enzyme and chemical structures. Carbohydrate destruction would remove vital energy sources for life. Lipid destruction would cause leakage through the cell membrane and consequent death. A case can even be made for eliminating the nucleic acids since this would disrupt the chromosome and genes, thereby preventing protein synthesis.

11. The bottle with the lower pH should be handled more carefully because it is more acidic and hence, more caustic to the skin.

12. Carbon atoms have four electrons in their outer shell. They cannot lose their electrons easily, nor can they attract electrons from other atoms. Hence, they enter into covalent bonds, sharing their four electrons with four other atoms or radicals. This leads to enormous variations of chemical combination.

13. Proteins are denatured by heat, thereby losing their tertiary structure and assuming a secondary structure. Heating contaminated food may denature the toxin, although the bacteria may remain alive. In botulism, the toxin is the key element in disease.

14. Water, the universal solvent of life, functions as the solution in which organic molecules are dissolved. It is also a participant in dehydration synthesis and digestions of proteins, lipids, carbohydrates, and nucleic acids. Other uses for water can also be found.

15. The proteins are not acidic because the acid (carboxyl) groups of the amino acid are used up to form the peptide bonds.

CHAPTER 3

1. The student should recall that protozoa are eukaryotes and use the characteristics of eukaryotes to describe the protozoa. A surprisingly large amount of information about protozoa can be obtained this way, especially if the characteristics are supplemented with some information from a previous biology course.

2. A New York newspaper actually had the audacity to print *E. coli* as "eecoli." The word bacteria should be bacterium; the "e" should be capitalized; the name should be italicized or underlined. Incidentally, the newspaper is no longer in existence.

3. In order to locate the properties and nature of the bacterium, microbiologists used many of the processes described in the section *Bacterial Taxonomy*. Identification of the Lyme disease spirochete was finally announced in 1985 by researchers at the State University of New York at Stony Brook. The spirochete was named *Borrelia burgdorferi* to honor Willy Burgdorfer who had previously observed it in the tick vector.

4. If water is substituted for alcohol, no decolorization will take place, since water is unable to decolorize the bacteria. Both Gram-positive and Gram-negative bacteria will remain blue-purple. Incidentally, you will note that I am capitalizing "Gram" throughout the text since it was originally a person's name. Some authors use the lower case, but I believe that retaining the identity of Christian Gram as the discoverer is important.

5. The modern nomenclature establishes "meter" as the basic unit of measurement and uses prefixes for multiples of fractions. The older term "micron" was used as a basic unit even though it was in fact a fraction of a meter. I recall using the word "millimicron" for one-thousandth of a micron. We now use the word nanometer.

6. Judging by the use of the word "protist" in general biology books, it is reasonable to believe that the chapter will probably cover eukaryotic microorganisms. However, some authors use protist as an umbrella expression for all microorganisms.

7. The water is probably slimy, in addition to having an eerie blue-green color. Cyanobacteria are the likely cause. Some algicide would be advisable.

8. No doubt an order for *Bergey's Manual of Systematic Bacteriology* will shortly be placed since the identification of unknown bacteria depends on information from this book.

9. You will receive an order of mushrooms because the Italian fungi (Italian pronunciation "foon-jee") translates to mushroom.

10. I suspect the owner of the car is a microbiologist. If you travel New York highways, keep your eye out for such a license plate.

11. This question is an application of the text discussion on the fluorescent microscope. The fluorescein would be tagged to the plague antibodies and combined with the blood. If plague bacilli were in the blood, the tagged antibodies would be attracted to the bacilli and the fluorescein would cause the bacilli to glow. If no bacilli were present, no attraction could take place and no organisms could be visualized.

12. It is rather common to see this mistake. Actually, the species is a conceptual entity whose name is *Homo sapiens*. Several species (e.g.,

Homo erectus, Homo habilis, etc.) are organized into the genus *Homo*. The binomial name thus consists of the genus name and the species modifier (not the species). If you still cannot understand this concept, ask yourself what a *sapiens* is; then ask what a *Homo sapiens* is.

13. Contrary to the belief of many, oil does not increase the magnification of the light microscope. It merely allows the gathering of enough light to make using the instrument possible. The magnification would be the same without the oil, but the resolution would be much reduced.

14. It would take almost 12 days of nonstop counting to tally each bacterium. To help students appreciate the enormity of this figure, begin at an arbitrary number and count along as, for instance, 231,573; 231,574; 231,575; 231,576; 231,577; and so on.

15. A class of artists or pseudoartists could have a field day with this project. Imagine all the shapes taken by protozoa, bacteria, fungi, algae, and the rest, and a rather diverse group of "patrons" would emerge. I am reminded of the famous bar scene in *Star Wars*.

CHAPTER 4

1. There are various advantages which derive from being able to form a spore, capsule, or flagellum. The answers may help you to see that the bacterium's life is not dramatically different than that of other creatures.

2. Incubation periods may be affected by several factors, one of which is the generation time. For tubercle bacilli the generation time is a long 18 hours, much longer than for streptococci. You should speculate on other factors.

3. Stuffing is made with various ingredients that have probably come in contact with bacteria, including bread cubes and spices, and it is often mixed by hand. The bacterial content is therefore high. It takes a long time for the viscera of a turkey to cool to refrigerator temperature, and bacteria have the opportunity to grow profusely before the cold affects them. It is best, therefore, to make the stuffing, refrigerate it to cool quickly, and stuff the turkey in the morning.

4. Toothbrushes dry quickly and bacteria cannot grow on dry environments. Also, toothbrushes have few nutrients to support bacterial growth, and toothpaste often contains inhibitory substances. Toothbrushes are not sterile, however, and they may transmit small numbers of bacteria.

5. Boiling may kill the majority of bacteria, but bacterial spores survive 2 or more hours of boiling. It is wrong to believe the water is sterile after a few minutes of boiling. It has been disinfected, however, and is perfectly safe to drink under normal circumstances.

Appendix E *(continued)*

Answers to Questions for Thought and Discussion

6. This description may be deciphered from descriptions in the chapter. The organism is surrounded by flagella, lives in an environment free of oxygen, utilizes preformed organic matter, exists at moderate temperature environments, and consists of a chain of bacilli. Cultivation may be performed on laboratory media, in live animals, in tissue cultures, or other environments that contain organic matter.

7. An enjoyable discussion may ensue by relating a bacterial growth curve to the growth of the United States population for the past 300 years. It is interesting to speculate in which phase the American population finds itself at present.

8. Arguments can be offered for each mode of life. While parasites appear to be in a dominant position during infection, the death of the host pushes them to the side as other decomposers take over. Also, parasites are dependent upon the host for life. On the other hand, saprobes may have to eke out an existence in a harsh environment, but they live independently. Additional outlooks for each mode should also be discussed.

9. In the pre-electron microscope days, the chapter on bacterial anatomy would probably be very brief. Certainly there would be no discussion of pili, plasmids, magnetosomes, ribosomes, or fluid mosaic model.

10. Consider the flagellum, pilus, capsule, spore, and other structures and note how they help pathogens overcome body defenses.

11. I find it interesting that after 3.5 billion years of evolution, only three major shapes have emerged in bacteria. There are no cubes, triangles, or other interesting geometric shapes. I cannot account for this observation. Perhaps you can.

12. Enzyme systems are adapted to a certain mode of existence, and growth in one system does not necessarily mean it will grow well in another system. You might speculate on conditions that might favor or inhibit growth.

13. Bacteria may produce toxins that interfere with metabolic patterns in the body. Thus, the physical growth of bacteria may not be a necessary prerequisite to disease, but consuming food that contains toxins may be dangerous.

14. Nitrogen is a relatively inert gas that will not support the energy-producing or synthesis metabolism of bacteria (Chapter 5). Without bacterial growth, decay will not take place.

15. The stem "methano-" refers to "methane-loving" and "-halophilus" refers to "salt-loving." See the section on archaeobacteria for the conditions under which these organisms live.

CHAPTER 5

1. The cartoon in question appeared in a syndicated newspaper and was authored by Athelstan Spilhaus, former President of the AIBS. Spilhaus probably conjectured that the cellulose of newspapers would be broken down to units of glucose. The glucose would be metabolized by bacteria into metabolic intermediates which would then be converted to amino acids by adding amino groups to carbon skeletons in a reversal of deamination. The amino acids are fed to cattle which would combine them to produce proteins.

2. Apparently this company is interested in producing citric acid because citric acid will accumulate in the absence of the enzyme citrase. Chapter 25 explores this industrial process in more detail and includes some uses for citric acid including beverages and clot prevention.

3. The lack of bacterial growth may be traced to lead in the water used to make the medium. Lead combines with the sulfhydryl groups of enzyme molecules and inhibits their activity. Lead pipes are rare but lead solders are sometimes used to seal them.

4. Fermentation is an anaerobic process and therefore absence of oxygen is essential. If oxygen were admitted, the metabolism of the yeast would shift to an aerobic mode with production of water and CO_2 rather than ethyl alcohol. The sealing of the vat excludes oxygen, but of equal importance is the carbon dioxide given off initially in aerobic respiration and then in fermentation. This carbon dioxide takes up all the air space in the wine and prohibits any oxygen from entering while the fermentation proceeds.

5. This question takes you to the origins of life on Earth. As the text indicates, photosynthetic organisms use the sun's energy to produce carbohydrates. In the process, water breaks down to release oxygen to the atmosphere. Organisms could now evolve to combine the oxygen with carbohydrate to release the energy for their own use.

6. One reason that ATP is not supplied to the growth medium is that it probably could not be absorbed to the cytoplasm of a bacterial cell. You should speculate on other possible reasons.

7. Actually both students are right since enzymes are protein molecules. Enzymes function in the protein synthesis process, as for example in the knitting together of mRNA molecules. But proteins (enzymes) also function to produce more enzymes (proteins).

8. A science fiction writer could have a field day with a story about how chlorophyll was altered and photosynthesis stopped on Earth. Perhaps you might like to try your hand at such a story.

9. Doubtlessly, the bacterium would choose the aerobic respiration route because 38 molecules of ATP arise as compared with 2 by fermentation. Also, fermentation yields waste products such as acids that are inhibitory, while CO_2 and H_2O are the waste products of aerobic respiration.

10. You should recognize the "flavin" part of riboflavin and guess that it is used to make flavin adenine dinucleotide (FAD). Here is an example of how recognizing word stems helps one to make an intelligent guess. Without FAD, oxidative phosphorylation grinds to a halt, energy production ceases, and death is imminent.

11. The individual probably assumes that the green sulfur bacteria will use the hydrogen sulfide as a hydrogen source in photosynthesis, thus liberating elemental sulfur. Whether it will work is a point for conjecture, and you might like to discuss the pros and cons of such a scheme.

12. By binding to messenger RNA, the streptomycin effectively inhibits protein synthesis. This discussion illustrates how chemicals can interfere with the process.

13. Stopping glycolysis would mean that the pyruvic acid fuel for the Krebs cycle would cease being made from carbohydrate. However, pyruvic acid could be produced from amino acids by deamination of alanine, and fatty acids might convert to acetyl-CoA for entry to the cycle.

14. Phosphate is used in ATP, in DNA and RNA, as ions in glycolysis, in NAD and FAD, and in numerous other instances.

15. This incident actually happened to me. I began writing at 9:00 A.M. and did not finish until 5:00 P.M. Essentially, the answer summarizes the chapter by showing how carbohydrates, fats, and proteins are digested and synthesized, and how the processes relate to one another. It is an excellent exercise in metabolism.

CHAPTER 6

1. Some ingenuity is required to explain the transformation as observed by Griffith in 1928. You might begin by assuming that a mistake was made and guessing how Griffith eliminated this possibility. From there you will have to let your intuition work overtime.

2. The huge amount of antibiotic unknowingly expelled into the environment is a factor in natural selection. Susceptible bacteria are killed, and populations of resistant bacteria emerge. These mutants may then infect hospital patients, with disastrous results.

3. Transposon movement is called "illegitimate recombination" because it happens in a single cell without the intervention of any other cell. Moreover, there is no passage of DNA into the cell from an outside source. Nevertheless, the cell is genetically different after the transposon has moved, so a case can be made for a recombination having occurred.

4. Arguments can be offered for any of the three processes to be most common. For example, transformation might be common because local debris from dead bacteria is often encountered, but the ability of DNA to pass through the

Appendix E *(continued)*

Answers to Questions for Thought and Discussion

recipient's wall and membrane would mitigate against a high incidence of transformation. Pluses and minuses likewise exist for the other two types.

5. It is interesting to note that a mutation in the direction of resistance is a foregone conclusion. One possible origin might be ultraviolet light from the sun. Another might be nitrites in the soil. If we accept the truth of the statement, then all antibiotics will eventually become useless. You might speculate on whether and how this trend can be short-circuited.

6. Recombination processes result in new genetic species arising from new biochemical characteristics and therefore, a rearrangement of the classification system will be required. Apparently, taxonomists will never want for work so long as recombination continues.

7. Several discoveries in this chapter have opened doors to other discoveries. Some examples: Avery's pointing to DNA as the transforming material; Berg's synthesis of a recombinant DNA molecule; Lederberg's work with conjugation. You should pick out other discoveries and understand why they were pivotal.

8. Mutation offers the opportunity for an organism to develop new genetic codes. The codes are expressed as biological characteristics that permit an organism to survive environmental pressures and, in doing so, evolve. Biologists discuss evolution from a population standpoint, but microbiologists can see the biochemical basis at the fundamental level. Mutation may be the driving force of evolution and transposable genetic elements (insertion sequences and transposons) may be the key to mutations.

9. Perhaps the organism was one which had existed previously but as a harmless species. A recombination process may have permitted it to assume a parasitic mode and thus cause disease. Another possibility is the acquisition of genes that allowed it to produce toxic substances, which lead to tissue destruction.

10. Probably the most appealing characteristic of the bacterial chromosome is that it is singular in nature. There is no second chromosome and thus, no dominance or recessiveness as in eukaryotic genetics. Also, the chromosome lies free in the cytoplasm without modifying protein. You can probably come up with other features which add to its appeal.

11. The label "second Industrial Revolution" has been widely appended to the development of genetic engineering and, to many observers, the label is justified because the process yields products that cannot be manufactured by other means. The manufacture of insulin, interferon, vaccines, and other products cited in the chapter are examples. Chapter 25 outlines multiple other offshoots of the process and delineates numerous products.

12. The increase in antibiotic-resistant *H. influenzae* is yet another example of the emergence of drug-resistant strains of bacteria. These strains probably exist in nature, having arisen by mutation and recombination, and are selected out by antibiotic use. The evolution demonstrates Darwin's principle of natural selection in nature. It shows how the fittest survive.

13. Why genetic variation exists in influenza viruses depends on the segments of RNA that make up the genome, but it is also conceivable that during replication, the viruses may acquire fragments of host DNA, as in generalized transduction, and express these bits of DNA as new antigens unrecognized by circulating antibodies.

14. Although the terms "reproduction" and "recombination" are sometimes confused, it is good to remember that reproduction leads to "more" cells, while recombination leads to "different" cells.

15. Broad spectrum drugs kill numerous different organisms in the environment. Should a resistant organism be present, the antibiotic will permit it to emerge as the dominant organism. Recombination may bring the genes to pathogenic species resulting in an antibiotic-resistant pathogen.

CHAPTER 7

1. Probably the most important factor was the introduction of the vaccine in 1887. Other factors might have included a strong surveillance system, the willingness of doctors to report cases, the establishment of a case definition to ensure that reports were accurate, the uses of tests to assess continued vaccine effectiveness, and the willingness of health departments to spend the necessary funds to help the eradication effort. Can you suggest any other public health measures that might have been used?

2. Many microbiologists believe that the antibiotics killed *Legionella pneumophila* in the tissue samples and thereby prevented its identification until 1976. The assumption of a viral pneumonia also precluded the search for *Legionella*. The outbreak of Pontiac fever illustrates that *Legionella* was indeed present before 1976, but detection methods were lacking, and antibiotic use stifled attempts to locate the bacillus.

3. This question assumes that a pathogen is in an advantageous condition, a situation with which you might take issue. Perhaps the virus places the bacillus in a difficult position because pathogenicity is not always desirable, especially after the host dies. On the other hand, the virus permits active growth in the tissues, a situation not possible without the toxin.

4. This is a tough one, and I wouldn't be surprised if you didn't guess correctly. At the turn

of the century, people believed that microorganisms hid in corners so they built this hospital with rounded edges, carefully avoiding any corners. It was actually one of the first attempts at preventive medicine.

5. Dead streptococci yield streptococcal antigens which may lead to immune complex formation after streptococcal antibodies are produced by the immune system. Rheumatic heart disease may be one result. Glomerulonephritis may be another. A fruitful discussion may be developed on alternative vaccines such as that used for *Haemophilus* meningitis.

6. Presumably, the purer air high in the mountains places less stress on the lungs than air in the crowded urban slums, and the respiratory system may be able to deal with the disease more effectively. Such spas no longer exist in the United States, where prevention and therapy are the keys to tuberculosis control. One might argue, however, that 25,000 cases per year does not reflect effective prevention or control of tuberculosis.

7. Rickettsiae and chlamydiae have no Gram designation because they cannot be seen under the light microscope, and part of the definition of the Gram reaction is the color assumed by the organism after Gram staining and on observation by light microscopy. *Mycobacterium* species have cell walls that cannot be penetrated by the stains used or the procedures employed in the process of Gram staining. However, *Mycobacterium* species stain by the acid-fast technique. Thus, they are neither Gram-positive nor Gram-negative; they are acid-fast.

8. The key to diagnosis in this case was the dropping red blood cell count. Apparently the patient was producing antibodies that at cold temperatures (e.g., February) react with and clump red blood cells, thereby causing them to disintegrate, thus, the dropping cell count. Another insight was the resistance to penicillin, a notable characteristic of mycoplasmas. The final diagnosis was primary atypical pneumonia due to *Mycoplasma pneumoniae*.

9. Pneumonia is generally not a "catchable" disease because the organisms are usually within the respiratory tract already. However, the parent is correct (although, indirectly) because lowering the body's resistance will encourage the organisms to proliferate, and pneumonia may develop.

10. Tuberculosis is a disease of crowded populations where people live in less-than-ideal settings. Poor nutrition and medical care, ghetto living, and a generally low quality of life contribute to its development. Once established, tuberculosis must be treated for months to rid the body of infection. Public health workers have a difficult time locating victims before the symptoms become acute and must keep

Appendix E *(continued)*

Answers to Questions for Thought and Discussion

after patients to take isoniazid for many weeks. Also, relapses are very common. The numbers thus remain high.

11. One reason is that infants produce few antibodies and must rely on those obtained by placental transfer from the mother. It is interesting to note that as women marry and have children at a later age, they have fewer antibodies to give their unborn child, a possible problem for the future. Another reason is a child's exposure to other newborns in the nursery. Also, the child cannot control its respiratory tract and cough up the mucus, as an adult can.

12. It would probably be a good idea to change the name to avoid confusion with the influenza virus. One choice is *Haemophilus meningitidis*. Perhaps you can come up with a more novel name.

13. The spread of diphtheria in the former Soviet Union was probably due to the population movements and migrations associated with the breakup of the country, and the re-establishment of new countries. At such a time, concern for one's life is primary, and the opportunity for receiving immunizations is minimal. A large population of children had probably emerged during the previous years with virtually no protection to diphtheria. The major effort to stop the epidemic were to have mass vaccinations of children and adults.

14. Modern microbiological discoveries are the fruits of labors of many investigators, so it is unlikely that one or two will be immortalized in a common bacterial name. However, part of the scientific name may honor an investigator. For example, *Borrelia burgdorferi*, the Lyme disease spirochete, is named for Willy Burgdorfer, the investigator who isolated the spirochete from the tick in the early 1980s.

15. Any animal-related profession would qualify, including veterinarian, zoo worker, pet shop owner, animal researcher, butcher, and so on. Many cases of Q fever resemble influenza and have probably been confused with this disease, thus keeping the number of reported cases low.

CHAPTER 8

1. The doctor was a carrier of typhoid fever. His fingers were contaminated with the bacilli, and each time he touched the patient's tongue, he transmitted the disease. It took several months before fellow physicians realized the secret of his "success."

2. Yogurt contains live bacteria that will supplement those already in the intestine and provide competition for any pathogens that might be consumed in food and/or water. Pasteurized yogurt has been heated and the bacterial

population has been diminished. Yogurt made from pasteurized milk has not been heated and the population has not been reduced.

3. There are several keys to solving this problem. Twelve hours incubation indicates a food intoxication rather than a food infection. "Roast beef" points to protein-rich foods, and the absence of diarrhea probably rules out staphylococcal food poisoning. A reasonable guess, based on the information at hand, would be *Clostridium perfringens*.

4. During summer months, people visit the local park and often stop to feed the ducks. Pieces of bread or other food sink to the bottom of the pond and accumulate with the excrement from ducks and any other organic waste that may be present. The water in mid-summer is often stagnant. In the anaerobic mud and slime at the bottom, *Clostridium botulinum* proliferates. It liberates its toxin to the water, the ducks drink the water, and limberneck follows. You might like to speculate on how this problem can be avoided.

5. Botulism can be a deadly disease, so an immunizing agent might be warranted. On the other hand, relatively few cases occur in the United States annually and most occur in people who can or otherwise preserve their own food. In view of this, immunization might not be warranted.

6. Chicken barbecues are much more common in summertime. What other poultry-related products are related more to summer than other seasons?

7. To say "salmonella food poisoning" is to violate at least two rules of microbiology. First, the word "salmonella" is incorrectly written; second, the disease is an "infection," not a "poisoning" because bacteria grow and multiply in the intestine (vs. staphylococcal food poisoning).

8. This actually happened to me some years ago. I had intended to show how soilborne organisms contaminate food and cause the can to swell. A colleague pointed out that I would probably be cultivating *Clostridium botulinum* from the soil and asked whether I wanted to expose students to this organism. I cancelled the experiment.

9. So far as I know, mushrooms are cultivated in dark, humid caves on trays of fresh manure and other organic matter. It is probable that clostridial spores enter the manure from the animal's intestine and cling to the mushrooms as they grow tall. Although the spores do not revert to vegetative cells in the mushrooms, they do remain with the mushrooms. When the latter are bottled, anaerobic conditions can be established following a failed sterilization procedure with steam. Now the toxin is produced

and, assuming the mushrooms are not boiled, the toxin will pass to the consumers. The important cultivation practice is the use of fresh manure for the process.

10. Clams use their filtering mechanisms to trap food particles (and microorganisms). As the population of typhoid bacilli accumulates in the "combs" of the filter, the clam becomes a lethal meal if eaten raw.

11. Chickens and other forms of poultry may be infected with *Salmonella* serotypes. Evidence of this problem is often difficult to ascertain, and poultry manufacturers may unknowingly send infected chickens off to the market. If the chicken is cut up on a carving board, the *Salmonella* cells may be deposited there and picked up by the salad items. Old, wooden carving boards are a particular problem because of numerous cracks and fissures. Thorough cooking kills the *Salmonella* in the poultry, but the salad is eaten raw and is a source of live bacteria. In this situation, the salad should be prepared before the chicken is cut up if the same carving board is used.

12. The public health official is probably asking a bit much to expect that doctors will recognize botulism symptoms. With about 50 cases per year, botulism is quite rare in the United States, and doctors who deal with botulism will probably be seeing their first and last case. Usually there is little to distinguish botulism from paralysis-associated disease such as a stroke. When the eating history of the patient is examined, however, the possibility of botulism arises.

13. The disease was brucellosis. All recovered and returned to work. The health department also made several recommendations to prevent further outbreaks, including the use of rubber gloves, face shields, and the use of negative air pressure on the floor.

14. My knee-jerk reaction would be to select the frozen hamburger since there would be less opportunity for staphylococci or other bacteria to grow during the past two days. However, if you cooked the hamburgers for equal amounts of time, the frozen one would absorb less heat than the fresh one and fewer internal bacteria would be destroyed. Therefore the fresh one might be safer. Then again . . .

15. In the cartoon, the illustrator showed chicken salad, mushrooms, and ham, all of which had been implicated in recent weeks in food poisoning episodes. You might like to suggest a menu of easily contaminated foods from examples in the chapter.

16. During peak lunchtime hours, harried workers in fast food restaurants are constantly tossing hamburgers onto the grill, flipping them quickly, and serving them before they're fully cooked. Bacteria survive the cooking process and

Appendix E *(continued)*

Answers to Questions for Thought and Discussion

infect the patrons. More extensive cooking at off-peak hours prevents such infection.

17. My "culinary experiences" in the kitchen have often involved preparing stuffed shells and lasagna (my Italian heritage helps considerably). In both cases, to make the cheese filling one uses ricotta, mozzarella, and Parmesan cheeses with appropriate spices (lots of basil and oregano), and a couple of fresh eggs to "hold things together," as my mother would say. The association of *Salmonella* with eggs is well-established, and it is quite likely that the eggs used by the manufacturer were contaminated. At the time, the inspectors focused their attention on the eggs in the pasta products and hit paydirt. Chalk up another one to bad eggs.

CHAPTER 9

1. This incident actually happened. The bipolar staining characteristic of the plague bacillus makes it appear as a safety pin, and the technician apparently thought diplococci were present. Also, the plague bacillus is a Gram-negative rod which indicates that the Gram stain technique was performed poorly. The erroneous conclusions appeared to have dire consequences.

2. The medical upheaval was bubonic plague. The town marks the farthest reach of the vast graveyard that the suburbs of London became after the plague.

3. Tetanus toxins are produced in the human body tissues as compared with botulism toxins which must be consumed, pass through the gastrointestinal tract, and then be absorbed into the bloodstream for distribution. From this standpoint, the tetanus toxin would appear to pose the greater threat.

4. Epidemic typhus is a disease spread by lice, and lice are common when armies engage in warfare. Personal hygiene and sanitation are lacking, and as soldiers huddle close together against the cold, lice pass easily among them and spread rickettsiae.

5. This incident also happened as stated. The story was broadcast by UPI and the intent seemed to be to minimize a serious situation. You should construct a letter as I did back in 1976.

6. From the viewpoint of costs alone, it is clear that achieving total immunization to tetanus in the American population is highly desirable. Based on the costs noted, 45,000 doses of vaccine could have been administered in the state for the single case treated. Without being too objective in your thought process, explore methods so that a wider immunization can be achieved.

7. Lyme disease is spread by deer ticks. Curtailing deer hunting provides more hosts for the ticks and an explosion of the tick population. More

ticks mean more possibilities for Lyme disease transmission.

8. Endemic typhus is caused by rickettsiae transmitted by the fleas of rodents. When the rat poison was used, the rats died and the fleas quickly left the dead animal bodies, only to infest the nearby human bodies. Five members of the household were subsequently infected.

9. Good quality boots, preferably hip length, would be a wise suggestion.

10. It was Rocky Mountain spotted fever. The progression of the rash from extremities to body trunk is the key.

11. The festering boils on humans and cattle are probably those of anthrax. Anthrax may be caused by airborne spores settling on the skin. It is fairly certain that ancient peoples were familiar with the disease since anthrax has been a concern in veterinary medicine for centuries. The French originally called it "charboneuse," a reference to the black encrusted boils that form.

12. The numerous soilborne diseases discussed in the chapter provide ample possibilities. Certainly tetanus and gas gangrene must be prime candidates for serious disease, but the list can also include anthrax, melioidosis, and leptospirosis. Since each disease has a different pathology, you should explore how each can be deadly in warfare.

13. When leaves and debris pile at the curbside, soilborne arthropods flourish, including ticks that commonly occur on grass and leaves. A child's chance playing in the leaves may bring it in contact with the ticks and any of the three diseases indicated may follow.

14. By ridding the community of its population of cats, the people would give the rodent population an opportunity to multiply rapidly. If infected fleas were to enter the large rodent population, bubonic plague could spread among the rodents and then, to the community. The chances of plague would be less if the rodent population was normally small.

15. Lice transmit typhus. Therefore, by reducing the louse population via head shaving, the people also reduced the opportunity for transfer of the rickettsiae of typhus. Accumulations of lice eggs are called nits. In past generations, nits were so common in the hair that people would habitually pick them out, a practice that has come to our time in the expression "nit-picking."

CHAPTER 10

1. I suspect you could debate the issue of a vaccine for many hours.

2. The STDs open the epithelial tissue to penetration to HIV deposited there especially since the virus is present in the infected person's semen. The genital ulcers in syphilis and chancroid are

typical portals of entry for HIV, and the eroded urethra provides an entryway for a person having gonorrhea, chlamydia, or other forms of NGU.

3. Impetigo is a skin disease often due to staphylococci and occurring most commonly among children. Children tend to have more contact with one another during the summer months than during any other time of the year, and the skin is often unclothed at this time.

4. Leprosy and tuberculosis are both caused by species of *Mycobacterium*. The immune system's response to leprosy results in some protection to tuberculosis (much like the syphilis-yaws relationship). Evidence for this is in the use of BCG for immunization against leprosy. Therefore, where the incidence of leprosy is high, the number of tuberculosis cases is likely low.

5. During douching, fluid is forcibly expelled into the vaginal tract to cleanse the tract. This fluid can force microorganisms in the vaginal tract up into the uterus and/or Fallopian tubes where infection can occur. Pelvic inflammatory disease (PID) may result.

6. The disease was leprosy.

7. The student was one of the few people in the United States that year to experience rat bite fever. *Streptobacillus moniliformis* was observed in tissue samples from her finger.

8. I can see pluses and minuses in this method. What do you think?

9. The World Health Organization attempts to restrict the spread of trachoma by maintaining squads of trachoma nurses whose duty is to visit remote villages and treat the eyes of all residents with antibiotics. This treatment is given to trachoma patients as well as potential victims of the disease.

10. Whenever I assist a student who is using a microscope with such a tube, I remove the tube. The practical benefit does not seem to outweigh the danger of contracting conjunctivitis, particularly when so many different students use the microscope. Disinfection is a useful alternative to removal, but this is not always possible.

11. Organizations that assist in control of leprosy point out that "leprosy" is a pejorative term and should be replaced by "Hansen's disease." This will probably take many years to accomplish in view of the long history of leprosy. You might wish to focus on the modern methods for detection and treatment of leprosy, and indicate that many diseases pose a greater threat to life. You might be interested to know that thousands of leprosy patients are treated at centers throughout the U.S., often on an outpatient basis.

Appendix E *(continued)*

Answers to Questions for Thought and Discussion

12. The person has probably acquired cat scratch disease. This type of lesion, low fever, swollen lymph nodes on one side, and the cat's involvement in the wound all point in this direction.

13. The woman probably had syphilis. Congenital syphilis often leads to miscarriage after the fourth month. The symptoms in the newborn, including Hutchinson's triad, also point to syphilis.

14. The high incidence of ectopic pregnancy reflects infection of the Fallopian tube that accompanies many STDs. As the tubes become blocked by infection and scar tissue, the fertilized egg cannot pass into the uterus, and development begins in the tube. Eventually, an ectopic or "tubal" pregnancy will result and a life-threatening situation may confront the woman.

15. The disease is dental caries. You may be surprised that caries has a disease connotation, but a close look reveals that dental caries fulfills the prerequisites for an infectious disease.

CHAPTER 11

1. The chapter contents should help considerably.

2. The concept certainly seems to be true. Then again, are viruses "organisms" or replicating entities that possess one but not all the properties of a living thing? And is "much" of their behavior or "all" of their behavior directed toward reproduction? And is it really "reproduction" as we generally mean the term? What do you think? Can you rephrase the concept to better suit the virus?

3. Oncogenes appear to function in the production of substances that are involved in the metabolism of body cells. When these substances are overproduced, however, they may transform the cell to a cancer cell. Hence, the "Jekyll and Hyde" connotation is applied to illustrate the good and bad effects.

4. The body's immune system is normally stimulated by proteins but not by nucleic acid. A proteinless particle would therefore be a poor stimulant to the body's defensive network. Furthermore, destruction of viral particles by the body is usually directed against protein, not nucleic acids. Outside the body, the physical and chemical methods of virus control (heat, antiseptics, etc.) are commonly directed at the protein portion. A nucleic acid strand would conceivably be more difficult to control and might require new methods of destruction.

5. By definition, a virus contains either DNA or RNA, but not both. A particle containing both would not be considered a virus, to begin with. Then there is the problem of why both are present. Is there some metabolic activity taking place? If so, then it is certainly not a virus. I would think that the finding would create a

substantial stir. Incidentally, chlamydiae were once thought to be viruses, but the group was reclassified with the discovery of both RNA and DNA in chlamydiae.

6. The pure culture techniques of Koch set the stage for the isolation of many bacterial agents of disease while lending credence to the germ theory. The test tube cultivation of viruses was an equally auspicious event because test tubes were significantly easier to deal with than animals or fertilized eggs. Many different viruses could now be cultivated in massive quantities. Not only was there an isolation of new viruses but also the development of many viral vaccines.

7. Is it possible that viruses have been holding the bacterial population in check all these millennia?

8. For bacterial disease, it is treatment; for viral disease, it is prevention. Think of all the vaccines available for various viral diseases and compare that number with the limited number available for bacterial diseases.

9. Viruses reproduce very quickly and destroy their host cells in the process. A single virus, for example, may replicate to 1000 viruses in an hour (1 cell lost); each of the 1000 viruses can then infect a new cell in the immediate vicinity and destroy it (1000 cells lost), while producing 1000 new viruses per cell in the process. Now there are 1,000,000 viruses . . .

10. It's not a bad way of showing how a tiny virus or two can be lost within a mass of antibody molecules. Thus surrounded, the viruses cannot reach their host cells, cannot replicate, and cannot continue the infection. It may be an oversimplification, but this is essentially how antibodies interact with viruses.

11. It's simple—"prion" sounds better than "proin."

12. Their purpose is to "weaken" the impact. The word attenuate comes from Latin stems that mean to thin out or weaken something.

13. Dmitri Iwanowski did not discover viruses by any stretch of the imagination. He merely showed that something smaller than a bacterium was capable of causing disease. At the time it was a momentous discovery because bacteria were believed to be the ultimate in simplicity. Iwanowski pointed out that a *filterable* virus caused tobacco mosaic disease. His key word was filterable to show that whatever the agent was, it was able to pass through a filter that trapped bacteria.

14. Viroids and prions are of special interest because they appear to lack protein and nucleic acid, respectively. Even at the level of viruses, both components seem to be necessary. In future years, it will be interesting to see how the research develops on viroids and prions. Viruses do or do not conform to other living things, depending on how one defines living things. Discussions as these are fruitful

because they get to the root of the student's perception of a living thing.

15. Reproduction has a connotation in biology that implies the generation of new individuals by asexual or sexual processes. Although new individuals are generated in viral replication, the process is neither asexual nor sexual, but a completely separate process seen nowhere else in biology. Hence, virologists prefer "replicate" to "reproduce."

CHAPTER 12

1. A generation ago, in 1965–1966, the last great rubella epidemic swept through the United States. Over 50,000 cases resulted of stillborn or deformed children born to pregnant women who contracted the disease. Hearing loss was a major defect in newborns. The current generation of deaf students reflects the tragedy of that era.

2. Here is what transpired: Twice-daily reports were collected from delegations on whether any members had measles symptoms. All visitors to medical stations were observed for measles symptoms. Letters were sent to all participants, volunteers, and staff advising them of the situation as well as control measures, signs, and symptoms. Daily telephone calls were made to all local hospital emergency rooms. State health departments of the competition participants were notified of the outbreak. Can you think of anything the epidemiologists missed? P.S. No additional cases occurred.

3. The symptoms point to Reye's syndrome. The child should be rushed to a hospital where supportive treatment such as a blood transfusion may be rendered.

4. A modern Thomas Syndeham should look for distinguishing markers among the diseases. Some examples are teardrop-shaped, fluid-filled lesions occurring in chickenpox, Koplik spots followed by the hairline rash of measles, mild measlelike symptoms of rubella, "slapped cheek" appearance in fifth disease, salivary swelling in mumps, deep pustules and pock formation in smallpox, and characteristic lesions of molluscum contagiosum.

5. Children are easier to reach at 15 months for immunization, but after 20 years or so, the immunity has probably worn thin. By contrast, teenagers may be harder to reach and certify for immunization, but the level of immunity will be much higher. Additional discussion can lead from here.

6. Answer #2 should give you a head start on this one. The problem to confront is how to convince the Amish population that vaccination is a good idea.

Appendix E *(continued)*

Answers to Questions for Thought and Discussion

7. There are several similarities between the diseases. The symptoms and methods of transmission (sexual contact) are similar, as are the viruses (DNA and icosahedral). In both cases, there is a possibility of cancer as a complication. There are many differences (types of virus, other areas of infection, treatments). You should be able to continue the list from here.

8. The key problem is the eight segments of RNA that make up the genome of the influenza virus. With eight RNA segments encoding proteins, a different variation of proteins may wind up in the new virus than was present in the original, and the new virus may thus have an antigenic combination different than the original one. Influenza vaccines are commonly prepared by cultivating viruses in chicken eggs, and the protein from the latter may be allergenic.

9. Herpes simplex viruses are associated with tumors of the cervix so there is great reluctance to use these viruses in a vaccine. However, genetically engineered viral fragments (such as used for hepatitis B) or genetically engineered viral antigens remain viable alternatives for a vaccine.

10. To test for influenza, a rather involved series of diagnostic tests would have to be formed. These include viral isolation from tissue in the laboratory and demonstration of agglutination of laboratory red blood cells. Antibodies for influenza would also be sought by such tests as hemagglutination-inhibition. Penicillin might be useful in retarding secondary bacterial infections that tend to accompany influenza.

11. The evidence points to chickenpox. By the way, you will note that I use "chickenpox" as a single word. This is to be consistent with smallpox and cowpox, both terms traditionally expressed as a single word.

12. It will certainly lead to a smelly shoe. It will also stop people from coming too close to you, which will decrease your possibility of coming in contact with disease organisms. Chapter 22 has a box on garlic that you might enjoy reading.

13. The incident helps prove that the agents of disease are transmissible. Pasteur and his supporters were right-on-target.

14. The change from monkeypox virus to smallpox virus could occur by any of several mechanisms. Generalized transduction, explored in Chapter 6, could account for the acquisition of new genes. Mutation is another possibility also explored in Chapter 6. Interestingly, the AIDS virus is believed to have existed in monkeys before an adaptation to humans took place. A monkeypox to smallpox virus adaptation is equally conceivable.

15. The virus causing the man's shingles is the chickenpox virus. If he were to come in contact with children, he might initiate an epidemic.

The advice is therefore justified. Indeed, in some hospitals shingles patients are placed in isolation for the duration of their illness.

CHAPTER 13

1. 1700 residents of New York most certainly did not experience rabies in 1994. The number most likely refers to the number of people who received rabies immunizations due to a possible exposure to rabies viruses from an animal. There was in fact only one case of rabies in all of New York that year.

2. Infectious mononucleosis can be spread by contact with the saliva of infected persons, so the advice is well-taken.

3. The answer was "breakbone fever."

4. Care to be a part of history? Then watch the continuing battle to eradicate polio from the world. Four strategies have been employed: (1) maintaining high vaccine coverage among children with at least three doses of oral vaccine; (2) developing sensitive systems of surveillance; (3) administering supplementary doses of vaccine during National Immunization Days; and (4) instituting "mopping up" vaccine campaigns at high-risk areas where wild polio virus is believed to exist.

5. In an early 1980s stage appearance, the rock singer Ozzy Osbourne is reputed to have bitten a bat's head. The teenagers were apparently emulating their "hero." For their trouble, they received a series of rabies immunizations.

6. Monkeys and other primates probably carry viruses that have adapted to their bodies without harm. When transferred to a new species such as the human body, the viruses may be extremely pathogenic. An epidemic of a previously unknown disease may ensue. You might reread the section on Lassa fever and other "remote" viruses to understand the possibilities of this happening.

7. With the gradual disappearance of polio as a public health problem, the need for a universal polio vaccination has become less acute, and epidemiologists are willing to accept that fewer people will be vaccinated by injection. Moreover, and of substantial importance, the only cases of polio in the United States in the years since 1993 have been related to the oral vaccine. When these cases occur, there is the possibility of an epidemic breaking out. Therefore, eliminating the oral vaccine from use effectively eliminates this possibility.

8. The letters indicate that the person is a carrier of the cytomegalovirus (CMV). CMV could pose a hazard if the blood were transfused to a person who was not previously a carrier of the virus and who had an immune system problem (such as AIDS). Knowing that the blood was CMV (+) would avoid this possibility. Also,

pregnant women would not receive this blood in a transfusion either because of the congenital significance of CMV.

9. Incidents such as these are the bane of laboratory instructors. The dangers that come to mind include AIDS and hepatitis B, but a case can be made for numerous other dangers such as staphylococcal disease. At this juncture, you might like to consider whether the pretreatment with alcohol is an effective safeguard against disease transfer.

10. This headline really happened. It meant that the Centers for Disease Control and Prevention has related hantavirus pulmonary syndrome to the Muerto Canyon virus.

11. This incident took place in 1971 as described in the question. The reporter was obviously given misinformation on the mode of transmission of hepatitis B, and the report made it appear that narcotics were being used.

12. Certainly, you should accept the offer. In fact if you work in any profession where blood is encountered (health-care or laboratory worker, mortician, police officer, corrections officer, firefighter) you should consider obtaining a hepatitis B immunization. Ask around. Your hospital, doctor's office, school, or other employer may offer it as a perk.

13. This is a palindrome, a phrase which can be read forwards or backwards. It is a unique palindrome because the only vowel used is A. Another palindrome which you might find interesting is one which reflects the dilemma of Napoleon: "ABLE WAS I ERE I SAW ELBA." This palindrome is a rare perfect palindrome because each word can be reversed.

14. The Sicilian barbers in the study used traditional razors that are nondisposable and unsterilized. They probably shaved themselves with the razors after shaving their patrons and transmitted hepatitis C viruses in bits of blood remaining on the razor. No doubt they also contributed to a high incidence of hepatitis C in their patrons. Perhaps the shaving parlor, the last bastion of male luxury, is doomed to extinction.

15. Disease reflects a competition between host and parasite. If the parasite overcomes the body defenses, disease ensues; but if the body defenses overcome the parasite, the latter is driven away. In each case, the parasite and defenses compete until one wins out. AIDS is dramatically different. The HIV virus eliminates body defenses by destroying portions of its immune system. Left defenseless, the body is subjected to marauding parasites in the form of opportunistic microorganisms. Few other diseases take this approach.

Appendix E (continued)

Answers to Questions for Thought and Discussion

CHAPTER 14

1. *Cryptococcus neoformans* is known for its ability to grow in pigeon droppings where there is an abundant supply of creatinine. The organisms are then airborne by wind gusts and inhaled to the lung where respiratory disease may develop. The action of the residents would therefore appear to be justified, at least microbiologically.

2. The amateur baker is smelling ethyl alcohol from the dough. During the night the yeasts used up all the oxygen in the dough, and as carbon dioxide from the Krebs cycle took up the airspace, an anaerobic environment developed. Yeast metabolism then shifted from respiration to fermentation, and ethyl alcohol evolved.

3. There are several possibilities as to why intestinal diseases of fungal origin are so few. Perhaps the spores are destroyed by stomach acid, or perhaps intestinal nutrients for fungal growth are lacking, or perhaps receptor sites for tissue attachment are not present, or perhaps the oxygen supply is limited. Discussions as these are valuable because they help you focus on the requirements for infection to take place.

4. Diagnosing oneself may not be a good idea because a misdiagnosis may result. If so, then the wrong medication will be taken and the infection will become progressively worse. This is the down side of over-the-counter sales of drugs.

5. Yeasts possess a variety of enzymes that may digest the organic materials in the anaerobic sludge at the bottom of a cesspool or septic tank. The metabolic schizophrenia in yeasts results from their ability to metabolize organic materials aerobically as well as anaerobically.

6. The affectionate cats probably have ringworm. By rubbing across the woman's legs, the cats transmit the fungi to her shins where the disease develops. Ringworm remains among the most transmissible fungal diseases.

7. This student has an intriguing idea. The chitin will encourage chitin-digesting bacteria to emerge, and she may then isolate them to develop a natural fungicide.

8. A pizza with mushrooms.

9. Lime raised the pH of Mr. A's soil and prevented fungal disease in the turf. The mushrooms in Mr. B's lawn in June are a signal that fungi can grow in the acidic environment, and the brown spots are proof. Mr. B should invest in lime next year.

10. Lots of luck.

11. Tests were performed for various fungi. The results pointed to *Histoplasma capsulatum* and histoplasmosis. All were treated and recovered.

12. It will be interesting to see which pronunciation your instructor prefers. I learned the second pronunciation and thought it was the only one until my colleagues put me on to the other pronunciations. I suspect they evolved from various attempts to pronounce the word as it was written.

13. Composting is the conversion of plant material to simpler compounds that can be recycled as "compost." The principle behind composting is that microorganisms, primarily bacteria and fungi, perform the chemical breakdowns. The *Aspergillus* was probably one of those organisms involved, and its spores filled the local air. The fungus could conceivably be dangerous because certain *Aspergillus* species are causes of lung infections. Also, fungal spores tend to cause allergic reactions, which also could pose a hazard to health if they are severe.

14. According to public health epidemiologists, the outbreak of coccidioidomycosis was most likely due to dust made airborne by the earthquake. The disease is not transmitted from person to person, but from inhalation of airborne *Coccidioides immitis*.

15. The aspergillum sprays water about, just as the fungus *Aspergillus* spreads its mycelium about quickly. Moreover, close inspection of the fruiting body of *Aspergillus*, as in the electron micrographs in this chapter, shows a radiating head with spores spraying out in all directions. Both words are derived from the Latin *aspergere*, meaning "to sprinkle."

CHAPTER 15

1. Sounds like a good idea for a contest.

2. Malaria remains the most widespread global health problem. Trichomoniasis affects over 2.5 million Americans annually. Giardiasis is among the most prevalent parasitic intestinal diseases in the United States. Pneumocystosis causes over half the deaths associated with AIDS. Toxoplasmosis should be a concern to anyone who owns a cat. I would offer that more attention to parasitology is warranted.

3. The British found they could improve the taste of quinine by mixing it with gin. The drink came to be their "gin and tonic."

4. *Leishmania* species often penetrate the white blood cells and remain there. Therefore it was deemed advisable to reject the service personnel as blood donors.

5. Antidrought measures usually include pools of water in ditches that transect the agricultural lands. Mosquitoes thrive in these ditches and since mosquitoes transmit the agents of malaria, the incidence of this disease increases.

6. Giardiasis and balantidiasis are two candidates.

7. It will probably depend on whether the residents have suffered from intestinal disease due to protozoa. The residents of Milwaukee, Wisconsin, would probably favor a proposal, but other taxpayers would probably reject it until . . .

8. This is quite a bold statement but it reflects the fact that dirty diapers are as capable of transmitting infectious disease as the open sewers of generations ago. Day-care center workers should take special note.

9. To avoid toxoplasmosis, the student might give the friend two helpful pieces of advice: have the hamburger well-done, and send the cats to someone else's house until the baby is born. Toxoplasmosis is a significant threat to a pregnant woman—the T in TORCH is toxoplasmosis.

10. As the borders of Rome expanded, Romans ventured into far-off lands where mosquitoes thrived and where the malaria parasite was prevalent. Infected Romans probably brought the disease back to Rome, and the mosquito populations of the aqueducts and pools propagated the parasite and spread the disease.

11. Here's an example of how a name can reveal much: *Entamoeba histolytica* means intestinal-amoeba (Ent-amoeba) that digests (lytic) tissue (histo-). The ameba lives in the intestine where it penetrates the tissue. Tissue penetration yields sharp appendicitislike pain as the nerves are encountered and blood as the blood vessels are reached. We therefore expect bloody stools and sharp pain. For diagnosis, the physician will ask for a stool specimen (intestine) and the laboratory technician will hunt for amebas. Transmission will probably be by fecal contamination of food or water.

12. The disease was probably trypanosomiasis (sleeping sickness).

13. It's the animals. Wild forest animals harbor the protozoa and spread them among the waters in the hills, mountains, and lakes. Human sanitation methods are adequate, but animal sanitation methods leave something to be desired.

14. Among the alternatives might be a new generation of water disinfectants, as well as finer filtration methods and filters on home water taps to trap *Cryptosporidium* species. Chapter 25 contains an overview of methods of water purification currently in use. You may locate other possibilities here.

15. African herdsmen know that tsetse flies are active only during the daytime. They have adapted this bit of knowledge into a method of preventing attacks of disease in their herds.

CHAPTER 16

1. Raw fish may lead to infection by *Diphyllobothrium latum*, bare feet would result in hookworm disease, snails could yield any of the fluke diseases, chestnuts might also carry intermediate forms of the flukes, mosquitoes which bear filaria worms would survive where there were no pesticides. The diplomat's report might recommend that a visit to the country might be above and beyond the call of duty.

Appendix E *(continued)*

Answers to Questions for Thought and Discussion

2. On a global scale, diseases as these "... impede national and individual development, make fertile land inhospitable, impair intellectual and physical growth, and exact a huge cost in treatment and control programs." Solutions are straightforward: develop new drugs, vaccines, diagnostic tests, and control methods. Can you suggest any novel approaches?

3. Many people have the impression that evolution always yields organisms that are more complex than their predecessors. The tapeworm illustrates the reverse. By adapting to its host, the tapeworm gradually lost its digestive tract and evolved into a less complex creature than its predecessor.

4. I would consider the situation dangerous because *Trichinella spiralis* is still quite prevalent in pork, and the preventative measures are minimal and the possibility of contracting trichinosis remains real. One day it may be possible to screen pork for trichinosis.

5. The incident in Egypt is offered by ecologists as an example of what happens if the balance of nature is upset. Microbiologically, the increase in the snail population was accompanied by increased levels of *Schistosoma* since the snail is the intermediary in the parasite's life cycle.

6. This question illustrates how knowing the stems of scientific words helps in deciphering their meaning. *Diphyllobothrium latum* is a very thin, broad tapeworm. It is associated with insanitary water which the worm enters via human feces.

7. Fish tapeworm (*Diphyllobothrium*) disease is an example. Others should be clear from this chapter.

8. The animal liver in the food might carry *Fasciola hepatica*, the liver fluke, which would then be transmitted to the dog or cat and possibly, the owner.

9. It appears that this patient may be suffering infection with Chinese liver flukes. The flukes invade the gallbladder and liver. Gallbladder blockage may lead to lack of bile in the intestine and thus, poor digestion of fats and fatty foods. The fact that Chinese liver flukes are transmitted by raw fish adds credence to the possibility.

10. The threadlike body may have been an eyeworm and the warmth of the fire may have brought it to the surface. Figure 16.14 in this chapter portrays an eyeworm in the eye.

11. It should be apparent from this statistic that cancer research is much better supported than parasitology. Multiple reasons can be offered for this, as the class might enjoy exploring.

12. Sausages made from pork are popular in Polish, German, and Italian communities, but pork sausage is not readily available in Jewish areas. Other pork products may also be used

as examples. Butchers should be sure to advise thorough cooking of pork products; they should purchase their meats from reputable dealers; and they should be careful of cross contamination with other meats via knives, aprons, cloths, and similar means.

13. Improved sanitation methods would reduce mosquito populations and reduce the possibility for transmission of *Wuchereria*. By contrast, clearing forest lands for agriculture and installing drainage ditches would encourage mosquito populations and increase the possibility for filariasis. Now it's your turn to continue the comparisons.

14. Steak tartare can be a vector for *Taenia saginata*, the beef tapeworm. After suffering from tapeworm infection, the person might switch to well-cooked hamburgers.

15. The concept of studying parasitology to appreciate the web of life is intriguing and worthy of note. Hundreds of thousands of individuals are infected with multicellular parasites. The relationship is benign in huge numbers of cases, but parasitical in many others as this chapter shows. Perhaps a geographical summary of the parasites might help you appreciate how worldwide the parasites are.

CHAPTER 17

1. Consider how many people have handled a stamp and the conditions under which it was stored before it comes to you. Then consider whether you would want to place it in your mouth.

2. The antibiotic has destroyed the bacteria in the urinary tract but has also eliminated the bacteria in the vaginal tract. These bacteria normally control the proliferation of *Candida albicans*, and now, *Candida* has overgrown the vaginal tract. The doctor might prescribe replacement of the bacteria by recommending tablets or suppositories that contain *Lactobacillus*. Many different types are commercially available.

3. This disease would have symptoms that developed slowly. The disease would linger for a long period of time, symptoms might not be obvious, parasites would have disseminated to the deeper organs and systems, the disease would not be transmissible to another individual, and the disease would probably have begun as a complication of an earlier infection.

4. The salad bar remains one of the most common possibilities for disease transmission in the restaurant business. You should discuss how to limit microbial transmission through chilling of foods, selection of foods which are to be offered at the salad bar, thorough washing of foods, cautions to patrons, turnover of foods, disinfection of salad bar areas, dust control, and other means. Reference may be made to the Note to the Student in Chapter 8.

5. Cholera bacilli are susceptible to stomach acid and thus, the bacilli may never have reached von Pettenkofer's intestine to cause disease. It is also conceivable that von Pettenkofer's anxiety caused his stomach to put out a higher-than-normal amount of acid, which would further reduce the bacterial population. Perhaps the laboratory isolate was less infectious than one directly from a diseased patient.

6. Bacteria have had more time to live and metabolize organic materials in a "quiet" mouth while you slept (as opposed to the very "busy" mouth of daytime). Therefore, the variety and quantity of gases are great. Controlling morning breath is another good reason for brushing the teeth before retiring.

7. This incident happened to me in 1982. At the emergency room, I was given a "tetanus shot," a preparation of tetanus toxoid to induce my immune system to produce tetanus antitoxins. These antitoxins would protect me against tetanus toxins since tetanus spores had probably entered the wound from the soil.

8. On the surface it would appear to be a good idea. But how would world ecology and population dynamics change if infectious disease did not limit populations? What other diseases would emerge to take their place (e.g., genetic disease, physiological disease, biochemical disease)? You can continue the discussion from here.

9. Droplets transfer a huge variety of microbial diseases, and this question may serve as a review if the diseases have been previously covered or as a prelude to diseases to be studied later in the course. The list will be rather extensive in either case.

10. Is it conceivable that a fever is meant to heat up the body so that the body defenses operate more efficiently? Or that diarrhea helps flush bacteria out of the intestinal tract? Or that a cough expels organisms from the respiratory tract? Or that a skin rash identifies infected individuals and causes others to stay away? Now it's your turn—identify another symptom and try to give a positive spin to it. And consider the implications the next time you reach for the Tylenol, Immodium, or Nyquil.

11. The low dose needed for the establishment of shigellosis simply means that relatively few bacilli are needed to bring on the disease. This factor makes controlling the disease quite difficult because it can spread quickly within the community.

12. Can you imagine how many people have handled a dollar bill before it gets into your hands? Wetting the fingers before touching the dollar bill brings whatever was on the bill into your mouth.

13. Most will probably consider the epidemic disease to be the greater threat because of its explosiveness but a case can probably be

Appendix E *(continued)*

Answers to Questions for Thought and Discussion

made for the endemic disease since it is often hidden from detection and strikes without warning. In the case of acute versus chronic disease, a strong argument can be offered for either.

14. Few people realize how many bacteria are present under the fingernails. When the fingers are brought to the mouth, infection can take place. In a day-care center (or any situation) where fecal matter or blood or tissue is contacted, it is a good idea to brush under the fingernails regularly.

15. It's simple. If a virus kills its victims quickly, then there is no way it will be around long enough to spread to new hosts. Of course, this theory assumes that the virus does not exist anywhere in nature where it can be contracted easily, and that it relies on human-to-human transfer to remain in existence. In this particular case, the more virulent the virus, the less likely it is to be slate-wiper (i.e., able to kill huge numbers of victims). You might like to contrast this episode with the outbreak of plague that killed a third of the European population in the 1300s. Here was another slate-wiper, but the conditions were very different.

CHAPTER 18

1. In many diseases, actual invasion of the tissues does not take place, but instead, the microorganisms grow along the body surfaces. In giardiasis, for example, the protozoa colonize the surface of the gastrointestinal tract and inhibit water absorption. In pertussis, the bacilli multiply along the epithelial surface of the respiratory tract and mucous plugs soon accumulate to cause suffocation. Other examples should emerge from the chapters on disease.

2. Begin with the chemical signals that stimulate phagocytosis and go from there. Remember that the heart and other organs are constantly "at work" but that the immune system lies at rest until it receives a signal to act. The signal can be an epitope; other signals are MHC molecules, interleukins, and antibody molecules.

3. Apparently the problems in developing commercially available lysozyme outweigh its usefulness. You should conjecture on possible problems. One suggestion is that lysozyme is a simple protein that the body may break down rapidly. A delivery system to the infection site might constitute another problem. It would be interesting to see whether genetic engineering techniques could produce a useful lysozyme product.

4. I think the individual would fare poorly. Modern humans have evolved an immune system that responds to myriad antigens from microorganisms, and our immune system is prepared for the appearance of these antigens (clonal

selection). The ancient immune system probably had no contact with these antigens and was primitive by modern standards (it did not have the immune system cells required for response). The individual's susceptibility to disease would probably be very high and I doubt whether survival would be possible. (Even if they managed to survive the microorganisms, I cannot see how ancient individuals could survive the food.)

5. The simple fact is that there are no natural openings to the blood system or the internal organs. Inward extensions of the skin form the respiratory and urinary tracts. Even the eye is a closed socket. This concept demonstrates the body's natural resistance to disease and illustrates why the deadly diseases are those in which invasion to the interior is made. Superficial diseases like pertussis or diphtheria are due to long-term effects indirectly associated with bacteria.

6. Here is another intriguing possibility for microbiology research. The obvious advantage is freedom from tooth decay, but many problems must be solved: is *S. mutans* the only agent of caries? Could the immune system be stimulated to produce enough IgA for protection? Would the IgA be delivered to and survive in the oral cavity? Would new organisms emerge as caries agents? Could IgA be used in a mouthwash or toothpaste?

7. The cockroach does appear to have an immune system, and the system apparently is a factor in its ability to survive. Far from being immunologically primitive, as suspected, the arthropods are able to produce antibodylike proteins. The next steps would be to identify and characterize these substances. How would you continue the experiments?

8. The brucellosis example exemplifies species immunity. Humans resist the serious effects of brucellosis because a key nutrient is lacking for the bacteria and growth on the placenta is retarded. You should conjecture on the source of species immunities in other animals. The temperature difference in chickens and anthrax is an example.

9. The botulism toxin is formed in canned food. In order for the toxin to affect the body, it must pass through the gastrointestinal tract where protein-digesting factors are at work. It must then pass through the intestinal wall to the bloodstream for distribution. The tetanus toxin is produced in the anaerobic tissue of a wound from which it easily enters the bloodstream. The natural barriers are therefore bypassed.

10. The tears contain lysozyme, which is active on the cell walls of Gram-positive bacteria. It is not surprising that bacterial conjunctivitis is due to a Gram-negative bacillus.

11. It's not necessarily unusual. How many times does an organism emit signals that stimulate the "wrong" response?

12. The immune system is centered in the lymph glands, and phagocytes congregate here. It is expected that early response to infection will take place in this tissue, and the swollen glands would attest to that.

13. When antigenic determinants arrive in the lymphoid tissue, they must seek out receptor sites on one of perhaps a million different cells. Only after that interaction has taken place will the immune system be stimulated. The analogy would appear correct.

14. You can have a field day with this one. Would you write a letter correcting the statement?

15. You are encouraged to exercise imagination on your trip and consider the perils of phagocytosis, mechanical barriers, lysozyme, T-lymphocytes, and other protective measures in the body. An appreciation for resistance mechanisms should evolve from this discussion.

CHAPTER 19

1. In herd immunity, the immunizing agent can be spread in numerous ways. For example, daycare workers changing the diaper of a child recently immunized to polio may pick up the viruses on their fingertips and inoculate themselves and other children they contact. Body secretions such as saliva or respiratory droplets may spread measles, mumps, or rubella viruses from children who recently received the MMR vaccine. These examples may serve as a starting point for the discussion.

2. Immunity tends to wear away with age as antibody levels decrease. Older women thus have fewer antibodies to donate to their offspring, and the child will be born with less protection to disease than a child born to a younger woman.

3. Blood scrapings would be mixed separately with antibodies prepared against human albumin, dog albumin, fish albumin, and chicken albumin. A precipitation reaction between the known antibodies and the unknown blood scrapings would identify the source of the blood. This is one example where immunology is of value in forensic medicine.

4. The approach would seem favorable. In effect, the woman is being used as an "immunological factory" to produce antibodies for her newborn as well as herself. I am hard-pressed to think of any negative effects other than the normal hazard of using a vaccine.

5. It would seem to be a good idea but it will take lots of money to implement. And the willingness for the American taxpayer to part with the funds will depend in part on how serious the threat of disease is perceived to be. You should continue the debate from here.

Appendix E *(continued)*

Answers to Questions for Thought and Discussion

6. IgM is not used in gamma globulin preparations possibly because the expense of obtaining it is prohibitive. Perhaps storage is a problem. The possibilities, however, are intriguing, and you might enjoy predicting the conditions that might make IgM use commonplace in the future.

7. With some insight and imagination, a case can be made for each type of immunity as being safest to obtain. Similarly, reasons may be offered for each type as being most helpful. Discussions such as these put immunity into perspective and help you make choices supported by what you have learned. Ultimately, I would think that the individual situation would dictate the choice.

8. Cells so-infected and displaying viral proteins would stimulate cell-mediated immunity and induce cytotoxic T-lymphocytes to be formed. The cells would then provide long-term immunity to disease. (Traditional vaccines stimulate antibody-mediated immunity and rely upon B-lymphocyte stimulation and antibody formation.) Thus far, the research is still in very rudimentary stages.

9. Lysis of the red blood cells would occur because without syphilis antigen, no activity would take place in the test system, and the complement would be left over for the indicator system. Without the mistake, the expected result would be no lysis.

10. The increased incidence of pertussis in adolescents and adults probably reflects the reduced use of the whole-cell killed vaccine in previous years. The reduced use resulted in less protection because fewer antibodies were formed. The acellular vaccine for pertussis should reduce the concerns associated with vaccine use and, with increased use, higher levels of immunity should result.

11. The one with the higher titer may have had a clinical infection, a subclinical infection, a mumps vaccination, a blood transfusion containing mumps antibodies, or an injection of antiserum. Any other possibilities?

12. Children between the ages of five and fifteen generally eat well and are well-clothed. Their cells are actively metabolizing, and their tissues are growing by leaps and bounds. Their immune systems are fully functional. They enjoy exercise for physical fitness and are relatively free of cares for mental fitness. At first signs of disease, they are given home or hospital care. These and so many other factors add up to a "golden age of resistance." It might be worthwhile to consider how these resistance factors gradually are compromised with age.

13. The meningococcal meningitis vaccine typifies a subunit vaccine, or second generation vaccine. It is far safer than a vaccine containing whole, attenuated microorganisms, which

may induce illness. Such a problem exists with the old pertussis vaccine, the polio vaccine, and others.

14. Keeping a body alive for vaccine production is certainly possible, but the ethical implications must be considered. Strong arguments can be made pro and con. You may wish to set up a debate panel on a topic such as this.

15. The original requirement for a blood test was to detect syphilis before it was transferred in sexual intercourse following marriage. The change in sexual practices during the last generation has made this somewhat obsolete. However, some health practitioners would prefer to keep the requirement as a way of detecting syphilis in people who might not otherwise visit a doctor. This group of practitioners appear to be losing the battle to keep the blood test requirement intact.

CHAPTER 20

1. Here is one view to the implication of Barnard's work: The rejection mechanism of the immune system appeared to stand in the way of successful transplants so attention turned to this system. With government money and support, concentrated study unlocked new avenues of immunology research and provided new insights to immunological mechanisms. Within a few years, the science of immunology blossomed.

2. No, the officer should be stopped. His blood cells contain B antigens which will agglutinate with b antibodies in the recipient's serum. The transfusion reaction will probably kill the recipient.

3. Formaldehyde used for preservation of the animals is probably causing a contact dermatitis. The sensitivity began to develop during the first two exposures and now, during the third exposure, it is manifesting itself. Rubber gloves would be helpful for protecting the hands.

4. The first thing you would have to do is think fast. Locate an "Anaphylaxis Kit" and follow the instructions with dispatch. The MicroFocus feature in this chapter will tell you about some of the things you will be injecting.

5. In the blood transfusion, the person acquired IgE from the blood donor who regularly had hay fever. Another possibility is that IgE-associated basophils entered from the donor's blood.

6. In the first animal, the raw egg will be digested by gastrointestinal system to amino acids and small peptides which are not antigenic. However, no digestion will take place in the second animal, and the immune system will probably produce antibodies against the egg protein. An allergic reaction is likely.

7. The "he" is William J. Clinton; the condition is hay fever.

8. First off, athlete's foot usually occurs in the webs of the toes while contact dermatitis occurs where the foot comes in contact with the shoe. Laboratory testing will reveal fungal cells if the condition is athlete's foot, but no cells if it is dermatitis. A patch test can be performed by the allergist to determine if a contact allergy exists.

9. She is suffering an allergic reaction. Insufficient sensitization had taken place after the previous injections but now, after the fifth injection, she was fully sensitized and basophils and mast cells began to degranulate. Most allergists recommend that a patient spend several minutes in the office after injections to be available in case allergic reaction takes place.

10. Tissue must be closely matched genetically for a transplant to be successful. Siblings are genetically close since they have the same parents, but the parents come from different backgrounds and probably have little genetic similarity.

11. The major histocompatibility complex codes for histocompatibility molecules, which exist on the cell surface. Over 50 versions of each gene have already been discovered, and the number of gene combinations and histocompatibility molecules is enormous. Immunologically, there is probably no one else like you.

12. The key to Cardan's success was removing the feathers from the mattress since they were allergenic. The other parts of the prescription probably helped too.

13. Blood typing indicates whether the man could have been the father, but it tells little else. Determining the major histocompatibility complex molecules of mother, child, and father would give a wealth of genetic information since it would show whether the maternal and paternal genes show up in the child.

14. The most obvious way of obtaining the Rh antibodies is to obtain blood from a woman who has had an Rh problem in a recent pregnancy. With the declining rate of Rh disease, this has presented a problem. The less obvious but more available way is to inject Rh-negative human males with Rh antigens and remove their blood at periodic intervals. The serum is used as a source of Rh antibodies. It is possible that human tissue cultures will be used as an antibody source in the future.

15. It is interesting to question whether the immune system is actually protecting the body during immune disorders. Allergies can be interpreted as protective devices to rid the body of antigens, and the theory can be extended to other types of hypersensitivity, as well as to transplants and tumors. Autoimmune diseases stand in stark contrast because the body appears to be attacking itself. An oxymoron is two terms that do not fit together (the textbook

Appendix E *(continued)*

Answers to Questions for Thought and Discussion

definition is "two mutually exclusive juxtaposed words"). "Immune disorder" appears to be an oxymoron because "immune" means "free of" and a disorder is a problem. Therefore, how can you have a problem that you don't have? (For another oxymoron, see the MicroFocus in Chapter 3.)

CHAPTER 21

1. One of the myths of everyday living is that "sterilized" materials are really sterilized—that is, free from all traces of life forms. Unfortunately, we are encouraged to believe that they are sterilized (why do you think so?) when in fact they are not. Treating the baby formula with heat in the "sterilizer" reduces the microbial population, but does not eliminate it completely. Some microorganisms remain, and if the bottles are left at room temperature, they will probably grow, and, although not pathogenic, will produce off colors and tastes. It's probably best to refrigerate the bottles even though having to warm the bottle makes feedings in the middle of the night a tad more difficult. Indeed, the manufacturer suggests refrigeration because he knows full well that the machine has not achieved complete sterilization.

2. Bringing water to a boil and then cooling it will effectively destroy most microorganisms, but up to thirty minutes may be required if large concentrations of hepatitis viruses, protozoal cysts, or other heat-resistant forms are present. Over two hours may be necessary for the destruction of anthrax spores, but the possibility of their presence is extremely remote. In the great majority of cases, the recommendation that water be boiled is merely a precaution by health officials, and the implication is that bringing water to a rolling boil is sufficient. Indeed, milk pasteurization is conducted at 71.6° C for 15 seconds, and this is deemed suitable for the destruction of pathogens in the milk.

3. A suspicious person might inquire what will happen within thirty days. Will spontaneous generation take place? Is it possible that the contents were sterilized at the manufacturing plant, but that the porous container is now permitting airborne microorganisms to enter? If so, then the product was once sterilized but is now contaminated, and evidence of contamination will appear by the expiration date.

4. Broiling of meat is done at extremely high heat. This heat would kill any surface organisms rapidly and the meat would be safe for consumption.

5. A five-minute total immersion of a scrupulously clean jar would be sufficient. Admittedly this would not sterilize the jar, but any remaining bacterial spores would probably not interfere with the accuracy of the specimen.

6. Cold temperatures in frozen ice are an excellent preservative, as the case of the wooly mammoth shows. Another factor would be the low number of decaying microorganisms in the soil of this region. Even the mammoth's hair and thick skin might have provided insulation from microorganisms.

7. Some possibilities for sterilizing agents in the normal household might be a pressure cooker, boiling water, wads of sterilized cotton used for wounds, an ultrasonic device for cleaning dental plates, a microwave oven, filtering material from an aquarium or swimming pool, or a steam sterilizer used for an infant's formula.

8. The budding microbiologist could sterilize disposable contaminated material by incinerating it.

9. Ultraviolet light would have a greater effect on DNA than RNA because DNA contains thymine bases not found in RNA. Formation of dimers by combining thymine residues could not take place in RNA. In DNA, however, the activity quickly leads to cellular death.

10. In the laboratory, the Bunsen burner is used to sterilize loops, needles, and the tips of tubes and flasks. Some imagination might be in order to identify other uses.

11. The process of fractional sterilization has many drawbacks, including the extensive time involved, the need for a steam apparatus, the expense of energy to run the machine, and the need for an incubator between steam exposures. Other drawbacks will probably surface during the discussion.

12. Milk's taste apparently changes as the temperature becomes too high. Also the protein will coagulate. The butterfat is another problem inasmuch as it will not pass through a filter. It should be recalled that milk is a suspension rather than a true solution. A fruitful discussion could take place on the economic implications of a useful sterilization method for milk.

13. One component of sunlight is ultraviolet light. The ultraviolet light kills numerous microorganisms in the atmosphere and in the upper layers of the soil. Dust particles, gases, and debris, however, deflect much of the energy so that sterilization is not achieved in most cases.

14. The autoclave is undoubtedly the first word to come to mind when we think of "sterilizer." However, the autoclave is merely a pressurized steam apparatus that can be used to achieve sterilization only under specific conditions. Unless these conditions are established in the machine, sterilization is not assured. Therefore, it would be incorrect to believe that "autoclave" and "sterilizer" are synonymous words.

15. Pasteurization merely implies the destruction of pathogenic microorganisms from milk or other liquid. The process has no effect on bacterial spores, and may be tolerated by protozoal or

worm cysts. Therefore a pasteurized product cannot be considered sterilized. By contrast, a sterilized product contains no life form of any type and may thus be considered pasteurized as well as sterilized.

CHAPTER 22

1. In selecting a laboratory disinfectant or sanitizing agent, it would be advisable to ask whether the agent was bactericidal or bacteriostatic under the desired conditions. Inquiry should also be made about its toxicity, solubility in water, shelf life, use in diluted form, penetrating ability, corrosiveness, temperature and pH of use, and cost. Many other avenues of inquiry are noted in the chapter.

2. Hot tubs present a problem to users because the water is warm enough (40° C to 42.2° C) to encourage the growth of *Pseudomonas*, and the water often remains stagnant between uses. Chlorination of the system and periodic checks of the chlorine level would be the most effective way of reducing the chance of contamination. Checks of the pH level and regular changes of the filter would also be helpful. Other problems associated with hot tubs are the unusually heavy bather load and the agitation and aeration of the water, both of which deplete residues of disinfectants.

3. To obtain a disinfectant, a bit of shopping might be in order. From the grocery store, a bottle of bleach could be purchased and used at a rate of a half-teaspoon per gallon of water, with 30 minutes contact time before consuming the water. From the pharmacy, a bottle of tincture of iodine would be useful at a rate of 5 drops per quart of water, with 30 minutes contact time before consuming the water. Alas, the post office would not yield anything of value.

4. Chlorhexidine is often used by dentists during treatments, but this chemical has not yet made its way into toothpastes. The up side is that it may reduce periodontal problems by reducing bacterial populations. The down side is that over a long period of time, the chemical may prove toxic to the tissues. Can you add to the list?

5. "Have you washed your hands?"

6. Some suggestions: a spoonful of bleach in a pail of water and several rinses with the diluted bleach; if the water reservoir is detachable, a scrubbing with Comet or some other chlorine cleanser; a wipe-down with Lysol brand cleaner.

7. Some ways of reducing the spread of microorganisms in day-care centers include: disinfection of table tops used for diaper changes, regular furniture and toy disinfection, and hand washing and use of an antiseptic after each diaper change. You should suggest other methods that are practical and useful.

Appendix E   *(continued)*

Answers to Questions for Thought and Discussion

8. A high-level germicide is one that achieves sterilization under established conditions. Ethylene oxide, glutaraldehyde, and formaldehyde fall into this category. Germicides that do not achieve sterilization, such as most antiseptics, are low-level germicides.

9. Bacteriostatic is an adjective derived from the noun bacteriostasis. The adjective means essentially the same as the noun coined by Churchman, that is, it applies to an agent which inhibits the further growth of bacteria without necessarily destroying them.

10. The dye that yielded the Gram-positive mouth was gentian violet, also known as crystal violet. The latter is used in Gram staining and gives bacteria a purple color. Antibiotics have largely replaced gentian violet for trench mouth, although many pharmacies still sell the drug.

11. This is an interesting dilemma. I would think that washing followed by disinfection would be most appropriate since the washing would mechanically remove microorganisms and the disinfectant would continue and complete the process on the residue.

12. The application of ethyl alcohol does not sterilize the skin. It mechanically removes skin oils, debris, and microorganisms; it dries the skin quickly; it smells "medical"; it is far better than doing nothing or washing with soap and water. But it does not sterilize the skin.

13. This advertisement was a nightmare. Some questions that occurred to me: Why wipe down or dunk a raw chicken? It's the inside that's a problem, not the outside—the cooking heat will take care of the outside. Why clean a raw pepper with Clorox? The skin is so smooth that ordinary water will probably wash away anything that's a problem. Why worry about the outsides of carrots? I always thought you peeled carrots before cooking them. Clean a red onion? Gimme' a break! Exactly how big is a sinkful? And why worry about being so precise about the amount of Clorox (1/8 cup) when you have no idea of the sink size? Can you spot the other problems? How about "... germs and bacteria"; "salmonella..."; "*like* chicken ..."?

14. One effective way of treating the thermometer would be to wash it well in hot soapy water, rinse it thoroughly, and immerse it in a tray of ethyl alcohol or rubbing alcohol for a minimum of ten minutes. Admittedly, this would require an interval of time before taking the next child's temperature, but the wait might be worth it, especially if the second child were not already sick.

15. Tincture of iodine is a solution of 2 percent iodine in ethyl alcohol. After some weeks or months, the alcohol may evaporate and the concentration of iodine will rise. Eventually it may be quite caustic to the skin and therefore dangerous to use. Hydrogen peroxide may contain traces of catalase which cause it to break down after some weeks or months. It will thus become ineffective.

CHAPTER 23

1. Pasteur and Joubert were probably observing the effects of an antibiotic produced by the contaminating microorganisms. It is clear that the concept of antibiosis was known a hundred years ago. You might enjoy speculating whether Pasteur and Joubert envisioned a therapeutic compound in the mixture of bacteria.

2. Observations as that in China illustrate how people were enjoying the effects of antibiotics without realizing it. In this case, the mold in the soybean curd was probably the source of the antibiotic. MicroFocus 23.5 details two other uses of antibiotics from microorganisms.

3. Sulfanilamide may have an effect on kidney tissues as the kidneys remove this foreign drug material from the bloodstream. Drinking a full canteen of water dilutes the sulfa drug and provides water to assist its removal from the kidney tissues into the urine.

4. Sounds like an interesting idea. What do you think?

5. The coming of antibiotics and modern medical practices to Nepal typifies how medical advances can disrupt the lives of a population. With more mouths to feed, forests had to be cleared, and great pressure was placed in preparing the land for crops. Living space soon became a premium, and natural resources, such as water supplies, were tapped to their limit. Greater populations also meant greater sanitation problems and, consequently, more opportunity for disease spread. Discussions such as these help us understand the negative aspects of medical advances.

6. The antibiotic was rifampin.

7. The soil is a highly volatile and complex environment where organisms compete for survival. Being able to produce an antibiotic gives an organism a selective advantage in this competition. *Streptomyces* species may owe their survival to their ability to destroy other organisms via antibiotic production.

8. Here are some origins of antibiotic names: erythromycin is named after *Streptomyces erythraeus*, a red-pigmented actinomycete that produces it; tetracycline is named for the four benzene rings found in the molecule; and nystatin is named for New York State because it was discovered at the Department of Health laboratory in New York. Use of a dictionary and some imagination should help reveal the origins of other antibiotic names in this chapter.

9. A synthetically produced antibiotic does not depend upon the whim and good nature of a microorganism for its production, nor does the manufacturer have to worry about a mutation in the organism or loss of the culture. On the other hand, chemical syntheses tend to be complex, and separating the final product may be tedious and less cost-effective. There is also the possibility that residual chemicals may be toxic in the patient.

10. The antibiotics came on the scene five years after the social security age was established at 65. As deaths from bacterial disease diminished in the elderly, and as improved health practices came into common use, the life expectancy began to rise until it reached the upper 70s, as it has today. I seriously doubt that if a social security system were established today, the age would be set at 65. You might enjoy speculating on what a suitable age might be.

11. It would seem likely that many genera of microorganisms could produce antibiotics and yet, only five genera have been pinpointed in this chapter. Perhaps this has to do with the soil habitat in which these genera normally exist. It is probable that many other genera of microorganisms produce antibiotics, but the chemicals are generally too toxic for use in the body.

12. Nobel Prizes are generally awarded for basic research that affects a broad cohort of people. The discovery of the therapeutic effects of chemical compounds would fit this concept. It might be interesting to survey the other chapters and examine the inside front cover of the text to discover other trends in research that are worthy of Nobel Prizes. You might conjecture on fundamental discoveries that failed to be recognized by the Nobel Prize. In addition, a list of "The Prizes Not Given" might be illuminating.

13. The super rats were not necessarily large rats, but species that were resistant to pesticides. The rats probably emerged after decades of pesticide applications as susceptible species were eliminated. The appearance of PPNG is reminiscent of the super rat emergence because, as susceptible gonococci disappeared, the resistant strain came forth. Both examples are elegant illustrations of Darwin's theory of evolution by natural selection. The fact that they have occurred in so few years is remarkable.

14. An antibiotic not absorbed from the gastrointestinal tract is useful for infections in the gastrointestinal tract. One that is rapidly expelled in the urine would be valuable for urinary tract infections. In these examples you can understand that the characteristics of a drug can be adapted to the use of the drug.

15. The antibiotic issue is one that can be argued for hours. Perhaps this might be a good question for a debate panel. The side effects, emergence of antibiotic-resistant bacteria, effect on human population growth, and human dependence on antibiotics would stimulate the

Appendix E *(continued)*

Answers to Questions for Thought and Discussion

anti-antibiotic side. The misery and death from disease, rising food costs, job impact in the pharmaceutical industry, and generally better quality of life might be discussed by the pro-antibiotic side.

CHAPTER 24

1. Ultraviolet light would reduce the level of spoilage organisms in the air and on surfaces exposed to the light. However, the UV light would not penetrate the food nor would it reach the side or bottom surfaces of the food. Also, there is the matter of public acceptance of irradiated foods. The idea is appealing, but not likely to be coming to fruition.

2. In the incident at the university facility, the health officials recommended cracking and using the eggs immediately, rather than storing them for the next morning. The outbreak occurred on the campus of Princeton University.

3. The correct sequence would be to do the salad before the chicken. *Salmonella* serotypes may be present in the chicken and cross-contamination to the salad may take place. Cooking will eliminate *Salmonella* in chicken, but salad is eaten raw and is potentially dangerous. You should explore other precautions. Thorough cleaning of the knife and cutting board is a good place to begin.

4. Here is another novel idea of researchers. The theory is that the harmless bacteria will crowd out and prevent the development of *Salmonella* serotypes. The process is called "flora replacement." It has been shown to be effective in research trials.

5. This is another example of truth being stranger than fiction. The incident occurred as described, and the baked potato was superb. (I had it with crab cakes and a veggie.) The sour cream was preserved primarily by the acid present, and no symptoms of food poisoning or any other intestinal illness developed in the following days. This incident points up that the expiration date is merely a statistical guess on when the product will look or taste bad. It is not necessarily the date on which the product will begin posing a peril to health.

6. Other factors notwithstanding, the preferred sequence would be to drop off the clothes, deliver the kids, pick up the broiler, buy the hamburgers and, finally, stop by for the eclairs.

7. The pasteurized yogurt has been heated after preparation and most of the helpful bacteria have been destroyed. I would recommend passing up on this type and taking the yogurt made with pasteurized milk.

8. "Bacteria put the holes in Swiss cheese." Describing to a child the nature of bacteria might be considerably harder. Blue cheese might pose a threat because the *Penicillium*

might form a penicillin-like end-product in the cheese. When consumed, the end-product might induce an allergic reaction.

9. One might be inclined to avoid such things as tuna, egg, or chicken salad and should be skeptical of ham slices or frankfurters that have not been well refrigerated. Leftover cream pies or warm salad dressings might also be hazardous.

10. Some examples to ponder: Custard egg products, ice cream, fresh mayonnaise, meat loaf mixtures, many sauces, prepared cheese products, and on and on.

11. Mathematical calculations reveal that between 10:00 A.M. and 2:30 P.M., the 100 bacilli became 51,200 bacilli. Thus the effects of a short generation time.

12. Vinegar is acidic, olive oil has chemical preservatives, the sugar in brown sugar creates an osmotic pressure, tea bags, cocoa mix, and spaghetti are dry, pancake syrup has sugar for osmotic pressure, soy sauce is very salty, and rice is dry.

13. The calves liver would probably be the better choice because it will spoil more rapidly than the steak. Liver is an organ meat, with a looser tissue consistency and richer blood supply than muscle tissue. Contamination is therefore more probable in the liver.

14. Orange.

15. The basket was overflowing with microbial cheer. Included was a jar of olives, can of sauerkraut, several cheeses, jar of mushrooms, bottles of wine and beer, containers of yogurt and sour cream, jar of pickles, a sausage, bottle of vinegar, and numerous other goodies that made me deliriously happy.

CHAPTER 25

1. This idea of connecting storm and sanitary sewers was a poor one because it resulted in a huge body of water where storm runoff polluted sanitary waste and vice versa. Separation probably would have made handling the waste considerably easier. However, you might present arguments for connecting the systems, the most obvious argument being the economic benefit.

2. The owner of the bumper sticker might have in mind such microbial contributions as the breakdown of sewage and their roles in the carbon, sulfur, and nitrogen cycles. Perhaps the microorganisms can also be thanked for their contributions to food and dairy product manufacture (Chapter 24) and industrial products (Chapter 25).

3. More carbohydrate breakdown means less residual carbohydrate and therefore, fewer calories. The product is "lite" beer.

4. The student apparently intends to cultivate large amounts of bacteria for use as a fat-digesting

cesspool treatment. (Of such ideas are fortunes made.) Most hardware stores sell bottles of "enzymes" for unclogging cesspools, but these act on the waste and then disappear. A culture of bacteria thriving in the cesspool would give a longer lasting effect. The options afforded by genetic engineering should not be overlooked.

5. Summertime pollution can be traced to many factors in addition to those cited: animals are more active in the summertime and they tend to drink from and defecate in the water; summer thunderstorms cause substantial soil runoff of organisms from animal remains; bathers introduce intestinal organisms to the water; algae invade the water inducing fish kills and bacterial growth during decomposition; and the heat, especially in a stagnant bay, leads to bacterial proliferation.

6. Despite their notorious reputations, the pathogens of disease are probably ill-equipped to compete with vast populations of other organisms in the soil environment. Possibly they do not possess the enzyme systems for soil metabolites nor the structural components to contend with environmental fluctuations in the soil. Antibiotic-producing soil organisms are also present.

7. Despite extensive reassurances, many microbiologists remain skeptical about the widespread use of *Bacillus thuringiensis*. What would happen, for example, if the organism mutated or underwent recombination to a pathogenic form? Ecologically speaking, is it possible that the insecticide is killing off insects that provide natural controls to other insects? These thoughts should provide a basis for a discussion.

8. To counter the adages, the following may be offered: 1) Soil runoff may contaminate the water despite the absence of large animals or humans, and small animals such as beavers or others may be sources. 2) Microorganisms survive in ice and snow, either of which may be contaminated with animal excrement. 3) Water may pick up contaminants during its journey out of the soil or over rocks; also it may have run into the ground farther up the hill. 4) Rapidly moving water may dilute contaminants more efficiently than still water, but it presents no assurance of being germ-free.

9. Why so much vitamin? Perhaps a defect in the enzyme system prevents the metabolism of the vitamin to the next step. Perhaps the organisms live in a symbiotic or synergistic relationship with other organisms that utilize the vitamin. As Perlman suggests in the Note to the Student: "There are not stupid microorganisms."

10. Ethyl alcohol is an antiseptic and disinfectant that kills a broad variety of microorganisms. Thus, the chances that pathogens can survive

Appendix E *(continued)*

Answers to Questions for Thought and Discussion

within beer, wine, or spirits are slim. It may be pointed out that beer and wine have a low alcohol content compared to the 70 percent ethyl alcohol discussed in Chapter 22.

11. The last few pages of the chapter delineate several areas in which genetic engineering can have substantial impact on technology in the years ahead. Questions have arisen, however, on several fronts: is the release of genetically engineered organisms to the environment safe to the ecosystem? Will genetic engineering processes be more economical than conventional processes? Can theory translate to practical value without knowing all the variables? Has the rush to computerization unrealistically fueled a rush to genetic engineering? Will commercialism impede scientific progress? The makings of a good discussion revolve about some of these concerns.

12. The grass is greener over the septic tank because the organic-rich water flowing from it provides a natural lawn fertilizer. In the wintertime, the hot water from a cesspool or septic tank warms the soil and melts the snow over it. To locate the cesspool or septic tank, one need only watch to see where the snow melts first.

13. The list of microbial products should be formidable and should include food, drink, and numerous industrial products.

14. The greater hazard probably exists in the smaller wading pool. This is where diaper-clad toddlers spread microorganisms to the water, where children are more likely to take a mouthful of water, where flatulence may eliminate microorganisms from the intestine, and where greater stagnation of the water may occur. Park officials may seek to limit disease transmission by providing constant water exchange and aeration, keeping vigilance on the chlorine content, watching for babies with soiled diapers, and posting signs advising parents not to allow children in the water if children have diarrhea.

15. The story on the inside back cover has been with me for years. Originally it involved cholera, a dervish, and the city of Bagdad. The idea is simple: an understanding of microorganisms can reduce fear of microorganisms and allow individuals to better cope with disease. In writing this book, one of my hopes was that you would come to know the microorganisms and understand the role they play in disease. Fear arising from ignorance can be a terrifying experience. I trust that you have acquired sufficient knowledge to help dispel some of the fear.

Appendix F
Answers to Review Questions

CHAPTER 1

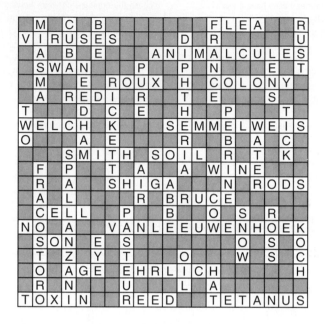

CHAPTER 2

1. IONIC 2. ORGANIC 3. HYDROGEN 4. GLUCOSE 5. FATS 6. ENZYMES
7. NUCLEOTIDES 8. PRIMARY 9. DEHYDRATION 10. ACIDITY 11. ELEMENT
12. ISOTOPES 13. CARBOXYL 14. MALTOSE 15. HYDROGEN

CHAPTER 3

1. G 2. T 3. K 4. L 5. X 6. U 7. N 8. W 9. D 10. A 11. V 12. C 13. P 14. F 15. O
16. I 17. B 18. J 19. Z 20. Q

CHAPTER 4

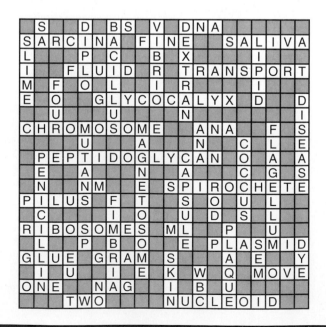

Appendix F (continued)

Answers to Review Questions

CHAPTER 5

1. metabolism; digestion; catabolism 2. protein; speed up; substrate 3. glycolysis; energy; ATP 4. fermentation; oxygen; glucose; alcohol 5. electrons; cytochromes; energy; ATP 6. pyruvic acid; carbon; carbon dioxide; NAD 7. amino; amino; deamination 8. carbon dioxide; photosynthesis; glucose 9. chemical reactions; carbohydrates; *Nirosomonas* 10. genetic; DNA; mRNA; transcription 11. ribosome; anticodon; an amino acid 12. repressor; operator; structural

CHAPTER 6

1. PLASMIDS 2. SEXDUCTION 3. MUTATION 4. SALMONELLA
5. COMPETENCE 6. FERTILITY 7. BACTERIOPHAGE 8. ENDONUCLEASE
9. PNEUMOCOCCUS 10. PILI 11. LYSOGENY 12. CHIMERA 13. LIGASE
14. INTERFERON 15. TRANSPOSON 16. GRIFFITH 17. CONJUGATION
18. MUTAGEN 19. VIRULENT 20. CHROMOSOME

CHAPTER 7

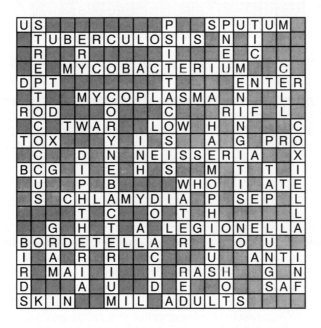

CHAPTER 8

1. ANTITOXIN 2. CHOLERA 3. UNDULATING 4. DYSENTERY 5. ROSE
6. STAPHYLOCOCCAL 7. CHICKENS 8. CARRIERS 9. NEGATIVE 10. CEREUS
11. CAMPYLOBACTER 12. ABORTION 13. PARALYSIS 14. NOSE 15. ACID
16. FLUID 17. RARE 18. VIBRIO 19. SALMONELLA 20. HEAT

CHAPTER 9

1. G 2. B 3. H 4. L 5. O 6. N 7. C 8. J 9. M 10. G 11. A 12. N 13. L 14. F 15. I
16. A 17. G 18. E 19. M 20. C 21. L 22. F 23. B 24. M 25. G

Appendix F *(continued)*

Answers to Review Questions

CHAPTER 10

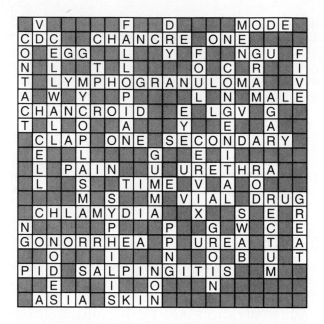

CHAPTER 11

1. CAPSID 2. HELIX 3. BACTERIOPHAGE 4. ANTIBODIES 5. NEGRI
6. ACYCLOVIR 7. INTERFERON 8. ATTENUATED 9. VIROID 10. TUMOR
11. ULTRAVIOLET 12. ONCOGENE 13. RECEPTOR 14. ICOSAHEDRON
15. GENOME 16. ENDERS 17. VIRION 18. LYSOGENY 19. AMANTADINE
20. ENVELOPE 21. STANLEY 22. FORMALDEHYDE 23. PRION
24. CAPSOMERES 25. INACTIVATED

CHAPTER 12

1. RNA; icosahedral; nose; mild 2. contact; blisters; emotional stress; acyclovir
3. in crops; teardrops; shingles; red; painful 4. measles; a skin rash; airborne
droplets 5. Reye's; testes; orchitis; liver; SSPE 6. measles; Koplik spots; mouth; red
rash 7. fifth, B19; parvovirus; DNA virus 8. airborne droplets; mumps; salivary
9. smallpox; cowpox; Jenner 10. influenza; vaccine; bacteria 11. RNA; lungs;
children; clump together; syncytia 12. papilloma viruses; sexual contact; molluscum
contagiosum 13. DNA; granules; common colds; eye; meninges 14. icosahedral;
million; thin; weeks; stress 15. infectious; transplacental passage; newborns; rubella;
herpes simplex 16. attenuated; children; long-term; measles; mumps; German measles

CHAPTER 13

1. F (RNA) 2. F (*Pneumocystis carinii*) 3. True 4. True 5. F (intestine) 6. True 7. F
(rabies) 8. True 9. F (lymphocyte) 10. True 11. True 12. F (gastroenteritis) 13. F
(eyes) 14. True 15. True 16. True 17. F (polio) 18. F (retrovirus) 19. F (Coxsackie)
20. True 21. F (liver) 22. F (inactivated viruses) 23. F (rotavirus) 24. True
25. True 26. F (infectious mononucleosis) 27. F (high) 28. F (hantaviruses)
29. F (ctyomegalovirus) 30. F (arm)

CHAPTER 14

1. I 2. Q 3. W 4. M 5. K 6. U 7. A 8. S 9. C 10. E 11. P 12. S 13. U 14. G 15. O
16. Q 17. B 18. P 19. H 20. M 21. S 22. Z 23. M 24. K 25. P

Appendix F *(continued)*

Answers to Review Questions

CHAPTER 15

```
C A T       M   P V I V A X       O N E
    O   B R U C E           T       Y
B O X   A   D   T       C D C   U R I N E
    O   L       N   G O         I
C   P L A S M O D I U M   P   W A T E R
O   L   N       A   A   N F   T       B
U   A   T I C K   R     E   H O T   C
G A S   I   C   D   P   U   M
H   M   D   T R I C H O M O N A S
    A   I   T   A       O   Y     M
L   I   U S   S   W B C       T E N
E N T A M O E B A       Y     A R
I   W     T   R   M O S Q U I T O
S   O   C N S   C   P   T   D   Z
H     M   E   O   C I L I A   S   O
M O V E     D     S         A I R
A     A   C   I     S       L   T
N O   T R Y P A N O S O M A   F E V E R
I     S   A     I           A
A   C Y S T       S A L I V A   F O U R
```

CHAPTER 16

1. a, b 2. c 3. a, c 4. c 5. a, b, c 6. b 7. a, c 8. a 9. b 10. a, b, c 11. a 12. a, b
13. c 14. b, c 15. c

CHAPTER 17

1. F (endemic) 2. True 3. True 4. F (infection) 5. F (liver) 6. True 7. F (low)
8. F (acme) 9. F (mechanical) 10. True 11. F (carrier) 12. F (antitoxins) 13. True
14. F (coagulase) 15. True 16. True 17. F (direct) 18. True 19. F (acute)
20. F (incubation)

CHAPTER 18

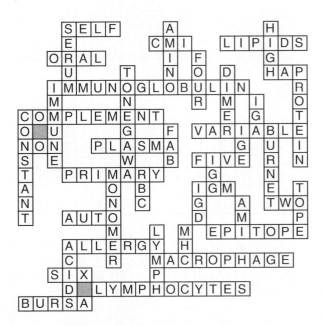

CHAPTER 19

1. innate, acquired 2. active; passive 3. measles; mumps; rubella 4. diphtheria;
pertussis, tetanus 5. natural; artificial 6. alum; mineral oil 7. hyperimmune serum;

Appendix F *(continued)*

Answers to Review Questions

antiserum 8. IgG; IgM; IgA, IgE 9. transplacental passage; breast feeding
10. labored breathing; swollen joints 11. capsule polysaccharides, pilus protein
12. hepatitis B; foot-and-mouth disease 13. genetic factors; physiological factors;
biochemical factors 14. inactivated; attenuated 15. diphtheria; tetanus
16. injection; oral consumption; nasal spray 17. gastrointestinal; respiratory
18. hepatitis B; chickenpox 19. therapy; prophylaxis 20. tetanus; diphtheria

CHAPTER 20

1. HISTAMINE 2. DEGEORGE'S 3. INDURATION 4. JOINTS 5. GRAVES
6. CYTOTOXIC 7. ATOPIC 8. ANTIGEN 9. PLASMA 10. TUBERCULOSIS
11. BASOPHILS 12. LYSIS 13. POSITIVE 14. KIDNEY 15. MHC 16. RASH
17. AUTOIMMUNE 18. CELLULAR 19. FACE 20. SMOOTH

CHAPTER 21

1. AUTOCLAVE 2. MEMBRANE 3. POWDER 4. DENATURATION 5. TYNDALL
6. DRYING 7. ULTRASONIC 8. GAMMA 9. SPORE 10. OXIDATION 11. PRESSURE
12. THIRTY 13. HOLDING 14. INSTRUMENTS 15. DIATOMS 16. ULTRAVIOLET
17. PLASTIC 18. TOXINS 19. SOLIDS 20. OSMOSIS 21. MICROWAVES 22.
BACILLUS 23. BUNSEN 24. CLEANING 25. TUBERCULOSIS

CHAPTER 22

1. F 2. B 3. J 4. I 5. C 6. D 7. A 8. K 9. N 10. E 11. A 12. C 13. J 14. E 15. I
16. C 17. J 18. I 19. B 20. E 21. K 22. F 23. K 24. F 25. C

CHAPTER 23

```
P E N I C I L L I N   R I N G S   N
H O   H           N         B E T A
    L O W   R I F A M P I N   W   M
  N   O     B       N Y     M I
  T E T R A C Y C L I N E   U S   S   N
G   O   A       N     I   T   T W O
R M   A     C     G E N   A   R
A Y   P   E Y E   R   J   T E E N S
M I C H   R P     A   E   I   P
N I   E R Y T H R O M Y C I N   T
E   N       A       T     L O C K
G     I   V   L I V E R   B S   M   I
A     C   I S O     Y     E Y   D
T B F O U R   S T R E P T O M Y C I N
I     L   U P     R     I   E   E
V T     S L O W   M O L D     S   Y
E A R S     R       T       S
  A T E N   I   C R E A M     K
  C E   O N E     I       N A I L
  S T A P H     C A N D I D A   N
```

CHAPTER 24

1. SALMONELLA 2. LACTOBACILLUS 3. CLOSTRIDIUM 4. SERRATIA
5. CHLAMYDIA 6. PROTEUS 7. STREPTOCOCCUS 8. BACILLUS
9. ACETOBACTER 10. LEUCONOSTOC 11. CLAVICEPS 12. CLOSTRIDIUM
13. PSEUDOMONAS 14. ENTEROBACTER 15. ASPERGILLUS

CHAPTER 25

1. a, c 2. b 3. b, c 4. a, b 5. a, b, c 6. a, b 7. b 8. a 9. b, c 10. a, b 11. a, b, c
12. a, c 13. c 14. b, c 15. b

Appendix G
Supplementary Readings

CHAPTER 1

Bardell, D. 1988. "The discovery of microorganisms by Robert Hooke." *ASM News* **54**:182–185.

Brock, T.D. 1961. *Milestones in Microbiology*. Prentice-Hall, Inc., Englewood Cliffs, NJ.

Cartwright, F.F. 1972. *Disease and History*. Thomas Y. Crowell Company, New York.

DeKruif, P. 1952. *Microbe Hunters*. Harcourt, Brace, Jovanovich, Inc., New York.

Hendrick, R. 1991. "Biology, history and Louis Pasteur." *American Biology Teacher* **53**:467–477.

Hesse, W. 1992. "Walther and Angeline Hesse—early contributors to bacteriology." *ASM News* **58**:425–429.

Lechevalier, H.A., and M. Solotorovsky. 1965. *Three Centuries of Microbiology*. McGraw-Hill Book Company, New York.

Root-Bernstein, R.S. 1988. "Setting the stage for discovery." *The Sciences*, May and June issues.

Zinsser, H. 1935. *Rats, Lice, and History*. Bantam Books, Inc., New York.

CHAPTER 2

Baker, J.J., and G.E. Allen. 1981. *Matter, Energy, and Life*, 4th ed. Addison-Wesley Publishing Co., Reading, MA.

Dickson, T.R. 1991. *Introduction to Chemistry*. John Wiley & Sons, New York.

Feigl, D.M. and J.W. Hill. 1991. *Foundations of Life: An Introduction to General, Organic, And Biological Chemistry*, 3rd ed. Macmillan, New York.

Malone, L.J. 1989. *Basic Concepts of Chemistry*, 3rd ed. John Wiley & Sons, New York.

Miller, G. 1987. *Chemistry: A Basic Introduction*, 4th ed. Wadsworth Publishing Company, Belmont, CA.

Multiple authors. 1985. "The molecules of life." *Scientific American* **253**(4).

Oxtoby, D.W., and N.H. Nachtrieb. 1990. *Principles of Modern Chemistry*. Saunders College Pub., Philadelphia.

Stryer, L. 1988. *Biochemistry*, 3rd ed. W.H. Freeman and Company, San Francisco, CA.

Timberlake, K. 1983. *Chemistry*, 3rd ed. Harper & Row, Publishers, New York.

CHAPTER 3

Anonymous, 1985. "Nomenclature." *ASM News* **51**:viii–x.

Barer, R. 1974. "Microscopes, microscopy, and microbiology." *Ann. Rev. Microbiol.* **28**:371–389.

Goodaa, G.W., D. Lloyd, and A.P.J. Trinici (eds.). 1980. "The eukaryotic microbial cell." In the 30th Symposium of the Society of General Microbiology, Cambridge University Press, New York.

Margulis, L. 1981. "How many kingdoms? Current views of biological classification." *American Biology Teacher* **43**:482–489.

Schwartz, R.M., and M.O. Dayhoff. 1978. "Origins of prokaryotes, eukaryotes, mitochondria, and chloroplasts." *Science* **199**:395.

Weaver, K.F. 1977. "Electronic voyage through an invisible world." *National Geographic* **151**:274–290.

Whittaker, R.H. 1969. "New concepts of kingdoms of organisms." *Science* **163**:150–160.

Woese, C.R. 1987. "Bacterial evolution." *Microbiol. Rev.* **51**:221–227.

Appendix G *(continued)*

Supplementary Readings

CHAPTER 4

Anonymous, 1992. "One pinch of soil." *Discover*, September 1992.

Belas, R. 1992. "The swarming phenomenon on *Proteus mirabilis.*" *ASM News* **58**:15–23.

Berg, H.C. 1975. "How bacteria swim." *Scientific American* **235**:36–45.

Costerton, J.W. G.G. Geesey, and K.J. Cheng. 1978. "How bacteria stick." *Scientific American* **245**(6):58–67.

Evans, R.H. 1983. "Archaebacteria: a new primary kingdom for our classrooms." *Amer. Biol. Teach.* **45**(3):139–143.

Hancock, R.E.W. 1991. "Bacterial outer membranes: evolving concepts." *ASM News* **57**:175–186.

Margulis, L., and K. Schwartz. 1988, *Five Kingdoms.* W.H. Freeman and Company. New York.

Razin, S. 1978. "The mycoplasmas." *Microbiol. Revs.* **42**:414–462.

Shapiro, J.A. 1991. "Multicellular behavior of bacteria." *ASM News* **57**:247–252.

Silverman, M., and M.I. Simon. 1977. "Bacterial flagella." *Ann. Rev. Microbiol.* **36**:407–428.

CHAPTER 5

Armstrong, F.B. 1989. *Biochemistry.* Oxford Univ. Press, New York.

Avers, C.J. 1986. *Molecular Cell Biology.* Benjamin Cummings Publishing Company, Redwood City, CA.

Ayala, F.J., and J.A. Kiger, Jr. 1984. *Modern Genetics*, 2nd ed. Benjamin Cummings Publishing Company, Redwood City, CA.

Freedman, D.H. 1991. "Bytes of life." *Discover*, November.

Holum, J.R. 1989. *Elements of General, Organic, and Biological Chemistry*, 3rd ed. John Wiley & Sons, New York.

Jacob, F., and J. Monod. 1961. "Genetic regulatory mechanisms in the synthesis of proteins." *J. Mol. Biol.* **3**:318–331.

Lechtman, M.D., B.L. Roohk, and R.J. Egan. 1978. *The Games Cells Play: Basic Concepts of Cellular Metabolism.* Benjamin Cummings Publishing Company, Redwood City, CA.

Lehninger, A.L. 1982. *Principles of Biochemistry.* Worth Publishers Inc., New York.

Vogel, S. 1988. "Breaking the code." *Discover*, August 1988.

Watson, J.D. 1987. *Molecular Biology of the Gene*, 4th ed. Benjamin Cummings Publishing Company, Redwood City, CA.

CHAPTER 6

Alcamo, I.E. 1994. *DNA Technology: The Awesome Skill.* William C. Brown, Dubuque, IA.

Birge, E.A. 1981. *Bacterial and Bacteriophage Genetics.* Springer-Verlag, New York.

Brock, W.E. 1987. "Bozeman chainsaw massacre." *Discover*, November 1987.

Cohen, S.N., and J.A. Shapiro, 1980. "Transposable genetic elements." *Scientific American* **242**:40–49.

Drlica, K. 1992. *Understanding DNA and Gene Cloning*, 2nd ed. John Wiley & Sons, New York.

Multiple authors. 1981. *Genetics: Readings from Scientific American.* W.H. Freeman and Company, San Francisco.

Multiple authors. 1984. "Special focus: genetic engineering." *American Biology Teacher* **46**: October and November 1984.

Novick, R.P. 1980. "Plasmids." *Scientific American* **243**:103–127.

Appendix G *(continued)*

Supplementary Readings

Sonea, S. 1988. "The global organism." *The Sciences*, July/August 1988.

Tangley, L. 1985. "New biology enters a new era." *Bioscience* **35**:270–275.

Watson, J.W., et al. 1992. *Recombinant DNA*, 2nd ed. W.H. Freeman and Company, New York.

Weaver, R.F. 1984. "Beyond supermouse: changing life's genetic blueprint." *National Geographic* **166**:818–847.

CHAPTER 7

Anonymous. 1991. "Resurgence in virulent streptococcal disease." *ASM News* **57**:619–620.

Bloom, B.R., and C. Murray. 1992. "Tuberculosis: commentary on a reemergent killer." *Science* **257**:1055–1066.

Caldwell, M. 1992. "Resurrection of a killer." *Discover*, December 1992.

Dajer, T. 1990. "Saint Vitus's dance." *Discover*, March 1990.

Ezzell, C. 1993. "Captain of the men of death." *Science News* **143**:90–94.

Frasch, C.E., and L.F. Mocca. 1982. "Strains of *Neisseria meningitidis* isolated from patients and their close contacts." *Infect. Imm.* **37**:155–159.

Fraser, D.W., and J.E. McDade. 1979. "Legionellosis." *Scientific American* **241**:82–99.

Hall, S. 1989. "The case of the poker players' plague." *Hippocrates*, September 1989.

Klass, P. 1993. "Spinal trap." *Discover*, January 1993.

Pitchenik, A.E., D. Fertel, and A.B. Bloch. 1988. "Mycobacterial disease: epidemiology, diagnosis, treatment, and prevention." *Clin. Chest Med.* **9**:425–441.

Veasy, L.G., et al. 1987. "Resurgence of acute rheumatic fever in the intermountain area of the United States." *New Eng. J. Med.* **316**:421–427.

CHAPTER 8

Anonymous. 1992. "Peru's cholera epidemic: health system's baptism by fire." *ASM News* **58**:178–181.

Burros, M. 1982. "Trying to solve the botulism mystery." *New York Times*, April 28, 1982.

Diamond, J. 1992. "The return of cholera." *Discover*, February 1992.

Franklin, D. 1992. "The case of the paralyzed travelers." *In Health*, December 1992.

Holmgren, J. 1981. "Actions of cholera toxin and the prevention and treatment of cholera." *Nature* **292**:413–417.

MacDonald, K.L., M.L. Cohen, and P.A. Blake. 1986. "The changing epidemiology of adult botulism in the United States." *Am. J. Epidemiol.* **124**:794–799.

St. Louis, M.E., et al. 1988. "The emergence of Grade A eggs as a major source of *Salmonella enteritidis*." *JAMA* **259**:2103–2107.

Shandra, W., C. Tackert, and P. Blake. 1983. "*Clostridium perfringens* in the United States." *J. Inf. Dis.* **147**(4):771–775.

Waters, T. 1992. "The fine art of making poison." *Discover*, August 1992.

CHAPTER 9

Anonymous. 1993. "The plague corners cats." *Health*, January 1993.

Barbour, A.G., and D. Fish. 1993. "The biological and social significance of Lyme disease." *Science* **260**:1610–1616.

Berger, B.W., O.J. Clemmensen, and A.B. Ackerman. 1983. "Lyme disease is a spirochetosis. A review of the disease and evidence for its cause." *Am. J. Dermatopathol.* **5**:111–124.

Dowell, V.R. 1984. "Botulism and tetanus: selected epidemiologic and microbiologic aspects." *Rev. Inf. Dis.* **6**:202–207.

Appendix G   *(continued)*
Supplementary Readings

Evans, M.E. 1981. "Tularemia and the tomcat." *JAMA* **246**:1343–1346.

Hendricks, M. 1991. "Biological weapons treaty review." *ASM News* **57**:358–363.

Klass, P. 1993. "An independent diagnosis." *Discover*, February 1993.

Leavesky, J.H. 1979. "The plague and its place in history." *Austral. Fam. Phys.* **8**:1005–1009.

Lovett, J., et al. 1987. "*Listeria monocytogenes* in raw milk: detection, incidence, and pathogenicity." *J. Food Protection.* **50**:188–192.

Marx, R.S., C.E. McCall, and J.S. Abramson. 1982. "Rocky Mountain spotted fever." *Amer. J. Dis. Child* **136**:16–18.

Patamuscucon, P., et al. 1982. "Melliodosis." *J. Pediatrics* **100**(2):175–182.

Rork, P.E., et al. 1982. "Gangrene of a fingertip from a mitten thread." *JAMA* **248**(8):924–925.

Sanford, J.M. 1984. "Leptospirosis: time for a booster." *N. Eng. J. Med.* **310**(8):524–525.

Schmitz, A. 1989. "After the bite." *Hippocrates*, May 1989.

Seligman, J. 1989. "Tiny tick, big worry." *Newsweek*, May 22, 1989, 66–72.

Steere, A.C., M.S. Grodzicki, and A.N. Kornblatt. 1983. "The spirochetal etiology of Lyme disease." *N. Eng. J. Med.* **308**(13):733–740.

CHAPTER 10

Anonymous, 1992. "Scratching the surface of the cat scratch disease controversy." *ASM News* **58**:655–657.

Arndt, K.A. 1983. "Battling sexually transmitted diseases." *Patient Care* **17**:63–68.

Cherry, L. 1985. "A hospital is no place for a sick person to be." *Discover*, October 1985.

Cowley, G. 1991. "Sleeping with the enemy." *Newsweek*, December 9, 1991.

Doberneck, R.C. 1982. "Pelvic actinomycosis associated with the use of IUDs." *Am. Surgeon* **48**:25–27.

Dolnick, E. 1988. "Good news for people with teeth." *Hippocrates*, July/August 1988.

Franklin, D. 1992. "The case of the lucky blues brother." *Health*, November 1992.

Friedland, S. 1982. "Reproductive ills of women are on the increase." *New York Times*, November 14, 1982.

Gibbons, W. "Clueing in on chlamydia." *Science News*, April 20, 1991.

Hadfield, T.L., et al. 1985. "Stalking the cause of cat-scratch disease." *Diag. Med.* **36**:23–27.

Lord, L.J. 1986. "Sex, with care." *U.S. News & World Report*, June 2, 1986.

McDonald, M.I., et al. 1982. "*Ureaplasma urealyticum* in patients with acute symptoms of urinary tract infections." *J. Urol.* **128**(3):517–523.

Perez-Stable, E.J. 1983. "Urethritis in men." *West J. Med.* **138**(3):426–429.

Radetsky, P. 1985. "The rise and (maybe not the) fall of toxic shock syndrome." *Science 85*, January/February 1985.

Sharp, D. 1992. "If you want to keep your teeth." *In Health*, December 1992.

Stine, G.S. 1992. *The Biology of Sexually Transmitted Diseases*. William C. Brown, Dubuque, IA.

CHAPTER 11

Butler, P., and A. Klug. 1978. "The assembly of a virus." *Scientific American* **129**:62–68.

Diener, T.O. 1983. "The viroid—a subviral pathogen." *American Scientist* **71**:481–488.

Fink, L. 1984. "Unraveling the molecular biology of cancer." *Bioscience* **34**(2):75–77.

Fraenkel-Conrat, H., et al. 1988. *Virology*, 2nd ed. Prentice-Hall, Englewood Cliffs, NJ.

Appendix G *(continued)*
Supplementary Readings

Freedman, D.H. 1992. "High seas lowlife." *Discover*, February 1992.

Henshaw, N.G. 1988. "Identification of viruses by methods other than electron microscopy." *ASM News* **54**:482–485.

Karpas, A. 1982. "Viruses and leukemia." *American Scientist* **70**:277–280.

Laughlin, C.A., et al. 1991. "Resistance to antiviral drugs." *ASM News* **57**: 514–518.

Levine, A. 1992. *The Viruses*. W.H. Freeman and Company, New York.

Luftig, R.B., et al. 1990. "Update on viral pathogenesis." *ASM News* **56**: 366–369.

Luria, S.E. 1984. *A Slot Machine, A Broken Test Tube*. Harper & Row, New York.

Miller, S.E. 1988. "Diagnostic virology by electron microscopy." *ASM News* **54**:475–480.

Prusiner, S. 1984. "Prions." *Scientific American* **251**:50–59.

Swierkosz, E.M. 1992. "Antiviral susceptibility testing: coming of age." *ASM News* **58**:83–88.

Wallis, C. 1986. "Viruses." *Time*, November 3, 1986.

CHAPTER 12

Adler, M.W. 1984. "Genital warts and molluscum contagiosum." *Br. Med. J.* **288**:213–215.

Behbehani, A.M. 1991. "The smallpox story: historical perspective." *ASM News* **57**:571–581.

Caldwell, M. 1992. "Vigil for a doomed virus." *Discover*, March 1992.

Cochi, S.L., et al. 1988. "Perspectives on the relative resurgence of mumps in the United States." *Am. J. Dis. Child.* **142**:499–505.

Fuller, L. 1983. "Unusual aspects of Kawasaki syndrome." *Am. Fam. Phys.* **28**:219–224.

Goodfield, J. 1985. "The last days of smallpox." *Science 85*, October 1985.

Gregg, C. T. 1983. *A Virus of Love and Other Tales of Medical Detection*. Charles Scribner's Sons, New York.

Gwaltney, J.M., P.B. Moskalski, and J.O. Hendley. 1978. "Hand-to-hand transmission of rhinovirus colds." *Ann. Intern. Med.* **88**:463–467.

Heller, L. 1992. "Hot stuff for a stuffy nose." *Health*, February 1992.

Henig, R.M. 1992. "Flu pandemic: a once and future menace." *New York Times Magazine*, November 29, 1992.

Markowitz, L.E., et al. 1989. "Patterns of transmission in measles outbreaks in the United States 1985–1986." *N. Eng. J. Med.* **320**:75–81.

Preblud, S., et al. 1984. "Varicella: Clinical manifestations, epidemiology, and health impact on children." *Ped. Inf. Dis.* **3**(6): 505–509.

Raeburn, P. 1985. "The Houdini virus (RSV)." *Science 85*, December 1985.

CHAPTER 13

Alcamo, I.E. 1993. *AIDS: The Biological Basis*. William C. Brown, Dubuque, IA.

Alter, M.J. 1989. "Importance of heterosexual activity in the transmission of hepatitis B and non-A, non-B hepatitis." *JAMA* **262**:1201–1205.

Boly, W. 1987. "Raggedy Ann town." *Hippocrates*, July/August 1987.

Caldwell, M. 1991. "Mad cows and wild proteins." *Discover*, April 1991.

Chasan, D.J. 1986. "The polio paradox." *Science 86*, April 1986.

Cowley, G. 1993. "A deadly desert illness." *Newsweek*, June 14, 1993.

Cowley, G. 1990. "Chronic fatigue syndrome: a modern medical mystery." *Newsweek*, November 12, 1990.

Dolin, R., J.J. Treanor, and H.P. Madore. 1987. "Novel agents of viral gastroenteritis in humans." *J. Infect. Dis.* **155**:365–376.

Appendix G *(continued)*
Supplementary Readings

Eron, C. 1981. *The Virus That Ate Cannibals: Six Great Medical Detective Stories.* Macmillan Publishing Company, New York.

Fan, H., R. Conner, and L. Villarreal. 1992. *The Biology of AIDS*, 2nd ed. Jones & Bartlett, Boston, MA.

Gajdusek, D.C. 1977. "Unconventional viruses and the origin and disappearance of kuru." *Science* **197**:943–945.

Langone, J. 1990. "Emerging viruses." *Discover*, December 1990.

Morse, S. 1990. "Stirring up trouble." *The Sciences*, September 1990.

Moser, P.W. 1985. "Rocky times for Rocky." *Discover*, May 1985.

Preston, R. 1992. "Crisis in the hot zone." *New York Times Magazine*, October 26, 1992.

Radetsky, P. 1989. "The ultimate parasite." *Discover*, August 1989.

Stein, R. 1993. "The ABCs of hepatitis." *American Health*, June 1993.

Weiss, R. 1989. "The viral advantage." *Science News.* **136**:200–206.

CHAPTER 14

Angier, N. 1986. "A stupid cell with all the answers." *Discover*, November 1986.

Anonymous. 1993. "Blowing away valley fever." *Health*, May 1993.

Caporael, L.R. 1976. "Ergotism: The satan loose in Salem?" *Science* **192**:21–26.

Large, E.C. 1962. *The Advance of the Fungi.* Dover Publications, New York.

Miller, J.A. 1987. "Fighting fungi with fungi." *Bioscience* **37**(4):248–250.

Moore-Landecker, E. 1982. *Fundamentals of the Fungi*, 2nd ed. Prentice-Hall, Englewood Cliffs, NJ.

Rippon, J.W. 1982. *Medical Mycology.* W.B. Saunders Company, Philadelphia.

Sacks, J.J., L. Ajello, and L.K. Crocket. 1986. "An outbreak and review of cave-associated histoplasmosis capsulati." *J. Med. Vet. Mycol.* **24**:313–325.

Stone, J. 1991. "Wild—very wild—kingdom." *Discover*, May 1991.

Stone, R. 1993. "Surprise! A fungus factory for taxol?" *Science*, **260**:154–156.

Walsh, T.J. 1988. "Recent advances in the treatment of systemic fungal infections." *ASM News* **54**:240–243.

West, S. 1989. "The case of the Arbor Day outbreak." *Hippocrates*, May 1989.

CHAPTER 15

Altman, L.K. 1993. "Outbreak of disease in Milwaukee undercuts confidence in water." *New York Times*, April 20, 1993.

Beck, J.W. and J.E. Davies. 1981. *Medical Parasitology*, 3rd ed. C.V. Mosby Company, St. Louis, MO.

Beck, M. 1993. "The endless plague." *Newsweek*, January 11, 1993.

Current, W. 1988. "The biology of *Cryptosporidium*." *ASM News* **54**:605–611.

Dajer, T. 1992. "Blackwater fever." *Discover*, May 1992.

Donelson, J.E., and A.C. Rice-Ficht. 1985. "Molecular biology of trypanosome antigenic variation." *Microbiol. Rev.* **49**:107–125.

Moser, P.W. 1991. "Danger in diaperland." *Discover*, September 1991.

Nash, M. 1993. "The waterworks flu." *Time*, April 19, 1993.

Tangley, L. 1987. "Malaria: fighting the African scourge." *Bioscience* **37**(2):94–98.

CHAPTER 16

Barclay, W. 1981. "Diphyllobothriasis." *JAMA* **246**(21):2483.

Beaver, P.C., et al. 1984. *Clinical Parasitology*, 9th ed. Lea and Febiger, Philadelphia.

Brown, H.W., and F.A. Neva. 1983. *Parasitology.* Appleton-Century-Crofts, Norwalk, CT.

Appendix G *(continued)*
Supplementary Readings

Fishman, J.A., and R.L. Perrone. 1984. "Colonic obstruction and perforation due to *T. trichiura*." *Am. J. Med.* **77**(1):154–156.

Gilman, R. H. 1982. "Hookworm disease: host-pathogen biology." *Rev. Infect. Dis.* **4**(4):824–829.

Iarotski. L.S., and A. Davis. 1982. "The schistosomiasis problem in the world." *Bull. WHO* **59**:115–121.

Kaimal, K., and B. Beyt. 1982. "Cardiac dysfunction in trichinosis." *N. Eng. J. Med.* **307**:374–375.

Loeb, L.J. 1982. "Gefilte fish and diphyllobothriasis." *JAMA* **247**(11):1566.

Most, H. 1984. "Treatment of parasitic infections of travelers and immigrants." *N. Eng. J. Med.* **310**(5):298–306.

Schieven, B.C. 1991. "*Echinococcus granulosus* hydatid disease." *ASM News* **57**:407–411.

Weller, P.F. 1983. "Parasitic infections of concern to primary care physicians." *Med. Times* **111**(8):32–36.

CHAPTER 17

Alper, J. 1993. "Ulcers as an infectious disease." *Science* **260**:159–160.

Boroson, W. 1984. "Learning from Typhoid Mary." *Science Digest*, March 1984.

Diamond, J. 1992. "The arrow of disease." *Discover*, October 1992.

Hall, S.S. 1987. "The invader." *Hippocrates*, September/October 1987.

Hedgpeth, J.W. 1993. "Foreign invaders." *Science* **261**:34–36.

Krause, R.M. 1992. "Origin of plagues: old and new." *Science* **257**:1073–1077.

Lowenstein, J.M. 1992. "Can we wipe out disease?" *Discover*, November 1992.

McNeill, W.H. 1976. *Plagues and Peoples*. Anchor Press, Doubleday, Garden City, NY.

Marples, M.J. 1969. "Life on the human skin." *Scientific American* **220**(1)108–115.

Miller, V.L. 1992. "*Yersinia* invasion genes and their products." *ASM News* **58**:26–31.

Rosebury, T. 1969. *Life on Man*. Berkeley Publishing Corporation, New York.

Stuller, J. 1990. "A question of bathing." *In Health*, March 1990.

Thomas, L. 1983. *The Youngest Science*. Bantam Books, New York.

CHAPTER 18

Bellanti, J.A. 1985. *Immunology III*. W.B. Saunders Company, Philadelphia.

Boehmer, H., and P. Kisielow. 1991. "How the immune system learns about self." *Scientific American*, October 1991.

Caldwell, M. 1991. "The immune challenge." *Discover*, August 1991.

Edelman, G.M. 1970. "The structure and function of antibodies." *Scientific American* **223**(2): 34–42.

Fox, J.L. 1990. "Bounty of immunity." *ASM News* **56**:87–89.

Jaret, P. 1992. "Mind over malady." *Health*, November 1992.

Jaret, P. 1986. "The wars within." *National Geographic*, June 1986.

Jaroff, L. 1988. "Stop that germ!" *Time*, May 23, 1988.

Kimball, J.W. 1983. *Introduction to Immunology*. Macmillan Publishing Company, New York.

Milstein, C. 1980. "Monoclonal antibodies." *Scientific American* **243**(4):66–74.

Schmeck, H.M., Jr. 1984. "Immune system: great mystery is solved after long quest." *New York Times*, June 19, 1984.

Silverstein, S.C. 1981. "The militant macrophage." *The Sciences* **213**:18–21.

Stites, D.P., and A. I. Terr. 1991. *Basic and Clinical Immunology*. Appleton & Lange, Norwalk, CA.

Todd, J. 1990. "A most intimate foe." *The Sciences*. March 1990.

Zwilling, B.S. 1992. "Stress affects disease outcomes." *ASM News* **58**:23–29.

Appendix G *(continued)*
Supplementary Readings

CHAPTER 19

Alcamo, I.E. 1993. "A vaccine for AIDS." *American Biology Teacher* **55**:198–202.
Cohen, J. 1993. "Naked DNA points way to vaccines." *Science* **259**:1691–1693.
Engleberg, N.C. 1991. "Nucleic acid probe tests for clinical diagnosis—where do we stand?" *ASM News* **57**:183–186.
Fox, J.L. 1988. "In search of a better typhoid vaccine." *ASM News* **54**:551–554.
Hanna, K.E. 1990. "Rubella and pertussis vaccines: convincingly safe?" *ASM News* **56**:470–476.
Monmaney, T. 1987. "Vaccines for adult diseases." *Newsweek*, October 12, 1987.
Norton, C. 1989. "Not just for kids." *Hippocrates*, September 1989.
Overbye, D. 1982. "Rosalyn Yalow: lady laureate of the Bronx." *Discover*, June 1982.
Roitt, I.M., J. Brostoff, and D.K. Male. 1985. *Immunology*. C.V. Mosby Co., St. Louis, MO.
Rytel, M.W. 1980. "Counterimmunoelectrophoresis: a diagnostic adjunct in clinical microbiology." *Lab. Med.* **11**:655–659.
Weiss, R. 1988. "Change the U.S. polio vaccine or leave it alone?" *ASM News* **54**:560–562.

CHAPTER 20

Clay-Poole, S.T., and I.L. Slesnick, 1983. "The beauty and biology of pollen." *Am. Biol. Teach.* **45**(7):366–370.
Cowley, G. 1992. "The quest for a cancer vaccine." *Newsweek*, October 19, 1993.
Dausset, J. 1981. "The major histocompatibility complex in man." *Science* **213**:1469–1473.
Franklin, D. 1989. "A nose for allergies." *Hippocrates*, March/April 1989.
Hall, S.S. 1993. "Allergic to the 20th century." *Discover*, May 1993.
Jaret, P. 1988. "The big itch." *Hippocrates*, July/August 1988.
Jaroff, L. 1992. "Allergies: nothing to sneeze at." *Time*, June 22, 1992.
Katzenstein, L. 1993. "Allergies: nothing to sneeze at." *American Health*, May 1993.
Kennel, S.J., et al. 1984. "Monoclonal antibodies in cancer detection and therapy." *Bioscience* **34**(3):150–154.
Newman, C. 1984. "Pollen: breath of life and sneezes." *National Geographic*, October 1984.
Rosenthal, E. 1993. "Burning down the house." *Discover*, May 1993.
Todd, J. 1990. "A most intimate foe." *The Sciences*, March/April 1990.

CHAPTER 21

Brandt, S.L., and P. Benner, 1980. "Infection control in hospitals. What are the challenges?" *Am. J. Nurs.* **80**:432–437.
Castle, M. 1980. *Hospital Infection Control*. John Wiley & Sons, New York.
Favero, M.S. 1980. "Sterilization, disinfection, and antisepsis in the hospital." In Lennette, F.H. et al., eds. *Manual of Clinical Microbiology*. American Society for Microbiology, Washington, D.C.
Roman, M. 1989. "The little waves that could." *Discover*, November 1989.
Russell, A.D., W.B. Hugo, and G.A. J. Ayliffe. 1982. *Principles and Practice of Disinfection, Preservation, and Sterilisation*. Blackwell Scientific Publications, Oxford, England.

CHAPTER 22

Ainamo, J. 1977. "Control of plaque by chemical agents." *J. Periodontol.* **4**:23–27.
Apostolov, K. 1980. "The effects of iodine on the biological activities of myxoviruses." *J. Hyg.* **84**:381–388.

Appendix G *(continued)*

Supplementary Readings

Block. S.S. 1983. *Disinfection, Sterilization, and Preservation*, 3rd ed. Lea and Febiger, Philadelphia.

Bryan, R.M., and L.A. Bland. 1981. "Occupational exposure to ethylene oxide: its effect and control." *J. Environ. Health* **43**:254–259.

Gorman, S.P., E.M. Scott, and A.D. Russell, 1980. "A review. Antimicrobial activity, uses, and mechanism of action of glutaraldehyde." *J. Appl. Bacteriol.* **48**:161–190.

Mallison, G.F. 1980. "Decontamination, disinfection, and sterilization." *Nurs. Clin. N. A.* **15**:757–761.

Nadolny, M.D. 1980. "Infection control in hospitals. What does the infection control nurse do?" *Am. J. Nurs.* **80**:430–440.

CHAPTER 23

Abraham, E.P. 1981. "The beta-lactam antibiotics." *Scientific American* **244**:76–85.

Cohen, M.L. 1992. "Epidemiology of drug resistance: implications for a post-antimicrobial era." *Science* **257**:1050–1055.

Davies, J. 1986. "Life among the aminoglycosides." *ASM News* **52**(12):620–624.

Garrod, L.P., H.P. Lambert, and F. O'Grady. 1981. *Antibiotics and Chemotherapy*. Churchill Livingston, New York.

Hudson, R.P. 1983. *Disease and Its Control*. Greenwood Press, Westport, CT.

Levy, S.B. 1988. "Tetracycline resistance determinants are widespread." *ASM News* **54**(8):418–421.

Mizuguchi, Y., et al. 1985. "Morphological changes induced by beta-lactam antibiotics in *Mycobacterium avium-intracellulare* complex." *Antimicro. Agents Chemoth.* **27**(4):541–547.

Neu, H.C. 1992. "The crisis in antibiotic resistance." *Science* **257**:1064–1072.

Sanders, C.C. 1991. "A problem with antibiotic susceptibility tests." *ASM News* **57**: 187–190.

Swierkosz, E.M. 1992. "Antiviral susceptibility testing: coming of age." *ASM News* **58**:83–87

Tenover, F.C. 1992. "Bauer and Kirby meet Watson and Crick." *ASM News* **58**:669–673.

Van der Meer, A. 1991. "Are you overdosing on antibiotics?" *Redbook*, December 1991.

CHAPTER 24

Ayres, J.C., J.O. Mundt, and W.E. Sandine. 1980. *Microbiology of Foods*. W.H. Freeman and Company, San Francisco.

Bibel, D.J. 1988. "Elie Metchnikoff's bacillus of long life." *ASM News* **54**(12):660–665.

Crowley, G., and J. McCormick. 1993. "How safe is our food?" *Newsweek*, May 24, 1993.

Dixon, B. 1979. *Magnificent Microbes*. Atheneum Press, New York.

Frazier, W.C., and D.C. Westhoff. 1978. *Food Microbiology*, 3rd ed. McGraw–Hill Book Company, New York.

Heller, L. 1991. "Yogurt: bring it back alive." *In Health*, July 1991.

Imperato, P., and G. Mitchell. 1985. "Bombarding bananas, zapping zucchini." *The Sciences*, January/February 1985.

Jaroff, L. 1992. "Is your fish really foul?" *Time*, June 29, 1992.

Jay, J.M. 1978. *Modern Food Microbiology*, 3rd ed. D. Van Nostrand Company, New York.

Katzenstein, L. 1992. "Food irradiation: the story behind the scare." *American Health*, December 1992.

Appendix G *(continued)*

Supplementary Readings

Long, P. 1987. "Guess what's coming to dinner." *Hippocrates*, November/December 1987.

Marth, E.H., ed. 1978. *Standard Methods for the Examination of Dairy Products*, 14th ed. American Public Health Association, Washington, D.C.

Mountney, G.J., and W. Gould, 1988. *Practical Food Microbiology and Technology*, 3rd ed. Van Nostrand Reinhold, New York.

Potter, M.E., et al. 1983. "Human *Campylobacter* infection associated with certified raw milk." *Am. J. Epidemiol.* **117**:473–483.

Rossmoore, H.W. 1976. *The Microbes, Our Unseen Friends*. Wayne State University Press, Detroit.

Sheagren, J.N. 1984. "*Staphylococcus aureus*—the persistent pathogen." *N. Eng. J. Med.* **310**(21):1368–1373.

Stone, J. 1990. "The dairy case." *Discover*, June 1990.

CHAPTER 25

Alcamo, I.E. 1994. *DNA Technology: The Awesome Skill*. William C. Brown, Dubuque, IA.

Armstrong, J.L., G.F. Rohrmann, and G.S. Beaudreau. 1985. "Delta endotoxin of *Bacillus thuringiensis* subsp. *israelensis*." *J. Bact.* **161**(1):39–46.

Blouston, G. 1985. "The biochemistry of Bacchus." *Science 85*, October 1985.

Ciferri, O. 1983. "*Spirulina*, the edible microorganism." *Microbiol. Rev.* **47**(4):551–578.

Culotta, E. 1992. "Red menace in the world's oceans." *Science* **257**:1476–1479.

Edmond, J.M., and K.V. Damm. 1983. "Hot springs on the ocean floor." *Scientific American* **284**:78–93.

Franklin, D. 1983. "Biotechnologists build a better bug." *Bioscience* **33**(11):678–680.

Greenberg, A. (ed.). 1985. *Standard Methods for the Examination of Water and Wastewater*, 16th ed. American Public Health Association, Washington, D.C.

Gustafson, T.L., et al. 1983. "*Pseudomonas folliculitis*: an outbreak and review." *Rev. Infect. Dis.* **5**:1–8.

Knight, P. 1991. "Baculovirus vectors for making proteins in insect cells." *ASM News* **57**:567–571.

McGaughey, G.I., and M.E. Whalon. 1992. "Managing insect resistance to *Bacillus thuringiensis* toxins." *Science* **258**:1451–1454.

Monmaney, T., et al. 1985. "Microbes for hire." *Science 85*, July/August 1985.

Multiple authors. 1981. "Industrial microbiology." *Scientific American* **245**(3) (entire issue).

Odom, J. 1990. "Industrial and environmental concerns with sulfate-reducing bacteria." *ASM News* **56**:473–476.

Oliwenstein, L. 1991. "Drugs by design." *Discover*, November 1991.

Payne, W.J. 1983. "Bacterial denitrification: asset or defect?" *Bioscience* **33**(5):319–323.

Stone, J. 1991. "Big brewhaha of 1800 B.C." *Discover*, January 1991.

Weaver, J.M. 1984. "Beyond supermouse: changing life's genetic blueprint." *National Geographic* **166**:818–847.

Weniger, B.G. et al. 1983. "An outbreak of waterborne giardiasis associated with heavy water runoff due to warm weather and volcanic ashfall." *Am. J. Pub. Health* **73**:868–872.

GLOSSARY

This glossary contains concise definitions of approximately 1000 microbiological terms, together with pronunciations where appropriate. The numbers in parentheses indicate the chapters in which the terms are discussed.

A

abscess A circumscribed pus-filled lesion characteristic of staphylococcal skin disease. (10, 18)

abyssal zone The environment at the bottom of oceanic trenches. (25)

acid-fast technique A process in which certain bacteria resist decolorization with acid alcohol after staining with a primary dye. (7)

acidophilus (as"i-dof'i-lus) **milk** Milk in which *Lactobacillus acidophilus* has been cultivated, or milk to which the bacterium has been added. (25)

acquired immune deficiency syndrome (AIDS) A serious viral disease caused by human immunodeficiency virus (HIV) in which the T-lymphocytes are destroyed and opportunistic infections occur in the patient. (13)

Actinomyces (ak"ti-no-mi'sēz) *israelii* A Gram-positive funguslike bacterial rod that causes actinomycosis. (10)

actinomycetes (ak"tĭ-no-mi-se'tēz) A group of soil microorganisms that exhibit funguslike properties when cultivated in the laboratory (23, 25)

actinomycosis An endogenous bacterial disease caused by *Actinomyces israelii* and accompanied by draining sinuses of the face and other organs. (10)

acute disease A disease that develops rapidly, shows substantial symptoms, and then comes to a climax. (17)

acyclovir (a-si'klo-vir) **(Zovirax)** A drug used as a topical ointment for herpes simplex and injected for herpes encephalitis. (11, 12)

adenosine triphosphate (ATP) A molecule that stores energy for use in chemical reactions; 1 mole of ATP liberates 7300 calories of energy when digested to adenosine diphosphate (ADP) and phosphate ions. (5)

adenovirus An icosahedral DNA virus involved in respiratory infections, viral meningitis, and viral conjunctivitis. (12)

adjuvant (ad'ju-vant) A substance attached to a vaccine component that increases the efficiency of a vaccine. (19)

Aeromonas hydrophila (a"er-o-mo'nas hi-drof'i-lah) A Gram-negative bacterial rod transmitted by food and water and involved in disease of the intestine. (8)

aflatoxins (af"lah-tok'sinz) Toxins produced by *Aspergillus flavus* that may pass from contaminated grain to humans in grains, milk, or meat products and may induce tumors. (14, 24)

agar (ahg'ar) A derivative of marine seaweed used as a solidifying agent in many microbiological media. (1, 4)

agglutination (ah-gloo"tĭ-na'shun) A type of antigen-antibody reaction that results in visible clumps of organisms or other material. (18, 19)

agranulocytosis (a-gran"u-lo-si-to'sis) The destruction of neutrophils (granulocytes) resulting from the reaction of antibodies with antigens on the neutrophil surface; a form of type II hypersensitivity. (20)

AIDS *See* acquired immune deficiency syndrome.

algae Any plantlike organisms that practice photosynthesis and differ structurally from mosses, ferns, and seed plants. (3)

alginate (al'gin-ate) A carbohydrate thickening agent used in ice cream, soups, and other foods; industrially produced by microorganisms. (25)

allergen An antigenic substance that stimulates an allergic reaction in the body. (20)

alloantigens Antigens that exist in certain but not all members of a given species; examples are the A, B, and Rh factors in humans. (18)

allograft A tissue graft between two members of the same species, such as between two humans. (20)

alpha-hemolytic streptococci Streptococci that partially destroy red blood cells; when cultivated in blood agar, an olive green color forms around colonies of these streptococci. (7)

amantadine (ah-man'tah-dēn) A synthetic drug thought to block the penetration of influenza viruses to host cells. (11, 12)

amebiasis (am"e-bi'ah-sis) A protozoal disease of the intestine caused by *Entamoeba histolytica*, transmitted by food and water, and accompanied by intestinal ulcers and involvement of multiple internal organs. (15)

Ames test A diagnostic procedure used to detect cancer agents by their ability to cause mutations in *Salmonella* cells. (6)

aminoglycosides (am"ĭ-no-gli'ko-sīdz) A group of antibiotics that contain amino groups bonded to carbohydrate groups; examples are gentamicin, streptomycin, and neomycin. (23)

amoxicillin (ah-moks"ĭ-sil'in) A semisynthetic penicillin antibiotic that is related to ampicillin. (23)

amphitrichous (am"fi-trik'us) **bacteria** Bacteria that possess flagella at the opposite poles of the cell. (4)

amphotericin (am"fo-ter'ĭ-sin) **B** An antifungal drug used to treat serious fungal diseases such as cryptococcosis, histoplasmosis, and blastomycosis. (14, 23)

ampicillin A semisynthetic penicillin derivative active against Gram-positive bacteria and certain Gram-negative bacteria. (23)

anabolism A chemical process involving the synthesis of organic compounds; usually an energy-utilizing process. (5)

anaerobic organism An organism that grows in an atmosphere free of oxygen. (4, 24)

analog A compound closely related to a naturally occurring compound. (11)

anaphylatoxins (an″ah-fĭ′lah-tok′sinz) Chemical substances that set into motion a series of events leading to smooth muscle contraction. (18)

anaphylaxis (an″ah-fĭ-lak′sis) A life-threatening allergic reaction in which a series of mediators cause contractions of smooth muscle throughout the body. (20)

Ancylostoma duodenale (an″ki-los′to-mah du-od-in-al′e) A multicellular roundworm parasite of the intestine and other organs transmitted by moist vegetation and commonly known as the Old World hookworm. (16)

anionic detergent A detergent that yields negatively charged ions in solution. (22)

anorexia (an″o-rek′se-ah) Loss of appetite.

anthrax A serious soilborne bacterial disease caused by *Bacillus anthracis* and accompanied by severe blood hemorrhaging. (9)

antibiotic A product of the metabolism of a microorganism that is inhibitory to other microorganisms. (23)

antibody A highly specific protein molecule produced by plasma cells in the immune system in response to a specific chemical substance; antibodies function in antibody-mediated immunity. (11, 18, 19)

anticodon A three-base sequence on the tRNA molecule that binds to the codon on the mRNA molecule during protein synthesis. (5)

antigen Any chemical substance that elicits a response by the body's immune system. (18)

antigenic determinant A geographical location on an antigen molecule that stimulates a specific response and to which the response is directed; often consists of several amino acids or monosaccharides. (18)

antigenic variation A process in which chemical changes occur periodically in antigens; known to take place in influenza viruses. (12)

antiglobulin antibody An antibody that reacts with human antibodies. (19)

antiseptic A chemical used to kill pathogenic microorganisms on a living object, such as the surface of the human body. (22)

antiserum Serum rich in a particular type or types of antibodies. (19)

antitoxins Antibodies that circulate in the bloodstream and provide protection against toxins by neutralizing them. (7, 17, 19)

ANUG Acute necrotizing ulcerative gingivitis, a bacterial infection of the mucous membranes in the oral cavity; sometimes called Vincent's angina or trench mouth. (10)

aplastic (a-plas′tik) **anemia** A side effect of chloramphenicol therapy in which red blood cells are produced with little or no hemoglobin. (23)

arboviral encephalitis Encephalitis due to a viral agent transmitted by an arthropod. (13)

archaeobacteria (ar″ke-o-bac-tēr′e-ah) A group of bacteria believed to be of ancient origin. (4)

arsphenamine (ars-fen′ah-min) An arsenic–phenol compound synthesized by Ehrlich and Hata for use against syphilis spirochetes; the first modern chemotherapeutic agent. (23)

Arthropoda A large phylum of animals having jointed appendages and a segmented body; includes insects such as lice, mosquitoes, and fleas and spiderlike arachnids such as ticks and mites. (17)

arthrospores Fungal spores formed by fragmentation of the hypha. (14)

artificially acquired active immunity Immunity resulting from the production of antibodies in response to antigens in a vaccine or toxoid. (19)

artificially acquired passive immunity Immunity resulting from an exposure to or injection of antibodies. (19)

Ascaris lumbricoides (as′ka-ris lum-bri-koid′ēz) A multicellular roundworm parasite of the intestine commonly known as the "roundworm." (16)

Ascomycetes (as″ko-mi-se′tēz) A group (or class) of fungi in which the sexual form of reproduction includes ascospore formation. (14)

ascospore A sexually produced spore formed by members of the Ascomycetes class of fungi. (14)

ascus A saclike structure that contains ascospores and is formed by members of the Ascomycetes class of fungi. (14)

aseptic (a-sep′tik) Free of microorganisms. (22)

aseptic meningitis A type of meningitis in which no bacterium or other agent can be readily identified; usually refers to meningitis due to a virus. (13)

aspergilloma (as″per-jil-o′mah) A dense, round ball of mycelium often occurring in the lungs and commonly due to *Aspergillus fumigatus*. (14)

Aspergillus flavus The fungus of the Ascomycetes class that may produce toxins in certain consumable foods. (14, 24)

Aspergillus fumigatus The fungus of the Ascomycetes group that may infect the lung and form a mass of hyphae called an aspergilloma. (14)

asthma (az′mah) A period of wheezing and stressed breathing resulting from a type I hypersensitivity reaction taking place in the respiratory tract. (20)

asymptomatic Without symptoms.

athlete's foot A fungal disease of the webs of the toes caused by various species of fungi. (14)

atopic (a-top′ik) **disease** A type I hypersensitivity characterized by limited production of IgE and a localized reaction in the body; also called common allergy. (20)

attenuated (ah-ten″u-a′ted) **virus** A weakened variant of a virus that emerges during successive transfers of the virus in tissue cultures; used in immunizations because of its reduced virulence; sometimes called "live virus." (11, 19)

autoantigens A person's own proteins and other organic compounds that elicit a specific response in the body. (18)

autoclave A laboratory instrument that sterilizes microbiological materials by means of steam under pressure. (21)

autograft Tissue taken from one part of the body and grafted to another. (20)

autoimmune disease A disease in which antibodies react with an individual's own chemical substances and cells. (18, 20)

avirulent organism An organism that is normally without pathogenicity. (17)

axial filaments Submicroscopic fibers located along the cell wall in certain species of spirochetes; contractions of the filaments yield undulating motion in the cell. (9)

azathioprine (a″za-thī′o-prēn) An antimitotic drug used to treat cancer. (20)

B

B19 disease An alternative name for fifth disease. (12)

Babesia microti (bah-be′ze-ah mi-cro′ti) A protozoan of the Sporozoa group that causes babesiosis. (15)

babesiosis A tickborne protozoal disease of the red blood cells caused by *Babesia microti* and accompanied by periods of high fever, anemia, and blood clotting. (15)

bacille Calmette–Guérin (BCG) (bah-sil′ā kal-met′ ga-ran′) A strain of attenuated *Mycobacterium bovis* used for immunization to tuberculosis and, on occasion, to leprosy. (7, 19, 20)

bacillus A bacterial rod. (4)

Bacillus cereus A Gram-positive sporeforming bacterial rod transmitted by food and water and a cause of intestinal disease. (8)

bacitracin (bas″ĭ-tra′sin) An antibiotic derived from a *Bacillus* species and effective against Gram-positive bacteria when used topically. (23)

bacterial dysentery An alternate name for shigellosis. (8)

bactericidal agent An agent that kills bacteria. (22)

bacteriochlorophyll (bak-te″re-o-klo′ro-fil) A pigment located in bacterial membrane systems that upon excitement by light loses electrons and initiates photosynthetic reactions. (5)

bacteriocins (bak-te″re-o′sinz) A group of bacterial proteins toxic to other bacteria. (6)

bacteriophage (bak-te″re-o-faj′) A type of virus that attacks and replicates within bacteria. (6, 11)

bacteriorhodopsin (bak-te″re-o-ro-dop′sin) A photosynthetic pigment found in archaeobacteria. (5)

bacteriostatic agent An agent that prevents the multiplication of bacteria without killing them. (22)

balantidiasis A waterborne and foodborne protozoal disease of the intestine caused by *Balantidium coli* and accompanied by mild to severe diarrhea. (15)

Balantidium (bal″an-tid′e-um) *coli* The ciliated protozoan that causes balantidiasis. (15)

Bang's disease An alternate name for brucellosis. (7)

barophilic (bar″o-fil′ik) **microorganisms** Microorganisms that live under conditions of high pressure. (25)

Bartonella bacilliformis The rickettsia that causes bartonellosis.

bartonellosis A relatively mild sandfly-transmitted disease of the blood caused by *Bartonella bacilliformis* and accompanied by fever and a skin rash.

Basidiomycetes A group (or class) of fungi in which the sexual form of reproduction includes basidiospore formation. (14)

basidiospore A sexually produced spore formed by members of the Basidiomycetes class of fungi. (14)

basidium A clublike structure that contains basidiospores and is formed by members of the Basidiomycetes class of fungi. (14)

basophil (ba′so-fil) A type of leukocyte that functions in allergic reactions and, possibly, other reactions not yet established. (17, 20)

bejel A mild syphilislike disease occurring in remote parts of the world and due to a species of *Treponema*. (10)

benthic zone The environment of the ocean floor. (25)

Bergey's Manual of Systematic Bacteriology The official manual of bacteriology that lists the names and characteristics of the known bacteria and presents a classification scheme for these organisms. (3)

beta-hemolytic streptococci Streptococci that destroy red blood cells completely; when cultivated in blood agar, a clearing forms around the colonies of the streptococci. (7)

beta-lactam nucleus The chemical group central to all penicillin antibiotics. (23)

beta-lactamase An alternative name for penicillinase, the enzyme that converts penicillin to penicilloic acid. (23)

bilharziasis (bil″har-zi′ah-sis) Another term for schistosomiasis. (16)

binary fission An asexual process by which a cell divides to form two new cells; specifics of the process vary among organisms. (4)

binomial system The system of nomenclature that uses the genus and specific modifier to refer to organisms. (3)

biochemical oxygen demand (BOD) A number referring to the amount of oxygen utilized by the microorganisms in a sample of water during a five-day period of incubation. (25)

biological vector An infected living organism that transmits disease agents. (17)

biphasic (bi-fāz′ik) **fungus** A fungus that grows in two different patterns depending on growth conditions; also known as diphasic fungus. (14)

bipolar staining A characteristic of *Yersinia* and *Francisella* species in which stain gathers at the poles of the cells giving cells the appearance of safety pins. (9)

bisphenols (bĭs-phe′nolz) Combinations of two phenol molecules used in disinfection. (22)

blackwater fever An alternative name for malaria. (15)

blanching A process in which food is subjected to steam for three to five minutes in order to destroy cellular enzymes and enhance preservation. (24)

blastospore A fungal spore formed by budding. (14)

B-lymphocyte A lymphocyte formed in the bursa of Fabricius in embryonic chicks and possibly in the fetal liver or bone marrow in humans; the lymphocyte is responsible for the system of antibody-mediated immunity; also called the B-cell. (18)

boil A raised pus-filled lesion. (8, 10, 18)

Bordetella pertussis A Gram-negative bacterial rod transmitted by droplets and the cause of pertussis (whooping cough). (7)

Borrelia burgdorferi (bo-rel′e-ah burg-dorf′er-i) A spirochete transmitted by ticks and the cause of Lyme disease. (9)

Borrelia recurrentis A spirochete transmitted by arthropods and the cause of relapsing fever. (9)

botulism A foodborne disease of the nervous system accompanied by paralysis. (8)

boutonneuse (boo-ten-ez′) **fever** A tickborne rickettsial disease due to *Rickettsia conori* and accompanied by fever and skin rash. (9)

bovine spongiform encephalopathy A viral disease of cows and other large animals in which infected brain tissue becomes soft and spongy; accompanied by neurological defects; also called "mad cow disease." (13)

bradykinin (brad″e-ki′nin) A peptide that functions in type I hypersensitivity reactions by contracting smooth muscles. (20)

breakbone fever A term applied to dengue fever reflecting sensations that the bones are breaking. (13)

broad-spectrum antibiotic An antibiotic useful for treating many groups of microorganisms including Gram-positive and Gram-negative bacteria, rickettsiae, fungi, and protozoa. (23)

bronchopneumonia Scattered patches of pneumonia, especially in the bronchial tree. (7)

Brucella abortus A Gram-negative bacterial rod transmitted by food and water and the cause of brucellosis in animals and undulant fever in humans. (7)

brucellosis A foodborne bacterial disease of the blood caused by *Brucella* species and accompanied in humans by blood involvement and undulating fever; in large animals, a disease of the reproductive organs; also known as Malta fever, Bang's disease, and undulant fever; called contagious abortion in animals. (7)

Bruton's agammaglobulinemia (a-gam″ah-glob″u-lĭ-ne′me-ah) An immune deficiency disease in which the body fails to produce plasma cells from B-lymphocytes. (20)

bubo A swelling of the lymph nodes. (9)

budding An asexual process of reproduction in the fungi, in which a new cell forms at the border of the parent cell and then breaks free to live independently. (14)

Burkitt's lymphoma A type of lymphoid cancer occurring in the connective tissues of the jaw; described by Dennis Burkitt in 1956. (11, 20)

bursa of Fabricius The lymphoid organ of the gastrointestinal tract of the embryonic chick in which B-lymphocytes are formed. (18)

C

Calymmatobacterium (kah-lim″mah-to-bak-te′re-um) *granulomatis* A Gram-negative rod and the cause of granuloma inguinale. (10)

Campylobacter (kam′-pi-lo-bak″ter) *jejuni* A Gram-negative curved bacterial rod transmitted by food and water and involved in campylobacteriosis of the intestine. (8)

campylobacteriosis (kam″pi-lo-bac″ter-i-o′sis) A foodborne and waterborne bacterial disease of the intestine caused by *Campylobacter jejuni* and accompanied by diarrhea. (8)

cancer A condition characterized by the radiating spread of cells that reproduce at an uncontrolled rate. (11, 20)

Candida albicans The fungus that causes candidiasis; also an opportunistic fungus that infects immunocompromised patients such as those with AIDS. (14)

candidiasis (kan-di-di′ah-sis) Infection due to *Candida albicans* and manifesting itself as a yeast infection of the intestine, the vaginal tract ("yeast disease"), or the mucous membranes of the mouth (thrush). (14)

capsid The surrounding layer of protein that encloses the genome of a virus. (11)

capsomere A protein subunit of the capsid found on the surface of a virus; the number of capsomeres varies among different viruses. (11)

capsule A layer of polysaccharides and small proteins that adheres to the surface of certain bacteria; serves as a buffer between the cell and its environment. (4)

carbenicillin (kar″ben-ĭ-sil′in) A semisynthetic penicillin antibiotic used to treat urinary tract infections caused by certain Gram-negative bacteria. (23)

carbuncle An enlarged abscess formed from the union of several smaller abscesses. (18)

carcinogens Cancer-causing substances. (11)

cardiolipin (kar″de-o-lip′in) An alcoholic extract of beef heart used as an antigen in the VDRL procedure. (19)

carrier One who has recovered from a disease but retains live organisms in the body and continues to shed them. (8, 17)

Carrion's disease An alternative name for bartonellosis. (10)

casein (ka'sēn) A mixture of three long chains of amino acids; the major protein in milk. (24)

catabolism A chemical process in which organic compounds are digested; usually an energy-liberating process. (5)

cationic detergent A detergent that yields positively charged ions in solution. (22)

cat scratch disease An infectious disease of the lymph channels probably of bacterial origin and accompanied by lymph node swellings. (10)

cavitation The formation and implosion of bubbles in a liquid in which ultrasonic vibrations are propagated. (21)

cefoxitin (se-foks'ĭ-tin) A semisynthetic derivative of a *Streptomyces* species used as a penicillin substitute where penicillin resistance is encountered. (23)

cellular immunity Immunity arising from the activity of T-lymphocytes on or near the body cells; also called tissue immunity and cell-mediated immunity. (18, 20)

cephalosporins (sef"ah-lo-spōr'inz) A group of antibiotics derived from the mold *Cephalosporium* and used against Gram-positive bacteria and certain Gram-negative bacteria. (23)

cercaria (ser-ka're-ah) A tadpolelike intermediary stage in the life cycle of a fluke. (16)

cestodes A group of flatworms commonly known as tapeworms. (16)

Chagas' disease An alternative name for American trypanosomiasis. (15)

chancre (shag'ker) A circular, purplish hard ulcer with a raised margin that occurs during primary syphilis. (10)

chancroid A sexually transmitted bacterial disease of the external genital organs caused by *Haemophilus ducreyi* and accompanied by soft, spreading chancres and swollen lymph nodes. (10)

Chediak–Higashi syndrome An immune disorder characterized by delayed killing of phagocytized microorganisms. (20)

chemically defined medium A medium in which the nature and quantity of each component are identified; also called a synthetic medium. (4)

chemoautotrophs (ke"mo-aw'to-trophz) Organisms that derive energy from chemical reactions and utilize the energy to synthesize nutrients from carbon dioxide. (5)

chemoheterotrophs (ke"mo-het'er-o-trophz) Organisms that derive energy from chemical reactions and utilize the energy to synthesize nutrients from carbon compounds other than carbon dioxide. (5)

chemotactic factor (CF) A lymphokine that draws phagocytes to the antigen site in cellular immunity. (18)

chemotaxis A chemical attraction. (18)

chemotherapeutic agent A synthetic or semisynthetic chemical compound that is inhibitory to microorganisms and is used in the body for therapeutic purposes; includes antibiotics synthetically manufactured. (23)

chickenpox A communicable skin disease caused by a DNA icosahedral virus and accompanied by teardrop-shaped, highly infectious, skin lesions. (12)

chimera (ki-mer'-ah) A plasmid engineered to contain a fragment of foreign DNA. (6)

chitin (kī'tin) A polymer of acetylglucosamine units found in the cell walls of fungi; chitin adds rigidity to the cell wall. (14)

"chlamydia" (klah-mid'e-ah) A sexually transmitted bacterial disease of the urethra and reproductive organs caused by *Chlamydia trachomatis* and accompanied by urinary tract symptoms and possible pelvic inflammatory disease leading to sterility. (10)

chlamydiae (klah-mid'e-e) A subgroup of rickettsiae seen only with the electron microscope and cultivated within living tissue medium. (3, 7, 10)

chlamydial ophthalmia (kla-mid'e-al of-thal'me-ah) Infection of the eye tissues by *Chlamydia trachomatis*; may occur in newborns from exposure to chlamydiae during birth. (10)

chlamydial pneumonia An influenzalike disease of the lungs caused by *Chlamydia pneumoniae*. (7)

Chlamydia pneumoniae A chlamydia and the cause of chlamydial pneumonia; formerly known as TWAR. (7)

Chlamydia psittaci (sit'a-si) A chlamydia and the cause of psittacosis. (7)

Chlamydia trachomatis (tra-ko'mah-tis) A chlamydia and the cause of "chlamydia" as well as trachoma. (10)

chloramines (klo'rah-mēnz) A group of chlorine derivatives formed by adding a chlorine to an amino group on a carrier molecule. (22)

chloramphenicol (klo"ram-fen'ĭ-kol) A broad-spectrum antibiotic derived from a *Streptomyces* species and used to treat typhoid fever and various other diseases; chloramphenicol interferes with protein synthesis. (23)

chlorhexidine (klor-hexs'ĭ-dēn) A bisphenol compound widely used as an antiseptic and disinfectant. (22)

chlorophyll A pigmented molecule that functions in photosynthesis; exists free in the cytoplasm of prokaryotes and within chloroplasts of eukaryotes. (3, 5, 15)

chloroquine (klor'o-kwin) A synthetic drug used to treat malaria in the clinical phase. (15)

chocolate agar A bacteriological medium consisting of a nutritious base and whole blood; the medium is heated to disrupt the blood cells and release the hemoglobin. (4)

cholera A foodborne and waterborne bacterial disease of the intestine caused by *Vibrio cholerae* and accompanied by massive diarrhea, fluid and electrolyte imbalance, and severe dehydration. (8)

chronic disease A disease that develops slowly, tends to linger for a long time, and requires long convalescence. (17)

cilia Hairlike appendages in eukaryotic cells; cilia assist motion in certain protozoa and are used for filtering air by respiratory epithelial cells. (15, 17)

Ciliophora (sil-e-of'o-rah) The class or group of protozoa whose members move by means of cilia. (15)

Claviceps purpurea A fungus that infects rye and other grains and produces toxins with hallucinatory properties that may be consumed. (24)

clindamycin An antibiotic used as a penicillin substitute and for certain anaerobic bacterial diseases. (23)

clinical disease A disease in which the symptoms are apparent. (17)

clonal selection hypothesis A theory that shows how certain lymphocytes are selected out from the mixed population of B-lymphocytes when stimulated by processed antigens. (18)

clone A collection, or colony, of identical cells arising from a single cell. (18)

Clonorchis sinensis (klo-nor'kis si-nen'sis) A multicellular flatworm parasite of the liver; commonly known as the Chinese liver fluke. (16)

clostridial myonecrosis (mi″o-ně-kro'sis) An alternative expression for gas gangrene referring to the death of muscle cells following invasion of certain *Clostridium* species. (9)

Clostridium botulinum A Gram-positive anaerobic sporeforming rod that causes botulism. (8)

Clostridium perfringens A Gram-positive anaerobic sporeforming rod that causes gas gangrene and a form of food poisoning. (8, 9)

Clostridium tetani A Gram-positive anaerobic sporeforming rod that causes tetanus. (9)

clotrimazole (klo-trī'mah-zōl) An imidazole drug often useful in the treatment of fungal disease such as candidiasis. (23)

coagulase An enzyme that catalyzes the formation of a fibrin clot; produced by virulent staphylococci. (17)

Coccidioides (kok-sid″e-oi'dēz) ***immitis*** A fungus transmitted by dust and the cause of coccidioidomycosis. (14)

coccidioidomycosis (kok-sid″e-oi″do-mi-ko'sis) A fungal disease of the lungs caused by *Coccidioides immitis* and accompanied by cough, malaise, and other respiratory symptoms. (14)

coccobacillus A form of bacteria characterized by rods with rounded edges. (4)

coccus A spherical bacterium. (4)

codon A three-base sequence on the mRNA molecule that specifies a particular amino acid in the protein molecule. (5)

coenocytic fungi (se″no-sit'ik) Fungi containing no septa in the hyphae. (14)

coenzyme An organic molecule that forms the nonprotein part of an enzyme molecule. (5)

cold agglutinin screening test (CAST) A laboratory procedure in which *Mycoplasma* antibodies agglutinate human red blood cells at cold temperatures. (7)

cold sores Herpes-induced blisters occurring on the lips, gums, nose, and adjacent areas. (12)

coliform (kol'ĭ-form) **bacteria** Gram-negative nonspore-forming bacilli usually found in the human and animal intestine; coliform bacteria ferment lactose to acid and gas. (25)

Colorado tick fever A tickborne disease caused by an RNA virus, occurring in the western United States, and accompanied by high fever and joint pains. (13)

colostrum (kŏ-los'trum) The first milk secreted from the mammary gland of animals or humans. (18, 24)

commensalism A close and permanent association between two populations of organisms in which only one population benefits. (4, 17)

common allergy An allergic reaction taking place in a localized area of the body; synonymous with atopic disease. (20)

communicable disease A disease that is transmissible among various hosts. (17)

competent cell A bacterium that can take up DNA in the recombination process of transformation. (6)

complement A group of proteins that functions in a cascading series of reactions during the response by the body to certain antigens; the complement cascade is stimulated by antigen–antibody activity. (18)

congenital rubella syndrome A condition in which rubella viruses pass across the placenta of an infected woman and cause damage in the fetus. (12)

conidia (ko-nid'e-ah) Asexually produced fungal spores formed on a supportive structure without an enclosing sac. (14)

conidiophore (ko-nid'e-o-fōr) The supportive structure on which conidia form. (14)

conjugation A type of bacterial recombination in which genetic material passes from a live donor cell into a live recipient cell during a period of contact. (6)

conjunctivitis (kon-junk″tĭ-vi'tis) A general expression for disease of the conjunctiva, the thin mucous membrane that covers the cornea and forms the inner eyelid. (10)

contact dermatitis A type IV hypersensitivity in which the immune system responds to allergens such as clothing materials, metals, and insecticides; the reaction is usually characterized by an induration. (20)

contagious abortion An alternate name for brucellosis in animals. (7)

contagious disease A communicable disease whose agent passes with particular ease among hosts. (17)

continuous flow technique An industrial process in which medium is continually added to a fermentation tank to replace that which has been used. (25)

convalescent serum Antibody-rich serum obtained from a convalescing patient. (19)

copepod (ko'pě-pod) A small aquatic arthropod that serves as an intermediary host for the fish tapeworm. (16)

Corynebacterium (ko-ri-ne″bac-te're-um) ***diphtheriae*** A Gram-positive club-shaped bacterial rod that causes diphtheria. (7)

covalent bond A chemical bond created by the sharing of electrons between atoms. (2, 5)

Coxiella (kok"se-el'lah) *burnetii* A rickettsia transmitted by food, water, arthropods, or contact and the cause of Q fever. (7)

Coxsackie (cook-sak'e) **virus** An RNA virus of the enterovirus group transmitted by food and water and involved in disease of the intestine; also the cause of disease of the heart muscle (myocarditis) and the chest wall (pleurodynia). (13)

cresols (kre'solz) Derivatives of phenol that contain methyl groups on the benzene ring. (22)

croup (kroop) A name given to upper respiratory tract infections often caused by adenoviruses; accompanied by hoarse coughing. (12)

cryptococcosis (krip"to-kok-o'sis) A fungal disease of the lungs and spinal cord that occurs as an opportunistic disease in persons with compromised immune systems such as by AIDS. (14)

Cryptococcus neoformans (krip"to-kok'us ne-o-form'anz) The fungus that causes cryptococcosis; also, an opportunistic fungus that infects immunocompromised patients such as those with AIDS. (14)

cryptosporidiosis (krip"to-spor-id"e-o'sis) A waterborne protozoal disease of the intestine caused by *Cryptosporidium coccidi* and *C. parvum*, accompanied by intense diarrhea, and often occurring in immunocompromised individuals such as AIDS patients. (15)

Cryptosporidium (krip"to-spor-id'e-um) *coccidi* An opportunistic protozoan that infects the intestines and causes cryptosporidiosis. (15)

Cryptosporidium parvum An opportunistic protozoan that infects the intestines and causes cryptosporidiosis. (15)

cyanobacteria (si"ah-no-bak-tēr'e-ah) A group of pigmented microorganisms occurring in unicellular and filamentous forms; formerly called blue-green algae. (3, 5, 25)

cyclosporin A A drug that suppresses cellular immunity and encourages acceptance of transplanted tissue. (20)

cyst A dormant and very resistant form of a microorganism such as a protozoan or multicellular parasite. (15, 16)

cytomegalovirus (si"to-meg"ah-lo-vi'rus) An icosahedral DNA virus that causes infected cells to enlarge. (13)

cytomegalovirus (CMV) disease A viral disease of multiple organs accompanied by nonspecific symptoms and malaise; transmissible to the fetus of a pregnant woman; an opportunistic disease in AIDS patients. (13)

cytotoxic hypersensitivity A cell-damaging or cell-destroying hypersensitivity that develops when IgG reacts with antigens on the surfaces of cells. (20)

D

dalton A unit of weight equal to the mass of one hydrogen atom; used to measure molecular weights. (2)

dander Particles of animal skin, hair, or feathers that may contain materials that cause allergic reactions. (20)

dapsone A chemotherapeutic agent used to treat leprosy patients. (10, 23)

Darling's disease An alternative name for histoplasmosis. (14)

deamination (de-am"ĭ-na'shun) The biochemical process in which amino groups are enzymatically removed from amino acids and hydroxyl groups are substituted; this allows the carbon skeleton to be used for energy purposes. (5)

decline phase The final portion of a bacterial growth curve in which environmental factors overwhelm the population and induce death. (4)

degerm To mechanically remove organisms from a surface. (22)

dehydration synthesis A process of bonding two molecules together by removing the products of water and joining the open bonds. (2)

denaturation A process in which proteins change from the tertiary structure to the secondary structure; heat and certain chemicals may induce denaturation. (2, 21)

dengue (deng'e) **fever** A viral disease transmitted by the *Aedes aegypti* mosquito and accompanied by bone-breaking sensations. (13)

Dermacentor andersoni The tick that transmits Rocky Mountain spotted fever and other infectious diseases. (9)

dermatomycosis (der-mah'to-mi-ko'sis) A fungal disease of the skin tissues. (14)

desensitization A process in which minute doses of antigens are used to remove antibodies from the body tissues to prevent a later allergic reaction. (20)

desquamation (des"kwah-ma'shun) Peeling of the skin about the fingers and toes; associated with toxic shock syndrome and Kawasaki disease. (12)

diarrhea Excessive loss of fluid from the gastrointestinal tract.

diatomaceous (di"ah-to-ma'shus) **earth** Filtering material composed of the remains of diatoms. (21, 25)

diatoms (di'ah-tomz) Eukaryotic marine microorganisms that practice photosynthesis; a type of unicellular algae. (3, 21, 25)

differential medium A growth medium in which different species of microorganisms can be distinguished. (4)

differential technique A staining or other procedure intended to separate organisms into different categories. (3)

DiGeorge's syndrome An immune deficiency disease in which the thymus fails to develop, thereby leading to a reduced number of T-lymphocytes. (20)

dimorphic fungus A fungus that takes a yeast form in the human body, and a hyphal form when cultivated in the laboratory; synonymous with biphasic. (14)

dinoflagellates A group of photosynthetic marine flagellates that form one of the foundations of the food chain in the ocean. (3, 25)

diphtheria (dif-the're-ah) A bacterial disease of the respiratory tract caused by toxin-producing *Corynebacterium diphtheriae* and accompanied by tissue destruction and the accumulation of pseudomembranes. (7)

Diphyllobothrium (di-fil″o-both're-um) *latum* A multicellular flatworm parasite of the intestine and other organs transmitted by contaminated seafood and commonly known as the fish tapeworm. (16)

dipicolinic (di″pik-o-lin'ik) **acid** An organic substance that helps stabilize the proteins in a bacterial spore and thereby increases spore resistance. (4, 21)

diplococcus A pair of cocci. (4, 7)

disease Any change from the general state of good health. (17)

disinfectant A chemical used to kill pathogenic microorganisms on a lifeless object such as a table top. (22)

DNA ligase An enzyme that binds together DNA fragments; important in genetic engineering experiments. (6)

DNA polymerase An enzyme that forms DNA by combining fragments during DNA replication. (6)

double helix The spiral-staircase arrangement of the chromosome in which two strands of DNA oppose each other with the nitrogenous bases forming the rungs of the staircase. (2, 5)

Downey cells Swollen lymphocytes with foamy cytoplasm and many vacuoles that develop as a result of infection with infectious mononucleosis viruses. (11, 13)

Dracunculus medinensis (drah-kung'ku-lus med-i-nen'sis) A multicellular roundworm parasite of the intestine and other organs transmitted by food and commonly known as the Guinea worm. (16)

droplets Airborne particles of mucus and sputum from the respiratory tract that contain disease organisms. (7, 12, 17)

dysentery A condition marked by frequent, watery stools, often with blood and mucus. (8, 15)

E

Eaton agent An alternative name for *Mycoplasma pneumoniae*. (7)

Ebola virus An RNA virus existing in Africa and causing a form of a type of hemorrhagic fever called Ebola fever. (13)

Echinococcus granulosus (e-ki″-no-kok'us gran-u-lo'sis) A multicellular roundworm parasite of the liver commonly known as the dog tapeworm and the cause of hydatid disease. (16)

echovirus An RNA virus transmitted by food and/or water and involved in disease of the intestine as well as the skin tissues. (13)

***E. coli* 0157:H7** A virulent strain of *Escherichia coli* that causes bloody diarrhea and kidney hemorrhaging after transmission in contaminated food and water. (3)

edema A swelling of the tissues brought about by an accumulation of fluid. (16, 20)

Ehrlichia (er-lik'e-ah) *canis* The rickettsia that causes ehrlichiosis. (9)

ehrlichiosis (er-lik-e-o'sis) A tickborne rickettsial disease caused by *Ehrlichia canis* and characterized by fever, headache, and malaise; occurs primarily in dogs. (9)

electrophoresis (e-lek″tro-fo-re'sis) A laboratory procedure characterized by the movement of charged organic molecules through an electrical field. (19)

elementary body An infectious form of a chlamydia in the early stage of reproduction. (10)

elephantiasis (el″ah-fan-tī'ah-sis) Swelling and distortion of the tissues, especially the legs and scrotum; commonly due to infection with *Wuchereria bancrofti*. (16)

encephalitis Inflammation of the tissue of the brain or infection of the brain. (13)

endemic typhus A relatively mild fleaborne disease of the blood caused by *Rickettsia typhi* and accompanied by fever and a skin rash. (9)

endogenous (en-doj'ĕ-nus) **disease** A disease caused by organisms commonly found within the body.

endonuclease An enzyme that cleaves a DNA molecule at the sugar–phosphate bond; used in genetic-engineering experiments. (6)

endotoxin A metabolic poison produced chiefly by Gram-negative bacteria; endotoxins are part of the bacterial cell wall and, consequently, are released on cell disintegration; they are composed of lipid-polysaccharide-peptide complexes. (17)

Entamoeba histolytica An ameboid protozoan transmitted by food and water and the cause of amebic dysentery. (15)

enteritis A synonym for gastrointestinal illness; commonly, an alternative name for salmonellosis. (8)

Enterobius vermicularis (en″ter-o'be-us ver″mik-u-la'ris) A multicellular roundworm parasite of the intestine commonly known as the pinworm. (16)

enterotoxin A toxin that is active in the gastrointestinal tract of the host. (8, 17)

enterovirus A virus that infects intestinal cells. (13)

envelope The flexible membrane of protein and lipid that surrounds many types of viruses. (11)

enzyme A reusable protein molecule that brings about a chemical change while remaining unchanged itself; the molecule may include a nonprotein part. (5)

enzyme-linked immunosorbent assay (ELISA) A serological test in which an enzyme system is used to detect test material linked to antigens or antibodies on a solid surface. (19)

eosinophil (e″o-sin'o-fil) A type of leukocyte whose functions are not clearly established but may involve phagocytosis. (17)

epidemic parotitis An alternate name for mumps. (12)

epidemic typhus A relatively serious louseborne disease of the blood caused by *Rickettsia prowazecki* and accompanied by high fever and a skin rash beginning on the body trunk. (9)

Epidermophyton (ep"e-der-mof'i-ton) **species** A fungus spread by contact and one of the causes of athlete's foot and ringworm. (14)

episome A plasmid attached to the chromosome of a bacterium. (6)

Epstein–Barr virus An icosahedral DNA virus thought to cause infectious mononucleosis; the virus is also associated with Burkitt's lymphoma. (13)

ergot disease A toxemia transferred to humans from rye grain; the toxin is produced by the fungus *Claviceps purpurea*. (14, 24)

erythema A zone of redness in the skin due to accumulation of blood.

erythema chronicum migrans (ECM) The expanding circular red rash that occurs on the skin of patients with Lyme disease. (9)

erythema infectiosum An alternative name for fifth disease. (7)

erythrogenic toxin A streptococcal toxin that leads to the rash in scarlet fever. (7)

erythromycin (ĕ-rith"ro-mi'sin) An antibiotic derived from a *Streptomyces* species; used against Gram-positive bacteria, mycoplasmas, and certain other organisms. (23)

Escherichia (esh-er-ik'e-ah) *coli* A Gram-negative bacterial rod transmitted by food and/or water and the cause of travelers' diarrhea as well as infantile diarrhea (8); also used widely in genetic engineering techniques (6) and as an indicator of water pollution (25).

estivo-autumnal malaria Malaria in which the attacks occur at widely spaced intervals. (15)

eukaryote (u-kar'e-ōt) A relatively complex organism whose cells contain organelles as well as a nucleus with multiple chromosomes and a nuclear membrane; reproduction involves mitosis. (3, 14)

exanthem (eg-zan'them) A maculopapular rash occurring on the skin surface. (12)

exotoxin A metabolic poison produced chiefly by Gram-positive bacteria; exotoxins are released to the environment on production; they are composed of protein and affect various organs and systems of the body. (7, 17)

F

F factor A fragment of DNA in the cyptoplasm of the F⁺ bacterial cell that may pass to a recipient bacterial cell in conjugation and change the recipient into an F⁺ cell. (6)

Fab fragment The portion of the antibody molecule that combines with the determinant site of the antigen. (18)

facultative organism An organism that grows in the presence or absence of oxygen. (4)

farmer's lung A condition that results from a type III immune complex hypersensitivity following exposure to antigens. (20)

Fasciola (fas-e-o'lah) *hepatica* A multicellular flatworm parasite of the liver commonly known as the liver fluke. (16)

Fasciolopsis (fas"e-o-lop'sis) *buski* A multicellular flatworm parasite of the intestine and other organs transmitted by water and commonly known as the intestinal fluke. (16)

Fc fragment The portion of the antibody molecule that combines with phagocytes, viral receptor sites, and complement. (18)

fermentation Anaerobic respiration in which intermediaries in the process are used as electron acceptors; also refers to the industrial use of microorganisms. (5, 25)

fifth disease A communicable skin disease caused by a DNA virus and accompanied by a fiery red rash especially on the cheeks of the face; also called B19 disease. (12)

filariform (fĭ-lār'ĭ-form) **larva** A hairlike intermediate form of the hookworm. (16)

fimbriae (fim'bre-ā) Short, hairlike structures used by bacteria for attachment; sometimes used as an alternative expression for pili. (4)

flaccid (flak'sid) **paralysis** Paralysis in which the limbs have little tone and become flabby. (8)

flagellum A long hairlike appendage composed of protein and responsible for motion in microorganisms; found in bacteria and protozoa. (4, 15)

flare A spreading zone of redness around a wheal; occurs during an allergic reaction. (20)

flash pasteurization A process of pasteurization in which milk is heated at 71.6° C for 15 to 17 seconds and then cooled rapidly; also known as HTST method for "high temperature, short time." (21, 24)

flatworms A common expression for the worms belonging to the phylum Platyhelminthes. (16)

flavin adenine dinucleotide (FAD) A coenzyme that functions in electron transport during oxidative phosphorylation. (5)

flocculation (flok"u-la'shun) The formation of jellylike masses of coagulated material in the water purification process; also, a serological reaction in which particulate antigens react with antibodies to form visible aggregates of material. (19, 25)

flucytosine (flu-si'to-sēn) An antifungal drug that interrupts nucleic acid synthesis in cells. (23)

fluid mosaic model The model for the cell membrane in microorganisms where protein globules "float" within two parallel layers of phospholipid. (4)

flukes A group of flatworms belonging to the phylum Platyhelminthes; also known as trematodes. (16)

folic acid The organic compound in bacteria whose synthesis is blocked by sulfonamide drugs. (5, 23)

fomites (fo′mitz) Inanimate objects such as clothing or utensils that carry disease organisms. (17)

foraminifera (fo-ram″ĭ-nif′er-ah) A group of shell-containing ameboid protozoa having chalky skeletons with windowlike openings between sections of the shell. (15)

formalin A solution of formaldehyde used as embalming fluid, in the inactivation of viruses, and as a disinfectant. (22)

Francisella tularensis A Gram-negative rod that displays bipolar staining and causes tularemia. (9)

Friedländer's bacillus An alternative name for *Klebsiella pneumoniae*. (7)

fruiting body The general name for an asexual reproductive structure of a fungus. (14)

functional group A group of atoms functioning as a unit. (2)

fungemia Dissemination of fungi through the circulatory system. (17)

Fungi (fun′jī) A kingdom in the Whittaker classification of living things composed of nongreen eukaryotic microorganisms; also the name for the mold and yeast microorganisms within the kingdom.

fungicidal agent An agent that kills fungi. (22)

fusospirochetal (fu″so-spi″ro-ke′tal) **disease** An alternative name for trench mouth. (10)

G

gamma globulin A general term for antibody-rich serum. (19)

gamma-hemolytic streptococci Streptococci that have no effect on red blood cells. (7)

gangrene The physiological process in which the enzymes from wounded tissue digest the surrounding layer of cells, whose enzymes digest the next layer of cells, and so on, thereby inducing a spreading death to the tissue cells; often called "dry gangrene." (9)

Gardnerella (gard-ner-el′ah) *vaginalis* A Gram-negative bacterial rod transmitted by sexual contact and the cause of vaginitis. (10)

gas gangrene A serious soilborne bacterial disease caused by *Clostridium perfringens* and accompanied by gas accumulation in the muscle tissues and gangrene; often called "wet gangrene." (9)

gastroenteritis Infection of the intestinal tract often due to a virus. (13)

gene A segment of a DNA molecule that provides the biochemical information for protein synthesis and inherited traits. (6)

generalized transduction A transduction in which the prophage accidentally incorporates bacterial DNA into its own DNA while replicating during the lytic cycle; the bacterial DNA is then carried into the next cell by the virus. (6)

generation time The time interval between bacterial divisions; varies among bacteria and can be as brief as 20 minutes. (4)

genome (je′nōm) The nucleic acid core of the virus. (11)

gentamicin (jen″tah-mi′sin) An aminoglycoside antibiotic often used to treat infections caused by Gram-negative bacteria. (23)

genus A rank in the classification system composed of two or more species; a collection of genera constitute a family. (3)

germ theory of disease The theory that holds that microorganisms are responsible for infectious diseases. (1)

germicidal agent An agent that kills microorganisms. (22)

Giardia (je-ar′de-ah) *lamblia* A flagellated protozoan transmitted by food and/or water and the cause of giardiasis. (15)

giardiasis (je-ar-di′ah-sis) A foodborne and waterborne protozoal disease of the intestine caused by *Giardia lamblia* and accompanied by mild to severe diarrhea. (15)

gibberellins (gib-ber-el′inz) A series of plant hormones that promote growth by stimulating cell elongation in the stem. (25)

Gilchrist's disease An alternative name for blastomycosis. (14)

glomerulonephritis (glo-mer″u-lo-nĕ-fri′tis) A complication of streptococcal disease involving inflammation of blood vessels in the kidney due to reactions between antigens and antibodies. (7, 20)

glycocalyx (gli″ko-ka′liks) A loose layer of slime around certain bacteria that assists attachment to a surface. (4)

glycolysis (gli-kol′ĭ-sis) A series of enzyme-catalyzed chemical reactions in which glucose is broken down into two molecules of pyruvic acid with a net gain of two ATP molecules. (5)

Golden Age of Microbiology The approximately 60-year period from 1857 to 1914 during which microorganisms were related to infectious disease. (1)

gonococcal ophthalmia (off-thal′mē-ah) Infection of the eye by *Neisseria gonorrhoeae*; may occur in newborns exposed during birth. (10)

gonococcus A colloquial expression for *Neisseria gonorrhoeae*. (10)

gonorrhea A sexually transmitted bacterial disease of the urethra and reproductive organs caused by *Neisseria gonorrhoeae* and accompanied by urinary tract symptoms and possible pelvic inflammatory disease leading to sterility. (10)

Goodpasture's syndrome A type II hypersensitivity in which antibodies are directed against antigens on the membranes of kidney cells, often resulting in kidney failure. (20)

graft versus host rejection (GVHR) A phenomenon in which a tissue graft produces immune substances against the host. (20)

Gram stain technique A technique used to divide bacteria into two groups, Gram-positive and Gram-negative, depending upon their ability to retain a crystal violet–iodine complex on treatment with an alcohol solution. (3)

granuloma inguinale A sexually transmitted bacterial disease of the external genital organs caused by *Calymmatobacterium granulomatis* and accompanied by bleeding ulcers and swollen lymph nodes. (10)

Graves' disease A type II hypersensitivity in which antibodies react with receptors on thyroid gland cells resulting in an overabundant secretion of thyroxine. (20)

gray syndrome A side effect of chloramphenicol therapy characterized by a sudden breakdown of the cardiovascular system. (23)

green rot Spoilage in eggs due to the production of a green, fluorescent pigment by *Pseudomonas* species. (24)

griseofulvin (gris″e-o-ful′vin) An antifungal drug used against tinea infections of the skin, hair, and nails. (14, 23)

Guillain–Barré syndrome (ge-yan′bar-ra′) A complication of influenza and chickenpox characterized by nerve damage and poliolike paralysis. (12)

Guinea worm The common name for *Dracunculus medinensis* and a multicellular roundworm parasite of the human blood and subcutaneous tissues. (16)

gumma (gum′ah) A soft, granular lesion that forms in the cardiovascular and/or nervous systems during tertiary syphilis. (10)

H

Haemophilus (he-mof′i-lus) **aegypticus** A Gram-negative bacterial rod transmitted by contact and involved in bacterial conjunctivitis. (10)

Haemophilus ducreyi A Gram-negative bacterial rod transmitted by sexual contact and the agent of chancroid. (10)

Haemophilus influenzae b A Gram-negative bacterial rod transmitted by respiratory droplets and the agent of *Haemophilus* meningitis; known as Hib. (7)

Haemophilus (he-mof′i-lus) **meningitis** An airborne bacterial disease caused by *Haemophilus influenzae* b and accompanied by respiratory distress followed by inflammation of the meninges. (7)

halogen A chemical element whose atoms have seven electrons in their outer shell; examples are iodine and chlorine. (22)

halophilic (hal″o-fil′ik) **microorganisms** Microorganisms that live in environments that have high concentrations of salt. (25)

Hansen's disease An alternative name for leprosy. (10)

hapten A small molecule that combines with tissue proteins or polysaccharides to form an antigen. (18)

Hashimoto's disease A type II hypersensitivity in which antibodies react with thyroid gland cells leading to a deficiency of thyroxine. (20)

hay fever A type I hypersensitivity reaction resulting from the inhalation of tree and grass pollens. (20)

helix A figure resembling a tightly wound coil such as a corkscrew or spring; one of the major shapes taken by the viral capsid. (3, 11)

helminth A term referring to a multicellular parasite; includes roundworms and flatworms. (16)

helper T-lymphocyte A T-lymphocyte that enhances the activity of B-lymphocytes. (18, 20)

hemagglutination (hem″ah-gloo″tĭ-na′shun) The agglutination of red blood cells. (11, 19)

hemagglutinin (hem″ah-gloo′tin-in) An enzyme on the surface spikes of certain influenza viruses that allows the viruses to bind to red blood cells. (12)

hemoglobin The red oxygen-carrying pigment in erythrocytes. (19)

hemolysins (he-mol′ĭ-sinz) Enzymes that dissolve red blood cells; produced by streptococci, staphylococci, gas gangrene, bacilli, and other microorganisms. (17)

hemolytic disease of the newborn A disease in which Rh antibodies from a pregnant woman combine with Rh antigens on the surface of fetal erythrocytes and destroy the latter; also known as Rh disease and erythroblastosis fetalis; a form of type II hypersensitivity. (20)

hemolytic uremic syndrome A set of symptoms that include hemorrhaging and inflammation of the kidney tissues; associated with *E. coli* 0157:H7. (3, 8)

hemorrhagic fever Any of a series of viral diseases characterized by high fever and hemorrhagic lesions of the throat and internal organs. (13)

hepatitis A A foodborne and waterborne disease of the liver caused by a highly resistant RNA virus and accompanied by jaundice, abdominal pain, and degeneration of the liver tissue. (13)

hepatitis B A bloodborne disease of the liver caused by a fragile DNA virus and accompanied by jaundice, abdominal pain, and degeneration of the liver tissue. (13)

hepatitis B core antigen (HBcAG) An antigen located in the inner lipoprotein coat enclosing the DNA of a hepatitis B virus. (13)

hepatitis B surface antigen (HBsAg) An antigen located in the outer surface coat of a hepatitis B virus; previously known as the Australia antigen. (13)

hepatitis C A bloodborne disease of the liver caused by a virus and accompanied by jaundice, abdominal pain, and degeneration of the liver tissue; also called non-A non-B hepatitis. (13)

hepatocarcinoma (hep-at″o-car-cin-o′mah) Cancer of the liver tissue. (13)

hermaphroditic (her-maf′ro-dit′ik) **organism** An organism that possesses both male and female reproductive organs. (16)

herpes simplex A viral disease of the skin and nervous system, often characterized by blisterlike sores. (11)

heterophile (het′er-o-fīl) **antigen** An antigen that occurs in apparently unrelated species of organisms. (9, 13, 18)

heterotrophic (het"er-o-trof'ik) **organism** An organism that feeds on preformed organic matter and obtains energy from this matter. (4)

hexachlorophene (hek"sah-klo'ro-fēn) A bisphenol compound containing six chlorine atoms; used in disinfectants and antiseptics. (22)

histamine (his'tah-mēn) A mediator in type I allergic reactions; histamine is released from the granules in mast cells and basophils and causes the contraction of smooth muscles. (2)

histocompatibility (his"to-kom-pat"ĭ-bil'ĭ-te) **molecules** Molecules on the surface of animal tissue cells; involved in transplant acceptance and rejection. (20)

Histoplasma capsulatum A fungus often found in the human lung and the cause of histoplasmosis. (14)

histoplasmosis A fungal disease of the lungs and other systemic organs caused by *Histoplasma capsulatum* often occurring in immunocompromised individuals. (14)

His-Werner disease An alternate name for trench fever. (9)

holding method A process of pasteurization in which milk is heated at 62.9° C for 30 minutes; also known as LTLT method for "low temperature, long time." (21, 24)

humoral immunity Immunity arising from the activity of antibodies directed against antigens in the bloodstream; antibody-mediated immunity. (18)

Hutchinson's triad Deafness, impaired vision, and notched, peg-shaped teeth; may accompany congenital syphilis. (10)

hyaluronidase (hi"ah-lu-ron'ĭ-dās) An enzyme that digests hyaluronic acid and thereby permits the penetration of parasites through the tissues; known as the spreading factor. (17)

hydatid cyst A thick-walled body formed in the human liver by *Echinococcus granulosus*. (16)

hydrogen bond A weak chemical bond that forms between protons and adjacent pairs of electrons. (2, 5)

hydrophobia "Fear of water," an emotional condition arising from the inability to swallow as a consequence of rabies. (13)

Hymenolepis (hi"me-nol'e-pis) *nana* A multicellular flatworm parasite of the intestine commonly known as the dwarf tapeworm. (16)

hyperimmune serum Serum that contains a higher than normal amount of a particular antibody. (19)

hypha (hi'fah) A microscopic filament of cells that represents the basic unit of a fungus. (14)

I

icosahedron A symmetrical figure composed of 20 triangular faces and 12 points; one of the major shapes taken by the viral capsid. (11)

idoxuridine (i-doks-ur'ĭ-dēn) The drug 5-iodo-2-deoxyuridine (IDU) that replaces thymine in DNA molecules and is useful in treating herpes simplex. (12)

IgA An antibody in humoral immunity; present in the respiratory and gastrointestinal tracts and in body secretions such as milk. (18)

IgD An antibody of uncertain function believed to act as a receptor site on B-lymphocytes. (18)

IgE The antibody involved in type I hypersensitivity reactions; fixes to the surface of mast cells and basophils. (18, 20)

IgG The major circulating antibody in antibody-mediated immunity; passes across the placental barrier; principal component of the secondary antibody response. (18, 19)

IgM The largest antibody formed in humoral immunity; remains in the circulation; principal component of the primary antibody response. (18, 19)

imidazoles (im-id'ah-zolz) A group of antifungal drugs that interfere with sterol synthesis in fungal cell membranes; includes miconazole and ketoconazole. (23)

immune adherence phenomenon Increased phagocytosis of a cell resulting from the attachment of certain complexes to the cell surface. (18)

immune complex hypersensitivity Type III hypersensitivity in which antigens combine with antibodies to form aggregates that are deposited in blood vessels or on tissue surfaces. (20)

immune complexes Aggregates of antigen–antibody material that are deposited in body tissues; characteristic of type III hypersensitivity. (20)

immunoelectrophoresis (im"mu-no-e-lek"tro-fo-re'sis) A laboratory procedure in which antigen molecules move through an electric field and then diffuse to meet antibody molecules to form a precipitation line. (19)

immunoglobulin An alternative term for antibody. (18)

immunologic tolerance The phenomenon in which the body tolerates itself but responds to nonself. (18)

imperfect fungi Fungi that are known to multiply only by asexual processes. (14)

impetigo contagiosum Infectious skin disease accompanied by abscesses and boils and generally due to staphylococci or streptococci. (10)

inactivated virus A weakened virus that results from treatment with physical or chemical agents; used in immunizating agents because of its reduced virulence; sometimes called a "dead virus." (11, 19)

inclusion A granulelike body that forms in cellular cytoplasm infected with certain organisms; the body represents accumulations of organisms such as chlamydiae or viruses. (10, 11)

incubation period The time that elapses between entry of a parasite to the host and the appearance of symptoms. (17)

induced mutation A mutation arising from a mutagenic agent used under controlled laboratory conditions. (6)

induration A thickening and drying of the skin tissue that occurs in type IV hypersensitivity reactions. (20)

infarction A blockage of the blood vessels.

infection The relationship between two organisms and the competition for supremacy that takes place between them. (17)

infection allergy A type IV hypersensitivity in which the immune system responds to the presence of certain microbial agents. (20)

infectious mononucleosis A disease of the white blood cells caused by an icosahedral DNA virus and accompanied by sore throat, mild fever, and malaise. (13)

inflammation A nonspecific defensive response to injury; usually characterized by red color from blood accumulation, warmth from the heat of blood, swelling from fluid accumulation, and pain from injury to local nerves. (18)

influenza A disease of the lungs caused by a helical RNA virus and accompanied by cough and malaise. (13)

initial body A noninfectious form of chlamydia in the later stage of reproduction. (10)

innate immunity An inborn capacity for resisting disease. (19)

insect An arthropod having six legs; examples are fleas, mosquitoes, and lice.

insertion sequence A segment of DNA that forms a copy of itself, after which the copy moves into areas of gene activity to interrupt the genetic coding sequence. (6)

interferon (in"ter-fēr'on) An antiviral protein produced by body cells on exposure to viruses; interferon triggers production of a second protein that binds to mRNA coded by the virus and thereby inhibits viral replication. (11)

interleukins (in"ter-loo'kinz) Lymphokines produced by white blood cells that act on other white blood cells; important in cellular immunity. (18)

interleukin 2 A lymphokine found to have value in cancer therapy by activating natural killer cells. (20)

intermediary host The host in which the larval or other intermediary stage of a multicellular parasite is found. (16)

invasiveness The ability of a parasite to invade the tissues of the host and cause structural damage to the tissues. (17)

iodophors (i-o'do-forz) Complexes of iodine and detergents that release iodine over a long period of time; used as antiseptics and disinfectants. (22)

ionizing radiations Radiations such as gamma rays and X rays that cause the formation of ions. (21)

isograft Tissue taken from an identical twin and grafted to the other twin. (20)

isomers (i'so-merz) Molecules with the same molecular formula but different structural formulas. (2)

isoniazid (i"so-ni'ah-zid) A chemotherapeutic agent effective against the tubercle bacillus. (7, 23)

isopropyl alcohol A two-carbon alcohol compound widely used as a disinfectant; known as rubbing alcohol. (22)

isotopes Variants of an atom in which the numbers of neutrons differ. (2)

Ixodes dammini (iks-o'dēz dam'in-i) The species of tick that transmits Lyme disease. (9)

J

jaundice A condition in which bile seeps into the circulatory system, causing a dull yellow color of the complexion. (13, 17)

Job's syndrome An immune disorder characterized by defective chemotaxis between phagocyte and microorganism. (20)

K

kala-azar An alternative name for leishmaniasis. (15)

kanamycin An aminoglycoside antibiotic derived from a *Streptomyces* species; used to treat infections caused by Gram-negative bacteria. (23)

Kaposi's sarcoma A type of skin cancer that affects immunocompromised patients such as those with AIDS. (13)

kappa factors Nucleic acid particles produced by species of *Paramecium*; appear responsible for the production of toxins by these protozoa. (15)

Kawasaki disease A disease of undetermined origin accompanied by a skin rash and desquamation and often involving the cardiovascular system. (12)

kefir (ke-fir') A fermentation product of goat's milk that contains acid and alcohol. (24)

keratitis Infection of the cornea of the eye.

keratoconjunctivitis (ker"ah-to-kon-junk"tĭ-vi'tis) Eye inflammation accompanied by infection of the cornea and conjunctiva; the condition is accompanied by tearing, swelling, and sensitivity to light. (12)

ketoconazole (ke-te-kon'ah-zōl) An imidazole drug often used in the treatment of various fungal diseases. (23)

killer T-lymphocyte A type of T-lymphocyte that attacks and destroys cells altered by the presence of antigens; important in the destruction of cancer cells; also called a killer cell. (18, 20)

Kirby–Bauer test An agar diffusion test used to determine the antibiotic concentration effective against a test organism. (23)

Klebs–Löffler bacillus A common term for *Corynebacterium diphtheriae*. (7)

Klebsiella (klebs-e-el'ah) ***pneumoniae*** A Gram-negative encapsulated bacterial rod transmitted by contact and involved in disease of the respiratory tract and intestine. (7, 10)

Koch's postulates A set of procedures by which a specific organism can be related to a specific disease. (1)

Koch–Weeks bacillus An alternative name for *Haemophilus aegypticus*. (10)

Koplik spots Red patches with white central lesions that form on the gums and walls of the pharynx during the early stages of measles. (11, 12)

Krebs cycle A cyclic series of enzyme-catalyzed reactions in which carbon from acetyl-CoA is released as carbon dioxide; the reactions also yield protons and high-energy electrons that are transported among coenzymes and cytochromes as their energy is released. (5)

kumiss (koo′mis) A fermentation product of mare's milk that contains acid and alcohol. (25)

kuru A slow virus disease characterized by slow degeneration of the brain tissue. (13)

L

lactalbumin A protein occurring in minor quantities in cow's milk. (24)

lactose A milk sugar composed of one molecule of glucose and one molecule of galactose. (2, 5, 24)

lag phase A portion of a bacterial growth curve encompassing the first few hours of the population's history; minimal reproduction occurs. (4)

larva A preadult immature stage in the life cycle of certain eukaryotic organisms; also, a tiny worm in the life cycle of a multicellular parasite. (16)

Lassa fever A hemorrhagic fever disease of the blood caused by an RNA virus and accompanied by throat lesions. (13)

lecithinase (les′ĭ-thĭ-nās) A toxin produced by gas gangrene bacilli that dissolves the membranes of tissue cells. (9)

Legionella (lēg-on-el′ah nu-mof″i-lah) **pneumophila** A Gram-negative bacterial rod transmitted by droplets and the cause of Legionnaires' disease. (7)

legionellosis An alternate name for Legionnaires' disease. (7)

Legionnaires' disease A bacterial disease of the lungs caused by *Legionella pneumophila* and accompanied by cough, fever, and malaise. (7)

legumes Plants that bear their seeds in pods; examples are beans, soybeans, and alfalfa. (26)

Leishmania (lēsh-ma′ne-ah) **species** Flagellated protozoa transmitted by sandflies and the causes of leishmaniasis; include *L. donovani* and *L. tropica*. (15)

leishmaniasis (lēsh-ma-ni′ah-sis) A protozoal disease of the white blood cells caused by *Leishmania* species, transmitted by sandflies, and accompanied by fever, sores, and emaciation; occurs in visceral and cutaneous forms. (15)

leproma A tumorlike growth on the skin associated with leprosy. (10)

lepromin (lep-ro′min) **test** A skin test used in the screening and diagnosis of leprosy. (10)

leprosy A bacterial disease transmitted by contact and caused by *Mycobacterium leprae*; accompanied by destruction of the skin tissues and the peripheral nerves leading to local anesthesia. (10)

Leptospira interrogans A spirochete and the cause of leptospirosis. (9)

leptospirosis A soilborne bacterial disease caused by *Leptospira interrogans* and accompanied by mild fever symptoms. (9)

leukemia A cancer of the white blood cells.

leukocidin (loo″ko-si′din) An enzyme that destroys phagocytes thereby preventing phagocytosis of the parasite. (17)

leukocyte An alternative name for the white blood cell. (17)

leukopenia A condition characterized by a drop in the normal number of white blood cells.

leukotrienes (loo′ko-trēnz) A class of organic compounds, which appear when allergens combine with IgE on mast cell surfaces and which increase vascular permeability, contract smooth muscles, and attract neutrophils. (20)

lichen An organism composed of a fungal mycelium within which are embedded photosynthetic algae. (14)

lincomycin An antibiotic used as a penicillin substitute for diseases caused by Gram-positive bacteria. (23)

lipase A fat-digesting enzyme produced by certain contaminating bacteria in milk. (24)

lipids A group of organic compounds that dissolve in organic solvents; composed of carbon, hydrogen, and oxygen; include fats; used in energy metabolism and structural compounds. (2, 4, 5)

Lipschütz bodies Inclusions that form in the nucleus of cells infected with herpesviruses. (11)

Listeria monocytogenes (lis-ter′e-ah mon″o-si-toj′e-nez) A small Gram-positive rod that causes listeriosis. (9)

listeriosis A soilborne and foodborne bacterial disease caused by *Listeria monocytogenes* and accompanied by mild symptoms except in pregnant women where miscarriage may occur. (9)

littoral zone The environment along the shoreline of an ocean. (25)

Loa loa A multicellular roundworm parasite of the eye commonly known as the eyeworm. (16)

lobar pneumonia Pneumonia that involves an entire side or lobe of the lung. (7)

local disease A disease restricted to a single area of the body, usually the skin. (17)

lockjaw A colloquial expression for tetanus based on the spasms of the jaw muscles. (9)

locus An individual site on the bacterial chromosome where genetic activity can be located. (6)

logarithmic (log) phase The second portion of a bacterial growth curve, in which active growth leads to a rapid rise in numbers of the population. (4)

long incubation hepatitis An alternate name for hepatitis B. (13)

louse A type of insect of the genus *Pediculus* that transmits epidemic typhus and other diseases. (9)

lumpy jaw The name for actinomycosis in the jaw; characterized by hard nodules in the tissue. (10)

Lyme disease A serious arthropodborne bacterial disease transmitted by the tick, caused by *Borrelia burgdorferi*, and accompanied by skin rash (ECM) and malaise and at a later time swelling and degeneration of the large joints. (9)

lymph nodes Bean-shaped organs along the lymph vessels that are involved in body response to disease; the major cells of these organs are phagocytes and lymphocytes. (17)

lymphoblast The young cell to which the T-lymphocyte reverts; lymphoblasts secrete lymphokines. (18)

lymphocyte A type of leukocyte that functions in the immune system. (17, 18, 20)

lymphocytic choriomeningitis (lim-fo-sit′ik kor″e-o-men-in-gi′tis) A disease of the brain tissue possibly due to a virus and accompanied by headache, drowsiness, and stupor and large numbers of lymphocytes in the meninges. (13)

lymphogranuloma (lim″fo-gran-u-lo′mah) **venereum** A sexually transmitted bacterial disease of the external genital organs caused by *Chlamydia trachomatis* and accompanied by swollen lymph nodes. (10)

lymphokines (lim′fo-kīnz″) Proteins that increase the efficiency of phagocytosis at the antigen site in cellular immunity. (18)

lymphopoietic (lim-fo″poi-et′ik) **cells** Primitive cells that arise from stem cells and are modified to form B-lymphocytes or T-lymphocytes. (18)

lyophilization (li-of″ĭ-li-za′shun) A process in which food or other material is deep frozen, after which its liquid is drawn off by a vacuum. (24)

lysogenic bacterium A bacterium that carries a prophage. (6)

lysogeny (li-soj′e-ne) The phenomenon in which a virus remains in the cell cytoplasm as a fragment of DNA or attaches to the chromosome, but fails to replicate in or destroy the cell. (6, 11)

lysosome A submicroscopic organelle found in eukaryotic cells; contains digestive enzymes. (3, 15, 18)

lysozyme (li′so-zīm) A nonspecific enzyme found in the tears and saliva that digests the peptidoglycan of Gram-positive bacteria and leads to their destruction. (5, 11, 18, 24)

lytic cycle The process wherein a virus replicates within a host cell and destroys the host cell in the process of replicating.

M

Machupo A type of viral hemorrhagic fever occurring primarily in South America. (13)

"mad cow disease" A common term for bovine spongiform encephalopathy. (10)

macrophages Large cells derived from monocytes and found within the tissues; macrophages actively phagocytize foreign bodies and comprise the reticuloendothelial system (RES). (17, 18)

macules Pink-red skin spots. (9, 12)

maculopapular (mak″u-lo-pap′u-lar) **rash** A rash consisting of pink-red spots that later become dark red before fading; occurs in rickettsial diseases. (9)

Madura foot Substantial swelling of the tissues of the foot due to *Nocardia asteroides*. (10)

magnetosome A cytoplasmic body in certain bacteria that assists orientation to the environment by aligning with the magnetic field. (4)

major histocompatibility complex A set of genes that controls expression of the histocompatibility molecules and is involved in transplant rejection. (20)

malaria A serious protozoal disease of the red blood cells caused by *Plasmodium* species, transmitted by mosquitoes, and accompanied by periods of high fever, anemia, and blood clotting. (15)

Malta fever An alternate name for brucellosis. (9)

mannitol salt agar An enriched medium that encourages the growth of staphylococci; the medium contains a high percentage of salt, a feature that makes it inhibitory to most other organisms. (4)

Marburg disease A viral disease caused by an RNA virus and accompanied by fever and hemorrhaging. (13)

mast cells Connective tissue cells to which IgE fixes in type I hypersensitivity reactions; the cells degranulate and release histamine during allergic attacks. (20)

Mastigophora (mas″ti-gof′o-rah) The class or group of protozoa whose members move by means of flagella. (15)

maternal antibodies Type G antibodies that cross the placenta from the maternal to the fetal circulation and protect the newborn for the first few months of life. (18)

measles A communicable respiratory disease caused by an RNA helical virus and accompanied by respiratory symptoms and a blushlike skin rash. (12)

mechanical vector A living organism or an object that transmits disease agents on its surface. (17)

melioidosis (me″le-oi-do′sis) A soilborne bacterial disease caused by *Pseudomonas pseudomallei* and accompanied by lung abscesses. (9)

memory cells Cells derived from B-lymphocytes or T-lymphocytes that react rapidly upon the future recurrence of antigens in the tissues. (18)

meninges The covering layer or layers of the brain and spinal cord; the individual meninges are the dura mater, arachnoid, and pia mater. (7, 13)

meningitis A general term for infection of the meninges due to any of several bacteria, fungi, viruses, or protozoa. (7, 13, 14, 15)

meningococcemia (mĕ-ning″go-kok-se′me-ah) A type of endotoxic shock due to *Neisseria meningitidis*. (7)

meningococcus A colloquial expression for *Neisseria meningitidis*. (7)

merozoite (mer″o-zo′ĭt) A stage in the life cycle of *Plasmodium* species; parasites invade the red blood cells in this form. (15)

mesophiles Organisms that grow at the temperature range of 20–40° C. (4)

metabolism The sum of all biochemical processes taking place in a living cell. (5)

metacercaria (met"ah-cer-cār′e-ah) An encysted intermediary stage in the life cycle of a fluke. (16)

metachromatic granules Phosphate-storing granules that stain deeply with methylene blue; commonly found in *Corynebacterium diphtheriae*; also called volutin granules. (4, 7)

metronidazole (me"tro-ni′dah-zōl) **(Flagyl)** A chemotherapeutic agent used to treat trichomoniasis and other protozoal diseases. (14, 23)

miasma (mi-az′mah) An ill-defined entity generally referring to an altered chemical quality of the atmosphere; believed to cause disease before establishment of the germ theory of disease. (1)

miconazole (mi-kon′ah-zōl) An imidazole drug often used in the treatment of topical and systemic fungal disease. (23)

microaerophilic (mi"kro-a-rō-fil′ik) **organism** An organism that grows best in an oxygen-reduced environment. (4)

microbe An alternative expression for microorganism.

microfilariae Intermediate eellike forms of *Wuchereria bancrofti*. (16)

micrometer A unit of measurement equivalent to one millionth of a meter; the unit is designated as μm and is commonly used in measuring microorganisms. (3)

microorganism A microscopic form of life including bacteria, viruses, fungi, protozoa, and some multicellular parasites.

miliary tuberculosis Tuberculosis that spreads through the body. (7)

miracidium (mi-rah-sid′e-um) A ciliated larva representing an intermediary stage in the life cycle of a fluke. (16)

mite A spiderlike arthropod that transmits such diseases as rickettsialpox and tsutsugamushi. (9)

MMR A three-component vaccine providing immunity to measles, mumps, and rubella. (12)

mold A type of fungus that consists of chains of cells and appears as a fuzzy mass. (14)

molecule The smallest part of a compound that retains the properties of the compound. (2)

molluscum (mŏ-lus′kum) **body** A cytoplasmic inclusion that occurs in cells infected with the viruses of molluscum contagiosum. (12)

molluscum contagiosum A viral skin disease caused by a DNA virus and accompanied by firm, waxy, wartlike lesions with a depressed center. (13)

Monera A kingdom in the Whittaker classification of living things; composed of prokaryotes such as bacteria and cyanobacteria. (3)

monoclonal antibodies Antibodies produced by a clone of hybridoma cells consisting of antigen-stimulated plasma cells fused to myeloma cells. (18)

monocyte A leukocyte with a large bean-shaped nucleus; functions in phagocytosis. (17)

monotrichous (mon"o-trik′us) **bacteria** Bacteria that possess a single flagellum. (4)

mortality rate The percentage of victims of a disease who succumb to the disease.

most probable number (MPN) A laboratory test in which a statistical evaluation is used to estimate the number of bacteria in a sample of fluid; often employed in determinations of coliform bacteria in water. (25)

moxalactam (mox-ah-lac′tam) A monobactam antibiotic used against Gram-negative bacteria. (23)

M protein A protein that enhances the pathogenicity of streptococci by allowing organisms to resist phagocytosis and adhere firmly to tissue. (7)

μm The symbol for micrometer, a millionth of a meter. (4)

mumps A communicable disease caused by an icosahedral DNA virus and accompanied by swelling of the parotid and other salivary glands. (12)

mutation A change in the characteristics of an organism arising from alteration of the chromosome. (6)

mutualism A close and permanent association between two populations of organisms in which both benefit from the association. (4)

myasthenia gravis A type II hypersensitivity in which antibodies react with acetylcholine receptors on the membranes of muscle fibers. (20)

mycelium (mi-se′le-um) A visible mass of tangled filaments of fungal cells. (14)

Mycobacterium avium-intracellulare An acid-fast bacterial rod that causes lung infection in immunocompromised persons such as those with AIDS. (7)

Mycobacterium leprae An acid-fast bacterial rod and the cause of leprosy. (10)

Mycobacterium tuberculosis An acid-fast bacterial rod and the cause of tuberculosis. (7)

mycologist One who studies the fungi. (14)

mycology The study of fungi. (14)

Mycoplasma hominis A mycoplasma that causes mycoplasmal urethritis. (10)

Mycoplasma pneumoniae A mycoplasma that causes mycoplasmal pneumonia (7).

mycoplasmal pneumonia An airborne primary pneumonia transmitted by droplets, caused by *Mycoplasma pneumoniae*, and accompanied by lung tissue destruction and respiratory symptoms. (7)

mycoplasmal urethritis A sexually transmitted disease caused by *Mycoplasma hominis* and accompanied by urinary tract discomfort. (10)

mycoplasmas A group of tiny bacteria that lack cell walls and are seen only with the electron microscope. (3, 7, 10)

mycotoxin A fungal toxin. (14)

myeloma A mass of cancerous cells. (18)

myocarditis Infection of the heart muscle often due to Coxsackie virus. (13)

N

Naegleria fowleri (na-gle′ri-ah fow-ler′i) A flagellated protozoan transmitted by water and the cause of primary amebic meningoencephalitis. (15)

nalidixic (nal-ĭ-diks′ik) **acid** A chemotherapeutic agent that blocks protein synthesis in certain Gram-negative bacteria that cause urinary tract infections. (23)

NANB hepatitis Non-A non-B hepatitis; a form of hepatitis caused by a virus unrelated to the viruses of hepatitis A and hepatitis B; the former term for hepatitis C and now used for any form of infectious hepatitis in which the virus remains unidentified. (13)

nanometer A unit of measurement equivalent to one billionth of a meter; the unit is designated as nm and is often used in measuring viruses and the wavelength of energy. (3)

narrow-spectrum antibiotic An antibiotic that is useful for a restricted group of microorganisms. (23)

naturally acquired active immunity Immunity resulting from the immune system's response to disease antigens. (9)

naturally acquired passive immunity Immunity resulting from the passage of antibodies to the fetus via the placenta or the milk of a nursing mother; synonymous with congenital immunity. (19)

Necator americanus A multicellular roundworm parasite of the intestine and other organs transmitted by moist vegetation and commonly known as the New World hookworm. (16)

necrosis Cell death.

negative stain technique The staining process that results in clear bacteria on a stained background when viewed with the light microscope. (3)

Negri bodies Cytoplasmic inclusions in brain cells infected with rabies viruses. (11)

Neisseria gonorrhoeae (ni-se′re-ah gon″o-re′ā) A Gram-negative bacterial diplococcus transmitted by sexual contact and the agent of gonorrhea. (10)

Neisseria meningitidis (ni-sc′rc-ah men-in″gi-ti′dis) A Gram-negative bacterial diplococcus transmitted by droplets and the cause of meningococcal meningitis. (7)

nematodes A common expression for roundworms belonging to the phylum Nemathelminthes. (16)

neomycin An aminoglycoside antibiotic derived from a *Streptomyces* species; used topically for infections caused by Gram-negative bacteria, especially in the eye. (23)

neoplasm An uncontrolled growth of cells; often called a tumor. (11)

Neufeld quellung reaction A laboratory test used to distinguish strains of *Streptococcus pneumoniae*. (7)

neuraminidase (nūr-ah-min′ĭ-dās) An enzyme on the surface spikes of certain viruses that dissolves the cell membrane when the virus attaches to a cell during replication. (12)

neurotoxin A toxin that is active in the nervous system of the host. (17)

neutralization A type of antigen–antibody reaction in which the activity taking place between reactants is not visible. (18, 19)

neutrophil A type of polymorphonuclear cell that functions chiefly as a phagocyte. (17)

nicotinamide adenine dinucleotide (NAD) A coenzyme that transports electrons during oxidative phosphorylation and fermentation reactions. (5)

night soil Human feces sometimes used as an agricultural fertilizer. (16)

nitrogen fixation The general term for the chemical process in which organisms trap atmospheric nitrogen and use it to form organic compounds. (25)

nitrogenous (ni-troj′en-us) **base** Any of five nitrogen-containing compounds found in nucleic acids including adenine, guanine, cytosine, thymine, and uracil. (2, 4, 5)

Nocardia asteroides An acid-fast funguslike bacterial rod that causes endogenous infections of the lungs and other internal organs accompanied by abscesses; the cause of Madura foot. (10)

noncommunicable disease A disease whose causative agent is acquired from the environment and is not easily transmitted to the next host. (17)

normal flora The populations of organisms that infect various parts of the body without usually causing disease. (17)

Norwalk virus A virus transmitted by food and/or water and involved in disease of the intestine. (13)

nosocomial (nos″o-ko′me-al) **disease** A disease acquired during an individual's stay at a hospital. (10)

nucleocapsid The combination of genome and capsid in a virus. (11)

nucleoid The chromosomal region of a bacterium. (4)

nucleotide A component of a nucleic acid consisting of a carbohydrate molecule, a phosphate group, and a nitrogenous base. (2, 5)

nutrient agar A common bacteriological growth medium consisting of beef extract, peptone, water, and agar. (4)

nutrient broth A common bacteriological growth medium consisting of beef extract, peptone, and water. (4)

nystatin (nis′tah-tin) An antifungal drug effective against *Candida albicans*. (14, 23)

O

Okazaki (o-ka-zak′e) **fragments** Segments of DNA that combine with one another to form a DNA molecule during chromosomal duplication. (6)

oncofetal antigen An antigen in tumor cells thought to be expressed during differentiation of the tissues in the embryonic stage. (20)

oncogene (ong′ko-jēn) A region of DNA in human cells thought to induce uncontrolled growth of the cell if permitted to function. (11)

oncology The study of tumors and cancers. (11, 20)

oocyst (o′o-sist) An oval body in the reproduction cycle of certain protozoa that develops by a complex series of asexual and sexual processes. (15)

oospore (o′o-spōr) A sexually produced spore that is formed by members of the Oomycetes class of fungi. (14)

operon The unit of gene activity that expresses a particular trait; also, the unit that controls protein synthesis. (5)

ophthalmia Severe inflammation of the eyes. (10)

opportunists Organisms that invade the tissues when the body defenses are suppressed. (13, 17)

opsonins (op′so-ninz) Antibodies or complement components that encourage phagocytosis. (18)

opsonization Enhanced phagocytosis due to the activity of antibodies or complement. (18)

orchitis (or-ki′tis) A condition caused by the mumps virus in which the virus damages the testes. (12)

organelles Membrane-bound compartments in eukaryotic cells; sites of various cellular functions. (3, 14, 15)

Oriental sore A cutaneous form of leishmaniasis characterized by skin sores. (15)

ornithosis An alternative name for psittacosis. (7)

osmosis The flow of water from a region of low concentration of chemical substance through a semipermeable membrane to a region of high concentration of the same chemical substance. (24)

oxidases Enzymes that catalyze oxidation–reduction reactions. (5)

oxidation A chemical change in which electrons are lost by an atom; also, the union of oxygen with a chemical substance. (2)

oxidative phosphorylation A series of sequential steps in which energy is released from electrons as they pass among coenzymes and cytochromes, and ultimately, to oxygen; the energy is used to combine phosphate ions with ADP molecules to form ATP molecules. (5)

P

pandemic A worldwide epidemic. (17)

papilloma (pap″ĭ-lo′mah) A tumor of the skin tissue. (11)

papilloma virus An icosahedral DNA virus and one of the causes of skin warts. (13)

papules Pink pimples on the skin. (12)

Paragonimus westermani (par″ah-gon′i-mus wes′ter-man-i) A multicellular flatworm parasite of the lung transmitted by seafood and commonly known as the lung fluke. (16)

parasite A type of heterotrophic organism that feeds on live organic matter such as another organism. (4, 17)

parasitemia Spread of protozoa and multicellular worms through the circulatory system. (15, 16)

parasitism A close and permanent association between two organisms in which one feeds on the other and may cause injury to the other organism. (4, 17)

parasitology The discipline of biology concerned with pathogenic protozoa. (15)

paromomycin (par′o-mo-mi″sin) A drug used in the treatment of amebiasis and balantidiasis. (15)

paroxysm (par-ok′sizm) A sudden intensification of symptoms such as a severe bout of coughing. (7)

parvovirus An icosahedral DNA virus that causes disease in dogs and is the cause of fifth disease in humans. (12)

passive agglutination An immunological procedure in which antigen molecules are adsorbed to the surface of latex spheres or other carriers that agglutinate when combined with antibodies. (19)

Pasteurella multocida (pas-tur-el′ah mul-toc′i-dah) A Gram-negative bacterial rod and the cause of pasteurellosis. (10)

pasteurellosis (pas″tur-el-lo′sis) A bacterial disease transmitted by animal bites, caused by *Pasteurella multocida*, and accompanied by abscess formation at the bite site. (10)

pasteurization A heating process that destroys pathogenic bacteria in a fluid such as milk and lowers the overall number of bacteria in the fluid. (21, 24)

pathogen A type of parasite that causes disease in the host organism. (4, 17)

pathogenicity The ability of a parasite to gain entry to a host and bring about a physiological or anatomical change interpreted as disease. (17)

Paul–Bunnell test A diagnostic procedure for infectious mononucleosis in which a sample of patient's serum is combined with sheep red blood cells; the cells agglutinate in a positive test. (13)

pebrine (pa-brēn′) A protozoal disease of silkworms studied by Louis Pasteur. (1)

Pediculus (ped-ik′u-lus) The genus of lice that transmit epidemic typhus and other diseases. (9)

pertussis A bacterial disease of the upper respiratory tract in which an accumulation of mucus causes a narrowing of the tubes and a characteristic "whoop" on inspiration. (7)

pellicle A rigid covering layer of certain protozoa composed in part of chitinlike material. (15)

pelvic inflammatory disease Disease of the pelvic organs; often a complication of a sexually transmitted disease. (10)

penicillin Any of a group of antibiotics derived from *Penicillium* species or produced synthetically; the antibiotics are effective for Gram-positive bacteria and several Gram-negative bacteria; they interfere with cell wall synthesis. (23)

penicillinase An enzyme produced by certain microorganisms that converts penicillin to penicilloic acid and thereby confers resistance against penicillin. (23)

peptidoglycan (pep″tĭ-do-gli′kan) A complex molecule of the bacterial cell wall composed of alternating units of N-acetylglucosamine and N-acetylmuramic acid; formation of the molecule is prevented by penicillin antibiotics. (4, 23)

period of acme The period in a disease at which specific symptoms occur and the disease is at its height. (17)

period of convalescence The period in a disease in which the body's systems return to normal. (17)

period of decline The period of a disease in which the symptoms subside. (17)

period of incubation *See* incubation period.

period of prodromal symptoms The period during which general symptoms occur in the body. (17)

peritrichous bacteria (per″e-trik′us) Bacteria that possess flagella over the entire surface of the cell. (4)

pH An abbreviation for the negative logarithm of the amount of hydrogen ions in 1 liter of solution; the pH scale extends from 1 to 14 and indicates the degree of acidity or alkalinity of a solution. (2)

phagocyte A cell that practices phagocytosis. (18)

phagocytosis (fag″o-sī-to′sis) A process in which solid particles are taken into the cell; important in nutritional processes and in defense against disease. (18)

phagosome A vesicle that contains particles of phagocytized material. (18)

phenol coefficient A number that indicates the effectiveness of an antiseptic or disinfectant as compared to phenol. (22)

phosphatase (fos′fah-tās) An enzyme normally found in milk; phosphatase is destroyed by pasteurization processes; its absence indicates that the pasteurization has been successful. (24)

photoautotrophs (fo″to-aw′to-trophz) Organisms that utilize light energy to synthesize nutrients from carbon dioxide. (5)

photoheterotrophs (fo″to-het′er-o-trophz) Organisms that utilize light energy to synthesize nutrients from carbon compounds other than carbon dioxide. (5)

photophobia Sensitivity to bright light.

photosynthesis The biochemical process in which light energy is converted to chemical energy used in carbohydrate synthesis. (5)

picornavirus (pi-kor′nah-vir″us) A small virus containing RNA in its genome. (13)

pigeon fancier's disease A condition that develops from a type III hypersensitivity reaction following exposure to antigens from a pigeon. (20)

pili (pil′i) Short, hairlike appendages of bacteria that anchor the cell to a surface; pili are also involved in conjugations between bacteria. (4, 6, 8)

pinkeye A disease of the conjunctival membranes of the eye usually due to bacteria or viruses and accompanied by red, swollen eyes. (10)

pinocytosis A type of phagocytosis in which materials dissolved in fluid are taken into the cell. (18)

pinta A mild syphilislike disease of the skin occurring in remote parts of the world and due to a species of *Treponema*. (10)

pinworm The common name for *Enterobius vermicularis*, a multicellular roundworm parasite that infects the intestine. (16)

plague A serious arthropodborne bacterial disease transmitted by the flea, caused by *Yersinia pestis* and accompanied by severe blood hemorrhaging. (9)

plaque A clear area on a lawn of bacteria where viruses have destroyed the bacteria; also, the gummy layer of gelatinous material consisting of bacteria and organic matter on the teeth. (10, 11)

plasma The fluid portion of blood remaining after the cells have been removed; serum plus the clotting agents. (17)

plasma cell The cell derived from B-lymphocyte; plasma cells produce antibodies. (18)

plasmid A small, closed loop molecule of DNA apart from the chromosome; plasmids carry genes for drug resistance and pilus formation and are used in genetic engineering experiments. (4, 6, 25)

Plasmodium species Protozoa of the sporozoa group that are transmitted by mosquitoes and infect human red blood cells and cause malaria; include *P. vivax*, *P. malariae*, and *P. falciparum*. (15)

pleomorphic (ple″o-mor′fik) **organism** An organism that appears in a variety of shapes. (9)

pleurodynia (ploo″o-din′e-ah) Infection of the chest wall commonly due to Coxsackie virus. (13)

pneumococcus A colloquial expression for *Streptococcus pneumoniae*. (7)

Pneumocystis carinii (nu″mo-sis′tis car-in′e-e) An opportunistic protozoan that infects the lungs and causes pneumonia in immunocompromised patients such as those with AIDS. (15)

Pneumocystis carinii pneumonia (PCP) A lung infection caused by *Pneumocystis carinii* and accompanied by consolidation of the lung and suffocation; occurs primarily in immunocompromised individuals such as AIDS patients; also called pneumocystosis. (15)

pneumonia Infectious disease of the lower respiratory tract. (7)

pneumonic plague A form of plague in which *Yersinia pestis* invades the lung tissues and causes severe respiratory symptoms. (9)

polymorphonuclear (pol″e-mor″fo-nu′kle-ar) **cell** A type of white blood cell with a multilobed nucleus. (17)

polyvalent serum Serum that contains a mixture of antibodies. (19)

Pontiac fever A form of Legionnaires' disease. (7)

portal of entry The site at which a parasite enters the host. (17)

Pott's disease Tuberculosis of the spine. (7)

pox Pitted scars remaining in recoverers from smallpox. (12)

precipitation A type of antigen–antibody reaction in which thousands of molecules of antigen and antibody cross-link to form particles of precipitate. (18, 19)

prednisone (pred′nih-sōn) A steroid hormone that suppresses the inflammatory response; used to retard transplant rejection. (20)

primaquine (prim′a-kwin) A synthetic drug used to treat malaria in the dormant phase. (15)

primary antibody response The initial response by the immune system to an antigen; characterized by an outpouring of IgM. (18)

primary atypical pneumonia (PAP) An airborne bacterial disease of the lungs caused by *Mycoplasma pneumoniae* and accompanied by degeneration of the lung tissue; also called mycoplasmal pneumonia and walking pneumonia. (7)

primary disease A disease that develops in an otherwise healthy individual. (17)

prions (prē′onz) Infectious particles of protein possibly involved in human diseases of the brain. (11)

proctitis Infection of the rectum. (10)

proglottid (pro-glot′id) One of a series of segments that make up the body of a tapeworm. (16)

prokaryote (pro-kar′e-ōt) A relatively simple organism composed of single cells having a single chromosome but no intracellular organelles, nucleus, or nuclear membrane; reproduction does not involve mitosis. (3, 5)

prontosil (pron′to-sil) A red dye found by Domagk to have significant antimicrobial activity when tested in live animals, and from which sulfanilamide was later isolated. (23)

properdin (pro′per-din) A protein that functions in the alternative pathway of complement activation. (18)

prophage The DNA fragment of a temperate phage. (6)

prophylactic serum (pro″fi-lak′tic) Antibody-rich serum used to protect against the development of a disease. (19)

proteins Chains of amino acids used as structural materials and enzymes in living cells; all proteins have primary and secondary structures, and some have a tertiary structure. (2, 5)

Protista One of the five kingdoms of Whittaker containing the protozoa and various other simple forms; also, the term covered by Haeckel to denote single-celled organisms of microscopic size. (3)

proto-oncogene A region of DNA in the chromosome of human cells; altered by carcinogens into oncogenes that transform cells. (11)

pruritus Itching sensations. (12)

pseudomembrane An accumulation of mucus, leukocytes, bacteria, and dead tissue in the respiratory passages of diphtheria patients. (7)

pseudomembranous colitis A condition of the intestinal wall characterized by membranous lesions; believed due to *Clostridium difficile*. (10, 23)

Pseudomonas aeruginosa (soo″do-mon′as a″er-jin-o′sa) A Gram-negative bacterial rod transmitted by contaminated materials and involved in disease of burnt tissue and the urinary tract. (10)

Pseudomonas pseudomallei A Gram-negative bacterial rod transmitted by soilborne materials and the cause of melioidosis. (9)

pseudopodia (soo″do-po′de-ah) Projections of the cell membrane that allow movement in members of the Sarcodina class of protozoa. (15)

psittacosis An airborne bacterial disease of the lung caused by *Chlamydia psittaci* and accompanied by respiratory discomfort and influenzalike symptoms; occurs in birds of the psittacine group (parrots, etc.) and in humans; also called ornithosis. (7)

psychrophiles (si′kro-fīlz) Organisms that grow at cold temperature ranges of 0° C to 20° C. (4, 24, 25)

psychrotrophic (si″kro-troph′ik) **organisms** Organisms that normally grow at medium temperatures but tolerate and grow at cold temperatures. (4, 24)

ptomaines (tō′mānz) Nitrogen compounds having a strong odor; once thought responsible for food poisoning. (8, 24)

pure culture A culture or colony of microorganisms of one type. (1)

pus A mixture of serum, dead tissue cells, leukocytes, and bacteria that accumulates at the site of infection. (18)

Q

Q fever A rickettsial disease characterized by flulike symptoms. (17)

quartan malaria A type of malaria in which the attacks occur at four-day intervals. (15)

quinine A drug derived from the bark of the *Cinchona* tree; used to treat malaria. (14, 23)

R

R factors Plasmids that occur frequently in Gram-negative bacteria and carry genes for drug resistance. (6, 8)

rabbit fever An alternative name for tularemia. (9)

rabies A serious nervous system disease due to a helical RNA virus, transmitted by an animal bite, and characterized by destruction of the brain tissue leading to paralysis and death. (13)

racial immunity Immunity present in one race of people but not in another. (18)

radioallergosorbent test (RAST) A type of radioimmunoassay in which antigens for the unknown antibody are attached to matrix particles. (19)

radioimmunoassay (RIA) An immunological procedure that uses radioactive-tagged antigens to determine the identity and amount of antibodies in a sample. (19)

rapid plasma reagin (RPR) test A screening procedure for syphilis in which serum from the patient is mixed with syphilis antigen; precipitation occurs in a positive test. (10)

rat bite fever A bacterial disease of the skin and blood due to either *Spirillum minor* or *Streptobacillus moniliformis* and transmitted by a rat bite. (10)

reagin An alternative name for the IgE that stimulates anaphylaxis in the body. (20)

real image An image that can be projected onto a screen; formed by the objective lens in light microscopy. (3)

recombinant DNA molecule A DNA molecule that carries foreign genes. (6)

recombination An alteration in the genetic information in a microorganism arising from the acquisition of DNA; in bacteria recombination occurs by transformation, conjugation, and transduction. (6, 15)

red rot Spoilage in milk and eggs due to the production of red pigment by *Serratia marcescens*. (24)

redia (re'de-ah) An intermediary stage in the life cycle of a fluke. (16)

relapsing fever A serious arthropodborne bacterial disease transmitted by ticks and lice, caused by *Borrelia recurrentis*, and accompanied by periods of fever. (9)

rennin An enzyme that accelerates the curdling of protein in milk. (24)

reservoir A human or other animal that retains disease organisms in the body but has not experienced disease and shows no evidence of illness. (17)

resolving power The numerical value of a lens system that indicates the size of the smallest object that can be seen clearly when using that system. (3, 11)

respiration Any biochemical process in which energy is liberated; respiration may occur in the presence of oxygen (aerobic respiration) or in its absence (anaerobic respiration). (5)

respiratory syncytial (RS) disease An airborne viral disease occurring in young children and accompanied by pneumonia in the lung tissues. (12)

respiratory syncytial (RS) virus A helical RNA virus that causes lung infections primarily in young children. (12)

restriction enzyme A type of endonuclease that splits open a DNA molecule at a specific restricted point; important in genetic engineering experiments. (6)

reticuloendothelial system (RES) A collection of large, monocyte-derived cells that practice phagocytosis within the tissues; also known as the mononuclear phagocyte system. (18)

retrovirus An RNA virus that uses reverse transcriptase to synthesize DNA using RNA as a template. (13)

reverse transcriptase An enzyme that synthesizes a DNA molecule from the code supplied by an RNA molecule. (6, 11, 13)

Reye's syndrome A complication of influenza and chickenpox characterized by vomiting and convulsions as well as liver and brain damage. (12)

rhabditiform larva (rab-dit'i-form) An elongated intermediate form of the hookworm. (16)

rheumatic fever A complication of streptococcal disease in which damage to the heart valves develops from the reactions between antigens and antibodies. (7, 20)

rheumatoid arthritis An autoimmune disease characterized by immune complex formation in the joints. (20)

rhinovirus An icosahedral RNA that causes diseases of the upper respiratory tract commonly called head colds. (12)

ribosomes Ultramicroscopic bodies of RNA and protein that participate in protein synthesis. (3, 4, 5, 11)

rice-water stools Particles of intestinal tissue released in the stools during episodes of cholera. (7)

rickettsiae (rik-et'se-e) A group of small bacteria generally transmitted by arthropods; most rickettsiae are cultivated only within living tissue medium. (3, 7, 9)

Rickettsia prowazecki The rickettsia transmitted by the louse and the cause of epidemic typhus. (9)

Rickettsia rickettsii The rickettsia transmitted by the tick and the cause of Rocky Mountain spotted fever. (9)

Rickettsia tsutsugamushi (soot"soo-ga-moosh'e) The rickettsia transmitted by the mite and the cause of tsutsugamushi. (9)

Rickettsia typhi The rickettsia transmitted by fleas and the cause of endemic typhus. (9)

rickettsialpox A relatively mild miteborne disease of the blood caused by *Rickettsia akari* and accompanied by fever and a skin rash. (9)

rifampin (rif-am'pin) An antibiotic prescribed for tuberculosis and leprosy patients and for carriers of *Neisseria* and *Haemophilus* species. (23)

Rift Valley fever A disease caused by an RNA virus, occurring primarily in the Rift Valley of Africa and accompanied by intense fever and joint pains. (13)

ringworm A fungal disease caused by numerous species of fungus and characterized by scaly, circular patches of infection, usually on the head; ringworm of the feet is athlete's foot; ringworm of the groin, nails, face, and other body regions is also possible. (14)

Ritter's disease Impetigo contagiosum of the skin due to *Staphylococcus aureus*. (10)

Rocky Mountain spotted fever (RMSF) A relatively serious tickborne disease of the blood caused by *Rickettsia rickettsii* and accompanied by high fever and a skin rash beginning on the body extremities and proceeding toward the body trunk. (9)

rolling circle mechanism A type of DNA replication in which the broken strand of DNA "rolls off" the loop and serves as a template for synthesis of a complementary strand of DNA. (6)

ropy bread Bread that has become soft and stringy due to capsular material deposited by a bacterium such as *Bacillus subtilis*. (4, 24)

ropy milk Milk that has become thick and viscous due to accumulation of capsular material deposited by a bacterium such as *Alcaligenes viscolactis*. (4, 24)

rose spots Bright red skin spots associated with diseases such as typhoid fever and relapsing fever. (8, 9)

rotavirus An RNA virus transmitted by food and/or water and involved in disease of the intestine. (13)

rubella A communicable skin disease caused by an icosahedral RNA virus and accompanied by mild respiratory symptoms and a measleslike rash; can cause damage in fetus if contracted by a pregnant woman. (12)

rubeola An alternate name for measles. (12)

S

Sabin vaccine A type of polio vaccine prepared with attenuated viruses; the vaccine is taken orally. (13)

Sabouraud dextrose agar (sab′oo-rō) A growth medium for fungi. (14)

saddleback fever A condition characterized by fluctuations of fever. (13)

Salk vaccine A type of polio vaccine prepared with viruses inactivated with formaldehyde; the vaccine is injected into the body. (13)

Salmonella enteritidis (en″ter-it′i-dis) A Gram-negative bacterial rod transmitted by food and/or water and involved in disease of the intestine. (8)

Salmonella typhi A Gram-negative bacterial rod transmitted by food and/or water and the cause of typhoid fever, a disease of the intestine. (8)

salpingitis (sal″pin-ji′tis) Blockage of the Fallopian tubes; a possible complication of a sexually transmitted disease. (10)

sandfly fever A sandfly-transmitted viral disease occurring in Mediterranean regions and accompanied by high fever and joint pains. (13)

sanitize To reduce the microbial populations to a safe level as determined by public health standards. (22)

saprobe a type of heterotrophic organism that feeds on dead organic matter such as rotting wood or compost; formerly called saprophyte. (4)

sarcina (sar′sĭ-nah) A cubelike packet of eight cocci. (4)

Sarcodina (sar″ko-di′nah) The class or group of protozoa whose members move by means of pseudopodia. (15)

sarcoma A tumor of the connective tissues. (11)

scalded skin syndrome A staphylococcal skin disease in infants characterized by red, wrinkled surface with a sandpaper texture. (10)

scarification A method of inoculating an immunizing agent by scratching the skin.

scarlatina A mild case of scarlet fever.

Schick test A skin test used to determine the effectiveness of diphtheria immunization. (7)

Schistosoma (shis-to-so′mah) **species** A group of multicellular flatworm parasites of the blood transmitted by contact with water and commonly known as blood flukes; include *S. mansoni* and *S. japonicum*. (16)

schistosomiasis A waterborne blood disease caused by *Schistosoma* species and characterized by fever, chills, and liver damage. (16)

schmutzdecke (shmoots′dek-ĕ) A slimy layer of microorganisms that develops in a slow sand filter. (25)

sclerotium (skle-ro′she-um) A hard purple body forming in grains contaminated with *Claviceps purpurea*. (14)

scolex The head region of a tapeworm wherein the attachment organ is located. (16)

scrapie A slow virus disease of sheep and other animals in which nerve damage causes the animal to scratch the skin. (10)

scrofula a bacterial disease of the lymph nodes of the neck caused by *Mycobacterium scrofulaceum*. (7)

secondary anamnestic response A vigorous immune response stimulated by a second or subsequent entry of antigens to the body. (19)

secondary antibody response The second response by the immune system to an antigen; characterized by an outpouring of IgG. (19)

secondary disease A disease that develops in a weakened individual. (17)

Seitz filter (sĭtz) A filter composed of a pad of porcelain or ground glass. (21)

selective medium A growth medium that contains ingredients to inhibit certain microorganisms while encouraging the growth of others. (4)

septicemia A generalized bacterial infection of the bloodstream due to any of several organisms including streptococci and staphylococci; once known as blood poisoning. (7, 17)

septum A cross wall occurring in the hypha of a fungus. (14)

serological reaction An antigen–antibody reaction studied under laboratory conditions involving serum. (19)

serology The branch of immunology that studies serological reactions. (19)

serotonin (ser″o-to′nin) A derivative of tryptophan that functions in type I hypersensitivity reactions to contract smooth muscles. (20)

serotype A rank of classification below the species level based on an organism's reaction with antibodies in serum; used for several bacteria, especially *Salmonella*. (3, 8)

Serratia marcescens (se-ra′she-ah mar-ses′ens) A Gram-negative red-pigmented bacterial rod involved as an opportunist in disease of the respiratory and urinary tracts. (7)

serum The fluid portion of the blood consisting of water, minerals, salts, proteins, and other organic substances; contains no clotting agents. (9, 17, 19)

serum sickness A type of hypersensitivity reaction in which the body responds to proteins contained in foreign serum. (19, 20)

severe combined immunodeficiency (SCID) An immune disease in which the lymph nodes lack both B-lymphocytes and T-lymphocytes. (20)

sexduction A process of recombination in which chromosomal genes pass from a donor cell to a recipient cell while attached to the F factor. (6)

***Shigella* species** A group of Gram-negative bacterial rods transmitted by food and/or water and the cause of shigellosis. (7)

shigellosis A foodborne and waterborne bacterial disease of the intestine caused by *Shigella* species and accompanied by extensive diarrhea, often with blood and mucus. (7)

shingles An alternative name for herpes zoster; a condition of the skin due to varicella–zoster virus. (12)

short incubation hepatitis An alternative name for hepatitis A. (13)

silage A type of animal feed produced by fermenting grains and other plants in silos, the huge cylindrical structures that often stand next to barns. (24)

"slapped cheek disease" A common name for fifth disease. (12)

slow virus disease A slow-growing viral disease in which the effects appear after a long time period. (13)

smallpox A now-extinct viral disease of the skin and body organs caused by a complex DNA virus and accompanied by bleeding skin pustules, disfigurements, and multiple organ involvements. (12)

smut A fungal disease of agricultural crops so named because of the sooty black appearance of infected plants. (14)

sodium hypochlorite (NaOCl) A derivative of chlorine used in disinfection practices; also known as bleach. (22)

sodoku (so'do-koo) An alternative name for rat-bite fever caused by a spirillum. (10)

soft chancre An alternative name for chancroid. (10)

specialized transduction A transduction in which the prophage carries some bacterial genes when it breaks free from the chromosome; the bacterial genes are then replicated and carried into the next cell by the virus. (6)

species The fundamental rank of the classification system; two or more species are grouped together as a genus. (3)

species immunity Immunity present in one species of organisms but not another species. (18)

specific immunological tolerance A phenomenon in which a person's own proteins and polysaccharides contact and inactivate cells that might later respond to them immunologically. (18)

spectinomycin An antibiotic used as a substitute for penicillin in cases of gonorrhea that are caused by penicillinase-producing *Neisseria gonorrhoeae*. (23)

spherule (sfer'ūl) A stage in the life cycle of the fungus *Coccidioides immitis*. (14)

spike A functional projection of the viral envelope. (11, 12)

spirillum A form of bacteria characterized by twisted or curved rods, generally with a rigid cell wall and flagella. (4)

Spirillum minor A spiral bacterium and one of the causes of rat bite fever. (10)

spirochete (spi'ro-kēt) A twisted bacterial rod with a flexible cell wall containing axial filaments for motility. (4)

spontaneous generation A theory suggesting that lifeless objects give rise to living things in their present form. (1)

spontaneous mutation A mutation that arises from chance events in the environment. (6)

sporangiospores Asexually produced fungal spores formed within a sporangium. (14)

sporangium (spo-ran'je-um) A protective sac that contains asexually produced fungal spores. (14)

spore A highly resistant structure formed from vegetative cells in several genera of bacteria including *Bacillus* and *Clostridium*; also, a reproductive structure formed by a fungus. (4, 14)

sporicidal agent An agent that kills bacterial spores. (22)

sporocyst An intermediary stage in the life cycle of a fluke; usually occurs in the snail. (16)

Sporothrix schenkii The fungus that causes sporotrichosis. (14)

sporotrichosis A soilborne fungal disease of the lymph channels caused by *Sporothrix schenkii* and characterized by knotlike growths under the skin surface and occasional skin lesions. (14)

Sporozoa The class or group of protozoa whose members have no means of locomotion in the adult form. (15)

sporozoite A stage in the life cycle of *Plasmodium* species; the parasite enters the human body in this form. (15)

sputum (spu'tum) Thick, expectorated matter from the lower respiratory tract. (7)

staphylococcus (staf"ĭ-lo-kok'us) A form of bacteria characterized by spheres in a grapelike cluster. (4, 24)

Staphylococcus aureus A Gram-positive grapelike cluster of cocci that can be the cause of food poisoning and/or infections of the skin (boils, abscesses), lungs, meninges, or other organs. (8, 10)

stationary phase The third portion of a bacterial growth curve in which the reproductive and death rates of cells are equal. (4)

stem cell A primordial cell of the bone marrow from which hematopoietic and lymphopoietic cells develop. (18)

sterilization The removal of all life forms, especially bacterial spores. (21)

stormy fermentation Fermentation and curdling of milk accompanied by gas accumulation that forces the curds apart. (24)

streptobacillus A chain of bacterial rods. (4)

Streptobacillus moniliformis A rod-shaped bacterium in chains and one of the causes of rat bite fever. (10)

streptococcus A chain of bacterial cocci. (4, 24)

Streptococcus mutans A Gram-positive chain of cocci and an important cause of dental caries. (10)

Streptococcus pneumoniae A Gram-positive chain of cocci and an important cause of bacterial pneumonia; also called the pneumococcus. (7)

Streptococcus pyogenes A Gram-positive chain of cocci and an important cause of streptococcal diseases such as scarlet fever. (7)

streptokinase (strep″to-ki′nās) An enzyme that dissolves fibrin clots; produced by virulent streptococci. (17)

streptomycin An antibiotic derived from *Streptomyces griseus* that is effective against Gram-negative bacteria and the tubercule bacillus; streptomycin interferes with protein synthesis. (23)

Strongyloides stercoralis (stron″ji-loi′dēz ster-ko-ral′is) A multicellular roundworm parasite of the intestine and lung. (16)

subacute sclerosing panencephalitis (SSPE) A rare complication of measles that affects the brain with loss of cortical functions and decrease in intellectual skills. (12)

subclinical disease A disease in which there are few or inapparent symptoms. (17)

substrate The substance upon which an enzyme acts. (5)

subunit vaccine A vaccine that contains parts of microorganisms such as capsular polysaccharides or purified pili. (19)

sulfamethoxazole (sul″fah-meth-oks′ah-zōl) A sulfonamide compound used for vaginal infections, conjunctivitis, and other diseases. (23)

sulfanilamides (sul″fah-nil′ah-mīds) A group of sulfurcontaining compounds used as antimicrobial agents. (23)

sulfur granules Small, hard granules found in tissue infected with *Actinomyces israelii*; resemble the sulfur granules used in pharmacy. (10)

suppressor T-lymphocyte A T-lymphocyte that interferes with the activity of B-lymphocytes. (18, 20)

swimmer's itch Dermatitis of the skin caused by the body's reaction to certain species of schistosomes. (16)

symbiosis (sim″bi-o′sis) An interrelationship between two populations of organisms where there is a close and permanent association. (4)

syncytia (sin-sish′ah) Giant tissue cells in culture formed by the fusion of cells infected with respiratory syncytial viruses. (12)

syndrome A collection of symptoms.

synergism (sin′er-jizm) A close and permanant association between two populations of organisms such that the populations are able to accomplish together what neither could otherwise accomplish alone. (4)

synthetic vaccine A vaccine that contains chemically synthesized parts of microorganisms, such as proteins normally found in viral capsids. (19)

syphilis A sexually transmitted bacterial disease of multiple organs caused by *Treponema pallidum* and occurring in three stages accompanied by extensive skin lesions, multiple organ involvements, and complications of the nervous and cardiovascular systems. (10)

systemic disease A disease that has disseminated to the deeper organs and systems of the body. (17)

systemic lupus erythematosus (SLE) An autoimmune disease in which antibodies form against nuclear components of the individual's cells and then unite with the antigens to form immune complexes in the skin and body organs. (20)

T

tabardillo (tab″ar-dēl′yo) An alternative name for endemic typhus. (9)

Taenia saginata A multicellular flatworm parasite of the intestine transmitted by contaminated beef and commonly known as the beef tapeworm. (16)

Taenia solium A multicellular flatworm parasite of the intestine transmitted by contaminated pork and commonly known as the pork tapeworm. (16)

tapeworms A group of ribbonlike flatworms belonging to the phylum Platyhelminthes; also called cestodes. (16)

taxonomy The science dealing with the systematized arrangements of related living things in categories. (3)

teichoic (ti-ko′ik) **acid** A polysaccharide found in the cell wall of Gram-positive bacteria. (4)

temperate phage A bacteriophage that enters a bacterium but does not replicate; the phage DNA may remain in the bacterial cytoplasm or attach to the bacterial chromosome. (6)

tertian malaria A type of malaria in which the attacks occur every 48 hours. (15)

tetanospasmin An exotoxin produced by *Clostridium tetani*; inhibits the removal of acetylcholine at the synapse and thereby stimulates muscle contractions. (9)

tetanus A soilborne bacterial disease of the muscles and nerves caused by toxins produced by *Clostridium tetani* and accompanied by uncontrolled muscle spasms. (9)

tetracyclines A group of antibiotics each of which is characterized by four benzene rings with attached side groups; used for many diseases caused by Gramnegative bacteria, rickettsiae, and chlamydiae; inhibit protein synthesis in bacteria. (9, 23)

therapeutic serum Antibody-rich serum used to treat a specified condition. (19)

thermal death point The temperature required to kill an organism in a given length of time. (21, 24)

thermal death time The length of time required to kill an organism at a given temperature. (21, 24)

thermoduric (ther"mo-du'rik) **organisms** Organisms that normally grow at medium temperatures but may tolerate and grow at high temperatures. (24)

thermophiles Organisms that grow at high temperature ranges of 40° C to 90° C. (4, 24)

thioglycollate (thi"o-gli'ko-lāt) **medium** A bacteriological medium that contains thioglycollic acid; the latter binds oxygen from the atmosphere and creates an environment suitable for anaerobic growth. (4)

thrombocytopenia (throm"bo-si"to-pe'ne-ah) A reduced count of blood platelets (thrombocytes) resulting from the reaction of antibodies with antigens on the platelet surface; a form of type II hypersensitivity. (20)

thrush Oral candidiasis. (14)

thymus A flat bilobed organ that lies in the neck below the thyroid; the T-lymphocyte is produced in this organ. (18)

tincture of iodine An iodine solution consisting of 2 percent iodine plus iodide in ethyl alcohol. (22)

tinea Any of a group of fungal infections of the skin; includes tinea pedis (athlete's foot) and tinea capitis (ringworm of the scalp). (14)

tinidazole (ti-nid'ah-zōl) A chemotherapeutic agent used to treat trichomoniasis. (15)

tissue immunity An alternative expression for cell-mediated immunity. (18)

titer (ti'ter) The most dilute concentration of antibody that will yield a positive reaction with specific antigen; a method of expressing the amount of antibody in a sample of serum. (19)

T-lymphocyte A lymphocyte that is modified in the thymus gland and is associated with the system of cell-mediated immunity; also called a T-cell. (13, 18)

TORCH An acronym for four diseases that pass from the mother to the unborn child; includes Toxoplasmosis, Rubella, Cytomegalovirus disease, and Herpes simplex; the O stands for other diseases. (12, 13, 15)

toxic shock syndrome A bacterial disease of the blood caused by toxins produced by *Staphylococcus aureus* and accompanied by shock and circulatory collapse. (10)

toxin A poisonous substance produced by a species of microorganism; bacterial toxins are classified as exotoxins or endotoxins. (17)

toxoid An immunizing agent produced from an exotoxin that elicits antitoxin production by the body. (7, 9, 17, 19)

Toxoplasma gondii A protozoan of the Sporozoa group and the cause of toxoplasmosis. (15)

toxoplasmosis A protozoal disease of the blood caused by *Toxoplasma gondii*, transmitted by contact with cats or consumption of beef, and accompanied by malaise and nonspecific symptoms; serious disease in immunocompromised patients; fetal injury in pregnant women. (15)

trachoma A contact disease of the eye caused by *Chlamydia trachomatis* and accompanied by tiny, pale nodules on the conjunctiva and possibly leading to blindness. (10)

transcription The biochemical process in which RNA is synthesized according to a code supplied by the bases of the DNA molecule. (5)

transduction A type of bacterial recombination in which a virus transports fragments of DNA from a donor cell to a recipient cell. (6)

transfer RNA (tRNA) A molecule of RNA that unites with amino acids and transports them to the ribosome in protein synthesis. (5)

transformation A type of bacterial recombination in which competent bacteria acquire fragments of DNA from disintegrated donor cells and incorporate the DNA into their chromosomes. (6)

translation The biochemical process in which the code on the mRNA molecule is translated into a sequence of amino acids in the protein molecule. (5)

transposon A segment of DNA that moves from one site on a DNA molecule to another site; transposons carry information for protein synthesis; also known as jumping genes. (6)

travelers' diarrhea A foodborne and waterborne bacterial disease of the intestine often caused by *Escherichia coli* and accompanied by diarrhea. (8)

trematodes A group of flatworms commonly known as flukes. (16)

trench fever A louseborne rickettsial disease due to *Rochalimaea quintana* and accompanied by fever and a skin rash. (9)

Treponema pallidum A spirochete and the cause of syphilis. (10)

Treponema pertenue A spirochete and the cause of yaws. (10)

Trichinella spiralis (trik"i-ne'ah spir-al'is) A multicellular roundworm parasite of the muscles transmitted by contaminated pork and the cause of trichinosis. (16)

trichinosis A foodborne disease of the muscles caused by the parasite *Trichinella spiralis*, contracted from poorly cooked pork, and accompanied by muscle pains. (16)

trichocysts (trik'o-sistz) Organelles of *Paramecium* species that discharge filaments used to trap the organism's prey. (15)

Trichomonas vaginalis A flagellated protozoan transmitted by sexual contact and the cause of trichomoniasis. (15)

trichomoniasis A protozoal disease of the reproductive tract caused by *Trichomonas vaginalis*, transmitted by sexual contact, and accompanied by discomfort in the urinary and lower reproductive tracts. (15)

Trichophyton (tri-kof'i-ton) **species** A fungus spread by contact and one of the causes of athlete's foot and ringworm. (14)

Trichuris trichiura (trik-u'ris trik-e-u'rah) A multicellular roundworm parasite of the intestine commonly known as the whipworm. (16)

trivalent vaccine A vaccine consisting of three components each of which stimulates immunity; examples are the diphtheria–pertussis–tetanus vaccine (DPT), the measles–mumps–rubella vaccine (MMR), and the trivalent oral polio vaccine (TOP). (13, 19)

Trombicula (trom-bik'u-lah) **species** Species of mites, a spiderlike arthropod. (9)

trophozoite (trof"o-zo'īt) The feeding form of a microorganism such as a protozoan. (15)

Trypanosoma brucei (tri-pan-o-so'mah bru'ce-i) A flagellated protozoan transmitted by tsetse flies and the cause of African trypanosomiasis. (15)

Trypanosoma cruzi A flagellated protozoan transmitted by the triatomid bug and the cause of South American trypanosomiasis. (15)

trypanosomiasis A protozoal disease of the blood caused by *Trypanosoma* species, transmitted by arthropods, and accompanied by periods of high fever, and coma; also called sleeping sickness. (15)

tsutsugamushi (soot"soo-ga-moosh'e) A relatively mild miteborne disease of the blood caused by *Rickettsia tsutsugamushi* and accompanied by fever and a skin rash; also called scrub typhus. (9)

tubercle A hard nodule that develops in tissue infected with *Mycobacterium tuberculosis*. (7)

tuberculin test A skin test used for early detection of tuberculosis; performed by applying PPD to the skin and noting a thickening of the skin with a raised vesicle in a few days. (7, 20)

tuberculosis An airborne bacterial disease of the lungs caused by *Mycobacterium tuberculosis* and accompanied by degeneration of the lung tissue and spread to other organs. (7)

tularemia A mild arthropodborne bacterial disease transmitted by the tick and by contact, caused by *Francisella tularensis* and accompanied by mild fever and vague symptoms of malaise. (9)

tumor An abnormal functionless mass of cells. (11, 20)

tyndallization (tyn"dal-i-za'shun) A sterilization method in which materials are heated in free-flowing steam for 30 minutes on each of three successive days. (21)

typhoid fever A foodborne and waterborne bacterial disease of the intestine caused by *Salmonella typhi* and accompanied by intestinal ulcers, severe fever, and blood involvement. (7)

U

ultrapasteurization A pasteurization process in which milk is heated at 82° C for 3 seconds. (24)

undulant fever The name applied to brucellosis in humans due to the undulating nature of the fever. (8)

Ureaplasma urealyticum (u-re"ah-plaz'mah u-re"ah-lit'i-kum) The mycoplasma that causes ureaplasmal urethritis. (10)

ureaplasmal urethritis A sexually transmitted bacterial disease of the urinary and genital organs caused by *Ureaplasma urealyticum* and accompanied by urinary tract symptoms. (10)

urticaria (ur"tĭ-ka're-ah) A hivelike rash of the skin. (20)

urushiol (u-roo'she-ol) A chemical substance on the leaves of certain plants that induces poison ivy. (20)

V

vaccination Originally the process by which Jenner introduced vaccinia (cowpox) to volunteers; currently, the term refers to any immunization procedure involving an injection.

vaccinia (vak-sin'e-ah) The alternative name for cowpox. (12)

vaginitis A general term for disease of the vagina. (10)

vancomycin An antibiotic used in therapy of diseases caused by Gram-positive bacteria, especially staphylococci. (23)

varicella An alternative name for chickenpox; means "little vessel," a reference to small chickenpox lesions. (12)

varicella–zoster immune globulin (VZIG) A preparation of purified antibodies from blood donors that gives some protection against chickenpox. (12)

variola The alternate name for smallpox. (12)

VDRL test A precipitation test used in the detection of syphilis antibodies. (19)

vector A living organism that transmits the agents of disease. (9, 17)

vesicle A fluid-filled skin lesion such as that occurring in chickenpox. (12)

vibrio A form of bacterium occurring as a curved rod; resembles a comma. (4, 7)

Vibrio cholerae A Gram-negative curved bacterial rod transmitted by food and/or water and the cause of cholera. (8)

Vibrio parahaemolyticus (par"ah-he-mo-lit'i-kus) A Gram-negative curved bacterial rod transmitted by food and/or water and involved in disease of the intestine. (8)

Vibrio vulnificus A marine bacterium known to cause intestinal infection when ingested in water. (25)

vidarabine (Vira-A) A drug useful in treating herpes zoster and herpes encephalitis. (11, 12)

Vincent's angina Bacterial infection of the tonsil and soft palate areas due to several species of bacteria; a form of ANUG. (10)

viremia Spread of viruses through the circulatory system. (17)

virion A completely assembled virus outside its host cell. (11)

viroids Tiny fragments of nucleic acids associated with certain plant diseases; possibly associated with animal disease. (11)

virucidal agent An agent that inactivates viruses. (22)

virulence The degree of pathogenicity of a parasite. (17)

virulent phage A bacteriophage that replicates within a bacterium and destroys the bacterium. (6)

viruses Particles of nucleic acid (either DNA or RNA) surrounded by a protein sheath; neither prokaryotic nor eukaryotic; highly infectious. (3, 6, 11)

V–Z virus The name given to the virus that causes varicella (chickenpox) and herpes zoster (shingles). (12)

W

walking pneumonia A colloquial expression for a mild case of pneumonia. (7, 12)

wart A small, usually benign skin growth commonly due to a virus. (12)

Waterhouse–Friderichsen syndrome The formation of hemorrhagic lesions in the adrenal glands; thought to be an allergic reaction with immune complex formation; associated with meningococcal meningitis. (7, 20)

Weil–Felix test A diagnostic procedure for rickettsial diseases performed by combining serum from a patient and *Proteus OX-19* on a slide; in a positive test, the *Proteus* cells clump together. (9)

wetting agents Agents that emulsify and solubilize particles clinging to a surface; examples are the soaps. (22)

wheal An enlarged hivelike zone of puffiness on the skin often due to an allergic reaction. (20)

whey The clear liquid remaining after protein has curdled out of milk. (24)

whipworm The common name for *Trichuris trichiura*. (16)

whooping cough The alternate name for pertussis. (7)

Widal (ve-dahl') **test** An immunological procedure used for diagnostic purposes in which *Salmonella* antibodies agglutinate *Salmonella* antigens. (8, 19)

woolsorter's disease An alternative name for anthrax, derived from the fact that people who work with wool inhale the spores of the causative agent. (9)

Wuchereria bancrofti A multicellular roundworm parasite of the lymph channels transmitted by the mosquito and the cause of filariasis or elephantiasis. (16)

X

xenograft (zen'o-graft) A tissue graft between members of different species such as between an animal and a human. (20)

Xenopsylla (zen"op-sil'ah) ***cheopis*** The rat flea, the vector in plague. (9)

Y

yaws A mild syphilislike disease occurring in remote parts of the world and due to a species of *Treponema*. (10)

yeast A type of fungus that is unicellular and resembles bacteria in culture. (5, 14)

yellow fever A highly fatal viral disease transmitted by the *Aedes aegypti* mosquito and accompanied by intense fever and jaundice. (13)

Yersinia enterocolitica A Gram-negative bacterial rod transmitted by food and/or water and involved in disease of the intestine. (8)

Yersinia pestis A Gram-negative rod that displays bipolar staining and causes plague. (9)

Z

zoonosis (zo"o-no'sis) An animal disease that may be transmitted to humans. (9, 17)

zoospore (zo'o-spōr) A flagellated spore produced asexually by members of the Oomycetes class of fungi. (14)

Zygomycetes A group (or class) of fungi in which the sexual form of reproduction includes zygospore formation. (14)

zygospore A sexually produced spore formed by members of the Zygomycetes class of fungi. (14)

Photograph Acknowledgments

Note: CDC = Centers for Disease Control and Prevention, Public Health Service, U.S. Department of Health and Human Services, Atlanta, Georgia.

PART 1 OPENER

Courtesy Alastair Pringle, Anhauser-Busch.

PART 1 MICROBIOLOGY PATHWAYS

© UPI/Corbis-Bettmann.

CHAPTER 1

1.1: From Parke-Davis Series "Great Moments in Medicine." 1.4: Courtesy of L. Tao from *J. Bacteriol.* **169**:2543 (1987) with permission of American Society for Microbiology. 1.5: Courtesy of Institut Pasteur, Paris. 1.7a: Courtesy of National Library of Medicine. 1.7b: The Bettmann Archive. 1.9: Courtesy of Institut Pasteur, Paris. 1.10a and b: Courtesy of National Library of Medicine.

CHAPTER 2

MicroFocus 2.2: Courtesy Dr. Z. Skobe, Forsyth Dental Center.

CHAPTER 3

3.2: From R.E. Mueller, et al. *AEM* **47**:715 (1984). 3.3a: From John, Cole, & Marciano-Cabral. *AEM* **47**:12–14 (1984). 3.3b: Scanning micrograph by L. Tetley, from K. Vickeman, *Nature* **273**:613 (1978). 3.3c: Courtesy of Robert L. Owen, *Gastroenterology* **76**:759–769 (1979). 3.4a: Courtesy of E.M. Peterson, R.J. Hawlay, and R.A. Calerone. 3.4b: Courtesy of Ilan Chet, from *J. Bacteriol.* **154**:1431 (1983). 3.8: Courtesy of Carl Zeiss, Inc., New York. 3.12: Courtesy of J. Robert Waaland/BPS. 3.13: Courtesy of C.L. Hathaway, CDCP. 3.15: Courtesy of N. Gotoh, from *J. Bacteriol.* **171**:983 (1989) with permission of American Society for Microbiology. MicroFocus 3.4: Courtesy of N. Pace and E. Angert, from *ASM News* **58**:419 (1992) with permission of American Society for Microbiology.

PART 2 OPENER

Courtesy of C. Nombela, from *J. Bacteriol.* (1990) with permission of American Society for Microbiology.

PART 2 MICROBIOLOGY PATHWAYS

© Hank Morgan/Photo Researchers, Inc.

CHAPTER 4

4.2a: Courtesy of D.L. Shungu, J.B. Cornett, and G.D. Schockman, *J. Bacteriol.* **138**:601 (1979), with permission from ASM. 4.2b: Courtesy of Drs. Z. Yoshii, J. Tokunaga, and J. Tawara, *Atlas of Scanning Electron Microscopy in Microbiology*, IGAKU-SHOIN, Ltd., 1976. 4.2c: Courtesy of Dr. M. Matsuhashi. From Yamada et al., *J. Bacteriol.* **129**:1513–1517 (1977), Figure 2A. 4.3a and b: D.L. Balkwill, D. Maratean, and R.P. Blakemore, reprinted from *J. Bacteriol.* **141**:1399–1408 (1980), with copyright permission from ASM. 4.4d: Photograph by Gary Gaard. Courtesy of Dr. A. Kelman (Department of Plant Pathology, University of Wisconsin-Madison). 4.5a: Courtesy of G. Biwas, from *J. Bacteriol.* **171**:657 (1989) with permission of American Society for Microbiology. 4.5b: Courtesy of Dr. B. Sugarman, Houston Veterans Administration Medical Center. 4.7: Courtesy of S.D. Acres, VIDO,

Photograph Acknowledgments *(continued)*

University of Saskatchewan, and J.W. Costeron, University of Calgary. From *Infect. Immun.* **37**(3):1170 (1982). 4.9: With permission of Victor Lorian, M.D., Bronx Lebanon Hospital Center, Bronx, NY. 4.10: Courtesy of E. Kellenberger, from *J. Bacteriol.* **173**:3149 (1992) with permission of American Society for Microbiology. 4.12a: Courtesy of Dr. C. Robinow. From *The Bacteria*, Vol. 1, Academic Press, New York (1960), Fig. 7, p. 214. 4.12b: Reproduced with permission from ASM and Gerhardt, Pankrants, and Scherrer, *Appl. Environ. Microbiol.* **32**:438–439 (1976). 4.12c: Courtesy of Drs. Yoshii, J. Tokunaga, and J. Tawara, *Atlas of Scanning Electron Microscopy in Microbiology*, IGAKU-SHOIN, Ltd., 1976. 4.14a: Courtesy of P. Scherer, from *ASM NEWS* **56**:569 (1990) with permission of American Society for Microbiology. 4.14b: Courtesy of Drs. Z. Yoshii, J. Tokunaga, and J. Tawara, *Atlas of Scanning Electron Microscopy in Microbiology*, IGAKU-SHOIN, Ltd., 1976. 4.16: Courtesy of R.E. Mueller et al., *AEM* **47**:715 (1984). MicroFocus 4.2: Courtesy of D. Bazlinski, from *ASM News* **55**:409 (1989) with permission of American Society for Microbiology.

CHAPTER 5

5.2: Courtesy of R. Belas, from *J. Bacteriol.* **173**:6280 (1991) with permission of American Society for Microbiology. 5.8: Courtesy of L. Tao, from *J. Bacteriol.* **169**:2543 (1987) with permission of American Society for Microbiology. 5.9: Courtesy of J.M. Sieburth, *Microbial Seascape's*, University Park Press, Baltimore (1979). 5.10: Courtesy of H.W. Jannasch, from *Appl. Environ. Microbiol.* **41**:528–538 (1981). 5.11: Courtesy of Phillip Sharp. MicroFocus 5.3: UPI/Bettmann Newsphotos.

CHAPTER 6

6.3: Courtesy of C. Nombela, from *J. Bacteriol.* **172**:2384 (1990) with permission of the American Society for Microbiology. 6.10: Courtesy of CDC. 6.11a: Courtesy of CDC. 6.11b: Courtesy of Dr. S.C. Holt, University of Massachusetts/BPS. 6.14a: Courtesy of Dr. S.N. Cohen, Stanford University.

PART 3 OPENER

Courtesy of V. Fischetti, from ASM News **57**:619 (1991) with permission of American Society for Microbiology.

PART 3 MICROBIOLOGY PATHWAYS

Courtesy of Janssen-Ortho, Inc.

CHAPTER 7

7.1a: Courtesy of Drs. Z. Yoshii, J. Tokunaga, and J. Tawara, *Atlas of Scanning Electron Microscopy in Microbiology*, IGAKU-SHOIN, Ltd., 1976. 7.1b: Courtesy of V. Fischetti, from *ASM News* **57**:619 (1991) with permission of American Society for Microbiology. 7.2: Courtesy of Armed Forces Institute of Pathology, Neg. No. 58-6180-2. 7.8: "Portrait of a Woman," by Franz Hals. The Detroit Institute of Arts, City of Detroit Purchase. 7.9a: Courtesy of Gourley, Leach, and Howard, *J. Gen. Microbiol.* **81**:475 (1974), with permission from ASM. 7.9b and c: Courtesy of Respiratory Disease Laboratory Section, DCDP, from *Appl. Environ. Microbiol.* **47**(3):467 (1984). 7:10: Courtesy of Dr. B. Sugarman, Houston Veterans Administration Medical Center. 7.11a: S. Ito and A.R. Flesher, Harvard Medical School. 7.11b: Courtesy of Dr. Barry Fields. 7.11c: Courtesy of Frederick Quinn, CDC. 7.13: Courtesy of Dr. Edgar Ribi and William R. Brown, Ribi ImmunoChem Research, Inc., Hamilton, Montana. 7.14: Courtesy of C.C. Kuo, from *J. Bacteriol.* **169**:3757 (1987) with permission of American Society for Microbiology.

Photograph Acknowledgments *(continued)*

CHAPTER 8

8.2a: Courtesy of CDC. 8.2b: Courtesy of K. Amako, Fukuoka University, Japan. 8.3: Courtesy of CDC. 8.6a: Courtesy of D.J. Thomas and T.A. McMeekin, University of Tasmania, Tasmania, Australia. 8.6b: Courtesy of Dr. W.L. Dentler. 8.8a and b: Courtesy of Drs. Z. Yoshii, J. Tokunaga, and J. Tawara, *Atlas of Scanning Electron Microscopy in Microbiology*, IGAKU-SHOIN, Ltd., 1976. 8.10: Courtesy of N. Schifferli, from *J. Bacteriol.* **173**:1230 (1991) with permission of American Society for Microbiology. 8.11 a and b: Courtesy of Diane E. Taylor, from *J. Bacteriol.* **164**(1):338–343 (1985). 8.13: Courtesy of V. Miller, from *ASM News* **58**:26 (1992) with permission of American Society for Microbiology.

CHAPTER 9

9.1a: Courtesy of Drs. Z. Yoshii, J. Tokunaga, and J. Tawara, *Atlas of Scanning Electron Microscopy in Microbiology*. IGAKU-SHOIN, Ltd., 1976. 9.2: Courtesy of CDC. 9.3: Courtesy of The Royal College of Surgeons, Edinburgh. 9.4: Courtesy of CDC. 9.5a and b: Courtesy of Drs. Z. Yoshii, J. Tokunaga, and J. Tawara, *Atlas of Scanning Electron Microscopy in Microbiology*, IGAKU-SHOIN, Ltd., 1976. 9.8a and b: Courtesy of Armed Forces Institute of Pathology (a) Neg. No. N-85837-2, and (b) Neg. No. AN1147-1A. 9.9a: Courtesy of A. Garon, Rocky Mountain Laboratory, Hamilton, Montana. 9.9b: Courtesy of J. Radolf, from *J. Bacteriol.* **173**:8004 (1991) with permission of American Society for Microbiology. 9.10: Courtesy of Allen C. Steere, from *Annals, Int. Med.* **86**:685 (June 1977). 9.12a, b, and c: Courtesy of CDC. MicroFocus 9.1: S & G Press Agency, Ltd.

CHAPTER 10

10.1a: Carroll H. Weiss 1981/Camera M.D. Studios. 10.1b: Courtesy of Armed Forces Institute of Pathology, Neg. No. 78-1413. 10.1c and d: Courtesy of CDC. 10.4a and b: Courtesy of Drs. Z. Yoshii, J. Tokunaga, and J. Tawara, *Atlas of Scanning Electron Microscopy in Microbiology*, IGAKU-SHOIN, Ltd., 1976. 10.5a and b: Courtesy of Dr. L.A. Page. From *Bergey's Manual of Determinative Bacteriology*, 8th ed., Williams and Wilkins Co.: Baltimore (1974). 10.6: Courtesy of D. Krause from *J. Bacteriol.* **173**:1041 (1991) with permission of American Society for Microbiology. 10.7a and b: Courtesy of American Leprosy Mission, 1 Broadway, Elmwood Park, NJ. 10.8: Courtesy of Armed Forces Institute of Pathology, Neg. No. 56-140750-2. 10.9: Courtesy of CDC. 10.10: Courtesy of CDC. 10.11: Courtesy of R.E. Muller et al., *AEM* **47**:715 (1984). 10.12b: Courtesy of the Naval Dental Research Institute, Naval Training Center, Great Lakes, Illinois. 10.13: Courtesy of CDC. 10.14: Courtesy of T.J. Marrie and J.W. Costeron, from *Appl. Environ. Microbiol.* **42**(6):1093–1102 (1981). 10.15: Courtesy of N. Gotoh, from *J. Bacteriol.* **171**:983 (1989) with permission of American Society for Microbiology. MicroFocus 10.1: Courtesy of Fitzgerald, Cleveland, Johnson, Miller, and Syukes, from *J. Bacteriol.* **130**:1337. MicroFocus 10-2: Courtesy of National Hansen's Disease Center, Carville, LA.

PART 4 OPENER

Courtesy of K.J. Kwon-Chung, from *J. Clin. Microbiol.* **30**:3290 (1992) with permission of American Society for Microbiology.

PART 4 MICROBIOLOGY PATHWAYS

© Michael Rosenfeld/Tony Stone Images.

CHAPTER 11

11.1a: Courtesy of National Library of Medicine. 11.1b: Courtesy of USDA photographs. 11.2: Courtesy of CDC. 11.4: Courtesy of C.K.Y. Fong, from *J. Clin.*

Photograph Acknowledgments *(continued)*

Microbiol. **30**:1612 (1992) with permission of American Society for Microbiology. 11.5b: Courtesy of Dr. Glenn Howard, Pall Corporation, Glen Cove, NY. 11.8: Courtesy of Dr. Sara Miller, Duke University Medical Center. 11.9a and b: Courtesy of Drs. Z. Yoshii, J. Tokunaga, and J. Tawara, *Atlas of Scanning Electron Microscopy in Microbiology*, IGAKU-SHOIN, Ltd., 1976. 11.12a and b: Courtesy of Drs. Z. Yoshii, J. Tokunaga, and J. Tawara, *Atlas of Scanning Electron Microscopy in Microbiology*, IGAKU-SHOIN, Ltd., 1976. 11.14c: Courtesy of Fisher Scientific Company. 11.15a: Courtesy of J. Smit, from *J. Bacteriol.* **173**:5677 (1991) with permission of American Society for Microbiology. 11.15b: Courtesy of J. Smit, from *J. Bacteriol.* **173**:5568 (1991) with permission of American Society for Microbiology.

CHAPTER 12

12.2: Fumio Uno, *J. Electron Microscopy* **28**:83–92 (1979). Figs. 7a and 8a. 12.3: Courtesy of Dr. Sara Miller, Duke University Medical Center. 12.4: Courtesy of Spotswood L. Spruance, from *J. Clin. Microbiol.* **22**(3):366–368 (1985). 12.6: Courtesy of CDC. 12.7: Courtesy of J. Vreeswijk, from *J. Clin. Microbiol.* **30**:2487 (1992) with permission of American Society for Microbiology. 12.8: Courtesy of CDC. 12.11: Courtesy of V. Murphy, from *ASM News* **57**:579 (1991) with permission of American Society for Microbiology. 12.13: Courtesy of Parke-Davis, Division of Warner-Lambert Company, Morris Plains, NJ. MicroFocus 12.3: Courtesy of Gertrude B. Elion.

CHAPTER 13

13.1a: Courtesy of Alyne K. Harrison, Viral Pathology Division, CDC. 13.1b: Courtesy of CDC. 13.3: Courtesy of Dr. E.H. Cook, Jr., Hepatitis and Viral Enteritis Division, CDC, Phoenix, AZ. 13.4 a and b: Courtesy of Fred B. Williams, Jr., U.S. Environmental Protection Agency, Cincinnati, OH 45268. 13.9: Courtesy of Bernard J. Poiesz, from *SUNY Research '84*, State University of New York. 13.10: Courtesy of Drs. F.A. Murphy and A.K. Harrison, in *Rhabdoviruses*, D.H.L. Bishop, ed., Vol. I, pp. 66–106, CRC Press, 1979. MicroFocus 13.2: Courtesy of Parke-Davis, Division of Warner-Lambert Company, Morris Plains, NJ.

CHAPTER 14

14.1: Courtesy of R. Simmons, from *ASM News* **57**:400 (1991) with permission of American Society for Microbiology. 14.2: Courtesy of C. Nombela, from *J. Bacteriol.* **172**:2384 (1990) with permission of American Society for Microbiology. 14.3: Courtesy of Drs. L.F. Ellis and S.W. Queener. Reproduced by permission of the National Research Council of Canada, from the *Canadian Journal of Microbiology*, Vol. 21, pp. 1982–1996. 14.4a and b: Courtesy of Drs. Z. Yoshii, J. Tokunaga, and J. Tawara, *Atlas of Scanning Electron Microscopy in Microbiology*, IGAKU-SHOIN, Ltd., 1976. 14.5a–f: Courtesy of Northern Regional Research Center, Peoria, IL 61604. 14.6: Courtesy of K.J. Kwon-Chung, from *J. Clin. Microbiol.* **30**:3290 (1992) with permission of American Society for Microbiology. 14.7: Courtesy of T. Sewall, from *ASM News* **54**:657 (1988) with permission of American Society for Microbiology. 14.9a: Courtesy of Joseph Schlitz Brewery. 14.9b: From F.G. Kessel and C.Y. Shih, *Scanning Electron Microscopy in Biology. A Students' Atlas on Biological Organization.* Springer-Verlag, New York, 1976. 14.10: Courtesy of Wine Institute of California. 14.11: Courtesy of M.A. Kydstra, from *Infect. Immun.* **16**:129 (1977). 14.12: Photograph by Marion J. Balish. Courtesy of Edward Balish, from *AEM* **47**(4):647 (1984). 14.13a and b: Courtesy of CDC. 14.16a: Courtesy of J.D. Raj, California State University, Long Beach, and S.S. Sekhon, Veterans Administration Hospital, Long Beach, from *ASM News* **49**(4), April 1983. 14.16b: Courtesy of CDC. MicroFocus 14.5: Courtesy of Dr. Gary Strobel, University of Montana.

Photograph Acknowledgments   *(continued)*

CHAPTER 15

15.1: Courtesy of D. White, from *ASM News* **54**:583 (1988) with permission of American Society for Microbiology. 15.3: From John, Cole, & Marciano-Cabral. *AEM* **47**:12–14 (1984). 15.4a–f: Courtesy of Carolina Biological Supply Company. 15.5: Courtesy of E. Bottone, from *J. Clin. Microbiol.* **30**:2447 (1992) with permission of American Society for Microbiology. 15.6: Courtesy of British Tourist Authority, New York. 15.9a–c: Courtesy of Robert L. Owen and the editors of *Gastroenterology* **76**:759–769 (1979). 15.10a: Courtesy of Armed Forces Institute of Pathology, Neg. No. 74-5195. 15.10b: Courtesy of Steven T. Brentans and John E. Donelson, University of Iowa. 15.11: Courtesy of T. Minnick, from *ASM News* **51**:239 (1985) with permission of American Society for Microbiology. 15.12: Courtesy of CDC. 15.14a and b: Courtesy of Chiappino, Nichols, and O'Connor, *J. Protozool.* **31**:228 (1984). 15.17: Courtesy of Saul Tzipori, from *Microbiological Review* **47**(1):84–96 (March 1983).

CHAPTER 16

16.1a: Courtesy of Armed Forces Institute of Pathology, Neg. No. 69-77765-2. 16.1b: Courtesy of Dr. Clive E. Bennet, Jr., *Parasitology* **61**:892 (1975). 16.4a–c: Courtesy of Dr. J. Jujino. 16.7: Courtesy of B. Schieven, from *ASM News* **57**:566 (1991) with permission of American Society for Microbiology. 16.8: Courtesy of Dr. K.A. Wright, Jr., *J. Parasitology* **65**(3):441 (1979). 16.10a and b: Courtesy of Armed Forces Institute of Pathology, (a) Neg. No. 69-3584, and (b) Neg. No. 60-3583. 16.12: Courtesy of Armed Forces Institute of Pathology, Neg. No. 74-11349. 16.13: Courtesy of Armed Forces Institute of Pathology, Neg. No. 4440-1. 16.14a and b: Courtesy of Armed Forces Institute of Pathology, (a) Neg. No. 73-6654, and (b) Neg. No. 75-11789-4.

PART 5 OPENER

© 1990 Dennis Kunkel.

PART 5 MICROBIOLOGY PATHWAYS

Benjamin/Cummings Publishing Co.

CHAPTER 17

17.1: Courtesy of Drs. Z. Yoshii, J. Tokunaga, and J. Tawara, *Atlas of Scanning Electron Microscopy of Microbiology*, IGAKU-SHOIN, Ltd., 1976. 17.2: Courtesy of Jose Galan, SUNY, Stony Brook. 17.6a and b: Courtesy of Drs. D.C. Savage and R.V.H. Blumershine, *Infect. Immun.* **10**:240–250 (1974). 17.7: Courtesy of B. Finlay, from *ASM News* **58**:486 (1992) with permission of American Society for Microbiology. 17.8a: Courtesy of J.A. Dowsett and J. Reid, from *Mycologia* **71**:329 (1979). 17.8b: Courtesy of D. Corwin and S. Falkow, Rocky Mountain Laboratory, NIAID, NIH, Hamilton, Montana. 17.9: Courtesy of Howard J. Faden.

CHAPTER 18

18.2: Courtesy of Drs. D.F. Bainton and M.G. Farquhar. 18.11: Courtesy of Dr. Juneann W. Murphy, from *Microbiology-84*, ASM.

CHAPTER 19

19.4a: Courtesy of CDC. 19.4b: From Razin et al., *Infect. Immun.* **30**:538–546 (1980). MicroFocus 19.1: The Bettmann Archive.

Photograph Acknowledgments *(continued)*

CHAPTER 20

20.3a and b: Photos by Scott T. Clay-Poole. 20.6a and b: Courtesy of Dr. Robert A. Marcus. 20.7. Courtesy of Armed Forces Institute of Pathology, Neg. No. 55-12646. 20.9: The Bettmann Archive. 20.11: Courtesy of American Cancer Society.

PART 6 OPENER

Courtesy of H. Konishin, from *J. Bacteriol.* **168**:1476 (1986) with permission of American Society for Microbiology.

PART 6 MICROBIOLOGY PATHWAYS

Courtesy of Amgen.

CHAPTER 21

21.1: Courtesy of AMSCO/American Sterilizer Company, Erie, PA 16512. 21.2a: The Bettmann Archive. 21.3: The Bettmann Archive. 21.6: Courtesy of AMSCO/American Sterilizer Company, Erie, PA 16512. 21.7: Dr. Jean-Noël Barbotin. 21.8a-d: Courtesy of Millipore Corporation, Bedford, MA.

CHAPTER 22

22.1: The Bettmann Archive. MicroFocus 22.1: Courtesy of Parke-Davis, Division of Warner-Lambert Company, Morris Plains, NJ. 22.5: Courtesy of Dr. Roger L. Anderson, CDC. 22.7: Courtesy of T.J. Marrie and J.W. Costerton, from *Appl. Environ. Microbiol.* **42**(6):1093 (1981).

CHAPTER 23

23.1: Courtesy of the Bettmann Archive. 23.2: Courtesy of Parke-Davis, Division of Warner-Lambert Company, Morris Plains, NJ. 23.4a and b: Courtesy of the National Library of Medicine. 23.5: Courtesy of Parke-Davis, Division of Warner-Lambert Company, Morris Plains, NJ. 23.7: Photo from Norton, 2nd. ed. Fig. 24.2, p. 727. 23.9a and b: Courtesy of H. Konishi, from *J. Bacteriol.* **168**:1476 (1986) with permission of American Society for Microbiology. 23.10: Courtesy of Dr. D.R. Zusman. 23.11a: Courtesy of Pfizer, Inc. 23.14: Courtesy of J.C. Nichel and J.W. Costeron, from *Antimicrobial Agents and Chemotherapy* **27**(4):619 (1985). MicroFocus 23.4: Courtesy of S. Horinuchi, from *J. Bacteriol.* **172**:3003 (1990) with permission of American Society for Microbiology.

PART 7 OPENER

© Manfred Kage/Peter Arnold, Inc.

PART 7 MICROBIOLOGY PATHWAYS

Courtesy of WHO, photo by J. & P. Hubley.

CHAPTER 24

24.5a and b: Courtesy of the National Food Processors Association, Washington, DC. 24.7: Courtesy of C.J. Thomas and R.A. McMeekin, University of Tasmania, Tasmania, Australia, from *Appl. Environ. Microbiol.* **41**(2):492–503. 24.8: Courtesy of The VirtisCo., Inc., Gardiner, NY. 24.10: Courtesy of Paul Gustafson. 24.11: Courtesy of M.A. Dawschel and H.P. Flemming, USDA-ARS, Raleigh, NC. 24.12: Courtesy of Kikkoman Foods, Inc., Walworth, WI.

Photograph Acknowledgments   *(continued)*

CHAPTER 25

25.1a and b: Courtesy of Tomio Kawata. 25.3: Reprinted with permission of the present publisher, Jones and Bartlett Publishers, Inc. from Shih and Kessel: *Living Images*, Science Books International, 1982. 25.4: Photo provided by A.C. Scott, Central Veterinary Laboratory, Wexbridge, U.K., British Crown Copyright 1982. 25.7: Courtesy of M. Dworkin, from *J. Bacteriol.* **173**:7810 (1991) with permission of American Society for Microbiology. 25.11: Allan R.J. Eaglesham, Boyce Thompson Institute. 25.13: Photo ACL, Bruxelles. Prière de verser 1000 FB au compte 000-0343629-55 du Patrimoine des Muscès royaux d/art et d/histoire. 10 parc du Cinquantenaire B-1040 Bruxelles. 25.15a and b: Courtesy of Wine Institute of California. 25.16: Courtesy of Pfizer, Inc., 25.17: Courtesy of Dr. P.C. Fitz-James, University of Western Ontario.

THE MICROBES IN COLOR

Plate 1a, 1b, 1d: Courtesy of Barry L. Batzing, Department of Biological Sciences, SUNY Cortland. 1c: Courtesy of Carole Rehkugler, Cornell University. Plate 2a: Courtesy of Dr. Stephen Zinder, Dept. of Microbiology, Cornell University, Ithaca, NY 14854. 2b: Courtesy of Dr. William Ghiorse, Cornell University. 2c, 2d: Courtesy of Barry L. Batzing, Department of Biological Sciences, SUNY Cortland. Plate 3a, 3b: Courtesy of Carole Rehkugler, Cornell University. 3c: Courtesy of Dr. Maurice Sheppard, Camp Lejeune, NC 28542. 3d: Courtesy of Barry L. Batzing, Department of Biological Sciences, SUNY Cortland. 3e, 3f: Courtesy of Dr. J.F. Timoney, Cornell University, Ithaca, NY 14853. Plate 4a: Courtesy of Susan Hughes, Milk Marketing Board, Madison, WI. 4b: Courtesy of Dr. Norman Dondero, Cornell University. 4c, 4d: Courtesy of Barry L. Batzing, Department of Biological Sciences, SUNY Cortland.

MICROBES AND BIOTECHNOLOGY

Plate 1a, 1b: Courtesy of Mr. John Doorley, Corporate Communications, Hoffman-LaRoche, Inc., Nutley, NJ 07110. 1c: Courtesy of Dr. Tom Currier, Dept. of Microbiology, Univ. of Washington, Seattle, WA 98195. Plate 2: Courtesy of Cornell University Biotechnology Program. Plate 3a: Courtesy of Mr. John Doorley, Corporate Communications, Hoffman-LaRoche, Inc., Nutley, NJ 07110. 3b: Courtesy of Cornell University Biotechnology Program. 3c: Courtesy of Dr. Georgia Lee Helmes, CIBA-GEIGY. Plate 4a, 4b: Courtesy of Mr. John Doorley, Corporate Communications, Hoffman-LaRoche, Inc., Nutley, NJ 07110. 4c: Manville Technical Celite® Division, Lompoc, CA.

INDEX

B refers to MicroFocus Box; F refers to figure; T refers to table.